Remote Sensing for Precision Nitrogen Management

Remote Sensing for Precision Nitrogen Management

Editors

Yuxin Miao

Raj Khosla

David J. Mulla

MDPI • Basel • Beijing • Wuhan • Barcelona • Belgrade • Manchester • Tokyo • Cluj • Tianjin

Editors

Yuxin Miao
Precision Agriculture Center,
University of Minnesota,
St. Paul, MN, USA

Raj Khosla
Department of Soil and Crop
Sciences, Colorado
State University,
Fort Collins, CO, USA

David J. Mulla
Precision Agriculture Center,
University of Minnesota,
St. Paul, MN, USA

Editorial Office
MDPI
St. Alban-Anlage 66
4052 Basel, Switzerland

This is a reprint of articles from the Special Issue published online in the open access journal *Remote Sensing* (ISSN 2072-4292) (available at: https://www.mdpi.com/journal/remotesensing/special_issues/Precision_N).

For citation purposes, cite each article independently as indicated on the article page online and as indicated below:

LastName, A.A.; LastName, B.B.; LastName, C.C. Article Title. *Journal Name* **Year**, *Volume Number*, Page Range.

ISBN 978-3-0365-5709-0 (Hbk)
ISBN 978-3-0365-5710-6 (PDF)

Cover image courtesy of Yuxin Miao.

Contents

About the Editors ix

Preface to "Remote Sensing for Precision Nitrogen Management" xi

Rui Dong, Yuxin Miao, Xinbing Wang, Zhichao Chen, Fei Yuan, Weina Zhang and Haigang Li
Estimating Plant Nitrogen Concentration of Maize Using a Leaf Fluorescence Sensor across Growth Stages
Reprinted from: *Remote Sens.* **2020**, *12*, 1139, doi:10.3390/rs12071139 1

Taifeng Dong, Jiali Shang, Jing M. Chen, Jiangui Liu, Budong Qian, Baoluo Ma, Malcolm J. Morrison, Chao Zhang, Yupeng Liu, Yichao Shi, Hui Pan and Guisheng Zhou
Assessment of Portable Chlorophyll Meters for Measuring Crop Leaf Chlorophyll Concentration
Reprinted from: *Remote Sens.* **2019**, *11*, 2706, doi:10.3390/rs11222706 23

Ke Zhang, Xiaokang Ge, Pengcheng Shen, Wanyu Li, Xiaojun Liu, Qiang Cao, Yan Zhu, Weixing Cao and Yongchao Tian
Predicting Rice Grain Yield Based on Dynamic Changes in Vegetation Indexes during Early to Mid-Growth Stages
Reprinted from: *Remote Sens.* **2019**, *11*, 387, doi:10.3390/rs11040387 43

Shanyu Huang, Yuxin Miao, Fei Yuan, Qiang Cao, Huichun Ye, Victoria I.S. Lenz-Wiedemann and Georg Bareth
In-Season Diagnosis of Rice Nitrogen Status Using Proximal Fluorescence Canopy Sensor at Different Growth Stages
Reprinted from: *Remote Sens.* **2019**, *11*, 1847, doi:10.3390/rs11161847 67

Romina de Souza, M. Teresa Peña-Fleitas, Rodney B. Thompson, Marisa Gallardo and Francisco M. Padilla
Assessing Performance of Vegetation Indices to Estimate Nitrogen Nutrition Index in Pepper
Reprinted from: *Remote Sens.* **2020**, *12*, 763, doi:10.3390/rs12050763 89

Cadan Cummings, Yuxin Miao, Gabriel Dias Paiao, Shujiang Kang and Fabián G. Fernández
Corn Nitrogen Status Diagnosis with an Innovative Multi-Parameter Crop Circle Phenom Sensing System
Reprinted from: *Remote Sens.* **2021**, *13*, 401, doi:10.3390/rs13030401 107

Min Jia, Jie Zhu, Chunchen Ma, Luis Alonso, Dong Li, Tao Cheng, Yongchao Tian, Yan Zhu, Xia Yao and Weixing Cao
Difference and Potential of the Upward and Downward Sun-Induced Chlorophyll Fluorescence on Detecting Leaf Nitrogen Concentration in Wheat
Reprinted from: *Remote Sens.* **2018**, *10*, 1315, doi:10.3390/rs10081315 125

Jian Yang, Shalei Song, Lin Du, Shuo Shi, Wei Gong, Jia Sun and Biwu Chen
Analyzing the Effect of Fluorescence Characteristics on Leaf Nitrogen Concentration Estimation
Reprinted from: *Remote Sens.* **2018**, *10*, 1402, doi:10.3390/rs10091402 143

Huichun Ye, Wenjiang Huang, Shanyu Huang, Bin Wu, Yingying Dong and Bei Cui
Remote Estimation of Nitrogen Vertical Distribution by Consideration of Maize Geometry Characteristics
Reprinted from: *Remote Sens.* **2018**, *10*, 1995, doi:10.3390/rs10121995 159

Bei Cui, Qianjun Zhao, Wenjiang Huang, Xiaoyu Song, Huichun Ye and Xianfeng Zhou
A New Integrated Vegetation Index for the Estimation of Winter Wheat Leaf Chlorophyll Content
Reprinted from: *Remote Sens.* **2019**, *11*, 974, doi:10.3390/rs11080974 . . . . **177**

Fenling Li, Li Wang, Jing Liu, Yuna Wang and Qingrui Chang
Evaluation of Leaf N Concentration in Winter Wheat Based on Discrete Wavelet Transform Analysis
Reprinted from: *Remote Sens.* **2019**, *11*, 1331, doi:10.3390/rs11111331 . . . . **195**

Ning Liu, Zizheng Xing, Ruomei Zhao, Lang Qiao, Minzan Li, Gang Liu and Hong Sun
Analysis of Chlorophyll Concentration in Potato Crop by Coupling Continuous Wavelet Transform and Spectral Variable Optimization
Reprinted from: *Remote Sens.* **2020**, *12*, 2826, doi:10.3390/rs12172826 . . . . **215**

Peng Zhou, Wei Yang, Minzan Li and Weichao Wang
A New Coupled Elimination Method of Soil Moisture and Particle Size Interferences on Predicting Soil Total Nitrogen Concentration through Discrete NIR Spectral Band Data
Reprinted from: *Remote Sens.* **2021**, *13*, 762, doi:10.3390/rs13040762 . . . . **237**

Nora Tilly and Georg Bareth
Estimating Nitrogen from Structural Crop Traits at Field Scale—A Novel Approach Versus Spectral Vegetation Indices
Reprinted from: *Remote Sens.* **2019**, *11*, 2066, doi:10.3390/rs11172066 . . . . **257**

Zhenhai Li, Xiuliang Jin, Guijun Yang, Jane Drummond, Hao Yang, Beth Clark, Zhenhong Li and Chunjiang Zhao
Remote Sensing of Leaf and Canopy Nitrogen Status in Winter Wheat (*Triticum aestivum* L.) Based on N-PROSAIL Model
Reprinted from: *Remote Sens.* **2018**, *10*, 1463, doi:10.3390/rs10091463 . . . . **281**

Jiale Jiang, Weidi Cai, Hengbiao Zheng, Tao Cheng, Yongchao Tian, Yan Zhu, Reza Ehsani, Yongqiang Hu, Qingsong Niu, Lijuan Gui and Xia Yao
Using Digital Cameras on an Unmanned Aerial Vehicle to Derive Optimum Color Vegetation Indices for Leaf Nitrogen Concentration Monitoring in Winter Wheat
Reprinted from: *Remote Sens.* **2019**, *11*, 2667, doi:10.3390/rs11222667 . . . . **299**

Hongjun Li, Yuming Zhang, Yuping Lei, Vita Antoniuk and Chunsheng Hu
Evaluating Different Non-Destructive Estimation Methods for Winter Wheat (*Triticum aestivum* L.) Nitrogen Status Based on Canopy Spectrum
Reprinted from: *Remote Sens.* **2020**, *12*, 95, doi:10.3390/rs12010095 . . . . **317**

Hengbiao Zheng, Tao Cheng, Dong Li, Xiang Zhou, Xia Yao, Yongchao Tian, Weixing Cao and Yan Zhu
Evaluation of RGB, Color-Infrared and Multispectral Images Acquired from Unmanned Aerial Systems for the Estimation of Nitrogen Accumulation in Rice
Reprinted from: *Remote Sens.* **2018**, *10*, 824, doi:10.3390/rs10060824 . . . . **333**

Lucas Prado Osco, Ana Paula Marques Ramos, Danilo Roberto Pereira, Érika Akemi Saito Moriya, Nilton Nobuhiro Imai, Edson Takashi Matsubara, Nayara Estrabis, Maurício de Souza, José Marcato Junior, Wesley Nunes Gonçalves, Jonathan Li, Veraldo Liesenberg and José Eduardo Creste
Predicting Canopy Nitrogen Content in Citrus-Trees Using Random Forest Algorithm Associated to Spectral Vegetation Indices from UAV-Imagery
Reprinted from: *Remote Sens.* **2019**, *11*, 2925, doi:10.3390/rs11242925 . . . . **351**

Hainie Zha, Yuxin Miao, Tiantian Wang, Yue Li, Jing Zhang, Weichao Sun, Zhengqi Feng and Krzysztof Kusnierek
Improving Unmanned Aerial Vehicle Remote Sensing-Based Rice Nitrogen Nutrition Index Prediction with Machine Learning
Reprinted from: *Remote Sens.* **2020**, *12*, 215, doi:10.3390/rs12020215 . . . . **369**

Lucas Prado Osco, José Marcato Junior, Ana Paula Marques Ramos, Danielle Elis Garcia Furuya, Dthenifer Cordeiro Santana, Larissa Pereira Ribeiro Teodoro, Wesley Nunes Gonçalves, Fábio Henrique Rojo Baio, Hemerson Pistori, Carlos Antonio da Silva Junior and Paulo Eduardo Teodoro
Leaf Nitrogen Concentration and Plant Height Prediction for Maize Using UAV-Based Multispectral Imagery and Machine Learning Techniques
Reprinted from: *Remote Sens.* **2020**, *12*, 3237, doi:10.3390/rs12193237 . . . . **391**

Ali Moghimi, Alireza Pourreza, German Zuniga-Ramirez, Larry E. Williams and Matthew W. Fidelibus
A Novel Machine Learning Approach to Estimate Grapevine Leaf Nitrogen Concentration Using Aerial Multispectral Imagery
Reprinted from: *Remote Sens.* **2020**, *12*, 3515, doi:10.3390/rs12213515 . . . . **409**

Xingang Xu, Lingling Fan, Zhenhai Li, Yang Meng, Haikuan Feng, Hao Yang and Bo Xu
Estimating Leaf Nitrogen Content in Corn Based on Information Fusion of Multiple-Sensor Imagery from UAV
Reprinted from: *Remote Sens.* **2021**, *13*, 340, doi:10.3390/rs13030340 . . . . **429**

Hengbiao Zheng, Wei Li, Jiale Jiang, Yong Liu, Tao Cheng, Yongchao Tian, Yan Zhu, Weixing Cao, Yu Zhang and Xia Yao
A Comparative Assessment of Different Modeling Algorithms for Estimating Leaf Nitrogen Content in Winter Wheat Using Multispectral Images from an Unmanned Aerial Vehicle
Reprinted from: *Remote Sens.* **2018**, *10*, 2026, doi:10.3390/rs10122026 . . . . **447**

Tyler J. Nigon, Ce Yang, Gabriel Dias Paiao, David J. Mulla, Joseph F. Knight and Fabián G. Fernández
Prediction of Early Season Nitrogen Uptake in Maize Using High-Resolution Aerial Hyperspectral Imagery
Reprinted from: *Remote Sens.* **2020**, *12*, 1234, doi:10.3390/rs12081234 . . . . **463**

Rajasheker R. Pullanagari, Gabor Kereszturi and Ian Yule
Integrating Airborne Hyperspectral, Topographic, and Soil Data for Estimating Pasture Quality Using Recursive Feature Elimination with Random Forest Regression
Reprinted from: *Remote Sens.* **2018**, *10*, 1117, doi:10.3390/rs10071117 . . . . **487**

James Brinkhoff, Brian W. Dunn, Andrew J. Robson, Tina S. Dunn and Remy L. Dehaan
Modelling Mid-Season Rice Nitrogen Uptake Using Multispectral Satellite Data
Reprinted from: *Remote Sens.* **2019**, *11*, 1837, doi:10.3390/rs11151837 . . . . **501**

Junjun Lu, Yuxin Miao, Wei Shi, Jingxin Li, Xiaoyi Hu, Zhichao Chen, Xinbing Wang and Krzysztof Kusnierek
Developing a Proximal Active Canopy Sensor-based Precision Nitrogen Management Strategy for High-Yielding Rice
Reprinted from: *Remote Sens.* **2020**, *12*, 1440, doi:10.3390/rs12091440 . . . . **523**

Beatriz Moreno-García, Mª Auxiliadora Casterad, Mónica Guillén and Dolores Quílez
Agronomic and Economic Potential of Vegetation Indices for Rice N Recommendations under Organic and Mineral Fertilization in Mediterranean Regions
Reprinted from: *Remote Sens.* **2018**, *10*, 1908, doi:10.3390/rs10121908 . . . . **545**

Marta Aranguren, Ander Castellón and Ana Aizpurua
Crop Sensor-Based In-Season Nitrogen Management of Wheat with Manure Application
Reprinted from: *Remote Sens.* **2019**, *11*, 1094, doi:10.3390/rs11091094 . . . . **567**

About the Editors

Yuxin Miao

Dr. Yuxin Miao is an Associate Director at the Precision Agriculture Center, and Associate Professor of Precision Agriculture and Nutrient Management in the Department of Soil, Water and Climate at University of Minnesota. He earned his Ph.D. in Soil Science with a specialization in Precision Agriculture from the University of Minnesota in 2005. From 2007 to 2017, Dr. Miao worked as an Associate Professor at China Agricultural University in Beijing. In 2017, he joined the University of Minnesota. He received the Pierre C. Robert Precision Agriculture Award from the International Society of Precision Agriculture in 2012, Outstanding Chinese Alumni Award from University of Minnesota in 2014, and the Kingenta Agricultural Science Award from the American Society of Agronomy in 2022. His research aim is to develop integrated precision nutrient and crop management systems for food security and sustainable development. A combination of field measurements, active canopy sensors, UAV and aerial as well satellite remote sensing, crop growth modeling, geospatial analysis, machine learning, and other new technologies are used to develop innovative nutrient and crop management solutions to improve crop yield, quality, nutrient use efficiency, economic returns and protect the environment. He is the founding leader of the Precision Nitrogen Management Community in International Society of Precision Agriculture. He has worked on different crops (wheat, corn, rice and potato) and different scales of farming systems. He has authored and coauthored over 100 publications. He serves as Associate Editor for several journals, including *Precision Agriculture*, *Remote Sensing*, and *Agronomy*.

Raj Khosla

Prof. Raj Khosla is the Head of the Agronomy Department at Kansas State University and is a globally recognized authority on Precision Agriculture. He has been engaged in precision agriculture since its inception and has made significant contributions in the development and spread of Precision Agriculture worldwide. He is the Founder and former President of the International Society of Precision Agriculture. Most recently, he served as a member of the National Academy of Science Executive Committee on Science Breakthrough 2030. In 2012, Dr. Khosla was named the Jefferson Science Fellow by the National Academy of Sciences and was appointed as the Senior Science Advisor on Food Security to the U.S. Department of State. In 2011, he was inducted by NASA to the US "Presidential Advisory Board on Positioning, Navigation and Timing". Prof. Khosla's research specializes in harnessing spatial and temporal heterogeneity in managed agro-ecosystems and translating those into better decision models. His group has extensively used remote sensing and other geo-spatial tools to enhance the production, resource use efficiency, profitability, and sustainability of managed agro-ecosystems. He has co-authored over 100 publications (book chapters, refereed journal articles, and others) and has been invited globally to over 30 countries for Keynotes at major international conferences, research collaborations and other invited talks. Prof. Khosla is a Fellow of American Society of Agronomy; Fellow of Soil Science Society of America; Fellow of Soil and Water Conservation Society and Honorary Life Fellow of International Society of Precision Agriculture.

David J. Mulla

Dr. Mulla received their Ph.D. degree in Agronomy from Purdue University with an emphasis on soil physics (1983). From 1983 to 1995, he was Assistant to Full Professor in the Dept. of Crop and Soil Sciences at Washington State University. Since 1995, he has been Professor and Larson Chair for Soil and Water Resources in the Dept. of Soil, Water, and Climate at the Univ. of Minnesota. Since January 2004, he has served as the Director of the Precision Agriculture Center at the University of Minnesota. Dr. Mulla's research emphasizes precision agriculture and precision conservation, land management and water quality interactions, ecosystem services, and international agricultural development. Dr. Mulla and his coauthors (including 40 MS and PhD students) have produced over 210 publications, and have received funding amounting to over $30 million. Dr. Mulla and his colleagues have extensive experience in applying geostatistics, remote sensing, machine learning and GIS terrain analysis in agricultural, biological and ecological systems, including applications for the development of targeted nutrient BMP placement and "management zones" in precision agriculture and precision conservation. Dr. Mulla's pioneering research on precision agriculture contributed significantly to its adoption in the US and around the world, fostering business opportunities, job growth and greater economic and environmental efficiency in agriculture. Dr. Mulla is an internationally recognized researcher and scholar. His research has taken him to over 20 countries. His peers elected him as a Fellow in the Soil Science Society of America (SSSA), and as a Fellow in the Agronomy Society of America. In 2012, he received the Pierre C. Robert Precision Agriculture Award from the International Society for Precision Agriculture. In 2013, he received the SSSA Soil Science Applied Research Award. He served as Associate and Technical Editor for the Soil Science Society of America Journal, and as Associate Editor for the *Journal of Precision Agriculture*.

Preface to "Remote Sensing for Precision Nitrogen Management"

Nitrogen is the most widely used macro nutrient in the world. Agriculture is a major source of N2O emissions and nitrate-nitrogen leaching. Precision nitrogen management aims to match nitrogen supply with crop nitrogen demand in both space and time to ensure high crop yields while increasing nitrogen use efficiency and protecting the environment. The effective and efficient non-destructive estimation and diagnosis of crop nitrogen status and in-season nitrogen recommendations is crucially important for the success of precision nitrogen management. Remote sensing is one of the key supporting technologies for precision nitrogen management and significant progress has been made in the last decade to develop proximal and remote sensing-based in-season crop nitrogen status diagnosis and recommendation strategies. To help readers keep up to date with the progress in the field, this Special Issue focuses on fundamental and applied research on the non-destructive estimation and diagnosis of crop leaf and plant nitrogen status and in-season nitrogen management strategies based on leaf sensors, proximal canopy sensors, unmanned aerial vehicle remote sensing, manned aerial remote sensing and satellite remote sensing technologies. Statistical and machine learning methods are used to predict plant-nitrogen-related parameters with sensor data or sensor data together with soil, landscape, weather and/or management information. Different sensing technologies or different modelling approaches are compared and evaluated. Strategies are developed to use crop sensing data for in-season nitrogen recommendations to improve nitrogen use efficiency and protect the environment. We greatly appreciate the support and contributions from all the authors and the editorial staff of Remote Sensing which made the development of this Special Issue a great success.

Yuxin Miao, Raj Khosla, and David J. Mulla
Editors

remote sensing

MDPI

Article

Estimating Plant Nitrogen Concentration of Maize Using a Leaf Fluorescence Sensor across Growth Stages

Rui Dong [1], Yuxin Miao [1,2,*], Xinbing Wang [1], Zhichao Chen [3], Fei Yuan [4], Weina Zhang [1] and Haigang Li [5]

1. College of Resources and Environment Sciences, China Agricultural University, Beijing 100193, China; BS20183030296@cau.edu.cn (R.D.); xbwang2020@cau.edu.cn (X.W.); bs20163030268@cau.edu.cn (W.A.)
2. Precision Agriculture Center, Department of Soil, Water and Climate, University of Minnesota, St. Paul, MN 55108, USA
3. School of Surveying and Land Information Engineering, Henan Polytechnic University, Jiaozuo 454000, China; czc@hpu.edu.cn
4. Department of Geography, Minnesota State University, Mankato, MN 56001, USA; fei.yuan@mnsu.edu
5. College of Grassland, Resources and Environment, Inner Mongolia Agricultural University, Hohhot 010011, Inner Mongolia, China; haigangli@cau.edu.cn

* Correspondence: ymiao@umn.edu

Received: 28 February 2020; Accepted: 30 March 2020; Published: 2 April 2020

Abstract: Nitrogen (N) is one of the most essential nutrients that can significantly affect crop grain yield and quality. The implementation of proximal and remote sensing technologies in precision agriculture has provided new opportunities for non-destructive and real-time diagnosis of crop N status and precision N management. Notably, leaf fluorescence sensors have shown high potential in the accurate estimation of plant N status. However, most studies using leaf fluorescence sensors have mainly focused on the estimation of leaf N concentration (LNC) rather than plant N concentration (PNC). The objectives of this study were to (1) determine the relationship of maize (Zea mays L.) LNC and PNC, (2) evaluate the main factors influencing the variations of leaf fluorescence sensor parameters, and (3) establish a general model to estimate PNC directly across growth stages. A leaf fluorescence sensor, Dualex 4, was used to test maize leaves with three different positions across four growth stages in two fields with different soil types, planting densities, and N application rates in Northeast China in 2016 and 2017. The results indicated that the total leaf N concentration (TLNC) and PNC had a strong correlation (R^2 = 0.91 to 0.98) with the single leaf N concentration (SLNC). The TLNC and PNC were affected by maize growth stage and N application rate but not the soil type. When used in combination with the days after sowing (DAS) parameter, modified Dualex 4 indices showed strong relationships with TLNC and PNC across growth stages. Both modified chlorophyll concentration (mChl) and modified N balance index (mNBI) were reliable predictors of PNC. Good results could be achieved by using information obtained only from the newly fully expanded leaves before the tasseling stage (VT) and the leaves above panicle at the VT stage to estimate PNC. It is concluded that when used together with DAS, the leaf fluorescence sensor (Dualex 4) can be used to reliably estimate maize PNC across growth stages.

Keywords: nitrogen status diagnosis; Dualex sensor; precision nitrogen management; leaf position; proximal sensing; nitrogen balance index

---

1. Introduction

Maize (*Zea mays* L.) is one of the three major grain crops in the world along with rice (*Oryza sativa* L.) and wheat (*Triticum aestivum* L.) [1]. Nitrogen (N) is one of the most essential nutrients that significantly affect maize yield, biomass, and grain quality. However, over-application of N is common in Chinese agriculture, resulting in many environmental problems [2,3]. Precision N management strategies aiming to apply the optimal amount of N fertilizer at the right time and place can help increase crop yield and N use efficiency, as well as reduce N surplus and environmental impacts [4–6].

The development and implementation of proximal and remote sensing technologies has provided new opportunities for non-destructive and real-time crop N status estimation on different scales [7–9]. Since leaf pigment concentrations, especially the chlorophyll concentration (Chl), can affect leaf reflectance properties and are highly correlated with N status, many optical sensors have been used to measure the canopy reflectance of a specific area or the entire crop field to estimate plant N status [7,10–12]. However, reflectance sensors are usually influenced by light conditions, soil and water background, and often saturate under high biomass conditions. Proximal fluorescence sensing is insensitive to soil backgrounds, environmental light, or biomass conditions and may overcome some of the problems in N status diagnosis [7,13]. As a traditional and standard indicator of plant N status, plant N concentration (PNC) is closely related to yield performance [14–16]. Therefore, PNC has been widely estimated through various sensing methods, and used as a reference in different N diagnostic methods, with critical PNC values established for different crops and growth stages [7,17,18]. Most importantly, unlike reflectance indices, fluorescence signals have stronger relationships with PNC as they are mainly affected by leaf Chl concentration but not by biomass or leaf area index (LAI) [13,19–21].

Dualex 4 (Force-A, Orsay, France) is a portable leaf fluorescence sensor and has been used in the past few years to monitor crop physiology and study N status diagnosis [22]. Apart from measuring Chl concentration through leaf transmittance, Dualex 4 can also measure leaf epidermal flavonoids (Flav) by comparing the Chl fluorescence induced by ultra-violet (UV) excitation at 375 nm to that induced by red light at 650 nm wavelength [23–25], and provides a new Chl/Flav ratio called N balance index (NBI). Numerous recent studies have focused on utilizing Dualex 4 to estimate N status in a variety of crops, and have revealed a significant relationship between Dualex 4 readings and N indicators. For example, Dualex 4-based Chl readings were found to be highly related to leaf Chl concentrations in four crops including corn, soybean (*Glycine max* L. Merr.), spring wheat (*Triticum aestivum* L.), and canola (*Brassica napus* L.) (R^2 = 0.69–0.90) [26]. Cartelat et al. [27] showed a strong linear relationship between phenolics (Phen) measured by Dualex sensor and leaf N concentration (LNC) (R^2 = 0.76) and further displayed the correlation between Phen and N nutrition index (NNI) (r = −0.60) for wheat. It was found that NBI calculated using a chlorophyll metercombined with an older version Dualex sensor could predict PNC and NNI most accurately for muskmelon (*Cucumis melo* L. cv. Tezac) (R^2 = 0.79–0.93 and 0.80–0.95) [28].

However, it is sometimes difficult to successfully use various proximal and remote sensors, including fluorescence sensors, because spectral data may vary due to different factors, such as soil conditions, crop growth stages, and leaf positions [7,9,21,29–32]. The variations of different soils in terms of quality, water, nutrition, and temperature often affect crop growth and lead to changes in crop properties [33–36]. Stress events resulting from environmental situations may eventually change the content of some compounds like leaf pigments [37–39]. This could modify leaf optical and fluorescence properties and be monitored by proximal or remote sensing technologies.

It has been a great challenge to use proximal and remote sensing technologies to reliably estimate PNC across growth stages [40]. The active canopy sensor GreenSeeker-based vegetation indices (VIs) could be used to predict LAI and aboveground biomass well (R^2 = 0.83–0.89), but had a poor performance for PNC estimation (R^2 = 0.47) across V5–V10 growth stages for spring maize in Northeast China [41]. Different VIs and prediction models will be needed to estimate PNC at different growth stages [40,42]. PNC can be more reliably estimated at later growth stages, but at early growth stages before crop canopy closure, the performance of most prediction models has been quite poor, even with hyperspectral remote sensing data [40,42]. Gabriel et al. [43] compared two leaf chlorophyll sensors (SPAD-502 and Dualex) for estimating maize LNC on two different dates, and demonstrated similar performance of the two sensors, with R^2 = 0.43 and 0.62 for SPAD and R^2 = 0.42 and 0.68 for Dualex.

Tremblay et al. [29] found that Dualex could be used to diagnose maize N status successfully within 21 days after topdressing but lost sensitivity at later stages. A hand-held canopy fluorescence sensor Multiplex has been used to detect N status in early growth stages of maize and it was proven that the Multiplex parameters were strongly influenced by N dose [20,44]. Moreover, strong relationships between fluorescence indices and N indicators of rice at different growth stages were revealed by Huang et al. [45]. To overcome the influence of growth stage or other factors, N sufficiency index (NSI) or response index (RI) is generally calculated by using a well-fertilized area as the reference to diagnose crop N status [41,45,46], but such approach will require a well-fertilized area or N rich plot, and generally cannot improve the prediction of PNC across growth stages [45]. Therefore, more studies are still needed to develop methods for reliable prediction of PNC across growth stages.

Several studies have reported the use of fluorescence sensing technology to evaluate crop N status. For example, Yang et al. [47] showed that there were consistent positive correlations between fluorescence parameters and LNC for different rice cultivars with R^2 varying from 0.70 to 0.90. Another study demonstrated that NBI and NBI1 based on the Multiplex 2 fluorescence sensor were linearly related to LNC with a high coefficient of determination for two turfgrass cultivars (R^2 = 0.85–0.87 and R^2 = 0.75–0.78, respectively) [48]. A study using a Dualex sensor calibrated three optical indices (Chl, Flav and NBI) against LNC of grapevine (*Vitis vinifera* L.) and verified that NBI was optimal for estimating LNC with a root-mean-square error (RMSE) smaller than 2 mg of N g^{-1} dry weight. The threshold values of NBI from 11 to 18 at flowering and 8 to 11 at bunch closure were proposed [17]. Recently, Zhang et al. [49] found Dualex 4 sensor parameters were significantly correlated with rice PNC at different growth stages (R^2 = 0.43–0.77) or across growth stages (R^2 = 0.52–0.69).

To date, few studies have reported on how to use leaf fluorescence sensors to accurately estimate crop PNC, especially developing strategies to effectively overcome the influence of different growth stages. Therefore, the objectives of this study were to (1) determine the relationship of maize LNC and PNC, (2) evaluate the main factors influencing the variations of leaf fluorescence sensor parameters, and (3) develop a practical strategy to reliably estimate maize PNC using leaf fluorescence sensor across different growth stages.

2. Materials and Methods

2.1. Study Site and Soil Description

The study was conducted in Lishu County (43°02′–43°46′N, 123°45′–124°53′E), Jilin Province in Northeast China from 2016 to 2017. Located in North Temperate Zone with four distinct seasons, this region has a semi-humid continental monsoon climate. The mean annual average temperature is 6.6 °C, and the annual average precipitation is 556 mm, about 80% of which occurs during the crop growing season from May to September. Figure 1 shows the precipitation distribution and mean temperature during the growing season in 2016–2017 in Lishu.

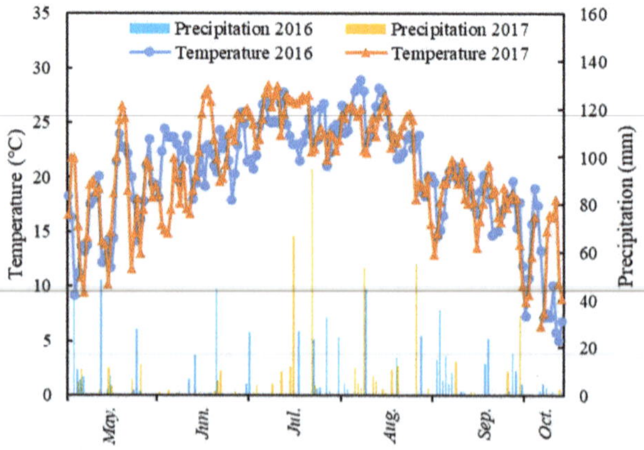

Figure 1. Daily precipitation (mm) and mean temperature (°C) during the growing season at the study site in 2016 and 2017.

Two sites with different soil types were selected for this study. Site 1 has Aeolian sandy soil (loamy sand), which is classified as typic Cryopsamments. Site 2 has Black soil (loamy clay), which is classified as typic Haploboroll in the United States Department of Agriculture (USDA) Soil Taxonomy.

2.2. Experimental Design

All the treatments of the experiment were the same in both fields. Using a split-plot design with three replications, a local maize cultivar Liangyu 66 was sown in early May each year with planting densities of 55,000, 70,000 and 85,000 plant ha^{-1} as the main plots, and six N treatments were established as the subplots, with total N doses of 0, 60, 120, 180, 240, and 300 kg ha^{-1}, respectively. The N fertilizers were applied in two split applications: 1/3 was broadcasted and incorporated into the soil with rotary tillage as basal N using ammonium sulfate before sowing, and 2/3 was banded as side-dress N using urea at around V8–V9 growth stage in early July. The subplots in a wide-narrow row planting pattern were 9 × 12 m with 1 m wide alley between the subplots. Sufficient phosphate (90 kg P_2O_5 ha^{-1}) and potash (90 kg K_2O ha^{-1}) fertilizers were applied before sowing to make sure P and K nutrients were not limiting for each plot.

There was no irrigation at Site 2, while about 50 mm of water was irrigated into the soil around mid-July at Site 1 in each year due to water stress in sandy soil. All plots were kept free of weeds, insects, and diseases with pesticides based on local standard practices. Detailed information about the experiments conducted in this study in 2016 and 2017 is listed in Table 1.

Table 1. Experimental set-up, plant sampling, and sensing stages in the experiment conducted in 2016 and 2017.

Site	Planting Date	Side Dressing Date	Harvest Date	Irrigation Date	Sampling and Sensing Stage
2016					
Site 1	May 7th	Jul. 3rd (57 DAS)	Oct. 6th	Jul. 13–16th (70–73 DAS)	V8 (49 DAS *, 50 DAS), V12 (70 DAS *, 73 DAS), VT (78 DAS *, 81 DAS)
Site 2	May. 5th	Jul. 4th (60 DAS)	Sep. 29th	No irrigation	V8 (50 DAS, 51 DAS *), V13 (72 DAS *), VT (78 DAS, 80 DAS *)
2017					
Site 1	May. 4th	Jul. 3rd (60 DAS)	Oct. 3rd	Jul. 11–13th (69–71 DAS)	V4 (30 DAS *), V6 (40 DAS), V8 (56 DAS), V11 (65 DAS *), VT (84 DAS, 86 DAS *)
Site 2	May. 3th	Jul. 2nd (60 DAS)	Oct. 2nd	No irrigation	V4 (29 DAS *), V6 (38 DAS), V8 (52 DAS), V11 (64 DAS *), VT (83 DAS, 85 DAS *)

* Data acquired at these stages were used to plot the relationships between SLNC and TLNC or PNC.

2.3. Dualex 4 Sensor Data Collection, Plant Sampling, and Measurements

Three representative plants located in the center rows of each plot of six N treatments at the 70,000 plants ha^{-1} density in the two fields were selected to be cut at the ground level at V8, V12, and VT in 2016, and V4, V11, and VT in 2017. Each single leaf was separated from the whole plant, and leaves in the same position of the three plants in each plot were mixed together. The stems were also mixed together.

The Dualex 4 sensor (Force-A, Orsay, France) was used in this study for proximal sensing. Three representative plants located in the inner rows of each plot of six N treatments at three densities were selected to be sampled and measured by the sensor. In particular, this sensor measures a leaf surface area of 20 mm^2. The plant samples and sensor readings were obtained at V6, V8, V12, and VT growth stages in 2016 and at V6, V8, and VT growth stages in 2017 in each field. Dualex 4 values were measured at around the leaf blade midpoint to avoid midribs or physical damage on the adaxial side (upper side) of the uppermost, second, and third fully expanded leaves before VT stage, as well as the leaf above the panicle, panicle leaf, and the leaf below the panicle at the VT growth stage for each plant in each plot. After sensing, all plant samples in each plot were separated into leaves and stems except for the V6 growth stage.

All samples were oven-dried at 105 °C for 30 min, then dried at 70 °C to a constant weight, and ground into fine powders to determine N concentration using a modified Kjeldahl digestion method [50]. The total NLC (TLNC) is the sum of the product of the N concentration of each leaf and its proportional weight, while PNC is calculated by adding the product of the N concentration of each organ and its proportional weight. As the leaves were the main component of the plant at the V6 growth stage, the TLNC at this stage was also used as PNC in this study.

The data acquisition dates and days after sowing (DAS) are shown in Table 1. For the convenience of discussion, the first, second, and third leaves counted from the top of the maize plant selected for sensor measurement in this research were abbreviated as Leaf 1, Leaf 2, and Leaf 3, respectively. The three Dualex 4 parameters (Chl, Flav, and NBI) at the same leaf position of three plants in each plot were averaged and used as the mean reading of that leaf position for the plot. The details of the Dualex 4 parameters are shown in Table 2.

Table 2. The details of Dualex 4-based parameters.

Parameters	Abbreviation	Algorithm
Chlorophyll	Chl	FRF_R/RF_R
Flavonoids	Flav	Log (FRF_R/FRF_{UV})
Nitrogen balance index	NBI	Chl/Flav
Modified chlorophyll	mChl	Chl/DAS
Modified flavonoids	mFlav	Flav × DAS
Modified nitrogen balance index	mNBI	NBI/DAS^2

2.4. Statistical Analysis

The data collected were pooled together to establish the relationships between single LNC (SLNC) and TLNC and PNC. In addition, the data collected by fluorescence sensing for each site, year, growth stage, and N rate were divided into a calibration dataset (two replications of the observations) and a validation dataset (one replication of the observations). The mean, standard deviation (SD), coefficient of variation (CV, %) of SLNC, TLNC, and PNC were calculated using Microsoft Excel (Microsoft Corporation, Redmond, WA, USA).

Besides three raw Dualex 4 parameters, three modified parameters derived from them based on DAS were used in this study: modified Chl (mChl), modified Flav (mFlav), and modified NBI (mNBI) were computed as the ratio of Chl to DAS, the product of Flav and DAS, and the ratio of NBI to the square of DAS (Table 2). The coefficient of determination (R^2) values for all relationships were calculated using SPSS 21.0 (SPSS Inc., Chicago, IL, USA), and the model with the highest R^2 between the Dualex parameters and the two N indices (TLNC and PNC) was selected and listed in this paper, and was used for further estimation of PNC. The performance of the established relationship models for predicting PNC was evaluated by comparing R^2, RMSE, and relative error (RE). The higher the R^2 and the lower the RMSE and RE, the higher the accuracy of the prediction models. Agronomic and proximal sensing data were subjected to the least significant difference (LSD) test at a 5% significance level to assess differences between the means of treatments using SPSS 21.0.

3. Results

3.1. Interrelationships of SLNC, TLNC, and PNC

We plotted SLNC against TLNC and PNC in Figure 2. The SLNC was highly related to both TLNC and PNC with R^2 varying from 0.91 to 0.98. The best relationships appeared in the results for Leaf 1 (R^2 = 0.98 and 0.96), followed by Leaf 2 and finally Leaf 3. As the sum of each single leaf, the TLNC was more related to SLNC than that of the whole plant, because the PNC was also affected by the stem N concentration. However, the strong correlations of TLNC and PNC to SLNC indicated the high probability to use information acquired from a single leaf, especially Leaf 1, to understand the N status of maize through the prediction of TLNC or PNC.

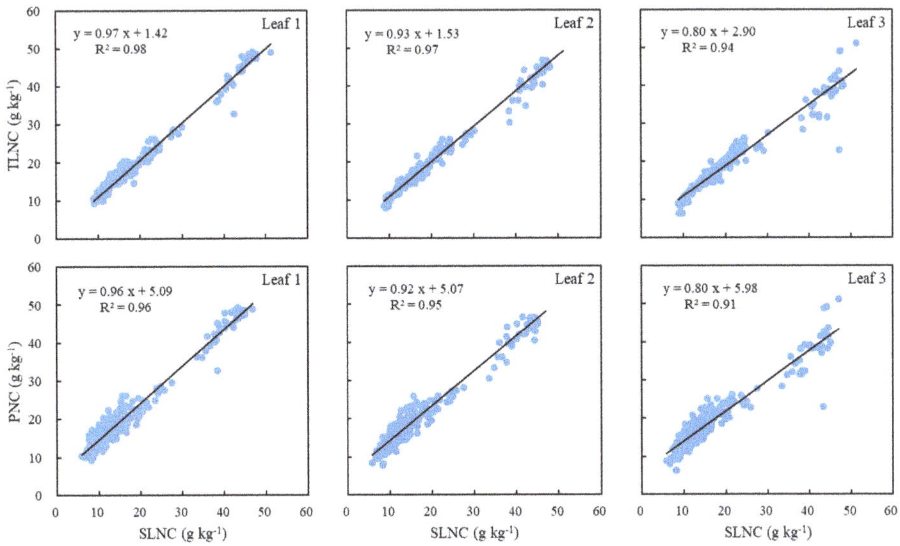

Figure 2. Relationships between single leaf N concentration (SLNC) obtained from three different leaves of a plant and total leaf N concentration (TLNC) and plant N concentration (PNC).

3.2. *Effects of Soil Type, Growth Stage, and N Rate on Maize TLNC and PNC*

Figure 3 indicated that PNC was slightly lower than TLNC, and both were significantly affected by the growth stage and N rate. Different soil types only resulted in different TLNC values (mean = 22.91 and 23.93 g kg^{-1} for Aeolian sandy soil field (Site 1) and Black soil field (Site 2), respectively). For PNC, there was no significant difference between the two fields, with 20.07 and 20.69 g kg^{-1} for Aeolian sandy soil field (Site 1) and Black soil field (Site 2), respectively (Figure 3a). In contrast, TLNC of four growth stages differed significantly, from 17.06 g kg^{-1} at the V12 growth stage to 31.03 g kg^{-1} at the V6 growth stage, while PNC ranged from 12.10 g kg^{-1} at the VT growth stage to 31.28 g kg^{-1} at the V6 growth stage, without a significant difference between the V12 and VT growth stages (Figure 3b). The mean values of TLNC and PNC increased with N rates from the lowest under 0 kg ha^{-1} N rate treatment (18.10 g kg^{-1} for TLNC and 16.25 g kg^{-1} for PNC) to the highest under 300 kg ha^{-1} N rate treatment (27.09 g kg^{-1} for TLNC and 23.44 g kg^{-1} for PNC). The values of TLNC were significantly different under low N treatments (0 and 60 kg ha^{-1}), but not under high N treatments (240 and 300 kg ha^{-1}). However, the values of PNC were not significantly different under either low or high N treatments (Figure 3c).

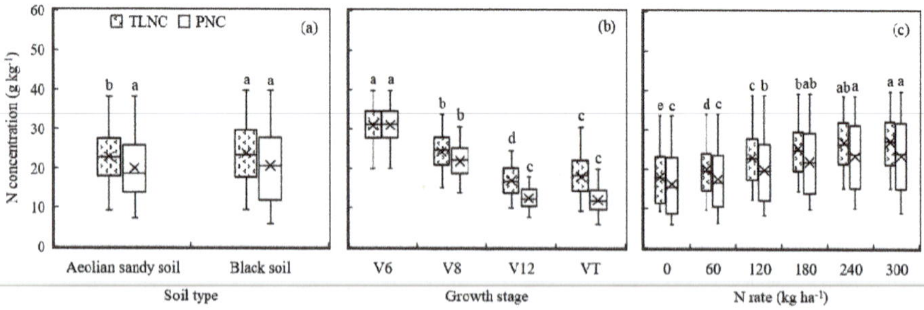

Figure 3. Total leaf N concentration (TLNC) and plant N concentration (PNC) as affected by different soil types (**a**), growth stages (**b**), and N rates (**c**). The different letters above the boxes indicate that the TLNC and PNC values differed significantly according to the least significant difference test at $p \leq 0.05$.

3.3. Effects of Soil Type, Growth Stage, and N Rate on Dualex 4 Parameters

The differences in Chl, Flav, and NBI measured with a Dualex 4 sensor as influenced by soil types, growth stages, and N rates were analyzed and displayed in Figure 4. The Dualex 4 sensor-based fluorescence parameter values obtained from the three differently positioned leaves did not exhibit significant variability under the two soil types, except for Chl values of Leaf 2, with 41.70 for Aeolian sandy soil field (Site 1) and 43.14 for Black soil field (Site 2) (Figure 4a,d,g). Fluorescence parameter values changed less significantly with growth stages than TLNC and PNC (Figures 3b and 4b,e,h). For Chl, there was no consistent trend as maize grew, although there was a significant difference at late growth stages for Leaf 1 and at early stages for Leaf 2 and Leaf 3 (Figure 4b). Flav values showed a constant decreasing trend with the growth of maize, similar to the changes of TLNC and PNC, but with less variation. Comparatively, the variation trend of Leaf 1 was the same as that of PNC, while the values of the other two leaves changed significantly with growth stages (Figures 3b and 4e). NBI also showed significant differences among different growth stages, particularly for Leaf 2 and Leaf 3, but NBI and TLNC and PNC exhibited opposite trends from Flav, increasing with the increase of DAS (Figures 3b and 4h). Besides, the Chl and NBI values increased with N rate up to 240 kg ha^{-1} for all three differently positioned leaves (Figure 4c,i). There was a negative relationship between Flav and N rate, with higher levels of Flav found in low N treatments and lower levels of Flav found in high N treatments. However, Flav did not vary much from 180 to 300 kg ha^{-1} N rates (Figure 4f). Moreover, the readings of the three fluorescence parameters of the three differently positioned leaves were very similar in different fields and under different growth stages and N rates (Figure 4).

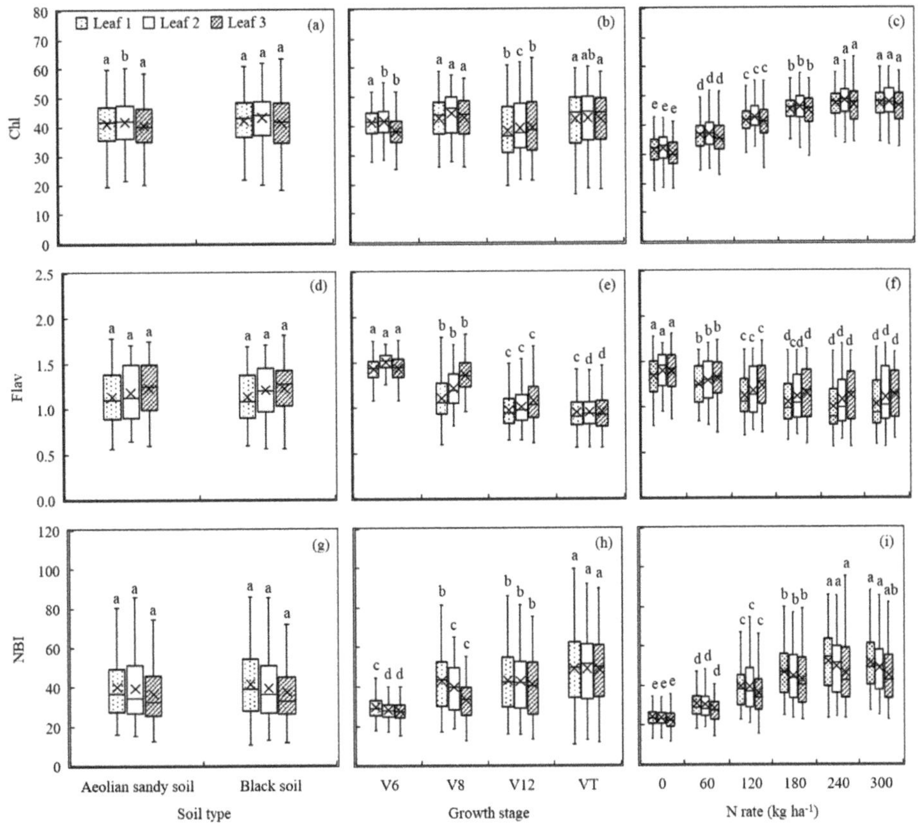

Figure 4. Dualex 4 parameters (chlorophyll concentration (Chl), leaf epidermal flavonoids (Flav), and N balance index (NBI)) as affected by different soil types, growth stages, and N rates. The different letters above the boxes indicate that the Dualex 4 parameter values differed significantly according to the least significant difference test at $p \leq 0.05$.

The results above indicated that growth stage was the main factor that influences the general model's estimation of maize N concentration. To overcome the problem of the high inconsistency between the Dualex 4 parameters and the TLNC and PNC across growth stages, the modified Dualex 4 parameters incorporating the information of DAS were computed and the variations of these new parameters are displayed in Figure 5. It was evident that changes in these new parameters of differently positioned leaves varied markedly among different growth stages after combining with DAS, although a reverse trend was observed in mFlav compared with the original Flav. The mNBI showed a similar trend as mChl with the growth stage.

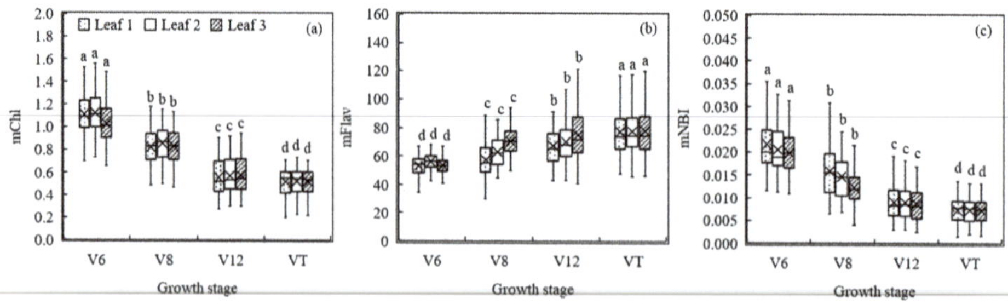

Figure 5. Modified Dualex 4 parameters (modified Chl (mChl) (**a**), modified Falv (mFlav) (**b**), and modified (mNBI)) (**c**) as affected by growth stage. Different letters above the boxes indicate that the modified Dualex 4 parameter values differed significantly according to the least significant difference test at $p \leq 0.05$.

3.4. Relationships between Dualex 4 Parameters and TLNC or PNC

All data acquired from the two study sites across growth stages were gathered together and divided into calibration and validation datasets to establish a general model for PNC estimation (Table 3).

Table 3. Descriptive statistics for TLNC and PNC for calibration and validation datasets.

Dataset	TLNC (g kg^{-1})			PNC (g kg^{-1})		
	Mean	**SD**	**CV (%)**	**Mean**	**SD**	**CV (%)**
Calibration dataset $n = 504$	23.59	7.24	31	20.61	8.73	42
Validation dataset $n = 252$	23.08	7.07	31	19.91	8.68	44

The mean values of TLNC (23.08–23.59 g kg^{-1}) were larger than those of PNC (19.91–20.61 g kg^{-1}) in both datasets. For the calibration dataset, across all growth stages, the PNC was more variable (CV = 42%) than the TLNC (CV = 31%), and the validation dataset had similar variability as the calibration dataset with a CV of 44% for PNC and of 31% for TLNC.

The original Dualex 4 readings, Chl, Flav, and NBI, demonstrated poor relationships with TLNC or PNC across growth stages, with R^2 ranging from 0.01 to 0.34 (Table 4). The mean values of Chl, Flav, and NBI for the three differently positioned leaves were calculated and their relationships with TLNC and PNC at four growth stages are displayed in Figure 6. The impact of growth stages on these relationships were significant.

Table 4. Coefficients of determination (R^2) for the relationships between Dualex 4-based parameters of three differently positioned leaves and two N status indicators (TLNC and PNC).

N Concentration (g kg^{-1})	Leaf Position	Chl		Flav		NBI		mChl		mFlav		mNBI	
		Model	**R^2**	**Model**	**R^2**	**Model**	**R^2**	**Model**	**R^2**	**Model**	**R^2**	**Model**	**R^2**
TLNC	Leaf 1	Q	0.34	Q	0.11	P	0.03	P	0.80	P	0.53	P	0.77
	Leaf 2	Q	0.34	Q	0.16	P	0.01	P	0.78	P	0.49	P	0.76
	Leaf 3	P	0.18	Q	0.10	Q	0.01	P	0.75	E	0.50	P	0.74
PNC	Leaf 1	Q	0.20	Q	0.22	Q	0.03	Q	0.84	P	0.56	P	0.80
	Leaf 2	Q	0.18	Q	0.33	Q	0.07	P	0.83	P	0.49	P	0.79
	Leaf 3	Q	0.10	Q	0.25	Q	0.09	P	0.78	P	0.50	P	0.75

Note: Q: Quadratic model; P: Power model; E: Exponential model.

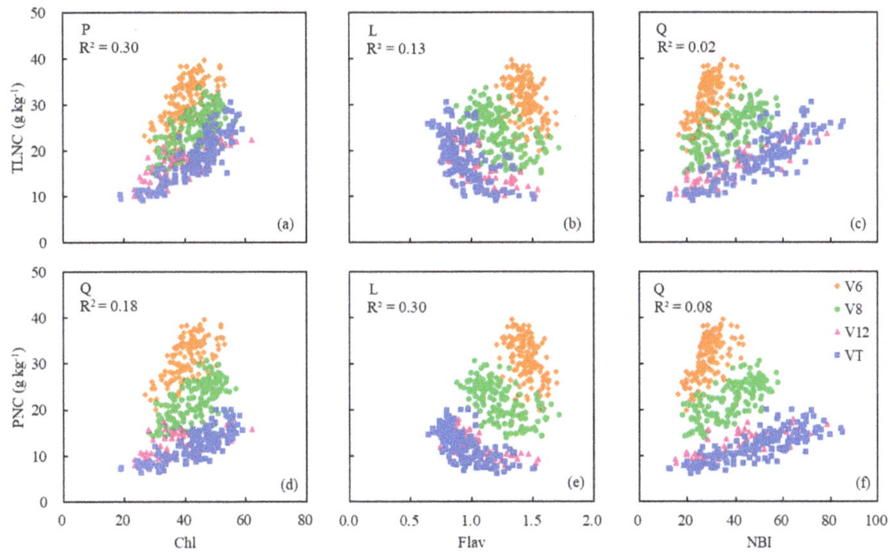

Figure 6. Relationships between original Dualex 4 parameters (Chl (**a**,**d**), Flav (**b**,**e**), and NBI (**c**,**f**)) for the mean values of three differently positioned leaves and two N status indicators (TLNC (**a**–**c**)) and PNC (**d**–**f**)) across growth stages. The models with the highest R^2 are displayed, with L (**e**), P (**a**), and Q (**d**) indicating linear, power, and quadratic models, respectively.

However, after combining with DAS, which was directly connected to the growth stage, the modified Dualex 4 parameters (mChl, mFlav, and mNBI) were highly related to TLNC and PNC (Table 4 and Figure 7). As maize grew, the values of mChl and mNBI decreased with the increase of TLNC and PNC (Figure 7a,c,d,f), while the values of mFlav increased with the growth stages and exhibited negative relationships with both TLNC and PNC accordingly (Figure 7b,e).

The mChl was most related to TLNC and PNC with R^2 ranging from 0.75 to 0.84 for single-position leaves. mNBI also performed very well, with R^2 ranging from 0.74 to 0.80. Nevertheless, the result for mFlav (R^2 = 0.49–0.56) was not as good as the above-mentioned two parameters, but it was much better than that of the original Flav (Table 4). A power function relationship was found for the models established based on the values measured on every single leaf for mChl, mFlav, and mNBI. Leaf 1 showed the best potential to estimate PNC with the highest R^2 values (Table 4). Moreover, by calculating the mean values of each parameter of the three leaves, the correlations between the modified Dualex 4 parameters and TLNC or PNC were improved to a certain extent, especially for mFlav (Figure 7b,e and Table 4). Compared with TLNC, PNC was more related to the modified Dualex 4 parameters (mChl, mFlav, and mNBI). This indicated that the modified Dualex 4 parameters had greater potential for direct estimation of PNC accurately (Figure 7 and Table 4).

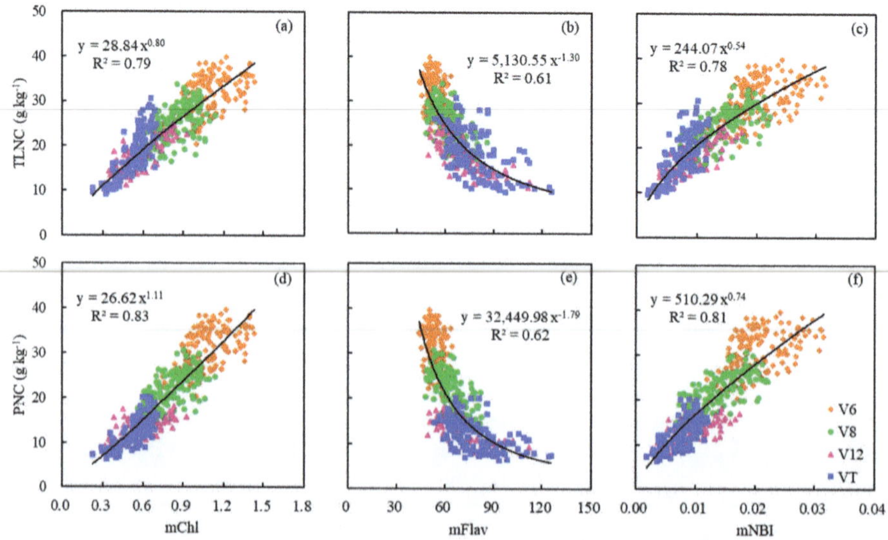

Figure 7. Relationships between modified Dualex 4 parameters (mChl, mFlav, and mNBI) for the mean values of three differently positioned leaves and two N status indicators (TLNC and PNC) across growth stages.

3.5. The Estimation of PNC Using Different Modified Dualex 4 Parameters

In order to establish good relationships between modified Dualex 4 parameters and PNC as well as TLNC, the above models describing their relationships were further evaluated with the validation dataset. The validation results of mChl for estimating TLNC and PNC were the best with the highest R^2 (0.71–0.79 and 0.73–0.84) and the lowest RMSE (3.27–0.81 and 3.46–4.48) and RE (14%–16% and 17–23%) (Table 5). Although the mFlav showed a better relationship with TLNC and PNC than the original Flav, it still did not perform as well as the other two modified Dualex parameters in predicting TLNC and PNC, with the lowest R^2 (0.42–0.53 and 0.40–0.53) and the highest RMSE (4.83–5.38 and 5.96–6.73) and RE (21%–23% and 30%–34%) (Table 5). For single leaf, mChl obtained from Leaf 1 predicted TLNC and PNC most accurately, while mFlav and mNBI obtained from Leaf 3 showed the highest accuracy among the three positioned leaves (Table 5). Besides, the results were further improved by averaging the mFlav and mNBI readings measured from three differently positioned leaves (Table 5 and Figure 8b,c,e,f). The data distribution for the V6 growth stage deviated from the 1:1 line more significantly than the other growth stages, which would influence the estimation results to some extent (Figure 8). Above all, the modified Dualex 4 parameters showed a better performance for direct PNC estimation across growth stages, but not for TLNC estimation, which implied that it was needless to estimate maize PNC by estimating TLNC using sensor readings acquired primarily from single leaves.

Table 5. Validation results for the estimation of TLNC and PNC using modified Dualex 4 parameters for single-position leaves by the general estimation models across growth stages.

N Concentration (g kg^{-1})	Leaf Position	mChl			mFlav			mNBI		
		R^2	RMSE	RE(%)	R^2	RMSE	RE(%)	R^2	RMSE	RE(%)
TLNC	Leaf 1	0.79	3.27	14	0.45	5.22	23	0.72	3.76	16
	Leaf 2	0.76	3.45	15	0.42	5.38	23	0.71	3.77	16
	Leaf 3	0.71	3.81	16	0.53	4.83	21	0.73	3.65	16
PNC	Leaf 1	0.84	3.46	17	0.41	6.66	33	0.73	4.52	23
	Leaf 2	0.82	3.68	18	0.40	6.73	34	0.74	4.37	22
	Leaf 3	0.73	4.48	23	0.53	5.96	30	0.75	4.29	22

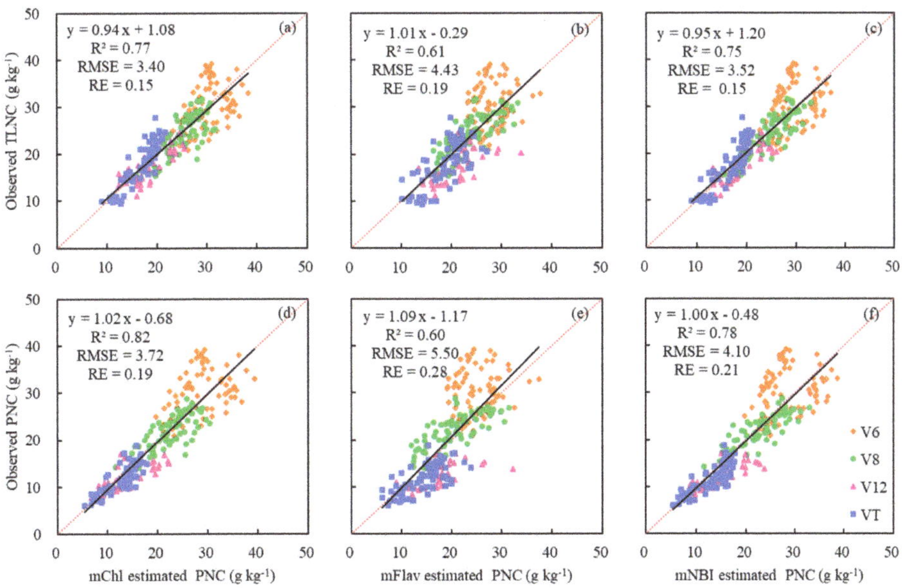

Figure 8. Validation results for the prediction of TLNC (**a–c**) and PNC (**d–f**) using the mean values of modified Dualex 4 parameters for three single-position leaves. Black and red lines indicate the regression and the 1:1 line, respectively.

4. Discussion

4.1. Feasibility of Estimating Maize N Status Using Single Leaf-based Dualex 4 Parameters

The SLNC of the fully expanded three leaves selected in our study was linearly related to both TLNC and PNC, with high R^2 values. Considering the good relationships between SLNC and TLNC and PNC (Figure 2), and the significant change trend of Dualex 4 parameters with the increase of N rate (Figures 3c and 4c,f,i), this study verified that it would be possible to predict maize N status using the information from a single leaf of maize and be possible to predict PNC directly or indirectly by predicting TLNC first.

The Chl values elevated with an increasing N rate (Figure 4c), which conforms to the finding that leaf chlorophyll is strongly affected by leaf N [51–53]. Leaf N contributed a large proportion to plant photosynthesis [54]. However, as an indicator of epidermal flavonoids, a carbon-based secondary metabolite, Flav exhibited a reverse trend with the changes of N rate (Figure 4f) because the carbon-based metabolite would be produced under low N level and would consequently cause the enhanced synthesis of flavonoids [55,56]. As the ratio of Chl to Flav, NBI increased with N rate

(Figure 4i) and has been considered to be a sensitive indicator of crop N status [13,28,44,57]. The results of this study indeed showed that the range of NBI variation was greater than that of Chl and Flav under different N rates (Figure 4c,f,i).

The Dualex 4 parameters' good discrimination ability of different N rates and the saturation at high N levels (240 and 300 kg ha^{-1}) (Figure 4) was in agreement with other studies using the fluorescence-based method [58]. The suggested optimal N rate for the experimental sites in our study was approximately 180–200 kg ha^{-1} [59], so the N rates of 240 and 300 kg ha^{-1} exceeded the optimal level. Some previous studies have concluded that fluorescence indices were insensitive to high crop N content in the range of optimal to excessive N levels [58,60]. In this study, as shown in Figure 3c, both TLNC and PNC varied clearly with the changes of N rate from 0 through 240 kg ha^{-1}, but there was no significant difference between 240 and 300 kg ha^{-1} N rates. Thus, in this study, the saturation of Dualex 4 parameters on high N rates was closely related to the N status of maize, and we could not draw a similar conclusion that the fluorescence indices were insensitive to high N content here.

4.2. Main Factor(s) Affecting the Establishment of the General Model and the Best Parameter(s) for PNC Estimation

Compared with stems, leaves are more important for photosynthesis. However, maize plants grown in the Aeolian sandy soil field (Site 1) were much more vulnerable to water deficiency than maize plants grown in the Black soil field (Site 2). Due to the N accumulation in stems, the PNC obtained in the Aeolian sandy soil field increased when the N concentration of stems was taken into account. Therefore, the difference in PNC between the two fields vanished (Figure 3a). The significant variation observed in TLNC values but the similarity observed in PNC values in the two study sites (Figure 3a) implies that it may be more difficult to evaluate TLNC from different soil fields, and it would be more feasible to establish a PNC predicting model neglecting the effect of soil types. Nevertheless, it is necessary to distinguish soil types when estimating PNC indirectly by estimating TLNC first (Figure 4a,d,g).

The changes of TLNC or PNC as affected by growth stages (Figure 3b) have been verified in diverse crops in other studies [28,61,62]. This is why growth stages must be strictly defined and distinguished for N status evaluation in most studies. Thus, the growth stage plays an important role in establishing general models for PNC estimation according to the findings of this study.

Chl measured by Dualex 4 has been calibrated in μg cm^{-2} units [22], which means Chl is a proxy of surface-based N. However, the TLNC and PNC mentioned in this study were both calculated based on mass. The area-based LNC changes very little during the growing season due to the expression of area-based LNC forces changes in specific leaf areas [63]. The time we selected to conduct the sensing was before the reproductive phase, and there was no obvious redistribution of N from the leaves to the ears that would result in the significant reduction of LNC. Furthermore, the Dualex 4 only measures a green leaf surface area of 20 mm^2 that would not exhibit too much variation of N concentration. Thus, the surface-based parameter Chl is relatively more stable than the mass-based TLNC and PNC (Figures 3b and 4b).

Different from Chl, the epidermal Flav estimated by Dualex 4, which is the representative of total leaf Flav [27,64], is considered as a surrogate of dry leaf mass per area with no units [22]. Meyer et al. [23] reported the positive correlations between mass-based polyphenols (mainly Flav) and leaf mass per area for woody plants, which revealed the regularity of the decrease in Flav was caused by the increase in leaf dry mass. Thicker leaves had a larger leaf mass per area, and the accumulation of organic matter in leaves played the role of Flav dilution. In addition, a previous study has reported that the accumulation of Flav was highly sensitive to light intensity [65]. As the maize grows, the leaf area and canopy closure of the uppermost leaves gradually increase, resulting in increased degree of shading. This may limit the expression of Flav synthetic genes and lead to a reduction in flavonoid content in these leaves during later growth stages. More studies are still needed on the periodic changes of Flav.

As the ratio of Chl to Flav, NBI is equivalent to mass-based Chl (Chl g^{-1} dry mass). Since Chl is relatively stable and Flav showed a downward trend as the maize grew, it is easy to understand the reverse trend of NBI to Flav (Figure 4h). The inconsistent relationships between the original Dualex 4 parameters (Chl, Flav, and NBI) and TLNC or PNC at different growth stages led to poor model performance across the growth stages (Table 4 and Figure 6). This makes the original Dualex 4 parameters unfeasible for estimating TLNC and PNC across growth stages.

Varvel et al. [46] pointed out that it is effective to evaluate crop N status by calculating growing degree days (GDD) when a specific growth stage cannot be determined. Following the idea of Teal et al. [66], who predicted maize yield using normalized difference vegetation index (NDVI) and GDD, DAS was used together with Dualex 4 parameters in this study to explore the potential of real-time PNC estimation across growth stages. The results of this study indicated the relationships between the DAS-based modified Dualex 4 parameters and TLNC and PNC across growth stages were significantly improved and were simulated by power function models. Furthermore, this study revealed a stronger relationship between the modified parameters and PNC than TLNC. In addition, the good relations between SLNC and PNC implied that PNC could be directly estimated well using a leaf sensor instead of indirect estimation through TLNC (Table 5 and Figure 8). This indicates the great potential of leaf-based Dualex 4 for assessing plant N status, not just maize leaf N status.

It has been reported that polyphenols (Phen, including Flav) may be a more specific indicator of crop N status than Chl because of the sensitivity of Chl to sulfur stress [67]. Padilla et al. [58] demonstrated the strong linear relationships between Flav, measured by a Multiplex sensor, and cucumber (*Cucumis sativus* 'Strategos') N concentration (R^2 = 0.87–0.95). Besides, some studies pointed out the superiority of NBI (or Chl/Phen) in N status estimation in a variety of crops, including maize [28,29,31,44,48,57,68]. However, mChl (Chl) performed best while mFlav (Flav) performed worst in this study.

The relatively poor performance of mFlav indicated the difficulty of estimating PNC accurately using Flav, while the difficulty for estimating Phen in leaves has been attributed to different solubility, distribution locations, and species varieties [23]. Flav has previously been shown to perform well in N status estimation for other crops such as vegetable, rice, and sweet cherry (*Prunus avium* L.) [49,52,68,69]. However, the relatively thick maize leaves could affect the transmittance of the light emitted by the sensor and then lead to the change of the parameter [22]. Measurements taken from both the adaxial (top) and abaxial (bottom) sides of the leaves may help improve maize PNC estimation by Flav, although the flavonoid contents in the adaxial and abaxial sides of the leaves are highly correlated in some crops [27,57]. Moreover, the change of Flav content is influenced by various factors in addition to N such as leaf thickness, light condition, water stress, pathogen attack, low temperature, and the onset of senescence [17,30,65,70]. Although affected by the poor performance of mFlav, the estimation results of mNBI-based models showed similar R^2 values with mChl, indicating the good performance of NBI for N indices estimation. These results conform to the findings of some previous studies [27–29,31,44,49,57].

4.3. The Most Suitable Leaf Position for Sensing Measurements and PNC Estimation

We tested three differently positioned leaves at each growth stage in this study in order to understand their effects on evaluating maize N status. The changes of the Dualex 4 based parameters with leaf positions were relatively smaller than that with the growth stage and N rate. Nevertheless, there were slight differences among the three leaves, which consequently displayed different performance in PNC prediction.

It was noted that leaf Chl and Flav contents depended on leaf age (leaf position) induced by light conditions [71,72]. The photosynthetic photon flux density would decrease with increasing canopy depth during crop growth and canopy development [73]. The priority supplement of N in younger leaves (upper leaves) re-translocated from older leaves (lower leaves) leads to the preferential distribution of N to the upper leaves [74] and consequently leads to higher rates of canopy photosynthesis, which needs a higher content of Chl. The finding of higher Chl values in Leaf 1 (the uppermost leaf) than in Leaf 3 (the lowest leaf) was similar to the study results for rice [74,75]. Higher Flav contents in older leaves have been reported [71,76,77], similar to the results of this study. Besides, the decrease of leaf mass per area along with leaf thickness as leaf position declines, the limited sunlight in the lower leaves mentioned above caused the change of Flav with the leaf position. However, there was a slightly increasing tendency of Chl measurement in Leaf 2, which may be caused by the close locations of the three leaves. As the uppermost leaf (Leaf 1) was newly expanded soon after Leaf 2, the N concentration of these two leaves would not differ too much. In addition, the average values used in this analysis may help reduce some variations. The Flav is an N-free compound, so its continuous change with leaf positions was more distinct.

To diagnose crop N status quickly and effectively, the test frequency of a leaf sensor must be reduced. Thus, selecting the most appropriate leaf rather than several leaves of a plant for sensing measurement with high accuracy and reliability is important. The measurements taken from Leaf 1 for all three modified Dualex parameters provided the highest model R^2 in TLNC and PNC predictions. The best performance of mFlav appeared in Leaf 3 in PNC estimation, and the mean values of three leaves improved the results significantly, but the results were not as good as the other two parameters. For mNBI, although the best result appeared in Leaf 3 too, it did not show a significant superiority over the results of the other two leaves. Thus, in conclusion, the measurement of Leaf 1 may be sufficient to estimate maize N status using Dualex 4-based mChl and mNBI whereas Leaf 3 is preferred to accurately estimate PNC using mFlav.

4.4. Implications for Practical Application and Future Research Needs

This study found that combining with DAS, the Dualex 4 leaf fluorescence sensor could be used to reliably estimate maize PNC across growth stages, even if only one leaf (uppermost) was sampled. The estimated PNC can then be compared with threshold or critical PNC values to diagnose maize N status by calculating N nutrition index (NNI) [41]. For this purpose, maize aboveground biomass will be needed and can be reliably estimated using active canopy sensors, like GreenSeeker or Crop Circle sensors [41,78]. Research is needed to develop methods to use Dualex 4 sensor to estimate NNI directly, as demonstrated for SPAD chlorophyll meter [79]. Based on the relationship between SPAD chlorophyll meter readings and PNC and the critical N concentration curve for maize, Yang et al. [80] established the critical or optimal chlorophyll meter (SPAD) reading curve. Measured chlorophyll meter readings can then be compared with critical chlorophyll meter readings at specific biomass to calculate NNI, without the need to estimate PNC. This is an innovative idea and should be tested with the Dualex sensor.

The relationship between leaf and canopy fluorescence parameters could be further investigated through the use of a simple model taking into account leaf fluorescence profile inside the canopy, structural variations of the canopy, and background reflection [81]. Such information can be useful to evaluate the possibility to use the proposed method in conjunction with remote sensing fluorescence measurements obtained from an unmanned aerial vehicle (UAV) [82], aerial or satellite remote sensing [83,84].

In theory, NBI combining both Chl and Flav should be more sensitive to crop N status than using Chl or Flav alone, as found by previous research [28,29,31,44,48,57,68]. However, NBI did not perform better than Chl in this study. Those previous studies either used an earlier model of Dualex sensor (cannot measure Chl) and chlorophyll meter to calculate NBI or used canopy fluorescence sensor Multiplex or studied other crops. Therefore, more studies are needed to confirm our results with maize and further evaluate the potential of improving corn N status diagnosis using both Chl and Flav information, as well as other related soil, weather, and management data.

5. Conclusions

This study demonstrated the reliability of maize PNC estimation by fluorescence parameters obtained from single leaves using a leaf sensor Dualex 4. The fluorescence parameter values did not exhibit significant variability under the two different soil conditions, which indicated the needlessness of establishing a soil-specific predicting model in this study, while the variations between the PNC and Dualex 4 parameters at different growth stages make it difficult to estimate PNC across all growth stages using the original Dualex 4 parameters. Nevertheless, the modified Dualex 4 parameters using the information of DAS overcame the problem caused by growth stage changes and allowed accurate estimation of PNC using a general model across growth stages. Among the three parameters obtained by Dualex 4, mChl and mNBI were more reliable indicators for PNC estimation. It was sufficient to take the fluorescence measurement from the uppermost leaf of maize. Further research needs to focus on improving the accuracy of PNC estimation in early growth stages and practical methods for maize N status diagnosis using proximal leaf fluorescence sensors.

Author Contributions: Y.M. and H.L. designed the experiment. R.D. conducted the experiment, performed the analysis, and wrote the original paper, X.W., Z.C., and W.Z. assisted in the experiment, plant and soil sampling, and sample processing. Y.M. and F.Y. reviewed and revised the manuscript. All authors have read and agreed to the published version of the manuscript.

Funding: This research was funded by the Key National Research and Development Program (2016YFD0200600; 2016YFD0200602), Norwegian Ministry of Foreign Affairs (SINOGRAIN II, CHN-17/0019), the Internationalization Training and Promotion Project of Graduate Students in China Agricultural University, and the UK Biotechnology and Biological Sciences Research Council (BB/P004555/1).

Acknowledgments: We would like to thank Guohua Mi, Zheng Fang, Xuezhi Yue, and Hainie Zha for their assistance during this study. We also would like to thank the local farmers for their cooperation in this research.

Conflicts of Interest: The authors declare no conflict of interest.

References

1. FAO. Cultured Aquatic Species Information Programme, Epinephelus Coioides. Available online: http://www.fao.org (accessed on 28 June 2019).
2. Gu, B.; Ju, X.; Wu, Y.; Erisman, J.W.; Bleeker, A.; Reis, S.; Smith, R.I. Cleaning up nitrogen pollution may reduce future carbon sinks. *Glob. Environ. Chang.* **2018**, *48*, 56–66. [CrossRef]
3. Zhang, F.; Chen, X.; Vitousek, P. An experiment for the world. *Nature* **2013**, *497*, 33–35. [CrossRef] [PubMed]
4. Jin, Z.; Archontoulis, S.V.; Lobell, D.B. How much will precision nitrogen management pay off? An evaluation based on simulating thousands of corn fields over the US corn-belt. *Field Crop. Res.* **2019**, *240*, 12–22. [CrossRef]
5. Wang, X.; Miao, Y.; Dong, R.; Chen, Z.; Guan, Y.; Yue, X.; Fang, Z.; Mulla, D.J. Developing active canopy sensor-based precision nitrogen management strategies for corn in Northeast China. *Sustainability* **2019**, *11*, 706.
6. Kyveryga, P.M.; Blackmer, A.M.; Zhang, J. Characterizing and classifying variability in corn yield response to nitrogen fertilization on subfield and field scales. *Agron. J.* **2009**, *101*, 269–277. [CrossRef]
7. Muñoz-Huerta, R.F.; Guevara-Gonzalez, R.G.; Contreras-Medina, L.M.; Torres-Pacheco, I.; Prado-Olivarez, J.; Ocampo-Velazquez, R.V. A review of methods for sensing the nitrogen status in plants: Advantages, disadvantages and recent advances. *Sensors* **2013**, *13*, 10823–10843. [CrossRef] [PubMed]
8. Corti, M.; Cavalli, D.; Cabassi, G.; Gallina, P.M.; Bechini, L. Does remote and proximal optical sensing successfully estimate maize variables? A review. *Eur. J. Agron.* **2018**, *99*, 37–50. [CrossRef]
9. Pinter, P.J.; Hatfield, J.L.; Schepers, J.S.; Barnes, E.M.; Moran, M.S.; Daughtry, C.S.T.; Upchurch, D.R. Remote sensing for crop management. *Photogramm. Eng. Remote. Sens.* **2003**, *69*, 647–664. [CrossRef]
10. Mulla, D.J. Twenty-five years of remote sensing in precision agriculture: Key advances and remaining knowledge gaps. *Biosyst. Eng.* **2013**, *114*, 358–371. [CrossRef]
11. Nigon, T.J.; Mulla, D.J.; Rosen, C.J.; Cohen, Y.; Alchanatis, V.; Rud, R. Evaluation of the nitrogen sufficiency index for use with high resolution, broadband aerial imagery in a commercial potato field. *Precis. Agric.* **2014**, *15*, 202–226. [CrossRef]
12. Delloye, C.; Weiss, M.; Defourny, P. Retrieval of the canopy chlorophyll content from Sentinel-2 spectral bands to estimate nitrogen uptake in intensive winter wheat cropping systems. *Remote. Sens. Environ.* **2018**, *216*, 245–261. [CrossRef]

13. Tremblay, N.; Wang, Z.; Cerovic, Z.G. Sensing crop nitrogen status with fluorescence indicators. A review. *Agron. Sustain. Dev.* **2012**, *32*, 451–464. [CrossRef]
14. Kovács, P.; Vyn, T.J. Relationships between ear-leaf nutrient concentrations at silking and corn biomass and grain yields at maturity. *Agron. J.* **2017**, *109*, 2898–2906. [CrossRef]
15. Gaju, O.; Allard, V.; Martre, P.; Le Gouis, J.; Moreau, D.; Bogard, M.; Hubbart, S.; Foulkes, M.J. Nitrogen partitioning and remobilization in relation to leaf senescence, grain yield and grain nitrogen concentration in wheat cultivars. *Field Crop Res.* **2014**, *155*, 213–223. [CrossRef]
16. Isfan, D.; Zizka, J.; D'Avignon, A.; Deschênes, M. Relationships between nitrogen rate, plant nitrogen concentration, yield and residual soil nitrate-nitrogen in silage corn. *Commun. Soil Sci. Plant Anal.* **1995**, *26*, 2531–2557. [CrossRef]
17. Cerovic, Z.G.; Ghozlen, N.B.; Milhade, C.; Obert, M.; Debuisson, S.; Moigne, M.L. Nondestructive diagnostic test for nitrogen nutrition of grapevine (*Vitis vinifera* L.) based on Dualex leaf-clip measurements in the field. *J. Agric. Food Chem.* **2015**, *63*, 3669–3680. [CrossRef] [PubMed]
18. Duan, D.; Zhao, C.; Li, Z.; Yang, G.; Yang, W. Estimating total leaf nitrogen concentration in winter wheat by canopy hyperspectral data and nitrogen vertical distribution. *J. Integr. Agric.* **2019**, *18*, 1562–1570. [CrossRef]
19. Huang, S.; Miao, Y.; Zhao, G.; Yuan, F.; Ma, B.; Tan, C.; Yu, W.; Gnyp, M.L.; Lenz-Wiedemann, V.I.S.; Rascher, U.; et al. Satellite remote sensing-based in-season diagnosis of rice nitrogen status in Northeast China. *Remote Sens.* **2015**, *7*, 10646–10667. [CrossRef]
20. Longchamps, L.; Khosla, R. Early detection of nitrogen variability in corn using fluorescence. *Agron. J.* **2014**, *106*, 511–518. [CrossRef]
21. Heege, H.J.; Reusch, S.; Thiessen, E. Prospects and results for optical systems for site-specific on-the-go control of nitrogen-top-dressing in Germany. *Precis. Agric.* **2008**, *9*, 115–131. [CrossRef]
22. Cerovic, Z.G.; Masdoumier, G.; Ghozlen, N.B.; Latouche, G. A new optical leaf-clip meter for simultaneous non-destructive assessment of leaf chlorophyll and epidermal flavonoids. *Physiol. Plant.* **2012**, *146*, 251–260. [CrossRef] [PubMed]
23. Meyer, S.; Cerovic, Z.G.; Goulas, Y.; Montpied, P.; Demotes-Mainard, S.; Bidel, L.P.; Moya, I.; Dreyer, E. Relationships between optically assessed polyphenols and chlorophyll contents, and leaf mass per area ratio in woody plants: A signature of the carbon-nitrogen balance within leaves? *Plant Cell Environ.* **2006**, *29*, 1338–1348. [CrossRef] [PubMed]
24. Cerovic, Z.G.; Ounis, A.; Cartelat, A.; Latouche, G.; Khosla, R. The use of chlorophyll fluorescence excitation spectra for the non-destructive in situ assessment of UV-absorbing compounds in leaves. *Plant Cell Environ.* **2002**, *25*, 1663–1676. [CrossRef]
25. Barnes, P.W.; Searles, P.S.; Ballaré, C.L.; Ryel, R.J.; Caldwell, M.M. Non-invasive measurements of leaf epidermal transmittance of UV radiation using chlorophyll fluorescence: Field and laboratory studies. *Physiol. Plant.* **2000**, *109*, 274–283. [CrossRef]
26. Dong, T.; Shang, J.; Chen, J.; Liu, J.; Qian, B.; Ma, B.; Morrison, M.J.; Zhang, C.; Liu, Y.; Shi, Y.; et al. Assessment of portable chlorophyll meters for measuring crop leaf chlorophyll concentration. *Remote Sens.* **2019**, *11*, 2706. [CrossRef]
27. Cartelat, A.; Cerovic, Z.G.; Goulas, Y.; Meyer, S.; Lelarge, C.; Prioul, J.L.; Barbottin, A.; Jeuffroy, M.H.; Gate, P.; Agati, G.; et al. Optically assessed contents of leaf polyphenolics and chlorophyll as indicators of nitrogen deficiency in wheat (*Triticum aestivum* L.). *Field Crop Res.* **2005**, *91*, 35–49. [CrossRef]
28. Padilla, F.M.; Peña-Fleitas, M.T.; Gallardo, M.; Thompson, R.B. Evaluation of optical sensor measurements of canopy reflectance and of leaf flavonols and chlorophyll contents to assess crop nitrogen status of muskmelon. *Eur. J. Agron.* **2014**, *58*, 39–52. [CrossRef]
29. Tremblay, N.; Wang, Z.; Bélec, C. Evaluation of the Dualex for the assessment of corn nitrogen status. *J. Plant Nutr.* **2007**, *30*, 1355–1369. [CrossRef]
30. Goulas, Y.; Cerovic, Z.G.; Cartelat, A.; Moya, I. Dualex: A new instrument for field measurements of epidermal ultraviolet absorbance by chlorophyll fluorescence. *Appl. Opt.* **2004**, *43*, 4488–4496. [CrossRef]
31. Tremblay, N.; Wang, Z.; Bélec, C. Performance of Dualex in spring wheat for crop nitrogen status assessment, yield prediction and estimation of soil nitrate content. *J. Plant Nutr.* **2009**, *33*, 57–70. [CrossRef]
32. Apostol, S.; Viau, A.A.; Tremblay, N.; Briantais, J.-M.; Prasher, S.; Parent, L.-E.; Moya, I. Laser-induced fluorescence signatures as a tool for remote monitoring of water and nitrogen stresses in plants. *Can. J. Remote Sens.* **2003**, *29*, 57–65. [CrossRef]

33. Xu, J.; Cai, H.; Wang, X.; Ma, C.; Lu, Y.; Ding, Y.; Wang, X.; Chen, H.; Wang, Y.; Saddique, Q. Exploring optimal irrigation and nitrogen fertilization in a winter wheat-summer maize rotation system for improving crop yield and reducing water and nitrogen leaching. *Agric. Water. Manage.* **2020**, *228*, 105904. [CrossRef]
34. Carolina, S.P.; Crossa, J.L.; Bonnett, D.; Yamaguchi-Shinozaki, K.; Reynolds, M.P. Phenotyping transgenic wheat for drought resistance. *J. Exp. Bot.* **2012**, *63*, 1799–1808. [CrossRef]
35. Scharf, P.C.; Kitchen, N.R.; Sudduth, K.A.; Davis, J.G. Spatially variable corn yield is a weak predictor of optimal nitrogen rate. *Soil Sci. Soc. Am. J.* **2006**, *70*, 2154–2160. [CrossRef]
36. Power, J.F.; Willis, W.O.; Grunes, D.L.; Reichman, G.A. Effect of soil temperature, phosphorus and plant age on growth analysis of barley. *Agron. J.* **1967**, *59*, 231–234. [CrossRef]
37. Sarker, U.; Oba, S. Response of nutrients, minerals, antioxidant leaf pigments, vitamins, polyphenol, flavonoid and antioxidant activity in selected vegetable amaranth under four soil water content. *Food Chem.* **2018**, *252*, 72–83. [CrossRef]
38. Lea, U.S.; Slimestad, R.; Smedvig, P.; Lillo, C. Nitrogen deficiency enhances expression of specific MYB and bHLH transcription factors and accumulation of end products in the flavonoid pathway. *Planta* **2007**, *225*, 1245–1253. [CrossRef]
39. Shapiro, C. Using a chlorophyll meter to manage nitrogen applications to corn with high nitrate irrigation water. *Commun. Soil Sci. Plant Anal.* **1999**, *30*, 1037–1049. [CrossRef]
40. Li, F.; Miao, Y.; Hennig, S.D.; Gnyp, M.L.; Chen, X.; Jia, L.; Bareth, G. Evaluating hyperspectral vegetation indices for estimating nitrogen concentration of winter wheat at different growth stages. *Precis. Agric.* **2010**, *11*, 335–357. [CrossRef]
41. Xia, T.; Miao, Y.; Wu, D.; Shao, H.; Khosla, R.; Mi, G. Active optical sensing of spring corn for in-season diagnosis of nitrogen status based on nitrogen nutrition index. *Remote Sens.* **2016**, *8*, 605. [CrossRef]
42. Li, F.; Miao, Y.; Feng, G.; Yuan, F.; Yue, S.; Gao, X.; Liu, Y.; Liu, B.; Ustin, S.L.; Chen, X. Improving estimation of summer maize nitrogen status with red edge-based spectral vegetation indices. *Field Crops Res.* **2014**, *157*, 111–123. [CrossRef]
43. Gabriel, J.L.; Quemada, M.; Alonso-Ayuso, M.; Lizaso, J.; Martín-Lammerding, D. Predicting N status in maize with clip sensors: Choosing sensor, leaf sampling point, and timing. *Sensors.* **2019**, *19*, 3881. [CrossRef] [PubMed]
44. Zhang, Y.; Tremblay, N.; Zhu, J. A first comparison of Multiplex®for the assessment of corn nitrogen status. *J. Food. Agric. Environ.* **2012**, *10*, 1008–1016.
45. Huang, S.; Miao, Y.; Yuan, F.; Cao, Q.; Ye, H.; Lenz-Wiedemann, V.I.S.; Bareth, G. In-Season diagnosis of rice nitrogen status using proximal fluorescence canopy sensor at different growth stages. *Remote Sens.* **2019**, *11*, 1847. [CrossRef]
46. Varvel, G.E.; Wilhelm, W.W.; Shanahan, J.F.; Schepers, J.S. An algorithm for corn nitrogen recommendations using a chlorophyll meter-based sufficiency index. *Agron. J.* **2007**, *99*, 701–706. [CrossRef]
47. Yang, J.; Gong, W.; Shi, S.; Du, L.; Sun, J.; Song, S.; Chen, B.; Zhang, Z. Analyzing the performance of fluorescence parameters in the monitoring of leaf nitrogen content of paddy rice. *Sci. Rep.* **2016**, *6*, 28787. [CrossRef]
48. Agati, G.; Foschi, L.; Grossi, N.; Guglielminetti, L.; Cerovic, Z.G.; Volterrani, M. Fluorescence-based versus reflectance proximal sensing of nitrogen content in Paspalum vaginatum and Zoysia matrella turfgrasses. *Eur. J. Agron.* **2013**, *45*, 39–51. [CrossRef]
49. Zhang, K.; Liu, X.; Ma, Y.; Zhang, R.; Cao, Q.; Zhu, Y.; Cao, W.; Tian, Y. A comparative assessment of measures [illegible]
50. Nelson, D.W.; Sommers, L.E. Determination of total nitrogen in plant material. *Agron. J.* **1973**, *65*, 109–112. [CrossRef]
51. Roca, L.F.; Romero, J.; Bohórquez, J.M.; Alcántara, E.; Fernández-Escobar, R.; Trapero, A. Nitrogen status affects growth, chlorophyll content and infection by Fusicladium oleagineum in olive. *Crop Prot.* **2018**, *109*, 80–85. [CrossRef]
52. Padilla, F.M.; de Souza, R.; Peña-Fleitas, M.T.; Gallardo, M.; Giménez, C.; Thompson, R.B. Different responses of various chlorophyll meters to increasing nitrogen supply in sweet pepper. *Front Plant Sci.* **2018**, *9*, 1752. [CrossRef] [PubMed]

53. Schlemmer, M.; Gitelson, A.; Schepers, J.; Ferguson, R.; Peng, Y.; Shanahan, J.; Rundquist, D. Remote estimation of nitrogen and chlorophyll contents in maize at leaf and canopy levels. *Int. J. Appl. Earth Obs.* **2013**, *25*, 47–54. [CrossRef]
54. Evans, J.R. Photosynthesis and nitrogen relationships in leaves of C3 plants. *Oecologia* **1989**, *78*, 9–19. [CrossRef] [PubMed]
55. Bragazza, L.; Freeman, C. High nitrogen availability reduces polyphenol content in Sphagnum peat. *Sci. Total Environ.* **2007**, *377*, 439–443. [CrossRef]
56. Liu, W.; Zhu, D.W.; Liu, D.H.; Geng, M.J.; Zhou, W.B.; Mi, W.J.; Yang, T.W.; Hamilton, D. Influence of nitrogen on the primary and secondary metabolism and synthesis of flavonoids in Chrysanthemum morifolium Ramat. *J. Plant Nutr.* **2010**, *33*, 240–254. [CrossRef]
57. Li, J.; Zhang, J.; Zhao, Z.; Lei, X.; Xu, X.; Weng, D.; Gao, Y.; Cao, L. Use of fluorescence-based sensors to determine the nitrogen status of paddy rice. *J. Agric. Sci.* **2013**, *151*, 862–887. [CrossRef]
58. Padilla, F.M.; Peña-Fleitas, M.T.; Gallardo, M.; Thompson, R.B. Proximal optical sensing of cucumber crop N status using chlorophyll fluorescence indices. *Eur. J. Agron.* **2016**, *73*, 83–97. [CrossRef]
59. Cui, Z.L.; Zhang, H.Y.; Chen, X.P.; Zhang, C.C.; Ma, W.Q.; Huang, C.D.; Zhang, W.F.; Mi, G.H.; miAo, Y.X.; Li, X.L.; et al. Pursuing sustainable productivity with millions of smallholder farmers. *Nature* **2018**, *555*, 363–366. [CrossRef]
60. Richardson, A.D.; Duigan, S.P.; Berlyn, G.P. An evaluation of noninvasive methods to estimate foliar chlorophyll content. *New Phytol.* **2002**, *153*, 185–194. [CrossRef]
61. Romero, I.; García-Escudero, E.; Martín, I. Leaf blade versus petiole analysis for nutritional diagnosis of Vitis vinifera L. cv. Tempranillo. *Am. J. Enol. Vitic.* **2012**, *64*, 50–64. [CrossRef]
62. Romero, I.; García-Escudero, E.; Martín, I. Effects of leaf position on blade and petiole mineral nutrient concentration of Tempranillo grapevine (*Vitis vinifera* L.). *Am. J. Enol. Vitic.* **2010**, *61*, 544–550. [CrossRef]
63. Ziadi, N.; Bélanger, G.; Gastal, F.; Claessens, A.; Lemaire, G.; Tremblay, N. Leaf nitrogen concentration as an indicator of corn nitrogen status. *Agron. J.* **2009**, *101*, 947–957. [CrossRef]
64. Kolb, C.A.; Pfündel, E.E. Origins of non-linear and dissimilar relationships between epidermal UV absorbance and UV absorbance of extracted phenolics in leaves of grapevine and barley. *Plant Cell Environ.* **2005**, *28*, 580–590. [CrossRef]
65. Agati, G.; Cerovic, Z.G.; Pinelli, P.; Tattini, M. Light-induced accumulation of ortho-dihydroxylated flavonoids as non-destructively monitored by chlorophyll fluorescence excitation techniques. *Environ. Exp. Bot.* **2011**, *73*, 3–9. [CrossRef]
66. Teal, R.K.; Tubana, B.; Girma, K.; Freeman, K.W.; Arnall, D.B.; Walsh, O.; Raun, W.R. In-season prediction of corn grain yield potential using normalized difference vegetation index. *Agron. J.* **2006**, *98*, 1488–1494. [CrossRef]
67. Samson, G.; Tremblay, N.; Dudelzak, A.E.; Babichenko, S.M.; Dextraze, L.; Wollring, J. Nutrient stress of corn plants: Early detection and discrimination using a compact multiwavelength fluorescent lidar. In Proceedings of the 4th EARSeL Workshop Lidar Remote Sensing of Land and Sea held during the 20th EARSeL Symposium, Dresden, Germany, 14–16 June 2000.
68. Overbeck, V.; Schmitz, M.; Tartachnyk, I.; Blanke, M. Identification of light availability in different sweet cherry orchards under cover by using non-destructive measurements with a Dualex™. *Eur. J. Agron.* **2018**, *93*, 50–56. [CrossRef]
69. Padilla, F.M.; Gallardo, M.; Peña-Fleitas, M.T.; De Souza, R.; Thompson, R.B. Proximal optical sensors for nitrogen management of vegetable crops: A review. *Sensors* **2018**, *18*, 2083. [CrossRef]
70. Agati, G.; Brunetti, C.; Di Ferdinando, M.; Ferrini, F.; Pollastri, S.; Tattini, M. Functional roles of flavonoids in photoprotection: New evidence, lessons from the past. *Plant Physiol. Biochem.* **2013**, *72*, 35–45. [CrossRef]
71. Louis, J.; Meyer, S.; Maunoury-Danger, F.; Fresneau, C.; Meudec, E.; Cerovic, Z.G. Seasonal changes in optically assessed epidermal phenolic compounds and chlorophyll contents in leaves of sessile oak (*Quercus petraea*): Towards signatures of phenological stage. *Funct. Plant Biol.* **2009**, *36*, 732–741. [CrossRef]
72. Louis, J.; Genet, H.; Meyer, S.; Soudani, K.; Montpied, P.; Legout, A.; Dreyer, E.; Cerovic, Z.G.; Dufrêne, E. Tree age-related effects on sun acclimated leaves in a chronosequence of beech (*fagus sylvatica*) stands. *Funct. Plant Biol.* **2012**, *39*, 323–331. [CrossRef]
73. Yang, H.; Li, J.; Yang, J.; Wang, H.; Zou, J.; He, J. Effects of nitrogen application rate and leaf age on the distribution pattern of leaf SPAD readings in the rice canopy. *PLoS ONE* **2014**, *9*, e92509. [CrossRef] [PubMed]

74. Wang, S.; Zhu, Y.; Jiang, H.; Cao, W. Positional differences in nitrogen and sugar concentrations of upper leaves relate to plant N status in rice under different N rates. *Field Crop Res.* **2006**, *96*, 224–234. [CrossRef]
75. Lin, F.F.; Qiu, L.F.; Deng, J.S.; Shi, Y.Y.; Chen, L.S.; Wang, K. Investigation of SPAD meter-based indices for estimating rice nitrogen status. *Compu. Electron. Agric.* **2010**, *71S*, 60–65. [CrossRef]
76. Abdallah, S.B.; Rabhi, M.; Harbaoui, F.; Zar-kalai, F.; Lachâal, M.; Karray-Bouraoui, N. Distribution of phenolic compounds and antioxidant activity between young and old leaves of *Carthamus tinctorius* L. and their induction by salt stress. *Acta Physiol. Plant.* **2013**, *35*, 1161–1169. [CrossRef]
77. Vagiri, M.; Conner, S.; Stewart, D.; Andersson, S.C.; Verrall, S.; Johansson, E.; Rumpunen, K. Phenolic compounds in blackcurrant (*Ribes nigrum* L.) leaves relative to leaf position and harvest date. *Food Chem.* **2014**, *172*, 135–142. [CrossRef] [PubMed]
78. Mulla, D.J.; Miao, Y. Precision Farming. In *Land Resources Monitoring, Modeling, and Mapping with Remote Sensing*; Thenkabail, P.S., Ed.; CRC Press: Boca Raton, FL, USA, 2016.
79. Ziadi, N.; Brassard, M.; Bélanger, G.; Claessens, A.; Tremblay, N.; Cambouris, A.N.; Nolin, M.C.; Parent, L.E. Chlorophyll measurements and nitrogen nutrition index for the evaluation of corn nitrogen status. *Agron. J.* **2008**, *100*, 1264–1273. [CrossRef]
80. Yang, Y.; Timlin, D.J.; Fleisher, D.H.; Lokhande, S.B.; Chun, J.A.; Kim, S.H.; Staver, K.; Reddy, V.R. Nitrogen concentration and dry-matter accumulation in maize crop: Assessing maize nitrogen status with an allometric function and a chlorophyll meter. *Commun. Soil Sci. Plant Anal.* **2012**, *43*, 1563–1575. [CrossRef]
81. Olioso, A.; Méthy, M.; Lacaze, B. Fluorescence as a Function of Canopy Structure and Leaf Fluorescence. *Remote. Sens. Environ.* **1992**, *41*, 239–247. [CrossRef]
82. Garzonio, R.; di Mauro, B.; Colombo, R.; Cogliati, S. Surface reflectance and sun-induced fluorescence spectroscopy measurements using a small hyperspectral UAS. *Remote Sens.* **2017**, *9*, 472. [CrossRef]
83. Coppo, P.; Taiti, A.; Pettinato, L.; Francois, M.; Taccola, M.; Drusch, M. Fluorescence imaging spectrometer (FLORIS) for ESA FLEX mission. *Remote Sens.* **2017**, *9*, 649. [CrossRef]
84. Miao, Y.; Mulla, D.J.; Randall, G.W.; Vetsch, J.A.; Vintila, R. Combining chlorophyll meter readings and high spatial resolution remote sensing images for in-season site-specific nitrogen management of corn. *Precis. Agric.* **2009**, *10*, 45–62. [CrossRef]

remote sensing

MDPI

Article

Assessment of Portable Chlorophyll Meters for Measuring Crop Leaf Chlorophyll Concentration

Taifeng Dong [1], Jiali Shang [1,*], Jing M. Chen [2,*], Jiangui Liu [1], Budong Qian [1], Baoluo Ma [1], Malcolm J. Morrison [1], Chao Zhang [1], Yupeng Liu [1], Yichao Shi [1], Hui Pan [1] and Guisheng Zhou [3]

1 Ottawa Research and Development Centre, Agriculture and Agri-Food Canada, Ottawa, ON K1A 0C6, Canada; taifeng.dong@canada.ca (T.D.); jiangui.liu@canada.ca (J.L.); Budong.Qian@canada.ca (B.Q.); baoluo.ma@canada.ca (B.M.); Malcolm.Morrison@Canada.ca (M.J.M.); Zhangc1700@yzu.edu.cn (C.Z.); Ykevin.Liu@mail.utoronto.ca (Y.L.); Yichao.Shi@Canada.ca (Y.S.); Pan.Hui@agr.gc.ca (H.P.)

2 Department of Geography and Planning, University of Toronto, Toronto, ON M5S 3G3, Canada

3 College of Agriculture, Yangzhou University, Yangzhou 225009, China; gszhou@yzu.edu.cn

* Correspondence: jiali.shang@canada.ca (J.S.); jing.chen@utoronto.ca (J.M.C.)

Received: 16 October 2019; Accepted: 13 November 2019; Published: 19 November 2019

Abstract: Accurate measurement of leaf chlorophyll concentration (LChl) in the field using a portable chlorophyll meter (PCM) is crucial to support methodology development for mapping the spatiotemporal variability of crop nitrogen status using remote sensing. Several PCMs have been developed to measure LChl instantaneously and non-destructively in the field, however, their readings are relative quantities that need to be converted into actual LChl values using conversion functions. The aim of this study was to investigate the relationship between actual LChl and PCM readings obtained by three PCMs: SPAD-502, CCM-200, and Dualex-4. Field experiments were conducted in 2016 on four crops: corn (*Zea mays* L.), soybean (*Glycine max* L. Merr.), spring wheat (*Triticum aestivum* L.), and canola (*Brassica napus* L.), at the Central Experimental Farm of Agriculture and Agri-Food Canada in Ottawa, Ontario, Canada. To evaluate the impact of other factors (leaf internal structure, leaf pigments other than chlorophyll, and the heterogeneity of LChl distribution) on the conversion function, a global sensitivity analysis was conducted using the PROSPECT-D model to simulate PCM readings under different conditions. Results showed that Dualex-4 had a better performance for actual LChl measurement than SPAD-502 and CCM-200, using a general conversion function for all four crops tested. For SPAD-502 and CCM-200, the error in the readings increases with increasing LChl. The sensitivity analysis reveals that deviations from the calibration functions are more induced by non-uniform LChl distribution than leaf architectures. The readings of Dualex-4 can have a better ability to restrict these influences than those of the other two PCMs.

Keywords: leaf chlorophyll concentration; portable chlorophyll meter; crop; PROSPECT-D; sensitivity analysis; remote sensing; radiative transfer model

1. Introduction

Estimation of plant traits using remote sensing data, such as leaf nitrogen concentration, leaf chlorophyll concentration (LChl) and leaf area index (LAI), is important for mapping the spatiotemporal variability of crop and soil conditions, and modeling crop nutrient balance, and crop productivity [1–3]. LChl is the main light-harvesting pigment that determines leaf photosynthetic capacity, and it is highly influenced by nitrogen fertilization [4–6]. Furthermore, incorporating LChl into process-based crop models could improve model performance [7,8]. LChl varies with leaf positions, species, crop types, crop growth stages and crop managements [7,9,10]; thus, knowledge on the spatiotemporal variability of LChl is important to understand the status of crop growth condition and productivity [8,11,12].

A number of studies have investigated the potential of remote sensing data in estimating LChl using statistical or physical based approaches [13–15]. There has been rapid development in new satellite sensors, such as multispectral satellite sensors with red-edge (680–750 nm) reflectance measurements (e.g., Sentinel-2 and VENµS) [16,17], the VNIR-SWIR hyperspectral satellite sensors (e.g., HyspIRI and EnMAP) [18,19], and the multi- and hyperspectral imaging systems mounted on a UAV system [19]. In particular, these sensors possess the red-edge or hyperspectral reflectance that is highly sensitive to changes in LChl [4]. This allows for improved accuracy for LChl estimation from remote sensing data at different spatial scales. Accurate in situ LChl measurements are essential for developing and validating remote-sensing LChl estimation models.

Destructive and non-destructive methods are often used for LChl measurement. Both methods rely on measured light absorption/transmission to determine LChl [2,20,21]. Conventionally, destructive measurement is conducted using a wet-chemical method in a lab setting [4,22]. Leaves are harvested from the plant and chlorophyll is extracted using organic solvents (e.g., acetone, methanol, ethanol, dimethyl sulphoxide (DMSO), or N-dimethyl formamide (DMF) [4,22,23]. A spectrophotometer, a fluorometer, or a high-performance liquid chromatography (HPLC) is often used to measure light absorptions at a few wavelength ranges [4,23], which are then used to determine LChl. The lab-based approach is costly, labour intensive and time consuming. In addition, destructive sampling does not allow for tracking the temporal dynamics of LChl of the same leaves [5].

Non-destructive methods provide a cost-efficient way for frequent measurement of LChl over a large area [2,9,10]. Studies have found that spectral indices derived from light absorbance or reflectance at the visible and near infrared (NIR) regions have good correlations with LChl [24–27] and can be used to develop non-destructive methods for LChl measurements [27,28]. Portable chlorophyll meters (PCMs), such as the SPAD-502/501 (Soil Plant Analysis Development (SPAD) chlorophyll meter, Konica–Minolta, Inc., Osaka, Japan), the CCM-200 (CCM-200 plus Chlorophyll Content Meter, Opti-Sciences, Inc., Hudson, NH), and the Dualex-4 (Dualex Scientific^{+TM} Polyphenol & Chlorophyll Meter, FORCE-A, Orsay, France), have been developed for non-destructive measurements of LChl and nitrogen in the field [2,29,30]. The readings from the PCMs (meter reading) are relative quantities that need to be converted to actual LChl. The transformation equations are usually established using meter readings and lab-measured LChl of the same leaf area [21,29–32]. For instance, Markwell et al. [21] developed a widely used exponential equation to estimate LChl from SPAD-502 readings, and Cerovic et al. [29] subsequently developed a generic conversion function for SPAD-502 readings based on more data collected in different studies (e.g., Markwell et al. [21], Richardson et al. [30] and Marenco et al. [33]).

It should be noted that factors other than LChl may also influence the light transmittance of a leaf, such as leaf structure, water content and leaf pigment distribution [2,34,35]. Environmental factors such as light intensity can also affect light transmittance of a leaf, resulting in measurement errors of LChl [36,37]. Influences on light transmittance can be categorized into two groups [38–40]. The first is the detour effect (light scattering), caused primarily by non-chlorophyll components (e.g., leaf architecture and dry matter), which can result in an increase in the path length of light inside a leaf [41]. The sieve effect occurs when light passes through leaf tissues without being absorbed, thereby decreasing total absorption [39,40,42]. The distribution of chlorophyll molecules within a leaf is usually non-uniform, associated with the structural organization of the grana within the chloroplasts, chloroplasts within the cells, and cells within the tissue layers [43,44]. Furthermore, the influences on light transmittance vary with wavelength. Since different PCMs are developed based on different wavelengths, they may be impacted differently by different factors. Large uncertainties have been reported when converting meter readings into LChl using a general conversion function for different crops [29,39]. An in depth understanding of the mechanisms for the PCMs is useful for improving protocols to obtain high-quality in situ LChl measurements. However, it is difficult to consider the impacts from all leaf/canopy and environmental factors through field experiments. Using a leaf radiative transfer model to simulate the complex light transmission processes inside a leaf may provide a solution [14,35,43]. This study, therefore, was designed to address the following: (1) the performances of different PCMs in estimating actual LChl, (2) the relationships between PCM reading and the actual LChl, (3) the sources of errors in PCM measurements based on simulations of light transmission in a leaf using radiative transfer model, and (4) the potential of deriving a generic conversion equation for a specific PCM. To address these, an experiment was conducted to collect PCM readings of corn (*Zea mays* L.), spring wheat (*Triticum aestivum* L.), soybean (*Glycine max* L. Merr.) and canola (*Brassica napus* L.) using the three aforementioned instruments (SPAD-502, CCM-200 and Dualex-4) at the Central Experimental Farm (CEF) of Agriculture and Agri-Food Canada (AAFC) in Ottawa, during the 2016 growing season.

2. Materials and Methods

2.1. Theoretical Basis of Portable Chlorophyll Meters

A portable chlorophyll meter measures the light transmittance of a leaf at two different wavelengths, the index band and the reference band. The index band resides in a chlorophyll absorption region, whereas the reference band is located in the NIR region. There is generally no light absorption in the reference band, which is used to compensate for mechanical differences caused by leaf structure, such as leaf thickness or/and leaf density [24,27,45].

The SPAD-502 (Konica–Minolta, Inc., Osaka, Japan) is the earliest and most widely used PCM. It quantifies LChl based on the difference of light transmittance between the NIR band (centered at 940 nm with a full-width at half magnitude (FWHM) of about 10 nm) and the red band (centered at 650 nm with FWHM of about 30 nm). The reading is formulated as [21,40]:

$$SPAD = k\left[\log\left(\frac{I'_{940}}{I_{940}}\right) - \log\left(\frac{I'_{650}}{I_{650}}\right)\right] + c = k[\log(T_{940}) - \log(T_{650})] + c \quad (1)$$

where k and c are calibration coefficients, I'_{940} and I'_{650} are transmitted light intensities at respective wavelengths, I_{940} and I_{650} are light intensities of the LED light sources, and T_{940} and T_{650} are light transmittances through the leaf. SPAD-502 has a 6 mm^2 (2 mm × 3 mm) measurement aperture. The claimed accuracy of the SPAD-502 reading is within ±1.0 unit for a range of 0 to 50 under normal conditions. The value may be less accurate when the reading is greater than 50.

The CCM-200 PCM (Opti-Sciences, Inc., Hudson, NH, USA, Apogee Instruments 2011) measures a chlorophyll content index (CCI)—the ratio of leaf light transmittance between the wavelength of 931 and 653 nm (T_{931} and T_{653}) ([39]:

$$CCI = \frac{I'_{931}/I_{931}}{I'_{653}/I_{653}} = \frac{T_{931}}{T_{653}} \quad (2)$$

where, I'_{931} and I'_{653} are the measured leaf light transmission intensities at respective wavelengths, and I_{931} and I_{653} are the light intensities of the LED light source centered at 931 (FWHM about 25 nm) and 653 nm (FWHM about 50 nm), respectively. Calibration is required every time the instrument is turned on. The CCM-200 has a sensing aperture of 71 mm^2 (9.5 mm diameter). The readings of CCM-200 range from 0 to 200, with a resolution of ±1.0 CCI units.

The Dualex-4 Scientific (FORCE-A, Orsay, France) is a new-generation polyphenol and chlorophyll meter that measures the leaf chlorophyll index (Chl), the flavonol index (Flav), the anthocyanin index (Anth) and nitrogen balance index (NBI) [29,46]. Flav and Anth are relative measures of flavonol and anthocyanins' concentration, respectively [46–48]. NBI is the ratio between Chl and Flav, corresponding to LChl corrected by dry leaf mass per unit area. Calculation of Chl uses a red-edge band centered at 710 nm and an NIR band centered at 850 nm. The reason for using a red-edge band is that indices based on the red-edge and the NIR wavelength have better sensitivity to chlorophyll concentration than the indices based on the red and NIR wavelengths [28,31,49]. Its circular sensing aperture has a diameter of 5 mm (~20 mm^2). The Chl readings range from 5 to 80 µg cm^{-2}. The function of Chl [29] is:

$$Chl = k\,(\frac{I'_{850}/I_{850}}{I'_{710}/I_{710}} - 1) + c = k(\frac{T_{850}}{T_{710}} - 1) + c \quad (3)$$

where k is the calibration coefficient to obtain leaf chlorophyll concentration in the unit of µg cm^{-2}, and c is a constant for correcting the potential bias of the model. Instrument calibration is required prior to use and is done with no sample in the measuring head.

2.2. Leaf Chlorophyll Concentration Measurement

Field experiments were conducted at the Central Experimental Farm (45.38° N, 75.71° W, Figure 1) in Ottawa, Ontario, Canada, which included a field experiment and a greenhouse experiment. Field sampling for corn, soybean, canola and spring wheat was conducted three times during the growing season on 21 June, 7 July and 4 August 2016. Measurements of LChl for corn and soybeans were collected from a rotation experiment field with eight rotation patterns. The field was divided into 48 strips, each about 9 m wide and 17 m long. All the corn plots received three levels of nitrogen (N) fertilizer (0, 100 and 200 kg N ha^{-1}) with six replications for each level of N application. Soybean was not fertilized. LChl were measured at the plots of continuous cropping with corn or soybean. Canola was planted in a field adjacent to the rotation field (Figure 1). The field was divided into two sections for two different experiments. The first experiment was designed to test eight rates of N application on two canola varieties (InVigor L140P and InVigor 5440), each with four replicates. Each plot was 4.0 m wide by 6.0 m long. The second experiment was designed to test 12 combinations of different rates of N and Sulphur application, with three rates of N (0, 75, 150 kg ha^{-1}) and four rates of Sulphur (0, 10, 20, 40 kg ha^{-1}). Each treatment had four replicates, resulting in a total of 48 plots with the same size as the first experiment. In this study, LChl was measured at plots receiving five levels of N application (0, 50, 100, 150 and 200 kg N ha^{-1}) in the first experiment, and plots receiving 75 kg N ha^{-1} and four rates of Sulphur (0, 10, 20, 40 kg ha^{-1}) in the second experiment. The spring wheat field was adjacent to the canola field (Figure 1). The field was treated with uniform nitrogen fertilization (150 kg N ha^{-1}). LChl was measured at 8 randomly selected plots.

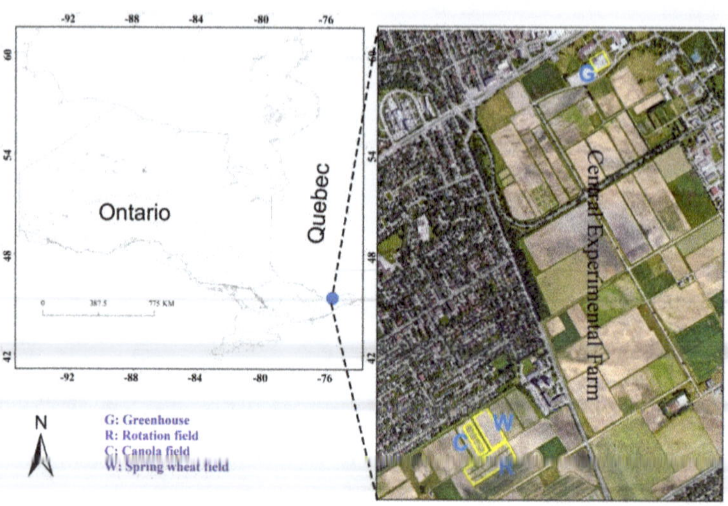

Figure 1. Location of the four sampling fields in the Central Experimental Farm, Ottawa, Ontario, Canada; background satellite imagery (9 June 2018) obtained from the Google Earth Pro.

In the greenhouse, canola was seeded on 2 May 2016. Five rates of N (0, 50, 75, 100, and 150 kg N ha^{-1}) were applied with six replications per treatment to examine the response of crop growth to N application. There were a total of 30 pots with a size of 15 cm in diameter and 15 cm in depth. The soil used was a mixture of one-half sand and one-half potting soil to create soil N deficient. The pots were placed on a bench 1 m above the ground, with a distance of 10 cm from each other. The greenhouse was controlled at 25–15 °C day/night temperature and with a 16 h photoperiod. Plants received both natural light, and light from fluorescence lamps with an intensity of 300 µmol m^{-2} s^{-1} during cloudy periods. Chlorophyll measurements using PCMs and leaf sampling were conducted on 30 May and 16 June 2016, corresponding to the plant's growth stage, with 5–6 leaves and early flowering, respectively.

A total of 220 plants were selected from the two experiments and from different dates (Table 1). For each plant, a fully expanded leaf from the top, usually the second leaf, was selected for measurement. Multiple readings were taken for a leaf depending on its size, and the average reading was used as the representative value for the meter reading of each leaf. In detail, four to five readings were taken for soybean and canola at the middle portion of the leaf and on two sides of the main rib (avoiding the midrib and veins). For spring wheat, six to eight readings were taken at the two sides of the main rib. Measurements with eight to ten readings for corn leaves were taken between the midrib and the leaf margin about 20 cm from the stalk. Readings were taken at the same area on the leaf for the three instruments.

Table 1. Description of the total number of samples for chlorophyll content measurement in each crop type.

	Crop Type	Seeding Date	May 30	June 16	June 21	July 07	August 04	Total
Experiment 1	Corn	May 18	-	-	19	18	18	55
	Soybean	May 12			6	9	10	25
	Spring wheat	April 27	-	-	7	9	6	23
	Canola	May 6			30	27	Harvested	57
Experiment 2 (greenhouse)	Canola	May 2	30	30	-	-	-	60
	Total	-	30	30	62	64	34	220

For destructive sampling, four to six discs (1.09 cm^2) were clipped from the same leaf at the same area where readings were taken using the three PCMs. Leaf discs were placed into a 15 mL plastic tube and kept cool. In the laboratory, the samples were stored in liquid nitrogen at −80 °C before further processing. For chemical determination of pigment concentration, the samples from each site were placed in 10 mL of ethanol solution (96%, *v/v*) and incubated at room temperature in the dark for four days until leaf samples turned white completely. The solution for each site was put in three cuvettes, and light absorption was measured for each cuvette using a Varian Cary 100 Bio UV-Visible Spectrophotometer (Thermo Electron Corporation, Madison, WI, USA) at three wavelengths: 665, 649 and 470 nm. Chlorophyll a (Chl_a), chlorophyll b (Chl_b) and total carotenoid concentrations (C_{ar}) were calculated using the following equations (Lichtenthaler et al. [22]):

$$Chl_a(\mu g\ cm^{-2}) = (13.95A_{665} - 6.88A_{649}) \times V/TLA \quad (4)$$

$$Chl_b(\mu g\ cm^{-2}) = (24.96A_{649} - 7.32A_{665}) \times V/TLA \quad (5)$$

$$LChl(\mu g\ cm^{-2}) = (6.63A_{665} + 18.08A_{649}) \times V/TLA \quad (6)$$

$$Car(\mu g\ cm^{-2}) = (1000A_{470} - 2.05Chl_a - 114.8Chl_b)/245 \times V/TLA \quad (7)$$

where *A* is the measured absorbance at different wavelengths given by the subscript (in nm); TLA is the total leaf area (cm^2) used, and *V* is the amount of ethanol (solvent, mL). The total chlorophyll concentration (LChl) was the sum of Chl_a and Chl_b. The composition of LChl (Chl_a/Chl_b ratio) was also calculated. Values for the three cuvettes of a site were averaged to obtain pigment concentrations for the site.

2.3. Statistical Analysis

Among the 220 samples collected, 194 samples were retained for analysis. The other 26 samples were excluded, as their measured absorbance at 649 nm was wrong (>1 absorbance unit). To identify the difference in leaf pigment traits among the four crops, statistical indicators, including the average, the minimum, the maximum, and the coefficient of variance (CV), of the meter readings and leaf pigments' concentrations were derived separately. Flav, Anth, and NBI, derived from the Dualex-4, were also analyzed. Correlations among leaf pigment concentrations determined from the lab, and linear or nonlinear regressions between meter readings and leaf pigment concentration (LChl and C_{ar}) were then analyzed, respectively. The purpose of this analysis was to investigate whether other pigments impacted on the conversion functions from optical reading to the actual LChl.

2.4. Sensitivity Analysis

Leaf reflectance and transmittance can be simulated using a leaf radiative transfer mode, by taking into consideration LChl as well as interference factors (e.g, leaf structure and leaf pigments other than LChl) [14,34,35,43]. PCM is developed based on sensing technologies through spectral measurements. The PROSPECT-D [14], the newest version of the PROSPECT leaf optical model, was used. The model simulates leaf directional–hemispherical reflectance and transmittance in the spectral range 400–2500 nm using a set of leaf parameters, such as leaf anthocyanins (C_{anth}), chlorophyll and carotenoid concentration, leaf water concentration (C_w), leaf dry matter concentration (C_m), and leaf structure parameter (N_s) [14,35].

To evaluate the detour effect, leaf–light transmittance was simulated using the PROSPECT-D model. Leaf brown pigment (C_{bp}) was given a value of zero, as the analysis was performed on green leaves only. Except for a constant value assigned to LChl, the other variables (N_s, C_w, C_{ar}, C_m and C_{anth}) were varied in a range determined from field measurements and the literature studies [7,14,35], following a uniform distribution (Table 2). Light transmittances at wavelengths centered at 940, 931, 850, 710, 653, and 650 nm were then simulated. The spectral response functions of the LEDs used in the PCMs [50–52] were also considered. A global sensitivity analysis (GSA) was conducted to determine the contribution variation for each leaf parameter, using the SIMLAB (Simulation Environment for Sensitivity and Uncertainty Analysis, http://simlab.jrc.ec.europa.eu/) software. Detailed information on the variance-based GSA and its application in the simulation of PROSPECT model can be found in the studies of Dong et al. [53] and Liu et al. [54].

Table 2. Summary of PROSPECT-D parameters used to simulate leaf transmittance.

Variable	Constant	Range	Step	Reference
Leaf structure parameter, N_s	1.55	1.0–2.8	0.2	[35]
Leaf chlorophyll concentration, LChl (µg cm^{-2})	48.39	10–80	5	Field collection
Leaf carotenoid concentration, C_{ar} (µg cm^{-2})	8.04	3.6–12.6	1.0	Field collection
Leaf water concentration, C_w (g cm^{-2})	0.0113	0.004–0.04	0.004	[35]
Leaf dry matter concentration, C_m (g cm^{-2})	0.0053	0.0017–0.0137	0.00133	[35]
Leaf anthocyanin concentration, C_{anth} (µg cm^{-2})	1.0	0–14.0	1.4	[14]
Leaf brown pigment, C_{bp}	0.0	-	-	[55]

To investigate the sieve effects, the approach proposed by Uddling et al. [38] was used in the PROSPECT-D simulations. It is assumed that the variation in actual LChl within the measured area follows a normal distribution around the average value of LChl (μ). The standard deviation (σ) of LChl was set within the range 10–50% of μ in steps of 10% of μ. The average LChl (μ) was varied from 10 to 80 µg cm^{-2} with steps of 5 µg cm^{-2}. For simplicity, other variables were given constant values obtained from field measurements (e.g., C_{ar}) or the literature studies (e.g., N_s, C_m and C_{anth})

(Table 2). For instance, C_{ar} was assigned as 8.04 g cm^{-2}, as that is the mean value of C_{ar} obtained in our in situ measurements, and Ns was 1.55 as an average value of cereal crops [35,54]. The variations in the log (T_{940}/T_{650}) (Equation (1)), the (T_{931}/T_{653}) (Equation (2)) and the $(T_{850}/T_{710}-1)$ (Equation (3)) responses to different degrees of heterogeneity of leaf chlorophyll concentration distribution were then analyzed.

3. Results

3.1. Variability of LChl

Statistics of meter readings and lab pigment concentration measurements for four crops together are given in Table 3. Readings by CCM-200 (CCM-200-CCI) had a wider range and larger variation (CV = 46.3%, 8.7–80.3) than those of SPAD-502 (CV = 18.3%, 25.5–70.3) and Dualex-4 (Dualex-4-Chl, CV = 22.5%, 22.1–61.0). Large variations were also observed in Flav (CV = 42.5%), Anth (CV = 46.3%) and NBI (CV = 65.3%) of Dualex-4. For lab chemical measurements, the absolute pigment concentration ranged from 25.6 to 83.6 μg cm^{-2} for LChl (μ = 48.4 μg cm^{-2} and CV = 25.3%) and 3.6 to 13.3 μg cm^{-2} for C_{ar} (CV = 28.0%). These suggest that the variability of the SPAD-502 readings and the Dualex-4-Chl readings was closer to the variability of actual LChl than that of the CCM-200-CCI. For the LChl composition, Chl_a (19.9 to 62.8 μg cm^{-2}) was generally greater than Chl_b (4.6 to 21.1 μg cm^{-2}), with a ratio (Chl_a/Chl_b) between 2.2 and 4.6 (μ = 3.3, CV = 14.4%). C_{ar} generally constituted about 14% of the sum of LChl and C_{ar}, as all selected leaves were in dark-green.

Table 3. Statistics of leaf pigment concentration measurements of four crops (corn, soybean, spring wheat and canola), using three portable chlorophyll meters and lab chemical methods.

LChl	Types	Mean	CV (%) [a]	Min [b]	Max [b]
Portable chlorophyll meter	SPAD-502	46.6	18.3	25.5	67.8
	CCM-200-CCI	33.1	46.3	8.7	75.8
	Dualex-4-Chl	37.4	22.6	22.1	61.0
	Dualex-4-Flav	1.2	42.5	0.2	2.0
	Dualex-4-Anth	0.1	46.3	0.0	0.2
	Dualex-4-NBI	41.1	65.3	11.6	122.5
Lab chemical measurement	C_{ar} (μg cm^{-2})	8.0	28.0	3.6	13.3
	Chl_a (μg cm^{-2})	37.1	26.0	19.9	62.8
	Chl_b (μg cm^{-2})	11.3	25.9	4.6	21.1
	Chl_a/Chl_b ratio	3.3	14.4	2.2	4.6
	LChl (μg cm^{-2})	48.4	25.3	25.6	83.6

[a] CV is the coefficient of variation, as the ratio of the standard error to the average value (n = 195); [b] Min and Max are the minimum and the maximum values, respectively.

Leaves of the four crops were different, according to the statistical characteristics of lab pigment concentration measurements in each crop (Figure 2). Chl_a/Chl_b of corn, with a mean value of 4.0 and range of 3.2–4.6, was higher than that of the other three crops, and Chl_a/Chl_b for canola, spring wheat and soybean crops were close to each other. Spring wheat had a larger mean value of pigment concentrations (LChl, Chl_a, Chl_b and C_{ar}) than other crops (Figure 2). Soybean had lower variation (CV = 15.4% for LChl and 11.5% for C_{ar}) than the other three crops (Table A1). Pigment concentrations of canola in the greenhouse were apparently lower than that of canola in the field. Similar differences can be found in the Flav index of Dualex-4 between the greenhouse canola (Flav = 0.3) and the field canola (Flav = 1.7) (Table A1). This may reflect the great difference in light conditions between field and greenhouse, as the Flav index of Dualex-4 is an indicator of light intensity [36,47,56].

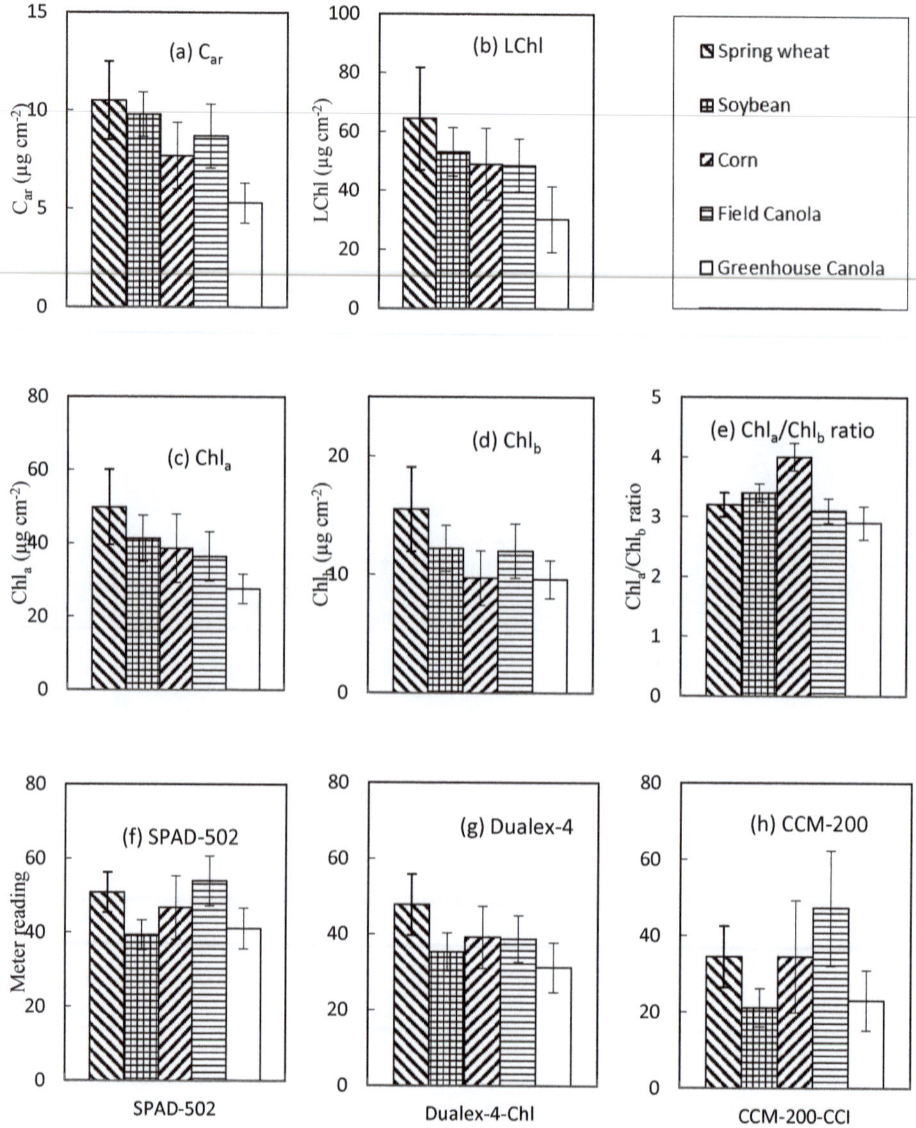

Figure 2. Mean value and standard value in four crops for C_{ar} (**a**), LChl (**b**), Chl_a (**c**), Chl_b, (**d**) Chl_a/Chl_b ratio (**e**), SPAD-502 (**f**), Dualex-4-CCI (**g**) and CCM-200 (**h**).

CCM-200 readings showed the largest variability for any specific crop (Table A1 and Figure 2). Figure 2f–h show that average values of meter readings by each PCM were different among the four crops. The relative differences of average values among the four crops for the SPAD-502 readings were similar to those of the CCM-200-CCI, but were different from those of Dualex-4-Chl. The averages of Dualex-4-Chl for the four crops showed similar relative differences to actual LChl (Figure 2b).

3.2. Correlation among Leaf Pigment Concentrations

Examination of the correlation between Chl_a, Chl_b, LChl and C_{ar} is helpful to understand the potential impact of other pigments on the conversion of PCM readings into actual LChl. Figure 3 shows that there were strong linear correlations among leaf pigment concentrations determined by lab chemical method. The linear regression between Chl_a and Chl_b for corn (C4 plant) was different from that of the other three crops (C3 plant), showing a greater Chl_a/Chl_b. In comparison, the relationship between the C_{ar} and LChl was more universal among the four crops, and both Chl_a and Chl_b had a strong correlation with C_{ar}.

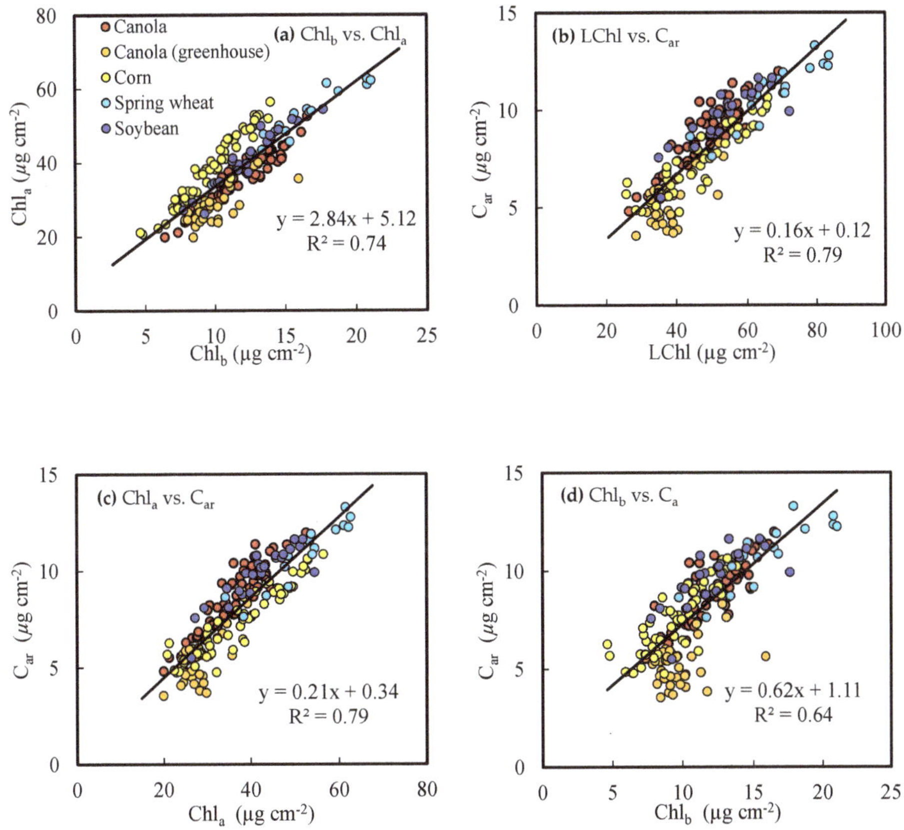

Figure 3. Scatterplots showing the linear relationship between leaf pigments in four different crops for (**a**) leaf chlorophyll a (Chl_a) vs. leaf chlorophyll b (Chl_b), (**b**) leaf chlorophyll concentration (Chl_a+Chl_a, LChl) vs. leaf carotenoid concentration (C_{ar}), (**c**) Chl_a vs. C_{ar} and (**d**) Chl_b vs. C_a.

3.3. Estimation of Pigment Concentration from PCM Readings

Figure 4 shows that both C_{ar} and LChl had a strong linear or nonlinear correlation with PCM readings. In general, the best regression model is nonlinear for SPAD-502 and CCM-200, but is linear for Dualex-4. It was observed that different crops had different relationships between SPAD-502 and CCM-200 readings and actual pigment concentrations (Table 4). Except for Dualex-4, errors of estimates in both LChl and C_{ar} increased with the increasing value of meter readings. For the same level of actual LChl (and C_{ar}), SPAD-502 and CCM-200 had lower readings for spring wheat and soybean than for corn and canola. The R^2 values for SPAD-502 (R^2 = 0.48 for LChl and R^2 = 0.40 for C_{ar}) were larger than those for CCM-200-CCI (R^2 = 0.40 for LChl and R^2 = 0.33 for C_{ar}), suggesting that the readings of SPAD-502 were better at restricting interference from other factors (e.g., leaf architectures) than CCM-200-CCI. The actual LChl values higher than 60 $\mu g\ cm^{-2}$ for corn and canola could not be well estimated from CCM-200-CCI. Dualex-4-Chl was the best PCM for consistent measurements of LChl for all the four crops ($R^2 = 0.74$). A generic function was possible for converting Dualex-4-Chl readings into actual LChl for the four crops, with an average accuracy of 87%.

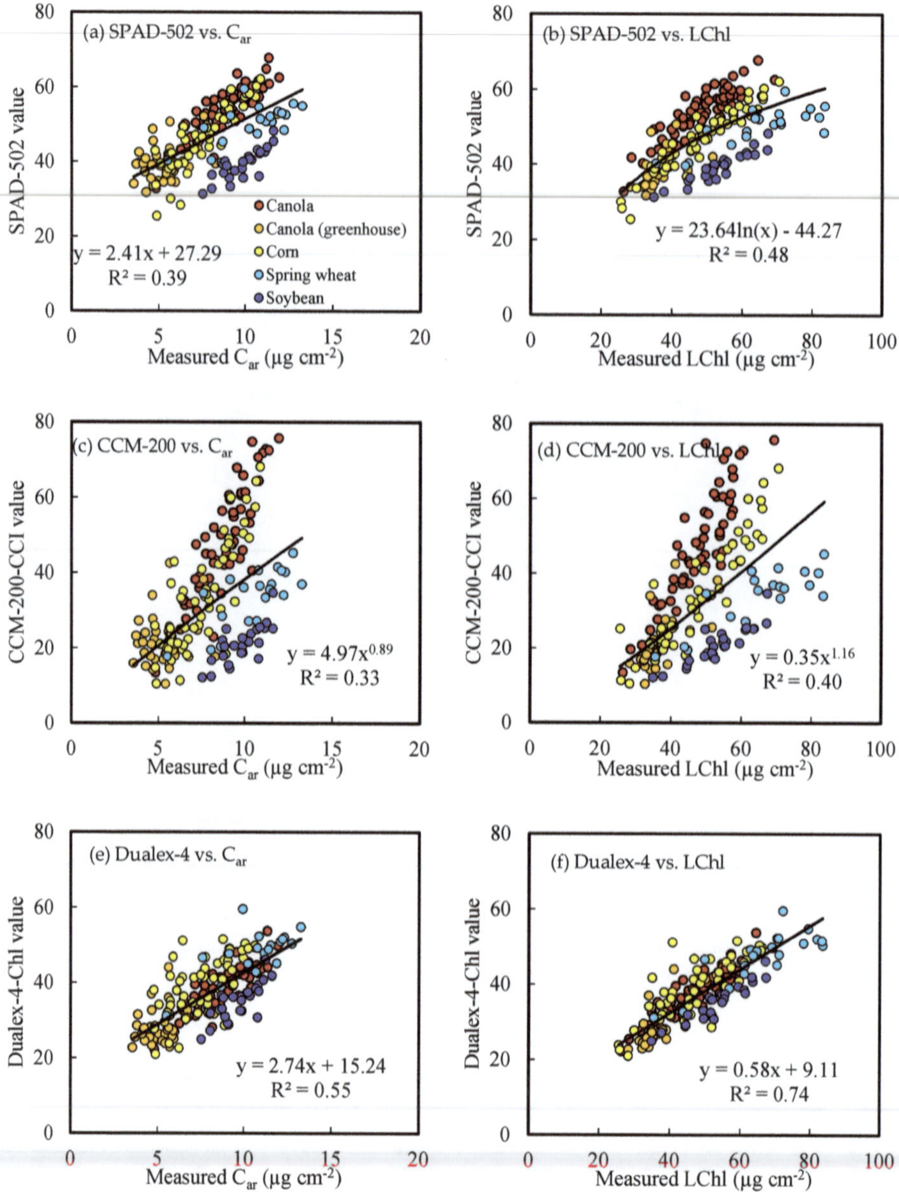

Figure 4. Scatterplots showing linear relationships between leaf carotenoid concentration (Car, μ cm^{-2}), leaf chlorophyll concentration (LChl, μ cm^{-2}) and meter readings for SPAD-502, Dualex-4 and CCM-200 in four different crops.

Table 4. Regression analysis between readings from the handheld chlorophyll meters (meter reading, (x) and lab-measured leaf carotenoid and chlorophyll concentration (y); RMSE (μg cm^{-2}) is the root–mean–square error.

	Handheld Chlorophyll Meter	Leaf Chlorophyll Concentration			Leaf Carotenoid Concentration		
		Regression	R^2	RMSE	Regression	R^2	RMSE
Canola	SPAD-502	y = 0.88x + 1.55	0.77	4.51	y = 0.21x − 2.78	0.80	0.99
	CCM-200-CCI	y = 0.45x + 27.25	0.75	5.93	y = 0.11x + 3.30	0.78	1.30
	Dualex-4-Chl	y = 1.14x + 4.28	0.83	3.86	y = 0.26x − 1.52	0.75	1.11
Corn	SPAD-502	y = 1.31x − 13.02	0.90	3.68	y = 0.17x − 0.47	0.74	0.93
	CCM-200-CCI	y = 0.72x + 23.18	0.81	5.04	y = 0.10x + 4.30	0.68	0.99
	Dualex-4-Chl	y = 1.21x + 0.37	0.69	6.33	y = 0.14x + 1.94	0.46	1.27
Soybean	SPAD-502	y = 1.91x − 21.71	0.88	2.79	y = 0.23x + 0.71	0.68	0.63
	CCM-200-CCI	$y = 8.53x^{0.60}$	0.84	3.44	$y = 3.22x^{0.37}$	0.59	0.72
	Dualex-4-Chl	y = 1.55x − 1.20	0.90	2.56	y = 0.18x + 3.45	0.64	0.66
Spring wheat	SPAD-502	$y = 10.71e^{0.04x}$	0.66	8.88	$y = 2.54e^{0.03x}$	0.48	1.48
	CCM-200-CCI	$y = 4.58x^{0.76}$	0.74	7.57	$y = 1.20x^{0.62}$	0.58	1.28
	Dualex-4-Chl	y =1.52x − 5.57	0.72	5.12	y = 0.19x + 1.76	0.52	1.01
All crops	SPAD-502	$y = 18.29e^{0.02x}$	0.48	9.31	y = 0.16x + 0.47	0.39	1.75
	CCM-200-CCI	$y = 14.49x^{0.34}$	0.40	10.12	$y = 2.18x^{0.37}$	0.33	1.88
	Dualex-4-Chl	y = 1.27x + 1.11	0.74	6.25	y = 0.20x + 0.51	0.55	1.48

3.4. Relationship of Meter Reading Averages and Deviations

Multiple readings were taken per plant sample using the PCMs, from which the average and the standard deviation can be derived for each sample plant. Figure 5 shows the relationship between the averages and the standard deviations of the PCM readings. The results of the greenhouse canola are shown as an example, using SPAD-502 measurements. The standard deviation generally increased for both SPAD-502 and CCM-200-CCI readings which suggests that the sources of measured error of the two instruments increased with actual LChl. However, the standard deviation in the Dualex-4-Chl readings was smaller and did not show an apparent increasing trend.

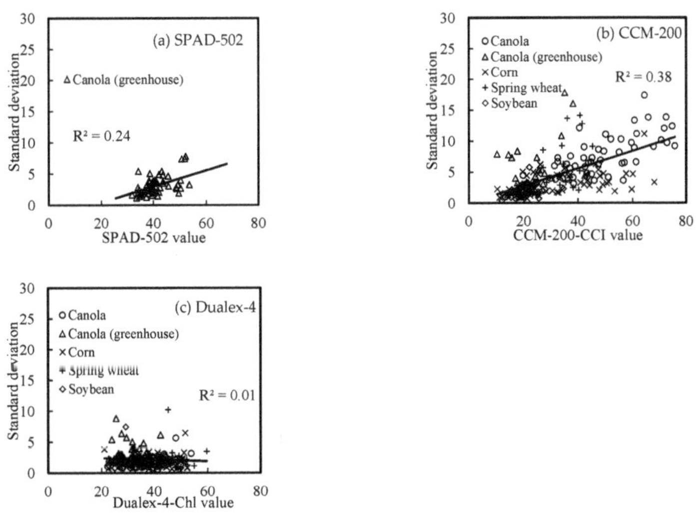

Figure 5. Relationship between meter readings and standard deviation error of measurements for SPAD-502, CCM-200 and Dualex-4; the meter reading was the average value of 4–5 readings of each leaf sample and the standard deviation value was derived from these readings; the standard deviation for SPAD-502 was only recorded for the canola in the Greenhouse.3.5. Factors Affecting Meter Readings.

3.4.1. Influence of Leaf Parameters

Results from the global sensitivity analysis (GSA) in Table 5 indicate that variability of light transmittance is primarily affected by LChl and N_s for the index band and by N_s and C_m for the reference band. Compared with the index band used in Dualex-4-CCI (710 nm), light transmittance in the index band used by both SPAD-502 (650 nm) and CCM-200-CCI (653 nm) is more sensitive to LChl, especially low LChl. N_s contributed >90% of the light transmittance variability for the reference band. Increasing N_s could increase light interaction probability within a leaf, and thereby boost light absorption and reduce light transmittance [14,40,57,58]. Increasing C_m would yield similar results but at a lower level. The impact of other parameters on light transmittance variability, including C_{ar}, C_{anth} and C_w, were relatively small.

Table 5. Results of the global sensitivity analysis (GSA). The first order index derived from the GSA represents the contribution of a parameter (Ns, LChl, C_{ar}, C_w, C_m and C_{anth} in the first row) to a variable (T650, T653 etc. in the second column), and the interaction effects of all the parameters.

		N_s	LChl	C_{ar}	C_w	C_m	C_{anth}	Interactions
Index band	T650 (SPAD-502)	11.36	79.73	0.00	0.00	0.01	0.00	8.90
	T653 (CCM-200)	12.00	79.31	0.00	0.00	0.02	0.00	8.67
	T710 (Dualex-4)	42.57	55.96	0.00	0.00	0.33	0.00	1.14
Reference band	T940 (SPAD-502)	95.74	0.00	0.00	0.35	3.53	0.00	0.38
	T931 (CCM-200)	95.72	0.00	0.00	0.38	3.52	0.00	0.38
	T850 (Dualex-4)	95.85	0.00	0.00	0.01	3.75	0.00	0.39
Ratio	T940/T650	11.37	74.13	0.00	0.01	0.05	0.00	14.44
	Log(T940/T650) (SPAD-502)	7.50	91.68	0.00	0.01	0.03	0.00	0.78
	(T931/T653) (CCM-200)	10.81	79.96	0.00	0.02	0.05	0.00	9.16
	(T850/T710) (Dualex-4)	10.91	84.39	0.00	0.00	0.17	0.00	4.53

Note: Ns: leaf structure parameter; LChl (μ cm^{-2}): leaf chlorophyll concentration; C_{ar} (μ cm^{-2}): leaf carotenoid concentration; C_w (μ cm^{-2}): leaf water concentration; C_m (μ cm^{-2}): leaf dry matter concentration and C_{anth} (μ cm^{-2}): leaf anthocyanin concentration.

Table 5 also shows that the ratio between NIR and VIS transmittance can suppress the interaction effects of leaf structure (e.g., N_s and C_m) on meter readings. The effect is more apparent for T850/T710 in Dualex-4-Chl. The influence of leaf parameters on T940/T650 used in SPAD-502 is not different from that on T931/T653 used in CCM-200. This is because the center wavelengths of the two bands used in SPAD-502 are close to that used in the CCM-200. However, it is important to note that the logarithmic transformation of T940/T650 used in the SPAD-502 instrument reduces the influence of N_s and other parameters, and thereby improves its sensitivity to LChl measurement, compared with the ratio of T940/T650 and the ratio of T931/T653. Compared to CCM-200 and Dualex-4, this increases the sensitivity of SPAD-502 to LChl. The meter readings of Dualex-4 are more sensitive to LChl compared with that of CCM-200.

Further analysis (Figure 6) showed that the variability of PCM readings caused by other interference factors (N_s, C_w, C_{ar}, C_m and C_{anth}) increases with LChl, consistent with the studies by Uddling et al. [38] and Nauš et al. [40]. The result in Figure 6(d) shows that log(T940/T650) used in SPAD-502 could have the best performance in reducing the influences of light scattering caused by interference factors (e.g., leaf structure), especially at a high LChl. The change in CV was more stable for Dualex-4 (T850/T710; 12.21–21.35%) and SPAD-502 (log (T940/T650); 7.66–12.21%). However, the change in CV for SPAD-502 (log (T940/T650)) was generally lower than for Dualex-4 (T850/T710).

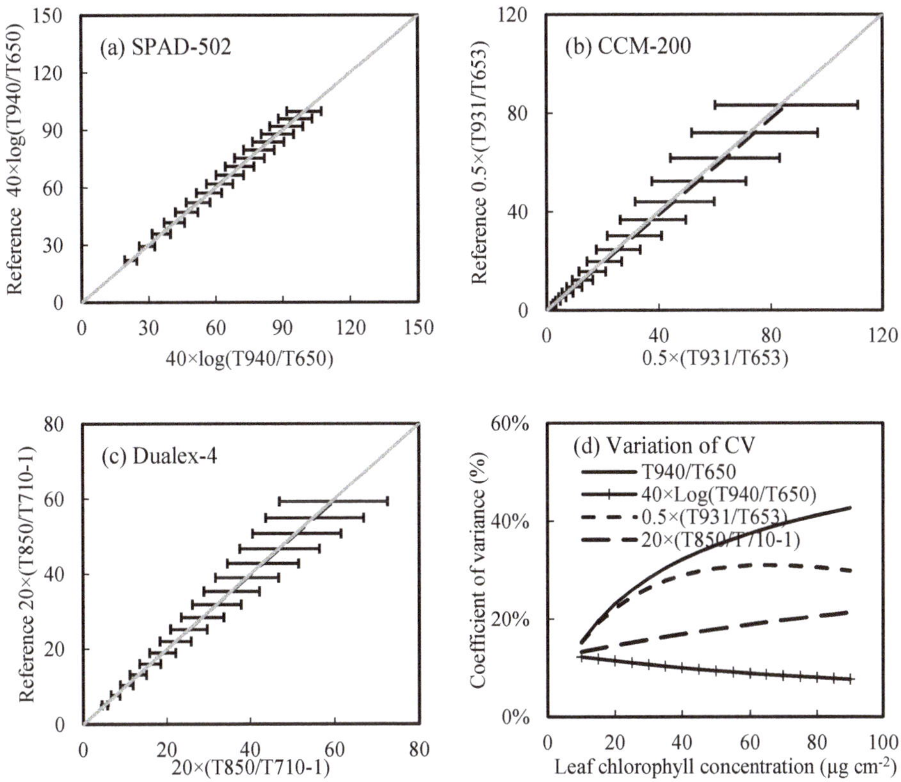

Figure 6. Uncertainty of the readings (shown by the error bars) due to interference factors other than leaf chlorophyll for the three instruments (**a–c**) and the relative error as a function of leaf chlorophyll concentration (**d**). The solid gray lines represent the 1:1 relationship (no impact from other factors). Parameters of the PROSPECT-D simulation are given in Table 2.

3.4.2. The Influence of Non-Uniform LChl Distribution

Figure 7 shows that the impact of non-uniform distribution of LChl on meter readings becomes more apparent with increasing LChl. Meter readings decrease with increasing heterogeneity of LChl distribution, in particular at high LChl values. This is more apparent for CCM-200 and less apparent for Dualex-4. The readings of Dualex-4 were the least affected by the non-uniform distribution of LChl, as the sensitivity of LChl to light transmission at 710 nm was weaker than at the red wavelengths (Table 5).

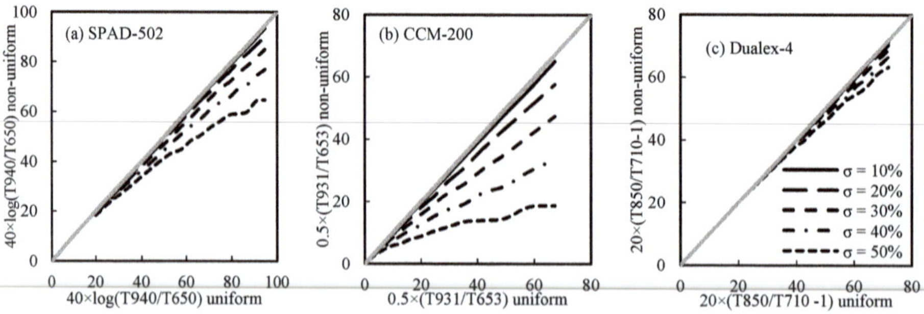

Figure 7. Influence of non-uniform distribution of leaf chlorophyll concentration on the readings of portable chlorophyll meters: (**a**) SPAD-502; (**b**) CCM-200; (**c**) Dualex-4; the solid gray lines (1:1 line) represent the change of light transmittance under the same amount of leaf chlorophyll concentration with uniform distribution. Parameters of the PROSPECT-D simulation are shown in Table 2.

4. Discussion

The four crops were selected to explore the capability of the three PCMs for crop LChl measurement. Leaf pigment contents were different among the four crops (Figure 2 and Table A1). In particular, the composition of LChl (i.e., Chl_a/Chl_b ratio, Table A1) was obviously different between C4 plants (corn) and C3 plants (spring wheat, canola and soybean) because of their difference in photosynthesis pathways [39,56,59,60]. Actual LChl, especially high LChl, had a larger linear correlation coefficient with the Dualex-4-Chl readings than with the SPAD-502 and the CCM-200-CCI for an individual crop and for the four crops combined (Table 4). Both SPAD-502 and CCM-200-CCI tend to have a nonlinear relationship with actual LChl, as they are sensitive to LChl at low LChl rates but are easily saturated at high LChl rates [61]. These results are consistent with the results from Cerovic et al. [29] and Casa et al. [32], in which Dualex-4-Chl performed better than the other two instruments. More importantly, there is a greater potential to develop a generic calibration function for LChl estimates of the four crops using Dualex-4-Chl than using SPAD-502 and CCM-200-CCI. Dualex-4 could be more accurate and applicable than the other two PCMs in measuring LChl over a wide, dynamic range using a generic conversion function. Time and effort taken to recalibrate the concversion fuction for different crops would be largely reduced by using the Dualex-4 compared to the other two PCMs. Moreover, Dualex-4 has lower uncertainties at high LChl rates (Figures 5 and 7). LChl generally shows an increasing trend during the vegetative stage, and reaches its maximum at the peak growing stage, which is an improtant inditor for asessing crop nitrogen uptake and crop yield [7,62]. Thus, Dualex-4 might be a better PCM to accuartely measure LChl at the peak growth stage.

Sensitivity analysis using the PROSPECT-D model simulation revealed that the leaf structure parameter (N_s) had a strong impact on the variability of PCM readings. The literature showed similar results; PCM readings were significantly affected by leaf internal architecture such as leaf thickness, specific leaf mass, and leaf succulence [12,27,33,58,63]. Variability in the leaf structure parameter greatly influenced light interactions within the leaf, resulting in significant changes in light transmittance in the VIS and NIR ranges [27,35,57]. Our study showed that the influence of leaf structure parameters was different on both the reference and the index bands. The influence of the parameters on the index band (710 nm) for Dualex-4-Chl was much greater than for SPAD-502 (650 nm) and CCM-200-CCI (653 nm), while the influence of the parameters on the reference band was comparable for all three PCMs (Table 5). The influence of multiple scattering of light by leaf tissues could be further reduced by taking a simple ratio of NIR to VIS transmittance [26,27,39,58,64]. The ratio of T850/T710 was better to restrict the influence of leaf structure parameters than T940/T650 and T937/T653. The logarithmic function applied to the ratio of 940/650 can restrict the influence of leaf structure parameters, especially at high LChl, and improve the sensitivity to C_{ab}. However, the overall ability of the three PCMs to reduce the impacts was limited (<15%), and their difference was small. The non-uniform distribution of

LChl within the measured area was another important factor influencing the variation of PCM readings. The sensing area of CCM-200 (71 mm^2) is much larger than the sensing area of SPAD-502 (6 mm^2) and Dualex-4 (20 mm^2), hence CCM-200 can be more susceptible to greater non-uniformity in LChl distribution. In particular, CCM-200-CCI had larger variations at high LChl, compared with SPAD-200 and Dualex-4 (Figure 4), which is consistent with the observations of Padilla et al. [65]. Our simulation, using the method by Uddling et al. [38], showed that CCM-200-CCI was most sensitive to the degree of non-uniformity of the LChl distribution, especially when LChl was high (Figure 7). Increased heterogeneity of LChl distribution across leaf area can result in an increased light transmittance and decreased light absorption at the red wavelengths [38]. Although similar wavelengths are employed in both SPAD-502 and CCM-200-CCI, the variability of SPAD-502 readings is not largely influenced by the heterogeneity distribution of LChl within the measured area. This could be attributed to the logarithmic transformation used in the SPAD-502, which helps reduce the divergence of non-uniform distribution of LChl [39]. Dualex-4-Chl was the best at restricting the influence of the non-uniform LChl distribution. The data in this study showed that the meter readings for both SPAD-502 and CCM-200 had crop-specific relationships with LChl (Figure 4). In particular, the deviations apparently increased at high rates of LChl. In addition, the errors of meter readings within the measured area increased with increasing LChl (Figure 5). The results in both Figures 4 and 5 are more consistent with Figure 7 than with Figure 6, suggesting that uncertainty in PCM readings was more due to non-uniform distribution of LChl than to leaf structure parameters. The studies of Parry et al. [39] and Richardson et al. [30] also showed that the non-uniform distribution of LChl at high LChl had an apparent influence on both SPAD-502 and CCM-200 readings. Previous studies [20,63] found that the heterogeneity of the pigment distribution was greater for leaves with a higher pigment content, as leaves with high chlorophyll concentration tend to have a high chlorophyll *density* in chloroplasts, rather than develop more chloroplasts [20].

The variations of light intensity can greatly influence chloroplast movement inside a leaf [37,40,43,66]. This can lead to variations of leaf optical properties (reflectance, absorbance and fluorescence) for the same amount of LChl, and therefore difference in PCM readings. Nauš et al. [40] found that the movement of chloroplast from the cell walls perpendicular to the incident light (face position) to the cell walls parallel to the incident light (the side position) could induce approximately 35% of the difference in SPAD-502 readings. Padilla et al. [36] reported that measurement time in day determines light intensity and can have a strong impact on SPAD-502 and CCM-200 readings. The ratio of Chl_a/Chl_b was considered to be an indicator of light intensity that has strong influence on chloroplast movement [56,67]. Strong relationships between the Chl_a/Chl_b ratio and SPAD-502 readings were observed by Netto et al. [67] and Li et al. [63]; however, we did not find similar relationships between the Chl_a/Chl_b ratio and PCM readings in the present study, similar to the result obtained by Parry et al. [39]. The large difference in light conditions between the field and the greenhouse for canola was revealed by the Flav index of Dualex-4 measurements, which was another indicator of light intensity [36,47,56]. In addition, the chloroplast movement is closely linked with the combined effects of light scattering (detour effect) and non-uniform chlorophyll distribution (sieve effect) [40,43,66]. For instance, Nauš et al. [40] reported that the impact of chloroplast movement on the relationship between SPAD readings and actual LChl was different between old and young tobacco leaves with different leaf structures (e.g., leaf area mass and leaf thickness). Our study results indirectly support this finding. In particular, the variability of the CCM-200-CCI readings was greatly affected by both light scattering and non-uniform LChl distribution. More recently, Stuckens et al. [68] developed a Dorsiventral Leaf Model (DLM) to simulate leaf radiative transfer by considering the influence of leaf asymmetry that is modeled by assigning non-uniform distributions of pigments, water and dry matter to palisade and mesophyll layers and by simulating different amounts of light diffusion for adaxial and abaxial leaf surfaces. Baránková et al. [43] developed a Simple Explicitly Non-Linear Empirical model for Leaf Optical Properties (SENLELOP model) to investigate the influence of chloroplast movement on the optical properties of green tobacco leaves. In future studies, a specially designed field experiment integrated with the SENLELOP model, DLM, or other, similar models, could lead to an improved understanding of the mechanistic relationship between optical readings and LChl.

5. Conclusions

In this study, we evaluated the performances of three commonly used portable chlorophyll meters (SPAD-502, CCM-200 and Dualex-4) in measuring the leaf chlorophyll concentration (LChl) of four different crops. Analyses were conducted based on field measurements of four crops (corn, soybean, spring wheat and canola), to explore the relationships between the actual LChl measured in the lab and readings from the portable chlorophyll meters (PCM). Simulation of leaf transmittance using the PROSPECT-D model was used to further explore the driving factors of light transmission on the used wavelengths, including leaf pigments other than chlorophyll, leaf internal structure, and the heterogeneity of LChl distribution. Major conclusions can be drawn as follows:

(1) SPAD-502 and CCM-200 readings of this study had larger dynamic ranges than Dualex-4 readings. The sources of error for both SPAD-502 and CCM-220 readings increased with increasing LChl, whereas they were relatively stable for Dualex-4;
(2) Relationships between SPAD-502 and CCM-200 readings and the actual LChl were more sensitive to crop type than the relationships between Dualex-4 and LChl;
(3) The sieve effect (caused by the heterogeneity of LChl distribution) would have more influence on PCM readings than the detour effect (caused by leaf parameters, such as leaf pigments and leaf internal structure) does. The ratio of light transmittance between the index and reference bands used in the Dualex-4-Chl was generally better at minimizing the interference factors;
(4) Our results suggest that Dualex-4 is a better choice for collecting LChl measurements for different crops in the field, compared with the SPAD-502 and the CCM-200.

Author Contributions: Conceptualization, J.S. and J.L.; methodology, T.D., J.S., J.M.C., and J.L.; formal analysis, T.D., J.S., J.M.C., J.L. and B.Q.; investigation, T.D., J.S., J.M.C., J.L. and B.Q.; writing—original draft preparation, T.D.; writing—review and editing, all authors (T.D., J.S., J.M.C., J.L., B.Q., B.M., M.J.M., C.Z., Y.L., Y.S., H.P., G.Z.); supervision, J.S., J.M.C., and J.L.; project administration, J.S., and J.M.C.

Funding: This work was supported by Agriculture and Agri-Food Canada's Lang Productivity (#1130) and Crop Harvest Monitoring (#1313) projects, as well as the Canadian Space Agency grant (#17SUSOARTO).

Acknowledgments: The authors thank the anonymous reviewers for the constructive criticism to improve the quality of the manuscript.

Conflicts of Interest: The authors declare no conflict of interest.

Appendix A

Table A1. Statistics of leaf pigment content measurements by crop type (canola, corn, soybean and spring wheat).

Crop		Types	Mean	CV (%) [a]	Range (Min–Max) [b]
Canola (Field, n = 57)	Chlorophyll meter	SPAD-502	53.9	12.5	32.8–67.8
		CCM-200-CCI	47.2	32.1	13.4–75.8
		Dualex-4-Chl	38.7	15.8	22.3–58.8
		Dualex-4-Flav	1.7	13.3	1.1–2.0
		Dualex-4-Anth	0.1	50.4	0.0–0.1
		Dualex-4-NBI	24.1	23.7	11.6–36.6
	Lab chemical measurement	Car (μg cm^{-2})	8.7	18.7	4.8–12.0
		Chla (μg cm^{-2})	[illegible]	[illegible]	[illegible]
		Chlb (μg cm^{-2})	12.0	19.2	6.4–16.6
		Chla/Chlb ratio	3.1	6.6	2.7–3.8
		LChl (μg cm^{-2})	48.6	18.4	26.4–69.2

Table A1. *Cont.*

Crop		Types	Mean	CV (%) [a]	Range (Min–Max) [b]
Canola (Greenhouse, n = 41)	Chlorophyll meter	SPAD-502	41.1	13.4	31.8–53.6
		CCM-200-CCI	23.0	34.3	8.7–46.9
		Dualex-4-Chl	31.2	21.1	22.8–52.9
		Dualex-4-Flav	0.3	19.3	0.2–0.5
		Dualex-4-Anth	0.1	31.9	0.0–0.1
		Dualex-4-NBI	90.7	12.2	65.6–122.5
	Lab chemical measurement	Car ($\mu g\ cm^{-2}$)	5.3	19.4	3.6–8.3
		Chla ($\mu g\ cm^{-2}$)	27.6	14.9	19.9–40.2
		Chlb ($\mu g\ cm^{-2}$)	9.6	16.6	7.8–15.9
		Chla/Chlb ratio	2.9	9.5	2.2–3.3
		LChl ($\mu g\ cm^{-2}$)	30.3	37.0	11.4–53.2
Corn (n = 52)	Chlorophyll meter	SPAD-502	46.7	18.4	25.5–62.2
		CCM-200-CCI	34.5	42.6	20.4–68.2
		Dualex-4-Chl	39.2	20.9	21.1–52.4
		Dualex-4-Flav	1.4	22.4	0.7–1.8
		Dualex-4-Anth	0.1	33.4	0.0–0.2
		Dualex-4-NBI	30.3	29.5	12.4–53.8
	Lab chemical measurement	Car ($\mu g\ cm^{-2}$)	7.7	22.1	4.8–10.9
		Chla ($\mu g\ cm^{-2}$)	38.6	24.1	20.8–56.5
		Chlb ($\mu g\ cm^{-2}$)	9.7	24.0	4.6–14.0
		Chla/Chlb ratio	4.0	5.8	3.2–4.6
		LChl ($\mu g\ cm^{-2}$)	49.0	25.1	25.6–70.5
Soybean (n = 25)	Chlorophyll meter	SPAD-502	39.3	10.2	31.4–48.3
		CCM-200-CCI	21.1	23.8	12.1–34.7
		Dualex-4-Chl	35.3	14.2	25.0–46.2
		Dualex-4-Flav	1.47	8.2	1.2–1.7
		Dualex-4-Anth	0.1	40.7	0.0–0.1
		Dualex-4-NBI	24.1	12.4	16.8–29.9
	Lab chemical measurement	Car ($\mu g\ cm^{-2}$)	9.8	11.5	7.6–11.6
		Chla ($\mu g\ cm^{-2}$)	41.3	15.3	27.2–51.8
		Chlb ($\mu g\ cm^{-2}$)	12.2	15.9	7.7–15.9
		Chla/Chlb ratio	3.4	4.4	3.0–3.8
		LChl ($\mu g\ cm^{-2}$)	53.2	15.4	34.9–67.3
Spring wheat (n = 20)	Chlorophyll meter	SPAD-502	50.8	10.8	39.7–59.6
		CCM-200-CCI	34.5	23.2	17.7–47.5
		Dualex-4-Chl	47.8	16.9	31.2–61.0
		Dualex-4-Flav	1.2	8.2	1.1–1.5
		Dualex-4-Anth	0.1	48.4	0.0–0.1
		Dualex-4-NBI	38.9	19.6	25.1–56.6
	Lab chemical measurement	Car ($\mu g\ cm^{-2}$)	10.5	18.9	5.5–13.3
		Chla ($\mu g\ cm^{-2}$)	49.8	20.6	26.3–62.8
		Chlb ($\mu g\ cm^{-2}$)	15.5	22.9	9.2–21.1
		Chla/Chlb ratio	3.2	6.3	2.9–3.6
		LChl ($\mu g\ cm^{-2}$)	64.4	27.0	32.3–97.8

[a] CV (%) is the coefficient of variation, as the ratio of the standard error to the average value (n = 195); [b] Min and Max are the minimum and the maximum values, respectively.

References

1. Corti, M.; Cavalli, D.; Cabassi, G.; Marino Gallina, P.; Bechini, L. Does remote and proximal optical sensing successfully estimate maize variables? A review. *Eur. J. Agron.* **2018**, *99*, 37–50. [CrossRef]
2. Padilla, F.M.; Gallardo, M.; Peña-Fleitas, M.T.; De Souza, R.; Thompson, R.B. Proximal optical sensors for nitrogen management of vegetable crops: A review. *Sensor* **2018**, *18*, 2083. [CrossRef] [PubMed]
3. Dong, T.; Shang, J.; Liu, J.; Qian, B.; Jing, Q.; Ma, B.; Huffman, T.; Geng, X.; Sow, A.; Shi, Y.; et al. Using RapidEye imagery to identify within-field variability of crop growth and yield in Ontario, Canada. *Precis. Agric.* **2019**, *20*, 1231–1250. [CrossRef]
4. Croft, H.; Chen, J.M. Leaf Pigment Content. In *Comprehensive Remote Sensing*; Liang, S., Ed.; Elsevier: Oxford, UK, 2018; pp. 117–142.

5. Blackburn, G.A. Hyperspectral remote sensing of plant pigments. *J. Exp. Bot.* **2007**, *58*, 855–867. [CrossRef] [PubMed]
6. Houlès, V.; Guérif, M.; Mary, B. Elaboration of a nitrogen nutrition indicator for winter wheat based on leaf area index and chlorophyll content for making nitrogen recommendations. *Eur. J. Agron.* **2007**, *27*, 1–11. [CrossRef]
7. Houborg, R.; Cescatti, A.; Migliavacca, M.; Kustas, W. Satellite retrievals of leaf chlorophyll and photosynthetic capacity for improved modeling of GPP. *Agric. For. Meteorol.* **2013**, *177*, 10–23. [CrossRef]
8. Houborg, R.; McCabe, M.F.; Cescatti, A.; Gitelson, A.A. Leaf chlorophyll constraint on model simulated gross primary productivity in agricultural systems. *Int. J. Appl. Earth Obs. Geoinf.* **2015**, *43*, 160–176. [CrossRef]
9. Gitelson, A.A.; Peng, Y.; Viña, A.; Arkebauer, T.; Schepers, J.S. Efficiency of chlorophyll in gross primary productivity: A proof of concept and application in crops. *J. Plant Physiol.* **2016**, *201*, 101–110. [CrossRef]
10. Zhang, Q.; Cheng, Y.-B.; Lyapustin, A.I.; Wang, Y.; Gao, F.; Suyker, A.; Verma, S.; Middleton, E.M. Estimation of crop gross primary production (GPP): fAPARchl versus MOD15A2 FPAR. *Remote Sens. Environ.* **2014**, *153*, 1–6. [CrossRef]
11. Escobar-Gutiérrez, A.J.; Combe, L. Senescence in field-grown maize: From flowering to harvest. *Field Crop. Res.* **2012**, *134*, 47–58. [CrossRef]
12. Shah, S.H.; Houborg, R.; McCabe, M.F. Response of Chlorophyll, Carotenoid and SPAD-502 Measurement to Salinity and Nutrient Stress in Wheat (*Triticum aestivum* L.). *Agronomy* **2017**, *7*, 61. [CrossRef]
13. Daughtry, C.; Walthall, C.; Kim, M.; De Colstoun, E.B.; McMurtrey, J. Estimating corn leaf chlorophyll concentration from leaf and canopy reflectance. *Remote Sens. Environ.* **2000**, *74*, 229–239. [CrossRef]
14. Féret, J.-B.; Gitelson, A.; Noble, S.; Jacquemoud, S. PROSPECT-D: Towards modeling leaf optical properties through a complete lifecycle. *Remote Sens. Environ.* **2017**, *193*, 204–215. [CrossRef]
15. Houborg, R.; McCabe, M.F. Adapting a regularized canopy reflectance model (REGFLEC) for the retrieval challenges of dryland agricultural systems. *Remote Sens. Environ.* **2016**, *186*, 105–120. [CrossRef]
16. Drusch, M.; Del Bello, U.; Carlier, S.; Colin, O.; Fernandez, V.; Gascon, F.; Hoersch, B.; Isola, C.; Laberinti, P.; Martimort, P. Sentinel-2: ESA's optical high-resolution mission for GMES operational services. *Remote Sens. Environ.* **2012**, *120*, 25–36. [CrossRef]
17. Shang, J.; Liu, J.; Ma, B.; Zhao, T.; Jiao, X.; Geng, X.; Huffman, T.; Kovacs, J.M.; Walters, D. Mapping spatial variability of crop growth conditions using RapidEye data in Northern Ontario, Canada. *Remote Sens. Environ.* **2015**, *168*, 113–125. [CrossRef]
18. Transon, J.; d'Andrimont, R.; Maugnard, A.; Defourny, P. Survey of hyperspectral earth observation applications from space in the sentinel-2 context. *Remote Sens.* **2018**, *10*, 157. [CrossRef]
19. Aasen, H.; Honkavaara, E.; Lucieer, A.; Zarco-Tejada, P.J. Quantitative Remote Sensing at Ultra-High Resolution with UAV Spectroscopy: A Review of Sensor Technology, Measurement Procedures, and Data Correction Workflows. *Remote Sens.* **2018**, *10*, 1091. [CrossRef]
20. Castelli, F.; Contillo, R.; Miceli, F. Non-destructive Determination of Leaf Chlorophyll Content in Four Crop Species. *J. Agron. Crop Sci.* **1996**, *177*, 275–283. [CrossRef]
21. Markwell, J.; Osterman, J.C.; Mitchell, J.L. Calibration of the Minolta SPAD-502 leaf chlorophyll meter. *Photosynth. Res.* **1995**, *46*, 467–472. [CrossRef]
22. Lichtenthaler, H.K.; Wellburn, A.R. Determinations of total carotenoids and chlorophylls a and b of leaf extracts in different solvents. *Analysis* **1983**, *603*, 142–196. [CrossRef]
23. Minocha, R.; Martinez, G.; Lyons, B.; Long, S. Development of a standardized methodology for quantifying total chlorophyll and carotenoids from foliage of hardwood and conifer tree species. *Can. J. For. Res.* **2009**, *39*, 849–861. [CrossRef]
24. Gitelson, A.A.; Gritz, Y.; Merzlyak, M.N. Relationships between leaf chlorophyll content and spectral reflectance and algorithms for non-destructive chlorophyll assessment in higher plant leaves. *J. Plant Physiol.* **2003**, *160*, 271–282. [CrossRef] [PubMed]
25. Lichtenthaler, H.K.; Gitelson, A.; Lang, M. Non-destructive determination of chlorophyll content of leaves of a green and an aurea mutant of tobacco by reflectance measurements. *J. Plant Physiol.* **1996**, *148*, 483–493. [CrossRef]
26. Datt, B. Remote Sensing of Chlorophyll a, Chlorophyll b, Chlorophyll a+b, and Total Carotenoid Content in Eucalyptus Leaves. *Remote Sens. Environ.* **1998**, *66*, 111–121. [CrossRef]
27. Sims, D.A.; Gamon, J.A. Relationships between leaf pigment content and spectral reflectance across a wide range of species, leaf structures and developmental stages. *Remote Sens. Environ.* **2002**, *81*, 337–354. [CrossRef]

28. Ciganda, V.; Gitelson, A.; Schepers, J. Non-destructive determination of maize leaf and canopy chlorophyll content. *J. Plant Physiol.* **2009**, *166*, 157–167. [CrossRef]
29. Cerovic, Z.G.; Masdoumier, G.; Ghozlen, N.B.; Latouche, G. A new optical leaf-clip meter for simultaneous non-destructive assessment of leaf chlorophyll and epidermal flavonoids. *Physiol. Plant.* **2012**, *146*, 251–260. [CrossRef]
30. Richardson, A.D.; Duigan, S.P.; Berlyn, G.P. An evaluation of noninvasive methods to estimate foliar chlorophyll content. *New Phytol.* **2002**, *153*, 185–194. [CrossRef]
31. Steele, M.R.; Gitelson, A.A.; Rundquist, D.C. A comparison of two techniques for nondestructive measurement of chlorophyll content in grapevine leaves. *Agron. J.* **2008**, *100*, 779–782. [CrossRef]
32. Casa, R.; Castaldi, F.; Pascucci, S.; Pignatti, S. Chlorophyll estimation in field crops: An assessment of handheld leaf meters and spectral reflectance measurements. *J. Agric. Sci.* **2015**, *153*, 876–890. [CrossRef]
33. Marenco, R.A.; Antezana-Vera, S.A.; Nascimento, H.C.S. Relationship between specific leaf area, leaf thickness, leaf water content and SPAD-502 readings in six Amazonian tree species. *Photosynthetica* **2009**, *47*, 184–190. [CrossRef]
34. Jacquemoud, S.; Baret, F. PROSPECT: A model of leaf optical properties spectra. *Remote Sens. Environ.* **1990**, *34*, 75–91. [CrossRef]
35. Féret, J.-B.; François, C.; Asner, G.P.; Gitelson, A.A.; Martin, R.E.; Bidel, L.P.; Ustin, S.L.; Le Maire, G.; Jacquemoud, S. PROSPECT-4 and 5: Advances in the leaf optical properties model separating photosynthetic pigments. *Remote Sens. Environ.* **2008**, *112*, 3030–3043. [CrossRef]
36. Padilla, F.M.; de Souza, R.; Peña-Fleitas, M.T.; Grasso, R.; Gallardo, M.; Thompson, R.B. Influence of time of day on measurement with chlorophyll meters and canopy reflectance sensors of different crop N status. *Precis. Agric.* **2019**, *20*, 1087–1106. [CrossRef]
37. Higa, T.; Wada, M. Chloroplast avoidance movement is not functional in plants grown under strong sunlight. *PlantCell Environ.* **2016**, *39*, 871–882. [CrossRef] [PubMed]
38. Uddling, J.; Gelang-Alfredsson, J.; Piikki, K.; Pleijel, H. Evaluating the relationship between leaf chlorophyll concentration and SPAD-502 chlorophyll meter readings. *Photosynth. Res.* **2007**, *91*, 37–46. [CrossRef]
39. Parry, C.; Blonquist, J.; Bugbee, B. In situ measurement of leaf chlorophyll concentration: Analysis of the optical/absolute relationship. *Plant Cell Environ.* **2014**, *37*, 2508–2520. [CrossRef]
40. Nauš, J.; Prokopová, J.; Řebíček, J.; Špundová, M. SPAD chlorophyll meter reading can be pronouncedly affected by chloroplast movement. *Photosynth. Res.* **2010**, *105*, 265–271. [CrossRef]
41. McClendon, J.H.; Fukshansky, L. On the interpretation of absorption spectra of leaves-II. The non-absorbed ray of the sieve effect and the mean optical pathlength in the remainder of the leaf. *Photochem. Photobiol.* **1990**, *51*, 211–216. [CrossRef]
42. McClendon, J.H.; Fukshansky, L. On the interpretation of absorption spectra of leaves–I. Introduction and the correction of leaf spectra for surface reflection. *Photochem. Photobiol.* **1990**, *51*, 203–210. [CrossRef]
43. Baránková, B.; Lazár, D.; Nauš, J. Analysis of the effect of chloroplast arrangement on optical properties of green tobacco leaves. *Remote Sens. Environ.* **2016**, *174*, 181–196. [CrossRef]
44. Barton, C.V. A theoretical analysis of the influence of heterogeneity in chlorophyll distribution on leaf reflectance. *Tree Physiol.* **2001**, *21*, 789–795. [CrossRef] [PubMed]
45. Ustin, S.L.; Gitelson, A.A.; Jacquemoud, S.; Schaepman, M.; Asner, G.P.; Gamon, J.A.; Zarco-Tejada, P. Retrieval of foliar information about plant pigment systems from high resolution spectroscopy. *Remote Sens. Environ.* **2009**, *113*, S67–S77. [CrossRef]
46. Goulas, Y.; Cerovic, Z.G.; Cartelat, A.; Moya, I. Dualex: A new instrument for field measurements of epidermal ultraviolet absorbance by chlorophyll fluorescence. *Appl. Opt.* **2004**, *43*, 4488–4496. [CrossRef]
47. Overbeck, V.; Schmitz, M.; Tartachnyk, I.; Blanke, M. Identification of light availability in different sweet cherry orchards under cover by using non-destructive measurements with a Dualex™. *Eur. J. Agron.* **2018**, *93*, 50–56. [CrossRef]
48. Meyer, S.; Cerovic, Z.G.; Goulas, Y.; Montpied, P.; Demotes-Mainard, S.; Bidel, L.P.R.; Moya, I.; Dreyer, E. Relationships between optically assessed polyphenols and chlorophyll contents, and leaf mass per area ratio in woody plants: A signature of the carbon–nitrogen balance within leaves? *Plant Cell Environ.* **2006**, *29*, 1338–1348. [CrossRef]

49. Schepers, J.S.; Blackmer, T.M.; Wilhelm, W.W.; Resende, M. Transmittance and Reflectance Measurements of CornLeaves from Plants with Different Nitrogen and Water Supply. *J. Plant Physiol.* **1996**, *148*, 523–529. [CrossRef]
50. Apogee Instruments, I. *CCM-200 plus Chlorophyll Meter, Product Manual*; Apogee Instruments, Inc.: Logan, UT, USA, 2011.
51. Konica, M. *Spad 502 Plus Chlorophyll Meter, Product Manual*; Konica Minolta: Chiyoda, Japan, 2011.
52. Raymond Hunt, E.; Daughtry, C.S. Chlorophyll meter calibrations for chlorophyll content using measured and simulated leaf transmittances. *Agron. J.* **2014**, *106*, 931–939. [CrossRef]
53. Dong, T.; Liu, J.; Shang, J.; Qian, B.; Ma, B.; Kovacs, J.M.; Walters, D.; Jiao, X.; Geng, X.; Shi, Y. Assessment of red-edge vegetation indices for crop leaf area index estimation. *Remote Sens. Environ.* **2019**, *222*, 133–143. [CrossRef]
54. Liu, J.; Pattey, E.; Jégo, G. Assessment of vegetation indices for regional crop green LAI estimation from Landsat images over multiple growing seasons. *Remote Sens. Environ.* **2012**, *123*, 347–358. [CrossRef]
55. Houborg, R.; Anderson, M.; Daughtry, C. Utility of an image-based canopy reflectance modeling tool for remote estimation of LAI and leaf chlorophyll content at the field scale. *Remote Sens. Environ.* **2009**, *113*, 259–274. [CrossRef]
56. Fritschi, F.B.; Ray, J.D. Soybean leaf nitrogen, chlorophyll content, and chlorophyll a/b ratio. *Photosynthetica* **2007**, *45*, 92–98. [CrossRef]
57. Jacquemoud, S.; Verhoef, W.; Baret, F.; Bacour, C.; Zarco-Tejada, P.J.; Asner, G.P.; François, C.; Ustin, S.L. PROSPECT+ SAIL models: A review of use for vegetation characterization. *Remote Sens. Environ.* **2009**, *113*, S56–S66. [CrossRef]
58. Serrano, L. Effects of leaf structure on reflectance estimates of chlorophyll content. *Int. J. Remote Sens.* **2008**, *29*, 5265–5274. [CrossRef]
59. Larcher, W. *Physiological Plant Ecology: Ecophysiology and Stress Physiology of Functional Groups*; Springer: Berlin, Germany, 2003.
60. Kouril, R.; Ilík, P.; Naus, J.; Schoefs, B. On the limits of applicability of spectrophotometric and spectrofluorimetric methods for the determination of chlorophyll a/b ratio. *Photosynth. Res.* **1999**, *62*, 107–116. [CrossRef]
61. Gitelson, A.; Solovchenko, A. Non-invasive quantification of foliar pigments: Possibilities and limitations of reflectance- and absorbance-based approaches. *J. Photochem. Photobiol. B Biol.* **2018**, *178*, 537–544. [CrossRef]
62. Dong, T.; Liu, J.; Qian, B.; Jing, Q.; Croft, H.; Chen, J.; Wang, J.; Huffman, T.; Shang, J.; Chen, P. Deriving maximum light use efficiency from crop growth model and satellite data to improve crop biomass estimation. *IEEE J. Sel. Top. Appl. Earth Obs. Remote Sens.* **2016**, *10*, 104–117. [CrossRef]
63. Li, J.; Yang, J.; Fei, P.; Song, J.; Li, D.; Ge, C.; Chen, W. Responses of rice leaf thickness, SPAD readings and chlorophyll a/b ratios to different nitrogen supply rates in paddy field. *Field Crop. Res.* **2009**, *114*, 426–432.
64. Gitelson, A.A.; Merzlyak, M.N. Signature analysis of leaf reflectance spectra: Algorithm development for remote sensing of chlorophyll. *J. Plant Physiol.* **1996**, *148*, 494–500. [CrossRef]
65. Padilla, F.M.; de Souza, R.; Peña-Fleitas, M.T.; Gallardo, M.; Giménez, C.; Thompson, R.B. Different responses of various chlorophyll meters to increasing nitrogen supply in sweet pepper. *Front. Plant Sci.* **2018**, *9*, 1752. [CrossRef] [PubMed]
66. Davis, P.A.; Caylor, S.; Whippo, C.W.; Hangarter, R.P. Changes in leaf optical properties associated with light-dependent chloroplast movements. *PlantCell Environ.* **2011**, *34*, 2047–2059. [CrossRef] [PubMed]
67. Netto, A.T.; Campostrini, E.; Oliveira, J.G.d.; Bressan-Smith, R.E. Photosynthetic pigments, nitrogen, chlorophyll a fluorescence and SPAD-502 readings in coffee leaves. *Sci. Hortic.* **2005**, *104*, 199–209. [CrossRef]
68. Stuckens, J.; Verstraeten, W.W.; Delalieux, S.; Swennen, R.; Coppin, P. A dorsiventral leaf radiative transfer model: Development, validation and improved model inversion techniques. *Remote Sens. Environ.* **2009**, *113*, 2560–2573. [CrossRef]

remote sensing

MDPI

Article

Predicting Rice Grain Yield Based on Dynamic Changes in Vegetation Indexes during Early to Mid-Growth Stages

Ke Zhang, Xiaokang Ge, Pengcheng Shen, Wanyu Li, Xiaojun Liu, Qiang Cao, Yan Zhu, Weixing Cao and Yongchao Tian *

National Engineering and Technology Center for Information Agriculture, Key Laboratory of Crop System Analysis and Decision Making, Ministry of Agriculture, Jiangsu Key Laboratory for Information Agriculture, Nanjing Agricultural University, 1 Weigang Road, Nanjing 210095, China; 2017201080@njau.edu.cn (K.Z.); gexk1992@163.com (X.G.); 15062276966@163.com (P.S.); lwany1996@163.com (W.L.); liuxj@njau.edu.cn (X.L.); qiangcao@njau.edu.cn (Q.C.); yanzhu@njau.edu.cn (Y.Z.); caow@njau.edu.cn (W.C.)

* Correspondence: yctian@njau.edu.cn; Tel.: +86-25-84399050; Fax: +86-25-84396672

Received: 22 January 2019; Accepted: 12 February 2019; Published: 14 February 2019

Abstract: Predicting the grain yield during early to mid-growth stages is important for initial diagnosis of rice and quantitative regulation of topdressing. In this study, we conducted four experiments using different nitrogen (N) application rates (0–400 kg N·ha^{-1}) in three Japonica rice cultivars (Wuyunjing24, Ningjing4, and Lianjing7) grown in Jiangsu province, Eastern China, from 2015–2016. Spectral reflectance data were collected multiple times during early to mid-growth stages using an active mounted sensor (RapidScan CS-45, Holland Scientific Inc., Lincoln, NE, USA). Data were then used to calculate optimal vegetation indexes (normalized difference red edge, NDRE; normalized difference vegetation index, NDVI; ratio vegetation index, RVI; red-edge ratio vegetation index, RERVI), which were used to develop a dynamic change model and in-season grain yield prediction model. The NDRE index was more stable than other indexes (NDVI, RVI, RERVI), showing less standard deviation at the same N fertilizer rate. The R^2 of the relationships between leaf area index (LAI), plant nitrogen accumulation (PNA), and NDRE also increased compared to other indexes. These findings suggest that NDRE is suitable for analysis of paddy rice N nutrition. According to real-time series changes in NDRE, the resulting dynamic model followed a sigmoid curve, with a coefficient of determination (R^2) >0.9 and relative root-mean-square error <5%. Moreover, the feature platform value (saturation value, SV) of the NDRE-based model accurately predicted the differences between treatments and the final grain yield levels. R^2 values of the relationship between SV and yield were >0.7. For every 0.1 increase in SV, grain yield increased by 3608.1 kg·ha^{-1}. Overall, our new dynamic model effectively predicted grain yield at stem elongation and booting stages, providing real-time crop N nutrition data for management of N fertilizer topdressing in rice production.

Keywords: red-edge; NDRE; dynamic change model; sigmoid curve; grain yield prediction

1. Introduction

Rice (*Oryza sativa* L.) is a staple food for billions of people worldwide. China is the largest rice producer in the world, with approximately 31 million hectares of paddy rice cultivation [1]. To ensure sustainable food production, agricultural producers invest in a significant amount of nitrogen (N) fertilizer [1]. Zhang et al. reported that the N use efficiency (NUE) of China's major food crops was less than 30% [2], representing 33% of the global NUE, and 40% of that in developed countries in Europe and America [3]. Scientific management of N fertilizer use is, therefore, crucial in realizing the goals of high yield, quality, and efficiency in rice production in eastern China.

Methods based on appropriate indicators, such as leaf area index (LAI), soil and plant analyzer development (SPAD), and N nutrition index (NNI), and yield forecasts are often used for crop growth diagnostics and fertilizer quantification [4]. Site-specific nutrient management (SSNM) of N fertilizer topdressing, which is based on comparisons of SPAD and threshold values at key growth stages, is one such method [5]. However, different varieties have different thresholds. For example, SPAD thresholds of rice cultivars Shanyou63 and Liangyoupei9 were found to be 36 and 38, respectively [6,7]. During rice-wheat rotation, SPAD values should be more than 37.5 to ensure high yield; if lower, 30 kg·ha^{-1} N fertilizer topdressing is applied [8]. Meanwhile, Qiong et al. recommended an SPAD threshold of 39–41 for the application of N fertilizer at critical growth stages in paddy rice [9] (Table 1). Moreover, Peng et al. suggested that, under SSNM management in China, a plot requires just 68% of the N fertilizer applied under traditional fertilization methods, resulting in a 5% increase in N use efficiency (NUE) [5]. However, SPAD thresholds are affected by a number factors including the species, regional climates, soil type, and N fertilizer management measures [7], resulting in significantly different thresholds.

Table 1. Summary of threshold distributions under different conditions.

Author	Eco-site	Variety	Threshold
Huang et al. [6]	Hubei, China	Shanyou63, Liangyoupei9	36, 38
Singh et al. [7]	Indo-Gangetic plains, India	PR118, PAU201, etc.	37.5
Qiong et al. [8]	Hubei, China	Peiliangyou3076, Yangliangyou6	39–41
He et al. [3]	Hubei, China	Liangyoupei9, Shanyou63	38, 39, 35–37
Fan et al. [10]	Helongjiang, China	Songjing98-128	38–40

The critical N dilution curve can also be used to describe in-season plant N status [11]. Based on this model, the N nutrition index (NNI), a widely applicable diagnostic indicator, is used for N diagnosis [12]. When NNI = 1, N nutrition is optimal, while NNI > 1 and NNI < 1 indicate excess and deficient N nutrition, respectively [13]. NNI can accurately predict the relative yield of rice at different stages, with coefficients of determination (R^2) greater than 0.74 [14]. Similar results were also obtained using the normalized difference vegetation index (NDVI) of GreenSeeker (Trimble Navigation Ltd. Sunnyvale, CA, USA) in barley and maize. NNI was previously found to account for 70% of the variation in crop N nutrition [15,16]; however, in practical applications, in different regions and with different rice varieties, NNI is less effective. Calculated NNI values were found to have a large range [17]. For example, Hu et al. reported NNI values ranging from 0.4–1.6 [18], while Houlès et al. suggested that 0.25–1.25 was the optimal range [19]. Moreover, Debaeke et al. revealed NNI values ranging from 0.5–1.5 in wheat [20]. Significant differences between studies were also revealed [21]. These differences in critical N concentrations, therefore, reduced the popularity and widespread use of NNI.

Yield prediction methods use crop conditions at early to mid-growth stages to predict the potential yield of that season, thereby determining the amount of fertilizer topdressing required [4]. In traditional methods, agronomic parameters such as the leaf area index (LAI) and plant dry matter (PDM) at different growth stages are commonly used to predict potential yield. The quantitative relationships between these indicators and final yield is then determined, allowing N fertilizer application rates to be recommended [22]. Portable spectral sensors are also used to collect crop spectral information, which is then used to determine vegetation indexes and the relationship with final yield. Moreover, Spitkó et al. revealed a relationship between NDVI and yield after anthesis; however, the correlation coefficient was less than 0.5 (r) [23]. Tubaña et al. also established a relationship between relative NDVI and relative yield (relative to the respective maximum) at panicle differentiation and for two weeks thereafter [24]. Similarly, the determination coefficient was less than 0.5. Meanwhile, Liu et al.

confirmed a relatively stable relationship between NDVI and grain yield in double-cropping rice during major growth periods, with R^2 values ranging from 0.6 to 0.65 [25].

Xue et al. also described the relationship between NDVI and N accumulation from tillering to early grain filling stages, and used the obtained sufficiency index (SI) to predict potential yield and guide topdressing nitrogen management [26]. The R^2 value was 0.8, resulting in effective improvements in NUE. Furthermore, Evert et al. studied the relationship between the weighted difference vegetation index (WDVI) and the N status of potato using MSR87 (CropScan Inc., Rochester, MN, USA) [27], revealing a significant correlation between WDVI and N uptake. Accordingly, they were able to establish a potential yield model for potato. Similarly, Morier et al. used FieldSpec HandHeld (Analytical Spectral Devices (ASD), Inc., Boulder, CO) to obtain hyperspectral data (325–1075 nm and a spectral resolution of 3.5 nm at 700 nm) for monitoring of potato populations, revealing that the best relationship existed between the index incorporating the red-edge wave (CIred-edge) and yield. As a result, they were able to explain 76% of the variability in total tuber yield at 55 days after planting [28].

Lukina et al. and Raun et al. also proposed an N fertilization optimization algorithm based on spectral index and the corresponding time of the growing degree days (GDD), which allowed calculations of the estimated yield coefficient in-season estimated yield (INSEY) [29,30]. This method eliminated the effect of different growth stages, compared with approaches based on vegetation indexes, and improved accuracy by 30% in terms of grain yield predictions in winter wheat [31]. However, studies also showed that NDVI and other spectral indexes have a significant saturation effect under high vegetation coverage [32]. In rice, accuracy decreases when the plant dry matter is greater than 3736 kg·ha^{-1} [33]. As a result, the prediction accuracy of the algorithm also decreased, with an R^2 of only 0.5 [34]. Moreover, in rice crops in particular, R^2 was only about 0.4–0.5 at stem elongation and booting stages [8,35]. However, Thompson et al. revealed that red-edge bands based on normalized difference red edge (NDRE) could effectively identify crop populations, with a higher correlation with indicators such as plant nitrogen accumulation, helping solve the problem of saturation [36]. Cao et al. used multispectral data of Crop Circle-ACS470 (Holland Scientific, Lincoln, NE, USA) to fit a variety of vegetation indexes including the normalized NIR index (NNIR), green soil-adjusted vegetation index (GSAVI), and modified enhanced vegetation index (MEVI); then, based on these indexes, they determined the relationship between INSEY and rice yield at stem elongation and booting [37]. These results revealed an R^2 value of 0.8 at stem elongation, 21–26% higher than that based on INSEY-NDVI and INSEY-RVI [38]. However, few studies examined the use of NDRE-related models in accurate management of crop N fertilizer topdressing.

Previous studies suggested problems of saturation with models using vegetation indexes such as the single spectral index, resulting in low prediction accuracy when used to predict nutrition or grain yield in paddy rice [38]. Screening of effective vegetation indexes and development of a stable accurate method for predicting rice grain yield is, therefore, required. This paper, therefore, aimed to (1) evaluate spectral index data obtained using the active multi-spectrometer RapidScan CS-45 (Holland Scientific Inc., Lincoln, NE, USA) during rice growth; (2) construct a dynamic model using the vegetation index data; and (3) develop a prediction algorithm of rice grain yield based on the dynamic model.

2. Materials and Methods

2.1. Experiment Design

Japonica rice is the main variety grown in Jiangsu Province, which has the highest average rice production in all China [2]. This study was carried out in two typical Japonica rice-producing areas in Jiangsu Province, the lower reaches of the Yangtze River (Rugao) and the Huaihe River Basin (Huai'an and Sihong) in northern Jiangsu. In Rugao, the predominant soil type is loam, with a total N content of 1.71 g·kg^{-1}, olsen phosphorus (P) content of 13.3 mg·kg^{-1}, and available potassium (K) content of 95.7 mg·kg^{-1}. The soil types in Huai'an and Sihong are yellow-brown and lime concretion black soil, respectively, with a total N content of 1.35 and 1.28 g·kg^{-1}, olsen P content of 32 and 27.6 mg·kg^{-1}, and available K content of 85.3 and 75.2 mg·kg^{-1}, respectively. The climate data of each experimental site (Exp. 1, Exp. 2, and Exp. 3) is shown in Figure 1. Four experiments were carried out in total using different N application rates (0–400 kg·N·ha^{-1}) and three Japonica rice varieties, Wuyunjing24 (WYJ24), Ningjing4 (NJ4), and Lianjing7 (LJ7), as detailed in Table 2.

Experiment 1 (Exp. 1): Experiment 1 was conducted over a single season from June to October, 2015, in Rugao (118.26°E, 33.37°N). Rice variety WYJ24 was sown on 15 May and transplanted on 15 June at a density of 15 × 30 cm. The plot was 7 by 5 m and covered a total area of 35 m^2. Four basal N fertilization rates (0 (N0), 120 (N5), 180 (N7), and 240 (N8) kg·ha^{-1}) were applied in the form of urea at a rate of 50% at pre-planting and 50% at tillering, and four topdressing N fertilizer rates (0 (T0), 80 (T3), 120 (T6), and 160 (T7) kg·ha^{-1}) were also applied in the form of urea at a rate of 50 at stem elongation and 50% at booting. A no-topdressing treatment (T0) was also examined at each N gradient. In each treatment, 127 kg·ha^{-1} P_2O_5 was applied as supplemental phosphate fertilizer and 225 kg·ha^{-1} K_2O (50% at pre-planting, 50% at stem elongation) was applied as potash fertilizer.

Experiment 2 (Exp. 2): Experiment 2 was conducted over a single season from June to October, 2015, in Sihong (118.26°E, 33.37°N). Varieties WYJ24, NJ4, and LJ7 were transplanted on 14 and 20 June. Four basal N fertilization rates (0 (N0), 36 (N1), 72 (N3), and 108 (N4) kg·ha^{-1}) were applied in the form of urea at a rate of 50% at pre-planting and 50% at tillering, and four topdressing N fertilizer rates (0 (T0), 84 (T4), 168 (T8), and 252 (T9) kg·ha^{-1}) were applied in the form of urea at a rate of 50% at stem elongation and 50% at booting. Other factors were as in Exp. 1.

Experiment 3 (Exp. 3): Experiment 3 was conducted in 2016 in Rugao (118.26°E, 33.37°N). Varieties WYJ24 and NJ4 were transplanted on 25 June. Four basal N fertilization rates (0 (N0), 60 (N2), 150 (N6), and 240 (N8) kg·ha^{-1}) were applied in the form of urea at a rate of 50% at pre-planting and 50% at tillering. For WYJ24, six topdressing N fertilizer rates (0 (T0), 40 (T1), 50 (T2), 80 (T3), 100 (T5), and 160 (T7) kg·ha^{-1}) were also applied in the form of urea at a rate of 50% at stem elongation and 50% at booting, and four topdressing N fertilizer rates (0 (T0), 40 (T1), 100 (T5), and 160 (T7) kg·ha^{-1}) were applied for NJ4. Other factors were as in Exp. 1.

Experiment 4 (Exp. 4): Experiment 4 was a repeat of Exp. 2, carried out in 2016. Detailed information on each experiment is provided in Table 2. Data from Exp. 1 and Exp. 3 were mainly used to develop the model, while data from Exp. 2 and Exp. 4 were used for model testing. Exp. 2 and 4 were performed to maintain uniformity across different study years.

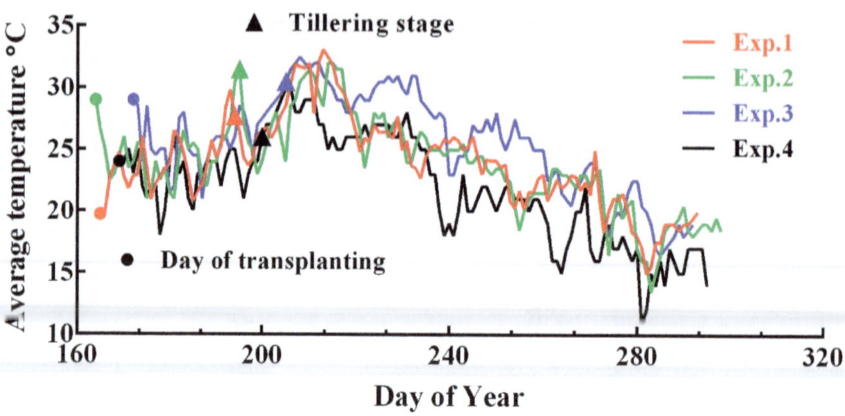

Figure 1. Daily temperatures during the rice growing seasons in 2015–2016.

Table 2. Basic characteristics of experiments 1, 2, 3 and 4.

Experiment	Location	Transplanting and Harvest Date	Cultivar	Treatment	Nitrogen fertilizer (N, kg·ha^{-1})				
					Basal and tillering fertilizer		Panicle fertilizer		Total fertilizer
Exp. 1 2015	Rugao 118.26°E, 33.37°N	14 June 25 Oct.	WYJ24	N0T0	0	(N0)	0	(T0)	0
				N5T0	120	(N5)	0	(T0)	120
				N5T3	120	(N5)	80	(T3)	200
				N7T0	180	(N7)	0	(T0)	180
				N7T6	180	(N7)	120	(T6)	300
				N8T0	240	(N8)	0	(T0)	240
				N8T7	240	(N8)	160	(T7)	400
Exp. 2 2015	SiHong 118.26°E, 33.37°N	14 June 25 Oct.	WYJ24, NJ4, LJ7	N0T0	0	(N0)	0	(T0)	0
				N1T4	36	(N1)	84	(T4)	120
				N3T8	72	(N3)	168	(T8)	240
				N4T9	108	(N4)	252	(T9)	360
Exp. 3 2016	RuGao 120.76°E, 32.27°N	25 June 26 Oct.	WYJ24	N0T0	0	(N0)	0	(T0)	0
				N2T0	60	(N2)	0	(T0)	60
				N2T1	60	(N2)	40	(T1)	100
				N2T3	60	(N2)	80	(T3)	140
				N6T0	150	(N6)	0	(T0)	150
				N6T2	150	(N6)	50	(T2)	200
				N6T5	150	(N6)	100	(T5)	250
				N8T0	240	(N8)	0	(T0)	240
				N8T3	240	(N8)	80	(T3)	320
				N8T7	240	(N8)	160	(T7)	400
			NJ4	N0T0	0	(N0)	0	(T0)	0
				N2T1	60	(N2)	40	(T1)	100
				N6T5	150	(N6)	100	(T5)	250
				N8T7	240	(N8)	160	(T7)	400
Exp. 4 2016	SiHong 118.26°E, 33.37°N	18 June 22 Oct.	WYJ24, NJ4, LJ7	N0T0	0	(N0)	0	(T0)	0
				N1T4	36	(N1)	84	(T4)	120
				N3T8	72	(N3)	168	(T8)	240
				N4T9	108	(N4)	252	(T9)	360

Note: "WYJ24" is "Wunyunjing24", "NJ4" is "Ningjing4", "LJ7" is "Lianjing7"; "N0–8" represents the different basal nitrogen fertilizer rates; "T0–9" represents the different topdressing N fertilizer rates.

2.2. Data Acquisition and Determination

Spectral reflectance data were collected using the active portable sensor RapidScan CS-45 (Holland Scientific Inc., Lincoln, NE, USA), and contained red (R, 670 nm), red-edge (Re, 730 nm), and near-infrared (NIR, 780 nm) wavebands. It has its own light source; thus, dependency on sunlight was avoided, preventing sensitivity to weather conditions. All spectral measurements were performed during cloud-free periods between 9:30 and 10:30 a.m. with the sensor placed 0.7 m above the canopy. Three representative lines were selected in each plot. The sensor records one spectral reflectance value every second along the line at uniform velocity, giving approximately 70–80 values per line. The average spectral index and reflectance in each plot was then calculated.

Exp. 1 and Exp. 3 were carried out from mid-tillering, with measurements every seven days until the end of tillering to flowering, when testing was conducted every three days. In Exp. 2 and Exp. 4, measurements were carried out simultaneously with sampling at each growth stage.

In each experiment, three hills from each plot were sampled for growth analysis at different stages during vegetative growth. Plants were manually uprooted and then cut at ground level for determination of N concentration. Fresh plants were separated into green leaf blade (leaf) and culm plus sheath (stem) samples, heated for 30 min at 105 °C to halt metabolic processes, and then dried at 80 °C in a forced-draft oven until reaching a constant weight. Plant dry matter (PDM) and leaf dry matter (LDM) were then determined before grinding the samples and passing them through a 1-mm sieve in a Wiley mill. The green fresh leaf samples were immediately scanned using an LI-3000A (Li-Cor, Lnr, Lincoln, NE) and the green leaf area of each layer was obtained to calculate the green LAI for each plot (the sum of different layers). The samples were then stored in plastic bags at room temperature until further chemical analysis. Whole-plant and leaf N concentrations (PNC and LNC) were determined using an elemental analyzer (vario MACRO cube; Elementar, Hanau, Germany); then, plant N accumulation (PNA) and leaf N accumulation (LNA) were calculated using the PDM/LDM and PNC/LNC values. Grain yield was determined in a 2-m^2 area in each plot and adjusted to a moisture content of 14.5%.

2.3. Data Analysis

2.3.1. Calculation of Relative Accumulated Growing Degree Days (RAGDD)

Meteorological data including temperature were collected using the automated weather station Dynameta-1K (Dynamax Inc., Houston, TX, USA), which was installed at each test site, and recorded using the EM50 data acquisition system (Decagon Devices Inc., Washington, USA) every 5 min. Data were then used to calculate the accumulation growing degree days (AGDD) and RAGDD. In this paper, RAGDD was used as a time variable and was calculated from the AGDD, while AGDD represents the sum of growing degree days (GDD) throughout each experiment [39]. AGDD can be predicted using GDD or time [40], and in this study, it was calculated as follows:

$$GDD = \frac{T_{\text{Max}} + T_{\text{Min}}}{2} - T_{BASE}, \tag{1}$$

where T_{Max} and T_{Min} represent the maximum and minimum temperatures on a specific day, respectively, and TBASE is the base temperature, which is usually set at 12.5 °C for Japonica rice [41]. RAGDD was then calculated as follows:

$$RAGDD = \frac{AGDD_i}{AGDD_{harvest}}, \tag{2}$$

where AGDDi represents the AGDD on a specific day, and $\text{AGDD}_{\text{harvest}}$ is the AGDD at harvest, representing the AGDD of the entire growth period.

2.3.2. Statistical Analyses

Mapping, profiling, curve fitting, and model building were carried out using OriginPro 9.0 (OriginLab Corporation, Northampton, MA, USA). Four commonly used spectral indexes were determined in this study: NDVI, NDRE, the ratio vegetation index (RVI), and red-edge ratio vegetation index (RERVI). To do so, three wavebands tests were carried out using RapidSCAN during N nutrition monitoring. The specific equations are shown in Table 3. Data-fitting processes were performed using Origin 9.0, choosing different equations based on convergence. Linear, quadratic, logarithmic, exponential, and rational models were evaluated; then, the model with the highest coefficient of determination (R^2) was adopted. The accuracy of the model was evaluated using the R^2 and relative root-mean-square error (RRMSE), and the 1:1 relationship between the experimental observations and model predictions were plotted as follows:

$$RRMSE = \sqrt{\frac{1}{n} \times \sum_{i=1}^{n} (P_i - O_i)^2} \times \frac{100}{\overline{O_i}}, \tag{3}$$

where n is the number of test samples, P_i is the model estimate, O_i is the observed value, and $\overline{O_i}$ is the average observed value.

Table 3. Equations of each vegetation index.

Spectral Index	Equation	Sensitivity Indicator	Reference
Normalized difference vegetation index (NDVI)	$\frac{NIR-R}{NIR+R}$	Grain yield, HI, N status, RUE, LAI, Biomass, Grain protein	[36,42]
Normalized difference red edge (NDRE)	$\frac{NIR-Re}{NIR+Re}$	LAI, Biomass, N status	[6,33]
Ratio vegetation index (RVI)	$\frac{NIR}{R}$	Grain yield, Biomass, LAI, Grain protein, N status	[37,42]
Red-edge vegetation index (RERVI)	$\frac{NIR}{Re}$	LAI, Biomass, N status	[37]

Note: NIR, Re, and R refer to the reflectance of near-infrared, red-edge, and red wavelengths, respectively. LAI: leaf area index, HI: harvest index (unit-less), RUE: radiation use efficiency.

3. Results

3.1. Dynamic Changes in Agronomic Parameters and Spectral Indexes before Flowering

The dynamic changes in rice agronomic parameters (LAI, PDM, and PNA) in relation to N nutrition were examined in Exp. 1–4 (Figure 2). Similar dynamic characteristics were observed between varieties under different N rates. The results showed that LAI first increased then remained stable (Figure 2A–D). During booting, LAI continued to increase, with the biggest growth rate appearing at the panicle initiation stage. After entering the booting stage, LAI stabilized and remained essentially unchanged thereafter. PDM increased at an invariable rate from tillering to flowering (Figure 2E–H), while PNA was consistent with LNA, continually increasing until flowering (Figure 2M–T). PNC differed slightly from the other N indicators, decreasing up until flowering at an increasing rate (Figure 2I–L).

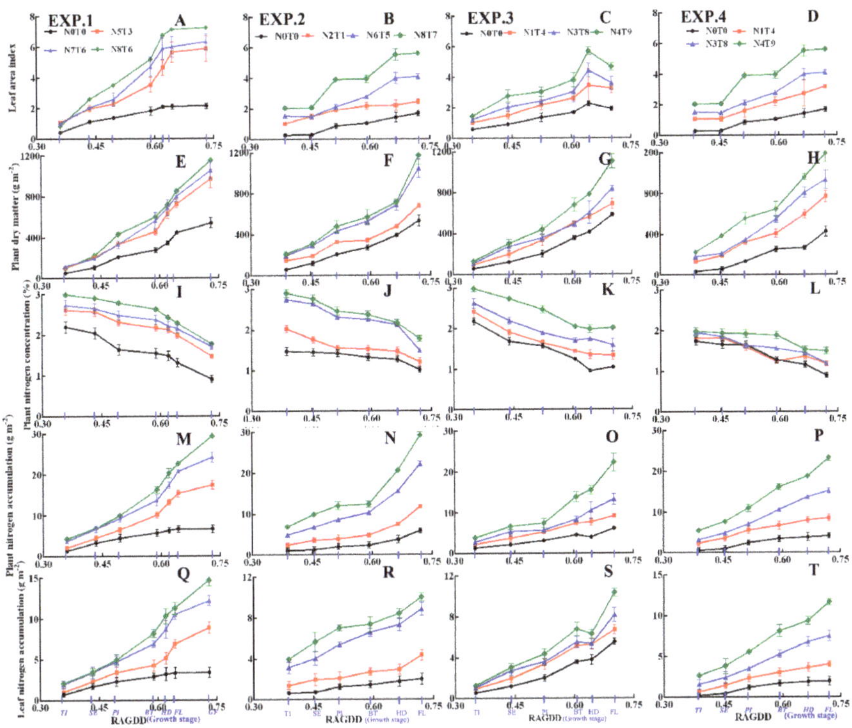

Figure 2. Dynamic changes in agronomic parameters before the grain filling stage. (**A–D**) Leaf area index, (**E–H**) plant dry matter, (**I–L**) nitrogen concentration, (**M–P**) whole-plant nitrogen accumulation, and (**Q–T**) leaf nitrogen accumulation in Exp. 1 and Exp. 4. "N0T0, N1T4, N2T0, N3T8, N6T5, N8T7, N5T3, N7T6, N3T8, and N4T9" represent basal and topdressing N fertilizer rates (0, 36 + 84, 60 + 0, 72 + 168, 150 + 100, 240 + 160, 120 + 80, 180 + 120, 72 + 168, and 108 + 252 kg·ha^{-1}, respectively). RAGDD: relative accumulated growing degree days; TI: tillering, SE: stem elongation, PI: panicle initiation, BT: booting, HD: heading, FL: flowering, and GF: grain filling.

As shown in Figure 3A–H, NDVI and NDRE increased slowly before stem elongation followed by a rapid rise till booting then a slow increase or static period. This was similar to the changes in LAI. Changes in the RVI and RERVI were similar to those of PDM, PNA, and LNA (Figure 3I–P); however, growth rates differed. The growth rate of PDM and PNA accelerated at the end of the stem elongation stage, while RVI remained relatively stable. PNC showed an opposite trend to all spectral indexes and consistency was also poor. All vegetation indexes showed different values under different N gradients. The red-edge-based vegetation indexes NDRE and RERVI showed higher distinction under different N fertilizer rates, while NDVI showed slight overlap between fertilizer rates. Compared with NDRE, RERVI showed a smaller range before flowering, making it hard to distinguish between growth stages and, therefore, less useful. In each single test, NDRE errors and fluctuations were smaller than those of NDVI, making it more stable. Moreover, ranges of all agronomic parameters were also small, suggesting that the environmental effect on NDVI was much bigger than the effect on NDRE.

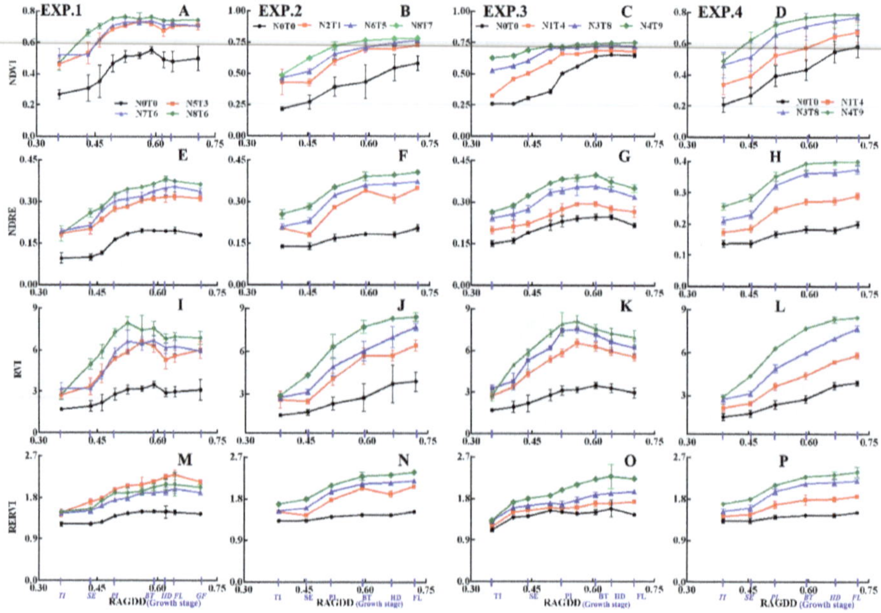

Figure 3. Dynamic changes in the vegetation indexes (NDRE, NDVI) before grain filling. "N0T0, N1T4, N2T0, N3T8, N6T5, N8T7 N5T3, N7T6, N3T8, and N4T9" represent different basal and topdressing N fertilizer rates (0, 36 + 84, 60 + 0, 72 + 168, 150 + 100, 240 + 160, 120 + 80, 180 + 120, 72 + 168, and 108 + 252 kg·ha^{-1}, respectively). RAGDD: relative accumulated growing degree days; TI: tillering, SE: stem elongation, PI: panicle initiation, BT: booting, HD: heading, FL: flowering, GF: grain filling; NDVI: normalized difference vegetation index, NDRE: normalized difference red edge; RERVI: red-edge ratio vegetation index, and RVI: ratio vegetation index.

3.2. Relationship between the Spectral Indexes and Agronomic Parameters

Using the results of Exp. 1–4 (Table 4), the relationships between each vegetation index and agronomic parameter were also determined, revealing differing results. The relationships between the spectral indexes and LAI and leaf N accumulation were strongest, while those with N concentration were lowest. Moreover, all correlations were low at the tillering stage, increasing with growth until reaching a maximum at stage. During the entire pre flowering period, the R^2 of the red-edge band-based indexes were higher than bidirectional reflectance simulation of the red-band-based vegetation indexes. Using the NDRE and NDVI findings in Exp. 1–4, NDVI (Figure 4A) saturation and an intermittent phenomenon were obvious, while the NDRE (Figure 4B) data were more clearly distributed and had a better relationship with LAI. This is also consistent with results in other crops [43]. The relationships between red-edge band-based vegetation indexes and agronomic parameters were, therefore, better than those indexes based on the red band.

Table 4. Coefficients of determination (R^2) of the correlations between vegetation indexes at different stages and the agronomic parameters.

Agronomic Parameter	Spectral Index	Mid-Tillering				Stem Elongation				Panicle Initiation				Booting				Pre-Flowering			
		Exp. 1	Exp. 2	Exp. 3	Exp. 4	Exp. 1	Exp. 2	Exp. 3	Exp. 4	Exp. 1	Exp. 2	Exp. 3	Exp. 4	Exp. 1	Exp. 2	Exp. 3	Exp. 4	Exp. 1	Exp. 2	Exp. 3	Exp. 4
LAI	NDRE	0.37	**0.7[illegible]**	**0.67**	**0.70**	0.51	**0.72**	0.69	0.71	0.65	0.67	0.77	0.75	0.70	0.75	0.83	0.79	0.67	0.64	0.74	0.78
	NDVI	0.21	0.5[illegible]	0.50	0.56	0.48	0.69	0.64	0.65	0.58	0.69	0.74	0.64	0.67	0.66	0.74	0.68	0.44	**0.66**	0.63	0.69
	RERVI	0.37	0.6[illegible]	0.66	0.70	**0.57**	0.71	**0.71**	**0.72**	**0.66**	0.70	**0.80**	**0.77**	**0.71**	**0.77**	**0.85**	**0.81**	**0.76**	0.61	**0.78**	**0.81**
	RVI	0.15	0.5[illegible]	0.43	0.44	0.46	0.70	0.62	0.58	0.63	**0.74**	0.78	0.66	0.66	0.74	0.77	0.67	0.61	0.57	0.67	0.70
PNC	NDRE	0.49	0.5[illegible]	0.46	0.13	**0.31**	**0.74**	**0.46**	0.19	0.27	0.70	0.52	0.10	0.34	**0.77**	**0.60**	0.37	<0.01	**0.26**	**0.28**	**0.59**
	NDVI	0.30	0.3[illegible]	0.29	0.07	0.21	0.63	0.39	**0.23**	0.25	0.58	0.46	0.16	0.32	0.57	0.49	0.35	<0.01	0.20	0.23	0.53
	RERVI	**0.50**	**0.5[illegible]**	**0.47**	**0.14**	0.19	0.74	0.40	0.16	0.23	**0.73**	**0.55**	**0.17**	**0.43**	0.76	0.62	**0.39**	<0.01	0.24	0.27	0.58
	RVI	0.21	0.3[illegible]	0.26	0.10	0.31	0.71	0.42	0.11	**0.32**	0.62	0.48	0.10	0.35	0.66	0.56	0.39	0.01	0.16	0.22	0.51
PNA	NDRE	**0.52**	**0.6[illegible]**	0.66	0.59	0.56	0.64	**0.68**	0.71	0.58	0.60	0.70	0.70	0.65	0.49	0.67	0.74	0.63	0.51	0.67	0.76
	NDVI	0.33	0.5[illegible]	0.53	0.53	0.48	**0.70**	0.63	0.59	0.56	0.54	0.64	0.64	0.61	0.53	0.63	0.60	0.40	**0.56**	0.56	0.62
	RERVI	0.52	0.6[illegible]	**0.68**	**0.70**	**0.59**	0.62	0.68	**0.72**	**0.63**	**0.62**	**0.73**	**0.72**	**0.77**	0.51	0.72	**0.77**	**0.74**	0.50	**0.72**	**0.79**
	RVI	0.26	0.5[illegible]	0.54	0.63	0.55	0.63	0.64	0.60	0.61	0.58	0.69	0.69	0.72	**0.67**	**0.77**	0.72	0.73	0.51	0.69	0.70
LNA	NDRE	**0.53**	**0.6[illegible]**	**0.71**	0.72	**0.64**	**0.75**	**0.75**	**0.71**	**0.62**	0.43	0.62	**0.74**	**0.70**	0.69	0.74	0.64	**0.63**	0.61	**0.72**	**0.79**
	NDVI	0.34	0.4[illegible]	0.51	0.57	0.54	0.72	0.70	0.69	0.55	0.41	0.56	0.64	0.62	0.62	0.65	0.55	0.60	**0.64**	0.67	0.65
	RERVI	0.52	0.6[illegible]	0.71	**0.73**	0.61	0.75	0.74	0.70	0.53	**0.46**	**0.60**	0.71	0.63	0.70	0.74	**0.66**	0.63	0.58	0.71	0.78
	RVI	0.28	0.4[illegible]	0.49	0.57	0.57	0.74	0.69	0.62	0.51	0.46	0.57	0.62	0.64	**0.74**	**0.75**	0.64	0.40	0.55	0.54	0.59

Notes: LAI: leaf area index, PNC: plant nitrogen concentration, PNA: plant nitrogen accumulation, LNA: leaf nitrogen accumulation, NDVI: normalized difference vegetation index, NDRE: normalized difference red edge, RERVI: red-edge ratio vegetation index, RVI: ratio vegetation index. The bold word means higher R^2 value among four experiments.

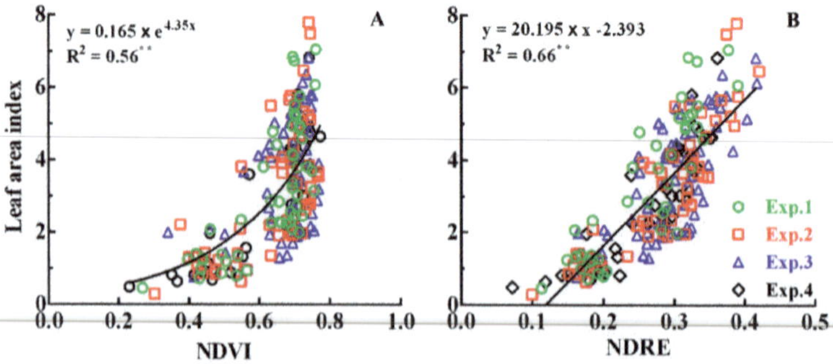

Figure 4. Relationship between (**A**) the normalized difference vegetation index (NDVI) and (**B**) normalized difference red edge (NDRE) and leaf area index (LAI) before flowering.

We also compared the relationships between agronomic parameters and final yield during the main growth period (Table 5). As pointed out previously, the relationship between agronomic parameters and yield differs at different growth stages [44,45]. This study yielded similar results. The relationship between LAI or N accumulation and yield was best at certain growth stages. Moreover, LAI had the highest coefficient and was more stable than N accumulation, highlighting its potential use in yield predictions.

Table 5. Coefficients of determination (R^2) of the correlations between the agronomic parameters at different stages and yield.

Growing stage	LAI	PNC	PNA	LNA
Tillering	0.82	0.16	0.73	0.76
Stem elongation	0.72	0.23	0.70	0.67
Panicle initiation	0.73	0.09	0.56	0.74
Booting	0.76	0.42	0.75	0.65

Notes: LAI: leaf area index, PNC: plant nitrogen concentration, PNA: plant nitrogen accumulation, LNA: leaf nitrogen accumulation.

As mentioned earlier, there were differences in the ranges of NDRE and RERVI before flowering. Stability and distinction between the N gradients of NDRE were better than those of RERVI (Figure 3). Moreover, the dynamic changes in LAI and NDRE were the most consistent (Figures 2 and 3). These results suggest that NDRE can be used to effectively reflect the growth status of rice under different N treatments, and in different eco-sites and different varieties, highlighting its potential use in growth diagnosis and dynamic model construction.

3.3. Construction of a Dynamic Model of NDRE

In this study, NDRE values were stable at first before showing a period of rapid growth throughout the entire pre-flowering stage, followed by a second stable, consistent with sigmoid-curve variation (Figure 5). Therefore, we chose a sigmoid curve to construct the dynamic change model of NDRE (Equation (4)).

$$y = \frac{A_1 - A_2}{1 + e^{(x - x_0)/dx}} + A_2, \quad (4)$$

where x represents the time-variable RAGDD, y is the spectral index NDRE value, A_1 is the previous platform value corresponding to the baseline value (BV) in each treatment, A_2 is the medium platform value corresponding to processing of the saturation value (SV), and x_0 is the midpoint of the sigmoid curve, that is, the point at which the model reaches inflection, indicating the different stages of dynamic change (Figure 5). Before x_0, rice is considered to be in the rapid growth stage, while, after x_0, the growth rate decreases and the population reaches a peak. Moreover, when $x = x_0$, $x = \frac{A_1+A_2}{2}$, the growth rate of the model is at its highest. Finally, dx represents the time constant of the model.

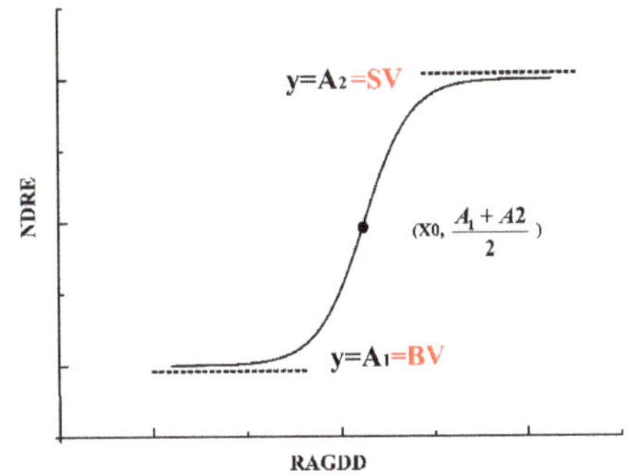

Figure 5. Dynamic characteristics of the sigmoid curve. NDRE: normalized difference red edge, RAGDD: relative accumulated growing degree day, A1: the platform value corresponding to the baseline value (BV) in each treatment, A2: the medium platform value corresponding to processing of the saturation value (SV), x_0: the midpoint of dynamic change, that is, the point at which the model reaches its inflection point, indicating the different stages of dynamic change.

At early and mid-tillering stages, resource requirements including fertilizer and water were relatively small, resulting in a small growing base. Moreover, vegetation coverage was also small, causing the canopy spectra to contain a lot of water and soil background information. As a result, the NDRE values were low and fluctuated around BV. In the mid-stage of growth, particularly early stem elongation, the growth rate increased rapidly, leading to a rapid growth period. After this, in the late stem elongation and booting stages, growth gradually reached a peak, and spectral index growth slowed until flowering.

Results of low (N2T1, 60 + 40 kg·ha^{-1}), medium (N5T3, 120 + 80 kg·ha^{-1}), and high (N8T7, 240 + 160 kg·ha^{-1}) N fertilizer averages are shown in Figure 6. At early growth stages, the differences in NDRE were not significant; however, SV values differed during mid-growth stages. The SV of N2T1 was low at approximately 0.23, while the values of N5T3 and N8T7 were 0.29 and 0.38. Meanwhile, NDVI arrived at its threshold earlier than NDRE as shown in Figure 6, and it was hard to distinguish between N2T1, N5T3, and N8T7 treatments.

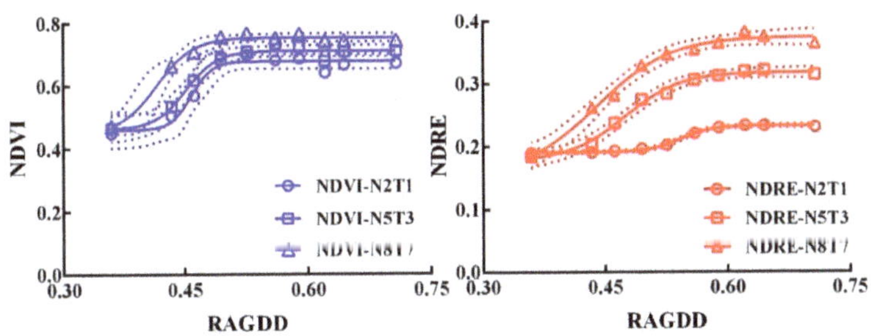

Figure 6. Dynamic fitting curves of each vegetation index (NDRE and NDVI) under high (N8T7, 240 + 160 kg·ha^{-1}), medium (N5T3, 120 + 80 kg·ha^{-1}), and low (N2T1, 60 + 40 kg·ha^{-1}) nitrogen treatment. The solid lines denote the logistic regression and the dotted lines represent the confidence bands (p = 0.95). Solid points denote measured values.

The model parameters were mainly obtained by fitting the measured values obtained at a high frequency testing, that is, under Exp. 1 and 3. In addition, BV and SV parameters were also estimated from NDRE values obtained at corresponding growth stages. For example, in the stem elongation to booting stages, NDRE entered the mid-stage of growth in the SV period. At this time, regardless of measurement errors, the measured NDRE values were approximately equal to the SV value. The SVs of Exp. 2 and Exp. 4 were subsequently obtained using this method. At the latest, SV values are determined at the early date.

3.4. Testing of Each Vegetation Index-Based Dynamic Model and Parameter Analysis

3.4.1. Verification of Model Construction Correctness

To confirm the accuracy of the dynamic models, they were fitted using each experiment and N treatment (Table 6). The results of model fitting based on NDRE showed that the R^2 values of different N gradients were greater than 0.9, while the values of RRMSE were less than 5%, suggesting that model values showed very small differences compared to the measured values, also indicating that the modeling effect of the different gradients was good. The normal distribution test of standard residuals under each treatment model showed that the probability of normal distribution was greater than 0.05, revealing that the standard residuals of each treatment were within normal distribution and that the model errors were random and generated by the model itself. Model fitting based on NDVI resulted in a high R^2 (0.87–0.96); however, the range of RRMSE was 7.5–21% higher than that of NDRE. Overall, these findings suggest that the model was applicable under different N levels, with different varieties, in different eco-sites, and in different years, confirming the accuracy of the model.

3.4.2. Changes in BV and SV under Different Treatments and Their Relationship with Agronomic Parameters

The NDRE values changed little at early stages of growth, fluctuating around the BV value. The BV under each treatment was less than 0.25 and there were no differences between treatments (Figure 7, NDRE), while BV and SV estimated using NDVI showed no obvious regularity. SV calculated using NDRE increased with increasing N fertilizer rates, and the trends between treatments were consistent. The differences with topdressing under the same basal fertilizer also differed, and the SV of T0 decreased by 15–25%. Moreover, a 10% increase in SV was observed between N0T0 and N5T0, N5T0 and N7T0, and N7T0 and N8T0. The differences between N application and topdressing levels also differed, suggesting that SV has obvious potential in growth status analysis in rice.

The relationship between SV and agronomic parameters was subsequently examined using the data from Exp. 1 and Exp. 3 (Figure 8). The relationships between SV and LAI, PDM, PNA, and LNA were good at both stem elongation (fourth leaf from the top stage) and booting stage (second leaf from the top stage). For NDRE, the R^2 of each of these relationships was more than 0.6. At stem elongation, there were slight differences but essentially similar correlations between SV and LAI, PDM, PNA, and LNA. Correlations at booting stage were better than those at stem elongation. Overall, LAI had the best relationship with SV. However, with NDVI, poor correlation was observed between all parameters.

Variation in the critical point x_0 under each treatment was also examined. As shown in Table 7, the model reached a critical point of dynamic change earlier in Exp. 1 than Exp. 3. This was especially true under identical N fertilizer treatments, namely, 0, 200, 240, and 320 kg·N·ha^{-1} N fertilizer topdressing. Moreover, the earliest critical point was observed under N8T0 (240 kg·N·ha^{-1}) treatment in both experiments (Exp. 1 and Exp. 3). Meanwhile, 100–200 kg·N·ha^{-1} plots showed the highest average RAGDD values.

Table 6. Summary of fitting analysis of the NDRE dynamic model under each treatment.

NDRE Exp. 1				NDRE Exp. 3				NDVI Exp. 1				NDVI Exp. 3			
Treatments	R^2	RRMSE (%)	p-Value of Residual's Normality	Treatments	R^2	RRMSE (%)	p-Value of Residual's Normality	Treatments	R^2	RRMSE (%)	p-Value of Residual's Normality	Treatments	R^2	RRMSE (%)	p-Value of Residual's Normality
N0T0	0.94	4.83	0.89	N0T0	0.94	2.43	0.9	N0T0	0.82	14.5	0.64	N0T0	0.87	21.0	0.72
N5T0	0.94	2.52	0.65	N2T0	0.9	1.55	0.69	N5T0	0.9	13.0	0.96	N2T0	-	-	-
N5T3	0.97	2.26	0.62	N2T1	0.93	2.17	0.79	N5T3	0.92	13.5	0.81	N2T1	0.89	19.0	0.9
N7T0	0.94	2.35	0.39	N6T0	0.95	1.07	0.49	N7T0	0.91	15.0	0.89	N6T0	0.91	17.0	0.84
N7T6	0.94	3.68	0.74	N6T5	0.96	1.9	0.51	N7T6	0.93	17.0	0.9	N6T5	0.92	16.0	0.97
N8T0	0.96	2.34	0.5	N8T0	0.9	1.93	0.29	N8T0	0.94	11.0	0.95	N8T0	0.83	15.0	0.73
N8T7	0.97	2.44	0.56	N8T7	0.91	2.59	0.44	N8T7	0.95	7.5	0.79	N8T7	0.9	16.0	0.81

Notes: RRMSE: relative root-mean-square error; "N0T0, N5T0, N5T3, N6T0, N7T0, N7T6, N8T0, and N8T7" represent different basal and topdressing N fertilizer rates (0 + 0, 120 + 0, 120 + 80, 150 + 0, 180 + 0, 180 + 120, 240 + 0, and 2400 + 160 kg·ha^{-1}, respectively). The normality test was conducted using the Shapiro–Wilk method at 0.05 significance. Reason for the fitting's failure was due to mutual dependency existence between parameters.

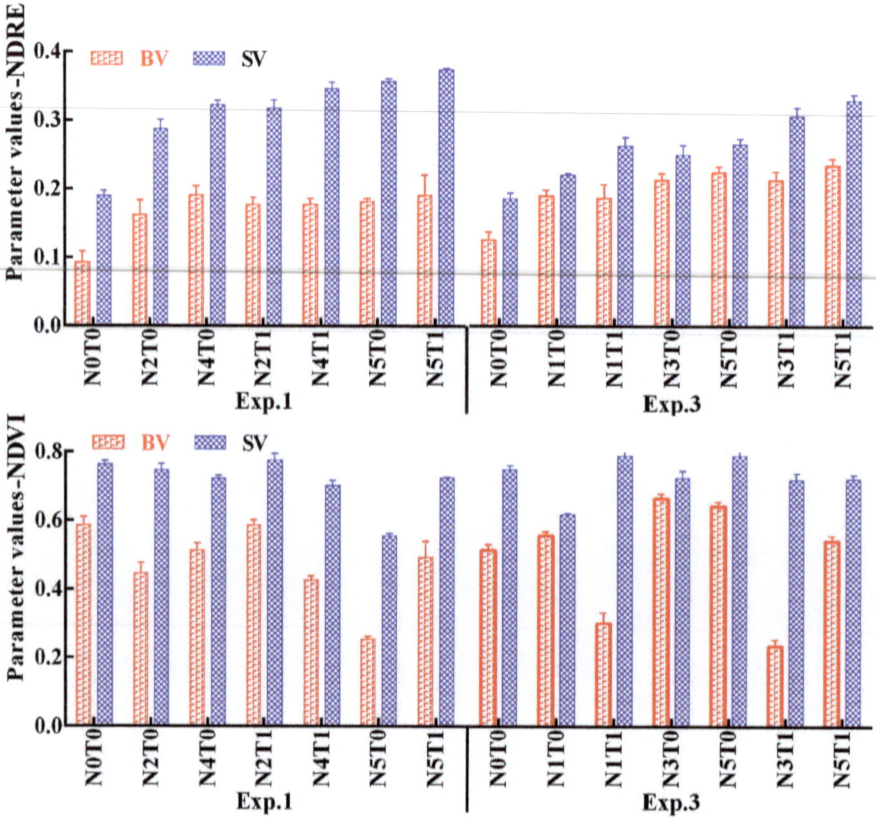

Figure 7. Baseline values (BV) and saturation values (SV) calculated using different vegetation indexes (NDRE, NDVI) under each treatment according to the total N rate. "N0T0, N5T0, N5T3, N7T0, N7T6, N8T0, and N8T7" represent different basal and topdressing N fertilizer rates (0 + 0, 120 + 0, 120 + 0, 180 + 0, 180 + 120, 240 + 0, and 240 + 160 kg·ha^{-1}, respectively).

Table 7. Distribution of the critical point of dynamic change under a single time series.

Experiment	Exp. 1						
Treatment (kg·ha^{-1})	N0T0 0	N5T0 120	N7T0 180	N5T3 200	N8T0 240	N7T6 300	N8T3 320
RAGDD	0.366	0.374	0.351	0.357	0.303	0.345	0.314
	Exp. 3						
Treatment (kg·ha^{-1})	N0T0 0	N2T0 60	N2T1 100	N6T0 150	N6T2 200	N8T0 240	N8T3 320
RAGDD	0.516	0.55	0.524	0.559	0.538	0.494	0.542

Notes: "N8T0, N8T3, N7T6, N7T0, N5T3, N0T0, N5T0, N2T1, N6T2, N2T0, and N6T0" represent different basal and topdressing N fertilizer rates (240 + 0, 240 + 80, 180 + 120, 180 + 0, 120 + 80, 0 + 0, 120 + 0, 60 + 40, 150 + 50, 60 + 0, and 150 + 0 kg·ha^{-1}, respectively).

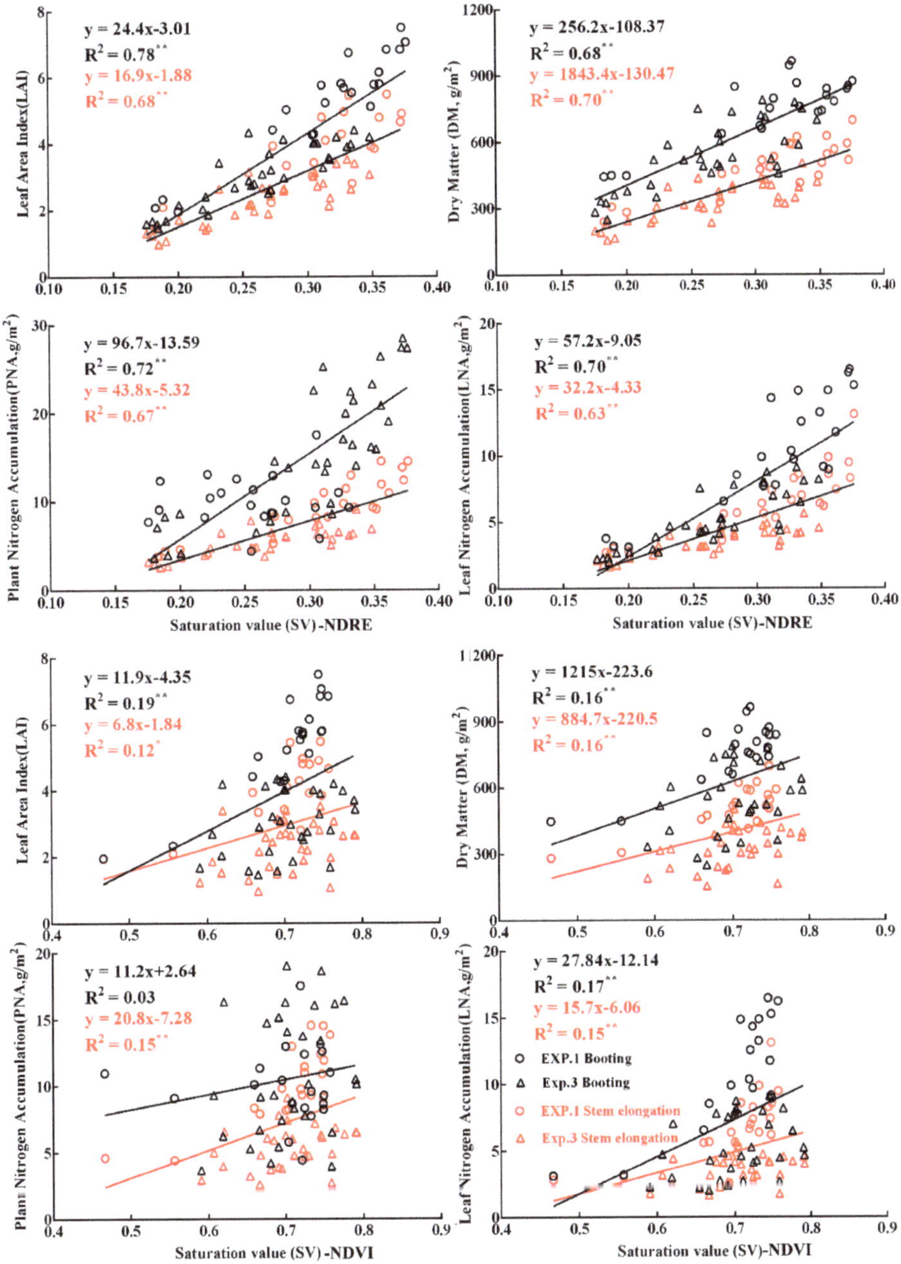

Figure 8. Relationships between saturation values (SV, calculated using NDRE or NDVI) and leaf area index (LAI), plant dry matter, plant nitrogen accumulation, and leaf nitrogen accumulation at stem elongation and booting.

3.5. Prediction Algorithm for Rice Grain Yield

The relationship between BV, SV, and x_0 with grain yield was also studied (Figure 9). The results showed a poor relationship between BV and yield (Figure 9A,E), while x_0 value was negatively correlated with yield, and, although the correlation was poor, differences were observed among cultivars (Figure 9B,F). There were no differences in SV between experiments, suggesting that SV showed the similar correlations with yield between different years, eco-sites, and varieties (Figure 9C). Meanwhile, the NDVI-based SV was poorly correlated with grain yield.

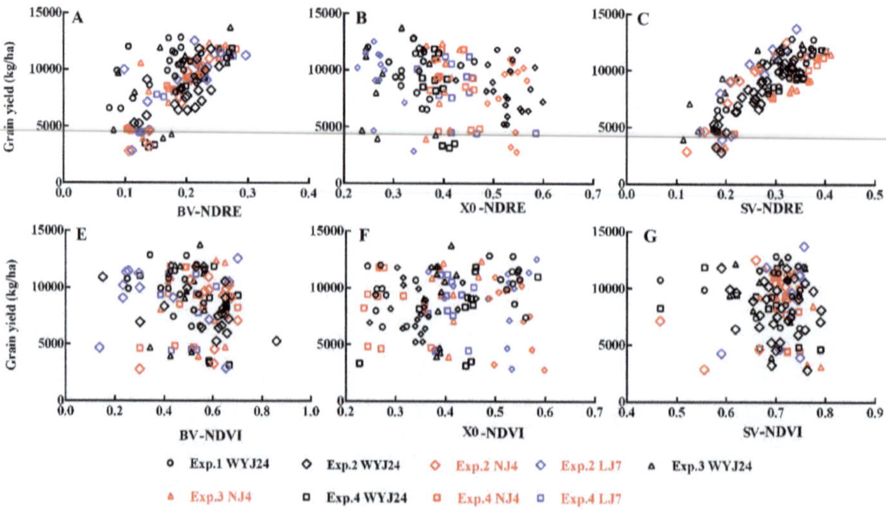

Figure 9. Relationships between each vegetation index (NDRE, NDVI) based on different model parameters (SV, BV, and x_0) and grain yield. BV: baseline value, SV: saturation value, and x_0: the midpoint of the sigmoid curve. (**A**,**E**) show the BV, (**B**,**F**) show the x_0, and (**C**,**G**) show the SV.

The quantitative relationship between SV and grain yield was developed (Figure 10) using the data from Exp. 1 and Exp. 3. As shown in Figure 10, grain yield was forecast using linear regression analysis based on the SV values. Grain yield was evenly distributed from 6000 to 13000 kg·ha^{-1} and the model was good. The slope of the linear model was 36081 kg·ha^{-1}, suggesting that with every 0.1 increase in SV, yield increased by 3608.1 kg·ha^{-1}.

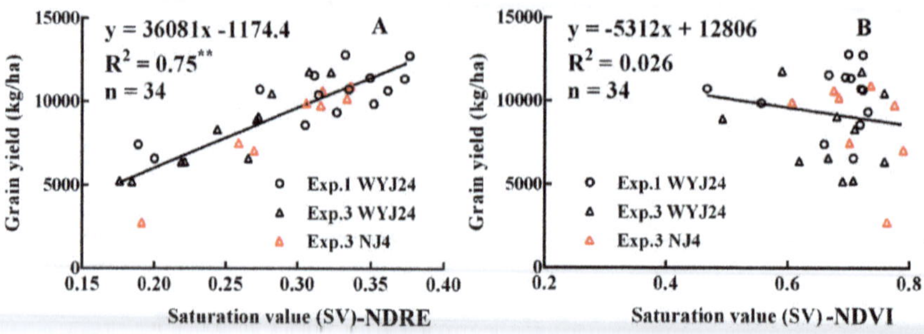

Figure 10. Yield prediction linear model based on saturation values (SV). SV in (**A**) was calculated using NDRE, and in (**B**) was calculated by NDVI.

The data from Exp. 2 and Exp. 4 were subsequently used to test the accuracy of the SV based yield prediction model (Figure 11). As shown, in Figure 11A, SV was obtained from the fitted sigmoid curve model, while, in Figure 11B, values were estimated using measured NDRE values from the stem elongation to booting stages. Both methods gave good prediction results. The predicted yields in both study years showed no obvious abnormalities compared with the observed yield, and the overall prediction error was approximately 20%. These findings suggest that the model is applicable in different varieties and different rice planting regions under different SV acquisition methods.

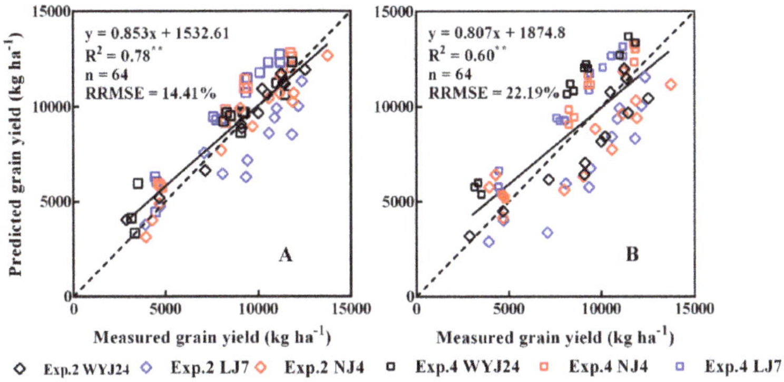

Figure 11. Test results of the yield prediction models based on saturation values (SV). SV in (**A**) was obtained from the fitted sigmoid curve model, and in (**B**) was estimated using measured NDRE (normalized difference red edge) values from stem elongation to booting stages.

We also used the same data from the same growth stages to estimate yield using INSEY (Figure 12) [46]. Of all stages, the best correlation was obtained in booting stage. As shown in Figure 12A, the R^2 of the INSEY-NDVI model was low with poor accuracy. Meanwhile, the INSEY-NDRE model had 70% prediction accuracy, but this was still less accurate than the SV-based yield prediction algorithm by about 10% (Figure 12B).

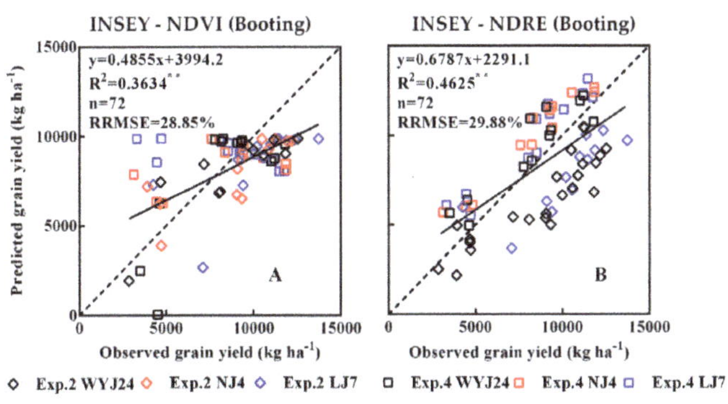

Figure 12. Test results of the yield prediction models based on in-season estimated yield (INSEY). (**A**) The normalized difference vegetation index (NDVI)-INSEY model and (**B**) the normalized difference red edge (NDRE)-INSEY model.

4. Discussion

4.1. Spectral Indexes for Rice Growth Diagnosis

NDVI is the most widely used spectral index. However, because of the characteristics of red light in NDVI itself [47,48], saturation occurs when the population or coverage reaches a certain level, thereby affecting the monitoring effect [41]. Muñoz-Huerta et al. revealed a positive linear relationship between RVI and N uptake in winter wheat, as well as insensitivity to growth stage and crop variety. In addition, RVI was previously used to estimate N status in over-fertilized fields [49]. It was also suggested that the red-edge-based vegetation indexes (RVI, NDRE) perform well in estimating NNI and plant N accumulation [37]. In this study, serious overlap and a cross between obtained values and the error ranges of different gradients in spectral indexes containing red bands was observed after stem elongation, preventing clarification of the differences between N gradients. Studies showed that the absorption of chlorophyll on the red-edge waveband is weaker than that of red light, with the red-edge region having stronger transmission ability with the crop canopy and leaf [25]. Use of red-edge rather than red-light bands can, therefore, reduce the saturation phenomenon, enhancing monitoring and diagnosis of crop N nutrition.

In this study, NDVI did not distinguish between different N treatments during mid-growth and late growth stages, similar to the findings of Liu et al. [41]. It was also previously revealed that, at an early growth stage (before stem elongation), NDVI resulted in a better relationship with rice grain yield; however, on approaching maturity, NDRE performed better [50]. Red-edge-based NDRE reflects this phenomenon via differences in saturation values. Our results suggest that the shortcoming of NDVI is that saturation at an early stage leads to no differences in SV with varying N rates. Yumiko et al. also reported that NDVI reached saturation earlier than NDRE under different N rates [50]. NDVI can, therefore, be used to determine low-land rice at early growth stages, but is less useful at mid-growth to late growth stages. Moreover, using vegetation indexes with the red-light band, correlations with agronomic parameters were poor, while the red-edge band-based indexes (NDRE and RERVI) showed a linear relationship with agronomic parameters, allowing differences between N fertilizer levels to be distinguished. However, changes in RERVI throughout the entire growth period were small, thereby preventing effective analysis between different stages. In the same plot, NDRE showed lower variability than RERVI. These findings suggest that NDRE can be used to reflect the growth status of different treatments, making it a suitable index for diagnosis of rice growth.

4.2. The Dynamic Model and Parameters of Rice Spectral Indexes

The results of this study suggest that NDRE conforms to a sigmoid curve. In early stages, NDRE fluctuated around background BV values at a low level. However, values began to rise rapidly after tillering. The critical point (x_0) of the dynamic model was then reached on entering stem elongation. Meanwhile, at flowering stage, the dynamic model became stable again and approached the feature platform SV.

Previous monitoring methods such as the saturation index (SI) method and NNI are based mainly on data collected at each growth stage [8]. These methods mostly lack biophysical mechanisms, often fail to obtain indicators, and consist of a complex calculation method [25]. Moreover, deviations in identification of growth stages, environmental factors, time selection, and so on have a significant impact on these single tests. Furthermore, single measured values are treated as a reflection of an entire stage in the calculations, resulting in greater deviation from the actual values [51].

In this paper, the parameters in the model have practical biological meaning. The range of x_0 reflects growth and development of rice at early stages. The findings also suggest that higher x_0 values occur under abnormally high temperatures in early stages compared with normal growing conditions. The x_0 value can be used as an indicator of the growth environment during early stages of growth; however, the poor correlation between x_0 and grain yield prevented quantification of yield predictions.

BV was also able to reflect growth status at early stages. However, when rice coverage was less than 50%, including the depth of the water layer, turbidity, soil background color, and so on, there was an impact on the spectral reflectance test [52,53]. In this study, BV values were low under all treatments, and no obvious differences between gradients were observed. Meanwhile, BV did not distinguish between differences in basal fertilizers levels in early stages of growth, suggesting that water and soil background information under low coverage conditions during this period obscure growth information contained in the spectral data. BV is, therefore, not applicable for direct monitoring and diagnosis of rice growth, although differences in LAI and other agronomic parameters were obvious. In the future, we aim to increase accuracy by adding variables such as coverage, soil, and water to improve the signal-to-noise ratio.

In the middle of the growing season, SV values overcame the above problems in x_0 and BV. SV values differed between different rice populations under different treatments and in different groups, and there was a good linear relationship with agronomic parameters such as LAI and aboveground N accumulation. SV values could be obtained not only from the fitted model, but also from measured values. Overall, the accuracy and availability of the model were high, confirming applicability for monitoring and diagnosis.

The dynamic model was successfully used to determine growth status using different rice varieties, in different eco-sites, and under different N rates before flowering with 90% variation. Bonfil et al. also reported similar dynamic changes in wheat in 2016 [52]. Moreover, our dynamic model also determined the growth status of paddy rice using RapidScan CS-45 at different grain yield levels. Of course, it can also be used to predict yield before flowering without any specific calibration. It should also be used on non-rainy days, since water droplets on the blade affect monitoring. In the future, the growth status after flowering stage will be examined further, and additional vegetation indexes will be used in developing a dynamic model.

4.3. SV-Based Yield Prediction Algorithm

In previous studies, spectral indexes were used to predict yield in crops such as wheat [54] and sugarcane [55] with good results. Bonfil used RapidScan CS-45 to monitor the dynamic changes in wheat, revealing a good linear relationship between NDRE and wheat yield in different varieties [54]. In this study, the dynamic pattern of NDRE at pre-flowering stages was consistent, and predicted yield was achieved under optimal growing conditions other than limited N fertilizer. Kanke et al. also revealed that vegetation indexes containing the red-edge band had a better linear relationship with agronomic parameters and yield at main growth stages in rice [55]. A linear model of yield prediction using different vegetation indexes as input variables was also established, within which NDRE and RERVI gave best correlations, with R^2 values greater than 0.8. However, the prediction equation was predicted using data from a single growth stage, and the parameters differed between stages, making it of no use between stages or deviations [56]. Similarly, Cowley et al. established an NDVI-based model for forecasting of rapeseed yield [57], while Liu et al. proved the dynamic model could be continuously used to monitor N nutrition status in a single growing season [41]. However, it could only be used when GDD was between 210 and 320 °C.

Thus, the problem of low prediction accuracy resulting from NDVI saturation when using traditional algorithms in rice under high yield conditions remains unsolved. In this study, SV showed a good linear relationship with agronomic parameters and yield. Moreover, SV values were the feature platform of each treatment, eliminating deviations between tests and different environments, thereby increasing applicability of the model, improving stability of the algorithm, and widening the available time range. In addition, SV values could be obtained not only from the fitted model, but also from measured values. Moreover, measurement and processing are relatively simple and do not require normalization or other transformations to maintain accuracy, simplifying our method compared with other algorithms [58].

Satellite remote sensing (RS) data were extensively applied to a wide range of research problems and practical applications, including yield predictions [59]. Zhu et al. used Moderate Resolution Imaging Spectroradiometer Enhanced Vegetation Index (MODIS EVI) time-series data to predict yield in winter wheat, revealing a high R^2 (0.70) and low RMSE (343.34 kg·hm^{-2}) [60]. Furthermore, Fletcher reported continuity of multispectral high-resolution optical observations over global terrestrial surfaces using Sentinel 2 [61], providing detailed information within field variability [62]. Our newly developed SV-based yield prediction algorithm must now be tested and applied under different RS platforms, such as unmanned aerial vehicles and Sentinel 2.

4.4. Interval System Errors between Experiments and Uncertainties of the Model

In different experiments, trends in indicators were basically the same, although the ranges of change were quite different, especially in Exp. 1 (2015) and Exp. 3 (2016). This was possibly due to the differences in meteorological conditions in 2015 and 2016 (Figure 13). From June to early July 2016, the temperature was slightly higher than in 2015; thus, growth was relatively fast in early and mid-tillering. However, in late July to August, unusually high temperatures and continuous rain occurred, which is not conducive to rice growth and development, thereby inhibiting growth. As a result, the agronomic parameters, spectral indexes, and other indicators decreased sharply in 2015.

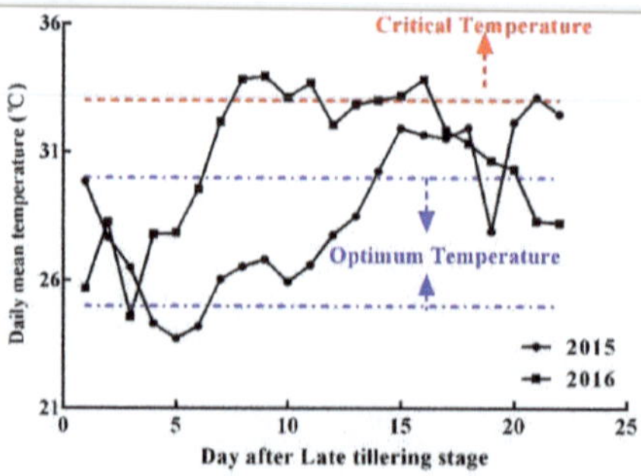

Figure 13. Daily average temperatures in the study site from late tillering to early stem elongation in 2015 and 2016.

This phenomenon was also reflected in another way. In both years, x_0 values remained relatively concentrated at the end of the tillering stage; however, there were differences between N rates. Moreover, the critical point shifted in 2016 compared to 2015. The average daily temperature was higher than 33 °C at the end of tillering and early stem elongation stage in Exp. 3, 2016. The optimum temperature of paddy rice at tillering is 25~30 °C, while the maximum temperature is 32 °C [42] (Figure 13). As a result, these high temperatures caused an abnormal increase in the RAGDD time variable, even though growth was inhibited and x_0 increased.

Model uncertainty received considerable attention in recent studies [63]. There is growing pressure on crop field production to conform to quality standards, which require evaluation and expression of the uncertainty of measurement results [64]. It is, therefore, important to determine the degree of uncertainty associated with a model and its application in predicting crop N status and yield. In this study, the dynamic model was developed using a large database, diverse eco-sites, different rice varieties, and differing N rates; however, errors were revealed when the model output was compared with observations. Moreover, it is difficult to evaluate uncertainty in practice. For example, the uncertainty of certain components is often unknown in crop field experiments, such as those associated with environmental effects or different sensors. During the 2015–2017 growing season, the sensors were calibrated each year to avoid errors; however, additional field experiments are needed to expand the database and therefore improve the robustness of our dynamic model.

5. Conclusions

We used the active spectrometer RapidScan CS-45 to determine the dynamic changes in different spectral indexes and agronomic parameters in rice. The results suggest that NDRE is more stable than other indexes, with the distinctions between different treatment gradients most obvious. NDRE also had the best relationship with agronomic parameters. The dynamic changes in NDRE and LAI were similar, and followed a "slow–fast–slow" trend during the pre-flowering period.

A sigmoid curve was used in NDRE dynamic model construction, with high accuracy and an R^2 of 0.9 and RRMSE of 5%. The model contained critical point x_0, background value BV, and feature platform value SV parameters. The relationships with yield showed that each of these parameters was correlated with yield, although SV was better than the other two parameters. SV showed a good linear relationship with in-season potential grain. Based on these findings, we constructed a yield prediction model. With every 0.1 increase in SV, grain yield increased by 3608.1 kg·ha^{-1}. The overall prediction accuracy of the model was greater than 80%. It can be used at stem elongation and booting, improving timeliness compared with existing forecasting algorithms. Our model is, therefore, useful for rice production management, purchasing and storage preparation, rice spot price trend forecasts, and food policy decision-making.

Author Contributions: X.G. and K.Z. performed the experiments with support from P.S., W.L., and Q.C.; Y.T., X.L., Y.Z., and W.C. provided advice and edited the manuscript; X.G. planned the experiments; and K.Z. wrote the manuscript. All authors read and approved the final manuscript.

Funding: This work was supported by the National Key R&D Program (2018YFD0300805), and the Science and Technology Support Program of Jiangsu [grant number BE2016375].

Acknowledgments: The authors would like to thank the anonymous reviewers for their constructive comments and suggestions.

Conflicts of Interest: The authors declare no conflicts of interest.

References

1. Lundström, C.; Lindblom, J.; Hansen, J.W.; Thornton, P.K.; Berentsen, P.B.M. Considering farmers' situated knowledge of using agricultural decision support systems (AgriDSS) to Foster farming practices: The case of CropSAT. *Agric. Syst.* **2018**, *159*, 9–20. [CrossRef]
2. Zhang, H.; Zhang, J.; Gong, J.; Yong, C.; Min, I.; Hui, G.; Dai, Q.; Huo, Z.; Ke, X.; Wei, H. The productive advantages and formation mechanisms of "Indica Rice to Japonica Rice". *Sci. Agric. Sinica* **2013**, *46*, 686–704.
3. He, J.; Xiao, H.; Huang, X.; Zhou, Q.; Hu, D. Effects of real-time and site-specific nitrogen management on rice yield and nitrogen use efficiency. *Southwest China J. Agric. Sci.* **2010**, *23*, 1132–1136.
4. Diacono, M.; Rubino, P.; Montemurro, F. Precision nitrogen management of wheat. A review. *Agron. Sustain. Dev.* **2013**, *33*, 219–241. [CrossRef]
5. Peng, S.; Buresh, R.J.; Huang, J.; Zhong, X.; Zou, Y.; Yang, J.; Wang, G.; Liu, Y.; Hu, R.; Tang, Q. Improving nitrogen fertilization in rice by Site-Specific N Management. *Agron. Sustain. Dev.* **2010**, *30*, 649–656. [CrossRef]
6. Huang, J.; He, F.; Cui, K.; Buresh, R.J.; Xu, B.; Gong, W.; Peng, S. Determination of optimal nitrogen rate for rice varieties using a chlorophyll meter. *Field Crops Res.* **2008**, *105*, 70–80. [CrossRef]
7. Peng, S.; Garcia, F.V.; Laza, R.C.; Sanico, A.L.; Visperas, R.M.; Cassman, K.G. Increased N-use efficiency using a chlorophyll meter on high-yielding irrigated rice. *Field Crops Res.* **1996**, *47*, 243–252. [CrossRef]
8. Bijay-Singh; Varinderpal-Singh; Yadvinder-Singh; Thind, H.S.; Kumar, A.; Gupta, R.K.; Kaul, A.; Vashistha, M. Fixed-time adjustable dose site-specific fertilizer nitrogen management in transplanted irrigated rice (*Oryza sativa* L.) in South Asia. *Field Crops Res.* **2012**, *126*, 63–69. [CrossRef]
9. Yi, Q.; Zhao, S.; Zhang, X.; Yang, L.; Xiong, G.; He, P. Yield and nitrogen use efficiency as influenced by real time and site specific nitrogen management in two rice cultivars. *Plant Nutr. Fertil. Sci.* **2012**, *18*, 777–785.
10. Fan, L.; Peng, X.; Liu, Y.; Song, T. Study on the site-specific nitrogen management of rice in cold area of northeastern China. *Sci. Agric. Sinica* **2005**, *38*, 1761–1766.
11. Justes, E.; Mary, B.; Meynard, J.M.; Machet, J.M.; Thelierhuche, L. Determination of a critical nitrogen dilution curve for winter wheat crops. *Ann. Bot.* **1994**, *74*, 397–407. [CrossRef]
12. Greenwood, D.J.; Neeteson, J.J.; Draycott, A. Quantitative relationships for the dependence of growth rate of arable crops on their nitrogen content, dry weight and aerial environment. *Plant Soil* **1986**, *91*, 281–301. [CrossRef]
13. Lemaire, G.; Jeuffroy, M.H.; Gastal, F. Diagnosis tool for plant and crop N status in vegetative stage: Theory and practices for crop N management. *Eur. J. Agron.* **2008**, *28*, 614–624. [CrossRef]
14. Ata-Ul-Karim, S.T.; Liu, X.; Lu, Z.; Yuan, Z.; Zhu, Y.; Cao, W. In-season estimation of rice grain yield using critical nitrogen dilution curve. *Field Crops Res.* **2016**, *195*, 1–8. [CrossRef]

15. Xia, T.; Miao, Y.; Wu, D.; Shao, H.; Khosla, R.; Mi, G. Active optical sensing of spring maize for in-season diagnosis of nitrogen status based on nitrogen nutrition index. *Remote Sens.* **2016**, *8*, 605. [CrossRef]
16. Zhao, B.; Liu, Z.; Ata-Ul-Karim, S.T.; Xiao, J.; Liu, Z.; Qi, A.; Ning, D.; Nan, J.; Duan, A. Rapid and nondestructive estimation of the nitrogen nutrition index in winter barley using chlorophyll measurements. *Field Crops Res.* **2016**, *185*, 59–68. [CrossRef]
17. Ata-Ul-Karim, S.T.; Liu, X.; Lu, Z.; Zheng, H.; Cao, W.; Zhu, Y. Estimation of nitrogen fertilizer requirement for rice crop using critical nitrogen dilution curve. *Field Crops Res.* **2017**, *201*, 32–40. [CrossRef]
18. Hu, Y.; Yang, J.P.; Lv, Y.M.; He, J.J. SPAD values and nitrogen nutrition index for the evaluation of rice nitrogen status. *Plant Prod. Sci.* **2014**, *17*, 81–92.
19. Houlès, V.; Guérif, M.; Mary, B. Elaboration of a nitrogen nutrition indicator for winter wheat based on leaf area index and chlorophyll content for making nitrogen recommendations. *Eur. J. Agron.* **2007**, *27*, 1–11. [CrossRef]
20. Debaeke, P.; Rouet, P.; Justes, E. Relationship between the normalized SPAD index and the nitrogen nutrition index: Application to durum wheat. *J. Plant Nutr.* **2006**, *29*, 75–92. [CrossRef]
21. Ata-Ul-Karim, S.T.; Zhu, Y.; Liu, X.; Cao, Q.; Tian, Y.; Cao, W. Comparison of different critical nitrogen dilution curves for nitrogen diagnosis in rice. *Sci. Rep.* **2017**, *7*, 42679. [CrossRef] [PubMed]
22. Bogdan, K.; Andrzej, L.; Andrzej, O.; Marek, K. The effect of tillage system and forecrop on the yield and values of LAI and SPAD indices of spring wheat. *Eur. J. Agron.* **2010**, *33*, 43–51.
23. Spitkó, T.; Nagy, Z.; Zsubori, T.; Szőke, C.; Berzy, T.; Pintér, J.; Márton, L. Connection between normalized difference vegetation index and yield in maize. *Plant Soil Environ.* **2016**, *62*, 293–298. [CrossRef]
24. Tubaña, B.S.; Harrell, D.L.; Walker, T.; Teboh, J.; Lofton, J.; Kanke, Y. In-season canopy reflectance-based estimation of rice yield response to nitrogen. *Agron. J.* **2012**, *104*, 1604–1611. [CrossRef]
25. Liu, K.; Yazhen, L.I.; Huiwen, H.U. Estimating the effect of urease inhibitor on rice yield based on NDVI at key growth stages. *Front. Agric. Sci. Eng.* **2014**, *1*, 150–157. [CrossRef]
26. Xue, L.; Li, G.; Qin, X.; Yang, L.; Zhang, H. Topdressing nitrogen recommendation for early rice with an active sensor in south China. *Precis. Agric.* **2014**, *15*, 95–110. [CrossRef]
27. Evert, F.K.V.; Booij, R.; Jukema, J.N.; Berge, H.F.M.T.; Uenk, D.; Meurs, E.J.J.; Geel, W.C.A.V.; Wijnholds, K.H.; Slabbekoorn, J.J. Using crop reflectance to determine sidedress N rate in potato saves N and maintains yield. *Eur. J. Agron.* **2012**, *43*, 58–67. [CrossRef]
28. Morier, T.; Cambouris, A.N.; Chokmani, K. In-season nitrogen status assessment and yield estimation using hyperspectral vegetation indices in a potato crop. *Agron. J.* **2015**, *107*, 1295–1309. [CrossRef]
29. Lukina, E.V.; Freeman, K.W.; Wynn, K.J.; Thomason, W.E.; Mullen, R.W.; Stone, M.L.; Solie, J.B.; Klatt, A.R.; Johnson, G.V.; Elliott, R.L. Nitrogen fertization optimization algorithm based on in-season estimates of yield and plant nitrogen uptake. *J. Plant Nutr.* **2001**, *24*, 885–898. [CrossRef]
30. Raun, W.R.; Solie, J.B.; Johnson, G.V.; Stone, M.L.; Lukina, E.V.; Thomason, W.E.; Schepers, J.S. In-season prediction of potential grain yield in winter wheat using canopy reflectance. *Agron. J.* **2001**, *93*, 131–138. [CrossRef]
31. Cao, Q.; Miao, Y.; Shen, J.; Yu, W.; Yuan, F.; Cheng, S.; Huang, S.; Wang, H.; Yang, W.; Liu, F. Improving in-season estimation of rice yield potential and responsiveness to topdressing nitrogen application with Crop Circle active crop canopy sensor. *Precis. Agric.* **2016**, *17*, 136–154. [CrossRef]
32. Shaver, T.M.; Khosla, R.; Westfall, D.G. Evaluation of two ground-based active crop canopy sensors in maize: Growth stage, row spacing, and sensor movement speed. *Soil Sci. Soc. Am. J.* **2010**, *74*, 2101–2108. [CrossRef]
33. Li, F.; Etc, Y.M. Estimating winter wheat biomass and nitrogen status using an active crop sensor. *Intell. Autom. Soft Comput.* **2010**, *16*, 1221–1230.
34. Lofton, J.; Tubana, B.S.; Kanke, Y.; Teboh, J.; Viator, H.; Dalen, M. Estimating sugarcane yield potential using an in-season determination of normalized difference vegetative index. *Sensors* **2012**, *12*, 7529. [CrossRef] [PubMed]
35. Yao, Y.; Miao, Y.; Jiang, R.; Khosla, R.; Gnyp, M.L.; Bareth, G. Evaluating different active crop canopy sensors for estimating rice yield potential. In Proceedings of the Second International Conference on Agro-Geoinformatics, Fairfax, VA, USA, 12–16 August 2013; pp. 538–542.
36. Thompson, L.J.; Ferguson, R.B.; Kitchen, N.; Frazen, D.W.; Mamo, M.; Yang, H.; Schepers, J.S. Model and sensor-based recommendation approaches for in-season nitrogen management in corn. *Agron. J.* **2015**, *107*, 2020–2030. [CrossRef]

37. Cao, Q.; Miao, Y.; Wang, H.; Huang, S.; Cheng, S.; Khosla, R.; Jiang, R. Non-destructive estimation of rice plant nitrogen status with Crop Circle multispectral active canopy sensor. *Field Crops Res.* **2013**, *154*, 133–144. [CrossRef]
38. Russelle, M.P.; Wilhelm, W.W.; Olson, R.A.; Power, J.F. Growth analysis based on degree days. *Crop Sci.* **1984**, *24*, 28–32. [CrossRef]
39. Su, L.; Wang, Q.; Bai, Y. An analysis of yearly trends in growing degree days and the relationship between growing degree day values and reference evapotranspiration in Turpan area, China. *Theor. Appl. Climatol.* **2013**, *113*, 711–724. [CrossRef]
40. Liu, X.; Ferguson, R.B.; Zheng, H.; Cao, Q.; Tian, Y.; Cao, W.; Zhu, Y. Using an active-optical sensor to develop an optimal NDVI dynamic model for high-yield rice Production (Yangtze, China). *Sensors* **2017**, *17*, 672. [CrossRef]
41. Li, H.; Zhao, C.; Yang, G.; Feng, H. Variations in crop variables within wheat canopies and responses of canopy spectral characteristics and derived vegetation indices to different vertical leaf layers and spikes. *Remote Sens. Environ.* **2015**, *169*, 358–374. [CrossRef]
42. Kanke, Y.; Raun, W.; Solie, J.; Stone, M.; Taylor, R. Red edge as a potential index for detecting differences in plant nitrogen status in winter wheat. *J. Plant Nutr.* **2012**, *35*, 1526–1541. [CrossRef]
43. Prasad, B.; Carver, B.F.; Stone, M.L.; Babar, M.A.; Raun, W.R.; Klatt, A.R. Genetic analysis of indirect selection for winter wheat grain yield using spectral reflectance indices. *Crop Sci.* **2007**, *47*, 1416–1425. [CrossRef]
44. Gutierrez, M.; Reynolds, M.P.; Raun, W.R.; Stone, M.L.; Klatt, A.R. Spectral water indices for assessing yield in elite bread wheat genotypes under well-irrigated, water-stressed, and high-temperature conditions. *Crop Sci.* **2010**, *50*, 197–214. [CrossRef]
45. Yao, Y.; Miao, Y.; Huang, S.; Gao, L.; Ma, X.; Zhao, G.; Jiang, R.; Chen, X.; Zhang, F.; Yu, K. Active canopy sensor-based precision N management strategy for rice. *Agron. Sustain. Dev.* **2012**, *32*, 925–933. [CrossRef]
46. Nguyrobertson, A.; Gitelson, A.; Peng, Y.; Viña, A.; Arkebauer, T.; Rundquist, D. Green leaf area index estimation in maize and soybean: Combining vegetation indices to achieve maximal sensitivity. *Agron. J.* **2012**, *104*, 1336. [CrossRef]
47. Gnyp, M.L.; Miao, Y.; Yuan, F.; Ustin, S.L.; Yu, K.; Yao, Y.; Huang, S.; Bareth, G. Hyperspectral canopy sensing of paddy rice aboveground biomass at different growth stages. *Field Crops Res.* **2014**, *155*, 42–55. [CrossRef]
48. Beck, P.S.A.; Atzberger, C.; Høgda, K.A.; Johansen, B.; Skidmore, A.K. Improved monitoring of vegetation dynamics at very high latitudes: A new method using MODIS NDVI. *Remote Sens. Environ.* **2006**, *100*, 321–334. [CrossRef]
49. Ferencz, Cs.; Bognár, P.; Lichtenberger, J.; Hamar, D.; Tarcsai, Gy.; Timár, G.; Molnár, G.; Pásztor, Sz.; Steinbach, P.; Székely, B.; et al. Crop yield estimation by satellite remote sensing. *Int. J. Remote Sens.* **2004**, *25*, 4113–4149. [CrossRef]
50. Muñoz-Huerta, R.F.; Guevara-Gonzalez, R.G.; Contreras-Medina, L.M.; Torres-Pacheco, I.; Prado-Olivarez, J.; Ocampo-Velazquez, R.V. A review of methods for sensing the nitrogen status in plants: Advantages, disadvantages and recent advances. *Sensors* **2013**, *13*, 10823–10843. [CrossRef] [PubMed]
51. Vaesen, K.; Gilliams, S.; Nackaerts, K.; Coppin, P. Ground-measured spectral signatures as indicators of ground cover and leaf area index: The case of paddy rice. *Field Crops Res.* **2001**, *69*, 13–25. [CrossRef]
52. Vinciková, H.; Hanuš, J.; Pechar, L. Spectral reflectance is a reliable water-quality estimator for small, highly turbid wetlands. *Wetl. Ecol. Manag.* **2015**, *23*, 933–946. [CrossRef]
53. Pradhan, S.; Bandhypadyay, K.K.; Sahoo, R.N.; Sehgal, V.K.; Singh, R.; Joshi, D.K.; Gupta, V.K. Prediction of wheat (Triticum astivum) grain biomass yield under different irrigation and nitrogen management practices using canopy reflectance spectra model. *Indian J. Agric. Sci.* **2013**, *83*, 1136–1143.
54. Amaral, L.R.; Molin, J.P.; Portz, G.; Finazzi, F.B.; Cortinove, L. Comparison of crop canopy reflectance sensors used to identify sugarcane biomass and nitrogen status. *Precis. Agric.* **2015**, *16*, 15–28. [CrossRef]
55. Bonfil, D.J. Wheat phenomics in the field by RapidScan: NDVI vs. NDRE. *Israel J. Plant Sci.* **2017**, *64*, 41–45. [CrossRef]
56. Kanke, Y.; Tubaña, B.; Dalen, M.; Harrell, D. Evaluation of red and red-edge reflectance-based vegetation indices for rice biomass and grain yield prediction models in paddy fields. *Precis. Agric.* **2016**, *17*, 507–530. [CrossRef]

57. Ali, A.M.; Thind, H.S.; Sharma, S.; Varinderpal-Singh. Prediction of dry direct-seeded rice yields using chlorophyll meter, leaf color chart and GreenSeeker optical sensor in northwestern India. *Field Crops Res.* **2014**, *161*, 11–15. [CrossRef]
58. Cowley, R.B.; Luckett, D.J.; Moroni, J.S.; Diffey, S. Use of remote sensing to determine the relationship of early vigour to grain yield in canola (*Brassica napus* L.) germplasm. *Crop Pasture Sci.* **2014**, *65*, 1288–1299. [CrossRef]
59. Schmidt, J.; Beegle, D.; Zhu, Q.; Sripada, R. Improving in-season nitrogen recommendations for maize using an active sensor. *Field Crops Res.* **2011**, *120*, 94–101. [CrossRef]
60. Zhu, Z.; Chen, L.; Zhang, J.; Pan, Y.; Zhu, W.; Hu, T. Division of winter wheat yield estimation by remote sensing based on MODIS EVI time series data and spectral angle clustering. *Spectrosc. Spectr. Anal.* **2012**, *32*, 1899–1904.
61. Verrelst, J.; Muñoz, J.; Alonso, L.; Delegido, J.; Rivera, J.P.; Camps-Valls, G.; Moreno, J. Machine learning regression algorithms for biophysical parameter retrieval: Opportunities for sentinel-2 and -3. *Remote Sens. Environ.* **2012**, *118*, 127–139. [CrossRef]
62. Sánchez, B.; Rasmussen, A.; Porter, J.R. Temperatures and the growth and development of maize and rice: A review. *Glob. Chang. Biol.* **2014**, *20*, 408–417. [CrossRef] [PubMed]
63. Clyde, M.; George, E.I. Model Uncertainty. *Stat. Sci.* **2004**, *19*, 81–94.
64. Hansen, L.P.; Sargent, T.J. Robust Control and Model Uncertainty. *Am. Econ. Rev.* **2001**, *91*, 60–66. [CrossRef]

remote sensing

MDPI

Article

In-Season Diagnosis of Rice Nitrogen Status Using Proximal Fluorescence Canopy Sensor at Different Growth Stages

Shanyu Huang [1,2,3], Yuxin Miao [1,4,*], Fei Yuan [5], Qiang Cao [6], Huichun Ye [7], Victoria I.S. Lenz-Wiedemann [1,2] and Georg Bareth [1,2]

1 International Center for Agro-Informatics and Sustainable Development (ICASD), College of Resources and Environmental Sciences, China Agricultural University, Beijing 100193, China
2 Institute of Geography, University of Cologne, 50923 Köln, Germany
3 Chinese Academy of Agricultural Engineering Planning & Design, Beijing 100125, China
4 Precision Agriculture Center, Department of Soil, Water and Climate, University of Minnesota, St. Paul, MN 55108, USA
5 Department of Geography, Minnesota State University, Mankato, MN 56001, USA
6 National Engineering and Technology Center for Information Agriculture, Nanjing Agricultural University, Nanjing 210095, China
7 Key Laboratory of Digital Earth Science, Institute of Remote Sensing and Digital Earth, Chinese Academy of Sciences, Beijing 100094, China
* Correspondence: ymiao@cau.edu.cn or ymiao@umn.edu; Tel.: +86-138-1104-8358 or +1-612-625-4731

Received: 7 July 2019; Accepted: 6 August 2019; Published: 8 August 2019

Abstract: Precision nitrogen (N) management requires an accurate and timely in-season assessment of crop N status. The proximal fluorescence sensor Multiplex®3 is a promising tool for monitoring crop N status. It performs a non-destructive estimation of plant chlorophyll, flavonol, and anthocyanin contents, which are related to plant N status. The objective of this study was to evaluate the potential of proximal fluorescence sensing for N status estimation at different growth stages for rice in cold regions. In 2012 and 2013, paddy rice field experiments with five N supply rates and two varieties were conducted in northeast China. Field samples and fluorescence data were collected in the leaf scale (LS), on-the-go (OG), and above the canopy (AC) modes using Multiplex®3 at the panicle initiation (PI), stem elongation (SE), and heading (HE) stages. The relationships between the Multiplex indices or normalized N sufficient indices (NSI) and five N status indicators (above-ground biomass (AGB), leaf N concentration (LNC), plant N concentration (PNC), plant N uptake (PNU), and N nutrition index (NNI)) were evaluated. Results showed that Multiplex measurements taken using the OG mode were more sensitive to rice N status than those made in the other two modes in this study. Most of the measured fluorescence indices, especially the N balance index (NBI), simple fluorescence ratios (SFR), blue–green to far-red fluorescence ratio (BRR_FRF), and flavonol (FLAV) were highly sensitive to N status. Strong relationships between these fluorescence indices and N indicators, especially the LNC, PNC, and NNI were revealed, with coefficients of determination (R^2) ranging from 0.40 to 0.78. The N diagnostic results indicated that the normalized N sufficiency index based on NBI under red illumination (NBI_R_{NSI}) and FLAV achieved the highest diagnostic accuracy rate (90%) at the SE and HE stages, respectively, while NBI_R_{NSI} showed the highest diagnostic consistency across growth stages. The study concluded that the Multiplex sensor could be used to reliably estimate N nutritional status for rice in cold regions, especially for the estimation of LNC, PNC, and NNI. The normalized N sufficiency indices based on the Multiplex indices could further improve the accuracy of N nutrition diagnosis by reducing the influences of inter-annual variations and different varieties, as compared with the original Multiplex indices.

Keywords: Multiplex®3 sensor; nitrogen balance index; nitrogen nutrition index; nitrogen status diagnosis; precision nitrogen management

1. Introduction

Nitrogen (N) is an essential nutrient for plant growth and development. However, excessive N fertilizer applications have led to severe environmental impacts in China [1,2]. Therefore, there has been a growing interest in developing precision N management strategies in agricultural research for many years. This requires the development of efficient and timely crop N status diagnosis strategies and technologies [3].

Plant or leaf N concentration is an indication of N nutritional status. The traditional N testing method in the laboratory is time-consuming and impractical for characterizing spatial and temporal variability in crop N status in precision N management. Alternatively, it is known that there is a strong relationship between plant chlorophyll content and N content, although other stress factors, such as water deficit, cold, heat, excess light, leaf or root pathogens, etc., can also influence chlorophyll content directly or indirectly [4,5]. Therefore, various instruments based on measuring chlorophyll have been developed to provide indirect, nondestructive, and real-time estimations of leaf N content [3,6]. For example, the Soil Plant Analysis Development chlorophyll meter (SPAD) developed by Minolta Camera Co. (Osaka, Japan) is a widely used portable instrument for measuring chlorophyll in leaves. The SPAD meter measures the difference in absorption between the red (660 nm) and near-infrared (940 nm) wavelengths [7,8]. Leaf chlorophyll absorbs red light but not infrared, therefore, the SPAD readings indicate plant chlorophyll concentration and N content [9–11]. However, the reliability of SPAD results is affected by factors such as growth stage, irradiance, water status, and leaf thickness [3,5,9].

Optical non-destructive remote sensing methods based on canopy reflectance measurements have also been widely used [12,13]. The high measuring efficiency of reflectance spectroscopy sensors and the strong correlation between their measurements and crop physiological and biochemical parameters offer a high potential for N management [14,15]. Proximal active sensors, such as GreenSeeker (NTech Industries, Inc., Ukiah, CA) and Crop Circle (Holland Scientific, NE, USA), have been used to diagnose N nutritional status in real-time and to guide in-season precision management for rice N fertilization [16,17]. However, the results based on the canopy reflectance are affected by various factors, such as soil characteristics, crop growth stages, and saturation under high biomass conditions [16,18]. In addition, it is more difficult to estimate chlorophyll or N status using optical remote sensing methods as the contribution of leaf area index and biomass to canopy reflectance is much greater than that of chlorophyll or N concentration [19,20].

Unlike reflectance indices, the fluorescence spectra are less affected by biomass or leaf area index [15,21]. At different N nutritional levels, the fluorescence intensities of leaves are significantly different near the 440 nm (Blue, B), 525 nm (Green, G), 685–690 nm (Red, R), and 735–740 nm (Near-infrared, NIR) wavelengths [22,23]. Studies have shown that the fluorescence ratio of NIR and R bands is highly correlated with chlorophyll concentration [24,25] and leaf N concentration [26]. Because the fluorescence ratio is only related to chlorophyll concentration or photosynthetic activities, soil background does not affect the vegetation spectra at the early growth stages. Longchamps and Khosla [27] observed that N supply levels in corn could be differentiated as early as the V5 phenological stage using a proximal fluorescence sensor. In contrast, the test results only became reliable starting from the V8 growth stage based on reflective sensors [28,29]. Therefore, chlorophyll fluorescence sensing is a powerful solution to the shortcomings of proximal reflectance sensors in crop N status monitoring.

In addition to chlorophyll content, polyphenols (mainly flavonols) can also be used to estimate plant N status. When N is deficient, polyphenols increase significantly due to carbon and N balance regulation mechanisms [30], although an increase in polyphenol concentration may also be related to their photoprotective roles [31]. These compounds are mainly concentrated in epidermal cells and have typical absorption peaks in the ultraviolet region [31–33]. Thus, N status diagnosis is improved by combining the polyphenol and chlorophyll fluorescence [34,35]. Lejealle et al. [36] demonstrated that the N balance index (NBI), the ratio of chlorophyll to flavonol, had a better and more stable correlation with leaf N concentration. Leaf fluorescence sensor Dualex (FORCE-A, Orsay, Paris, France) and canopy fluorescence sensor Multiplex (FORCE-A, Orsay, Paris, France) can be used to estimate plant polyphenol contents as well as chlorophyll content. Thus, in addition to N, these fluorescence sensors can detect physiological and biochemical plant parameters such as anthocyanins [37], or plant diseases [38]. Furthermore, Multiplex is a canopy sensor that is more efficient than a leaf sensor such as Dualex. It allows rapid large-area measurements with simultaneous GPS data recording for field map generation [39,40]. However, studies based on Multiplex are still limited, especially for rice N status diagnosis and precision N management. Zhang et al. [41] identified FERARI (fluorescence excitation ratio anthocyanin relative index), SFR_G (simple fluorescence ratio under green excitation), SPAD/DUAD (Dualex reading), and SPAD reading as valuable indicators for monitoring corn N status at early stages. In addition, they found leaf-scale (LS) Multiplex measurements were better related to N treatments than the Multiplex readings made from a small distance above the canopy (AC). Li et al. [42] reported that both leaf-based SPAD and canopy Multiplex indices could be used to predict rice leaf N contents. The on-the-go (OG) measurement mode means placing the sensor probe in the canopy close to the blade and continuously collecting data during the move. This mode is most efficient in data collection. In a recent study, Diago et al. [39] confirmed the capability of the Multiplex sensor using the OG mode to estimate key nutritional parameters in grapevine leaves in motion by calibrating Multiplex against the leaf-clip Dualex sensor. However, further systematic and comprehensive study is necessary to investigate the application potential of canopy fluorescence sensing for rice N status diagnosis in motion.

Hence, the main objectives of this study were to: (1) Determine the feasibility of using canopy multispectral fluorescence sensing system such as the Multiplex sensor to estimate N status in paddy rice by comparing the results of three measurement modes (LS, AC, and OG); and (2) establish and validate the estimation models for N indicators based on the optimal Multiplex indices. In addition, to reduce the influences of varieties, years, sites, and other factors, the normalized N sufficiency index (NSI) was calculated and included in the analysis of the fluorescence indices. Well-fertilized reference plots were used to normalize reflectance measurements as more stable rice N diagnostic results might be obtained when calculating the NSI [43].

2. Materials and Methods

2.1. Experimental Design

Two field trials were conducted at the Jiansanjiang Experiment Station of China Agricultural University (47°15′N, 132°39′E), Sanjiang Plain, Heilongjiang Province, northeast China. The field experiments in 2012 and 2013 included five different N rates (0, 70, 100, 130, 160 kg N ha^{-1}) and two Japonica rice varieties, Kongyu 131 (KY 131) and Longjing 21 (LJ 21). These two represent the main varieties in this region: KY131 has 11 leaves, four elongation nodes and about 127 maturity days, while LJ 21 has 12 leaves and needs 133 maturity days. Planting density was approximately 30 hills m^{-2} for KY 131 and 28 hills m^{-2} for LJ 21, with an identical row spacing of 0.3 m. The plot size was 4.5 m × 9 m. The N fertilizer applications were split into 40%, 30%, 30%, and applied before the transplanting, at the active tillering stage, and at the stem elongation (SE) stage, respectively. Phosphate (50 kg P_2O_5 ha^{-1}) fertilizers were applied before transplanting, and potash (100 kg K_2O ha^{-1}) fertilizers were applied as two splits, 50% as basal fertilizer and 50% as panicle fertilizer at the SE stage. The two experiments were carried out in a randomized complete block design with three replicates. Table 1 lists the details of the two experiments.

Table 1. Details of nitrogen (N) rate experiments with two rice cultivars conducted during 2012–2013.

Experiment	Year	Cultivar	Transplanting Date	Sampling Date and Stage
1	2012	KY 131	18 May	21 June (PI), 29 June (SE), 23 July (HE)
1	2012	LJ 21	18 May	25 June (PI), 2 July (SE), 23 July (HE)
2	2013	KY 131	17 May	23 June (PI), 2 July (SE), 22 July (HE)
2	2013	LJ 21	17 May	28 June (PI), 6 July (SE), 27 July (HE)

PI: Panicle Initiation; SE: Stem Elongation; HE: Heading.

2.2. Fluorescence Measurements

The portable fluorescence sensor Multiplex®3 was used in this study. It is an active sensor involving four emission light sources (UV_A, green, red or blue) to excite the fluorescence in plant tissues. Generally, the UV_A (375 nm), green (530 nm), and red (630 nm) emission light sources were used for plant monitoring while the blue (470 nm) emission light source was used for calibration. The sensor has three filtered detectors for fluorescence recording including blue–green fluorescence (447 nm) (BGF), red fluorescence (665 nm) (RF) and far-red fluorescence (735 nm) (FRF). The fluorescence measurements were not influenced by ambient light conditions [41].

All the variables provided by the Multiplex sensor and their explanations are listed in Table 2. There are nine measured single fluorescence variables under three excitations and ten calculated indices. The Simple Fluorescence Ratio (SFR) index is the ratio of the FRF and RF emission under red (SFR_R) or green (SFR_G) illumination. SFR is related to leaf chlorophyll content. Due to the chlorophyll absorption waveband overlapping with its fluorescence emission red band, the chlorophyll re-absorption occurs at the shorter red wavelength rather than at the far-red wavelength. Therefore, using the FRF as a reference, the absorption of the RF reflects the content of chlorophyll [24,44]. Accordingly, SFR increases as chlorophyll content increases. The Flavonols (FLAV) index compares the fluorescence emission density of the far-red fluorescence under ultraviolet (FRF_UV) and red excitation (FRF_R). It is related to the flavonoid concentration of the epidermal layer [45,46]. The N Balance Index (NBI) is defined as the ratio of SFR and FLAV [47]. Therefore, NBI is proportional to both chlorophyll and flavonoid concentrations. Blue–green to far-red fluorescence ratio (BRR_FRF) index is the ratio of BGF and FRF under UV excitation. The ratio of fluorescence emission at 440 nm and 740 nm (F_{440}/F_{740}) wavelength has been shown to be sensitive to environmental changes and growth conditions, and can detect plant stress before visible symptoms occur [18,48]. The Anthocyanins (ANTH) index and FERARI are both proven to be correlated with skin anthocyanin content [49].

Measurements were taken in motion with the sensor bottom along and just touching the crop leaves on two representative rows in the center of each plot. The data collected from the two rows were averaged to represent the plot [50]. This approach was named "measuring in motion" or "on-the-go" (OG) mode. Notably, in the OG mode, the Multiplex sensor was placed right above the crop leaves manually while walking forward as it is impractical to use a motorized vehicle or to mount the sensor on a vehicle in paddy fields. For comparison purposes, data were also measured in the LS and AC modes. The leaf-borne measurements in the LS mode were collected in the laboratory by taking ten leaves in the second position from the top. In the AC mode, the measurements were collected in the field by randomly selecting ten representative hills and placing the Multiplex sensor at a distance of approximately 10 cm above each selected hill. The average reading was then used to represent the plot.

2.3. Plant Sampling and Measurements

At the PI, SE, and HE stages, for each plot three to six plant samples were collected in each plot from the same rows where fluorescence sensor measurements were acquired. Various N status indicators, including above-ground biomass (AGB), plant N concentration (PNC), leaf N concentration (LNC), plant N uptake (PNU), and N nutrition index (NNI), were determined. The detailed sampling dates and related information are listed in Table 1. Roots from all the plant samples were removed and the samples were cleaned with water and then separated into leaves, stems, and panicles (e.g., HE stage). All samples were dried at 105 °C for half an hour to reduce plant metabolic activities. After being dried in an oven at 70–80 °C to constant weight, the samples were weighed. N concentrations for leaves, stems, and panicles were determined using the standard Kjeldahl method. The PNC is the sum of the products of each organ's N concentration and its proportional weight. The PNU was determined by multiplying PNC by AGB. The NNI is defined as the ratio of the actual PNC (N_a) and the critical N concentration (N_c), with N_c being calculated following the equation developed for rice in this region [51].

Table 2. Description of the variables and indices recorded by the Multiplex sensor (modified from Table 1 by Zhang et al. [41]).

Variables	Formula	Explanation
BGF_UV	/	Blue–green Fluorescence under UV excitation
RF_UV	/	Red Fluorescence under UV excitation
FRF_UV	/	Far-Red Fluorescence under UV excitation
BGF_G	/	Reflected Blue–Green light under Green excitation
RF_G	/	Red Fluorescence under Green excitation
FRF_G	/	Far-Red Fluorescence under Green excitation
RF_R	/	Red Fluorescence under Red excitation
FRF_R	/	Far-Red Fluorescence under Red excitation
SFR_G	FRF_G/RF_G	Simple Fluorescence Ratio under Green excitation
SFR_R	FRF_R/RF_R	Simple Fluorescence Ratio under Red excitation
BRR_FRF	BGF_UV/FRF_UV	Blue–green to Far-Red Fluorescence Ratio under UV excitation
FER_RUV	FRF_R/FRF_UV	Flavonols under Red and UV excitation
FLAV	Log (FRF_R/FRF_UV)	Flavonols under Red and UV excitation
FER_RG	FRF_R/FRF_G	Anthocyanins under Red and Green excitation
ANTH	Log (FRF_R/FRF_G)	Anthocyanins under Red and Green excitation
NBI_G	FRF_UV/RF_G	Nitrogen Balance Index under UV and Green excitation
NBI_R	FRF_UV/RF_R	Nitrogen Balance Index under UV and Red excitation
FERARI#	Log (5000/FRF_R)	Fluorescence Excitation Ratio Anthocyanin Relative Index

the variable is not measured in the "on-the-go" mode.

2.4. Statistical Analysis

The Multiplex data of the three measurement modes at each sampling stage, year, and cultivar obtained under the varied N supply were subjected to analysis of variance (ANOVA) using SAS software (SAS Institute, Cary, NC, USA). Moreover, the means for each treatment were compared using the least significant difference (LSD) test at the 95% level of significance. Relationships between the Multiplex indices and N status indicators were determined. All of the in-situ samples were divided into two groups by a stratified random sampling method, with approximately 2/3 of the data used for model calibration and the remaining for model validation. Simple linear regression analyses were performed with SPSS 20.0 (SPSS Inc., Chicago, Illinois, USA). The coefficient of determination (R^2) was calculated for comparison. The relationships between Multiplex indices and N status indicators established at different growth stages were validated. The root mean square error (RMSE) and the relative error (RE), shown in Equations (1) and (2), between the predicted and observed values were used for evaluation.

$$\mathrm{RMSE} = \sqrt{\frac{1}{n}\sum_{i=1}^{n}(y_i - \hat{y}_i)^2} \tag{1}$$

$$\mathrm{RE}(\%) = \frac{RMSE}{\overline{y}} \times 100 \tag{2}$$

where y_i, $\hat{y}_i$, and $\overline{y}$ were the observed, predicted, and mean value, respectively.

In order to evaluate normalized vegetation indices for improving the estimation of N nutrition indicators, the well-fertilized plots were used as N sufficient reference to calculate the NSI index. The NSI index equals to the ratio of Multiplex indices of the plots receiving normal N rates and the well-fertilized plots. In this study, the treatment with the largest shoot dry matter was defined as the well-fertilized plot, corresponding to the treatment of 130 or 160 kg N ha^{-1}.

Finally, the NNI diagnostic results of validation data using Multiplex indices were compared to the observed NNI by areal agreement and Kappa statistics [52]. Both used the same diagnostic criteria: N was deficient when NNI < 0.95, N was optimal when NNI is between 0.95–1.05, and N was in surplus when NNI > 1.05. The areal agreement (%) and Kappa statistics were used to determine the accuracy of the diagnostic results. The areal agreement indicates the percentage of two groups sharing a common category or diagnostic class [53]. The Kappa statistic is a more robust measure of the agreement of two classifications by correcting the agreement that occurs by chance. When Kappa equals 1, it indicates that the two categorization systems are identical. Kappa ≥ 0.60 indicates a satisfactory agreement, while the Kappa < 0.4 indicates weak agreement [54].

3. Results

3.1. Comparison of the Three Measurement Modes

The three aforementioned measurement modes were applied for each treatment plot. The results were compared to determine the best measurement mode. The abbreviation of the measurement mode is added to the variable as a prefix. For example, AC_SFR_G represents the Multiplex index SFR_G obtained from above the canopy.

Figure 1 shows box plots of Multiplex indices obtained from the three measuring modes at two phenological stages. Since FLAV and ANTH are the log transformation values of FER_RUV and FER_RG, only FLAV and ANTH were selected for the analysis to avoid duplication. During both the PI and SE growth stages, the mean values of the Multiplex indices (except for the BRR_FRF) obtained using the OG mode were significantly higher ($P \leq 0.05$) than the leaf-borne indices measured in the LS mode, whereas the latter were significantly higher than the measurements obtained in the AC mode. In addition, substantial large data ranges were shown in the NBI_G and NBI_R indices measured in motion and the leaf-borne BRR_FRF value.

The ANOVA results are listed in Table 3, indicating the sensitivities of the Multiplex indices to five N rates (0, 70, 100, 130, and 160 kg N ha^{-1}). N fertilization treatment effects were significant for most of the Multiplex variables measured in the OG mode whereas the opposite was true for the leaf-borne measurements. Comparatively, the N treatment effects were more significant for the readings obtained in the OG mode than those collected in the AC mode, while the measurements collected using the LS mode showed the minimum sensitivity to N rate changes. As an example, Figure 2 further demonstrates the comparison of the selected Multiplex indices at each N application rate for cultivar KY 131 in 2013. From Figure 2 we can also see that the indices measured using the OG mode could better distinguish the effects of N supply compared to the other two modes. Especially, the Multiplex indices measured in the OG mode performed consistently better than the others for differentiating high N-application treatment (≥100 kg ha^{-1}) effects. Similar results were achieved for the variety LJ 21, or for the experiments conducted in 2012.

In conclusion, the ANOVA analysis results showed that most of the Multiplex variables obtained in the OG mode were more sensitive to N supply, followed by the AC modes, while measurements made in the LS mode were the least influenced by N supply (Table 3, Figure 2). Thus, only the results of the OG mode were used for further analysis and discussion.

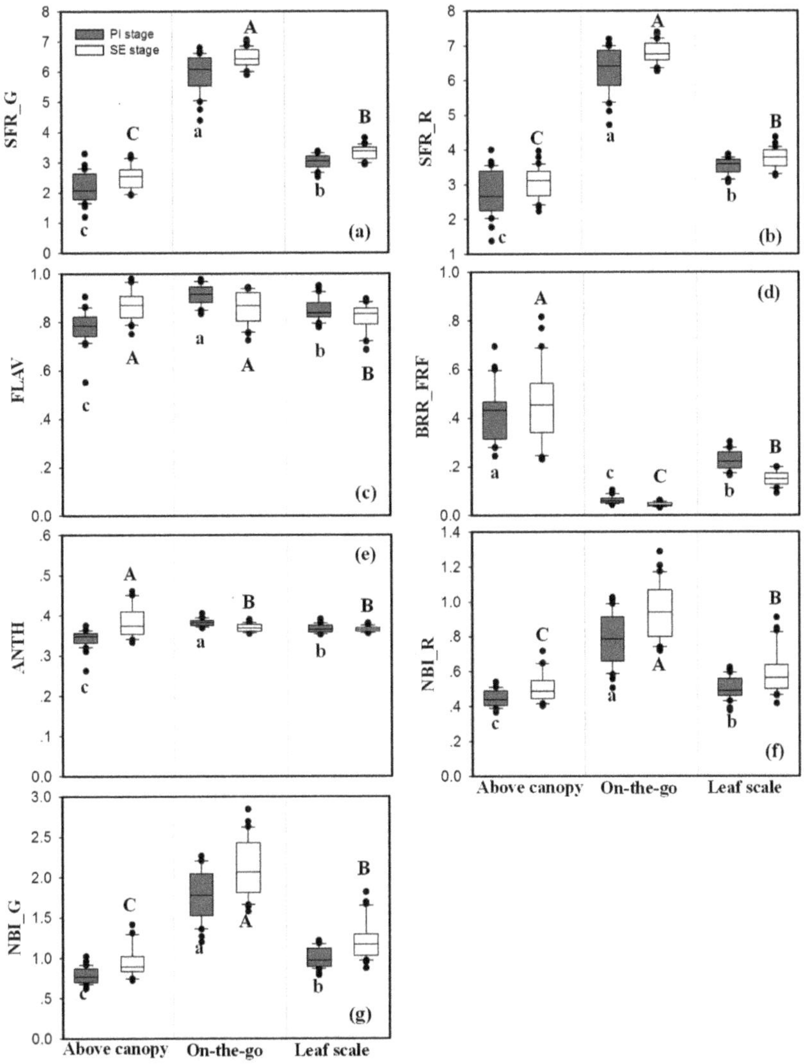

Figure 1. Box plots of selected Multiplex index values for the above canopy, on-the-go, and leaf scale measuring modes at the panicle initiation and stem elongation stages in 2013: Simple fluorescence ratio under green excitation (SFR_G) (**a**), simple fluorescence ratio under red excitation (SFR_R) (**b**), flavonol (FLAV) (**c**), blue–green to far-red fluorescence ratio (BRR_FRF) (**d**), anthocyanins (ANTH) (**e**), nitrogen balance index under UV and red excitation (NBI_R) (**f**), and nitrogen balance index under UV and green excitation (NBI_G) (**g**). Within the same growth stage, different lowercase (panicle initiation stage) or uppercase letters (stem elongation stage) above or below the boxes indicate that the Multiplex index values differed significantly according to the least significant difference test at $P \leq 0.05$.

Table 3. Significance test (ANOVA) of Multiplex variables measured in three modes across 5 N rates: Above canopy (AC), "on-the-go" (OG), and leaf scale (LS) at the panicle initiation (PI), stem elongation (SE), and heading (HE) growth stages for the rice varieties Kongyu 131 (KY 131) and Longjing 21 (LJ 21) in 2013.

Variety	Stage	BGF_UV	RF_UV	FRF_UV	BGF_G	RF_G	FRF_G	RF_R	FRF_R	SFR_G	SFR_R	BRR_FRF	FLAV	ANTH	NBI_G	NBI_R	FERARI
								AC measurement mode									
KY 131	PI	NS	*	*	NS	*	*	*	*	**	**	NS	NS	NS	NS	NS	**
	SE	NS	NS	NS	NS	NS	NS	NS	NS	NS	*	NS	NS	NS	*	NS	NS
								OG measurement mode									
	PI	***	**	**	**	***	***	***	***	***	***	**	*	*	**	**	
	SE	**	***	***	NS	**	***	**	**	**	**	**	**	**	**	**	
	HE	NS	***	***	NS	**	**	*	**	**	**	**	**	**	***	***	
								LS measurement mode									
	PI	NS	NS	NS	NS	NS	NS	NS	*	*	*	NS	NS	NS	NS	NS	NS
	SE	NS	NS	NS	NS	NS	*	NS	*	***	**	NS	NS	NS	**	*	*
	HE	NS	NS	NS	**	NS	*	NS	*	**	*	*	NS	NS	*	*	*
								AC measurement mode									
LJ 21	PI	NS	*	*	NS	*	*	*	*	**	**	**	*	*	**	*	**
	SE	NS	*	*	NS	*	*	*	*	*	*	**	NS	NS	NS	NS	*
								OG measurement mode									
	PI	NS	*	*	NS	NS	*	NS	*	**	**	*	*	*	**	**	
	SE	NS	**	**	NS	NS	**	NS	**	***	***	*	*	*	**	**	
	HE	NS	***	***	NS	**	***	NS	***	***	***	***	***	NS	***	***	
								LS measurement mode									
	PI	NS	NS	NS	NS	NS	NS	NS	NS	*	**	NS	NS	NS	*	NS	NS
	SE	NS	**	**	NS	NS	*	NS	NS	*	NS	**	NS	NS	*	*	*
	HE	NS	NS	NS	NS	NS	NS	NS	NS	NS	NS	NS	NS	NS	NS	NS	NS

*** Significant at the 0.001 level; ** Significant at the 0.01 level; * Significant at the 0.05 level; NS Not significant.

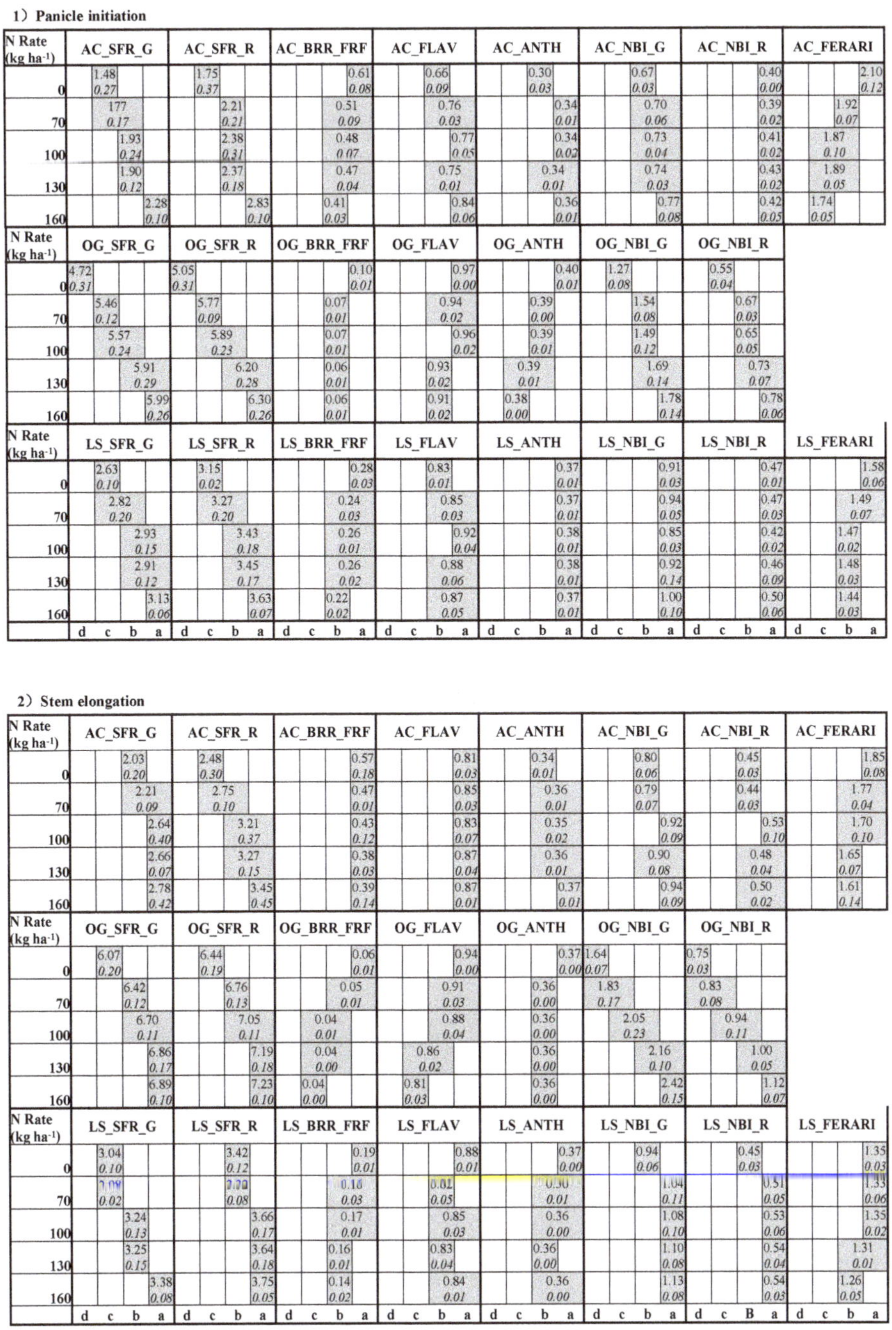

Figure 2. *Cont.*

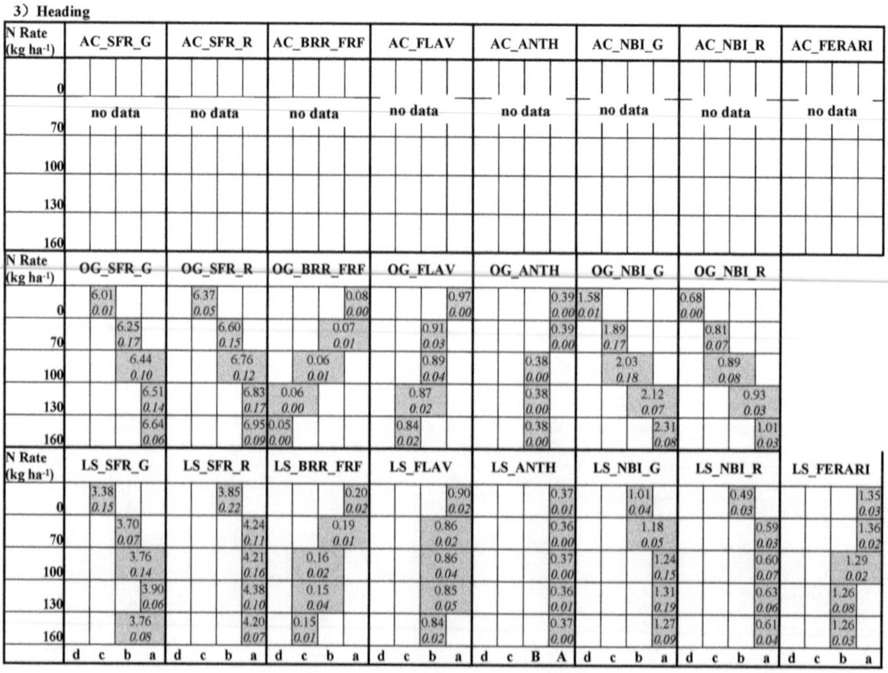

Figure 2. Mean value comparisons for each Multiplex index at the three growth stages and each N application rate (kg N ha^{-1}) for variety KY 131 in 2013. Means and standard errors are shown in each cell for N rate treatments with each measurement mode (Above Canopy, AC; On-the-go, OG; Leaf Scale, LS). Different lowercase letters at the bottom of the plot at each growth stage indicate significant differences according to the least significant difference test at $P \leq 0.05$.

3.2. Changes in Multiplex Indices ("On-The-Go" Mode) over Growth Stages under Different N Supplies

The SFR_G, SFR_R, NBI_G, and NBI_R indices demonstrated an increasing trend as N rate increased, while a decreasing trend was shown for FLAV (Figure 2). Comparatively, the ANTH and BRR_FRF values were less sensitive to the changes in N rates. The values of these SFR_G, SFR_R, NBI_G, and NBI_R indices increased from the PI to SE stage but decreased slightly from the SE to HE stage, because the panicle formation decreased the chlorophyll/N concentration in the upper layer at the HE stage. The opposite was true for the BRR_FRF, FLAV, and ANTH. NBI_G and NBI_R could differentiate different N application rates the best regardless of the growth stages, followed by SFR_G, SFR_R, BRR_FRF, and FLAV. The performance of ANTH was the worst (Figure 2).

3.3. Correlations between Multiplex Indices ("On-The-Go" Mode) and N Status Indicators

The linear regression results of the seven Multiplex indices and the five N status indicators at three growth stages across the two rice varieties are shown in Table 4. The SFR_G, SFR_R, NBI_G, and NBI_R indices were positively correlated with the N indicators whereas the BRR_FRF, ANTH, and FLAV were inversely associated with them. The R^2 of the regression models based on these indices varied from 0.03 to 0.78. The best performing index varied at different stages, but NBI_G and NBI_R showed steady high correlations with all five N status indicators. The second-best performing indices were BRR_FRF and FLAV. The SFR_G and SFR_R indices displayed high or moderate correlations with the N indicators during the PI or HE stage, respectively.

Table 4. The coefficients of determination (R^2) for the linear relationships between standard and normalized Multiplex indices and N status indicators (leaf N concentration (LNC), plant N concentration (PNC), N nutrition index (NNI), aboveground biomass (AGB) and plant N uptake (PNU)) for two varieties at the panicle initiation (PI), stem elongation (SE), and heading (HE) growth stages.

Multiplex Indices	LNC (g kg^{-1})			PNC (g kg^{-1})			NNI			AGB (t ha^{-1})			PNU (kg ha^{-1})		
	PI	SE	HE	PI	SE	HE	PI	SE	HE	PI	SE	HE	PI	SE	HE
Standard indices															
SFR_G	0.63 **	0.30 **	0.49 **	0.64 **	0.34 **	0.46 **	0.72 **	0.31 **	0.59 **	0.60 **	0.14 *	0.41 **	0.66 **	0.21 **	0.58 **
SFR_R	0.59 **	0.28 **	0.42 **	0.58 **	0.34 **	0.38 **	0.66 **	0.29 **	0.54 **	0.56 **	0.13 *	0.45 **	0.61 **	0.19 **	0.57 **
BRR_FRF	0.53 **	0.52 **	0.67 **	0.47 **	0.48 **	0.66 **	0.57 **	0.48 **	0.72 **	0.50 **	0.26 **	0.33 **	0.54 **	0.39 **	0.59 **
FLAV	0.40 **	0.64 **	0.55 **	0.39 **	0.64 **	0.55 **	0.58 **	0.73 **	0.67 **	0.55 **	0.50 **	0.38 **	0.59 **	0.68 **	0.59 **
ANTH	0.38 **	0.12 *	0.27 **	0.41 **	0.14 *	0.33 **	0.60 **	0.10 *	0.47 **	0.60 **	0.03NS	0.36 **	0.61 **	0.06NS	0.48 **
NBI_G	0.54 **	0.68 **	0.62 **	0.52 **	0.71 **	0.61 **	0.69 **	0.78 **	0.76 **	0.63 **	0.50 **	0.47 **	0.68 **	0.71 **	0.71 **
NBI_R	0.52 **	0.67 **	0.58 **	0.52 **	0.71 **	0.56 **	0.70 **	0.77 **	0.74 **	0.64 **	0.47 **	0.51 **	0.69 **	0.68 **	0.72 **
Normalized indices															
SFR_G_{NSI}	0.58 **	0.39 **	0.67 **	0.65 **	0.42 **	0.70 **	0.55 **	0.54 **	0.69 **	0.35 **	0.45 **	0.24 **	0.43 **	0.50 **	0.52 **
SFR_R_{NSI}	0.57 **	0.42 **	0.62 **	0.61 **	0.46 **	0.67 **	0.52 **	0.57 **	0.68 **	0.33 **	0.45 **	0.25 **	0.40 **	0.52 **	0.52 **
BRR_FRF_{NSI}	0.49 **	0.34 **	0.63 **	0.48 **	0.41 **	0.74 **	0.41 **	0.56 **	0.76 **	0.26 **	0.50 **	0.28 **	0.33 **	0.56 **	0.58 **
$FLAV_{NSI}$	0.42 **	0.51 **	0.70 **	0.44 **	0.60 **	0.76 **	0.41 **	0.74 **	0.82 **	0.26 **	0.55 **	0.34 **	0.33 **	0.70 **	0.64 **
$ANTH_{NSI}$	0.51 **	0.40 **	0.57 **	0.64 **	0.40 **	0.65 **	0.54 **	0.57 **	0.56 **	0.34 **	0.52 **	0.11 *	0.42 **	0.58 **	0.35 **
NBI_G_{NSI}	0.59 **	0.53 **	0.69 **	0.61 **	0.63 **	0.75 **	0.55 **	0.76 **	0.78 **	0.35 **	0.55 **	0.31 **	0.43 **	0.71 **	0.61 **
NBI_R_{NSI}	0.60 **	0.55 **	0.69 **	0.65 **	0.64 **	0.75 **	0.58 **	0.77 **	0.79 **	0.36 **	0.56 **	0.31 **	0.46 **	0.72 **	0.61 **

** Correlation is significant at the 0.01 level; * Correlation is significant at the 0.05 level; NS Correlation is not significant.

Compared to the counterpart of the standard indices, the normalized sufficiency indices SFR_G$_{NSI}$, SFR_R$_{NSI}$, and ANTH$_{NSI}$ exhibited better linear relationships with LNC, PNC, and NNI in most of the cases, especially at the SE and HE stages. The NBI_G$_{NSI}$ and NBI_R$_{NSI}$ displayed enhanced relationships with the LNC and PNC at the PI and HE stages, and with the NNI at the HE stage. The BRR_FRF$_{NSI}$, FLAV$_{NSI}$ showed improved associations with PNC at the PI and HE stages, and with NNI at the SE and HE stages. All the standard indices showed moderate–high relationships with the AGB and PNU during the PI and HE stages, while at the HE stage, greatly improved R^2 values were obtained using the normalized indices.

3.4. Validation of the Estimation Models for N Status Indicators

In order to diagnose rice N status, linear regression models between the Multiplex indices and N indicators were established. The regression models varied across growth stages. Table 5 lists the best performing models at the PI, SE, and HE stages. The best performing indices differed across the stages. However, the relationships of NBI_G and NBI_R with N indicators were relatively more stable. After normalization, the NBI_R$_{NSI}$ showed an absolute advantage for N status estimation at the PI and SE growth stages, while the FLAV$_{NSI}$ demonstrated to be optimal for estimating most of the N indicators.

Table 5. Equations and coefficients of determination of linear regression models (n = 40) at different growth stages based on the best performing Multiplex index and crop N indicators (LNC, PNC, NNI, PNU, and AGB).

Growth Stage	Standard Indices	Model	R^2	Normalized Indices	Model	R^2
PI	SFR_G	LNC = 4.468x + 5.932	0.63	NBI_R$_{NSI}$	LNC = 23.918x + 10.413	0.60
PI	SFR_G	PNC = 2.912x + 4.961	0.64	NBI_R$_{NSI}$	PNC = 15.323x + 8.247	0.65
PI	SFR_G	NNI = 0.2442x-0.5188	0.72	NBI_R$_{NSI}$	NNI = 1.1412x − 0.1116	0.58
PI	NBI_R	PNU = 88.184x-33.56	0.69	NBI_R$_{NSI}$	PNU = 85.908x − 43.67	0.46
PI	NBI_G	AGB = 1.5268x-1.1565	0.64	NBI_R$_{NSI}$	AGB = 2.905x − 1.1184	0.36
SE	NBI_G	LNC = 8.707x + 14.352	0.68	NBI_R$_{NSI}$	LNC = 17.96x + 16.279	0.55
SE	NBI_G	PNC = 5.544x + 9.082	0.71	NBI_R$_{NSI}$	PNC = 12.317x + 9.542	0.64
SE	NBI_G	NNI = 0.5003x + 0.0582	0.78	NBI_R$_{NSI}$	NNI = 1.1571x + 0.0601	0.77
SE	NBI_G	PNU = 51.494x-40.873	0.71	NBI_R$_{NSI}$	PNU = 120.8x − 42.157	0.72
SE	NBI_G	AGB = 1.7391x-0.4502	0.50	NBI_R$_{NSI}$	AGB = 4.1975x − 0.5961	0.56
HE	BRR_FRF	LNC = -210.31x + 47.452	0.67	NBI_G$_{NSI}$	LNC = 21.646x + 15.473	0.69
HE	BRR_FRF	PNC = -131.79x + 24.313	0.66	FLAV$_{NSI}$	PNC = -31.59x + 49.591	0.76
HE	NBI_G	NNI = 0.5942x-0.054	0.76	FLAV$_{NSI}$	NNI = -3.0631x + 4.3956	0.82
HE	NBI_R	PNU = 213.07x-67.623	0.72	FLAV$_{NSI}$	PNU = -462.81x + 612.19	0.64
HE	NBI_R	AGB = 8.3363x + 0.0609	0.51	FLAV$_{NSI}$	AGB = -15.729x + 24.131	0.34

Figure 3 shows the RE values of the validation models for six Multiplex indices (SFR_G, SFR_R, BRR_FRF, FLAV, NBI_G, and NBI_R) and the N status indicators. The RE values for AGB and PNU estimations based on these indices decreased steadily with advancing growth stages, while a slightly increasing trend was observed for the LNC and PNC estimation models from the SE to HE stage. The RE values for LNC (4.50%–10.24%) and PNC (5.87%–10.87%) models were much smaller than those for AGB (15.49%–30.18%) and PNU (19.31%–31.25%), while the REs of NNI remained similar during the three growth stages. At the earlier to middle growth stages, NBI_R and NBI_G presented a lower RE than the other four indices for all the five N indicators. At the HE stage, however, the prediction accuracies of the six indices were similar.

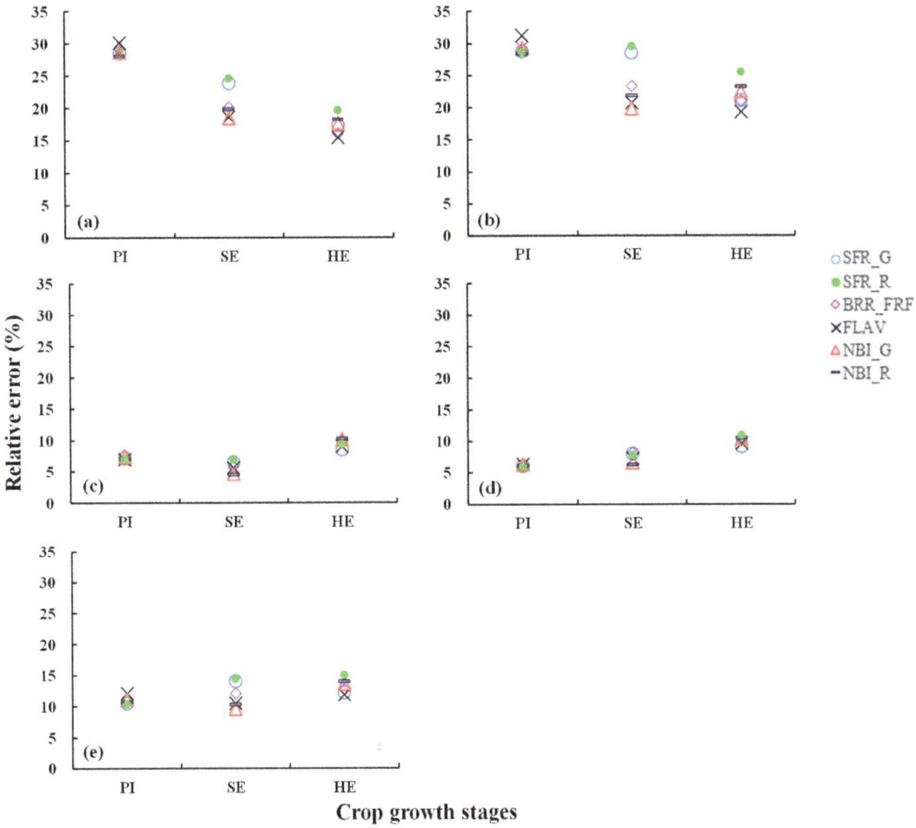

Figure 3. The relative error (RE) values of the validation analysis based on the regression models of the six Multiplex indices and the N status indicators for (**a**) above-ground biomass (AGB), (**b**) plant N uptake (PNU), (**c**) leaf N concentration (LNC), (**d**) plant N concentration (PNC), and (**e**) nitrogen nutrition index (NNI) at the panicle initiation (PI), stem elongation (SE), and heading (HE) stages.

3.5. Rice N Status Diagnosis

The best performing indices including SFR_G, BRR_FRF, NBI_G, NBI_R, NBI_G_{NSI}, NBI_R_{NSI} and $FLAV_{NSI}$ were validated using independent data sets (Table 5). Moderate–high model performance with R^2 ranging from 0.34 to 0.82 was observed, especially for the LNC, PNC, and NNI estimations. The areal agreement and Kappa statistics were compared at the critical N fertilizer application stages (SE and HE) to evaluate the N diagnostic accuracies of the indices. Results confirmed that the NNI models based on NBI_R and NBI_G performed consistently well at the SE and HE growth stages, and their corresponding NSI indices further improved the results (Table 6). At the SE stage, the NBI_R_{NSI} achieved the highest diagnostic accuracy (areal agreement = 90%; Kappa = 0.84), while the best accuracy was achieved by FLAV at the HE stage (areal agreement = 90%; Kappa = 0.76). In addition, across the two growth stages, the NBI_R_{NSI} showed the highest diagnostic consistency, followed by the BRR_FRF_{NSI}.

Table 6. Agreement and Kappa statistics for different indices (SFR_G, SFR_R, BRR_FRF, FLAV, ANTH, NBI_G, and NBI_R) and corresponding normalized indices (SFR_G$_{NSI}$, SFR_R$_{NSI}$, BRR_FRF$_{NSI}$, FLAV$_{NSI}$, ANTH$_{NSI}$, NBI_G$_{NSI}$, and NBI_R$_{NSI}$) regarding diagnostic results (Nitrogen Nutrition Index) at different growth stages.

Comparison	Agreement (%)		Kappa statistics		Comparison	Agreement (%)		Kappa statistics	
	SE	HE	SE	HE		SE	HE	SE	HE
SFR_G and NNI	75	65	0.554 ***	0.310 *	SFR_G$_{NSI}$ and NNI	70	85	0.494 **	0.661 ***
SFR_R and NNI	70	70	0.510 ***	0.322 *	SFR_R$_{NSI}$ and NNI	75	85	0.583 ***	0.661 ***
BRR_FRF and NNI	60	70	0.363 *	0.409 **	BRR_FRF$_{NSI}$ and NNI	80	80	0.655 ***	0.538 ***
FLAV and NNI	75	90	0.605 ***	0.763 ***	FLAV$_{NSI}$ and NNI	75	80	0.558 ***	0.570 ***
ANTH and NNI	55	75	0.283NS	0.355 *	ANTH$_{NSI}$ and NNI	80	65	0.669 ***	0.227NS
NBI_G and NNI	75	80	0.595 ***	0.590 ***	NBI_G$_{NSI}$ and NNI	75	85	0.673 ***	0.698 ***
NBI_R and NNI	75	80	0.595 ***	0.590 ***	NBI_R$_{NSI}$ and NNI	90	85	0.840 ***	0.698 ***

*** Significant at the 0.001 level; ** Significant at the 0.01 level; * Significant at the 0.05 level; NS Not significant.

4. Discussion

4.1. Multiplex Measurement Modes and Estimation of Crop N Indicators by Fluorescence Indices

The N treatment effects were more significant for the readings obtained in the OG mode than those collected in the AC mode (Table 3, Figure 2), which is different from the finding by Diago et al. [39] who reported a 20% loss of information occurred when using the Multiplex on-the-go (compared to the AC mode) for N assessment of grapevine. This is because the OG measurements in this study were taken manually by placing the Multiplex sensor right on the top of the rice rather than a small distance above the rice canopy while passing through the rice paddy. In contrast, in the study by Diago et al. [39], the Multiplex sensor was mounted onto an all-terrain vehicle and placed 1.5 m above the ground so that the leaves on the mid-part of the canopy were automatically measured at a 20 cm distance, the same measuring distance as their AC mode. In addition, this study revealed that measurements made using the LS mode were the least sensitive to N supply, contrasting to the result by Zhang et al. [41] who found Multiplex measurements made from corn leaves were more capable of distinguishing plant N status than those made from above the plants. While the leaf scale measurements made in this study were collected in the laboratory by taking ten leaves in the second position from the top, the leaf-borne measurements by Zhang et al. [41] were made on 20 representative plants in the center two rows of each plot in the field, which is more similar to the OG rather than the LS method in this study. Another possible reason for their better results with the LS method is that the OG measurements do not give much time to choose leaves and result in more random leaf choosing than the LS method, which may have an unwanted tendency to choose "good" leaves. This is particularly true for maize, because systematic use of a representative leaf is easier, as it is well known which is the most representative leaf for each growth stage, given its determined growth. For rice, the individual leaves are quite small and the signal obtained during the measurements taken on a leaf is relatively weak and can be easily affected by other factors. One advantage the "measuring in motion" or OG mode has is efficiency, especially when the sensor is mounted on a vehicle or other automatic devices, which might make practical applications of such non-destructive technology over large areas possible. Bringing the sensor close to or even touching the leaves of the crop in OG mode may help reduce information loss. However, further well-designed studies are needed to confirm this finding.

Strong relationships between the Multiplex indices (SFR_G, SFR_R, BRR_FRF, FLAV, NBI_G, and NBI_R) measured in the OG mode and the five N indicators were achieved with low RE and high R^2 values (Table 4, Figure 3). This finding conforms to previous research results in this field [26,42,55,56]. Many studies confirmed SFR was a good fluorescence index for chlorophyll content monitoring [24,28,57]. However, in this study, it was found that the R^2 of the SFR_G, SFR_R for LNC, PNC, and NNI estimation decreased steadily from the early-stage to later stages, while an opposite trend was observed for FLAV. Padilla et al. [55] found that the relationships between the NNI and SFR_G changed with the phenological stages of cucumber (*Cucumis sativus* L.). Firstly, the consistency of the relationship between chlorophyll content and N concentration varied with crop development, leading to different performances of SFR for N concentration estimation. For example, the linear correlation between LNC and chlorophyll meter readings of rice was weaker at the SE stage than at other growth stages [58]. Secondly, the weaker differentiation ability of SFR under the unlimited N conditions may also be a reason [55]. The performance of FLAV increased from the PI to HE stages, which was confirmed by Padilla et al. [59], who found the relationship of FLAV and NNI increased at the middle to late growth stages. The better performance of FLAV at the later stage may be attributed to the accumulation of the flavonols content in leaves under light radiation [44,60]. The NBI_G and NBI_R indices were shown as the best indices for estimating the N indicators (Table 5). Many studies have proven that the NBI indices appeared to be the most efficient in estimating the N status [47,56,59,61]. This is because the NBI is a ratio of SFR and FLAV, which makes it more robust than using FLAV or SFR alone to reduce the effects of leaf age or other factors [34,36,47]. The NBI_G and NBI_R had similar performance in this study, as demonstrated by Longchamps and Khosla [27]. Moreover, Longchamps and Khosl [27] found that SFR was less sensitive to N application than NBI, which conforms to our results, as shown in Figure 2. In most cases, the SFR_G and SFR_R indices could not distinguish between the 100 and 130 kg N ha^{-1} treatments, but NBI could. The BRR_FRF index was significantly correlated with the N nutritional status and was especially sensitive to N deficiency in this research. When there is N stress, the fluorescence ratio of blue–green/far-red will increase after exposure to elevated UV radiation to avoid or alleviate the damage of the photosynthetic apparatus [62]. Generally, the UV-protection response takes place before the chlorophyll damage can be seen, so the BRR_FRF can also be considered as a potential index that can realize early N deficiency detection [63]. The BRR_FRF was also very sensitive to environmental stresses, such as disease and drought [38,63,64]. The ANTH index provided by Multiplex is commonly used to reflect anthocyanin content, which corresponds to the maturation degree of fruit [49,65]. In this research, the low values of ANTH were due to the low anthocyanin content in the rapid vegetative growth phase for rice [66]. Nevertheless, ANTH was also found to be closely related to the leaf chlorophyll concentration in some studies [38,41]. This study revealed that ANTH was significantly related to N status indicators in PI and HE growth stages with moderate R^2 values (Table 4).

4.2. Normalized Nitrogen Sufficiency Fluorescence Indices

Our research involved two years and two varieties of experiments. In these experiments, N fertilization rate is the main variable, and the variation is so high that it will probably override any other source of variation. In a commercial field, many factors can influence N availability, N and Chlorophyll relationship, or Chlorophyll (+FLAV and ANTH) relations to fluorescence indices, including biotic or abiotic stresses at the moment of measurement or in the history of the crop or even the field. The normalized N sufficiency index approach has been suggested to reduce the influence of the varieties, developmental stages, and other variables on SPAD values or spectral data [3,11,67]. From the results of this study, in most cases, the normalized NSIs were better associated with the LNC, PNC, and NNI (Table 4). The R^2 of the $ANTH_{NSI}$ was improved the most, followed by the NBI_G_{NSI} and NBI_R_{NSI}. However, the improvement in R^2 for BRR_FRF_{NSI} was minimal. The variance analysis of this study showed consistent results, which demonstrated that NSI indices could reduce the influence of inter-annual and growth stage differences. Since NNI itself is a diagnostic criterion, it represents an optimal N status when NNI is equal to one [68]. Most of the NSI indices greatly improved the NNI diagnostic accuracy at the critical topdressing (SE and HE) stages (Table 6). Similarly, Lu et al. [43] observed that the NNI inversion through the normalized vegetation indices further improved the N nutrition diagnostic results of rice.

Hussain et al. [69] proposed a critical NSI value of 0.90 for rice. However, in this study, when the NSI indices were 0.90, different optimal NNI values, ranging from 0.85 to 2.14, were derived by different indices at different developmental stages. Only the corresponding optimal NNI values for the NBI_G_{NSI} and NBI_R_{NSI} indices were close to one (ranging from 0.91 to 1.19). Therefore, to avoid the risk of misdiagnosis, the NSI threshold was not used as a diagnostic criterion directly. A possible reason for this is that the N fertilizer application rate in this study was only 1.3–1.6 times higher than the optimal amount instead of 1.8–2.0 times higher than recommended for the well-fertilized N plot as Hussain et al. [69] suggested. Furthermore, all of the Multiplex indices were divided by the readings of the N rate with the largest shoot dry matter at each sampling date to obtain a sufficiency index. However, Varvel et al. [70] suggested that the maximum readings within each cropping system, variety, and year should be considered as the normalized criterion. Obviously, with different normalization criteria, different sufficiency indices will be obtained, which will affect the corresponding analysis results. Another limitation of the NSI approach is that well-fertilized reference plots need to be established in each farmer's field for practical application purposes, and some farmers may not be willing to do this. More in-depth and systematic research is expected in the future.

4.3. The Application Potential of the Multiplex Sensor

The Multiplex indices presented good R^2 values for LNC and PNC estimations at the earlier growth stages (Table 4). In particular, the validation data showed that the RE values for LNC and PNC estimations were as low as 6%–7% (Figure 3c,d). This is consistent with the results of Cerovic et al. [71] and Agati et al. [56], who have shown a high correlation between the fluorescence index and LNC. NBI and LNC had a fairly linear relationship. Therefore, the NBI indices can be used to more accurately estimate a wider range of LNC. Agati et al. [56] also found the results based on reflectance imaging (camera picture) are less sensitive to N application than fluorescence-based indices. Research by Stroppiana et al. [72] and Yu et al. [19] on rice showed unsatisfactory results for the estimation of LNC and PNC based on reflectance spectroscopy. This is possibly due to the fact that the effect of N on the leaf area index and biomass is much greater than its effect on chlorophyll content. Second, near-infrared radiation is hardly absorbed in the canopy and is highly transmissive, so its correlation with leaf area index or biomass is extremely high; while visible light, especially the blue and red radiation, is easily absorbed by chlorophyll and its transmittance is low, so it is highly correlated with chlorophyll content [15,72]. On the other hand, changes in plant metabolism indicators are fast or slow due to changes in response to the environment. However, the sensitivity of reflectance-based parameters does not always provide satisfactory monitoring results [73]. Demotes-Mainard et al. [74] observed that changes in N concentration took precedence over changes in biomass. Thus, fluorescence-based techniques that are highly sensitive to plant N status information may address the limitation of reflectance-based methods [27,73]. Similarly, the Multiplex indices, especially the NBI_G and NBI_R, presented accurate estimation for NNI, with R^2 reaching a maximum of 0.72–0.78, and the validation results also showed a low inversion error for NNI (RMSE $\leq$ 0.16, RE $\leq$ 15%) (Table 4). Many studies have confirmed that NBI has a strong estimation potential for NNI [47,55,59]. This is because NBI is the ratio of SFR to FLAV. The SFR index was considered to be an important parameter for estimating chlorophyll concentration, which was often used as an index of surface-based N [75], while the FLAV parameter directly reflects flavonol content, which is controlled by light as well as leaf mass per area, and has a strong correlation with leaf mass [76]. Therefore, NBI as the SFR/FLAV ratio is the best N nutrition diagnostic index.

The fluorescence-based indices are more sensitive to chlorophyll or N content than the reflectance-based indices and can detect the difference in N nutrition status earlier. However, the difference of the stage-based models between the indices and the N nutrition indicators based on the canopy reflectance instrument is smaller than that based on fluorescence [59]. The surface area of the crop involved in each test when using the canopy reflectance spectroscopy sensors is larger than the fluorescence sensors [59]. Therefore, canopy reflectance measurements are more representative, while fluorescence instruments require increasing the number of tests to obtain sufficiently representative data. Although the performance of the fluorescence sensor was quite good for estimating LNC, PNC, and NNI, 22%–60% of their variability was still not explained. In addition, the fluorescence sensor did not perform very well for estimating plant biomass in the middle to late stages (Table 4). It has been suggested to combine the fluorescence and reflectance data to improve the estimation of plant N status [3,73]. This may be one of the important research directions in the future.

5. Conclusions

This research compared the LS, OG, and AC measurement modes of the fluorescence instrument Multiplex®3 and determined that the OG mode was best suited for this rice N status study. Using the OG mode, stable test results and crop growth information were derived. The results revealed that the fluorescence indices of NBI, SFR, BRR_FRF, and FLAV were significantly correlated to all five N status indicators from the PI through HE growth stages. Among them, NBI_G and NBI_R were the best performing indices and highly correlated to LNC (R^2 = 0.52–0.68), PNC (R^2 = 0.52–0.71), NNI (R^2 = 0.69–0.78), AGB (R^2 = 0.47–0.64), and PNU (R^2 = 0.68–0.72) at the three growth stages. The normalized sufficiency indices of the Multiplex parameters could greatly improve the LNC, PNC, and NNI estimation ability, especially at the HE stage. The N diagnostic results indicated that the NBI_R_{NSI} and FLAV achieved the highest diagnostic accuracy rate (90%) at the SE and HE stage, respectively, while NBI_R_{NSI} showed the highest analytical consistency across growth stages. The results suggest that the Multiplex sensor can be used to reliably estimate N nutritional status for rice in cold regions, especially for the estimation of LNC, PNC, and NNI. The normalized sufficiency indices based on Multiplex indices may further improve the accuracy of N nutrition diagnosis by reducing the differences between years and varieties.

Author Contributions: Y.M. and G.B. conceived and guide the study. S.H. and Q.C. conducted the field experiments. S.H. and H.Y. performed the data analysis. S.H. wrote the original manuscript. Y.M., F.Y. and V.I.S.L.-W. revised the manuscript. G.B. and H.Y. reviewed and edited the manuscript. All authors read and approved the final manuscript.

Funding: This research was financially supported by the National Key Research and Development Program of China (2016YFD0200600, 2016YFD0200602), National Basic Research Program (2015CB150405), and the Norwegian Ministry of Foreign Affairs (SINOGRAIN II, CHN-17/0019).

Acknowledgments: We would like to thank the supports by Wen Yang, Huamin Zhu, and Fengyan Liu at the Jiansanjiang Institute of Agricultural Sciences. We also would like to thank Jianning Shen, Weifeng Yu, and Shanshan Cheng for their fieldwork and contributions in spectral data collection.

Conflicts of Interest: The authors declare no conflict of interest.

References

1. Guo, J.H.; Liu, X.J.; Zhang, Y.; Shen, J.L.; Han, W.X.; Zhang, W.F.; Christie, P.; Goulding, K.W.; Vitousek, P.M.; Zhang, F.S. Significant acidification in major Chinese croplands. *Science* **2010**, *327*, 1008–1010. [CrossRef]
2. Miao, Y.; Stewart, B.A.; Zhang, F. Long-term experiments for sustainable nutrient management in China. A review. *Agron. Sustain. Dev.* **2011**, *31*, 397–414. [CrossRef]
3. Samborski, S.M.; Tremblay, N.; Fallon, E. Strategies to make use of plant sensors-based diagnostic information for nitrogen recommendations. *Agron. J.* **2009**, *101*, 800–816. [CrossRef]
4. Evans, J.R. Nitrogen and photosynthesis in the flag leaf of wheat (*Triticum aestivum* L.). *Plant Physiol.* **1983**, *72*, 297–302. [CrossRef] [PubMed]
5. Blackmer, T.M.; Schepers, J.S. Use of a chlorophyll meter to monitor nitrogen status and schedule fertigation for corn. *J. Prod. Agric.* **1995**, *8*, 56–60. [CrossRef]
6. Schlemmer, M.R.; Francis, D.D.; Shanahan, J.F. Remotely measuring chlorophyll content in corn leaves with differing nitrogen levels and relative water content. *Agron. J.* **2005**, *97*, 106–112. [CrossRef]
7. Schröder, J.J.; Neeteson, J.J.; Oenema, O.; Struik, P.C. Does the crop or the soil indicate how to save nitrogen in maize production? Reviewing the state of the art. *Field Crop Res.* **2000**, *66*, 151–164. [CrossRef]
8. Schepers, J.S.; Blackmer, T.M.; Francis, D.D. Predicting N fertilizer needs for corn in humid regions: Using chlorophyll meters. In *Predicting N Fertilizer Needs for Corn in Humid Regions*; Bock, B.R., Kelley, K.R., Eds.; National Fertilizer and Environmental Research Center: Muscle Shoals, AL, USA, 1992; pp. 105–114.
9. Schepers, J.S.; Francis, D.D.; Vigil, M.; Below, F.E. Comparison of corn leaf nitrogen concentration and chlorophyll meter readings. *Commun. Soil Sci. Plan.* **1992**, *23*, 2173–2187. [CrossRef]

10. Markwell, J.; Osterman, J.C.; Mitchell, J.L. Calibration of the Minolta SPAD-502 leaf chlorophyll meter. *Photosynth. Res.* **1995**, *46*, 467–472. [CrossRef]
11. Lin, F.F.; Qiu, L.F.; Deng, J.S.; Shi, Y.Y.; Chen, L.S.; Wang, K. Investigation of spad meter-based indices for estimating rice nitrogen status. *Comput. Electron. Agric.* **2010**, *71*, S60–S65. [CrossRef]
12. Ali, M.M.; Al-Ani, A.; Eamus, D.; Tan, D.K.Y. Leaf nitrogen determination using non-destructive techniques—A review. *J. Soil Sci. Plant Nutr.* **2017**, *40*, 928–953.
13. Mulla, D.J.; Miao, Y. Precision Farming. In *Land Resources Monitoring, Modeling, and Mapping with Remote Sensing*; Thenkabail, P.S., Ed.; CRC Press: Boca Raton, FL, USA, 2016.
14. Gitelson, A.A.; Gritz, Y.; Merzlyak, M.N. Relationships between leaf chlorophyll content and spectral reflectance and algorithms for non-destructive chlorophyll assessment in higher plant leaves. *J. Plant Physiol.* **2003**, *160*, 271–282. [CrossRef] [PubMed]
15. Heege, H.J.; Reusch, S.; Thiessen, E. Prospects and results for optical systems for site-specific on-the-go control of nitrogen-top-dressing in Germany. *Precis. Agric.* **2008**, *9*, 115–131. [CrossRef]
16. Yao, Y.; Miao, Y.; Huang, S.; Gao, L.; Ma, X.; Zhao, G.; Jiang, R.; Chen, X.; Zhang, F.; Yu, K.; et al. Active canopy sensor-based precision n management strategy for rice. *Agron. Sustain. Dev.* **2012**, *32*, 925–933. [CrossRef]
17. Cao, Q.; Miao, Y.; Shen, J.; Yu, W.; Yuan, F.; Cheng, S.; Huang, S.; Wang, H.; Yang, W.; Liu, F. Improving in-season estimation of rice yield potential and responsiveness to topdressing nitrogen application with Crop Circle active crop canopy sensor. *Precis. Agric.* **2016**, *17*, 136–154. [CrossRef]
18. Olfs, H.W.; Blankenau, K.; Brentrup, F.; Jasper, J.; Link, A.; Lammel, J. Soil-and plant-based nitrogen-fertilizer recommendations in arable farming. *J. Soil Sci. Plant Nutr.* **2005**, *168*, 414–431. [CrossRef]
19. Yu, K.; Li, F.; Gnyp, M.L.; Miao, Y.; Bareth, G.; Chen, X. Remotely detecting canopy nitrogen concentration and uptake of paddy rice in the Northeast China Plain. *ISPRS J. Photogramm. Remote Sens.* **2013**, *78*, 102–115. [CrossRef]
20. Huang, S.; Miao, Y.; Zhao, G.; Yuan, F.; Ma, X.; Tan, C.; Yu, W.; Gnyp, M.; Lenz-Wiedemann, V.; Rascher, U. Satellite remote sensing-based in-season diagnosis of rice nitrogen status in Northeast China. *Remote Sens.* **2015**, *7*, 10646–10667. [CrossRef]
21. Bredemeier, C.; Schmidhalter, U. Laser-induced chlorophyll fluorescence sensing to determine biomass and nitrogen uptake of winter wheat under controlled environment and field condition. In Proceedings of the 5th European Conference on Precision Agriculture, Uppsala, Sweden, 9–12 June 2005; Wageningen Academic Publishers: Wageningen, The Netherlands, 2005; pp. 273–280.
22. McMurtrey, J.E., III; Chappelle, E.W.; Kim, M.S.; Meisinger, J.J.; Corp, L.A. Distinguishing nitrogen fertilization levels in field corn (*Zea mays* L.) with actively induced fluorescence and passive reflectance measurements. *Remote Sens. Environ.* **1994**, *47*, 36–44. [CrossRef]
23. Langsdorf, G.; Buschmann, C.; Sowinska, M.; Babani, F.; Mokry, M.; Timmermann, F.; Lichtenthaler, H.K. Multicolour fluorescence imaging of sugar beet leaves with different nitrogen status by flash lamp UV-excitation. *Photosynthetica* **2000**, *38*, 539–551. [CrossRef]
24. Gitelson, A.A.; Buschmann, C.; Lichtenthaler, H.K. The chlorophyll fluorescence ratio F-735/F-700 as an accurate measure of the chlorophyll content in plants. *Remote Sens. Environ.* **1999**, *69*, 296–302. [CrossRef]
25. Cerovic, Z.G.; Goutouly, J.P.; Hilbert, G.; Destrac-Irvine, A.; Martinon, V.; Moise, N. Mapping winegrape quality attributes using portable fluorescence-based sensors. *Frutic* **2009**, *9*, 301–310.
26. Yang, J.; Wei, G.; Shi, S.; Lin, D.; Sun, J.; Song, S.; Chen, B.; Zhang, Z. Analyzing the performance of fluorescence parameters in the monitoring of leaf nitrogen content of paddy rice. *Sci. Rep.* **2016**, *6*, 28787. [CrossRef] [PubMed]
27. Longchamps, L.; Khosla, R. Early detection of nitrogen variability in maize using fluorescence. *Agron. J.* **2014**, *106*, 511–518. [CrossRef]
28. Teal, R.K.; Tubana, B.; Girma, K.; Freeman, K.W.; Arnall, D.B.; Walsh, O.; Raun, W.R. In-season prediction of corn grain yield potential using normalized difference vegetation index. *Agron. J.* **2006**, *98*, 1488–1494. [CrossRef]
29. Martin, K.L.; Girmaa, K.; Freemana, K.W.; Teala, R.K.; Tubańaa, B.; Arnalla, D.B.; Chunga, B.; Walsha, O.; Solieb, J.B.; Stoneb, M.L.; et al. Expression of variability in corn as influenced by growth stage using optical sensor measurements. *Agron. J.* **2007**, *99*, 384–389. [CrossRef]
30. Jones, C.G.; Hartley, S.E. A protein competition model of phenolic allocation. *Oikos* **1999**, *86*, 27–44. [CrossRef]

31. Burchard, P.; Bilger, W.; Weissenbock, G. Contribution of hydroxycinnamates and flavonoids to epidermal shielding of UV-A and UV-B radiation in developing rye primary leaves as assessed by ultraviolet-induced chlorophyll fluorescence measurements. *Plant Cell Environ.* **2000**, *23*, 1373–1380. [CrossRef]
32. Knogge, W.; Weissenböck, G. Tissue-distribution of secondary phenolic biosynthesis in developing primary leaves of *Avena sativa* L. *Planta* **1986**, *167*, 196–205. [CrossRef]
33. Cerovic, Z.G.; Ounis, A.; Cartelat, A.; Latouche, G.; Goulas, Y.; Meyer, S.; Moya, I. The use of chlorophyll fluorescence excitation spectra for the non-destructive in situ assessment of UV-absorbing compounds in leaves. *Plant Cell Environ.* **2002**, *25*, 1663–1676. [CrossRef]
34. Tremblay, N.; Wang, Z.; Bélec, C. Evaluation of the Dualex for the assessment of corn nitrogen status. *J. Soil Sci. Plant Nutr.* **2007**, *30*, 1355–1369. [CrossRef]
35. Tremblay, N.; Wang, Z.; Belec, C. Performance of Dualex in spring wheat for crop nitrogen status assessment, yield prediction and estimation of soil nitrate content. *J. Soil Sci. Plant Nutr.* **2009**, *33*, 57–70. [CrossRef]
36. Lejealle, S.; Evain, S. Multiplex: A new diagnostic tool for management of nitrogen fertilization of turgrass. In Proceedings of the 10th International Conference on Precision Agriculture, Denver, CO, USA, 18–21 July 2010.
37. Cerovic, Z.G.; Moise, N.; Agati, G.; Latouche, G.; Ghozlen, N.B.; Meyer, S. New portable optical sensors for the assessment of winegrape phenolic maturity based on berry fluorescence. *J. Food Compos. Anal.* **2008**, *21*, 650–654. [CrossRef]
38. Yu, K.; Leufen, G.; Hunsche, M.; Noga, G.; Chen, X.; Bareth, G. Investigation of leaf diseases and estimation of chlorophyll concentration in seven barley varieties using fluorescence and hyperspectral indices. *Remote Sens.* **2013**, *6*, 64–86. [CrossRef]
39. Diago, M.P.; Rey-Carames, C.; Le Moigne, M.; Fadaili, E.M.; Tardaguila, J.; Cerovic, Z.G. Calibration of non-invasive fluorescence-based sensors for the manual and on-the-go assessment of grapevine vegetative status in the field. *Aust. J. Grape Wine R.* **2016**, *22*, 438–449. [CrossRef]
40. Song, X.; Yang, G.; Yang, C.; Wang, J.; Cui, B. Spatial variability analysis of within-field winter wheat nitrogen and grain quality using canopy fluorescence sensor measurements. *Remote Sens.* **2017**, *9*, 237. [CrossRef]
41. Zhang, Y.P.; Tremblay, N.; Zhu, J.J. A first comparison of Multiplex® for the assessment of corn nitrogen status. *J. Food Agric. Environ.* **2012**, *10*, 1008–1016.
42. Li, J.W.; Zhang, J.X.; Zhao, Z.; Lei, X.D.; Xu, X.L.; Lu, X.X.; Weng, D.L.; Gao, Y.; Cao, L.K. Use of fluorescence-based sensors to determine the nitrogen status of paddy rice. *J. Agric. Sci.* **2013**, *151*, 862–871. [CrossRef]
43. Lu, J.; Miao, Y.; Wei, S.; Li, J.; Yuan, F. Evaluating different approaches to non-destructive nitrogen status diagnosis of rice using portable RapidSCAN active canopy sensor. *Sci. Rep.* **2017**, *7*, 14073. [CrossRef]
44. Pedrós, R.; Goulas, Y.; Jacquemoud, S.; Louis, J.; Moya, I. FluorMODleaf: A new leaf fluorescence emission model based on the PROSPECT model. *Remote Sens. Environ.* **2010**, *114*, 155–167. [CrossRef]
45. Ounis, A.; Cerovic, Z.G.; Briantais, J.M.; Moya, I. Dual-excitation FLIDAR for the estimation of epidermal UV absorption in leaves and canopies. *Remote Sens. Environ.* **2001**, *76*, 33–48. [CrossRef]
46. Agati, G.; Cerovic, Z.G.; Pinelli, P.; Tattini, M. Light-induced accumulation of ortho-dihydroxylated flavonoids as non-destructively monitored by chlorophyll fluorescence excitation techniques. *Environ. Exp. Bot.* **2011**, *73*, 3–9. [CrossRef]
47. Cartelat, A.; Cerovic, Z.G.; Goulas, Y.; Meyer, S.; Lelarge, C.; Prioul, J.L.; Barbottin, A.; Jeuffroy, M.H.; Gate, P.; Agati, G. Optically assessed contents of leaf polyphenolics and chlorophyll as indicators of nitrogen deficiency in wheat (*Triticum aestivum* L.). *Field Crop. Res.* **2005**, *91*, 35–49. [CrossRef]
48. Lichtenthaler, H.K. Vegetation stress: An introduction to the stress concept in plants. *J Plant Physiol.* **1996**, *148*, 4–14. [CrossRef]
49. Ghozlen, N.B.; Cerovic, Z.G.; Germain, C.; Toutain, S.; Latouche, G. Non-destructive optical monitoring of grape maturation by proximal sensing. *Sensors* **2010**, *10*, 10040–10068. [CrossRef] [PubMed]
50. Huang, S.; Miao, Y.; Yuan, F.; Cao, Q.; Ye, H.; Lenz-Wiedemann, V.; Khosla, R.; Bareth, G. Proximal fluorescence sensing for in-season diagnosis of rice nitrogen status. *Adv. Anim. Biosci.* **2017**, *8*, 343–348. [CrossRef]
51. Huang, S.; Miao, Y.; Cao, Q.; Yao, Y.; Zhao, G.; Yu, W.; Shen, J.; Yu, K.; Bareth, G. Critical nitrogen dilution curve for rice nitrogen status diagnosis in Northeast China. *Pedosphere* **2018**, *28*, 814–822. [CrossRef]
52. Xia, T.; Miao, Y.; Wu, D.; Shao, H.; Khosla, R.; Mi, G. Active optical sensing of spring maize for in-season diagnosis of nitrogen status based on nitrogen nutrition index. *Remote Sens.* **2016**, *8*, 605. [CrossRef]

53. Bausch, W.C.; Khosla, R. QuickBird satellite versus ground-based multi-spectral data for estimating nitrogen status of irrigated maize. *Precis. Agric.* **2010**, *11*, 274–290. [CrossRef]
54. Landis, R.J.; Koch, G.G. The measurement of observer agreement for categorical data. *Biometrics* **1977**, *33*, 159–174. [CrossRef]
55. Padilla, F.M.; Peña-Fleitas, M.T.; Gallardo, M.; Thompson, R.B. Proximal optical sensing of cucumber crop N status using chlorophyll fluorescence indices. *Eur. J. Agron.* **2016**, *73*, 83–97. [CrossRef]
56. Agati, G.; Foschi, L.; Grossi, N.; Volterrani, M. In field non-invasive sensing of the nitrogen status in hybrid bermudagrass (*Cynodon dactylon*, × *C. transvaalensis*, Burtt Davy) by a fluorescence-based method. *Eur. J. Agron.* **2015**, *63*, 89–96. [CrossRef]
57. Buschmann, C. Variability and application of the chlorophyll fluorescence emission ratio red/far-red of leaves. *Photosynth. Res.* **2007**, *92*, 261–271. [CrossRef] [PubMed]
58. Yuan, Z.; Ata-Ul-Karim, S.T.; Cao, Q.; Lu, Z.; Cao, W.; Zhu, Y.; Liu, X. Indicators for diagnosing nitrogen status of rice based on chlorophyll meter readings. *Field Crop. Res.* **2016**, *185*, 12–20. [CrossRef]
59. Padilla, F.M.; Peña-Fleitas, M.T.; Gallardo, M.; Thompson, R.B. Evaluation of optical sensor measurements of canopy reflectance and of leaf flavonols and chlorophyll contents to assess crop nitrogen status of muskmelon. *Eur. J. Agron.* **2014**, *58*, 39–52. [CrossRef]
60. Barthod, S.; Cerovic, Z.; Epron, D. Can dual chlorophyll fluorescence excitation be used to assess the variation in the content of UV-absorbing phenolic compounds in leaves of temperate tree species along a light gradient? *J. Exp. Bot.* **2007**, *58*, 1753–1760. [CrossRef] [PubMed]
61. Goulas, Y.; Cerovic, Z.G.; Cartelat, A.; Moya, I. Dualex: A new instrument for field measurements of epidermal ultraviolet absorbance by chlorophyll fluorescence. *Appl. Opt.* **2004**, *43*, 4488–4496. [CrossRef]
62. Lichtenthaler, H.K.; Schweiger, J. Cell wall bound ferulic acid, the major substance of the blue-green fluorescence emission of plants. *J. Plant Physiol.* **1998**, *152*, 272–282. [CrossRef]
63. Buschmann, C.; Lichtenthaler, H.K. Principles and characteristics of multi-colour fluorescence imaging of plants. *J. Plant Physiol.* **1998**, *152*, 297–314. [CrossRef]
64. Bürling, K.; Cerovic, Z.G.; Cornic, G.; Ducruet, J.M.; Noga, G.; Hunsche, M. Fluorescence-based sensing of drought-induced stress in the vegetative phase of four contrasting wheat genotypes. *Environ. Exp. Bot.* **2013**, *89*, 51–59. [CrossRef]
65. Agati, G.; Pinelli, P.; Cortés, E.S.; Romani, A.; Cartelat, A.; Cerovic, Z.G. Nondestructive evaluation of anthocyanins in olive (*Olea europaea*) fruits by in situ chlorophyll fluorescence spectroscopy. *J. Agric. Food Chem.* **2005**, *53*, 1354–1363. [CrossRef] [PubMed]
66. Chen, Y.; Chen, J.; Yan, B.; Cui, L.; Pan, J.; Kai, G. Agronomic traits of a new characteristic rice line 'Huxuan 102'. *Acta Agric. Shanghai* **2015**, *2*, 60–64.
67. Zubillaga, M.; Urricariet, S. Assessment of nitrogen status in wheat using aerial photography. *Commun. Soil Sci. Plan.* **2005**, *36*, 1787–1798. [CrossRef]
68. Lemaire, G.; Jeuffroy, M.H.; Gastal, F. Diagnosis tool for plant and crop N status in vegetative stage: Theory and practices for crop N management. *Eur. J. Agron.* **2008**, *28*, 614–624. [CrossRef]
69. Hussain, F.; Bronson, K.F.; Peng, S. Use of chlorophyll meter sufficiency indices for nitrogen management of irrigated rice in Asia. *Agron. J.* **2000**, *92*, 875–879.
70. Varvel, G.E.; Wilhelm, W.W.; Shanahan, J.F.; Schepers, J.S. An algorithm for corn nitrogen recommendations using a chlorophyll meter based sufficiency index. *Agron. J.* **2007**, *99*, 701–706. [CrossRef]
71. Cerovic, Z.G.; Ghozlen, N.B.; Milhade, C.; Obert, M.; Debuisson, S.; Le Moigne, M. Nondestructive diagnostic test for nitrogen nutrition of grapevine (*Vitis vinifera* L.) based on Dualex leaf-clip measurements in the field. *J. Agr. Food Chem.* **2015**, *63*, 3669–3680. [CrossRef] [PubMed]
72. Stroppiana, D.; Boschetti, M.; Brivio, P.A.; Bocchi, S. Plant nitrogen concentration in paddy rice from field canopy hyperspectral radiometry. *Field Crop. Res.* **2009**, *111*, 119–129. [CrossRef]
73. Tremblay, N.; Wang, Z.; Cerovic, Z.G. Sensing crop nitrogen status with fluorescence indicators. a review. *Agron. Sustain. Dev.* **2012**, *32*, 451–464. [CrossRef]
74. Demotes-Mainard, S.; Boumaza, R.; Meyer, S.; Cerovic, Z.G. Indicators of nitrogen status for ornamental woody plants based on optical measurements of leaf epidermal polyphenol and chlorophyll contents. *Sci. Hortic.* **2008**, *115*, 377–385. [CrossRef]

75. Agati, G.; Foschi, L.; Grossi, N.; Guglielminetti, L.; Cerovic, Z.G.; Volterrani, M. Fluorescence-based versus reflectance proximal sensing of nitrogen content in *Paspalum vaginatum* and *Zoysia matrella* turfgrasses. *Eur. J. Agron.* **2013**, *45*, 39–51. [CrossRef]
76. Meyer, S.; Cerovic, Z.G.; Goulas, Y.; Montpied, P.; Demotesmainard, S.; Bidel, L.P.; Moya, I.; Dreyer, E. Relationships between optically assessed polyphenols and chlorophyll contents, and leaf mass per area ratio in woody plants: A signature of the carbon-nitrogen balance within leaves? *Plant Cell Environ.* **2006**, *29*, 1338–1348. [CrossRef] [PubMed]

remote sensing

MDPI

Article

Assessing Performance of Vegetation Indices to Estimate Nitrogen Nutrition Index in Pepper

Romina de Souza [1,*], M. Teresa Peña-Fleitas [1], Rodney B. Thompson [1,2], Marisa Gallardo [1,2] and Francisco M. Padilla [1,2]

1 Department of Agronomy, University of Almeria, Carretera de Sacramento s/n, 04120 La Cañada, Almería, Spain; mtpena.fl@ual.es (M.T.P.-F.); rodney@ual.es (R.B.T.); mgallard@ual.es (M.G.); f.padilla@ual.es (F.M.P.)

2 CIAIMBITAL Research Centre for Mediterranean Intensive Agrosystems and Agrifood Biotechnology, University of Almeria, 04120 La Cañada, Almería, Spain

* Correspondence: sdi672@ual.es; Tel.: +34-950-014-101

Received: 30 January 2020; Accepted: 25 February 2020; Published: 26 February 2020

Abstract: Vegetation indices (VIs) can be useful tools to evaluate crop nitrogen (N) status. To be effective, VIs measurements must be related to crop N status. The nitrogen nutrition index (NNI) is a widely accepted parameter of crop N status. The present work evaluates the performance of several VIs to estimate NNI in sweet pepper (*Capsicum annuum*). The performance of VIs to estimate NNI was evaluated using parameters of linear regression analysis conducted for calibration and validation. Three different sweet pepper crops were grown with combined irrigation and fertigation, in Almería, Spain. In each crop, five different N concentrations in the nutrient solution were frequently applied by drip irrigation. Proximal crop reflectance was measured with Crop Circle ACS470 and GreenSeeker handheld sensors, approximately every ten days, throughout the crops. The relative performance of VIs differed between phenological stages. Relationships of VIs with NNI were strongest in the early fruit growth and flowering stages, and less strong in the vegetative and harvest stages. The green band-based VIs, GNDVI, and GVI, provided the best results for estimating crop NNI in sweet pepper, for individual phenological stages. GNDVI had the best performance in the vegetative, flowering, and harvest stages, and GVI had the best performance in the early fruit growth stage. Some of the VIs evaluated are promising tools to estimate crop N status in sweet pepper and have the potential to contribute to improving crop N management of sweet pepper crops.

Keywords: canopy reflectance; crop N status; *Capsicum annuum*; proximal optical sensors

1. Introduction

Vegetable crops production is characterized by nitrogen (N) losses and the associated environmental problems [1–3]. The most common environmental problems include ground and surface water contamination, eutrophication of surface water, and nitrous oxide (N_2O) emission [4,5]. These problems are often a consequence of the high use of N fertilizer as a way to ensure optimal growth and production [6], which generally exceeds the demand of the crops [3,7,8]. Knowing crop N requirements and matching N supply to crop demand are requirements to reduce N contamination of water bodies by intensive vegetable production [3,9,10]. Various tools are available for monitoring crop N status [3,11]. A traditional tool is leaf nutrient analysis, which requires laborious and time-consuming laboratory work, and which generally cannot characterize the temporal and spatial variability of N status [12,13]. These are major drawbacks, because knowledge of temporal and spatial variability of crop N status appreciably assists the matching of N supply to crop N requirements [14]. Optical sensors are devices that provide rapid, effective, and nondestructive assessment of crop N status, in the field [3,15]. They enable frequent assessment throughout a crop, and assessment of spatial variability.

Amongst the proximal optical sensors, canopy reflectance sensors have two very positive features in that they can measure large areas of a crop and have on-the-go measurement capability [16].

Crop reflectance measurements can be used to assess N status of field crops [11]. These measurements are based on the differential reflection of wavelengths of radiation [3], which are absorbed and reflected by the crop in different proportions, depending on crop N status [15]. Generally, the light wavelengths used are in red, green, and near-infrared ranges [11]. More recently, the red-edge has been proposed to overcome the reported saturation of the red band [17,18]. Using reflectance data of different wavelengths, vegetation indices are calculated, which commonly combine reflectance data from 2–3 wavelengths [19].

Measurements of crop reflectance can be made with proximal sensors positioned relatively close to the canopy, from several centimeters to a few meters away [15]. Due to the field of view of the sensors, each individual measurement can integrate a large area of crop canopy [3,20]. Depending on the sensor, continuous measurements can be made as the sensor passes along the crop canopy ("on-the-go" measurement), thereby integrating large surface areas of crop canopy [15].

In order to use vegetation indices, calculated from canopy reflectance measurement, as a proxy of crop N status, calibration is required. A commonly-used approach is to determine the relationship between values of a given vegetation index and a measure of crop N status, such as the nitrogen nutrition index (NNI) [13,21]. NNI is calculated by dividing the actual crop N content by the critical crop N content [22,23], the latter being the lowest crop N content necessary for nonlimiting growth. Values of NNI equal to 1 indicate optimal N nutrition [24], and any deviation from 1 indicates excess N (i.e., NNI>1) or deficient N (i.e., NNI<1) crop status.

Numerous studies have reported that vegetation indices, obtained with canopy reflectance sensors, are strongly related to crop biomass and yield [11,25,26]. Appreciably, fewer studies have assessed the capability of vegetation indices, measured with proximal reflectance sensors, to assess crop N status [27]. Most studies have been conducted in cereal crops such as wheat [12,27,28] and rice [29,30]; very few with vegetable crops such as sweet pepper. To use vegetation indices as estimators of crop N status, it is necessary to derive a regression equation between the measured vegetation index (independent variable) and crop NNI (dependent variable) [31]. This procedure requires firstly fitting a regression equation between the vegetation index and crop NNI with a calibration dataset [27], and secondly, it requires validation of this regression equation with an independent, validation dataset.

Environmental problems associated with the high use of N fertilizer in vegetable production systems have been reported for diverse regions [5], such as southeastern (SE) Spain [32], SE United States [6], and China [7,33]. Greenhouse production systems are major sources of vegetables [34]. Within greenhouse-based vegetable production systems, sweet pepper is one of the most important vegetable crop [35]. In SE Spain, approximately 40,000 ha [36] of highly-concentrated greenhouses are used for intensive vegetable production; 30,000 ha are located in the Almería province. This system is characterized by high rates of N fertilizer and an excessive N supply [2,37] that are associated with nitrate contamination of underlying aquifers [32]. There is increasingly strong pressure to improve crop N management to reduce aquifer contamination from this vegetable production system. In Almeria, sweet pepper is one of the most important crops; each year, it is grown on 8000 ha [38].

Given the pressure to improve N management in greenhouse-based vegetable production [2,7] and that sweet pepper is a major crop, information is required of tools and sensors that inform of the N status of sweet pepper crops grown in greenhouses. Such tools will provide vital information of the adequacy of ongoing N management, enabling optimal N fertilizer use and ensuring less environmentally harmful N losses [3].

In the present work, eight vegetation indices, calculated from canopy reflectance measurements obtained with two different proximal sensors, were evaluated to estimate crop N status of sweet pepper. Firstly, calibration regression equations of each vegetation index to crop NNI were fitted. Secondly, these regression equations were subsequently validated using a different dataset. Thirdly, using the validated equations between vegetation indices and crop NNI, sufficiency values were derived for each vegetation index for optimal N nutrition, for the major phenological stages of sweet pepper crops.

2. Material and Methods

2.1. Site and Experimental Design

Three experiments with sweet pepper (*Capsicum annuum* cv. Melchor) were carried out in a plastic greenhouse in Almería, southeast Spain. The greenhouses were located in the experimental station of the University of Almería (36° 51′ 51″ N, 2° 16′ 56″ W, 92 m altitude). The first crop was grown in 2014–2015, the second in 2016–2017, and the third in 2017–2018 (Table 1). All crops spanned a summer-winter cycle. Further details of the greenhouse are in Padilla et al. 2014 [39] and 2017 [21]. The crops were grown in an artificial layered soil, known as "enarenado", typical of commercial greenhouse crops in SE Spain [2].

Table 1. Duration, beginning of nitrogen (N) treatments, concentration of mineral N (NO_3^-–N + NH_4^+–N) applied in nutrient solutions, and mineral N amount applied in fertigation, in the three sweet pepper crops. DAT: Days after transplanting.

Crop	Cycle	Duration (Days)	Beginning of N Treatments	Mineral N Concentration of Treatments (mmol N L^{-1})	Amount of Mineral N Applied (kg N ha^{-1})
2014–2015	12 August – 29 January	170	1 DAT	N1: 2.4 N2: 6.2 N3: 12.6 N4: 16.1 N5: 20.0	N1: 64 N2: 189 N3: 516 N4: 804 N5: 990
2016–2017	19 July – 24 March	248	9 DAT	N1: 2.0 N2: 5.3 N3: 9.7 N4: 13.5 N5: 17.7	N1: 88 N2: 302 N3: 561 N4: 1052 N5: 1320
2017–2018	21 July – 20 February	214	10 DAT	N1: 2.0 N2: 5.7 N3: 9.7 N4: 13.1 N5: 16.7	N1: 86 N2: 304 N3: 519 N4: 870 N5: 1198

The crops were established by transplanting 35-day old seedlings, in twin rows (0.8 m between twin rows and 1.2 m between twin rows) and 0.5 m distance between plants within each line, with a plant density of 2 plants m^{-2}. Each experimental plot measured 6 by 6 m, giving a total of 72 plants per replicate plot. There were three twin rows of plants, 6 m in length in each plot. The middle twin row was used for canopy reflectance measurements.

Water and fertilizers were applied combined through fertigation, by using an above-ground drip irrigation system. Each plant was planted close to an emitter. The fertigation was applied every two–three days, depending on crop demand. The three experiments consisted of a fully randomized block design, with five N treatments and four replicates per treatment. The N treatments were applied by fertigation by using different nutrient solutions with increasing N concentration. All other macro and micronutrients were applied in the nutrient solution to ensure they were not limiting. The treatments were: Very deficient N (N1), deficient N (N2), conventional N (N3), excessive N (N4), and very excessive N (N5) (Table 1). N was applied mostly (90%) as nitrate (NO_3^-), the rest as ammonium (NH_4^+). The crop was physically supported which is typical for pepper production in this system. Crop management followed local practices.

2.2. Canopy Reflectance with Optical Sensors

Two active proximal reflectance sensors were used to measure canopy reflectance information throughout the three crops. In the first crop, reflectance measurements were made weekly and in the second and third crops every two weeks. The sensors used were the GreenSeeker handheld sensor (Trimble Navigation Limited, Sunnyvale, CA, USA) and the Crop Circle ACS-470 (Holland Scientific Inc., Lincoln, NE, USA). The measurements were made by positioning both sensors vertically and parallel to the crop rows, so that the upper limit of the field of view was at the height of the most recently fully expanded leaf [15].

The GreenSeeker includes two light sources, visible (660 nm—red light) and near-infrared (NIR) (780 nm). This sensor measures the fraction of emitted lights reflected from the crop to calculate the vegetation index NDVI which is explained in Table 2. The GreenSeeker handheld sensor was positioned at 60 cm horizontal distance to the foliage; the field of view was an oval with a height of ≈ 25 cm. The measuring mode was the individual measurement ("one-shot"). For each date of measurement, eight marked plants were measured per replicate plot, and the mean value was determined.

Table 2. Vegetation indices calculated in the present study.

Index	Acronym	Equation	Reference
Normalized Difference Vegetation Index	NDVI	$\frac{NIR-RED}{NIR+RED}$	Sellers [40]
Green Normalized Difference Vegetation Index	GNDVI	$\frac{NIR-Green}{NIR+Green}$	Ma et al. [41]
Red Ratio of Vegetation Index	RVI	$\frac{NIR}{Red}$	Birth and McVey [42]
Green Ratio of Vegetation Index	GVI	$\frac{NIR}{Green}$	Birth and McVey [42]
Red Edge Normalized Difference Vegetation Index	RENDVI	$\frac{NIR-Red\ Edge}{NIR+Red\ Edge}$	Gitelson and Merzlyak [43]
Chlorophyll Index	CI	$\frac{NIR}{Red\ Edge}$	Gitelson et al. [44]
Canopy Chlorophyll Content Index	CCCI	$\frac{RENDVI-RENDVImin}{RENDVImax-RENDVImin}$	Fitzgerald et al. [28]
MERIS Terrestrial Chlorophyll Index	MTCI	$\frac{NIR-Red\ Edge}{Red\ Edge-Red}$	Dash and Curran [45]

The Crop Circle ACS-470 used filters at 550 nm (green), 670 nm (red), 760 nm (near infrared; NIR), and 730 nm (red edge). The sensor was positioned at a 45 cm horizontal distance. The field of view was a rectangle of ≈ 26 (vertical) × 5 (horizontal) cm. The measurements were made in two separate passes. Each pass consisted of a 4 m transect in each line of plants in the middle twin row of each plot. In the first pass, green, red, and NIR filters were used; in the second pass, red edge, red, and NIR filters were used. Measurements were collected at a frequency of 10 readings per second. On-the-go measurements were made by walking at approximately 1.5 km h^{-1}. In total, 200 individual measurements were collected per plot. Data were stored in a portable GeoSCOUT GLS-400 data logger (Holland Scientific, Inc.). The vegetation indices shown in Table 2 were calculated from reflectance values of individual wavelengths.

2.3. Crop Sampling and NNI Determination

In each of the crops, periodical aboveground biomass samplings (approximately every 14 days) were made to determine dry matter (DM). For each replicate plot in each sampling, two complete plants were selected and removed. The dry weights of different components of the plants (stem, leaf, and fruit) were recorded by oven-drying until constant weight at 65 °C. In each replicate plot, the fruit production and pruned material were recorded throughout the crop in eight marked plants. Subsamples of dry material were ground prior to analysis of N content (%N) in a Dumas-type elemental analyzer (Rapid N, Elementar, Analysensysteme GmbH, Hanau, Germany). The amount of N was calculated by multiplying the %N by the dry matter mass of the corresponding component.

The NNI was calculated using the critical N curve derived for greenhouse-grown sweet pepper crop: Critical N = $4.71{\cdot}DM^{-0.22}$ (Alejandra Rodríguez, University of Almeria, unpublished data). The NNI was calculated by dividing the N content measured in the crop by the critical N content. The NNI value for each reflectance measurement day was by interpolating DM and crop N content values between two consecutive biomass samplings [46].

2.4. Data Analysis

Data of reflectance measurements and NNI were grouped and analyzed for phenological stage. Four main phenological stages were considered for sweet pepper, according to de Souza et al. [47], as: (1) Vegetative, (2) flowering, (3) early fruit growth, and (4) harvest. The definition of the phenological stages is in de Souza et al. [47]. Within each phenological stage, several canopy reflectance measurements and biomass samplings were conducted. To integrate data of the various measurements within each phenological stage, integrated NNI and vegetation indices values were calculated, according to Lemaire and Gastal [24] and Padilla et al. [21], as:

$$\text{Integrated index} = \frac{1}{D}\cdot\sum(V.ds) \tag{1}$$

where D was the duration of the phenological stage, V was the value of NNI or vegetation index for each day of measurement, and ds was the duration between two successive measurements [47].

Predictive regression functions were evaluated to estimate NNI of sweet pepper for each of the eight vegetation indices assessed in the current work. For each phenological stage of the three crops; data of integrated vegetation indices and the corresponding integrated NNI were pooled. This created a single pooled data set of 60 data points for each vegetation index in each phenological stage considering the three crops together. The 60 data points, for each phenological stage were randomly separated into two groups: Forty data points (2/3 of total data) for the calibration dataset, and 20 data points (1/3 of total data) for the validation dataset. With the calibration dataset, simple linear regression analyses were conducted with the integrated vegetation index as the independent variable (x variable) and NNIi as the dependent variable (y variable). The software CurveExpert Professional®2.2.0 software (Daniel G. Hyams, MS, USA) was used. Validation of the equations that related each integrated vegetation index with NNIi, for each phenological stage, was then conducted with the validation dataset. Validation consisted of calculating the predicted NNIi from the calibration equation for each combination of vegetation index and phenological stage. Predicted NNIi values were then compared with the NNIi values of the validation dataset. Linear regression analysis was made between observed NNIi (independent variable) and predicted NNIi (dependent variable) and the root mean square error (RMSE) of the NNI estimation was determined. The RMSE was calculated as:

$$\text{RMSE} = \sqrt{\sum_{1}^{n}\frac{(E_i - O_i)^2}{n}} \tag{2}$$

where n is the number of samples, Ei is the estimated value of the relationship, and Oi is the observed value [48].

The performance of the different vegetation indices was evaluated according to Xin-feng et al., 2013 [49]; this procedure considers both the calibration and validation results. Coefficient of determination (R^2) and RMSE values of the linear regression of the calibration dataset, and the R^2 and RMSE values, absolute values of slope-1, and absolute values of intercept of the linear regression of the validation dataset were used [49]. Slope-1 is the absolute value of the slope after subtracting one from the slope of the linear regression. The use of this parameter effectively normalizes slope values and enabled ranking of all integrated vegetation indices from lowest to highest values.

The performance of each vegetation index was calculated (i) by sorting R^2 in decreasing order and RMSE in ascending order for the calibration and validation datasets separately, and (ii) the sorting of absolute slope-1 and absolute intercept values in ascending order for the validation dataset [49]. The best performing vegetation index was that which had the lowest sum of these six factors [49]. Additionally, the performance of the validation regression equation of the different vegetation indices, in each phenological stage, was assessed by comparing the relative error (RE) between observed and estimated NNIi values. The relative error was calculated as:

$$\text{RE: } \frac{RMSE}{Oi} \tag{3}$$

where Oi is the average of observed values.

Sufficiency values of each vegetation index, for each phenological stage, were calculated from the regression equations of the calibration datasets. The equations of calibration for each phenological stage were solved for NNI = 1, according to Lemaire et al. [23].

3. Results

3.1. Phenological Relationships Between Integrated Vegetation Indices and Integrated NNI (NNIi), for Calibration Dataset

For the calibration data, the relationships between most of the integrated vegetation indices and NNIi, in each phenological stage, were highly significant (Table 3, Figure 1). In the vegetative stage, the coefficients of determination (R^2) of these relationships were generally low; averaged across all vegetation indices, the R^2 of the vegetative stage was 0.45 ± 0.05. In the vegetative stage, the R^2 ranged from 0.19 to 0.50 for most of the vegetation indices, except the GNDVIi which had a R^2 value of 0.63 (Table 3). In the flowering stage, the coefficients of determination were slightly higher than in the vegetative stage, with an average R^2 value for all vegetation indices of 0.52 ± 0.03 and a range from 0.38 to 0.65 (Table 3). The highest R^2 values were obtained in phenological stage corresponding to early fruit growth, where the average R^2 value across all vegetation indices was 0.71 ± 0.04, with a range from 0.52 to 0.84 (Table 3). The harvest stage had the lowest R^2 values of all phenological stages for all vegetation indices considered together, the average was 0.27 ± 0.02, with a range from 0.19 to 0.42 (Table 3).

Comparing the performance of different vegetation indices to estimate NNI throughout the crops, the integrated vegetation indices that were based on the green band (GNDVI and GVI) had higher and more consistent R^2 values in the first three phenological stages, which were the vegetative, flowering, and early fruit growth stages. The R^2 values for GNDVIi were 0.63, 0.65, and 0.62 for the vegetative, flowering, and early fruit growth stage, respectively. For GVIi, the R^2 values were 0.56, 0.60, and 0.63, respectively, for the same phenological stages (Table 3). The R^2 values of RVIi, CIi, and CCIi vegetation indices were low ($R^2 < 0.50$) and very similar in the vegetative and flowering stages but increased in the early fruit growth stage (Table 3). For the rest of the integrated vegetation indices evaluated, the R^2 values increased from the vegetative to early fruit growth stages, being lowest in the harvest stage (Table 3). Sufficiency values of each integrated vegetation index, for each phenological stage, were calculated from the regression equations of the calibration datasets. The equations of calibration for each phenological stage were solved for NNI = 1.

Table 3. Equations, coefficient of determination (R^2), and root mean square error (RMSE) of linear regressions between integrated vegetation indices (x variable) and integrated nitrogen nutrition index (NNIi, y variable) at different phenological stages, for calibration data (n = 40). Significance of regressions are indicated with asterisks close to R^2 values. ***, $p < 0.001$; **, $p < 0.01$). Abbreviations for vegetation indices are in Table 2.

Index	Vegetative			Flowering			Early Fruit Growth			Harvest		
	Equation	R^2	RMSE	Equation	R^2	RMSE	Equation	R^2	RMSE	Equation	R^2	RMSE
$NDVIi_{CS}$	NNIi = 1.314x - 0.024	0.27***	0.139	NNIi = 7.225x - 5.368	0.63***	0.121	NNIi = 7.393x - 5.530	0.52***	0.151	NNIi = 1.443x - 0.260	0.42***	0.135
NDVIi	NNIi = 2.268x - 0.726	0.48***	0.118	NNIi = 6.347x - 4.495	0.54***	0.137	NNIi = 7.277x - 5.340	0.65***	0.131	NNIi = 1.395x - 0.233	0.27***	0.151
GNDVIi	NNIi = 2.431x - 0.562	0.63***	0.099	NNIi = 3.872x - 1.817	0.65***	0.118	NNIi = 4.135x - 2.117	0.62***	0.135	NNIi = 1.294x + 0.014	0.32***	0.146
RVIi	NNIi = 0.061x + 0.422	0.48***	0.117	NNIi = 0.069x + 0.002	0.46***	0.147	NNIi = 0.088x - 0.319	0.68***	0.125	NNIi = 0.026x + 0.608	0.22**	0.156
GVIi	NNIi = 0.144x + 0.290	0.56***	0.108	NNIi = 0.167x - 0.091	0.60***	0.126	NNIi = 0.170x - 0.236	0.63***	0.135	NNIi = 0.067x + 0.510	0.27***	0.151
RENDVIi	NNIi = 3.891x - 0.288	0.51***	0.114	NNIi = 4.308x - 0.659	0.38***	0.151	NNIi = 7.820x - 2.170	0.82***	0.094	NNIi = 1.718x + 0.279	0.19**	0.175
CIi	NNIi = 0.923x - 0.855	0.48***	0.118	NNIi = 1.055x - 1.356	0.50***	0.142	NNIi = 1.517x - 2.609	0.83***	0.092	NNIi = 0.425x + 0.005	0.30***	0.148
CCCIi	NNIi = 1.655x + 0.124	0.45***	0.122	NNIi = 1.445x + 0.193	0.44***	0.150	NNIi = 2.484x - 0.680	0.83***	0.092	NNIi = 0.726x + 0.522	0.23**	0.155
MTCIi	NNIi = 0.473x + 0.269	0.19**	0.147	NNIi = 0.998x - 0.469	0.46***	0.147	NNIi = 1.533x - 1.538	0.84***	0.088	NNIi = 0.440x + 0.285	0.23**	0.156

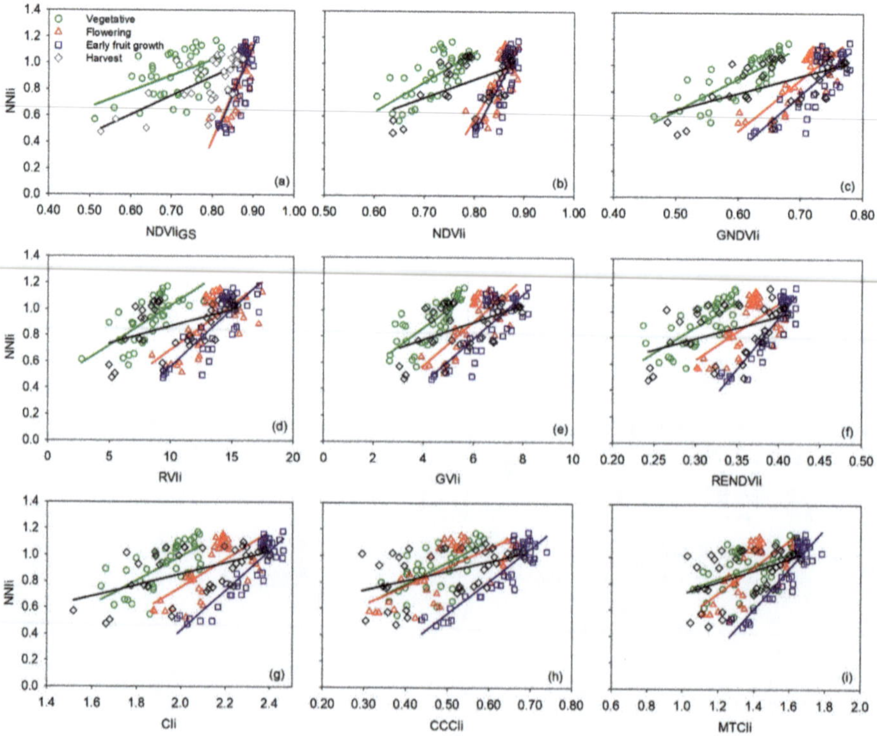

Figure 1. Linear regressions between each integrated vegetation index and integrated nitrogen nutrition index (NNIi) for the four vegetative stages, for calibration data (n = 40). Circle: Vegetative; Triangle: Flowering; Square: Early fruit growth; and Diamond: Harvest stage. Panel (**a**) shows normalized index vegetation index (NDVI) measured with GreenSeeker sensor and the other panels (**b–i**) show indices calculated with the Crop Circle sensor. Results of regression are in Table 3. Abbreviations for vegetation indices are in Table 2.

3.2. Validation of the Phenological Relationships Between Vegetation Indices and NNIi

Validation of the relationships established with the calibration dataset was made with an independent and different dataset. For all of the vegetation indices analyzed, the vegetative stage had the worst validation results. In this stage, the average R^2 and RMSE values for all indices were 0.46 ± 0.05 and 0.123 ± 0.006, respectively (Table 4). The validation results, for all vegetation indices, improved in the flowering and early fruit growth phenological stages. In these stages, the average R^2 and RMSE values for all indices were 0.63 ± 0.04 and 0.127 ± 0.006, for the flowering stage, and were 0.87 ± 0.02 and 0.120 ± 0.008, for the fruit growth stage, respectively (Table 4). In the harvest stage, validation results were intermediate (Table 4), with average R^2 and RMSE values for all indices of 0.59 ± 0.09 and 0.125 ± 0.002, respectively.

Table 4. Results of validation analysis for each vegetation index at different phenological stages. Equations, coefficient of determination (R^2), and root mean square error (RMSE) of linear regression between observed NNIi values (x variable) and predicted NNIi values (y variable), for validation dataset (n = 20). Significance of regressions are indicated with asterisks close to R^2 values. ***, p < 0.001; **, p < 0.01; ns: Not significant. Abbreviations for vegetation indices are in Table 2.

Index	Vegetative			Flowering			Early Fruit Growth			Harvest		
	Equation	R^2	RMSE	Equation	R^2	RMSE	Equation	R^2	RMSE	Equation	R^2	RMSE
$NDVIi_{GS}$	NNIi_Pred = 0.[illegible]3x + 0.468	0.55***	0.109	NNIi_Pred = 0.895x + 0.061	0.67***	0.132	NNIi_Pred = 1.088x - 0.146	0.70***	0.147	NNIi_Pred = 0.532x + 0.432	0.58***	0.119
NDVIi	NNIi_Pred = 0.[illegible]9x + 0.468	0.33**	0.135	NNIi_Pred = 0.946x + 0.017	0.69***	0.132	NNIi_Pred = 1.342x - 0.398	0.89***	0.130	NNIi_Pred = 0.439x + 0.522	0.57***	0.124
GNDVIi	NNIi_Pred = 0.8[illegible]6x + 0.193	0.55***	0.121	NNIi_Pred = 0.905x + 0.092	0.78***	0.099	NNIi_Pred = 1.215x - 0.240	0.89***	0.094	NNIi_Pred = 0.483x + 0.489	0.62***	0.118
RVIi	NNIi_Pred = 0.5[illegible]0x + 0.431	0.49**	0.117	NNIi_Pred = 0.730x + 0.255	0.65***	0.120	NNIi_Pred = 1.256x - 0.292	0.87***	0.111	NNIi_Pred = 0.402x + 0.571	0.64***	0.124
GVIi	NNIi_Pred = 0.6[illegible]6x + 0.323	0.59***	0.106	NNIi_Pred = 0.826x + 0.185	0.76***	0.100	NNIi_Pred = 1.189x - 0.190	0.91***	0.079	NNIi_Pred = 0.452x + 0.527	0.62***	0.121
RENDVIi	NNIi_Pred = 0.5[illegible]8x + 0.404	0.55***	0.109	NNIi_Pred = 0.518x + 0.461	0.41**	0.153	NNIi_Pred = 1.427x - 0.468	0.86***	0.145	NNIi_Pred = 0.37x + 0.565	0.58***	0.139
CIi	NNIi_Pred = 0.5[illegible]2x + 0.471	0.46**	0.122	NNIi_Pred = 0.716x + 0.275	0.59***	0.131	NNIi_Pred = 1.418x - 0.466	0.90***	0.132	NNIi_Pred = 0.444x + 0.525	0.60***	0.121
CCCIi	NNIi_Pred = 0.2[illegible]4x + 0.661	0.17ns	0.159	NNIi_Pred = 0.612x + 0.379	0.54***	0.136	NNIi_Pred = 1.296x - 0.337	0.87***	0.120	NNIi_Pred = 0.377x + 0.581	0.55***	0.129
MTCIi	NNIi_Pred = 0.2[illegible]0x + 0.718	0.49**	0.131	NNIi_Pred = 0.652x + 0.338	0.55***	0.136	NNIi_Pred = 1.305x - 0.356	0.89***	0.120	NNIi_Pred = 0.398x + 0.561	0.58***	0.126

Generally, for the vegetative stage, the slope of linear regression between observed and predicted NNIi values was appreciably different to one for all of the vegetation indices evaluated; the average slope value for all indices was 0.513 ± 0.057 (Table 4). The exception was the GNDVIi that had a slope of 0.806. Compared to the 1:1 line, there was a tendency for all vegetation indices except GNDVI to overestimate NNI values for NNI values < 0.9, and a tendency to underestimate NNI, for NNI values >0.9 (Figure 2, green circles). In the flowering stage, the slopes of the regression between observed and predicted NNIi values were closer to one for all of the vegetation indices evaluated (average value of 0.756± 0.049), and particularly so for NDVI (0.946) and GNDVI (0.905) (Table 4). Compared to the 1:1 line, all of the vegetation indices except for NDVI and GNDVI tended to overestimate NNI values at NNI values <1, and underestimate NNI at NNI values >1 (Figure 2, red triangles). In the early fruit growth stage, slopes of linear regression between observed and predicted NNIi values were slightly above 1 for all of the vegetation indices evaluated (average value of 1.282 ± 0.036) (Table 4). Compared to the 1:1 line, all vegetation indices underestimated NNI values at the whole range of NNI observed in the early fruit growth stage (Figure 2, blue squares). In the harvest stage, the slopes of the regression between observed and predicted NNIi values were close to 0.5 for all vegetation indices evaluated (average value of 0.433± 0.018) (Table 4). Compared to the 1:1 line, all vegetation indices overestimated NNI values, at NNI values < 0.9, and underestimate NNI, at NNI values > 0.9 during the harvest stage (Figure 2, grey diamonds).

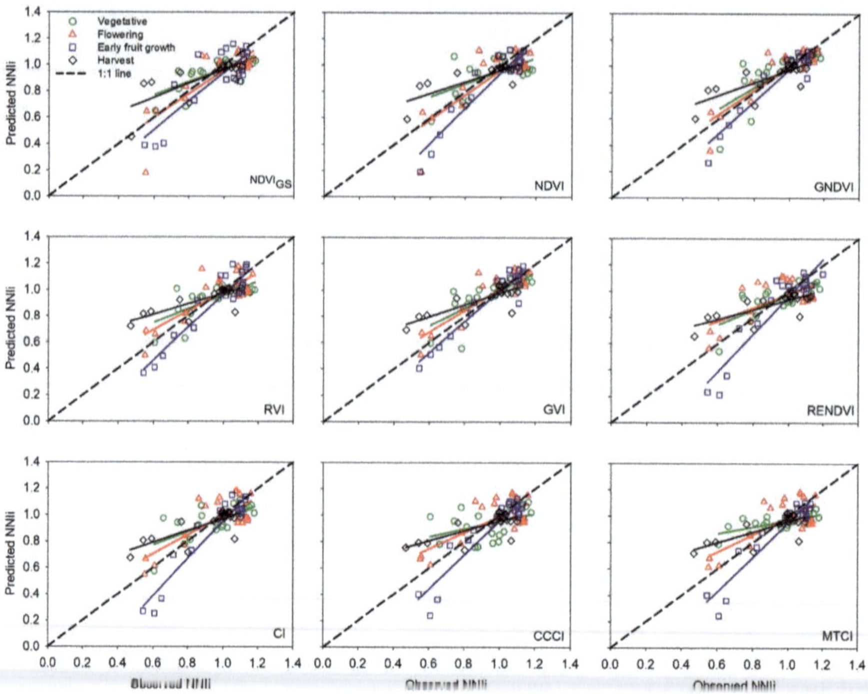

Figure 2. Relationships between observed integrated Nitrogen Nutrition Index (NNIi) and predicted NNIi for the four phenological stages, for validation data (n = 20). Circle: Vegetative; Triangle: Flowering; Square: Early fruit growth; and Diamond: Harvest stage. Panel (**a**) shows NDVI measured with GreenSeeker sensor and the other panels (**b–i**) show indices calculated with Crop Circle sensor. Dotted line represents the 1:1 line. Results of regression are in Table 4. Abbreviations for vegetation indices are in Table 2.

The relative error (RE) of the validation analysis for all vegetation indices evaluated in each phenological stage are presented in Figure 3. For the vegetative, flowering, and early fruit growth stages, RE ranged from 8% to 17%, with average RE values, for all indices, of 12.9% ± 0.59%, 13.2% ± 0.66% and 12.4% ± 0.78%, respectively. In the harvest stage, the RE ranged from 12% to 15%, with an average value of 13.5% ± 0.21% for all indices. The vegetation indices GVI and GNDVI had consistently lower RE values in most of the phenological stages (average RE values across phenological stages of 10.7% ± 1.03% and 11.3% ± 0.81%, respectively), followed by the RVI index (averaged RE values across phenological stages of 12.4% ± 0.40%). The GVI had the lowest RE in the vegetative, flowering, and early fruit growth stages (Figure 3). In contrast, the RENDVI index was the vegetation index with highest RE values throughout the four phenological stages (average value of 14.5% ± 1.09%).

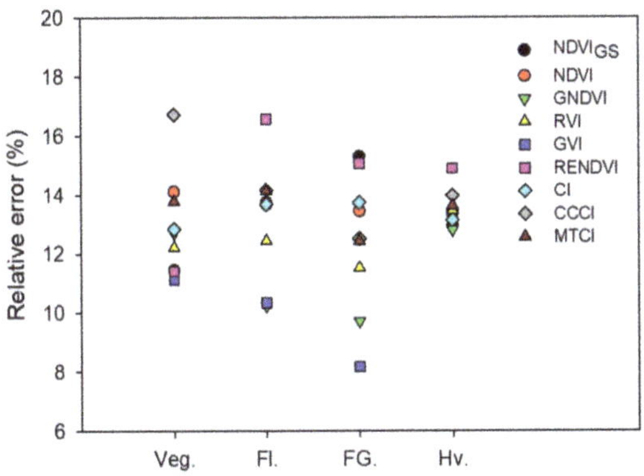

Figure 3. Relative error of linear relationships between observed integrated Nitrogen Nutrition Index (NNIi) values and predicted NNIi for each vegetation index at different phenological stages, for validation data (n = 20). Veg: Vegetative stage; Fl: Flowering stage; FG: Early fruit growth stage; Hv: Harvest stage. Abbreviations for vegetation indices are in Table 2.

3.3. Performance of Vegetation Indices

The classification of vegetation indices based on R^2 and RMSE of linear regression analysis of the calibration and validation datasets, and on the slope and intercept values of linear regressions of the validation dataset, showed that six (NDVI, RVI, RENDVI, CI, CCCI, and MTCI) of the nine vegetation indices evaluated had their best performance in the early fruit growth stage (Table 5). For NDVI, RVI, CI, CCCI, and MTCI, the flowering stage, was the phenological stage in which the second best results were obtained, for these indices, which were only slightly inferior to those in the early fruit growth stage. The RENDVI index was the exception, where the second best performance was in the vegetative stage (Table 5). For the three remaining vegetation indices ($NDVI_{GS}$ measured with GreenSeeker, GNDVI, and GVI), the best performance was in the flowering stage, followed by the early fruit growth stage. The worst performance for most of the vegetation indices occurred in the harvest stage, followed by the vegetative stage (Table 5).

Table 5. Ranking of best performing phenological stage for each vegetation index. Performance was evaluated using R^2 and RMSE of linear regression of calibration and validation datasets, and slope and intercept of linear regression of validation dataset. Numbers in brackets show the performance of each phenological stage. The best performance is the one which has the lowest value.

Best Performance	$NDVI_{GS}$	NDVI	GNDVI	RVI	GVI	RENDVI	CI	CCCI	MTCI
1st	Flowering (10)	Early fruit growth (10)	Flowering (9)	Early fruit growth (8)	Flowering (10) Early fruit growth (10)	Early fruit growth (10)	Early fruit growth (11)	Early fruit growth (6)	Early fruit growth (7)
2nd	Early fruit growth (14)	Flowering (12)	Early fruit growth (14)	Flowering (14)	Vegetative (17)	Vegetative (11)	Flowering (13)	Flowering (16)	Flowering (14)
3rd	Harvest (16)	Vegetative (18)	Vegetative (15)	Vegetative (15)	Harvest (23)	Flowering (19)	Vegetative (19)	Harvest (18)	Harvest (17)
4th	Vegetative (20)	Harvest (20)	Harvest (22)	Harvest (23)		Harvest (20)	Harvest (20)	Vegetative (20)	Vegetative (22)

For each phenological stage, the performance of the vegetation indices was compared to one another in Table 6. In the vegetative, flowering, and harvest stages, the best performing index was GNDVI. In early fruit growth stage, the best performing index was GVI. The performance of various indices, in this ranking was not constant in the different stages (Table 6). For example, in the vegetative stage, the second and third best performing vegetation indices were GVI and RENDVI, respectively, but RENDVI was one of the worst performing indices in the other three stages. Similar results were obtained for MTCI, which was the second-best performing index in the early fruit growth stage but was amongst the last positions in the other three stages. Overall, considering the four stages together, the best performing vegetation index was GNDVI, followed by GVI. The vegetation indices that performed worse were CCCI and MTCI (Table 6).

Table 6. Ranking of best performing indices for each phenological stage. Performance was evaluated using R^2 and RMSE of linear regression of calibration and validation datasets, and slope and intercept of linear regression of validation dataset. Numbers in brackets show the performance of each index. The best performance index is the one which has the lowest value.

Best Performance	Vegetative	Flowering	Early Fruit Growth	Harvest	Whole Crop
1st	GNDVI (12)	GNDVI (9)	GVI (20)	GNDVI (12)	GNDVIi (60)
2nd	GVI (13)	$NDVI_{GS}$ (18) GVI (18)	MTCI (22)	$NDVI_{GS}$ (13)	GVIi (73)
3rd	RENDVI (20)	NDVI (19)	CCCI (26)	CI (21)	$NDVIi_{GS}$ (98)
4th	RVI (25)	RVI (32) CI (32)	GNDVI (27) RVI (27)	GVI (22)	CIi (117)
5th	$NDVI_{GS}$ (29)	MTCI (40)	CI (31)	NDVI (33)	RVIi (121)
6th	CI (33)	CCCI (48)	NDVI (37)	RVI (37)	NDVIi (131)
7th	NDVI (42)	RENDVI (54)	$NDVI_{GS}$ (38)	MTCI (38)	MTCIi (147)
8th	MTCI (47)		RENDVI (42)	CCCI (46)	RENDVIi (164)
9th	CCCI (49)			RENDVI (48)	CCCIi (169)

3.4. *Sufficiency Values of Vegetation Indices*

Sufficiency values of each vegetation index for each phenological stage were derived from the calibration equations. Figure 4 shows the dynamics of sufficiency values of each vegetation index throughout the four phenological stages. The sufficiency values of the best performing vegetation indices ranged 0.64–0.76 for GNDVI, and 4.93–7.36 for GVI. The largest difference between sufficiency values, for most of the vegetation indices evaluated, was between vegetative and flowering stages. On average, the relative increase in sufficiency values from the vegetative to the flowering stage was approximately 10% for NDVI, measured both with Crop Circle and GreenSeeker sensors, and for GNDVI, RENDVI, and CI. There were much larger relative increases for RVI and GVI, which were 34% and 25% higher in the flowering stage compared to the vegetative stage. In contrast, the smallest differences in sufficiency values between these two stages were for the CCCI and MTCI indices, with values of approximately 5%. Generally, the sufficiency values for the early fruit growth stage were similar to those for the flowering and harvest stages; except for GVI, CCCI, and MTCI which were on average 13% higher in the early fruit growth stage compared to the flowering stage.

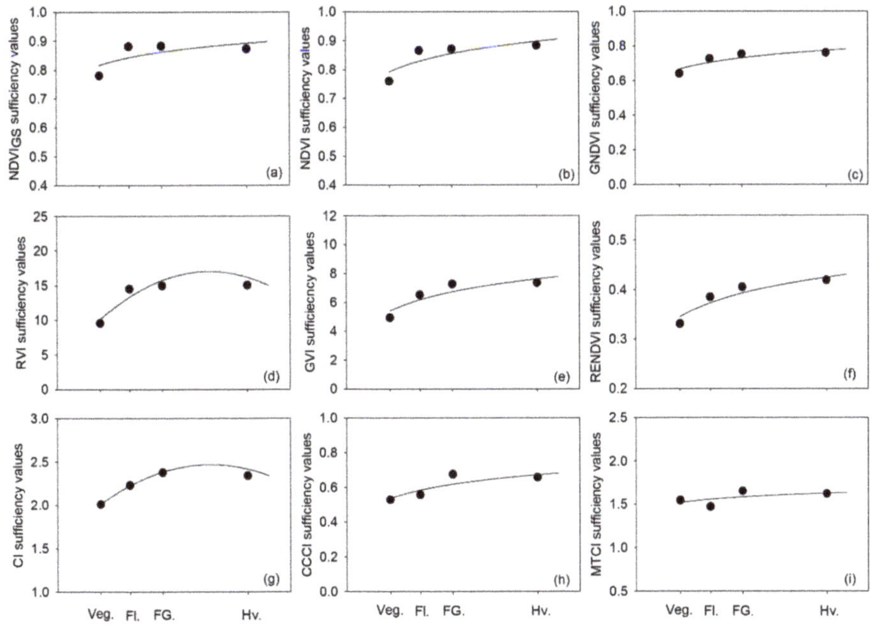

Figure 4. Sufficiency values of integrated vegetation indices calculated for NNI = 1 over different phenological stages: Veg: Vegetative; Fl: Flowering; FG: Early fruit growth; and Hv: Harvest. Abbreviations for vegetation indices are in Table 2. Panel (**a**) shows NDVI measured with GreenSeeker sensor and the other panels (**b–i**) show indices calculated with Crop Circle sensor.

4. Discussion

In sweet pepper, the R^2 and RMSE of linear regressions between vegetation indices and crop NNI were variable between phenological stages throughout the crop and between the eight vegetation indices evaluated. The best performance of vegetation indices for estimation of crop NNIi, in terms of R^2 and RMSE values, was in the early fruit growth stage. Using these criteria, the worst performance for estimating NNIi was in the harvest stage, followed by the vegetative stage, for most of the vegetation indices evaluated. Similar variability of performance of vegetation indices, throughout a crop, was reported by Hatfield and Prueger [50], in maize and soybean. In that research, the relative performance of different vegetation indices for estimating leaf chlorophyll content differed as the growing season progressed. Similarly, Yu et al. [29] found in rice that some red edge-band based vegetation indices had better performance to estimate plant N concentration after the heading stage. In wheat, performance of vegetation indices varied across growth stages, with better results after flowering was reported by Li et al. [51].

According to the analysis of R^2 and RMSE values of linear relationships between vegetation indices and NNIi, for the calibration dataset, vegetation indices based on reflectance of the green band (i.e., GNDVI and GVI) had consistently higher R^2 values and lower RMSE values throughout most of the phenological stages of the crop (average R^2 value of 0.62 in the three first phenological stages). These results indicate that vegetation indices based on reflectance of the green band estimated crop NNIi with more accuracy than the rest of vegetation indices evaluated. In rice, Cao et al. [30] also found that the best indices to estimate NNI, in the different parts of the crop cycle, were mainly green band based indices. In the present study, there was an exception in the early fruit growth stage, where vegetation indices based on reflectance of red edge band (i.e., RENDVI, CI, CCCI, and MTCI) had higher R^2 values (average R^2 value across these four indices of 0.83) than green band based vegetation indices such as GNDVI and GVI. These results are consistent with Yu et al. [29], who found that the red edge based vegetation indices were more sensitive to plant N concentration particularly after the heading stage in rice, whereas green based vegetation indices were more sensitive to plant N concentration in the rest of stages.

Vegetation indices based on reflectance in the green band and in the red edge band are very sensitive to leaf and crop greenness [41,52,53]. They have been preferred over red band based reflectance indices as indicators of crop N status [11,21,25] because of higher sensitivity, particularly of the red edge band at high chlorophyll levels contents [43]. In the present study, vegetation indices based on reflectance in the green band were more sensitive to estimate crop N status than vegetation indices based on the red edge band in all phenological stages, except for the early fruit growth stage when green pepper fruits developed and enlarged. It is possible that the abundance of green tissues in this stage, formed by green pepper fruits and leaves, caused some degree of saturation of the green band but not of the red edge band.

To validate the calibration linear regression equations (derived from the calibration data set) that estimated crop NNIi values from integrated vegetation index measurements, the same calibration linear regression equations were used to estimate NNIi values from the validation data set for each vegetation index. The NNIi values estimated, with this procedure, were compared with the integrated measured NNI values, by linear regression analysis. For most of the vegetation indices evaluated, there was a deficient validation of the regression equations in the vegetative and harvest stages, and more successful validation in the flowering and early fruit growth stages. This interpretation is based on the slope of linear regression between observed and predicted NNIi values, and on the calculated relative error.

In the vegetative and harvest stages, the slopes of linear regression diverged appreciably from one and the relative errors were very different to 0% which would represent perfect validation of regression equations [54–56]. In contrast, in the flowering and early fruit growth stages, all indices had slopes and relative errors closer to 1 and 0%, respectively. The reason for this poor calibration in the vegetative and harvest stages may be associated to characteristics of the crop canopy in these two stages. In the vegetative stage, the plants are small and have low foliage density, which could affect reflectance measurements by the inclusion of background noise [11,15]. Johansen and Tømmervik [57] and Wang et al. [58] reported a lack of precision with NDVI until the plant canopy achieved adequate coverage. In the harvest stage, mottling and discoloration of older leaves, because of crop age, can affect reflectance measurements, as has been reported for cucumber [46].

Validation results (slopes of linear regression between observed and predicted NNIs values, and relative errors) were consistent with the evaluation conducted taking into account results of regression of both calibration and validation datasets [49]. There was a clear tendency for better performance of most vegetation indices in the early fruit growth stage, followed by the flowering stage. The performance of most vegetation indices was worse in the vegetative and harvest stages, most likely due to the characteristics of the crop canopy in these two stages, insufficient foliage density in the vegetative stage, and aging foliage in the harvest stage, as discussed previously.

Analyzing the performance of each vegetation index to estimate NNI within each phenological stage, GNDVI was the best index in three of four phenological stages (vegetative, flowering, and harvest), and GVI was the best performing vegetation index in the other stage (early fruit growth). This is in agreement with Padilla et al. [21] and de Souza et al., [59] who reported these two indices (i.e., GNDVI and GVI) to be more strongly related to NNI in cucumber. Likewise, green vegetation indices of processing tomato were more strongly related to leaf N content than red vegetation indices [60]. Very similar results were also obtained in broccoli [61].

The RENDVI and CCCI were the worst performing indices. The poor performance of CCCI index, in the present study, in the vegetative stage is inconsistent with the results obtained in maize by Li et al. [62]. Li et al. [62] reported that CCCI successfully excluded the effect of soil reflectance when crop cover was low. The different results with CCCI in our study and that of Li et al. [62] may be due to the measurement procedures and the structure of the different crops. In the current study, measurements were made from the side of the crop, and in maize from above. Moreover, the pepper crops in the current study were vertically supported. It is possible that CCCI was influenced by the small areas of empty background, between adjacent pepper plants, that were exposed by vertically supporting the plants.

Comparing the sufficiency values obtained, for maximum dry matter production between the different phenological stages, it was possible to derive a unique sufficiency value for the complete crop cycle for CCCI of 0.61 and for MTCI of 1.57 because the relative differences between phenological stages were small (Figure 4). For the other indices, such as RVI and GVI, the relative difference between the sufficiency value of vegetative and flowering stages were too large (around 30%) to be able to derive a unique sufficiency value for the complete crop cycle. For the rest of the indices evaluated (NDVIGS, NDVI, GNDVI, RENDVI, and CI), the relative difference between vegetative and flowering stages was approximately 10%. In the current work, it was considered not possible to calculate a unique sufficiency value for the whole crop cycle for sweet pepper for the NDVI, GNDVI, RVI, GVI, and RENDVI indices. However, in cucumber, it was possible to calculate a unique sufficiency value for the entire crop cycle because of the relative constancy of sufficiency values throughout the cycle [21]. Overall, the sufficiency values derived for sweet pepper were higher than those derived for cucumber [21], for equivalent indices. This difference may be due to the relatively high chlorophyll content and greenness of sweet pepper crops compared to other vegetable and cereal species [63].

5. Conclusions

The present work evaluated the capacity of different vegetation indices to estimate crop NNI in the vegetative, flowering, early fruit growth, and harvest phenological stages of sweet pepper. There were differences in the performance of the indices within individual phenological stages and between stages. The best performance of all indices was in the early fruit growth and the flowering stages. The best performing indices to assess crop N in sweet pepper were the green band based indices GNDVI and GVI which had the best results for all phenological stages.

Author Contributions: Data curation, R.d.S. and M.T.P.-F.; Funding acquisition, R.B.T., M.G., and F.M.P.; Investigation, R.d.S. and M.T.P.-F.; Methodology, R.d.S., M.T.P.-F., R.B.T., M.G., and F.M.P.; Writing—original draft, R.d.S. and F.M.P.; Writing—review and editing, R.d.S., R.B.T., and F.M.P. All authors have read and agreed to the published version of the manuscript.

Funding: The Spanish Ministry of Economy and Competitiveness (AGL2015-67076-R) provided funding for this research. The Spanish Ministry of Economy and Competitiveness supported RdS and FMP through a Formación de Profesorado Investigador (BES-2016-076706) and a Ramón y Cajal grant (RYC-2014-15815), respectively.

Acknowledgments: We thank the staff and students of the Research Station of the UAL-Anecoop Foundation.

Conflicts of Interest: The authors declare no conflict of interest.

References

1. Hartz, T.K. Vegetable production best management practices to minimize nutrient loss. *Horttechnology* **2006**, *16*, 398–403. [CrossRef]
2. Thompson, R.B.; Martínez-Gaitan, C.; Gallardo, M.; Giménez, C.; Fernández, M.D. Identification of irrigation and N management practices that contribute to nitrate leaching loss from an intensive vegetable production system by use of a comprehensive survey. *Agric. Water Manag.* **2007**, *89*, 261–274. [CrossRef]
3. Thompson, R.B.; Tremblay, N.; Fink, M.; Gallardo, M.; Padilla, F.M. Tools and strategies for sustainable nitrogen fertilisation of vegetable crops. In *Advances in Research on Fertilization Management in Vegetable Crops*; Tei, F., Nicola, S., Benincasa, P., Eds.; Springer: Berlin/Heidelberg, Germany, 2017; pp. 11–63.

4. Congreves, K.A.; Van Eerd, L.L. Nitrogen cycling and management in intensive horticultural systems. *Nutr. Cycl. Agroecosystems* **2015**, *102*, 299–318. [CrossRef]
5. Padilla, F.M.; Gallardo, M.; Manzano-Agugliaro, F. Global trends in nitrate leaching research in the 1960–2017 period. *Sci. Total Environ.* **2018**, *643*, 400–413. [CrossRef] [PubMed]
6. Zotarelli, L.; Dukes, M.D.; Scholberg, J.M.S.; Muñoz-Carpena, R.; Icerman, J. Tomato nitrogen accumulation and fertilizer use efficiency on a sandy soil, as affected by nitrogen rate and irrigation scheduling. *Agric. Water Manag.* **2009**, *96*, 1247–1258. [CrossRef]
7. Ju, X.T.; Kou, C.L.; Zhang, F.S.; Christie, P. Nitrogen balance and groundwater nitrate contamination: Comparison among three intensive cropping systems on the North China Plain. *Environ. Pollut.* **2006**, *143*, 117–125. [CrossRef] [PubMed]
8. Soto, F.; Gallardo, M.; Thompson, R.B.; Peña-Fleitas, M.T.; Padilla, F.M. Consideration of total available N supply reduces N fertilizer requirement and potential for nitrate leaching loss in tomato production. *Agric. Ecosyst. Environ.* **2015**, *200*, 62–70. [CrossRef]
9. Zhu, J.H.; Li, X.L.; Christie, P.; Li, J.L. Environmental implications of low nitrogen use efficiency in excessively fertilized hot pepper (*Capsicum frutescens* L.) cropping systems. *Agric. Ecosyst. Environ.* **2005**, *111*, 70–80. [CrossRef]
10. Meisinger, J.J.; Schepers, J.S.; Raun, W.R. Crop Nitrogen Requirement and Fertilization. *Nitrogen Agric. Syst.* **2008**, 563–612. [CrossRef]
11. Fox, R.H.; Walthall, C.L. Crop monitoring technologies to assess nitrogen status. In *Nitrogen in Agricultural Systems, Agronomy Monograph No. 49*; Schepers, J.S., Raun, W.R., Eds.; American Society of Agronomy: Madison, WI, USA; Crop Science Society of America: Madison, WI, USA; Soil Science Society of America: Madison, WI, USA, 2008; pp. 647–674.
12. Yao, X.; Yao, X.; Jia, W.; Tian, Y.; Ni, J.; Cao, W.; Zhu, Y. Comparison and intercalibration of vegetation indices from different sensors for monitoring above-ground plant nitrogen uptake in winter wheat. *Sensors* **2013**, *13*, 3109–3130. [CrossRef]
13. Mistele, B.; Schmidhalter, U. Estimating the nitrogen nutrition index using spectral canopy reflectance measurements. *Eur. J. Agron.* **2008**, *29*, 184–190. [CrossRef]
14. Hansen, P.M.; Schjoerring, J.K. Reflectance measurement of canopy biomass and nitrogen status in wheat crops using normalized difference vegetation indices and partial least squares regression. *Remote Sens. Environ.* **2003**, *86*, 542–553. [CrossRef]
15. Padilla, F.M.; Gallardo, M.; Peña-Fleitas, M.T.; de Souza, R.; Thompson, R.B. Proximal Optical Sensors for Nitrogen Management of Vegetable Crops: A Review. *Sensors* **2018**, *18*, 2083. [CrossRef] [PubMed]
16. Samborski, S.M.; Tremblay, N.; Fallon, E. Strategies to make use of plant sensors-based diagnostic information for nitrogen recommendations. *Agron. J.* **2009**, *101*, 800–816. [CrossRef]
17. Cammarano, D.; Fitzgerald, G.J.; Casa, R.; Basso, B. Assessing the robustness of vegetation indices to estimate wheat N in mediterranean environments. *Remote Sens.* **2014**, *6*, 2827–2844. [CrossRef]
18. Basso, B.; Fiorentino, C.; Cammarano, D.; Schulthess, U. Variable rate nitrogen fertilizer response in wheat using remote sensing. *Precis. Agric.* **2015**, *17*, 168–182. [CrossRef]
19. Bannari, A.; Morin, D.; Bonn, F.; Huete, A.R. A review of vegetation indices. *Remote Sens. Rev.* **1995**, *13*, 95–120. [CrossRef]
20. Schepers, J.S.; Blackmer, T.M.; Wilhelm, W.W.; Resende, M. Transmittance and reflectance measurements of corn leaves from plants with different nitrogen and water supply. *J. Plant Physiol.* **1996**, *148*, 523–529. [CrossRef]
21. Padilla, F.M.; Peña-Fleitas, M.T.; Gallardo, M.; Thompson, R.B. Determination of sufficiency values of canopy reflectance vegetation indices for maximum growth and yield of cucumber. *Eur. J. Agron.* **2017**, *84*. [CrossRef]
22. Greenwood, D.J.; Gastal, F.; Lemaire, G.; Draycott, A.; Millard, P.; Neeteson, J.J. Growth rate and % N of field grown crops: Theory and experiments. *Ann. Bot.* **1991**, *67*, 181–190. [CrossRef]
23. Lemaire, G.; Jeuffroy, M.H.; Gastal, F. Diagnosis tool for plant and crop N status in vegetative stage. Theory and practices for crop N management. *Eur. J. Agron.* **2008**, *28*, 614–624. [CrossRef]
24. Lemaire, G.; Gastal, F.N. Uptake and Distribution in Plant Canopies. In *Diagnosis of the Nitrogen Status in Crops*; Springer: Berlin/Heidelberg, Germany, 1997; pp. 3–43. ISBN 978-3-642-64506-8.
25. Hatfield, J.L.; Gitelson, A.A.; Schepers, J.S.; Walthall, C.L. Application of spectral remote sensing for agronomic decisions. *Agron. J.* **2008**, *100*, S117–S131. [CrossRef]

26. Haboudane, D.; Miller, J.R.; Pattey, E.; Zarco-Tejada, P.J.; Strachan, I.B. Hyperspectral vegetation indices and novel algorithms for predicting green LAI of crop canopies: Modeling and validation in the context of precision agriculture. *Remote Sens. Environ.* **2004**, *90*, 337–352. [CrossRef]
27. Chen, P. A comparison of two approaches for estimating the wheat nitrogen nutrition index using remote sensing. *Remote Sens.* **2015**, *7*, 4527–4548. [CrossRef]
28. Fitzgerald, G.; Rodriguez, D.; O'Leary, G. Measuring and predicting canopy nitrogen nutrition in wheat using a spectral index-The canopy chlorophyll content index (CCCI). *Field Crops Res.* **2010**, *116*, 318–324. [CrossRef]
29. Yu, K.; Li, F.; Gnyp, M.L.; Miao, Y.; Bareth, G.; Chen, X. Remotely detecting canopy nitrogen concentration and uptake of paddy rice in the Northeast China Plain. *ISPRS J. Photogramm. Remote Sens.* **2013**, *78*, 102–115. [CrossRef]
30. Cao, Q.; Miao, Y.; Wang, H.; Huang, S.; Cheng, S.; Khosla, R.; Jiang, R. Non-destructive estimation of rice plant nitrogen status with Crop Circle multispectral active canopy sensor. *Field Crops Res.* **2013**, *154*, 133–144. [CrossRef]
31. Chen, P.; Haboudane, D.; Tremblay, N.; Wang, J.; Vigneault, P.; Li, B. New spectral indicator assessing the efficiency of crop nitrogen treatment in corn and wheat. *Remote Sens. Environ.* **2010**, *114*, 1987–1997. [CrossRef]
32. Pulido-Bosch, A.; Bensi, S.; Molina, L.; Vallejos, A.; Calaforra, J.M.; Pulido-Leboeuf, P. Nitrates as indicators of aquifer interconnection. Application to the Campo de Dalias (SE—Spain). *Environ. Geol.* **2000**, *39*, 791–799. [CrossRef]
33. Cui, M.; Sun, X.; Hu, C.; Di, H.J.; Tan, Q.; Zhao, C. Effective mitigation of nitrate leaching and nitrous oxide emissions in intensive vegetable production systems using a nitrification inhibitor, dicyandiamide. *J. Soils Sediments* **2011**, *11*, 722–730. [CrossRef]
34. Boulard, T.; Raeppel, C.; Brun, R.; Lecompte, F.; Hayer, F.; Carmassi, G.; Gaillard, G. Environmental impact of greenhouse tomato production in France. *Agron. Sustain. Dev.* **2011**, *31*, 757–777. [CrossRef]
35. Abd-El-Baky, H.M.; Ali, S.A.; El-Haddad, Z.; El-Ansary, Z. Some Environmental Parameters Affecting Sweet Pepper Growth and Productivity Under Different Greenhouse Forms in Hot and Humid Climatic Conditions. *J. Soil Sci. Agric. Eng.* **2010**, *1*, 225–247.
36. Junta de Andalucía Cartografía de Invernaderos en el Litorall de Andalucía Oriental. Available online: https://www.juntadeandalucia.es/export/drupaljda/estudios_informes/16/12/CartografiainvernaderosenellitoraldeAndaluíaOriental_v161201.pdf (accessed on 26 November 2019).
37. Jadoski, S.; Thompson, R.B.; Peña-Fleitas, M.T.; Gallardo, M. Regional N Balance for an Intensive Vegetable Production System in South-Eastern Spain. In Proceedings of the Abstracts of Nev 2013 International Workshop on Nitrogen, Environment and Vegetables, Turin, Italy; 2013; pp. 50–51.
38. Valera, D.L.; Belmonte, L.J.; Molina-Aiz, F.D.; López, A.; Camacho, F. The greenhouses of Almería, Spain: Technological analysis and profitability. *Acta Hortic.* **2017**, *1170*, 219–226. [CrossRef]
39. Padilla, F.M.; Peña-Fleitas, M.T.; Gallardo, M.; Thompson, R.B. Evaluation of optical sensor measurements of canopy reflectance and of leaf flavonols and chlorophyll contents to assess crop nitrogen status of muskmelon. *Eur. J. Agron.* **2014**, *58*, 39–52. [CrossRef]
40. Sellers, P.J. Canopy reflectance, photosynthesis and transpiration. *Int. J. Remote Sens.* **1985**, *6*, 1335–1372. [CrossRef]
41. Ma, B.L.; Morrison, M.J.; Dwyer, L.M. Canopy light reflectance and field greenness to assess nitrogen fertilization and yield of maize. *Agron. J.* **1996**, *88*, 915–920. [CrossRef]
42. Birth, G.S.; McVey, G.R. Measuring the color of growing turf with a reflectance spectrophotometer. *Agron. J.* **1968**, *60*, 640–643. [CrossRef]
43. Gitelson, A.; Merzlyak, M.N. Spectral reflectance changes associated with autumn senescence of Aesculus hippocastanum L. and Acer platanoides L. leaves. Spectral features and relation to chlorophyll estimation. *J. Plant Physiol.* **1994**, *143*, 286–292. [CrossRef]
44. Gitelson, A.A.; Gritz, Y.; Merzlyak, M.N. Relationships between leaf chlorophyll content and spectral reflectance and algorithms for non-destructive chlorophyll assessment in higher plant leaves. *J. Plant Physiol.* **2003**, *160*, 271–282. [CrossRef]
45. Dash, J.; Curran, P.J. The MERIS terrestrial chlorophyll index. *Int. J. Remote Sens.* **2004**, *25*, 5403–5413. [CrossRef]

46. Padilla, F.M.; Peña-Fleitas, M.T.; Gallardo, M.; Giménez, C.; Thompson, R.B. Derivation of sufficiency values of a chlorophyll meter to estimate cucumber nitrogen status and yield. *Comput. Electron. Agric.* **2017**, *141*, 54–64. [CrossRef]
47. De Souza, R.; Peña-fleitas, M.T.; Thompson, R.B.; Gallardo, M.; Grasso, R.; Padilla, F.M. The Use of Chlorophyll Meters to Assess Crop N Status and Derivation of Sufficiency Values for Sweet Pepper. *Sensors* **2019**, *19*, 2949. [CrossRef] [PubMed]
48. Zhao, B.; Ata-Ul-Karim, S.T.; Liu, Z.; Zhang, J.; Xiao, J.; Liu, Z.; Qin, A.; Ning, D.; Yang, Q.; Zhang, Y.; et al. Simple Assessment of Nitrogen Nutrition Index in Summer Maize by Using Chlorophyll Meter Readings. *Front. Plant Sci.* **2018**, *9*, 11. [CrossRef] [PubMed]
49. Xin-feng, Y.; Xia, Y.; Yong-chao, T.; Jun, N.I.; Xiao-jun, L.; Wei-xing, C.; Yan, Z. A New Method to Determine Central Wavelength and Optimal Bandwidth for Predicting Plant Nitrogen Uptake in Winter Wheat. *J. Integr. Agric.* **2013**, *12*, 788–802. [CrossRef]
50. Hatfield, J.L.; Prueger, J.H. Value of using different vegetative indices to quantify agricultural crop characteristics at different growth stages under varying management practices. *Remote Sens.* **2010**, *2*, 562–578. [CrossRef]
51. Li, F.; Mistele, B.; Hu, Y.; Chen, X.; Schmidhalter, U. Reflectance estimation of canopy nitrogen content in winter wheat using optimised hyperspectral spectral indices and partial least squares regression. *Eur. J. Agron.* **2014**, *52*, 198–209. [CrossRef]
52. Daughtry, C.S.T.; Walthall, C.L.; Kim, M.S.; de Colstoun, E.B.; McMurtrey, J.E. Estimating corn leaf chlorophyll concentration from leaf and canopy reflectance. *Remote Sens. Environ.* **2000**, *74*, 229–239. [CrossRef]
53. Raper, T.B.; Varco, J.J. Canopy-scale wavelength and vegetative index sensitivities to cotton growth parameters and nitrogen status. *Precis. Agric.* **2015**, *16*, 62–76. [CrossRef]
54. Yang, J.M.; Yang, J.Y.; Liu, S.; Hoogenboom, G. An evaluation of the statistical methods for testing the performance of crop models with observed data. *Agric. Syst.* **2014**, *127*, 81–89. [CrossRef]
55. Gallardo, M.; Thompson, R.B.; Giménez, C.; Padilla, F.M.; Stöckle, C.O. Prototype decision support system based on the VegSyst simulation model to calculate crop N and water requirements for tomato under plastic cover. *Irrig. Sci.* **2014**, *32*, 237–253. [CrossRef]
56. Piñeiro, G.; Perelman, S.; Guerschman, J.P.; Paruelo, J.M. How to evaluate models: Observed vs. predicted or predicted vs. observed? *Ecol. Modell.* **2008**, *216*, 316–322. [CrossRef]
57. Johansen, B.; Tømmervik, H. The relationship between phytomass, NDVI and vegetationcommunities on Svalbard. *Int. J. Appl. Earth Obs. Geoinf.* **2014**, *27*, 20–30. [CrossRef]
58. Wang, Y.W.; Mao, P.S.; Dunn, B.L.; Arnall, D.B. Use of an active canopy sensor and SPAD chlorophyll meter to quantify geranium nitrogen status. *HortScience* **2012**, *47*, 45–50. [CrossRef]
59. De Souza, R.; Grasso, R.; Teresa Peña-Fleitas, M.; Gallardo, M.; Thompson, R.B.; Padilla, F.M. Effect of Cultivar on Chlorophyll Meter and Canopy Reflectance Measurements in Cucumber. *Sensors* **2020**, *20*, 509. [CrossRef]
60. Gianquinto, G.; Orsini, F.; Fecondini, M.; Mezzetti, M.; Sambo, P.; Bona, S. A methodological approach for defining spectral indices for assessing tomato nitrogen status and yield. *Eur. J. Agron.* **2011**, *35*, 135–143. [CrossRef]
61. El-Shikha, D.M.; Waller, P.; Hunsaker, D.; Clarke, T.; Barnes, E. Ground-based remote sensing for assessing water and nitrogen status of broccoli. *Agric. Water Manag.* **2007**, *92*, 183–193. [CrossRef]
62. Li, F.; Miao, Y.; Feng, G.; Yuan, F.; Yue, S.; Gao, X.; Liu, Y.; Liu, B.; Ustin, S.; Susan, L.; et al. Improving estimation of summer maize nitrogen status with red edge-based spectral vegetation indices. *Field Crops Res.* **2014**, *157*, 111–123. [CrossRef]
63. Parry, C.; Blonquist, J.M.; Bugbee, B. In situ measurement of leaf chlorophyll concentration: Analysis of the optical/absolute relationship. *Plant. Cell Environ.* **2014**, *37*, 2508–2520. [CrossRef]

remote sensing

Article

Corn Nitrogen Status Diagnosis with an Innovative Multi-Parameter Crop Circle Phenom Sensing System

Cadan Cummings, Yuxin Miao *, Gabriel Dias Paiao, Shujiang Kang and Fabián G. Fernández

Precision Agriculture Center, Department of Soil, Water and Climate, University of Minnesota, Saint Paul, MN 55108, USA; cummi428@umn.edu (C.C.); gdiaspai@umn.edu (G.D.P.); skang2010apr@gmail.com (S.K.); fabiangf@umn.edu (F.G.F.)
* Correspondence: ymiao@umn.edu

Abstract: Accurate and non-destructive in-season crop nitrogen (N) status diagnosis is important for the success of precision N management (PNM). Several active canopy sensors (ACS) with two or three spectral wavebands have been used for this purpose. The Crop Circle Phenom sensor is a new integrated multi-parameter proximal ACS system for in-field plant phenomics with the capability to measure reflectance, structural, and climatic attributes. The objective of this study was to evaluate this multi-parameter Crop Circle Phenom sensing system for in-season diagnosis of corn (*Zea mays* L.) N status across different soil drainage and tillage systems under variable N supply conditions. The four plant metrics used to approximate in-season N status consist of aboveground biomass (AGB), plant N concentration (PNC), plant N uptake (PNU), and N nutrition index (NNI). A field experiment was conducted in Wells, Minnesota during the 2018 and the 2019 growing seasons with a split-split plot design replicated four times with soil drainage (drained and undrained) as main block, tillage (conventional, no-till, and strip-till) as split plot, and pre-plant N (PPN) rate (0 to 225 in 45 kg ha^{-1} increment) as the split-split plot. Crop Circle Phenom measurements alongside destructive whole plant samples were collected at V8 +/−1 growth stage. Proximal sensor metrics were used to construct regression models to estimate N status indicators using simple regression (SR) and eXtreme Gradient Boosting (XGB) models. The sensor derived indices tested included normalized difference vegetation index (NDVI), normalized difference red edge (NDRE), estimated canopy chlorophyll content (eCCC), estimated leaf area index (eLAI), ratio vegetation index (RVI), canopy chlorophyll content index (CCCI), fractional photosynthetically active radiation (fPAR), and canopy and air temperature difference (ΔTemp). Management practices such as drainage, tillage, and PPN rate were also included to determine the potential improvement in corn N status diagnosis. Three of the four replicated drained and undrained blocks were randomly selected as training data, and the remaining drained and undrained blocks were used as testing data. The results indicated that SR modeling using NDVI would be sufficient for estimating AGB compared to more complex machine learning methods. Conversely, PNC, PNU, and NNI all benefitted from XGB modeling based on multiple inputs. Among different approaches of XGB modeling, combining management information and Crop Circle Phenom measurements together increased model performance for predicting each of the four plant N metrics compared with solely using sensing data. The PPN rate was the most important management metric for all models compared to drainage and tillage information. Combining Crop Circle Phenom sensor parameters and management information is a promising strategy for in-season diagnosis of corn N status. More studies are needed to further evaluate this new integrated sensing system under diverse on-farm conditions and to test other machine learning models.

Keywords: precision nitrogen management; active canopy sensing; integrated sensing system; machine learning; nitrogen nutrition index

Citation: Cummings, C.; Miao, Y.; Paiao, G.D.; Kang, S.; Fernández, F.G. Corn Nitrogen Status Diagnosis with an Innovative Multi-Parameter Crop Circle Phenom Sensing System. *Remote Sens.* **2021**, *13*, 401. https://doi.org/10.3390/rs13030401

Received: 31 December 2020
Accepted: 19 January 2021
Published: 24 January 2021

1. Introduction

Agricultural nutrient management has been historically guided using grower knowledge of cultivated land and soil supply of essential nutrients such as nitrogen (N), phosphorous, and potassium. Over the past forty years, the development of precision agriculture has offered an alternative method of guiding nutrient management leveraged on using proximal and remote sensing, data analysis, and smart machinery to optimize fertilizer application timing and rate to match nutrient supply with crop demand [1]. Within commercial crop production, N is frequently the primary limiting nutrient for plant growth [2]. Limiting conditions are often attributable to N mobility within the soil horizon and susceptibility for losses through leaching, denitrification, and volatilization processes [3,4]. Improving N management is critical to protection of water resources and reduction of atmospheric greenhouse gas levels [5]. Centered on matching N supply with crop N demand in both space and time, precision N management (PNM) has the potential to increase N use efficiency by reducing N losses while maintaining crop yields [6,7].

For corn (*Zea mays* L.) production, N fertilizer timing and rate are critical aspects to mitigating N loss [8]. Physiologically, corn plant N concentration (PNC) is highest earlier in its vegetative growth and decreases until plant senescence; however, plant N demand is greatest midway through the growing season when the plant is rapidly increasing in biomass. Historically, N fertilizer is applied in full around the time of planting with the expectation that sufficient N will persist throughout the season to facilitate optimal plant growth. This practice is viable for growing seasons with low early season N loss and ideal weather conditions. However, it is not conductive for field seasons with high N loss potential from heavy or frequent rain events. For this reason, optimal in-season N management must develop tools which determine plant N status accurately and non-destructively [9]. Corn plants predominantly exhibit N deficiency symptoms of stunting due to decreased cell division and leaf chlorosis of older leaves [10]. Plénet and Lemaire [11] established an empirical allometric critical N dilution curve, which calculates the minimum PNC needed to optimally grow as "critical" N concentration (Nc) depending on aboveground biomass (AGB). Corn N status can be determined by calculating N nutrition index (NNI), which is defined as the ratio of actual PNC to Nc. Since the development of corn NNI, subsequent studies have evaluated its efficacy and utilized it as a tool to improve corn N status diagnosis and to guide side-dress N application [12–15].

To apply NNI in commercial agriculture, there are several methods to determine corn AGB and PNC. Traditional destructive sampling and analysis is not only time consuming and expensive but also cannot adequately capture spatial or temporal variability because it is a snapshot of crop health at a specific location and day of year [16,17]. As a result, proximal and remote sensing technologies have been developed for real-time non-destructive N status estimation. Canopy sensors are more efficient than destructive sampling because they can be quickly collected and return instantaneous estimations of plant health. Additionally, active instruments are superior and more repeatable compared to passive sensors because their measurements are independent of environmental light conditions.

Three of the most frequently utilized active canopy sensors (ACS) for corn N management are the two-band GreenSeeker (Trimble Inc., Sunnyvale, CA, USA), the three-band Crop Circle ACS-430 (Holland Scientific, Lincoln, NE, USA), and the three-band RapidScan CS-45 (Holland Scientific, Lincoln, NE, USA). Researchers have developed empirical techniques to estimate in-season N status through correlating multispectral band reflectance measurements or calculated vegetative indices (VIs) with crop N status indicators. Xia et al. [13] used a GreenSeeker sensor to predict corn NNI and found the sensor derived VIs could moderately predict NNI directly (R^2 between 0.56–0.65) at V7–V10 growth stage when used with N-rich plots as reference to calculate response index. However, the GreenSeeker sensor did not perform well when solely using VIs to predict NNI (R^2 between 0.33–0.55) without using N rich plots. Paiao et al. [18] evaluated GreenSeeker and RapidSCAN sensors for corn plant N status estimation from V1 to R1 in Minnesota. The study found that optimum N rates did not correlate well with proximal sensor measurements prior to V12 stage, which could limit their values for determining side-dress N needs around V8–V9 stages.

The Crop Circle Phenom is a new integrated multi-parameter ACS, which measures spectral reflectance of red, red-edge, and near-infrared wavelengths to calculate normalized difference vegetation index (NDVI) [19] and the normalized difference red edge (NDRE) [20] as well as to provide several other variables, including estimated canopy chlorophyll content (eCCC), estimated leaf area index (eLAI), atmospheric pressure, relative humidity, reflected and incoming photosynthetically active radiation (PAR), and canopy and air temperatures. These additional metrics can be used to calculate physiological metrics such as fractional PAR (fPAR) and canopy-air temperature difference (ΔTemp). Previous research indicated that PAR [21] and canopy temperature [22] could be used to estimate biomass and crop N stress. Therefore, through measuring spectral, estimated structural characteristics, and climatic variables, the Crop Circle Phenom sensor system is hypothesized to be able to improve corn N status estimation and diagnosis compared to only using vegetation indices such as NDVI and NDRE. To date, no study has been reported for the evaluation of this new integrated sensor system for in-season corn N status estimation. Therefore, the objective of this research was to evaluate the potential of the Crop Circle Phenom sensor system for in-season diagnosis of corn N status across different drainage and tillage systems under variable N supply conditions.

2. Materials and Methods

2.1. Study Site

The study was conducted in southcentral Minnesota near Wells, MN (43°51′15.7″ N 93°43′47.2″ W) in the 2018 and the 2019 growing seasons. The predominant soil types at the site are Marna silty clay loam (fine, smectitic, mesic Vertic Endoaquolls) and Nicollet silty-clay loam (fine-loamy, mixed, superactive, mesic Aquic Hapludolls). The experiment was conducted in a randomized complete-block design with a split-splitplot arrangement and four replications in a corn-soybean (Glycine max L.) rotation where both crops are present every year. The main plot was set up in 2011 with subsurface tile drainage where half of the blocks are fully closed (undrained) and the other half are fully open (drained). For more details, see Fernández et al. [23]. The sub-plot includes three tillage treatments established in 2017: no-tillage (NT), strip-tillage (ST), and conventional-tillage (CT). The sub-sub-plot is six pre-plant N (PPN) rate treatments (0, 45, 90, 135, 180, 225 kg-N ha^{-1}) initiated in 2017 (Figure 1). The trials are part of a larger experiment with N timing also being evaluated at various early growth stages, but only the PPN treatments were used for this project. Each treatment plot was composed of four planted rows approximately nine meters in length with 76 cm row spacing and approximately 83,000 plants ha^{-1} density. Nitrogen was applied as urea+Agrotain (46-0-0) (urea with N-(n-butyl) thiophosphoric triamide (NBPT)) (Koch Fertilizer LLC, Wichita, KS, USA) in mid-May within a week of planting the crop. The Pioneer hybrid P9929AMXT was used in this study.

2.2. Proximal Sensor Collection

Proximal sensing data were collected around V8 growth stages in the 2018 and the 2019 growing seasons using a Crop Circle Phenom canopy sensor (Holland Scientific, Lincoln, NE, USA). This sensor fuses the instrument capabilities of a Crop Circle ACS-430 and a Crop Circle DAS43X sensor using a GeoScout X controller, which simultaneously geotags and timestamps each unique measurement (Figure 2). Analogous to prior studies which have utilized the Crop Circle ACS-430, the Phenom sensor collects reflectance data in red (670 nm), red-edge (RE, 730 nm), and near-infrared (NIR, 780 nm) wavelengths as well as automatically calculated NDVI and NDRE. Furthermore, the Crop Circle Phenom sensor system also calculates eLAI and eCCC using empirical relationships with spectral bands. In addition to spectral data, this sensor system collects environmental information from a DAS43X sensor that measures atmospheric pressure, relative humidity, incoming and reflected PAR, canopy temperature, and air temperature. Supplemental vegetation indices were selected based on their previously published ability to approximate plant N metrics, including canopy chlorophyll content index (CCCI) and ratio vegetation index (RVI). Canopy and air temperature difference (ΔTemp) and fPAR were also calculated (Table 1).

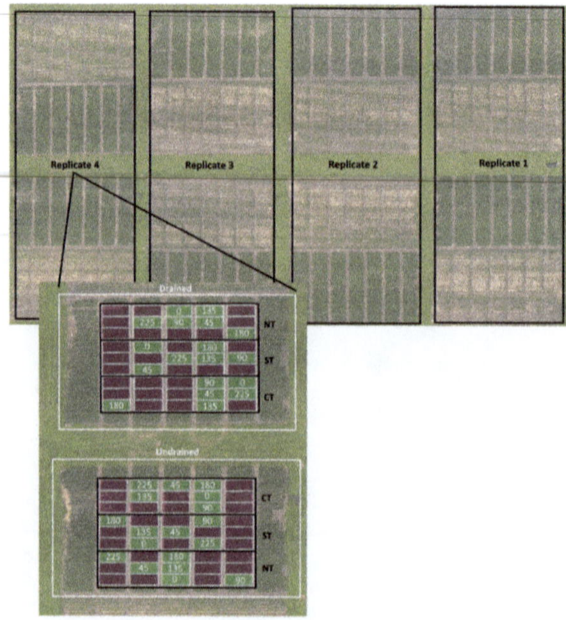

Figure 1. Wells research site experimental design with four replicates of block resolution drainage treatments and sub-plot tillage and sub-sub plot pre-plant N treatments. Green plots signify pre-plant N treatments while purple plots are timing treatments outside the realm of this study. NT, ST, and CT stand for no-till, strip-tillage, and conventional-tillage, respectively. The numbers for the pre-plant N treatment plots indicate the N rates (kg ha^{-1}).

Table 1. List of sensor parameters calculated using the Crop Circle Phenom.

Vegetation Index	Abbreviation	Formula	Reference
Normalized Difference Vegetation Index	NDVI	$\frac{(NIR-RED)}{(NIR+RED)}$	[19]
Normalized Difference Red Edge	NDRE	$\frac{(NIR-RE)}{(NIR+RE)}$	[20]
Estimated Canopy Chlorophyll Content	eCCC	$\frac{(a*NIR-b*RE)}{(c*RE-d*R)}$ *where a, b, c, d are scaling constants*	[24]
Estimated Leaf Area Index	eLAI	$k*\ln(1-NDVI)$ *where k is a scaling constant*	[25]
Ratio Vegetation Index	RVI	$\frac{NIR}{R}$	[26]
Canopy Chlorophyll Content Index	CCCI	$\frac{(NDRE)}{(NDVI)}$	[27]
Delta Temperature	ΔTemp	Canopy Temp (C)—Air Temp (C)	[28]
Fractional Photosynthetically Active Radiation	fPAR	$\frac{Reflected\ PAR}{Incoming\ PAR}$	[28]

The Crop Circle Phenom system was fitted to a custom mount and handle to enable the user to hold the sensor level at nadir approximately 30 cm above the canopy and approximately a meter ahead of the operator to avoid casting a shadow on the area of interest. Two measurements were collected in each plot from the center two treatment rows, and the readings were averaged to represent each plot. The sensor metadata provide estimated distance between sensor and canopy derived from the spectral band observations and the inverse square law. The estimated distance to canopy occasionally varied within plot, and rapidly changing sensor readings (>50 cm) were removed.

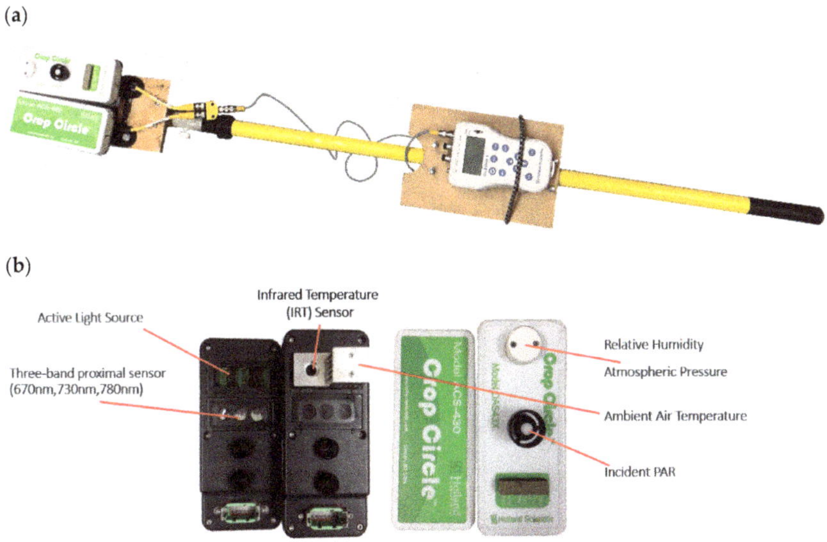

Figure 2. Crop Circle Phenom sensor (**a**) custom assembly with extendable pole and (**b**) close up view of ACS-430 and DAS43X sensor components.

2.3. Plant Sampling and Analysis

Following sensor measurements, six whole plant samples were collected at V8 growth stage, oven-dried at 60 °C to a constant weight, weighed for AGB determination, ground, and analyzed for total N by combustion [29]. Stand count measurements were collected from each plot around V8 growth stage from 12.2 m of crop rows from the two center rows. Total dried AGB was calculated using plot stand counts and average dried biomass weight per plant for each plot. Furthermore, PNU (kg ha^{-1}) was calculated using AGB and PNC. Plant N status was evaluated by calculating N_c and NNI using the critical N dilution curve developed by Plénet and Lemaire [11] (Equations (1) and (2)). The allometric function estimates Nc at different dried AGB weight (W). The authors observed the relationship was best utilized between 1 Mg ha^{-1} and 22 Mg ha^{-1} but recommend a constant N_c of 3.4% be applied under 1 Mg ha^{-1} dried AGB.

$$N_c = 3.4 * W^{-0.37} \tag{1}$$

$$NNI = \frac{PNC}{N_c} \tag{2}$$

2.4. Data Analysis

The dataset consisted of 275 unique plot observations representing the 2018 and the 2019 growing seasons across drainage, tillage, and PPN treatment variables. A handful of plots (n = 13) were accidentally not collected or were removed due to irregular sensor readings, which reduced the measurement count from the overall 288 unique plots. Training and testing datasets were produced by randomly selecting three of the four drained and undrained experimental blocks as a training dataset (n = 208) and using the remaining block data as a testing dataset (n = 67). This methodology was selected to maintain an approximately equal distribution of drainage, tillage, and PPN treatments in both the training and the testing datasets. Using Crop Circle Phenom derived spectral and climatic parameters and destructively sampled corn N indicators, simple regression (SR) and eXtreme Gradient Boosting (XGB) machine learning-based approaches were investigated to predict AGB, PNC, PNU, and NNI. The training dataset was used to fit each of the SR and the XGR regression models, while the testing dataset was solely utilized to validate the final performance of each of the models.

Each of the selected Crop Circle Phenom measured parameters was individually evaluated for predicting AGB, PNC, PNU, and NNI using the SciPy curve_fit Python function [30]. In addition to fitting linear models, exponential, power, and quadratic models were also evaluated and compared to create best fit for each sensor metric. The model with the lowest training mean absolute error (MAE) and root mean square error (RMSE) calculated using scikit-learn package [31] was selected as the optimal model.

To evaluate the benefit of fusing multiple sensor parameters alongside management data, XGB regression models were constructed and compared to SR. Drainage and tillage treatments were hypothesized to influence in-season N status, yet neither could be easily included in SR modeling. The XGB machine learning package was examined to allow categorical variables to be evaluated in conjunction with the quantitative proximal sensor data. Three distinct levels of input variables were investigated for XGB modeling, which consisted of (1) default vegetation indices of NDVI and NDRE automatically calculated by the Crop Circle Phenom sensor system, (2) NDVI and NDRE plus additional Crop Circle Phenom collected variables, and (3) Crop Circle Phenom sensor data plus management information (drainage, tillage, and PPN).

The XGB regression model was adopted as a machine learning strategy to improve plant N status prediction due to its ease of use and ability to be tuned towards small datasets to avoid overfitting through altering the hyperparameter inputs [32]. This valuable characteristic is primarily due to its ability to be tuned for learning rate and size of decision trees.

Machine learning models to predict N status variables were constructed using the Python package XGBoost Regressor [32]. Tuning the machine learning hyperparameters was performed using the XGBoost built-in cross-validation function, which was only utilized within the training dataset. To perform cross-validation for each plant growth parameter, a Python function was constructed, which utilized three k-folds within the training dataset to test various max depth, minimum child weights, and learning rates. Hyperparameter tuning is critical to machine learning model performance because they together govern the performance of the model through minimizing overall loss versus risk of model overfitting [33]. Since a tree based XGBoost model is used, max depth and minimum child weight decide depth of tree and number of samples per node, respectively, whereas learning rate controls how successive trees weigh input features (Figure 3). To avoid overfitting training data during tuning, an early stopping parameter was used to halt subsequent boosting rounds after five iterations where MAE did not improve. The parameter set that returned the lowest MAE was used as the starting parameters for the XGB regression model.

The model performance was evaluated using mean absolute error (MAE) and root mean squared error (RMSE) (Equations (3) and (4)) alongside coefficient of determination (R^2). Both error metrics calculate the average difference between predicted and observed variables where n is the number of measurements, y_i is the i-th observed measurement, and $\hat{y}_i$ is the corresponding predicted measurement.

$$MAE = \frac{1}{n}\sum_{i=1}^{n}|y_i - \hat{y}_i| \tag{3}$$

$$RMSE = \sqrt{\frac{1}{n}\sum_{i=1}^{n}(y_i - \hat{y}_i)^2} \quad (4)$$

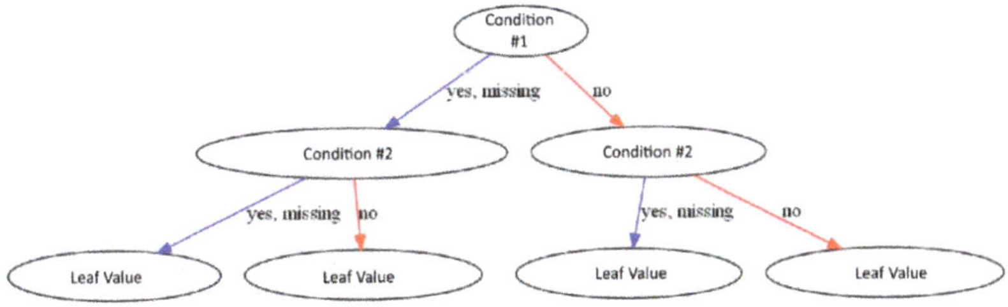

Figure 3. Example eXtreme Gradient Boosting (XGB) XGBoost regression tree composed of a series of conditional statements that test each observation with successive branches and leaf nodes deciding the predicted split value of a target variable.

2.5. Corn N Status Diagnosis

The NNI values were used to diagnose corn N status using the following threshold values: NNI < 0.95, 0.95 ≤ NNI ≤ 1.05, NNI > 1.05 for deficient, optimum, and surplus N status, respectively [13,34]. Using measured and predicted NNI values by SR and XGB models from the test dataset, the accuracy of corn N status diagnosis was evaluated using areal agreement and kappa statistics [13,35]. The areal agreement is the percentage of predicted and measured diagnostic results sharing a common classification, while kappa statistics is a more robust indicator of the agreement of the two diagnostic results that is adjusted for random chance classification [36]. The kappa statistics values < 0.4, 0.4–0.6, and > 0.6 indicate weak, moderate, and strong agreement [37].

3. Results

3.1. Corn N Status Indicator Variability

Across the experiment treatments and two site years, PNU demonstrated the greatest amount of variability (coefficient of variation (CV) around 40%) with a range of 3.95 to 101.68 kg ha^{-1} (Table 2). NNI fluctuated comparably less between 0.34 and 1.40 with a CV around 30%. The PNC and the AGB statistics show similar variability, with CV of 26–27%. Random selection of three of the four drainage replicates into training data and one drainage replicate block into testing data resulted in comparable statistics to construct and validate N status models. The large variabilities in N status indicators (CV = 25.96–40.11%) indicated the suitability of the datasets for evaluating the Crop Circle Phenom sensor system.

Table 2. Descriptive statistics of aboveground biomass (AGB), plant N concentration (PNC), plant N uptake (PNU), and N nutrition index (NNI) at V7–V8 growth stage for training and testing datasets across drainage, tillage, N treatments, and site years.

	Training Set (n = 208)				Testing Set (n = 67)			
	Max	Min	Mean	CV(%)	Max	Min	Mean	CV(%)
AGB (Mg ha^{-1})	3.27	0.59	2.03	26.89	2.95	0.85	1.88	25.96
PNC (g kg^{-1})	3.86	0.95	2.48	26.54	3.68	1.14	2.45	27.04
PNU (kg ha^{-1})	101.68	3.95	51.79	39.27	86.98	7.79	46.82	40.11
NNI	1.38	0.28	0.95	29.34	1.40	0.34	0.91	30.22

3.2. Crop Circle Phenom Sensor Inter-Parameter Correlation

Several of the Crop Circle Phenom sensor parameter combinations were strongly related (Figure 4). One such example of a strongly correlated parameter paring was eCCC and eLAI with a nearly linear relationship (R^2 = 1) (Figure 4). Overall correlations between spectral sensor metrics were moderate to strong (R^2 = 0.70–0.98), whereas environmental temperature and PAR metrics were less correlated (R^2 = 0.12–0.46).

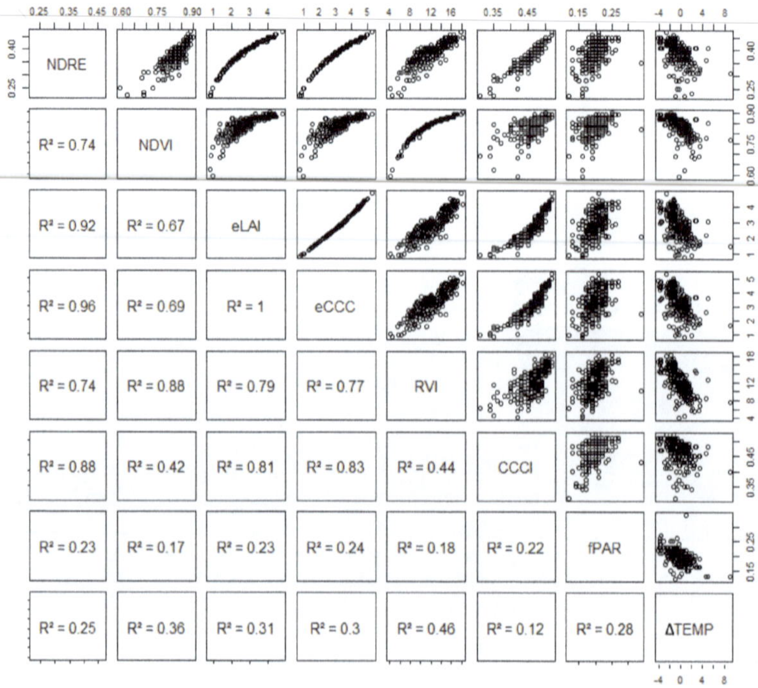

Figure 4. Correlation pairs between the Crop Circle Phenom metrics. Upper panel displays the relationship between sensor metrics and lower corner panel shows coefficient of determination (R^2).

3.3. *Simple Regression Analysis*

Simple regression models showed significant variation in prediction ability between the Crop Circle Phenom parameters and the four plant N status indicators (Table 3). Across the sensor parameters, NDVI (MAE = 0.23 Mg ha^{-1}), NDRE (MAE = 0.24 Mg ha^{-1}), and RVI (MAE = 0.24 Mg ha^{-1}) performed the best for predicting AGB. Conversely, CCCI outperformed the other sensor metrics for predicting PNC (MAE = 0.41 g N 100g DM^{-1}) and NNI (MAE = 0.16 g N 100g DM^{-1}). The eCCC parameter was the best performing sensor parameter for predicting PNU (MAE = 11.12 kg ha^{-1}). PNU was overall the most difficult N status indicator for the sensor parameters to predict (MAE range 11.21 to 14.24 kg ha^{-1}). Compared to spectral parameters, fPAR and ΔTemp both performed poorly for all N status indicators. In several instances, a suitable model could not be fit for all sensor metrics, and, therefore, SR results were not reported.

The best performing metric for each testing SR model was plotted in Figure 5. Training models suggested that AGB, PNU, and NNI were best fit using a non-linear model because their MAE and RMSE decreased compared to linear models.

Table 3. The performance of simple regression (SR) models using Crop Circle Phenom sensor parameters for predicting corn N status indicators across years and treatments. NDVI: normalized difference vegetation index; NDRE: normalized difference red edge; eCCC: estimated canopy chlorophyll content; eLAI: estimated leaf area index; RVI: ratio vegetation index; CCCI: canopy chlorophyll content index; fPAR: fractional photosynthetically active radiation; MAE: mean absolute error; RMSE: root mean squared error.

Parameter	Regression Model	Training			Testing		
		R^2	MAE	RMSE	R^2	MAE	RMSE
Aboveground Biomass (AGB)							
NDVI	$y = 20.56x^2 - 24.76x + 8.29$	0.46	0.31	0.40	0.66	0.23	0.28
NDRE	$y = 0.35e^{4.43x}$	0.45	0.30	0.40	0.60	0.24	0.31
eLAI	$y = 0.44x + 0.76$	0.45	0.30	0.40	0.58	0.25	0.32
eCCC	$y = 0.39x + 0.72$	0.45	0.30	0.40	0.58	0.25	0.31
RVI	$y = 0.13x + 0.44$	0.45	0.31	0.40	0.65	0.24	0.29
CCCI	$y = 24.07x^2 - 12.40x + 2.53$	0.34	0.33	0.44	0.36	0.31	0.39
fPAR	$y = 8.09x + 0.42$	0.15	0.40	0.50	0.16	0.34	0.44
ΔTemp	$y = -0.18x + 1.96$	0.25	0.37	0.47	0.26	0.32	0.42
Plant Nitrogen Concentration (PNC)							
NDRE	$y = 5.88x + 0.16$	0.16	0.49	0.60	0.23	0.48	0.58
eLAI	$y = -0.25x^2 + 1.64x - 0.05$	0.21	0.48	0.58	0.27	0.47	0.56
eCCC	$y = -0.18x^2 + 1.40x - 0.01$	0.23	0.47	0.58	0.29	0.47	0.55
CCCI	$y = 9.58x - 2.00$	0.27	0.45	0.56	0.41	0.41	0.50
Plant Nitrogen Uptake (PNU)							
NDVI	$y = 106.79x^{4.21}$	0.26	14.42	17.46	0.26	13.85	16.09
NDRE	$y = 276.23x - 56.97$	0.38	12.48	15.95	0.48	11.51	13.48
eLAI	$y = -4.07x^2 + 37.35x - 19.11$	0.38	12.44	15.99	0.49	11.21	13.29
eCCC	$y = -2.68x^2 + 30.05x - 16.11$	0.39	12.36	15.91	0.50	11.12	13.22
RVI	$y = 3.52x + 8.33$	0.24	14.69	17.68	0.23	14.24	16.35
CCCI	$y = 353.23x^2 + 34.6x - 42.08$	0.38	12.30	16.01	0.46	11.25	13.76
Nitrogen Nutrition Index (NNI)							
NDVI	$y = 0.13e^{2.34x}$	0.12	0.22	0.26	0.10	0.22	0.26
NDRE	$y = 3.32x - 0.36$	0.30	0.19	0.23	0.41	0.18	0.21
eLAI	$y = 0.17x + 0.46$	0.25	0.20	0.24	0.34	0.19	0.22
eCCC	$y = 0.16x + 0.43$	0.27	0.20	0.24	0.37	0.19	0.22
RVI	$y = 0.03x + 0.57$	0.10	0.23	0.26	0.08	0.23	0.26
CCCI	$y = 6.16x^{2.48}$	0.38	0.18	0.22	0.53	0.16	0.19
fPAR	$y = 3.36x + 0.28$	0.10	0.22	0.26	0.08	0.23	0.26

3.4. Machine Learning Modeling Using eXtreme Gradient Boosted (XGB) Regression

The XGB regression models with NDVI and NDRE performed relatively well. Although adding additional sensor variables as inputs improved the model performance with training dataset for all the four N status indicators, the testing results were not improved (Table 4). The XGB models with all Crop Circle Phenom metrics combined with management information performed the best with both training and testing datasets, except AGB for training.

Validation models using testing dataset observations resulted in N status indicator estimation with model accuracy of $R^2 > 0.6$ and RMSE < 0.40 for all AGB, PNC, and PNU, but lower model accuracy was present for PNU (Figure 6). Model performance for the training and the testing datasets suggested a considerable difference between including traditional vegetation indices compared to using all sensor and management information. Comparing the performance of models using NDVI and NDRE verses models which utilized all Crop Circle Phenom parameters, the NDVI and the NDRE-based models matched or outperformed the full parameter models for all four N status indicators when validated using the testing dataset (Table 4).

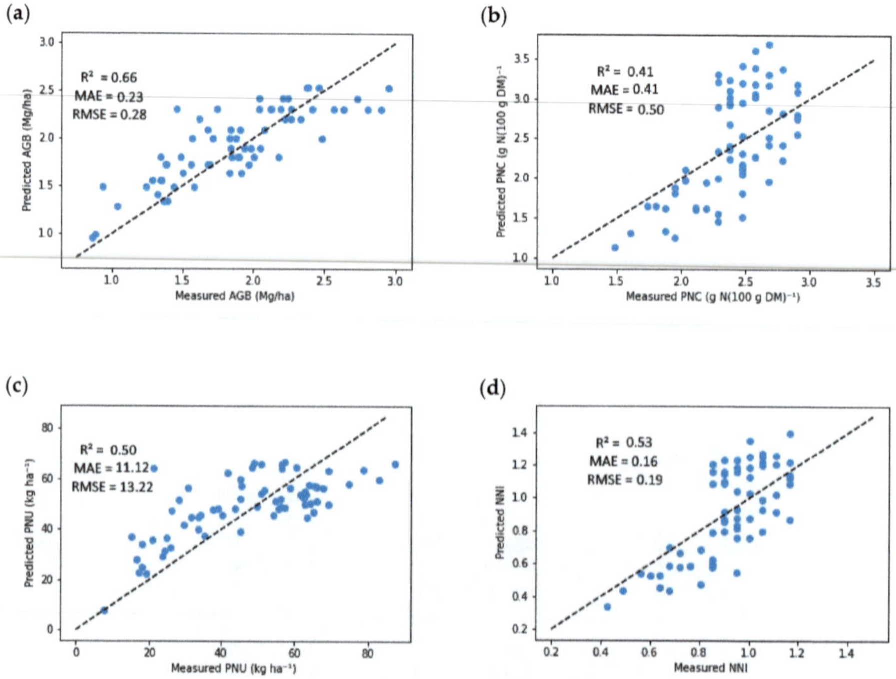

Figure 5. Measured versus predicted (**a**) aboveground biomass (AGB) using SR NDVI, (**b**) plant N concentration (PNC) using SR CCCI, (**c**) plant N uptake (PNU) using SR eCCC, and (**d**) N nutrition index (NNI) using SR CCCI.

Table 4. eXtreme gradient boosted (XGB) model performance using different levels of Crop Circle Phenom sensor and management variables for predicting aboveground biomass, plant N concentration, plant N uptake, and N nutrition index.

Plant Variables	Input Variables	Training			Testing		
		R^2	MAE	RMSE	R^2	MAE	RMSE
Aboveground Biomass(Mg ha^{-1})	NDRE + NDVI	0.61	0.26	0.34	0.54	0.26	0.33
	All Phenom Sensor Metrics	0.83	0.17	0.23	0.50	0.28	0.34
	Phenom Metrics + Management	0.70	0.23	0.30	0.60	0.24	0.30
Plant N Concentration	NDRE + NDVI	0.64	0.32	0.40	0.59	0.33	0.42
	All Phenom Sensor Metrics	0.82	0.21	0.28	0.50	0.38	0.46
	Phenom Metrics + Management	0.88	0.18	0.23	0.66	0.27	0.38
Plant N Uptake	NDRE + NDVI	0.51	11.13	14.18	0.43	11.80	14.10
	All Phenom Sensor Metrics	0.61	9.76	12.59	0.35	12.18	15.01
	Phenom Metrics + Management	0.80	7.08	9.05	0.44	10.83	14.00
N Nutrition Index	NDRE + NDVI	0.65	0.13	0.16	0.55	0.15	0.18
	All Phenom Sensor Metrics	0.85	0.08	0.11	0.52	0.15	0.19
	Phenom Metrics + Management	0.96	0.04	0.06	0.65	0.13	0.16

Note: Management data included drainage, tillage, and pre-plant N rate.

Since the Crop Circle Phenom is a new sensor system, only two site years of data are available. This limitation was mitigated through hyperparameter tuning of max depth, minimum child weight, and learning rate. No overall patterns of greater max depth, min child weight, or learning rate were observed by adding additional sensor or management parameters (Table 5). Cross-validation models using three k-folds were also employed to tune the hyperparameters using the training dataset. In the case of PNU, manual tuning was instead performed because the cross-validation model did not converge on suitable parameters.

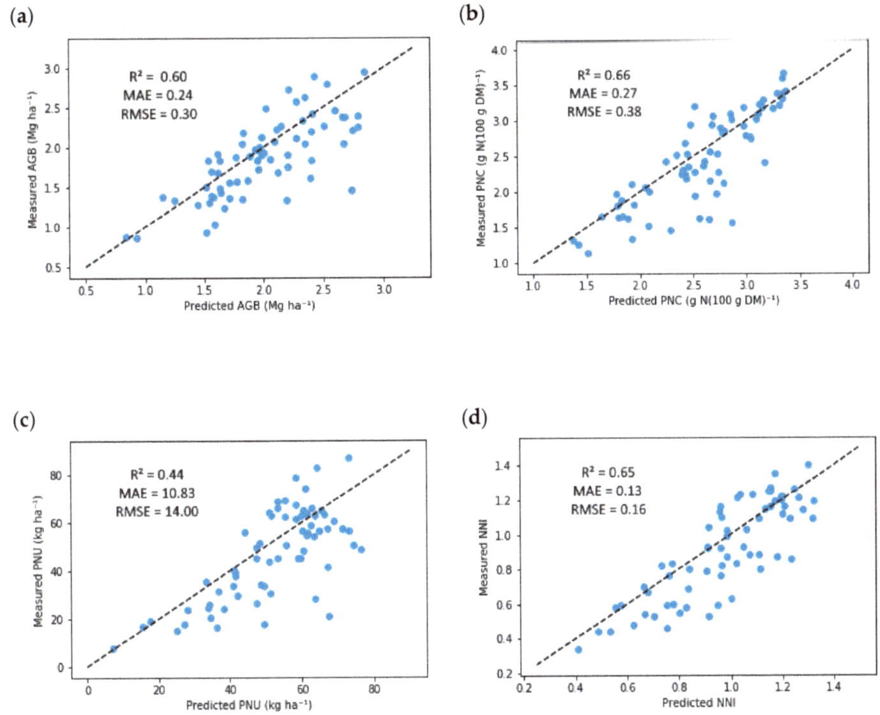

Figure 6. Measured versus predicted (**a**) aboveground biomass (AGB), (**b**) plant N concentration (PNC), (**c**) plant nitrogen uptake (PNU), and (**d**) nitrogen nutrition index (NNI) by XGB regression models using testing dataset and all phenom metrics and management data.

Table 5. XGB cross-validation hyperparameters. Mean absolute error was minimized for ABG, PNC, and NNI using built-in function and was manually tuned for PNU.

Plant Variables	Input Variables	Hyperparameter Parameters		
		Max Depth	Min Child Weight	Learning Rate
Aboveground Biomass	NDRE + NDVI	2	5	0.10
	Phenom Sensor Metrics	4	5	0.05
	Sensor Metrics + Management	[illegible]	[illegible]	[illegible]
Plant N Concentration	NDRE + NDVI	3	1	0.10
	All Phenom Sensor Metrics	2	4	0.15
	All Sensor Metrics + Management	3	3	0.05
Plant N Uptake	NDRE + NDVI	2	3	0.05
	All Phenom Sensor Metrics	2	3	0.05
	All Sensor Metrics + Management	2	3	0.10
N Nutrition Index	NDRE + NDVI	4	1	0.05
	All Phenom Sensor Metrics	4	5	0.05
	All Sensor Metrics + Management	3	3	0.15

3.5. Relative Importance of Input Variables

The importance values were calculated for each N status prediction model to indicate relative worth of sensing and management parameters using the XGB plot_importance tool. Average gain value per model split was selected as the parameter used to measure a feature's F score. This metric computed the average split value that each Crop Circle Phenom or management parameter was selected in and averaged their value for each N status indicator. The resulting model suggests PPN was the most important input variable for predicting PNC, PNU, and NNI (Figure 7b–d). However, sensing parameters NDRE, RVI, and NDVI were the most important parameters for predicting AGB (Figure 7a). CCCI was the most important sensor parameter to be included for estimating PNC, PNU or NNI, however, it was one of the lower importance sensor metrics for AGB prediction.

Tillage and drainage variables were not rated highly for predicting PNC, PNU, or NNI. An exception was that no-till (NT) was the fourth highest ranked metric for predicting AGB, although its F score was significantly lower compared to the top sensor metrics (Figure 7a). Drainage was predicted to have a high importance for predicting plant N status indicators due to its correlation with N loss processes, however, it consistently had a lower feature importance compared to sensing parameters and PPN.

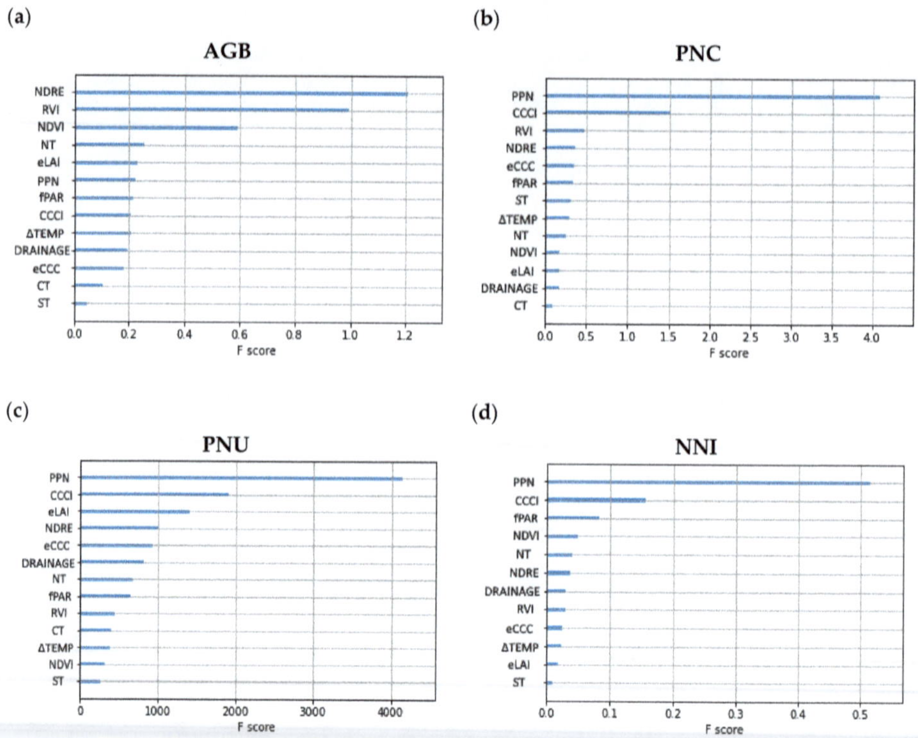

Figure 7. Relative importance of sensor metrics and management variables for predicting (**a**) AGB, (**b**) PNC, (**c**) PNU, and (**d**) NNI as represented by the F score values from XGB regression.

3.6. Diagnosis of In-Season N Status Using NNI

The areal agreement and the kappa coefficient statistics for evaluating the efficacy of each sensor modeling technique to diagnose corn N status (NNI < 0.95 = deficient, 0.95 < NNI < 1.05 = optimum, NNI > 1.05 surplus) are given in Table 6. Among the 67 measurements, 37 plots were deficient, 4 were optimum, and 26 were surplus. Using the best performing NDRE and CCCI single sensor parameters to estimate corn N status demonstrated acceptable diagnostic accuracy for deficient conditions based on testing data (62–70%), however, both parameters performed poorly when diagnosing surplus corn N condition (<42%). Comparing XGB modeling techniques, only the model combining Crop Circle Phenom sensor parameters with management data achieved kappa statistics of over 0.4, with the overall areal agreement of 72%. Although the XGB models using NDVI and NDRE or all Crop Circle Phenom sensor data both improved overall corn N status diagnostic accuracy compared with SR models using NDRE or CCCI, neither of them achieved moderate agreement based on kappa statistics (0.4–0.6).

Table 6. Corn N status diagnosis accuracy based on NNI prediction using SR and XGB regression results. Model precision was assessed using areal agreement (%) and kappa statistics (NNI < 0.95 = deficient, 0.95 < NNI < 1.05 = optimum, NNI > 1.05 surplus).

	Areal Agreement (%)				Kappa Statistics
	Deficient (n = 37)	Optimum (n = 4)	Surplus (n = 26)	Overall (n = 67)	
NDRE	**70**	**25**	23	49	0.22
CCCI	62	50	42	54	0.26
XGB NDVI+NDRE	70	0	50	58	0.31
XGB All Phenom Metrics	68	25	46	57	0.29
XGB Phenom + Management	68	50	81	72	0.54

4. Discussion

4.1. Crop Circle Phenom Comparison to Similar Proximal Active Canopy Sensors

This research was conducted to evaluate the potential of the new multi-parameter Crop Circle Phenom sensor system, which has traditional spectral band reflectance and vegetation indices as well as climatic and physiological metrics. Compared with the commonly used three-band Crop Circle ACS 430 or RapidSCAN CS-45 sensors that calculate NDVI and NDRE, the Crop Circle Phenom system also provides eLAI and eCCC. These additional estimated parameters proved beneficial for estimating PNC and PNU in SR models, as both outperformed NDVI and NDRE. Regarding AGB estimation, eLAI and eCCC performed similarly to NDVI and NDRE. This is not a surprise, as NDVI has been extensively used for AGB estimation [38,39]. Similarly, the commonly used three band active canopy sensors and the Crop Circle Phenom sensor system would have comparable performance estimating NNI or PNU, since NNI was best predicted using CCCI, and PNU was estimated similarly well using eCCC, eLAI, NDRE, or CCCI, which can all be calculated by all these sensors.

Aside from the estimated LAI and CCC metrics, the key potential advantage of the Crop Circle Phenom sensor system is the derivation of ΔTemp and fPAR. The ΔTemp parameter has been commonly used to identify crop water stress [40,41], however, limited research has been conducted to investigate how crop N status influences canopy temperature. Tan et al. [42] found that rice canopy temperature responded to N rate, with N stress causing higher temperatures. Similarly, Alzaben, Fraser, and Swanton [22] used thermal imagery to investigate the relation between canopy temperature and N status. The study observed that both corn leaf and whirl temperatures statistically responded to N treatment, with optimal N corresponding to lower canopy temperature. For this study, ΔTemp calculated from the sensor's air and canopy temperature readings showed a poor relationship with all four plant N metrics using SR. This result contrasts with previous research but could be explained by the inability of the proximal sensor to separate soil and plant signals. Alzaben, Fraser, and Swanton [22] were able to separate plants from soil background using a segmentation algorithm, which they indicated considerably changed their measured plot temperatures.

As with ΔTemp, fPAR has not been thoroughly studied in crop nutrient management. Although PAR information can be used as a component to estimate crop biomass, yield, and primary productivity [21], no published article has used it as a metric to estimate plant N status. For this reason, fPAR was investigated in this study and compared with traditional vegetative indices. The results of this study indicated that fPAR was marginally related to other sensor metrics (R = 0.17–0.28) and was not as important as vegetation indices for predicting corn N status based on SR analysis, including AGB.

Although fPAR did not perform well using SR, it showed more potential when used in XGB regression models as it was ranked as one of the most important variables for predicting NNI. This result indicated that fPAR was not an important predictor of N status individually but could provide important information complementary to spectral vegetation indices. However, ΔTemp did not rank highly in any of the N indicator models. The ΔTemp information may be beneficial to help differentiate different stress factors since it has already been shown to detect water stress, as demonstrated by Jensen et al. [43] and DeJonge et al. [41].

4.2. Modelling Strategies for In-Season Corn N Status Prediction and Diagnosis

The SR modeling was evaluated as a simplistic approach to model N nutrition metrics; however, limitations were discovered when including categorical field management variables. Additionally, determining the correct model fit for each sensor metric is difficult since most relationships are non-linear. Therefore, machine learning methods that can include categorical variables may be a better approach to model non-linear relationships.

The XGB regression was investigated as a machine learning method to predict in-season N metrics using three distinct levels: NDVI and NDRE, all selected Crop Circle Phenom sensor parameters, and all sensor parameters as well as drainage, tillage, and PPN management information. The results indicated that corn N status indicators were best predicted when sensor data and management information were utilized together. The PPN rate information was highly important for predicting PNC, PNU, and NNI, however, it was not as important for predicting AGB. Compared to PPN, drainage and tillage information did not contribute as strongly to the prediction of any of the plant N metrics since their F scores were significantly lower than most sensor metrics. This low feature importance was unexpected since both drainage and tillage were found to be significant factors for grain yield (data not shown).

Models using all Crop Circle parameters overall performed better than models only using NDVI and NDRE using the training dataset, although this did not translate into improved performance of the testing dataset that was at best comparable to using only NDVI and NDRE. Comparing the importance of each sensor metric, CCCI was the most informative sensor index for PNC, PNU, and NNI. This supports previous studies [44–46]. It should be noted that the CCCI used in this study is a simplified index calculated as NDRE/NDVI, while the original CCCI was based on the theory of two-dimensional planar domain involving both NDRE and NDVI [47,48]. More studies are needed to further evaluate the simplified and the original CCCI for applications in crop N status prediction and diagnosis.

The N status diagnosis results also indicated that XGB models using two or more variables outperformed SR models using one variable. The XGB models using NDVI and NDRE or all selected Crop Circle Phenom sensor derived variables performed similarly, with the same areal agreement (57%) and slight difference in kappa statistics (0.32 vs. 0.36). Adding management information further improved the N status diagnostic accuracy, with areal agreement of 72% and kappa statistics of 0.54. This result highlighted the importance of combining management information with crop sensor data.

Few previous studies have attempted to combine sensor data with soil and climate data to improve in-season N recommendations [49–51], however, limited studies have been reported to combine management practice information with crop sensor data for in-season N status prediction and diagnosis. Countless machine learning models have been used for predicting crop N status indicators [52–55], however, XGB regression was selected for this project because it includes self-contained cross validation modules to perform hyperparameter tuning and the ability to define an early stopping parameter to mitigate overfitting. Specifically, the ability to tune for learning rate was important within our limited dataset because it further mitigated the risk of overfitting our training models. Nevertheless, the results in this study indicated that overfitting was still a problem. More studies are needed to broaden the dataset and evaluate different machine learning methods [51,54].

4.3. Implications for On-Farm Applications

Proximal sensing systems are beneficial for on-farm use because they require minimal training to collect data and fewer processing resources than aerial or satellite imagery. The Crop Circle Phenom sensor system is designed to be mounted on a vehicle or tractor, which makes it more difficult to be carried by hand for small plot research. To deploy it in small plot experiments, a custom pole was constructed to mount the two sensors and the GeoScout data logger. Another difference compared to similar proximal sensors is the Phenom requires an external 12 volt battery to power its active sensor light for calculating reflectance. Although the Crop Circle Phenom requires modifications for small plot research, adapting the sensor system for commercial field applications would be much easier because the mounting hardware and the electrical wiring were designed for use on a field implement. This ease of use for commercial applications is also due to its GPS connectivity and ability to quickly swap out the sensor across a range of field implements from sprayers to fertilizer spreaders, which enables whole field resolution readings to be collected throughout the growing season.

Another way in which the Crop Circle Phenom can set itself apart as a proximal sensing system is through its multi-parameter spectral, environmental, and physiological metrics. Utilizing biophysical relationships between spectral features and temperature, the Crop Circle Phenom can be used to estimate ΔTemp and fPAR. Although utilized in this study to investigate N status, these metrics have the potential to differentiate various stress factors such as water status and pathological issues. However, both these management considerations were outside the scope of this research and should be investigated in the future.

The PPN information was an important factor to use with crop sensor data for in-season N status prediction and diagnosis. Such data can be easily obtained from as-applied maps and should be included in in-season N status diagnosis, especially when variable rate PPN is applied.

5. Conclusions

The Crop Circle Phenom sensing system possesses multi-parameter indices that can be used to measure crop canopy reflectance, eLAI, eCCC, and calculate ΔTemp and fPAR. The eLAI and eCCC indices performed slightly better than NDVI and NDRE for predicting PNC and PNU using SR models. As a result, these indices warrant inclusion in future sensor-based diagnosis methods alongside traditional vegetation indices. In contrast, SR models using ΔTemp or fPAR did not perform well for predicting plant N status indicators. This poorer model performance could be due to inability to segment soil from plant reflectance, as is possible with imagery or potted plant experiments. Nonetheless, both ΔTemp and fPAR parameters were useful for N status prediction when used alongside reflectance parameters with machine learning models, such as XGB regression. The CCCI parameter was found to be an important vegetation index for predicting PNC, PNU, and NNI in both SR and XGB modeling. This improvement over NDVI and NDRE indicates CCCI should be included in future sensor guided management research.

The Crop Circle Phenom sensor system shows promise as a tool for in-season corn N status prediction and diagnosis across different drainage, tillage, N supply, and site year conditions. Combining management information, especially PPN, with Crop Circle

Phenom sensor data using machine learning can improve corn N status prediction and diagnosis compared to only using sensor data. Additional studies are needed to further evaluate this new multi-parameter Crop Circle Phenom sensing system with more site year data using additional tree based supervised models.

Author Contributions: Y.M. conceived the study. F.G.F. designed and implemented the field experiment. C.C., G.D.P., S.K. collected plant and sensor data. C.C. performed data analysis, wrote the original paper, S.K. assisted with data analysis, Y.M., F.G.F. and G.D.P. reviewed and revised the manuscript, Y.M. and F.G.F. secured funding. All authors have read and agreed to the published version of the manuscript.

Funding: This research was funded by the Minnesota Department of Agriculture/Agricultural Fertilizer Research and Education Council (MDA/AFREC R2018-25, R2019-20 and R2020-32), the Minnesota Soybean Research and Promotion Council (project 00071830), the USDA National Institute of Food and Agriculture (State project 1016571), and the Startup Fund.

Institutional Review Board Statement: Not applicable.

Informed Consent Statement: Not applicable.

Data Availability Statement: The data presented in this study are available on request from the corresponding author. The data are not yet publicly available.

Acknowledgments: We would like to thank the Field Crew at the Department of Soil, Water and Climate, Hui Shao for their assistance with field and sensor data collection, and Seth Naeve for drainage design and long-term management of the study site.

Conflicts of Interest: The authors declare no conflict of interest.

References

1. Gebbers, R.; Adamchuk, V. Precision Agriculture and Food Security. *Science* **2010**, *327*, 828–831. [CrossRef] [PubMed]
2. Subedi, K.D.; Ma, B.L. Assessment of Some Major Yield-Limiting Factors on Maize Production in a Humid Temperate Environment. *Field Crops Res.* **2009**, *110*, 21–26. [CrossRef]
3. Andraski, T.W.; Bundy, L.G.; Brye, K.R. Crop Management and Corn Nitrogen Rate Effects on Nitrate Leaching. *J. Environ. Qual.* **2000**, *29*, 1095–1103. [CrossRef]
4. Ma, B.L.; Wu, T.Y.; Tremblay, N.; Deen, W.; McLaughlin, N.B.; Morrison, M.J.; Stewart, G. On-Farm Assessment of the Amount and Timing of Nitrogen Fertilizer on Ammonia Volatilization. *Agron. J.* **2010**, *102*, 134–144. [CrossRef]
5. Balafoutis, A.; Beck, B.; Fountas, S.; Vangeyte, J.; van der Wal, T.; Soto, I.; Gómez-Barbero, M.; Barnes, A.; Eory, V. Precision Agriculture Technologies Positively Contributing to Ghg Emissions Mitigation, Farm Productivity and Economics. *Sustainability* **2017**, *9*, 1339. [CrossRef]
6. Diacono, M.; Rubino, P.; Montemurro, F. Precision Nitrogen Management of Wheat. A Review. *Agron. Sustain. Dev.* **2013**, *33*, 219–241. [CrossRef]
7. Cao, Q.; Miao, Y.; Feng, G.; Gao, X.; Liu, B.; Liu, Y.; Li, F.; Khosla, R.; Mulla, D.J.; Zhang, F. Improving Nitrogen Use Efficiency with Minimal Environmental Risks Using an Active Canopy Sensor in a Wheat-Maize Cropping System. *Field Crops Res.* **2017**, *214*, 365–372. [CrossRef]
8. Cassman, K.; Dobermann, A.; Walters, D. Agroecosystems, Nitrogen-Use Efficiency, and Nitrogen Management. *Biogeochemistry* **2006**, *79*, 132–140. [CrossRef]
9. Mistele, B.; Schmidhalter, U. Estimating the Nitrogen Nutrition Index Using Spectral Canopy Reflectance Measurements. *Eur. J. Agron.* **2008**, *29*, 184–190. [CrossRef]
10. Silva, J.; Uchida, R. Essential Nutrients for Plant Growth. In *Plant Nutrient Management in Hawaii's Soils: Approaches for Tropical and Subtropical Agriculture*; Silva, J.A., Uchida, R., Eds.; College of Tropical Agriculture and Human Resources, University of Hawaii at Manoa: Manaoa, HI, USA, 2000; pp. 31–55.
11. Plénet, D.; Lemaire, G. Relationships between Dynamics of Nitrogen Uptake and Dry Matter Accumulation in Maize Crops. Determination of Critical N Concentration. *Plant Soil* **1999**, *216*, 65–82. [CrossRef]
12. Cilia, C.; Panigada, C.; Rossini, M.; Meroni, M.; Busetto, L.; Amaducci, S.; Boschetti, M.; Picchi, V.; Colombo, R. Nitrogen Status Assessment for Variable Rate Fertilization in Maize through Hyperspectral Imagery. *Remote Sens.* **2014**, *6*, 6549–6565. [CrossRef]
13. Xia, T.; Miao, Y.; Wu, D.; Shao, H.; Khosla, R.; Mi, G. Active Optical Sensing of Spring Maize for In-Season Diagnosis of Nitrogen Status Based on Nitrogen Nutrition Index. *Remote Sens.* **2016**, *8*, 605. [CrossRef]
14. Ziadi, N.; Brassard, M.; Bélanger, G.; Claessens, A.; Tremblay, N.; Cambouris, A.N.; Nolin, M.C.; Parent, L.É. Chlorophyll Measurements and Nitrogen Nutrition Index for the Evaluation of Corn Nitrogen Status. *Agron. J.* **2008**, *100*, 1264–1273. [CrossRef]

15. Zhao, B.; Duan, A.; Ata-Ul-Karim, S.T.; Liu, Z.; Chen, Z.; Gong, Z.; Zhang, J.; Xiao, J.; Liu, Z.; Qin, A.; et al. Exploring New Spectral Bands and Vegetation Indices for Estimating Nitrogen Nutrition Index of Summer Maize. *Eur. J. Agron.* **2018**, *93*, 113–125. [CrossRef]
16. Basso, B.; Cammarano, D.; Grace, P.R.; Cafiero, G.; Sartori, L.; Pisante, M.; Landi, G.; de Franchi, S.; Basso, F. Criteria for Selecting Optimal Nitrogen Fertilizer Rates for Precision Agriculture. *Ital. J. Agron.* **2009**, *4*, 147–158. [CrossRef]
17. Miao, Y.; Mulla, D.J.; Randall, G.W.; Vetsch, J.A.; Vintila, R. Combining Chlorophyll Meter Readings and High Spatial Resolution Remote Sensing Images for In-Season Site-Specific Nitrogen Management of Corn. *Precis. Agric.* **2009**. [CrossRef]
18. Paiao, G.D.; Fernández, F.F.; Spackman, J.A.; Kaiser, D.E.; Weisberg, S. Ground-Based Optical Canopy Sensing Technologies for Corn–Nitrogen Management in the Upper Midwest. *Agron. J.* **2020**, *112*, 2998–3011. [CrossRef]
19. Rouse, J.W.; Haas, R.H.; Schell, J.A.; Deering, D.W. Monitoring Vegetation Systems in the Great Plains with ERTS. In *Proceedings of the Third Earth Resources Technology Satellite—1 Symposium*; Held by Goddard Space Flight Center at Washington, DC on 10–14 December 1973: Prepared at Goddard Space Flight Center; Scientific and Technical Information Office, National Aeronautics and Space Administration: Washington, DC, USA, 1974; Volume 351, pp. 309–317.
20. Barnes, E.M.; Clarke, T.R.; Richards, S.E. Coincident Detection of Crop Water Stress, Nitrogen Status and Canopy Density Using Ground Based Multispectral Data. In Proceedings of the Fifth International Conference on Precision Agriculture, Madison, WI, USA, 16–19 July 2000; Robert, P.C., Rust, R.H., Larson, W.E., Eds.; American Society of Agronomy (CD-ROM): Madison, WI, USA, 2000.
21. Serrano, L.; Filella, I.; Peñuelas, J. Remote Sensing of Biomass and Yield of Winter Wheat under Different Nitrogen Supplies. *Crop Sci.* **2000**, *40*, 723–731. [CrossRef]
22. Alzaben, H.; Fraser, R.; Swanton, C. An Inverse Correlation between Corn Temperature and Nitrogen Stress: A Field Case Study. *Agron. J.* **2019**, *111*, 3207–3219. [CrossRef]
23. Fernández, F.G.; Fabrizzi, K.P.; Naeve, S.L. Corn and Soybean's Season-Long in-Situ Nitrogen Mineralization in Drained and Undrained Soils. *Nutr. Cycl. Agroecosyst.* **2017**, *107*, 33–47. [CrossRef]
24. Holland, K.H.; Schepers, J.S. Active Proximal Sensing: Review of Waveband Selection, Vegetation Indices, Scientific Trump Cards, Etc. In Proceedings of the ASA CSSA SSSA 2011 International Annual Meetings, San Antonio, TX, USA, 16–19 October 2011.
25. Jones, H.G.; Vaughan, R.A. *Remote Sensing of Vegetation: Principles, Techniques, and Applications*; Oxford University Press: New York, NY, USA, 2010; ISBN 9780199207794.
26. Jordan, C.F. Derivation of Leaf-Area Index from Quality of Light on the Forest Floor. *Ecology* **1969**, *50*, 663–666. [CrossRef]
27. Long, D.S.; Eitel, J.U.H.; Huggins, D.R. Assessing Nitrogen Status of Dryland Wheat Using the Canopy Chlorophyll Content Index. *Crop Manag.* **2009**, *8*, 1–8. [CrossRef]
28. Holland Scientific. *Crop Circle Phenom User's Guide*; Holland Scientific: Lincoln, NE, USA, 2016.
29. Horneck, D.A.; Miller, R.O. Determination of total nitrogen in plant tissue. In *Handbook of Reference Methods for Plant Analysis*; CRC Press: Boca Raton, FL, USA, 1998; pp. 75–84.
30. Virtanen, P.; Gommers, R.; Oliphant, T.E.; Haberland, M.; Reddy, T.; Cournapeau, D.; Burovski, E.; Peterson, P.; Weckesser, W.; Bright, J.; et al. SciPy 1.0: Fundamental Algorithms for Scientific Computing in Python. *Nat. Methods* **2020**, *17*, 261–272. [CrossRef] [PubMed]
31. Pedregosa, F.; Varoquaux, G.; Gramfort, A.; Michel, V.; Thirion, B.; Grisel, O.; Blondel, M.; Prettenhofer, P.; Weiss, R.; Dubourg, V.; et al. Scikit-Learn: Machine Learning in Python. *J. Mach. Learn. Res.* **2011**, *12*, 2825–2830.
32. Chen, T.; Guestrin, C. XGBoost: A Scalable Tree Boosting System Tianqi. In Proceedings of the 22nd ACM SIGKDD International Conference on Knowledge Discovery and Data Mining, San Francisco, CA, USA, 13–17 August 2016; pp. 785–794.
33. Schratz, P.; Muenchow, J.; Iturritxa, E.; Richter, J.; Brenning, A. Hyperparameter Tuning and Performance Assessment of Statistical and Machine-Learning Algorithms Using Spatial Data. *Ecol. Model.* **2019**, *406*, 109–120. [CrossRef]
34. Huang, S.; Miao, Y.; Zhao, G.; Yuan, F.; Ma, X.; Tan, C.; Yu, W.; Gnyp, M.L.; Lenz-Wiedemann, V.I.S.; Rascher, U.; et al. Satellite Remote Sensing-Based in-Season Diagnosis of Rice Nitrogen Status in Northeast China. *Remote Sens.* **2015**, *7*, 10646–10667. [CrossRef]
35. Lu, J.; Miao, Y.; Shi, W.; Li, J.; Yuan, F. Evaluating Different Approaches to Non-Destructive Nitrogen Status Diagnosis of Rice Using Portable RapidSCAN Active Canopy Sensor. *Sci. Rep.* **2017**. [CrossRef]
36. Cohen, J. A Coefficient of Agreement for Nominal Scales. *Educ. Psychol. Meas.* **1960**, *20*, 37–46. [CrossRef]
37. Landis, J.R.; Koch, G.G. The Measurement of Observer Agreement for Categorical Data. *Biometrics* **1977**, *33*, 159. [CrossRef]
38. Freeman, K.W.; Girma, K.; Arnall, D.B.; Mullen, R.W.; Martin, K.L.; Teal, R.K.; Raun, W.R. By-Plant Prediction of Corn Forage Biomass and Nitrogen Uptake at Various Growth Stages Using Remote Sensing and Plant Height. *Agron. J.* **2007**, *99*, 530–536. [CrossRef]
39. Wang, X.; Miao, Y.; Guan, Y.; Xia, T.; Lu, J.; Mulla, D.J. An evaluation of two active sensor systems for non-destructive estimation of spring maize biomass. In Proceedings of the Fifth International Conference on Agro-Geoinformatics (Agro-Geoinformatics 2016), Tianjin, China, 18–20 July 2016; pp. 1–6. [CrossRef]
40. Jackson, R.D.; Idso, S.B.; Reginato, R.J.; Pinter, P.J. Canopy Temperature as a Crop Water Stress Indicator. *Water Resour. Res.* **1981**, *17*, 1133–1138. [CrossRef]
41. DeJonge, K.C.; Taghvaeian, S.; Trout, T.J.; Comas, L.H. Comparison of Canopy Temperature-Based Water Stress Indices for Maize. *Agric. Water Manag.* **2015**, *156*, 51–62. [CrossRef]

42. Yan, C.; Ding, Y.; Wang, Q.; Liu, Z.; Li, G.; Muhammad, I.; Wang, S. The Impact of Relative Humidity, Genotypes and Fertilizer Application Rates on Panicle, Leaf Temperature, Fertility and Seed Setting of Rice. *J. Agric. Sci.* **2010**, *148*, 329–339. [CrossRef]
43. Jensen, H.E.; Svendsen, H.; Jensen, S.E.; Mogensen, V.O. Canopy-Air Temperature of Crops Grown under Different Irrigation Regimes in a Temperate Humid Climate. *Irrig. Sci.* **1990**, *11*, 181–188. [CrossRef]
44. Cammarano, D.; Fitzgerald, G.; Basso, B.; O'Leary, G.; Chen, D.; Grace, P.; Fiorentino, C. Use of the Canopy Chlorophyl Content Index (CCCI) for Remote Estimation of Wheat Nitrogen Content in Rainfed Environments. *Agron. J.* **2011**, *103*, 1597–1603. [CrossRef]
45. Perry, E.M.; Fitzgerald, G.J.; Nuttall, J.G.; O'Leary, G.J.; Schulthess, U.; Whitlock, A. Rapid Estimation of Canopy Nitrogen of Cereal Crops at Paddock Scale Using a Canopy Chlorophyll Content Index. *Field Crops Res.* **2012**, *134*, 158–164. [CrossRef]
46. Li, F.; Miao, Y.; Feng, G.; Yuan, F.; Yue, S.; Gao, X.; Liu, Y.; Liu, B.; Ustin, S.L.; Chen, X. Improving Estimation of Summer Maize Nitrogen Status with Red Edge-Based Spectral Vegetation Indices. *Field Crops Res.* **2014**, *157*, 111–123. [CrossRef]
47. Clarke, T.R.; Moran, M.S.; Barnes, E.M.; Pinter, P.J.; Qi, J. Planar Domain Indices: A Method for Measuring a Quality of a Single Component in Two-Component Pixels. *Int. Geosci. Remote Sens. Symp.* **2001**, *3*, 1279–1281. [CrossRef]
48. Fitzgerald, G.; Rodriguez, D.; O'Leary, G. Measuring and Predicting Canopy Nitrogen Nutrition in Wheat Using a Spectral Index-The Canopy Chlorophyll Content Index (CCCI). *Field Crops Res.* **2010**, *116*, 318–324. [CrossRef]
49. Bushong, J.T.; Mullock, J.L.; Miller, E.C.; Raun, W.R.; Brian Arnall, D. Evaluation of Mid-Season Sensor Based Nitrogen Fertilizer Recommendations for Winter Wheat Using Different Estimates of Yield Potential. *Precis. Agric.* **2016**, *17*, 470–487. [CrossRef]
50. Bean, G.M.; Kitchen, N.R.; Camberato, J.J.; Ferguson, R.B.; Fernandez, F.G.; Franzen, D.W.; Laboski, C.A.M.; Nafziger, E.D.; Sawyer, J.E.; Scharf, P.C.; et al. Improving an Active-Optical Reflectance Sensor Algorithm Using Soil and Weather Information. *Agron. J.* **2018**, *110*, 2541–2551. [CrossRef]
51. Ransom, C.J.; Camberato, J.J.; Carter, P.R.; Ferguson, R.B. Statistical and Machine Learning Methods Evaluated for Incorporating Soil and Weather into Corn Nitrogen Recommendations. *Comput. Electron. Agric.* **2019**, *164*, 104872. [CrossRef]
52. Yao, X.; Huang, Y.; Shang, G.; Zhou, C.; Cheng, T.; Tian, Y.; Cao, W.; Zhu, Y. Evaluation of Six Algorithms to Monitor Wheat Leaf Nitrogen Concentration. *Remote Sens.* **2015**, *7*, 14939–14966. [CrossRef]
53. Chlingaryan, A.; Sukkarieh, S.; Whelan, B. Machine Learning Approaches for Crop Yield Prediction and Nitrogen Status Estimation in Precision Agriculture: A Review. *Comput. Electron. Agric.* **2018**, *151*, 61–69. [CrossRef]
54. Zheng, H.; Li, W.; Jiang, J.; Liu, Y.; Cheng, T.; Tian, Y.; Zhu, Y.; Cao, W.; Zhang, Y.; Yao, X. A Comparative Assessment of Different Modeling Algorithms for Estimating Leaf Nitrogen Content in Winter Wheat Using Multispectral Images from an Unmanned Aerial Vehicle. *Remote Sens.* **2018**, *10*, 2026. [CrossRef]
55. Zha, H.; Miao, Y.; Wang, T.; Li, Y.; Zhang, J.; Sun, W. Sensing-Based Rice Nitrogen Nutrition Index Prediction with Machine Learning. *Remote Sens.* **2020**, *12*, 215. [CrossRef]

remote sensing

MDPI

Article

Difference and Potential of the Upward and Downward Sun-Induced Chlorophyll Fluorescence on Detecting Leaf Nitrogen Concentration in Wheat

Min Jia [1,2,3,4], Jie Zhu [1,2,3,4], Chunchen Ma [1,2,3,4], Luis Alonso [5], Dong Li [1,2,3,4], Tao Cheng [1,2,3,4], Yongchao Tian [1,2,3,4], Yan Zhu [1,2,3,4], Xia Yao [1,2,3,4,*] and Weixing Cao [1,2,3,4,*]

1 National Engineering and Technology Center for Information Agriculture, Nanjing Agricultural University, Nanjing 210095, Jiangsu, China; 2016201075@njau.edu.cn (M.J.); 2016101051@njau.edu.cn (J.Z.); 2016101050@njau.edu.cn (C.M.); lidongmath@163.com (D.L.); tcheng@njau.edu.cn (T.C.); yctian@njau.edu.cn (Y.T.); yanzhu@njau.edu.cn (Y.Z.)

2 Key Laboratory for Crop System Analysis and Decision Making, Ministry of Agriculture, Nanjing 210095, Jiangsu, China

3 Jiangsu Key Laboratory for Information Agriculture, Nanjing 210095, Jiangsu, China

4 Jiangsu Collaborative Innovation Center for Modern Crop Production, Nanjing 210095, Jiangsu, China

5 Department of Earth Physics and Thermodynamics, University of Valencia, 46010 València, Spain; luis.alonso@uv.es

* Correspondence: yaoxia@njau.edu.cn (X.Y.); caow@njau.edu.cn (W.C.); Tel.: +86-25-84396565 (X.Y.); Fax: +86-25-84396565 (X.Y.)

Received: 9 June 2018; Accepted: 10 August 2018; Published: 20 August 2018

Abstract: Precise detection of leaf nitrogen concentration (LNC) is helpful for nutrient diagnosis and fertilization guidance in farm crops. Numerous researchers have estimated LNC with techniques based on reflectance spectra or active chlorophyll fluorescence, which have limitations of low accuracy or small scale in the field. Given the correlation between chlorophyll and nitrogen contents, the response of sun-induced chlorophyll fluorescence (SIF) to chlorophyll (Chl) content reported in a few papers suggests the feasibility of quantifying LNC using SIF. Few studies have investigated the difference and power of the upward and downward SIF components on monitoring LNC in winter wheat. We conducted two field experiments to evaluate the capacity of SIF to monitor the LNC of winter wheat during the entire growth season and compare the differences of the upward and downward SIF for LNC detection. A FluoWat leaf clip coupled with a ASD spectrometer was used to measure the upward and downward SIF under sunlight. It was found that three (↓FY687, ↑FY687/↑FY739, and ↓FY687/↓FY739) out of the six SIF yield (FY) indices examined were significantly correlated to the LNC (R^2 = 0.6, 0.51, 0.75, respectively). The downward SIF yield indices exhibited better performance than the upward FY indices in monitoring the LNC with the ↓FY687/↓FY739 being the best FY index. Moreover, the LNC models based on the three SIF yield indices are insensitive to the chlorophyll content and the leaf mass per area (LMA). These findings suggest the downward SIF should not be neglected for monitoring crop LNC at the leaf scale, although it is more difficult to measure with current instruments. The downward SIF could play an increasingly important role in understanding of the SIF emission for LNC detection at different scales. These results could provide a solid foundation for elucidating the mechanism of SIF for LNC estimation at the canopy scale.

Keywords: sun-induced chlorophyll fluorescence (SIF); SIF yield indices; upward; downward; leaf nitrogen concentration (LNC); wheat (*Triticum aestivum* L.)

1. Introduction

Nitrogen, an essential element in chlorophyll and in enzymes needed for photosynthesis, plays an important role in maintaining crop growth and enhancing grain yield [1]. However, over-fertilization will result in lower nitrogen-use efficiency and environmental pollution. Leaf nitrogen concentration (LNC) can be used to diagnose the nutritional status and guide precise fertilization [2–4]. Therefore, many studies have been devoted to accurately monitoring crop LNC [5–7]. The close correlation between chlorophyll (Chl) and nitrogen content makes it possible to quantify the crop LNC with empirical methods [8]. Existing studies have illustrated that reflectance-based parameters/vegetation indices could be used to monitor the LNC [9–11], which are based on the absorption characteristics of chemical components. However, until now, these studies have some shortcomings, such as mixed signals from the plants and the soil, lack of specificity of the nitrogen stress, and limitations to the specific ranges of biomass, leaf area, and chlorophyll [12]. In the past decades, chlorophyll fluorescence (ChlF), the light emitted by chlorophyll has proven to be highly related to crop physiology and sensitive to plant nitrogen status when compared with reflectance signals [13].

ChlF, used to probe many aspects of photosynthesis of plants and other photosynthetic organisms, is emitted by Chl, independent of soil interference and biomass. Govindjee [14] suggested that ChlF could provide abundant information about photosynthetic characteristics, including pigment system composition, de-excitation energy, rates of electron transfer reactions on Photosystem II (PS II), actual photochemical quantum yields and coefficients of photochemical quenching. Since nitrogen is the main element of Chl and enzymes, plant nitrogen content affects the Photosystem I (PS I) and Photosystem II (PS II) functions, and then influences the photosynthetic characteristics by affecting the photosynthetic pigment content and physical changes in pigment-protein complexes. Therefore, plant nitrogen content could affect the photosynthetic function leading to the changes of ChlF emission. Meanwhile, the strong link between LNC and ChlF provides an empirical basis for detecting plant LNC.

ChlF can be measured with active and passive ChlF techniques depending on the type of excitation light source. The active ChlF measurements have been proposed as possible species-specific approach to monitor the LNC and indentify nutrient deficiency of crops by several scientists [15–20]. For example, laser-induced fluorescence parameters (F685, F740: fluorescence intensity at 685 nm and 740 nm; F740/F685: ratio of fluorescence intensity at 740 nm and 685 nm) are reported as a potential method for non-destructively monitoring paddy rice LNC [21]. It seems that much progress has been made to detect LNC using ChlF [22–24]. To date, there still exist some limitations to the application of active ChlF. First, due to the artificial light source used to excite the leaf fluorescence emission, the active ChlF is mainly used for individual leaves and small plants. Moreover, it is unrealistic to be applied at the large scale with the limitation of laser pulse energy and background interference [25,26]. Additionally, it is difficult to extrapolate the result because the shape and intensity of the active ChlF varies with the excitation light source [14].

Sun-induced fluorescence (SIF), also known as passive chlorophyll fluorescence, has been widely used in recent years as a promising approach to probing plant physiology, net photosynthesis, stress status at different scales, i.e., leaf, canopy, region and global [27,28]. SIF, a bimodal spectrum ranging from 650 nm to 850 nm, is emitted directly by Chls under the excitation of sun-light. It is composed of two peaks, with the first peak (685 nm) located in the red region, which is mainly attributable to Photosystem II (PS II), and the second peak (740 nm) located in the far-red region, which is attributed to Photosystem I (PS I) and PS II [29,30]. Compared with the values of the reflected and transmitted radiation, though the signal of leaf SIF is relatively small (just about 2–5% in the near-infrared), it plays an important role in characterizing the photosynthetic process [29,31]. SIF has been employed as an effective means not only for detecting plant photosynthetic capacity [32], light-use efficiency [33,34], stress, and injury [35,36], but also for other physiological parameters related to nitrogen fertility conditions. The ChlF peak ratio is known to be an accurate estimator of leaf Chl content [23,24], which is an indirect association between ChlF and LNC, mediated by chlorophyll. Tubuxin et al. [37] has estimated the Chl content using SIF at various growth stages of paprika (*Capsicum annuum cv.* 'Sven') plants. Moreover, Du et al. [38] reported a high relationship between SIF at the canopy level and photosynthetically active radiation absorbed by chlorophyll, although it is affected by species-specific, bio-chemical components and canopy structure, particularly at the O_2-B band. The model-based analysis has shown that the slope of gross primary production and SIF tends to be smaller with increasing Chlorophyll a + b content (Cab). The slope is only sensitive when Cab is <20 μg·cm^{-2} and is stable when Cab is >20 μg·cm^{-2} [39]. The studies mentioned above provide the experimental basis to probe the LNC utilizing SIF, which is closely linked with the management of nitrogen fertility. However, few studies so far have explored the feasibility and potential of SIF to detect the LNC in agronomic crops.

Concerning a typical bifacial leaf, SIF is emitted from both sides [13,40]. Although the upward and downward SIF are generated by the same incident light, there are differences between them due to the internal pigment distribution and structural factors [13]. Descriptions of the upward and downward SIF characteristics are helpful to interpret the remote sensing signal. Understanding and comparing the contribution of the upward and downward SIF in the total SIF helps in recognizing the change of SIF in the propagation process in a remote sensing manner. Moreover, it should also be noted that two SIF emission peaks are affected by strong internal absorption, which could affect the percentage of the upward and downward SIF and change the ratio between red and far-red peaks for both sides. So far, no studies have investigated the capacity and difference of upward and downward SIF in the LNC detection. Few researchers have compared the ability of the two SIF peaks to estimate LNC, which would be beneficial to understanding the mechanism of monitoring the LNC based on SIF.

The overall goal of this study is to estimate LNC in wheat, a major food crop, using SIF related parameters. To fulfill this goal, four main objectives are pursued: (1) to understand the variation of upward and downward SIF spectra under different LNC levels; (2) to compare the differences of the correlations between the upward and downward SIF spectra and LNC; (3) to construct an empirical model for estimation of LNC based on the upward and downward SIF-related parameters; (4) to evaluate the performance of the LNC models under various Chl content and leaf mass per area (LMA) levels in wheat.

2. Materials and Methods

2.1. Experimental Design

Two completely randomized block design field experiments were carried out over one growing season with different nitrogen application rates, planting densities and different winter wheat cultivars replicated three times, in Jiangsu province of East China. During two experiments, 50% nitrogen fertilizer was applied at the pre-planting stage and 50% at the jointing period for N+ treatments. During Experiment 1, 120 kg·hm^{-2} of monocalcium phosphate (P_2O_5) and 135 kg·ha^{-1} of potassium chloride (KCl) were applied prior to seeding for all treatments. During Experiment 2, 105 kg hm^{-2} P_2O_5 was applied at sowing and 135 kg·ha^{-1} KCl was split 50% at sowing and 50% at jointing. Crop management followed local standard practices in wheat production. The detailed information is provided in Table 1.

Table 1. Design of two field experiments and data acquisition.

Experiment (Exp.)	Year	Plot Size (m × m)	Wheat Cultivar	Planting Density	N Application Rate (Kg·ha^{-1})	Sampling Date	Number of Samples
1 Rugao (32°15′N, 120°38′E)	2016–2017	5 × 6	Yangmai 15 (V1) Yangmai 16 (V2)	25 cm 40 cm	0, 150, 300	Jointing, Booting, Heading, Anthesis	29 30 30 30
2 Sihong (33°27′N, 118°13′E)	2016–2017	6 × 7	Huaimai 20 (V3) Xumai 30 (V4)	25 cm	0, 90, 180, 270, 360	Booting, Heading, Anthesis	30 30 30

2.2. Measurements of Sun-Induced Fluorescence at the Leaf Scale

2.2.1. Acquisition of the Upward (↑F) and Downward (↓F) SIF Spectra at the Leaf Scale

Reflectance (R), transmittance (T) without fluorescence contribution, and chlorophyll fluorescence (F) datasets were measured under natural illumination with clear sky conditions using an ASD FieldSpec Pro FR2500 spectrometer (ASD) (Analytical Spectral Devices, Boulder, CO, USA) coupled with the FluoWat leaf clip (Producción por mecanizados villanueva S.L.U, Spain) from 10:00–11:30 at each growth stage [13,41,42]. The ASD recorded data at 1 nm intervals in the region of 350–2500 nm. The spectral data were collected with a sampling interval of 1.4 nm and a full width at half maximum (FWHM) spectral resolution of 3 nm in 350–1000 nm. It recorded reflectance with a sampling interval of 2 nm and a FWHM spectral resolution of 10 nm in 1000–2500 nm.

Due to the flexible design of the FluoWat leaf clip, fiber can be vertically positioned onto the adaxial and abaxial leaf (Figure 1A). The incoming sun radiance (I) was measured as the reflected radiance of a Spectralon reflectance standard (ODM-98, Gigahertz-Optik GmbH, Türkenfeld, Germany). When the leaf was clamped into the FluoWat leaf clip, the incident solar beam could be manually aligned into the open aperture with the direction of 45° relative to the leaf surface, and the R-T-F datasets were measured. Then, a high performance low pass filter (<650 nm, Producción por mecanizados villanueva S.L.U, Spain) was used to cut off the light above 650 nm, the upward and downward fluorescence emission (↑F and ↓F) were obtained separately (Figure 1B) [13,42]. Upward and downward sun-induced fluorescence emission (↑F and ↓F) were measured at the same point of the upper and lower epidermis of the wheat leaf, respectively. The point of the measurement was at two-thirds of the distance from the leaf base. The first, second, and third fully-expanded leaves from the top were selected randomly from a plant in every plot. The leaf veins were avoided to appear in the detection area during the measurements. Additional details regarding the data acquisition are provided in Figures 1 and 2.

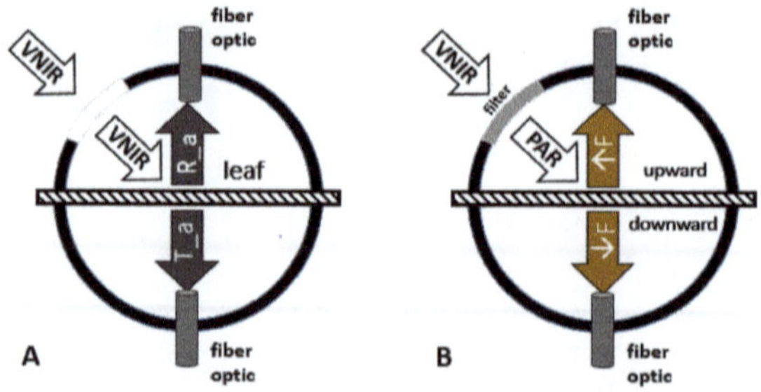

Figure 1. Scheme of the FluoWat leaf clip during measurement. Reflectance and transmittance with the contribution of SIF are measured by inserting a fiber into the upward and downward position of the FluoWat leaf clip (**A**); with the short-pass filter (wavelength shorter than 650 nm), the upward and downward SIF are collected (**B**).

2.2.2. Sun-Induced Fluorescence (SIF) Yield Indices

Since the intensity of the incident light was different on every testing day during the whole wheat growth stages, SIF yield indices (FY, unitless) were calculated by normalizing the SIF for the absorbed incoming photosynthetic active radiance (APAR). APAR equals the integration of incoming sun radiance in the photosynthetic active radiation (PAR) region (400–700 nm) multiplying with the fraction of the light absorbed in the PAR region of (fAPAR) [42] (Equations (1)–(5)). The total measured Chl fluorescence yield (FY_{tot}, unitless) equals the sum of the upward and downward Chl fluorescence yield (FY_{tot} = ↑FY + ↓FY). Additionally, several SIF yield indices constructed by the red and the far-red emission peaks also were calculated to track the characteristics of the SIF spectra (Table 2). More details about the measurements and the formulas of SIF yield indices used in the paper, can be seen in [13,42].

$$\mathrm{PAR} = \int_{400}^{700} \mathrm{I} \cdot d\lambda \tag{1}$$

$$\mathrm{fAPAR} = (1 - \mathrm{R} - \mathrm{T}) \tag{2}$$

$$\mathrm{APAR} = \int_{400}^{700} [\mathrm{I} \times \mathrm{fAPAR}] d\lambda \tag{3}$$

$$\uparrow \mathrm{FY} = \frac{\uparrow \mathrm{F}}{\mathrm{APAR}} \tag{4}$$

$$\downarrow \mathrm{FY} = \frac{\downarrow \mathrm{F}}{\mathrm{APAR}} \tag{5}$$

Figure 2. Drawing of the SIF measurement.

The distribution of SIF peaks in the red region and far-red region was close to the normal distribution, as shown in Figure 3. The data collected from the two ecological sites showed that the peak in the red region was located in the range of 685–690 nm, among which a 40% SIF peak appeared at 687 nm (Figure 3A). The peak in the far-red region was situated around 735–741 nm, of which a 58% SIF peak was discovered at 739 nm (Figure 3B). Therefore, the SIF yield indices calculated in this study were based on the peak emission positions of 687 nm and 739 nm.

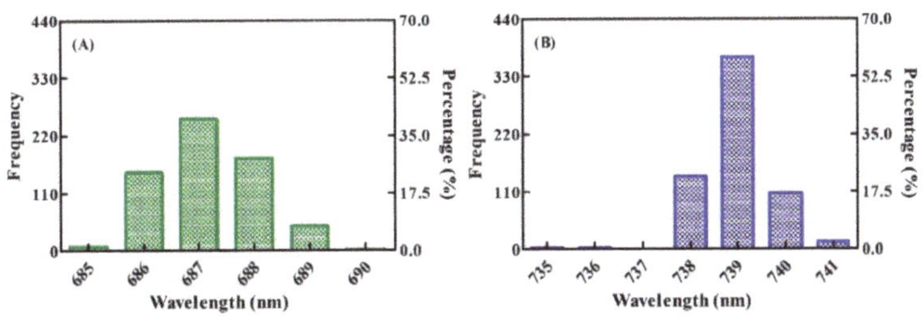

Figure 3. Distributions of the SIF peak emission positions in winter wheat ((**A**) the red region, (**B**) the far-red region).

Table 2. SIF yield indices used in this study.

	SIF Yield Indices	Definition	Formula
Upward	↑FY687 (%)	Upward SIF emission at 687 nm normalized by APAR	↑F687/APAR
	↑FY739 (%)	Upward SIF emission at 739 nm normalized by APAR	↑F739/APAR
	↑FY687/↑FY739 (%)	The ratio of upward SIF emission peaks	↑FY687/↑FY739
Downward	↓FY687 (%)	Downward SIF emission at 687 nm normalized by APAR	↓F687/APAR
	↓FY739 (%)	Downward SIF emission at 739 nm normalized by APAR	↓F739/APAR
	↓FY687/↓FY739 (%)	The ratio of downward SIF emission peaks	↓FY687/↓FY739

2.3. Measurements of Leaf Biochemical Parameters

All leaves were detached for scanning and weighing after the measurement of SIF. Leaf area was determined with a leaf area meter LI-3000 (LI-COR, Inc., Lincoln, NE, USA). Finally, leaves were oven-dried at 105 °C for 30 min, and then at 80 °C for 48 h until a constant dry weight (DW) was obtained. LMA was calculated as grams of dry mass per square centimeter. The LNC in the leaf tissues (% or $Gn \cdot g^{-1}$ DW) was determined using the micro-Kjeldahl method in this study. The leaf Chl content was estimated by the PROCWT model which couples PROSPECT with continuous wavelet transform [43].

2.4. Calculation of Vegetation Indices

To compare the differences in detection of LNC between the SIF and the vegetation indices, some widely used vegetation indices from previous studies are cited, such as a normalized difference vegetation index (NDVI), enhanced vegetation index (EVI2), red edge inflection point (REP), green NDVI, green chlorophyll index (CI_{green}), and red edge chlorophyll index ($CI_{red\ edge}$). NDVI, EVI2, and green NDVI are generally applied for the remote estimation of canopy LAI, while REP, CI_{green}, and $CI_{red\ edge}$ are widely used to estimate the Chl content. The detailed information is summarized in Table 3.

Table 3. Vegetation indices used in this study.

Index	Equation	Reference
Normalized difference vegetation index (NDVI)	(R810 − R690)/(R810 + R690)	[44]
Enhanced vegetation index (EVI2)	2.5 × (R810 − R690)/(R810 + 2.4 × R690 + 1)	[45]
Red edge inflection point (REP)	R700 + 40 × [(R670 + R780)/2 − R700)/(R740 − R700)]	[46]
Green NDVI	(R800 − R550)/(R800 + R550)	[47]
Green chlorophyll index (CI_{green})	(R800/R550) − 1	[48,49]
Red edge chlorophyll index ($CI_{red\ edge}$)	(R800/R720) − 1	[48,49]

2.5. Statistical Analysis

The constructed models of combined wheat LNC data were evaluated in leave-one-out cross validation (LOOCV). The predictive performance of LNC models on SIF yield indices was evaluated using different statistical parameters: The coefficient of determination (R^2); root mean square error (RMSE); relative root mean square error (RRMSE).

The fitness between the predicted and observed values were evaluated by the square of correlation coefficient (R^2), the root mean square error (RMSE) [50], and relative root mean square error (RRMSE) [51].

3. Results

3.1. Characteristics of SIF Spectra at the Leaf Scale under Varied Nitrogen Rates

We took the Experiment 2 data as the example to demonstrate the differences of the downward and upward SIF in the region of 665–850 nm at different nitrogen levels (Figure 4), and SIF in the region of 650–665 nm was deleted due to the noise of partial transmittance of the filter in the region. Generally, Figure 4A–C show that, with the increase of LNC, SIF yield is decreasing in the red region, while it rises in the near-infrared region, which was observed on both the upward and downward SIF yield. The red and far-red peaks are clearly visible at the upward SIF spectra, with a lower peak in the red region than that in the far-red region (Figure 4A); but, in the downward SIF yield spectrum, the red peak is inconspicuous, especially for the leaves with high LNC (Figure 4B).

Figure 4D–F show how the signal of the downward SIF is generally weaker than that of the upward SIF for a given LNC. Both the signal of two peaks in the red region and the far-red region for the downward SIF is significantly smaller than that in the upward SIF. However, the peak in the far-red region is only a little smaller in downward SIF than that in the upward SIF. The changing trend was consistent under the different nitrogen levels.

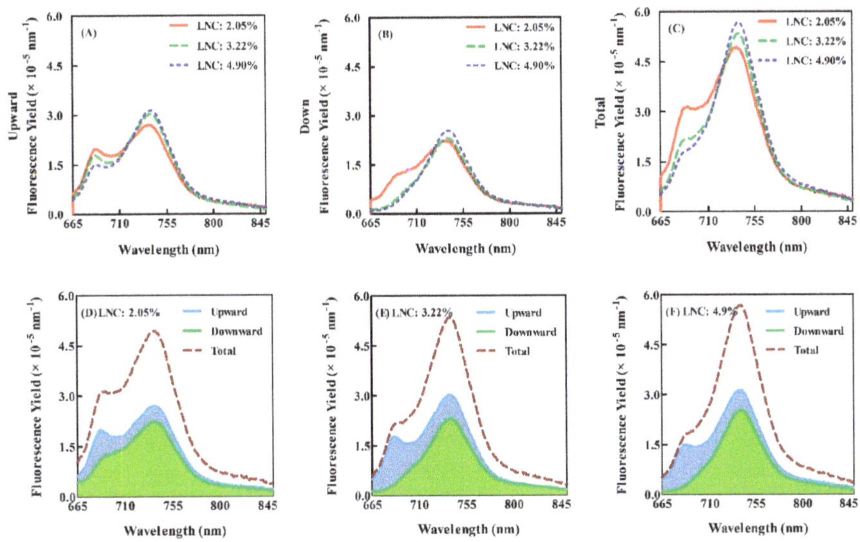

Figure 4. The upward, downward and total SIF yield spectra under different nitrogen levels in Experiment 2. Top row, each figure comparing different LNC contents for a given SIF component (**A**–**C**). Bottom row, each figure comparing the different SIF contributions for a given LNC (**D**–**F**).

3.2. Correlations between the Upward and Downward SIF Yield and Three Given LNC Ranges for the Winter Wheat

Correlation coefficients between the upward and downward SIF yield spectra and the three different ranges of LNC are shown in Figure 5. It shows that the sensitive band to the LNC is approximately at 686 nm in the visible spectral range. There is a quite flat curve in the region of 730–770 nm in the far-red region, with the close to highest correlation with LNC at 758 nm for both upward and downward SIF, as shown in Figure 5. The LNC is negatively correlated to both upward and downward SIF of wheat leaves in the region of 650–720 nm, which shows similar trends in different bands under all the LNC ranges. However, the correlation coefficient in the region of 720–830 nm is positive when the LNC was <3.8% for bidirectional SIF, nevertheless, it is negative in the region of 650–850 nm when 3.8% < LNC < 5% for the downward SIF (Figure 5B). Overall, the downward SIF exhibits a closer relationship with the LNC than that of the upward SIF for the whole SIF region, especially when the LNC is >3.8%. Meanwhile, the correlation coefficient between the LNC and SIF is significantly higher in the red region than that in the far-red region, especially when LNC is <3.8%.

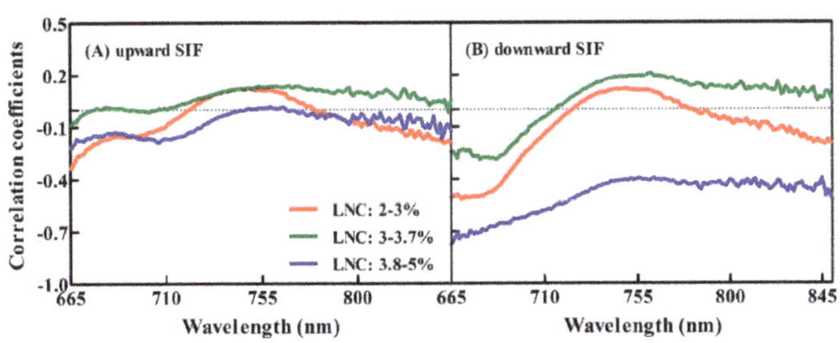

Figure 5. Correlation coefficients between the upward, downward SIF spectra and the three different ranges of LNC.

3.3. Constructing the LNC Estimation Models on SIF Yield Indices in Wheat

The scatter plots for the 210 pairs of the LNC and the upward and downward SIF yield indices samples for two experiments (Experiment 1 and Experiment 2) are shown in Figure 6. The best-fit function for the relationships between SIF yield indices and the LNC are all nonlinear. It shows that ↓FY687 performs much better than ↑FY687 with R^2 as 0.58, 0.72 and 0.60 for Experiment 1, Experiment 2 and combined datasets, respectively (Figure 6A,D). Neither the upward nor downward FY739 indices present any correlation (Figure 6B,D). The upward red/far-red peak ratios (↑FY687/↑FY739) yield the performance with R^2 values of 0.45, 0.65, and 0.51 for Experiment 1, Experiment 2, and combined data, respectively. The downward red/far-red peak ratio ↓FY687/↓FY739 exhibits the strongest relationship with a non-linear character in each dataset (R^2 = 0.72, 0.80 and 0.75 for Experiment 1 dataset, Experiment 2 dataset, and combined, respectively) (Figure 6F).

Figure 6A–C show that, for all upward SIF yield indices, the upward red/far-red peak ratio index ↑FY687/↑FY739 has better fit with the LNC, although with an R^2 of 0.65 it is not very strong. Among the downward SIF yield indices, the ratio ↓FY687/↓FY739 has the best ability to estimate the LNC due to the highest fit, in this case, with an R^2 between 0.72 and 0.80. Briefly, ↓FY687, ↑FY687/↑FY739 and ↓FY687/↓FY739 yield better results with LNC than other SIF yield indices. However, it seems that these three SIF yield indices lack sensitivity to the LNC at low LNC values.

To conclude, the best-fit functions for the relationships between SIF yield indices and LNC were mostly nonlinear with better performance in downward SIF yield indices than that of the upward ones. The red/far-red peak ratio indices showed higher correlation with LNC than that of the single peak SIF yield indices, particularly for ↓FY687/↓FY739 on which the nonlinear exponent prediction function was built with the best goodness of fit. Moreover, it was also found that the differences of the ↓FY687/↓FY739 models between the two data sets were the smallest.

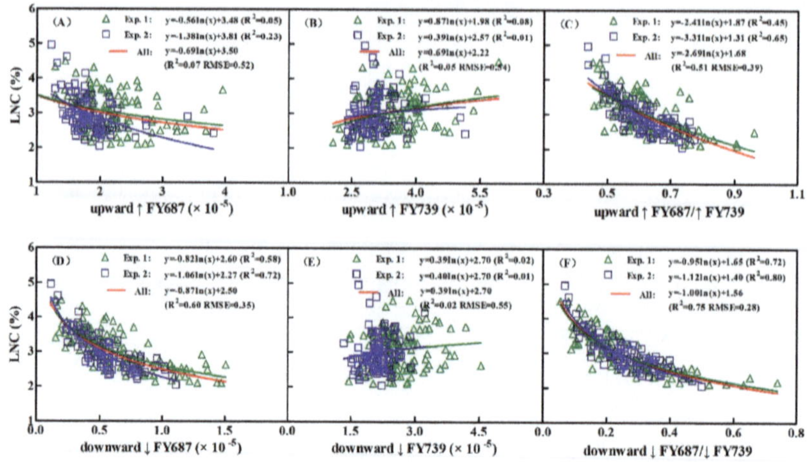

Figure 6. LNC plotted against SIF yield indices (Top row is the upward; Bottom row is the downward). (**A**) the upward ↑FY687 ($\times10^{-5}$); (**B**) the upward↑FY739 ($\times10^{-5}$); (**C**) the upward ↑FY687/↑FY739; (**D**) the downward ↓FY687 ($\times10^{-5}$); (**E**) the downward ↓FY739 ($\times10^{-5}$); (**F**) the downward ↓FY687/↓FY739. Note: green, blue, and red lines are the best-fit function for the Experiment 1 (Exp. 1), Experiment 2 (Exp. 2), and the two datasets combined, respectively.

3.4. Validation of the Estimated LNC Model on SIF Yield Indices in Wheat

The LNC models constructed on the upward and downward SIF yield indices group of FY687, FY739 and red/far-red peak ratio indices (FY687/FY739) were validated by the two combined datasets using LOOCV with three statistical parameters of the coefficients of determination (R^2), the root mean square error (RMSE) and relative root mean square error (RRMSE) (Figure 7). Among them, the estimation models built on ↓FY687 and red/far-red peak ratio indices group (↑FY687/↑FY739 and ↓FY687/↓FY739) performed well overall (Figure 7C,D,F). Additionally, ↓FY687/↓FY739 did best with the highest correlation (0.74) and the lowest RRMSE (9.25%), followed by ↑FY687/↑FY739 (R^2 = 0.50, RMSE = 0.39, RRMSE = 12.88%). The regression line was close to y = x line. The 1:1 plotting with the observed and predicted values exhibited the reliability and accuracy of the derived models, as shown in Figure 7C,F.

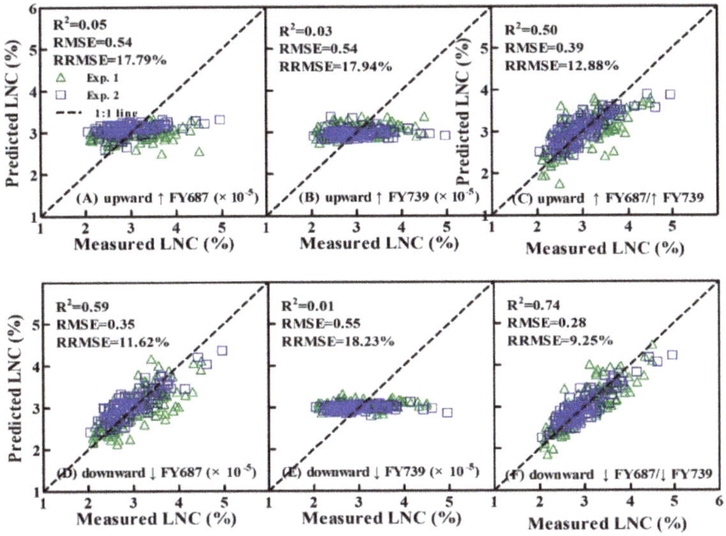

Figure 7. Comparisons between measured and predicted LNC. (**A**) the upward ↑FY687 ($\times 10^{-5}$); (**B**) the upward↑FY739 ($\times 10^{-5}$); (**C**) the upward ↑FY687/↑FY739; (**D**) the downward ↓FY687 ($\times 10^{-5}$); (**E**) the downward ↓FY739 ($\times 10^{-5}$); (**F**) the downward ↓FY687/↓FY739. Data points from the Experiment 1 and Experiment 2 data sets are shown in green (triangle) and blue (square), respectively.

Table 4 illustrates the performance of monitoring models for LNC on vegetation indices used in previous studies for 210 combined samples. Based on the statistical parameters of the calibration and validation sets, $CI_{red\ edge}$ shows the highest correlation to LNC, followed by CI_{green}, Green NDVI, and REP, which yielded better precision than NDVI and EVI for LNC detection. Consistent with the performance in calibration, NDVI and EVI show poorer accuracy than the other four vegetation indices. The vegetation indices NDVI and EVI used for estimation of LAI did not perform well since they were not sensitive when the LNC was high. Compared with the SIF yield indices, only downward fluorescence ratios (↓FY687/↓FY739) appeared to perform better than $CI_{red\ edge}$, but not significantly.

Table 4. Better performing LNC models based on vegetation indices in the calibration and validation.

Vegetation Index	Calibration		Validation			Reference
	Equation	R^2	R^2	RMSE	RRMSE	
EVI	y = 6.74x − 0.67	0.25	0.23	0.49		[45]
NDVI	y = 8.67x − 2.62	0.35	0.33	0.45	14.9%	[44]
Green NDVI	$y = 0.91e^{2.39x}$	0.64	0.61	0.38	11.30%	[47]
REP	y = 0.18x − 125.03	0.65	0.63	0.33	10.991%	[46]
CI_{green}	$y = 1.62e^{0.30x}$	0.67	0.63	0.36	11.10%	[48,49]
$CI_{red\ edge}$	$y = 1.70e^{1.30x}$	0.71	0.68	0.30	10.54%	[48,49]
↓FY687/↓FY739	y = −ln(x) + 1.56	0.75	0.74	0.28	9.25%	This study

3.5. Assessing the LNC Models on SIF Yield Indices under Individual Stage, Different LNC, Chl Content, and Leaf Structure LMA

It can be seen that ↓FY687, ↑FY687/↑FY739, and ↓FY687/↓FY739 yield stable relationships with LNC under various stages with broader ranges of Chl content, in Figure 6. Due to the more complete data acquisition period in Experiment 1 than that of Experiment 2, we took Experiment 1 as the example to compare the performance of these three LNC models in individual stages (Figure 8). They all had the best performance during the anthesis stage, followed by the heading stage, booting stage, and jointing stage. This might be caused by the large difference among the samples at the anthesis stage, which led to the wide range of LNC, Chl content and leaf-structure properties. Meanwhile, the range of ↓FY687/↓FY739 and ↓FY687 were larger than ↑FY687/↑FY739, therefore, the LNC models based on ↓FY687/↓FY739 and ↓FY687 have stronger applicability.

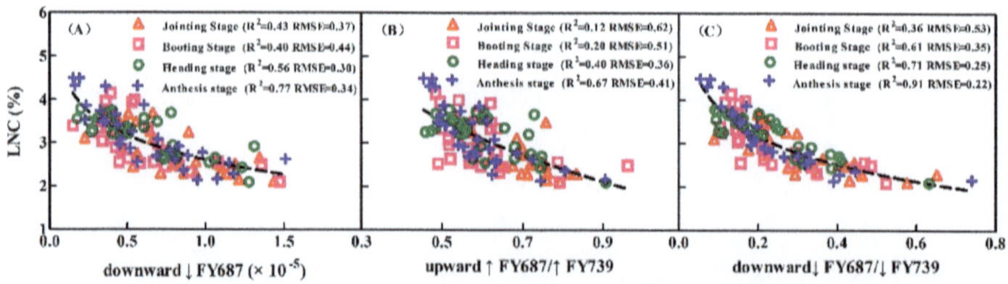

Figure 8. LNC plotted against SIF yield indices at different growth stages: ↓FY687 (**A**); ↑FY687/↑FY739 (**B**) and ↓FY687/↓FY739 (**C**). The data are shown in red (triangle) for jointing stage samples, magenta (square) for booting stage samples, green (circle) for heading stage samples, and blue (plus) for anthesis stage samples. All regressed lines are statistically significant ($p < 0.001$).

SIF emission is influenced by the re-absorption of Chl and leaf structure. The data of the LNC and SIF were divided into five groups according to the value of Chl content and LMA in this study. The objective was to evaluate LNC models on the conditions of practical growth status with varied Chl content and LMA. Figure 9 shows the scatter diagram of SIF yield indices and LNC at different Chl content levels. Observing the Chl content at different groups, the distribution of SIF–LNC scatter changed a little with the change of Chl content. Especially when Chl content is about 55 μg·cm^{-2}, the ↑FY687/↑FY739 and ↓FY687/↓FY739 were not sensitive to LNC. The SIF yield indices ↓FY687 and ↓FY687/↓FY739 still were sensitive to LNC under the low Chl concentrations, just the model changed a little. Generally, the changes of the SIF–LNC models in five groups were not significant. As shown in Figure 10A, although Chl was highly related to LNC, Chl content exerted influence on the relationships between ↓FY687, ↑FY687/↑FY739, and ↓FY687 and the LNC.

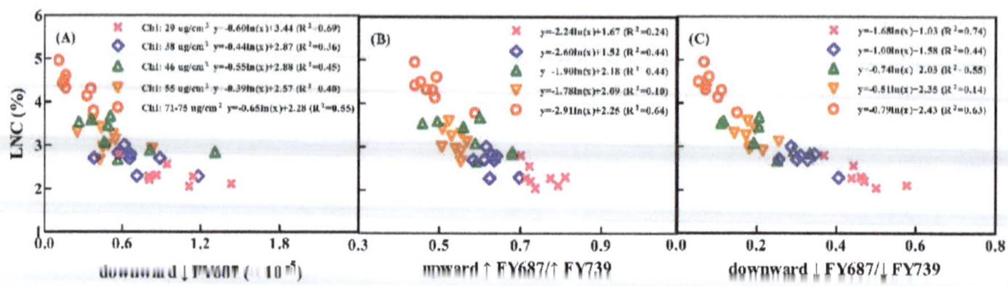

Figure 9. Effects of different Chl content on the relationships between SIF yield indices and LNC. (**A**) ↓FY687 (×10^{-5}); (**B**) ↑FY687/↑FY739; (**C**) ↓FY687/↓FY739.

We explored the LMA impact on the relationship between SIF yield indices and the LNC with a sensitivity analysis using some measured data with different LMA categories (Figure 11). All the SIF yield indices showed less variability to LMA (Figure 11) for LMA values in each range (14–97 g/m^2) and all LNC ranges (1–5%). SIF were sensitive to LNC all the time and the relationships between SIF and LNC at five groups hardly changed. Therefore, the relationships between all SIF yield indices and LNC were almost independent of the LMA. Meanwhile, Figure 10B shows that the LMA is not relevant to LNC in any case. It can be concluded that the SIF emission also is affected by other leaf properties.

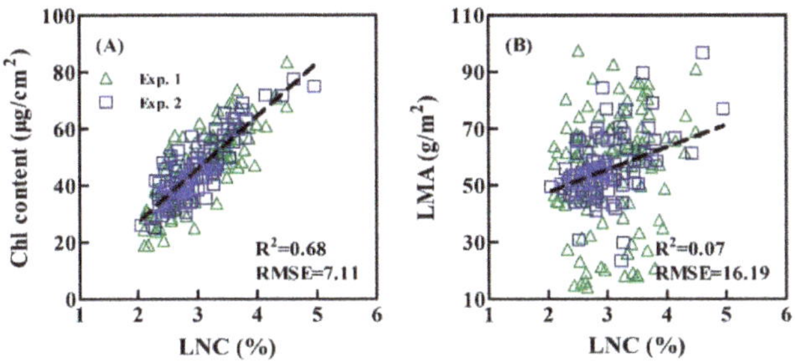

Figure 10. Chl content (**A**) and LMA (**B**) versus LNC for two ecological datasets. The data of Experiment1 and Experiment 2 are shown in green (triangle) and blue (square), respectively.

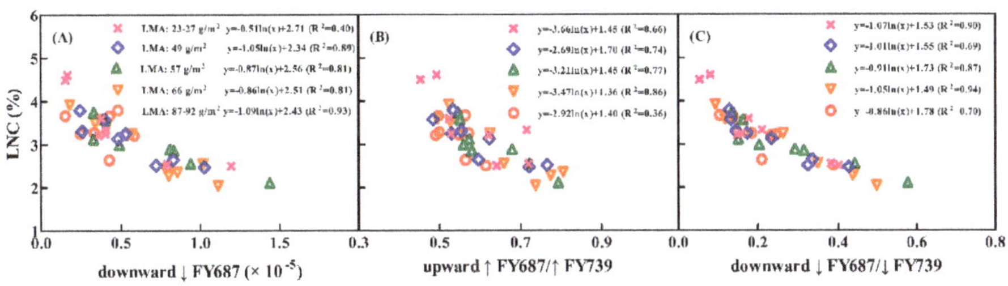

Figure 11. Effects of varied LMA on SIF yield indices to LNC. (**A**) ↓FY687 ($\times 10^{-5}$); (**B**) ↑FY687/↑FY739; (**C**) ↓FY687/↓FY739.

4. Discussion

4.1. Power of the Upward and Downward SIF Yield Indices (↑FY and ↓FY) in LNC Detection

This study showed that the absolute value of upward SIF emission was higher than that of the downward SIF for all leaves. Figure 12A,B,C also show that the value of the downward SIF yield indices (↓FY687 and ↓FY739) are generally lower than that in the upward fluorescence (↑FY687 and ↑FY739). Due to the stronger absorption, scattering effect for the SIF emission [52], and more chlorophyll content around the upper epidermis, it was observed that the upward SIF radiance and SIF yield indices were higher compared with that of the downward SIF, especially for the SIF peak in the red region, which is consistent with [40,53]. Much of the light is absorbed by the palisade parenchyma, in the view of light propagation in the leaves. The SIF emitted downward possibly was self-absorbed a second time by Chl in the leaf, reducing the downward SIF, which was weaker with the spongiform parenchyma acting as a diffuser [54].

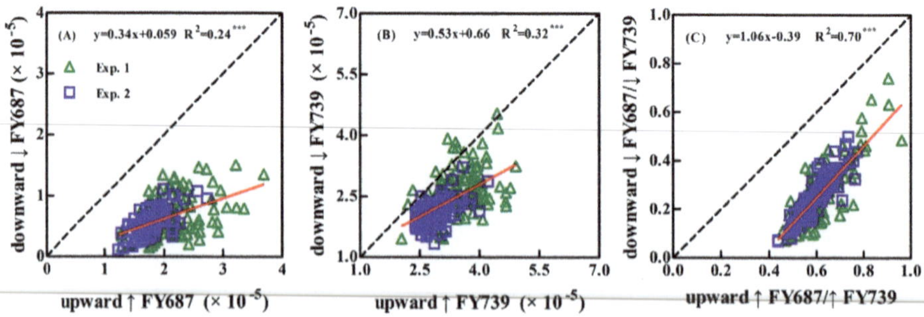

Figure 12. Linear relationship between upward and downward SIF yield indices: A (upward ↑FY687, downward ↓FY687), B (upward ↑FY739, downward ↓FY739) and C (upward ↑FY687/ ↑FY739, downward ↓FY687/ ↓FY739) for two datasets of Experiment 1 and Experiment 2 shown in green (triangle) and blue (square), respectively. Statistical significance is shown as * $p < 0.05$; ** $p < 0.01$; *** $p < 0.001$.

Since the LNC is directly linked to Chl content, it is to be expected that the emission of the peak in the red region decreased with the increasing of the LNC, however, increased in the far-red region. This phenomenon of the transmittance and reflectance characteristics for wheat leaves was consistent with the upward and downward SIF, which is the same as the previous results [55–57]. The reason for the low accuracy of ↑FY739 and ↓FY739 for the LNC is the strong self-absorption of chlorophyll in the red region rather than in the far-red region. The red peak of SIF is more likely to be closely correlated to the LNC than the far-red peak, which agrees with previous study [52]. Although the downward SIF signal was weaker than that of the upward SIF, the correlation coefficient between the LNC and the downward SIF was higher than that with the upward SIF. Bidirectional observations revealed that downward SIF yield indices (↓FY687 and ↓FY687/↓FY739), ↓FY687/↓FY739 especially has a closer relationship with the LNC than that with the upward SIF yield indices (↑FY687 and ↑FY687/↑FY739) (Figures 5 and 6). Zhao et al. [53] also indicated that the downward SIF was more sensitive to Chl content than the upward SIF in the red region by the sensitivity analysis. The cause for these phenomena might be the absorption of Chl, which was more obvious for the downward SIF than that for the upward SIF [58,59]. Moreover, this effect can also be influenced by the leaf thickness, which will affect the photon's path length within the leaf [53]. Van Wittenberghe [13] concluded that the downward SIF, as an important part of SIF emission, should be taken into consideration when interpreting the SIF signal at the scales of leaf, canopy, and landscape, which validates the results of this study.

4.2. Reason for Better Performance of Peak Ratio Indices in LNC Detection

Compared with the single peak SIF yield indices, the red/far-red peak ratio indices (↑FY687/↑FY739, ↓FY687/↓FY739) reflected the proportion of two photosystems. Regarding the two peaks in the SIF, the peak located in the red spectral region (around 687 nm) mainly originated from Photosystem II (PS II), and the other peak (about 739 nm) in the far-red region attributed to both Photosystem I (PS I) and PS II [29,30]. The reason for the better performance of red/far-red peak ratio in LNC estimation is not only the stacking of the thylakoid membranes, but also the associated changes in spillover [60], led to the red/far-red peak ratio performing better in detecting several types of growth status. According to the relevant studies over the last decades, it has been verified that the red/far-red peak ratio index (F687/F739) from active fluorescence technology is correlated with the maximum photochemical efficiency of PS II, largely due to reabsorption in the red region for ChlF, which was suggested to be an indicator to Chl content [61,62], especially in diagnosing the plant stress status [63]. Van Wittenberghe et al. [13] found that both the red/far-red peak ratio indices (↓FY (687)/↓FY (741) and ↑FY (687)/↑FY (739)) have a high correlation with Chl content, and they decrease with the increasing of Chl content. According to Tubuxin et al. [37], solar-induced and artificial light-induced Chl fluorescence yield ratios of 686 nm and 760 nm were both highly correlated with the Chl content. Due to the strong relationship between Chl content and LNC, the red/far-red peak ratio indices (↑FY687/↑FY739, ↓FY687/↓FY739) could be used to estimate the LNC, especially for the downward SIF yield indices ↓FY687/↓FY739, which also corresponded well with the result drawn by [22–24].

It can be seen that ↓FY687, ↑FY687/↑FY739, and ↓FY687/↓FY739 all lacked sensitivity to the LNC at low LNC values. Since the LNC models constructed in this study were empirical, inevitably there would be a problem on the high value or low value. Figure 6 shows that, although the SIF yield indices might lack sensitivity to those parameters of the LNC at low LNC values, it displays that the relationships between ↓FY687, ↓FY687/↓FY739, and LNC—shown in Figure 9A,B—were better than expected. The SIF yield indices ↓FY687 and ↓FY687/↓FY739 still were sensitive to the LNC under the low Chl concentrations, which changed a little compared to the model built on all data.

Generally, simpler methods (SPAD, Dualex) are preferred to estimate the LNC [64,65], however, neither can be used at the leaf scale. Although the reflectance-based parameters are widely used to monitor LNC [66], they lack specificity to nitrogen stress [12]. Compared with the performance of vegetation indices, only the result yielded by the downward fluorescence ratio (↓FY687/↓FY739) was better than $CI_{red\ edge}$, while the relationships between other SIF yield indices and the LNC was not as good as $CI_{red\ edge}$. This might be because the LNC used in this paper refers to the total nitrogen, including the nitrogen involved in photosynthesis and nitrogen in other forms. SIF mainly monitors the nitrogen which is involved in photosynthesis. The mechanism of SIF for the LNC estimation is different from that of hyperspectral remote sensing, which is based on the spectral absorption characteristics caused by the chemical bonds in the chemical composition under certain light intensities. The RubisCo and the Chl are two major storage for nitrogen, both of which are involved in photosynthesis. Evans [67] found that, within species, there are strong linear relationships between nitrogen and both RuBP carboxylase and chlorophyll. Since Chl can be measured easily, in this study only the Chl was taken into consideration, which could be the reason the monitoring accuracy was not very high.

4.3. Performance of the Relationships between SIF Yield Indices and LNC under the Varied Chl Content and Leaf Structure LMA

The influence of the Chl content and the LMA on the LNC models was explored using the field-measured datasets. Chl content is a significant factor in linking SIF with LNC, so the conclusion of this study can provide a reference for LNC estimation using SIF yield indices. The SIF yield indices seemed to be less accurate at a lower LNC content. Regarding low Chl content, the slopes of ↓FY687, ↓FY687/↓FY739, and LNC were still sensitive, however, the SIF–LNC model changed a little, including the slope and intercept. Zhang et al. [39] also found that the slope of gross primary production and SIF also was insensitive when Cab was >20 $\mu g \cdot cm^{-2}$ using the model-based analysis.

We did not record leaf angle and leaf thickness, so the LMA was the alternative for leaf structure. This study showed that the relationships between SIF yield indices and LNC were relatively insensitive to the LMA. Van Wittenberghe et al. [13] illustrated there was not a significant influence of specific leaf area on ↓F755/↑F755. The result probably was caused by the emission fluorescence escaped from the leaf being affected by the other properties, such as the pigments and leaf structure., To reduce the influence of variable light intensity, the SIF yield indices used in the study were normalized by APAR, Therefore, regardless of saturated or unsaturated light conditions, the effects of light were minimized, meanwhile, the kinetic effects of the non-saturating light and the variability of chloroplast movement were not considered for the SIF measurement collected under variable light intensities. Therefore, in the ongoing research, we would focus on improving the machine for SIF measurement at the leaf scale and providing a more practical SIF index to estimate the LNC combining the result of canopy scale and assessing the influence of canopy structure properties (leaf area, leaf angle, and more) on the SIF yield indices.

5. Conclusions

We assessed the capacity of upward and downward SIF yield indices for LNC estimation at the leaf scale under various growth stages for winter wheat. The signal of the upward SIF was higher than that of the downward SIF due to the re-absorption and scattering effect. Bidirectional observations revealed that downward SIF had the higher fitness with the LNC than that of upward SIF. Downward SIF also played an important role for predicting the LNC of winter wheat. The downward red/far-red peak ratio index (↓FY687/↓FY739) showed the highest correlation to LNC with stable performance. The relationships between LNC and these three SIF yield indices (↓FY687, ↑FY687/↑FY739, and ↓FY687/↓FY739) were hardly influenced by the LMA and Chl content. Therefore, SIF can be used as a new priority with higher accuracy than vegetation indices to detect nitrogen content directly for future study in the scales of field and region. SIF can be affected by the canopy structure, background, and atmospheric absorption, but, in this study we simply assessed the LNC using SIF yield indices under the varied ranges of Chl content and LMA. We should consider the impact of those factors on LNC estimation by SIF yield indices in their on-going research.

Author Contributions: M.J., W.C., and X.Y. conceived and designed the research. M.J., J.Z., and C.M. contributed significantly to the field experiments and data collection. M.J., L.A., W.C., and X.Y. made important contributions to the research method, data analysis, and manuscript revision. D.L. provided the model PROCWT and contributed to retrieve chlorophyll content. T.C., Y.T., and Y.Z. contributed in providing suggestions for the research method, and data analysis.

Funding: This research was funded by the National Key Research and Development Program of China (2016YFD0300601), the National Natural Science Foundation of China (31671582), Jiangsu Distinguished Professor Program, Jiangsu Collaborative Innovation Center for Modern Crop Production, the Priority Academic Program Development of Jiangsu Higher Education Institutions (PAPD), and Qinghai Project of Transformation of Scientific and Technological Achievements (2018-NK-126), China.

Acknowledgments: We are thankful to Jose Moreno of the Image Process Laboratory in Valencia for providing the FluoWat leaf clip. We would also like to express our appreciation for the advice given by Liangyun Liu of Chinese Academy of Science, Yongguang Zhang of Nanjing University, and Feng Zhao of Beihang University. We are grateful to the reviewers for their suggestions and comments which significantly improved the quality of this paper.

Conflicts of Interest: The authors declare no conflict of interest.

References

1. Clevers, J.; Gitelson, A.A. Remote estimation of crop and grass chlorophyll and nitrogen content using red-edge bands on Sentinel-2 and -3. *Int. J. Appl. Earth Obs. Geoinf.* **2013**, *23*, 344–351. [CrossRef]
2. Miao, Y.; Mulla, D.J.; Hernandez, J.A.; Wiebers, M.; Robert, P.C. Potential impact of precision nitrogen management on corn yield, protein content, and test weight. *Soil Sci. Soc. Am. J.* **2007**, *71*, 1490–1499. [CrossRef]
3. Diacono, M.; Rubino, P.; Montemurro, F. Precision nitrogen management of wheat: A review. *Agron. Sustain. Dev.* **2013**, *33*, 219–241. [CrossRef]
4. Yang, J.; Gong, W.; Shi, S.; Du, L.; Sun, J.; Song, S. Estimation of nitrogen content based on fluorescence spectrum and principal component analysis in paddy rice. *Plamt Soil Environ.* **2016**, *62*, 178–183. [CrossRef]
5. Inoue, Y.; Sakaiya, E.; Zhu, Y.; Takahashi, W. Diagnostic mapping of canopy nitrogen content in rice based on hyperspectral measurements. *Remote Sens. Environ.* **2012**, *126*, 210–221. [CrossRef]
6. Yao, X.; Ren, H.; Cao, Z.; Tian, Y.; Cao, W.; Zhu, Y.; Cheng, T. Detecting leaf nitrogen content in wheat with canopy hyperspectrum under different soil backgrounds. *Int. J. Appl. Earth Obs. Geoinf.* **2014**, *32*, 114–124. [CrossRef]
7. Yao, X.; Huang, Y.; Shang, G.; Zhou, C.; Cheng, T.; Tian, Y.; Cao, W.; Zhu, Y. Evaluation of Six Algorithms to Monitor Wheat Leaf Nitrogen Concentration. *Remote Sens.* **2015**, *7*, 14939–14966. [CrossRef]
8. Schlemmer, M.; Gitelson, A.A.; Schepers, J.; Ferguson, R.; Peng, Y.; Shanahan, J.; Rundquist, D. Remote estimation of nitrogen and chlorophyll contents in maize at leaf and canopy levels. *Int. J. Appl. Earth Obs. Geoinf.* **2013**, *25*, 47–54. [CrossRef]

9. Chen, P.; Haboudane, D.; Tremblay, N.; Wang, J.; Vigneault, P.; Li, B. New spectral indicator assessing the efficiency of crop nitrogen treatment in corn and wheat. *Remote Sens. Environ.* **2010**, *114*, 1987–1997. [CrossRef]
10. Nguy-Robertson, A.; Gitelson, A.A.; Peng, Y.; Viña, A.; Arkebauer, T.; Rundquist, D. Green Leaf Area Index Estimation in Maize and Soybean: Combining Vegetation Indices to Achieve Maximal Sensitivity. *Agron. J.* **2012**, *104*, 1336–1347. [CrossRef]
11. Clevers, J.G.P.W.; Kooistra, L. Using hyperspectral remote sensing data for retrieving canopy chlorophyll and nitrogen content. *IEEE J. Sel. Top. Earth Obs. Remote Sens.* **2012**, *5*, 574–583. [CrossRef]
12. Tremblay, N.; Wang, Z.; Cerovic, Z.G. Sensing crop nitrogen status with fluorescence indicators: A review. *Agron. Sustain. Dev.* **2011**, *32*, 451–464. [CrossRef]
13. Van Wittenberghe, S.; Alonso, L.; Verrelst, J.; Moreno, J.; Samson, R. Bidirectional sun-induced chlorophyll fluorescence emission is influenced by leaf structure and light scattering properties—A bottom-up approach. *Remote Sens. Environ.* **2015**, *158*, 169–179. [CrossRef]
14. Govindjee, G. *Chlorophyll Fluorescence: A Bit of Basics and History*; Springer: Dordrecht, The Netherlands, 2004; pp. 1–41.
15. Kuckenberg, J.; Tartachnyk, I.; Noga, G. Detection and differentiation of nitrogen-deficiency, powdery mildew and leaf rust at wheat leaf and canopy level by laser-induced chlorophyll fluorescence. *Biosyst. Eng.* **2009**, *103*, 121–128. [CrossRef]
16. Cendrero-Mateo, M.P.; Moran, M.S.; Papuga, S.A.; Thorp, K.R.; Alonso, L.; Moreno, J.; Ponce-Campos, G.; Rascher, U.; Wang, G. Plant chlorophyll fluorescence: Active and passive measurements at canopy and leaf scales with different nitrogen treatments. *J. Exp. Bot.* **2016**, *67*, 275–286. [CrossRef] [PubMed]
17. Kalaji, H.M.; Oukarroum, A.; Alexandrov, V.; Kouzmanova, M.; Brestic, M.; Zivcak, M.; Samborska, I.A.; Cetner, M.D.; Allakhverdiev, S.I.; Goltsev, V. Identification of nutrient deficiency in maize and tomato plants by invivo, chlorophyll a, fluorescence measurements. *Plant Physiol. Biochem.* **2014**, *81*, 16–25. [CrossRef] [PubMed]
18. Živčák, M.; Olšovská, K.; Slamka, P.; Galambošová, J.; Rataj, V.; Shao, H.B.; Brestič, M. Application of chlorophyll fluorescence performance indices to assess the wheat photosynthetic functions influenced by nitrogen deficiency. *Plant Soil Environ.* **2014**, *60*, 210–215. [CrossRef]
19. Živčák, M.; Olšovská, K.; Slamka, P.; Galambošová, J.; Rataj, V.; Shao, H.B.; Kalaji, H.M.; Brestič, M. Measurements of chlorophyll fluorescence in different leaf positions may detect nitrogen deficiency in wheat. *Zemdirbyste-Agriculture* **2014**, *101*, 437–444. [CrossRef]
20. Yang, J.; Gong, W.; Shi, S.; Du, L.; Sun, J.; Song, S.; Chen, B.; Zhang, Z. Analyzing the performance of fluorescence parameters in the monitoring of leaf nitrogen content of paddy rice. *Sci. Rep.* **2016**, *6*, 28787. [CrossRef] [PubMed]
21. Cartelat, A.; Cerovic, Z.G.; Goulas, Y.; Meyer, S.; Lelarge, C.; Prioul, J.L.; Barbottin, A.; Jeuffroy, M.H.; Gate, P.; Agati, G.; et al. Optically assessed contents of leaf polyphenolics and chlorophyll as indicators of nitrogen deficiency in wheat (*Triticum aestivum* L.). *Field Crops Res.* **2005**, *91*, 35–49. [CrossRef]
22. Buschmann, C. Variability and application of the chlorophyll fluorescence emission ratio red/far-red of leaves. *Photosynth. Res.* **2007**, *92*, 261–271. [CrossRef] [PubMed]
23. Gitelson, A.; Buschmann, C.; Lichtenthaler, H.K. The chlorophyll fluorescence ratio F735/F700 as an accurate measure of the chlorophyll content in plants. *Remote Sens. Environ.* **1999**, *69*, 296–302. [CrossRef]
24. Lichtenthaler, H.K.; Hak, R.; Rinderle, U. The chlorophyll fluorescence ratio F690/F730 in leaves of different chlorophyll content. *Photosynth. Res.* **1990**, *25*, 295–298. [CrossRef] [PubMed]
25. Rosema, A.; Zahn, H. Laser Pulse Energy Requirements for Remote Sensing of Chlorophyll Fluorescence. *Remote Sens. Environ.* **1997**, *62*, 101–108. [CrossRef]
26. Zhang, Y.J.; Zhao, C.J.; Liu, L.Y.; Wang, J.H.; Wang, R.C. Chlorophyll Fluorescence Detected Passively by Difference Reflectance Spectra of Wheat (*Triticum aestivum* L.) Leaf. *J. Integr. Plant Biol.* **2005**, *47*, 1228–1235. [CrossRef]
27. Porcar-Castell, A.; Tyystjarvi, E.; Atherton, J.; Van der Tol, C.; Flexas, J.; Pfundel, E.E.; Moreno, J.; Frankenberg, C.; Berry, J.A. Linking chlorophyll a fluorescence to photosynthesis for remote sensing applications: Mechanisms and challenges. *J. Exp. Bot.* **2014**, *65*, 4065–4095. [CrossRef] [PubMed]

28. Rascher, U.; Alonso, L.; Burkart, A.; Cilia, C.; Cogliati, S.; Colombo, R.; Damm, A.; Drusch, M.; Guanter, L.; Hanus, J.; et al. Sun-induced fluorescence—A new probe of photosynthesis: First maps from the imaging spectrometer HyPlant. *Glob. Chang. Biol.* **2015**, *21*, 4673–4684. [CrossRef] [PubMed]
29. Papageorgiou, G.C.; Govindjee, G. *Chlorophyll a Fluorescence—A Signature of Photosynthesis*; Springer: Dordrecht, The Netherlands, 2004; p. 818.
30. Baker, N.R. Chlorophyll Fluorescence: A Probe of Photosynthesis In Vivo. *Annu. Rev. Plant Biol.* **2008**, *59*, 89–113. [CrossRef] [PubMed]
31. Franck, F.; Juneau, P.; Popovic, R. Resolution of the Photosystem I and Photosystem II contributions to chlorophyll fluorescence of intact leaves at room temperature. *BBA Bioenerg.* **2002**, *1556*, 239–246. [CrossRef]
32. Guanter, L.; Zhang, Y.; Jung, M.; Joiner, J.; Voigt, M.; Berry, J.A.; Frankenberg, C.; Huete, A.R.; Zarco-Tejada, P.; Lee, J.E.; et al. Global and time-resolved monitoring of crop photosynthesis with chlorophyll fluorescence. *Proc. Natl. Acad. Sci. USA* **2014**, *111*, E1327–E1333. [CrossRef] [PubMed]
33. Liu, L.; Zhang, Y.; Jiao, Q.; Peng, D. Assessing photosynthetic light-use efficiency using a solar-induced chlorophyll fluorescence and photochemical reflectance index. *Int. J. Remote. Sens.* **2013**, *34*, 4264–4280. [CrossRef]
34. Wagle, P.; Zhang, Y.; Jin, C.; Xiao, X. Comparison of solar-induced chlorophyll fluorescence, light-use efficiency, and process-based GPP models in maize. *Ecol. Appl.* **2016**, *26*, 1211–1222. [CrossRef] [PubMed]
35. Ni, Z.; Liu, Z.; Huo, H.; Li, Z.L.; Nerry, F.; Wang, Q.; Li, X. Early Water Stress Detection Using Leaf-Level Measurements of Chlorophyll Fluorescence and Temperature Data. *Remote Sens.* **2015**, *7*, 3232–3249. [CrossRef]
36. Zhao, F.; Guo, Y.; Huang, Y.; Reddy, K.N.; Zhao, Y.; Molin, W.T. Detection of the onset of glyphosate-induced soybean plant injury through chlorophyll fluorescence signal extraction and measurement. *J. Appl. Remote Sens.* **2015**, *9*, 097098. [CrossRef]
37. Tubuxin, B.; Rahimzadeh-Bajgiran, P.; Ginnan, Y.; Hosoi, F.; Omasa, K. Estimating chlorophyll content and photochemical yield of photosystem II (PhiPSII) using solar-induced chlorophyll fluorescence measurements at different growing stages of attached leaves. *J. Exp. Bot.* **2015**, *66*, 5595–5603. [CrossRef] [PubMed]
38. Du, S.; Liu, L.; Liu, X.; Hu, J. Response of Canopy Solar-Induced Chlorophyll Fluorescence to the Absorbed Photosynthetically Active Radiation Absorbed by Chlorophyll. *Remote Sens.* **2017**, *9*, 911. [CrossRef]
39. Zhang, Y.; Guanter, L.; Berry, J.A.; Van der Tol, C.; Yang, X.; Tang, J.; Zhang, F. Model-based analysis of the relationship between sun-induced chlorophyll fluorescence and gross primary production for remote sensing applications. *Remote Sens. Environ.* **2016**, *187*, 145–155. [CrossRef]
40. Louis, J.; Cerovic, Z.G.; Moya, I. Quantitative study of fluorescence excitation and emission spectra of bean leaves. *J. Photochem. Photobiol. B* **2006**, *85*, 65–71. [CrossRef] [PubMed]
41. Alonso, L.; Gomez-Chova, L.; Vila-Frances, J.; Amoros-Lopez, J.; Guanter, L.; Calpe, J.; Moreno, J. Sensitivity analysis of the fraunhofer line discrimination method for the measurement of chlorophyll fluorescence using a field spectroradiometer. In Proceedings of the IEEE International Geoscience and Remote Sensing Symposium (IGARSS), Barcelona, Spain, 23–28 July 2007; pp. 3756–3759.
42. Van Wittenberghe, S.; Alonso, L.; Verrelst, J.; Hermans, I.; Delegido, J.; Veroustraete, F.; Valcke, R.; Moreno, J.; Samson, R. Upward and downward solar-induced chlorophyll fluorescence yield indices of four tree species as indicators of traffic pollution in Valencia. *Environ. Pollut.* **2013**, *173*, 29–37. [CrossRef] [PubMed]
43. Li, D.; Cheng, T.; Jia, M.; Zhou, K.; Lu, N.; Yao, X.; Tian, Y.; Zhu, Y.; Cao, W. PROCWT: Coupling PROSPECT with continuous wavelet transform to improve the retrieval of foliar chemistry from leaf bidirectional reflectance spectra. *Remote Sens. Environ.* **2018**, *206*, 1–14. [CrossRef]
44. Rouse, J.W. *Monitoring the Vernal Advancement and Retrogradation (Greenwave Effect) of Natural Vegetation*; NASA/GSFCT Technical Report; NTRS: Chicago, IL, USA, 1974.
45. Jiang, Z.; Huete, A.; Didan, K.; Miura, T. Development of a two-band enhanced vegetation index without a blue band. *Remote Sens. Environ.* **2008**, *112*, 3833–3845. [CrossRef]
46. Guyot, G.; Baret, F. Utilisation de la Haute Resolution Spectrale pour Suivre L'etat des Couverts Vegetaux. *Spectr. Signat. Objects Remote Sens.* **1988**, *287*, 279.
47. Gilabert, M.A.; Gandía, S.; Meliá, J. Analyses of spectral-biophysical relationships for a corn canopy. *Remote Sens. Environ.* **1996**, *55*, 11–20. [CrossRef]

48. Gitelson, A.A.; Gritz, Y.; Merzlyak, M.N. Relationships between leaf chlorophyll content and spectral reflectance and algorithms for non-destructive chlorophyll assessment in higher plant leaves. *J. Plant Physiol.* **2003**, *160*, 271–282. [CrossRef] [PubMed]
49. Gitelson, A.A.; Viña, A.; Ciganda, V.; Rundquist, D.C.; Arkebauer, T.J. Remote estimation of canopy chlorophyll content in crops. *Geophys. Res. Lett.* **2005**, 32. [CrossRef]
50. Cheng, T.; Riaño, D.; Ustin, S.L. Detecting diurnal and seasonal variation in canopy water content of nut tree orchards from airborne imaging spectroscopy data using continuous wavelet analysis. *Remote Sens. Environ.* **2014**, *143*, 39–53. [CrossRef]
51. Gnyp, M.L.; Miao, Y.; Yuan, F.; Ustin, S.L.; Yu, K.; Yao, Y.; Huang, S.; Bareth, G. Hyperspectral canopy sensing of paddy rice aboveground biomass at different growth stages. *Field Crops Res.* **2014**, *155*, 42–55. [CrossRef]
52. Zhao, F.; Guo, Y.; Huang, Y.; Verhoef, W.; Van der Tol, C.; Dai, B.; Liu, L.; Zhao, H.; Liu, G. Quantitative Estimation of Fluorescence Parameters for Crop Leaves with Bayesian Inversion. *Remote Sens.* **2015**, *7*, 14179–14199. [CrossRef]
53. Vogelmann, T.C.; Han, T. Measurement of gradients of absorbed light in spinach leaves from chlorophyll fluorescence profiles. *Plant Cell Environ.* **2000**, *23*, 1303–1311. [CrossRef]
54. Vogelman, T.C.; Nishio, J.N.; Smith, W.K. Leaves and light capture: Light propagation and gradients of carbon fixation within leaves. *Trends Plant Sci.* **1996**, *1*, 65–70. [CrossRef]
55. Thomas, J.R.; Gausman, H.W. Leaf Reflectance vs. Leaf Chlorophyll and Carotenoid Concentrations for Eight Crops1. *Agron. J.* **1977**, *69*, 799–802. [CrossRef]
56. Bauerle, W.L.; Weston, D.J.; Bowden, J.D.; Dudley, J.B.; Toler, J.E. Leaf absorptance of photosynthetically active radiation in relation to chlorophyll meter estimates among woody plant species. *Sci. Hortic.* **2004**, *101*, 169–178. [CrossRef]
57. Wang, J.F.; He, D.X.; Song, J.X.; Dou, H.J.; Du, W.F. Non-destructive measurement of chlorophyll in tomato leaves using spectral transmittance. *Int. J. Agric. Biol. Eng.* **2015**, *8*, 73–78.
58. Fournier, A.; Daumard, F.; Champagne, S.; Ounis, A.; Goulas, Y.; Moya, I. Effect of canopy structure on sun-induced chlorophyll fluorescence. *ISPRS J. Photogramm. Remote Sens.* **2012**, *68*, 112–120. [CrossRef]
59. Middleton, E.M.; Cheng, Y.B.; Corp, L.A.; Campbell, P.K.E.; Huemmrich, K.F.; Zhang, Q.; Kustas, W.P. Canopy level Chlorophyll Fluorescence and the PRI in a cornfield. In Proceedings of the IEEE International Geoscience and Remote Sensing Symposium (IGARSS), Munich, Germany, 22–27 July 2012; pp. 7117–7120.
60. Butler, W.L.; Kitajima, M. Energy transfer between Photosystem II and Photosystem I in chloroplasts. *Biochim. Biophys. Acta* **1975**, *396*, 72–85. [CrossRef]
61. Hák, R.; Lichtenthaler, H.K.; Rinderle, U. Decrease of the chlorophyll fluorescence ratio F690/F730 during greening and development of leaves. *Radiat. Environ. Biophys.* **1990**, *29*, 329–336. [CrossRef] [PubMed]
62. Pedrós, R.; Goulas, Y.; Jacquemoud, S.; Louis, J.; Moya, I. FluorMODleaf: A new leaf fluorescence emission model based on the PROSPECT model. *Remote Sens. Environ.* **2010**, *114*, 155–167. [CrossRef]
63. Rossini, M.; Meroni, M.; Celesti, M.; Cogliati, S.; Julitta, T.; Panigada, C.; Rascher, U.; Van der Tol, C.; Colombo, R. Analysis of Red and Far-Red Sun-Induced Chlorophyll Fluorescence and Their Ratio in Different Canopies Based on Observed and Modeled Data. *Remote Sens.* **2016**, *8*, 412. [CrossRef]
64. Errecart, P.M.; Agnusdei, M.G.; Lattanzi, F.A.; Marino, M.A. Leaf nitrogen concentration and chlorophyll meter readings as predictors of tall fescue nitrogen nutrition status. *Field Crops Res.* **2012**, *129*, 46–58. [CrossRef]
65. Li, J.W.; Zhang, J.X.; Zhao, Z.; Lei, X.D.; Xu, X.L.; Lu, X.X.; Weng, D.L.; Gao, Y.; Cao, L.K. Use of fluorescence-based sensors to determine the nitrogen status of paddy rice. *J. Agric. Sci.* **2013**, *151*, 862–871. [CrossRef]
66. Li, F.; Miao, Y.; Hennig, S.D.; Gnyp, M.L.; Chen, X.P.; Jia, L.L.; Bareth, G. Evaluating hyperspectral vegetation indices for estimating nitrogen concentration of winter wheat at different growth stages. *Precis. Agric.* **2010**, *11*, 335–357. [CrossRef]
67. Evans, J.R. Photosynthesis and nitrogen relationships in leaves of C3 plants. *Oecologia* **1989**, *78*, 9–19. [CrossRef] [PubMed]

remote sensing

MDPI

Article

Analyzing the Effect of Fluorescence Characteristics on Leaf Nitrogen Concentration Estimation

Jian Yang [1], Shalei Song [2,*], Lin Du [1], Shuo Shi [3,4], Wei Gong [3,4], Jia Sun [3] and Biwu Chen [3]

1 Faculty of Information Engineering, China University of Geosciences, Wuhan 430074, Hubei, China; yangjian@cug.edu.cn (J.Y.); dulin@cug.edu.cn (L.D.)

2 Wuhan Institute of Physics and Mathematics, Chinese Academy of Sciences, Wuhan 430071, Hubei, China

3 State Key Laboratory of Information Engineering in Surveying, Mapping and Remote Sensing, Wuhan University, Wuhan 430079, Hubei, China; shishuo@whu.edu.cn (S.S.); liesmars_lidar@foxmail.com (W.G.); sunjia@whu.edu.cn (J.S.); cbw_think@whu.edu.cn (B.C.)

4 Collaborative Innovation Center of Geospatial Technology, Wuhan 430079, Hubei, China

* Correspondence: songshalei@gmail.com (S.S.); Tel.: +86-134-7608-2677

Received: 24 July 2018; Accepted: 30 August 2018; Published: 3 September 2018

Abstract: Leaf nitrogen concentration (LNC) is a significant indicator of crops growth status, which is related to crop yield and photosynthetic efficiency. Laser-induced fluorescence is a promising technology for LNC estimation and has been widely used in remote sensing. The accuracy of LNC monitoring relies greatly on the selection of fluorescence characteristics and the number of fluorescence characteristics. It would be useful to analyze the performance of fluorescence intensity and ratio characteristics at different wavelengths for LNC estimation. In this study, the fluorescence spectra of paddy rice excited by different excitation light wavelengths (355 nm, 460 nm, and 556 nm) were acquired. The performance of the fluorescence intensity and fluorescence ratio of each band were analyzed in detail based on back-propagation neural network (BPNN) for LNC estimation. At 355 nm and 460 nm excitation wavelengths, the fluorescence characteristics related to LNC were mainly located in the far-red region, and at 556 nm excitation wavelength, the red region being an optimal band. Additionally, the effect of the number of fluorescence characteristics on the accuracy of LNC estimation was analyzed by using principal component analysis combined with BPNN. Results demonstrate that at least two fluorescence spectral features should be selected in the red and far-red regions to estimate LNC and efficiently improve the accuracy of LNC estimation.

Keywords: laser-induced fluorescence; leaf nitrogen concentration; back-propagation neural network; principal component analysis; fluorescence characteristics

---

1. Introduction

Chlorophyll is an essential factor in crop photosynthesis, and nitrogen (N), a main element in chlorophyll, can favorably affect the growth and quality of crops. Leaf nitrogen concentration (LNC) is a key indicator of the N nutrition in crops, and scientific N fertilization measurements can be obtained by monitoring the LNC [1,2]. Thus, estimating leaf nitrogen concentration (LNC) accurately and nondestructively is important for the accurate diagnosis and quality evaluation of plant growth status [3,4]. The development of remote sensing has made it a significant tool for monitoring plant growth at the leaf, canopy, and landscape levels [5–8]. Many researchers have investigated hyperspectral remote sensing and found a certain difference among the sensitive bands of the LNC for different crops [9–11]. Moreover, the optimal bands will vary at different growth stages of the same crops [12]. Therefore, chlorophyll fluorescence was proposed and utilized for monitoring crop growth status. Chlorophyll fluorescence has shown to be a promising technology for monitoring crop growth status.

Variable chlorophyll fluorescence (or Kautsky kinetics) posits that fluorescence intensity changes with time at a certain wavelength. These fluorescence characteristics are mainly used to monitor the photosynthesis of crops [13–15]. However, Kautsky kinetics are seldom utilized in the monitoring of N status or other agricultural application due to limitations such as the requirement of a fixed distance of measurement and the need for dark-adaption 15 min before measurement [16]. Thus, Tartachnyk and Rademacher [17] suggested that laser-induced chlorophyll fluorescence (LICF) is more conducive to monitoring the N status of crops compared with Kautsky kinetics. Laser-induced fluorescence (LIF) spectra contain abundant spectral information. Fluorescence intensity will be related to different nutrient stresses [14]. Owing to its rapid, non-destructive, and high sensitivity properties, LICF has been widely utilized in monitoring N fertilizer levels in crops [8,18–20].

Subhash and Mohanan [21] pointed out that the fluorescence intensity ratios F690/F725nm and F690/F705nm have great potential for remote sensing-based monitoring of the effect of nutrient stress on paddy rice growth status. Some research analyzed the monitoring ability of LICF for crop nutrition stress and proposed that the fluorescence ratio F685/F730 is mainly related to foliar chlorophyll concentration when no significant reduction in chlorophyll concentration has occurred [22,23]. Gu et al. [24] analyzed the fluorescence characteristics of rice, which has suffered flooding and waterlogging, and demonstrated that LICF can also be used to analyze water stress in crops. Anderson et al. attempted to estimate the cowpea (*Vigna unguiculata* (L.) *Walp*) yield based on the LICF characteristics and found that the fluorescence characteristics were sensitive to change in photosynthetic activity [25]. Yang et al. [26,27] discussed the performance of the LICF for LNC estimation in paddy rice with the support vector machine and back-propagation neural network (BPNN) models. However, studies about the performance of chlorophyll fluorescence intensity or fluorescence ratios of each band for LNC estimation are still sparse. In addition, LIF contains large amounts of spectral information, and studies are lacking on the correct number of fluorescence characteristics that should be selected from the fluorescence spectrum to estimate LNC. Thus, the main target of the present study is to analyze the performance of the fluorescence intensity and fluorescence ratio of each band for LNC estimation in paddy rice with different excitation light wavelengths. Additionally, the effect of the number of fluorescence characteristics on the accuracy of LNC estimation was discussed by using principal component analysis (PCA) combined with BPNN.

2. Materials and Methods

2.1. Materials and Experimental Design

The paddy rice variety Yangliangyou 6 was cultivated in Huazhong Agricultural University, Wuhan City in the province of Hubei, China (Figure 1). The rice was seeded on 30 April 2018, and transplanted to the field on 27 May 2015. Four levels of urea fertilizer (0, 120, 180, 240 kg/ha) were used and divided into three splits (60% at seeding, 20% at tillering and 20% at shooting). In addition, the experimental field had a randomized complete block design with three replications for each treatment under the same cultivation conditions. In each plot, nine fully expanded second leaves from the top were gathered randomly with three replicates for each experimental field on 26 July 2018, providing a total of 324 samples. The fresh leaf samples were sealed in plastic bags, stored in ice chests, and then transported to the laboratory for fluorescence measurement. During the fluorescence measurement, the samples were held on a black paperboard, which is a non-fluorescent material [28].

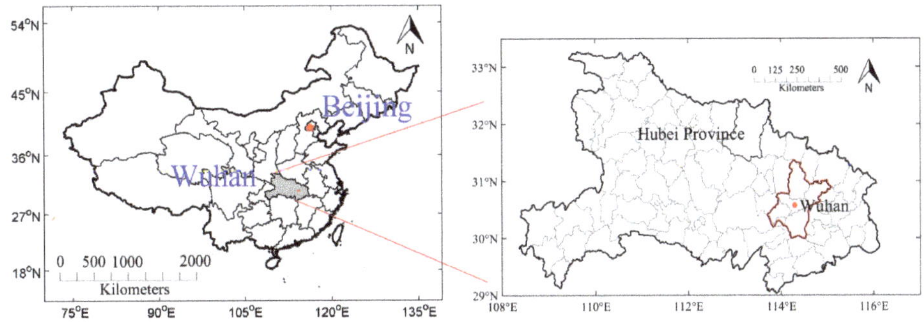

Figure 1. Location of experimental fields, Huazhong Agricultural University in Wuhan, Hubei province, China.

2.2. Measurement of Fluorescence Spectra

The system utilized for LIF measurement consists of three main parts (Figure 2), including the laser emission, optical receiver, and fluorescence signal detection [26]. Generally, the fluorescence is emitted in a longer wavelength region than the excitation wavelength and chlorophyll fluorescence is mainly located in 600–800 nm region [29]. Therefore, 355 nm (ultraviolet), 460 nm (blue), and 556 nm (red) excitation lights were used in the present study to induce plant fluorescence. The 355 nm excitation light source is a neodymium-doped yttrium aluminum garnet laser. The 460 nm and 556 nm lasers used were made by Spectra-Physics. The excitation light was irradiated on the surface of the samples at a near 90° angle, and the emitted fluorescence was collected using the convex lens at a near 90° angle on the same side. In addition, a long-pass filter (Semrock BLP01-355R-25 with the edge of 361 nm and the 93% transmittance at 364.9–900 nm for 355 nm excitation light; and LP02-633RE-25 with the edge of 633 nm and the 93% transmittance at 636.9–1427.4 nm for 460 and 556 nm excitation light sources) was placed before the optical fiber probe and was utilized to reduce the reflected light from the laser entering system. Then, the fluorescence signal entered the spectrometer (Princeton Instrument SP2500i with the spectral resolution of 0.5 nm) by using the single-mode optical fiber with a diameter of 200 μm and was detected by an intensified charge-coupled device camera. Fluorescence data was stored in a personal computer for subsequent analysis.

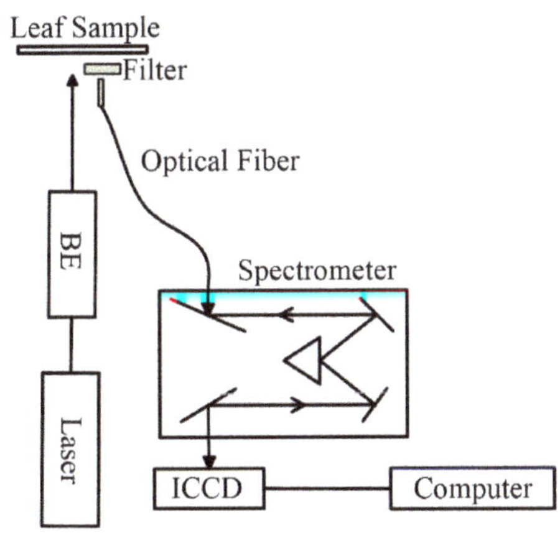

Figure 2. The schematic of the LIF system. BE, beam expander (5 times at 355 nm, 460 nm and 556 nm); ICCD, intensified charge-coupled device.

The fluorescence spectral regions were 360–800 nm, 640–790 nm, and 640–800 nm for the 355 nm, 460 nm and 556 nm excitation lights. The sample interval was 0.5 nm. After the fluorescence spectra was measured, all samples were immediately carried to the Wuhan Academy of Agricultural Science and Technology for LNC analysis. The standard Kjeldahl method was utilized to determine LNC in the present study [3]. Firstly, these samples were cut into pieces and oven-dried at 105 °C for 30 min and then at 80 °C until constant weight for chemical analysis. Then, those pigments soluble in organic solvents were extracted from leaves in acetone for 12 h in the dark at room temperature. The detailed description of standard Kjeldahl method can be found in reference [30].

2.3. Back-Propagation Neural Network

BPNN is a kind of feedforward network with the advantage of self-adaption and self-learning with good performance and has been widely utilized in solving various nonlinear problems. A BPNN model usually consists of three layers, namely, input, hidden, and output layers. The weight of the neurons can be adjusted based on the errors between the target output values and measured values. The training process of BPNN model involves updating the weights until the average sum squared error of the training dataset is minimized within the specified tolerance. A brief introduction about BPNN can be referred to in a previous study [31,32]. The fluorescence characteristics of each excitation light were randomly divided into two datasets, namely, 70% as the training dataset and another 30% as the validation set for LNC prediction [32,33]. In the BPNN model, the fluorescence characteristics served as the input parameter to train the model, and the LNC responded to the output parameters. The coefficient of determination (R^2) between the predicted and measured LNC, the root mean square error (RMSE), and the relative error (RE) were utilized in this study to assess the performance of fluorescence characteristics for LNC estimation. Each characteristic setting was repeated 100 times based on the BPNN model, and the average was obtained.

2.4. Principal Component Analysis

The fluorescence spectra contain spectral information with hundreds of bands that provide high dimensionality that is usually much higher than the number of available training samples. Furthermore, the high-dimensional fluorescence spectra have significant redundancy, because large autocorrelation between adjacent variables is observed. PCA as a tool for dimensionality analysis can efficiently extract the main characteristic variables without any notable loss of information from hundreds or thousands spectral bands [34]. In the process of spectral analysis by using the PCA, the eigen values and eigen vectors of covariance matrix composed of fluorescence spectra were first calculated. Then, multi-dimensional data vectors were mapped from the spectral space to a new orthogonal space using the principal components (PCs). The new variables were calculated based on the combination of the original spectra [35].

$$w_i = \sum_{j=1}^{k} p^2 \left(X_j, Y_i\right) \tag{1}$$

where, X_j is the PC, Y_i represents the original values at ith bands, w_i corresponds to the sum of the kth PC for the ith wavelength, and P is the loading weight of the latent variables. Thus, the analysis can be efficiently simplified using fewer calculated variables than the original ones [33].

3. Results

3.1. Fluorescence Spectrum

Figure 3 shows the fluorescence spectrum of paddy rice leaf excited by 355 nm laser.

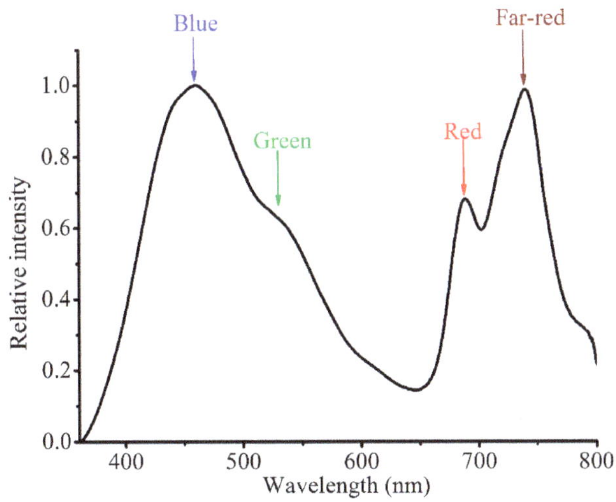

Figure 3. Paddy rice foliar fluorescence spectrum excited by 355 nm excitation light.

The fluorescence spectrum displayed three main fluorescence peaks at 460, 685, and 740 nm, and a peak shoulder at 525 nm (Figure 3). According to previous research works, chlorophyll fluorescence mainly focuses on the wavelength ranges from 600 and 800 nm and peak centering at 685 and 740 nm. The fluorescence peak at 740 nm is attributed to the antenna chlorophyll of Photosystems I and II. Another fluorescence peak at 685 nm corresponds to Chlorophyll *a*, which is associated with Photosystem II [36]. The fluorescence peak shoulder at 525 nm and the peak at 460 nm are attributed to riboflavin and nicotinamide adenine dinucleotide, respectively [37,38]. Figure 3 also shows that the chlorophyll fluorescence at 685 nm is lower than that at 740 nm [18,39]. As the chlorophyll fluorescence is related to the biochemical content of the leaf, it can be applied to remote sensing monitoring and has been identified as a promising technology in the quantitative monitoring of remote sensing.

3.2. LNC Estimation Based on Fluorescence Spectra

To analyze the predictive ability of the fluorescence spectra for monitoring LNC in paddy rice, the BPNN algorithm was used to inversely predict LNC based on the fluorescence spectra. The relationship between the measured and predicted LNC were established and illustrated in Figure 4.

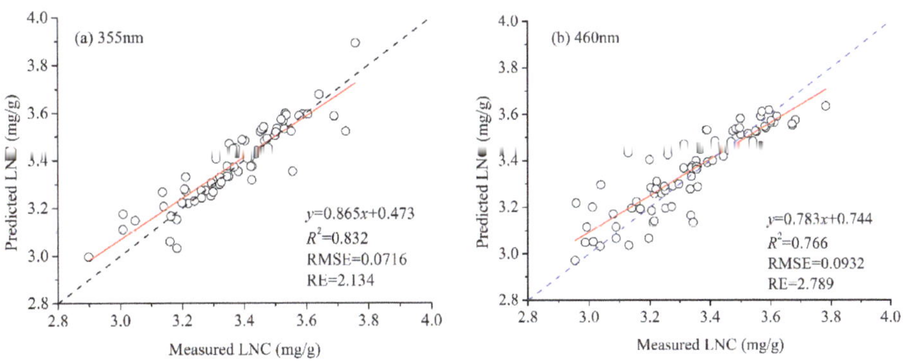

Figure 4. *Cont.*

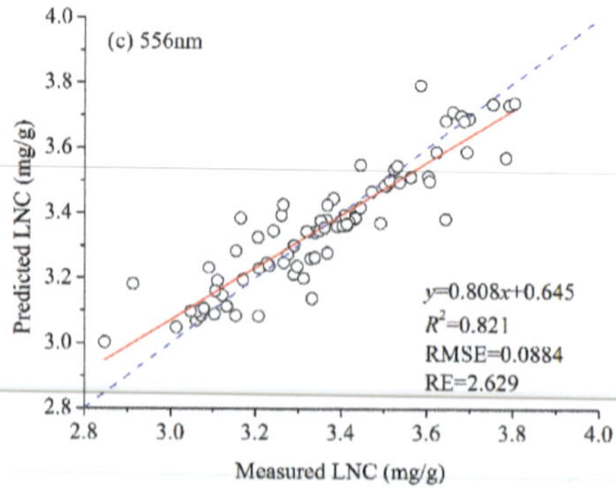

Figure 4. Relationship between the measured and predicted LNC by using BPNN based on fluorescence spectra with different excitation light wavelengths. (**a**) 355 nm, (**b**) 460 nm, (**c**) 556 nm. The blue dotted line is the 1:1 line. The red solid line is the linear regression.

The results of the comparison among the R^2 values in Figure 4 shows that the fluorescence spectra exhibit a promising potential for revising LNC, and the overall R^2 exceeds 0.76. The red solid line denotes the linear regression analysis results between the predicted and measured LNC. For 355 nm (R^2 = 0.832) and 556 nm (R^2 = 0.821) excitation lights, the inversion results show better predictive performance, having higher R^2 and lower RMSE and RE than those for 460nm (R^2 = 0.766) excitation light.

3.3. *Performance of Each Band's Fluorescence Intensity for LNC Estimation*

The sampling interval of the fluorescence spectrum is 0.5 nm for the three excitation light wavelengths. The fluorescence spectra contain a large number of spectral bands, which may be autocorrelated between adjacent bands. To discuss the performance of the fluorescence intensity of each band for estimating LNC, the fluorescence characteristics of each band were used as the input parameter to train the BPNN model to estimate LNC. Each setting was repeated 100 times and the average of R^2 was obtained for every band performance assessment in the LNC estimation. This method can eliminate the local optimum effect of the internal parameters on the BPNN model. The R^2 between the measured and predicted LNC based on the fluorescence characteristic of a single wavelength with different excitation light wavelengths is shown in Figure 5.

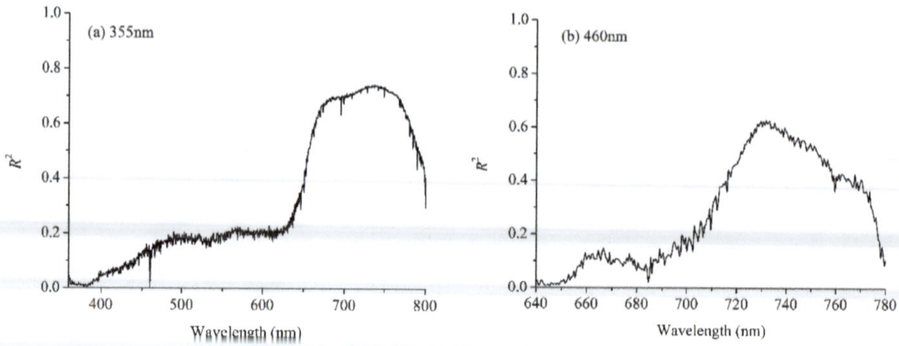

Figure 5. *Cont.*

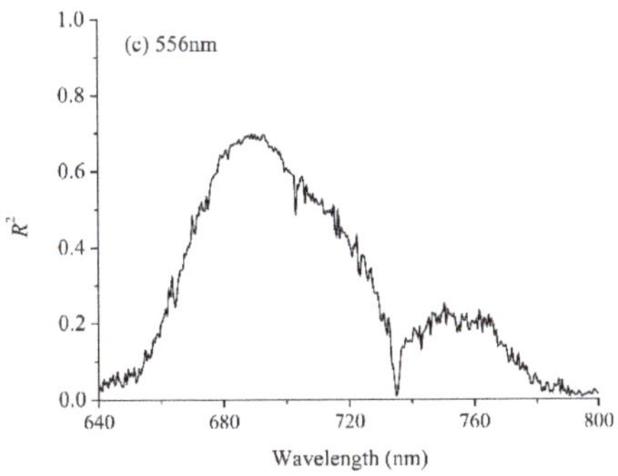

Figure 5. R^2 between the measured and predicted LNC with single wavelength fluorescence served as the input parameter to train the BPNN model for different excitation light wavelengths. (**a**) 355 nm, (**b**) 460 nm, (**c**) 556 nm.

As shown in Figure 5, different bands exhibited different performances in the LNC estimation. For 355 nm excitation light, the chlorophyll fluorescence from 670 nm to 750 nm displayed better performance ($R^2 > 0.70$) in LNC estimation based on the BPNN model than the other bands. For 460 nm excitation light, the optimal fluorescence bands mainly focus on the near 735 nm with the R^2 more than 0.65. For 556 nm excitation light, the optimal fluorescence bands are located near 685 nm, and the R^2 is more than 0.7. Thus, the fluorescence characteristics of paddy rice, which can be used to estimate LNC, were mainly located in the red and far-red fluorescence spectral regions.

3.4. *Performance of Fluorescence Ratio for LNC Estimation*

To analyze the optimal fluorescence characteristics band ratios for estimating paddy rice LNC, further analysis was done on the performance of fluorescence ratios for LNC estimation based on the BPNN model by using datasets with different excitation light wavelengths (Figure 6).

Figure 6 shows the equipotential graphs of R^2 between the measured and predicted LNC with the two wavelengths on the vertical axis and the abscissa with different excitation light wavelengths. An overview of the statistical consequence for the performance of all fluorescence ratios was also provided. For 355 nm excitation light, the fluorescence ratio between red and blue wavelengths displayed higher R^2 than other ratios for estimating LNC. For the 460 and 556 nm excitation lights, the fluorescence ratios between far-red and red wavelengths exhibited higher R^2 than other ratios for estimating LNC. Thus, the chlorophyll fluorescence characteristics ratio located at the red and far-red regions showed better performance for LNC estimation than other regions.

3.5. *LNC Estimation Based on PCA*

3.5.1. Accumulative Variance Analysis

PCA was utilized for the analysis of the internal correlation and reduction of dimensionality of the fluorescence spectra. The most significant characteristic variables were extracted, because the fluorescence spectra contain large amounts of information, which may influence its performance in the LNC estimation. The cumulative variance changes with PC are shown in Figure 7. Apparently, the cumulative variance with additional PC was increased slightly when the number of PCs exceeded four (Table 1).

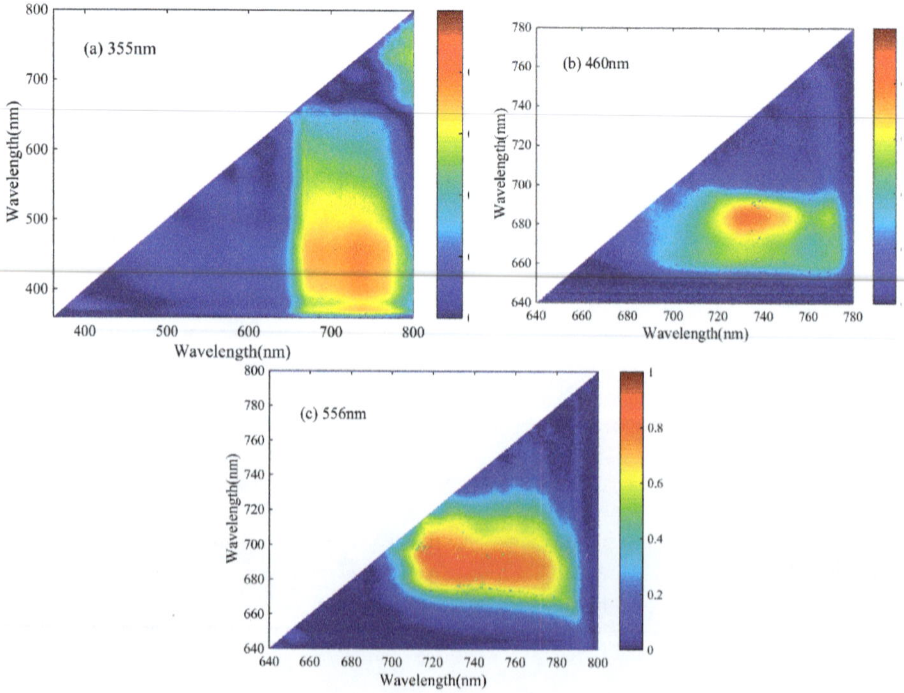

Figure 6. Equipotential graphs of R^2 between the measured and predicted LNC based on the BPNN model with fluorescence ratio of each band for different excitation light wavelengths. (**a**) 355 nm, (**b**) 460 nm, (**c**) 556 nm.

Table 1. Percentages of explained variance for the first four principal components (PCs) with different excitation wavelengths.

	355 nm		460 nm		556 nm	
	Eigen Values	**Explained Variance**	**Eigen Values**	**Explained Variance**	**Eigen Values**	**Explained Variance**
PC1	11.05	80.95%	4.08	66.24%	4.98	75.88%
PC2	1.26	15.94%	1.08	17.18%	0.52	14.24%
PC3	0.13	1.02%	0.58	9.37%	0.28	4.71%
PC4	0.08	0.61%	0.18	2.94%	0.12	1.64%

According to previous research, the first four PCs were mainly attributed to the red and far-red regions [26]. For 355 nm excitation light, the PC3 contains the blue region, but the rate of contribution was less than 2%. Additionally, other PCs contained less spectral information, which is related to LNC that can be ignored and will not influence the performance of the extracted characteristics for estimating LNC.

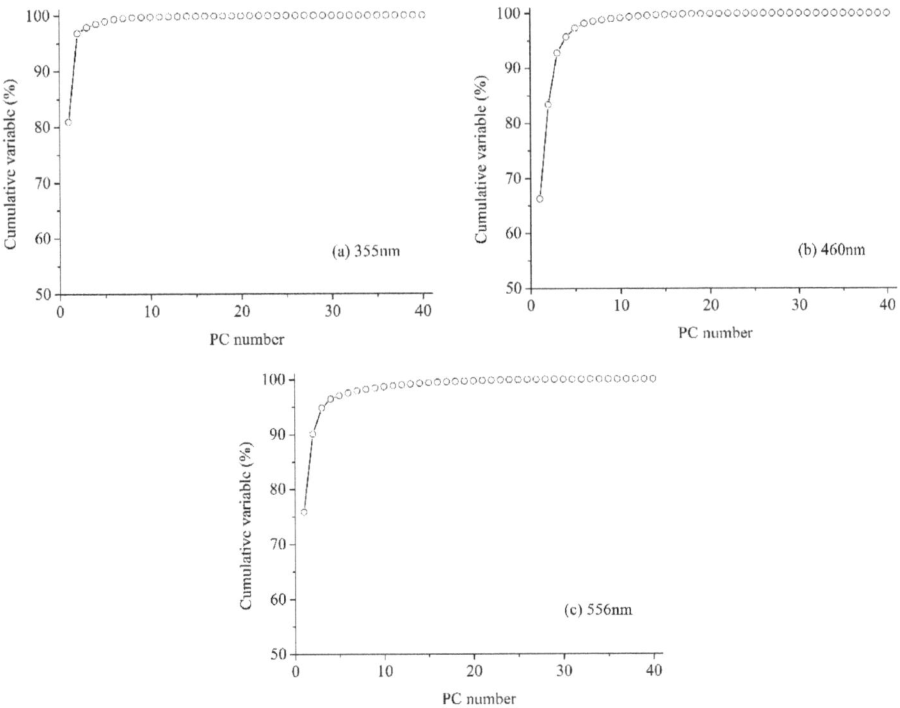

Figure 7. Cumulative explained variances of PC for different excitation light wavelengths. (**a**) 355 nm, (**b**) 460 nm, (**c**) 556 nm.

3.5.2. Performance of New Variables for LNC Estimation

Then, the new variables were calculated based on each PC and used to estimate LNC based on the BPNN model. The R^2 between the measured and predicted LNC changed with the number of new variables (Figure 8).

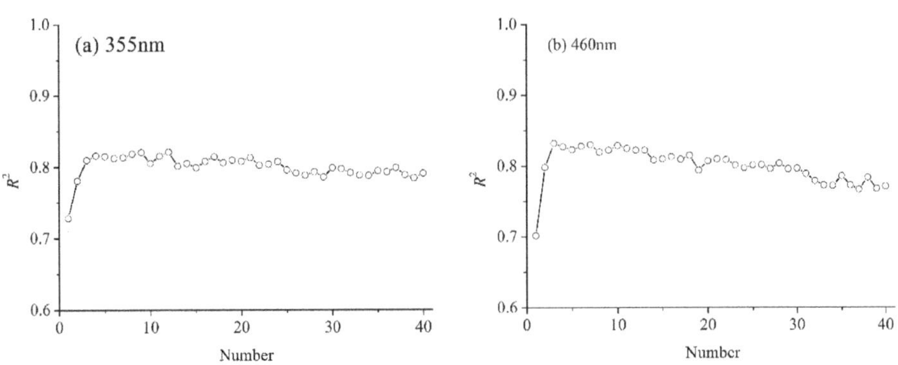

Figure 8. *Cont.*

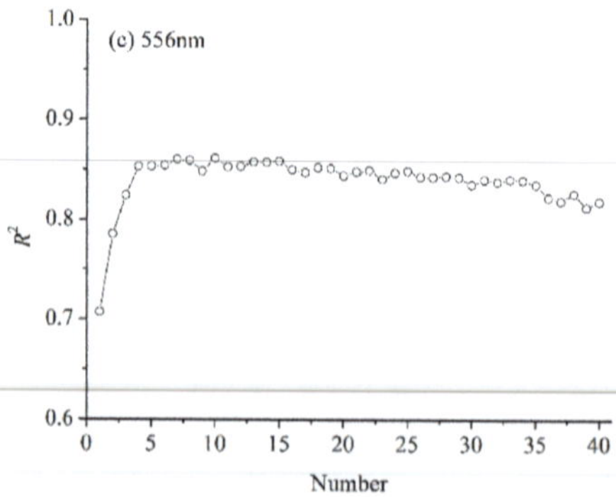

Figure 8. R^2 between the measured and predicted LNC changes with PC number based on the BPNN model at different excitation light wavelengths. (**a**) 355 nm, (**b**) 460 nm, (**c**) 556 nm.

As shown in Figure 8, the trend in R^2 changes consistently with the number of variables at different excitation light wavelengths. R^2 first increased with increasing the number of variables, and then decreased when the number of variables was over four or five. For 355 nm excitation light, the variable number is four when the R^2 reached maximum. For 460 nm and 556 nm excitation light, the R^2 exhibited the same tendency. Thus, the increase of explained variance with additional PC will be reduced to less than 1%, which means that the new calculated variable contains less spectral information than the raw variables. In addition, the results also demonstrated that the fluorescence spectra contain a large amount of information, which will influence the LNC estimation.

3.5.3. Estimation of LNC Based on Calculated Variables

The first four PCs were utilized for further study. The new variables calculated based on PC were provided as input parameters for training the BPNN model. The scatter plots between the measured and predicted LNC were established via linear regression analysis (Figure 9).

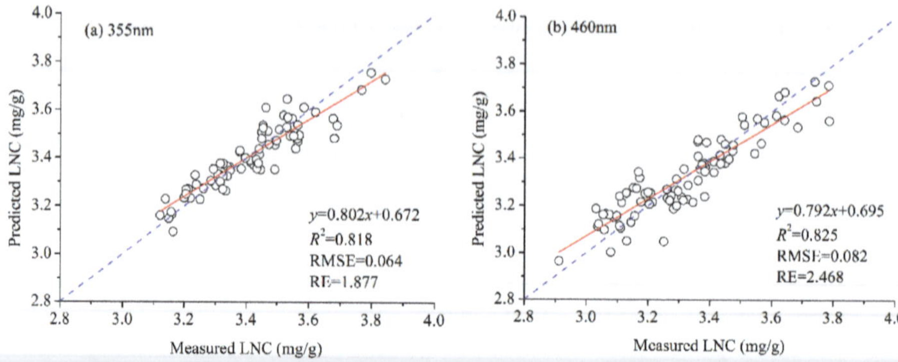

Figure 9. *Cont.*

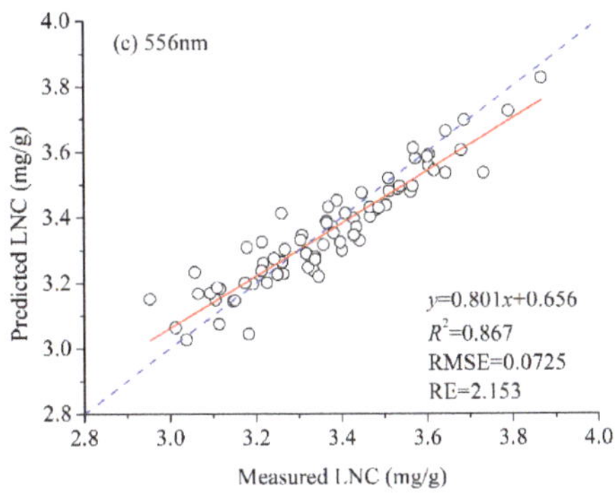

Figure 9. Relationship between the predicted LNC by using PCA combined with BPNN and the measured LNC for different excitation light wavelengths. (**a**) 355 nm, (**b**) 460 nm, (**c**) 556 nm. The blue dotted line is the 1:1 line. The red solid line is the linear regression.

As shown in Figure 9, the performance of new variables calculated through PCA for monitoring LNC was analyzed for different excitation light wavelengths. The solid line represents the linear regression analysis that denotes the correlation between the predicted and measured LNC. The inversion results demonstrated that PCA can efficiently extract the fluorescence spectral characteristics for LNC estimation with high R^2 and low RMSE, and RE values. All the R^2 values exceed 0.80 and can reach up to 0.86. Therefore, the extracted fluorescence characteristics located at red and far-red region can be effectively utilized to estimate LNC.

4. Discussion

LNC is a significant indicator for estimating crop growth status. A large number of correlational studies have been done by using passive and active remote sensing technologies. Plant fluorescence emitted by the chlorophyll in the leaf, which is related to photosynthetic pigments, can be affected by LNC. In addition, the molecular structure of chlorophyll contains a porphyrin ring structure which is consisted of carbon©, oxygen (O), and N. Thus, fluorescence can be utilized in the estimation of the biochemical content of the leaf [40,41]. Most recently, related studies have been conducted on the application of LIF [42] to monitor the biochemical content of crops. In the present study, we mainly discussed the effect of fluorescence characteristics on the LNC estimation and provided a guide for the selection of fluorescence characteristics.

Chlorophyll fluorescence displayed two main fluorescence peaks, and the fluorescence peak at 685 nm is less intensive than that at 740 nm (Figure 3). The reason is that the fluorescence emitted between 680 and 695 nm was more strongly reabsorbed by the chlorophyll pigment in the upper layer leaf cells than the fluorescence emitted between 730 and 750 nm. Thus, chlorophyll pigment had little influence on the fluorescence peak at 740 nm [39]. The fluorescence spectra excited by different excitation light wavelengths exhibited different performance for the LNC estimation (Figure 4). The results showed that 355 nm and 556 nm excitation lights are superior to 460 nm excitation light in the LNC estimation. The possible interpretation may be that fluorescence spectrum excited by 355 nm excitation light contains more fluorescence characteristics (360–800 nm), and that excited by 556 nm excitation light can penetrate deeper in the leaf than other excitation lights [26,43]. Furthermore, the absorption of foliar pigment may also influence the excitation light in the inner leaf influencing the performance of fluorescence characteristics for the LNC estimation.

The fluorescence spectrum contained large amounts of spectral information, and the red and far-red regions exhibited good performance in estimating LNC (Figure 5). This result is mainly because a large autocorrelation between different bands exists, and the two adjacent bands carry similar fluorescence information. Thus, fluorescence characteristics can be selected from the near fluorescence peaks at 685 nm and 740 nm to estimate LNC and are not just limited to the fluorescence characteristics peaks. For 355 nm and 460 nm excitation lights, the fluorescence can be chosen from the far-red region for LNC estimation, and for 556 nm excitation light, the red region may be the optimal band. The results of PCA showed that the first three PCs can explain more than 90% of the spectral information with different excitation light wavelengths (Figure 7). However, the trend of R^2 changes with variable number by increasing and then decreasing when the number of fluorescence characteristics was over four or five (Figure 8). PCA can extract significant spectral information, and the increase of explained variance will be small with an increase of PC number. In addition, when additional PCs cannot provide more information than raw spectral information, the performance of the extracted fluorescence characteristics will decrease in their ability to estimate LNC. Thus, when the number of PCs exceeded four or five, the R^2 will decrease with an increase of variable number. However, to improve the inversion accuracy of the LNC, selecting at least two fluorescence spectral features in the red and far-red region, respectively, is necessary. Related investigation showed that the extracted fluorescence characteristics were mainly located at the red and far-red regions, which can be effectively applied to LNC estimation (Figure 9) [26]. In addition, the selection of fluorescence characteristics is not just limited in the fluorescence characteristic peaks at 685 nm and 740 nm for LNC estimation. The fluorescence characteristics, which are located in the red and far-red regions (660–770 nm), can be also efficiently applied to LNC estimation [44].

In this study, a detailed analysis of the performance of the fluorescence spectral characteristics of each band for LNC estimation was conducted based on BPNN. However, some limitations should be considered for further studies. For the BPNN, although each setting was repeated one hundred times and the average was obtained to eliminate the local optimum effect of the internal parameters of the BPNN model, the optimal network architecture also needs to be analyzed. Furthermore, the effect of the number of fluorescence characteristics on the reversion performance was discussed in detail. Selecting the number of optimal fluorescence characteristics for LNC estimation also needs to be addressed in further studies. In addition, the effect of the spectral sampling interval and the type of crops on the performance of fluorescence characteristics for estimating LNC will be a promising study direction.

5. Conclusions

In this study, the performance of the fluorescence characteristics and fluorescence ratio of each band for the LNC estimation were analyzed in detail based on the BPNN model combined with PCA. Furthermore, the effect of the number of fluorescence characteristics on the accuracy of LNC estimation was also analyzed. The results demonstrated that the fluorescence characteristic, which is related to the LNC, is mainly located in the red and far-red regions, and the latter is superior to the former for LNC monitoring. For 355 nm and 460 nm excitation lights, the fluorescence characteristics can be chosen from the far-red region for LNC estimation, and for 556 nm excitation light, the red region is optimal. Thus, the selection of fluorescence characteristics is not just limited in the fluorescence peaks for LNC estimation. What's more, this study found that selecting at least two fluorescence spectral features in the red and far-red regions is necessary for LNC estimation. Additional research will be conducted to extend the results to different crop varieties.

Author Contributions: This experiment was conducted and designed by J.Y., W.G., S.S. (Shalei Song), L.D., S.S. (Shuo Shi), J.S. and B.C. The data processing was performed by J.Y. The manuscript written by J.Y. In addition, S.S. (Shalei Song) has provided significant contributions for manuscript preparation. All authors reviewed the manuscript.

Funding: This research was funded by the National Key Research and Development Program of China (Grant No. 2018YFB0504500), the National Natural Science Foundation (Grant No. 41571370), the Natural Science Foundation of Hubei Province (Grant No. 2018CFB272), the Open Fund of State Laboratory of Information Engineering in Surveying, Mapping and Remote Sensing, Wuhan University (Grant No. 17R05), Fundamental Research Funds for the Central Universities, China University of Geosciences (Wuhan) (Grant No. CUG170661).

Acknowledgments: The authors wish to thank College of Plant Science & Technology of Huazhong Agricultural University for providing the experimental samples and wish to thank Wuhan Academy of Agricultural Science & Technology for providing the LNCs of samples.

Conflicts of Interest: The authors declare no conflict of interest.

References

1. Li, F.; Mistele, B.; Hu, Y.; Chen, X.; Schmidhalter, U. Reflectance estimation of canopy nitrogen content in winter wheat using optimised hyperspectral spectral indices and partial least squares regression. *Eur. J. Agron.* **2014**, *52*, 198–209. [CrossRef]
2. Gameiro, C.; Utkin, A.; Cartaxana, P.; da Silva, J.M.; Matos, A. Rapid and nondestructive estimation of the nitrogen nutrition index in winter barley using chlorophyll measurements. *Agric. Water Manag.* **2016**, *164*, 127–136. [CrossRef]
3. Tian, Y.C.; Yao, X.; Yang, J.; Cao, W.X.; Hannaway, D.B.; Zhu, Y. Assessing newly developed and published vegetation indices for estimating rice leaf nitrogen concentration with ground- and space-based hyperspectral reflectance. *Field Crop. Res.* **2011**, *120*, 299–310. [CrossRef]
4. Feng, W.; Yao, X.; Zhu, Y.; Tian, Y.; Cao, W. Monitoring leaf nitrogen status with hyperspectral reflectance in wheat. *Eur. J. Agron.* **2008**, *28*, 394–404. [CrossRef]
5. Cecchi, G.; Mazzinghi, P.; Pantani, L.; Valentini, R.; Tirelli, D.; De Angelis, P. Remote sensing of chlorophyll a fluorescence of vegetation canopies: 1. Near and far field measurement techniques. *Remote Sens. Environ.* **1994**, *47*, 18–28. [CrossRef]
6. Zarco-Tejada, P.J.; Rueda, C.A.; Ustin, S.L. Water content estimation in vegetation with modis reflectance data and model inversion methods. *Remote Sens. Environ.* **2003**, *85*, 109–124. [CrossRef]
7. Stroppiana, D.; Boschetti, M.; Brivio, P.A.; Bocchi, S. Plant nitrogen concentration in paddy rice from field canopy hyperspectral radiometry. *Field Crop. Res.* **2009**, *111*, 119–129. [CrossRef]
8. Cendrero-Mateo, M.P.; Moran, M.S.; Papuga, S.A.; Thorp, K.; Alonso, L.; Moreno, J.; Ponce-Campos, G.; Rascher, U.; Wang, G. Plant chlorophyll fluorescence: Active and passive measurements at canopy and leaf scales with different nitrogen treatments. *J. Exp. Bot.* **2016**, *67*, 275–286. [CrossRef] [PubMed]
9. Song, S.; Gong, W.; Zhu, B.; Huang, X. Wavelength selection and spectral discrimination for paddy rice, with laboratory measurements of hyperspectral leaf reflectance. *ISPRS J. Photogramm.* **2011**, *66*, 672–682. [CrossRef]
10. Diacono, M.; Rubino, P.; Montemurro, F. Precision nitrogen management of wheat. A review. *Agron. Sustain. Dev.* **2013**, *33*, 219–241. [CrossRef]
11. He, L.; Song, X.; Feng, W.; Guo, B.B.; Zhang, Y.S.; Wang, Y.H.; Wang, C.Y.; Guo, T.C. Improved remote sensing of leaf nitrogen concentration in winter wheat using multi-angular hyperspectral data. *Remote Sens. Environ.* **2016**, *174*, 122–133. [CrossRef]
12. Osborne, S.L.; Schepers, J.S.; Francis, D.D.; Schlemmer, M.R. Detection of phosphorus and nitrogen deficiencies in corn using spectral radiance measurements. *Agron. J.* **2002**, *94*, 1215–1221. [CrossRef]
13. Kalaji, H.M.; Oukarroum, A.; Alexandrov, V.; Kouzmanova, M.; Brestic, M.; Zivcak, M.; Samborska, I.A.; Cetner, M.D.; Allakhverdiev, S.I.; Goltsev, V. Identification of nutrient deficiency in maize and tomato plants by in vivo chlorophyll a fluorescence measurements. *Plant Physiol. Biochem.* **2014**, *81*, 16–25. [CrossRef] [PubMed]
14. Živcak, M.; Olsovska, K.; Slamka, P.; Galambošová, J.; Rataj, V.; Shao, H.; Brestič, M. Application of chlorophyll fluorescence performance indices to assess the wheat photosynthetic functions influenced by nitrogen deficiency. *Plant Soil Environ.* **2014**, *60*, 210–215. [CrossRef]
15. Zivcak, M.; Brestic, M.; Kunderlikova, K.; Olsovska, K.; Allakhverdiev, S.I. Effect of photosystem I inactivation on chlorophyll a fluorescence induction in wheat leaves: Does activity of I photosystem play any role in OJIP rise? *J. Photochem. Photobiol. B Biol.* **2015**, *152*, 318–324. [CrossRef] [PubMed]
16. Mauromicale, G.; Ierna, A.; Marchese, M. Chlorophyll fluorescence and chlorophyll content in field-grown potato as affected by nitrogen supply, genotype, and plant age. *Photosynthetica* **2006**. [CrossRef]
17. Tartachnyk, I.; Rademacher, I. Estimation of nitrogen deficiency of sugar beet and wheat using parameters of laser induced and pulse amplitude modulated chlorophyll fluorescence. *J. Appl. Bot.* **2003**, *77*, 61–67.

18. Apostol, S.; Viau, A.A.; Tremblay, N. A comparison of multiwavelength laser-induced fluorescence parameters for the remote sensing of nitrogen stress in field-cultivated corn. *Can. J. Remote Sens.* **2007**, *33*, 150–161. [CrossRef]
19. Yang, J.; Gong, W.; Shi, S.; Du, L.; Sun, J.; Ma, Y.Y.; Song, S.L. Accurate identification of nitrogen fertilizer application of paddy rice using laser-induced fluorescence combined with support vector machine. *Plant Soil Environ.* **2015**, *61*, 501–506. [CrossRef]
20. Kalaji, H.M.; Schansker, G.; Brestic, M.; Bussotti, F.; Calatayud, A.; Ferroni, L.; Goltsev, V.; Guidi, L.; Jajoo, A.; Li, P. Frequently asked questions about chlorophyll fluorescence, the sequel. *Photosynth. Res.* **2017**, *132*, 13–66. [CrossRef] [PubMed]
21. Subhash, N.; Mohanan, C.N. Laser-induced red chlorophyll fluorescence signatures as nutrient stress indicator in rice plants. *Remote Sens. Environ.* **1994**, *47*, 45–50. [CrossRef]
22. Günther, K.; Dahn, H.G.; Lüdeker, W. Remote sensing vegetation status by laser-induced fluorescence. *Remote Sens. Environ.* **1994**, *47*, 10–17. [CrossRef]
23. Campbell, P.K.E.; Middleton, E.M.; Corp, L.A.; Kim, M.S. Contribution of chlorophyll fluorescence to the apparent vegetation reflectance. *Sci. Total Environ.* **2008**, *404*, 433–439. [CrossRef] [PubMed]
24. Gu, X.; Xu, P.; Qiu, H.; Feng, H. *Monitoring the Chlorophyll Fluorescence Parameters in Rice under Flooding and Waterlogging Stress Based on Remote Sensing*; IEEE: Waikoloa, HI, USA, 2014.
25. Anderson, B.; Buah-Bassuah, P.K.; Tetteh, J.P. Using violet laser-induced chlorophyll fluorescence emission spectra for crop yield assessment of cowpea (*Vigna unguiculata* (L.) walp) varieties. *Meas. Sci. Technol.* **2004**, *15*, 1255–1265. [CrossRef]
26. Yang, J.; Sun, J.; Du, L.; Chen, B.; Zhang, Z.; Shi, S.; Gong, W. Effect of fluorescence characteristics and different algorithms on the estimation of leaf nitrogen content based on laser-induced fluorescence lidar in paddy rice. *Opt. Express* **2017**, *25*, 3743–3755. [CrossRef] [PubMed]
27. Yang, J.; Gong, W.; Shi, S.; Du, L.; Sun, J.; Song, S.; Chen, B.; Zhang, Z. Analyzing the performance of fluorescence parameters in the monitoring of leaf nitrogen content of paddy rice. *Sci. Rep.* **2016**. [CrossRef] [PubMed]
28. Saito, Y.; Kanoh, M.; Hatake, K.I.; Kawahara, T.D.; Nomura, A. Investigation of laser-induced fluorescence of several natural leaves for application to lidar vegetation monitoring. *Appl. Opt.* **1998**, *37*, 431–437. [CrossRef] [PubMed]
29. Krause, G.H.; Weis, E. Chlorophyll fluorescence as a tool in plant physiology. *Photosynth. Res.* **1984**, *5*, 139–157. [CrossRef] [PubMed]
30. Kjeldahl, J. Neue methode zur bestimmung des stickstoffs in organischen körpern. *Z. Anal. Chem.* **1883**, *22*, 366–382. [CrossRef]
31. Keiner, L.E.; Yan, X.H. A neural network model for estimating sea surface chlorophyll and sediments from thematic mapper imagery. *Remote Sens. Environ.* **1998**, *66*, 153–165. [CrossRef]
32. Samborska, A.I.; Alexandrov, V.; Sieczko, L.; Kornatowska, B.; Goltsev, V.; Magdalena, D.C.; Kalaji, H.M. Artificial Neural Networks and Their Application in Biological and Agricultural Research. Available online: http://signpostejournals.com/ejournals/Portals/12/v22.pdf (accessed on 3 September 2018).
33. Yi, Q.X.; Huang, J.F.; Wang, F.M.; Wang, X.Z.; Liu, Z.Y. Monitoring rice nitrogen status using hyperspectral reflectance and artificial neural network. *Environ. Sci. technol.* **2007**, *41*, 6770–6775. [CrossRef] [PubMed]
34. Galvão, L.S.; Pizarro, M.A.; Epiphanio, J.C.N. Variations in reflectance of tropical soils: Spectral-chemical composition relationships from aviris data. *Remote Sens. Environ.* **2001**, *75*, 245–255. [CrossRef]
35. Bro, R.; Smilde, A.K. Principal component analysis. *Anal. Methods* **2014**, *6*, 2812–2831. [CrossRef]
36. Chappelle, E.W.; McMurtrey, J.E.; Kim, M.S. Identification of the pigment responsible for the blue fluorescence band in the laser induced fluorescence (lif) spectra of green plants, and the potential use of this band in remotely estimating rates of photosynthesis. *Remote Sens. Environ.* **1991**, *36*, 213–218. [CrossRef]
37. Tremblay, N.; Wang, Z.; Cerovic, Z.G. Sensing crop nitrogen status with fluorescence indicators. A review. *Agron. Sustain. Dev.* **2012**, *32*, 451–464. [CrossRef]
38. Chappelle, E.W.; Wood, F.M.; Wayne Newcomb, W.; McMurtrey, J.E. Laser induced fluorescence of green plants. 3: LIF spectral signatures of five major plant types. *Appl. Opt.* **1985**, *24*, 74–80. [CrossRef] [PubMed]
39. Malenovsky, Z.; Mishra, K.B.; Zemek, F.; Rascher, U.; Nedbal, L. Scientific and technical challenges in remote sensing of plant canopy reflectance and fluorescence. *J. Exp. Bot.* **2009**, *60*, 2987–3004. [CrossRef] [PubMed]

40. Pieruschka, R.; Klimov, D.; Kolber, Z.S.; Berry, J.A. Monitoring of cold and light stress impact on photosynthesis by using the laser induced fluorescence transient (LIFT) approach. *Funct. Plant Biol.* **2010**, *37*, 395–402. [CrossRef]
41. Kalaji, H.M.; Jajoo, A.; Oukarroum, A.; Brestic, M.; Zivcak, M.; Samborska, I.A.; Cetner, M.D.; Łukasik, I.; Goltsev, V.; Ladle, R.J. Chlorophyll a fluorescence as a tool to monitor physiological status of plants under abiotic stress conditions. *Acta Physiol. Plant* **2016**, *38*, 1–11. [CrossRef]
42. Hoge, F.E.; Swift, R.N.; Yungel, J.K. Feasibility of airborne detection of laser-induced fluorescence emissions from green terrestrial plants. *Appl. Opt.* **1983**, *22*, 2991–3000. [CrossRef] [PubMed]
43. Schweiger, J.; Lang, M.; Lichtenthaler, H.K. Differences in fluorescence excitation spectra of leaves between stressed and non-stressed plants. *J. Plant Physiol.* **1996**, *148*, 536–547. [CrossRef]
44. Meroni, M.; Rossini, M.; Guanter, L.; Alonso, L.; Rascher, U.; Colombo, R.; Moreno, J. Remote sensing of solar-induced chlorophyll fluorescence: Review of methods and applications. *Remote Sens. Environ.* **2009**, *113*, 2037–2051. [CrossRef]

remote sensing

MDPI

Article

Remote Estimation of Nitrogen Vertical Distribution by Consideration of Maize Geometry Characteristics

Huichun Ye [1,2], Wenjiang Huang [1,3,*], Shanyu Huang [4], Bin Wu [1], Yingying Dong [1] and Bei Cui [1,3]

1 Key Laboratory of Digital Earth Science, Institute of Remote Sensing and Digital Earth, Chinese Academy of Sciences, Beijing 100094, China; yehc@radi.ac.cn (H.Y.); speed12344@163.com (B.W.); dongyy@radi.ac.cn (Y.D.); cuibei@radi.ac.cn (B.C.)
2 Key Laboratory of Agricultural Big Data, Ministry of Agriculture, Beijing 100081, China
3 Key Laboratory of Earth Observation, Hainan Province, Sanya 572029, China
4 Institute of Geography, University of Cologne, 50923 Köln, Germany; ivy331919@163.com
* Correspondence: huangwj@radi.ac.cn; Tel.: +86-10-82178169

Received: 8 October 2018; Accepted: 6 December 2018; Published: 9 December 2018

Abstract: The vertical leaf nitrogen (N) distribution in the crop canopy is considered to be an important adaptive response of crop growth and production. Remote sensing has been widely applied for the determination of a crop's N status. Some studies have also focused on estimating the vertical leaf N distribution in the crop canopy, but these analyses have rarely considered the plant geometry and its influences on the remote estimation of the N vertical distribution in the crop canopy. In this study, field experiments with three types of maize (*Zea mays* L.) plant geometry (i.e., horizontal type, intermediate type, and upright type) were conducted to demonstrate how the maize plant geometry influences the remote estimation of N distribution in the vertical canopy (i.e., upper layer, middle layer, and bottom layer) at different growth stages. The results revealed that there were significant differences among the three maize plant geometry types in terms of canopy architecture, vertical distribution of leaf N density (LND, g m^{-2}), and the LND estimates in the leaves of different layers based on canopy hyperspectral reflectance measurements. The upright leaf variety had the highest correlation between the lower-layer LND (R^2 = 0.52) and the best simple ratio (SR) index (736, 812), and this index performed well for estimating the upper (R^2 = 0.50) and middle (R^2 = 0.60) layer LND. However, for the intermediate leaf variety, only 25% of the variation in the lower-layer LND was explained by the best SR index (721, 935). The horizontal leaf variety showed little spectral sensitivity to the lower-layer LND. In addition, the growth stages also affected the remote detection of the lower leaf N status of the canopy, because the canopy reflectance was dominated by the biomass before the 12th leaf stage and by the plant N after this stage. Therefore, we can conclude that a more accurate estimation of the N vertical distribution in the canopy is obtained by canopy hyperspectral reflectance when the maize plants have more upright leaves.

Keywords: nitrogen; vertical distribution; plant geometry; remote sensing; maize

1. Introduction

Nitrogen (N) is an essential nutrient and the most limiting nutrient for crop growth and yield [1–3]. N deficiency in crops causes the older and lower leaves on the plant to turn yellow and wilt and ultimately impact leaf photosynthesis, crop yield, and quality [4–7]. The overapplication or misapplication of N fertilizers on farmland can also result in crop lodging, reduced efficiency in N use, and increased risk of environmental contamination [8–10]. Fortunately, precision N management has the potential to solve these problems by matching the N supply with the crop's N requirements at the correct rate, place, and time [11]. Additionally, remote sensing provides an alternative to large-area crop N diagnosis. Various methods have been presented for the estimation of canopy total N or canopy mean N concentration using hyperspectral reflectance information. However, these methods seldom consider the non-uniformity of the N distribution [12–16].

Various studies reported that the vertical leaf N distribution of a plant canopy is non-uniform; i.e., the shaded lower leaves generally have a lower N content than the upper illuminated leaves [17–21]. It has also been shown that the lower leaves are more sensitive to N deficiency than the upper leaves [22]. The transmission of N in response to N stress is generally from the bottom, old leaves to the upper, new leaves in crops [23]. Thus, N deficiencies usually occur first in the bottom leaves, whereas excess N affects the upper leaves [9]. Therefore, N fertilization should ideally be applied according to the vertical leaf N distribution, which is critical for an early assessment of the crop growth status.

A few studies have focused on estimating the vertical leaf N distribution in the crop canopy using remote sensing technology. The existing studies can be grouped into two categories based on the acquisition type of the hyperspectral data; i.e., the leaf N content of different vertical layers can be estimated using top-view spectral data observations [24,25] or multi-angular canopy reflectance data can be obtained to select the optimal observation zenith angle of different vertical layers [9,16].

Canopy structure is an important factor influencing the vertical N distribution in plant canopies [26–29] because it determines the light distribution within the canopy [30]. Plants allocate N based on the light distribution pattern within the canopy, and the illuminated upper leaves accumulate more N than the shaded lower leaves in an effort to maximize total canopy photosynthesis [21,31]. Other factors that affect canopy structure, such as the plant geometry (i.e., horizontal type, intermediate type, or upright type), also result in changes in the vertical N distribution of the crop canopy [21]. In densely planted stands, plants with relatively horizontal leaves increase the leaf shade and decrease the light transmittance rate, whereas plants with relatively upright leaves have a plant architecture that allows the lower leaves to capture more light [32–35]. Maize (*Zea mays* L.) is a large-grain plant grown across the world. During the last 50 years, whether selection for morphological traits of maize was unintentional in the United States [36] or intentional in China [37,38], the leaves of maize have become more upright, increasing their tolerance to high planting densities and achieving genetic gains [36,37,39,40]. However, until recently, few studies have considered the plant geometry (i.e., horizontal, intermediate, or upright leaf types) and its influences on the remote estimation of the N vertical distribution in the canopy of maize [21].

The objectives of this study are to (i) demonstrate how plant geometry influences the remote estimation of the N vertical distribution in the canopy of maize at different growth stages, (ii) evaluate the performance of published hyperspectral vegetation indices (VIs) for estimating the N vertical distribution in the maize canopy, and (iii) determine the optimum published vegetation index to obtain a high-precision estimate of the N content in the leaves of the lower canopy.

2. Materials and Methods

2.1. Experimental Site

Field experiments were conducted in 2002 and 2003 in the area of the Xiaotangshan Precision Agriculture Experimental Base (40°11′N, 116°27′E) in Beijing City, China. The soil type at the field site is Eutric Cambisols according to the International Union of Soil Sciences (IUSS) Working Group World Reference Base (WRB) soil classification system [41], and the soil texture is a silty clay loam according to the United States Department of Agriculture (USDA) soil texture triangle [42]. The nutrient contents of the top soil (0–30 cm) were as follows: 1.9–2.2 g kg^{-1} of organic matter, 10.2–12.3 mg kg^{-1} of ammonium N, 16.2–18.0 mg kg^{-1} of nitrate N, 15.2–17.6 mg kg^{-1} of available phosphorus, and 225–230 mg kg^{-1} of available potassium. The cropping system is winter wheat (*Triticum aestivum* L.) summer maize rotation, which is most popular in North China. The field site is located in a warm temperate climate zone in a semi-moist continental monsoon region. The annual average temperature, precipitation and sunlight hours are 10–12 °C, 600–700 mm and 2700–2800 h, respectively. The seasonal distribution of precipitation is uneven, with 70% of the precipitation occurring from July to September. The climatic characteristics are suitable for many field crops, including wheat, maize, and soybeans (*Glycine max* L.), which can only be harvested twice a year.

2.2. Experimental Design

In this study, the treatments consisted of three maize plant geometry types. The classification of the maize plant geometry types was based on the leaf orientation value (LOV) that was proposed by Pepper et al. (1977) [43]. Specifically, maize varieties with an LOV $\geq 60°$ were categorized as a horizontal leaf type, those with $30° < \text{LOV} < 60°$ were categorized as an intermediate leaf type, and those with LOV $\leq 30°$ were categorized as an upright leaf type. The LOV was calculated as follows:

$$\text{LOV} = 1/n\sum(90-\theta) \times \left(L_f/L\right) \quad (1)$$

where θ is the measured leaf angle, L_f is the length from the leaf collar to the flagging point of the measured leaves, L is the leaf length, and n is the number of measured leaves [43]. Two experiments were performed from June to September in 2002 and 2003 at the Xiaotangshan Precision Agriculture Experimental Base.

The experiment in 2002 was a split-plot design. The main plot consisted of three maize plant geometry treatments, including one horizontal variety (Nongda 80), one intermediate variety (Jingyu 7), and one upright variety (Tangyu 10). The subplot consisted of three N application rates: 0, 150, and 300 kg N ha^{-1} with urea. Initially, 50% of the urea was applied at the 6th leaf stage (V6), and the rest at the 12th leaf stage (V12). The treatment plot size was 15 m by 7 m and the planting distance was 70 cm by 30 cm (planting density of 48,000 plants ha^{-1}). The field management, such as weeding, irrigation and pesticide applications, followed the local standard practices.

The experiment in 2003 involved three plant geometry treatments with 15 varieties, including five horizontal varieties (Nongda 80, 96–3, Zhengdan 958, Yuyu 22 and Nongda 108), five intermediate varieties (Jingyu 7, Zhongyuandan 32, Zhongdan 9409, Gaoyou 115 and Zhongnuo 2), and five upright varieties (Tangyu 10, Hudan 2000, Jingshibai 1, Tangkang 5 and Jiangzao 13). The treatment plot size was 15 m by 7 m and the planting distance was 70 cm by 30 cm. All the treatments were conducted under the same management practices according to the local standard practices for summer maize production.

2.3. Hyperspectral Reflectance Measurement

The canopy spectral reflectance was measured using an ASD FieldSpec® Pro FR spectroradiometer (Analytical Spectral Devices, Boulder, CO, USA) between 10:00 AM and 2:00 PM (Beijing local time) under clear sky conditions. The ASD spectroradiometer was configured with a spectral range from 350 nm to 2500 nm, a 1.4 nm sampling interval between 350 nm and 1000 nm, a 2 nm sampling interval between 1000 nm and 2500 nm, and with 3 nm spectral resolution at 700 nm, and 10 nm spectral resolution at 1400 nm and 2100 nm. The hyperspectral data were subdivided into 1 nm bandwidths by using a self-driven interpolation method of the ASD spectroradiometer, and they were then saved to the connected PC. In this study, we used data from the common spectral region of 350–1050 nm. The field of view of the ASD spectroradiometer was 25°. A fiber-optic probe was fixed at the top end of a height-adjustable inverted-L pole. All canopy reflectance measurements were obtained from a distance of 1.6 m above the maize canopy. The reflectance measurements were acquired randomly at three sites in each plot and then averaged to represent the canopy reflectance of the plot. The spectral measurements and plant samples were obtained at the major growth stages of the maize plant, including the 6th leaf stage (V6), 12th leaf stage (V12), 14th leaf stage (V14), tasseling stage (VT), silking stage (R1), and milk stage (R3) for both the 2002 and 2003 experiments.

2.4. Plant Vertical Layered Sampling and Measurements

After measuring the canopy reflectance, three maize plants near the locations of the spectral measurement were cut at ground level and quickly transported to the nearby laboratory. The plants were divided into three layers based on 1/3 position of the total number of leaves from top to bottom of the canopy (i.e., the upper layer is the first layer, the middle layer is the second layer, and the bottom layer is the third layer). The stamens after the VT were not considered in this study. All green leaves were separated from the stems, and were cleaned and oven-dried for 15 min at 105 °C and later at 80 °C for 8 h to a constant weight. The leaf area index (LAI) of each layer was determined using the gravimetric method [44]. This method correlates the dry weight of leaves and leaf area using the ratio of leaf dry mass to leaf area (specific leaf weight, g m^{-2}), which is predetermined from a sub sample extracted from collected leaves. The leaf N concentration (%) of each layer was determined using the Kjeldahl method following concentrated sulphuric acid (H_2SO_4) digestion as described by Helrich (1990) [45].

The crop N concentration in plants in dense canopies declines as they grow and the dry matter increases during the growth period, even when there is an ample supply of N [46]. This "dilution effect" exists in most crops, such as rice (*Oryza sativa* L.) [1], wheat [2], maize [3], and potato (*Solanum tuberosum* L.) [47]. It was demonstrated that the rate of the above-ground biomass production exceeds the rate of N uptake by plants before heading, when the biomass dominates the canopy reflectance [48]. In contrast, the rate of increase in biomass decreases and the "dilution effect" culminates after heading, when the plant N dominates the canopy reflectance. The leaf N density (LND, g m^{-2}), which embodies the N concentration, biomass, and LAI, is defined as the accumulation of N in the plant leaves per unit area [9]. In this study, the LND was used as the indicator of the N status. The LND of each layer is defined as

$$\mathrm{LND}_i = \%\mathrm{N}_i \times \mathrm{SLW}_i \times \mathrm{LAI}_i \quad (2)$$

where i denotes the position of the layer of the maize canopy; %N is the leaf N concentration (%); SLW denotes the specific leaf weight (g m^{-2}).

The canopy structure, namely the LAI vertical distribution in a canopy, affects the light interception that greatly impacts the vertical distribution patterns of the canopy N [18,30,49,50]. In most studies, the leaf N distribution in canopies was expressed as a function of the relative cumulative LAI (a ratio of the cumulative LAI from the top to the bottom of the canopy to the total LAI of the canopy, LAI_C/LAI_T) of a given depth [21,26,51]. In this study, we used the relative LAI of a given layer, which is a ratio of the subtotal leaf area of a given layer to the total leaf area index of the canopy (LAI_i/LAI_T, where i = 1, 2 and 3 represents the upper layer, middle layer and bottom layer of the canopy, respectively) to allow for the comparison of the maize varieties. The sum of the LAI_i/LAI_T of all layers is one.

2.5. Hyperspectral Vegetation Indices and Data Analysis

The selected published hyperspectral VIs that have been used to estimate the crop N status are listed in Table 1. In order to determine the effects of the band combinations and growth stages on the relationships and the performance of the VIs for deriving the N status of each layer of different plant types, new VIs were calculated. These consisted of simple ratio indices (SR = B1/B2) involving all possible 2-band combinations from 350 to 1050 nm. The growth stages were separated into the V6 to V12 stages (when the canopy was not closed) and the V14 to R3 stages (when the canopy was fully closed) to evaluate the effect of the growth stages on the relationship and performance of the VIs. In addition, the analyses were combined across all growth stages. The correlation and regression analyses were performed using the MATLAB 7.0 software (The MathWorks, Inc., Natick, MA, USA).

Table 1. Published vegetation indices used in this study.

Index	Formula	Reference
Simple ratio (SR) 1	B810/B560	[13]
SR 2	B750/B710	[52]
Normalized difference vegetation index (NDVI)	(B800 − B680)/(B800 + B680)	[53]
Green NDVI	(B750 − B550)/(B750 + B550)	[54]
Normalized difference red edge (NDRE) Index	(B790 − B720)/(B790 + B720)	[55]
Optimized soil-adjusted vegetation index (OSAVI)	1.16 × (B800 − B670)/(B800 + B670 + 0.16)	[56]
MERIS * terrestrial chlorophyll index (MTCI)	(B754 − B709)/(B709 − B681)	[57]
Double-peak nitrogen index (NDDA)	(B680 + B756 − 2 × B718)/(B756 − B680)	[58]
Modified red-edge normalized difference vegetation Index (mND705)	(B750 − B705)/(B750 + B705 − 2 × B445)	[59]
New double difference (DDn) Index	2 × B710 − B660 − B760	[60]
Modified chlorophyll absorption ratio index (MCARI)	(B750 − B705 − 0.2 × (B750 − B550)) × (B750/B705)	[61]
MCARI/OSAVI	MCARI/OSAVI	[62]

* MERIS = Medium Resolution Imaging Spectrometer.

3. Results

3.1. Canopy Structure Characteristics

As shown in Figure 1, the upper layer LAI_1/LAI_T of the horizontal, intermediate, and upright leaf varieties ranged from 0.30 to 0.62, 0.24 to 0.62, and 0.22 to 0.64, respectively, decreasing with maize growth. Prior to the V12 stage, there were small differences in the LAI_1/LAI_T among the different plant geometry types, whereas, after the V12 stage, the LAI_1/LAI_T of the upright varieties decreased the most, followed by the intermediate varieties and horizontal varieties. In contrast, the LAI_3/LAI_T of the bottom layer of all plant types initially increased prior to the V14 stage and after that, it decreased slightly as the crop grew. The LAI_3/LAI_T ranged from 0.06 to 0.29 for the horizontal varieties, from 0.07 to 0.33 for the intermediate varieties, and from 0.06 to 0.44 for the upright varieties. The LAI_2/LAI_T values of the middle layer of the three plant types were relatively stable during the entire growth period.

Furthermore, the differences in the LAI_i/LAI_T among the different varieties became more apparent after the V14 stage. The coefficients of variation (CV) of the LAI_1/LAI_T of the different plant types after the V14 stage ranged from 13.3 to 20.7% with a mean of 16.7%, whereas it varied from 1.4 to 4.4% with a relatively small mean of 2.7% prior to the V12 stage. Similar to the upper layer, the variation of the LAI_3/LAI_T of the different plant types was greater after the V14 stage (CV ranged from 21.8 to 35.0% with a mean of 26.7%) than before the V12 stage (CV ranged from 9.3 to 22.1% with a mean of 14.4%). The difference in the LAI_2/LAI_T between the different plant types was relatively small.

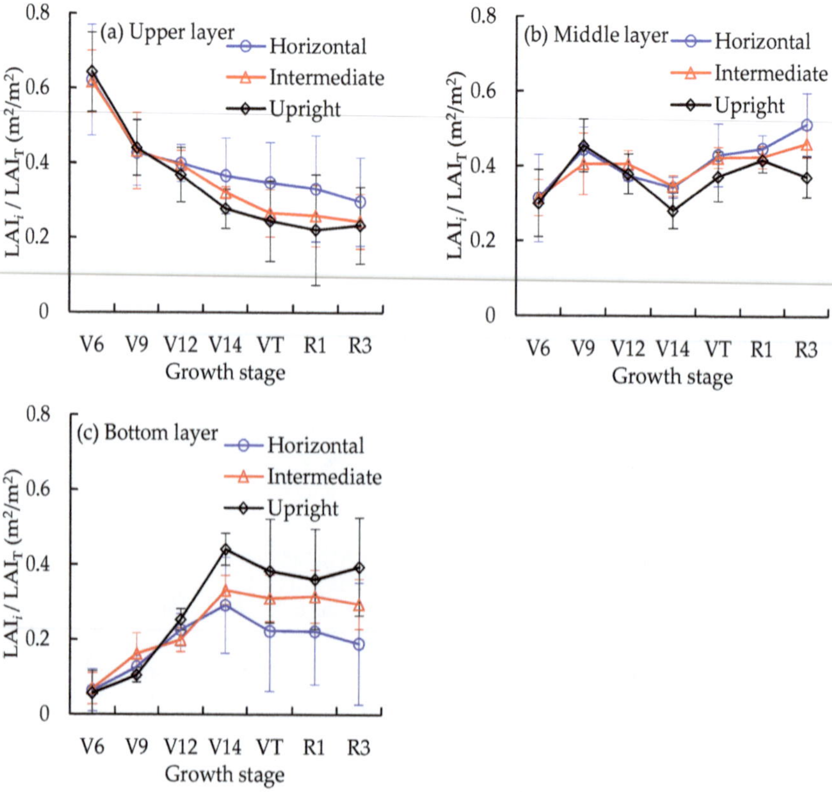

Figure 1. Comparison of the relative leaf area index (LAI) (LAI_i/LAI_T) of the vertical canopy layers for different maize plant geometry types. LAI_i/LAI_T represents the ratio of the subtotal leaf area of a given layer (LAI_i) to the total LAI of the canopy (LAI_T). (**a**) Upper layer; (**b**) middle layer; (**c**) bottom layer. Error bars represent standard error of the mean.

3.2. Vertical Distribution Characteristics of Leaf N Density

The LND values of the vertical canopy layers of the different plant geometry types are listed in Table 2. From top to bottom within the canopies, the LND showed a decreasing trend. For the horizontal leaf varieties across the growth stages, the LND at the bottom layer ranged between 0.13 and 2.66 g m^{-2} (mean = 0.93 g m^{-2}, and CV = 80.0%) and was lower but more variable than that in the upper and middle layers, where the LND was from 1.30 to 3.71 g m^{-2} (mean = 2.22 g m^{-2}, and CV = 28.0%) and 0.50 to 2.91 g m^{-2} (mean = 1.82 g m^{-2}, and CV = 29.1%), respectively. This trend was also true for both the intermediate and upright leaf varieties. The variation in LND was larger in the bottom layer than that in the upper and middle layers across all the stages, and the CV values in the upper, middle and bottom layers of the intermediate leaf varieties were 27.8%, 33.2% and 68.3%, respectively, and for the upright leaf varieties, the values were 25.3%, 34.8% and 81.4%, respectively. The average LND values in the bottom layer of the intermediate and upright leaf varieties were 1.08 and 1.18 g m^{-2}, respectively, and both were significantly lower than those in the upper and middle layers. This is consistent with previous studies, which reported that the lower shaded leaves generally have a lower N content than the upper illuminated leaves [21,28,29,63,64]. The LND gradient between the upper and bottom layers was 1.29 g m^{-2} for the horizontal varieties, 1.09 g m^{-2} for the intermediate varieties, and 0.85 g m^{-2} for the upright varieties, respectively, indicating that the vertical N gradients decreased as the canopy leaves changed to a more upright geometry. This occurs because the upright leaf variety has a high light transmittance rate, and leaves growing under better light conditions accumulate more N than shaded leaves [31]. In addition, the average LND values in the bottom layer were lower during the V6 to V12 stages than during the V14 to R3 stages (0.57 vs. 1.18 g m^{-2} for the horizontal varieties, 0.65 vs. 1.14 g m^{-2} for the intermediate varieties, and 0.64 vs. 1.56 g m^{-2} for the upright varieties, respectively).

Table 2. Descriptive statistics of the leaf N density (g m^{-2}) in the vertical canopy layers of different maize plant geometry types.

Growth Stage	Canopy Layer	n	Min	Max	Mean	SD [1]	CV [2] (%)
Horizontal leaf varieties							
All stages	Upper layer	49	1.30	3.71	2.22	0.62	28.0
	Middle layer	49	0.50	2.91	1.82	0.53	29.1
	Bottom layer	49	0.13	2.66	0.93	0.74	80.0
V6 to V12 stages	Upper layer	21	1.41	3.71	2.37	0.74	31.2
	Middle layer	21	0.50	2.44	1.59	0.56	35.4
	Bottom layer	21	0.13	1.33	0.57	0.38	66.4
V14 to R3 stages	Upper layer	28	1.30	2.80	2.06	0.43	21.1
	Middle layer	28	1.39	2.91	2.02	0.42	20.9
	Bottom layer	28	0.17	2.66	1.18	0.84	70.6
Intermediate leaf varieties							
All stage	Upper layer	49	0.88	3.74	2.17	0.72	33.2
	Middle layer	49	0.71	3.24	2.18	0.61	27.8
	Bottom layer	49	0.25	3.01	1.08	0.74	68.3
V6 to V12 stages	Upper layer	21	1.33	3.74	2.59	0.74	28.7
	Middle layer	21	0.71	3.24	2.02	0.61	30.1
	Bottom layer	21	0.25	1.29	0.65	0.35	53.4
V14 to R3 stages	Upper layer	28	0.88	2.72	1.86	0.53	28.7
	Middle layer	28	1.25	3.15	2.30	0.59	25.8
	Bottom layer	28	0.37	3.01	1.40	0.79	56.8
Upright leaf varieties							
All stage	Upper layer	49	0.81	3.68	2.03	0.71	34.8
	Middle layer	49	0.93	2.96	2.09	0.53	25.3
	Bottom layer	49	0.12	3.38	1.18	0.96	81.4
V6 to V12 stages	Upper layer	21	1.32	3.68	2.36	0.70	29.8
	Middle layer	21	0.93	2.94	1.98	0.66	33.2
	Bottom layer	21	0.12	1.91	0.64	0.48	75.0
V14 to R3 stages	Upper layer	28	0.81	3.35	1.79	0.62	34.5
	Middle layer	28	1.28	2.96	2.18	0.41	18.7
	Bottom layer	28	0.18	3.38	1.56	1.04	66.5

[1] Standard deviation. [2] Coefficient of variation.

3.3. Performance of Published Vegetation Indices

The correlation analysis results between the published indices and the LND in the vertical canopy layers of the different plant geometry types are shown in Table 3. All 12 published indices performed well for the upper layer ($p < 0.05$) for all varieties and growth stages. As the depth of the canopy increased, the correlations between the LND and the published indices gradually weakened, but they differed among the plant types. For the bottom layer of the canopy, the five best performing published indices for the upright leaf varieties were SR 1, SR 2, normalized differentiated red-edge index (NDRE), MERIS terrestrial chlorophyll index (MTCI), and double-peak nitrogen index (NDDA) with R^2 values ranging from 0.42 to 0.48, whereas those for intermediate leaf varieties were SR 1, SR 2, green-normalized differentiated vegetation index (NDVI), NDRE, and modified chlorophyll absorption ratio index (MCARI) with R^2 values ranging from 0.18 to 0.20. The horizontal leaf varieties had low correlations between the bottom layer LND and the spectral indices ($R^2 < 0.02$, $p > 0.05$).

Table 3. Coefficients of determination (R^2) for the relationship between the vegetation indices and the leaf N density.

Index	All Stages			V6 to V12 Stages			V14 to R3 Stages		
	Upper Layer	Middle Layer	Bottom Layer	Upper Layer	Middle Layer	Bottom Layer	Upper Layer	Middle Layer	Bottom Layer
Horizontal leaf varieties									
SR 1	0.63 **	0.23 **	0.00	0.68 **	0.06	0.20	0.41 **	0.38 **	0.09
SR 2	0.67 **	0.23 **	0.00	0.78 **	0.14	0.27 *	0.34 **	0.27 **	0.11
NDVI	0.75 **	0.14 *	0.01	0.83 **	0.09	0.24	0.57 **	0.19 *	0.17
Green NDVI	0.69 **	0.22 **	0.02	0.75 **	0.08	0.22	0.42 **	0.39 **	0.08
NDRE	0.60 **	0.30 **	0.02	0.73 **	0.15	0.31	0.21 *	0.31 **	0.10
OSAVI	0.75 **	0.14 *	0.01	0.83 **	0.09	0.24	0.58 **	0.20 *	0.16
MTCI	0.23 **	0.30 **	0.02	0.38 *	0.33 *	0.30 *	0.10	0.19 *	0.03
NDDA	0.25 **	0.34 **	0.02	0.40 *	0.29 *	0.31 *	0.06	0.24 *	0.05
mND705	0.69 **	0.23 **	0.01	0.83 **	0.19	0.30 *	0.22 *	0.16	0.15
DDn	0.39 **	0.07	0.00	0.54 **	0.02	0.31 *	0.33 **	0.36 **	0.13
MCARI	0.46 **	0.06	0.00	0.64 **	0.03	0.29 *	0.37 **	0.30 **	0.15
MCARI/OSAVI	0.40 **	0.05	0.01	0.58 **	0.03	0.29 *	0.33 **	0.30 **	0.14
Intermediate leaf varieties									
SR 1	0.59 **	0.50 **	0.20 **	0.55 **	0.50 **	0.50 **	0.38 **	0.51 **	0.07
SR 2	0.52 **	0.54 **	0.20 **	0.51 **	0.60 **	0.50 **	0.32 **	0.50 **	0.15 *
NDVI	0.45 **	0.46 **	0.15 **	0.57 **	0.53 **	0.35 *	0.19	0.60 **	0.22 *
Green NDVI	0.55 **	0.50 **	0.18 **	0.52 **	0.56 **	0.40 *	0.37 **	0.53 **	0.11
NDRE	0.49 **	0.49 **	0.20 **	0.43 *	0.64 **	0.49 **	0.20 *	0.41 **	0.05
OSAVI	0.46 **	0.46 **	0.15 *	0.56 **	0.53 **	0.35 *	0.21 *	0.59 **	0.22 *
MTCI	0.41 **	0.43 **	0.18 **	0.22	0.63 **	0.49 **	0.30 **	0.25 *	0.03
NDDA	0.35 **	0.36 **	0.16 **	0.18	0.61 **	0.43 *	0.12	0.16	0.00
mND705	0.45 **	0.48 **	0.17 **	0.46 **	0.63 **	0.37 *	0.27 **	0.41 **	0.16 *
DDn	0.18 *	0.42 **	0.13 *	0.13	0.50 **	0.47 **	0.06	0.43 **	0.07
MCARI	0.26 **	0.47 **	0.13 *	0.28	0.57 **	0.54 **	0.17	0.50 **	0.10
MCARI/OSAVI	0.20 **	0.44 **	0.12 *	0.19	0.54 **	0.52 **	0.15	0.47 **	0.09
Upright leaf varieties									
SR 1	0.58 **	0.51 **	0.44 **	0.60 **	0.50 **	0.54 **	0.47 **	0.53 **	0.34 **
SR 2	0.61 **	0.56 **	0.42 **	0.64 **	0.57 **	0.53 **	0.55 **	0.54 **	0.39 **
NDVI	0.53 **	0.39 **	0.24 **	0.54 **	0.33 *	0.28 *	0.59 **	0.50 **	0.32 **
Green NDVI	0.59 **	0.50 **	0.37 **	0.59 **	0.48 **	0.44 **	0.50 **	0.51 **	0.30 **
NDRE	0.63 **	0.56 **	0.45 **	0.63 **	0.64 **	0.52 **	0.52 **	0.50 **	0.36 **
OSAVI	0.53 **	0.39 **	0.24 **	0.54 **	0.33 *	0.27 *	0.59 **	0.49 **	0.32 **
MTCI	0.53 **	0.56 **	0.48 **	0.49 **	0.76 **	0.61 **	0.50 **	0.48 **	0.39 **
NDDA	0.52 **	0.56 **	0.45 **	0.48 **	0.77 **	0.56 **	0.44 **	0.45 **	0.33 **
mND705	0.61 **	0.58 **	0.36 **	0.65 **	0.59 **	0.43 *	0.58 **	0.52 **	0.36 **
DDn	0.43 **	0.43 **	0.27 **	0.56 **	0.53 **	0.55 **	0.35 **	0.34 **	0.20 *
MCARI	0.47 **	0.46 **	0.29 **	0.61 **	0.47 **	0.54 **	0.42 **	0.46 **	0.28 **
MCARI/OSAVI	0.43 **	0.45 **	0.28 **	0.59 **	0.50 **	0.59 **	0.40 **	0.42 **	0.26 **

Analyses were performed using Pearson correlation. * Significant at 0.05 level based on a two-tailed test. ** Significant at 0.01 level based on a two-tailed test.

For the different growth stages, the correlation between the upper-layer LND and the published indices was higher before the V12 stage than after the V12 stage. The possible reason is that the spectral information of the upper leaves is influenced by the maize tassels at the top of the canopy during the later stages. Similar to the results for the upper layer, the bottom-layer LND of all varieties had good correlations with the published indices during the V6 to V12 stages. In comparison, during the V14 to R3 stages, the correlations varied greatly for the different varieties during the V14 to R3 stages. The upright leaf varieties exhibited the best correlations between the LND and all the published indices and were significant at the 0.05 significance level (R^2 ranging from 0.20 to 0.39). The intermediate leaf varieties exhibited significant correlations at the 0.05 significance level (R^2 ranging from 0.15–0.22), whereas the horizontal leaf varieties had non-significant correlations with the published indices ($R^2 < 0.17$, $p > 0.05$).

3.4. Identification of New Vegetation Indices

The analysis of the published VIs shows that the SR index generally provided a good estimate of the LND vertical distribution in the maize canopy. Therefore, we further examined all the possible 2-band combinations from 350 to 1150 nm and correlated them with the LND in the different canopy layers. The matrix plots of the coefficients of determination (R^2) between the SRs and the LND show that the relationships varied in terms of canopy layers, plant geometry types, and growth stages (Figures 2–4). Consistent with the results of the published vegetation indices, the R^2 for the SRs was higher for the upper layer than the lower layers. The R^2 values of the lower layers exhibited greater differences in the contour plots for the different maize plant geometry types (Figure 2). When the data for all growth stages were combined, the performances of the SRs were significantly affected by the plant geometry types, especially in the middle and the bottom layers. For instance, in the bottom layers of both the intermediate and upright leaf varieties, the highest R^2 values were observed for the red-edge bands (700–760 nm) paired with the red-edge to near-infrared (NIR) bands (700–1050 nm), followed by the green bands (550–580 nm) paired with the red-edge to NIR bands (700–1050 nm). The best new SR index (736, 812) produced the highest R^2 value of 0.52 for all stages for modeling the bottom-layer LND of the upright leaf varieties. The same SR index (736, 812) was also suitable for estimating the upper and middle-layer LND of the upright leaf varieties, with R^2 values of 0.50 and 0.60, respectively. For the intermediate leaf varieties, the highest R^2 value generated by the SR index (721, 935) was only 0.25. For the horizontal leaf varieties, however, few bands showed sensitivity to the bottom-layer LND.

For the different growth stages, there were good correlations between the SRs and the LND for all canopy layers except for the horizontal leaf varieties during the V6 to V12 stages. The sensitive bands in the middle and bottom layers were centered at (i) the red-edge bands (700–760 nm) paired with the red-edge to NIR bands (700–1050 nm) and (ii) the blue bands (440–490 nm) paired with the orange bands (590–640 nm) (Figure 3). For the bottom layer, for example, the best new SRs yielded large R^2 values than the published indices. The R^2 values were 0.71 for the SR index (753, 812), 0.63 for the SR index (748, 793), and 0.38 for the SR index (740, 775) for the upright, intermediate, and horizontal leaf varieties, respectively. In comparison, during the V14 to R3 stages, the R^2 values for the SRs were considerably lower for the bottom layer. The sensitive bands for the bottom layer were shifted to (i) the orange to red bands (590–710 nm) paired with the red-edge to NIR bands (710–1050 nm) and (ii) green bands (520–600 nm) paired with the red bands (600–700 nm) (Figure 4). The SR index (524, 583) and SR index (591, 611) were the best new SRs for the upright ($R^2 = 0.54$) and intermediate ($R^2 = 0.30$) leaf varieties, respectively. Additionally, there were also very poor relationships between the SRs and the bottom-layer LND of the horizontal leaf varieties.

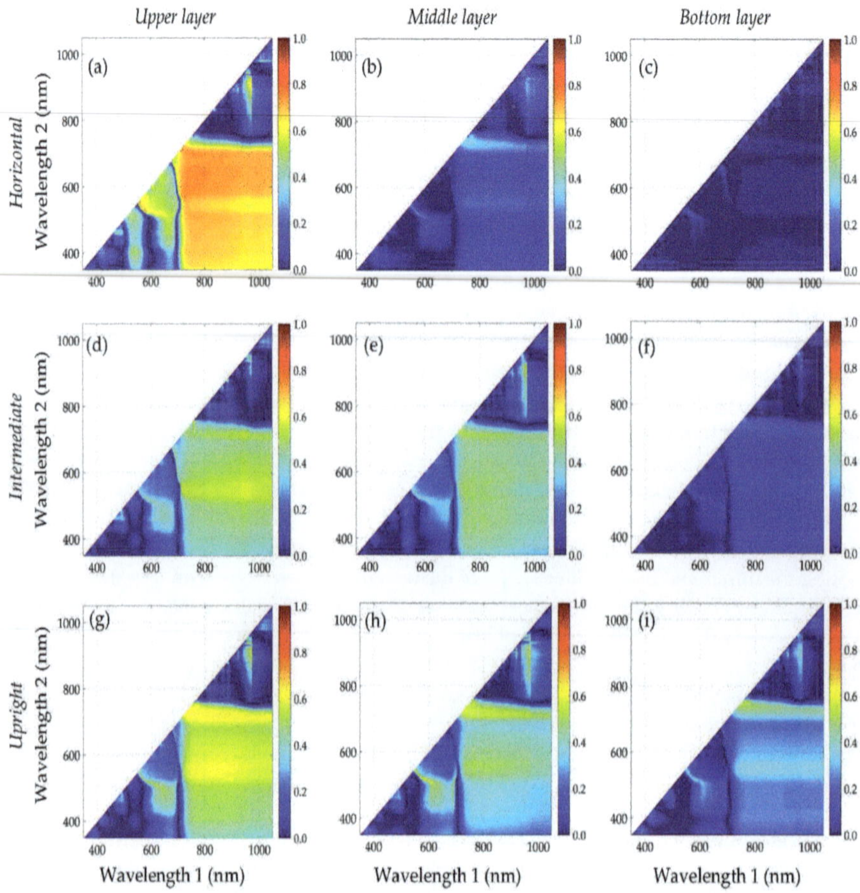

Figure 2. Maps showing the coefficients of determination (R^2) for the relationships between the leaf N density and SR index calculated from all possible 2-band combinations in the range of 35–1050 nm across all stages. (**a**) The upper layer of horizontal leaf varieties; (**b**) the middle layer of the horizontal leaf varieties; (**c**) the bottom layer of the horizontal leaf varieties; (**d**) the upper layer of the intermediate leaf varieties; (**e**) the middle layer of the intermediate leaf varieties; (**f**) the bottom layer of the intermediate leaf varieties; (**g**) the upper layer of the upright leaf varieties; (**h**) the middle layer of the upright leaf varieties; and (**i**) the bottom layer of the upright leaf varieties.

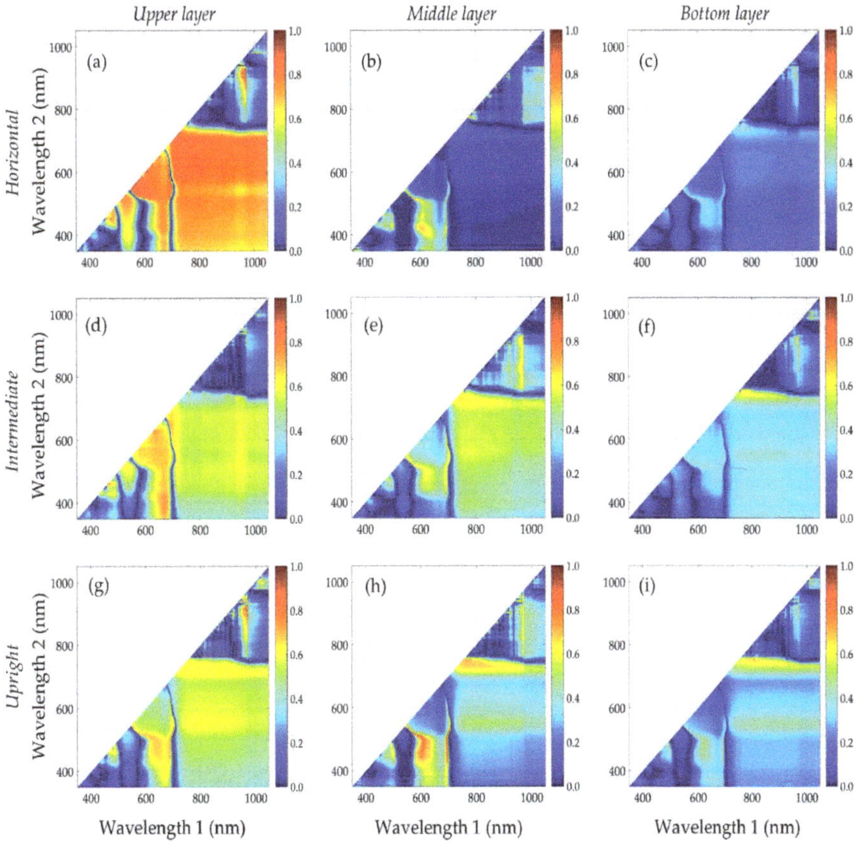

Figure 3. Maps showing the R^2 for the relationships between the leaf N density and SR index calculated from all possible 2-band combinations in the range of 350–1050 nm during the V6 to V12 stages. (**a**) The upper layer of the horizontal leaf varieties; (**b**) the middle layer of the horizontal leaf varieties; (**c**) the bottom layer of the horizontal leaf varieties; (**d**) the upper layer of the intermediate leaf varieties; (**e**) the middle layer of the intermediate leaf varieties; (**f**) the bottom layer of the intermediate leaf varieties; (**g**) the upper layer of the upright leaf varieties, (**h**) the middle layer of the upright leaf varieties; and (**i**) the bottom layer of the upright leaf varieties.

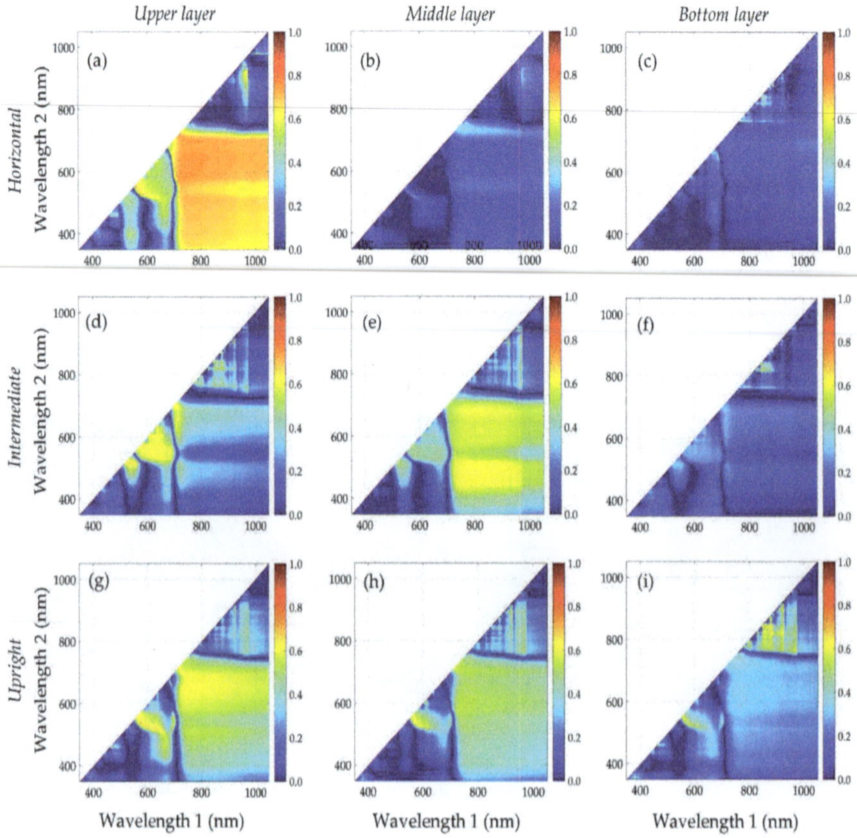

Figure 4. Maps showing the R^2 for the relationships between the leaf N density and SR index calculated from all possible 2-band combinations in the range of 350–1050 nm during the V14 to R3 stages. (**a**) The upper layer of the horizontal leaf varieties; (**b**) the middle layer of the horizontal leaf varieties; (**c**) the bottom layer of the horizontal leaf varieties; (**d**) the upper layer of the intermediate leaf varieties; (**e**) the middle layer of the intermediate leaf varieties; (**f**) the bottom layer of the intermediate leaf varieties; (**g**) the upper layer of the upright leaf varieties; (**h**) the middle layer of the upright leaf varieties; and (**i**) the bottom layer of the upright leaf varieties.

4. Discussion

The non-uniformity of the vertical leaf N distribution in the canopy of plants has been demonstrated in many studies; i.e., the lower shaded leaves generally have a lower N content than the upper illuminated leaves [21,28,29,63,64]. Our results revealed that there were significant differences in the LAI_i/LAI_T between the different maize plant geometry types (i.e., the order of the upper-layer LAI_1/LAI_T was horizontal leaf varieties > intermediate leaf varieties > upright leaf varieties, whereas the order of the bottom-layer LAI_3/LAI_T was upright leaf varieties > horizontal leaf varieties > intermediate leaf varieties). This resulted in a decrease in the vertical N gradients between the upper layer and the bottom layer as the plant type changed to a more upright geometry (i.e., 1.29, 1.09, and 0.85 g m^{-2} for the horizontal, intermediate, and upright varieties, respectively). In addition, the variation in the bottom-layer LND was larger than that of the upper-layer LND across the entire growth period (i.e., CV: 80.0% vs. 28.0% for the horizontal leaf varieties, 68.3% vs. 33.2% for the intermediate leaf varieties, 81.4% vs. 34.8% for the upright leaf varieties). This is consistent with the results of Lu (1994) [22], who reported that the lower leaves of the crop canopy are more sensitive to N deficiency or N translocation than the upper leaves.

The most encouraging result, however, was that the maize geometry plant type had a significant effect on the detectability of the vertical leaf N distribution in the canopy using canopy spectral reflectance, as well as the sensitivity of the bands. As shown in Table 2 and Figure 3, the correlations are higher between the lower leaf N and the spectral indices (both for the published indices and the new SRs) for maize plants with upright leaves than those with relatively flat leaves. For the bottom layers of the upright leaf varieties, the red edge and NIR bands performed better than the visible bands, which is in agreement with the results of Feng et al. (2008) [65] for wheat and those of Yu et al. (2013) [14] for rice. Actually, 52% of the variation in the bottom layer LND of upright leaf varieties was explained by the best performing newly proposed SR index (736, 812), whereas only 25% of the variation in the bottom layer LND of the intermediate leaf varieties was explained by the best performing newly proposed SR index (721, 935). Few bands were sensitive to the LND of the bottom layer of the horizontal leaf varieties. This is attributed to the fact that, in plants with upright leaves, the bottom layer contributes more to the canopy spectral reflectance than in plants with relatively flat leaves [30,32–34,37].

The results also showed that the growth stages had a large influence on the remote estimation of the N vertical distribution in the canopy. During the V6 to V12 stages, the sensitive bands for detecting the bottom leaf N were the red-edge bands (700–760 nm) paired with the red-edge to NIR bands (700–1050 nm), and these were shifted to the orange to red bands (590–710 nm) paired with the red-edge to NIR bands (710–1050 nm) during the V14 to R3 stages. This occurred because the canopy reflectance was dominated by the biomass before the V12 stage and by the plant N after this stage [14]. In addition, many studies have shown that the green duration and the senescence speed of the lower leaves of the crop are closely related to the amount of light quanta that is received [34,66]. When the canopy is fully closed during the later growth stage (i.e., after the V14 stage in this study), the bottom leaves begin to senesce and wilt.

The timely and accurate diagnosis of maize N nutritional status is essential for precise N management. The vertical leaf N distribution is very important for N status diagnosis because when N is deficient, lower leaves will show deficiency symptoms first. The results of this study indicated that it was feasible to estimate the N vertical distribution in the maize canopy using canopy hyperspectral reflectance before the V12 stage. These results provide a theoretical basis for early diagnosis of maize N status using remote sensing for precision N management, especially for upright leaf type varieties. According to the local standard practices, the topdressing of N for maize is mainly applied between V6 and V12 stages. The N vertical distribution information in the canopy of maize before the V12 stage can be used to guide in-season N management. More studies are needed to develop corresponding N recommendation algorithms based on such information. However, there are still some limitations in this study. The factors affecting the reflectivity should be considered in future research; for example, the sun angle, background, LAI, leaf angle distribution, and pigment content [67,68]. Moreover, the hyperspectral reflectance data obtained from the top of the canopy by ASD FieldSpec® Pro FR spectroradiometer represents spectrally mixed information because the upper, middle, and bottom-layer leaves cannot be distinguished. In the future, we plan to use hyperspectral imaging acquired from different zenith and azimuth angles, which may be more suitable for estimating the vertical N distribution in the crop canopy. A fusion of LiDAR and hyperspectral (or multispectral) data collected from unmanned aerial vehicle (UAV) platforms also has great potential in estimating N vertical distribution in the crop canopies at field and farm scales for precision agriculture applications.

5. Conclusions

This study evaluated the influences of the maize plant geometry (i.e., horizontal type, intermediate type, and upright type) on the remote estimation of the vertical N distribution in a canopy at different growth stages. The results showed that there were significant differences between the three maize plant geometry types in terms of canopy architecture and N vertical distribution. The vertical LND gradients between the upper layer and bottom layer decreased as the leaves of the maize changed to a more upright geometry (i.e., 1.29, 1.09, and 0.85 g m^{-2} for the horizontal, intermediate, and upright varieties, respectively). In comparison, the lower-layer LND was more sensitive than the upper-layer LND for the combined growth stages. The results also showed that the maize plant geometry had a significant impact on the estimation of the vertical N distribution based on canopy spectral reflectance. The upright leaf varieties had the highest correlation between the lower-layer LND (R^2 = 0.52) and the new SR index (736, 812), and this index also performed well for estimating the upper (R^2 = 0.50) and middle (R^2 = 0.60) layer LND. However, for the intermediate leaf variety, only 25% of the variation in the bottom-layer LND was explained by the best new SR index (721, 935) and few bands showed sensitivity to the bottom-layer LND of the horizontal leaf varieties. Furthermore, the growth stages also influenced the remote detection of the lower-leaf N status of the canopy, because the canopy reflectance was dominated by the biomass before the V12 stage and by the plant N after this stage. Therefore, we can conclude that a more accurate estimation of the N vertical distribution in the canopy can be obtained by canopy hyperspectral reflectance for maize plants with more upright leaves.

Author Contributions: H.Y. performed the data analysis and wrote the manuscript. W.H. conducted the field experiments and guided the study. S.H., B.W., Y.D. and B.C. provided suggestions for the study, reviewed and edited the manuscript. All authors read and approved the manuscript.

Funding: This research was funded by the National Natural Science Foundation of China (41871339, 41501468, 41571354, 41601466), the Key Laboratory of Agricultural Big Data, Ministry of Agriculture and Rural Affairs, Beijing, China, and Hainan Provincial Department of Science and Technology under Grant (ZDKJ2016021, ZDKJ2017009).

Acknowledgments: We would like to thank the field campaign at the National Experimental Station for Precision Agriculture, Beijing, China.

Conflicts of Interest: The authors declare no conflict of interest.

References

1. Sheehy, J.E.; Dionora, M.J.A.; Mitchell, P.L.; Peng, S.; Cassman, K.G.; Lemaire, G.; Williams, R.L. Critical nitrogen concentrations: Implications for high-yielding rice (*Oryza sativa* L.) cultivars in the tropics. *Field Crop Res.* **1998**, *59*, 31–41. [CrossRef]
2. Ziadi, N.; Bélanger, G.; Claessens, A.; Lefebvre, L.; Cambouris, A.N.; Tremblay, N.; Nolin, M.C.; Parent, L.É. Determination of a critical nitrogen dilution curve for spring wheat. *Agron. J.* **2010**, *102*, 241–250. [CrossRef]
3. Yue, S.C.; Sun, F.L.; Meng, Q.F.; Zhao, R.F.; Li, F.; Chen, X.P.; Zhang, F.S.; Cui, Z.L. Validation of a critical nitrogen curve for summer maize in the North China Plain. *Pedosphere* **2014**, *24*, 76–83. [CrossRef]
4. Scholberg, J.; McNeal, B.L.; Boote, K.J.; Jones, J.W.; Locascio, S.J.; Olson, S.M. Nitrogen stress effects on growth and nitrogen accumulation by field-grown tomato. *Agron. J.* **2000**, *92*, 159–167. [CrossRef]
5. Zhao, D.; Reddy, K.R.; Kakani, V.G.; Reddy, V.R. Nitrogen deficiency effects on plant growth, leaf photosynthesis, and hyperspectral reflectance properties of sorghum. *Eur. J. Agron.* **2005**, *22*, 391–403. [CrossRef]
6. Boussadia, O.; Steppe, K.; Zgallai, H.; El Hadj, S.B.; Braham, M.; Lemeur, R.; Van Labeke, M.C. Effects of nitrogen deficiency on leaf photosynthesis, carbohydrate status and biomass production in two olive cultivars 'Meski'and 'Koroneiki'. *Sci. Hortic.* **2010**, *123*, 336–342. [CrossRef]
7. Zhang, Y.; Tan, L.; Zhu, Z.; Yuan, L.; Xie, D.; Sun, C. TOND1 confers tolerance to nitrogen deficiency in rice. *Plant J.* **2015**, *81*, 367–376. [CrossRef]
8. Ju, X.T.; Xing, G.X.; Chen, X.P.; Zhang, S.L.; Zhang, L.J.; Liu, X.J.; Cui, Z.L.; Yin, B.; Christie, P.; Zhu, Z.L.; et al. Reducing environmental risk by improving N management in intensive Chinese agricultural systems. *Proc. Natl. Acad. Sci. USA* **2009**, *106*, 3041–3046. [CrossRef]
9. Huang, W.; Yang, Q.; Pu, R.; Yang, S. Estimation of Nitrogen Vertical Distribution by Bi-Directional Canopy Reflectance in Winter Wheat. *Sensors* **2014**, *14*, 20347–20359. [CrossRef]
10. Qiu, S.J.; He, P.; Zhao, S.C.; Li, W.J.; Xie, J.G.; Hou, Y.P.; Grant, C.A.; Zhou, W.; Jin, J.Y. Impact of Nitrogen Rate on Maize Yield and Nitrogen Use Efficiencies in Northeast China. *Agron. J.* **2015**, *107*, 305–313. [CrossRef]

11. Li, F.; Miao, Y.; Hennig, S.D.; Gnyp, M.L.; Chen, X.; Jia, L.; Bareth, G. Evaluating hyperspectral vegetation indices for estimating nitrogen concentration of winter wheat at different growth stages. *Precis. Agric.* **2010**, *11*, 335–357. [CrossRef]
12. Hansen, P.M.; Schjoerring, J.K. Reflectance measurement of canopy biomass and nitrogen status in wheat crops using normalized difference vegetation indices and partial least square regression. *Remote Sens. Environ.* **2003**, *86*, 542–553. [CrossRef]
13. Xue, L.; Cao, W.; Luo, W.; Dai, T.; Zhu, Y. Monitoring leaf nitrogen status in rice with canopy spectral reflectance. *Agron. J.* **2004**, *96*, 135–142. [CrossRef]
14. Yu, K.; Li, F.; Gnyp, M.L.; Miao, Y.; Bareth, G.; Chen, X. Remotely detecting canopy nitrogen concentration and uptake of paddy rice in the Northeast China Plain. *ISPRS-J. Photogramm. Remote Sens.* **2013**, *78*, 102–115. [CrossRef]
15. Huang, S.; Miao, Y.; Zhao, G.; Yuan, F.; Ma, X.; Tan, C.; Yu, W.; Gnyp, M.L.; Lenz-Wiedemann, V.I.S.; Rascher, U.; et al. Satellite remote sensing-based in-season diagnosis of rice nitrogen status in Northeast China. *Remote Sens.* **2015**, *7*, 10646–10667. [CrossRef]
16. He, L.; Song, X.; Feng, W.; Guo, B.B.; Zhang, Y.S.; Wang, Y.H.; Wang, C.Y.; Guo, T.C. Improved remote sensing of leaf nitrogen concentration in winter wheat using multi-angular hyperspectral data. *Remote Sens. Environ.* **2016**, *174*, 122–133. [CrossRef]
17. Dreccer, M.F.; Slafer, G.A.; Rabbinge, R. Optimization of vertical distribution of canopy nitrogen. *J. Crop Prod.* **1997**, *1*, 47–77. [CrossRef]
18. Dreccer, M.F.; Van Oijen, M.; Schapendonk, A.H.C.M.; Pot, C.S.; Rabbinge, R. Dynamics of vertical leaf nitrogen distribution in a vegetative wheat canopy. Impact on canopy photosynthesis. *Ann. Bot.* **2000**, *86*, 821–831. [CrossRef]
19. Huang, W.; Wang, Z.; Huang, L.; Lamb, D.W.; Ma, Z.; Zhang, J.; Wang, J.; Zhao, C. Estimation of vertical distribution of chlorophyll concentration by bi-directional canopy reflectance spectra in winter wheat. *Precis. Agric.* **2011**, *12*, 165–178. [CrossRef]
20. Wang, Z.; Wang, J.; Zhao, C.; Zhao, M.; Huang, W.; Wang, C. Vertical distribution of nitrogen in different layers of leaf and stem and their relationship with grain quality of winter wheat. *J. Plant Nutr.* **2012**, *28*, 73–91. [CrossRef]
21. Li, H.; Zhao, C.; Huang, W.; Yang, G. Non-uniform vertical nitrogen distribution within plant canopy and its estimation by remote sensing: A review. *Field Crop Res.* **2013**, *142*, 75–84. [CrossRef]
22. Lu, J.L. *Plant Nutrition*, 1st ed.; the First Volume; Beijing Agricultural University Press (continuing China Agricultural University Press): Beijing, China, 1994. (In Chinese)
23. Wang, Z.; Wang, J.; Liu, L.; Huang, W.; Zhao, C.; Lu, Y. Estimation of nitrogen status in middle and bottom layers of winter wheat canopy by using ground-measured canopy reflectance. *Commun. Soil Sci. Plant* **2005**, *36*, 2289–2302. [CrossRef]
24. Wang, J.H.; Huang, W.J.; Lao, C.L.; Zhang, L.D.; Luo, C.B.; Wang, T.; Liu, L.Y.; Song, X.Y.; Ma, Z.H. Inversion of winter wheat foliage vertical distribution based on canopy reflected spectrum by partial least squares regression method. *Spectrosc. Spect. Anal.* **2007**, *27*, 1319–1322, (In Chinese with English abstract).
25. Ciganda, V.S.; Gitelson, A.A.; Schepers, J. How deep does a remote sensor sense? expression of chlorophyll content in a maize canopy. *Remote Sens. Environ.* **2012**, *126*, 240–247. [CrossRef]
26. Hirose, T.; Werger, M.J.A.; Pons, T.L.; van Rheenen, J.W.A. Canopy structure and leaf nitrogen distribution in a stand of Lysimachia vulgaris L. as influenced by stand density. *Oecologia* **1988**, *77*, 145–150. [CrossRef]
27. Anten, N.P.R.; Schieving, F.; Werger, M.J.A. Patterns of light and nitrogen distribution in relation to whole canopy carbon gain in C3 and C4 mono- and dicotyledonous species. *Oecologia* **1995**, *101*, 504–513. [CrossRef] [PubMed]
28. Lötscher, M.; Stroh, K.; Schnyder, H. Vertical leaf nitrogen distribution in relation to nitrogen status in grassland plants. *Ann. Bot.* **2003**, *92*, 679–688. [CrossRef]
29. Bertheloot, J.; Martre, P.; Andrieu, B. Dynamics of light and nitrogen distribution during grain filling within wheat canopy. *Plant Physiol.* **2008**, *148*, 1707–1720. [CrossRef]
30. Monsi, M.; Saeki, T. On the factor light in plant communities and its importance for matter production. *Ann. Bot.* **2005**, *95*, 549–567. [CrossRef]
31. Hirose, T.; Werger, M.J.A. Maximizing daily canopy photosynthesis with respect to the leaf nitrogen allocation pattern in the canopy. *Oecologia* **1987**, *72*, 520–526. [CrossRef]

32. Maddonni, G.A.; Otegui, M.E.; Cirilo, A.G. Plant population density, row spacing and hybrid effects on maize canopy architecture and light attenuation. *Field Crop. Res.* **2001**, *71*, 183–193. [CrossRef]
33. Lambert, R.J.; Johnson, R.R. Leaf angle, tassel morphology, and the performance of maize hybrids. *Crop Sci.* **1978**, *18*, 499–502. [CrossRef]
34. Azumi, Y.; Watanabe, A. Evidence for a senescence-associated gene induced by darkness. *Plant Physiol.* **1991**, *95*, 577–583. [CrossRef] [PubMed]
35. Ku, L.X.; Zhao, W.M.; Zhang, J.; Wu, L.C.; Wang, C.L.; Wang, P.A.; Zhang, W.Q.; Chen, Y.H. Quantitative trait loci mapping of leaf angle and leaf orientation value in maize (*Zea mays* L.). *Theor. Appl. Genet.* **2010**, *121*, 951–959. [CrossRef] [PubMed]
36. Duvick, D.N.; Smith, J.S.C.; Cooper, M. Long-term selection in a commercial hybrid maize breeding program. *Plant Breed. Rev.* **2004**, *24*, 109–151.
37. Ma, D.L.; Xie, R.Z.; Niu, X.K.; Li, S.K.; Long, H.L.; Liu, Y.E. Changes in the morphological traits of maize genotypes in China between the 1950s and 2000s. *Eur. J. Agron.* **2014**, *58*, 1–10. [CrossRef]
38. Tong, P. The course and the achievement of corn variety improvement in the20th century in China. *China Hist. Mater. Sci. Technol.* **2001**, *22*, 113–127.
39. Meghji, M.R.; Dudley, J.W.; Lambert, R.J.; Sprague, G.F. Inbreeding depression, inbred and hybrid grain yields, and other traits of maize genotypes representing three eras. *Crop Sci.* **1984**, *24*, 545–549. [CrossRef]
40. Russell, W.A. Genetic improvement of maize yields. *Adv. Agron.* **1991**, *46*, 245–298.
41. IUSS Working Group WRB. *World Reference Base for Soil Resources 2006*; World Soil Resources Reports No. 103; FAO: Rome, Italy, 2006.
42. Soil Survey Division Staff. *Soil Survey Manual*; United States Department of Agriculture: Washington, DC, USA, 1993.
43. Pepper, G.E.; Pearce, R.B.; Mock, J.J. Leaf orientation and yield of maize. *Crop Sci.* **1977**, *17*, 883–886. [CrossRef]
44. Jonckheere, I.; Fleck, S.; Nackaerts, K.; Muys, B.; Coppin, P.; Weiss, M.; Baret, F. Methods for leaf area index determination. Part I: Theories, techniques and instruments. *Agric. For. Meteorol.* **2004**, *121*, 19–35. [CrossRef]
45. Helrich, K. *Official Methods of Analysis of the AOAC*, 15th ed.; AOAC: Arlington, VA, USA, 1990; pp. 16–17.
46. Greenwood, D.J.; Neeteson, J.J.; Draycott, A. Quantitative relationships for the dependence of growth rate of arable crops on their nitrogen content, dry weight and aerial environment. In *Fundamental, Ecological and Agricultural Aspects of Nitrogen Metabolism in Higher Plants*; Springer: Dordrecht, The Netherlands, 1986; Volume 91, pp. 281–301.
47. Giletto, C.M.; Echeverría, H.E. Critical nitrogen dilution curve for processing potato in Argentinean humid pampas. *Am. J. Potato Res.* **2012**, *89*, 102–110. [CrossRef]
48. Mistele, B.; Schmidhalter, U. Estimating the nitrogen nutrition index using spectral canopy reflectance measurements. *Eur. J. Agron.* **2008**, *29*, 184–190. [CrossRef]
49. Hirose, T.; Werger, M.J.A.; van Rheenen, J.W.A. Canopy development and leaf nitrogen distribution in a stand of Carex acutiformis. *Ecology* **1989**, *70*, 1610–1618. [CrossRef]
50. Archontoulis, S.V.; Vos, J.; Yin, X.; Bastiaans, L.; Danalatos, N.G.; Struik, P.C. Temporal dynamics of light and nitrogen vertical distributions in canopies of sunflower, kenaf and cynara. *Field Crop Res.* **2011**, *122*, 186–198. [CrossRef]
51. Schieving, F.; Pons, T.L.; Werger, M.J.A.; Hirose, T. The vertical distribution of nitrogen and photosynthetic activity at different plant densities in Carex acutiformis. *Plant Soil* **1992**, *14*, 9–17. [CrossRef]
52. Zarco-Tejada, P.J.; Miller, J.R.; Noland, T.L.; Mohammed, G.H.; Sampson, P.H. Scaling-up and model inversion methods with narrowband optical indices for chlorophyll content estimation in closed forest canopies with hyperspectral data. *IEEE Trans. Geosci. Remote Sens.* **2001**, *39*, 1491–1507. [CrossRef]
53. Blackburn, G.A. Quantifying chlorophylls and carotenoids at leaf and canopy scales: An evaluation of some hyperspectral approaches. *Remote Sens. Environ.* **1998**, *66*, 273–285. [CrossRef]
54. Gitelson, A.; Merzlyak, M.N. Quantitative estimation of chlorophyll-a using reflectance spectra: Experiments with autumn chestnut and maple leaves. *J. Photochem. Photobiol. B-Biol.* **1994**, *22*, 247–252. [CrossRef]
55. Fitzgerald, G.J.; Rodriguez, D.; Christensen, L.K.; Belford, K.; Sadras, V.O.; Clarke, T.R. Spectral and thermal sensing for nitrogen and water status in rainfed and irrigated wheat environments. *Precis. Agric.* **2006**, *7*, 233–248. [CrossRef]

56. Rondeaux, G.; Steven, M.; Baret, F. Optimization of soil-adjusted vegetation indices. *Remote Sens. Environ.* **1996**, *55*, 95–107. [CrossRef]
57. Dash, J.; Curran, P.J. The MERIS terrestrial chlorophyll index. *Int. J. Remote Sens.* **2004**, *25*, 5403–5413. [CrossRef]
58. Feng, W.; Guo, B.B.; Wang, Z.J.; He, L.; Song, X.; Wang, Y.H.; Guo, T.C. Measuring leaf nitrogen concentration in winter wheat using double-peak spectral reflection remote sensing data. *Field Crop. Res.* **2014**, *159*, 43–52. [CrossRef]
59. Sims, D.A.; Gamon, J.A. Relationships between leaf pigment content and spectral reflectance across a wide range of species, leaf structures and developmental stages. *Remote Sens. Environ.* **2002**, *81*, 331–354. [CrossRef]
60. Le Maire, G.; François, C.; Soudani, K.; Berveiller, D.; Pontailler, J.Y.; Bréda, N.; Genet, H.; Davi, H.; Dufrêne, E. Calibration and validation of hyperspectral indices for the estimation of broadleaved forest leaf chlorophyll content, leafmass per area, leaf area index and leaf canopy biomass. *Remote Sens. Environ.* **2008**, *112*, 3846–3864. [CrossRef]
61. Wu, C.; Niu, Z.; Tang, Q.; Huang, W. Estimating chlorophyll content from hyperspectral vegetation indices: Modeling and validation. *Agric. For. Meteorol.* **2008**, *148*, 1230–1241. [CrossRef]
62. Daughtry, C.S.T.; Walthall, C.L.; Kim, M.S.; De Colstoun, E.B.; McMurtrey, J.E. Estimating corn leaf chlorophyll concentration from leaf and canopy reflectance. *Remote Sens. Environ.* **2000**, *74*, 229–239. [CrossRef]
63. Drouet, J.L.; Bonhomme, R. Do variations in local leaf irradiance explain changes to leaf nitrogen within row maize canopies? *Ann. Bot.* **1999**, *84*, 61–69. [CrossRef]
64. Shiratsuchi, H.; Yamagishi, T.; Ishii, R. Leaf nitrogen distribution to maximize the canopy photosynthesis in rice. *Field Crop Res.* **2006**, *95*, 291–304. [CrossRef]
65. Feng, W.; Yao, X.; Zhu, Y.; Tian, Y.C.; Cao, W.X. Monitoring leaf nitrogen status with hyperspectral reflectance in wheat. *Eur. J. Agron.* **2008**, *28*, 394–404. [CrossRef]
66. Ottman, M.J.; Welch, L.F. Supplemental radiation effects on senescence, plant nutrients, and yield of field-grown corn. *Agron. J.* **1988**, *80*, 619–626. [CrossRef]
67. Schlemmer, M.; Gitelson, A.; Schepers, J.; Ferguson, R.; Peng, Y.; Shanahan, J.; Rundquistb, D. Remote estimation of nitrogen and chlorophyll contents in maize at leaf and canopy levels. *Int. J. Appl. Earth Obs. Geoinf.* **2013**, *25*, 47–54. [CrossRef]
68. Feng, W.; Qi, S.; Heng, Y.; Zhou, Y.; Wu, Y.; Liu, W.; He, L.; Li, X. Canopy vegetation indices from in situ hyperspectral data to assess plant water status of winter wheat under powdery mildew stress. *Front. Plant Sci.* **2017**, *8*, 1219. [CrossRef] [PubMed]

remote sensing

MDPI

Article

A New Integrated Vegetation Index for the Estimation of Winter Wheat Leaf Chlorophyll Content

Bei Cui [1,2,3], Qianjun Zhao [4], Wenjiang Huang [1,3,*], Xiaoyu Song [5], Huichun Ye [1,3] and Xianfeng Zhou [6]

1 Aerospace Information Research Institute, Chinese Academy of Sciences, Beijing 100094, China; cuibei@aircas.ac.cn (B.C.); yehc@aircas.ac.cn (H.Y.)
2 University of Chinese Academy of Sciences, Beijing 100049, China
3 Key Laboratory of Earth Observation, Sanya 572029, Hainan Province, China
4 Bureau of Science & Technology for Development, Chinese Academy of Sciences, Beijing 100864, China; zhaoqj@radi.ac.cn
5 Beijing Research Center for Information Technology in Agriculture, Beijing Academy of Agriculture and Forestry Sciences, Beijing 100097, China; songxy@nercita.org.cn
6 College of Life Information Science and Instrument Engineering, Hangzhou Dianzi University, Hangzhou 310018, China; zhouxianfeng@hdu.edu.cn
* Correspondence: huangwj@aircas.ac.cn; Tel.: +86-0898-88597739

Received: 28 March 2019; Accepted: 20 April 2019; Published: 23 April 2019

Abstract: Leaf chlorophyll content (LCC) provides valuable information about the nutrition and photosynthesis statuses of crops. Vegetation index-based methods have been widely used in crop management studies for the non-destructive estimation of LCC using remote sensing technology. However, many published vegetation indices are sensitive to crop canopy structure, especially the leaf area index (LAI), when crop canopy spectra are used. Herein, to address this issue, we propose four new spectral indices (The red-edge-chlorophyll absorption index (RECAI), the red-edge-chlorophyll absorption index/optimized soil-adjusted vegetation index (RECAI/OSAVI), the red-edge-chlorophyll absorption index/ the triangular vegetation index (RECAI/TVI), and the red-edge-chlorophyll absorption index/the modified triangular vegetation index(RECAI/MTVI2)) and evaluate their performance for LCC retrieval by comparing their results with those of eight published spectral indices that are commonly used to estimate LCC. A total of 456 winter wheat canopy spectral data corresponding to physiological parameters in a wide range of species, growth stages, stress treatments, and growing seasons were collected. Five regression models (linear, power, exponential, polynomial, and logarithmic) were built to estimate LCC in this study. The results indicated that the newly proposed integrated RECAI/TVI exhibited the highest LCC predictive accuracy among all indices, where R^2 values increased by more than 13.09% and RMSE values reduced by more than 6.22%. While this index exhibited the best association with LCC ($0.708^{**} \leq r \leq 0.819^{**}$) among all indices, RECAI/TVI exhibited no significant relationship with LAI ($0.029 \leq r \leq 0.167$), making it largely insensitive to LAI changes. In terms of the effects of different field management measures, the LCC predictive accuracy by RECAI/TVI can be influenced by erective winter wheat varieties, low N fertilizer application density, no water application, and early sowing dates. In general, the newly developed integrated RECAI/TVI was sensitive to winter wheat LCC with a reduction in the influence of LAI. This index has strong potential for monitoring winter wheat nitrogen status and precision nitrogen management. However, further studies are required to test this index with more diverse datasets and different crops.

Keywords: leaf chlorophyll content; red-edge reflectance; spectral index; winter wheat

---

1. Introduction

As winter wheat is one of the most important food crops in China, the timely and accurate monitoring of the growth and nutrition of this crop contributes to proper field management. The leaf chlorophyll content (LCC), which includes the contents of chlorophyll *a* and chlorophyll *b*, can provide crucial information for understanding vegetation stress [1,2], physiological status, and photosynthesis potential [3,4]. In addition, LCC is strongly related to the N content [5–8] and can be used as a close proxy for the N concentration at the leaf level [9,10]. The traditional measurement approach in the laboratory is relatively time- and labor-consuming, making it difficult to meet the practical demands of precise crop management in large fields. With the development of remote sensing techniques, remotely sensed data have been widely used to accurately and non-destructively monitor crop chlorophyll contents [2,11,12].

Currently, the semi-empirical index-based approach is commonly used to estimate crop chlorophyll content. According to the typical spectral absorption characteristics of chlorophyll pigments, red and near-infrared (NIR) spectral bands are primarily used to build chlorophyll content indices [4]. The red-edge spectrum has received much attention for many years for monitoring chlorophyll content, and the red edge has been identified to be more sensitive to chlorophyll contents than the red part of the spectrum [2,13]. The "red edge" refers to the steep part between the chlorophyll absorption valley in the red band and the high reflection shoulder in the NIR band. The importance of the red-edge spectra for estimating chlorophyll content was demonstrated by extensive studies based on the combined PROSPECT leaf optical properties model and SAIL canopy bidirectional reflectance model (PROSAIL)-simulated spectra or ground-measured spectra. Therefore, some red-edge parameters and chlorophyll indices were developed based on the red-edge band(s). Several scholars have developed red-edge parameters, such as the red-edge position, red-edge amplitude, red-edge width, red-edge kurtosis, minimum amplitude, and red-edge amplitude/minimum amplitude, to predict crop chlorophyll contents. Yao et al. [14] analyzed the relationship between the red-edge spectrum features of the winter wheat canopy and leaf chlorophyll content (SPAD) values in different growing periods and noted that the model-based red-edge kurtosis exhibited the highest SPAD predictive accuracy. Liu et al. [15] calculated seven red-edge parameters as inputs for a back-propagation (BP) neural network estimation model and studied the LCCs of *Pinus massoniana*. Gitelson and Merzlyak [2] proposed that the sensitivity of the normalized difference vegetation index (NDVI) to chlorophyll content can be improved by replacing the reflectance in the red band with the reflectance in the red-edge band at approximately 690–710 nm. Zillmann et al. [16] concluded that the normalized difference red edge (NDRE) index was strongly linearly related to the winter wheat chlorophyll content at the canopy level based on RapidEye images. Based on the strong absorption characteristics of chlorophyll in the red band, Kim et al. [17] built a new chlorophyll absorption ratio index (CARI) to reduce the effect from non-photosynthetic materials by using the ratio of the reflectances at 700 nm and 550 nm. Daughtry et al. [18] proposed the modified chlorophyll absorption ratio index (MCARI) by introducing the R700/R670 ratio to the CARI index to reduce background effects. However, MCARI was still sensitive to background properties. Then, Haboudane et al. [19] introduced the R_{700}–R_{550} term to MCARI to further reduce the effects from the background and proposed a new index called the transformed chlorophyll absorption in reflectance index (TCARI). Gitelson et al. [20] found that the reciprocal reflectance in the range from 695–705 nm is closely related to LCC and proposed a new index called $CI_{red\text{-}edge}$ that obviously improved the accuracy of chlorophyll content prediction. Many scholars have used the $CI_{red\text{-}edge}$ index to estimate the chlorophyll contents of different crop species [8,16,21,22]. Based on the band settings of medium resolution imaging spectrometer (MERIS) data, Dash and Curran [23] proposed the medium resolution imaging spectrometer (MERIS) terrestrial chlorophyll index (MTCI), which is strongly related to the red-edge position and has been used to successfully predict vegetation chlorophyll contents at the canopy level. Maire et al. [24] proposed the new double difference (DD) index to estimate tree chlorophyll contents according to the "peak jump" and the multiple-peak features existing on the first derivative of the spectral reflectance. Jin et al. [25]

built two new indices, the double-peak canopy nitrogen index I (DCNI I) and the ratio of the plant pigment ratio to the NDVI (PPR/NDVI), to estimate cotton LCCs with high predictive accuracy.

Most vegetation indices have been developed to estimate chlorophyll content at the canopy level. However, research on the estimation of LCCs using a vegetation index approach with crop canopy spectra is relatively limited and unsatisfactory, mainly because crop canopy spectra are affected by not only leaf biochemical parameters and leaf distribution but also crop canopy structure, soil nutrients, atmosphere, and other factors. In addition, red-edge spectra are influenced by both chlorophyll pigments and the leaf area index (LAI) [26,27]. As a result, the precision of leaf-scale chlorophyll content inversions based on crop canopy spectral data is often low. Therefore, the influence of LAI prevents the use of red-edge information as an LCC estimator. An optical chlorophyll index should be both sensitive to chlorophyll content and insensitive to other interference factors [28]. However, uncoupling the interplay of chlorophyll pigments and LAI on spectral reflectance is a challenging issue for the estimation of crop chlorophyll content at the leaf level by remote sensing. Only a few studies have explored this problem using specific crop types. Daughtry et al. [18] used simulated data to demonstrate that the combination of two groups of vegetation indices that minimize background reflectance contributions and strongly respond to leaf chlorophyll concentrations can be used to estimate the leaf chlorophyll concentration at the leaf level with minimal confounding effects from the LAI and soil background. Haboudane et al. [19] proposed that the transformed chlorophyll absorption in the reflectance index/optimized soil-adjusted vegetation index (TCARI/OSAVI) is both very sensitive to LCCs and very resistant to impacts from LAI and solar zenith angle. He also estimated corn LCCs using TCARI/OSAVI and achieved good results. Kooistra and Clevers [29] used TCARI/OSAVI, TCI/OSAVI (the triangular chlorophyll index/ optimized soil-adjusted vegetation index), and CVI (chlorophyll vegetation index) to estimate the LCC in potato crops using RapidEye images, and the best result was obtained using TCARI/OSAVI. Clevers et al. [30] also found that TCARI/OSAVI provided a good linear estimation of the LCC in potato crops using Sentinel-2 images. Cui et al. [31] also found that TCARI/OSAVI was the best index to predict LCC with strong anti-disturbance ability. Crop type had a clear influence on the predictions of LCCs using these combined vegetation indices. Haboudane et al. [32] found that the predictive accuracy of the wheat LCC was obviously lower than that of the corn LCC.

Therefore, the objective of this study was to develop a new approach for LCC estimation for winter wheat based on crop canopy reflectance with minimum sensitivity to LAI and consistent sensitivity to different crop growth conditions. Given this goal, the spectral response characteristics of the red-edge region of chlorophyll pigments were fully considered for creating the new chlorophyll indices. The sensitivity of the newly proposed indices to LCC and the insensitivity to LAI were analyzed using a large amount of field data. Finally, the consistent performance of the best spectral index on the estimation of LCC under various field management strategies (winter wheat variety, quantity of N fertilizer, quantity of water applied, sowing date) was evaluated.

2. Materials and Methods

2.1. Experimental Site and Experimental Design

The field experiments were performed at the National Experimental Station for Precision Agriculture, Changping District of Beijing, China (40°10.6′N, 116°26.3′E). The site has a temperate climate, with an average annual precipitation of 507.7 mm and a mean annual temperature of 13.8 °C. The field soil is silty clay loam. Winter wheat (*Triticum aestivum* L.) was planted at this site in the 2001–2002 and 2009–2010 growing seasons. To obtain a wide range of LCCs, various field treatments were implemented for these experiments.

In the 2001–2002 campaign, the study site was divided into 48 small plots, each of which was 32.4 m × 30 m, separated by a 1-m wide isolation strip from adjacent plots. Winter wheat was planted on 26–27 September 2001, and four N fertilization densities (0, 150, 300, and 450 kg ha^{-1}), four water treatment plans (0, 225, 450, and 675 m^3 ha^{-1}), and three winter wheat varieties (Zhongyou 9507, Jing 9428, and Jingdong 8) were used in these experiments. Table 1 shows the descriptions of the experimental designs. Zhongyou 9507 and Jing 9428 are horizontal varieties, and Jingdong 8 is an erective variety. For all plots, one-third of the total N fertilization was applied pre-planting, one-third was applied at the tillering stage (Zadoks scale 20, Z20), and the remainder was applied at the stem elongation stage (Z30). Half of the water was applied at the tillering stage (Z20), and the remainder was applied at the elongation stage (Z30). The wheat LAI, LCC, and canopy spectra were measured at tillering (Z25), stem elongation (Z31 and Z34), booting (Z41), head emergence (Z54), and pollination (Z68), and the crop was harvested on 20 June 2002. This procedure produced a total of 288 samples for the 2002 campaign.

Table 1. Descriptions of the winter wheat experimental design in 2001–2002.

N fertilizer \ Water	W1 (0 m^3 ha^{-1})	W2 (225 m^3 ha^{-1})	W3 (450 m^3 ha^{-1})	W4 (675 m^3 ha^{-1})	Variety
N1 (0 kg ha^{-1})	12	13	36	37	Zhongyou 9507
	11	14	35	38	Jing 9428
	10	15	34	39	Jingdong 8
N2 (150 kg ha^{-1})	9	16	33	40	Zhongyou 9507
	8	17	32	41	Jing 9428
	7	18	31	42	Jingdong 8
N3 (300 kg ha^{-1})	6	19	30	43	Zhongyou 9507
	5	20	29	44	Jing 9428
	4	21	28	45	Jingdong 8
N4 (450 kg ha^{-1})	3	22	27	46	Zhongyou 9507
	2	23	26	47	Jing 9428
	1 [1]	24	25	48	Jingdong 8

[1] The numbers represent the sequence numbers of different plots.

In the 2009–2010 campaign, winter wheat was planted in 36 plots (a total area of 5040 m^2). This campaign used three winter wheat cultivars (Nongda 195, Jing 9428, and Jingdong 13), four N fertilization densities (56, 82, 109, and 135 kg ha^{-1}), and three sowing dates (25 September and 5 and 15 October). Table 2 shows the descriptions of the experimental design. Nongda 195 and Jing 9418 are horizontal varieties, and Jingdong 13 is an erective variety. The nitrogen fertilizer was applied twice, with the first application on 23 September (56 kg ha^{-1} for each plot) and the second application on 21 April (0, 26, 53, and 79 kg ha^{-1}) for the sowing treatment on 25 September, and one N fertilizer level (53 kg ha^{-1}) was applied for the sowing treatments on 5 and 15 October. The seeding rates were 152, 217, and 279 kg ha^{-1} for the sowing dates of 25 September and 5 and 15 October, respectively. The other field management measures were consistent with conventional management by farmers. The wheat LAI, LCC, and canopy reflectance spectra were measured at stem elongation (Z36), booting (Z41), flowering (Z65), and milk (Z73 and Z75). Although winter wheat was planted on different dates, there were no major differences in development stages between the different treatments, especially after the head emergence stage. The winter wheat was harvested on 23 June 2010, and a total of 168 samples were produced for the 2010 campaign.

Table 2. Descriptions of the winter wheat experimental design in 2009–2010.

		Sowing Date						Variety
		25 September		5 October		15 October		
N fertilizer	56 kg ha^{-1}	2	1 [1]	/ [2]	/	/	/	Nongda 195
		10	9	/	/	/	/	Jing 9428
		18	17	/	/	/	/	Jingdong 13
	82 kg ha^{-1}	4	3	/	/	/	/	Nongda 195
		12	11	/	/	/	/	Jing 9428
		20	19	/	/	/	/	Jingdong 13
	109 kg ha^{-1}	6	5	26	25	32	31	Nongda 195
		14	13	28	27	34	33	Jing 9428
		22	21	30	29	36	35	Jingdong 13
	135 kg ha^{-1}	8	7	/	/	/	/	Nongda 195
		16	15	/	/	/	/	Jing 9428
		24	23	/	/	/	/	Jingdong 13

[1] The numbers stand for the sequence numbers of different plots. [2] / stands for no plot.

2.2. Field Measurements

2.2.1. Measurement of the Reflectance Spectrum from the Winter Wheat Canopy

During the 2002 and 2010 growing seasons, a 1-m^2 area of winter wheat was selected for canopy reflectance measurements using a portable field spectroradiometer (FieldSpec-FR2500, ASD, USA) with a spectral range from 350 to 2500 nm and spectral resolutions of 3 nm from 350 to 1050 nm and 10 nm from 1050 to 2500 nm. To ensure accurate measurements, the canopy spectral data were acquired under clear, blue skies between 10:00 and 14:00 h (Beijing Local Time) at a height of 1.3 m above the wheat canopy with a field of view of 25° to maintain the same viewing geometry. A total of 20 measurements were collected in each plot, and the average spectrum was retained as the spectrum for the plot. The measured radiance was converted into absolute reflectance using a calibration from a white Spectralon®(Labsphere, Inc., North Sutton, NH, USA) reference panel, as follows:

$$R_{target} = \frac{DN_{target}}{DN_{reference}} \times R_{reference} \quad (1)$$

where R_{target} is the spectral reflectance of the winter wheat canopy, DN_{target} and $DN_{reference}$ are the radiances of the winter wheat canopy and the white reference panel, respectively, and $R_{reference}$ is the reflectance of the white reference panel.

2.2.2. Measurement of Plant Parameters

After acquiring the canopy spectra, all plants in four 1-m long rows per plot (with a row spacing of 25 cm) were harvested on each investigation date, placed in a plastic bag and transported to the laboratory for measurement of LCC and green LAI.

The LCC was measured in the laboratory using standard procedures [illegible]. First, fresh winter wheat leaf samples from a certain area in each plot were mixed with a given volume of 80% alcohol solution. Each sample was placed in a cuvette and stored in the dark at 25 °C for 48 h. Next, the absorbance of pigments at 663 and 646 nm was measured using an L6 ultraviolet-visible spectrophotometer (INESA, China). The concentrations of chlorophyll *a* and chlorophyll *b* were calculated using

$$Chl_a\left(mg\ L^{-1}\right) = 12.21 \times A_{663} - 2.81 \times A_{646} \quad (2)$$

$$Chl_b\left(mg\ L^{-1}\right) = 20.13 \times A_{646} - 5.03 \times A_{663} \quad (3)$$

$$Chl_{a+b}\left(mg\,L^{-1}\right) = Chl_a\left(mg\,L^{-1}\right) + Chl_b\left(mg\,L^{-1}\right) \quad (4)$$

$$Chl_{a+b}\left(mg\,g^{-1}\right) = \left[Chl_{a+b}\left(mg\,L^{-1}\right) \times V_T(ml)\right] / \left[W(g) \times 1000\right] \quad (5)$$

where Chl_a and Chl_b are the chlorophyll *a* and chlorophyll *b* concentrations (mg L^{-1}), A_{663} and A_{646} are the absorbances of the extract solution at 663 and 646 nm, Chl_{a+b} is the chlorophyll *a* + *b* content per unit leaf weight (mg g^{-1}), V_T is the volume in mL of leaf chlorophyll extract solution, and W is the leaf weight (g).

The green LAI was measured using the dried-weight method [34]. In each plot, all green leaves from the samples were separated from the stems. Thirty leaves were randomly selected from all the green leaves to ensure that all ages and sizes of leaves were included. Then, a leaf segment of approximately 1 cm^2 was cut from the middle parts of the thirty leaves, and these leaf segments served as reference leaves. All green leaves, including the reference leaves, were oven dried at 70 °C to a constant weight. The reference leaves and the remaining leaves were weighed. The relationship between fresh leaf area and leaf dry weight for the reference leaves was used to convert the dry weight of all green leaves into fresh leaf area. The green LAI was calculated using

$$\text{LAI} = \frac{S_r W_t}{S_l W_r} \quad (6)$$

where S_r (m^2) is the area of the fresh reference leaves, W_t (g) is the total dry weight of all green leaves, S_l (m^2) is the sampled land area, and W_r (g) is the dry weight of the reference leaves.

The statistical analyses of the measured winter wheat LCC and LAI for the 2001–2002 and 2009–2010 datasets are shown in Table 3. For the 2002 dataset, the mean values of LCC and LAI are 3.102 and 2.311, respectively, and both LCC and LAI have moderate levels of variation. For the 2010 dataset, the mean values of LCC and LAI are 2.913 and 2.069, respectively, and both LCC and LAI also have moderate levels of variation.

Table 3. Results of the statistical analysis of the measured winter wheat LCC (mg g^{-1}) and LAI.

Datasets	Parameter	Mean	Min	Max	SD	CV (%)	*n*
2002	LCC	3.102	1.832	6.439	1.014	32.688	288
	LAI	2.311	0.434	4.859	1.017	43.995	288
2010	LCC	2.913	1.066	5.879	0.925	31.741	168
	LAI	2.069	0.394	5.374	0.869	42.030	168

LCC = leaf chlorophyll content; LAI = leaf area index; Min = the minimum value; Max = the maximum value; SD = the standard deviation; CV = the coefficient of variation; n = the number of samples.

2.3. Spectral Indices

2.3.1. New Spectral Index

The spectral characteristics of green vegetation at the canopy level with different LCCs, which were simulated by the combined PROSPECT + SAIL model (teledetection.ipgp.jussieu.fr/prosail/) via MATLAB R2015a software (MathWorks, Inc., Natick, MA, USA), are shown in Figure 1. To investigate the effect of chlorophyll content on canopy spectral reflectance, LCC was set to change from 10 to 70 μg cm^{-2} with a step of 10 μg cm^{-2}, LAI was fixed to a value of 3, carotenoid content was 8, cbrown content was 0, equivalent water thickness (Cw) was 0.0015 cm, dry matter content (Cm) was 0.0035 g cm^{-2}, and the leaf structure parameter (N) was 1.41. These input parameters were fixed with reasonable values based on field measurements and previous studies [35,36]. As shown in Figure 1, chlorophyll pigments mainly affect the visible spectral range; when LCC increases, the reflectance in the green, red, and red-edge bands decreases gradually. The reflectance in the NIR bands approaches an approximately constant value. Based on this phenomenon, we tried to construct a new spectral index that is thought of as a ratio of the difference between the reflectance in the NIR band and that in the red-edge band to the reflectance in the green band (chlorophyll absorption minimum). This ratio form can effectively enhance the spectral difference between different LCC levels. The higher the LCC is, the larger the value of the new index.

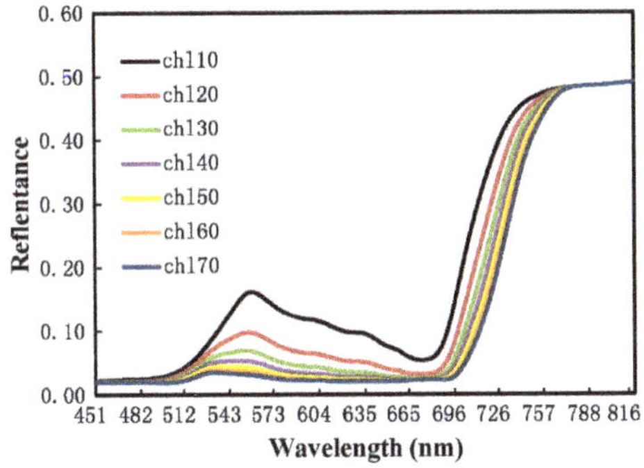

Figure 1. Crop canopy spectral reflectance with various LCCs, as simulated by the combined PROSPECT leaf optical properties model and SAIL canopy bidirectional reflectance model (PROSAIL) model.

Previous studies also demonstrated that the combined effects of chlorophyll concentration and LAI variation strongly influence the abrupt changes affecting the vegetation reflectance in the red-edge region, as shown in Figure 2, which uses spectra simulated by the SAIL radiative transfer model [32]. According to Figure 2, the wavelength regions that are most sensitive to leaf chlorophyll variability are centered at 550 nm in the green peak and 720 nm in the red edge. In contrast, the LAI generates weak variations in the reflectance spectrum at 550 and 720 nm. The reflectance in the NIR region does not vary with LCC. Therefore, we selected the bands at 550 nm (green band), 720 nm (red-edge band), and 800 nm (NIR band) to build the new spectral index. Moreover, Kim [17] showed that the ratio of reflectance at 700 nm and 550 nm (the bands corresponding to the minimum absorption of the photosynthetic pigments) is closely influenced by non-photosynthetic materials in the canopy. Therefore, this ratio was introduced to the new spectral index to reduce the effects of non-photosynthetic materials in the remote estimates of LCC. Finally, a new spectral index called the red-edge-chlorophyll absorption index (RECAI) is defined as follows:

$$(R_{800} - R_{720}) / R_{550} \times (R_{700} / R_{550}) \quad (7)$$

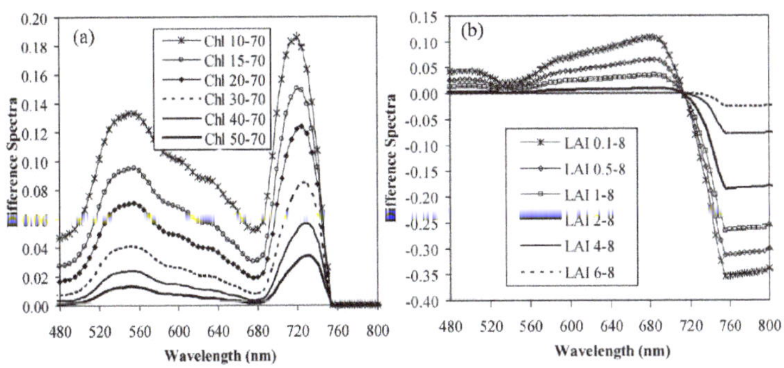

Figure 2. Relative difference in canopy reflectance [32]. (**a**) Difference between reflectance spectra corresponding to various chlorophyll contents and reflectance spectra corresponding to LCC = 70 μg cm^{-2}. In the legend, Chl 30–70 is the difference between the reflectance spectra corresponding to LCC = 30 and 70 μg cm^{-2}. (**b**) Difference between reflectance spectra representing various LAI values and the spectrum corresponding to LAI = 8. The legend follows the same format as in the left panel.

Due to the effect of LAI on LCC estimates, we developed three other integrated indices in this study: RECAI/OSAVI, RECAI/TVI, and RECAI/MTVI2 (see Table 4). These indices are based on previously published methods that take the ratio of two VIs: one is sensitive to the canopy chlorophyll content, and the other is sensitive to LAI [18,19,29,30,32]. The optimized soil-adjusted vegetation index (OSAVI) [37], the triangular vegetation index (TVI) [38], and the modified TVI (MTVI2) [39] are confirmed to be sensitive to LAI and were used to build integrated indices with chlorophyll-related indices effective for estimating LCC. Such ratios can minimize the influence of LAI and maximize the sensitivity to LCC. Because the value of RECAI/TVI is very small, it is scaled up by two orders of magnitude in practical calculations. OSAVI, TVI, and MTVI2 are calculated as follows:

$$\mathrm{OSAVI} = 1.16(\mathrm{R}_{800} - \mathrm{R}_{670})/(\mathrm{R}_{800} + \mathrm{R}_{670} + 0.16) \tag{8}$$

$$\mathrm{TVI} = 0.5[120(\mathrm{R}_{750} - \mathrm{R}_{550}) - 200(\mathrm{R}_{670} - \mathrm{R}_{550})] \tag{9}$$

$$\mathrm{MTVI2} = \frac{1.5[1.2(\mathrm{R}_{800} - \mathrm{R}_{550}) - 2.5(\mathrm{R}_{670} - \mathrm{R}_{550})]}{\sqrt{(2\mathrm{R}_{800} + 1)^2 - (6\mathrm{R}_{800} - 5\sqrt{\mathrm{R}_{670}}) - 0.5}} \tag{10}$$

2.3.2. Spectral Indices in this Study

For the present study, we selected eight spectral indices from the literature to evaluate their capacity and consistency in estimating chlorophyll content. The details of these indices are provided in Table 4. The eight published chlorophyll-related VIs based on discrete red-edge and green bands (chlorophyll absorption minimum) and/or the red band (chlorophyll absorption maximum) have been confirmed to be closely related to LCC [24,40–44].

Table 4. Spectral indices used in this study.

Index	Formula	Reference
Green chlorophyll index ($\mathrm{CI_{green}}$)	$\mathrm{R_{783}/R_{550}} - 1$	[11,20]
Red-edge chlorophyll index ($\mathrm{CI_{red\text{-}edge}}$)	$\mathrm{R_{783}/R_{705}} - 1$	[11,20]
Moderate-resolution imaging spectrometer terrestrial chlorophyll index (MTCI)	$(\mathrm{R_{750}} - \mathrm{R_{710}})/(\mathrm{R_{710}} - \mathrm{R_{680}})$	[23]
Red-edge model index (R-M)	$(\mathrm{R_{750}/R_{720}}) - 1$	[11]
Double-peak canopy nitrogen index I (DCNI I)	$[(\mathrm{R_{750}} - \mathrm{R_{670}}+0.09)(\mathrm{R_{750}} - \mathrm{R_{700}})]/(\mathrm{R_{700}} - \mathrm{R_{670}})$	[25]
The modified chlorophyll absorption ratio index/optimized soil-adjusted vegetation index (MCARI/OSAVI)	$[(\mathrm{R_{700}} - \mathrm{R_{670}}) - 0.2(\mathrm{R_{700}} - \mathrm{R_{550}})]$ $(\mathrm{R_{700}/R_{670}})/[1.16(\mathrm{R_{800}}-\mathrm{R_{670}})/(\mathrm{R_{800}}+\mathrm{R_{670}}+0.16)]$	[32]
The transformed chlorophyll absorption in the reflectance index/optimized soil-adjusted vegetation index (TCARI/OSAVI)	$3[(\mathrm{R_{700}} - \mathrm{R_{670}}) - 0.2(\mathrm{R_{700}} - \mathrm{R_{550}})(\mathrm{R_{700}/R_{670}})]/$ $[1.16(\mathrm{R_{800}} - \mathrm{R_{670}})/(\mathrm{R_{800}}+\mathrm{R_{670}}+0.16)]$	[19]
The triangular chlorophyll index/optimized soil-adjusted vegetation index (TCI/OSAVI)	$[1.2(\mathrm{R_{700}} - \mathrm{R_{550}}) - 1.5(\mathrm{R_{670}} - \mathrm{R_{550}})\sqrt{(\mathrm{R_{700}/R_{670}})}]/$ $[1.16(\mathrm{R_{800}} - \mathrm{R_{670}})/(\mathrm{R_{800}} + \mathrm{R_{670}} + 0.16)]$	[32]
The red-edge-chlorophyll absorption index (RECAI)	$(\mathrm{R_{800}} - \mathrm{R_{720}})/\mathrm{R_{550}}*(\mathrm{R_{700}/R_{550}})$	This study
The red-edge-chlorophyll absorption index/ optimized soil-adjusted vegetation index (RECAI/OSAVI)	RECAI/OSAVI	This study
The red-edge-chlorophyll absorption index/ the triangular vegetation index (RECAI/TVI)	100RECAI/TVI	This study
The red-edge-chlorophyll absorption index/ the modified triangular vegetation index (RECAI/MTVI2)	RECAI/MTVI2	This study

2.4. Analysis Method and Software

For this work, the measured dataset was used to estimate LCC through a semi-empirical vegetation index approach. The dataset was divided randomly into two subsets: 80% as the training dataset (365 samples) and 20% as the validation dataset (91 samples). Five regression models (linear, power, exponential, polynomial, and logarithmic) were used to model the relationships between LCC and different spectral indices. The coefficient of determination (R^2) and the root mean square error (RMSE) were selected as the accuracy indicators of the statistical models. The details of the R^2 and RMSE indicators are available in Richter [45]. The conventional statistical analysis of various parameters was performed using SPSS 18.0 software (SPSS Inc., Chicago, IL). The regression models were established, and the validation procedures were performed using MATLAB R2015a software (The Math Works, Inc., Natick, MA, USA).

3. Results

3.1. Prediction LCC by VIs

The LCC was predicted based on the published and newly proposed spectral indices using a ground-measured dataset. Approximately 365 samples from this dataset were used to build empirical regression models between LCC and VIs (linear, power, exponential, polynomial, and logarithmic). The scatterplots and estimation results are shown in Figure 3. As shown in this figure, RECAI/TVI exhibited the highest R^2 value (0.573) and the lowest RMSE value (0.663 mg g^{-1}), followed by TCARI/OSAVI (R^2 = 0.498 and RMSE = 0.707 mg g^{-1}), TCI/OSAVI (R^2 = 0.373 and RMSE = 0.795 mg g^{-1}), MCARI/OSAVI (R^2 = 0.301 and RMSE = 0.841 mg g^{-1}), and then the other indices. Unexpectedly, CI_{green} and $CI_{red\text{-}edge}$ did not exhibit good performance in this study. As determined by the fit lines, the prediction models of CI_{green}, $CI_{red\text{-}edge}$, R-M, DCNI I, RECAI, and RECAI/TVI were exponential models, while the other vegetation indices were power models. The scatterplots between LCC and RECAI/TVI, TCARI/OSAVI, TCI/OSAVI, and MCARI/OSAVI were regular, while the others were diverse. There was a near-linear relationship between RECAI/TVI and LCC, which was more linear than the relationships between LCC and TCARI/OSAVI, TCI/OSAVI, and MCARI/OSAVI.

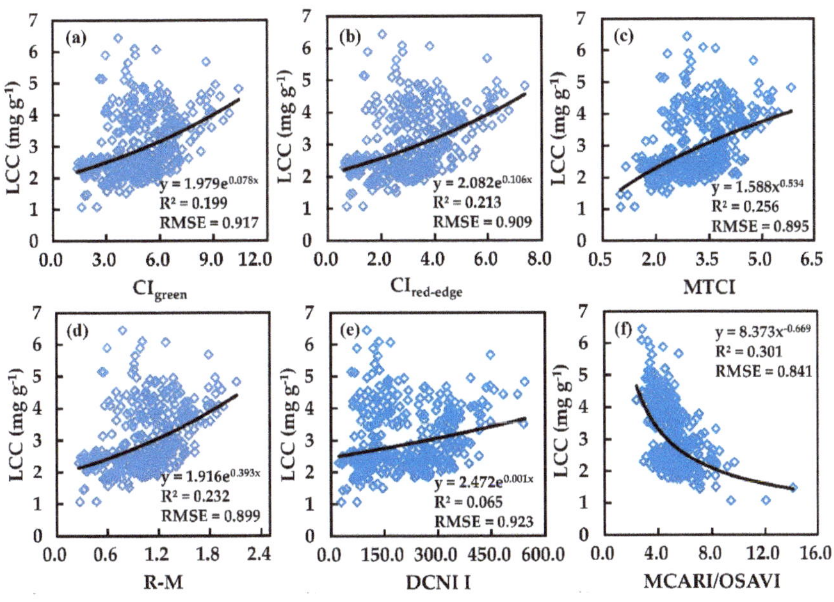

Figure 3. *Cont.*

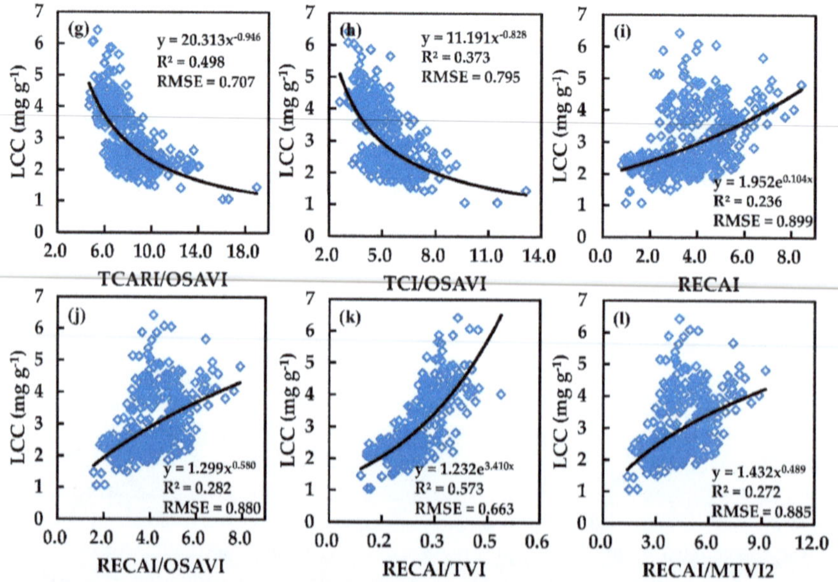

Figure 3. (**a–l**) The scatterplots between measured LCC versus CI_{green}, $CI_{red\text{-}edge}$, MTCI, R-M, DCNI I, MCARI/OSAVI, TCARI/OSAVI, TCI/OSAVI, RECAI, RECAI/OSAVI, RECAI/TVI, and RECAI/MTVI2 for the 2002 + 2010 combined calibration dataset (n = 365), respectively.

The predicted performances of the integrated vegetation indices (MCARI/OSAVI, TCARI/OSAVI, TCI/OSAVI, RECAI/TVI, RECAI/OSAVI, and RECAI/MTVI2 with $0.272 \leq R^2 \leq 0.573$) were superior to those of the single vegetation indices (CI_{green}, $CI_{red\text{-}edge}$, MTCI, R-M, and DCNI I with $0.065 \leq R^2 \leq 0.256$), which again proved that the ratio method was helpful for improving the estimation accuracy when using crop canopy spectral data.

3.2. Comparing the LCC Estimation Performances of the Indices

The predicted LCC values were plotted against field LCC measurements using the remaining samples (91 samples). The R^2 and RMSE values were selected to assess the confidence of the relationship (Figure 4). Figure 4 (RECAI/TVI) shows that there was very good agreement between the estimated LCC values and the field-measured LCC values, with the highest R^2 values (0.518) and the lowest RMSE values (0.658 mg g^{-1}), followed by TCARI/OSAVI (R^2 = 0.473 and RMSE = 0.680 mg g^{-1}), TCI/OSAVI (R^2 = 0.296 and RMSE = 0.787 mg g^{-1}), MCARI/OSAVI (R^2 = 0.217 and RMSE = 0.832 mg g^{-1}), R-M (R^2 = 0.204 and RMSE = 0.840 mg g^{-1}), and RECAI/OSAVI (R^2 = 0.201 and RMSE = 0.843 mg g^{-1}). The order of the validation accuracies of these models was similar to the order of the prediction accuracies. From Figure 4, the slope of the validation linear fit for RECAI/TVI was closer to unity than that for the other indices. The LCCs were overestimated for low values and underestimated for high values by RECAI/TVI, TCARI/OSAVI, and TCI/OSAVI, especially TCARI/OSAVI and TCI/OSAVI. The LCC prediction results evidently showed that RECAI/TVI had the greatest potential for estimating winter wheat LCC. The DCNI Iprovided the worst predictions in this study with the measured datasets, which was inconsistent with the results obtained by [25]. This discrepancy can be justified by the fact that DCNI I was initially proposed for LCC estimation in cotton, which has a completely different canopy structure than winter wheat.

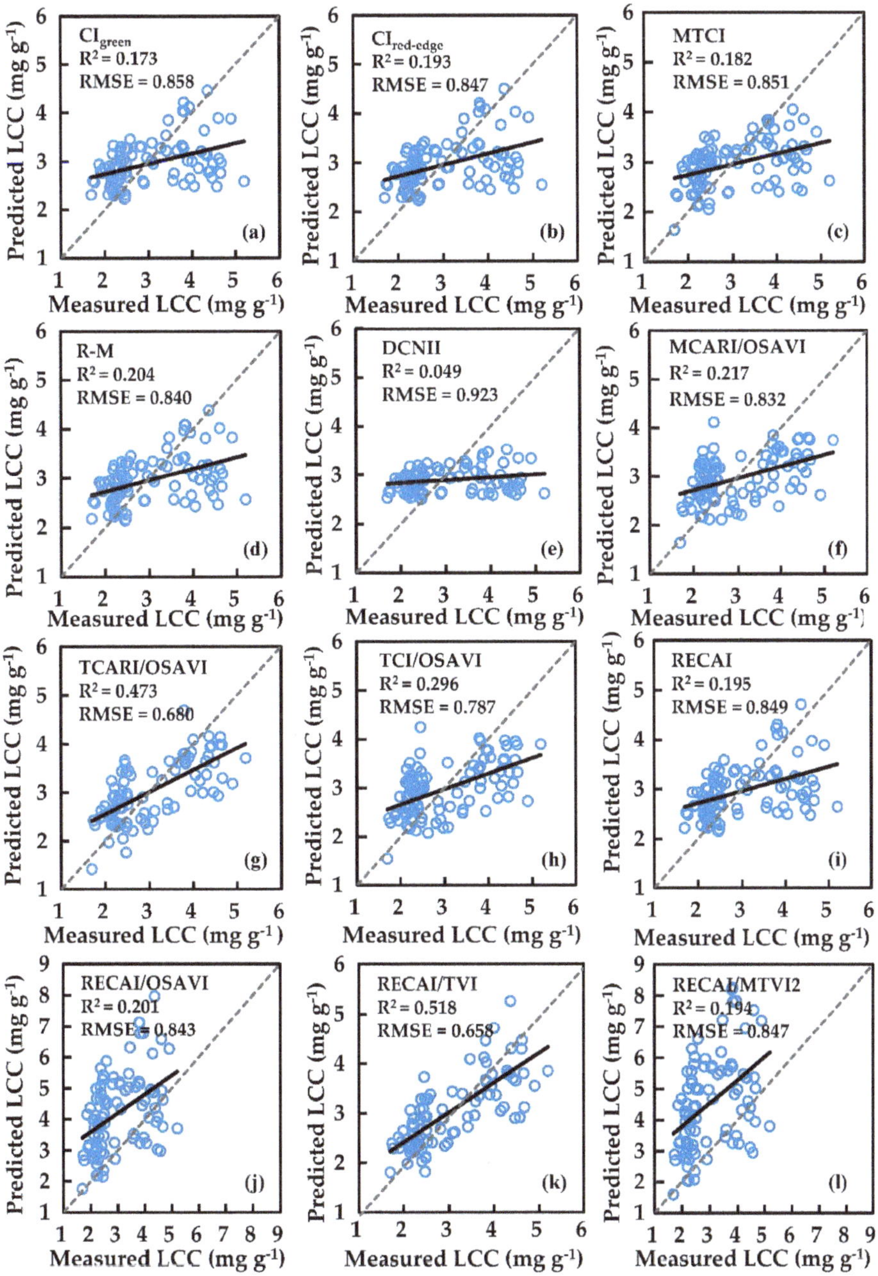

Figure 4. (**a**–**l**) The validation scatterplots between measured and predicted LCC by CI_{green}, $CI_{red\text{-}edge}$, MTCI, R-M, DCNI I, MCARI/OSAVI, TCARI/OSAVI, TCI/OSAVI, RECAI, RECAI/OSAVI, RECAI/TVI, and RECAI/MTVI2 for the dataset (n = 91), respectively.

3.3. Effect of LAI on the Assessment of LCC

To analyze the influence of LAI on the LCC estimation, the relationships between the spectral indices and LCC and LAI were analyzed under five LAI levels (0 < LAI < 1, 1 ≤ LAI < 2, 2 ≤ LAI < 3, 3 ≤ LAI < 4, LAI ≥ 4). The results are shown in Table 5.

Table 5. Pearson correlation coefficients (r values) between spectral indices and winter wheat LCC and LAI.

	0 < LAI < 1 (n[1] = 52)		1 ≤ LAI < 2 (n = 143)		2 ≤ LAI < 3 (n = 154)		3 ≤ LAI < 4 (n = 94)		LAI ≥ 4 (n = 13)	
	LCC	LAI	LCC	LAI	LCC	LAI	LCC	LAI	LCC	LAI
CI_{green}	0.665 **	0.422 **	0.468 **	0.469 **	0.426 **	0.366 **	0.692 **	0.014	0.613 *	0.116
$CI_{red\text{-}edge}$	0.661 **	0.422 **	0.465 **	0.481 **	0.464 **	0.346 **	0.725 **	−0.004	0.581 *	0.182
MTCI	0.526 **	0.233	0.432 **	0.453 **	0.454 **	0.293 **	0.758 **	−0.111	0.673 *	−0.043
R-M	0.664 **	0.426 **	0.486 **	0.497 **	0.485 **	0.347 **	0.746 **	−0.024	0.630 *	0.085
DCNI I	0.102	0.333 *	0.022	0.508 **	0.177 *	0.388 **	0.533 **	0.078	0.359	0.049
MCARI/OSAVI	−0.570 **	0.185	−0.526 **	−0.150	−0.524 **	0.146	−0.519 **	0.364 **	−0.523	0.249
TCARI/OSAVI	−0.668 **	0.054	−0.652 **	−0.246 **	−0.669 **	0.040	−0.676 **	0.264 *	−0.568 *	0.017
TCI/OSAVI	−0.608 **	0.155	−0.588 **	-0.148	−0.585 **	0.133	−0.568 **	0.338 **	−0.551	0.186
RECAI	0.688 **	0.398 **	0.542 **	0.427 **	0.454 **	0.357 **	0.698 **	−0.009	0.730 **	−0.103
RECAI/OSAVI	0.621 **	0.255	0.545 **	0.341 **	0.450 **	0.335 **	0.693 **	−0.041	0.750 **	−0.207
RECAI/TVI	0.777 **	0.029	0.819 **	0.109	0.708 **	0.053	0.722 **	−0.167	0.730 **	−0.106
RECAI/MTVI2	0.615 **	0.238	0.535 **	0.353 **	0.445 **	0.342 **	0.694 **	−0.035	0.742 **	−0.177

** and * indicate significance at the 0.01 and 0.05 levels, respectively. [1] n is the number of samples. CI_{green} = green chlorophyll index, $CI_{red\text{-}edge}$ = red-edge chlorophyll index, MTCI = moderate-resolution imaging spectrometer terrestrial chlorophyll index, R-M = red-edge model index, DCNI I = double-peak canopy nitrogen index I; MCARI/OSAVI = the modified chlorophyll absorption ratio index/optimized soil-adjusted vegetation index, TCARI/OSAVI = the transformed chlorophyll absorption in the reflectance index/optimized soil-adjusted vegetation index, TCI/OSAVI = the triangular chlorophyll index/ optimized soil-adjusted vegetation index, RECAI = the red-edge-chlorophyll absorption index, RECAI/OSAVI = the red-edge-chlorophyll absorption index/ optimized soil-adjusted vegetation index, RECAI/TVI = the red-edge-chlorophyll absorption index/ the triangular vegetation index, and RECAI/MTVI2 = the red-edge-chlorophyll absorption index/ the modified triangular vegetation index.

For 0 < LAI < 1, all indices except DCNI I were strongly sensitive to LCC at the 0.01 confidence level. MTCI, MCARI/OSAVI, TCARI/OSAVI, TCI/OSAVI, RECAI/OSAVI, RECAI/TVI, and RECAI/MTVI2 were not correlated with LAI, but the other indices were sensitive to LAI. The RECAI/TVI exhibited the highest correlation with LCC (r = 0.777**, significant at the 0.01 level) and no correlation with LAI (r = 0.029) among the indices. For 1 ≤ LAI < 2, all indices except DCNI I were significantly correlated with LCC (0.432 ≤ r ≤ 0.819). Only MCARI/OSAVI, TCI/OSAVI, and RECAI/TVI showed no correlation with LAI, whereas the other indices showed a strong correlation with LAI. The RECAI/TVI also showed the highest correlation coefficient with LCC (r = 0.819**) and no relationship with LAI (r = 0.109). For 2 ≤ LAI < 3, all indices were correlated with LCC (0.177 ≤ r ≤ 0.708). The RECAI/TVI exhibited the best correlation with LCC (r = 0.708**), followed by TCARI/OSAVI (r = 0.669**). Most indices were sensitive to LAI (0.293 ≤ r ≤ 0.388) except for MCARI/OSAVI (r = 0.146), TCARI/OSAVI (r = 0.040), TCI/OSAVI (r = 0.133), and RECAI/TVI (r = 0.053). For 3 ≤ LAI < 4, all indices were significantly related to LCC at the 0.01 level, and RECAI/TVI still showed a very strong correlation with LCC (r = 0.722 at the 0.01 level). Most of the vegetation indices showed no correlation with LAI, except for MCARI/OSAVI, TCARI/OSAVI, and TCI/OSAVI. For LAI ≥ 4, the relationships between vegetation indices and LCC were weakened for most indices, but RECAI, RECAI/OSAVI, RECAI/TVI, and RECAI/MTVI2 still showed strong correlations with LCC at the 0.01 level. No indices were related to LAI. In general, at both low and high LAI levels, RECAI/TVI was the most closely related to LCC at the 0.01 confidence level. This index was poorly related to LAI, indicating that it may be considered the best index for estimating LCC.

4. Discussion

4.1. Building a New Vegetation Index for Retrieving Winter Wheat Leaf Chlorophyll Content

The change in LCC contributes to the variation in reflectance in the visible spectral region, especially in the green, red, and red-edge bands. In this study, we adhered to the following three principles to select the optimal bands for the new spectral index RECAI. (1) To minimize the effect of LAI, the reflectance of the selected bands should be more sensitive to LCC and less sensitive to LAI. According to Figure 2, wavelength regions centered on 550 nm in the green peak and 720 nm in the red edge are most sensitive to leaf chlorophyll variability and least sensitive to LAI variability [32]. (2) To mitigate the potential saturation problem, we selected the green (550 nm) and red-edge (700 nm and 720 nm) bands instead of the red band (the maximum absorption of chlorophyll is near 670 nm). Some researchers found that even low chlorophyll content saturates the absorption in the red spectral region, whereas the reflectance in a wide range from 530 to 630 nm and near 700 nm remains sensitive to high chlorophyll contents [2,20,46]. (3) To reduce the effects of non-photosynthetic materials, we selected the ratio of reflectance at 550 and 700 nm, which corresponds to the minimum absorption of chlorophyll. This selection was based on the finding that the ratio of R_{700} and R_{550} is considered constant at the leaf level, despite the variability in chlorophyll content [17,19,35,47].

However, Table 5 indicates that RECAI was still strongly correlated with LAI, indicating that RECAI did not effectively minimize the effect of LAI variations on LCC estimation. To overcome this drawback, we utilized the method in this study by taking a ratio of two VIs: one that is sensitive to the canopy chlorophyll content and another that is sensitive to LAI effects [18,19,30–32]. Thus, TVI, MTVI2, and OSAVI were introduced to reduce the effect on RECAI on the LAI variation. According to the results from the ground-measured datasets (Figure 3 and Table 5), all three spectral indices (RECAI/TVI, RECAI/OSAVI, and RECAI/MTVI2) improved the LCC predictive accuracy and reduced the sensitivity to LAI compared with RECAI. The RECAI/TVI exhibited very good performance in terms of LCC estimation; RECAI/MTVI2 and RECAI/OSAVI did not perform as well as RECAI/TVI, which may be explained by the relationship between (i) TVI, MTVI2, OSAVI and (ii) LCC, LAI. As shown in Table 6, TVI, OSAVI, and MTVI2 were strongly correlated with LAI and weakly correlated (or even uncorrelated) with LCC, which met the demands of the abovementioned method for developing a ratio index to reduce the effect of LAI. The TVI was more sensitive to LAI and less sensitive to LCC than OSAVI and MTVI2, which was why RECAI/TVI performed better for LCC estimation than RECAI/OSAVI and RECAI/MTVI2.

Table 6. Correlation coefficients (*r*) between (i) TVI, OSAVI, MTVI2 and (ii) LCC, LAI for the 2002, 2010, and 2002+2010 datasets.

	LCC			LAI		
TVI	−0.193 **	0.167 *	−0.037	0.800 **	0.735 **	0.779 **
OSAVI	0.308 **	0.220 **	0.326 **	0.799 **	0.643 **	0.764 **
MTVI2	0.307 **	0.370 **	0.269 **	0.761 **	0.720 **	0.690 **

** Indicates significance at the 0.01 level; * indicates significance at the 0.05 level. N = 288, 168, and 456 for the 2002, 2010, and 2002+2010 datasets, respectively.

4.2. Effect of Field Management Measures on the RECAI/TVI Index

Field management measures strongly impact crop growth and lead to variations in LCC. In this study, we analyzed how winter wheat varieties, N fertilizer, irrigation volume, and sowing dates affect the newly proposed RECAI/TVI (Figure 5).

For different winter wheat varieties, Jing 9428, Zhongyou 9507 and Nongda 195 are horizontal winter wheat varieties, whereas Jingdong 8 and Jingdong 13 are erective varieties. From Figure 5a,b, RECAI/TVI performed better for the horizontal varieties than for the erective varieties on both the 2002 and 2010 datasets. In 2002, the R^2 values of the horizontal varieties (Jing 9428 and Zhongyou 9507) were 0.682 and 0.688, respectively, whereas the R^2 of the erective variety (Jingdong 8) was only 0.351. In 2010, the R^2 values of the horizontal varieties (Jing 9428 and Nongda 195) were 0.736 and 0.714, respectively, whereas the R^2 of the erective variety (Jingdong 13) was 0.590. For a given LAI, horizontal winter wheat can effectively increase the vegetation fraction and reduce the influence of soil compared with the erective variety. That is, the RECAI/TVI of horizontal varieties can obtain a more accurate estimate of LCC.

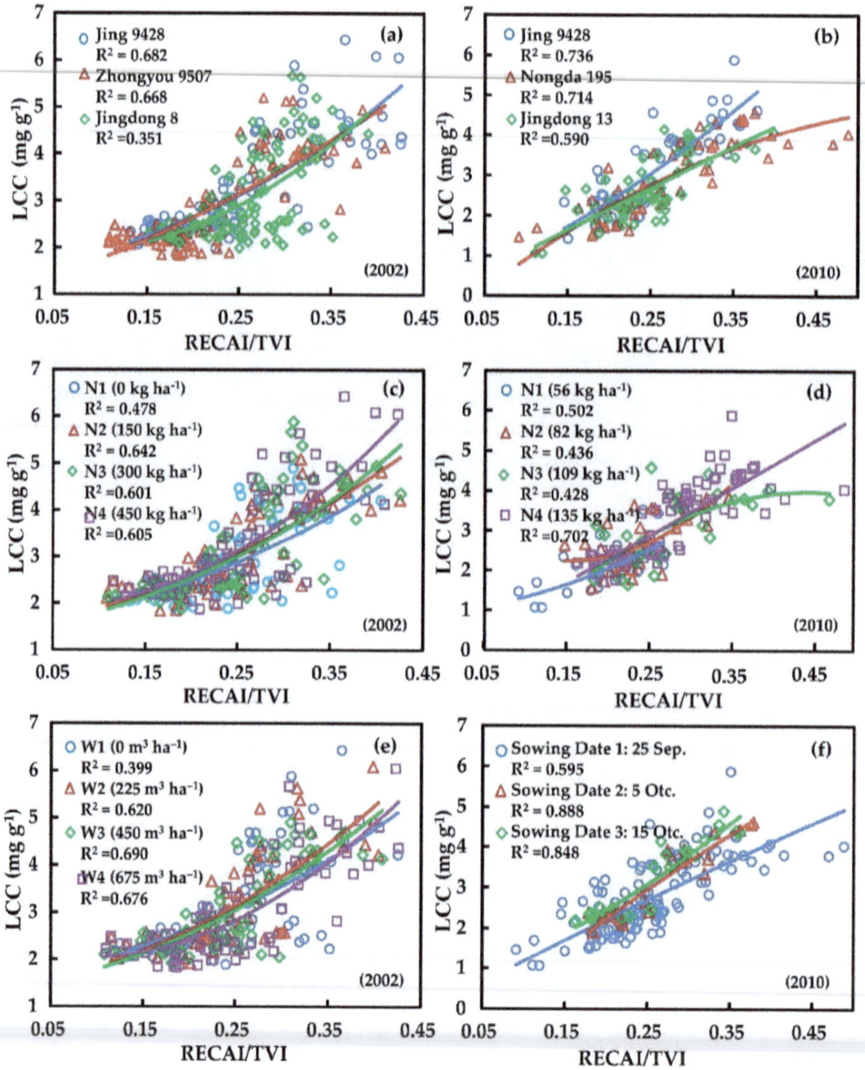

Figure 5. Influence of different field management measures on RECAI/TVI. (**a**), (**b**), (**c**), (**d**), (**e**), (**f**) stand for the scatter plots between RECAI/TVI and LCC under different crop varieties (2002), crop varieties (2010), nitrogen fertilization densities (2002), nitrogen fertilization densities (2010), water treatments, and bowing date treatments, respectively.

For different N fertilizer contents, for both the 2002 and 2010 datasets, high N fertilizer density can improve the accuracy of LCC estimates when using RECAI/TVI (Figure 5c,d). For example, low N application densities (0, 56, 82, and 109 kg ha^{-1}) provided LCC estimates with low accuracy by RECAI/TVI ($0.428 \leq R^2 \leq 0.502$). A high N application density (135, 150, 300, and 450 kg ha^{-1}) provides highly accurate LCC estimates by RECAI/TVI ($0.601 \leq R^2 \leq 0.702$). Another phenomenon is that overfertilization with nitrogen can reduce the accuracy of LCC estimates.

For different water treatments on the 2002 dataset, no water treatment (W1) led to the worst prediction accuracy of LCC with RECAI/TVI ($R^2 = 0.399$), whereas the other water treatments led to better prediction accuracy of LCC with RECAI/TVI ($R^2 \geq 0.620$) (see Figure 5e). No irrigation treatment (W1) had a strong impact on seed germination and seedlings, which led to worse growth and large soil background influence by the RECAI/TVI method. Due to natural precipitation, the different water treatments (W2, W3, and W4) had little effect on crop growth. Therefore, there were no obvious differences in the effect of soil background on LCC estimated by RECAI/TVI.

The different sowing dates in the 2010 dataset affected the LCC predictive performance of RECAI/TVI (see Figure 5f). For sowing date 1 (25 September), the R^2 between RECAI/TVI and LCC was 0.595, whereas for sowing dates 2 (5 October) and 3 (15 October), the R^2 values were 0.888 and 0.848, respectively. This result may be attributed to the tiller number for winter wheat in the different sowing date treatments. Because sowing late may lead to the capacity reduction in tillering, especially before winter, a larger seeding rate was applied in the field on sowing dates 2 (217 kg ha^{-1}) and 3 (279 kg ha^{-1}) than that applied on sowing date 1 (152 kg ha^{-1}) to obtain a sufficiently high tiller number. However, during the stem elongation stage, which is the end of the tillering period, the number of stems for sowing dates 2 (626 m^{-2}) and 3 (468 m^{-2}) was obviously lower than that for sowing date 1 (857 m^{-2}). This result indicated that an excessive tiller number may reduce the performance of RECAI/TVI on LCC estimation.

4.3. Comparison of the Performances of Different VIs

In this study, we found that LCC exhibited different relationships with different spectral indices. In general, the ratio vegetation indices, which comprise VIs for estimating canopy chlorophyll contents and VIs for estimating LAI, exhibited good predictive performance for LCC. For example, RECAI/TVI, TCARI/OSAVI, MCARI/OSAVI, TCI/OSAVI, RECAI/OSAVI, and RECAI/MTVI2 showed better predictive accuracies with large RMSE values between 0.658 mg g^{-1} and 0.847 mg g^{-1}, whereas CI_{green}, $CI_{red\text{-}edge}$, MTCI, R-M, DCNI I, and RECAI showed large RMSE values between 0.840 mg g^{-1} and 0.923 mg g^{-1}. This result is in agreement with the results of previous studies on LCC estimation [29]. The TCARI/OSAVI has been widely used to estimate LCC in published research. In this study, TCARI/OSAVI showed better performance for LCC retrieval, which is consistent with the results of other studies [28–30]. However, this index did not estimate the LCC of winter wheat as well as other crop species (corn, potato) as indicated by other published reports, and this index performed slightly worse than RECAI/TVI in this study. This result can be interpreted by the viewpoint that TCARI/OSAVI is still sensitive to the soil background and LAI variations, especially when LAI < 3 [32,48]. The chlorophyll indices (CI_{green} and $CI_{red\text{-}edge}$) did not perform well in LCC estimations in our study, which is in accordance with the results of a previous study [30] that proved that the chlorophyll indices were more suitable for the estimation of canopy chlorophyll content.

Overall, the newly proposed RECAI/TVI greatly improved the predictive accuracy, effectively overcame the saturation problem and reduced the effect of LAI when used to estimate winter wheat LCC. Thus, RECAI/TVI is considered the best spectral index for estimating LCC. However, some problems still need to be addressed in future work. Remote estimates of LCC always depend strongly on the growth stage because changes in vegetation fraction, plant type, and other factors may lead to variations in the canopy spectral reflectance. Furthermore, the performance of RECAI/TVI at a given growth stage remains to be addressed to help precision field management. Additionally, the performance of the index in estimating the LCC of other crops, such as rice or corn, should be examined.

In future research, we will put much more effort into confirming the capacity of RECAI/TVI to estimate LCC given different crop growth stages, different types of crops, different spatial scales, and different spectral data.

5. Conclusions

Chlorophyll is a vital pigment for photosynthesis and directly or indirectly reflects crop nutritional status, growth, vigor, etc. This study focused on testing a remote sensing method to accurately estimate the LCC of winter wheat over crop canopies with minimum effects from LAI. Based on the measured datasets ($n = 456$), we evaluated the performance of eight published and four newly proposed spectral indices for LCC retrieval. The results showed that the newly proposed RECAI/TVI performed excellently for LCC estimation, with R^2 values that increased by more than 13.09% and RMSE values that decreased by more than 6.22%. In addition, whether at low LAI or at high LAI, RECAI/TVI showed a significant relationship with LCC ($0.708^{**} \leq r \leq 0.819^{**}$) and revealed no relationship with LAI variation ($0.029 \leq r \leq 0.167$), indicating that it clearly reduced the influence of LAI. These results also indicate that RECAI/TVI may be considered the optimal index for estimating winter wheat LCC. The LCC predictive accuracy of RECAI/TVI can be influenced by erective winter wheat varieties, a low density of N fertilizer, no water application, and early sowing (excessive tiller number). Therefore, due to the complexity of crop growth conditions, the capability of RECAI/TVI to accurately estimate the LCC of winter wheat should be further verified by applying this index to a more varied range of field-measured data.

Author Contributions: W.H. and X.S. conceived and designed the experiments. B.C. analyzed the experimental data and drafted the manuscript. X.Z. provided guidance on data processing. Q.Z., W.H., X.S. and H.Y. checked the results and revised the manuscript. All authors read and approved the final version of the manuscript.

Funding: This study was supported by the National Key R&D Program of China (2016YFD0300601), the National Natural Science Foundation of China (41501468, 41601466), the Youth Innovation Promotion Association CAS (2017085), the Hainan Provincial Key R&D Program of China (ZDYF2018073), the Agricultural Science and Technology Innovation of Sanya (2015KJ04), the National special support program for high-level personnel recruitment (Wenjiang Huang), and the ten-thousand Talents Program (Wenjiang Huang).

Conflicts of Interest: The authors declare no conflicts of interest.

References

1. Carter, G.A. Ratios of leaf reflectances in narrow wavebands as indicators of plant stress. *Int. J. Remote Sens.* **1994**, *15*, 697–704. [CrossRef]
2. Gitelson, A.A.; Merzlyak, M.N. Remote estimation of chlorophyll content in higher plant leaves. *Int. J. Remote Sens.* **1997**, *18*, 2691–2697. [CrossRef]
3. Osborne, B.A.; Raven, J.A. Light absorption by plants and its implications for photosynthesis. *Biol. Rev.* **1986**, *61*, 1–60. [CrossRef]
4. Curran, P.J.; Dungan, J.L.; Gholz, H.L. Exploring the relationship between reflectance red edge and chlorophyll content in slash pine. *Tree Physiol.* **1990**, *7*, 33–48. [CrossRef] [PubMed]
5. Sage, R.F.; Robert, W.P.; Seemann, J.R. The Nitrogen Use Efficiency of C3 and C4 Plants. *Plant Physiol.* **1987**, *85*, 355–359. [CrossRef]
6. Miao, Y.; Mulla, D.J.; Randall, G.W.; Vetsch, J.A.; Vintila, R. Combining chlorophyll meter readings and high spatial resolution remote sensing images for in-season site-specific nitrogen management of corn. *Precis. Agric.* **2008**, *10*, 45–62. [CrossRef]
7. Clevers, J.G.P.W.; Kooistra, L. Using Hyperspectral Remote Sensing Data for Retrieving Canopy Chlorophyll and Nitrogen Content. *IEEE J. Sel. Top. Appl. Earth Obs. Remote Sens.* **2012**, *5*, 574–583. [CrossRef]
8. Schlemmer, M.; Gitelson, A.; Schepers, J.; Ferguson, R.; Peng, Y.; Shanahan, J.; Rundquist, D. Remote estimation of nitrogen and chlorophyll contents in maize at leaf and canopy levels. *Int. J. Appl. Earth Obs. Geoinf.* **2013**, *25*, 47–54. [CrossRef]
9. Baret, F.; Houles, V.; Guerif, M. Quantification of plant stress using remote sensing observations and crop models: The case of nitrogen management. *J. Exp. Bot.* **2007**, *58*, 869–880. [CrossRef]

10. Gitelson, A.A.; Peng, Y.; Arkebauer, T.J.; Schepers, J. Relationships between gross primary production, green LAI, and canopy chlorophyll content in maize: Implications for remote sensing of primary production. *Remote Sens. Environ.* **2014**, *144*, 65–72. [CrossRef]
11. Gitelson, A.A.; Viña, A.; Ciganda, V.; Rundquist, D.C.; Arkebauer, T.J. Remote estimation of canopy chlorophyll content in crops. *Geophys. Res. Lett.* **2005**, *32*. [CrossRef]
12. Ustin, S.L.; Gitelson, A.A.; Jacquemoud, S.; Schaepman, M.; Asner, G.P.; Gamon, J.A.; Zarco-Tejada, P. Retrieval of foliar information about plant pigment systems from high resolution spectroscopy. *Remote Sens. Environ.* **2009**, *113*, S67–S77. [CrossRef]
13. Li, F.; Miao, Y.X.; Feng, G.H.; Yuan, F.; Yue, S.; Gao, X.; Liu, Y.; Liu, B.; Ustin, S.L.; Chen, X. Improving estimation of summer maize nitrogen status with red edge-based spectral vegetation indices. *Field Crops Res.* **2014**, *157*, 111–123. [CrossRef]
14. Yao, F.Q.; Cai, H.J.; Li, Y.L.; Luo, W.B. Monitoring Winter Wheat SPAD Based on Red Edge Parameter Derived from Hyperspectral Reflectance. *China Rural Water Hydropower* **2015**, *3*, 84–87.
15. Liu, W.Y.; Pan, J. A hyperspectral assessment model for leaf chlorophyll content of Pinus massoniana based on neural network. *Chin. J. Appl. Ecol.* **2017**, *28*, 1128–1136.
16. Zillmann, E.; Schönert, M.; Lilienthal, H.; Siegmann, B.; Jarmer, T.; Rosso, P.; Weichelt, H. Crop Ground Cover Fraction and Canopy Chlorophyll Content Mapping using RapidEye imagery. *ISPRS Int. Arch. Photogramm. Remote Sens. Spat. Inf. Sci.* **2015**, *40*, 149–155. [CrossRef]
17. Kim, M.; Daughtry, C.; Chappelle, E.; McMurtrey, J., III; Walthall, C. The use of high spectral resolution bands for estimating absorbed photosynthetically active radiation (Apar). In Proceedings of the 6th Symposium on Physical Measurements and Signatures in Remote Sensing, Val D'Isere, France, 17–21 January 1994; pp. 299–306.
18. Daughtry, C.S.T.; Walthall, C.L.; Kim, M.S.; Colstoun, E.B.D.; Mcmurtrey, J.E.I. Estimating Corn Leaf Chlorophyll Concentration from Leaf and Canopy Reflectance. *Remote Sens. Environ.* **2000**, *74*, 229–239. [CrossRef]
19. Haboudane, D.; Millera, J.R.; Tremblay, N.; Zarco-Tejadad, P.J.; Dextraze, L. Integrated narrow-band vegetation indices for prediction of crop chlorophyll content for application to precision agriculture. *Remote Sens. Environ.* **2002**, *81*, 416–426. [CrossRef]
20. Gitelson, A.A.; Gritz, Y.; Merzlyak, M.N. Relationships between leaf chlorophyll content and spectral reflectance and algorithms for non-destructive chlorophyll assessment in higher plant leaves. *J. Plant Physiol.* **2003**, *160*, 271–282. [CrossRef] [PubMed]
21. Gitelson, A.A.; Keydan, G.P.; Merzlyak, M.N. Three-band model for noninvasive estimation of chlorophyll, carotenoids, and anthocyanin contents in higher plant leaves. *Geophys. Res. Lett.* **2016**, *33*, L11402. [CrossRef]
22. Ciganda, V.; Gitelson, A.; Schepers, J. Non-destructive determination of maize leaf and canopy chlorophyll content. *J. Plant Physiol.* **2009**, *166*, 157–167. [CrossRef]
23. Dash, J.; Curran, P.J. The MERIS terrestrial chlorophyll index. *Int. J. Remote Sens.* **2004**, *25*, 5403–5413. [CrossRef]
24. Maire, L.G.; François, C.; Dufrêne, E. Towards universal broad leaf chlorophyll indices using PROSPECT simulated database and hyperspectral reflectance measurements. *Remote Sens. Environ.* **2004**, *89*, 1–28. [CrossRef]
25. Jin, X.; Li, Z.; Feng, H.; Xu, X.; Yang, G. Newly Combined Spectral Indices to Improve Estimation of Total Leaf Chlorophyll Content in Cotton. *IEEE J. Sel. Top. Appl. Earth Obs. Remote Sens.* **2014**, *7*, 4589–4600. [CrossRef]
26. Lamb, D.W.; Steyn-Ross, M.; Schaare, P.; Hanna, M.M.; Silvester, W.; Steyn-Ross, A. Estimating leaf nitrogen concentration in ryegrass (Lolium spp.) pasture using the chlorophyll red-edge: Theoretical modelling and experimental observations. *Int. J. Remote Sens.* **2002**, *23*, 3619–3648. [CrossRef]
27. Liu, J.; Pattey, E.; Jégo, G. Assessment of vegetation indices for regional crop green LAI estimation from Landsat images over multiple growing seasons. *Remote Sens. Environ.* **2012**, *123*, 347–358. [CrossRef]
28. Liang, L.; Di, L.; Zhang, L.; Deng, M.; Qin, Z.; Zhao, S.; Lin, H. Estimation of crop LAI using hyperspectral vegetation indices and a hybrid inversion method. *Remote Sens. Environ.* **2015**, *165*, 123–134. [CrossRef]
29. Kooistra, L.; Clevers, J.G.P.W. Estimating potato leaf chlorophyll content using ratio vegetation indices. *Remote Sens. Lett.* **2016**, *7*, 611–620. [CrossRef]
30. Clevers, J.; Kooistra, L.; Marnix, V.D.B. Using Sentinel-2 Data for Retrieving LAI and Leaf and Canopy Chlorophyll Content of a Potato Crop. *Remote Sens.* **2017**, *9*, 405. [CrossRef]

31. Cui, S.; Zhou, K. A comparison of the predictive potential of various vegetation indices for leaf chlorophyll content. *Earth Sci. Inf.* **2017**, *10*, 169–181. [CrossRef]
32. Haboudane, D.; Tremblay, N.; Miller, J.R.; Vigneault, P. Remote Estimation of Crop Chlorophyll Content Using Spectral Indices Derived From Hyperspectral Data. *IEEE Trans. Geosci. Remote Sens.* **2008**, *46*, 423–437. [CrossRef]
33. Porra, R.J. The chequered history of the development and use of simultaneous equations for the accurate determination of chlorophylls a and b. *Photosynth. Res.* **2002**, *73*, 149–156. [CrossRef]
34. Wang, Z.; Wang, J.; Liu, L.; Huang, W.; Zhao, C.; Lu, Y. Estimation of Nitrogen Status in Middle and Bottom Layers of Winter Wheat Canopy by Using Ground-Measured Canopy Reflectance. *Commun. Soil Sci. Plant Anal.* **2005**, *36*, 2289–2302. [CrossRef]
35. Wu, C.; Niu, Z.; Tang, Q.; Huang, W. Estimating chlorophyll content from hyperspectral vegetation indices: Modeling and validation. *Agric. For. Meteorol.* **2008**, *148*, 1230–1241. [CrossRef]
36. Zhou, X.; Huang, W.; Kong, W.; Ye, H.; Dong, Y.; Casa, R. Assessment of leaf carotenoids content with a new carotenoid index: Development and validation on experimental and model data. *Int. J. Appl. Earth Obs. Geoinf.* **2017**, *57*, 24–35. [CrossRef]
37. Rondeaux, G.; Steven, M.; Baret, F. Optimization of soil-adjusted vegetation indices. *Remote Sens. Environ.* **1996**, *55*, 95–107. [CrossRef]
38. Broge, N.H.; Leblanc, E. Comparing prediction power and stability of broadband and hyperspectral vegetation indices for estimation of green leaf area index and canopy chlorophyll density. *Remote Sens. Environ.* **2000**, *76*, 156–172. [CrossRef]
39. Haboudane, D. Hyperspectral vegetation indices and novel algorithms for predicting green LAI of crop canopies: Modeling and validation in the context of precision agriculture. *Remote Sens. Environ.* **2004**, *90*, 337–352. [CrossRef]
40. Gates, D.M.; Keegan, H.J.; Schleter, J.C.; Weidner, V.R. Spectral Properties of Plants. *Appl. Opt.* **1965**, *4*, 11–20. [CrossRef]
41. Horler, D.N.H.; Dockray, M.; Barber, J. The red edge of plant leaf reflectance. *Int. J. Remote Sens.* **1983**, *4*, 273–288. [CrossRef]
42. Datt, B.A. New Reflectance Index for Remote Sensing of Chlorophyll Content in Higher Plants: Tests using Eucalyptus Leaves. *J. Plant Physiol.* **1999**, *154*, 30–36. [CrossRef]
43. Blackburn, G.A. Hyperspectral remote sensing of plant pigments. *J. Exp. Bot.* **2007**, *58*, 855–867. [CrossRef] [PubMed]
44. Hatfield, J.L.; Gitelson, A.A.; Schepers, J.S.; Walthall, C.L. Application of Spectral Remote Sensing for Agronomic Decisions. *Agron. J.* **2008**, *100*, S117–S131. [CrossRef]
45. Richter, K.; Atzberger, C.; Hank, T.B.; Mauser, W. Derivation of biophysical variables from Earth observation data: Validation and statistical measures. *J. Appl. Remote Sens.* **2012**, *6*, 063557. [CrossRef]
46. Croft, H.; Chen, J.; Zhang, Y. The applicability of empirical vegetation indices for determining leaf chlorophyll content over different leaf and canopy structures. *Ecol. Complex.* **2014**, *17*, 119–130. [CrossRef]
47. Brogea, N.; Mortensen, J. Deriving green crop area index and canopy chlorophyll density of winter wheat from spectral reflectance data. *Remote Sens. Environ.* **2002**, *81*, 45–57. [CrossRef]
48. Zarco-Tejada, P.J.; Berjón, A.; López-Lozano, R.; Miller, J.R.; Martín, P.; Cachorro, V.; González, M.R.; de Frutos, A. Assessing vineyard condition with hyperspectral indices: leaf and canopy reflectance simulation in a row-structured discontinuous canopy. *Remote Sens. Environ.* **2005**, *99*, 271–287. [CrossRef]

remote sensing

MDPI

Article

Evaluation of Leaf N Concentration in Winter Wheat Based on Discrete Wavelet Transform Analysis

Fenling Li [1,2], Li Wang [1], Jing Liu [1,2], Yuna Wang [1] and Qingrui Chang [1,2,*]

1 College of Natural Resources and Environment, Northwest A&F University, Yangling 712100, Shaanxi, China; fenlingli@nwafu.edu.cn (F.L.); wangli1990@nwsuaf.edu.cn (L.W.); jingliu@nwsuaf.edu.cn (J.L.); 2018050822@nwafu.edu.cn (Y.W.)

2 Key Laboratory of Plant Nutrition and the Agri-Environment in Northwest China, Ministry of Agriculture, Yangling 712100, Shaanxi, China

* Correspondence: changqr@nwsuaf.edu.cn; Tel.: +86-135-7183-5969

Received: 19 April 2019; Accepted: 31 May 2019; Published: 3 June 2019

Abstract: Leaf nitrogen concentration (LNC) is an important indicator for accurate diagnosis and quantitative evaluation of plant growth status. The objective was to apply a discrete wavelet transform (DWT) analysis in winter wheat for the estimation of LNC based on visible and near-infrared (400–1350 nm) canopy reflectance spectra. In this paper, in situ LNC data and ground-based hyperspectral canopy reflectance was measured over three years at different sites during the tillering, jointing, booting and filling stages of winter wheat. The DWT analysis was conducted on canopy original spectrum, log-transformed spectrum, first derivative spectrum and continuum removal spectrum, respectively, to obtain approximation coefficients, detail coefficients and energy values to characterize canopy spectra. The quantitative relationships between LNC and characteristic parameters were investigated and compared with models established by sensitive band reflectance and typical spectral indices. The results showed combining log-transformed spectrum and a sym8 wavelet function with partial least squares regression (PLS) based on the approximation coefficients at decomposition level 4 most accurately predicted LNC. This approach could explain 11% more variability in LNC than the best spectral index mSR_{705} alone, and was more stable in estimating LNC than models based on random forest regression (RF). The results indicated that narrowband reflectance spectroscopy (450–1350 nm) combined with DWT analysis and PLS regression was a promising method for rapid and nondestructive estimation of LNC for winter wheat across a range in growth stages.

Keywords: leaf nitrogen concentration; discrete wavelet transform; partial least squares; hyper-spectra

1. Introduction

Nitrogen (N) is one of the essential elements in plants. N deficiency seriously affects the photosynthesis process and physiological metabolism and results in poor wheat grain yield and quality [1,2]. Excessive N over fertilization not only fails to increase crop yield, but also causes the unnecessary purchase of fertilizer and results in environmental pollution [3,4]. Knowledge of in-season plant N status is the key to guiding N fertilization for farmers. Leaf nitrogen concentration (LNC) is an important indicator of N nutrition in crops and is desired to be obtained by a rapid, non-destructive method [5]. Hyperspectral remote sensing captures continuous and subtle spectral absorption features of crop canopy from narrow bands, which has been widely applied to differentiate and quantify the biophysical and biochemical parameters of agricultural crops [6]. Absorption characteristics of N itself are quite weak and often are expressed by amino acid absorption characteristics in protein. The sensitive absorption wavelength of N lies in short-wave-infrared (SWIR), which is easily obscured by water-vapor absorption characteristics [7,8]. Visible spectrum (VIS) and near-infrared (NIR) band reflectance are often used to estimate LNC indirectly due to the strong positive correlation with leaf chlorophyll content and pronounced sensitivity to canopy structures [9–12].

Among the estimating approaches, linear or non-linear regression models are typically analyzed based on individual input variable of sensitive waveband reflectance. However, the canopy reflectance spectra and sensitive wavebands of LNC are easily and strongly influenced by the soil background, vegetation canopy geometry and atmospheric conditions [13]. Spectral transformation techniques such as first derivative transformation, continuum removal and log-transformation processing techniques are applied to reduce the effects of the surroundings and to enhance the spectral sensitivity to crop N content [11,14,15]. However, it remains to be discussed about which spectral transformation is more effective in canopy N evaluation. Meanwhile, a number of spectral indices sensitive to chlorophyll and canopy structures with robustness have been developed to minimize spectral noise and to estimate N-related indicators [16–18]. These include the modified normalized difference (mND_{705}) and the modified simple ratio (mSR_{705}), which effectively reduce the impact of differences in leaf surface reflectance and improve the sensitivity of pigment and N content estimation [19]. Chen et al. [20] developed the three-band double-peak canopy nitrogen index (DCNI) to predict the LNC of maize and wheat during the critical N management stage. However, these indices are calculated by utilizing a limited number of wavelengths in specific spectral regions, which have not exploited the entire range in hyperspectral data [21] and are calibrated against a specific database, which cannot be generalized to other databases [22]. There is an urgent need to propose an approach that could take advantage of the entire canopy spectral information as well as diminish the impacts of band autocorrelation and data redundancy.

The wavelet transform (WT) is a multi-resolution analysis tool that has found several applications in signal processing and compression [23], pattern recognition and classification [24], and recently was involved in precision agriculture applications such as detection of crop-yield-reducing weeds [25], estimation of leaf chlorophyll content [22], crop residue management [26] and diagnosis of crop diseases [27]. Discrete wavelet transform (DWT) is capable of decomposing canopy original spectra into different DWT coefficients in fine-scale detail coefficient (DC) and coarse-scale approximation coefficient (AC) on the basis of mother wavelet functions. DC provides a detailed view of the input hyperspectral signal in response to noise and special information inhered in the signal. Low-frequency AC is an expression of global behavior of the signal, which corresponds to the main and large trend in a signal. The AC and DC together reflect the time-frequency properties of the canopy spectral signal at different scales [23–25]. It is considered as a productive tool for hyperspectral feature extraction [25], and has been successfully used in quantifying pigment concentrations [28], retrieving soil moisture [29], estimating crop residue mass [26] and leaf area index (LAI) mapping [30]. However, little application of DWT in crop agronomy parameter evaluation is reported in the literature. Moreover, the relationship between DWT features and LNC has not yet been studied.

In this study, we focused on the relationship between leaf N concentration and canopy spectral reflectance to explore the possibilities of using the entire VIS-NIR region (400–1350 nm) for winter wheat LNC assessment. The objectives of this study were (1) to analyze the impacts of the spectral transformation type, mother wavelet and decomposition level on feature extraction of the entire canopy spectra with a DWT analysis; (2) to construct DWT-based LNC estimating models with partial least squares (PLS) and random forest (RF) regression and (3) to compare the models in (2) with sensitive band reflectance-based and spectral index-based LNC estimation models (SR-LNC and SI-LNC) to find a promising LNC monitoring model across a range of wheat growth stages.

2. Materials and Methods

2.1. Data Acquisition

2.1.1. Experimental Design

The experiments were conducted in Guanzhong region, Shaanxi Province, China. The winter wheat was planted in mid-October and harvested in mid-June of the following year. In experiment 1, the commonly adopted wheat cultivar in this region, Xiaoyan 22 was cultivated in 2013–2015 at No.1 experiment station of Northwest Agriculture and Forestry University (108°03′ E, 34°17′ N; elevation: 454 m). A total of 24 plots were set and each plot size was 12 m^2 (3 m × 4 m) with a planting row spacing of 0.2 m and a plant density of 185 $kgha^{-1}$. Six N rates (0, 30, 60, 90, 120 and 150 kg ha^{-1}) and six P rates (0, 15, 30, 60, 75 and 90 kg ha^{-1}) were employed with two replications. 30 kg ha^{-1} P_2O_5 and 60 kg ha^{-1} N were applied as a basal fertilizer for N and P treatments, respectively. There was no K fertilizer application due to the K sufficiency in this area. Experiment 2 was conducted at Qian County (108°10′ E, 34°37′ N; elevation: 830 m) in Xianyang City during the years 2014 to 2015. A total of 36 plots were set and each plot was 36 m^2 (6 m × 6 m) with a similar cultivar as experiment 1. Six N rates (0, 45, 90, 135, 180 and 225 kg ha^{-1}), six P rates (0, 22.5, 45, 67.5, 90 and 112.5 kg ha^{-1}) and six K rates (0, 15, 30, 45, 60 and 75 kg ha^{-1}) were applied and replicated twice. 60 kg ha^{-1} K_2O and 45 kg ha^{-1} P_2O_5, 60 kg ha^{-1} K_2O and 90 kg ha^{-1} N and 90 kg ha^{-1} N and 45 kg ha^{-1} P_2O_5 were applied as a basal fertilizer for N, P and K treatments, respectively, before planting. For all the treatments, N, P_2O_5 and K_2O fertilizers were applied as urea, potassium chloride and superphosphate, respectively. All the plot crop planting and management patterns followed the local standard practices for wheat production.

2.1.2. Canopy Spectral Measurement

All canopy spectral measurements were collected by a field portable spectrometer (SVC HR-1024I, USA) whose sensor could collect the canopy spectrum from 350–2500 nm with a sampling interval of 3.5 nm for 350–1000 nm, 3.6 nm for 1000–1850 nm and 2.5 nm for the 1850–2500 nm spectral region. Each spectral measurement was obtained with a 25° field-of-view operating from a height of 1.3m above the ground under clear sky conditions between 10:00 and 14:00 local time. Before each measurement, a white $BaSO_4$ calibration panel was used to calculate the black and baseline reflectance. To minimize the effects caused by the surroundings, the canopy spectrum in each plot was obtained by randomly selecting three sampling sites, and then averaging these into a single spectral sample. Each sample consisted of an average of ten scans at an optimized integration time. Canopy spectral data were measured during the main growth stages in each growing season. A total of 84, 84, 74 and 73 samples were obtained in tillering, jointing, booting and filling growth stages.

2.1.3. Leaf Nitrogen Concentration Measurement

Samples for LNC determination were collected immediately after measurements of the canopy spectra. Wheat plants from an area of 0.08 m^2 (0.2 m × 0.4 m) and 0.25 m^2 (0.5 m × 0.5 m) of each plot in experiments 1 and 2 were cut respectively at ground level. All green leaves were separated from stems, sealed in plastic bags and transferred to the laboratory with ice chests. Then, the samples were oven-dried at 105 °C for 30 min, followed by oven drying at 80 °C until a constant weight was achieved. Finally, dried leaves were finely ground and a subsample of ground leaves was taken to analyze for LNC (g per 100 g dry weight, %) using the Kjeldahl method [31].

2.2. Spectral Transformation

The spectral response in VIS-NIR bands at 400–1350 nm was used to monitor the wheat LNC in this study. All the canopy spectral reflectance curves were resampled at 1 nm spectral interval, and then the Savitzky–Golay smoothing procedure [32] with a nine-point moving window and a second-order polynomial fitting was applied to each spectrum. The smoothed canopy spectrum was labeled as the original spectrum (OS). After that, three kinds of commonly used spectral transformations were calculated to compare with OS, including a log-transformed spectrum (LOGS), first derivative spectrum (FDS) and continuum removal spectrum (CRS). LOGS was determined by calculating a log function of the spectral reflectance's reciprocal [14]. FDS was derived through calculating differences in reflectance between adjacent wavebands [33]. CRS was obtained by normalizing the absorption valley in the spectral curve onto the continuum line of the absorption valley [34].

2.3. Analytical Methods

2.3.1. Discrete Wavelet Transform Analysis

The background and principle of discrete wavelet transform (DWT) can be found in the literature [25]. It can be described as a set of inner products between a finite-length signal and a discretized wavelet basis made by scaled and transformed versions of a mother wavelet. The result is known as the DWT coefficient. The mathematical expression is as follows,

$$W_{i,k} = \left\langle f(\lambda), \phi_{i,k}(\lambda) \right\rangle, \tag{1}$$

$$\phi_{i,k}(\lambda) = \frac{1}{\sqrt{2^i}} \psi\left(\frac{\lambda}{2^i} - k\right), \tag{2}$$

where $W_{i,k}$ is a DWT coefficient; $f(\lambda)$ is a signal; $\varphi_{i,k}(\lambda)$ is the discretized wavelet basis used to fit optimally the signal; i is the ith decomposition level or step and k is the kth wavelet coefficient at the ith level. With DWT analysis, signals are analyzed over a discrete set of scales, typically being dyadic (2^i, $i = 1, 2, 3, \ldots$ i is the decomposition level) [25,28,30].

In practice, the wavelet basis performs as a set of high-pass and low-pass filters to decompose the signal into low-scale, high-frequency detail coefficients (DCs) and high-scale, low-frequency approximation coefficients (ACs) according to scale. The length of the ACs and DCs is related to the type of mother wavelet and decomposition level. In a multi-level decomposition, the filtering process can be iterated with successive approximations being decomposed in turn, so that the signal is broken down into many lower resolution components (Figure 1). Through the DWT, not only can the detailed behavior be separated from the macroscopic behavior, but also the dimensionality of hyperspectral data is reduced. All the information in the original signal is contained in the ACs at a particular decomposition level i (L_i) plus the DCs at decomposition level 1 to level j (L_1–L_i).

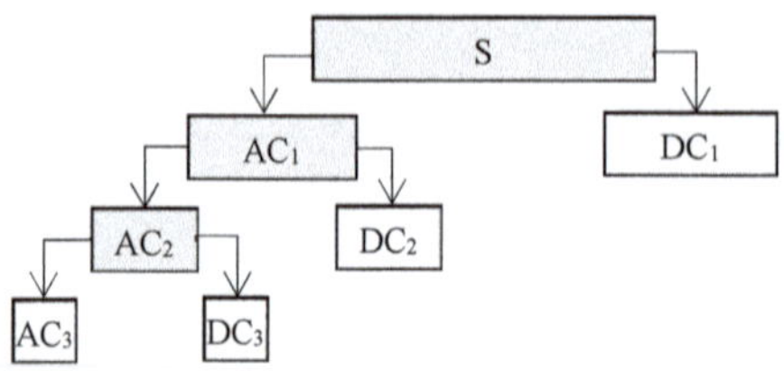

Figure 1. Process of multi-level discrete wavelet transform decomposition of signal S. AC_i and DC_i denote the approximation coefficient (AC) and detail coefficient (DC) at the ith decomposition level (L_i). Taking L_3 for example, the output wavelet decomposition vectors include AC_3, DC_1, DC_2 and DC_3. The size of each box demonstrates the length of the successive approximation and detail coefficients vectors.

Considering the canopy spectrum as a signal changing with wavelength, the ACs and DCs after multi-level DWT decomposition were investigated as feature parameters of canopy spectra to find whether it is a productive tool for LNC estimation in this paper. DWT analysis could be implemented with the function 'wavedec' in MATLAB Wavelet Toolbox. The function, 'wrcoef', was used to reconstruct the spectral signal so as to find out how ACs delineate the canopy spectral information. Energy value (EV) [30] is a set of compressed ACs and DCs, which tries to take advantage of ACs and

DCs to characterize the whole signal information. It also can be considered as a feature parameter of canopy hyper-spectral signal and can be obtained through the expression (3):

$$EV_i = \sqrt{\frac{1}{K}\sum_{k=1}^{K} {w_{i,k}}^2}, \tag{3}$$

where EV_i denotes the wavelet energy value of the *i*th decomposition level (L_i), $w_{i,k}$ is the *k*th wavelet coefficient at L_i and K represents the total number of wavelet coefficients under each level. As shown in Figure 1, EV_3 will be calculated with the AC_3, DC_1, DC_2 and DC_3 according to expression (3). Five mother wavelet functions including db10, sym8, coif5, bior6.8 and rbio6.8 from the Daubechies, Symlet, Coiflet, Biorthogonal and Reverse biorthogonal wavelet families, respectively, were assessed in this study, which are commonly tested wavelet families in the canopy spectra decomposition [25–28,30].

2.3.2. Existing Spectral Indices Calculation

A total of ten correlated hyperspectral indices from three categories were selected and examined for comparison with the DWT approach, including (1) chlorophyll indices: Modified red edge simple ratio index (mSR_{705}), MERIS terrestrial chlorophyll index (MTCI), structurally insensitive pigment index (SIPI) and normalized pigment chlorophyll index (NPCI); (2) nitrogen indices: Nitrogen reflectance index (NRI), normalized difference red-edge index (NDRE) and double-peak canopy nitrogen index (DCNI); (3) greenness indices: Green normalized difference vegetation index (GNDVI), optimized soil-adjusted vegetation index (OSAVI) and modified triangular vegetation index ($MTVI_2$). The definitions and reference sources for these ten spectral indices are summarized in Table 1.

Table 1. The definitions and reference sources of narrowband spectral indices tested in this study.

Category	Index	Formula	Developed by
Chlorophyll indices	mSR_{705}	$(R_{750} - R_{445})/(R_{705} - R_{445})$	[19]
	MTCI	$(R_{754} - R_{709})/(R_{709} - R_{681})$	[35]
	SIPI	$(R_{800} - R_{445})/(R_{800} - R_{680})$	[36]
	NPCI	$(R_{430} - R_{680})/(R_{430} + R_{680})$	[36]
Nitrogen indices	NRI	$(R_{570} - R_{670})/(R_{570} + R_{670})$	[37]
	NDRE	$(R_{790} - R_{720})/(R_{790} + R_{720})$	[38]
	DCNI	$(R_{720} - R_{700})/(R_{700} - R_{670})/(R_{720} - R_{670} + 0.03)$	[20]
Greenness indices	GNDVI	$(R_{750} - R_{550})/(R_{750} + R_{550})$	[39]
	OSAVI	$1.16(R_{800} - R_{670})/(R_{800} + R_{670} + 0.16)$	[40]
	$MTVI_2$	$1.5(1.2(R_{800} - R_{550}) - 2.5(R_{670} - R_{550}))/\mathrm{sqrt}((2R_{800} + 1)^2 - (6R_{800} - 5\mathrm{sqrt}(R_{670})) - 0.5)$	[41]

R_i is the reflectance at *i* nm wavelength

2.3.3. Modeling Method

The ordinary least squares (OLS) regression analysis was used to construct the LNC estimation model based on sensitive-band reflectance and spectral index. Two types of multivariate models, partial least squares (PLS) regression and random forest (RF) regression, were carried out to establish the estimation model on the basis of wavelet coefficients and energy values.

PLS regression is a bilinear multivariate regression method. It compresses the input data into a number of independent latent variables (LVs) and maximizes the covariance between the LV scores and dependent variables. The operation makes it possible to avoid high collinearity among multi-variables and shows better prediction performance when compared with stepwise regression or principal component regression [21,42]. The basic PLS algorithm could be obtained in Geladi and Kowalski [43]. The number of latent variables is selected on the basis of the standard error of leave-one-out cross-validation. Parameter optimization and modeling were implemented with the PLS Toolbox based on MATLAB® 7.0 (MathWorks, Inc., Natick, MA, USA).

RF regression is an ensemble machine learning algorithm based on regression trees [44,45]. It uses the bootstrap sampling method and randomized subspace method to build decision trees, in which only randomly selected predictors are used for each tree. The final prediction result is determined by the average of all decision trees. Two parameters need to be optimized in RF: 'ntree', the number of regression trees grown based on a bootstrap sample of the observations; 'mtry', the number of different predictors (independent variables) tested at each node. The random forest library developed in the R package (R Development Core Team 2008) was employed to implement the RF algorithm in this study. The parameter 'mtry' was set as 1/3 of the number of independent variables and the 'ntree' was set at 500 as suggested by Breiman [43].

2.3.4. Calibration and Validation

In order to ensure the range of LNC is represented in both datasets, all the observations were pooled together and then divided into calibration and validation dataset according to LNC values in an ascending sort order with a proportion of 4:1 [46]. The calibration and validation data sets were evenly distributed (Table 2). LNC ranged from 0.22% to 3.87% across all growth stages, with an average value of 1.47%. The coefficient of variation (CV) was 52.03% indicating a moderate temporal variation.

Table 2. The statistical description of leaf nitrogen concentration (%) across all growth stages.

Data Set	No. of Samples	Min	Max	Range	Mean	SD	Variance	Skewness	Kurtosis	CV (%)
Whole	315	0.22	3.87	3.64	1.47	0.77	0.59	0.76	2.98	52.03
Calibration	252	0.22	3.60	3.38	1.46	0.76	0.58	0.73	2.88	51.88
Validation	63	0.35	3.87	3.52	1.5	0.79	0.63	0.86	3.26	53.00

The coefficient of determination (R^2), root mean square error (RMSE), relative error (RE, %) and the ratio of prediction to deviation (RPD) were used to measure the predictive performance of each estimation model by different methods. Higher values of R^2 and RPD, and lower values of RMSE and RE indicate better dependability and accuracy of the regression model in predicting LNC [21,47,48]. RPD is a ratio of standard deviation to RMSE. RPD values greater than 2.0 indicate a stable and accurate predictive model, an RPD value between 1.4 and 2.0 indicates a fair model that could be improved by more accurate prediction techniques and a value less than 1.4 indicates poor predictive capacity [21]. Rc^2, Rv^2, RMSEc, RMSEv, REc, REv, RPD_c and RPD_v in this paper represented R^2, RMSE, RE and RPD in the calibration and validation data set, respectively. A 1:1 plot of observed vs. estimated values was drawn to demonstrate the degree of model fit.

3. Results

3.1. LNC Estimation Models (SR-LNC) Based on Sensitive-Band Reflectance

3.1.1. Correlations Between Canopy Spectra and LNC

In general, winter wheat canopy reflectance was more significantly correlated with LNC at VIS than NIR wavelengths (Figure 2). The absolute values of the correlation coefficient were greater than 0.6 at 400–750 nm. Correlations were similar from 614 to 640 nm, with the strongest negative value of −0.75. The correlation between FDS and LNC changed rapidly at 400–1300 nm. It was slightly better at 435–465 nm than other wavelengths, and the best correlation coefficient was −0.76 at 447 nm. A weak negative correlation between LOGS and LNC was found in the near-infrared region (750–1300 nm), whereas a strong positive correlation was observed at 400–750 nm, which significantly improved the relationship between OS and LNC at 600–700 nm. The highest correlations appeared at 642–648 nm with a correlation coefficient of 0.83. Correlation between CRS and LNC was better at 400–765 nm, 934–1050 nm, 1124–1290 nm and 1304–1350 nm than that between OS and LNC. Moreover, the correlation coefficients were greater than 0.6 at 400–760 nm and 1180–1270 nm, and the best correlation coefficient was −0.85 at 721–727 nm. As reflectance at 640 nm was affected by chlorophyll absorption [8], and 725 nm within the red-edge wavelength region was highly related with canopy nitrogen content [17], 640 nm and 725 nm were selected as the bands of OS and CRS most sensitive to LNC, respectively. According to the best correlation with canopy spectra, 447 nm and 645 nm (center part of 642–648 nm) were regarded as the bands of FDS and LOGS most sensitive to LNC, respectively. The spectral reflectance of sensitive bands is used to establish LNC estimation models.

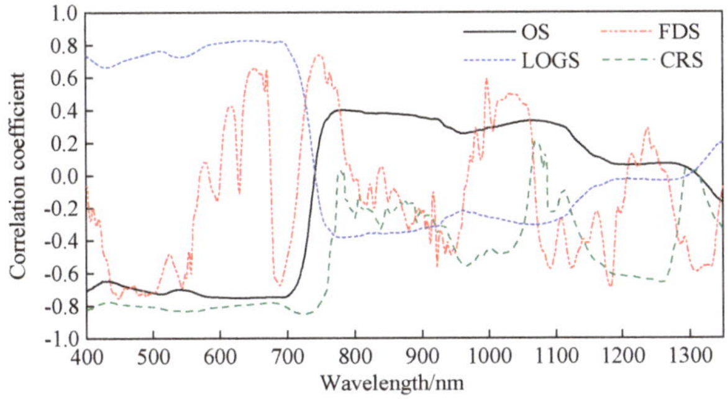

Figure 2. Correlation coefficients between leaf nitrogen concentration and transformed canopy spectra, including original spectrum (OS), first derivative spectrum (FDS), log-transformed spectrum (LOGS) and continuum removal spectrum (CRS).

3.1.2. Construction of SR-LNC Estimation Models

The calibration and validation accuracy of FDS at 447 nm and OS at 640 nm were similar to each other (Figure 3; Figure 4). LOGS at 645 nm and CRS at 725 nm significantly improved the LNC prediction accuracy relative to the FDS and OS models, and the R_c^2s in exponential prediction models were 0.73 and 0.79, respectively. The R_v^2s in validation samples were 0.72 and 0.85, $RMSE_V$s were 0.42 and 0.35 and RE_Vs were 28.38 and 20.65 for LOGS and CRS, respectively. Scatter plots between predicted and measured LNC values indicated that higher LNC values were underestimated (Figure 4). CRS was superior to other spectra in predicting LNC at sensitive reflectance bands, with predicted and measured LNC values falling close to the 1:1 line.

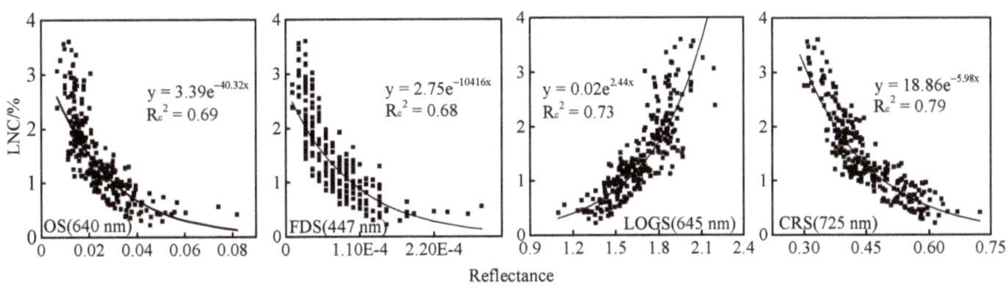

Figure 3. Leaf nitrogen concentration prediction models based on sensitive band reflectance. The solid line represents the exponential fitting.

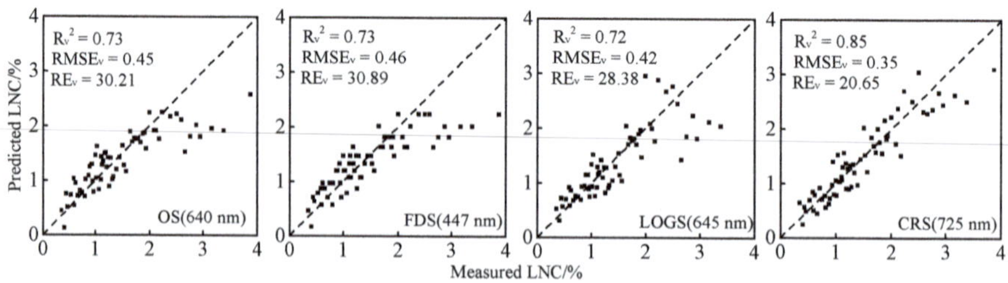

Figure 4. Scatter plots between the measured and predicted leaf nitrogen concentration based on sensitive band reflectance. The dash line is the 1:1 line.

3.2. *LNC Estimation Models (SI-LNC) Based on Spectral Indices*

All the spectral indices were significantly correlated with LNC as shown in Table 3. The mSR_{705} index was the best of ten spectral indices, which could explain 83% variability in LNC. The higher R_v^2 (0.86), lower $RMSE_v$ (0.28) and RE_v (18.81) also illustrated a better performance of mSR_{705}. The NDRE index was the best of three nitrogen indices, which could explain 80% of variability in LNC. The GNDVI was the best among the three greenness indices. Both GNDVI and NDRE were exponentially related to LNC, while the accuracy of GNDVI was slightly worse than NDRE.

Table 3. Estimation models and prediction errors of leaf nitrogen concentration based on the spectral indices. ** at 0.01 significance level.

Category	Index	Correlation Coefficient	Equation	R_c^2	R_v^2	$RMSE_v$	RE_v
Chlorophyll indices	mSR_{705}	0.91 **	$LNC = 0.2702x - 0.6773$	0.83	0.86	0.28	18.81
	MTCI	0.89 **	$LNC = 0.5454x - 1.0901$	0.78	0.84	0.31	20.94
	SIPI	0.79 **	$LNC = 1E - 06e^{15.28x}$	0.71	0.69	0.57	37.80
	NPCI	0.80 **	$LNC = 1.9583e^{4.13x}$	0.70	0.71	0.45	30.13
Nitrogen indices	NRI	0.70 **	$LNC = 6.2342x - 0.5199$	0.50	0.50	0.56	37.64
	NDRE	0.86 **	$LNC = 0.046e^{6.28x}$	0.80	0.85	0.30	20.27
	DCNI	0.79 **	$LNC = 0.039x - 0.8904$	0.63	0.74	0.41	27.06
Greenness indices	GNDVI	0.85 **	$LNC = 0.002e^{8.44x}$	0.81	0.82	0.33	21.90
	OSAVI	0.69 **	$LNC = 0.0099e^{6.61x}$	0.55	0.54	0.55	36.61
	$MTVI_2$	0.60 **	$LNC = 5.293x - 1.0551$	0.36	0.42	0.60	40.08

3.3. *LNC Estimation Models (DWT-LNC) Based on DWT Features*

3.3.1. Selection of Optimum Mother Wavelet and Decomposition Level

The number of DWT coefficients describes the extent of data compression. As shown in Table 4, the number of DWT coefficients changed with the mother wavelet and decomposition level, and was independent on spectral transformation types. It tended to drop off from L_1 to L_{12}, and the downward trend became stable at L_{10}. Among five mother wavelets, sym8 had the strongest data compression ability, while coif5 was the weakest. For example, the total number of wavebands in this study was 951 (from 400–1350 nm). After DWT analysis at decomposition level 10, the number of DWT coefficients with mother wavelet function sym8 was 15, while coif5 had 29, which was determined from the wavelet basis length [23].

Table 4. The number of wavelet coefficients under different mother wavelets and decomposition levels.

Mother Wavelet	L_1	L_2	L_3	L_4	L_5	L_6	L_7	L_8	L_9	L_{10}	L_{11}	L_{12}
bior6.8	484	250	133	75	46	31	24	20	18	17	17	17
coif5	490	259	144	86	57	43	36	32	30	29	29	29
db10	485	252	135	77	48	33	26	22	20	19	19	19
rbio6.8	484	250	133	75	46	31	24	20	18	17	17	17
sym8	483	249	132	73	44	29	22	18	16	15	15	15

Since the approximation coefficient (AC) was considered as an indicator of global information for the canopy spectrum, ACs of each transformed spectrum at decomposition level 1 to 12 were utilized to perform signal reconstruction in order to find out how ACs delineated the canopy reflectance spectra. The correlations between canopy spectral signals and reconstruction signals are shown in Figure 5. The correlation coefficient decreased from L_5 until leveling off at L_{10}, which indicated that the explanatory and signal restoring ability of ACs to canopy spectra declined gradually from L_5 to L_{10}. All the correlation coefficients were still above 0.7 at L_{10} except with FDS, which went down to less than 0.6 rapidly after decomposition level 6. The mother wavelet db10 was more labile compared to the others, especially poor was the large fluctuation in CRS correlations. Taking into account the data compression effectiveness, stability of mother wavelet and ability to maintain the information quality of the canopy spectra, a mother wavelet sym8 at decomposition level of L_1–L_{10} was chosen to conduct the DWT to analyze the correlation with LNC.

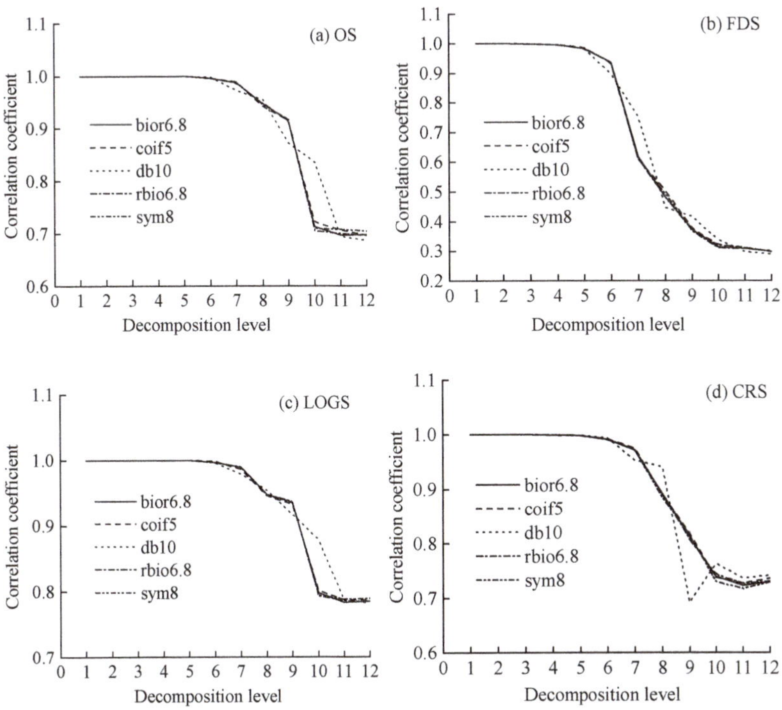

Figure 5. Correlations between reconstructed signals and transformed spectra for different mother wavelets at each decomposition level. (**a**) OS, (**b**) FDS, (**c**) LOGS and (**d**) CRS.

3.3.2. DWT-LNC Models Based on PLS Regression

PLS Regression Using Wavelet ACs

All the LNC estimation models established by calibration of sample measurements with ACs of sym8 at L_1 to L_{10} passed the 0.01 significance level test. The number of latent variables (LVs) extracted by PLS regression increased and then decreased with the decomposition level (Figure 6). The maximum number of latent variables emerged at L_5, which implied a lower convergence rate of the PLS regression at L_5. As a whole, all the estimating models had high prediction accuracy (R_c^2 was greater than 0.70). The general trend of R_c^2 in each predicting model increased and then decreased gradually with an increasing decomposition level (Figure 7). The ACs could explain 90%–93% of the variability in LNC at L_1 to L_5, except with CRS at L_1. The variation of R_v^2 was consistent with R_c^2, and all the R_v^2s were greater than 0.75. Table 5 shows the variation in RMSE and RE. LOGS had stronger prediction ability over other models, with the highest values of R_v^2 and R_c^2, and lowest RE_v and $RMSE_v$ values at L_1–L_{10}. All the $RMSE_v$s were below 0.30. The R_v^2s were 0.93, 0.93 and 0.91 respectively for the LOGS at L_3, L_4 and L_5. Measured and predicted LNC values with ACs from LOGS at L_3–L_5 closely approximated a 1:1 line (Figure 8), and the PLS model at L_4 yielded the optimal prediction accuracy ($RMSE_v$ = 0.20, RE_v = 13.47).

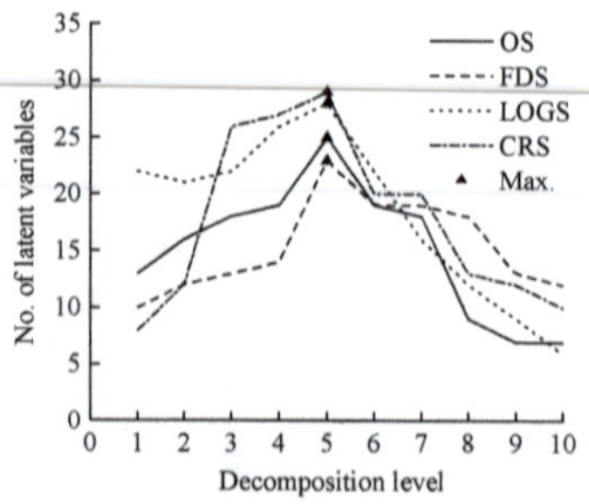

Figure 6. The number of latent variables in a partial least squares regression based on approximation coefficients.

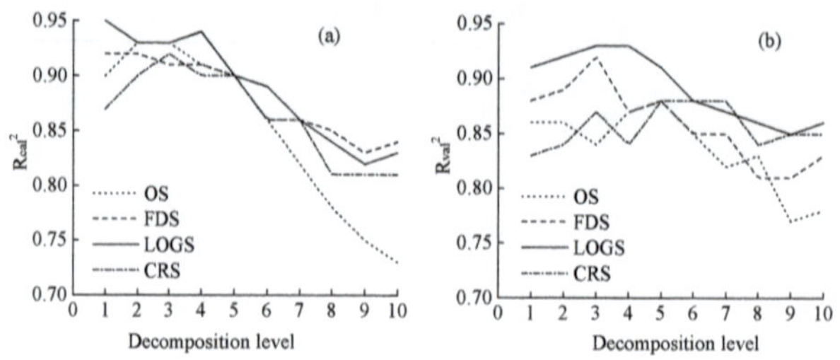

Figure 7. Relationships between determination coefficients and decomposition level of partial least squares regression models with approximation coefficients in the calibration (**a**) and validation set (**b**).

Table 5. The coefficient of determination, root mean square error and relative error of the validation set based on a partial least squares regression with wavelet approximation coefficients.

	OS			FDS			LOGS			CRS		
	R_v^2	$RMSE_v$	RE_v	R_v^2	$RMSE_v$	RE_v	R_v^2	$RMSE_v$	RE_v	R_v^2	$RMSE_v$	RE_v
AC_1	0.86	0.29	19.59	0.88	0.27	18.07	0.91	0.24	15.78	0.83	0.33	22.06
AC_2	0.86	0.30	19.68	0.89	0.26	17.22	0.92	0.23	15.19	0.84	0.33	22.13
AC_3	0.84	0.32	21.24	0.92	0.23	15.50	0.93	0.20	13.59	0.87	0.33	22.18
AC_4	0.87	0.29	19.09	0.87	0.27	18.18	0.93	0.20	13.47	0.84	0.33	22.20
AC_5	0.88	0.28	18.45	0.88	0.29	19.49	0.91	0.23	15.60	0.88	0.28	18.37
AC_6	0.85	0.31	20.98	0.85	0.31	20.76	0.88	0.27	17.95	0.88	0.27	18.24
AC_7	0.82	0.34	22.62	0.85	0.31	20.52	0.87	0.30	20.11	0.88	0.27	17.99
AC_8	0.83	0.33	22.28	0.81	0.31	20.84	0.86	0.30	19.83	0.84	0.31	20.87
AC_9	0.77	0.38	25.13	0.81	0.35	23.07	0.85	0.30	20.29	0.85	0.31	20.74
AC_{10}	0.78	0.37	24.90	0.83	0.32	21.64	0.86	0.29	19.59	0.85	0.30	20.14

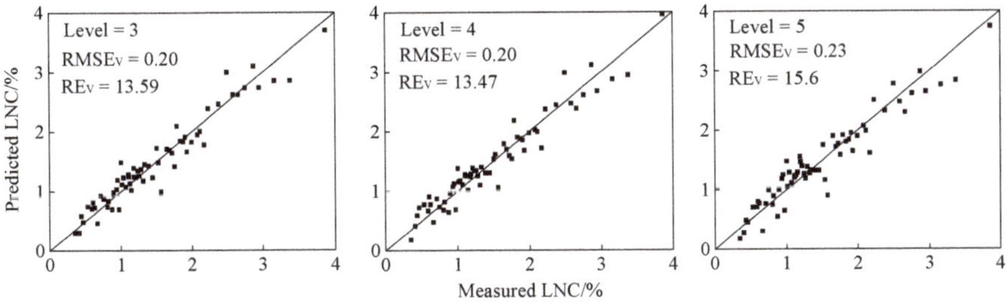

Figure 8. Relationships between measured and predicted leaf nitrogen concentration (%) based on the approximation coefficients of the log-transformed spectra (LOGS) at decomposition level 3, 4 and 5 in validation set.

PLS Regression Using Wavelet DCs

The high-frequency detail coefficient of DWT analysis represented noise or minor absorption in the canopy spectrum [21,28,49]. Figure 5 shows that the correlation coefficients between canopy spectral signals and reconstruction signals by ACs at decomposition level 1 to 5 were close to 1, which indicates that DCs at L_1 to L_5 were very small in amplitude (near zero) and could be removed without major loss in the information content of the signal. Table 6 summarizes the validation results of PLS models based on DCs at L_6–L_{10}. Prediction accuracy decreased with the decomposition level. The performance of LOGS was better than other spectral transformations, but it was still worse than PLS modeling with ACs (Table 5).

Table 6. The coefficient of determination, root mean square error and relative error of the validation set based on a partial least squares regression with detail coefficients at decomposition level 6 to level 10. DC_i denotes the detail coefficient (DC) at L_i.

DC	OS			FDS			LOGS			CRS		
	R_v^2	$RMSE_v$	RE_v	R_v^2	$RMSE_v$	RE_v	R_v^2	$RMSE_v$	RE_v	R_v^2	$RMSE_v$	RE_v
DC_6	0.84	0.32	21.43	0.81	0.35	23.02	0.89	0.26	17.53	0.85	0.30	20.25
DC_7	0.83	0.33	21.75	0.77	0.38	25.19	0.89	0.27	17.73	0.84	0.32	21.41
DC_8	0.83	0.33	22.17	0.57	0.52	34.54	0.87	0.29	18.38	0.81	0.35	23.13
DC_9	0.77	0.38	25.36	0.48	0.57	38.05	0.88	0.28	18.77	0.81	0.36	23.76
DC_{10}	0.76	0.39	26.17	0.39	0.62	41.63	0.88	0.28	19.55	0.77	0.38	25.53

PLS Regression Using EVs

Energy values achieve further compression of spectral signals. After multi-level one-dimensional wavelet analysis at decomposition level *n*, *n* + 1 variables were used to calculate the energy value according to Equation (3), and the relationship between energy value and LNC was analyzed by using PLS regression (Table 7). With an increase in decomposition level, R_v^2 increased and then decreased with OS and LOGS, while a general tendency of first increasing then decreasing and a conspicuous monotonic increase were found, respectively, with FDS and CRS (Table 7). However, all the R_v^2 reached the maximum at L_{10}. Energy value could explain over 80% of the variability of LNC at L_{10} with fewer variables (number of variables was 11). LOGS still gave the best performance over the other transformations (with R_c^2 of 0.85 and R_v^2 of 0.88), but the overall accuracy was still lower than those using approximate coefficients at L_3–L_5 (Figure 6). The RMSE and RE of validation set were 0.26 and 17.82 respectively, also a poorer correlation compared with the results in Table 5.

Table 7. The coefficient of determination, root mean square error and relative error of the validation set based on a partial least squares regression with wavelet energy values. EV_i denotes energy value (EV) at decomposition level *i*.

EV	OS			FDS			LOGS			CRS		
	R_v^2	$RMSE_v$	RE_v	R_v^2	$RMSE_v$	RE_v	R_v^2	$RMSE_v$	RE_v	R_v^2	$RMSE_v$	RE_v
EV_1	0.14	0.73	49.02	0.46	0.58	38.83	0.47	0.57	38.22	0.64	0.47	31.61
EV_2	0.25	0.69	46.14	0.43	0.59	39.80	0.72	0.42	28.67	0.71	0.43	28.43
EV_3	0.19	0.71	47.34	0.41	0.61	40.58	0.71	0.44	29.36	0.72	0.41	27.66
EV_4	0.19	0.71	47.34	0.36	0.63	42.08	0.65	0.46	31.14	0.74	0.40	26.74
EV_5	0.29	0.67	44.42	0.32	0.65	43.47	0.69	0.43	29.05	0.75	0.39	26.34
EV_6	0.74	0.42	27.93	0.65	0.47	26.41	0.81	0.34	22.85	0.75	0.39	26.41
EV_7	0.79	0.36	23.99	0.66	0.47	31.49	0.81	0.34	22.91	0.76	0.38	25.88
EV_8	0.87	0.29	19.22	0.74	0.41	27.61	0.83	0.32	21.56	0.80	0.35	23.56
EV_9	0.87	0.29	19.15	0.76	0.40	27.02	0.84	0.32	21.35	0.81	0.34	23.19
EV_{10}	0.87	0.29	19.28	0.82	0.35	23.23	0.88	0.26	17.82	0.83	0.33	22.17

3.3.3. DWT-LNC Based on RF Regression

ACs and EV_{10} were selected to build the RF regression models. As showed in Table 8, R^2s in the validation set were slightly lower than the calibration set and most of them were less than 0.90, except for results using ACs of LOGS in L_4 and L_5. After L_5, the $RMSE_v$ and RE_v tended to go up slightly for all transformations especially with FDS. In general, LOGS was better than other transformations in estimating LNC by ACs with RF regression. ACs at L_4 had the best $RMSE_v$ and RE_v, being 0.24 and 16.08, respectively, while the ACs at L_{10} were the worst in the RF regression models. The accuracy of RF models based on energy values of wavelet coefficients was improved compared with the PLS regression, but still poorer than using ACs.

Table 8. Validation of leaf nitrogen concentration estimation models based on a random forest regression with discrete wavelet transform features. AC_i denotes approximation coefficient (AC) at L*i*; EV_{10} is the energy value at L_{10}.

	OS			FDS			LOGS			CRS		
	R_v^2	$RMSE_v$	RE_v	R_v^2	$RMSE_v$	RE_v	R_v^2	$RMSE_v$	RE_v	R_v^2	$RMSE_v$	RE_v
AC_1	0.86	0.29	19.64	0.87	0.29	19.30	0.89	0.26	17.53	0.89	0.27	17.93
AC_2	0.86	0.30	19.86	0.87	0.29	19.25	0.89	0.27	17.91	0.89	0.28	18.38
AC_3	0.87	0.29	19.34	0.87	0.29	19.20	0.89	0.26	17.21	0.88	0.28	18.41
AC_4	0.86	0.29	19.52	0.86	0.29	19.52	0.91	0.24	16.08	0.89	0.27	17.78
AC_5	0.87	0.28	18.89	0.87	0.29	19.01	0.90	0.25	16.35	0.89	0.28	18.44
AC_6	0.86	0.28	18.91	0.86	0.29	19.44	0.86	0.28	18.40	0.88	0.29	19.25
AC_7	0.88	0.29	19.41	0.71	0.43	28.76	0.86	0.30	20.12	0.86	0.31	20.77
AC_8	0.83	0.33	21.84	0.69	0.45	29.85	0.85	0.31	20.84	0.84	0.32	21.57
AC_9	0.85	0.31	20.67	0.64	0.49	32.45	0.80	0.36	24.02	0.81	0.35	23.27
AC_{10}	0.83	0.32	21.49	0.51	0.56	37.15	0.74	0.41	27.17	0.76	0.39	26.23
EV_{10}	0.82	0.35	23.34	0.77	0.38	25.36	0.86	0.29	19.43	0.84	0.32	21.16

3.4. Estimation Accuracy Comparison

Compared with OS at 640 nm, the sensitive band reflectance of CRS at 725 nm provided a significant improvement in the accuracy of estimating LNC. The R_c^2 and R_v^2 increased to 0.79 and 0.85, and the $RMSE_v$ and RE_v decreased to 0.35 and 20.65, respectively (Figure 2). The accuracy also was better than the results of LOGS and FDS at 645 nm and 447 nm, respectively (Figures 3 and 4). However, it was still lower than the performance of some spectral indices especially mSR_{705} (R_c^2 = 0.83, R_v^2 = 0.86, $RMSE_v$ = 0.28 and RE_v = 18.81; Table 3). The LOGS was obviously distinguished from four transformed canopy spectra in the discrete wavelet transform analysis and exhibited a promising potential for revising LNC. For PLS modeling, LOGS combined with ACs at L_4 produced the best performance both in the calibration and validation sets. The prediction result of LNC in the high-value region was better than SR-based and SI-based LNC estimation models. DCs and EVs performed worse in LNC evaluation by using PLS regression. The best prediction accuracy of DCs was at the decomposition level 6 (R_v^2 = 0.89, $RMSE_v$ = 0.26 and RE_v = 17.53), which was slightly higher than the mSR_{705} index. EVs at L_{10} had similar prediction accuracy as DCs at L_6. With RF regression, LOGS at L_4 showed the highest accuracy in LNC prediction on the basis of ACs, with the R_v^2, $RMSE_v$ and RE_v being 0.91, 0.24 and 16.28, respectively, which were slightly worse than the PLS regression. LNC estimation result using energy values was improved by a RF regression, but it was still lower than the PLS and RF models with ACs at L_4.

The ration of prediction to deviation (RPD) is calculated by dividing the standard deviation (SD) of the reference data by the standard error of prediction. Since the SD values for calibration and validation data set are constants, the RPD value will not change the ranking of best performing indicators. However, the RPD is a dimensionless parameter and can be classified to different categories to evaluate the model accuracy. Table 9 exhibited the RPD values in some of the estimation models. All the RPDs in the calibration set and validation set exceeded 2.0 except the OLS regression model based on the CRS_{725}, which indicated all the estimation models in Table 9 had stable and accurate predictive abilities. The PLS model with AC_4 produced the best performance (RPD_c = 3.97 and RPD_v = 3.95), and followed by the RF regression model with AC_4 (RPD_c = 3.04 and RPD_v = 3.29). Overall, by comparing all the methods in this article with statistical indicators of R^2, RMSE, RE and RPD, an integrated approach using DWT ACs and PLS regression exhibited the highest stability and reliability in LNC estimation.

Table 9. The ratio of prediction to deviation (RPD) values in the calibration models and validation models of leaf nitrogen concentration.

Model	OLS Regression		PLS Regression			RF Regression	
	CRS_{725}	mSR_{705}	AC_4	DC_6	EV_{10}	AC_4	EV_{10}
RPD_c	1.95	2.43	3.97	2.81	2.61	3.04	2.38
RPD_v	2.26	2.82	3.95	3.04	3.04	3.29	2.72

4. Discussion

A large number of studies have been conducted with passive and active remote sensing technologies for the timely and non-destructive evaluation of LNC [9–11,13–17,50]. In the present study, we mainly discussed the effect of the entire range of canopy reflectance (400–1350 nm) on LNC estimation and provided a guide for feature extraction by DWT analysis and PLS regression.

4.1. Sensitive Band Reflectance and Spectral Transformation

This study demonstrated that spectral measurements were apparently useful for describing the N status of wheat canopies. It is known that pure chlorophyll a and b had absorption peaks at red and blue wavelength regions, respectively. The red edge (680–760 nm) caused by the strong absorption of pigments in the red spectrum and leaf scattering in the NIR spectrum has been found to be sensitive to crop growth [12,14,16–18]. As N concentration was linked to the plant photosynthetic pigments concentration especially chlorophyll, the correlation between leaf nitrogen concentration and spectral reflectance in visible light was better than in the near-infrared region as shown in Figure 2. All the sensitive wavebands of spectral transformations were located in visible light (OS at 640 nm, FDS at 447 nm, LOGS at 645 nm and CRS at 725 nm). In previous research, the first derivative was closely related to N concentration in corn and wheat [20], which was designed to eliminate background signals or noise and to resolve overlapping spectral features. The LOG transformation performed accurately compared with the original reflectance when estimating N concentrations [14]. CRS yielded the highest accuracy in estimating grass leaf nitrogen concentrations, followed by the LOGS [51]. Continuum removal enhanced the differences in absorption strength [34]. All indicated that spectral transformation could provide more sensitive features than canopy original spectrum and could be used to increase the accuracy of crop N estimation. We compared the LNC estimation accuracy of these four spectral transformations in this study. The result showed the performance of CRS (R_c^2 = 0.79) was higher than that of LOGS, FDS and OS (Figure 3) in winter wheat LNC prediction. The same conclusion was obtained in the grass foliar nitrogen retrieval reported by Ramoelo et al. [51], where an R^2 of 0.81 was based on a greenhouse experiment using continuum removal in combination with a PLS regression. As a whole, the exponential model was more suitable for delineating the quantitative relationship between sensitive band reflectance and LNC than a linear regression (Figure 3). This may be caused by the fact that the relationship between leaf N and chlorophyll concentration was not linear [10,52].

4.2. Relationship Between Spectral Indices and LNC

Canopy spectra obtained by remote sensing were affected by the canopy structure and surrounding conditions, while the spectral index could eliminate the impacts at a certain extent through the combination of characteristic bands [12]. Reflectance in the red-edge region (680–760 nm) was closely related to the chlorophyll content in plants as well as the nutritional status of plants, which had always been considered to be important in relationships with biochemical or biophysical parameters [53]. The mSR_{705} index of chlorophyll indices, NDRE index of nitrogen indices and GNDVI of greenness indices in this paper were constructed on the basis of red edge reflectance, could explain 83%, 80% and 81% of the variability in LNC, respectively (Table 3), and had better performance than other spectral indices. This was consistent with the report that the GNDVI performed similarly as NDRE in estimating maize N concentration [52]. Green and red edge reflectance were sensitive to a wider range of chlorophyll levels than red reflectance. The predictive ability of NDRE in the category of nitrogen indices was higher than with NRI and DCNI in this study (Table 3). NDRE is similar in form to NDVI, but with the red band being replaced by a red edge band. NRI, a normalized near-infrared over green waveband reflectance ratio, has been used to assess in-season corn N status and to develop N variability maps [37,54], but in our study, the predictive ability (R_c^2 = 0.50) of NRI was lower than NDRE (R_c^2 = 0.80). These results demonstrated the importance of red edge vegetation indices for estimating winter wheat N status.

Three-band spectral indices were proposed to solve the saturation problem associated with two-band indices [18,55,56]. Chen et al. [20] developed the three-band spectral index DCNI using the double-peak characteristics of the red edge to predict the nitrogen content of maize and wheat showed the determination coefficient of the prediction equation being 0.72 and 0.44, respectively. However, the DCNI (R_c^2 = 0.63) explained 20% less variability in LNC than mSR_{705} in this study (Table 3). The RMSE of mSR_{705} ($RMSE_v$ = 0.28) indicated an ordinary performance of the LNC prediction model, and an $RMSE_v$ of 0.41 indicated poor model performance of DCNI [48].

4.3. Features and Parameters Selection of the DWT Analysis

The mother wavelet function and decomposition level were two crucial parameters required for DWT analysis. The scaled and translated mother wavelet was used to fit the canopy spectra. Cocchi et al. [57] demonstrated the difficulty in developing a priori rules for identifying the most appropriate mother wavelet because the optimum mother wavelet changed with the particular task [58]. Five commonly used mother wavelets in vegetation spectra decomposition were tested in this study. Results showed that the mother wavelet sym8 was excellent in dimensionality reduction and signal reconstruction than other wavelets. Features based on the sym8 made a good LNC estimation accuracy (Tables 5–7), which indicated the shape of the wavelet sym8 could explicate the differences between the LNC.

Three DWT features (AC, DC and EV) were extracted to analyze the relationship with LNC. The wavelet low-frequency approximation coefficient (AC) was a reflection of global features in canopy spectra, while the high-frequency detail coefficient (DC) was a depiction of noise information, which together contained all of the information presented in the original spectrum. Energy value (EV) was a set of compressed ACs and DCs, which tried to take advantage of ACs and DCs to express the whole signal information. Blackburn and Ferwerda [59] used the approximation coefficients at level 8 along with detail coefficients from level 1 to 8 to estimate chlorophyll concentration. However, most of the characteristic information of canopy spectra was in the approximation coefficients at a lower decomposition scale. For their study, DCs were very small in amplitude and could be removed without major loss in the information content of the signal kept in ACs. After a certain decomposition level, more and more useful information would be eliminated, contributing to noisy signals and declining information content of canopy reflectance spectra [59]. In contrast, we found that ACs in decomposition level 1 to 5 preserved almost 100% of the information features for the canopy spectra (Figure 5). Our results also illustrated that LNC was much better correlated with the main information in ACs, not DCs (Table 5; Table 6), so the ACs in L_1 to L_5 could be used directly to predict the LNC, instead of using the whole reflectance spectra, while ignoring the high-frequency DCs.

Energy value achieved a further data compression of canopy spectral signals. Pu and Gong (2004) [28] indicated that the energy value features extracted by the WT method were the most effective way of mapping forest crown closure (CC) and leaf area index (LAI). The mapped accuracy for CC was 84.90% and for LAI was 75.39%. In this study, the EVs at AC_4 with PLS regression could explain 83% (R_c^2 = 0.83) of the variability in LNC. EVs had a poorer performance and ability to estimate LNC compared with the ACs. This might be due to excessive dimensionality reduction, resulting in a loss of some sensitive information in the canopy spectrum.

Our results showed that the wavelet coefficients were different depending upon whether they were derived from the reflectance spectra or transformed spectra. The differences indicated that transformed spectra were more sensitive to LNC than original reflectance spectra (Tables 5–8). The LOG-transformation was more useful in extracting additional information that was more difficult to obtain from other transforms of reflectance spectra.

4.4. Estimation Models of LNC

The results of this study indicated that the combination of multi-spectral bands generally improved the accuracy of LNC estimation (Table 4). The hyper-spectral narrow-band index (mSR_{705}) explained 4% more variability in LNC estimation than the best performing single sensitive band CRS_{725}. The advantage of multi-variable regression was obvious (Table 9). Approximation coefficients at decomposition level 4 using LOG-transformed spectra had the best prediction accuracy in PLS-LNC models. The prediction accuracy of the RF-LNC model with LOGS at L_4 was similar to the PLS-LNC model, while the R^2, RMSE, RPD and RE of validation set were slightly worse than the PLS regression (Tables 8 and 9). That is, the prediction and verification accuracy of the PLS model was more stable than that of the RF model. This is likely because the accuracy of the RF model was greatly influenced by the undefined input parameters, although it was efficient for large input variables and non-linear problems [42–45].

All the RPD values of validation models in Table 9 being greater than 2.0 indicated good LNC prediction models were built with different approaches. However, the OLS model based on the sensitive band reflectance should be noted that higher LNC values were underestimated (Figure 4). Considering the difficulty of feature extraction with wavelets, spectral index mSR_{705} might alternatively be suggested to predict LNC, especially in multi-spectral remote sensing applications. Wavelet analysis of a reflectance spectrum was performed by scaling and shifting the wavelet function to produce wavelet coefficients that were assigned to different frequency components. It made the DWT analysis to have the potential to capture much more of the information contained within the canopy hyper-spectra [25,28,30,57–59]. Our results showed that using DWT coefficients and PLS regression together could overcome the limitations of individual variable technology and offer a practical approach to LNC detection. The model produced by using AC_4 with PLS regression had the best performance (RPD_c = 3.97 and RPD_v = 3.95) and was recommended for LNC estimation across all growth stages.

4.5. Research Challenges

In this study, the performance of SR-LNC and VI-LNC models were analyzed in detail and compared with models by DWT analysis combing with PLS and RF regression. The LOGS combined with DWT ACs and PLS regression made good use of the full canopy reflectance spectra and produced good prediction accuracy of LNC, but it was difficult to interpret exactly which wavelength was contributing to the best performing models. It remains a particular challenge to test the performance of more mother wavelets. We also need to find whether this method could be successfully applied and whether it works well across various growth stages, varieties and eco-sites for estimation of LNC, and whether canopy spectra information can be used to detect the LNC status of the crop as precisely as the nitrogen nutrition index (NNI) approach.

5. Conclusions

Canopy spectra measurements were useful for estimating the nitrogen status of a wheat crop, thereby providing information to help decide on nitrogen fertilizer application in precision farming systems. The results of this study demonstrated that sensitive band reflectance of transformation canopy spectra and spectral indices gave a better correlation for LNC than the correlation using the original canopy spectra. DWT analysis accomplished feature extraction successfully from the narrow-band hyperspectral canopy spectrum across VIS and NIR wavelengths on the basis of keeping original spectrum information quality and reducing canopy spectral data space dimensions. Combining LOGS and the sym8 mother wavelet, approximation coefficients at the 4th decomposition level provided the best approach for estimating LNC by a PLS regression. This approach could explain 11% more variability in LNC than the corresponding best performing spectral index mSR_{705} and was more stable in LNC estimation than the RF regression.

Author Contributions: F.L., J.L. and Q.C. conceived and designed the experiment. L.W., Y.W., F.L. and J.L. conducted the experiment. F.L. conducted the data analysis and prepared the manuscript. Q.C. and J.L. revised the manuscript. All authors read and approved the final version.

Funding: This research was funded by the National Natural Science Foundation of China (41701398), the Fundamental Research Funds for the Central Universities (2452017108), and the Hi-Tech Research and Development Program (863) of China (2013AA102401-2).

Conflicts of Interest: The authors declare no conflict of interest.

References

1. Singh, B.; Singh, G. Effects of controlled irrigation on water potential, nitrogen uptake and biomass production in Dalbergia sissoo seedlings. *Environ. Exp. Bot.* **2006**, *55*, 209–219. [CrossRef]
2. Wang, H.; Guo, Z.; Shi, Y.; Zhang, Y.; Yu, Z. Impact of tillage practices on nitrogen accumulation and translocation in wheat and soil nitrate-nitrogen leaching in drylands. *Soil Till. Res.* **2015**, *153*, 20–27. [CrossRef]
3. Banedjschafie, S.; Bastani, S.; Widmoser, P.; Mengel, K. Improvement of water use and N fertilizer efficiency by subsoil irrigation of winter wheat. *Eur. J. Agron.* **2008**, *28*, 1–7. [CrossRef]
4. Munoz-Huerta, R.F.; Guevara-Gonzalez, R.G.; Contreras-Medina, L.M.; Torres-Pacheco, I.; Prado-Olivarez, J.; Ocampo-Velazquez, R.V. A review of methods for sensing the nitrogen status in plants: Advantages, disadvantages and recent advances. *Sensors* **2013**, *13*, 10823–10843. [CrossRef]
5. Vigneau, N.; Ecarnot, M.; Rabatel, G.; Roumet, P. Potential of field hyperspectral imaging as a non destructive method to assess leaf nitrogen content in Wheat. *Field Crop Res.* **2011**, *122*, 25–31. [CrossRef]
6. Miao, Y.; Mulla, D.J.; Randall, G.W.; Vetsch, J.A.; Vintila, R. Combining chlorophyll meter readings and high spatial resolution remote sensing images for in-season site-specific nitrogen management of corn. *Precis. Agric.* **2009**, *10*, 45–62. [CrossRef]
7. Curran, P.J. Remote sensing of foliar chemistry. *Remote Sens. Environ.* **1989**, *30*, 271–278. [CrossRef]

8. Fourty, T.; Baret, F.; Jacquemoud, S.; Schmuck, G.; Verdebout, J. Leaf optical properties with explicit description of its biochemical composition: direct and inverse problems. *Remote Sens. Environ.* **1996**, *56*, 104–117. [CrossRef]
9. Poorter, H.; Evans, J.R. Photosynthetic nitrogen-use efficiency of species that differ inherently in specific leaf area. *Oecologia* **1998**, *116*, 26–37. [CrossRef]
10. Schepers, J.S.; Francis, D.D.; Vigil, M.; Below, F.E. Comparison of corn leaf nitrogen concentration and chlorophyll meter readings. *Commun. Soil Sci. Plan.* **1992**, *23*, 2173–2187. [CrossRef]
11. Kokaly, R.F. Investigating a physical basis for spectroscopic estimates of leaf nitrogen concentration. *Remote Sens. Environ.* **2001**, *75*, 153–161. [CrossRef]
12. Gitelson, A.A.; Vina, A.; Ciganda, V.; Rundquist, D.C.; Arkebauer, T.J. Remote estimation of canopy chlorophyll content in crops. *Geophys. Res. Lett.* **2005**, *32*, 93–114. [CrossRef]
13. Feng, W.; Zhang, H.Y.; Zhang, Y.S.; Qi, S.L.; Heng, Y.R.; Guo, B.B.; Ma, D.Y.; Guo, T.C. Remote detection of canopy leaf nitrogen concentration in winter wheat by using water resistance vegetation indices from in-situ hyperspectral data. *Field Crop Res.* **2016**, *198*, 238–246. [CrossRef]
14. Yoder, B.J.; Pettigrew-Crosby, R.E. Predicting nitrogen and chlorophyll content and concentrations from reflectance spectra (400-2500 nm) at leaf and canopy scales. *Remote Sens. Environ.* **1995**, *53*, 199–211. [CrossRef]
15. Mutanga, O.; Skidmore, A.K.; van Wieren, S. Discriminating tropical grass (Cenchrus ciliaris) canopies grown under different nitrogen treatments using spectroradiometry. *ISPRS J. Photogramm.* **2003**, *57*, 263–272. [CrossRef]
16. Tian, Y.C.; Yao, X.; Yang, J.; Cao, W.X.; Hannaway, D.B.; Zhu, Y. Assessing newly developed and published vegetation indices for estimating rice leaf nitrogen concentration with ground-and space-based hyperspectral reflectance. *Field Crop Res.* **2011**, *120*, 299–310. [CrossRef]
17. Inoue, Y.; Sakaiya, E.; Zhu, Y.; Takahashi, W. Diagnostic mapping of canopy nitrogen content in rice based on hyperspectral measurements. *Remote Sens. Environ.* **2012**, *126*, 210–221. [CrossRef]
18. Feng, W.; Guo, B.B.; Wang, Z.J.; He, L.; Song, X.; Wang, Y.H.; Guo, T.C. Measuring leaf nitrogen concentration in winter wheat using double-peak spectral reflection remote sensing data. *Field Crop Res.* **2014**, *159*, 43–52. [CrossRef]
19. Sims, D.A.; Gamon, J.A. Relationships between leaf pigment content and spectral reflectance across a wide range of species, leaf structures and developmental stages. *Remote Sens. Environ.* **2002**, *81*, 337–354. [CrossRef]
20. Chen, P.; Haboudane, D.; Tremblay, N.; Wang, J.; Vigneault, P.; Li, B. New spectral indicator assessing the efficiency of crop nitrogen treatment in corn and wheat. *Remote Sens. Environ.* **2010**, *114*, 1987–1997. [CrossRef]
21. Yao, X.; Huang, Y.; Shang, G.; Zhou, C.; Cheng, T.; Tian, Y.; Cao, W.; Zhu, Y. Evaluation of six algorithms to monitor wheat leaf nitrogen concentration. *Remote Sens.* **2015**, *7*, 14939–14966. [CrossRef]
22. Wang, H.F.; Huo, Z.G.; Zhou, G.S.; Liao, Q.H.; Feng, H.K.; Wu, L. Estimating leaf SPAD values of freeze-damaged winter wheat using continuous wavelet analysis. *Plant Physiol. Bioch.* **2016**, *98*, 39–45. [CrossRef] [PubMed]
23. Quandt, V.I.; Pacola, E.R.; Pichorim, S.F.; Gamba, H.R.; Sovierzoski, M.A. Pulmonary crackle characterization: approaches in the use of discrete wavelet transform regarding border effect, mother-wavelet selection, and subband reduction. *Res. Biomed. Eng.* **2015**, *31*, 148–159. [CrossRef]
24. Chang, T.; Kuo, C.C. Texture analysis and classification with tree-structured wavelet transform. *IEEE Trans. Image Process.* **1993**, *2*, 429–441. [CrossRef] [PubMed]
25. Bruce, L.M.; Koger, C.H.; Li, J. Dimensionality reduction of hyperspectral data using discrete wavelet transform feature extraction. *IEEE Trans. Geosci. Remote* **2002**, *40*, 2331–2338. [CrossRef]
26. Sahadevan, A.S.; Shrivastava, P.; Das, B.S.; Sarathjith, M.C. Discrete wavelet transform approach for the estimation of crop residue mass from spectral reflectance. *IEEE J.-STARS* **2014**, *7*, 2490–2495. [CrossRef]
27. Shi, Y.; Huang, W.; Zhou, X. Evaluation of wavelet spectral features in pathological detection and discrimination of yellow rust and powdery mildew in winter wheat with hyperspectral reflectance data. *J. Appl. Remote Sens.* **2017**, *11*, 026025. [CrossRef]
28. Blackburn, G.A. Wavelet decomposition of hyperspectral data: a novel approach to quantifying pigment concentrations in vegetation. *Int. J. Remote Sens.* **2007**, *28*, 2831–2855. [CrossRef]

29. Peng, J.; Shen, H.; Wu, J.S. Soil moisture retrieving using hyperspectral data with the application of wavelet analysis. *Environ. Earth Sci.* **2013**, *69*, 279–288. [CrossRef]
30. Pu, R.; Gong, P. Wavelet transform applied to EO-1 hyperspectral data for forest LAI and crown closure mapping. *Remote Sens. Environ.* **2004**, *91*, 212–224. [CrossRef]
31. Bremner, J.M. Determination of nitrogen in soil by the Kjeldahl method. *J. Agric. Sci.* **1960**, *55*, 11–33. [CrossRef]
32. Savitzky, A.; Golay, M.J. Smoothing and differentiation of data by simplified least squares procedures. *Anal. Chem.* **1964**, *36*, 1627–1639. [CrossRef]
33. Dawson, T.P.; Curran, P.J.; Plummer, S.E. LIBERTY-Modeling the effects of leaf biochemical concentration on reflectance spectra. *Remote Sens. Environ.* **1998**, *65*, 50–60. [CrossRef]
34. Clark, R.N.; Roush, T.L. Reflectance spectroscopy: Quantitative analysis techniques for remote sensing applications. *J. Geophys. Res.-Solid Earth* **1984**, *89*, 6329–6340. [CrossRef]
35. Daughtry, C.S.T.; Walthall, C.L.; Kim, M.S.; De Colstoun, E.B.; McMurtrey Iii, J.E. Estimating corn leaf chlorophyll concentration from leaf and canopy reflectance. *Remote Sens. Environ.* **2000**, *74*, 229–239. [CrossRef]
36. Penuelas, J.; Baret, F.; Filella, I. Semi-empirical indices to assess carotenoids / chlorophyll a ratio from leaf spectral reflectance. *Photosynthetica* **1995**, *31*, 221–230.
37. Filella, I.; Serrano, L.; Serra, J.; Penuelas, J. Evaluating wheat nitrogen status with canopy reflectance indices and discriminant analysis. *Crop Sci.* **1995**, *35*, 1400–1405. [CrossRef]
38. Barnes, E.M.; Clarke, T.R.; Richards, S.E.; Colaizzi, P.D.; Haberland, J.; Kostrzewski, M.; Waller, P.; Choi, C.; Riley, E.; Thompson, T.; et al. Coincident detection of crop water stress, nitrogen status and canopy density using ground based multispectral data. In Proceedings of the 5th International Conference on Precision Agriculture, Bloomington, MN, USA, 16–19 July 2000.
39. Gitelson, A.A.; Kaufman, Y.J.; Merzlyak, M.N. Use of a green channel in remote sensing of global vegetation from EOS-MODIS. *Remote Sens. Environ.* **1996**, *58*, 289–298. [CrossRef]
40. Rondeaux, G.; Steven, M.; Baret, F. Optimization of soil-adjusted vegetation indices. *Remote Sens. Environ.* **1996**, *55*, 95–107. [CrossRef]
41. Haboudane, D.; Miller, J.R.; Pattey, E.; Zarco-Tejada, P.J.; Strachan, I.B. Hyperspectral vegetation indices and novel algorithms for predicting green LAI of crop canopies: Modeling and validation in the context of precision agriculture. *Remote Sens. Environ.* **2004**, *90*, 337–352. [CrossRef]
42. Ecarnot, M.; Compan, F.; Roumet, P. Assessing leaf nitrogen content and leaf mass per unit area of wheat in the field throughout plant cycle with a portable spectrometer. *Field Crop Res.* **2013**, *140*, 44–50. [CrossRef]
43. Geladi, P.; Kowalski, B.R. Partial least-squares regression: a tutorial. *Anal. Chim. ACTA* **1986**, *185*, 1–17. [CrossRef]
44. Breiman, L. Random forests. *Mach. Learn.* **2001**, *45*, 5–32. [CrossRef]
45. Liaw, A.; Wiener, M. Classification and regression by randomForest. *R News* **2002**, *2*, 18–22.
46. Wang, J.; Ding, J.; Abulimiti, A.; Cai, L. Quantitative estimation of soil salinity by means of different modeling methods and visible-near infrared (VIS–NIR) spectroscopy, Ebinur Lake Wetland, Northwest China. *PeerJ* **2018**, *6*, e4703. [CrossRef] [PubMed]
47. Nguyen, H.T.; Lee, B.W. Assessment of rice leaf growth and nitrogen status by hyperspectral canopy reflectance and partial least square regression. *Eur. J. Agron.* **2006**, *24*, 349–356. [CrossRef]
48. Zhu, Y.; Yao, X.; Tian, Y.; Liu, X.; Cao, W. Analysis of common canopy vegetation indices for indicating leaf nitrogen accumulations in wheat and rice. *Int. J. Appl. Earth Obs.* **2008**, *10*, 1–10. [CrossRef]
49. Jahn, B.R.; Brooksby, P.A.; Upadhyaya, S.K. Wavelet-based spectral analysis for soil nitrate content measurement. *Trans. ASAE* **2005**, *48*, 2065–2071. [CrossRef]
50. Cao, Q.; Miao, Y.; Feng, G.; Gao, X.; Li, F.; Liu, B.; Yue, S.; Cheng, S.; Ustin, S.L.; Khosla, R. Active canopy sensing of winter wheat nitrogen status: An evaluation of two sensor systems. *Comput. Electron Agric.* **2015**, *112*, 54–67. [CrossRef]
51. Ramoelo, A.; Skidmore, A.K.; Schlerf, M.; Mathieu, R.; Heitkonig, I.M. Water removed spectra increase the retrieval accuracy when estimating savanna grass nitrogen and phosphorus concentrations. *ISPRS J. Photogramm.* **2011**, *66*, 408–417. [CrossRef]

52. Li, F.; Miao, Y.; Feng, G.; Yuan, F.; Yue, S.; Gao, X.; Liu, Y.; Liu, B.; Ustin, S.L.; Chen, X. Improving estimation of summer maize nitrogen status with red edge-based spectral vegetation indices. *Field Crop Res.* **2014**, *157*, 111–123. [CrossRef]
53. Cho, M.A.; Skidmore, A.K. A new technique for extracting the red edge position from hyperspectral data: The linear extrapolation method. *Remote Sens. Environ.* **2006**, *101*, 181–193. [CrossRef]
54. Diker, K.; Bausch, W.C. Potential use of nitrogen reflectance index to estimate plant parameters and yield of maize. *Biosyst. Eng.* **2003**, *85*, 437–447. [CrossRef]
55. Hunt, E.R.; Daughtry, C.S.T.; Eitel, J.U.; Long, D.S. Remote sensing leaf chlorophyll content using a visible band index. *Agron. J.* **2011**, *103*, 1090–1099. [CrossRef]
56. Wang, W.; Yao, X.; Yao, X.; Tian, Y.; Liu, X.; Ni, J.; Cao, W.; Zhu, Y. Estimating leaf nitrogen concentration with three-band vegetation indices in rice and wheat. *Field Crop Res.* **2012**, *129*, 90–98. [CrossRef]
57. Cocchi, M.; Seeber, R.; Ulrici, A. Multivariate calibration of analytical signals by WILMA (wavelet interface to linear modelling analysis). *J. Chemometr.* **2003**, *17*, 512–527. [CrossRef]
58. Ferwerda, J.G.; Jones, S.D. Continuous wavelet transformations for hyperspectral feature detection. In *Progress in Spatial Data Handling*; Springer: Berlin/Heidelberg, Germany, 2006.
59. Blackburn, G.A.; Ferwerda, J.G. Retrieval of chlorophyll concentration from leaf reflectance spectra using wavelet analysis. *Remote Sens. Environ.* **2008**, *112*, 1614–1632. [CrossRef]

remote sensing

MDPI

Article

Analysis of Chlorophyll Concentration in Potato Crop by Coupling Continuous Wavelet Transform and Spectral Variable Optimization

Ning Liu [1,2], Zizheng Xing [1], Ruomei Zhao [1], Lang Qiao [1], Minzan Li [1,2], Gang Liu [2] and Hong Sun [1,*]

1 Key Laboratory of Modern Precision Agriculture System Integration Research, Ministry of Education, China Agricultural University, Beijing 100083, China; ningliu@cau.edu.cn (N.L.); S20183081362@cau.edu.cn (Z.X.); S20193081422@cau.edu.cn (R.Z.); b20193080667@cau.edu.cn (L.Q.); limz@cau.edu.cn (M.L.)

2 Key Laboratory of Agricultural Information Acquisition Technology, Ministry of Agriculture and Rural Affairs, China Agricultural University, Beijing 100083, China; pac@cau.edu.cn

* Correspondence: sunhong@cau.edu.cn; Tel.: +86-135-5272-6986

Received: 21 July 2020; Accepted: 30 August 2020; Published: 31 August 2020

Abstract: The analysis of chlorophyll concentration based on spectroscopy has great importance for monitoring the growth state and guiding the precision nitrogen management of potato crops in the field. A suitable data processing and modeling method could improve the stability and accuracy of chlorophyll analysis. To develop such a method, we collected the modelling data by conducting field experiments at the tillering, tuber-formation, tuber-bulking, and tuber-maturity stages in 2018. A chlorophyll analysis model was established using the partial least-square (PLS) algorithm based on original reflectance, standard normal variate reflectance, and wavelet features (WFs) under different decomposition scales (2^1–2^{10}, Scales 1–10), which were optimized by the competitive adaptive reweighted sampling (CARS) algorithm. The performances of various models were compared. The WFs under Scale 3 had the strongest correlation with chlorophyll concentration with a correlation coefficient of −0.82. In the model calibration process, the optimal model was the Scale3-CARS-PLS, which was established based on the sensitive WFs under Scale 3 selected by CARS, with the largest coefficient of determination of calibration set (R_c^2) of 0.93 and the smallest $R_c^2 - R_{cv}^2$ value of 0.14. In the model validation process, the Scale3-CARS-PLS model had the largest coefficient of determination of validation set (R_v^2) of 0.85 and the smallest root–mean–square error of cross-validation (RMSEV) value of 2.77 mg/L, demonstrating good prediction capability of chlorophyll concentration. Finally, the analysis performance of the Scale3-CARS-PLS model was measured using the testing data collected in 2020; the R^2 and RMSE values were 0.69 and 3.36 mg/L, showing excellent applicability. Therefore, the Scale3-CARS-PLS model could be used to analyze chlorophyll concentration. This study indicated the best decomposition scale of continuous wavelet transform and provided an important support method for chlorophyll analysis in the potato crops.

Keywords: standard normal variate (SNV); continuous wavelet transform (CWT); wavelet features optimization; competitive adaptive reweighted sampling (CARS); partial least square (PLS)

1. Introduction

Potato (*Solanum tuberosum*) is the world's fourth-largest food crop following rice, wheat, and maize [1,2]. Chlorophyll, as the essential photosynthetic pigment of potato leaves, reflects growth information about plant health [3] and photosynthetic rate [4], and its content is also significantly correlated with the concentration of nitrogen [5]. Therefore, the accurate analysis of the chlorophyll concentration of potato plants is of great importance for nitrogen management in precision agriculture.

Compared with the time-consuming and laborious chemical analysis of chlorophyll concentration [6,7], the modern spectroscopy analysis, as a non-destructive and rapid monitoring method, has advantages in the inversion of chlorophyll concentration of crops due to the principle of light absorption by molecular or chemical bonding [8,9]. At present, the crop analysis method based on spectroscopy primarily includes proximal spectroscopy analysis and remote sensing [10]. The former has advantages of high resolution and accurate data sampling [11]. Thus, it is suitable for the spectroscopy mechanism studies (e.g., the characteristic absorption bands of some material components) and the development of analysis algorithms, thereby laying a foundation for methods of large-scale and large-area remote sensing [12]. Thus, the motivation of this study is to accurately analyze the chlorophyll concentration in potato crops based on proximal spectroscopy.

Three major problems of crop chlorophyll content analysis based on proximal spectroscopy methods include spectral signal-noise reduction, characteristic variable analysis, and analysis model establishment [13,14]. Among them, the noise reduction of spectral data is the primary step to improve the spectral data performance. During spectral data collection, especially in the field environment, noises such as high-frequency noises [15] and scattering effects [16] are inevitably introduced [17]. Accordingly, previous studies have reported that the standard normal variate (SNV) can effectively correct the scattering effect resulting from different light reflection paths to improve the predictive capability of spectral data [18]. The Savitzky–Golay (S-G) smoothing method can reduce the high-frequency noise of spectral data resulting from instrument vibration or electromagnetic interference [19]. However, the main disadvantage of the S-G method is that the smoothing window size is not fixed, which requires complex optimization according to specific spectral data to select the optimal window size [20,21].

Regarding characteristic variable analysis, many methods have been developed to improve accuracy of chlorophyll concentration analysis [22]. One technique is to build a spectral reflectance index; for instance, the normalized difference vegetation index [23], chlorophyll index [24], weighted-difference vegetation index [25], and structural independent pigment index [26] are used to estimate the leaf chlorophyll concentration. Yu [27] reported that the reflectance ratio vegetation index could eliminate the influence of structural difference of wheat canopy on chlorophyll analysis. Another method is to select sensitive wavelengths; for instance, the red edge and characteristic absorption wavelengths are used to analyze chlorophyll concentration. Sun et al. [28] analyzed the spectral migration characteristics of winter wheat at jointing, booting, flowering, and milk-ripening stages. The wavelengths at red edge positions were extracted to establish the chlorophyll analysis models. Five sensitive wavelengths, namely, 680, 716, 1104, 1882, and 1920 nm, were selected to establish the model for chlorophyll and water-content detection.

However, the above methods cannot completely remove noises and present the features of the spectra [29]. Continuous wavelet transformation (CWT) has outstanding quality of time and frequency domain and can decompose a spectrum into numerous wavelet features (WFs) to effectively characterize spectral signals and eliminate high-frequency noises of spectral data [30]. Previous studies [31–33] have reported that continuous wavelet analysis achieves good performance on crop growth-parameter estimation. Li [32] indicated that WFs under 2^3, 2^4, and 2^5 (middle- and low-frequency) scales could reduce the phenomenon of "fingerprint spectrum" with serious vibration noises to improve the analysis accuracy for the leaf nitrogen concentration of wheat and rice crops. The analysis accuracy was higher than the normalized difference vegetation index. Lu et al. [34] indicated that the sensitive WFs of stripe rust and powdery mildew of wheat are distributed in the 2^2, 2^3, and 2^4 scales, and that the WFs could capture the pigment and water content in wheat leaf. These studies show that WFs under middle- and low-frequency scale factors can capture the peak and valley of an absorption feature of physical and chemical materials [33].

In terms of CWT application, some points remain unclear, especially chlorophyll concentration analysis of potato crop during different stages by using spectroscopy combined with the CWT method. Meanwhile, the partial least-square (PLS) regression model is used to explore and evaluate the relationship between scales of WFs and chlorophyll concentration [35,36], which can solve multicollinearity problems [37] among variables by executing a principal component analysis on the independent and dependent variable matrices. Occasionally, the PLS regression model contains uninformative variables, which result in poor prediction accuracy, overfitting phenomenon [38], and further lowering of the model stability. Some studies have reported that in contrast to the correlation analysis [39] and successive projection algorithm [40] methods, the competitive adaptive reweighted sampling (CARS) algorithm, serving as a sensitive wavelength-selection algorithm [41], can improve the performance of the PLS regression model by eliminating invalid variables [42,43]. Thus, we attempt herein to establish a high-performance chlorophyll content analysis model using the CARS-PLS method.

Accordingly, the study aimed to discuss the improvement in spectral data analysis performance by CWT. We focused on the effect of WFs under different decomposition scales on identifying valuable spectral variables and reducing high-frequency noise to enhance the analysis accuracy of potato chlorophyll concentration during growth periods. Combined with SNV correction, we proposed a CWT-CARS-PLS method to establish the high-performance chlorophyll content analysis model. In this model, CWT was used to eliminate the high-frequency noise and extract the valuable spectral variables, and CARS was applied to select WF variables. The effectiveness of CWT in the analysis ability of dynamic estimation of potato chlorophyll concentration was highlighted by comparing and analyzing the model results of original reflectance (Ref) and SNV reflectance (SNV).

The objectives of this study were to (1) determine the dynamic relationships between chlorophyll concentration and canopy spectra at different growth stages; (2) compare the chlorophyll concentration analysis capability of the WFs under different decomposition scales, Ref, and SNV reflectance; (3) use CARS to select sensitive variables and establish various CARS-PLS analysis models; and (4) validate and evaluate the performance of PLS and CARS-PLS models established by Ref, SNV, and different scale WFs.

2. Materials and Methods

2.1. Data Acquisition

2.1.1. Spectral Data Collection

Experiments were conducted in 2018 and 2020, respectively. Collected data in 2018 were used to develop the spectral variable optimization method and propose the chlorophyll analysis model by CWT. Measured data in 2020 were used to test the proposed method for potato chlorophyll analysis.

In 2018, field experiments were conducted at the National Precision Agriculture Experiment Station in Xiaotangshan, Beijing, China (40°16′25″ N, 116°44′03″ E). The potato crop was planted on 10 April 2018. According to management practices by farmers, the total N rate was 400 kg N ha^{-1}, with 12% applied at tillering stage, 33% at tuber formation stage, 38% at 127 tuber expansion stage, and the remaining 12% N at tuber maturation stage. Although the growth period of the Atlantic cultivar is about 90 days in Beijing, the canopy changes greatly from the tillering to the tuber maturation stages, after that the leaves of the crop turns to yellow. In order to establish a chlorophyll analysis model to analyze the chlorophyll concentration of potato canopy, as shown in Table 1, spectra data were collected at four growth stages on May 15, May 24, June 7 and June 19, respectively, which were the tillering stage with appearing flower buds (S1), tuber formation stage with flowers (S2), tuber expansion stage after flowers fell (S3), and tuber maturation stage during leaves turning yellow (S4). Eighty plots with a size of 1 m × 1 m were used. Figure 1 shows the location of the field and photos of potato crops for different growth stages. From each growth stage, 80 groups of data were collected, in which 6 groups were invalid because of the influence of low vegetation coverage. Thus, 74 groups of data were retained at S1. Thus, the modeling dataset had a total of 314 groups of reflectance spectra.

Table 1. Information of potato crop samples.

Growth Stage	Potato-Crop Characteristics	Samples	
		Modelling	Testing
S1	Appearing flower buds, having about 12 leaves	74	40
S2	Appearing flowers	80	40
S3	Flowers falling, stems and leaves aging	80	40
S4	Stems and leaves withering, upper leaves turning yellow	80	40

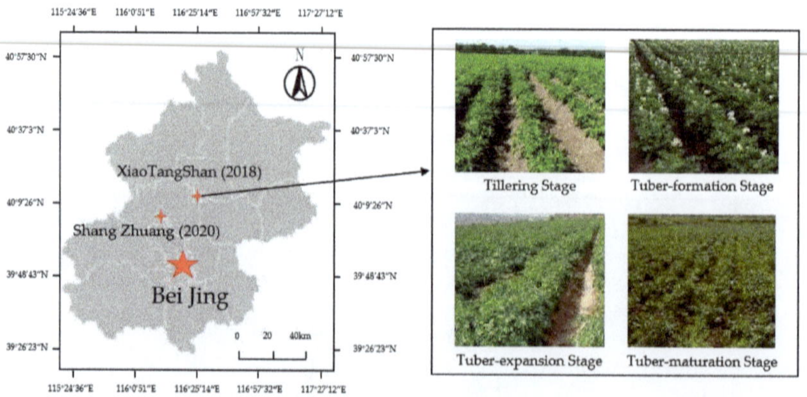

Figure 1. Location of the field and photos of potato crops for different growth stages.

In 2020, the testing experiment was conducted at the Shang Zhuang Experiment Station of China Agricultural University in Beijing, China (40°08′12″ N, 116°10′44″ E), as shown in Figure 1. The cultivar of the potato crop was Dutch. Due to the epidemic influence in the spring of 2020, the potato crop was planted on 5 June 2020, almost two months later than in 2018, and spectral data were collected on July 11, July 21, July 30 and August 12, respectively. N application and field management practices were similar to experiments in 2018. In addition, the sampling and data collection methods were also the same as in 2018. Although the growth stages of experiments might not exactly match the stages in 2018, collected data could be used to test the effectiveness of proposed methods on the analysis of chlorophyll concentration in potato canopy. Thus, a total of 160 samples were collected from four growth stages as a testing dataset in this paper. Details about the potato growth stages and sampling dates are given in Table 1.

Regarding spectral measurements and leaf sampling, one potato plant was randomly selected in each plot, for which canopy spectral data were collected three times and the average value was calculated to represent the canopy spectrum of the sample. The reflectance spectra were measured by using a ASD FieldSpec-HandHeld-2 spectrometer (Analytical Spectral Devices, Boulder, CO, USA), whose measured wavelength range is 325–1075 nm with step interval of 1 nm, spectral resolution < 3 nm, integration time ≥ 8.5 ms, and standard field-of-view of 25°. There were 751 wavelength variables per spectrum. During data collection, the ASD device was located directly above the sample plant canopy, and the vertical distance from sensor to canopy was about 30 cm. According to geometric operation, the sensor footprint on the potato plant canopy was about 0.02 m^2. The spectral reflectance was corrected by a standard calibration whiteboard (Spectralon Standard Correction Board, Labsphere Co., Ltd., North Sutton, NH, USA) every 10 min to eliminate the interference of variation in solar illumination intensity spectral data.

2.1.2. Chlorophyll Content Measurement

Three leaves in each sample plant canopy were randomly collected and were put into a freshness protection bag, which was numbered and stored in a portable thermal insulation box. Then, the chlorophyll concentration was determined based on the standard chemical methods in the laboratory [44]. Each potato leaf was cut into pieces. About 0.04 g pieces of each leaf were placed in a 25 mL mixture of acetone and anhydrous ethanol to extract chlorophyll. The volume ratio of acetone to anhydrous ethanol was 2:1. The extraction solution was placed in darkness for 24 h. The absorbance at 645 and 663 nm of extraction solution was then measured using a visible-infrared spectrophotometer (UV-752, Shimadzu, Kyoto, Japan) that could measure in the wavelength range of 200–1000 nm based on single beam optical system with step interval of 0.1 nm, optical system of a single beam, light source of a tungsten lamp and deuterium lamp, and spectral bandwidth of 4 nm. Chlorophyll concentration was calculated by the following equations:

$$C_a = 12.72A_{663} - 2.59A_{645} \tag{1}$$

$$C_b = 22.88A_{645} - 4.67A_{663} \tag{2}$$

$$C_t = C_a + C_b \tag{3}$$

where A_{645} and A_{663} are the absorbance at 645 and 663 nm, respectively; C_a and C_b are the concentrations of chlorophyll-a and chlorophyll-b, respectively; and C_t is the total chlorophyll concentration, whose unit is mg/L in the study.

2.2. Data Analysis

The main data-processing steps are shown in Figure 2. The first part was to convert original reflectance spectra (Ref), which included SNV reflectance (SNV) data obtained from original reflectance by standard normal variate correction, and the wavelet features (WFs) were obtained by continuous wavelet transform (CWT). The second part was to establish analysis models, including PLS models based on the full spectral wavelengths and CARS-PLS models based on the sensitive wavelength variables selected by the CARS algorithm. The third part was to compare the chlorophyll analysis performance of various models.

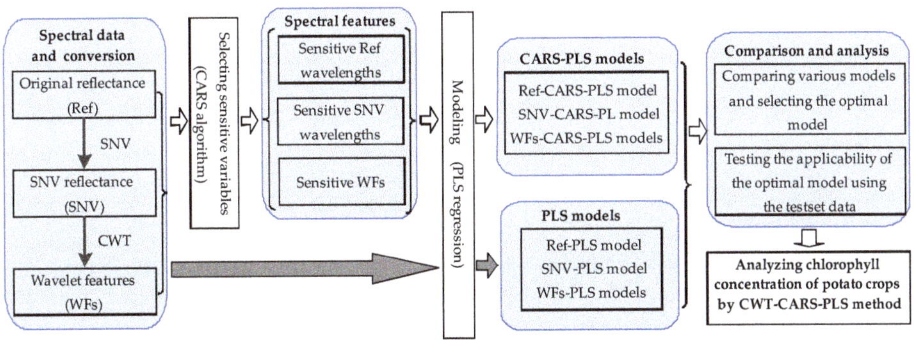

Figure 2. Flowchart of the main data processing.

2.2.1. SNV Correction

SNV is a certified method that can remove both additive and multiplicative effects in spectral data [45,46]. In SNV, each spectrum was being centered and then scaled by the corresponding standard deviation. It could be calculated with Equation (4):

$$z_i = \frac{x_i - \mu}{\sigma} \tag{4}$$

where x_i is the reflectance of the i nm, μ is the average reflectance of a spectrum, σ is the standard deviation of a spectrum, z_i is the reflectance after SNV of the i nm. In this work, the reflectance spectra corrected by SNV were denoted as SNV reflectance (SNV).

2.2.2. CWT

Mathematically, CWT is a liner operation that performs the convolution of reflectance spectrum with a scaled and shifted mother wavelet. The transform process is shown as Equation (5):

$$W_f(a,b) = \frac{1}{\sqrt{a}} \int_{-\infty}^{+\infty} f(\lambda) \bullet \psi(\frac{\lambda - b}{a}) \bullet d\lambda \tag{5}$$

where $\psi(\lambda)$ is the mother wavelet function, $f(\lambda)$ is the reflectance spectrum, and $W_f(a,b)$ is the wavelet coefficient (denoted as $WF_{a,b}$) for the scaling factor a and the shifting factor b. The scaling factor indicates the width of the scaled mother wavelet. The scaling factor used in this study was at dyadic scales $2^n (n = 1, 2, \cdots, 10)$, denoted as scale 1, scale 2, ... , scale 10, sequentially. The shifting factor was the central wavelength of the shifted mother wavelet. The physical and chemical components of crops had characteristic spectral absorption. b could be used to capture the peak and valley of an absorption feature, and the scaling factor a could be comparable with the width of an absorption feature. A crop leaf reflectance spectrum in the 325–1075 nm range consisted of a background continuum on which a number of absorption features attributable to pigments, water, and dry matter were superimposed [30]. Previous research had suggested that the shape of the absorption features is similar to that of the Gaussian function [47] or a combination of multiple Gaussian functions [48]. Thus, the second derivative of Gaussian, also known as the Mexican Hat, was used as the mother wavelet function in this study. All CWT operations were accomplished using the IDL 6.3 Wavelet Toolkit (ITT Visual Information Solutions, Boulder, CO, USA).

The one-dimensional SNV spectra were transformed into two-dimensional wavelet power map data composed of scaling (frequency scale) and shifting (spectral wavelength) factors by using the CWT. According to previous literature, the scaling factor from 1 to 3 belongs to low frequency, the scaling factor from 4 to 7 belongs to middle frequency, and the scaling factor from 8 to 10 belongs to high frequency [30–34]. The sensitive spectral variables of potato chlorophyll could be selected from these wavelet coefficients.

2.2.3. CARS

CARS, proposed by imitating the "survival of the fittest" principle of the Darwinian theory of evolution, is an efficient strategy to select sensitive variables depending on the absolute values of regression coefficients ($|\alpha|$) [43]. The steps of CARS can be summarized as follows [49,50]. First, $|\alpha|$ values are computed and used as indices to evaluate the importance of each variable. Second, the N subsets are selected by N Monte Carlo sampling runs based on the $|\alpha|$ of each variable. Third, a two-step procedure involving an exponentially decreasing function (EDF) and adaptive reweighted sampling (ASR) is used to select sensitive variables. In this step, EDF is utilized to remove the variables whose regression coefficients are relatively small in each sampling run. Following a decrease in EDF-based enforced variables, ARS is used to further eliminate the variables through a competitive way. Finally, the above three steps are repeated until the standard error of cross-validation is obtained, and then the optimal subset of variables is selected.

2.2.4. PLS Method

The PLS regression method proposed by Geladi [51] was used to solve multicollinearity problems among variables. PLS regression simultaneously executed principal component decomposition on the spectral reflectance matrix and the leaf chlorophyll concentration matrix [52], which were correlated in the decomposition process. A linear regression model was then established between them to analyze the chlorophyll concentration of potato leaves. To prevent model overfitting, internal interaction verification was performed by leave-one-out cross-validation (LOOCV), and the optimal latent variation was selected based on the largest coefficient of determination of the cross-validation set (R^2_{cv}). The program package of SNV, CARS, and PLS algorithms is available at the http://www.libpls.net/index.php.

2.2.5. CWT-CARS-PLS

A new spectral data analysis method named CWT-CARS-PLS was proposed in this study. The sensitive variables selected by CARS can remove the uninformative variables and enhance the PLS model performance. Thus, CARS combined with PLS regression (CARS-PLS) was an effective algorithm to establish the quantitative analysis model. CWT can also transform the one-dimensional SNV spectra into two-dimensional wavelet coefficients. Regarding decomposition, CWT can reduce the high-frequency noises of spectral data and extract the valuable spectral variables. Then, CWT combined with CARS-PLS (CWT-CARS-PLS) can deeply identify sensitive WFs and establish a high-performance analysis model. The proposed CWT-CARS-PLS algorithm is briefly introduced in Figure 3. All data calculations including SNV correction, PLS, CARS-PLS, and CWT-CARS-PLS were completed using MATLAB R2018a software.

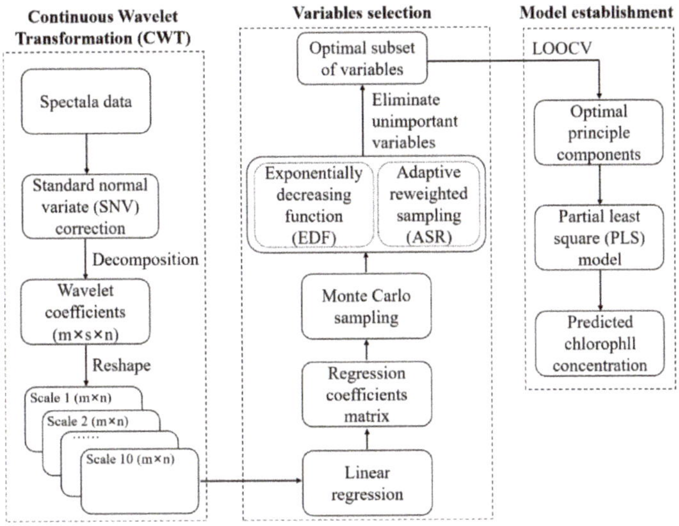

Figure 3. Flowchart of Continuous wavelet transformation-competitive adaptive reweighted sampling-partial least-square (CWT-CARS-PLS) algorithm.

2.3. Model Evaluation Indicators

To establish the analysis model, the modelling dataset was divided into a calibration and a validation set through sample-set partitioning based on the joint X-Y distance (SPXY) algorithm. This algorithm can comprehensively differentiate independent and dependent variables among samples [53,54].

The calibration set (200 samples) was used to train the PLS model. The validation set (114 samples) was used to verify the established analysis model's performance. The performance of the PLS model was evaluated with the determination coefficient of validation set (R^2) and the root mean square error (RMSE) as follows:

$$R^2 = 1 - \frac{\sum_{i=1}^{n} (y_i - y_i^*)^2}{\sum_{i=1}^{n} (y_i - \overline{y})^2} \quad (6)$$

$$RMSE = \sqrt{\frac{\sum_{i=1}^{n} (y_i - y_i^*)^2}{n}} \quad (7)$$

where y_i and y_i^* are the measured and predicted chlorophyll concentrations for sample i, respectively. $\overline{y}$ is the average value of measured chlorophyll, and n is the number of samples applied for the calibration or validation set. The difference value ($R_c^2 - R_{cv}^2$) between the R^2 of calibration set (R_c^2) and R^2 of cross-validation (R_{cv}^2) can be used as an indicator to judge the model stability, and a smaller value of $R_c^2 - R_{cv}^2$ value implies a more stable model. Furthermore, the R^2 of validation set (R_v^2) and the *RMSE* of validation set (*RMSEV*) can be utilized to evaluate the PLS model accuracy, and a higher R_v^2 and smaller *RMSEV* indicate a better model with stronger predictive capability.

3. Results

3.1. Statistics on Chlorophyll Concentration of Modeling Data

Chlorophyll concentrations were measured from S1 to S4. The average value at each stage was calculated and used to indicate the dynamic changes of potato growth. Results are shown in Figure 4. Chlorophyll concentration increased from 28.12 mg/L at S1 to 31.04 mg/L with the highest value at S2, and then decreased gradually to 15.36 mg/L at the S4 stage.

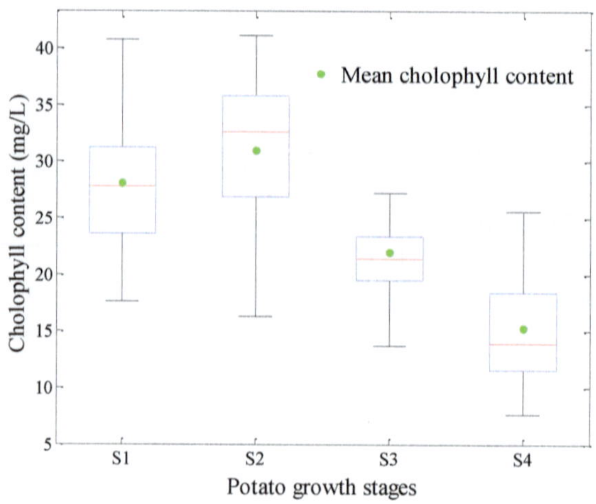

Figure 4. Statistical box line graph of chlorophyll concentration during potato growth stage for modeling dataset collected in 2018.

The results of the dataset partitioned by the SPXY algorithm were shown in Table 2, which shows the statistical description of the sample set for each growth stage and the combination of data from all four stages. Samples from all growth stages were combined to represent the changes in chlorophyll concentration. The modelling dataset for the chlorophyll concentration analysis model consisted of calibration and validation sets with 200 and 114 samples, respectively. The maximum value of the calibration set (41.20 mg/L) was larger than that of the validation set (37.46 mg/L), and the minimum value of the calibration set (7.66 mg/L) was smaller than that of the validation set (8.20 mg/L). The division result by SPXY was reasonable, and the calibration set could strongly represent the entire dataset.

Table 2. Chlorophyll concentration statistics of modeling dataset.

Samples	Data Set	Sample Number	Max (mg/L)	Min (mg/L)	Mean (mg/L)	STD (mg/L)
S1	All	74	40.77	17.64	28.12	5.05
	Calibration	50	40.77	17.64	28.27	5.31
	Validation	24	33.12	19.64	27.48	3.86
S2	All	80	41.20	16.30	31.04	5.81
	Calibration	50	41.20	16.30	30.23	6.29
	Validation	30	37.46	25.26	33.45	3.04
S3	All	80	35.63	13.70	22.00	4.18
	Calibration	50	35.63	13.70	22.04	4.65
	Validation	30	26.47	16.39	21.86	2.36
S4	All	80	32.25	7.66	15.36	5.45
	Calibration	50	32.25	7.66	15.73	5.93
	Validation	30	20.69	8.20	14.24	3.55
All stages	All	314	41.20	7.66	24.05	7.95
	Calibration	200	41.20	7.66	24.07	7.95
	Validation	114	37.46	8.20	24.00	8.00

3.2. Spectral Data Analysis

3.2.1. Analysis of Spectral Response during Growth

Figure 5a shows the Ref curves of the potato crop canopy. Serious scattering effects were observed in the Ref spectra among samples because of the different collection times and light reflection paths. After SNV correction, the noise caused by the scattering effects was significantly eliminated, and the dispersion among spectral curves was significantly reduced, as shown in Figure 5b. Accordingly, the SNV spectra were used for subsequent continuous wavelet transformation and modeling analysis.

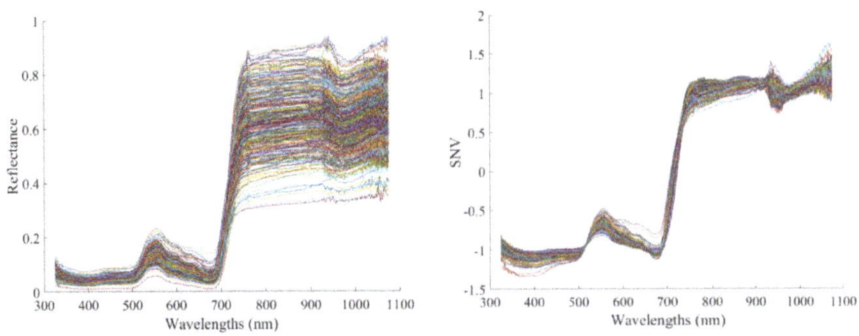

(a) (b)

Figure 5. Reflectance spectra of potato-crop canopy of four growth stages. (**a**) Original reflectance (Ref) spectra; (**b**) standard normal variate (SNV) reflectance spectra.

Furthermore, we examined the dynamic changes between different stages based on the average SNV spectrum of each stage. Figure 6 shows the reflectance of each stage. Their trends were similar in the visible (400–760 nm) and near-infrared (761–1000 nm) regions. In the visible region, the minimum reflectance appeared near 400 and 680 nm due to a strong absorption by the pigment. In the near-infrared region, the reflectance sharply increased from 711 nm to 760 nm because a reflective surface cavity existed in the spongy structure of the mesophyll. Although strong reflection existed in 761–1000 nm as a horizontal platform, a weak reflectance valley appeared near 970 nm because of the weak absorption of leaf water content.

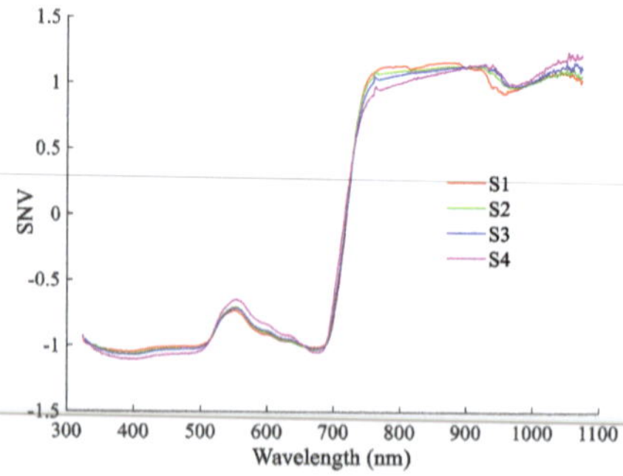

Figure 6. Average SNV spectral curve of potato crop canopy per growth stage.

However, significant changes were observed in some specific bands during growth. Within 530–640 nm, the SNV spectral reflectance increased with growth. The average SNV reflectance at S4 was significantly lower than that at the others, whereas the average SNV reflectance of S2 and S3 were very close. Within 740–880 nm, the SNV spectral reflectance decreased gradually. Small reflectance peaks were observed near 763 nm at S2–S4 stages. In the bands of 910–960 nm, the average value at S1 was significantly lower than those at the other stages.

3.2.2. Analysis of Wavelet Coefficient Curves under Different Decomposition Scales

The SNV spectral curves were decomposed into wavelet coefficients by CWT under 10 decomposition scales. The CWT results for some of the samples are shown in Figure 7. We observed that with increased scale, the wavelet coefficients gradually enlarged and the high-frequency noises were gradually reduced. Thus, the spectral curves were smoothed, and some characteristic absorption peaks were amplified under suitable decomposition scales, as shown in Scales 1–6 (Figure 7). However, when the decomposition scales were too large, the spectral curve became excessively smoothed and caused the the specific characteristic absorption peaks to disappear, which was not conducive to quantitative analysis, as shown in Scales 7–10 (Figure 7).

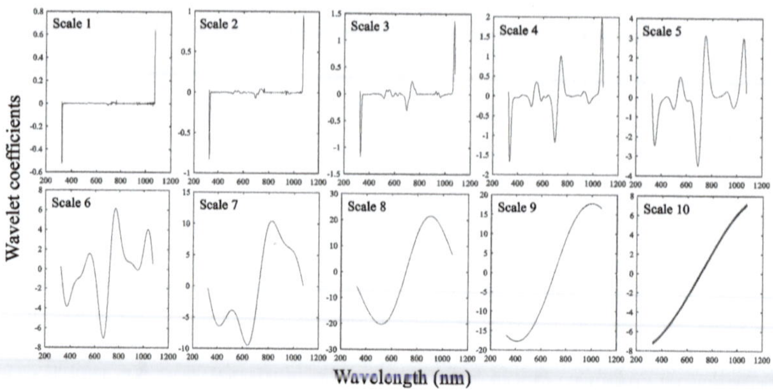

Figure 7. Wavelet coefficient curves under different decomposition scales.

3.3. *Correlation of Spectra and Wavelet Features with Chlorophyll Concentration*

3.3.1. Correlation Analysis between Chlorophyll Concentration and Spectra

Figure 8 shows the correlation coefficient curves between the chlorophyll concentration and Ref and SNV. Compared with Ref, the correlation coefficient between SNV and chlorophyll concentration was higher overall, illustrating that SNV correction reduced the noise of the original spectra and improved the analysis performance of spectral data. Furthermore, the correlation relationship between the chlorophyll concentration and SNV was analyzed. Within the ranges of 387–509, 519–633, and 744–844 nm, the absolute values of the correlation coefficient ($|r|$) were higher than 0.6. The peak value of the positive correlation occurred at 678 nm, and the r was 0.411. The peak value of negative correlation occurred at 702 nm, and the r was −0.715. Within 845–917 nm, the positive correlation gradually decreased before becoming a negative correlation, and then $|r|$ gradually increased.

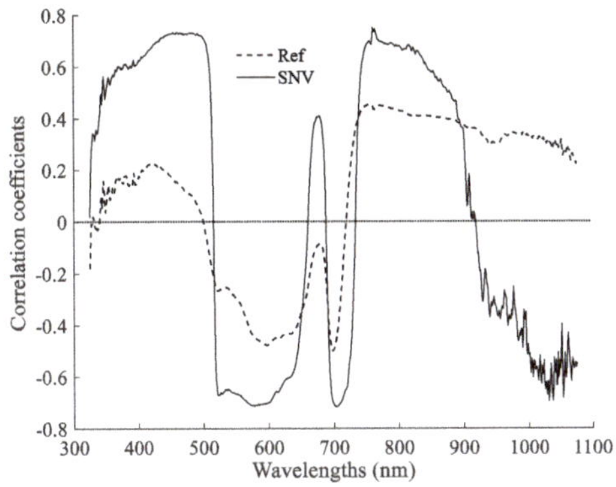

Figure 8. Correlation coefficient curve between chlorophyll concentration and spectra.

To further understand how the spectra changed with potato growth, correlation analysis was conducted between SNV and chlorophyll concentration from S1 to S4. Figure 9 shows the correlation coefficient curves. The chlorophyll concentration was correlated positively with the reflectance spectra within the range of 400–500 and 650–700 nm. However, a negative correlation existed between them within 510–630 and 701–750 nm. Furthermore, four band regions were highly correlated, including 400–510, 521–610, 701–740, and 761–920 nm. Overall, the correlation coefficients of S1–S4 had significant differences within 400–600, 601–620, and 700–902 nm. Conversely, the curve trend of the correlation coefficients of S2 and S3 was very similar.

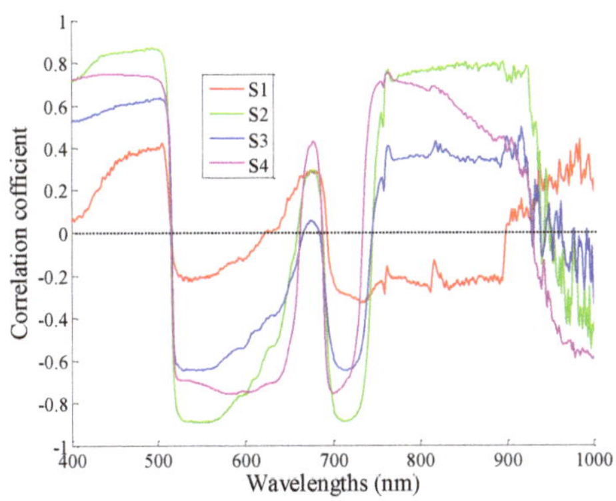

Figure 9. Correlation coefficient curve between chlorophyll concentration and SNV.

3.3.2. Correlation Analysis between Chlorophyll and Wavelet Features

The correlation coefficients between the chlorophyll and wavelet coefficients were calculated in the decomposition Scales 1–10 to draw the correlation coefficient distribution map, as shown in Figure 10. The correlation coefficient was represented by different colors and color values of each pixel in the map, which could help select the high correlation WFs. We observed that the correlation coefficients varied in different decomposition scales (scaling factors) and wavelength locations (shifting factors).

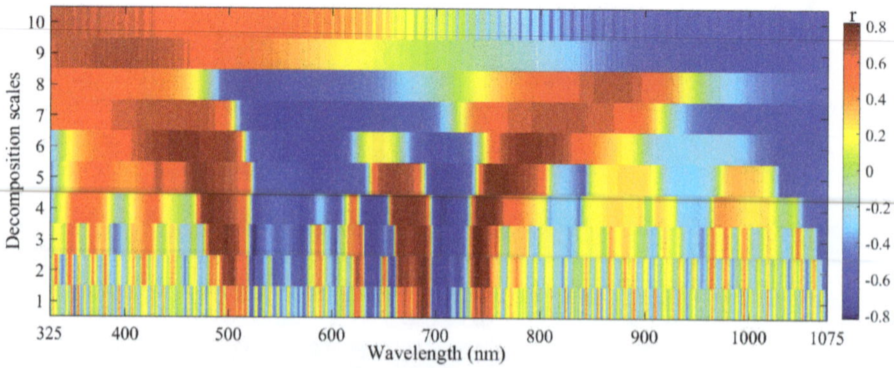

Figure 10. Correlation coefficient map between wavelet features and growth stages.

3.3.3. Comparison of Correlation Coefficient

The highest correlation coefficients of Ref, SNV, and WFs are shown in Table 3. We observed that the correlation coefficient of SNV (r = 0.75) was higher than Ref (r = 0.50), which revealed that SNV correction could effectively remove the noise of spectral data. For WFs, the correlation coefficient gradually increased form Scale 1 to Scale 3, and then the correlation coefficient gradually decreased. The strongest correlation was found in Scale 3 located in 524 nm (r = −0.82), and the Ref had the weakest correlation (r = −0.50) located in 698 nm.

Table 3. Correlation coefficient (*r*) between chlorophyll and original reflectance, SNV reflectance, and wavelet features.

	Feature Location	Highest *r*
	Wavelength (nm)	
Ref	698	−0.50
SNV	761	0.75
Scale 1	687	−0.78
Scale 2	739	0.81
Scale 3	524	−0.82
Scale 4	744	0.78
Scale 5	755	0.79
Scale 6	786	0.75
Scale 7	547	−0.74
Scale 8	515	−0.71
Scale 9	400	0.70
Scale 10	1038	−0.70

Moreover, the correlation coefficients of WFs in Scales 1–6 were higher than those of SNV, illustrating that CWT could enhance the correlation of chlorophyll by decomposing spectral data. The correlation coefficients of WFs in Scales 7–10 were also lower than those of SNV, further revealing that spectral data decomposing in too large scales were no longer helpful for quantitative analysis.

3.4. Establishment and Comparison of Chlorophyll Analysis Models

3.4.1. Sensitive Chlorophyll Variables Selected Using CARS

For the CWT-CARS-PLS, the sensitive WFs in each decomposition scale were selected, and the chlorophyll analysis PLS models were established for every scale. For comparison with CWT-CARS-PLS, the sensitive wavelengths were selected from Ref and SNV data to establish the Ref-CARS-PLS and SNV-CARS-PLS, respectively. The LOOCV was always operated to obtain the optimal principle components (PCs) in establishing the PLS models. The number of variables and PCs of various PLS models are shown in Table 4. For the chlorophyll analysis models, the maximal number of variables was 227 in Scale5-CARS-PLS model, and the minimal number of variables was 31 in Scale1-CARS-PLS. However, the minimal number of PCs was three in Scale3-CARS-PLS.

Table 4. Variable information of various CARS-PLS models.

Models	Variables Number	PCs	Models	Variables Number	PCs
Ref-CARS-PLS	61	21	Scale 5-CARS-PLS	227	21
SNV-CARS-PLS	64	17	Scale 6-CARS-PLS	48	17
Scale 1-CARS-PLS	31	12	Scale 7-CARS-PLS	33	15
Scale 2-CARS-PLS	61	13	Scale 8-CARS-PLS	54	18
Scale 3-CARS-PLS	57	3	Scale 9-CARS-PLS	57	17
Scale 4-CARS-PLS	178	19	Scale 10-CARS-PLS	33	28

The location of sensitive variables selected from Ref, SNV, and WFs in Scales 1–10 by using CARS algorithm are shown in Figure 11. All sensitive wavelengths selected by CARS were distributed in the visible and near-infrared regions. However, for the calibration model established by various sensitive variables, the predictive accuracy of Scale3-CARS-PLS model was the optimum, as shown in Figure 12a. Furthermore, the sensitive WFs of Scale3-CARS-PLS were analyzed through the leaf information. The number of variables of Scale3-CARS-PLS was 57. These sensitive WFs were located at 346, 389, 419, 425, 426, 431, 435, 436, 437, 520, 523, 535, 546, 547, 563, 579, 580, 590, 591, 620, 625, 661, 662, 667, 684, 685, 688, 690, 693, 698, 716, 717, 733, 739, 742, 751, 752, 767, 781, 811, 824, 825, 848, 857, 858, 875, 890, 909, 919, 929, 939, 948, 960, 963, 968, 973, and 985 nm. Among them, the WFs located in the visible region could reflect the leaf pigment. The WFs located in near-infrared regions could reflect the leaf structure and other leaf substance; for instance, the WF at 929 nm reflected the leaf fat, the WF at 973 nm near 970 nm reflected the leaf water content, and the WF at 985 nm reflected leaf starch.

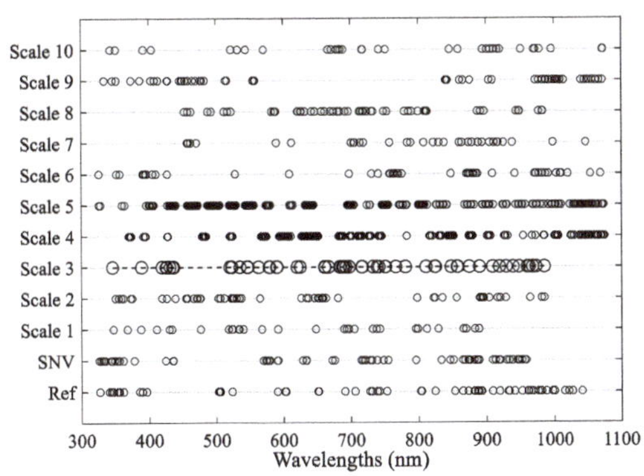

Figure 11. Location of sensitive variables selected from Ref, SNV, and wavelet features (WFs) in Scales 1–10 by using the CARS algorithm.

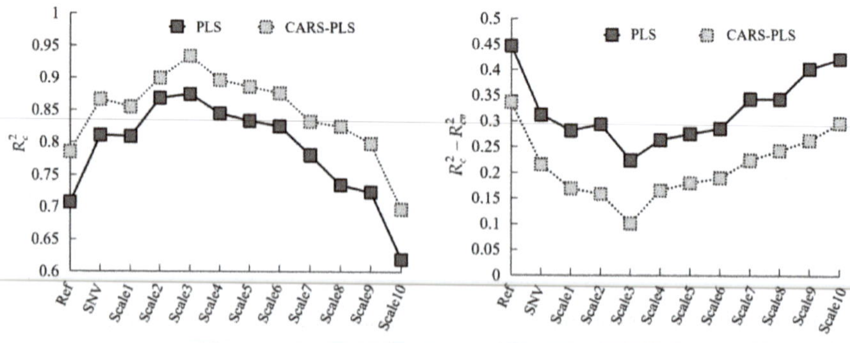

Figure 12. Comparison of model performance (R_c^2,$R_c^2 - R_{cv}^2$) between PLS and CARS-PLS established by Ref, SNV, and WFs in Scales 1–10. (**a**) R_c^2 of PLS and CARS-PLS; (**b**) $R_c^2 - R_{cv}^2$ of PLS and CARS-PLS.

3.4.2. Comparison of the Performance of PLS and CARS-PLS Models

The chlorophyll analysis models were established using the CARS-PLS method, and the modeling results (R_c^2 and $R_c^2 - R_{cv}^2$) are shown in Figure 12. To highlight the advantages of selecting sensitive variables by CARS, the analysis models were also established using the PLS method. Figure 12a shows that for all variable categories, the R_c^2 of CARS-PLS was higher than that of PLS, illustrating that CARS could effectively eliminate uninformative variables and improve model accuracy. The $R_c^2 - R_{cv}^2$ of CARS-PLS was lower than that of PLS, as shown in Figure 12b, which revealed that CARS could reduce model complexity and enhance model stability.

Furthermore, the R_c^2 of SNV-CARS-PLS was higher than that of Ref-CARS-PLS. The R_c^2 of CWT-CARS-PLS models established based on WFs was higher than those of models based on Ref and SNV. For CWT-CARS-PLS, the R_c^2 gradually increased from Scale 1 to Scale 3 and then R_c^2 gradually decreased. Based on the value of $R_c^2 - R_{cv}^2$, the stability of SNV-CARS-PLS was stronger than that of Ref-CARS-PLS. For CWT-CARS-PLS models, stability gradually strengthened from Scale 1 to Scale 3 and then gradually weakened. The stability of CARS-PLS models based on Scales 1–6 was stronger than those of Ref-CARS-PLS and SNV-CARS-PLS, which was consistent with the correlation analysis in Section 3.3. The above results demonstrated that the CWT could deeply identify spectral data to improve model performance.

3.5. Validation of Chlorophyll Analysis Models

The validation results of various chlorophyll analysis models are shown in Figure 13. The same as the calibration models, the CARS-PLS models had a higher determination coefficient of validation set (R_v^2) than the PLS models, and the CARS-PLS models had a smaller $RMSEV$ than the PLS models primarily because the invalid variables were removed by the CARS algorithm. Furthermore, CWT-CARS-PLS models under Scales 2–6 had higher R_v^2 values ($R_v^2 > 0.81$) and smaller RESEV values ($RMSEV < 3.34$ mg/L) than the Ref-CARS-PLS ($R_v^2 = 0.65$, $RMSEV = 4.11$ mg/L) and the SNV-CARS-PLS ($R_v^2 = 0.75$, $RMSEV = 3.55$ mg/L) models. Moreover, Scale3-CARS-PLS showed the highest R_v^2 value of 0.85 and the smallest root–mean–square error of cross-validation (RMSEV) value of 2.77 mg/L, as shown in Figure 14. These chlorophyll concentration values were evenly distributed on both sides of the 1:1 line, further illustrating that the proposed Scale3-CARS-PLS model had good stability.

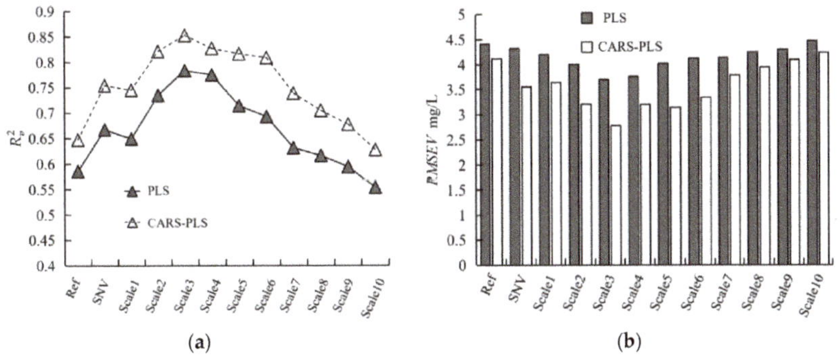

Figure 13. Validation results of PLS and CARS-PLS established by Ref, SNV, and WFs in Scales 1–10. (**a**) R_v^2 of PLS and CARS-PLS models; (**b**) root–mean–square error of cross-validation (RMSEV) of PLS and CARS-PLS.

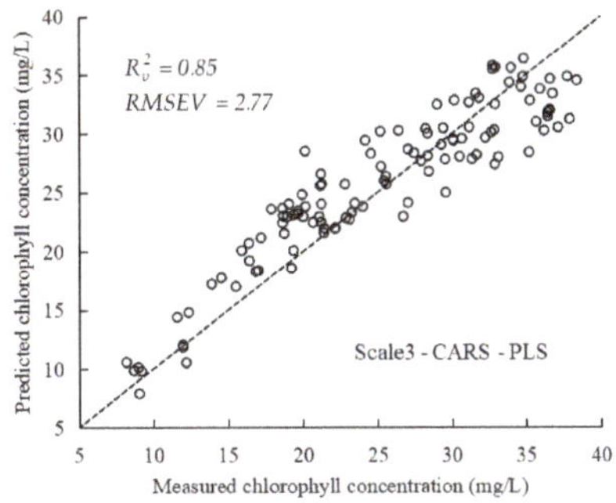

Figure 14. Predicted value of validation set for Scale3-CARS-PLS model.

3.6. Testing of the Developed Scale3-CARS-PLS Model

The testing set data collected in 2020 were used to test the stability and applicability of the developed Scale3-CARS-PLS model. The chlorophyll concentration of testing set data ranged from 8.81 mg/L to 39.59 mg/L, and the average content was 19.18 mg/L. The chlorophyll concentration range of the test set was smaller than that of the modeling set, which ranged from 7.66 mg/L to 41.20 mg/L.

The canopy reflectance spectra of the testing dataset (160 samples) were corrected by standard normal variate to obtain the SNV reflectance, then the CWT was performed on the SNV reflectance and the CARS algorithm was used to select the sensitive WFs under scale 3, and then the WFs were substituted into the Scale3-CARS-PLS model to predict chlorophyll concentration. In order to highlight the performance of Scale3-CARS-PLS, the reflectance spectra of the testing dataset were substituted into Ref-PLS and Ref-CARS-PLS models, and the SNV reflectance data were substituted into the SNV-CARS-PLS model. The scatter plot of 1:1 was created, as shown in Figure 15, to visually demonstrate the chlorophyll concentration prediction results. The performance of Ref-CARS-PLS was better than Ref-PLS, which showed that CARS could eliminate the valueless variables to improve the model analysis ability. The model performance of SNV-SCAR-PLS was further enhanced due to the SNV pre-processing by correcting the scattering effect. Then, the Scale3-CARS-PLS model showed the strongest R^2 of 0.69 and the smallest RMSE value of 3.36 mg/L, which illustrated that the Scale3-CARS-PLS model possessed good analysis capability, and the spectral analysis method had good applicability. Figure 15d shows that these chlorophyll values were evenly distributed on both sides of the 1:1 line, further illustrating that the proposed Scale3-CARS-PLS model had good stability.

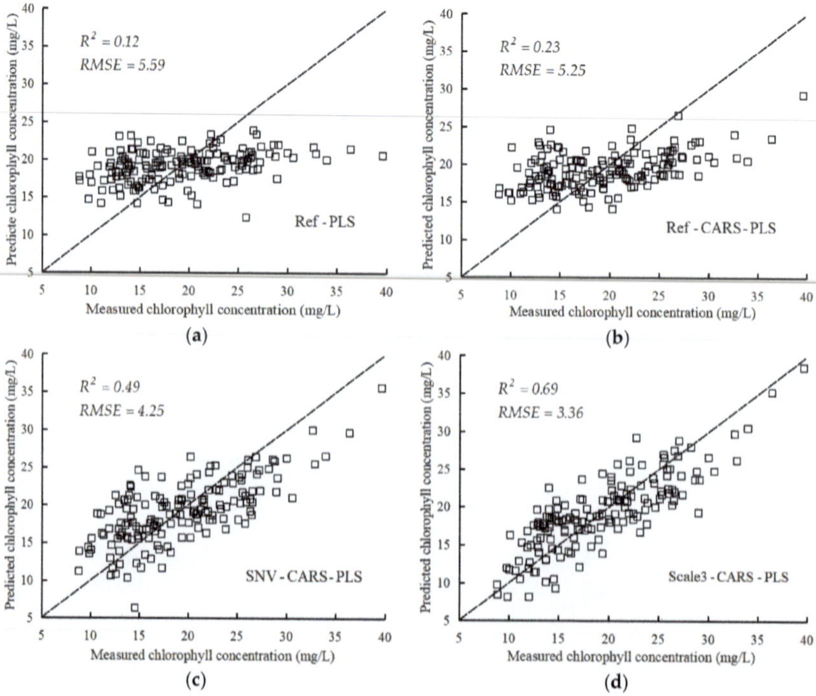

Figure 15. Testing results of the various models. (**a**) Ref-PLS; (**b**) Ref-CARS-PLS; (**c**) SNV-CARS-PLS; (**d**) Scale3-CARS-PLS.

4. Discussion

Spectroscopy is a rapid and non-destructive method of gathering crop-pigment information [55,56]. In this study, the spectral characteristic response and chlorophyll concentration change at different stages were analyzed and discussed. Results demonstrated that the average reflectance was close in S2 and S3, and that the correlation curves between the reflectance and chlorophyll concentration of S2 and S3 had similar change trends. According to the potato phenology, a new tuber forms by stolons after the plant flowers at S2 and the tuber expands at S3. Consequently, nutrient availability and balance are transferred from aboveground stems and leaves to underground tubers during these periods. This phenomenon may explain why some of the plants have similar physiology status and spectral responses than others [57].

4.1. *Abilities of Denoising and Sensitive-Variable Mining of CWT at Different Decomposition Scales*

SNV can effectively reduce scattering noise to enhance the analysis performance of spectral data [45,46]. After SNV correction, dispersion among spectral curves was significantly reduced (Figure 5), and the correlation between spectral data and chlorophyll concentration was enhanced (as shown in Figure 8). Accordingly, SNV spectra were used for further CWT and modeling analysis.

After CWT, the spectral reflectance was transformed into the wavelet coefficients, as shown in Figure 7. With increased decomposition scales from 1 to 6, the wavelet coefficient curve was smoothed, and some characteristic absorption peaks amplified. Then, the curve was excessively smoothed, resulting in the disappearance of the characteristic absorption location. The above content was consistent with previous literature reporting that WFs in the middle- and low-frequency scales could capture the absorption characteristics of the physical and chemical substances of crops [33,58] and effectively eliminate the high-frequency noise of spectral data [36,59]. High-frequency WFs could remove the absorption features and could not efficiently analyze the physiological and biochemical compositions [60].

The absolute value of the highest correlation coefficient between chlorophyll concentration and WFs under Scales 1–6 was higher than SNV (0.75), illustrating that the CWT could enhance the correlation of chlorophyll concentration by decomposing spectral data. Previous studies [30,34,59–61] have reported the same results, such as Wang [59] who indicated that the correlation between wavelet coefficients and pigments was significantly higher than that of vegetation index and sensitive wavelengths. Furthermore, with increased decomposition scales from 1 to 3, the absolute value of the highest correlation coefficient of WFs increased from 0.78 to 0.82 and then gradually decreased to 0.70, illustrating that high-frequency WFs were not conducive to quantitative analysis [32,33,60,61]. WFs under Scale 3 exhibited the strongest correlation relationship with chlorophyll concentration.

4.2. Uninformative Variable Elimination by CARS Algorithm

Given that a spectrometer collects reflectance data based on near-contiguous spectral bands, the selection of sensitive wavelengths or variables is one of key steps in the chlorophyll analysis to solve multiple mutual lineal problems of overfitting and redundancy [62]. Thus, wavelengths and WFs need be selected by effective algorithms to remove the uninformative variables and to enhance model performance [59,63]. The CARS developed based on the model population analysis strategy [64] can be used to consider the contribution of each variable to the analysis model to select informative spectral variables. Relative to the PLS models, the number of input variables of CARS-PLS models was reduced significantly, and the CARS-PLS models possessed more excellent prediction ability, as shown in Figures 12a and 13. Overfitting frequently occurred during the modeling process, caused by the increasing number of model variables, which affected the stability and accuracy of the PLS model [65]. Accordingly, internal cross-validation was performed in this study. The difference in the determination coefficient of calibration and cross-validation sets ($R_c^2 - R_{cv}^2$) was used as an indicator to determine the model stability [66]. As shown in Figure 12b, the $R_c^2 - R_{cv}^2$ values of the CARS-PLS models were lower than those in the PLS models, further illustrating that the CARS can effectively eliminate redundant variables and improve the analysis of the model's stability.

4.3. Chlorophyll Content Analysis Capability of WFs under Different Decomposition Scales

We further analyzed the performance of various CARS-PLS models. From the point of view of model stability, the $R_c^2 - R_{cv}^2$ values of the CARS-PLS models based on Scales 1–6 were smaller than those of Ref-CARS-PLS and SNV-CARS-PLS, and the $R_c^2 - R_{cv}^2$ value of Scale3-CARS-PLS model was the smallest, showing that Scale3-CARS-PLS model had the strongest stability. From the point of view of prediction capability of the model, the $RMSEV$ values of CARS-PLS models based on Scales 2–6 were smaller than those of SNV-CARS-PLS model, and Scale3-CARS-PLS showed the strongest R_v^2 value of 0.85 and the smallest RMSEV value of 2.77 mg/L, as shown in Figure 14. For WFs under Scale 3, 57 sensitive WFs were selected by the CARS algorithm, whose locations were evenly distributed in the visible (37 variables) and near-infrared (20 variables) region, as shown in Figure 11. Previous studies have reported that the spectral data in the visible region can analyze pigment content [67]. Moreover, the spectral data in the near-infrared region can reflect other substances' information and crop-canopy structure, which can improve the robustness of the chlorophyll analysis model [68].

Moreover, previous studies reported the detection of chlorophyll concentration in crops based on spectral wavelengths or/and spectral indices. Sun [28] selected 11 sensitive wavelengths for analyzing the chlorophyll concentration of potato leaf, with the R_v^2 of the model of 0.77. Tao [69] screened the red edge position using the linear extrapolation method for estimating the chlorophyll concentration of potato with R_c^2 of 0.87. However, the R_c^2 and R_v^2 of the analysis model developed by coupling CWT with CARS methods in this paper is 0.93 and 0.86, respectively. Above content demonstrated that CWT could deeply identify spectral data to improve model performance, and that the sensitive WFs under Scale 3 possessed the best excellent prediction capability for chlorophyll concentration of potato crops.

4.4. Generalizability of This Study to Future Works

A comprehensive analysis of testing results showed that spectral data could be processed using CWT. Sensitive variables were selected using CARS, which was suitable for model-variable optimization and prediction-capability improvement. Finally, the analysis performance of the Scale3-CARS-PLS model was tested using another variety of potato crop, the R^2, and RMSE was 0.69 and 3.36 mg/L, as shown in Figure 15, which demonstrated that the Scale3-CARS-PLS model possessed good stability and excellent applicability. Previous studies reported that the chlorophyll concentration is significantly correlated with the concentration of nitrogen [5,70]. Therefore, the study could provide a theoretical support for precision nitrogen management in the potato field, and a method reference for large-scale remote sensing analysis of potato chlorophyll concentration.

However, this method was based on specific spectral data for potato crops. The restrictions were based on the existence of other datasets or potato varieties [22,71]. Therefore, more datasets from wide-ranging potato varieties, planting patterns, and experimental fields should be collected to develop a stable and accurate classification model using CWT-CARS-PLS method.

5. Conclusions

We presented an effective method for analyzing the chlorophyll concentration of potato plants through canopy spectroscopy. The dynamic responses of canopy spectra at different growth stages were analyzed. The spectral characteristics were found to significantly differ between S1, S2–S3, and S4. However, the SNV spectral reflectance curves in S2 and S3 were similar. The performances of Ref, SNV, WFs under different decomposition scales, CARS-PLS, and CWT-CARS-PLS in analyzing chlorophyll concentration were compared based on the model results. The CARS-PLS model established by WFs under different scales obtained by CWT exhibited the most excellent analysis ability and reliability. Scale3-CARS-PLS model had fewer variables, smallest $R_c^2 - R_{cv}^2$ value, strongest R_v^2, and weakest $RMSEV$ for chlorophyll analysis. The analysis performance of the Scale3-CARS-PLS model was tested using another variety of potato crop with a satisfactory result. Based on spectral data, the WFs under Scale 3 showed excellent chlorophyll-content prediction capability. Thus, the proposed CWT-CARS-PLS was a potentially accurate and efficient method of analyzing the chlorophyll concentration of potato crops. This study could provide a method reference for large-scale remote sensing analysis of chlorophyll concentration and a theoretical support for precision nitrogen management of potato crops.

Author Contributions: Conceptualization, N.L., Z.X., R.Z., L.Q., M.L., G.L. and H.S.; data curation, N.L., Z.X., R.Z., L.Q., M.L., G.L. and H.S.; formal analysis, N.L., Z.X., R.Z., L.Q., M.L., G.L. and H.S.; funding acquisition, M.L. and H.S.; investigation, N.L., Z.X., R.Z., L.Q., M.L., G.L. and H.S.; methodology, N.L., Z.X., R.Z., L.Q., M.L., G.L. and H.S.; project administration, M.L., G.L. and H.S.; Resources, N.L., L.Q., G.L. and H.S.; software, N.L., Z.X., R.Z. and H.S.; supervision, M.L., G.L. and H.S.; validation, N.L., Z.X. and G.L.; visualization, N.L., Z.X., R.Z. and L.Q.; writing–original draft, N.L.; writing–review & editing, M.L., G.L. and H.S. All authors have read and agreed to the published version of the manuscript.

Funding: The project was supported by the National Natural Science Fund of China (Grant No. 31971785, 31501219), the Fundamental Research Funds for the Central Universities of China (Grant No. 2020TC036), and the Graduate Training Project of China Agricultural University (JG2019004 and YW2020007).

Acknowledgments: Thanks to Tao Zheng and Li Wu for their help in the field experiment.

Conflicts of Interest: The authors declare no conflict of interest.

References

1. Liu, N.; Zhao, R.; Qiao, L.; Zhang, Y.; Li, M.; Sun, H.; Xing, Z.; Wang, X. Growth stages classification of potato crop based on analysis of spectral response and variables optimization. *Sensors* **2020**, *20*, 3995. [CrossRef] [PubMed]
2. Shillito, R.M.; Timlin, D.J.; Fleisher, D. Yield response of potato to spatially patterned nitrogen application. *Agric. Ecosyst. Environ.* **2009**, *129*, 107–116. [CrossRef]
3. Vesali, F.; Omid, M.; Mobli, H.; Kaleita, A. Feasibility of using smart phones to estimate chlorophyll content in corn plants. *Photosynthetica* **2017**, *55*, 603–610. [CrossRef]

4. Zhenjiang, Z.; Mohamed, J.; Finn, P. Using ground-based spectral reflectance sensors and photography to estimate shoot N concentration and dry matter of potato. *Comput. Electron. Agric.* **2018**, *144*, 154–163. [CrossRef]
5. Clevers, J.G.P.W.; Kooistra, L. Using hyperspectral remote sensing data for retrieving canopy chlorophyll and nitrogen content. In Proceedings of the 2011 3rd Workshop on Hyperspectral Image and Signal Processing: Evolution in Remote Sensing (WHISPERS), Lisbon, Portugal, 6–9 June 2011; pp. 574–583. [CrossRef]
6. Sonobe, R.; Hirono, Y.; Oi, A. Non-destructive detection of tea leaf chlorophyll content using hyperspectral reflectance and machine learning algorithms. *Plants* **2020**, *99*, 368. [CrossRef] [PubMed]
7. Fairooz, J.; Mat-Jafri, M.Z.; Hwee-San, L.; Wan-Maznah, W.O. Laboratory measurement: Chlorophyll-a concentration measurement with acetone method using spectrophotometer. In Proceedings of the 2014 IEEE International Conference on Industrial Engineering and Engineering Management, Selangor Darul Ehsan, Malaysia, 9–12 December 2015; pp. 744–748. [CrossRef]
8. Igamberdiev, R.M.; Bill, R.; Schubert, H.; Lennartz, B. Analysis of cross-seasonal spectral response from Kettle Holes: Application of remote sensing techniques for chlorophyll estimation. *Remote Sens.* **2012**, *4*, 3481–3500. [CrossRef]
9. Wen, P.; He, J.; Fang, N. Estimating leaf nitrogen concentration considering unsynchronized maize growth stages with canopy hyperspectral technique. *Ecol. Indic.* **2019**, *107*, 1–16. [CrossRef]
10. Zhang, Y.; Guanter, L.; Berry, J.A.; Christiaan, T.; Yang, X.; Tang, J.; Zhang, F. Model-based analysis of the relationship between sun-induced chlorophyll fluorescence and gross primary production for remote sensing applications. *Remote Sens. Environ.* **2016**, *187*, 145–155. [CrossRef]
11. Thomas, C.A.; Binzel, R.P. Identifying meteorite source regions through near-Earth object spectroscopy. *Icarus* **2010**, *205*, 419–429. [CrossRef]
12. Reddy, V.; Gaffey, M.J.; Abell, P.A.; Hardersen, P.S. Constraining albedo, diameter and composition of near-Earth asteroids via near-infrared spectroscopy. *Icarus* **2012**, *219*, 382–392. [CrossRef]
13. Zhang, J.; Sun, H.; Gao, D.; Qiao, L.; Liu, N.; Li, M.; Zhang, Y. Detection of Canopy Chlorophyll Content of Corn Based on Continuous Wavelet Transform Analysis. *Remote Sens.* **2020**, *12*, 2741. [CrossRef]
14. Nigon, T.J.; Mulla, D.J.; Rosen, C.J. Evaluation of the nitrogen sufficiency index for use with high resolution, broadband aerial imagery in a commercial potato field. *Precis. Agric.* **2014**, *15*, 202–226. [CrossRef]
15. Liu, P.; Sohn, H.; Jeon, I. Nonlinear spectral correlation for fatigue crack detection under noisy environments. *J. Sound Vib.* **2017**, *400*, 305–316. [CrossRef]
16. Silalahi, D.; Midi, H.; Arasan, J.; Mustafa, M.S.; Caliman, J.P. Robust generalized multiplicative scatter correction algorithm on pretreatment of near infrared spectral data. *Vib. Spectrosc.* **2018**, *97*, 55–65. [CrossRef]
17. Zeaiter, M.; Roger, J.M.; Bellon-Maurel, V. Robustness of models developed by multivariate calibration: Part II: The influence of pre-processing methods. *TrAC Trends Anal. Chem.* **2005**, *24*, 437–445. [CrossRef]
18. Fearn, T.; Riccioli, C.; Garrido-Varo, A. On the geometry of SNV and MSC. *Chemom. Intell. Lab. Syst.* **2009**, *96*, 22–26. [CrossRef]
19. Luo, J.; Ying, K.; He, P.; Bai, J. Properties of Savitzky–Golay digital differentiators. *Digit. Signal Process.* **2005**, *15*, 122–136. [CrossRef]
20. Shen, Q.; Zhang, B.; Li, J.; Wu, D.; Song, Y.; Zhang, F.; Wang, G. Characteristic wavelengths analysis for remote sensing reflectance on water surface in Taihu Lake. *Spectrosc. Spectr. Anal.* **2011**, *31*, 1892–1897. [CrossRef]
21. Zhang, Z.; Ding, J.; Zhu, C.; Wang, J. Combination of efficient signal pre-processing and optimal band combination algorithm to predict soil organic matter through visible and near-infrared spectra. *Spectrochim. Acta Part. A Mol. Biomol. Spectrosc.* **2020**, *240*, 118553. [CrossRef]
22. Xu, S.; Wang, M.; Shi, X. Hyperspectral imaging for high-resolution mapping of soil carbon fractions in intact paddy soil profiles with multivariate techniques and variable selection. *Geoderma* **2020**, *370*, 114368. [CrossRef]
23. Rondeaux, G.; Steven, M.; Baret, F. Optimization of soil-adjusted vegetation indices. *Remote Sens. Environ.* **1996**, *55*, 95–107. [CrossRef]
24. Gitelson, A.; Merzlyak, M.N. Spectral reflectance changes associated with autumn senescence of aesculus hippocastanum l. and acer platanoidesl. Leaves. Spectral features and relation to chlorophyll estimation. *J. Plant Physiol.* **1994**, *143*, 286–292. [CrossRef]

25. Salas, E.A.L.; Henebry, G.M. A new approach for the analysis of hyperspectral data: Theory and sensitivity analysis of the moment distance method. *Remote Sens* **2014**, *6*, 20–41. [CrossRef]
26. Sukhova, E.; Sukhov, V. Connection of the photochemical reflectance index (PRI) with the photosystem II quantum yield and nonphotochemical quenching can be dependent on variations of photosynthetic parameters among investigated plants: A meta-analysis. *Remote Sens.* **2018**, *10*, 771. [CrossRef]
27. Yu, K.; Lenz-Wiedemann, V.; Chen, X. Estimating leaf chlorophyll of barley at different growth stages using spectral indices to reduce soil background and canopy structure effects. *ISPRS J. Photogramm. Remote Sens.* **2014**, *97*, 58–77. [CrossRef]
28. Sun, H.; Zheng, T.; Liu, N. Vertical distribution of chlorophyll in potato plants based on hyperspectral imaging. *Trans. Chin. Soc. Agric. Eng.* **2018**, *34*, 149–156. [CrossRef]
29. Sun, J.; Shi, S.; Yang, j.; Gong, W.; Qiu, F.; Wang, L.; Du, L.; Chen, B. Wavelength selection of the multispectral lidar system for estimating leaf chlorophyll and water contents through the PROSPECT model. *Agric. For. Meteorol.* **2019**, *266*, 43–52. [CrossRef]
30. Cheng, T.; Rivard, B.; Sánchez-Azofeifa, A. Spectroscopic determination of leaf water content using continuous wavelet analysis. *Remote Sens. Environ.* **2011**, *115*, 659–670. [CrossRef]
31. Li, D.; Cheng, T.; Zhou, K.; Zheng, H.; Yao, X.; Tian, Y.; Zhu, Y.; Cao, W. WREP: A wavelet-based technique for extracting the red edge position from reflectance spectra for estimating leaf and canopy chlorophyll contents of cereal crops. *ISPRS J. Photogramm. Remote Sens.* **2017**, *129*, 103–117. [CrossRef]
32. Li, D.; Wang, X.; Zheng, H. Estimation of area- and mass-based leaf nitrogen contents of wheat and rice crops from water-removed spectra using continuous wavelet analysis. *Plant Methods* **2018**, *14*, 76–96. [CrossRef]
33. Wang, H.; Huo, Z.; Zhou, G.; Liao, Q.; Feng, H.; Wu, L. Estimating leaf SPAD values of freeze-damaged winter wheat using continuous wavelet analysis. *Plant. Physiol. Biochem.* **2016**, *98*, 39–45. [CrossRef] [PubMed]
34. Lu, J.; Huang, W.; Zhang, J. Quantitative identification of yellow rust and powdery mildew in winter wheat based on wavelet feature. *Spectrosc. Spectr. Anal.* **2016**, *36*, 1854–1858. [CrossRef]
35. Metz, M.; Lesnoff, M.; Abdelghafour, F.; Akbarinia, R.; Masseglia, F.; Roger, J.M. A "big-data" algorithm for KNN-PLS. *Chemom. Intell. Lab. Syst.* **2020**, 104076. [CrossRef]
36. Huang, Y.; Cao, J.; Ye, S.; Duan, J.; Wu, L.; Li, Q.; Min, S.; Xiong, Y. Near-infrared spectral imaging for quantitative analysis of active component in counterfeit imidacloprid using PLS regression. *Optik* **2013**, *124*, 1644–1649. [CrossRef]
37. Fujiwara, K.; Sawada, H.; Kano, M. Input variable selection for PLS modeling using nearest correlation spectral clustering. *Chemom. Intell. Lab. Syst.* **2012**, *18*, 109–119. [CrossRef]
38. Cai, W.; Li, Y.; Shao, X. A variable selection method based on uninformative variable elimination for multivariate calibration of near-infrared spectra. *Chemom. Intell. Lab.* **2008**, *90*, 188–194. [CrossRef]
39. Mehmood, T.; Liland, K.H.; Snipen, L.; Sæbø, S. A review of variable selection methods in Partial Least Squares Regression. *Chemom. Intell. Lab. Syst.* **2012**, *118*, 62–69. [CrossRef]
40. Li, H.; Xu, Q.; Liang, Y. libPLS: An integrated library for partial least squares regression and linear discriminant analysis. *Chemom. Intell. Lab. Syst.* **2018**, *176*, 34–43. [CrossRef]
41. Devos, O.; Duponchel, L. Parallel genetic algorithm co-optimization of spectral pre-processing and wavelength selection for PLS regression. *Chemom. Intell. Lab. Syst.* **2011**, *107*, 50–58. [CrossRef]
42. Jiang, H.; Zhang, H.; Chen, Q.; Mei, C.; Liu, G. Identification of solid state fermentation degree with FT-NIR spectroscopy: Comparison of wavelength variable selection methods of CARS and SCARS. *Spectrochim. Acta Part A Mol. Biomol. Spectrosc.* **2015**, *149*, 1–7. [CrossRef]
43. Li, H.; Liang, Y.; Xu, Q.; Cao, D. Key wavelengths screening using competitive adaptive reweighted sampling method for multivariate calibration. *Anal. Chim. Acta.* **2009**, *648*, 77–84. [CrossRef] [PubMed]
44. Sumanta, N.; Imranul Haque, C. Spectrophotometric analysis of chlorophylls and carotenoids from commonly grown fern species by using various extracting solvents. *Res. J. Chem. Sci.* **2014**, *4*, 63–69. [CrossRef]
45. Haitao, S.; Peiqiang, Y. Comparison of grating-based near-infrared (NIR) and Fourier transform mid-infrared (ATR-FT/MIR) spectroscopy based on spectral preprocessing and wavelength selection for the determination of crude protein and moisture content in wheat. *Food Control* **2017**, *82*, 57–63. [CrossRef]

46. Barnes, R.J.; Dhanoa, M.S.; Lister, S.J. Standard normal variate transformation and de-trending of near-infrared diffuse reflectance spectra. *Appl. Spectrosc.* **1989**, *43*, 772–777. [CrossRef]
47. Miller, J.R.; Hare, E.W.; Wu, J. Quantitative characterization of the vegetation red edge reflectance 1. An inverted-Gaussian reflectance model. *Int. J. Remote Sens.* **1990**, *11*, 1755–1773. [CrossRef]
48. Maire, G.; François, C.; Dufrêne, E. Towards universal broad leaf chlorophyll indices using PROSPECT simulated database and hyperspectral reflectance measurements. *Remote Sens. Environ.* **2004**, *89*, 1–28. [CrossRef]
49. Zheng, K.; Li, Q.; Wang, J.; Geng, J.; Cao, P.; Sui, T.; Wang, X.; Du, Y. Stability competitive adaptive reweighted sampling (SCARS) and its applications to multivariate calibration of NIR spectra. *Chemom. Intell. Lab. Syst.* **2012**, *112*, 48–54. [CrossRef]
50. Li, Q.; Huang, Y.; Song, X.; Zhang, J.; Min, S. Moving window smoothing on the ensemble of competitive adaptive reweighted sampling algorithm. *Spectrochim. Acta Part A Mol. Biomol. Spectrosc.* **2019**, *214*, 129–138. [CrossRef] [PubMed]
51. Geladi, P.; Kowalski, B.R. Partial least-squares regression: A tutorial. *Anal. Chim. Acta* **1986**, *185*, 1–17. [CrossRef]
52. Wang, Z.; Sakuno, Y.; Koike, K. Evaluation of chlorophyll-a estimation approaches using iterative stepwise elimination partial least squares (ISE-PLS) regression and several traditional algorithms from field hyperspectral measurements in the Seto Inland Sea, Japan. *Sensors* **2018**, *18*, 2656. [CrossRef]
53. Galvão, R.K.; Araujo, M.C.; José, G.E. A method for calibration and validation subset partitioning. *Talanta* **2005**, *67*, 736–740. [CrossRef] [PubMed]
54. Tian, H.; Zhang, L.; Li, M. Weighted SPXY method for calibration set selection for composition analysis based on near-infrared spectroscopy. *Infrared Phys. Technol.* **2018**, *95*, 88–92. [CrossRef]
55. Meroni, M.; Rossini, M.; Guanter, L.; Alonso, L.; Rascher, U.; Colombo, R.; Moreno, J. Remote sensing of solar-induced chlorophyll fluorescence: Review of methods and applications. *Remote Sens. Environ.* **2009**, *113*, 2037–2051. [CrossRef]
56. Singhal, G.; Bansod, B.; Mathew, L.; Goswami, J.; Choudhury, B.U.; Raju, P.L.N. Chlorophyll estimation using multi-spectral unmanned aerial system based on machine learning techniques. *Remote Sens. Appl. Soc. Environ.* **2019**, *15*, 100235. [CrossRef]
57. Awgchew, H.; Gebremedhi, H.; Taddesse, G. Influence of nitrogen rate on nitrogen use efficiency and quality of potato (*Solanum tuberosum* L.) varieties at Debre Berhan, Central Highlands of Ethiopia. *Int. J. Soil Sci.* **2016**, *12*, 10–17. [CrossRef]
58. Guo, Y.; Zhou, Y.; Tan, J. Wavelet analysis of pulse-amplitude-modulated chlorophyll fluorescence for differentiation of plant samples. *J. Theor. Biol.* **2015**, *370*, 116–120. [CrossRef]
59. Wang, Z.; Chen, J.; Fan, Y. Evaluating photosynthetic pigment contents of maize using UVE-PLS based on continuous wavelet transform. *Comput. Electron. Agric.* **2020**, *169*, 105–160. [CrossRef]
60. Li, D.; Cheng, T.; Jia, M. PROCWT: Coupling PROSPECT with continuous wavelet transform to improve the retrieval of foliar chemistry from leaf bidirectional reflectance spectra. *Remote Sens. Environ.* **2018**, *206*, 1–14. [CrossRef]
61. Li, Y.; Brian, K.V.; Li, Y. Lifting wavelet transform for Vis-NIR spectral data optimization to predict wood density. *Spectrochim. Acta Part A Mol. Biomol. Spectrosc.* **2020**, 118566. [CrossRef]
62. Huang, X.; Luo, Y.; Xia, L. An efficient wavelength selection method based on the maximal information coefficient for multivariate spectral calibration. *Chemom. Intell. Lab. Syst.* **2019**, *194*, 103872. [CrossRef]
63. Sun, H.; Liu, N.; Wu, L.; Chen, L.; Yang, L.; Li, M.; Zhang, Q. Water content detection of potato leaves based on hyperspectral image. *IFAC PapersOnLine* **2018**, *51*, 443–448. [CrossRef]
64. Li, H.; Liang, Y.; Xu, Q.; Cao, D. Model population analysis for variable selection. *J. Chemom.* **2009**, *24*, 418–423. [CrossRef]
65. Andries, J.P.M.; Vander Heyden, Y.; Buydens, L.M.C. Improved variable reduction in partial least squares modelling based on Predictive-Property-Ranked Variables and adaptation of partial least squares complexity. *Anal. Chim. Acta* **2011**, *705*, 292–305. [CrossRef]
66. Gourvénec, S.; Fernández-Pierna, J.A.; Massart, D.L.; Rutledge, D.N. An evaluation of the PoLiSh smoothed regression and the Monte Carlo Cross-Validation for the determination of the complexity of a PLS model. *Chemom. Intell. Lab. Syst.* **2003**, *68*, 41–51. [CrossRef]

67. Qiao, L.; Zhang, Z.; Chen, L.; Sun, H.; Li, M.; Li, L.; Ma, J. Detection of chlorophyll content in Maize Canopy from UAV Imagery. *IFAC PapersOnLine* **2019**, *52*, 330–335. [CrossRef]
68. Li, L.; Ren, T.; Ma, Y.; Wei, Q.; Wang, S.; Li, X.; Cong, R.; Liu, S.; Lu, J. Evaluating chlorophyll density in winter oilseed rape (*Brassica napus* L.) using canopy hyperspectral red-edge parameters. *Comput. Electron. Agric.* **2016**, *126*, 21–31. [CrossRef]
69. Tao, Z.; Ning, L.; Li, W.; Li, M.; Sun, H.; Zhang, Q.; Wu, J. Estimation of Chlorophyll Content in Potato Leaves Based on Spectral Red Edge Position. *IFAC PapersOnLine* **2018**, *51*, 602–606. [CrossRef]
70. Ji, R.; Shi, W.; Wang, Y.; Zhang, H.; Min, J. Nondestructive estimation of bok choy nitrogen status with an active canopy sensor in comparison to a chlorophyll meter. *Pedosphere* **2020**, *30*, 769–777. [CrossRef]
71. Liu, N.; Liu, G.; Sun, H. Real-time detection on spad value of potato plant using an in-field spectral imaging sensor system. *Sensors* **2020**, *20*, 3430. [CrossRef]

remote sensing

MDPI

Article

A New Coupled Elimination Method of Soil Moisture and Particle Size Interferences on Predicting Soil Total Nitrogen Concentration through Discrete NIR Spectral Band Data

Peng Zhou, Wei Yang *, Minzan Li and Weichao Wang

Key Laboratory of Modern Precision Agriculture System Integration Research, Ministry of Education, China Agricultural University, Beijing 100083, China; zhoupeng@cau.edu.cn (P.Z.); limz@cau.edu.cn (M.L.); wwc1031@cau.edu.cn (W.W.)
* Correspondence: cauyw@cau.edu.cn

Abstract: Rapid and accurate measurement of high-resolution soil total nitrogen (TN) information can promote variable rate fertilization, protect the environment, and ensure crop yields. Many scholars focus on exploring the rapid TN detection methods and corresponding soil sensors based on spectral technology. However, soil spectra are easily disturbed by many factors, especially soil moisture and particle size. Real-time elimination of the interferences of these factors is necessary to improve the accuracy and efficiency of measuring TN concentration in farmlands. Although, many methods can be used to eliminate soil moisture and particle size effects on the estimation of soil parameters using continuum spectra. However, the discrete NIR spectral band data can be completely different in the band attribution with continuum spectra, that is, it does not have continuity in the sense of spectra. Thus, relevant elimination methods of soil moisture and particle size effects on continuum spectra do not apply to the discrete NIR spectral band data. To solve this problem, in this study, moisture absorption correction index (MACI) and particle size correction index (PSCI) methods were proposed to eliminate the interferences of soil moisture and particle size, respectively. Soil moisture interference was decreased by normalizing the original spectral band data into standard spectral band data, on the basis of the strong soil moisture absorption band at 1450 nm. For the PSCI method, characteristic bands of soil particle size were identified to be 1361 and 1870 nm firstly. Next, normalized index N_p, which calculated wavelengths of 1631 and 1870 nm, was proposed to eliminate soil particle size interference on discrete NIR spectral band data. Finally, a new coupled elimination method of soil moisture and particle size interferences on predicting TN concentration through discrete NIR spectral band data was proposed and evaluated. The six discrete spectral bands (1070, 1130, 1245, 1375, 1550, and 1680 nm) used in the on-the-go detector of TN concentration were selected to verify the new method. Field tests showed that the new coupled method had good effects on eliminating interferences of soil moisture and soil particle size.

Keywords: discrete NIR spectral band data; soil total nitrogen concentration; moisture absorption correction index; particle size correction index; coupled elimination

Citation: Zhou, P.; Yang, W.; Li, M.; Wang, W. A New Coupled Elimination Method of Soil Moisture and Particle Size Interferences on Predicting Soil Total Nitrogen Concentration through Discrete NIR Spectral Band Data. *Remote Sens.* **2021**, *13*, 762. https://doi.org/10.3390/rs13040762

Academic Editor: Won-Ho Nam

Received: 28 December 2020
Accepted: 17 February 2021
Published: 19 February 2021

1. Introduction

Excessive application of commercial fertilizers has been cited as a source of contamination of surface and groundwater [1–4]. As one of the important parts of precision agriculture (PA), based on the needs of the plant and the soil nutrients state, variable rate (VR) fertilization can reduce the overuse of manure in the farmlands to protect the soil and environment [5–8]. Rapid, real-time, and accurate acquisition of high-resolution soil parameter information is the basis of applying VR fertilization operation [9].

As the main nutrient for plant and soil, soil total nitrogen (TN) is one of the key factors that determine plant nutrient levels and soil fertility [10,11]. The estimation of TN concentration status is crucial from agricultural and environmental points of view [12–14].

Traditional wet chemical methods of measuring TN concentration are time-consuming, costly, have low efficiency, and also cause environmental pollution due to their consumption of harmful chemicals [15–17]. Spectral analysis technology is a fast, nondestructive, online, and free pollution method that shows an increasing development potential in TN concentration measurement [18–21]. Many researchers focused on exploring rapid TN detection methods and corresponding soil sensors based on spectral technology in the past four decades. In the 1980s, Dalal and Henry [22] showed it is possible to measure TN concentration through near-infrared (NIR) reflectance spectroscopy. Since then, many scholars have succeeded to predict TN concentration based on NIR reflectance spectroscopy in the laboratory [23–25].

However, collecting soil samples in the field requires a lot of human labor. It will also take a relatively long time to obtain the measurement results [17]. Thus, estimation of TN concentration through NIR reflectance spectroscopy in the laboratory does not satisfy the requirement of VR fertilization operation. Simultaneously, many scholars focus on developing soil parameter sensors to supply rapid and high-resolution acquisition of soil parameters. Sudduth and Hummel [26] designed and tested a portable, NIR spectrophotometer intended for in-field use. Mouazen et al. [27,28] developed a soil sensing device with the measuring range of the spectrometer from 306.5 to 1710.9 nm. Decreasing the capital is significant for the development of soil sensors. Thus, cheaper light-emitting diodes (LEDs), laser diodes, and tungsten halogen light sources are used as the alternative component of the spectrometer to develop soil sensors [29–31]. Tang et al. [32] developed a portable SOC analyzer using 850 nm as a sensitive wavelength. An et al. [31] developed a portable TN sensor based on six discrete NIR bands (940, 1050, 1100, 1200, 1300, and 1550 nm) to detect TN content, which used the LEDs as the detection light source. Zhou et al. [30] developed an in-situ TN detector with seven discrete NIR bands (1260, 1330, 1360, 1430, 1530, 1580, and 1660 nm) based on the laser diodes, which realized in-situ detection of TN concentration in the farmlands. Li et al. [33] developed the portable TN concentration detector with six discrete NIR bands (1108, 1248, 1336, 1450, 1537, and 1696 nm), which used the tungsten halogen light sources as the detection light source.

However, several problems remain through spectral technology to predict soil parameters. First, the soil spectra are easily affected by interactions with various soil constituents (e.g., minerals and soil nutrients) [34,35]. Especially, field soil samples have more variations in soil moisture and other physical conditions (i.e., soil particle size, and water). Water and hydroxyls (O-H) have strong influences in the NIR region. Overtones of O-H stretching absorb near 1450 nm, and it conceals the absorption information of soil N-H bonds in the NIR spectroscopy, resulting in the spectral curve only reflecting the changing trend of soil moisture. However, the absorption information of soil N-H bonds reflects the concentration of TN. Reflectance from a soil sample varies with its soil particle size in the NIR regions of the electromagnetic spectrum. As the soil particle size decreased, the reflectance also gradually increased. It also conceals the absorption information of soil N-H bonds in the NIR spectroscopy. Thus, the soil moisture and particle size effects on the NIR spectroscopy must be eliminated before performing the estimation of TN by the NIR spectroscopy [36–38].

Many scholars have analyzed the interferences of soil particle size and moisture on soil parameter prediction, and several studies have proposed methods to eliminate these disturbances. For the soil moisture, Tekin et al. [39] used NIR reflectance spectroscopy to study the effects of different soil moisture contents on soil organic matter (SOM) prediction. The root mean square error of validation (RMSEP) and the residual prediction deviation (RPD) for dry soil samples reached 1.26% and 2.83, respectively, and the soil samples with the highest moisture content had values of 1.55% and 2.29, respectively. Zhang et al. [23] studied the real-time NIR spectra of soil samples without artificial drying and sieving, and six sensitive bands were determined for predicting TN concentration through wavelet analysis and continuum removal techniques. Prediction models of TN concentration were established through partial least squares regression (PLSR) and support vector machine

(SVM), respectively. For the PLSR model, the coefficient of determination of calibration (R_c^2) was 0.602, and the coefficient of determination of prediction (R_v^2) was 0.634. For the SVM model, R_c^2 reached 0.823, and R_v^2 reached 0.810. It revealed that the soil moisture interference was removed from the real-time soil NIR reflectance spectroscopy during TN prediction using the proposed method of wavelet analysis and continuum removal. External parameter orthogonalization, direct standardization, global moisture modeling, slope bias correction, and selective wavelength modeling have also been applied to eliminate the effect of soil moisture on soil NIR reflectance spectroscopy [40]. For the soil particle size, Barthès et al. [41] analyzed the NIR reflectance spectroscopy of sandy soils with different particle sizes in Burkina Faso and Congo and established the prediction models of TN concentration under different soil particle sizes. The results showed that the model with less than 2 mm soil particle size was better than the model established using other soil particle sizes. Bogrekci and Lee [42] investigated the effects of soil particle size on the prediction of soil rapid available phosphorus (P) using ultraviolet, visible, NIR spectroscopy. They found that for the sandy soil, with the decrease in soil particle size, the soil reflectivity increased and the absorbance decreased. The individual prediction method for each soil particle size produced better soil P predictions with lower standard errors of prediction for coarse, medium, and fine sand particles than the method that removed the effect of particle size on absorbance spectra and predicted P concentration of the soil samples.

Summarizing the above research, there was no relevant research on the elimination of soil moisture and particle size effects on predicting soil parameters through the discrete NIR spectral band data. Soil parameter detectors developed based on discrete NIR spectral band data also face serious interferences caused by soil moisture and particle size. The methods used in the laboratory to eliminate the interference of soil particle size and moisture by sieving after grinding and drying cannot be applied to the discrete NIR spectral band data in the field sensing. The above elimination methods of soil particle size and moisture are not suitable for discrete NIR spectral band data acquisition in the field sensing. An et al. [36] studied the effects of soil moisture and particle size on the prediction of TN concentration using a portable TN detector developed based on six discrete NIR bands (940, 1050, 1100, 1200, 1300, and 1550 nm). The moisture absorption index and mixed calibration set were proposed to eliminate the soil moisture and particle size interferences, respectively. However, the mixed calibration set method entailed a complicated process and could not perform real-time elimination of the interference of soil particle size.

The combination effects and real-time elimination methods of soil moisture and particle size on predicting TN concentration through the discrete NIR spectral band data have rarely been explored. The discrete NIR spectral band data can be completely different in the band attribution with continuum spectra, it does not have continuity in the sense of spectra. Therefore, the study of methods for eliminating the effects of soil particle size and moisture on the discrete NIR spectral band data has important theoretical significance and practical application value for improving the performance of detectors based on the discrete NIR spectral band data. The main goal of the paper is to establish an on-line elimination method of soil moisture and particle size interferences, which could be used in the discrete NIR spectral band data to improve the measurement accuracy of soil parameters. Thus, in this study, six discrete NIR spectral bands (1070, 1130, 1245, 1375, 1550, and 1680 nm) used in the on-the-go detector of TN concentration were chosen as the object to perform the research [29]. A new coupled elimination method of soil moisture and particle size interferences on predicting TN concentration through discrete NIR spectral band data was proposed and evaluated.

2. Materials and Methods

2.1. Major Steps of the Study

Figure 1 shows the major procedures in this research, which included three stages. During stage 1, 60 soil samples were collected from Field 1. In order to research the effect of soil moisture on discrete NIR spectral band data, the soil particle size was eliminated by artificial sieving. Additionally, the moisture absorption correction index (MACI) method to eliminate the disturbance of soil moisture on discrete NIR spectral band data was proposed and validated. During stage 2, 24 soil samples were collected from Field 2. Before performing analysis, soil moisture disturbance was eliminated through drying by the oven. Next, four groups of soil samples with different particle sizes were obtained, each group contained six TN concentration levels, and each TN concentration level contained four soil samples. A total of 96 soil samples with different particle sizes and TN concentrations were generated from the 24 soil samples. Next, the effect of soil particle size on discrete NIR spectral band data was explored and the characteristic wavebands of soil particle size were determined. The particle size correction index (PSCI) method was proposed and validated to eliminate the effect of soil particle size on discrete NIR spectral band data.

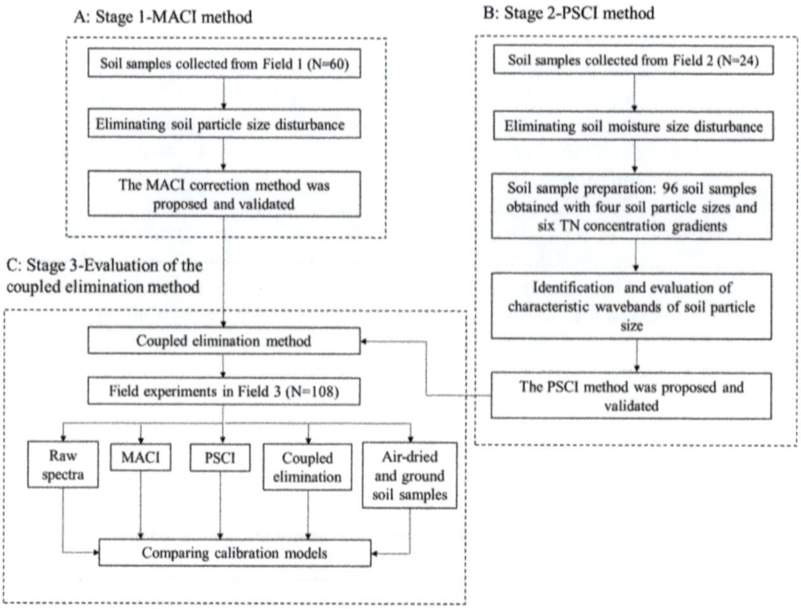

Figure 1. Flow chart of major steps adopted in this study. In the diagram, MACI refers to the moisture absorption correction index and PSCI is the particle size correction index.

During stage 3, a field experiment was performed in Field 3 to evaluate the coupled elimination method through the on-the-go detector, which included five processing methods. Finally, five models of TN concentration were calibrated and compared.

2.2. On-The-Go Detector of TN Concentration

An on-the-go detector of TN concentration was previously developed through discrete NIR spectral band data. This detector was reported in our early research [29]. Figure 2 shows the structure of the on-the-go detector, which consisted of mechanical, optical, and control units. The mechanical unit was the basis of the entire detector and supplied a downward force for the work of the subsoiler. Thus, the electrical control cabinet of the mechanical unit was designed as a three tier structure to integrate the optical and control units. The optical unit is used for sensing soil in the soil detection trench and a control unit embedded into an SVM model for obtaining real-time readings of TN concentration. A tungsten halogen light source was a better choice for the vehicle equipment than LED light and laser sources with respect to light intensity and stability, which provided a broadband source over the 300–2500 nm range. The selected performance parameters of the tungsten halogen light were as follows: <25% of radiation uniformity, <1% of unstable time, 20 × 20 mm radiation area, adjustable radiation intensity, 150 W of maximum power, and 2000 h of average life. The light had to be transmitted to the soil surface with minimal light loss. Therefore, an NIR optical fiber with a transmission range of 800–1900 nm was used. The diameter of the selected filter was 5 mm, and the average full width at half maximum (FWHM) bandwidth was 20 nm. The selected photoelectric sensor was InGaAs with a responding range of 800–1800 nm and a spectral responsiveness of 0–0.9. The sapphire glass attached to the bottom of the sensor mounting plate was used to ensure that the sensors and the filters were free from soil pollution. When the on-the-go detector performed detection of TN concentration in farmland, it was mounted on the tractor through a three-point suspension structure. The depth of the subsoiler that pushed into the soil exceeded 200 mm, which was used to loosen the soil and smash large soil particles. The flattener made the soil more flat and suitable for spectrum detection. The NIR light emitted by the tungsten halogen light source was transmitted through an NIR fiber to the flat detection soil surface. The diffused reflection light of the detected soil was changed into a single-band light by each optical filter, and each photoelectric detector converted the single-band light into an electrical signal. All detected electrical signals were amplified, filtered, and digitized into absorbance data, and then all data were collected and displayed in real-time by the data receiving software embedded with the whiteboard data, by which and detected spectral data the absorbance data were calculated at each sensitive wavelength. For this on-the-go detector, the group of the discrete NIR spectral bands of 1070, 1130, 1245, 1375, 1550, and 1680 nm was selected as the sensitive wavebands to predict TN concentration. Furthermore, the on-the-go detector was used to embed the new coupled elimination method of soil moisture and particle size interferences on predicting TN concentration through discrete NIR spectral band data.

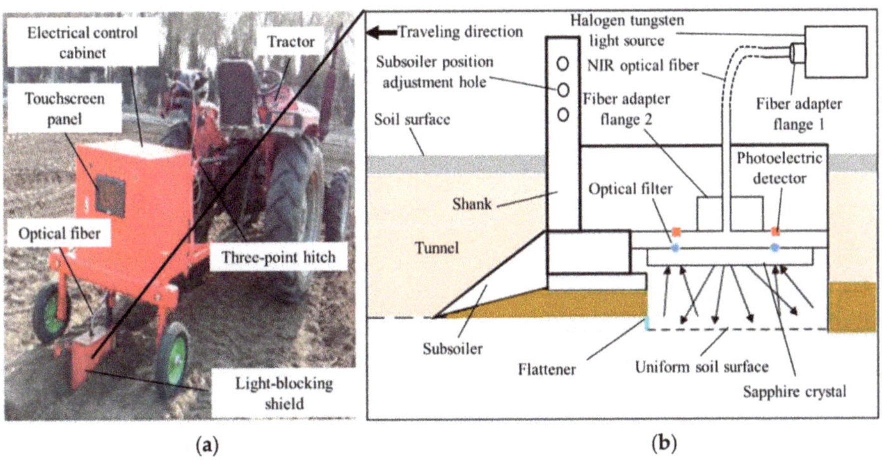

Figure 2. Structure of the on-the-go detector. (**a**) Working scene of the on-the-go detector; (**b**) subsoiler and optical path of the detection unit.

2.3. Experimental Materials and Methods

During stages 1 and 2 (Figure 1), soil samples were collected from the experimental farm of China Agricultural University on the outskirts of Beijing to perform the research of soil moisture and particle size interference elimination method. The soil type was common brown soil in northern China. The experimental area is in the range of north latitude 40.15395721–40.1534362 and east longitude 116.2155061–116.2150130. Since the fertilization usually occurs from the soil surface to the soil depth of 30 cm, soil samples were collected from 5 to 30 cm after removing 5 cm of the topsoil [43]. A total of 60 soil samples used to study the effect of soil moisture were collected at Field 1 (F1, 10 ha). Corn and wheat were cropped in F1 with tillage. In total, 24 soil samples were collected from the Field 2 (F2, 6 ha), which was a standard field without fertilization (the soil nutrient concentration is almost zero). The soil textural fractions (clay, silt, and sand) of F2 were the same with F1, both the two fields were sandy loam. In total, 2–2.5 kg of soil at each sampling point was collected and put it into double-layer cowhide bags for sealing to prevent the loss of soil moisture. A total of 84 soil samples were collected.

The first set of soil samples (N = 60) was used to study the effect of soil moisture. All soil samples were sieved under laboratory conditions using a 20-mesh sifter (0.9 mm) to produce pretreated samples with consistent particle sizes [36].

Another set of soil samples (N = 24) was used to study the effect of soil particle size. In order to obtain soil samples with different TN concentration, the standard soil samples were first dried in a drying box, the temperature was set to 85 °C, and the time for drying was 24 h [44,45]. The air-dried soil samples were randomly divided into six groups, four in each group. The soil nitrogen solution was configured with urea [46–48]. The concentration gradient of the soil nitrogen solution is divided into 1–6 grades, grade 1 nitrogen concentration is 0 $g \cdot kg^{-1}$, grade 2 nitrogen concentration is 0.04 $g \cdot kg^{-1}$, grade 3 nitrogen concentration is 0.08 $g \cdot kg^{-1}$, grade 4 nitrogen concentration is 0.12 $g \cdot kg^{-1}$, grade 5 nitrogen concentration is 0.16 $g \cdot kg^{-1}$, and grade 6 nitrogen concentration is 0.2 $g \cdot kg^{-1}$. When configuring the soil samples, we simulated the average summer soil moisture content (7%) of the cultivated layer of the Shangzhuang Experimental Station to configure the soil samples. After the soil configuration was completed, the drying process was carried out, and all the soil samples were sieved, respectively, 10-mesh sieves (2.0 mm), 20-mesh sieves (0.9 mm), 40-mesh sieves (0.45 mm), and 80-mesh sieves (0.2 mm). Finally, four groups of soil samples with different particle sizes were obtained, each group contained six TN concentration levels, and each TN concentration level contained four soil samples. A total of 96 soil samples with different particle sizes and TN concentrations were generated from the 24 soil samples.

During stage 3 (Figure 1), soil samples for verification were used to evaluate the new coupled elimination method. A verification experiment was conducted on the experimental farm of China Agricultural University. The soil type was common brown soil in northern China Field 3 (F3, 5.5 ha) which was one long-term research site located within 1.0 km of F1. Corn and wheat were cropped in F3 with tillage. The soil textural fractions (clay, silt, and sand) of F3 were the same as that of F1, which were sandy loam. Figure 3 shows the location and soil samples collection point in Field 3.

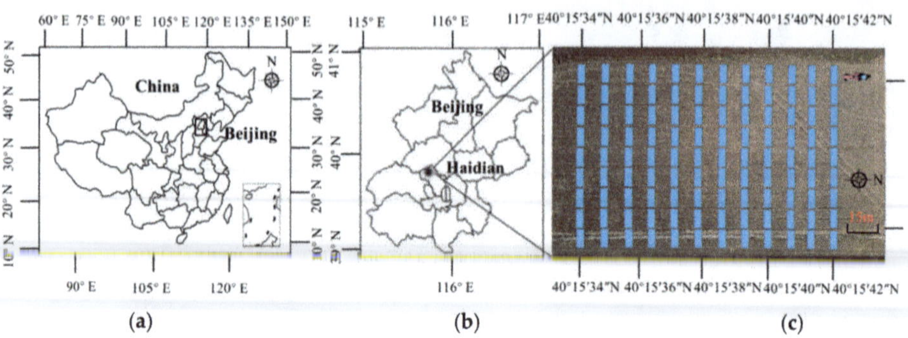

Figure 3. Soil sampling locations and treatments of the experiments at Field 3 (a) in Beijing (b) of China (c).

The experiment consisted of two parts which were data of discrete NIR spectral band data acquisition and soil samples collection. The experimental plot was 240 × 190 m, which was divided into 108 sampling cells. Each cell had a dimension of 5 × 20 m. Data and soil samples were collected every 10 m in each cell, blue cells in Figure 3 show the location of acquiring data and soil samples. All samples were placed in a double-thickness sampling bag to prevent moisture dispersion and then sent to the laboratory for optical and chemical measurements.

2.4. Laboratory Measurement

For each soil sample, it was divided into two parts, one was used for TN concentration measurement, and another was used for NIR reflectance spectroscopy. The TN concentration was tested with the FOSS Kjeltec™2300 Nitrogen Analyzer produced by FOSS, Sweden [49]. After grinding each soil sample, weighing 2.0 g, was put into a long test tube. The long test tube was also filled with 6.2 g catalyst. K_2SO_4:CuSO4.5H_2O was mixed and ground at 30:1. Finally, 20 mL of concentrated sulfuric acid was added to the long test tube for nitrification. The temperature of the nitration furnace was set to 420 °C for 2.5 h [50]. After nitrification, it was cooled down and the total nitrogen concentration was measured with the FOSS Kjeltec™2300 nitrogen analyzer. The NIR reflectance spectroscopy of soil samples was collected by the MATRIX-I Fourier Transform NIR Spectroscopy Analyzer produced by Bruker, Germany. Before soil spectra collection, the analyzer parameters needed to be set. The spectrum collection range, spectrum sampling interval, and scan times were set to 780–2550 nm, 3 nm, and 64 times, respectively. Approximately 20 g of soil samples were put into a quartz cuvette with a diameter of 50 mm, and put into the rotating sample cell of the analyzer for spectrum detection [17,51].

2.5. Model Accuracy and Methodology

In addition to the coefficient of determination (R^2), the root mean square error of validation (RMSEP), and the residual prediction deviation (RPD) were used to evaluate the prediction model of TN concentration [52]. Equations (1) and (2) present their calculation Equations.

$$\text{RMSEP} = \sqrt{\frac{\sum_{j=1}^{n_p}\left(y_j - \hat{y}_{pj}\right)^2}{n_p - 1}}, \tag{1}$$

$$\text{RPD} = \sqrt{\sum_{i=1}^{n_p}\left(y_j - \overline{y}_j\right)^2 / (n_p - 1)} \Big/ \sqrt{\sum_{i=1}^{n_p}\left(y_j - \hat{y}_{pj}\right)^2 / (n_p - 1)}, \tag{2}$$

where j is the number of soil samples, y_j is the TN concentration measured with the FOSS Kjeltec™2300 Nitrogen Analyzer, $\hat{y}_{cj}$ is the TN concentration predicted by the modeling set, $\hat{y}_{pj}$ is the TN concentration predicted by the validation set, $\overline{y}_j$ is the average value of the TN concentration measured with the FOSS Kjeltec™2300 Nitrogen Analyzer, n_c is the model set sample number, and n_p is the validation set sample number.

3. Results

3.1. Research on Eliminating the Interference of Soil Moisture

3.1.1. Interference of Soil Moisture on Discrete Spectral Band Data

The first set of 60 calibration samples with consistent particle sizes were measured through the developed on-the-go detector. To analyze the effect of soil moisture on soil discrete NIR spectral band data quantitatively, the spectral data of a soil sample with TN concentration of 0.1135 $g \cdot kg^{-1}$ was used as the research object. Figure 4 shows the absorbance values at seven sensitive wavelengths under different soil moisture contents. The absorbance curve obtained by the on-the-go detector from 1070 to 1680 nm showed a consistent trend, indicating that the curve was a good reflection of the soil spectral information [53]. Taking the soil absorbance at 1450 nm as an example, when the soil moisture content increased from 3.18% to 14.16%, the soil absorbance increased from 0.79 to 1.11, and the soil spectral absorbance increased by 40.51%. Absorbance data analysis indicated that along with the soil moisture content increase, the absorbance of the soil spectrum increased rapidly. Soil moisture had a serious influence on soil discrete NIR spectral band data [54–56]. The effects of soil moisture must be decreased when predicting TN concentration.

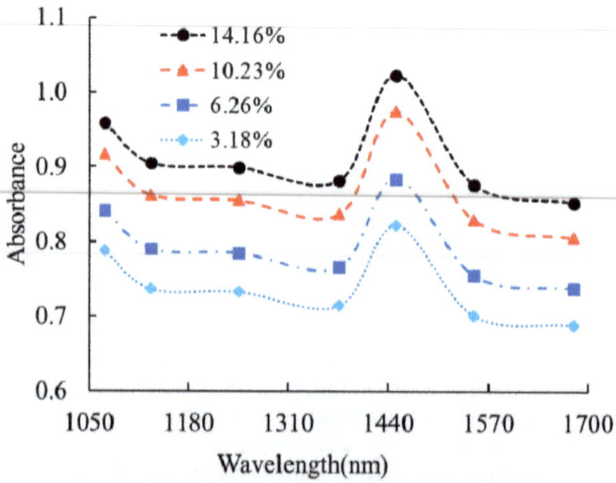

Figure 4. Absorbance curves of soil samples with different soil moisture contents.

Our goal is to eliminate the influence of soil moisture on the prediction of TN concentration through the on-the-go detector. Soil NIR spectroscopy indicates that three distinct moisture absorption peaks are present in the soil NIR spectroscopy at 1450, 1940, and 2210 nm [57,58]. However, in our research, we only focused on the absorption peak of O-H in H_2O at 1450 nm. O-H have strong influences in the NIR spectroscopy. Overtones of O-H stretching absorb near 1450 nm, and it conceals the absorption information of soil N-H bonds, resulting in the spectral curve only reflecting the change trend of soil moisture. However, the absorption information of soil N-H bonds reflects the concentration of TN. If the raw spectrum of the measured soil can be normalized to a standard spectrum with the same soil moisture content, the absorption information of soil N-H bonds on the NIR spectroscopy could be effectively extracted. Thus, the interference of soil moisture on predicting TN concentration could be decreased. However, when using the on-the-go detector to measure soil discrete NIR spectral band data, we were unable to directly obtain the soil moisture content in the fields. Thus, an alternative method must be explored to solve this problem.

3.1.2. Eliminating the Interference of Soil Moisture on Discrete NIR Spectral Band Data

To eliminate the interference of soil moisture, we need to classify different soil samples based on soil moisture content. The previous analysis revealed the difficulty of obtaining soil moisture content in real-time. Therefore, the MACI method was proposed to eliminate the effect of soil moisture on predicting TN concentration through the on-the-go detector. It was used to convert raw soil spectral data into standard spectral data. The MACI value of the on-the-go detector was calculated using Equation (3).

$$\text{MACI} = \frac{1 - A_{1450}}{1 + A_{1450}} \times 100, \tag{3}$$

where A_{1450} is the spectral absorbance value of the soil samples at 1450 nm and MACI is the soil moisture absorption index correlation. Soil samples with different moisture contents can be divided into groups based on MACI values; thus, we divided the detected soil samples into five types based on MACI values. Table 1 shows the five MACI values when the MACI value is used to represent the soil moisture level. Their corresponding ranges of soil moisture content and accuracy are also shown. When the soil moisture was between

0% and 3.0%, the MACI value was less than 8.66. When the soil moisture was 3.0–6.0%, 6.0–10.0%, and 10.0–13.0%, the MACI values were 8.66–12.66, 12.66–14.66, and 14.66–16.66, respectively. When the soil moisture content was higher than 13%, the MACI value was greater than 16.6.

Table 1. Accuracy of MACI classification and correction factors.

Serial Number	Soil Moisture Content Grade	MACI	Accuracy Rate (%)	W_j
1	Low level (0–3.0)	>9.11	100	1
2	Low-medium level (3.0–6.0)	9.11–6.05	88	0.92
3	Middle level (6.0–10.0)	6.05–2.03	86	0.86
4	High-medium level (10.0–13.0)	2.03–1.12	81	0.76
5	High level (>13.0)	<1.12	100	0.68

Note: MACI = moisture absorption correction index.

The classification of soil samples was the best according to the classification of MACI values. Although the soil moisture content could not be directly obtained in real-time when the on-the-go detector was used to measure soil in the fields, the soil samples could be classified by MACI.

Each classification group required a correction factor to normalize the spectral data, and Equation (5) was used to calculate the correction factor. In this equation, j (j = 1, 2,..., 5) is the serial number of the taxonomic group, W_j is the correction factor for the taxonomic group j, A_{avgj} is the soil sample average absorbance of group j, and A_{avgs} is the soil sample average absorbance of group 1. The soil sample average absorbance value of group 1 was used as the reference absorbance value. Table 1 also shows all five correction factors.

$$W_j = \frac{A_{avgj}}{A_{avgs}} \tag{4}$$

After determining the correction factor W_j ($j = 1, 2, \ldots, 5$), the raw absorbance data were corrected according to

$$A_{jc} = A_j \times W_j, \tag{5}$$

where A_j is the soil sample original absorbance value of group j, A_{jc} is the soil sample corrected normalized absorbance value, and W_j is the correction factor.

Absorbance data at 1070, 1130, 1245, 1375, 1450, 1550, and 1680 nm were detected using the on-the-go detector. Using the proposed MACI method to eliminate the interference of soil moisture, the wavelength absorbance at 1450 nm was selected to calculate the MACI value, and the raw spectral data were normalized to eliminate the interference caused by soil moisture content. Figure 5 shows the different absorbance curves for the change in soil moisture content after normalization. When the soil moisture content was 3.18% and 14.16%, the absorbance values at 1450 nm were 0.72 and 0.75, respectively; the absorbance value only increased by 4.16%. Compared with the soil spectrum without correction by the MACI method, the effect of soil moisture on soil discrete NIR spectral band data was greatly reduced, and the MACI method proposed in this study was proven to be effective in reducing soil moisture on the discrete NIR spectral band data.

3.2. Research on Eliminating the Interference of Soil Particle Size

3.2.1. Eliminating the Interference of Soil Moisture on Discrete NIR Spectral Band Data

Table 2 statistically analyzes the 96 soil samples prepared. The TN concentration in the soil was detected by the FOSS Kjeltec™2300 Nitrogen Analyzer [49]. The TN concentration value shows that the TN concentration of the prepared soil sample is distributed in the range of 0.003–0.206 g·kg^{-1}, with approximately 0.04 g·kg^{-1} as the TN concentration interval, divided into six TN concentration grades, the gradient distribution of TN concentration grade is reasonable.

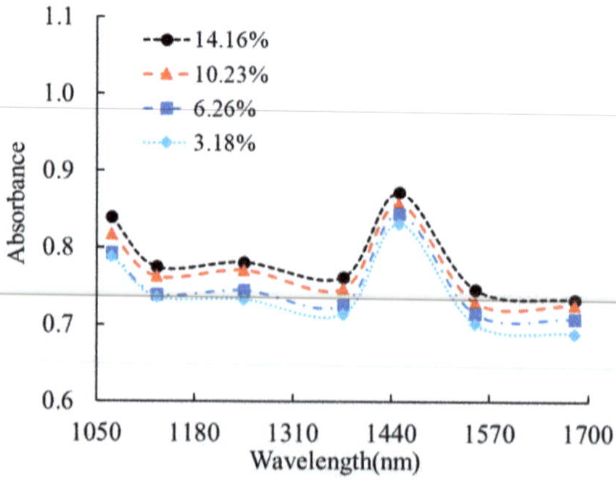

Figure 5. Corrected absorbance at seven discrete NIR spectral bands.

Table 2. Statistics of 96 soil samples prepared. The range of soil total nitrogen (TN) represents the measured TN of soil samples prepared.

Soil Particle Size	TN Level g·kg^{-1}	Range of TN g·kg^{-1}	Particle Size	TN Level g·kg^{-1}	Range of TN g·kg^{-1}
2.0 mm	0	0.005–0.012	0.9 mm	0	0.003–0.012
	0.04	0.036–0.047		0.04	0.029–0.037
	0.08	0.071–0.077		0.08	0.076–0.091
	0.12	0.109–0.123		0.12	0.119–0.133
	0.16	0.162–0.171		0.16	0.153–0.176
	0.2	0.186–0.194		0.2	0.191–0.206
0.45 mm	0	0.006–0.02	0.2 mm	0	0.009–0.018
	0.04	0.031–0.042		0.04	0.026–0.04
	0.08	0.063–0.071		0.08	0.060–0.067
	0.12	0.119–0.131		0.12	0.128–0.138
	0.16	0.142–0.153		0.16	0.153–0.169
	0.2	0.178–0.190		0.2	0.187–0.199

Note: TN = soil total nitrogen.

According to the analysis in Table 2, although the soil samples are collected from standard fields that are not fertilized all year, there is still TN in the soil samples, which will affect the TN concentration of the prepared soil samples.

Using the drying oven at 80 °C for 24 h, the second set of 96 soil samples was dried to eliminate the influence of soil moisture [17]. Subsequently, all soil samples in the second set were sieved by sifters of different meshes, and discrete NIR spectral band data were obtained by the on-the-go detector. Figure 6 shows the absorbance value of a single soil sample at six discrete NIR spectral bands with the TN concentration of 0.086 g·kg^{-1}, including five curves, which are the original soil spectrum, the soil spectra with soil particle sizes of 2.0, 0.9, 0.45, and 0.2 mm. When the soil particle size decreased from 2 to 0.2 mm, the corresponding soil samples' average absorbance also decreased from 0.74 to 0.63, and the overall absorbance changed by 14.86%. The results showed that soil particle size had a considerable effect on the soil discrete NIR spectral band data. When the soil particle size was between 0.45 and 0.9 mm, the absorbance of soil discrete NIR spectral band data was relatively stable. Larger and smaller particle sizes will cause a large difference on soil absorbance. Figure 6 also shows that the air-dried method could significantly decrease the interference of soil moisture on the spectral band of 1450 nm. Therefore, no absorption peak was observed at 1450 nm.

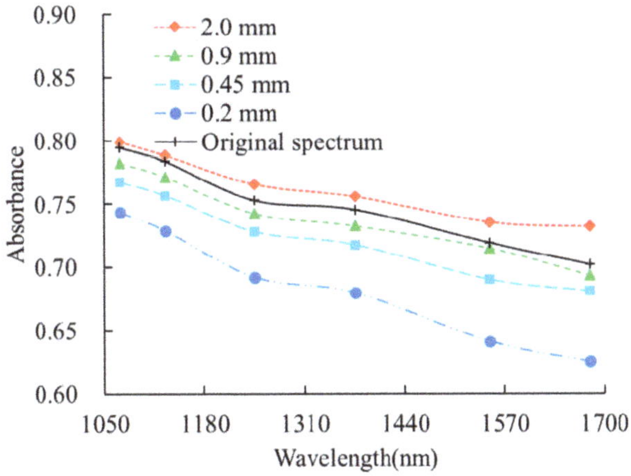

Figure 6. Absorbance curves of soil samples with different soil particle sizes.

3.2.2. Identification of Characteristic Wavebands of Soil Particle Size

Figure 7 shows the spectral standard deviation curves of four different soil particle sizes. It can be concluded from Figure 7 that the standard deviation value in the range of 850–1000 nm was smaller than in the spectral range of 1000–2500 nm. As each spectral standard deviation curve is under the same particle size, the interference of soil particle size is eliminated, and only the TN concentration is different. Thus, the 1000–2500 nm range is the sensitive spectral region of TN concentration. Simultaneously, the black curve in Figure 7 is the spectral standard deviation curve of all 96 soil samples, by the comparative analysis between the five standard deviation curves in the spectral range of 1000–2500 nm. After adding the difference in soil particle size, the black standard deviation curve in the spectral range between 1340–1420 nm and 1795–1955 nm changed up and down. In order to show the changes between different spectral standard deviation curves, spectral ranges 1340–1420 and 1795–1955 nm of the spectral standard deviation curve were extracted in Figure 7. Obviously, this change indicates that the above two spectral regions are closely related to the change of soil particle size and are sensitive spectral regions of soil particle size. Therefore, in this study, the extreme points 1361 and 1870 nm of the two spectral regions were selected as the characteristic wavebands of soil particle size.

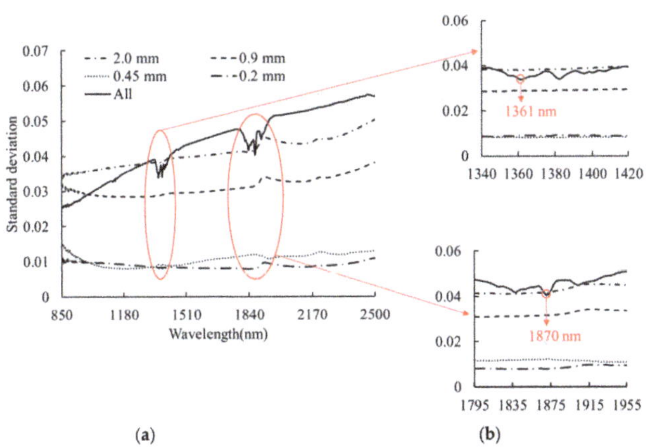

Figure 7. Standard deviation spectra. (**a**) Standard deviation spectra curve in the range 850–2500 nm; (**b**) standard deviation spectra curves in the ranges 1340–1420 nm and 1795–1955 nm.

3.2.3. Research on the Classification of Soil Particle Size

The 96 soil samples were divided into two groups, with 48 soil samples in each group. Each group of soil samples includes four soil particle size categories, each soil particle size category includes six TN concentration grades, and each soil concentration grade includes two soil samples. The two characteristic wavebands 1361 and 1870 nm were used to classify the soil particle size. The normalized index N_p was used as a single variable to predict the soil particle size, and the N_p is calculated by Equation (6). The soil classification model is constructed by the support vector machine (SVM) algorithm. After many attempts, the penalty parameter c of the SVM classification model is 2 and the kernel function is 1 to obtain the optimal classification result. Figure 8 shows the prediction results of soil particle size based on SVM. The overall classification accuracy rate of the four soil categories is 97.92%. Among the four soil categories, the classification accuracy rate when the soil particle sizes are 2.0, 0.45, and 0.2 mm is 100%, and the classification accuracy rate when the soil particle size is 0.90 mm is 91.67%.

$$N_p = \frac{A_{1870} - A_{1361}}{A_{1870} + A_{1361}}, \tag{6}$$

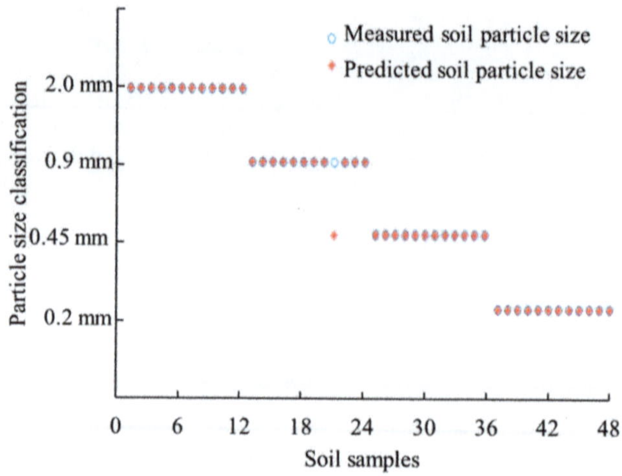

Figure 8. Classification of soil particle size based on support vector machine (SVM).

In Equation (6): A_{1870} and A_{1361} are the absorbance of the soil at 1870 and 1361 nm.

3.2.4. Eliminating the Interference of Soil Particle Size

Based on the soil particle size classification model established by SVM, the normalized index N_p was used as a single input to classify the soil particle size. The results show that the method is feasible. Based on the above results, a soil particle size correction coefficient P_c was proposed to correct the origin discrete NIR spectral band data. The soil particle size correction coefficient P_c is obtained by Equation (7),

$$P_c = \frac{\left(\frac{A_{1870} - A_{1361}}{A_{1870} + A_{1361}}\right)}{\overline{N}_p}, \tag{7}$$

In Equation (7): A_{1870} and A_{1361} are the absorbance of the corrected soil spectrum at 1870 and 1361 nm, and $\overline{N}_p$ is the normalized index of the absorbance of the 0.20 mm soil at 1870 and 1361 nm as the reference value. It is generally believed that the smaller soil particle size can eliminate the influence of soil particle size to the greatest extent, so the spectrum data with a particle size of 0.20 mm was determined as the reference spectrum.

The absorbance A_c of discrete NIR spectral band data corrected by the N_p was obtained by Equation (8),

$$A_c = A \times P_c, \tag{8}$$

In Equation (8): A is the soil absorbance of discrete NIR spectral band data before correction, and A_c is the soil absorbance after correction. In this study, the absorbance at six discrete NIR bands (1070, 1130, 1245, 1375, 1550, 1680 nm) of the on-the-go detector were used to verify the PSCI method. Figure 9 shows the absorbance of discrete NIR spectral band data after soil particle size correction. A comparative analysis of Figures 6 and 9 was computed. For example, at 1245 nm, the difference between the absorbance value with a soil particle size of 0.2 mm and the absorbance value with soil particle sizes 2.0, 0.9, 0.45 mm, and original spectrum were 0.0737, 0.0506, 0.0361, and 0.0607, respectively. After correction with the PSCI method, the difference was 0.0263, 0.0111, −0.0011, and 0.0228, and the difference was reduced by 64.31%, 78.06%, 103.04%, and 62.43% after correction. The results show that the soil particle size correction coefficient can significantly reduce the interference of soil particle size to a greater extent.

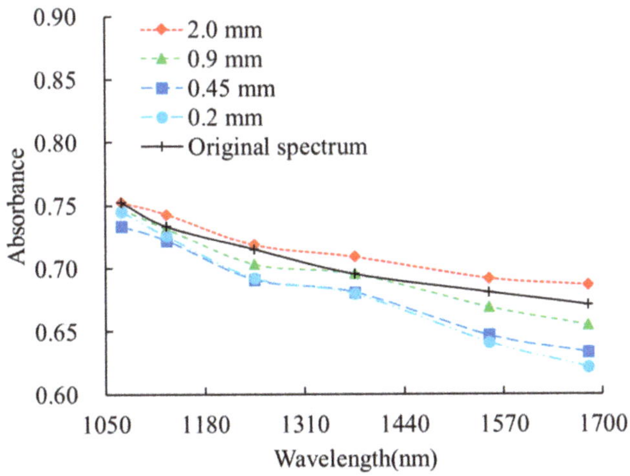

Figure 9. Corrected absorbance at six discrete NIR spectral bands.

3.3. Coupled Elimination Method of Soil Moisture and Particle Size Interferences

In order to achieve on-line coupled elimination of soil moisture and particle size interferences on predicting TN concentration through discrete NIR bands data, we performed detection of TN concentration through the soil sensors developed based on discrete spectral bands in the farmlands, MACI was firstly used to decrease the interference of soil moisture on discrete spectral band data. Then, PSCI was calculated to decrease the effect of soil particle size on the discrete spectral band data after corrected with the MACI method. Finally, TN concentration was obtained through the discrete spectral band data with coupled elimination of soil moisture and particle size interferences. Figure 10 shows the flowchart of the coupled elimination method.

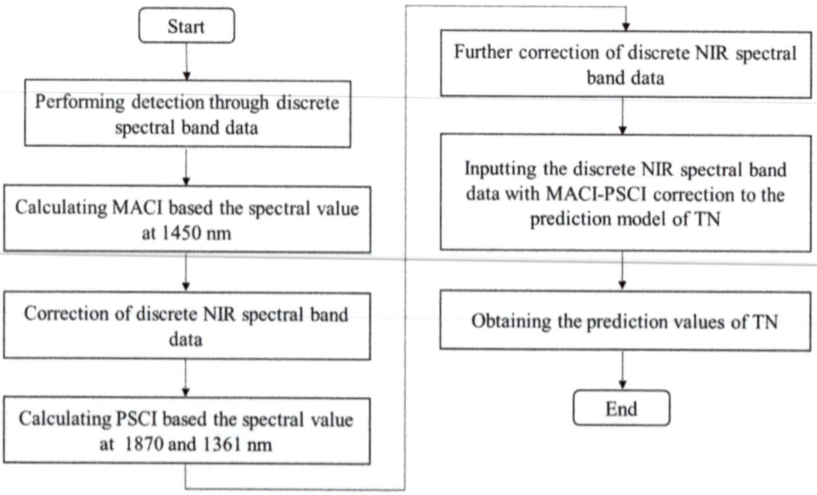

Figure 10. Flowchart of the coupled elimination method.

3.4. Evaluation of the New Coupled Elimination Method

Discrete NIR spectral band data was obtained used the on-the-go detector by in-field and laboratory measurement. The spectral data after different moisture and particle size removal processes were used to establish the TN prediction model. The elimination effects of different soil particle size and moisture removal methods on the prediction of TN concentration by discrete NIR spectral band data were evaluated. Figure 11 showed the comparison of elimination effects. The data used for prediction models included (a) raw discrete NIR spectral band data, model 1; (b) discrete NIR spectral band data using the proposed MACI correlation method to eliminate soil moisture disturbance, model 2; (c) discrete NIR spectral band data using the proposed PSCI method to eliminate soil particle size disturbance, model 3; (d) discrete NIR spectral band data using the proposed new coupled elimination method to eliminate soil moisture and particle size disturbances, model 4; (e) discrete NIR spectral band data on dried and ground soil samples, model 5. SVM was used to establish a prediction model for TN concentration based on data treated by different methods. In the established SVM model, the kernel function is the radial basis kernel function, three parameters affecting model regression results: penalty parameter C, radial basis kernel function γ parameter λ, and loss parameter ε, and they are 1.3, 0.25, and 0.01, respectively. Table 3 and Figure 10 show the calibration results. For the raw discrete NIR spectral band data, the prediction R^2 was 0.65. However, when the model was established by discrete NIR spectral band data after eliminating soil moisture and soil particle size interference, the RPD value of prediction model 4 reached 2.59, and the prediction R^2 reached 0.84, respectively. For models 2 and 3, there is only soil moisture or particle size disturbance eliminated by the MACI or PSCI method, respectively. However, both of them have an improvement of prediction R^2 for 12.3% and 9.2% compared with model 1. The above analysis shows that the effects of soil moisture and soil particle size on the discrete NIR spectral band were serious. Further, the TN concentration prediction model 5 was established under laboratory conditions through the discrete NIR spectral band data obtained from the soil samples after drying and grinding. The prediction accuracy of model 4 is still below that of model 5. Although, the proposed MACI and PSCI methods were quantitatively proven to have good effects on eliminating soil moisture and soil particle size interferences on predicting TN concentration through discrete NIR spectral band data. However, the 8.5% decrease of prediction was R^2 between models 4 and 5, which means that the effects of soil moisture and particle size were not eliminated using the new coupled elimination method. Thus, it is necessary to continue to explore other elimination methods of soil moisture and particle size effects on predicting TN concentration through discrete NIR spectral band data.

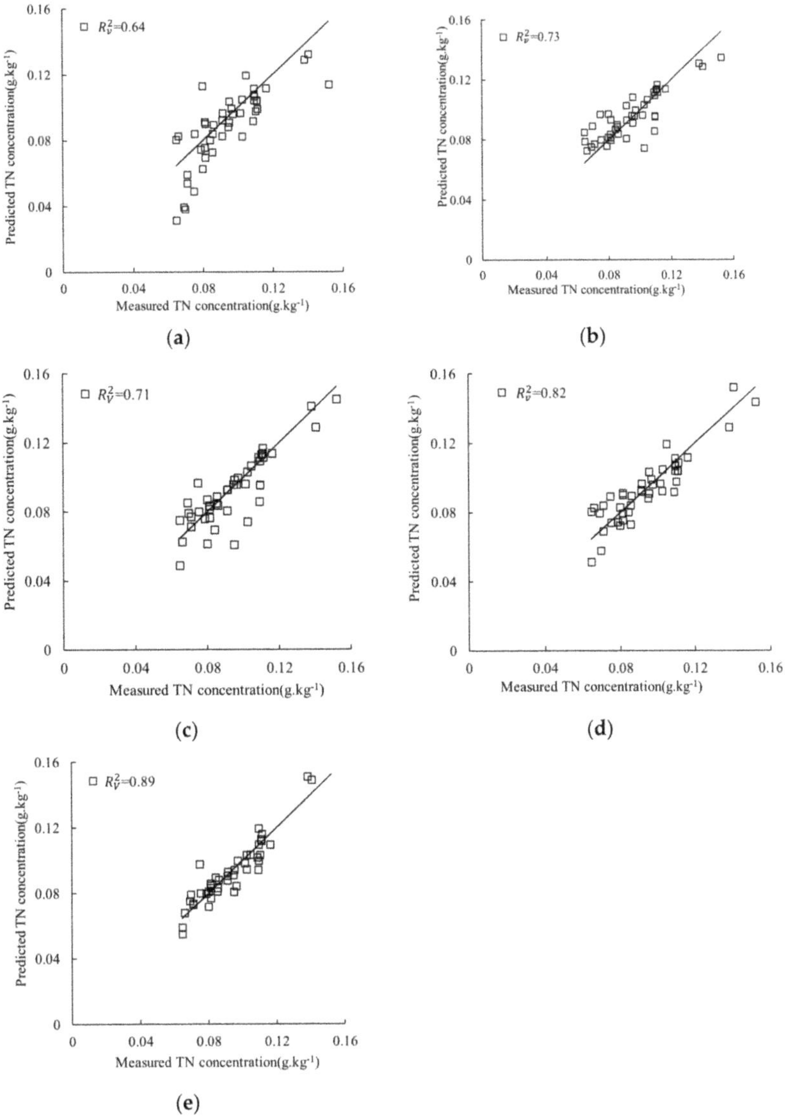

Figure 11. Comparison of elimination effects. (**a**) Estimation of TN concentration with raw discrete NIR spectral band data; (**b**) estimation of TN concentration with discrete NIR spectral band data through the proposed MACI correlation method to eliminate soil moisture disturbance; (**c**) estimation of TN concentration with discrete NIR spectral band data through the proposed PSCI method to eliminate soil particle size disturbance; (**d**) estimation of TN concentration with discrete NIR spectral band data through the proposed MACI and PSCI methods to eliminate soil moisture and particle size disturbances; (**e**) estimation of TN concentration with discrete NIR spectral band data on dried and ground soil samples.

Table 3. Comparison of elimination efforts.

Elimination Method	R_v^2	RMSEP(g·kg^{-1})	RPD
Model 1	0.64	0.278	1.59
Model 2	0.71	0.221	2.06
Model 3	0.73	0.202	2.12
Model 4	0.82	0.166	2.53
Model 5	0.89	0.121	2.72

Note: R_v^2 = coefficient of determination of validation; RMSEP = root mean square error of validation; RPD = residual prediction deviation; model 1 = raw discrete NIR spectral band data; model 2 = discrete NIR spectral band data using the proposed MACI correlation method to eliminate soil moisture disturbance; model 3 = discrete NIR spectral band data using the proposed PSCI method to eliminate soil particle size disturbance; model 4 = discrete NIR spectral band data using the proposed new coupled elimination method to eliminate soil moisture and particle size disturbances; model 5 = discrete NIR spectral band data on dried and ground soil samples.

4. Discussion

4.1. Role of New Coupled Elimination Method in Predicting TN through Discrete NIR Spectral Band Data

From the perspective of cost, the development of real-time rapid soil parameter detectors based on discrete NIR spectral band has become a current research hotspot [29–33]. However, for discrete NIR band data, adjacent spectral data can be completely different in the attribution of the bands, and they do not have continuity in the sense of spectrum. Thus, there is no theoretical basis for direct differentiation of a set of discrete NIR spectral band data that has no connection in the spectral sense [59]. In addition, due to the lack of discrete NIR spectral data, it does not meet the requirements of the Savitzky-Golay convolution derivation method for the number of window points, and its operation cannot be performed [60,61]. Soil particle size and moisture will cause serious interference with discrete NIR band data and affect its use [36]. Therefore, the study of effective correction methods to eliminate the interference of soil particle size and moisture on discrete NIR spectral band data has important theoretical significance and practical application value for improving the performance of instruments developed based on discrete NIR spectral band data.

There is no relevant research on the elimination of soil moisture and particle size effects on predicting soil parameters through the discrete NIR spectral band data. This study analyzes the effects of soil moisture and soil particle size on the discrete NIR spectral band data using qualitative and quantitative methods. The absorbance of soil samples increased with the increase in soil moisture content at the same TN concentration level. As the soil particle size decreased, the absorbance of soil samples decreased gradually at the same TN concentration level [36]. Correspondingly, we proposed a method to eliminate the effects of soil moisture and soil particle size on discrete NIR spectral band data. A field experiment also was conducted to evaluate the elimination effects of the proposed MACI and PSCI methods. When the model was established by discrete NIR spectral band data after eliminating soil moisture and soil particle size interference, the RPD value of prediction model 4 reached 2.59, and the prediction R^2 reached 0.84, respectively. The results showed that the new coupled elimination method could effectively reduce the interferences of soil particle size and moisture on the prediction of TN concentration.

From the perspective of actual use, the new coupled elimination method can be applied to other on-the-go and portable soil parameter testing equipment developed based on discrete NIR spectral band data. This study facilitates the development of soil sensors based on discrete NIR spectral band data.

4.2. Comparison of the New Coupled Elimination Method to the Similar

Table 4 shows the comparison results of the new coupled elimination method to the similar. According to the evaluation parameters (R_v^2 and RMSEP), our study obtains slightly high estimation accuracies compared to the similar in Table 4. Even though the new coupled elimination method was not applied to the on-the-go detector, the estimation accuracies of TN were higher than the portable detector and the Veris P4000. Applying the new coupled elimination method to the on-the-go detector, the estimation accuracy was improved significantly. The R_v^2 increased from 0.64 to 0.82, and the RMSEP decreased from 0.278 to 0.166. For the portable detector, the R_v^2 was 0.76 with the moisture absorbance correction method and mixed calibration set used in the portable detector. Compared to the similar, the new coupled elimination method achieved better elimination effects of soil particle size and moisture on NIR spectroscopy. These comparison results further verify the validity of the new coupled elimination method in eliminating the effects of soil particle size and moisture on discrete NIR spectral band data.

Table 4. Validation results of soil total nitrogen (TN) concentration.

N	Spectral Measuring Equipment	Spectral Range	Elimination Factor	Elimination Method	Calibration Algorithm	R_v^2	RMSEP	References
48	Portable detector	940, 1050, 1100, 1200, 1300, and 1550 nm	/	/	BPNN	0.45	0.215	[36]
48	Portable detector	940, 1050, 1100, 1200, 1300, 1450, and 1550 nm	Soil moisture and soil particle size	PMAI and mixed calibration set	BPNN	0.76	0.030	[36]
90	FT-NIR analyzer (MATRIX-I, Bruker corp., Germany)	800–2564 nm	Soil moisture	Wavelet de-compositions	SVM	0.81	0.053	[23]
140	AgroSpec portable VIS-NIR spectrophotometer (Tec5 Technology for Spectroscopy, Germany)	305–2200 nm	Soil moisture and soil particle size	FD transformation with 31 smoothing points and SNV	Cubist method	0.73	0.071	[62]
708	Veris P4000 (Veris Technologies, Inc., Salina, KS, USA)	343–2202 nm	Soil moisture	EPO	PLS	0.63	0.024	[63]
108	On-the-go detector	1070, 1130, 1245, 1375, 1550, and 1680 nm	/	/	SVM	0.64	0.278	This study
108	On-the-go detector	1070, 1130, 1245, 1361, 1375, 1450, 1550, 1680, and 1870 nm	Soil moisture and soil particle size	New coupled elimination method	SVM	0.82	0.166	This study

Note: N = number of samples; R_v^2 = coefficient of determination of validation; RMSEP = root mean scheme.

4.3. Uncertainty in Current Work and Future Work

Although the new coupled elimination method was quantitatively proven to have good effects on eliminating soil moisture and soil particle size interferences on predicting TN concentration using discrete NIR spectral band data, the 8.5% decrease of prediction R^2 was between models 4 and 5, which means that the effects of soil moisture and particle size were not eliminated using the new coupled elimination method. Thus, it is necessary to continue to explore other elimination methods of soil moisture and particle size effects on predicting TN concentration through discrete NIR spectral band data.

Additionally, the experimental conditions in fields are complex, and different regions and soil types also will affect the measurement of discrete NIR spectral band data of soil.

Moreover, field vehicle-mounted detectors are still limited in terms of coving large fields in a short time. The unmanned aerial vehicles (UAVs) have provided the potential alternative to the quick estimation of soil properties coving large fields in a short time. Thus, the hyperspectral cameras or the soil detectors based on the discrete NIR spectral band data mounted on the UAVs will be explored in future research. Therefore, further experiments are needed to verify the effectiveness and practicability of the proposed methods.

5. Conclusions

Commercialized soil nutrients detectors developed based on discrete NIR spectral band data cannot perform on-line elimination of soil moisture and particle size disturbances on predicting TN concentration. Bringing the detected data back to the laboratory for particle size and moisture interferences' elimination is necessary. This process severely restricts the efficiency of field testing and cannot achieve a quick estimation of TN concentration in the farmlands. This paper reports on the online coupled elimination method of soil moisture and particle size interferences on predicting TN concentration through the discrete NIR spectral band data. The field test also was performed to evaluate the proposed coupled elimination method through the on-the-go detector.

Author Contributions: P.Z. and W.W. designed experiments, conducted field data and sample collection, and supervised soil processing and laboratory analyses; P.Z. and W.Y. implemented spectral data processing and calibration modeling under the supervision of W.Y. and M.L.; the original draft of the manuscript was written by P.Z. with editorial contributions from W.Y., and M.L. All authors have read and agreed to the published version of the manuscript.

Funding: This study was supported by the National Key Research Projects (2017YFD0201500–2017YFD0201501, 2016YFD0700300–2016YFD0700304) and the National Natural Science Foundation of China (31801265).

Data Availability Statement: Not applicable.

Acknowledgments: The authors wish to thank teachers and students in the Key Laboratory of Modern Precision Agriculture System Integration Research for the help of Sample processing and chemical analysis. We would like to thank Weichao Wang for his help with field data collection. Additionally, great thanks to Sudduth from the USDA-ARS Cropping Systems and Water Quality Research Unit for his suggestions to improve this paper.

Conflicts of Interest: The authors declare no conflict of interest.

References

1. Bordoli, J.M.; Mallarino, A.P. Deep and shallow banding of phosphorus and potassium as alternatives to broadcast fertilization for no-till Corn. *Agron. J.* **1998**, *90*, 27–33. [CrossRef]
2. Domangue, R.J.; Mortazavi, B. Nitrate reduction pathways in the presence of excess nitrogen in a shallow eutrophic estuary. *Environ. Pollut.* **2018**, *238*, 599–606. [CrossRef]
3. Shi, P.; Zhang, Y.; Song, J.; Li, P.; Wang, Y.; Zhang, X. Response of nitrogen pollution in surface water to land use and social-economic factors in the Weihe River watershed, northwest China. *Sustain. Cities Soc.* **2019**, *50*, 101658. [CrossRef]
4. Staver, K.W.; Brinsfield, R.B. Patterns of soil nitrate availability in corn production systems: Implications for reducing groundwater contamination. *J. Soil Water Conserv.* **1990**, *45*, 318–323.
5. Martins, R.N.; Sárvio, D.; Valente, M.; Tadeu, J.; Rosas, F.; Santos, F.S.; Ferreira, F.; Dos, L.; Carolina, A.; Nascimento, C. Communications in Soil Science and Plant Analysis Site-specific Nutrient Management Zones in Soybean Field Using Multivariate Analysis: An Approach Based on Variable Rate Fertilization Site-specific Nutrient Management Zones in Soybean Field Using Multivariate Analysis: An Approach Based on Variable Rate. *Commun. Soil Sci. Plant. Anal.* **2020**, *51*, 687–700.
6. Yu, J.; Yin, X.; Raper, T.B.; Jagadamma, S.; Chi, D. Nitrogen Consumption and Productivity of Cotton under Sensor-based Variable-rate Nitrogen Fertilization. *Agron. J.* **2019**, *111*, 3320–3328. [CrossRef]
7. Qi, J.; Tian, X.; Li, Y.; Fan, X.; Yuan, H.; Zhao, J.; Jia, H. Design and experiment of a subsoiling variable rate fertilization machine. *Int. J. Agric. Biol. Eng.* **2020**, *13*, 118–124. [CrossRef]
8. Mouazen, A.M.; Kuang, B. Soil & Tillage Research On-line visible and near infrared spectroscopy for in-field phosphorous management. *Soil Tillage Res.* **2016**, *155*, 471–477.
9. Nawar, S.; Corstanje, R.; Halcro, G.; Mulla, D.; Mouazen, A.M. Delineation of Soil Management Zones for Variable-Rate Fertilization: A Review. *Adv. Agron.* **2017**, *143*, 175–245.

10. Batjes, N.H. Total carbon and nitrogen in the soils of the world. *Eur. J. Soil Sci.* **1996**, *47*, 151–163. [CrossRef]
11. Wang, S.; Zhuang, Q.; Jin, X.; Yang, Z.; Liu, H. Predicting Soil organic carbon and soil nitrogen stocks in topsoil of forest ecosystems in northeastern china using remote sensing data. *Remote Sens.* **2020**, *12*, 1115. [CrossRef]
12. Maeda, Y.; Tashiro, N.; Enoki, T.; Urakawa, R.; Hishi, T. Effects of species replacement on the relationship between net primary production and soil nitrogen availability along a topographical gradient: Comparison of belowground allocation and nitrogen use efficiency between natural forests and plantations. *For. Ecol. Manag.* **2018**, *422*, 214–222. [CrossRef]
13. Jin, Z.; Chen, C.; Chen, X.; Hopkins, I.; Zhang, X.; Han, Z.; Jiang, F.; Billy, G. The crucial factors of soil fertility and rapeseed yield–A five year field trial with biochar addition in upland red soil, China. *Sci. Total Environ.* **2019**, *649*, 1467–1480. [CrossRef]
14. Yoshida, H.; Takehisa, K.; Kojima, T.; Ohno, H.; Nakagawa, H. Modeling the effects of N application on growth, yield and plant properties associated with the occurrence of chalky grains of rice. *Plant. Prod. Sci.* **2016**, *1008*, 30–42. [CrossRef]
15. Kuang, B.; Mouazen, A.M. Non-biased prediction of soil organic carbon and total nitrogen with vis-NIR spectroscopy, as affected by soil moisture content and texture Keywords. *Biosyst. Eng.* **2013**, *114*, 249–258. [CrossRef]
16. Nocita, M.; Stevens, A.; Toth, G.; Panagos, P.; Montanarella, L. Prediction of soil organic carbon content by diffuse reflectance spectroscopy using a local partial least square regression approach. *Soil Biol. Biochem.* **2014**, *68*, 337–347. [CrossRef]
17. Zhang, Y.; Li, M.; Zheng, L.; Qin, Q.; Suk, W. Spectral features extraction for estimation of soil total nitrogen content based on modified ant colony optimization algorithm. *Geoderma* **2019**, *333*, 23–34. [CrossRef]
18. Bao, Y.; Meng, X.; Ustin, S.; Wang, X.; Zhang, X.; Liu, H. Vis-SWIR spectral prediction model for soil organic matter with diff erent grouping strategies. *Catena* **2020**, *195*, 104703. [CrossRef]
19. Debaene, G.; Nied, J.; Pecio, A.; Anna, Ż. Effect of the number of calibration samples on the prediction of several soil properties at the farm-scale. *Geoderma* **2014**, *215*, 114–125. [CrossRef]
20. Lin, L.; Gao, Z.; Liu, X. Estimation of soil total nitrogen using the synthetic color learning machine (SCLM) method and hyperspectral data. *Geoderma* **2020**, *380*, 114664. [CrossRef]
21. Johnson, J.; Vandamme, E.; Senthilkumar, K.; Sila, A.; Shepherd, K.D.; Saito, K. Near-infrared, mid-infrared or combined diff use reflectance spectroscopy for assessing soil fertility in rice fields in sub-Saharan Africa. *Geoderma* **2019**, *354*, 113840. [CrossRef]
22. Dalal, R.C.; Henry, R.J. Simultaneous determination of moisture, organic carbon, and total nitrogen by near infrared reflectance spectrophotometry. *Soil Sci. Soc. Am. J.* **1986**, *50*, 120–123. [CrossRef]
23. Zhang, Y.; Li, M.Z.; Zheng, L.H.; Zhao, Y.; Pei, X. Soil nitrogen content forecasting based on real-time NIR spectroscopy. *Comput. Electron. Agric.* **2016**, *124*, 29–36. [CrossRef]
24. Jiang, Q.; Li, Q.; Wang, X.; Wu, Y.; Yang, X.; Liu, F. Estimation of soil organic carbon and total nitrogen in different soil layers using VNIR spectroscopy: Effects of spiking on model applicability. *Geoderma* **2017**, *293*, 54–63. [CrossRef]
25. Conforti, M.; Matteucci, G.; Buttafuoco, G. Using laboratory Vis-NIR spectroscopy for monitoring some forest soil properties. *J. Soils Sediments* **2018**, *18*, 1009–1019. [CrossRef]
26. Sudduth, K.A.; Hummel, J.W. Portable, Near-infrared spectrophotometer for rapid soil analysis. *Trans. ASAE* **1993**, *36*, 185–193. [CrossRef]
27. Mouazen, A.M.; Alhwaimel, S.A.; Kuang, B.; Waine, T. Multiple on-line soil sensors and data fusion approach for delineation of water holding capacity zones for site specific irrigation. *Soil Tillage Res.* **2014**, *143*, 95–105. [CrossRef]
28. Mouazen, A.M.; Maleki, M.R.; De Baerdemaeker, J.; Ramon, H. On-line measurement of some selected soil properties using a VIS–NIR sensor. *Soil Tillage Res.* **2007**, *93*, 13–27. [CrossRef]
29. Zhou, P.; Li, M.; Yang, W.; Ji, R.; Meng, C. Development of vehicle-mounted in-situ soil parameters detector based on NIR diffuse reflection. *Spectrosc. Spect. Anal.* **2020**, *40*, 2856–2861.
30. Zhou, P.; Zhang, Y.; Yang, W.; Li, M.; Liu, Z.; Liu, X. Development and performance test of an in-situ soil total nitrogen-soil moisture detector based on near-infrared spectroscopy. *Comput. Electron. Agric.* **2019**, *160*, 51–58. [CrossRef]
31. An, X.; Li, M.; Zheng, L.; Liu, Y.; Sun, H. A portable soil nitrogen detector based on NIRS. *Precis. Agric.* **2014**, *15*, 3–16. [CrossRef]
32. Tang, N.; Li, M.Z.; Sun, J.Y.; Zheng, L.H.; Pan, L. Development of soil-organic-matter fast-determination instrument based on spectroscopy. *Spectrosc. Spectr. Anal.* **2007**, *27*, 2139.
33. Li, M.; Yao, X.; Yang, W.; Zhou, P.; Hao, Z.; Zheng, L. Design of New Portable Detector for Soil Total Nitrogen Content Based on High-power Tungsten Halogen Lamp and "One-Six" Special Optical Fiber. *Trans. Chinese Soc. Agric. Mach.* **2019**, *50*, 169–174.
34. Wang, Y.P.; Lee, C.K.; Dai, Y.H.; Shen, Y. Effect of wetting on the determination of soil organic matter content using visible and near-infrared spectrometer. *Geoderma* **2020**, *376*, 114528. [CrossRef]
35. Stenberg, B. Effects of soil sample pretreatments and standardised rewetting as interacted with sand classes on Vis-NIR predictions of clay and soil organic carbon. *Geoderma* **2010**, *158*, 15–22. [CrossRef]
36. An, X.; Li, M.; Zheng, L.; Sun, H. Eliminating the interference of soil moisture and particle size on predicting soil total nitrogen content using a NIRS-based portable detector. *Comput. Electron. Agric.* **2015**, *112*, 47–53. [CrossRef]
37. Rienzi, E.A.; Mijatovic, B.; Mueller, T.G.; Matocha, C.J.; Sikora, F.J.; Castrignanò, A. Prediction of Soil Organic Carbon under Varying Moisture Levels Using Reflectance Spectroscopy. *Soil Sci. Soc. Am. J.* **2014**, *78*, 958–967. [CrossRef]
38. Kaiser, M.; Kleber, M.; Berhe, A.A. How air-drying and rewetting modify soil organic matter characteristics: An assessment to improve data interpretation and inference. *Soil Biol. Biochem.* **2015**, *80*, 324–340. [CrossRef]
39. Tekin, Y.; Tumsavas, Z.; Mouazen, A.M. Effect of moisture content on prediction of organic carbon and pH using visible and near-infrared spectroscopy. *Soil Sci. Soc. Am. J.* **2012**, *76*, 188–198. [CrossRef]

40. Wijewardane, N.K.; Ge, Y.; Morgan, C.L. Morgan Moisture insensitive prediction of soil properties from VNIR reflectance spectra based on external parameter orthogonalization. *Geoderma* **2016**, *267*, 92–101. [CrossRef]
41. Barthès, B.G.; Brunet, D.; Hien, E.; Enjalric, F.; Conche, S.; Freschet, G.T.; d'Annunzio, R.; Toucet-Louri, J. Determining the distributions of soil carbon and nitrogen in particle size fractions using near-infrared reflectance spectrum of bulk soil samples. *Soil Biol. Biochem.* **2008**, *40*, 1533–1537. [CrossRef]
42. Bogrekci, I.; Lee, W.S. Improving phosphorus sensing by eliminating soil particle size effect in spectral measurement. *Trans. ASAE.* **2005**, *48*, 1971–1978. [CrossRef]
43. Lee, K.S.; Sudduth, K.A.; Drummond, S.T.; Lee, D.H.; Kitchen, N.R.; Chung, S.O. Calibration methods for soil property estimation using reflectance spectroscopy. *Trans. ASABE* **2010**, *53*, 675–684. [CrossRef]
44. Yao, X.; Yang, W.; Li, M.; Zhou, P.; Chen, Y.; Hao, Z.; Liu, Z. Prediction of Total Nitrogen in Soil Based on Random Frog Leaping Wavelet Neural Network. *IFAC-PapersOnLine* **2018**, *51*, 660–665. [CrossRef]
45. Yao, X.; Yang, W.; Li, M.; Zhou, P.; Liu, Z. Prediction of Total Nitrogen Content in Different Soil Types Based on Spectroscopy. *IFAC-PapersOnLine* **2019**, *52*, 270–276. [CrossRef]
46. Mcdowell, W.H.; Magill, A.H.; Aitkenhead-Peterson, J.A.; Aber, J.D.; Merriam, J.L.; Kaushal, S.S. Effects of Chronic Nitrogen Amendment on Dissolved Organic Matter and Inorganic Nitrogen in Soil Solution. *For. Ecol. Manag.* **2004**, *196*, 29–41. [CrossRef]
47. Jämtgård, S.; Näsholm, T.; Huss-danell, K. Soil Biology and Biochemistry Nitrogen Compounds in Soil Solutions of Agricultural Land. *Soil Biol. Biochem.* **2010**, *42*, 2325–2330. [CrossRef]
48. Stark, J.M.; Hart, S.C. Diffusion Technique for Preparing Salt Solutions, Kjeldahl Digests, and Persulfate Digests for Nitrogen-15 Analysis. *Soil Sci. Soc. Am. J.* **1996**, *60*, 1846–1855. [CrossRef]
49. Jia, S.; Zhang, J.; Li, G.; Yang, X. Predicting Soil Nitrogen and Organic Carbon Using Near Infrared SpectroscopyCoupled with Variable Selection. *Appl. Eng. Agric.* **2014**, *30*, 641–647.
50. Zheng, L.; Li, M.; Pan, L.; Sun, J.; Tang, N. Application of wavelet packet analysis in estimating soil parameters based on NIR spectra. *Spectrosc. Spectr. Anal.* **2009**, *29*, 1549–1552.
51. Zheng, L.; Li, M.; Pan, L.; Sun, J.; Tang, N. Estimation of soil organic matter and soil total nitrogen based on NIR spectroscopy and BP neural network. *Spectrosc. Spectr. Anal.* **2008**, *28*, 1160–1164.
52. Zhou, P.; Sudduth, K.A.; Veum, K.S.; Li, M. Selection of characteristic wavebands to minimize soil moisture effects with in-situ soil spectroscopy. In *2020 ASABE Annual International Virtual Meeting*; ASABE: Saint Joseph, MI, USA, 2020.
53. Ogen, Y.; Faigenbaum-golovin, S.; Granot, A.; Shkolnisky, Y. Removing Moisture Effect on Soil Reflectance Properties: A Case Study of Clay Content Prediction. *Pedosph. An. Int. J.* **2019**, *29*, 421–431. [CrossRef]
54. Jiang, Q.; Chen, Y.; Guo, L.; Fei, T.; Qi, K. Estimating Soil Organic Carbon of Cropland Soil at Different Levels of Soil Moisture Using VIS-NIR Spectroscopy. *Remote Sens.* **2016**, *8*, 755. [CrossRef]
55. Castaldi, F.; Palombo, A.; Pascucci, S.; Pignatti, S.; Santini, F. Reducing the Influence of Soil Moisture on the Estimation of Clay from Hyperspectral Data: A Case Study Using Simulated PRISMA Data. *Remote Sens.* **2015**, *7*, 15561–15582. [CrossRef]
56. Lobell, D.B.; Asner, G.P. Moisture Effects on Soil Reflectance. *Soil Sci. Soc. Am. J.* **2002**, *66*, 722–727. [CrossRef]
57. Li, M.Z.; Han, D.H.; Wang, X. *Spectral Analysis and Application*; Science Press: Beijing, China, 2006.
58. Lu, W.Z. *Modern Near Infrared Spectroscopy Analytical Technology*; China Petrochemical Press: Beijing, China, 2010.
59. Hu, A.Q.; Yuan, H.F.; Song, C.F.; Li, X.Y. A correction method of baseline drift of discrete spectrum of NIR. *Spectrosc. Spectr. Anal.* **2014**, *34*, 2606–2611.
60. Carle, C. Comments on Smoothing and Differentiation of Data by Simplified Least Square Procedure. *Anal. Chem.* **1972**, *44*, 1906–1909.
61. Savitzky, A.; Golay, M.J. Smoothing and Differentiation of Data by Simplified Least Squares Procedures. *Anal. Chem.* **1964**, *36*, 1627–1639. [CrossRef]
62. Morellos, A.; Pantazi, X.; Moshou, D.; Alexandridis, T.; Whetton, R.; Tziotzios, G.; Wiebensohn, J.; Bill, R.; Mouazen, A.M. Direct Special Issue: Proximal Soil Sensing Machine learning based prediction of soil total nitrogen, organic carbon and moisture content by using VIS-NIR spectroscopy. *Biosyst. Eng.* **2016**, *152*, 104–116. [CrossRef]
63. Veum, K.S.; Parker, P.A.; Sudduth, K.A.; Holan, S.H. Predicting Profile Soil Properties with Reflectance Spectra via Bayesian Covariate-Assisted External Parameter Orthogonalization. *Sensor* **2018**, *18*, 3869. [CrossRef]

remote sensing

MDPI

Article

Estimating Nitrogen from Structural Crop Traits at Field Scale—A Novel Approach Versus Spectral Vegetation Indices

Nora Tilly * and Georg Bareth

Institute of Geography, GIS & RS Group, University of Cologne, D-50923 Cologne, Germany
* Correspondence: nora.tilly@uni-koeln.de; Tel.: +49-221-470-6265; Fax: +49-221-470-4563

Received: 9 August 2019; Accepted: 22 August 2019; Published: 3 September 2019

Abstract: A sufficient nitrogen (N) supply is mandatory for healthy crop growth, but negative consequences of N losses into the environment are known. Hence, deeply understanding and monitoring crop growth for an optimized N management is advisable. In this context, remote sensing facilitates the capturing of crop traits. While several studies on estimating biomass from spectral and structural data can be found, N is so far only estimated from spectral features. It is well known that N is negatively related to dry biomass, which, in turn, can be estimated from crop height. Based on this indirect link, the present study aims at estimating N concentration at field scale in a two-step model: first, using crop height to estimate biomass, and second, using the modeled biomass to estimate N concentration. For comparison, N concentration was estimated from spectral data. The data was captured on a spring barley field experiment in two growing seasons. Crop surface height was measured with a terrestrial laser scanner, seven vegetation indices were calculated from field spectrometer measurements, and dry biomass and N concentration were destructively sampled. In the validation, better results were obtained with the models based on structural data ($R^2 < 0.85$) than on spectral data ($R^2 < 0.70$). A brief look at the N concentration of different plant organs showed stronger dependencies on structural data (R^2: 0.40–0.81) than on spectral data (R^2: 0.18–0.68). Overall, this first study shows the potential of crop-specific across-season two-step models based on structural data for estimating crop N concentration at field scale. The validity of the models for in-season estimations requires further research.

Keywords: terrestrial laser scanning; spectrometer; plant height; vegetation indices; biomass; nitrogen concentration; precision agriculture

1. Introduction

Nitrogen (N) is a fundamental component of proteins and thus essential for any kind of living. Its relevance for the nutrition of plants and for the photosynthesis process is indisputable [1]. Plants consist of a metabolic part with a high N concentration, such as leaves which are mostly responsible for photosynthetic processes, and a structural part with a low N concentration, such as the stem which is necessary for the plant architecture [2]. During early growing stages, crop N demand primarily results from the leaf area expansion to initiate growth [3]. When canopy closure is reached, plants compete for light and invest more N in stem elongation to place their leaves to the better-illuminated top layers [4,5]. Another role for the crop development can be attributed to the stem, as plants use stem N as source for grain N later in the growing season [3]. If this storage is insufficient and further sources are missing, such as N released by natural leaf senescence or soil N, plants let leaves die off for the required N [3]. This is likely to reduce the photosynthetic activity and inhibit growth. Hence, N plays the most important role in the fertilization of arable and forage cropping systems [6].

The rapidly growing world population requires an increasing food production, which goes along with a rising use of N fertilizers [7]. The global use of N fertilizer increased from ~10 Tg/year in the 1960s to ~80 Tg/year around 1990 [8] and might reach ~190 Tg in 2020 [9]. While aiming at increasing yield, it is frequently neglected that plants can assimilate N only up to a certain value. N losses into the environment has negative consequences, such as nitrous oxide emissions, nitrate leaching, or eutrophication [8,10,11]. A strong research interest can hence be observed on the general N cycle, the N use efficiency (NUE) of plants, and the improvement and optimization of fertilization practices [12–18].

A benchmark for quantifying the plant status at field scale is the crop-specific nitrogen nutrition index (NNI), defined as ratio between actual and critical N concentration [19,20]. The critical N concentration of dry biomass is defined as minimum N required for maximum growth [21]. If N supply is not limited, the N concentration generally decreases while dry biomass increases across a growing season. This allometric relation can be expressed with a negative power function [20]. A sequence of critical N concentration values can be plotted against dry biomass, which is then defined as crop-specific critical N dilution curve (NDC) [19,22,23]. The NNI and NDC concept hence requires knowledge about dry biomass and N. Both values can either be determined destructively [24] or, what is more common, estimated from proximal or remote sensing. Due to their non-destructive characteristics, these tools are highly attractive for precision agriculture or site-specific management and extensively investigated since the 1980s [25,26].

Remotely sensed spectral reflectance properties are known to be worthwhile for investigating vegetation [27]. Numerous studies investigated the usability for determining NNI [28,29] or calculated vegetation indices (VIs) for estimating crop N [30–34]. Dry biomass can also be estimated from VIs, but measurements are known to be affected by saturation effects at later growing stages [35–37]. Major disadvantages of all passive spectral measurements are the dependency on solar radiation and the influence through atmospheric conditions, which require repeated calibrations during a campaign [38,39]. These problems can be avoided by using bidirectional spectrometers or active sensors [30,40,41]. Further limiting factors are, however, the influence of differing soil properties, which is particularly important in the early growing period due to the low vegetation fraction [42], and plant properties, such as the leaf inclination angle [43]. Measurements with field spectrometers can also cover only small parts of the crop canopy. Based on their extensive research, Gastal et al. [4] concluded that the NNI is more a research than a management tool and non-invasive, cost-effective methods are required. They suggest chlorophyll measurements with a SPAD meter as promising tool for determining the N status. Various studies show its usability [44] and propose it as tool for fertilizer recommendations [45]. The amount of SPAD measurements is limited due to its handheld characteristics, which reinforces the general weakness of spectral measurements. An alternative robust approach for determining crop N at field scale is hence desirable.

The limitations of spectral sensing approaches, led to new research activities in crop monitoring, namely the sensing of structural properties such as crop height and crop density [46–48]. At field scale, these traits can be derived by sensing methods which produce 3D data such as terrestrial laser scanning (TLS) [49,50] or photogrammetric processing using Structure from Motion (SfM) and Multi-View Stereopsis (MVS) [51]. Other field studies measure with light curtains [52] or ultrasonic sensors [36,53]. In a comparative study on different sensors, best results were achieved with laser scanning in comparison to an ultrasonic sensor and drone-based imaging [54]. Several studies showed that dry biomass can be estimated from structural crop traits captured with TLS, drone-based imaging, or oblique stereo image acquisition [55–63]. Only a few studies compared biomass estimations based on different sensors so far [64]. From the existing ones, it can be summarized that structural estimators, such as crop height, outperform spectral ones [47,59,65]. TLS and SfM/MVS approaches allow capturing large areas in a high spatial resolution. Hence, in-field variabilities can be detected.

In summary, it is widely accepted that understanding processes and traits, which are involved in the N cycle of plants, are extremely important to optimize crop production [4,66]. Plant height is recognized as relevant trait [67,68], but until recent developments in proximal sensing, structural traits were hardly measurable in a sufficient spatial and temporal resolution at field scale. Manual measurements were laborious and prone to errors [54]. The literature shows that dry biomass is estimable either from 3D or spectral measurements, considering the limitations for the latter. In contrast, to the best of the authors' knowledge, crop N at field scale is only estimated from spectral information. Structural traits, determined from 3D data, are neglected so far. According to Seginer et al. [69] the light competition among individual plants occurs early in the growing season, as canopy closure is rapidly reached in typically dense agricultural fields. Hence, potential height differences among individual plants or in-field variations, which affect the amount of biomass, should be discernible. Along with the allometric relation between dry biomass and N concentration, the arising research question can be formulated as: Can the crop-specific N concentration at field scale be estimated from its indirect link to structural traits?

In a comprehensive study, Tilly et al. [65] found that TLS-derived plant height is a strong estimator for dry barley biomass ($R^2 < 0.85$) in contrast to VIs, which showed varying performance (R^2: 0.07–0.87). This new study further investigates this data set in terms of the stated research question as on the one hand, the quality and suitability of the data set was proven, but on the other hand, a so far unconsidered data set of destructively measured N was available. The overall aim of this study is to investigate whether the indirect link between structural traits and N concentration can be used for robust estimations at field scale. Two novel model designs are developed which investigate the interrelations between TLS-derived crop surface height, dry biomass, and N concentration. Another VI-based model is established for comparison. A brief look is thrown on the relation of the estimators to the dry biomass and N concentration of individual plant organs.

2. Materials and Methods

2.1. Data Acquisition

During the growing seasons 2013 and 2014, the data sets for this study were captured on a spring barley (*Hordeum vulgare* L.) field experiment, conducted at the experimental station Klein–Altendorf of the University of Bonn, Germany (50°37′N, 6°59′E, altitude 186 m). The exact locations of the fields varied slightly between the years due to crop rotation. Soil and climatic conditions were almost equal with a flat surface, a clayey silt luvisol, an average yearly precipitation of ~600 mm, and a daily average temperature of 9.3 °C [70,71]. The soil available N and organic matter was determined in early spring each year (2013: 42 kg/ha Nmin and 1.7% organic matter; 2014: 44 kg/ha Nmin and 1.8% organic matter). The field was subdivided into 36 small-scale plots (3 × 7 m). For half of the plots, a farmer's common rate of 80 kg/ha N fertilizer was applied, for the other half a reduced rate of 40 kg/ha. The fertilizer was applied 1 and 5 days after seeding in 2013 and 2014, respectively. In 2013, each fertilization scheme was carried out once for 18 cultivars. In 2014, the number of cultivars was reduced to 6 and each fertilization scheme was repeated three times. This study considers only these cultivars (Barke, Beatrix, Eunova, Trumpf, Mauritia, and Sebastian). Aasen and Bolten [72] present the randomized block design in the field plan of 2014. TLS and spectral measurements were carried out for monitoring plant development across the growing seasons. Plant height and dry biomass were manually measured as reference. The proximal sensing and reference measurements were carried out within a maximum timespan of three days per campaign. The main details are outlined in the following. For an extended description it is referred to Tilly et al. [65].

2.1.1. Terrestrial Laser Scanning

The time-of-flight scanner Riegl LMS-Z420i was used (near-infrared laser beam; beam divergence of 0.25 mrad; and measurement rate of up to 11,000 points/sec). Its field of view is up to 80° in the vertical and 360° in the horizontal direction and resolutions between 0.034° and 0.046° were used for this study. A Nikon D200 was mounted on top for colorizing the point clouds in the post-processing. The scanner was mounted on a hydraulic platform of a tractor (sensor height ~4 m above ground).

During each campaign, the field was scanned from its four corners to lower shadowing effects and to attain an almost uniform spatial coverage. The coordinates of all scan positions were measured with the RTK DGPS system Topcon HiPer Pro. The position of a highly reflective cylinder arranged on a ranging pole was measured with the DGPS system and detected in each scan. An exact georeferencing and co-registration can thus be achieved with the DGPS-derived coordinates of the scan positions.

The main steps of the TLS data processing can be summarized to merging and cleaning of the point clouds, filtering of the highest points, which are regarded as crop canopy, and spatially resolved calculation of CSH. Crop surface models (CSMs), which represent the crop surface, were established. Crop surface height (CSH) is then calculated by spatially subtracting the digital terrain model (DTM) height from the CSM. A detailed description is given by Tilly et al. [60]. The result is a raster data set for each campaign with pixel-wise stored CSH. The CSH was averaged plot-wise to achieve a common spatial base with the other measurements (Figure 1).

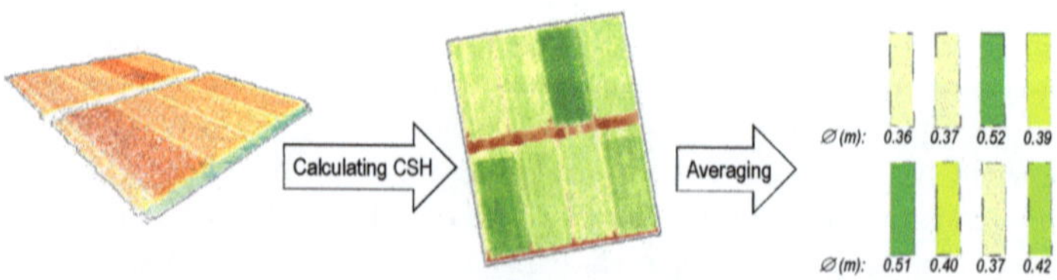

[A] [B] [C]

Figure 1. Steps from point cloud to plot-wise averaged crop surface height (CSH): **[A]** Filtered point cloud representing the crop surface; **[B]** Raster data set of CSH; **[C]** Plot-wise averaged CSH values.

2.1.2. Spectral Measurements

The ASD® FieldSpec3 was used (measurement range from 350 to 2500 nm; sampling interval of 1.4 nm and 2 nm in visible-near-infrared and short-wave infrared, respectively; spectra are resampled to 1 nm resolution). A cantilever with a pistol grip was used to avoid shadows and a water level ensured a nadir view. Samples were taken from 1 m above canopy and no fore optic was used. Thus, the field of view was 25°, resulting in a circular footprint area with a radius of ~22 cm on the canopy. Six positions were taken for each plot and averaged for the analysis. Ten measurements per position were carried out and instantly averaged. The spectrometer warmed up for at least 30 min prior to measurements. It was calibrated with a spectralon calibration panel every 10 min or after illumination change. Measurements were carried out around noon to ensure the sun was at its highest.

In the post-processing, six VIs were calculated by Tilly et al. [65] for estimating biomass. These VIs were now further investigated regarding their relation to N concentration. The NIR-based simple ratio index R_{760}/R_{730} was added for this study, as it was powerful for indicating the N status of wheat [40] and performed well for grain yield prediction with similar cultivars of spring barley as the here investigated [73]. The formulas of the VIs are given in Table 1. Different wavelength domains were covered by choosing two VIs in the near-infrared (NIR), three VIs in the visible-near-infrared (VISNIR), and one VI in the visible (VIS) domain.

Table 1. Vegetation indices. Table modified from Tilly et al. [65].

Wave-Length Domains	Vegetation Index	Formula	References
NIR	GnyLi	$(R900 \times R1050 - R955 \times R1220)/(R900 \times R1050 + R955 \times R1220)$	[74]
	NRI	$(R874 - R1225)/(R874 + R1225)$	[75]
	R_{760}/R_{730}	$R760/R730$	[40]
VISNIR	NDVI	$(R798 - R670)/(R798 + R670)$	[76]
	RDVI	$(R798 - R670)/\left(\sqrt{R798 + R670}\right)$	[77]
	REIP	$700 + 40 * \frac{\left(\frac{R670+R780}{2}\right) - R700}{R740 - R700}$	[78]
VIS	RGBVI	$(R_{green}{}^2 - R_{blue} \times R_{red})/(R_{green}{}^2 + R_{blue} \times R_{red})$	[59]

2.1.3. Reference Measurements

The BBCH scale was used to describe phenological stages and steps of plant development (Acronym BBCH is derived from the funding members: Biologische Bundesanstalt (German Federal Biological Research Centre for Agriculture and Forestry), Bundessortenamt (German Federal Office of Plant Varieties), and Chemical Industry) [79,80]. The campaign-wise averaged BBCH codes are plotted against day after seeding (DAS) in Figure 2. According to these codes, the campaigns covered the main vegetative phase, between stem elongation (Code 31) and end of fruit development (Code 79). In each campaign, the height of ten plants per plot was measured and averaged. The main stem height including ears was measured. Thin awns were excluded, as it is unlikely that they are captured by the laser scanner. In the defined sampling area of each plot, the aboveground biomass of a 0.2 × 0.2 m square was destructively taken each time and used to determine dry biomass in the laboratory. Each plant was previously divided into its individual organs (stem, leaf, and ear) and separately treated.

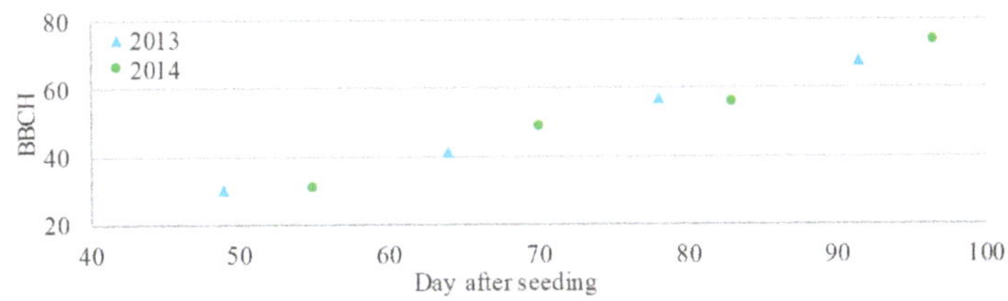

Figure 2. Campaign-wise averaged codes for the developmental steps (BBCH).

The dry biomass of the individual plant organs was further used for determining N concentration with the elemental analyzer vario EL cube [81]. For each plot, 2 to 3 mg dry biomass from stem, leaf, and ear were pulverized and homogenized. Each sample was catalytically combusted at 950 °C. The gas is separated through selective trap columns for determining the elemental composition by a thermoconductivity detector. The values of the stem, leaf, and ear samples were combined and averaged for each plot. The values of the entire plant were used in this first approach, as the entire crop surface and hence the whole plant is captured by the laser scanner. A brief overview about the values of individual plant organs will be given.

2.2. Estimation Models

This study aims at investigating whether the crop-specific indirect link between structural traits and crop N concentration (%N) can be used for robust estimations at field scale. The data set was split into four subsets for a leave-one-out cross validation. All possible combinations of the subsets for calibration and one subset for validation were calculated. For a first model run, subset 1 contained the data sets of all campaigns from 2013 (n = 48). Each other subset contained the measurements of all campaigns from one repetition from 2014 (each n = 48). For a second and third model run, the data sets were split according to the levels of N fertilization, reducing the number of values per subset for each of these runs (n = 24).

Dry biomass (DBio) was first investigated as potential estimator for %N, based on the allometric relation between DBio and %N. A two-step model was designed as shown in Figure 3 [A]. The first step of the model aims at determining DBio from crop surface height (CSH). An exponential biomass regression model (BRM) was calibrated from two subsets, since this performed best for the observed period of the growing season [65]. DBio of the other two subsets was then separately determined and validated. The determined values are hereinafter referred to as $DBio_{mod}$. In the second step, one subset of $DBio_{mod}$ and %N was used for calibrating a nitrogen regression model (NRM). As NRMs are newly investigated here, linear, exponential, and power function models were established. These NRMs were then applied to the remaining subset and validated. All calibrations were evaluated with the coefficient of determination ($R_C{}^2$). The formula is given in (1). For each validation, besides the coefficient of determination (R_V^2), the root mean square error (RMSE), and relative RMSE (rRMSE) were determined. The formulas are given in (2) to (4). All statistics are calculated for each run.

$$R_c^2 = \left(\frac{\sum_{i=1}^{n}(x-\overline{x})(y-\overline{y})}{\sqrt{\sum_{i=1}^{n}(x-\overline{x})^2 \sum_{i=1}^{n}(y-\overline{y})^2}} \right)^2 \quad (1)$$

where x refers to the estimator (CSH, $DBio_{mod}$, or VI) and y refers to the value which is estimated (DBio or %N).

$$R_v^2 = \left(\frac{\sum_{i=1}^{n}(x-\overline{x})(y-\overline{y})}{\sqrt{\sum_{i=1}^{n}(x-\overline{x})^2 \sum_{i=1}^{n}(y-\overline{y})^2}} \right)^2 \quad (2)$$

where x refers to the estimated values (DBio or %N) and y refers to the measured values (DBio or %N).

$$RMSE = \sqrt{\frac{\sum_{i=1}^{n}(x_i - y_i)^2}{n}} \quad (3)$$

where x refers to the estimated values, y refers to the measured values, and n is the number of samples.

$$rRMSE = \frac{RMSE}{\overline{M}} \quad (4)$$

where $\overline{M}$ is the across season mean value of the validation subset (DBio or %N).

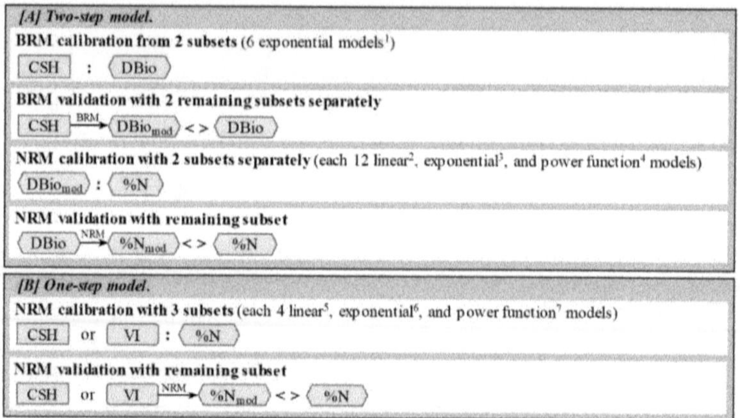

[1] $DBio_{mod} = a \cdot e^{b \cdot CSH}$ — a and b = factors, e = base of the natural logarithm
[2] $\%N_{mod} = a \cdot DBio_{mod} + b$ — a = slope, b = intercept
[3] $\%N_{mod} = a \cdot e^{b \cdot DBio_{mod}}$ — a and b = factors, e = base of the natural logarithm
[4] $\%N_{mod} = a \cdot DBio_{mod}^b$ — a and b = factors
[5] $\%N_{mod} = a \cdot VI \text{ or } CSH + b$ — a = slope, b = intercept
[6] $\%N_{mod} = a \cdot e^{b \cdot VI \text{ or } CSH}$ — a and b = factors, e = base of the natural logarithm
[7] $\%N_{mod} = a \cdot VI \text{ or } CSH^b$ — a and b = factors

Figure 3. Concept of the two-step (**A**) and one-step (**B**) models (BRM: biomass regression model; NRM: nitrogen regression model; CSH: crop surface height; VI: vegetation index; DBio: dry biomass; $DBio_{mod}$: modeled dry biomass; %N: nitrogen concentration; $\%N_{mod}$: modeled nitrogen concentration).

Considering the physiological principles of plant responses to light competition, CSH is suggested as potential estimator for crop N status. Similar to the two step model, linear, exponential, and power function NRMs were established as one-step model (Figure 3 [B]). As only one subset was needed for the validation, three subsets were used for the calibration. Again, three model runs were carried out, for the entire data set and for the common and reduced level of N fertilization. The calibrations were evaluated by the R_C^2 and the validations by the R_V^2, RMSE, and rRMSE.

In addition, NRMs based on spectral data were established. In a pre-test, the R^2 was calculated for each of the seven VIs vs. %N based on linear, exponential, and power function regressions. Based on the results, VIs were selected for the further considerations in the one-step model. The model design is equal to the NRMs based on CSH (Figure 3 [B]).

2.3. Relationship Between Crop Traits and Plant Organs

In this first approach DBio and %N of the entire plant were used for the estimation models. Nonetheless, a brief look is taken at the relationship between the estimators and the individual plant organs. As outlined in the introduction, N is required by different plant organs depending on the growing stage. The competing for light among neighboring plants influences the N dilution [19,69], which is observable in various environments [82,83]. Scatterplots of CSH or VI vs. DBio and %N per plant organ were established to get a first impression. Furthermore, ear DBio was plotted against stem %N and leaf %N, since these storages are important for the ear and grain development.

3. Results

3.1. Crop Traits

The following is a brief description of the captured crop traits. Table 2 shows the statistics of CSH, DBio, and %N. The values were calculated for the four campaigns of 2013, acting as subset 1 in the model design, and for the four campaigns of 2014 separately for subset 2 to 4. The subsets show similar patterns within each growing stage and comparable developments across the growing season. All values are additionally plotted against DAS in Figure 4 [A] to [C]. As a first approximation, the temporal development of all crop traits can quite good be explained by a linear trend (R^2: 0.82–0.86). Better results can be achieved with a logarithmic function for CSH (R^2: 0.88) and exponential functions for DBio (R^2: 0.86) and %N (R^2: 0.89). DBio is plotted against CSH in Figure 4 [D]. The relation between these traits was intensively discussed by Tilly et al. [65]. Considering the aim of this study, %N is plotted against DBio in Figure 4 [E] and against CSH in Figure 4 [F]. A linear regression explains the trend of DBio vs. %N quite well (R^2: 0.70), but a power function works better (R^2: 0.77). For CSH vs. %N the linear trend fits slightly better (R^2: 0.81) then the exponential one (R^2: 0.75). These values confirm the quality of the obtained data. The rather good link of %N to DBio and CSH should be noted, which supports the idea of using the indirect link for the estimation models.

Table 2. Statistics for the plot-wise averaged crop surface height (CSH), the destructively taken dry biomass (DBio), and nitrogen concentration (%N) measured in the laboratory. The values were calculated for subset 1 of 2013 and subsets 2, 3, and 4 of 2014 (DAS: Day after seeding; n: number of samples; mean: mean value; SD: standard deviation).

	DAS	Subset	n	CSH (m)		DBio (g/m^2)		%N (g/kg)	
				Mean	SD	Mean	SD	Mean	SD
2013	49	1	12	0.22	0.13	168.31	59.11	39.91	6.19
	64	1	12	0.46	0.17	415.31	152.51	20.56	3.73
	78	1	12	0.76	0.11	883.38	343.55	17.83	2.99
	92	1	12	0.78	0.07	1258.88	229.69	14.04	1.83
Mean				*0.56*	*0.12*	*681.47*	*196.22*	*23.09*	*3.68*
2014	41	2	12	0.17	0.03	86.69	31.22	39.94	3.82
	41	3	12	0.17	0.03	100.81	24.29	40.79	3.96
	41	4	12	0.17	0.03	79.54	26.14	39.37	3.78
	56	2	12	0.41	0.04	304.02	63.80	35.51	4.45
	56	3	12	0.42	0.05	290.90	77.10	35.69	3.94
	56	4	12	0.41	0.05	274.56	61.47	35.13	3.89
	70	2	12	0.63	0.03	487.73	82.10	22.02	3.11
	70	3	12	0.64	0.04	482.81	102.39	22.04	2.41
	70	4	12	0.62	0.04	425.95	107.53	23.07	2.69
	97	2	12	0.76	0.04	1247.56	285.68	11.13	2.13
	97	3	12	0.79	0.04	1138.10	296.94	11.33	1.86
	97	4	12	0.77	0.04	1113.50	263.13	10.73	1.62
Mean				*0.50*	*0.04*	*502.68*	*118.48*	*27.23*	*3.14*

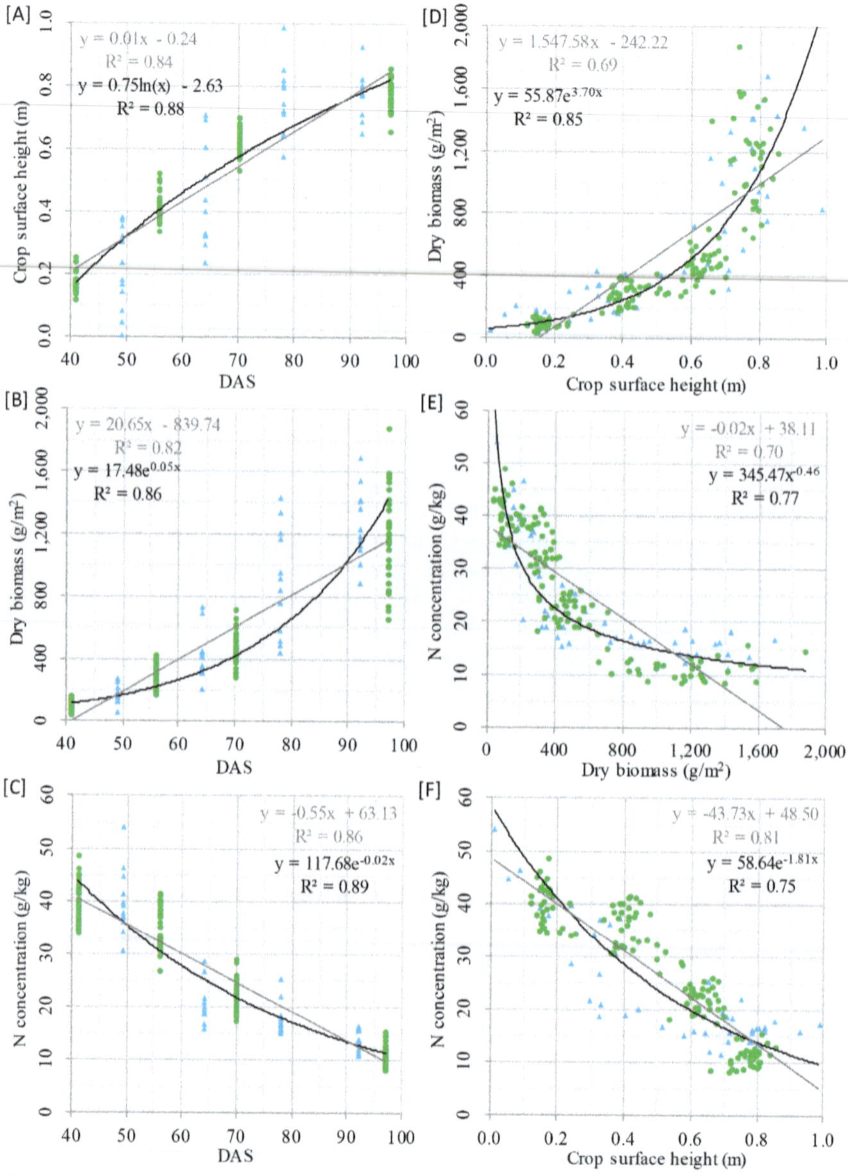

Figure 4. Blue triangles and green dots represent the values of 2013 and 2014, respectively, with linear (light grey) and the best fitting (black) regression lines. **[A]**–**[C]**: Scatterplots of the crop traits vs. day after seeding (DAS). **[D]**–**[F]**: Crop traits plotted against each other.

3.2. Nitrogen Estimation

A two-step model was designed for investigating whether the indirect link between CSH, DBio, and %N can be used for estimating %N from structural traits. The reason for this assumption was that %N is negatively related to DBio, which, in turn, can be estimated from CSH. Further one-step models based on CSH and VIs were established for comparison.

3.2.1. Two-Step Models

In the first step, two subsets of CSH and DBio were used for calibrating the BRM and applied to the other two subsets, separately. 18 exponential BRMs were established, as every possible subset combination was calculated in three runs. The mean values for the calibration and validation are given in Table 3, summarized per model run. The statistics for each possible subset combination are shown in the appendix (Table A1). As comparative value for the RMSE, it is referred to the across-season mean DBio of 681.47 g/m^2 and 502.68 g/m^2 for 2013 and 2014, respectively (Table 2). The differences between the model runs are negligible with quite high mean $R_C{}^2$ values of 0.86–0.87. In the validation, the run with the commonly fertilized plots performed slightly better ($R_V{}^2$: 0.81). The other runs yielded comparable good results ($R_V{}^2$: 0.74 and 0.72 for the entire data set and the reduced fertilized plots, respectively).

Table 3. Summarized mean values for biomass regression models (Calibration: $R_C{}^2$: coefficient of determination; Validation: $R_V{}^2$: coefficient of determination, RMSE: root mean square error (g/m^2), rRMSE: relative root mean square error; Entire: entire data set; Common: common N fertilization; Reduced: reduced N fertilization). The statistics for each possible subset combination are shown in the appendix (Table A1).

	$R_C{}^2$	$R_V{}^2$	RMSE	rRMSE
Entire	0.86	0.74	238.00	0.43
Common	0.86	0.81	213.81	0.40
Reduced	0.87	0.72	256.93	0.45

In the second step, one subset of $DBio_{mod}$ and %N was used for calibrating the NRM. The remaining subset was used for the validation. NRMs based on linear, exponential, and power function regressions were calculated for investigating the type of regression between DBio and %N. 36 NRMs were established, since this step is again performed for every possible subset combination in three runs. The mean values are given in Table 4. The statistics for each possible subset combination are shown in the appendix (Table A2). As comparative value for the RMSE, it is referred to the across-season mean %N values of 23.09 g/kg and 27.23 g/kg for 2013 and 2014, respectively (Table 2). The differences between the model runs are again almost negligible. The best results were achieved with the exponential NRM for the data set of the commonly fertilized plots ($R_C{}^2$: 0.84 and $R_V{}^2$: 0.80). The power function NRM for the entire data set performed weaker. However, these results are also far from poor with a mean $R_C{}^2$ of 0.76 and $R_V{}^2$ of 0.71. The RMSE is also similar for the linear and exponential NRM, but slightly higher for the power function NRM.

Table 4. Summarized mean values for nitrogen regression models (NRMs) based on two-step models (Calibration: $R_C{}^2$: coefficient of determination; Validation: $R_V{}^2$: coefficient of determination, RMSE: root mean square error (g/kg), rRMSE: relative root mean square error; Entire: entire data set; Common: common N fertilization; Reduced: reduced N fertilization). The statistics for each possible subset combination are shown in the appendix (Table A2).

	Linear NRM				Exponential NRM				Power Function NRM			
	$R_C{}^2$	$R_V{}^2$	RMSE	rRMSE	$R_C{}^2$	$R_V{}^2$	RMSE	rRMSE	$R_C{}^2$	$R_V{}^2$	RMSE	rRMSE
Entire	0.77	0.71	7.90	0.31	0.76	0.81	7.32	0.28	0.76	0.71	8.75	0.33
Common	0.82	0.69	8.10	0.30	0.84	0.80	6.96	0.25	0.80	0.70	10.05	0.36
Reduced	0.76	0.76	8.32	0.35	0.75	0.82	7.67	0.31	0.76	0.74	7.44	0.30

3.2.2. One-Step Models

Similar to the second step of the two-step model, NRMs based on linear, exponential, and power function regressions were established with CSH as estimator. 12 NRMs were established, since the models are again performed for every possible subset combination in three runs. The mean values are given in Table 5. The statistics for each possible subset combination are shown in the appendix (Table A3). As comparative value for the RMSE, it is again referred to Table 2. The differences between the three model runs are almost negligible. The best results were achieved with the linear NRMs (R_C^2: 0.81–0.83 and R_V^2: 0.82–0.85) followed by the exponential NRMs (R_C^2: 0.75–0.78 and R_V^2: 0.78–0.80). Slightly worse performed the power function NRMs (R_C^2: 0.58–0.66 and R_V^2: 0.58–0.68). The RMSE is again similar for the linear and exponential NRM, but slightly higher for the power function NRM.

Table 5. Summarized mean values for nitrogen regression models (NRMs) based on crop surface height (Calibration: R_C^2: coefficient of determination; Validation: R_V^2: coefficient of determination, RMSE: root mean square error (g/kg), rRMSE: relative root mean square error; Entire: entire data set; Common: common N fertilization; Reduced: reduced N fertilization). The statistics for each possible subset combination are shown in the appendix (Table A3).

	Linear NRM				Exponential NRM				Power function NRM			
	R_C^2	R_V^2	RMSE	rRMSE	R_C^2	R_V^2	RMSE	rRMSE	R_C^2	R_V^2	RMSE	rRMSE
Entire	0.81	0.82	5.05	0.19	0.76	0.78	5.84	0.22	0.60	0.58	15.81	0.58
Common	0.82	0.82	4.68	0.19	0.75	0.79	5.06	0.21	0.66	0.68	6.43	0.26
Reduced	0.83	0.85	4.89	0.18	0.78	0.80	6.17	0.23	0.58	0.63	21.53	0.84

As comparative data to the estimations from structural traits, the usability of the calculated VIs for estimating %N was investigated. The R^2 of each plot-wise averaged VI vs. %N was calculated for each of the seven VIs based on a linear, an exponential, and a power function regression as a pre-test. Based on these results (Table 6), four VIs were selected for further analysis. The GnyLi and NRI were chosen as they performed best (R^2: 0.47–0.68). Even though the REIP and the R_{760}/R_{730} yielded far poorer results (R^2: 0.04–0.18), they were further investigated, as these VIs are known from literature for its usefulness for estimating %N [30,32,40]. The NDVI, RDVI, and RGBVI were neglected due to their poor performance (R^2: 0.10–0.19).

Table 6. Pre-test: Coefficient of determination (R^2) between the plot-wise averaged vegetation indices and the nitrogen concentration.

	Linear	Exponential	Power Function
GnyLi	0.66	0.59	0.48
NDVI	0.11	0.06	0.06
NRI	0.68	0.61	0.47
RDVI	0.19	0.12	0.13
REIP	0.18	0.10	0.10
RGBVI	0.10	0.05	0.05
R_{760}/R_{730}	0.11	0.12	0.05

Similar to the one-step models based on CSH, NRMs based on linear, exponential, and power function regressions were established with GnyLi, NRI, REIP, and R_{760}/R_{730} as estimators. The mean values are given in Table 7. The structure of the table is equal to Table 5. The statistics for each possible subset combination are shown in the appendix (Table A4). As it could be assumed from the pre-test, the R_{760}/R_{730} and REIP showed the overall worst performance with mean R_C^2 of 0.02–0.18 and R_V^2 of 0.15–0.27 and R_C^2 of 0.06–0.28 and R_V^2 of 0.23–0.35, respectively. The GnyLi and NRI yielded better and among themselves comparable results. The differences between the model runs can again be neglected. Mean R_C^2 values between 0.45 and 0.68 were obtained with both VIs. The linear models performed best; closely followed by the exponential NRMs and the power function provided the worst results. A similar pattern can be stated for the validation. The linear NRMs showed the best performance with R_V^2 of 0.67–0.70. They are closely followed by the exponential ones (R_V^2: 0.58–0.61). The power function models performed slightly weaker (R_V^2: 0.47–0.63). For all VIs, the RMSE is again lower for the linear and exponential models than for the power function models.

Table 7. Summarized mean values for nitrogen regression models (NRMs) based on vegetation indices (Calibration: R_C^2: coefficient of determination; Validation: R_V^2: coefficient of determination, RMSE: root mean square error (g/kg), rRMSE: relative root mean square error; Entire: entire data set; Common: common N fertilization; Reduced: reduced N fertilization). The statistics for each possible subset combination are shown in the appendix (Table A4).

	Linear NRM				Exponential NRM				Power Function NRM			
	R_C^2	R_V^2	RMSE	rRMSE	R_C^2	R_V^2	RMSE	rRMSE	R_C^2	R_V^2	RMSE	rRMSE
GnyLi												
Entire	0.66	0.67	6.66	0.25	0.59	0.61	7.57	0.29	0.49	0.50	10.70	0.40
Common	0.66	0.67	6.73	0.24	0.61	0.59	8.12	0.29	0.46	0.47	11.46	0.41
Reduced	0.67	0.69	6.32	0.25	0.58	0.67	6.54	0.26	0.53	0.63	7.46	0.30
NRI												
Entire	0.68	0.68	6.47	0.24	0.61	0.62	7.47	0.28	0.47	0.48	9.87	0.37
Common	0.68	0.68	6.58	0.24	0.63	0.60	7.90	0.29	0.45	0.46	10.30	0.37
Reduced	0.69	0.70	6.07	0.24	0.60	0.66	6.60	0.26	0.52	0.59	8.40	0.33
REIP												
Entire	0.19	0.23	10.56	0.39	0.10	0.25	137.22	5.07	0.10	0.25	10.59	0.39
Common	0.28	0.33	10.25	0.37	0.18	0.35	88.71	3.47	0.19	0.35	10.12	0.37
Reduced	0.14	0.24	10.22	0.41	0.06	0.25	178.00	6.98	0.06	0.25	10.37	0.42
R_{760}/R_{730}												
Entire	0.11	0.15	10.96	0.42	0.05	0.16	11.25	0.43	0.06	0.18	11.11	0.42
Common	0.18	0.24	10.63	0.38	0.11	0.26	10.74	0.39	0.03	0.17	13.06	0.52
Reduced	0.07	0.15	10.58	0.43	0.02	0.16	10.83	0.44	0.12	0.27	10.63	0.38

3.3. *Relationship Between Crop Traits and Plant Organs*

DBio and %N of the entire plant were used for the models in this first approach, as the entire crop surface and hence the whole plant is captured by the laser scanner. At this point, a brief look is thrown on the relation between the used estimators and individual plant organs. Figure 5 shows DBio [A] and %N [B] per plant organ plotted against CSH. It is rather obvious that stem DBio depends on CSH (R^2: 0.84), whereas leaf DBio shows a weaker R^2 of 0.36. CSH and ear DBio show a moderate relation (R^2: 0.68). From a plant physiological point of view, the stem has a key role in the plant growth process. While competing for light with their neighboring plants, a large stem length alias high CSH, accompanied by high biomass, supports the plant to place their leaves to the better-illuminated top layers. Furthermore, plants use stem N for the grain development [3]. As it can be seen in Figure 5 [B], stem %N is rather strong related to CSH (R^2: 0.81). It is also visible that generally more N is allocated to the metabolic actives leaves in comparison to the stem. Even more interesting is, however, the similar negative slope of leaf %N and stem %N vs. CSH. All investigated VIs provided poorer results. As shown for the NRI in Figure 5 [C], slightly weaker values were achieved for stem DBio (R^2: 0.34) and leaf DBio (R^2: 0.83). Rather poor values can be stated for ear DBio (R^2: 0.24). Considerably lower values were achieved for NRI vs. %N of the individual organs as shown in Figure 5 [D]. Only R^2 of 0.57, 0.68, and 0.18 were reached for stem, leaf, and ear %N, respectively.

During the growing season, ear DBio can be used for quantitatively predicting the expected yield, which is most important for crop production. How well the grain develops depends on the available N, with stem N as main source, followed by leaf N, released through natural senescence, and soil N [3]. Since plants let their leaves die off if these sources are too low, a sufficient N storage can prevent this process, which would inhibit photosynthesis. During the early growing season, before ears are present, and where crop development can mainly be influenced, stem and leaf %N are hence most important. Ear DBio vs. stem %N and leaf %N are plotted in Figure 6 to investigate this relation. A slightly higher coefficient of determination was reached for stem %N (R^2: 0.78) compared to leaf %N (R^2: 0.77). Moreover, the scattering of the values is higher for leaf %N.

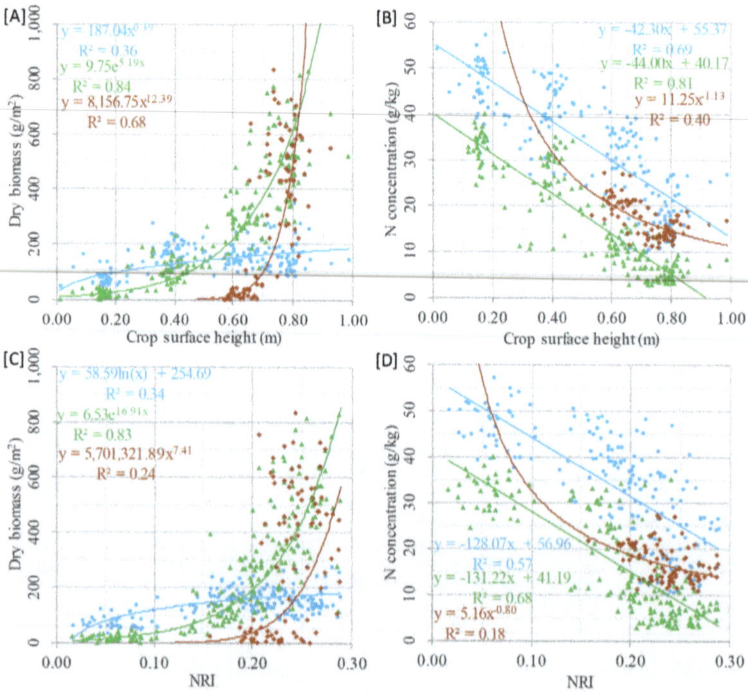

Figure 5. Scatterplots of dry biomass and N concentration per plant vs. crop surface height ([**A**] and [**B**]) or NRI ([**C**] and [**D**]) with best-fitting regression lines. Blue dots, green triangles, and orange rhombs represent the values of leaf, stem, and ear, respectively.

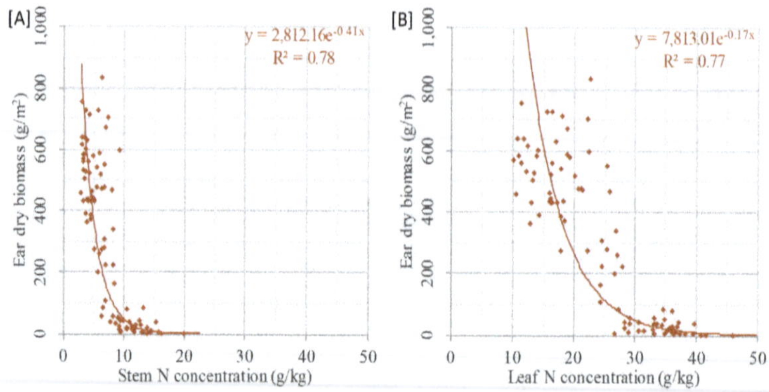

Figure 6. Scatterplots of the ear dry biomass vs. N concentration of stem [**A**] and leaf [**B**] with best-fitting regression lines.

4. Discussion

A first investigation of crop-specific models based on structural traits for indirectly estimating N concentration (%N) at field scale can be stated as overall aim of this study. Crop biomass and %N are two of the most important traits under investigation in precision agriculture. The usability of TLS-derived crop surface height (CSH) or rather plant height as estimator for biomass has been demonstrated by several authors [57–63]. An increasing number of studies on estimating biomass highlight the benefits of crop height as structural trait in comparison to spectral measurements [47,59,65]. In contrast, %N is commonly estimated from spectral data [4,31–33]. It is well known that N is negatively related to biomass during the growing season, which, in turn, can be estimated from CSH. Based on this indirect link, a two-step model was designed for determining DBio based on CSH in the first step, followed by estimating %N with a nitrogen regression model (NRM) from the determined DBio in the second step. In addition, CSH was examined as estimator in one-step NRMs and similar NRMs based on VIs were established.

The development of CSH, DBio, and %N across the growing season showed that each crop trait has a certain pattern, which can be expressed with mathematic functions (R^2: 0.82–0.89). The relation between DBio and %N can be expressed with a power function (R^2: 0.77). Similar patterns were shown for winter barley [24] and corn [29]. That the allometric relation of %N vs. DBio during the vegetative growth can be expressed with a negative power function has early been recognized [20] and is the fundament of the NNI and NDC concept. As shown in Figure 4 [F], CSH vs. %N appears to be linearly (R^2: 0.81) or exponentially (R^2: 0.75) related, which supports the idea of using this structural trait as estimator for the N status of crops. Unfortunately, no comparable studies at field scale were found.

The applicability of biomass regression models (BRMs) based on CSH was demonstrated by Tilly et al. [65]. In contrast, crop-specific NRMs are newly developed in this study and thus examined more closely with linear, exponential, and power functions. Since the NRMs of all approaches were similarly designed, the results can well be compared. A point in common of all NRMs is that only negligible differences between the three runs could be observed. This might be explained by the minor difference between the fertilizer treatments of only 40 kg/ha. In contrast, in the literature differences of 70 to 220 kg/ha can be found for common field experiments [30,41,73,84] or single experiments with extremely high rates of up to 420 kg/ha [30].

Spectral approaches for determining %N are presented in different studies [30–34]. In these studies, R^2 values between 0.50 and 0.80 were reached for different crops. The best performing VI in this study reached comparable results ($R^2 < 0.70$). Good results achieved with the R_{760}/R_{730} and REIP in other studies [30,32,40] could not be confirmed here ($R_V{}^2 < 0.35$). Even though these VIs will be excluded from the model comparison, some explanations for their weak performance should be formulated. First, all other studies investigated wheat not barley. Furthermore, Erdle et al. (2011) [40] and Baresel et al. [30] investigated measurements within one growing stage and used bidirectional radiometer or active sensors. Similar to this study, Li et al. [32] investigated different growing stages with a field spectrometer but obtained beside the quite good R^2 of 0.55 also weak results of 0.21 for the REIP. It can be summarized that the R_{760}/R_{730} and REIP might be suitable for estimating crop %N. It should, however, be further investigated how prone these VIs are to effects such as saturation [37], multiangular reflection [85], leaf inclination angle [43], or general atmospheric conditions. Further research should involve additional VIs, which were investigated for N content and N uptake, such as the red edge-based canopy chlorophyll content index [86] or the optimum multiple narrow band reflectance model [31]. Another approach that can be found in the literature are contour plot analyses. Conclusions of two studies are, however, that it is challenging to obtain the necessary information with two-band VIs due to the N dilution effect [31] and that it will be difficult to develop a simple sensor that can determine the N concentration of plants with two or three bands for different growth stages [32]. The presented approach based on structural data shall introduce an alternative method.

The power function produced the poorest results of the three function types, with the lowest R^2 and highest RMSE values. Furthermore, the weakest results were achieved with the two remaining VIs. For both VIs, best results were obtained with linear models, whereas the NRI performed slightly better ($R_V{}^2$: 0.68–0.70) than the GnyLi ($R_V{}^2$: 0.67–0.69). The NRMs based on $DBio_{mod}$ yielded better results and performed best with exponential functions ($R_V{}^2$: 0.81–0.82). The best and most robust results were obtained with the NRMs based on CSH ($R_V{}^2$: 0.78–0.85). It should be further investigated whether the fusion of CSH and VIs in multivariate models improves the estimation, as it was observed for the BRMs [65]. The benefit of the fusion of spectral and structural traits are also highlighted by other studies. The yield prediction of drought stressed barley for example could be improved [73]. Acquiring 3D and spectral data with one sensor could therefore be worthwhile, which might be realizable with recently emerging hyperspectral laser scanners [87]. First approaches of estimating rice leaf %N from multispectral and hyperspectral scanners can already be found [88,89]. These studies, however, only investigated the spectral information. Further research on the fusion of structural and spectral traits is required.

Quality, usability, and robustness of models can be assessed according to their accuracy and precision. Widely used metrics are the R_C^2, R_V^2, RMSE, and rRMSE. In case of the NRMs, R_C^2 shows if CSH-based DBio models, CSH or VIs can express variations of %N and hence if it is a suitable indicator. R_V^2 shows whether this relation is robust and transferable to independent data. It can well be used to compare the models among each other. RMSE and rRMSE are suitable for comparing the precision between the models. However, for evaluating the accuracy of NRMs for estimating %N, actual measurements must be considered. A major issue of across-season estimations is the very varying %N. The rRMSE refers to the across-season mean value, which can obviously hardly be used as reliable reference for the accuracy of the NRM at a certain point in time. As shown for the BRMs [90], more credible values can be obtained with campaign-wise separated analyses. Similar results are very assumable for the NRMs, which requires further research. The validity of the RMSE for time-specific models must be further investigated to evaluate the usability of the approach for site-specific management. Beyond that, studies are needed on the transferability of the models to other field locations, years, or crops. Good results have already been achieved for the transferability of BRMs for paddy rice [91]. However, it is known from the literature that the relationship between biomass and nitrogen is very different depending on the crop [24,92]. Accordingly, it can be assumed that NRMs must be calibrated crop-specific, as is the case for approaches with the nitrogen nutrition index (NNI). Novel machine learning approaches and algorithms, summarized as artificial intelligence, should also be taken into account for estimating crop traits [93–95]. Approaches of estimating grass sward biomass [96], quantifying rice N status [97], or monitoring wheat leaf %N [98] can already be found. As mentioned, an important point for site-specific management is the applicability of models at certain stages within the growing season. Analyses within one campaign were not very reliable with the available data set, due to the amount of data. Accordingly, further analyses with more extensive data sets are desirable for verifying the time-specific applicability of this approach.

The brief look at the relation between the estimators and individual plant organs suggests that CSH might be useful for estimating stem DBio and %N. The observed relation between CSH and leaf %N supports findings of Yin et al. [99]. They found a good relationship between plant height of corn and leaf %N. Beside the known fact that more N is allocated to the metabolic actives leaves than to the stem, a similar negative slope of leaf %N and stem %N vs. CSH was found. As the quantification of metabolic and structural N demand of individual plant organs is required for example for growth simulation models [5], further research on this is desirable. Moreover, this research can make an important contribution to understand the processes and traits, which are involved in the growth and N cycle of plants. For some crops, such as wheat, barley, and maize, the pre-anthesis ear and stem growth are somehow interlinked [100]. From the rather good relation between ear DBio and stem %N (R^2: 0.78) or leaf %N (R^2: 0.77), it can be interpreted that these individual plant organs might be useful for yield approximations during the growing season. Further research on this is required.

Considering finally the applicability in the field, the main benefits of capturing structural traits in comparison to spectral measurements were stated in the introduction. A main disadvantage of acquiring spectral data with handheld spectrometers is the limitation to several discrete measurements. It has to be mentioned that in recent years several robotic system were developed as high-throughput mobile field platform [101,102]. Such platforms can accelerate the capturing process or even enable an autonomous data acquisition. Furthermore, approaches of a spatially resolved acquisition of spectral data can be found [72,103]. As these attempts are all quite new, the applicability and validity of the results has still to be investigated. Besides the development of appropriate sensors, the emergence of unmanned aerial vehicles (UAVs) as platform was a major milestone. As reviewed by several authors [104–106], various studies attach digital cameras and multi- or hyperspectral imagers to UAVs for a nadir acquisition of surface properties. The size, payload, and stability of UAVs were rapidly improved through technical developments across recent years, which allows attaching heavier sensors, such as laser scanners. So far, such systems are occasionally used for capturing structural traits in forestry such as canopy height [107], tree stem diameter [108], or crown base height [109]. Major disadvantages are the high cost and large data processing [105]. Nevertheless, such UAV-based laser scanners would be beneficial for the acquisition of agricultural fields, as the nadir perspective should allow a deeper penetration of the vegetation and hence a more detailed capturing of the crop surface and structural traits. Crop density could also be captured from a nadir perspective. This is important in particular for the here investigated research question, as crop density influences the light competition process. This major process of plant growth can hence be examined more closely. Combining crop height and density information might allow precise and robust biomass and %N estimations. Finally, this research can have an influence on conventional agriculture, as sowing densities and fertilizer applications could be optimized.

5. Conclusions

Non-destructively estimating and monitoring crop biomass and nitrogen (N) concentration can support site-specific management. Studies show that dry biomass can be estimated by remotely sensing 3D or spectral information. In contrast, N estimations at field scale so far only base on spectral measurements, which are biased by illumination changes, vegetation fraction, plant properties, or saturation effects. This survey pursued a novel approach of estimating N concentration based on the indirect link to structural data. This link is based on the negative relationship between N and biomass, which can be estimated from crop height.

In two subsequent years, crop surface height was measured by terrestrial laser scanning (TLS), seven vegetation indices (VI) were calculated from field spectrometer measurements, and destructive measurements of dry biomass and N concentration were carried out. Based on the indirect link between crop surface height and N concentration, novel nitrogen regression models were established in three designs. The models based on modeled dry biomass and crop surface height reached R^2 values of up to 0.85 and 0.82, respectively, and thereby outperformed the VIs (best R^2: 0.70). The relation between crop surface height and N concentration of individual plant organs revealed that stem N showed a stronger dependency (R^2: 0.81) in comparison to leaf N (R^2: 0.69) and ear N (R^2: 0.40). With the best performing VI, the NRI, considerably weaker results were achieved (R^2 for stem, leaf, and ear N were 0.57, 0.68, and 0.18, respectively). This stem N is particularly important for plant growth, as it is one N source for the grain development.

In summary, these first results open perspectives for indirect models based on structural data as indicator for crop N concentration at field scale. Due to the crop-specific relationship between dry biomass and N concentration, it can be assumed that the models will have to be calibrated crop-specific. Furthermore, the time-specific applicability of the models must be further investigated. Capturing 3D data with proximal sensing can be regarded as being more robust and comfortable to carry out in comparison to spectral measurements. Beside the data acquisition, the analysis of 3D data and determination of structural traits could be designed more user-friendly than the interpretation of spectral data. Hence, beside the usability as a research tool, TLS or photogrammetric approaches should be considered for improving non-invasive and cost-effective N fertilizer management tools for farmers. For research purposes, such as monitoring plant-internal dynamics, investigating particular processes, or detecting diseases, the value of spectral information is indisputable. The fusion of 3D and spectral data can therefore be very beneficial.

Author Contributions: Conceptualization, N.T. and G.B.; methodology, writing—original draft preparation, N.T.; writing—review and editing, funding acquisition, G.B.

Funding: The field data acquisition was carried out within the CROP.SENSe.net project in context of the Ziel 2-Programms NRW 2007–2013 "Regionale Wettbewerbsfähigkeit und Beschäftigung", financially supported by the Ministry for Innovation, Science and Research (MIWF) of the state North Rhine Westphalia (NRW) and European Union Funds for regional development (EFRE) (005-1103-0018).

Acknowledgments: Thank you, Helge Aasen, for a great joint work and all constructive advices. We thank the student assistants for doing a great job in the field and during the post-processing. Dirk Hoffmeister and Martin Gnyp are to be thanked for their knowledgeable support of the TLS and spectral measurements, respectively. We extend our thanks to the staff of the Campus Klein–Altendorf for carrying out the experiment and the collaborators at the Forschungszentrum Jülich for supporting the laboratory work. Agim Ballvola is acknowledged for the coordination of the CROP.SENSe.net research group barley.

Conflicts of Interest: The authors declare no conflict of interest. The funders had no role in the design of the study; in the collection, analyses, or interpretation of data; in the writing of the manuscript, or in the decision to publish the results.

Appendix A

Table A1. Statistics for each run of the biomass regression models (Calibration: R_C^2: coefficient of determination; Validation: R_v^2: coefficient of determination, RMSE: root mean square error (g/m^2), rRMSE: relative root mean square error).

Calibration Subset	Validation Subset	R_C^2	R_V^2	RMSE	rRMSE
Entire data set					
1 and 2	3	0.83	0.79	189.74	0.38
1 and 2	4	0.83	0.75	204.09	0.43
1 and 3	2	0.84	0.81	232.62	0.44
1 and 3	4	0.84	0.75	207.14	0.44
1 and 4	2	0.81	0.81	229.79	0.43
1 and 4	3	0.81	0.79	190.90	0.38
2 and 3	1	0.90	0.61	322.25	0.47
2 and 3	4	0.90	0.75	202.88	0.43
2 and 4	1	0.89	0.60	343.49	0.50
2 and 4	3	0.89	0.79	191.64	0.38
3 and 4	1	0.89	0.61	313.03	0.46
3 and 4	2	0.89	0.81	228.41	0.43
Common N fertilization					
1 and 2	3	0.80	0.89	148.80	0.33
1 and 2	4	0.80	0.79	196.42	0.42
1 and 3	2	0.82	0.78	278.95	0.53
1 and 3	4	0.82	0.78	205.86	0.44
1 and 4	2	0.80	0.79	260.68	0.49
1 and 4	3	0.80	0.89	133.74	0.29
2 and 3	1	0.91	0.78	241.05	0.36
2 and 3	4	0.91	0.79	192.90	0.41
2 and 4	1	0.91	0.79	227.57	0.34
2 and 4	3	0.91	0.91	140.50	0.31
3 and 4	1	0.92	0.78	261.10	0.39
3 and 4	2	0.92	0.79	278.15	0.53
Reduced N fertilization					
1 and 2	3	0.85	0.77	220.80	0.40
1 and 2	4	0.85	0.72	203.87	0.43
1 and 3	2	0.86	0.85	171.58	0.32
1 and 3	4	0.86	0.72	204.06	0.43
1 and 4	2	0.83	0.85	184.17	0.35
1 and 4	3	0.83	0.77	231.74	0.42
2 and 3	1	0.92	0.53	412.87	0.60
2 and 3	4	0.92	0.71	210.76	0.44
2 and 4	1	0.88	0.52	371.16	0.54
2 and 4	3	0.88	0.76	322.87	0.58
3 and 4	1	0.89	0.53	388.46	0.56
3 and 4	2	0.89	0.86	160.78	0.30

Table A2. Statistics for each run of the nitrogen regression models (NRMs) based on two-step model. (Calibration: $R_C{}^2$: coefficient of determination; Validation: $R_V{}^2$: coefficient of determination, RMSE: root mean square error (g/kg), rRMSE: relative root mean square error).

Calibration Subset	Validation Subset	Linear NRM				Exponential NRM				Power Function NRM			
		$R_C{}^2$	$R_V{}^2$	RMSE	rRMSE	$R_C{}^2$	$R_V{}^2$	RMSE	rRMSE	$R_C{}^2$	$R_V{}^2$	RMSE	rRMSE
Entire data set													
3	4	0.88	0.79	5.97	0.22	0.88	0.86	6.86	0.25	0.80	0.65	12.90	0.48
4	3	0.88	0.77	6.40	0.23	0.87	0.85	6.11	0.22	0.77	0.73	7.71	0.28
2	4	0.88	0.79	7.18	0.27	0.87	0.86	5.49	0.20	0.77	0.62	15.14	0.56
4	2	0.88	0.74	8.61	0.32	0.87	0.86	6.35	0.23	0.77	0.60	13.84	0.51
2	3	0.88	0.77	7.46	0.27	0.87	0.87	6.00	0.22	0.77	0.73	7.57	0.28
3	2	0.88	0.74	7.49	0.28	0.88	0.85	6.10	0.22	0.80	0.62	11.61	0.43
1	4	0.44	0.79	6.98	0.26	0.46	0.86	8.94	0.33	0.72	0.72	6.34	0.23
4	1	0.87	0.54	9.66	0.42	0.87	0.65	7.00	0.30	0.77	0.79	5.08	0.22
1	3	0.42	0.77	7.84	0.29	0.44	0.85	10.30	0.38	0.72	0.79	7.52	0.27
3	1	0.86	0.54	8.58	0.37	0.87	0.65	6.70	0.29	0.80	0.79	4.87	0.21
1	2	0.44	0.74	7.30	0.27	0.46	0.85	9.62	0.35	0.72	0.69	7.25	0.27
2	1	0.87	0.54	11.34	0.49	0.87	0.73	8.37	0.36	0.77	0.79	5.22	0.23
Common N fertilization													
3	4	0.88	0.79	6.20	0.22	0.90	0.87	6.61	0.23	0.82	0.64	11.66	0.41
4	3	0.95	0.74	6.25	0.22	0.95	0.82	7.49	0.26	0.82	0.73	8.08	0.28
2	4	0.93	0.79	7.31	0.26	0.93	0.89	5.50	0.19	0.80	0.62	17.09	0.61
4	2	0.95	0.71	11.42	0.40	0.95	0.85	6.90	0.24	0.82	0.56	21.11	0.74
2	3	0.93	0.74	6.41	0.22	0.93	0.82	7.31	0.25	0.80	0.74	7.37	0.25
3	2	0.88	0.71	8.84	0.31	0.90	0.83	6.60	0.23	0.82	0.60	15.63	0.55
1	4	0.56	0.79	5.63	0.20	0.59	0.87	6.73	0.24	0.77	0.72	6.74	0.24
4	1	0.94	0.53	10.72	0.43	0.96	0.67	6.99	0.28	0.82	0.79	5.85	0.24
1	3	0.54	0.74	6.97	0.24	0.57	0.82	8.56	0.30	0.77	0.78	8.51	0.29
3	1	0.85	0.53	8.38	0.34	0.89	0.67	6.52	0.26	0.82	0.80	5.33	0.22
1	2	0.56	0.71	7.48	0.26	0.59	0.83	7.35	0.26	0.77	0.67	7.51	0.26
2	1	0.92	0.53	11.62	0.47	0.94	0.67	6.95	0.28	0.80	0.79	5.67	0.23
Reduced N fertilization													
3	4	0.93	0.80	5.58	0.21	0.92	0.85	7.19	0.28	0.83	0.64	11.43	0.44
4	3	0.85	0.81	6.78	0.26	0.82	0.89	4.77	0.18	0.75	0.74	7.37	0.28
2	4	0.90	0.80	6.64	0.26	0.89	0.84	5.32	0.21	0.79	0.63	12.71	0.49
4	2	0.85	0.84	5.25	0.20	0.82	0.89	5.28	0.21	0.75	0.68	7.97	0.31
2	3	0.89	0.81	9.60	0.37	0.89	0.89	5.66	0.22	0.79	0.73	7.73	0.30
3	2	0.93	0.84	5.35	0.21	0.92	0.88	6.06	0.24	0.83	0.68	7.66	0.30
1	4	0.38	0.80	7.98	0.31	0.38	0.76	17.15	0.66	0.65	0.72	5.85	0.23
4	1	0.84	0.58	8.05	0.38	0.82	0.69	6.41	0.30	0.75	0.85	3.81	0.18
1	3	0.37	0.81	7.90	0.30	0.37	0.89	10.66	0.41	0.65	0.80	8.40	0.32
3	1	0.92	0.58	17.41	0.81	0.92	0.69	6.87	0.32	0.83	0.85	5.45	0.25
1	2	0.37	0.84	8.38	0.33	0.37	0.87	9.33	0.36	0.65	0.75	6.90	0.27
2	1	0.89	0.58	10.92	0.51	0.89	0.76	7.33	0.34	0.79	0.85	3.98	0.19

Table A3. Statistics for each run of the nitrogen regression models (NRMs) based on crop height (Calibration: $R_C{}^2$: coefficient of determination; Validation: $R_V{}^2$: coefficient of determination, RMSE: root mean square error (g/kg), rRMSE: relative root mean square error).

Calibration Subset	Validation Subset	Linear NRM				Exponential NRM				Power Function NRM			
		$R_C{}^2$	$R_V{}^2$	RMSE	rRMSE	$R_C{}^2$	$R_V{}^2$	RMSE	rRMSE	$R_C{}^2$	$R_V{}^2$	RMSE	rRMSE
Entire data set													
1, 2, and 3	4	0.80	0.85	4.51	0.17	0.75	0.77	5.55	0.21	0.59	0.65	7.19	0.27
1, 2, and 4	3	0.80	0.87	4.54	0.17	0.74	0.79	5.69	0.21	0.58	0.65	7.48	0.28
1, 3, and 4	2	0.80	0.84	4.74	0.18	0.75	0.76	5.76	0.21	0.59	0.63	7.49	0.28
2, 3, and 4	1	0.85	0.73	6.41	0.24	0.78	0.81	6.36	0.24	0.63	0.41	41.10	1.52
Common N fertilization													
1, 2, and 3	4	0.81	0.89	3.98	0.14	0.77	0.80	5.20	0.18	0.56	0.68	7.54	0.27
1, 2, and 4	3	0.82	0.87	4.76	0.16	0.77	0.79	5.94	0.21	0.56	0.68	8.28	0.29
1, 3, and 4	2	0.82	0.86	4.30	0.15	0.77	0.76	5.72	0.20	0.56	0.63	7.77	0.27
2, 3, and 4	1	0.87	0.78	6.53	0.26	0.81	0.85	7.84	0.32	0.65	0.52	62.51	2.53
Reduced N fertilization													
1, 2, and 3	4	0.81	0.84	4.36	0.17	0.74	0.77	5.24	0.20	0.67	0.64	7.06	0.27
1, 2, and 4	3	0.78	0.92	3.37	0.13	0.72	0.84	4.52	0.17	0.66	0.68	7.14	0.28
1, 3, and 4	2	0.81	0.87	4.54	0.18	0.74	0.80	5.18	0.20	0.67	0.66	6.99	0.27
2, 3, and 4	1	0.88	0.65	6.46	0.30	0.79	0.74	5.31	0.25	0.64	0.75	4.53	0.21

Table A4. Statistics for each run of the nitrogen regression models (NRMs) based on vegetation indices (Calibration: R_C^2: coefficient of determination; Validation: R_V^2: coefficient of determination, RMSE: root mean square error (g/kg), rRMSE: relative root mean square error).

Calibration Subset	Validation Subset	Linear NRM				Exponential NRM				Power Function NRM			
		R_C^2	R_V^2	RMSE	rRMSE	R_C^2	R_V^2	RMSE	rRMSE	R_C^2	R_V^2	RMSE	rRMSE
GnyLi													
Entire data set													
1, 2, and 3	4	0.67	0.64	6.99	0.26	0.61	0.55	8.84	0.33	0.53	0.30	18.22	0.67
1, 2, and 4	3	0.66	0.68	6.77	0.25	0.59	0.60	7.75	0.28	0.48	0.51	8.75	0.32
1, 3, and 4	2	0.69	0.58	7.64	0.28	0.62	0.51	8.77	0.32	0.51	0.40	9.89	0.36
2, 3, and 4	1	0.63	0.76	5.24	0.23	0.54	0.79	4.91	0.21	0.43	0.77	5.94	0.26
Common N fertilization													
1, 2, and 3	4	0.68	0.64	7.31	0.26	0.64	0.52	9.64	0.34	0.52	0.28	19.32	0.68
1, 2, and 4	3	0.66	0.68	6.77	0.23	0.60	0.58	7.84	0.27	0.45	0.48	9.14	0.32
1, 3, and 4	2	0.70	0.55	7.55	0.26	0.64	0.45	9.28	0.32	0.48	0.36	9.97	0.35
2, 3, and 4	1	0.61	0.82	5.29	0.21	0.55	0.81	5.75	0.23	0.40	0.77	7.39	0.30
Reduced N fertilization													
1, 2, and 3	4	0.67	0.65	6.46	0.25	0.58	0.58	7.60	0.29	0.56	0.51	10.17	0.39
1, 2, and 4	3	0.66	0.68	6.48	0.25	0.58	0.63	7.26	0.28	0.54	0.55	7.99	0.31
1, 3, and 4	2	0.70	0.59	7.61	0.30	0.61	0.57	8.03	0.31	0.56	0.54	8.37	0.33
2, 3, and 4	1	0.63	0.83	4.74	0.22	0.53	0.89	3.27	0.15	0.47	0.90	3.31	0.15
NRI													
Entire data set													
1, 2, and 3	4	0.69	0.66	6.74	0.25	0.62	0.56	8.22	0.31	0.49	0.41	10.50	0.39
1, 2, and 4	3	0.67	0.71	6.45	0.24	0.61	0.61	7.68	0.28	0.47	0.46	9.11	0.34
1, 3, and 4	2	0.71	0.61	7.35	0.27	0.64	0.52	8.83	0.33	0.51	0.31	13.17	0.49
2, 3, and 4	1	0.65	0.74	5.33	0.20	0.58	0.78	5.14	0.19	0.42	0.75	6.69	0.25
Common N fertilization													
1, 2, and 3	4	0.69	0.66	6.83	0.25	0.65	0.55	8.09	0.29	0.46	0.36	9.76	0.35
1, 2, and 4	3	0.67	0.68	6.66	0.23	0.62	0.57	8.01	0.28	0.44	0.41	9.39	0.32
1, 3, and 4	2	0.70	0.59	7.27	0.25	0.66	0.49	9.47	0.33	0.49	0.32	14.22	0.50
2, 3, and 4	1	0.64	0.80	5.59	0.23	0.59	0.80	6.05	0.24	0.40	0.73	7.84	0.32
Reduced N fertilization													
1, 2, and 3	4	0.69	0.66	6.40	0.25	0.61	0.58	8.01	0.31	0.55	0.47	12.89	0.50
1, 2, and 4	3	0.67	0.73	5.96	0.23	0.59	0.65	6.91	0.27	0.51	0.55	7.96	0.31
1, 3, and 4	2	0.72	0.61	7.34	0.29	0.63	0.57	8.00	0.31	0.54	0.50	8.61	0.34
2, 3, and 4	1	0.66	0.78	4.60	0.21	0.56	0.85	3.46	0.16	0.46	0.86	4.14	0.19
REIP													
Entire data set													
1, 2, and 3	4	0.19	0.17	11.16	0.41	0.10	0.19	25.28	0.93	0.10	0.19	10.81	0.40
1, 2, and 4	3	0.20	0.15	10.94	0.40	0.11	0.17	12.91	0.48	0.11	0.17	11.57	0.43
1, 3, and 4	2	0.21	0.12	11.20	0.41	0.12	0.14	93.14	3.44	0.12	0.14	11.41	0.42
2, 3, and 4	1	0.14	0.47	8.94	0.33	0.06	0.49	417.55	15.42	0.06	0.50	8.58	0.32
Common N fertilization													
1, 2, and 3	4	0.30	0.21	10.46	0.37	0.21	0.23	27.47	0.97	0.21	0.23	10.20	0.36
1, 2, and 4	3	0.31	0.19	11.44	0.39	0.21	0.20	29.25	1.01	0.21	0.20	11.35	0.39
1, 3, and 4	2	0.30	0.19	10.19	0.36	0.20	0.22	27.72	0.97	0.21	0.22	10.24	0.36
2, 3, and 4	1	0.19	0.73	8.91	0.36	0.11	0.74	270.41	10.94	0.11	0.75	8.70	0.35
Reduced N fertilization													
1, 2, and 3	4	0.13	0.14	10.17	0.39	0.06	0.15	26.21	1.01	0.06	0.15	10.52	0.41
1, 2, and 4	3	0.14	0.14	10.93	0.42	0.07	0.15	11.07	0.43	0.07	0.15	11.37	0.44
1, 3, and 4	2	0.16	0.08	11.65	0.45	0.08	0.09	653.62	25.51	0.08	0.09	11.44	0.45
2, 3, and 4	1	0.11	0.60	8.15	0.38	0.04	0.62	21.11	0.98	0.04	0.61	8.15	0.38
R_{760}/R_{730}													
Entire data set													
1, 2, and 3	4	0.12	0.08	10.91	0.40	0.05	0.10	11.31	0.42	0.06	0.11	11.12	0.41
1, 2, and 4	3	0.13	0.08	11.48	0.42	0.06	0.09	12.09	0.44	0.07	0.10	11.97	0.44
1, 3, and 4	2	0.13	0.06	11.42	0.42	0.06	0.07	11.73	0.43	0.07	0.09	11.63	0.43
2, 3, and 4	1	0.07	0.39	10.04	0.43	0.02	0.40	9.87	0.43	0.03	0.42	9.72	0.42
Common N fertilization													
1, 2, and 3	4	0.20	0.12	10.84	0.38	0.13	0.14	10.75	0.38	0.14	0.16	10.59	0.38
1, 2, and 4	3	0.21	0.11	11.27	0.39	0.12	0.12	11.55	0.40	0.13	0.13	11.51	0.40
1, 3, and 4	2	0.21	0.10	10.76	0.38	0.12	0.12	10.82	0.38	0.13	0.14	10.72	0.37
2, 3, and 4	1	0.11	0.65	9.61	0.39	0.05	0.66	9.00	0.40	0.06	0.67	[illegible]	0.30
Reduced N fertilization													
1, 2, and 3	4	0.07	0.06	10.57	0.41	0.02	0.06	11.23	0.43	0.03	0.06	20.24	0.78
1, 2, and 4	3	0.08	0.06	11.02	0.42	0.02	0.06	11.59	0.45	0.03	0.07	11.58	0.45
1, 3, and 4	2	0.09	0.04	11.42	0.45	0.03	0.04	11.76	0.46	0.03	0.05	11.74	0.46
2, 3, and 4	1	0.05	0.46	9.34	0.44	0.01	0.47	8.75	0.41	0.01	0.49	8.66	0.40

References

1. Andrews, M.; Raven, J.A.; Lea, P.J. Do plants need nitrate? the mechanisms by which nitrogen form affects plants. *Ann. Appl. Biol.* **2013**, *163*, 174–199. [CrossRef]
2. Caloin, M.; Yu, O. Analysis of the time course of change in nitrogen content in *Dactylis glomerata* L. using a model of plant growth. *Ann. Bot.* **1984**, *54*, 69–76. [CrossRef]
3. Jamieson, P.D.; Semenov, M.A. Modelling nitrogen uptake and redistribution in wheat. *Field Crops Res.* **2000**, *68*, 21–29. [CrossRef]
4. Gastal, F.; Lemaire, G.; Durand, J.-L.; Louarn, G. Quantifying crop responses to nitrogen and avenues to improve nitrogen-use efficiency François. In *Crop Physiology*; Sadras, V.O., Calderini, D.F., Eds.; Elsevier Inc.: London, UK; Waltham, MA, USA; San Diego, CA, USA, 2015; pp. 161–206.
5. Lemaire, G.; van Oosterom, E.; Sheehy, J.; Jeuffroy, M.H.; Massignam, A.; Rossato, L. Is crop N demand more closely related to dry matter accumulation or leaf area expansion during vegetative growth? *Field Crops Res.* **2007**, *100*, 91–106. [CrossRef]
6. Greenwood, D.J.; Gastal, F.; Lemaire, G.; Draycott, A.; Millard, P.; Neeteson, J.J. Growth rate and %N of field grown crops: Theory and experiments. *Ann. Bot.* **1991**, *67*, 181–190. [CrossRef]
7. Vitousek, P.; Naylor, R.; Crews, T.; David, M.B.; Drinkwater, L.E.; Holland, E.; Johnes, P.J.; Katzenberger, J.; Martinelli, L.A.; Matson, P.A.; et al. Nutrient imbalances in agricultural development. *Science* **2009**, *324*, 1519–1520. [CrossRef]
8. Matson, P.A. Agricultural intensification and ecosystem properties. *Science* **1997**, *277*, 504–509. [CrossRef] [PubMed]
9. FAO. *World Fertilizer Trends and Outlook to 2020*; FAO: Rome, Italy, 2017.
10. Tilman, D.; Cassman, K.G.; Matson, P.A.; Naylor, R.; Polasky, S. Agricultural sustainability and intensive production practices. *Nature* **2002**, *418*, 671–677. [CrossRef]
11. Ju, X.-T.; Xing, G.-X.; Chen, X.-P.; Zhang, S.-L.; Zhang, L.-J.; Liu, X.-J.; Cui, Z.-L.; Yin, B.; Christie, P.; Zhu, Z.-L.; et al. Reducing environmental risk by improving N management in intensive Chinese agricultural systems. *Proc. Natl. Acad. Sci. USA* **2009**, *106*, 8077–8078. [CrossRef]
12. Schröder, J.J. The position of mineral nitrogen fertilizer in efficient use of nitrogen and land: A review. *Nat. Resour.* **2014**, *5*, 936–948. [CrossRef]
13. Nkebiwe, P.M.; Weinmann, M.; Bar-Tal, A.; Müller, T. Fertilizer placement to improve crop nutrient acquisition and yield: A review and meta-analysis. *Field Crops Res.* **2016**, *196*, 389–401. [CrossRef]
14. Azeem, B.; Kushaari, K.; Man, Z.B.; Basit, A.; Thanh, T.H. Review on materials & methods to produce controlled release coated urea fertilizer. *J. Control. Release* **2014**, *181*, 11–21.
15. Sanz-Cobena, A.; Lassaletta, L.; Aguilera, E.; del Prado, A.; Garnier, J.; Billen, G.; Iglesias, A.; Sánchez, B.; Guardia, G.; Abalos, D.; et al. Strategies for greenhouse gas emissions mitigation in Mediterranean agriculture: A review. *Agric. Ecosyst. Environ.* **2017**, *238*, 5–24. [CrossRef]
16. Zhang, M.; Yao, Y.; Tian, Y.; Ceng, K.; Zhao, M.; Zhao, M.; Yin, B. Increasing yield and N use efficiency with organic fertilizer in Chinese intensive rice cropping systems. *Field Crops Res.* **2018**, *227*, 102–109. [CrossRef]
17. Zhao, R.F.; Chen, X.P.; Zhang, F.S.; Zhang, H.; Schröder, J.; Römheld, V. Fertilization and nitrogen balance in a wheat-maize rotation system in North China. *Agron. J.* **2006**, *98*, 938–945. [CrossRef]
18. Greenwood, D.J. Modelling of crop response to nitrogen fertilizer. *Philos. Trans. R. Soc. Lond. Ser. B Biol.* **1982**, *296*, 351–362. [CrossRef]
19. Lemaire, G.; Gastal, F. N uptake and distribution in plant canopies. *J. Exp. Bot.* **1997**, *53*, 789–799.
20. Lemaire, G.; Salette, J.; Sigogne, M.; Terrasson, J.-P. Relation entre dynamique de croissance et dynamique de prélèvement d ' azote pour un peuplement de graminées fourragères. I.—Etude de l' effet du milieu. *Agronomie* **1984**, *4*, 423–430. [CrossRef]
21. Ulrich, A. Physiological bases for assessing the nutritional requirements of plants. *Annu. Rev. Plant Physiol.* **1952**, *3*, 207–228. [CrossRef]
22. Gislum, R.; Boelt, B. Validity of accessible critical nitrogen dilution curves in perennial ryegrass for seed production. *Field Crops Res.* **2009**, *111*, 152–156. [CrossRef]
23. Plénet, D.; Lemaire, G. Relationships between dynamics of nitrogen uptake and dry matter accumulation in maize crops. Determination of critical N concentration. *Plant Soil* **2000**, *216*, 65–82. [CrossRef]

24. Zhao, B. Determining of a critical dilution curve for plant nitrogen concentration in winter barley. *Field Crops Res.* **2014**, *160*, 64–72. [CrossRef]
25. Mulla, D.J. Twenty five years of remote sensing in precision agriculture: Key advances and remaining knowledge gaps. *Biosyst. Eng.* **2012**, *114*, 358–371. [CrossRef]
26. Liaghat, S.; Balasundram, S.K. A Review: The role of remote sensing in precision agriculture. *Am. Soc. Agric. Biol. Eng.* **2010**, *5*, 50–55. [CrossRef]
27. Atzberger, C. Advances in remote sensing of agriculture: Context description, existing operational monitoring systems and major information needs. *Remote Sens.* **2013**, *5*, 949–981. [CrossRef]
28. Mistele, B.; Schmidhalter, U. Estimating the nitrogen nutrition index using spectral canopy reflectance measurements. *Eur. J. Agron.* **2008**, *29*, 184–190. [CrossRef]
29. Tremblay, N.; Fallon, E.; Ziadi, N. Sensing of crop nitrogen status: Opportunities, tools, limitations, and supporting information requirements. *Horttechnology* **2011**, *21*, 274–281. [CrossRef]
30. Baresel, J.P.; Rischbeck, P.; Hu, Y.; Kipp, S.; Hu, Y.; Barmeier, G.; Mistele, B. Use of a digital camera as alternative method for non-destructive detection of the leaf chlorophyll content and the nitrogen nutrition status in wheat. *Comput. Electron. Agric.* **2017**, *140*, 25–33. [CrossRef]
31. Yu, K.; Li, F.; Gnyp, M.L.; Miao, Y.; Bareth, G.; Chen, X. Remotely detecting canopy nitrogen concentration and uptake of paddy rice in the Northeast China Plain. *ISPRS J. Photogramm. Remote Sens.* **2013**, *78*, 102–115. [CrossRef]
32. Li, F.; Miao, Y.; Hennig, S.D.; Gnyp, M.L.; Chen, X.; Jia, L.; Bareth, G. Evaluating hyperspectral vegetation indices for estimating nitrogen concentration of winter wheat at different growth stages. *Precis. Agric.* **2010**, *11*, 335–357. [CrossRef]
33. Stroppiana, D.; Boschetti, M.; Brivio, P.A.; Bocchi, S. Plant nitrogen concentration in paddy rice from field canopy hyperspectral radiometry. *Field Crops Res.* **2009**, *111*, 119–129. [CrossRef]
34. Hansen, P.M.; Schjoerring, J.K. Reflectance measurement of canopy biomass and nitrogen status in wheat crops using normalized difference vegetation indices and partial least squares regression. *Remote Sens. Environ.* **2003**, *86*, 542–553. [CrossRef]
35. Blackburn, G.A. Quantifying chlorophylls and caroteniods at leaf and canopy scales. *Remote Sens. Environ.* **1998**, *66*, 273–285. [CrossRef]
36. Reddersen, B.; Fricke, T.; Wachendorf, M. A multi-sensor approach for predicting biomass of extensively managed grassland. *Comput. Electron. Agric.* **2014**, *109*, 247–260. [CrossRef]
37. Thenkabail, P.S.; Smith, R.B.; De Pauw, E. Hyperspectral vegetation indices and their relationships with agricultural crop characteristics. *Remote Sens. Environ.* **2000**, *71*, 158–182. [CrossRef]
38. Adamchuk, V.I.; Hummel, J.W.; Morgan, M.T.; Upadhyaya, S.K. On-the-go soil sensors for precision agriculture. *Comput. Electron. Agric.* **2004**, *44*, 71–91. [CrossRef]
39. Psomas, A.; Kneubühler, M.; Huber, S.; Itten, K. Hyperspectral remote sensing for estimating aboveground biomass. *Int. J. Remote Sens.* **2011**, *32*, 9007–9031. [CrossRef]
40. Erdle, K.; Mistele, B.; Schmidhalter, U. Comparison of active and passive spectral sensors in discriminating biomass parameters and nitrogen status in wheat cultivars. *Field Crops Res.* **2011**, *124*, 74–84. [CrossRef]
41. Cao, Q.; Miao, Y.; Wang, H.; Huang, S.; Cheng, S.; Khosla, R.; Jiang, R. Non-destructive estimation of rice plant nitrogen status with crop circle multispectral active canopy sensor. *Field Crops Res.* **2013**, *154*, 133–144. [CrossRef]
42. Baret, F.; Guyot, G. Potentials and limits of vegetation indices for LAI and APAR assessment. *Remote Sens. Environ.* **1991**, *35*, 161–173. [CrossRef]
43. Zou, X.; Haikarainen, I.; Haikarainen, I.; Mäkelä, P.; Mõttus, M.; Pellikka, P. Effects of crop leaf angle on LAI-sensitive narrow-band vegetation indices derived from imaging spectroscopy. *Appl. Sci.* **2018**, *8*, 1435. [CrossRef]
44. Ziadi, N.; Bélanger, G.; Claessens, A.; Lefebvre, L.; Tremblay, N.; Cambouris, A.N.; Nolin, M.C.; Parent, L.-É. Plant-based diagnostic Tools for evaluating wheat nitrogen status. *Crop Sci.* **2010**, *50*, 2580–2590. [CrossRef]
45. Houles, V.; Guérif, M.; Mary, B. Elaboration of a nitrogen nutrition indicator for winter wheat based on leaf area index and chlorophyll content for making nitrogen recommendations. *Eur. J. Agron.* **2007**, *27*, 1–11. [CrossRef]

46. Bareth, G.; Bendig, J.; Tilly, N.; Hoffmeister, D.; Aasen, H.; Bolten, A. A comparison of UAV- and TLS-derived plant height for crop monitoring: Using polygon grids for the analysis of crop surface models (CSMs). *Photogramm. Fernerkund. Geoinf.* **2016**, *2016*, 85–94. [CrossRef]
47. Marshall, M.; Thenkabail, P. Developing in situ non-destructive estimates of crop biomass to address issues of scale in remote sensing. *Remote Sens.* **2015**, *7*, 808–835. [CrossRef]
48. Aasen, H.; Bareth, G. Spectral and 3D nonspectral approaches to crop trait estimation using ground and UAV sensing. In *Hyperspectral Remote Sensing of Vegetation*; Thenkabail, P.S., Lyon, G.J., Huete, A., Eds.; CRC Press Taylor & Francis Group: Boca Raton, FL, USA; London, UK; New York, NY, USA, 2018; pp. 103–131.
49. Hoffmeister, D.; Bolten, A.; Curdt, C.; Waldhoff, G.; Bareth, G. High resolution Crop Surface Models (CSM) and Crop Volume Models (CVM) on field level by terrestrial laser scanning. In Proceedings of the 6th International Symposium on Digital Earth, Beijing, China, 3 November 2010; Volume 7840.
50. Lumme, J.; Karjalainen, M.; Kaartinen, H.; Kukko, A.; Hyyppä, J.; Hyyppä, H.; Jaakkola, A.; Kleemola, J. Terrestrial laser scanning of agricultural crops. *Int. Arch. Photogramm. Remote Sens. Spat. Inf. Sci.* **2008**, *37*, 563–566.
51. Harwin, S.; Lucieer, A. Assessing the accuracy of georeferenced point clouds produced via multi-view stereopsis from Unmanned Aerial Vehicle (UAV) imagery. *Remote Sens.* **2012**, *4*, 1573–1599. [CrossRef]
52. Montes, J.M.; Technow, F.; Dhillon, B.S.; Mauch, F.; Melchinger, A.E. High-throughput non-destructive biomass determination during early plant development in maize under field conditions. *Field Crops Res.* **2011**, *121*, 268–273. [CrossRef]
53. Pittman, J.; Arnall, D.; Interrante, S.; Moffet, C.; Butler, T. Estimation of biomass and canopy height in bermudagrass, alfalfa, and wheat using ultrasonic, laser, and spectral sensors. *Sensors* **2015**, *15*, 2920–2943. [CrossRef]
54. Yuan, W.; Li, J.; Bhatta, M.; Shi, Y.; Baenziger, P.; Ge, Y. Wheat height estimation using LiDAR in comparison to ultrasonic sensor and UAS. *Sensors* **2018**, *18*, 3731. [CrossRef]
55. Brocks, S.; Bareth, G. Estimating barley biomass with crop surface models from oblique RGB imagery. *Remote Sens.* **2018**, *10*, 268. [CrossRef]
56. Näsi, R.; Viljanen, N.; Kaivosoja, J.; Alhonoja, K.; Hakala, T.; Markelin, L.; Honkavaara, E. Estimating biomass and nitrogen amount of barley and grass using UAV and aircraft based spectral and photogrammetric 3D features. *Remote Sens.* **2018**, *10*, 1082. [CrossRef]
57. Hämmerle, M.; Höfle, B. Mobile low-cost 3D camera maize crop height measurements under field conditions. *Precis. Agric.* **2018**, *19*, 630–647. [CrossRef]
58. Eitel, J.U.H.; Vierling, L.A.; Long, D.S. Simultaneous measurements of plant structure and chlorophyll content in broadleaf saplings with a terrestrial laser scanner. *Remote Sens. Environ.* **2010**, *114*, 2229–2237. [CrossRef]
59. Bendig, J.; Yu, K.; Aasen, H.; Bolten, A.; Bennertz, S.; Broscheit, J.; Gnyp, M.L.; Bareth, G. Combining UAV-based plant height from crop surface models, visible, and near infrared vegetation indices for biomass monitoring in barley. *Int. J. Appl. Earth Obs. Geoinf.* **2015**, *39*, 79–87. [CrossRef]
60. Tilly, N.; Hoffmeister, D.; Cao, Q.; Huang, S.; Lenz-Wiedemann, V.; Miao, Y.; Bareth, G. Multitemporal crop surface models: Accurate plant height measurement and biomass estimation with terrestrial laser scanning in paddy rice. *J. Appl. Remote Sens.* **2014**, *8*, 083671. [CrossRef]
61. Ehlert, D.; Adamek, R.; Horn, H.-J. Laser rangefinder-based measuring of crop biomass under field conditions. *Precis. Agric.* **2009**, *10*, 395–408. [CrossRef]
62. Bareth, G.; Schellberg, J. Replacing manual rising plate meter measurements with low-cost UAV-derived sward height data in grasslands for spatial monitoring. *PFG J. Photogramm. Remote Sens. Geoinf. Sci.* **2018**, *86*, 157–168. [CrossRef]
63. Hoffmeister, D.; Waldhoff, G.; Korres, W.; Curdt, C.; Bareth, G. Crop height variability detection in a single field by multi-temporal terrestrial laser scanning. *Precis. Agric.* **2016**, *17*, 296–312. [CrossRef]
64. Wijesingha, J.; Moeckel, T.; Hensgen, F.; Wachendorf, M. Evaluation of 3D point cloud-based models for the prediction of grassland biomass. *Int. J. Appl. Earth Obs. Geoinf.* **2019**, *78*, 352–359. [CrossRef]
65. Tilly, N.; Aasen, H.; Bareth, G. Fusion of plant height and vegetation indices for the estimation of barley biomass. *Remote Sens.* **2015**, *7*, 11449–11480. [CrossRef]
66. Cassman, K.G.; Dobermann, A.; Walters, D.T. Agroecosystems, nitrogen-use efficiency, and nitrogen management. *AMBIO A J. Hum. Environ.* **2002**, *31*, 132–140. [CrossRef] [PubMed]

67. Freeman, K.W.; Girma, K.; Arnall, D.B.; Mullen, R.W.; Martin, K.L.; Teal, R.K.; Raun, W.R. By-plant prediction of corn forage biomass and nitrogen uptake at various growth stages using remote sensing and plant height. *Agron. J.* **2007**, *99*, 530–536. [CrossRef]
68. Katsvairo, T.W.; Cox, W.J.; Van Es, H.M. Spatial growth and nitrogen uptake variability of corn at two nitrogen levels. *Agron. J.* **2003**, *95*, 1000–1011. [CrossRef]
69. Seginer, I. Plant spacing effect on the nitrogen concentration of a crop. *Eur. J. Agron.* **2004**, *21*, 369–377. [CrossRef]
70. Uni Bonn Soil Campus Klein-Altendorf. Available online: http://www.cka.uni-bonn.de/standort/copy_of_boden (accessed on 9 August 2019).
71. Uni Bonn Climate Campus Klein-Altendorf. Available online: http://www.cka.uni-bonn.de/standort/copy_of_klima (accessed on 9 August 2019).
72. Aasen, H.; Bolten, A. Multi-temporal high-resolution imaging spectroscopy with hyperspectral 2D imagers—From theory to application. *Remote Sens. Environ.* **2018**, *205*, 374–389. [CrossRef]
73. Rischbeck, P.; Elsayed, S.; Mistele, B.; Barmeier, G.; Heil, K.; Schmidhalter, U. Data fusion of spectral, thermal and canopy height parameters for improved yield prediction of drought stressed spring barley. *Eur. J. Agron.* **2016**, *78*, 44–59. [CrossRef]
74. Gnyp, M.L.; Bareth, G.; Li, F.; Lenz-Wiedemann, V.I.S.; Koppe, W.; Miao, Y.; Hennig, S.D.; Jia, L.; Laudien, R.; Chen, X.; et al. Development and implementation of a multiscale biomass model using hyperspectral vegetation indices for winter wheat in the North China plain. *Int. J. Appl. Earth Obs. Geoinf.* **2014**, *33*, 232–242. [CrossRef]
75. Koppe, W.; Li, F.; Gnyp, M.L.; Miao, Y.; Jia, L.; Chen, X.; Zhang, F.; Bareth, G. Evaluating multispectral and hyperspectral satellite remote sensing data for estimating winter Wheat GROWTH parameters at regional scale in the North China plain. *Photogramm. Fernerkund. Geoinf.* **2010**, *3*, 167–178. [CrossRef]
76. Rouse, J.W.; Haas, R.H.; Schell, J.A.; Deering, D.W. Monitoring vegetation systems in the great plains with ERTS. In *Third Earth Resources Technology Satellite-1 Symposium*; NASA: Washington, DC, USA, 1974; Volume 1, pp. 309–317.
77. Roujean, J.L.; Breon, F.M. Estimating PAR absorbed by vegetation from bidirectional reflectance measurements. *Remote Sens. Environ.* **1995**, *51*, 375–384. [CrossRef]
78. Guyot, G.; Baret, F. Utilisation de la haute resolution spectrale pour suivre l'etat des couverts vegetaux. *Spectr. Signat. Objects Remote Sens.* **1988**, *287*, 279.
79. Meier, U. *Growth Stages of Mono- and Dicotyledonous Plants*, 2nd ed.; Blackwell: Berlin, Germany, 2001.
80. Lancashire, P.D.; Bleiholder, H.; van den Boom, T.; Langelüddeke, P.; Strauss, R.; Weber, E.; Witzenberger, A. A uniform decimal code for growth stages of crops and weeds. *Ann. Appl. Biol.* **1991**, *119*, 561–601. [CrossRef]
81. Elementar Analysensysteme GmbH Vario EL Cube. Available online: https://www.elementar.de/en/products/organic-elemental-analysis/vario-el-cube.html (accessed on 9 August 2019).
82. Tremmel, D.C.; Bazzaz, F.A. Plant architecture and allocation in different neighborhoods: Implications for competitive success. *Ecology* **1995**, *76*, 262–271. [CrossRef]
83. Matsumoto, Y.; Oikawa, S.; Yasumura, Y.; Hirose, T.; Hikosaka, K. Reproductive yield of individuals competing for light in a dense stand of an annual, *Xanthium canadense*. *Oecologia* **2008**, *157*, 185–195. [CrossRef] [PubMed]
84. Miao, Y.-F.; Wang, Z.-H.; Li, S.-X. Relation of nitrate N accumulation in dryland soil with wheat response to N fertilizer. *Field Crops Res.* **2015**, *170*, 119–130. [CrossRef]
85. Burkart, A.; Aasen, H.; Alonso, L.; Menz, G.; Bareth, G.; Rascher, U. Angular dependency of hyperspectral measurements over wheat characterized by a novel UAV based goniometer. *Remote Sens.* **2015**, *7*, 725–746. [CrossRef]
86. Li, F.; Miao, Y.; Feng, G.; Yuan, F.; Yue, S.; Gao, X.; Liu, Y.; Liu, B.; Ustin, S.L.; Chen, X. Improving estimation of summer maize nitrogen status with red edge-based spectral vegetation indices. *Field Crops Res.* **2014**, *157*, 111–123. [CrossRef]
87. Vauhkonen, J.; Hakala, T.; Suomalainen, J.; Kaasalainen, S.; Nevalainen, O.; Vastaranta, M.; Holopainen, M.; Hyyppä, J. Classification of spruce and pine trees using active hyperspectral LiDAR. *IEEE Geosci. Remote Sens. Lett.* **2013**, *10*, 1138–1141. [CrossRef]
88. Sun, J.; Yang, J.; Shi, S.; Chen, B.; Du, L.; Gong, W.; Song, S. Estimating rice leaf nitrogen concentration: Influence of regression algorithms based on passive and active leaf reflectance. *Remote Sens.* **2017**, *9*, 15. [CrossRef]

89. Du, L.; Gong, W.; Shi, S.; Yang, J.; Sun, J.; Zhu, B.; Song, S. Estimation of rice leaf nitrogen contents based on hyperspectral LIDAR. *Int. J. Appl. Earth Obs. Geoinf.* **2016**, *44*, 136–143. [CrossRef]
90. Tilly, N. In-season estimation of barley biomass with plant height derived by terrestrial laser scanning. In Proceedings of the 13th International Conference on Precision Agriculture, St. Louis, MO, USA, 31 July–4 August 2016; p. 10.
91. Tilly, N.; Hoffmeister, D.; Cao, Q.; Lenz-Wiedemann, V.; Miao, Y.; Bareth, G. Transferability of models for estimating paddy rice biomass from spatial plant height data. *Agriculture* **2015**, *5*, 538–560. [CrossRef]
92. Lemaire, G.; Jeuffroy, M.-H.; Gastal, F. Diagnosis tool for plant and crop N status in vegetative stage. *Eur. J. Agron.* **2008**, *28*, 614–624. [CrossRef]
93. Kamilaris, A.; Prenafeta-Boldú, F.X. Deep learning in agriculture: A survey. *Comput. Electron. Agric.* **2018**, *147*, 70–90. [CrossRef]
94. Chlingaryan, A.; Sukkarieh, S.; Whelan, B. Machine learning approaches for crop yield prediction and nitrogen status estimation in precision agriculture: A review. *Comput. Electron. Agric.* **2018**, *151*, 61–69. [CrossRef]
95. Holloway, J.; Mengersen, K. Statistical machine learning methods and remote sensing for sustainable development goals: A review. *Remote Sens.* **2018**, *10*, 21. [CrossRef]
96. Viljanen, N.; Honkavaara, E.; Näsi, R.; Hakala, T.; Niemeläinen, O.; Kaivosoja, J. A novel machine learning method for estimating biomass of grass swards using a photogrammetric canopy height model, images and vegetation indices captured by a drone. *Agriculture* **2018**, *8*, 28. [CrossRef]
97. Shao, Y.; Zhao, C.; Bao, Y.; He, Y. Quantification of nitrogen status in rice by least squares support vector machines and reflectance spectroscopy. *Food Bioprocess Technol.* **2012**, *5*, 100–107. [CrossRef]
98. Yao, X.; Huang, Y.; Shang, G.; Zhou, C.; Cheng, T.; Tian, Y.; Cao, W.; Zhu, Y. Evaluation of six algorithms to monitor wheat leaf nitrogen concentration. *Remote Sens.* **2015**, *7*, 14939–14966. [CrossRef]
99. Yin, X.; Hayes, R.M.; McClure, M.A.; Savoy, H.J. Assessment of plant biomass and nitrogen nutrition with plant height in early-to mid-season corn. *J. Sci. Food Agric.* **2012**, *92*, 2611–2617. [CrossRef]
100. Ritchie, J.T.; Singh, U.; Godwin, D.C.; Bowen, W.T. Cereal growth, development and yield. In *Understanding Options for Agricultural Production. Systems Approaches for Sustainable Agricultural Development*; Tsuji, G.Y., Hoogenboom, G., Thornton, P.K., Eds.; Springer: Dordrecht, The Netherlands, 2013; Volume 7, pp. 79–98.
101. Deery, D.; Jimenez-Berni, J.; Jones, H.; Sirault, X.; Furbank, R. Proximal remote sensing buggies and potential applications for field-based phenotyping. *Agronomy* **2014**, *4*, 349–379. [CrossRef]
102. Barker, J.; Zhang, N.; Sharon, J.; Steeves, R.; Wang, X.; Wei, Y.; Poland, J. Development of a field-based high-throughput mobile phenotyping platform. *Comput. Electron. Agric.* **2016**, *122*, 74–85. [CrossRef]
103. Burkart, A.; Cogliati, S.; Schickling, A.; Rascher, U. A novel UAV-based ultra-light weight spectrometer for field spectroscopy. *IEEE Sens. J.* **2014**, *14*, 62–67. [CrossRef]
104. Aasen, H.; Honkavaara, E.; Lucieer, A.; Zarco-Tejada, P.J. Quantitative remote sensing at ultra-high resolution with UAV spectroscopy: A review of sensor technology, measurement procedures, and data correctionworkflows. *Remote Sens.* **2018**, *10*, 1091. [CrossRef]
105. Yang, G.; Liu, J.; Zhao, C.; Li, Z.; Huang, Y.; Yu, H.; Xu, B.; Yang, X.; Zhu, D.; Zhang, X.; et al. Unmanned aerial vehicle remote sensing for field-based crop phenotyping: Current status and perspectives. *Front. Plant Sci.* **2017**, *8*, 1111. [CrossRef] [PubMed]
106. Colomina, I.; Molina, P. Unmanned aerial systems for photogrammetry and remote sensing: A review. *ISPRS J. Photogramm. Remote Sens.* **2014**, *92*, 79–97. [CrossRef]
107. Brede, B.; Lau, A.; Bartholomeus, H.M.; Kooistra, L. Comparing Riegl Ricopter UAV LiDAR derived canopy height and DBH with terrestrial LiDAR. *Sensors* **2017**, *17*, 2371. [CrossRef] [PubMed]
108. Wieser, M.; Mandlburger, G.; Hollaus, M.; Otepka, J.; Glira, P.; Pfeifer, N. A case study of UAS borne laser scanning for measurement of tree stem diameter. *Remote Sens.* **2017**, *9*, 1154. [CrossRef]
109. Wallace, L.; Watson, C.; Lucieer, A. Detecting pruning of individual stems using airborne laser scanning data captured from an unmanned aerial vehicle. *Int. J. Appl. Earth Obs. Geoinf.* **2014**, *30*, 76–85. [CrossRef]

remote sensing

MDPI

Article

Remote Sensing of Leaf and Canopy Nitrogen Status in Winter Wheat (*Triticum aestivum* L.) Based on N-PROSAIL Model

Zhenhai Li [1,2,3], Xiuliang Jin [4], Guijun Yang [1,2], Jane Drummond [5], Hao Yang [1,2], Beth Clark [6], Zhenhong Li [3,*] and Chunjiang Zhao [1,2,*]

1 Beijing Research Center for Information Technology in Agriculture, Beijing Academy of Agriculture and Forestry Sciences, Beijing 100097, China; lizh323@126.com (Z.L.); yanggj@nercita.org.cn (G.Y.); yangh@nercita.org.cn (H.Y.)
2 National Engineering Research Center for Information Technology in Agriculture, Beijing 100097, China
3 School of Engineering, Newcastle University, Newcastle upon Tyne NE1 7RU, UK
4 UMR EMMAH, INRA, NAPV, 84914 Avignon, France; xiuliang.jin@inra.fr
5 School of Geographical and Earth Sciences, University of Glasgow, Glasgow G12 8QQ, UK; Jane.Drummond@glasgow.ac.uk or jane_drummond_nd_29@talktalk.net
6 School of Natural and Environmental Sciences, Newcastle University, Newcastle upon Tyne NE1 7RU, UK; Beth.clark@ncl.ac.uk
* Correspondence: Zhenhong.li@ncl.ac.uk (Z.L.); zhaocj@nercita.org.cn (C.Z.); Tel.: +44-756-509-1984 (Z.L.); +86-51503215 (C.Z.)

Received: 26 June 2018; Accepted: 10 September 2018; Published: 13 September 2018

Abstract: Plant nitrogen (N) information has widely been estimated through empirical techniques using hyperspectral data. However, the physical model inversion approach on N spectral response has seldom developed and remains a challenge. In this study, an N-PROSAIL model based on the N-based PROSPECT model and the SAIL model canopy model was constructed and used for retrieving crop N status both at leaf and canopy scales. The results show that the third parameter (3rd-par) retrieving strategy (leaf area index (LAI) and leaf N density (LND) optimized where other parameters in the N-PROSAIL model are set at different values at each growth stage) exhibited the highest accuracy for LAI and LND estimation, which resulted in R^2 and RMSE values of 0.80 and 0.69, and 0.46 and 21.18 $\mu g \cdot cm^{-2}$, respectively. It also showed good results with R^2 and RMSE values of 0.75 and 0.38% for leaf N concentration (LNC) and 0.82 and 0.95 $g \cdot m^{-2}$ for canopy N density (CND), respectively. The N-PROSAIL model retrieving method performed better than the vegetation index regression model (LNC: RMSE = 0.48 − 0.64%; CND: RMSE = 1.26 − 1.78 $g \cdot m^{-2}$). This study indicates the potential of using the N-PROSAIL model for crop N diagnosis on leaf and canopy scales in wheat.

Keywords: leaf nitrogen concentration; canopy nitrogen density; radiative transfer model; hyperspectral; winter wheat

1. Introduction

Nitrogen (N) is a critical nutrient element for maintaining photosynthesis, enhancing production, and improving grain quality in crops, but the excess use of N fertilizer also results in a series of plant and environmental problems (e.g., vigorous growth, and eutrophication) [1,2]. Precision farming, which considers the crop spatial N distribution, plays an important role in solving these problems, whereby the accurate crop N estimation by remote sensing technology has the potential to precisely manage N by supplying a crop's N requirement at the right place and right time [3,4].

Various sensitive spectral features and vegetation indices have been analyzed for crop N estimation. Kokaly proposed that leaf N concentration (LNC, %) in rice can be estimated by two reflectance absorptions at 2054 nm and 2172 nm [5]. Niu et al. showed that fresh LNC could be estimated using first derivative reflectance spectrum at 2120 nm and 1120 nm [6]. Cho and Skidmore reported on extracting two optimal red edge region equations (REP = $-(c_1 - c_2)/(m_1 - m_2)$, where c_1 and c_2 are the intercepts of a far-red line connected between 679.65 and 694.30 nm, and near-infrared line connected between 732.46 and 760.41 nm or 723.64 and 760.41 nm, respectively; m_1 and m_2 are the corresponding slope of c_1 and c_2, respectively) with high N sensitivity [7]. Previous studies demonstrated that spectral indices have a high correlation with crop N status [8–16], and many spectral indices centered on crop N were proposed. Normalized Difference N Index (NDNI = $[\log(1/R_{1510}) - \log(1/R_{1680})]/[\log(1/R_{1510}) + \log(1/R_{1680})]$ was used to estimate LNC in native shrub vegetation [9]. Elshikha et al. presented the Canopy Chlorophyll Content Index [CCCI = $(\text{NDRE} - \text{NDRE}_{min})/(\text{NDRE}_{max} - \text{NDRE}_{min})$, where NDRE = $(R_{790} - R_{720})/(R_{790} + R_{720})$] as an effective indicator to monitor cotton N when the canopy cover reached above 30% [17]. Yao et al. found the model for canopy N density (CND, g·m^{-2}) estimation based on NDSI (R_{860}, R_{720}), RSI (R_{990}, R_{720}), NDSI (FD_{736}, FD_{720}), and RSI (FD_{725}, FD_{516}) performed well in wheat [15]. Chen et al. developed a new index named Double-peak Canopy N Index [DCNI = $(R_{720} - R_{700})/(R_{700} - R_{670})/(R_{720} - R_{670} + 0.03)$] to estimate plant N concentration (PNC) in corn and wheat [13]. Feng et al. created a Water Resistance N Index [WRNI = $(R_{735} - R_{720}) \times R_{900}/R_{min(930\text{-}980)})/(R_{735} + R_{720})$] to improve the accuracy of the LNC model by minimizing water effects at different growth stages [16].

Moreover, the application of artificial intelligence methods in crop N estimation has been reported in many recent studies and has been proved to be a better predictor than using only sensitive spectral features or vegetation indices [3,4,18–24]. Among these, Hansen et al., Ecarnot et al., and Li et al. showed that partial least square regression (PLSR) could accurately predict LNC in winter wheat and winter oilseed rape [18–20]. Miphokasap et al. indicated that estimaing PNC by stepwise multiple linear regression (SMLR) performed a higher estimation than the model by vegetation indices [4]. Zhang et al. demonstrated that an artificial neural network (ANN) improved the prediction of LNC with consistently higher R^2 values, and was better than that by SMLR [21]. Xu et al. proposed that the optimal combination principle (OCP) method to monitor LNC in barley would exhibit better performance than the vegetation indices methods [22,23]. Yao et al. and Li et al. compared the method of SMLR, PLSR, ANN, and support vector machines regression (SVR), to monitor LNC, and showed PLSR and SVR to be preferred choices for estimating LNC, and ANN was also recommended when sufficient sample size was available [1,24].

Much research on crop N estimation has been reported, and some results were satisfactory. However, a statistical relationship between spectral information and crop N status cannot be expected everywhere and every time, even for a particular sensor [25,26]. The physical model inversion approach on N spectral response has rarely been developed, except for an N-based PROSPECT model (N-PROSPECT) which was extended from a PROSPECT model by replacing the specific absorption coefficients of chlorophyll in the model with equivalent N absorption coefficients, and which could accurately simulate and retrieve leaf N density (LND) at the leaf scale [27]. Thus, crop N status both at leaf and canopy scales could be retrieved through integrating the N-PROSPECT model and the SAIL canopy model [28], to be defined as the N-PROSAIL model in this study. In addition, an ill-posed problem, also called inaccurate inversion, existed and is unavoidable in model parameters inversion, which could be attributed to: (1) multiple solutions in the process of the inversion and (2) uncertainties from measurements and model assumptions [25,26]. Using prior information has been demonstrated as a very efficient solution to this problem [29,30].

To develop an N estimation model based on physical model and explore the suitable strategies for estimating leaf and canopy N status, the objectives of this research were: (1) to develop the N-PROSAIL model to simulate canopy reflectance responses to leaf N density (LND); (2) to assess N status both at leaf and canopy scales, i.e., LNC and canopy N density (CND, g·m^{-2} soil), in winter wheat using the N-PROSAIL model; (3) to reduce the ill-posed inversion and improve the accuracy of related N variables by setting prior parameters values in the N-PROSAIL model at different growth stages; and (4) to evaluate the performance of the N-PROSAIL model method by comparing it with LNC and CND estimated by vegetation index methods.

2. Materials and Methods

2.1. Experimental Design

The experiments were conducted over four growing seasons, 2012–2013, 2013–2014, 2014–2015, and 2015–2016, at the Xiaotangshan National Experimental Station for Precision Agriculture, (40.17°N, 116.43°E) in Beijing, China (Figure 1). The test variables included various cultivars, different N fertilization rates and irrigation amounts (Table 1). Experiment (Exp.) 1 was carried out in 2012–2013 as a completely randomized design with two replications of four wheat cultivars and four N fertilizer applications rates. Exp. 2 and 3 were conducted in 2013–2015 with an orthogonal experimental design with three replications of two wheat cultivars, four N fertilizer applications rates, and three irrigation amounts. Exp. 4 was designed in 2015–2016 as a completely randomized design with three replications of two wheat cultivars and four N fertilizer applications rates. Other management procedures such as pest management, weed control, and phosphate and potassium fertilizer followed local standard practices for winter wheat production.

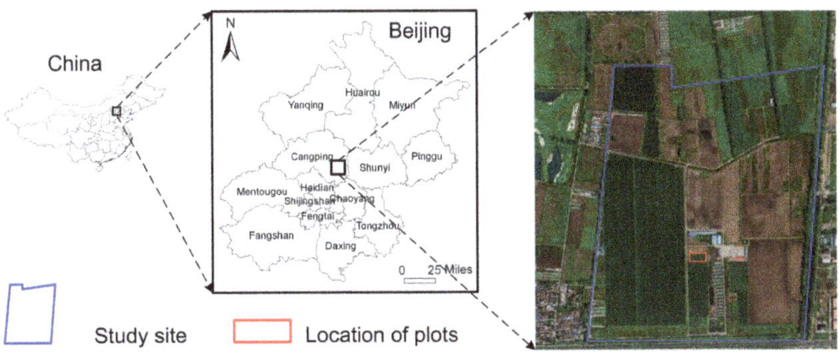

Figure 1. Study site and experiment plot.

Table 1. Summary of cultivar, soil characteristics and treatments for the four experiments.

No.	Season	Cultivar	Treatments	Dataset
Exp. 1	2012–2013	Nongda211, Zhongmai175, Jing9843, Zhongyou206	N rate (kg N ha^{-1}): 0, 110, 220, 440	Calibration
Exp. 2	2013–2014	Jing9843, Zhongmai175	N rate (kg N ha^{-1}): 0, 90, 180, 270; Irrigation rate (mm): 0, 146, 292	Calibration
Exp. 3	2014–2015	Jing9843, Zhongmai175	N rate (kg N ha^{-1}): 0, 90, 180, 270; Irrigation rate (mm): 0, 192, 384	Validation
Exp. 4	2015–2016	Lunxuan167, Jingdong18	N rate (kg N ha^{-1}): 18, 90, 180, 270;	Calibration

2.2. Data Acquisition

2.2.1. Canopy Spectral Data

Canopy hyperspectral reflectances were obtained at four key growth stages (Zadoks growth stage: 31, 47, 65, 75 of winter wheat) [31] (Table 2). Canopy reflectances were measured by an ASD FieldSpec Handheld spectrometer (Analytical Spectral Devices Inc., USA) with a spectral range of 350–2500 nm. Under clear sky conditions between 10:00 and 14:00 Beijing time, the spectrometer was held at a height of 1.0 m above the canopy to ensure the same corresponding at different growth stages. A 40 cm by 40 cm $BaSO_4$ calibration panel served as a black, baseline reflectance. To reduce the possible effects of sky and field conditions, spectral measurements were taken at four sites in each plot and then averaged to represent the canopy reflectance of each plot. Vegetation and panel radiance measurements were taken as the average of 20 scans at an optimized integration time, with a dark current correction for each spectrometry measurement. A panel radiance measurement was taken before and after each vegetation measurement by two scans. Bare soil refelectance was also acquired at each growth period so that it could be used as the input of the N-PROSAIL model.

Table 2. List of acquired experiment data in the four wheat experiments.

Growth Stage	Zadoks	Date	No.	LAI	Cm	LND	LNC	CND	Spectral
Exp. 1: 2012–2013									
Stem elongation	31	23 Apr.	32	✓	✓	✓	✓	✓	–
Booting	47	6 May	32	✓	✓	✓	✓	✓	✓
Anthesis	65	19 May	32	✓	✓	✓	✓	✓	✓
Milk development	75	1 Jun.	32	✓	✓	✓	✓	✓	✓
Exp. 2: 2013–2014									
Stem elongation	33	11 Apr.	48	✓	✓	✓	✓	✓	✓
Booting	45	20 Apr.	48	✓	✓	✓	✓	✓	✓
Anthesis	65	7 May	48	✓	✓	✓	✓	✓	✓
Milk development	75	20 May	48	✓	✓	✓	✓	✓	✓
Exp. 3: 2014–2015									
Stem elongation	31	14 Apr.	48	✓	✓	✓	✓	✓	✓
Booting	47	26 Apr.	48	✓	✓	✓	✓	✓	✓
Anthesis	65	12 May	48	✓	✓	✓	✓	✓	✓
Milk development	75	26 May	48	✓	✓	✓	✓	✓	✓
Exp. 4: 2015–2016									
Stem elongation	31	15 Apr.	24	✓	✓	✓	✓	✓	✓
Booting	47	29 Apr.	24	✓	✓	✓	✓	✓	✓
Anthesis	65	11 May	24	✓	✓	✓	✓	✓	✓
Milk development	75	26 May	24	✓	✓	✓	✓	✓	✓

2.2.2. Plant Measurements

The aboveground vegetation at spectral measurement positions was collected immediately by randomly cutting 0.25 m^2 in each plot, and the number of tillers counted. Then, plant samples of 20 representative wheat tillers were randomly selected from the collected cut plants. All green leaves were separated from the stems. A laser leaf area meter (CI-203, CID Bio-Science Inc., WA, USA) was used to measure the leaf area, and electronic scales (±0.01 g) and drying oven were used to get the leaf dry mass, wet matter weight and dry matter weight (C_m, g·m^{-2}) [26]. Equivalent water thickness (C_w, g·cm^{-2} or cm) was calculated based on wet matter weight and dry matter weight [32]. A Carlo-Erba NA 1500 dry combustion analyzer (Carlo Erba, Milan, Italy) [33] was used to measure leaf N concentration (LNC, %). Finally, leaf N density (LND, μg·cm^{-2} leaves) was calculated as the

LNC multiplied by the Cm, and canopy N density (CND, g·m^{-2} soil) was acquired by leaf area index (LAI) multiplied by LND [27,34].

2.3. Model and Methods

2.3.1. Inversion Procedure of LNC and CND Estimation

The inversion procedure of LNC and CND estimation with the N-PROSAIL model is shown in Figure 2. Specific steps were as follows:

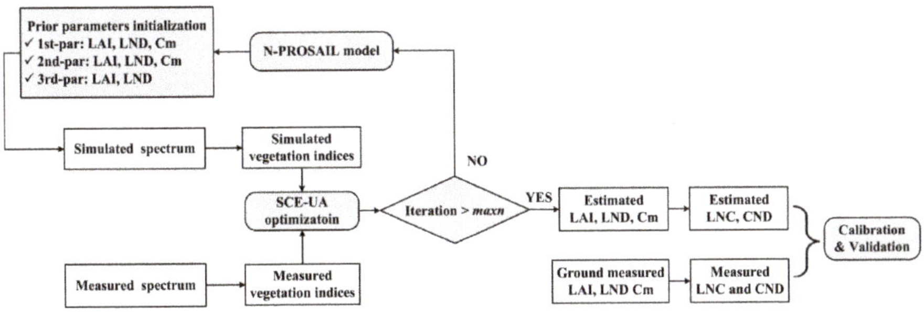

Figure 2. Flowchart of estimating leaf N concentration (LNC) and canopy N density (CND) with the N-PROSAIL model and the Shuffled Complex Evolution method developed at the University of Arizona (SCE-UA) method.

(1) Prior parameters initialization: parameters in the N-PROSAIL model (see Section 2.3.2) were initialized and simulated references can be simulated based on these set parameters. Three strategies of parameter (par) setting were tested and are as follows:

(i) 1st-par: LAI, LND and Cm were optimized and the other six parameters in the N-PROSAIL model (Table 3) were set as fixed values in the whole growth period;
(ii) 2nd-par: LAI, LND and Cm were optimized and the other six parameters in the N-PROSAIL model (Table 3) were set as fixed values at each growth stage;
(iii) 3rd-par: LAI and LND were optimized and Cm was also set as fixed values at each growth stage since it has low variation in specific crop cultivar at one growth stage [35,36]. The other six parameters in the N-PROSAIL model were set as 2nd-par.

(2) Vegetation indices calculation: simulated and measured vegetation indices (see Section 2.3.3) were computed and the optimal vegetation indices correlative with LAI, LND and Cm were chosen to construct cost function in optimization method.

(3) SCE-UA optimization: the SCE-UA method (see Section 2.3.4) was used to optimize parameters in the N-PROSAIL model. Parameters of the SCE-UA method were set and parameters in the N-PROSAIL model were updated until the number of optimization iterations was more than the maximum number of trials (*maxn*) allowed before optimization was terminated.

(4) Validation: the end LAI, LND and Cm were considered as optimized values and they were used to calculate LNC and CND. Three strategies were validated by comparing measured and simulated values. In addition, the inversion performance by the N-PROSAIL model was compared with LNC and CND estimation modeled by vegetation indices.

Table 3. Specific ranges and setting for input parameters in the N-PROSAIL model.

Parameters	1st-par Setting	2nd-par/3rd-par Setting			
		Z.S. 3.1	Z.S. 47	Z.S. 65	Z.S. 75
Leaf N density (LND, $\mu g \cdot cm^{-2}$)	60–230	75–230	70–200	80–200	60–200
Leaf area index (LAI, $m^2 \cdot m^{-2}$)	0.4–8.5	1.5–6.5	1.0–8.5	0.5–7.0	0.3–4.5
Dry matter content (C_m, $g \cdot m^{-2}$)	40	23–60 [#]; 40 [##]	28–45 [#]; 36 [##]	30–54 [#]; 39 [##]	30–70 [#]; 46 [##]
Equivalent water content (C_w, cm)	0.010	0.0138	0.0127	0.0098	0.0085
Leaf structure parameter (Ns)	1.5	1.24	1.07	1.03	1.16
Leaf inclination distribution (LID, deg)	50	50.5	47.2	45.0	43.8
Soil brightness parameter (R_{soil})	0.44	0.49	0.73	0.62	0.35
Hot spot parameter (S_L)	0.15	0.24	0.10	0.12	0.21
Solar zenith angle (θ_s, deg)	45	30	25	22	18

[#] The ranges of Cm at the 2nd-par setting according to calibration dataset; [##] the ranges of Cm at the 3rd-par setting according to calibration dataset.

2.3.2. The N-PROSAIL Model

The N-PROSAIL model is a combination of the N-PROSPECT leaf model [27] and SAILH canopy model [28]. At the leaf scale, PROSPECT uses a leaf structure parameter and leaf biochemical contents to simulate directional-hemispherical reflectance and transmittance of various leaves [37]. The N-PROSPECT model was developed from the PROSPECT model by replacing the specific absorption coefficients of chlorophyll in the PROSPECT model with equivalent N absorption coefficients [27,37]. LNC can be estimated according to LND and C_m. At canopy scale, SAILH considers canopy structures (LAI, leaf angle distribution (θs)), soil brightness, and other angle information to generate canopy reflectance [28]. LAI can be retrieved according to the SAIL model. Then, CND can be calculated by multiplying LAI and LND. During the inversion of the N-PROSAIL model to retrieve crop parameters, nine parameters needed to be determined (Table 3). These parameters were determined at different growth stages. LAI and LND were the main retrieval parameters and they were given intervals based on calibration dataset. C_w, C_m and Soil brightness parameter (R_{soil}) values were obtained by averaging measured values at different growth stages. The solar zenith angle (θ_s) was calculated for the time in the experiment when the hyperspectral data was measured. Leaf structure parameter (Ns), hot spot parameter (S_L) and leaf inclination distribution (LID) values were firstly optimized by the SCE-UA method using the 2012–2013, 2013–2014 and 2015–2016 wheat experiment data, where parameters of LAI, LND, C_w, C_m, R_{soil}, θs were known inputs. Then the mean values of Ns, LID at different growth stages were the final results. Finally, we got the parameter set of C_w, C_m, R_{soil}, θ_s, Ns, S_L and LID at each growth stage, respectively (Table 3).

2.3.3. Selection of Spectral Index

Fifteen vegetation indices correlated with agronomic parameters in previous results were calculated using the equations listed in Table 4. Using these vegetation indices has the following two purposes. (1) Selecting the best vegetation index correlation with LAI, LND and Cm to establish the cost function, an equation to evaluate the consistency between simulated and measured target, in the N-PROSAIL model (see Section 2.3.4). (2) Establishing regression equations of LNC and CND, which were used to validate the inversion performance using the N-PROSAIL model.

2.3.4. SCE-UA Algorithm for LNC and CND Estimation

The SCE-UA method (the abbreviation for the Shuffled Complex Evolution method developed at the University of Arizona) proposed by Duan et al. is a general purpose global optimization program [50]. The method has the advantages of search efficiency at high-parameter dimensionality, convergence speed and computational efficiency, and global searching stability [51,52], and it has been proved to be a useful and effective optimization method in past studies [53–55]. Duan et al. [51] and Wang et al. [54] give a detailed description of the steps of the SCE-UA method. There are many parameters in the SCE-UA method, but most of them were set as default in the method. The number of complexes in a sample population (*ngs*) and the maximum number of trials (*maxn*) were determined by the actual condition and they are 2 and 1000 in this study, respectively [54]. The cost function used to compare simulated with measured vegetation indices in this study was selected as follows:

$$J = \sum_{i=1}^{N} \frac{\sqrt{(VIm_i - VIs_i)^2}}{VIm_i} \quad (1)$$

where J is the value of cost function and N is the determined number of vegetation indices. *VIm* and *VIs* are the measured vegetation indices and simulated vegetation indices by the N-PROSAIL model, respectively. In this study, three vegetation indices respectively correlated with LAI, LND, and Cm, were selected into the cost function. In the process of iterative inversion, the minimum of J value (*minJ*) was given as 5% to avoid model overfitting. Thus, the terminal condition happens when the iterative number is larger than *maxn* or the J value is less than *minJ*.

Table 4. Summary of vegetation indices studied for the N-PROSAIL inversion and N estimation.

Vegetation Index	Formulation	Reference
$CI_{red\ edge}$	$(R_{750}/R_{720}) - 1$	Gitelson et al. [38]
GNDVI	$(R_{750} - R_{550})/(R_{750} + R_{550})$	Baret et al. [39]
MCARI/MTVI2	MCARI/MTVI2; MCARI: $(R_{700} - R_{670} - 0.2(R_{700} - R_{550}))(R_{700}/R_{670})$; MTVI2: $0.5(1.2(R_{800} - R_{500})\text{-}2.5(R_{670} - R_{550}))/$ sqrt $(2(R_{800} + 1)^2 - (6R_{800} - 5\text{sqrt}(R_{670})) - 0.5)$	Eitel et al. [40]
mND705	$(R_{750} - R_{705})/(R_{750} + R_{705} - 2R_{445})$	Sims et al. [41]
MSR	$(R_{800}/R_{670} - 1)/\text{sqrt}(R_{800}/R_{670} + 1)$	Chen [42]
ND705	$(R_{750} - R_{705})/(R_{750} + R_{705})$	Sims et al. [41]
NDVI	$(R_{890} - R_{670})/(R_{890} + R_{670})$	Pearson et al. [43]
ND_{LMA}	$(R_{2260} - R_{1490})/(R_{2260} + R_{1490})$	Le Maire et al. [44]
DCNI	$(R_{720} - R_{700})/(R_{700} - R_{670})/(R_{720} - R_{670} + 0.03)$	Chen et al. [13]
WDRVI	$(0.1R_{890} - R_{670})/(0.1R_{890} + R_{670})$	Gitelson et al. [45]
sLAIDI	$s(R_{1050} - R_{1250})/(R_{1050} + R_{1250})$, s = 5	Delalieux et al. [46]
GI	R_{551}/R_{677}	Zarco-Tejada et al. [47]
SPVI	$0.4(3.7(R_{800} - R_{670}) - 1.2\text{abs}(R_{550} - R_{670}))$	Vincini et al. [48]
$NDVI_{canste}$	$(R_{760} - R_{708})/(R_{760} + R_{708})$	Steddom et al. [49]
NDRE	$(R_{790} - R_{720})/(R_{790} + R_{720})$	Fitzgerald et al. [12]

2.3.5. Statistical Analysis

Data collected from 2012–2013, 2013–2014, and 2015–2016 ($n = 384$) were mainly used for analyzing the correlation between vegetation indices and agronomic variables, calibrating the parameters of N-PROSAIL model, and developing the regression models by vegetation indices. Data collected from 2014–2015 (n = 192) were used to validate the estimating performance by N-PROSAIL model and regression models.

Pearson Correlations (r) between vegetation indices and agronomic variables (LAI, LND, LNC, and CND) were analyzed using Microsoft Office Excel (Microsoft Corporation, Washington, DC, USA). The determination coefficient (R^2) and root mean square error (RMSE) were used to test the general performance of different models in this study. All calculations were made using the MATLAB (v2007, MathWorks Inc., Natick, MA, USA), and all graphs were made using the R statistical software RStudio (v1.0.44, RStudio Inc., Boston, MA, USA).

3. Results

3.1. Correlations among LAI, Cm, LND, LNC, and CND

Correlation coefficients between agronomic variables were analyzed using the calibration set (Table 5). The results showed highly significant differences (p-value < 0.01) between agronomic variables, but correlation (r) values showed high differences. CND had the highest correlation with LAI and LND, with r values of 0.90 and 0.84, respectively. LND calculated from LNC showed a strong correlation ($r = 0.73$) with LNC, while LAI also demonstrated a high correlation with LNC ($r = 0.66$) although the two variables were acquired separately. Cm exhibited negative correlations with LAI, LNC, and CND and a positive correlation with LND, with r values of −0.55, −0.19, −0.30, and 0.45, respectively.

Table 5. Correlations between agronomic variables ($n = 384$).

Variable	LAI	Cm	LND	LNC
Cm	−0.55 **			
LND	0.22 **	0.45 **		
LNC	0.66 **	-0.19 **	0.73 **	
CND	0.90 **	-0.30 **	0.52 **	0.84 **

** Model significance at the 0.01 probability level ($p < 0.01$).

3.2. Correlations between Agronomic Variables and Vegetation Indices

Fifteen vegetation indices correlated with agronomic values were analyzed (Table 6). The results showed that all spectral indices were highly significantly related to LAI (p-value < 0.01) except ND_{LMA} which had a correlation significant at the 0.05 level. The fourteen spectral indices except DCNI had correlations greater than 0.68, and MSR had the highest correlation with LAI, with r value of 0.80. Correlation coefficients between Cm and the spectral indices showed that fifteen vegetation indices, except DCNI, indicated highly significant differences (p-values < 0.01), but the absolute r values were only from 0.14 to 0.35, which were lower than the r values between LAI and the corresponding spectral indices. All spectral indices were highly significantly related to LND (p-value < 0.01). The maximum and minimum correlations were with MCARI/MTVI2 and ND_{LMA}, with r values of −0.56 and −0.19, respectively. According to the correlations analysis, MSR, GI, and MCARI/MTVI2 were first used to develop the cost function in the N-PROSAIL model.

Table 6. Correlations between agronomic variables and vegetation indices ($n = 384$).

Variable	LAI	Cm	LND	LNC	CND
$CI_{red\ edge}$	0.76 **	−0.22 **	0.49 **	0.70 **	0.81 **
GNDVI	0.77 **	−0.21 **	0.49 **	0.70 **	0.79 **
MCARI/MTVI2	−0.69 **	0.14 **	−0.56 **	−0.71 **	−0.75 **
mND705	0.74 **	−0.21 **	0.52 **	0.73 **	0.78 **
MSR	0.80 **	−0.28 **	0.41 **	0.67 **	0.81 **
ND705	0.76 **	−0.23 **	0.50 **	0.71 **	0.79 **
NDVI	0.74 **	−0.23 **	0.47 **	0.68 **	0.75 **
ND_{LMA}	0.12*	−0.26 **	−0.19 **	0.00 #	0.06 #
DCNI	0.40 **	−0.01 #	0.45 **	0.48 **	0.49 **
WDRVI	0.79 **	−0.27 **	0.44 **	0.68 **	0.80 **
sLAIDI	0.68 **	−0.20 **	0.38 **	0.56 **	0.69 **
GI	0.79 **	−0.35 **	0.30 **	0.61 **	0.76 **
SPVI	0.79 **	−0.28 **	0.44 **	0.71 **	0.82 **
$NDVI_{canste}$	0.76 **	−0.22 **	0.51 **	0.72 **	0.79 **
NDRE	0.74 **	−0.19 **	0.52 **	0.71 **	0.79 **

* Model significance at the 0.05 probability level ($p < 0.05$); ** Model significance at the 0.01 probability level ($p < 0.01$); # Model with no significance.

Correlations of LNC and CND to the fifteen vegetation indices were also analyzed (Table 6). The results showed that all vegetation indices, expect ND_{LMA}, were identified as significantly correlated with LNC and CND, respectively. In particular, $CI_{red\ edge}$, GNDVI, MCARI/MTVI2, mND705, ND705, SPVI, $NDVI_{canste}$, and NDRE showed relatively higher correlations with LNC ($r \geq 0.70$) than the others, while $CI_{red\ edge}$, GNDVI, MSR, ND705, WDRVI, SPVI, $NDVI_{canste}$, and NDRE exhibited higher correlations with CND ($r \geq 0.79$) than the others. Therefore, these vegetation indices could be used to further establish regression models with the purpose of comparing and evaluating the performance of LNC and CND estimation using the N-PROSAIL model.

3.3. LAI, LND, and Cm Estimation Using the N-PROSAIL Model Inversion

The N-PROSAIL model using the SCE-UA method was first applied to retrieve LAI, Cm, and LND in this study. Three parameter settings, 1st-par setting, 2nd-par setting, and 3rd-par setting, were tried in this process in order to get the best estimation of LNC and CND. The retrieved results of each agronomic variable with each parameter setting are shown in Figure 3 and Table 7.

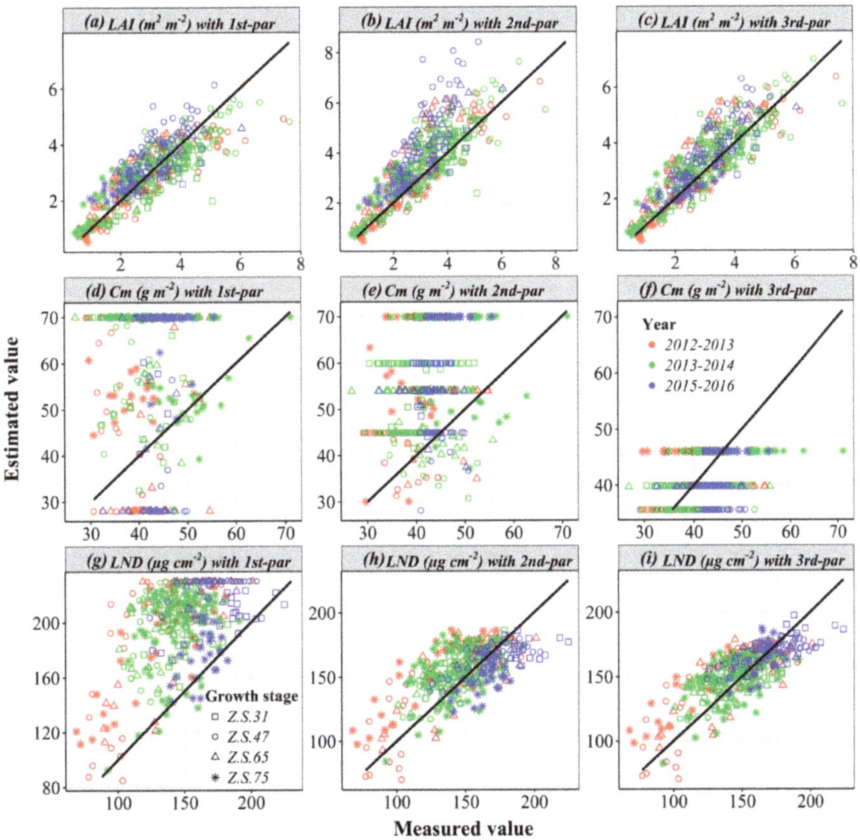

Figure 3. Comparison of measured and estimated values of leaf area index (LAI) (**a–c**), Cm (**d–f**), and LND (**g–i**) based on the N-PROSAIL model in winter wheat across the calibration set.

The results showed that a high consistency between the measured LAI and simulated LAI by the N-PROSAIL model inversion with the three parameter setting (Figure 3a–c). At different growth stages, the R^2 and RMSE values between the simulated LAI and measured LAI for the 1st-par setting, the 2nd-par setting, and the 3rd-par setting ranged from 0.45–0.69 and 0.56–0.93, 0.45–0.68 and 0.61–1.46, and 0.59–0.70 and 0.55–0.88, respectively (Table 7). The 3rd-par setting exhibited a relatively higher R^2 and lower RMSE than the other two parameter settings. The relationships between the simulated and measured LAIs at all growth stages were analyzed together, and the results showed that LAI estimation by the N-PROSAIL model inversion with the 3rd-par setting (R^2 = 0.75 and RMSE = 0.73) was superior to the LAI estimation with the 1st-par setting (R^2 = 0.67 and RMSE = 0.74) and the 2nd-par setting (R^2 = 0.67 and RMSE = 1.08). The independent data of 2014–2015 was used to test the estimation performance, and LAI estimation with the 3rd-par setting (R^2 = 0.80 and RMSE = 0.69) appeared stable compared with LAI estimation with the 1st-par setting (R^2 = 0.81 and RMSE = 0.64) and the 2nd-par setting (R^2 = 0.76 and RMSE = 0.93). These results indicated that the 3rd-par setting to retrieve LAI had the best estimation accuracy.

Table 7. Comparison of different agronomic parameters estimation with different parameter setting.

Parameter Setting	Statistical Categories	No.	LAI ($m^2{\cdot}m^{-2}$)		Cm ($g{\cdot}m^{-2}$)		LND ($\mu g{\cdot}cm^{-2}$)		LNC (%)		CND ($g{\cdot}m^{-2}$)	
			R^2	RMSE	R^2	RMSE	R^2	RMSE	R^2	RMSE	R^2	RMSE
1st-par	Z.S.31	72	0.45 **	0.76	0.02 #	26.43	0.29 **	42.24	0.00 #	1.10	0.36 **	1.66
	Z.S.47	104	0.56 **	0.93	0.07 #	25.50	0.48 **	70.07	0.49 **	1.99	0.69 **	3.19
	Z.S.65	104	0.69 **	0.66	0.00 #	21.84	0.42 **	64.42	0.47 **	2.32	0.74 **	2.61
	Z.S.75	104	0.67 **	0.56	0.00 #	21.86	0.25 **	48.33	0.11 **	0.64	0.68 **	1.78
	Calibration	384	0.67 **	0.74	0.00 #	23.79	0.30 **	58.49	0.21 **	1.69	0.66 **	2.45
	Validation	192	0.76 **	0.93	0.00 #	27.93	0.34 **	62.86	0.18 **	1.23	0.76 **	1.47
2nd-par	Z.S.31	72	0.45 **	0.82	0.00 #	18.51	0.34 **	18.53	0.03 #	1.24	0.20 **	2.16
	Z.S.47	104	0.50 **	1.46	0.06 #	8.21	0.61 **	18.60	0.38 **	0.52	0.68 **	2.57
	Z.S.65	104	0.68 **	1.15	0.04 #	11.85	0.51 **	17.57	0.21 **	0.65	0.73 **	2.15
	Z.S.75	104	0.67 **	0.61	0.00 #	22.68	0.25 **	26.55	0.02 #	1.00	0.68 **	1.24
	Calibration	384	0.67 **	1.08	0.05 #	16.12	0.45 **	20.80	0.14 **	0.87	0.67 **	2.08
	Validation	192	0.81 **	0.64	0.18 **	18.50	0.39 **	24.15	0.51 **	0.94	0.82 **	1.03
3rd-par	Z.S.31	72	0.59 **	0.57	0.00 #	4.81	0.57 **	15.16	0.34 **	0.41	0.44 **	1.31
	Z.S.47	104	0.67 **	0.88	0.00 #	6.35	0.65 **	17.29	0.64 **	0.65	0.76 **	1.52
	Z.S.65	104	0.70 **	0.81	0.00 #	6.09	0.56 **	16.14	0.74 **	0.47	0.76 **	1.40
	Z.S.75	104	0.66 **	0.55	0.00 #	6.64	0.48 **	20.08	0.67 **	0.30	0.69 **	1.00
	Calibration	384	0.75 **	0.73	0.13 **	6.10	0.59 **	17.43	0.62 **	0.48	0.75 **	1.32
	Validation	192	0.80 **	0.69	0.47 **	8.19	0.46 **	21.18	0.75 **	0.38	0.82 **	0.95

* Model significance at the 0.05 probability level ($p < 0.05$). ** Model significance at the 0.01 probability level ($p < 0.01$). # Model with no significance.

Cm estimations with the 1st-par setting and the 2nd-par setting in the N-PROSAIL model were achieved (Table 7 and Figure 3d–f). The results showed no significant difference between measured Cm and estimated Cm with the 2nd-par setting across the validation set and with the 3rd-par setting across the calibration and validation set. In the 1st-par setting, the interval of Cm at different growth period was not considered, and the limited range of Cm was set the same at all growth periods. Many estimation results were ranged on both sides of the interval, and there were relatively high deviations, with RMSE values of 23.79 and 27.93 $g{\cdot}m^{-2}$ for the calibration set and validation set, respectively (Figure 3d). The deviation between the measured Cm and estimated Cm was decreased to 16.12 and 18.50 $g{\cdot}m^{-2}$ for the calibration set and the validation set in the 2nd-par setting, where the interval of Cm values were based on the statistical results of the calibration set at different growth periods. However, many estimated values ranged on both sides of the interval as well (Figure 3e). Cm estimations with the 3rd-par setting in the N-PROSAIL model showed relatively lower RMSE than that of the 1st-par and 2nd-par setting, with RMSE values of 6.10 and 8.19 g m^{-2} for the calibration set and validation set, respectively. The results showed that Cm inversion using the optimizing algorithm of this study had a high deviation even given the limited ranges at different growth stages, and one Cm value at each growth stage had the lowest deviation in this study.

Finally, LND estimations with the three parameters setting in the N-PROSAIL model were compared (Table 7 and Figure 3g–i). The retrieval accuracy of LND was effectively improved after considering the optimizing parameters (the 2nd-par and 3rd-par settings), with R^2 and RMSE values of 0.45 and 20.80 $\mu g{\cdot}cm^{-2}$ for the 2nd-par setting, and 0.59 and 17.43 $\mu g{\cdot}cm^{-2}$ for the 3rd-par setting, respectively. However, the R^2 and RMSE values for the 1st-par setting were 0.30 and 58.49 $\mu g{\cdot}cm^{-2}$, respectively. The improvements at different growth stages were also significant compared with the 1st-par setting, with an increase in R^2 by 0–0.13 and 0.14–0.28, and a decrease in RMSE by 21.78–51.47 $\mu g{\cdot}cm^{-2}$ and 27.08–52.78 $\mu g{\cdot}cm^{-2}$ for the 2nd-par setting and the 3rd-par setting, respectively. The phenomenon of overestimation at LND estimation by the N-PROSAIL model with the 1st-par setting was clear (Figure 3g). LND estimation with the optimized parameters setting (the 2nd-par setting and the 3rd-par setting) reduced the problem of LND overestimation. According to the above comparison, LND estimation with the 3rd-par setting had a slightly better performance than that with the 2nd-par setting (Figure 3h,i). The validation results showed that the best performance was produced by the 3rd-par setting (R^2 = 0.46 and RMSE = 21.18 $\mu g \cdot cm^{-2}$), followed by the 2nd-par setting (R^2 = 0.39 and RMSE = 24.15 $\mu g \cdot cm^{-2}$), and finally the 1st-par setting (R^2 = 0.34 and RMSE = 62.86 $\mu g \cdot cm^{-2}$). Therefore, the LND estimation confirmed the operational potential of the N-PROSAIL model inversion with optimizing parameters, and the retrieval without considering Cm inversion had the best LND estimation.

3.4. LNC and CND Estimation Based on LAI, LND, and Cm

The following estimations were to acquire LNC calculated by LND and Cm, and CND calculated by LAI and LND, respectively. The retrieved results of each agronomic variable with default parameters and optimized parameters were also compared in Figure 4 and Table 7.

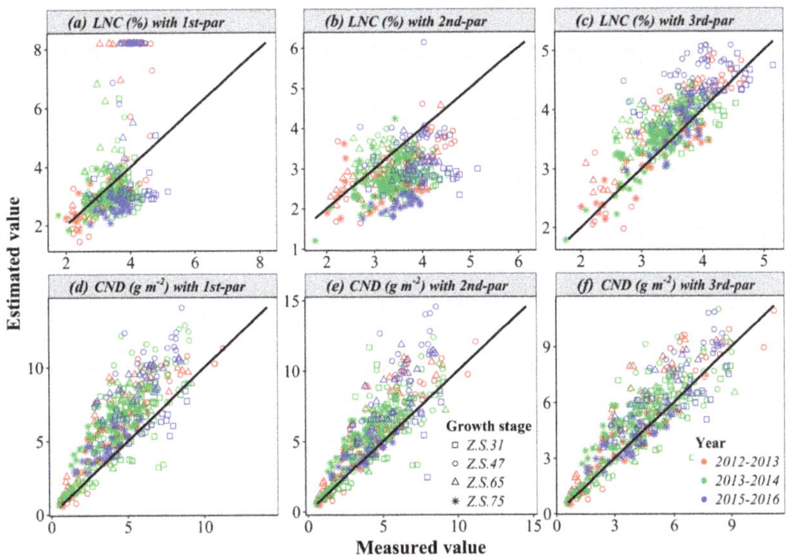

Figure 4. Comparison of measured and estimated values of LNC (**a–c**) and CND (**d–f**) based on the N-PROSAIL model in winter wheat across the calibration set.

Relative to the N-PROSAIL model inversion by the 1st-par and the 2nd-par setting, LNC estimation using the 3rd-par setting performed relatively better, with R^2 and RMSE values of 0.62% and 0.48% for the calibration set. The inversion of the N-PROSAIL model by the 3rd-par setting at different growth stages was also estimated better than by the 1st-par and the 2nd-par setting (Table 7). Many LNC estimations were overestimated, which resulted from the overestimation of LND and Cm (Figure 3d,g and Figure 4a). The 2nd-par setting considering the parameter settings at different growth stages mitigated the overestimation of LNC, but LNC estimation did not performed well (R^2 = 0.14 and RMSE = 0.87%). The independent data of 2014–2015 was used to validate the model stability, and the inversion of the N-PROSAIL model by the 3rd-par setting exhibited a superior result for LNC estimation with R^2 and RMSE values of 0.75 and 0.38%. This study further showed that calibrating parameters at different growth stages is necessary, and the retrieval without considering Cm inversion achieved a satisfactory estimation for LNC.

Together, using the N-PROSAIL model, it was able to get CND on the basis of LAI and LND (Figure 4d–f). The R^2 and RMSE values for the 1st-par setting at different growth stages ranged from

0.36 to 0.74 and 1.66 to 3.19 g·m^{-2}, respectively, and their values at all growth stages were 0.66 and 2.45 g·m^{-2} (Table 7). CND estimations for the 2nd-par setting across different growth stages ranged from 0.20 to 0.73 and 1.24 to 2.57 g·m^{-2}, respectively, and their values at all growth stages were 0.67 and 2.08 g·m^{-2} (Table 7). CND estimations for the 3rd-par setting showed a good performance, with a higher R^2 values and lower RMSE values across different stages, and the R^2 and RMSE values across different growth stages were 0.75 and 1.32 g·m^{-2}, respectively. Then, the validation dataset was used to validate the model stability, and the model inversion with the 3rd-par setting (R^2 = 0.82 and RMSE = 0.95 g·m^{-2}) also performed better than the inversions with the 1st-par setting (R^2 = 0.76 and RMSE = 1.47 g·m^{-2}) and the 2nd-par setting (R^2 = 0.82 and RMSE = 1.03 g·m^{-2}). The results of this study suggest that the 3rd-par setting in the N-PROSAIL model inversion is the best choice.

3.5. Comparison of the N-PROSAIL Model Method with the Vegetation Index Method

To evaluate the performance of LNC and CND estimation by the N-PROSAIL method with parameter optimization (estimations of LNC and CND with the 3rd-par setting were using in the following comparison), the estimation results using the N-PROSAIL model method were compared with the estimation results by the vegetation index method. Fourteen spectral indices with significant relationships were used to fit the regression models (Linear model, Power model, Exponential model, and logarithmic) of LNC, and the best regression model was selected as optimal regression models (Table 8). In these regression models, ten indices with R^2 values for LNC and vegetation indices were greater than 0.50, seven were greater than 0.55, and $CI_{red\ edge}$, mND705 and $NDVI_{canste}$ had the highest R^2 values of 0.58, 0.58, and 0.57, respectively (Figure 3a–c). Compared with LNC estimation using the N-PROSAIL model, the R^2 value between the estimated LNC and measured LNC was 0.62, which was better than all the regression models by vegetation indices. The validation set was used to test the model accuracy, the estimated values and measured values were compared based on the RMSE (Table 8 and Figure 4a–d). The regression model between LNC and MCARI/MTVI2 had the lowest RMSE value of 0.48% among the fourteen regression models with the RMSE values ranging from 0.48% to 0.65% (Figure 5). The RMSE value from measured LNC and predicted LNC by the N-PROSAIL model was 0.38%, which was lower than the values from all regression models by vegetation index (Tables 7 and 8). The results also showed that the consistency between the estimated values and the measured values of the N-PROSAIL method performed better than any vegetation index methods. The results showed that the N-PROSAIL model inversion could be a good method to estimate LNC.

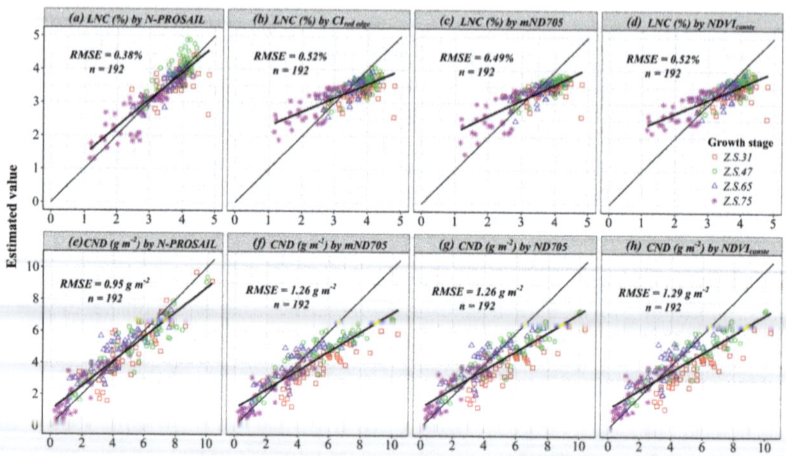

Figure 5. The relationships between measured and estimated values of LNC (**a**–**d**) and CND (**e**–**h**) in winter wheat by using data from 2014–2015 (n = 192).

Table 8. Relationships between LNC and vegetation indices (n = 384).

Vegetation Index	LNC Model #	R^2	RMSE (%)	CND Model	R^2	RMSE ($g{\cdot}m^{-2}$)
$CI_{red\ edge}$	$y = 3.224x^{0.274}$	0.58	0.52	$y = 3.005x^{1.107}$	0.79	1.34
GNDVI	$y = 4.654x^{0.837}$	0.54	0.51	$y = 13.583x^{3.464}$	0.78	1.32
MCARI/MTVI2	$y = 4.802e^{-5.938x}$	0.56	0.48	$y = 14.41e^{23.28x}$	0.73	1.38
mND705	$y = 4.320x^{0.589}$	0.58	0.49	$y = 0.206e^{4.230}$	0.80	1.26
MSR	$y = 2.550x^{0.274}$	0.53	0.55	$y = 1.121x^{1.134}$	0.78	1.34
ND705	$y = 1.870e^{0.977x}$	0.57	0.52	$y = 0.325e^{4.061x}$	0.80	1.26
NDVI	$y = 1.399e^{1.087x}$	0.52	0.55	$y = 0.0925e^{4.518x}$	0.77	1.34
DCNI	$y = 0.935x^{0.411}$	0.27	0.52	$y = 0.894e^{0.0604x}$	0.30	1.73
WDRVI	$y = 1.345x + 3.335$	0.48	0.55	$y = 3.181e^{1.738x}$	0.75	1.33
sLAIDI	$y = 2.648e^{0.415x}$	0.35	0.65	$y = 1.226e^{1.827x}$	0.59	1.78
GI	$y = 2.746x^{0.405}$	0.44	0.64	$y = 1.496x^{1.718}$	0.67	1.67
SPVI	$y = 4.954x^{0.426}$	0.57	0.62	$y = 16.354x^{1.675}$	0.75	1.63
$NDVI_{canste}$	$y = 1.889e^{1.004x}$	0.57	0.52	$y = 0.342e^{4.069x}$	0.80	1.29
NDRE	$y = 4.968x^{0.438}$	0.56	0.52	$y = 17.04x^{1.758}$	0.77	1.42

#: Linear and nonlinear regression (power regression, exponential regression and logarithmic regression) were conducted and listed are the optimal regression model of each vegetation indices. x: vegetation index; y: LNC or CND.

Similar comparisons were obtained for CND estimation between the N-PROSAIL model and vegetation indices methods. A highly significant regression relationship between measured CND and estimated CND was demonstrated both for the N-PROSAIL model method and the vegetation indices methods. The R^2 value between the estimated CND by the N-PROSAIL model and measured CND was 0.75. For the regression model by vegetation index, ten regression models with R^2 values for CND estimation were greater than 0.75 (Table 8), and three regression models by mND705, ND705 and $NDVI_{canste}$ were up to 0.80 (Figure 6d–f). The results of model validation from the validation set showed that the N-PROSAIL method (RMSE = 0.95 $g{\cdot}m^{-2}$) performed better than the vegetation indices methods (RMSE = 1.26 − 1.78 $g{\cdot}m^{-2}$) for estimation of CND (Table 8 and Figure 6). An advantage of stability using the N-PROSAIL model inversion was demonstrated. Furthermore, there are underestimations at high CND values using vegetation indices estimation, but the N-PROSAIL method could mitigate the phenomenon of underestimation. Overall, our results indicated that using the N-PROSAIL model with parameters optimizing has great potential for LNC and CND estimation in winter wheat.

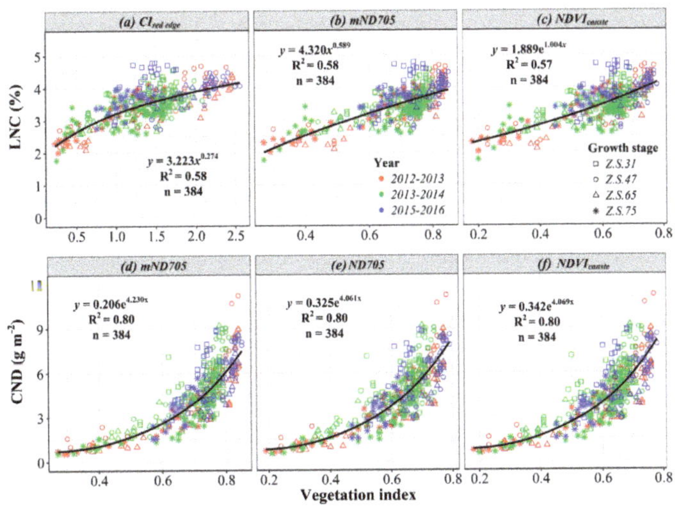

Figure 6. The best three regression models by vegetation index for LNC (**a–c**) and CND (**d–f**) estimation in winter wheat across the calibration set (n = 384).

4. Discussion

The N-PROSAIL model integrating the N-PROSPECT model and the SAILH model were developed to retrieve N status at leaf scale (LNC) and canopy scale (CND) [27,28]. LAI, Cm, and LND were retrieved from the N-PROSAIL model, and LNC and CND were calculated according to these relationships. LND had more variability on LNC than Cm. LNC showed a strong correlation with LND (r = 0.73) and relative low correlation with Cm (r = −0.19). This is because Cm is mainly determined by the plant cultivars and growth stage, and has little change when certain other conditions change [35]. But LND composed of two compartments, high N concentration in metabolic tissues and low N concentration in plant architecture, changes dynamically with plant growth [56]. Therefore, the results indicated the effect of LND on LNC estimation to be more sensitive to the effect from Cm. At the canopy scale, CND strongly reflects the variability of LAI (r = 0.90) as LND was relatively stable (r = 0.52). At different growth stages, canopy information, e.g., LAI, varied significantly, especially the variation between two growth stages [22,30]. This is explained by the lower coefficient of variation of LND compared to LAI (Table 5). This conclusion is in agreement with the study of Darvishzadeh [30], who found that canopy chlorophyll content was more related to LAI with an r value of 0.94.

According to the results of LAI, Cm, and LND estimations, the relationship between measured LAI and its estimation reflects more consistence and accuracy than LND and Cm (Figure 3). The main reasons are as follows: (1) LAI is the canopy characteristics and one of the variables most affected by canopy reflectance, while Cm and LND are variables at leaf level and their variations were lower than LAI; (2) the most correlated vegetation indices between these three variables were selected to build the cost function. LAI showed best correlation with MSR (r = 0.80) and was also highly correlated with MCARI/MTVI2 (r = −0.69) and GI (r = 0.79) (Table 6). The deviation was also relatively lower than with the other two variables. The study result is in line with the findings of previous studies by Feret et al. [57], Darvishzadeh et al. [30], and Li et al. [26].

In this study, three inversion strategies of estimating LNC and CND were tried in order to improve the estimation accuracy. In the 1st-par setting, LAI, Cm, and LND were considered for retrieval, and the other parameters in the N-PROSAIL model were set default values. The results showed that many Cm values were estimated at both sides of the interval and overestimations of LND were obvious. The estimations of these three variables were improved by prior parameters initialization, which is to limit the interval of the three values and assigning different values to the other parameters at different growth stages (the 2nd-par setting). The RMSE of Cm between estimated and measured values greatly decreased from 27.93 to 18.50 $g \cdot m^{-2}$, and the overestimation of LND was eliminated (Figure 3 and Table 7). These resulted in the improvement of LNC and CND estimation, with higher R^2 values of 0.51 and 0.82 and lower RMSE values of 0.94% and 1.03 $g \cdot m^{-2}$ for LNC and CND, respectively. The results indicated the necessity of giving different values of non-optimizing parameters in the N-PROSAIL model at different growth stages. As shown in this study, the statistical values of C_w and calibrating solar zenith angle varied as plant growth progressed, and N and LID also showed difference at different growth stages (Table 3). The uncertainties of model inversion were reduced through giving the different values to these parameters at different growth stages. In the 3rd-par setting, we attempted not to retrieve the Cm value and set the mean values of Cm at each growth stage as the model input. The 3rd-par setting showed more improvement than the 2nd-par setting. The lower RMSE, 8.19 $g \cdot m^{-2}$, for Cm between estimated and measured values demonstrated the lower certainty than the above two parameters setting, and the inversion of LAI and LND were also improved (Table 7). It has two explanations for this: (1) one parameter decreased can reduce the ill-posed problem in model parameters inversion (Figure 3d–f) [29,30]; (2) Cm is an important parameter in the crop growth model and has one initial value for specific crop cultivars and changes as growth goes on [35,36]. Jacquemoud indicated that Cm was first assumed a constant for one plant in the PROSPECT model construction [37]. Thus, the 3rd-par setting considered as a default value at each growth stage is reasonable. The estimated performance for every variable with the 3rd-par setting also exhibits the highest accuracy. Therefore, using the 3rd-par setting in the N-PROSAIL model is a better strategy for estimating plant N status.

$CI_{red\ edge}$, mND705 and $NDVI_{canste}$ were selected as the best three vegetation indices to estimate LNC. They all showed overestimation at low LNC and these samples were mainly measured at Z.S. 75. CND estimation by mND705, ND705, and $NDVI_{canste}$ demonstrated the same phenomenon, with CND estimation at a high value and at Z.S. 31 showed underestimation. Estimated LNC and CND using the N-PROSAIL model showed a higher accuracy than using vegetation indices. Two advantages of using the N-PROSAIL model can explain the situation. Firstly, estimating LAI and LND by the N-PROSAIL model are interactional. The two parameters are retrieved all at once and adjusted according to changes to each other, and are acquired relatively accurately in the end. So LNC and CND estimations were taken into account the results of LAI and LND. Secondly, different parameters, except LAI and LND in N-PROSAIL, were calibrated as different values at various growth stages, which greatly reduced the deviation of model. The improvement of this step was significant according to the compared results of 1st-par setting and 3rd-par setting (Figure 4).

The results showed the potential of a priori information (setting different parameters values at various growth stages) in using N-PROSAIL model for LNC and CND retrieval in winter wheat, which is also suitable to apply in other crops (rice, maize, cotton, etc.) by giving corresponding crop parameters values. However, it is still necessary to define this information as accurately as possible. Four critical growth stages in winter wheat were selected in this study and parameters in the N-PROSAIL model at different growth stages were set respectively. Further studies should focus on considering more growth stages or a high temporal resolution [58], and in this situation, the considered fixed parameters set across different growth stage will increase. The estimation accuracy will be influenced if the growth stage was determined inaccurately. Moreover, fixed parameters were determined in the study area. Further studies should verify whether these parameters change across different regions. Future research should also focus on validating the model using multi-platforms, such as unmanned aerial vehicle and satellite platform. The study should finally point that accurate plant N estimation is an important step towards precision N management, and more studies are needed to develop N-PROSAIL model-based in-season N recommendation algorithms and management strategies.

5. Conclusions

In this study, the N-PROSAIL model was established to retrieve winter wheat LNC and CND at different growth stages. The results suggested that:

(1) The 3rd-par setting retrieval strategies with LAI and LND optimized and other parameters in the N-PROSAIL model fixed at each growth stage exhibited the highest accuracy. The retrieved LAI (R^2 = 0.80 and RMSE = 0.69) and LND (R^2 = 0.46 and RMSE = 21.18 $\mu g \cdot cm^{-2}$) were consistent with the measured LAI and LND, respectively.

(2) LNC and CND were accurately estimated using the N-PROSAIL model, with R^2 and RMSE values of 0.75 and 0.38%, and 0.82 and 0.95 $g \cdot m^{-2}$, respectively.

(3) Compared with vegetation indices regression model, the N-PROSAIL model results reduced the problems of overestimation at low LNC and underestimation at high CND, and showed better performance than any vegetation index regression model (LNC: RMSE = 0.48–0.64%; CND: 1.26–1.78 $g \cdot m^{-2}$). The N-PROSAIL model shows a great potential to estimate canopy N status at leaf and canopy scales in winter wheat.

Author Contributions: G.Y., Z.L. (Zhenhai Li) and C.Z. conceived and designed the experiments; Z.L. (Zhenhai Li) and H.Y. performed the experiments; Z.L. (Zhenhai Li) and X.J. analyzed the data; Z.L. (Zhenhai Li), Z.L. (Zhenhong Li) and C.Z. discussed and drafted the manuscript. Z.L. (Zhenhong Li), J.D. and B.C. revised the manuscript and edited English language. All authors read and approved the final version.

Funding: This study was supported by the National Natural Science Foundation of China (Grant No. 61661136003, 41471285, 41601369), the UK Science and Technology Facilities Council through the PAFiC project (Ref:

ST/N006801/1), the Open Research Fund of Key Laboratory of Digital Earth Science, Institute of Remote Sensing and Digital Earth, Chinese Academy of Sciences (No. 2016LDE008), and the National Key Technologies of Research and Development Program (2016YFD0300602-04).

Acknowledgments: We are grateful to Weiguo Li, Hong Chang and Ling Kong for data collection.

Conflicts of Interest: The authors declare no conflict of interest.

References

1. Yao, X.; Huang, Y.; Shang, G.; Zhou, C.; Cheng, T.; Tian, Y.; Cao, W.; Zhu, Y. Evaluation of Six Algorithms to Monitor Wheat Leaf Nitrogen Concentration. *Remote Sens.* **2015**, *7*, 14939–14966. [CrossRef]
2. Li, F.; Miao, Y.; Hennig, S.D.; Gnyp, M.L.; Chen, X.; Jia, L.; Bareth, G. Evaluating hyperspectral vegetation indices for estimating nitrogen concentration of winter wheat at different growth stages. *Precis. Agric.* **2010**, *11*, 335–357. [CrossRef]
3. Karimi, Y.; Prasher, S.O.; Patel, R.M.; Kim, S.H. Application of support vector machine technology for weed and nitrogen stress detection in corn. *Comput. Electron. Agric.* **2006**, *51*, 99–109. [CrossRef]
4. Miphokasap, P.; Honda, K.; Vaiphasa, C.; Souris, M.; Nagai, M. Estimating Canopy Nitrogen Concentration in Sugarcane Using Field Imaging Spectroscopy. *Remote Sens.* **2012**, *4*, 1651–1670.
5. Kokaly, R.F. Investigating a Physical Basis for Spectroscopic Estimates of Leaf Nitrogen Concentration. *Remote Sens. Environ.* **2001**, *75*, 153–161. [CrossRef]
6. Niu, Z.; Chen, Y.; Sui, Z.; Zhang, Q.Y.; Zhao, C.J. Mechanism Analysis of Leaf Biochemical Concentration by High Spectral Remote Sensing. *J. Remote Sens.* **2000**, *4*, 125–130.
7. Cho, M.A.; Skidmore, A.K. A new technique for extracting the red edge position from hyperspectral data: The linear extrapolation method. *Remote Sens. Environ.* **2006**, *101*, 181–193. [CrossRef]
8. Barnes, E.M.; Clarke, T.R.; Richards, S.E.; Colaizzi, P.D.; Haberland, J.; Kostrzewski, M.; Waller, P.; Choi, C.; Riley, E.; Thompson, T.; et al. Coincident Detection of crop Water Stress, Nitrogen Status and Canopy Density Using Ground-Based Multispectral Data. Available online: https://naldc.nal.usda.gov/download/4190/PDF (accessed on 11 September 2018).
9. Serrano, L.; Peñuelas, J.; Ustin, S.L. Remote sensing of nitrogen and lignin in Mediterranean vegetation from AVIRIS data: Decomposing biochemical from structural signals. *Remote Sens. Environ.* **2002**, *81*, 355–364. [CrossRef]
10. Fitzgerald, G.J.; Rodriguez, D.; Christensen, L.K.; Belford, R.; Sadras, V.O.; Clarke, T.R. Spectral and thermal sensing for nitrogen and water status in rainfed and irrigated wheat environments. *Precis. Agric.* **2006**, *7*, 233–248.
11. Feng, W.; Yao, X.; Zhu, Y.; Tian, Y.C.; Cao, W.X. Monitoring leaf nitrogen status with hyperspectral reflectance in wheat. *Eur. J. Agron.* **2008**, *28*, 394–404. [CrossRef]
12. Fitzgerald, G.; Rodriguez, D.; O'Leary, G. Measuring and predicting canopy nitrogen nutrition in wheat using a spectral index—The canopy chlorophyll content index (CCCI). *Field Crop. Res.* **2010**, *116*, 318–324. [CrossRef]
13. Chen, P.; Haboudane, D.; Tremblay, N.; Wang, J.; Vigneault, P.; Li, B. New spectral indicator assessing the efficiency of crop nitrogen treatment in corn and wheat. *Remote Sens. Environ.* **2010**, *114*, 1987–1997. [CrossRef]
14. Tian, Y.C.; Yao, X.; Yang, J.; Cao, W.X.; Hannaway, D.B.; Zhu, Y. Assessing newly developed and published vegetation indices for estimating rice leaf nitrogen concentration with ground-and space-based hyperspectral reflectance. *Fuel Energy Abstr.* **2011**, *120*, 299–310. [CrossRef]
15. Yao, X.; Jia, W.; Si, H.; Guo, Z.; Tian, Y.; Liu, X.; Cao, W.; Zhu, Y. Exploring novel bands and key index for evaluating leaf equivalent water thickness in wheat using hyperspectra influenced by nitrogen. *PLoS ONE* **2014**, *9*, e96352. [CrossRef] [PubMed]
16. Feng, W.; Zhang, H.Y.; Zhang, Y.S.; Qi, S.L.; Heng, Y.R.; Guo, B.B.; Ma, D.Y.; Guo, T.C. Remote detection of canopy leaf nitrogen concentration in winter wheat by using water resistance vegetation indices from in-situ hyperspectral data. *Field Crop. Res.* **2016**, *198*, 238–246. [CrossRef]
17. Elshikha, D.M.; Barnes, E.M.; Clarke, T.R.; Hunsaker, D.J.; Haberland, J.A.; Pinter, P.J.; Waller, P.M.; Thompson, T.L. Remote Sensing of Cotton Nitrogen Status Using the Canopy Chlorophyll Content Index (CCCI). *Trans. Asabe* **2008**, *51*, 73–82.

18. Hansen, P.M.; Schjoerring, J.K. Reflectance measurement of canopy biomass and nitrogen status in wheat crops using normalized difference vegetation indices and partial least squares regression. *Remote Sens. Environ.* **2003**, *86*, 542–553. [CrossRef]
19. Ecarnot, M.; Compan, F.; Roumet, P. Assessing leaf nitrogen content and leaf mass per unit area of wheat in the field throughout plant cycle with a portable spectrometer. *Field Crop. Res.* **2013**, *140*, 44–50. [CrossRef]
20. Li, L.; Lu, J.; Wang, S.; Ma, Y.; Wei, Q.; Li, X.; Cong, R.; Ren, T. Methods for estimating leaf nitrogen concentration of winter oilseed rape (*Brassica napus* L.) using in situ, leaf spectroscopy. *Ind. Crop. Prod.* **2016**, *91*, 194–204. [CrossRef]
21. Zhang, C.; Kovacs, J.M.; Wachowiak, M.P.; Flores-Verdugo, F. Relationship between Hyperspectral Measurements and Mangrove Leaf Nitrogen Concentrations. *Remote Sens.* **2013**, *5*, 891–908. [CrossRef]
22. Xu, X.G.; Zhao, C.J.; Wang, J.H.; Li, C.J.; Yang, X.D. Associating new spectral features from visible and near infrared regions with optimal combination principle to monitor leaf nitrogen concentration in barley. *J. Infrared Millim. Waves* **2013**, *32*, 351. [CrossRef]
23. Xu, X.G.; Zhao, C.J.; Wang, J.H.; Zhang, J.C.; Song, X.Y. Using optimal combination method and in situ hyperspectral measurements to estimate leaf nitrogen concentration in barley. *Precis. Agric.* **2014**, *15*, 227–240. [CrossRef]
24. Li, Z.; Nie, C.; Wei, C.; Xu, X.; Song, X.; Wang, J. Comparison of Four Chemometric Techniques for Estimating Leaf Nitrogen Concentrations in Winter Wheat (*Triticum aestivum*) Based on Hyperspectral Features. *J. Appl. Spectrosc.* **2016**, *83*, 240–247. [CrossRef]
25. Atzberger, C. Object-based retrieval of biophysical canopy variables using artificial neural nets and radiative transfer models. *Remote Sens. Environ.* **2004**, *93*, 53–67. [CrossRef]
26. Li, Z.; Jin, X.; Wang, J.; Yang, G.; Nie, C.; Xu, X.; Feng, H. Estimating winter wheat (*Triticum aestivum*) LAI and leaf chlorophyll content from canopy reflectance data by integrating agronomic prior knowledge with the PROSAIL model. *Int. J. Remote Sens.* **2015**, *36*, 2634–2653. [CrossRef]
27. Yang, G.; Zhao, C.; Pu, R.; Feng, H.; Li, Z.; Li, H.; Sun, C. Leaf nitrogen spectral reflectance model of winter wheat (Triticum aestivum) based on PROSPECT: simulation and inversion. *J. Appl. Remote Sens.* **2015**, *9*. [CrossRef]
28. Verhoef, W. Light scattering by leaf layers with application to canopy reflectance modeling: the SAIL model. *Remote Sens. Environ.* **1984**, *16*, 125–141. [CrossRef]
29. Combal, B.; Baret, F.; Weiss, M.; Trubuil, A.; Mace, D.; Pragnere, A.; Myneni, R.; Knyazikhin, Y.; Wang, L. Retrieval of canopy biophysical variables from bidirectional reflectance: Using prior information to solve the ill-posed inverse problem. *Remote Sens. Environ.* **2002**, *84*, 1–15. [CrossRef]
30. Darvishzadeh, R.; Skidmore, A.; Schlerf, M.; Atzberger, C. Inversion of a radiative transfer model for estimating vegetation LAI and chlorophyll in a heterogeneous grassland. *Remote Sens. Environ.* **2008**, *112*, 2592–2604. [CrossRef]
31. Zadoks, J.C.; Chang, T.T.; Konzak, C.F. A decimal code for the growth stages of cereals. *Weed Res.* **1974**, *14*, 415–421. [CrossRef]
32. Liu, L.; Song, B.; Zhang, S.; Liu, X. A Novel Principal Component Analysis Method for the Reconstruction of Leaf Reflectance Spectra and Retrieval of Leaf Biochemical Contents. *Remote Sens.* **2017**, *9*. [CrossRef]
33. Schepers, J.S.; Francis, D.D.; Thompson, M.T. Simultaneous determination of total C, total N, and 15N on soil and plant material 1. *Commun. Soil Sci. Plant Anal.* **1989**, *20*, 949–959. [CrossRef]
34. Wang, J.; Zhao, C.; Huang, W. *Quantitative Remote Sensing of Agriculture: Theory and Application*; Science Press: Beijing, China, 2008. (In Chinese)
35. Jones, J.W.; Hoogenboom, G.; Porter, C.H.; Boote, K.J.; Batchelor, W.D.; Hunt, L.A.; Wilkens, P.W.; Singh, U.; Gijsman, A.J.; Ritchie, J.T. The DSSAT cropping system model. *Eur. J. Agron.* **2003**, *18*, 235–265. [CrossRef]
36. Boogaard, H.L.; Diepen, C.A.; Rotter, R.P.; Cabrera, J.A.; Laar, H.H. User's guide for the WOFOST 7.1 Crop Growth Simulation Model and WOFOST Control Center. Available online: http://library.wur.nl/WebQuery/wurpubs/309027 (accessed on 11 September 2018).
37. Jacquemoud, S.; Baret, F. PROSPECT: A model of leaf optical properties spectra. *Remote Sens. Environ.* **1990**, *34*, 75–91. [CrossRef]
38. Gitelson, A.A.; Vina, A.; Ciganda, V.; Rundquist, D.C.; Arkebauer, T.J. Remote estimation of canopy chlorophyll content in crops. *Geophys. Res. Lett.* **2005**, 32. [CrossRef]

39. Baret, F.; Guyot, G. Potentials and limits of vegetation indices for LAI and APAR assessment. *Remote Sens. Environ.* **1991**, *35*, 161–173. [CrossRef]
40. Eitel, J.U.H.; Long, D.S.; Gessler, P.E.; Smith, A.M.S. Using in-situ measurements to evaluate the new RapidEye™ satellite series for prediction of wheat nitrogen status. *Int. J. Remote Sens.* **2007**, *28*, 4183–4190. [CrossRef]
41. Sims, D.A.; Gamon, J.A. Relationships between leaf pigment content and spectral reflectance across a wide range of species, leaf structures and developmental stages. *Remote Sens. Environ.* **2002**, *81*, 337–354. [CrossRef]
42. Chen, J.M. Evaluation of vegetation indices and a modified simple ratio for boreal applications. *Can. J. Remote Sens.* **1996**, *22*, 229–242. [CrossRef]
43. Pearson, R.L.; Miller, L.D. Remote Mapping of Standing Crop Biomass for Estimation of the Productivity of the Shortgrass Prairie. Available online: http://adsabs.harvard.edu/abs/1972rse.conf.1355P (accessed on 11 September 2018).
44. Le Maire, G.; François, C.; Soudani, K.; Berveiller, D.; Pontailler, J.Y.; Bréda, N.; Genet, H.; Davi, H.; Dufrêne, E. Calibration and validation of hyperspectral indices for the estimation of broadleaved forest leaf chlorophyll content, leaf mass per area, leaf area index and leaf canopy biomass. *Remote Sens. Environ.* **2008**, *112*, 3846–3864. [CrossRef]
45. Gitelson, A.A. Wide dynamic range vegetation index for remote quantification of biophysical characteristics of vegetation. *J. Plant Physiol.* **2004**, *161*, 165–173. [CrossRef] [PubMed]
46. Delalieux, S.; Somers, B.; Hereijgers, S.; Verstraeten, W.W.; Keulemans, W.; Coppin, P.A. near-infrared narrow-waveband ratio to determine Leaf Area Index in orchards. *Remote Sens. Environ.* **2008**, *112*, 3762–3772. [CrossRef]
47. Zarco-Tejada, P.J.; Berjón, A.; López-Lozano, R.; Miller, J.R.; Martín, P.; Cachorro, V.; González, M.R.; De Frutos, A. Assessing vineyard condition with hyperspectral indices: Leaf and canopy reflectance simulation in a row-structured discontinuous canopy. *Remote Sens. Environ.* **2005**, *99*, 271–287. [CrossRef]
48. Vincini, M.; Frazzi, E.; D'Alessio, P. Angular Dependence of Maize and Sugar Beet VIs from Directional CHRIS/Proba Data. Available online: https://www.researchgate.net/publication/228413259_Angular_dependence_of_maize_and_sugar_beet_VIs_from_directional_CHRISProba_data (accessed on 11 September 2018).
49. Steddom, K.; Heidel, G.; Jones, D.; Rush, C.M. Remote detection of rhizomania in sugar beets. *Phytopathology* **2003**, *93*, 720–726. [CrossRef] [PubMed]
50. Duan, Q.Y.; Gupta, V.K.; Sorooshian, S. Shuffled complex evolution approach for effective and efficient global minimization. *J. Optim. Theory Appl.* **1993**, *76*, 501–521. [CrossRef]
51. Duan, Q.; Sorooshian, S.; Gupta, V.K. Optimal use of the SCE-UA global optimization method for calibrating watershed models. *J. Hydrol.* **1994**, *158*, 265–284. [CrossRef]
52. Zhu, X.; Zhao, Y.; Feng, X. A methodology for estimating Leaf Area Index by assimilating remote sensing data into crop model based on temporal and spatial knowledge. *Chin. Geogr. Sci.* **2013**, *23*, 550–561. [CrossRef]
53. Jin, H.; Wang, J.; Bo, Y.; Chen, G.; Xue, H. Data Assimilation of MODIS and TM Observations into CERES-Maize Model to Estimate Regional Maize Yield. Available online: https://doi.org/10.1117/12.860315 (accessed on 11 September 2018).
54. Wang, H.; Zhu, Y.; Li, W.; Cao, W.; Tian, Y. Integrating remotely sensed leaf area index and leaf nitrogen accumulation with RiceGrow model based on particle swarm optimization algorithm for rice grain yield assessment. *J. Appl. Remote Sens.* **2014**, *8*. [CrossRef]
55. Huang, J.; Ma, H.; Su, W.; Zhang, X.; Huang, Y.; Fan, J.; Wu, W. Jointly assimilating MODIS LAI and ET products into the SWAP model for winter wheat yield estimation. *IEEE J. Sel. Top. Appl. Earth Obs. Remote Sens.* **2015**, *8*, 4060–4071. [CrossRef]
56. Lemaire, G.; Jeuffroy, M.H.; Gastal, F. Diagnosis tool for plant and crop N status in vegetative stage: Theory and practices for crop N management. *Eur. J. Agron.* **2008**, *28*, 614–624. [CrossRef]
57. Feret, J.B.; François, C.; Asner, G.P.; Gitelson, A.A.; Martin, R.E.; Bidel, L.P.; Ustin, S.L.; Le Maire, G.; Jacquemoud, S. PROSPECT-4 and 5: Advances in the leaf optical properties model separating photosynthetic pigments. *Remote Sens. Environ.* **2008**, *112*, 3030–3043. [CrossRef]
58. Koetz, B.; Baret, F.; Poilvé, H.; Hill, J. Use of coupled canopy structure dynamic and radiative transfer models to estimate biophysical canopy characteristics. *Remote Sens. Environ.* **2005**, *95*, 115–124. [CrossRef]

remote sensing

MDPI

Article

Using Digital Cameras on an Unmanned Aerial Vehicle to Derive Optimum Color Vegetation Indices for Leaf Nitrogen Concentration Monitoring in Winter Wheat

Jiale Jiang [1,2,3,4], Weidi Cai [1,2,3,4], Hengbiao Zheng [1,2,3,4], Tao Cheng [1,2,3,4], Yongchao Tian [1,2,3,4], Yan Zhu [1,2,3,4], Reza Ehsani [5], Yongqiang Hu [6], Qingsong Niu [6], Lijuan Gui [6] and Xia Yao [1,2,3,4,*]

1 National Engineering and Technology Center for Information Agriculture, Nanjing Agricultural University, Nanjing 210095, China; jialejiang@njau.edu.cn (J.J.); 11115133@njau.edu.cn (W.C.); zhenghb@njau.edu.cn (H.Z.); tcheng@njau.edu.cn (T.C.); yctian@njau.edu.cn (Y.T.); yanzhu@njau.edu.cn (Y.Z.)

2 Key Laboratory for Crop System Analysis and Decision Making, Ministry of Agriculture, Nanjing Agricultural University, Nanjing 210095, China

3 Jiangsu Key Laboratory for Information Agriculture, Nanjing Agricultural University, Nanjing 210095, China

4 Jiangsu Collaborative Innovation Center for Modern Crop Production, Nanjing Agricultural University, Nanjing 210095, China

5 School of Engineering, University of California, 5200 Lake Road, Merced, CA 95343, USA; rehsani@ucmerced.edu

6 Qinghai Science and Technology Information Research Institute Co. LTD, Qinghai 810000, China; yqhu@qhinfo.net (Y.H.); nqs@qhinfo.net (Q.N.); glj@qhinfo.net (L.G.)

* Correspondence: yaoxia@njau.edu.cn; Tel.: +86-25-8439-6565

Received: 29 September 2019; Accepted: 13 November 2019; Published: 14 November 2019

Abstract: Commercially available digital cameras can be mounted on an unmanned aerial vehicle (UAV) for crop growth monitoring in open-air fields as a low-cost, highly effective observation system. However, few studies have investigated their potential for nitrogen (N) status monitoring, and the performance of camera-derived vegetation indices (VIs) under different conditions remains poorly understood. In this study, five commonly used VIs derived from normal color (RGB) images and two typical VIs derived from color near-infrared (CIR) images were used to estimate leaf N concentration (LNC). To explore the potential of digital cameras for monitoring LNC at all crop growth stages, two new VIs were proposed, namely, the true color vegetation index (TCVI) from RGB images and the false color vegetation index (FCVI) from CIR images. The relationships between LNC and the different VIs varied at different stages. The commonly used VIs performed well at some stages, but the newly proposed TCVI and FCVI had the best performance at all stages. The performances of the VIs with red (or near-infrared) and green bands as the numerator were limited by saturation at intermediate to high LNCs (LNC > 3.0%), but the TCVI and FCVI had the ability to mitigate the saturation. The results of model validations further supported the superiority of the TCVI and FCVI for LNC estimation. Compared to the other VIs derived using RGB cameras, the relative root mean square errors (RRMSEs) of the TCVI were improved by 8.6% on average. For the CIR images, the best-performing VI for LNC was the FCVI (R^2 = 0.756, RRMSE = 14.18%). The LNC–TCVI and LNC–FCVI were stable under different cultivars, N application rates, and planting densities. The results confirmed the applicability of UAV-based RGB and CIR cameras for crop N status monitoring under different conditions, which should assist the precision management of N fertilizers in agronomic practices.

Keywords: leaf nitrogen concentration; leaf nitrogen accumulation; unmanned aerial vehicle (UAV); digital camera; vegetation indices

1. Introduction

Nitrogen (N) is a component of many important compounds in plants, and thus plays an important role in plant growth [1,2]. Plant growth dominantly depends on the N supply [3]. A deficiency in N would reduce crop photosynthesis, whereas higher rates of N fertilization do not necessarily improve crop yield and can lead to serious water pollution [4–6]. Furthermore, leaf N concentration (LNC) is related to the photosynthetic capacity of leaves, and thus allows N fertilizer applications and grain quality to be modeled [7,8]. Therefore, timely quantification of LNC is a prerequisite for fertilization guidance and environmental quality [9,10].

Unmanned aerial vehicle (UAV) platforms have become a promising approach in precision agricultural assessments because they enable the non-destructive measurement of crop growth status, with a very high spatio-temporal resolution [11,12]. Due to their advantages of low cost, light weight, convenient operation, and simple data processing, digital cameras have been commonly deployed on UAVs in crop phenotype research [13]. Compared to other sensors, digital cameras can operate successfully in a range of working environments. Given that adequate image exposure can be set based on the weather conditions, data can be collected under both sunny and cloudy conditions [13]. Therefore, color images can be instantly acquired for researchers and farmers to monitor crop growth status [14].

A consumer-grade RGB camera is an "off-the-shelf" device, with red, green, and blue channels. Because each pixel value in color images can be calculated from the digital number (DN) values of specific bands, color indices can be extracted to accentuate a particular vegetation greenness and identify the vegetation features [14,15]. Hunt et al., [16] used the normalized green–red difference index (NGRDI) from RGB images to estimate the biomass of corn, alfalfa, and soybean, and found a linear correlation between the NGRDI and biomass. Kawashima and Nakatani [17] used a video camera to analyze the color of wheat leaves for estimating chlorophyll content. Woebbecke et al. [18] investigated the capability of several color indices to distinguish vegetation from the background, and found that the excess green vegetation index (ExG) could provide a near-binary intensity image outlining a plant region of interest. Moreover, color indices from RGB cameras containing a large amount of information regarding crop status can be used to estimate the vegetation fraction, plant height, biomass, and yield [19–21]. However, many vegetation indices (VIs) proposed for crop status monitoring contain the near infrared (NIR) bands [22–24]. Therefore, RGB cameras with a Bayer-pattern array of filters have been modified by replacing either the blue or red channel with a NIR channel to obtain color near-infrared (CIR) images [25]. Based on a newly-developed digital CIR camera system, Hunt et al., [25] found a strong correlation between the green normalized difference vegetation index (GNDVI) and leaf area index (LAI) in winter wheat. This CIR camera system has also been used to assess winter crop biomass [26]. Four VIs, the normalized difference vegetation index (NDVI), enhanced NDVI (ENDVI), GNDVI, and ExG, derived from UAV-based RGB and CIR images, have been shown to be reliable to assess experimental plots [27].

Previous studies have indicated that it is feasible to estimate crop growth status using RGB and CIR images, but few studies have investigated their usability for N status monitoring [12,28]. Firstly, the capability of digital cameras to monitor wheat LNC at different growth stages remains poorly understood. Since the composition of canopy components (e.g., leaves, tassels) and background materials (e.g., soil) varied sharply during whole growth stages of winter wheat [29], the performance of digital cameras on estimating LNC is necessary to be tested across different growth stages. Secondly, it is crucial to investigate the saturation problem of VIs, with varying LNC values. Because the application of N fertilizer has increased recently in China, it is important to effectively monitor LNC under middle to high application levels. Thirdly, the capability of digital cameras for LNC estimation under different conditions is unclear. The relationships among VIs and LNC in cereal crops have been investigated in terms of the mechanisms involved based on hyperspectral remote sensing. Many studies have proposed effective VIs that can be adjusted to variations in growth stage [30] and geographic location [31], and that can reduce the effects of the soil background [32]. Therefore, there is also a need to assess the capability of digital cameras for estimating wheat LNC under different conditions.

The overall objective of this study was to evaluate whether digital cameras mounted on UAVs could be applied to monitor LNC in winter wheat. Five typical VIs derived from RGB images and two widely used VIs derived from CIR images were selected to estimate LNC. Additionally, we developed the true color vegetation index (TCVI) and the false color vegetation index (FCVI) from RGB and CIR images, respectively. Experiments with different wheat varieties, planting densities, and N application rates were conducted in the field to: (1) quantify the relationship between LNC and the VIs from digital imagery at different growth stages, (2) evaluate the saturation sensitivity of the VIs under various LNC levels, and (3) validate the applicability of the LNC estimation models under different treatments.

2. Materials and Methods

2.1. Study Site and Experimental Design

The study site was located in Rugao City, Jiangsu Province, China (120°45′ E, 32°16′ N), as shown in Figure 1. The regional annual precipitation of this area is around 927.53 mm, with an average annual temperature of 16.59 °C. Two field experiments using winter wheat (*Triticum aestivum* L.) were designed that included three N application rates, two planting densities (D), and two varieties (V) in two growing seasons (Table 1). In each experiment, a split design was used with three replications and there were 36 plots, each with a plot area of 35 m^2 (Figure 1). The basal fertilizer included 120 kg/ha P_2O_5 and 120 kg/ha K_2O and there were three N application rates (0, 150, and 300 kg/ha as urea) applied at the end of October 2013/2014. Compound fertilizer was applied in early March 2014/2015, including N applications at the three different rates. The N fertilizers were applied in 50% as basal fertilizer at the sowing day and in 50% at the jointing stage. All other agronomic management was undertaken according to local wheat production practices.

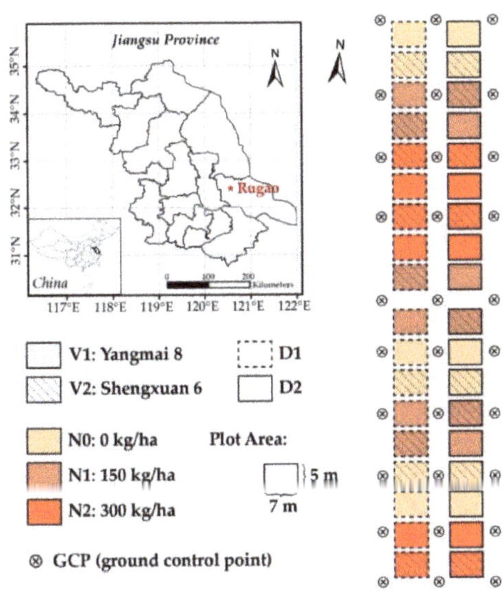

Figure 1. The study site of field experiments with two winter wheat varieties (V1, V2), two planting densities (D1, D2) and three nitrogen application rates (N0, N1, N2) in Rugao City, Jiangsu Province, China.

Table 1. Experimental Designs and Sampling Dates.

Experiment	Year	Wheat Varieties	N Application Rates (kg/ha)	Plot Area (m^2)	Planting Density (plants/ha)	Sampling Dates
Exp.1	2013–2014	V1: Yangmai 8 V2: Shengxuan 6	N0: 0 N1: 150 N2: 300	7 × 5	D1: 3.0×10^6 D2: 1.5×10^6	9/15/23 April 6 May
Exp.2	2014–2015	V1: Yangmai 8 V2: Shengxuan 6	N0: 0 N1: 150 N2: 300	7 × 5	D1: 2.4×10^6 D2: 1.5×10^6	8/17/25 April 6 May

2.2. *Data Acquisition and Processing*

2.2.1. Color Images from Unmanned Aerial Vehicle (UAV)

In this study, an eight-rotor ARF-MikroKopter UAV (Figure 2a) was used as the platform for the UAV-camera system and its specifications are listed in Table 2. A Canon 5D Mark III (Canon Inc., Japan) commercial digital camera (Figure 2b) was mounted on the UAV and took RGB images in continuous mode. The CIR camera was modified from a Canon SX260HS camera (Canon Inc.) by changing the original red channel to a near-infrared channel. The main parameters of the two cameras are described in Table 3. An MC-32 remote control module and a ThinkPad laptop were used to control the autonomous UAV flight. During each flight, the camera was fixed on a two-axis gimbal, with the lens positioned vertically downward at 50 m altitude. Considering the lighting conditions, the exposure time and shutter speed were fixed for each campaign. UAV campaigns were conducted at noon in clear weather under stable light conditions. The spatial resolution of the RGB and CIR imagery was 3 cm. The acquisition dates of these images are listed in Table 4.

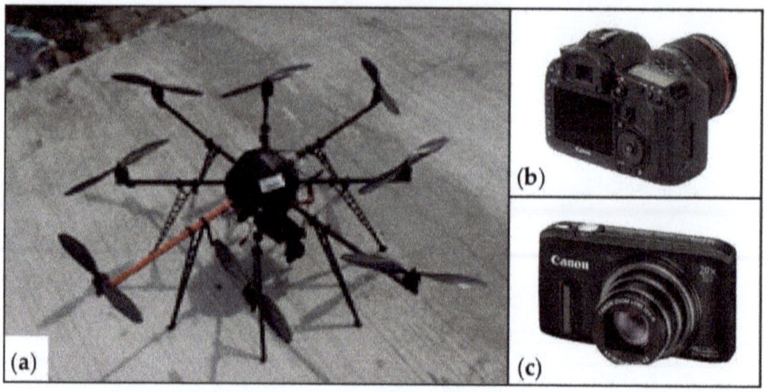

Figure 2. The unmanned aerial vehicle (UAV) camera system: (a) ARF-MikroKopter UAV, (b) normal color (RGB) camera (Canon 5D Mark III) and (c) color near-infrared (CIR) camera (Canon SX260HS).

Table 2. Specifications of ARF-MikroKopter Unmanned Aerial Vehicle (UAV).

Parameter	Value
Weight (without batteries)	2050 g
Size	73 (width) × 73 (length) × 36 (height) cm
Battery Wight (4s/5000)	520 g
Maximum payload	2500 g
Flight duration	8~41 min
Temperature range	−5 °C ~ 35 °C

Table 3. Main Parameters of Normal Color (RGB) and Near Infrared (NIR-G-B) Cameras.

Parameter	Value	
	RGB Camera	CIR Camera
Blue Channel	Visible blue light	Visible blue light
Green Channel	Visible green light	Visible green light
Red Channel	Visible red light	
NIR Channel		670–770 nm
Geometric Resolution	5760 × 3840 pixel	4000 × 3000 pixel
Focal Length	24 mm	4 mm

Table 4. Acquisition Dates of Unmanned Aerial Vehicle (UAV)-Based Digital Camera Images.

Date		Growth Stage	RGB Imagery	CIR Imagery
Exp. 1 (2014)	9 April	Booting	✘	✔
	15 April	Heading	✘	✔
	23 April	Anthesis	✘	✔
	6 May	Filling	✘	✔
Exp. 2 (2015)	8 April	Booting	✔	✔
	17 April	Heading	✔	✔
	25 April	Anthesis	✔	✔
	6 May	Filling	✔	✔

Before image pre-processing, the original digital images were screened. We selected images with a heading overlap rate of ~70% and a side overlap rate of ~30%, and excluded images with excessive repetition. The selected images were then pre-processed, including a lens distortion correction, image mosaicking, image registration, and ortho-rectification. First, lens distortion was corrected based on the Brown Model and the correction coefficients were calculated by an Agisoft Lens. Second, image mosaicking was conducted in Photoscan (Airsoft LLC, Russia). Third, image registration was referred to ground control points (GCPs) in the experimental area (see Figure 1). The GCPs were painted on a road surface as black annuluses, with inner and outer diameters of 10 and 50 cm, respectively. The geographic coordinates of the GCPs were determined using a real time kinematic (RTK) GPS system, with an error less than 2 cm in the horizontal direction and less than 3 cm in the vertical direction. Finally, ortho-rectification was automatically performed using Photoscan. Figure 3 shows the processed images from RGB and CIR cameras at the four growth stages.

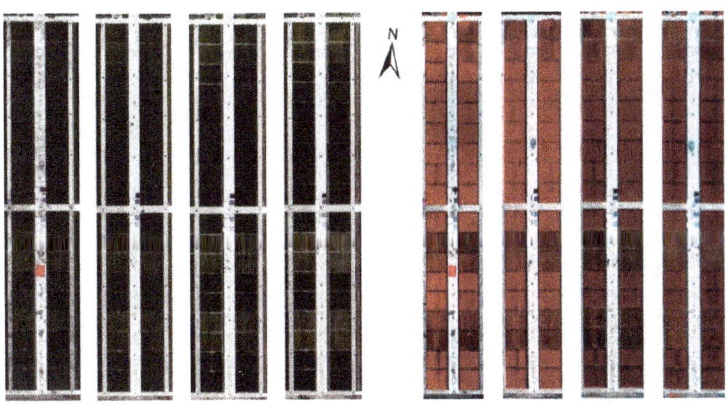

(a) (b) (c) (d) (e) (f) (g) (h)

Figure 3. The digital images from normal color (RGB) (**a–d**) camera and color near-infrared (CIR) (**e–h**) at the booting (**a**,**e**), heading (**b**,**f**), anthesis (**c**,**g**) and filling (**d**,**h**) stages. Red-green-blue and near infrared (NIR)-red-green channels are presented as RGB for RGB and CIR cameras.

2.2.2. Determination of Leaf N Concentration (LNC)

Common measures of crop LNC are either area-based (LNC_{area}, g/m^2) or mass-based (LNC_{mass}, %). LNC_{area} is the N mass per unit leaf area and LNC_{mass} is the ratio of N mass to leaf dry mass. These two measures can be converted between each other through the leaf mass per area (LMA), i.e., $LNC_{area} = LNC_{mass} \times LMA$ [33]. Therefore, LNC_{area} should be determined based on both LNC_{mass} and LMA measures, while LNC_{mass} can be measured directly in the laboratory. Given its strong connection with photosynthetic capacity [34] and its widespread use in fertilization management [35,36], LNC_{mass} has received more attention and has been estimated from remotely sensed data more often than its counterpart LNC_{area} [37–40]. The term LNC hereafter refers to LNC_{mass}.

Ground destructive samplings were taken at critical growth stages of winter wheat (Table 1) on the same dates as the UAV campaigns. Thirty hills of wheat plants were randomly cut above the ground surface of each plot and separated into leaves, stems, and panicles. All leaves were oven-dried at 105 °C for 30 min and then at 80 °C until a constant weight. We then grounded and stored all samples in plastic bags for chemical analysis. The LNC was determined based on the micro-Keldjahl method [41] with SEAL AutoAnalyzer 3 HR (SEAL Analytical, Ltd., German).

2.2.3. Derivation of Vegetation Indices (VIs)

The main VIs derived from digital camera images are summarized in Table 5. For RGB cameras, five widely used color indices were chosen in this study: the NGRDI [42], Kawashima index (IKaw) [17], red green ratio index (RGRI) [43], visible atmospherically resistance index (VARI) [42], and ExG [18]. For the CIR camera, we used the GNDVI [44] and enhanced normalized difference vegetation index (ENDVI), because GNDVI is related to chlorophyll concentration, while the ENDVI was recommended by the company that manufactured the modified cameras (www.maxmax.com) and also can monitor vegetation vigor [27]. The VIs could be categorized into two groups according to the number of channels used. Some were constructed by two channels, such as NGRDI, IKaw, RGRI, and GNDVI. The others (i.e., VARI, ExG, and ENDVI) were established using three channels.

Table 5. Formulas and References of Possible Vegetation Indices (VIs).

Camera	VI	Name	Formula
RGB	NGRDI	Normalized green red difference index	(G−R)/(G+R)
	IKaw	Kawashima index	(R−B)/(R+B)
	RGRI	Red green ratio index	R/G
	VARI	Visible atmospherically resistance index	(G−R)/(G+R−B)
	ExG	Excess green vegetation index	(2G−R−B)/(G+R+B)
	TCVI [1]	True Color Vegetation Index	1.4*(2R−2B)/(2R−G−2B+255*0.4)
CIR	GNDVI	Green normalized difference vegetation index	(NIR−G)/(NIR+G)
	ENDVI	Enhanced normalized difference vegetation index	(NIR+G−2B)/(NIR+G+2B)
	FCVI [2]	False Color Vegetation Index	1.5*(2NIR+B−2G)/(2G+2B−2NIR+255*0.5)

[1] TCVI is the True Color Vegetation Index derived from RGB images. [2] FCVI is the False Color Vegetation Index from CIR images. They are newly conducted in this paper.

Color vegetation index (CVI) derived from digital imagery is often calculated as ratios of DN values. However, the previous VIs listed in Table 5 were proposed without considering background material, such as soil in the field. Previous studies have suggested that the influence of the soil background can be reduced by adjusting the VIs with a term representing soil brightness [22,20]. Based on the ratio form and an equivalent soil adjustment, the commonly used CVIs from digital cameras can be expressed as:

$$CVI = \frac{(1+L)(a_1R + a_2G + a_3B)}{(a_4R + a_5G + a_6B + 255 * L)} \quad (1)$$

where G and B are the DN values of the green and blue channels, respectively; R represents the red component of RGB imagery or the near-infrared component of CIR imagery; a_i is the coefficient of each channel; and L is a soil background adjustment parameter.

To explore the capability of digital cameras for estimating LNC in wheat, we constructed new CVI from both RGB and CIR images. They were determined by optimizing the values of a_i and L based on the cost function J defined as:

$$J = 1 - \frac{\sum_{i=1}^{n} (LNC_{m,i} - LNC_{p,i})^2}{\sum_{i=1}^{n} (LNC_{m,i} - \overline{LNC_{m,i}})^2} \quad (2)$$

where LNC_m is the measured LNC. LNC_p was predicted by the best fitted function of CVI and $\overline{LNC_p}$ is the average value of LNC_p. According to the imaging principle of a Bayer filter, we set the value of a_i as an element of {−2, −1, 0, 1, 2}. Based on the setting rules of a soil background adjustment parameter in [22], L ranges from 0 to 1, with intervals of 0.1. The value of L varies by the amount or cover of green vegetation: in very high vegetation regions, L = 0; and in areas with no green vegetation, L = 1. Given that the maximum value of the DN was 255, L was multiplied by 255 in the denominator. All data from Exp. 2 were used in Equation (2). All possible combinations of variables (i.e., a_i and L) were traversed, and the corresponding values of J were compared. The values of a_i and L in Equation (1) that yield the best J were the optimal variables. Consequently, the TCVI from RGB images and the FCVI from CIR images were determined as follows:

$$TCVI = \frac{(1 + 0.4) * (2R - 2B)}{(2R - G - 2B + 255 * 0.4)} \quad (3)$$

$$FCVI = \frac{(1 + 0.5) * (2NIR + B - 2G)}{(2G + 2B - 2NIR + 255 * 0.5)} \quad (4)$$

2.3. Data Analysis and Evaluation

The channel information from digital cameras that constituted the VIs was first analyzed based on the data from Exp. 2. We analyzed how the DN values of different channels changed with the changes of LNC in wheat. To compare and evaluate the performances of different VIs for estimating LNC, the quantitative relationship between the VIs and LNC was analyzed at different growth stages. For all growth stages of winter wheat, the LNC–VI models from both RGB and CIR images were calibrated and validated with a 10-fold cross-validation procedure using the data from Exp. 2. The whole dataset was randomly divided into three equal-sized sub-datasets with two sub-datasets used as the calibration (training) dataset and the rest as the validation (testing) dataset. The process was repeated 10 times [45]. For the CIR cameras, the estimation models were also validated independently with the data from Exp. 1. The performance of the different VIs and models were evaluated using the determination coefficient (R^2) and the relative root mean square error (RRMSE).

$$R^2 = 1 - \frac{\sum_{i=1}^{n} (O_i - P_i)^2}{\sum_{i=1}^{n} (O_i - \overline{O_i})^2} \quad (5)$$

$$RRMSE = \sqrt{\frac{1}{n}\sum_{i-1}^{n} (O_i - P_i)^2} \times \frac{1}{\overline{O_i}} \times 100\% \quad (6)$$

where n is the number of samples; O_i is the observed LNC value; and P_i is the estimated value. The saturation sensitivity of the VI versus LNC was evaluated using the index of noise equivalent (NE∆LNC) [46,47].

$$NE\Delta LNC = \frac{RMSE\{VIvs.LNC\}}{d(VI)/d(LNC)} \quad (7)$$

where $RMSE\{VIvs.LNC\}$ is the root mean square error (RMSE) of the best fit function and the actual LNC value, and *d(VI)/d(LNC)* is the first derivative of VI with respect to LNC. A higher NE∆LNC indicates a lower sensitivity to LNC. The accuracies of the optimal estimation models from RGB and

CIR images were compared under different treatments (i.e., different varieties, N application rates, and planting densities) using RRMSE.

3. Results

3.1. Changes of Digital Number (DN) Values in Different Channels

Figure 4 shows the changes of DN values in different channels with the variation in LNC for images from the RGB and CIR cameras. The DN values of all channels from the RGB camera decreased when the LNC increased to 3.0%, and then became flatter as the LNC continually increased (Figure 4a). Conversely, the DN values of the near infrared channel from the CIR camera increased as the LNC increased (Figure 4b). The variations of the DN values in the blue and green channels from both cameras were similar, but the values were different. For the RGB camera, the values of the green channel were higher than those of the blue channel. For the CIR camera, the DN values of the green and blue channels were similar.

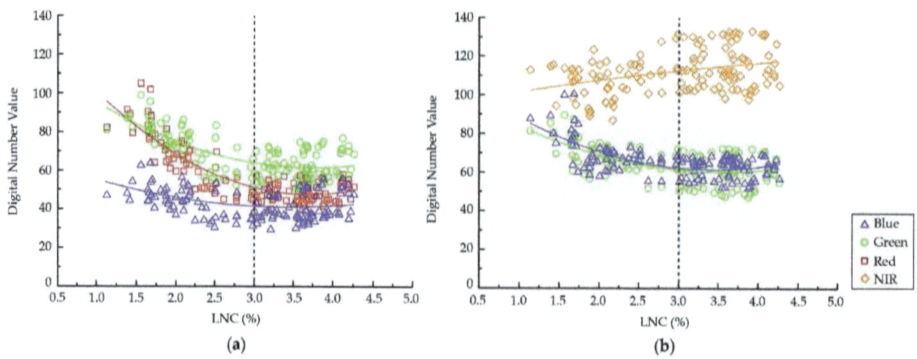

Figure 4. Changes of digital number values in different channels of (**a**) normal color (RGB) images and (**b**) color near-infrared (CIR) images with leaf N concentration (LNC) in winter wheat.

3.2. Leaf N Concentration (LNC) Estimation Model Constraction and Validation

3.2.1. Quantitative Relationships between Leaf N Concentration (LNC) and Vegetation Indices (VIs) at Different Growth Stages

Figure 5 shows the relationship between LNC and the VIs determined from both RGB and CIR images at different growth stages of winter wheat. For the VIs derived from RGB images, the results were quite different (Figure 5a–f). When the LNC increased, the NGRDI, VARI, and ExG increased, while the IKaw, RGRI, and TCVI decreased. The exponential relationship between the ExG and LNC was weak at all four stages (Figure 5e). The relationship between the NGRDI and LNC (Figure 5a) was similar to that between the VARI and LNC, but the fitting curves of the VARI and LNC at the booting and heading stages were stronger than those of the NGRDI and LNC (Figure 5d). Unlike the other VIs, the relationship between the TCVI and LNC was almost identical at all four stages (Figure 5f). For the VIs derived from CIR images, the GNDVI, ENDVI, and FCVI all increased as the LNC increased (Figure 5g–i). The results for the GNDVI and FCVI at each growth stage were similar (Figure 5g,i). The sample points for the relationship between the ENDVI and LNC were distributed along the fitting curves at the booting, anthesis, and filling stages, but at the heading stage the results were scattered (Figure 5h). At intermediate to high LNC levels, the NGRDI, RGRI, VARI, and GNDVI were not sensitive to changes in LNC > 3.0%, especially at the heading stage.

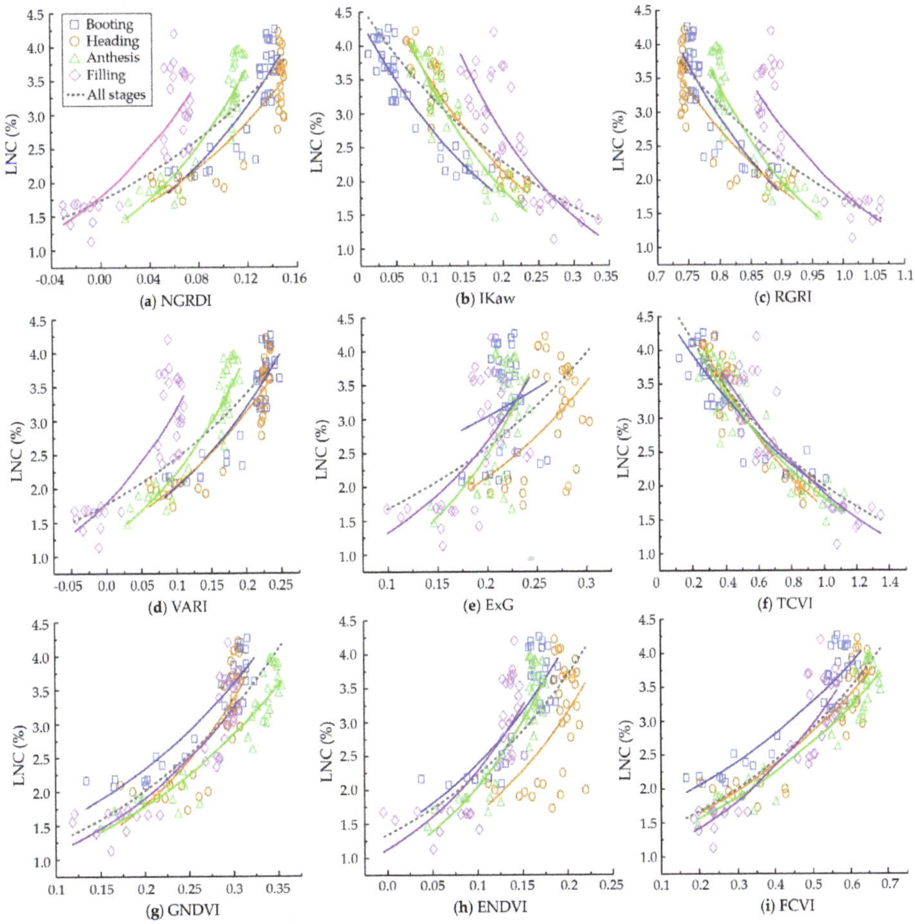

Figure 5. Leaf N concentration (LNC) (%) plotted against different vegetation indices (VIs) at different growth stages of winter wheat. (**a**) normalized green-red difference index (NGRDI), (**b**) Kawashima index (IKaw), (**c**) red green ratio index (RGRI), (**d**) visible atmospherically resistance index (VARI), (**e**) excess green vegetation index (ExG), (**f**) true color vegetation index (TCVI), (**g**) green normalized difference vegetation index (GNDVI), (**h**) enhanced normalized difference vegetation index (ENDVI), (**i**) false color vegetation index (FCVI). Statistics are given in Table 6.

Table 6. Relationship between leaf N concentration (LNC) and vegetation indices (VIs) at different growth stages of winter wheat. x and y are VI and LNC, respectively. R^2 is the determination coefficient. Relative root mean square error (RRMSE) (%) is the relative root mean square error. The bold values mean the most accurate result of normal color (RGB) and color near-infrared (CIR) cameras for each term.

Growth Stage		RGB Camera						CIR Camera		
		Normalized Green-Red Difference Index (NGRDI)	IKaw	Red Green Ratio Index (RGRI)	VARI	Excess Green Vegetation Index (ExG)	True Color Vegetation Index (TCVI)	Green Normalized Difference Vegetation Index (GNDVI)	Enhanced Normalized Difference Vegetation Index (ENDVI)	False Color Vegetation Index (FCVI)
Booting	Function	$y = 1.19e^{8x}$	$y = 4.42e^{-4.68x}$	$y = 141e^{-4.84x}$	$y = 1.28e^{4.57x}$	$y = 1.81e^{2.58x}$	$y = 4.73e^{-0.89x}$	$y = 0.99e^{4.3x}$	$y = 1.36e^{5.6x}$	$y = 1.5e^{1.56x}$
	R^2	0.752	**0.880**	0.743	0.818	0.032	0.854	0.853	0.664	**0.866**
	RRMSE	11.7	**8.3**	11.9	10.1	22.5	9.1	**9.1**	13.8	9.5
Heading	Function	$y = 1.31e^{6.56x}$	$y = 5.6e^{-4.79x}$	$y = 64.7e^{-3.95x}$	$y = 1.35e^{4.18x}$	$y = 0.79e^{5.01x}$	$y = 5.62e^{-1.21x}$	$y = 0.52e^{6.25x}$	$y = 0.87e^{6.42x}$	$y = 1.15e^{1.81x}$
	R^2	0.792	**0.918**	0.782	0.841	0.284	0.911	0.706	0.313	**0.822**
	RRMSE	12.9	**8.6**	13.1	11.2	23.8	8.9	14.1	22.9	**11.5**
Anthesis	Function	$y = 1.21e^{9.26x}$	$y = 6.19e^{-5.95x}$	$y = 237e^{-5.31x}$	$y = 1.25e^{5.73x}$	$y = 0.4e^{9.17x}$	$y = 5.11e^{-1.04x}$	$y = 0.71e^{4.63x}$	$y = 0.95e^{7.67x}$	$y = 1.1e^{1.79x}$
	R^2	0.875	0.825	0.868	**0.892**	0.458	0.855	0.889	0.811	**0.911**
	RRMSE	10.6	11.8	10.8	**10.0**	21.8	11.5	9.7	12.9	**9.0**
Filling	Function	$y = 1.79e^{8.34x}$	$y = 9.21e^{-6.11x}$	$y = 143e^{-4.37x}$	$y = 1.78e^{5.71x}$	$y = 0.66e^{6.98x}$	$y = 5.56e^{-1.08x}$	$y = 0.65e^{5.34x}$	$y = 1.13e^{7.02x}$	$y = 0.87e^{2.45x}$
	R^2	0.771	0.746	0.769	0.779	0.426	**0.813**	0.821	0.648	**0.849**
	RRMSE	18.5	18.6	18.4	18.1	27.5	**16.3**	15.2	20.0	**15.1**
All	Function	$y = 1.74e^{5.21x}$	$y = 4.61e^{-3.52x}$	$y = 34.4e^{-3x}$	$y = 1.76e^{3.27x}$	$y = 1.1e^{4.27x}$	$y = 5.09e^{-1.02x}$	$y = 0.77e^{4.77x}$	$y = 1.34e^{5.08x}$	$y = 1.15e^{1.85x}$
	R^2	0.631	0.659	0.634	0.651	0.252	**0.852**	0.744	0.507	**0.792**
	RRMSE	18.7	17.7	18.6	18.1	26.4	**12.1**	15.8	21.7	**14.0**

To quantitatively analyze the ability of the VIs to estimate LNC at different growth stages, the R^2 and RRMSE values at each growth stage and over all four stages are listed in Table 6. Generally, the results at the filling stage were much worse than at the other three stages, with an RRMSE higher than 15%. In the first three stages, the RRMSE values of most VIs were around 10%. However, the performances of the ExG and ENDVI were much poorer than those of the other VIs, especially at the heading and filling stages. The results for the TCVI and FCVI were not remarkably better than the other VIs at each growth stage, but they were obviously better for all four stages combined. For the RGB camera, the IKaw performed best at the booting and heading stages, and the VARI had the highest R^2 and lowest RRMSE values at the anthesis stage. Although the TCVI results were not the most accurate in the first three stages, they were superior to those of the other VIs derived from RGB images at the filling stage. For the CIR camera, the performance of the ENDVI was much poorer than that of the GNDVI and FCVI. The FCVI performed best at each stage, although the RRMSE of 9.5% at the booting stage was slightly higher than the value of 9.1% for the GNDVI. Compared to the VIs from RGB images, the FCVI results were less accurate than those of the IKaw at the first two stages, but were the most accurate at the anthesis and filling stages. For all four stages combined, the performance of the TCVI was remarkable, with an R^2 value of 0.852 and RRMSE of 12.1%, followed by the FCVI with an R^2 value of 0.792 and RRMSE of 14.0%.

3.2.2. Validation of the Leaf N Concentration (LNC) Models for Wheat

The validation statistics for the LNC models constructed from the different VIs are presented in Table 7. According to the results of the 10-fold cross-validation, the estimation models constructed from the TCVI had the highest accuracy. Compared to the other VIs from the RGB camera, the RRMSEs were improved by 6.32% to 15.76% for LNC based on the TCVI. For CIR images, the statistical values of the FCVI were the best for LNC, followed by the GNDVI. The independent validation indicated that the FCVI was capable of producing accurate LNC estimations at all stages.

Table 7. Validation statistics for the leaf N concentration (LNC) estimation models from different vegetation indices (VIs). The bold values mean the most accurate result of normal color (RGB) and color near-infrared (CIR) cameras for each term.

Camera	VI	Cross-Validation		Independent Validation	
		R^2	RRMSE (%)	R^2	RRMSE (%)
RGB	Normalized Green Red Difference Vegetation Index (NGRDI)	0.591	18.24		
	IKaw	0.618	17.79		
	Red green ratio index (RGRI)	0.608	18.66		
	VARI	0.603	18.37		
	Excess green vegetation index (ExG)	0.150	27.23		
	True color vegetation index (TCVI)	**0.848**	**11.47**		
CIR	Green normalized difference vegetation index (GNDVI)	0.720	16.13	0.523	23.66
	Enhanced normalized difference vegetation index (ENDVI)	0.492	20.62	0.207	37.99
	False color vegetation index (FCVI)	**0.756**	**14.18**	**0.627**	**13.61**

3.3. *Saturation Sensitivity of Vegetation Indices (VIs) at Different Leaf N Concentrations (LNCs)*

As shown in Figure 6, the saturation of the selected VIs was tested under the LNC range of 0 to 5%. The NE△LNC of ExG increased sharply as the LNC increased. The ExG was the VI with the fastest and largest increase, followed by the ENDVI. When LNC < 3%, the NE△LNC of the IKAW, RGRI, and VARI were similar. However, when LNC > 3%, the NE△LNC of the IKAW increased more slowly than that of the RGRI and VARI. Compared to the other VIs, the NE△LNC of the TCVI and FCVI were much lower, with that of the TCVI being the smallest. The NE△LNC of the TCVI was still the lowest of all the VIs at high LNCs.

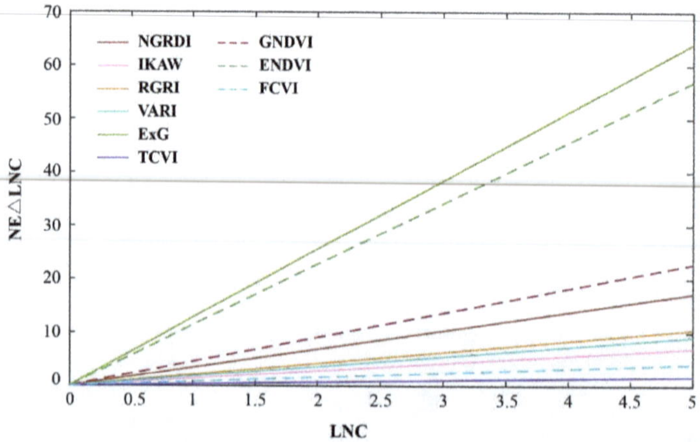

Figure 6. Sensitivity of the selected vegetation indices (VIs) to leaf N concentration (LNC).

3.4. *Applicability of the Leaf N Concentration (LNC) Models under Different Treatments*

According to the results in Table 6, the optimal LNC estimation models for the whole season from RGB and CIR images were constructed by the TCVI and FCVI, respectively. Table 8 shows a comparison of the optimal estimation models from RGB and CIR images under the different treatments. Generally, the RRMSE values of LNC estimations under the various treatments were lower than 15%.

Table 8. Relative root mean square error (RRMSE, %) values for leaf N concentration (LNC) estimations based on normal color (RGB) and color near-infrared (CIR) cameras under different treatments.

Treatment		RGB Imagery	CIR Imagery
Variety	Yangmai 8	13.95	14.69
	Shengxuan 6	9.88	13.34
N rates (kg/ha)	0	13.01	16.41
	150	11.23	12.61
	300	11.64	13.35
Planting Density (plants/ha)	3.0×10^6	9.01	14.17
	1.5×10^6	14.49	13.87

For the different wheat varieties, the RRMSE values of LNC estimations from RGB images were lower than those from CIR images, with the lowest RRMSE of 9.88% for Shengxuan 6. Under the different N application rates, the LNC models from RGB images performed better than those from CIR images. For both RGB and CIR images, the RRMSE values for N application rates of 150 kg/ha were lower than for the other N application rates. As planting density decreased, the accuracy of LNC estimation from RGB images decreased, while for CIR images it increased. The best results for the lowest planting density were obtained from CIR images, and the lowest RRMSE values for the highest planting density were obtained from RGB images.

4. Discussion

4.1. Performance of the Vegetation Indices (VIs) Derived from Digital Imagery in Estimating Wheat Leaf N Concentration (LNC)

The estimation of VIs is the most common and simplest way to extract information on crop growth status from digital images [27]. Different VIs calculated from different wavelengths highlight various vegetation properties [48–50]. Due to the different band combinations and different formulas used, the accuracy of N status estimation varies between the different VIs.

In this study, seven commonly used VIs (i.e., NGRDI, IKAW, RGRI, VARI, ExG, GNDVI, and ENDVI) and two new VIs (i.e., TCVI and FCVI) derived from both RGB and CIR images for estimating LNC were analyzed in wheat. As shown in Table 9, the numerator of the VIs from RGB images consisted of two channels, except ExG. Among them, the numerator of IKaw and TCVI contained the red and blue bands, while the red-green VIs (i.e., NGRDI, RGRI, and VARI) used the red and green bands as the numerator. The results in Figures 5 and 6 indicated that the ability to mitigate the saturation of the red-green VIs under high LNC levels was weakest for IKaw and TCVI. This might be because the red and blue bands are the chlorophyll and carotenoid absorption bands [51]. Given that crops respond to N status mainly by a change in chlorophyll concentration in the leaves [52], VIs using ratios or normalized differences of values acquired in the red and blue bands were significantly related to the N status of crops [53–55]. The GNDVI, which is calculated from the normalized difference between values of the green and NIR bands, was also limited by the saturation under intermediate to high LNC levels (Figures 5g and 6). Unlike the GNDVI, the proposed FCVI acquired information in the blue band, which reduced the saturation. Although the ENDVI also used all three bands, its performance in estimating LNC was relatively poor. Therefore, we cannot conclude that a VI incorporating more bands is able to produce more reliable LNC results. The accuracy is dependent on both the band configuration and VI formulation.

Table 9. Coefficients for Different Vegetation Indices (VIs).

	VI	Coefficients for Different Channels						Channels			L
		a_1	a_2	a_3	a_4	a_5	a_6	R	G	B	
Normal color (RGB)	Normalized green red difference vegetation index (NGRDI)	−1	1	0	1	1	0	DN_{red}	DN_{green}		0
	IKaw	1	0	−1	1	0	1	DN_{red}		DN_{blue}	0
	Red green ratio index (RGRI)	1	0	0	0	1	0	DN_{red}	DN_{green}		0
	VARI	−1	1	0	1	1	−1	DN_{red}	DN_{green}	DN_{blue}	0
	Excess green vegetation index (ExG)	−1	2	−1	1	1	1	DN_{red}	DN_{green}	DN_{blue}	0
	True color vegetation index (TCVI)	2	0	−2	2	−1	−2	DN_{red}	DN_{green}	DN_{blue}	0.4
Color near-infrared (CIR)	Green normalized difference vegetation index (GNDVI)	1	−1	0	1	1	0	DN_{nir}	DN_{green}		0
	Enhanced normalized difference vegetation index (ENDVI)	1	1	−2	1	1	2	DN_{nir}	DN_{green}	DN_{blue}	0
	False color vegetation index (FCVI)	1	−2	1	−2	2	2	DN_{nir}	DN_{green}	DN_{blue}	0.5

4.2. Accuracy and Universality of Leaf N Concentration (LNC) Estimation Models in Wheat

In this study, the LNC estimation models constructed with the TCVI and FCVI were sensitive under varying LNCs (from 0 to 5%) and had a better accuracy under the different treatments. Moreover, they were generalizable from the booting to filling stages of wheat. As the growth stage progressed, dramatic changes in the composition of canopy components and background materials can occur. These changes pose a critical challenge for the timely monitoring of crop N status. For the late stages (booting to filling) in the reproductive period, a single LNC–TCVI or LNC–FCVI model could be fitted, with a high efficiency and low error (Figure 5 and Table 6). Although ExG is mainly used to extract vegetation from different backgrounds and has been widely cited [12,42,56], its performance for estimating N status was very poor, which was consistent with previous results [12]. Previous studies have suggested that adjusting VIs with a term representing soil brightness could reduce the effect of the soil background [22,28]. Given that a soil background adjustment parameter (*L*) was added to the TCVI and FCVI, they had the potential to reduce the significant effect of soil background during the early stages of the vegetative period.

Due to the limited experiment, we only made an independent validation for the LNC–VI models from CIR images based on Exp. 1, and the LNC–VI models from both RGB images were calibrated and validated with a 10-fold cross-validation procedure using the data from Exp. 2. Since the FCVI and TCVI were developed based on the relationships between CVI and LNC from Exp.2, there is a need to validate their performance with independent measurements under different conditions (i.e., varied crop types, different site conditions and other cameras). Although the coefficients of FCVI and TCVI had been modified, the accuracy of LNC–TCVI or LNC–FCVI models was not always the highest (see Table 6). For example, the IKaw performed best at the booting and heading stages, and the VARI had the highest R^2 and lowest RRMSE values at the anthesis stage. However, these relationships may vary with crop type, site condition, and growth stage as influenced by the variation in physiological processes. Therefore, the universality of the LNC estimation models needs to be further verified during the early growth stages and under various conditions.

4.3. Capability of Commercial Digital Cameras for Wheat Leaf N Concentration (LNC) Estimation

The development of simple but efficient methods to monitor crop growth across a wide range of LNCs is urgently needed for precision agriculture in China. This is because the application of N fertilizer has increased in recent years in an attempt to boost production. Therefore, the timely monitoring of crop N status under intermediate to high N application levels is essential to maximize yield. In addition, easy-to-operate and low-cost equipment is required to help the owners of large farms or smallholders with fertilization management. Due to the low price and convenient operation of digital cameras, they have potential application prospects. In this study, UAV campaigns were conducted at noon in clear weather under stable lighting conditions. However, radiation correction should be considered in the future work, especially when experiencing changeable light intensity (e.g., for rice monitoring in summer).

To explore the capability of commercial digital cameras to estimate LNC in wheat, we constructed new VIs (i.e., TCVI and FCVI) from both RGB and CIR images. Because the TCVI and FCVI were based on large amounts of field experimental data under different conditions, they performed reliably in indicating the LNC in wheat. It is suggested that commercial digital cameras have the capability to derive optimum VIs for LNC monitoring in winter wheat. In addition, the performance of RGB images was generally better than that of CIR images (see Table 8). This study supported the widespread agreement that digital cameras are powerful tools for assessing crop growth status and further proved the applicability of UAV-based RGB and CIR cameras for the monitoring of crop N status.

5. Conclusions

The applicability of digital cameras mounted on UAVs for monitoring the LNC of winter wheat was evaluated in this study. Seven commonly used VIs (i.e., NGRDI, IKaw, RGRI, VARI, ExG, GNDVI, and ENDVI) and two proposed VIs (i.e., TCVI and FCVI) were used to estimate LNC at different growth stages of winter wheat. The performances of NGRDI, RGRI, VARI, and GNDVI were limited by saturation at intermediate to high LNCs (i.e., LNC > 3.0%). The accuracy of LNC estimation using ExG was the poorest among the VIs tested, while the optimal models were constructed by the TCVI and FCVI for all stages. The models were then cross-validated with datasets from different cultivars, N application rates, and planting densities. Compared to the other VIs derived from the RGB camera, the RRMSE values of the TCVI were improved by 6.32% to 15.76% for LNC. For the CIR camera, the statistical values of the FCVI were the best for determining LNC (R^2 = 0.756, RRMSE = 14.18%). The independent validation also indicated that the FCVI was capable of accurately estimating LNC at all growth stages. In summary, commercial digital cameras mounted on an UAV are feasible for monitoring wheat LNC at the farm-scale, especially under the high N fertilizer applications and different treatments typically used in fields in modern China.

In practical terms, commercial digital cameras are low-cost and easy to operate for researchers and farmers. Although the equipment is applicable, the LNC–VI relationship may vary with crop type, site condition, and growth stage, and may be influenced by variations in physiological processes. Therefore, additional calibrations are needed for different conditions before extending this method to other crops. For example, the use of UAV-based digital cameras for crop N status monitoring should be further investigated during the early stages of the vegetative period. Moreover, fluctuating ambient lighting conditions are an issue that should be addressed in future studies.

Author Contributions: X.Y. and T.C. conceived and designed the experiments. J.J. and H.Z. performed the experiments. J.J. analyzed the data. J.J. and X.Y. wrote the paper. All authors contributed to the interpretation of results and editing of the manuscript.

Funding: This work was supported by grants from the Key Projects (Advanced Technology) of Jiangsu Province (BE2019383), National Natural Science Foundation of China (31671582, 31971780), Jiangsu Qinglan Project, Jiangsu Collaborative Innovation Center for Modern Crop Production (JCICMCP), the Priority Academic Program Development of Jiangsu Higher Education Institutions (PAPD), Qinghai Project of Transformation of Scientific and Technological Achievements (2018-NK-126); Xinjiang Corps Great Science and Technology Projects (2018AA00403), and the 111 project (B16026).

Acknowledgments: We would like to thank former graduate students Yong Liu and Ni Wang for their help with field data collection.

Conflicts of Interest: The authors declare no conflict of interest.

References

1. Benedetti, R.; Rossini, P. On the use of NDVI profiles as a tool for agricultural statistics: The case study of wheat yield estimate and forecast in Emilia Romagna. *Remote Sens. Environ.* **1993**, *45*, 311–326. [CrossRef]
2. Rouse, J.W.J.; Haas, R.H.; Schell, J.A.; Deering, D.W. *Monitoring Vegetation Vegetation Systems in the Great Plains with ERTS*; NASA: Washington, DC, USA, 1973; Volume 351, p. 309.
3. Basnyat, B.M.P.; Moulin, A.; Pelcat, Y.; Lafond, G.P. Optimal time for remote sensing to relate to crop grain yield on the Canadian prairies. *Can. J. Plant Sci.* **2004**, *84*, 97–103.
4. Hatfield, J.L.; Gitelson, A.A.; Schepers, J.S.; Walthall, C.L.; Pearson, C.H. Application of spectral remote sensing for agronomic decisions. *Agron. J.* **2015**, *100*, 117–131. [CrossRef]
5. Ju, X.T.; Xing, G.X.; Chen, X.P.; Zhang, S.L.; Zhang, L.J.; Liu, X.J.; Cui, Z.L.; Yin, B.; Christie, P.; Zhu, Z.L.; et al. Reducing environmental risk by improving N management in intensive Chinese agricultural systems. *Proc. Natl. Acad. Sci. USA* **2009**, *106*, 3041–3046. [CrossRef]
6. Mulla, D.J. Twenty-five years of remote sensing in precision agriculture: Key advances and remaining knowledge gaps. *Biosyst. Eng.* **2013**, *114*, 358–371. [CrossRef]
7. Hansen, P.; Schjoerring, J.K. Reflectance measurement of canopy biomass and nitrogen status in wheat crops using normalized difference vegetation indices and partial least squares regression. *Remote Sens. Environ.* **2003**, *86*, 542–553. [CrossRef]
8. Tan, C.; Guo, W.; Wang, J. Predicting grain protein content of winter wheat based on landsat TM images and leaf nitrogen Content. In Proceedings of the International Conference on Remote Sensing, Environment and Transportation Engineering, Nanjing, China, 24–26 June 2011; pp. 5165–5168.
9. Zhang, F.; Cui, Z.; Fan, M.; Zhang, W.; Chen, X.; Jiang, R. Integrated Soil–Crop System Management: Reducing Environmental Risk while Increasing Crop Productivity and Improving Nutrient Use Efficiency in China. *J. Environ. Qual.* **2011**, *40*, 1051. [CrossRef]
10. Miao, Y.; Zhang, F. Long-term experiments for sustainable nutrient management in China. A review. *Agron. Sustain. Dev.* **2011**, *31*, 397–414. [CrossRef]
11. Bendig, J.; Yu, K.; Aasen, H.; Bolten, A.; Bennertz, S.; Broscheit, J.; Gnyp, M.L.; Bareth, G. Combining UAV-based plant height from crop surface models, visible, and near infrared vegetation indices for biomass monitoring in barley. *Int. J. Appl. Earth Obs. Geoinf.* **2015**, *39*, 79–87. [CrossRef]

12. Zheng, H.; Cheng, T.; Li, D.; Zhou, X.; Yao, X.; Tian, Y.; Cao, W.; Zhu, Y. Evaluation of RGB, Color-Infrared and Multispectral Images Acquired from Unmanned Aerial Systems for the Estimation of Nitrogen Accumulation in Rice. *Remote Sens.* **2018**, *10*, 824. [CrossRef]
13. Yang, G.; Liu, J.; Zhao, C.; Li, Z.; Huang, Y.; Yu, H.; Xu, B.; Yang, X.; Zhu, D.; Zhang, X. Unmanned Aerial Vehicle Remote Sensing for Field-Based Crop Phenotyping: Current Status and Perspectives. *Front. Plant Sci.* **2017**, *8*, 1111. [CrossRef] [PubMed]
14. Vadrevu, K.P. *Introduction to Remote Sensing (FIFTH EDITION)*; Campbell, J.B., Wynne, R.H., Eds.; Guilford Press: New York, NY, USA, 2013; Volume 28, pp. 117–118, ISBN 978-160918-176-5.
15. Du, M.; Noguchi, N. Monitoring of Wheat Growth Status and Mapping of Wheat Yield's within-Field Spatial Variations Using Color Images Acquired from UAV-camera System. *Remote Sens.* **2017**, *9*, 289. [CrossRef]
16. Hunt, E.R.; Cavigelli, M.; Daughtry, C.S.T.; McMurtrey, J.E.; Walthall, C.L. Evaluation of Digital Photography from Model Aircraft for Remote Sensing of Crop Biomass and Nitrogen Status. *Precis. Agric.* **2005**, *6*, 359–378. [CrossRef]
17. Nakatani, M.; Kawashima, S. An Algorithm for Estimating Chlorophyll Content in Leaves Using a Video Camera. *Ann. Bot.* **1998**, *81*, 49–54.
18. Woebbecke, D.M.; Meyer, G.E.; Von Bargen, K.; Mortensen, D.A. Color Indices for Weed Identification Under Various Soil, Residue, and Lighting Conditions. *Trans. ASAE* **1995**, *38*, 259–269. [CrossRef]
19. Golzarian, M.R.; Frick, R.A.; Rajendran, K.; Berger, B.; Roy, S.; Tester, M.; Lun, D.S. Accurate inference of shoot biomass from high-throughput images of cereal plants. *Plant Methods* **2011**, *7*, 2. [CrossRef] [PubMed]
20. Geipel, J.; Link, J.; Claupein, W. Combined Spectral and Spatial Modeling of Corn Yield Based on Aerial Images and Crop Surface Models Acquired with an Unmanned Aircraft System. *Remote Sens.* **2014**, *6*, 10335–10355. [CrossRef]
21. Cui, R.-X.; Liu, Y.-D.; Fu, J.-D. [Estimation of Winter Wheat Biomass Using Visible Spectral and BP Based Artificial Neural Networks]. *Guang pu xue yu guang pu fen xi = Guang pu* **2015**, *35*, 2596. [PubMed]
22. Huete, A.; Huete, A. A soil-adjusted vegetation index (SAVI). *Remote Sens. Environ.* **1988**, *25*, 295–309. [CrossRef]
23. Tucker, C.J. Red and photographic infrared linear combinations for monitoring vegetation. *Remote Sens. Environ.* **1979**, *8*, 127–150. [CrossRef]
24. Jordan, C.F. Derivation of Leaf-Area Index from Quality of Light on the Forest Floor. *Ecology* **1969**, *50*, 663–666. [CrossRef]
25. Hunt, E.R.; Hively, W.D.; Fujikawa, S.J.; Linden, D.S.; Daughtry, C.S.T.; Mccarty, G.W. Acquisition of NIR-Green-Blue Digital Photographs from Unmanned Aircraft for Crop Monitoring. *Remote Sens.* **2010**, *2*, 290–305. [CrossRef]
26. Hunt, E.R.; Hively, W.D.; Mccarty, G.W.; Daughtry, C.S.T.; Forrestal, P.J.; Kratochvil, R.J.; Carr, J.L.; Allen, N.F.; Fox-Rabinovitz, J.R.; Miller, C.D. NIR-Green-Blue High-Resolution Digital Images for Assessment of Winter Cover Crop Biomass. *GIScience Remote Sens.* **2011**, *48*, 86–98. [CrossRef]
27. Rasmussen, J.; Ntakos, G.; Nielsen, J.; Svensgaard, J.; Poulsen, R.N.; Christensen, S. Are vegetation indices derived from consumer-grade cameras mounted on UAVs sufficiently reliable for assessing experimental plots? *Eur. J. Agron.* **2016**, *74*, 75–92. [CrossRef]
28. Li, Y.; Chen, D.; Walker, C.; Angus, J. Estimating the nitrogen status of crops using a digital camera. *Field Crop. Res.* **2010**, *118*, 221–227. [CrossRef]
29. Li, F.; Gnyp, M.L.; Jia, L.; Miao, Y.; Yu, Z.; Koppe, W.; Bareth, G.; Chen, X.; Zhang, F. Estimating N status of winter wheat using a handheld spectrometer in the North China Plain. *Field Crop. Res.* **2008**, *106*, 77–85. [CrossRef]
30. Zhu, Y.; Tian, Y.; Yao, X.; Liu, X.; Cao, W. Analysis of Common Canopy Reflectance Spectra for Indicating Leaf Nitrogen Concentrations in Wheat and Rice. *Plant Prod. Sci.* **2007**, *10*, 400–411. [CrossRef]
31. Moharana, S.; Dutta, S. Spatial variability of chlorophyll and nitrogen content of rice from hyperspectral imagery. *ISPRS J. Photogramm. Remote Sens.* **2016**, *122*, 17–29. [CrossRef]
32. Yao, X.; Ren, H.; Cao, Z.; Tian, Y.; Cao, W.; Zhu, Y.; Cheng, T. Detecting leaf nitrogen content in wheat with canopy hyperspectrum under different soil backgrounds. *Int. J. Appl. Earth Obs. Geoinf.* **2014**, *32*, 114–124. [CrossRef]

33. Wright, I.J.; Reich, P.B.; Westoby, M.; Ackerly, D.D.; Baruch, Z.; Bongers, F.; Cavender-Bares, J.; Chapin, T.; Cornelissen, J.H.C.; Diemer, M.; et al. The worldwide leaf economics spectrum. *Nature* **2004**, *428*, 821–827. [CrossRef]
34. Evans, J.R. Photosynthesis and nitrogen relationships in leaves of C3 plants. *Oecologia* **1989**, *78*, 9–19. [CrossRef] [PubMed]
35. Filella, I.; Serrano, L.; Serra, J.; Peñuelas, J. Evaluating Wheat Nitrogen Status with Canopy Reflectance Indices and Discriminant Analysis. *Crop. Sci.* **1995**, *35*, 1400–1405. [CrossRef]
36. Houlès, V.; Guérif, M.; Mary, B. Elaboration of a nitrogen nutrition indicator for winter wheat based on leaf area index and chlorophyll content for making nitrogen recommendations. *Eur. J. Agron.* **2007**, *27*, 1–11. [CrossRef]
37. Yao, X.; Huang, Y.; Shang, G.; Zhou, C.; Cheng, T.; Tian, Y.; Cao, W.; Zhu, Y. Evaluation of Six Algorithms to Monitor Wheat Leaf Nitrogen Concentration. *Remote Sens.* **2015**, *7*, 14939–14966. [CrossRef]
38. Mutanga, O.; Adam, E.; Adjorlolo, C.; Abdel-Rahman, E.M. Evaluating the robustness of models developed from field spectral data in predicting African grass foliar nitrogen concentration using WorldView-2 image as an independent test dataset. *Int. J. Appl. Earth Obs. Geoinf.* **2015**, *34*, 178–187. [CrossRef]
39. Lepine, L.C.; Ollinger, S.V.; Ouimette, A.P.; Martin, M.E. Examining spectral reflectance features related to foliar nitrogen in forests: Implications for broad-scale nitrogen mapping. *Remote Sens. Environ.* **2016**, *173*, 174–186. [CrossRef]
40. Ecarnot, M.; Compan, F.; Roumet, P. Assessing leaf nitrogen content and leaf mass per unit area of wheat in the field throughout plant cycle with a portable spectrometer. *Field Crop. Res.* **2013**, *140*, 44–50. [CrossRef]
41. Bremner, J.M.; Sparks, D.L.; Page, A.L.; Helmke, P.A.; Loeppert, R.H.; Soltanpour, P.N.; Tabatabai, M.A.; Johnston, C.T.; Sumner, M.E. Nitrogen—Total. *Methods Soil Anal. Chem. Methods Part* **1996**, *72*, 532–535.
42. Gitelson, A.A.; Kaufman, Y.J.; Stark, R.; Rundquist, D. Novel algorithms for remote estimation of vegetation fraction. *Remote Sens. Environ.* **2002**, *80*, 76–87. [CrossRef]
43. Gamon, J.A.; Surfus, J.S. Assessing Leaf Pigment Content and Activity with a Reflectometer. *New Phytol.* **2010**, *143*, 105–117. [CrossRef]
44. Gitelson, A.A.; Kaufman, Y.J.; Merzlyak, M.N. Use of a green channel in remote sensing of global vegetation from EOS-MODIS. *Remote Sens. Environ.* **1996**, *58*, 289–298. [CrossRef]
45. Li, H.; Liu, G.; Liu, Q.; Chen, Z.; Huang, C. Retrieval of Winter Wheat Leaf Area Index from Chinese GF-1 Satellite Data Using the PROSAIL Model. *Sensors* **2018**, *18*, 1120. [CrossRef] [PubMed]
46. Govaerts, Y.M.; Verstraete, M.M.; Pinty, B.; Gobron, N. Designing optimal spectral indices: A feasibility and proof of concept study. *Int. J. Remote Sens.* **1999**, *20*, 1853–1873. [CrossRef]
47. Gitelson, A.A. Remote estimation of crop fractional vegetation cover: The use of noise equivalent as an indicator of performance of vegetation indices. *Int. J. Remote Sens.* **2013**, *34*, 6054–6066. [CrossRef]
48. Agapiou, A.; Hadjimitsis, D.G.; Alexakis, D.D. Evaluation of Broadband and Narrowband Vegetation Indices for the Identification of Archaeological Crop Marks. *Remote Sens.* **2012**, *4*, 3892–3919. [CrossRef]
49. Li, F.; Mistele, B.; Hu, Y.; Chen, X.; Schmidhalter, U. Reflectance estimation of canopy nitrogen content in winter wheat using optimised hyperspectral spectral indices and partial least squares regression. *Eur. J. Agron.* **2014**, *52*, 198–209. [CrossRef]
50. Hunt, E.R., Jr.; Doraiswamy, C.P.; McMurtrey, E.J.; Daughtry, T.C.S.; Akhmedov, B. A visible band index for remote sensing leaf chlorophyll content at the canopy scale. *Int. J. Appl. Earth Obs. Geoinf.* **2013**, *21*, 103–112. [CrossRef]
51. Chappelle, E.W.; Kim, M.S.; McMurtrey, J.E., III. Ratio analysis of reflectance spectra (RARS): An algorithm for the remote estimation of the concentrations of chlorophyll A, chlorophyll B, and carotenoids in soybean leaves. *Remote Sens. Environ.* **1992**, *39*, 239–247. [CrossRef]
52. Peñuelas, J.; Filella, I. Visible and near-infrared reflectance techniques for diagnosing plant physiological status. *Trends Plant Sci.* **1998**, *3*, 151–156. [CrossRef]
53. Peñuelas, J.; Gamon, J.; Fredeen, A.; Merino, J.; Field, C. Reflectance indices associated with physiological changes in nitrogen- and water-limited sunflower leaves. *Remote Sens. Environ.* **1994**, *48*, 135–146. [CrossRef]
54. Carter, G.A.; Miller, R.L. Early detection of plant stress by digital imaging within narrow stress-sensitive wavebands. *Remote Sens. Environ.* **1994**, *50*, 295–302. [CrossRef]

55. Peñuelas, J.A.; Gamon, J.; Griffin, K.L.; Field, C.B. Assessing community type, plant biomass, pigment composition, and photosynthetic efficiency of aquatic vegetation from spectral reflectance. *Remote Sens. Environ.* **1993**, *46*, 110–118. [CrossRef]
56. Lamm, R.D.; Slaughter, D.C.; Giles, D.K. Precision Weed Control System for Cotton. *Trans. ASAE* **2002**, *45*, 231–238.

remote sensing

Article

Evaluating Different Non-Destructive Estimation Methods for Winter Wheat (*Triticum aestivum* L.) Nitrogen Status Based on Canopy Spectrum

Hongjun Li [1], Yuming Zhang [1,*], Yuping Lei [1], Vita Antoniuk [2] and Chunsheng Hu [1]

1 Key Laboratory of Agricultural Water Resources, The Innovative Academy of Seed Design, Center for Agricultural Resources Research, Institute of Genetics and Developmental Biology, Chinese Academy of Sciences, 286 Huaizhong Road, Shijiazhuang 050021, China; lhj@sjziam.ac.cn (H.L.); leiyp@sjziam.ac.cn (Y.L.); cshu@sjziam.ac.cn (C.H.)

2 Department of Agroecology, Aarhus University, Blichers Allé 20, 8830 Tjele, Denmark; vita.antoniuk@agro.au.dk

* Correspondence: ymzhang@sjziam.ac.cn; Tel.: +86-311-8587-1620

Received: 18 November 2019; Accepted: 24 December 2019; Published: 26 December 2019

Abstract: Compared to conventional laboratory testing methods, crop nitrogen estimation methods based on canopy spectral characteristics have advantages in terms of timeliness, cost, and practicality. A variety of rapid and non-destructive estimation methods based on the canopy spectrum have been developed on the scale of space, sky, and ground. In order to understand the differences in estimation accuracy and applicability of these methods, as well as for the convenience of users to select the suitable technology, models for estimation of nitrogen status of winter wheat were developed and compared for three methods: drone equipped with a multispectral camera, soil plant analysis development (SPAD) chlorophyll meter, and smartphone photography. Based on the correlations between observed nitrogen status in winter wheat and related vegetation indices, green normalized difference vegetation index (GNDVI) and visible atmospherically resistant index (VARI) were selected as the sensitive vegetation indices for the drone equipped with a multispectral camera and smartphone photography methods, respectively. The correlation coefficients between GNDVI, SPAD, and VARI were 0.92 ** and 0.89 **, and that between SPAD and VARI was 0.90 **, which indicated that three vegetation indices for these three estimation methods were significantly related to each other. The determination coefficients of the 0–90 cm soil nitrate nitrogen content estimation models for the drone equipped with a multispectral camera, SPAD, and smartphone photography methods were 0.63, 0.54, and 0.81, respectively. In the estimation accuracy evaluation, the method of smartphone photography had the smallest root mean square error (RMSE = 9.80 mg/kg). The accuracy of the smartphone photography method was slightly higher than the other two methods. Due to the limitations of these models, it was found that the crop nitrogen estimation methods based on canopy spectrum were not suitable for the crops under severe phosphate deficiency. In addition, in estimation of soil nitrate nitrogen content, there were saturation responses in the estimation indicators of the three methods. In order to introduce these three methods in the precise management of nitrogen fertilizer, it is necessary to further improve their estimation models.

Keywords: canopy spectrum; non-destructive nitrogen status diagnosis; drone; multispectral camera; SPAD; smartphone photography

1. Introduction

Since the Green Revolution, chemical fertilizers have been widely applied in agricultural production to greatly increase the crop yield [1]. Nowadays, about half of the crop growth depends on nitrogen (N) fertilizer, and global agriculture consumes more than 50% of the total chemical N fixation each year [2]. However, the N use efficiency (NUE, the ratio of output N in yield to input N in fertilizers) in agriculture is generally low, and the NUEs of cereal crops are only about 33% [3]. This leads to an increased production cost. Furthermore, agricultural N pollution becomes a major problem and threatens the environment and human health [4].

As a management model of modern agriculture and an effective way for sustainable development of agriculture, precision agriculture has high potential in economic and ecological benefits of agricultural production through specific inputs of various elements, such as water, fertilizers, and pesticides. In order to implement precise management of N fertilizer, field sampling and laboratory testing methods are often adopted to determine the content of various forms of N in plants and soil, so that the amount of N fertilizer is recommended accordingly [5,6]. Although the laboratory testing methods are accurate, they are time-consuming, labor-intensive, and costly. It is a challenge to conduct large-scale sample monitoring while it is especially necessary for areas where farmland has large spatial variability [7]. Therefore, there is an urgent need to develop rapid, non-destructive, simple, and low-cost N estimation techniques to meet the demands for agricultural production.

When a crop is subjected to N deficiency stress, its canopy reflectance spectrum will change in certain bands. Consequently, using spectral diagnostic techniques, researchers have developed a variety of rapid and non-destructive methods to estimate N status of crops on three scales: space, sky, and ground. In space, satellite remote sensing can obtain spectral information of crops on a relatively extensive area. Satellites with high spatial resolution, such as QuickBird, IKONOS, and RapidEye, improved the ability to identify spatial heterogeneity of crop N status (CNS) [8,9]. It was found that the green normalized difference vegetation index (GNDVI) of wheat (*Triticum aestivum* L.) obtained by QuickBird image was significantly correlated with the N concentration of its flag leaves, and GNDVI can be used to monitor the N status of wheat [10]. Eitel et al. [11] verified the ability of RapidEye in wheat N estimation based on ground-based spectral observations. However, affected by various factors such as the temporal resolution of remote sensing, cloud interference, and the cost of commercial remote sensing images, satellite remote sensing often fails to meet the data needs for N fertilizer management. In the sky, high-resolution ground information can be quickly obtained at low cost and in real-time by using a multi-channel spectrometer or digital camera mounted on a drone, which compensates for the shortcomings of the satellite remote sensing [12–14]. Using the multispectral image or color image obtained by aerial photography, many vegetation indices (VIs) can be calculated. Using these VIs, the N status of the crop can be estimated [15,16]. On the ground, portable hyperspectral instruments and the soil plant analysis development (SPAD) chlorophyll meter are widely used in crop N diagnosis and yield prediction by monitoring the spectral characteristics of crops [17–20]. In recent years, digital cameras and smartphones have also been used in the research and application for the crop N status estimation [21,22]. By combining the camera function of a smartphone with the image analysis software installed on the smartphone, estimation of the crop N status and fertilization recommendation can be implemented on-site. The convenience of the smartphone gives this method a good application prospect [23,24].

The development of non-destructive estimation methods for CNS provides producers with a variety of techniques to choose from in implementing precise management for N fertilizer. However, these estimation methods differ in terms of equipment requirements, estimation accuracy, and application limitations. Studies have reported that some kinds of VI (e.g., normalized difference vegetation index, NDVI) can reach saturation in dense vegetation canopy, and the SPAD chlorophyll meter has a saturation response at higher leaf chlorophyll content [25]. Facing different non-destructive estimation methods, producers need to know how to make a smart choice suitable for a particular crop. In addition, the comparison of various non-destructive estimation methods is beneficial to know their limitations and meet different production demands through technical complementarity. Therefore, the objective of this research was to compare the three methods of non-destructive estimation of winter wheat N status, namely, a drone equipped with a multispectral camera (hereinafter referred to as UAVMC), SPAD, and smartphone photography (hereinafter referred to as PHONEP) methods. By performing this inter-comparison, it is expected to clarify the applicability and limitations of these three methods in the precision management of N fertilizer for winter wheat and provide a basis for the improvement for these methods and simplify the selection of estimation methods for the producer.

2. Materials and methods

2.1. Field Experiments

The fertilizer level experiment was conducted at the Luancheng Agro-Ecosystem Experimental Station (37.8897°N, 114.6935°E) of the Chinese Academy of Sciences. This area has loam soil and is situated in the temperate semi-humid monsoon climate. The main crop rotation system is winter wheat and maize (*Zea mays* L.), and all their straws after harvest are returned to the field. The experiment has been established since 1997. There were 4 levels of N fertilizer applications: 0, 200, 400, and 600 kg(N)·ha^{-1}·a^{-1} (marked as N0, N1, N2, and N3, respectively), 3 levels of phosphorus (P): 0, 32.5, and 64 kg(P)·ha^{-1}·a^{-1} (marked as P0, P1, and P2, respectively), and 2 levels of potassium (K): 0 and 50 kg(K)·ha^{-1}·a^{-1} (marked as K0 and K2, respectively). 1/4 of the total amount of N fertilizer was applied as winter wheat base fertilizer, and 1/4 was applied as a top dressing at the jointing stage of winter wheat, with the other 1/2 being applied as a top dressing for maize. 1/2 of the total K fertilizer was applied as the base fertilizer for winter wheat, and the other 1/2 was applied as the top dressing for maize. All P fertilizers were applied as base fertilizer for winter wheat. A total of 16 treatments were selected from the orthogonal incomplete design. A randomized 48 plots (8 × 14 m^2 each) with 3 replications of each treatment were arranged (as shown in Figure 1 and Table 1). The variety of the winter wheat used was "Kelong199".

Figure 1. The arrangement of the fertilizer level experiment (aerial photography on 4 April 2019).

Table 1. The treatments of N, P, and K.

Treatment	Plot Number in Figure 1	Treatment	Plot Number in Figure 1	Treatment	Plot Number in Figure 1	Treatment	Plot Number in Figure 1
N3P1K1	1, 2, 3	N2P2K1	4, 5, 6	N1P1K1	7, 8, 9	N0P0K1	10, 11, 12
N3P0K0	22, 23, 24	N2P0K0	19, 20, 21	N1P0K0	16, 17, 18	N0P0K0	13, 14, 15
N3P1K0	25, 26, 27	N2P1K0	28, 29, 30	N1P1K0	31, 32, 33	N0P1K0	34, 35, 36
N3P2K0	46, 47, 48	N2P2K0	43, 44, 45	N1P2K0	40, 41, 42	N0P2K0	37, 38, 39

2.2. Aerial Photography of UAVMC and the Reference VIs

All of the experimental campaigns were conducted during the standing stage of the winter wheat on 12 April 2018 and 4 April 2019, before topdressing in the jointing stage. The weather conditions were all suitable for acquiring aerial photography (cloud-free without strong winds). Time was selected from

10:00 a.m. to 2:00 p.m. with a relatively high solar altitude. Parrot Sequoia was used as a multispectral camera in this study, which had two sensors: a multispectral sensor and a sunlight sensor, with a total weight of 105 g. Spectral band parameters of the Parrot Sequoia are shown in Table 2. The sensors were mounted on an unmanned aerial vehicle DJI Phantom 4 quadrotor drone to acquire aerial photography.

In order to cover all the experimental area with enough overlaps, the aerial routes were programmed by the drone ground station software (DJI Ground Station Pro, an iPad app). The flight altitude was set to 50 m, and the frontal and side overlaps were all set to 80%. Before each flight, the multispectral images of a calibration target provided by the manufacturer were recorded for the radiometric calibration in post-processing. After uploading the aerial route to the drone, the drone completed the programmed flight and the spectrum camera took the images automatically. Pix4D software was used for the post-processing of the multispectral images to obtain an entire image of the experimental area and its reflectance data for the 4 bands. Figure 2 shows the pseudo color multispectral image of the experimental area.

Table 2. Specification of the Sequoia multispectral sensor used in this study.

Band	Band Width(nm)	Wave Width(nm)	Image Resolution	Field of ViewH° × V°
Green	40	550	1280 × 960	62.2 × 48.7
Red	40	660	1280 × 960	62.2 × 48.7
Red edge	40	735	1280 × 960	62.2 × 48.7
Near Infrared	40	790	1280 × 960	62.2 × 48.7

Figure 2. Pseudo color multispectral image of the experimental area on 4 April 2019 (Red-Green-Near Infrared).

Referring to the related research in remote sensing monitoring of CNS, 11 commonly used VIs (see Table 3) that can be retrieved using the multispectral image were selected as reference VIs. The most sensitive VI will be selected from them to establish the estimation models for N status for winter wheat.

Table 3. VIs derived from the multispectral image used in this study.

Name of VI	Abbreviation	Equation	Reference
Difference vegetation index	DVI	$\text{DVI} = R_{NIR} - R_{RED}$	[26]
Green normalized difference vegetation index	GNDVI	$\text{GNDVI} = (R_{NIR} - R_{GRE})/(R_{NIR} + R_{GRE})$	[27]
Modified non-linear vegetation index	MNLI	$\text{MNLI} = \left(1.5R_{NIR}^2 - 1.5R_{GRE}\right)/\left(R_{NIR}^2 + R_{RED} + 0.5\right)$	[28]
The second modified soil-adjusted vegetation index	MSAVI2	$\text{MSAVI2} = \frac{2R_{NIR}+1-\sqrt{(2R_{NIR}+2)^2-8(R_{NIR}-R_{RED})}}{2}$	[29]
Modified simple ratio	MSR	$\text{MSR} = \frac{\left(\frac{R_{NIR}}{R_{RED}} - 1\right)}{\left(\sqrt{\frac{R_{NIR}}{R_{RED}}}+1\right)}$	[30]

Table 3. *Cont.*

Name of VI	Abbreviation	Equation	Reference
Normalized vegetation index	NDVI	$NDVI = (R_{NIR} - R_{RED})/(R_{NIR} + R_{RED})$	[31]
Non-linear vegetation index	NLI	$NLI = \left(R_{NIR}^2 - R_{RED}\right)/\left(R_{NIR}^2 + R_{RED}\right)$	[32]
Optimized soil-adjusted vegetation index	OSAVI	$OSAVI = (R_{NIR} - R_{RED})/(R_{NIR} + R_{RED} + 0.16)$	[33]
Renormalized difference vegetation index	RDVI	$RNDVI = (R_{NIR} - R_{RED})/\sqrt{(R_{NIR} + R_{RED})}$	[34]
Ratio vegetation index	RVI	$RVI = R_{NIR}/R_{RED}$	[35]
Soil-adjusted vegetation index	SAVI	$SAVI = 1.5(R_{NIR} - R_{RED})/(R_{NIR} + R_{RED} + 0.5)$	[36]

Where R_{RED}, R_{GRE}, and R_{NIR} are reflectances of the red, green, and near-infrared bands, respectively.

2.3. Taking Photos of Winter Wheat with a Smartphone and the Reference Color-Based VIs

Studies have shown that there were no significant differences in the color parameters of the canopy photos obtained by different mobile phones [22]. Based on the previous research on the changes in canopy color parameters caused by different photographing methods, winter wheat was photographed by a smartphone at a height of 1 meter above the canopy and a 60° angle to the ground [22]. Using automatic exposure mode, photos were taken at three different locations in each plot and saved as JPG format. Figure 3 shows the winter wheat canopy photos from plots with different N fertilizer levels.

Figure 3. Winter wheat canopy photos. (**a–d**) Are from the plots with N3, N2, N1, and N0, respectively.

A variety of color-based VIs can be obtained by combining calculations using the Red (R), Green (G), and Blue (B) color channel values of the crop canopy photo. Referring to the related research on the estimation of CNS using canopy photo, 10 color-based VIs were selected as reference VIs for the establishment of estimation models (Table 4).

Since the photos of winter wheat in the field were taken at different time points, the values of the R, G, and B channels were susceptible to the intensity of the sun's illumination, suggesting that the absolute values of the single-color channels of the photos were not suitable for the estimation of CNS. To eliminate the effects of solar illumination changes, color-based VIs are often calculated using the ratio of different color channels or standardized form [37,38]. Therefore, this study normalized the original calculation equations of EXG and GMR in Table 4 by dividing (2G − R − B) and (G − R) by (R + G + B), respectively, in order to eliminate the influence of solar illumination changes.

Table 4. Color-based VIs used in this study.

Name of VI	Abbreviation	Equation	Reference
The dark green color index	DGCI	$DGCI = ((H-60)/60 + (1-S) + (1-B))/3$	[39]
Excess green index	EXG	$EXG = (2G-R-B)/(R+G+B)$	[40]
Green leaf index	GLI	$GLI = (2G-R-B)/(2G+R+G)$	[41]
The difference between green and red	GMR	$GMR = (G-R)/(R+G+B)$	[42,43]
Green-red vegetation index	GRVI	$GRVI = (G-R)/(G+R)$	[41]
Normalized blueness intensity	NBI	$NBI = B/(R+G+B)$	[44]
Normalized greenness intensity	NGI	$NGI = G/(R+G+B)$	[44]
Normalized redness intensity	NRI	$NRI = R/(R+G+B)$	[44]
SAVI green	SAVIGreen	$SAVI_{Green} = 1.5(G-R)/((G+R+0.5))$	[45]
Visible atmospherically resistant index	VARI	$VARI = (G-R)/(R+G-B)$	[46]
The dark green color index	DGCI	$DGCI = ((H-60)/60 + (1-S) + (1-B))/3$	[39]

Where, R, G, and B are the values of the red, green, and blue channels, respectively.

2.4. *Measurements of Nitrogen (N) Status of Winter Wheat*

Conventional laboratory testing methods were adopted to measure the N status of winter wheat on 12 April 2018 and 4 April 2019. The above-ground part of winter wheat was sampled from 1 m^2 in each plot, and total N (TN) was measured using the Kjeldahl method. Winter wheat root zone soil layers in each plot were collected at the depths of 0–30 cm, 30–60 cm, and 60–90 cm, respectively. These soil samples were extracted with 1 $mol \cdot L^{-1}$ KCl and the nitrate nitrogen content was measured using the ultraviolet (UV) spectrometry method. 30 winter wheat samples were randomly selected in each plot. The SPAD value of the first fully expanded leaf of each sample was measured in the field with a SPAD-502 chlorophyll meter, and the average value was recorded as the SPAD value for this plot.

2.5. *Analytical Methods*

Considering the wide application of soil testing and formula fertilization technology in precision agriculture, the estimation models for soil nitrate nitrogen content were developed for the methods of UAVMC, SPAD, and PHONEP in this study respectively, with the nitrate nitrogen content of winter wheat root zone soil layers being the inversion targets.

For the UAVMC method, the first step was to remove pixels of bare soil from the multispectral images to avoid the contamination of bare ground on the analysis. To achieve this goal, the NDVI of this area was calculated using the multispectral image according to the equation listed in Table 3. Winter wheat and bare soil were distinguished to obtain the mask files for winter wheat in 2018 and 2019 based on the threshold for bare soil NDVI. The 11 VIs of winter wheat listed in Table 3 were calculated using multispectral images and the mask files for winter wheat. Correspondingly, the average value of these VIs for each plot was obtained using the map for the 48 plots. Based on the correlation analysis between the 11 VIs, TN of winter wheat, and the soil nitrate nitrogen content of each layer (excluding the data for the P0 plots, as for the other two methods), the VI with the largest correlation coefficient was selected to establish the estimation models for the inversion of soil nitrate nitrogen content in root layers.

For the SPAD method, the correlations between SPAD values, TN of winter wheat, and the soil nitrate nitrogen content of each layer were analyzed. Then, the estimation models were established based on the regression equations of SPAD values and soil nitrate nitrogen content in root layers.

For the method of PHONEP, the analysis procedures were similar to the UAVMC method. To remove the non-vegetation pixels in each photo, $SAVI_{Green}$ and GMR were calculated using the R, G, and B of the winter wheat canopy photos referring to the equations in Table 4. According to the threshold ($SAVI_{Green} > 0$ and GMR > 0), the leaf mask for each photo was obtained [44,45]. Using the corresponding leaf mask, the average values of 10 VIs of winter wheat in Table 4 were calculated for each canopy photo. There were 3 photos for each plot, and the average value of the 10 VIs in the 3 photos was taken as the value for this plot. Similarly, based on the correlations between the 10 VIs, the TN of winter wheat, and the soil nitrate nitrogen content for each layer, the VI with the largest correlation coefficient was selected to establish estimation models of soil nitrate nitrogen content in root layers.

All the data from 2018 and half of the data from 2019 (data of plots numbered 1 to 24 as shown in Figure 1) were used for the development of estimation models. The remaining data (data of plots numbered 25 to 48 in 2019) were used to evaluate the estimation accuracy of the three methods by the parameters of determination coefficient (R^2) and root mean square error (RMSE). The RMSE was calculated from:

$$\text{RMSE} = \sqrt{\frac{1}{n}\sum\nolimits_{i=1}^{n}(\hat{y}_i - y_i)^2} \quad (1)$$

where n was the number of plots, and $\hat{y}$, y_i were the estimated and measured values, respectively. The correlations between the measured CNS (TN and nitrate nitrogen content at different soil layers) and reference VIs were analyzed using SPSS software. The calculation and spatial statistics of the VIs using multispectral images and canopy photos were implemented by the ERDAS IMAGINE software.

3. Results

3.1. Variation of CNS in the Fertilizer Level Experiment

The N supply for winter wheat was routinely determined by the nitrate content in root zone soil from the jointing stage to the harvest of winter wheat (0–90 cm) [47]. This practice was usually used in order to determine the amount of N fertilizer needed. The level of TN for winter wheat could also reflect the N status of the crop and was closely related to the N supply capacity of the root zone soil. Therefore, the nitrate nitrogen content in the 0–90 cm soil layer and the TN of winter wheat were measured in this study. As shown in Figure 4, due to the water, fertilizer, and other managements that were identical every year in the variable fertilizer level experiment, the two parameters measured during the same period in 2018 and 2019 were very close to each other. The biggest inter-annual differences of the nitrate nitrogen content of 0–90 cm soil and the TN of winter wheat were 28.20 mg/kg and 0.52%, respectively.

The distributions of the two parameters were consistent with the setting of the N fertilizer levels. With the increase of N fertilizer application, both parameters showed an upward trend. It is worth noting that the two parameters for the plots where P was 0 (numbered 10 to 24 in Figure 1). The content of soil nitrate nitrogen showed a rapid increase, while the TN of winter wheat in plots 16 to 24 stayed at around 3.05%. The severe shortage of P fertilizer inhibited the absorption of N fertilizer by winter wheat, and the accumulation of N fertilizer in the soil made the content of nitrate nitrogen of these plots significantly higher than that of other plots. Therefore, the data from these P0 plots were excluded during analysis.

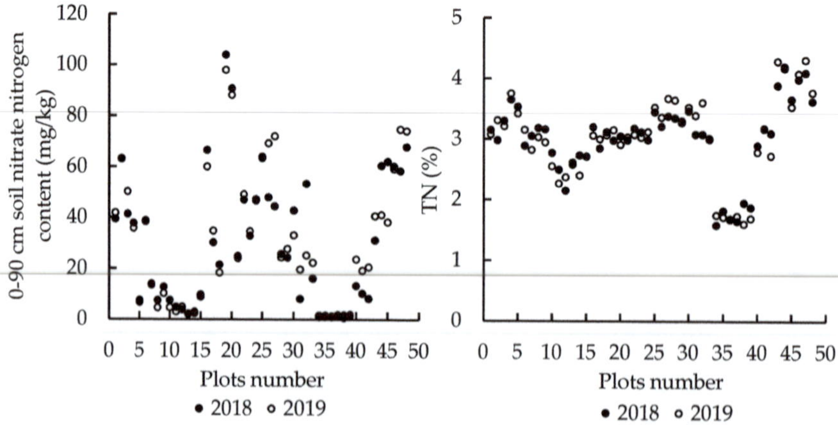

Figure 4. The distribution of TN of winter wheat and 0–90 cm soil nitrate nitrogen content in different fertilizer level experimental plots (2018 and 2019).

3.2. Estimation Models for the Method of UAVMC

The correlations between the 11 spectral VIs of winter wheat, TN of plant, and the nitrate nitrogen content in 0–30 cm, 30–60 cm, 60–90 cm, and 0–90 cm soil layers were analyzed, and all the correlation coefficients are listed in Table 5. All the 11 spectral VIs of winter wheat were significantly and positively correlated to the TN of their plants. Among these VIs, the correlation coefficient between GNDVI and TN of plants was the highest, reaching to 0.90 **. Although the correlation coefficient between RVI and TN of plants was the lowest, it still reached to 0.83 **. The N status of winter wheat was closely related to its spectral characteristics.

During the standing stage of winter wheat, 92.7% of its total root length was concentrated in the soil root layer of 0–40 cm [48]. Therefore, the N status of winter wheat should be closely related to the N supply capacity of the upper soil layer. The correlation coefficients between 11 spectral VIs and the 0–30 cm soil nitrate nitrogen content were the highest. With the increase of soil depth, the correlation coefficients decreased gradually. Among all the spectral VIs, the correlation coefficients between GNDVI and soil nitrate nitrogen content of the three layers were the highest. They were 0.52 **, 0.48 **, and 0.42 ** from the top to bottom layers. For the whole soil layer (0–90 cm), the correlation coefficient was 0.52 **.

Table 5. Correlation coefficients between spectral VIs and nitrogen status of plant and soil.

Spectral VIs	With TN of Plants	With Soil Nitrate Nitrogen Content			
		0–30 cm	30–60 cm	60–90 cm	0–90 cm
DVI	0.88 **	0.49 **	0.46 *	0.38 *	0.50 **
GNDVI	0.90 **	0.52 **	0.48 **	0.42 **	0.52 **
MNLI	0.87 **	0.51 **	0.47 **	0.38 *	0.51 **
$MSAVI_2$	0.87 **	0.50 **	0.46 **	0.37 *	0.51 **
MSR	0.89 **	0.51 **	0.48 **	0.39 **	0.52 **
NDVI	0.88 **	0.47 **	0.43 **	0.37 *	0.48 **
NLI	0.89 **	0.44 **	0.39 *	0.34 *	0.44 **
OSAVI	0.89 **	0.45 **	0.39 *	0.34 *	0.45 **
RDVI	0.89 **	0.51 **	0.43 **	0.37 *	0.50 **
RVI	0.83 **	0.50 **	0.38 *	0.26 *	0.46 **
SAVI	0.88 **	0.45 **	0.40 **	0.35 *	0.45 **

* Correlation is significant at the 0.05 level. ** Correlation is significant at the 0.01 level. (Data from 2018 and 2019).

Correlation analysis showed that GNDVI was the most sensitive spectral VI with N status of winter wheat. Therefore, GNDVI was selected as the estimation indicator, and the estimation models were established for the inversion of soil nitrate nitrogen content in different root depths of winter wheat. The models were:

Estimation model for 0–30 cm:

$$Y_{0\text{–}30} = 659.65\text{GNDVI}^{4.667}\ R^2 = 0.61 \tag{2}$$

Estimation model for 30–60 cm:

$$Y_{30\text{–}60} = 218.88\text{GNDVI}^{5.033}\ R^2 = 0.60 \tag{3}$$

Estimation model for 60–90 cm:

$$Y_{60\text{–}90} = 782.74\text{GNDVI}^{7.747}\ R^2 = 0.54 \tag{4}$$

Estimation model for 0–90 cm:

$$Y_{0\text{–}90} = 415.16\text{GNDVI}^{4.984}\ R^2 = 0.63 \tag{5}$$

3.3. Estimation Models for the SPAD Method

The correlation coefficients between the SPAD value of winter wheat leaves and TN of plants, nitrate nitrogen content in 0–30 cm, 30–60 cm, 60–90 cm, and 0–90 cm soil layers are shown in Table 6. SPAD and all winter wheat N status parameters were significantly correlated. The correlation coefficients between the SPAD value and soil nitrate nitrogen content decreased gradually with the increase of soil depth.

Table 6. Correlation coefficients between SPAD value and N status of plant and soil.

	With TN of Plants	With Soil Nitrate Nitrogen Content			
		0–30 cm	30–60 m	60–90 cm	0–90 cm
SPAD	0.85 **	0.57 **	0.50 **	0.43 **	0.55 **

** Correlation is significant at the 0.01 level (Data from 2018 and 2019).

Taking the SPAD value as the estimation indicator, the estimation models for soil nitrate nitrogen content in 0–30 cm, 30–60 cm, 60–90 cm, and 0–90 cm layers were:

Estimation model for 0–30 cm:

$$Y_{0\text{–}30} = 3.79 * 10^{-5}\text{SPAD}^{3.530}\ R^2 = 0.55 \tag{6}$$

Estimation model for 60–30 cm:

$$Y_{30\text{–}60} = 6.52 * 10^{-6}\text{SPAD}^{3.639}\ R^2 = 0.45 \tag{7}$$

Estimation model for 60–90 cm:

$$Y_{60\text{–}90} = 5.19 * 10^{-9}\text{SPAD}^{5.360}\ R^2 = 0.45 \tag{8}$$

Estimation model for 0–90 cm:

$$Y_{0\text{–}90} = 1.11 * 10^{-5}\text{SPAD}^{3.676}\ R^2 = 0.54 \tag{9}$$

3.4. Estimation Models for the PHONEP Method

The correlation coefficients between 10 color-based VIs of winter wheat, TN of plants, and the nitrate nitrogen content in 0–30 cm, 30–60 cm, 60–90 cm, and 0–90 cm soil layers are shown in Table 7. All the color-based VIs were significantly correlated with the TN of the plants. Among them, EXG, GLI, NGI, and NRI were negatively correlated, and the rest were positively correlated. Affected by the distribution characteristics of winter wheat roots, the correlation coefficients between color-based VIs and soil nitrate nitrogen content decreased with the increase of depth. Among the 10 color-based VIs, VARI had the best correlation with TN of plants, and its correlation coefficient was 0.91 **. The correlation coefficients between VARI and the soil nitrate content in 0–30 cm, 30–60 cm, and 0–90 cm soil layers were the largest. In the soil layers 60–90 cm, the correlation coefficient of VARI was slightly lower than that of DGCI. Considering that the correlation coefficients between VARI and TN of plant, 0–30 cm, 30–60 cm, and 0–90 cm soil nitrate nitrogen content were the largest, VARI was selected as the most sensitive color-based VI to establish the estimation model. Similarly, VARI was also selected by other researchers to develop the crop N estimation models [22,48].

Table 7. Correlation coefficients between color-based VIs and nitrogen status of plant and soil.

Color-Based VIs	With TN of Plants	With Soil Nitrate Nitrogen Content			
		0–30 cm	30–60 cm	60–90 cm	0–90 cm
DGCI	0.70 **	0.64 **	0.63 **	0.62 **	0.66 **
EXG	−0.71 **	−0.65 **	−0.64 **	−0.59 **	−0.64 **
GLI	−0.54 **	−0.64 **	−0.61 **	−0.56 **	−0.61 **
GMR	0.68 **	0.45 **	0.33 *	0.21 *	0.40 **
GRVI	0.83 **	0.55 **	0.55 **	0.53 **	0.57 **
NBI	0.65 **	0.67 **	0.60 **	0.60 **	0.64 **
NGI	−0.49 **	−0.57 **	−0.54 **	−0.50 **	0.54 **
NRI	−0.77 **	−0.64 **	−0.62 **	−0.54 **	0.65 **
$SAVI_{Green}$	0.68 **	0.54 **	0.48 **	0.38 *	0.52 **
VARI	0.91 **	0.72 **	0.67 **	0.60 **	0.72 **

* Correlation is significant at the 0.05 level. ** Correlation is significant at the 0.01 level (Data from 2018 and 2019).

Taking the VARI as the estimation indicator, the estimation models for soil nitrate nitrogen content in 0–30 cm, 30–60 cm, 60–90 cm, and 0–90 cm layers were:

Estimation model for 0–30 cm:

$$Y_{0\text{–}30} = 4311.7\text{VARI}^{2.1796}\ R^2 = 0.82 \quad (10)$$

Estimation model for 30–60 cm:

$$Y_{30\text{–}60} = 2580.3\text{VARI}^{2.4889}\ R^2 = 0.71 \quad (11)$$

Estimation model for 60–90 cm:

$$Y_{60\text{–}90} = 6750.4\text{VARI}^{3.286}\ R^2 = 0.67 \quad (12)$$

Estimation model for 0–90 cm:

$$Y_{0\text{–}90} = 2904.2\text{VARI}^{2.3097}\ R^2 = 0.81 \quad (13)$$

3.5. Validation

In view of the guiding role of root zone soil nitrate nitrogen content in the precise management of N fertilizer, the accuracy of the estimation models for the 0–90 cm soil nitrate nitrogen content by using the methods of UAVMC, SPAD, and PHONEP were evaluated. Figure 5 shows the linear correlations between nitrate nitrogen content in the 0–90 cm soil obtained by the laboratory testing method and that estimated by the three methods. The PHONEP method had the best estimation accuracy (R^2 = 0.93 and RMSE = 9.80 mg/kg). The SPAD method had the lowest estimation accuracy (R^2 = 0.61 and RMSE = 19.80 mg/kg). For the method of UAVMC, R^2 is 0.86 and RMSE is 12.40 mg/kg. As shown in Figure 5, all three methods had relatively high estimation accuracy in the low-value areas of soil nitrate nitrogen content. While in the high-value areas of soil nitrate nitrogen content, the estimated values of the three methods were significantly lower than the measured values.

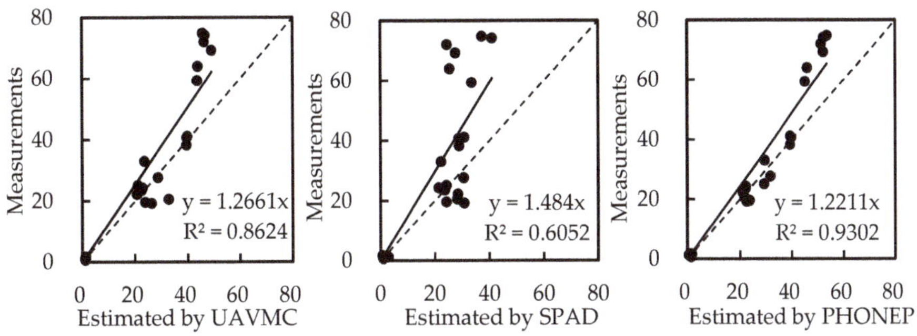

Figure 5. Correlations between nitrate nitrogen content in the 0–90 cm soil layers (mg/kg) obtained by the laboratory testing method and estimated by the methods of UAVMC, SPAD, and PHONEP.

4. Discussion

4.1. Comparison of the Three Estimation Methods

In this study, GNDVI, SPAD, and VARI were selected as estimation indicators for the three methods to establish estimation models for soil nitrate nitrogen content. The correlation coefficients between GNDVI, SPAD, and VARI were 0.92 ** and 0.89 ** respectively, and that between SPAD and VARI was 0.90 **. The significant correlation between these three indicators showed a high similarity of the three methods in the estimation of soil nitrate nitrogen content.

For the estimation models, the determination coefficients of the estimation equations for the PHONEP method were 0.82, 0.71, 0.67, and 0.81 for the soil layers of 0–30 cm, 30–60 cm, 60–90 cm, and 0–90 cm respectively, which were all slightly higher than the other two methods. The determination coefficients and RMSEs in the estimation accuracy evaluation indicated that the PHONEP method was the best, while the SPAD method was the worst method in terms of the estimation accuracy.

4.2. Effect of P Fertilizer Shortage on CNS Estimation

Considering the synergistic effects of N, P, and K in the nutrient intake of crops, the severe shortage of P fertilizer could inhibit the absorption of N by crops, and therefore, the P0 plots were excluded from the analyses. In order to analyze the impact of P fertilizer shortage on the diagnosis of CNS, taking VARI as an example, the estimation models with and without the data from P0 plots were compared.

In the fertilizer level experiment, plots with P0 contained four levels of N fertilizer, N0, N1, N2, and N3. As shown in Figure 6a, the points with P0 are concentrated at the lower left of the figure. This finding indicated that even if the N fertilizer was sufficient (such as N3), the TN of winter wheat plant was still generally low. If the points with P0 were removed, the correlation between VARI and TN of winter wheat plants was significantly improved (Figure 6b). The smaller VARI was correlated with the lower nitrate nitrogen content in the soil layer of 0–90 cm as shown in Equation (13). However, in Figure 6c, there were some points with low VARI and high soil nitrate nitrogen content in the 0–90 cm soil layer. This could be attributed to the shortage of P fertilizer. When this happened, the

NUE of winter wheat was reduced, leading to the retention of applied N in the soil. If these P0 points were excluded, the determination coefficient of the fitted equation for VARI and the 0–90 cm soil nitrate nitrogen content was significantly increased (Figure 6d). Results showed that the estimation method based on the canopy spectrum of winter wheat was not suitable for the estimation of N status of crops under a severe P fertilizer shortage. In actual production, the amount of P fertilizer applied was generally not too low to affect the absorption of N fertilizer by crops. Nevertheless, we should pay attention to the existence of this phenomenon.

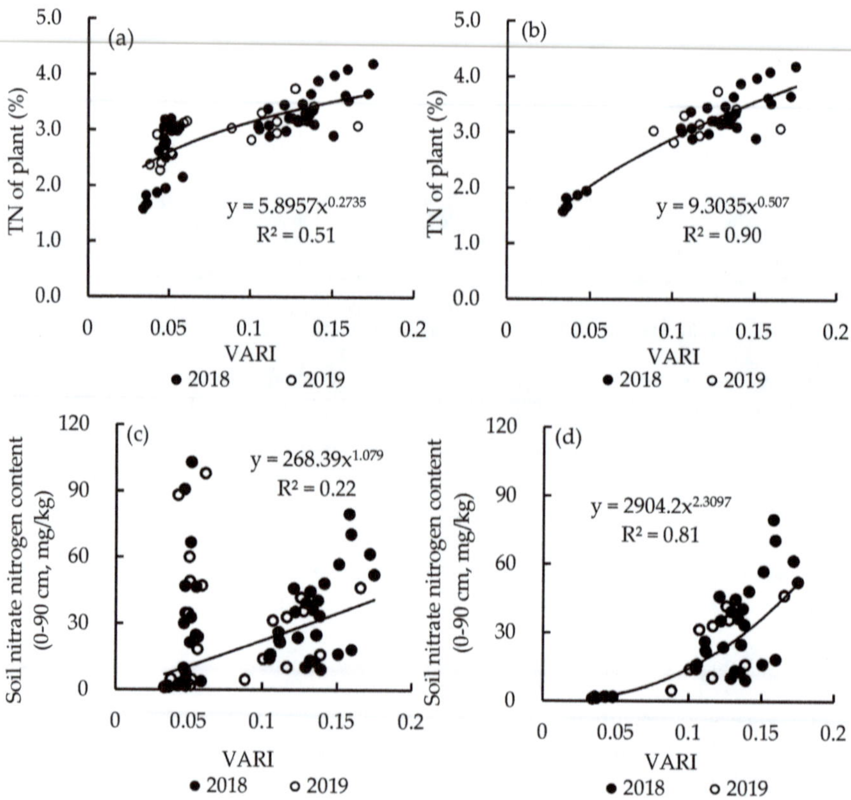

Figure 6. The impacts of P fertilizer shortage on N estimation model for the PHONEP method. **(a,b)** Are the distributions between VARI and TN of plant for the data including and excluding P0 plots, respectively. **(c,d)** Are the distributions between VARI and nitrate nitrogen content in the 0–90 cm soil layer for the data including and excluding P0 plots, respectively.

Due to the fact that more than 80% of the K absorbed by crops was stored in its straw, these K could be quickly released into the soil by straw returning [49]. In this way, plots with K0 in the fertilizer level experiment could get enough K to support the growth of winter wheat. Therefore, data from plots with K0 were involved in the development of estimation models.

4.3. The Saturation Response of the Estimation Indices

Some VIs (such as NDVI) might reach saturation in high-density vegetation areas [50], and the SPAD method had this limitation when N supply was sufficient for the crop as well [51]. Taking the measurements in this study as an example (excluding P0 plots), the saturation response of the estimation VIs (GNDVI, SPAD, and VARI) of the three methods in inversion of the 0–90 cm soil nitrate nitrogen content was analyzed and is shown in Figure 7. The distributions of the three VIs were similar to each other. In the low-value region of soil nitrate nitrogen content in the 0–90 cm depth, all the three VIs and the soil nitrate nitrogen content showed the quadratic polynomial correlations. With the increase of soil nitrate nitrogen content, the three VIs rapidly increased and reached their maximum values, i.e., their saturation points. After the saturation points, the three VIs did not significantly increase with the increase of nitrate nitrogen content in the soil. As shown in Figure 7a,b, the values of GNDVI and SPAD were hardly changed after their saturation points. Figure 7c showed that the value of soil nitrate nitrogen content at the saturation point of VARI was greater than that of GNDVI and SPAD. Moreover, after its saturation point, with the increase of soil nitrate nitrogen content, VARI tended to increase slowly. This indicated that the PHONEP method was relatively better than the other two methods. Therefore, it is necessary to pay attention to the saturation response of VIs for the estimation of CNS, or to improve the estimation accuracy by using the method of segmentation.

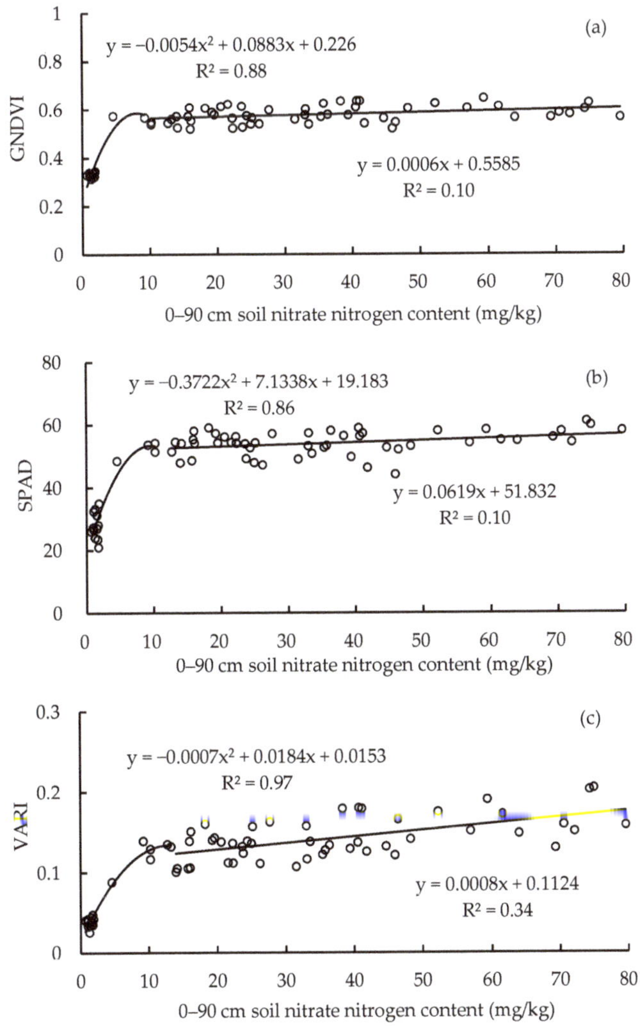

Figure 7. Analyses of the saturation response of GNDVI, SPAD, and VARI. (**a–c**) Show the saturation response for the method of GNDVI, SPAD, and VARI, respectively.

5. Conclusions

Fertilizers play irreplaceable roles in agricultural production, making it possible to provide more food from limited arable land for a growing population. However, excessive application for fertilizers and low use efficiency have increased the production cost and environmental pollution. It is crucial to implement rational use of chemical fertilizers under the premise of ensuring substantial grain yields. In order to achieve this goal, practical crop nutrient estimation methods and fertilization techniques are needed, especially for the N fertilizer that has the largest application amount. Compared to the laboratory testing method, rapid non-destructive estimation methods of CNS based on the canopy spectrum have more advantages in terms of timeliness, cost, practicality, and thus, have received extensive attention and research.

Based on the fertilizer level experiment conducted during 2018 and 2019, GNDVI and VARI were selected as the sensitive VIs for the methods UAVMC and PHONEP, respectively. The correlation coefficients between GNDVI and SPAD, VARI were 0.92 ** and 0.89 **, and the correlation coefficient between SPAD and VARI was 0.90 **. The estimation VIs for the three methods were significantly correlated with each other. The determination coefficients of the 0–90 cm soil nitrate nitrogen content estimation models for the UAVMC, SPAD, and PHONEP methods were 0.63, 0.54, and 0.81, respectively. Moreover, the PHONEP method had the smallest RMSE (9.8 mg/kg) in the estimation accuracy evaluation. Therefore, in terms of estimation accuracy, the PHONEP method was slightly higher than the other two methods. In addition, it was found that CNS estimation methods based on canopy spectrum were not suitable for crops under severe P deficiency due to the synergistic effects between different nutrients. At the same time, the estimation VIs of the three methods had a saturation response in the estimation of soil nitrate nitrogen content. The method of UAVMC is suitable for rapid CNS estimation for large-area crops, and SPAD and PHONEP are convenient for the random sampling and monitoring of crops. All three methods have good application prospects, but their estimation models still need to be further improved in order to achieve an efficient application in the precise management of N fertilizer.

Author Contributions: Conceptualization and the experiments design, H.L., Y.Z., and C.H.; conducted the field experiments, H.L., Y.L., and V.A.; analyzed the data, H.L. and Y.Z.; writing, H.L.; English language modification, V.A.; project administration, H.L. All authors have read and agreed to the published version of the manuscript.

Funding: This research was funded by the National Key Research and Development Program of China (2016YFD0800106) and (2016YFD0200307-06), and the National Natural Science Foundation of China (No. 41971262).

Acknowledgments: We would like to thank the Tianyao Company for their support of drone aerial photography technology.

Conflicts of Interest: The authors declare no conflict of interest.

References

1. Mogollón, J.; Lassaletta, L.; Beusen, A.; Van Grinsven, H.; Westhoek, H.; Bouwman, A. Assessing future reactive nitrogen inputs into global croplands based on the shared socioeconomic pathways. *Environ. Res. Lett.* **2018**, *13*. [CrossRef]
2. Hawkesford, M.J. Reducing the reliance on nitrogen fertilizer for wheat production. *J. Cereal Sci.* **2014**, *59*, 276–283. [CrossRef] [PubMed]
3. Raun, W.R.; Johnson, G.V. Improving nitrogen use efficiency for cereal production. *Agron. J.* **1999**, *91*, 357–363. [CrossRef]
4. Cameira, M.; Mota, M. Nitrogen related diffuse pollution from horticulture production—Mitigation practices and assessment strategies. *Horticulturae* **2017**, *3*, 25. [CrossRef]
5. Feng, W.; Zhu, Y.; Yao, X.; Tian, Y.; Zhuang, S.; Cao, W. Monitoring plant nitrogen accumulation dynamics with hyperspectral remote sensing in wheat. *Sci. Agric. Sin.* **2008**, *41*, 1937–1946.
6. Qi, Y.; Leng, Y.; Wang, M.; Hu, Y.; Bai, Y. Design of decision support system for soil testing and formula fertilization based on the intelligent agriculture. In Proceedings of the 2016 4th International Conference on Machinery, Materials and Information Technology Applications, Xi'an, China, 10–11 December 2016.

7. Li, G.; Zhu, L.; Li, J. Present status of research and application of non-destructive measurement of nitrogen nutrition diagnosis. *Heilongjiang Agric. Sci.* **2008**, *4*, 127–129.
8. Ali, M.; Al-Ani, A.; Eamus, D.; Tan, D.K. Leaf nitrogen determination using non-destructive techniques—A review. *J. Plant Nutr.* **2017**, *40*, 928–953. [CrossRef]
9. Shou, L.; Jia, L.; Cui, Z.; Chen, X.; Zhang, F. Using high-resolution satellite imaging to evaluate nitrogen status of winter wheat. *J. Plant Nutr.* **2017**, *30*, 1669–1680. [CrossRef]
10. Wright, D.L.; Rasmussen, V.P.; Ramsey, R.D.; Baker, D.J.; Ellsworth, J.W. Canopy reflectance estimation of wheat nitrogen content for grain protein management. *GISci. Remote Sens.* **2004**, *41*, 287–300. [CrossRef]
11. Eitel, J.; Long, D.; Gessler, P.; Smith, A. Using in-situ measurements to evaluate the new RapidEye™ satellite series for prediction of wheat nitrogen status. *Int. J. Remote Sens.* **2007**, *28*, 4183–4190. [CrossRef]
12. Jia, Y.; Su, Z.; Shen, W.; Yuan, J.; Xu, Z. UAV remote sensing image mosaic and its application in agriculture. *Int. J. Smart Home* **2016**, *10*, 159–170. [CrossRef]
13. Dash, J.; Pearse, G.; Watt, M. UAV multispectral imagery can complement satellite data for monitoring forest health. *Remote Sens.* **2018**, *10*, 1216. [CrossRef]
14. Hassan, M.A.; Yang, M.; Rasheed, A.; Yang, G.; Reynolds, M.; Xia, X.; Xiao, Y.; He, Z. A rapid monitoring of NDVI across the wheat growth cycle for grain yield prediction using a multi-spectral UAV platform. *Plant Sci.* **2019**, *282*, 95–103. [CrossRef] [PubMed]
15. Liu, C.; Wang, Z.; Chen, Z.; Zhou, L.; Yue, X.; Miao, Y. Nitrogen monitoring of winter wheat based on unmanned aerial vehicle remote sensing image. *Trans. Chin. Soc. Agric. Mach.* **2018**, *49*, 207–214.
16. Li, H.; Li, J.; Lei, Y.; Zhang, Y. Diagnosis of nitrogen nutrition of winter wheat and summer corn using images from digital camera equipped on unmanned aerial vehicle. *Chin. J. Eco Agric.* **2017**, *25*, 1832–1841.
17. Monostori, I.; Árendás, T.; Hoffman, B.; Galiba, G.; Gierczik, K.; Szira, F.; Vágújfalvi, A. Relationship between SPAD value and grain yield can be affected by cultivar, environment and soil nitrogen content in wheat. *Euphytica* **2016**, *211*, 103–112. [CrossRef]
18. Röll, G.; Hartung, J.; Graeff-Hönninger, S. Determination of plant nitrogen content in wheat plants via spectral reflectance measurements: Impact of leaf number and leaf position. *Remote Sens.* **2019**, *11*, 2794. [CrossRef]
19. Li, Z.; Jin, X.; Yang, G.; Drummond, J.; Yang, H.; Clark, B.; Li, Z.; Zhao, C. Remote sensing of leaf and canopy nitrogen status in winter wheat (*Triticum aestivum* L.) based on N-PROSAIL model. *Remote Sens.* **2018**, *10*, 1463. [CrossRef]
20. Liu, C.; Fang, Z.; Chen, Z.; Zhou, L.; Yue, X.; Wang, Z.; Wang, C.; Miao, Y. Nitrogen nutrition diagnosis of winter wheat based on ASD Field Spec3. *Trans. Chin. Soc. Agric. Eng.* **2018**, *34*, 162–169.
21. Jia, L.; Chen, X.; Zhang, F.; Buerkert, A.; Roemheld, V. Optimum nitrogen fertilization of winter wheat based on color digital camera images. *Commun. Soil Sci. Plant Anal.* **2007**, *38*, 1385–1394. [CrossRef]
22. Xia, S.; Zhang, C.; Li, H.; Zhang, Y.; Hu, C. Study on nitrogen diagnosis and fertilization recommendation of winter wheat using canopy digital images from cellphone camera. *Chin. J. Eco Agric.* **2018**, *26*, 538–546.
23. Kaur, N.; Singh, D. Android based mobile application to estimate nitrogen content in rice crop. *Int. J. Comput. Trends Technol. IJCTT* **2016**, *38*, 87–91. [CrossRef]
24. Intaravanne, Y.; Sumriddetchkajorn, S. Android-based rice leaf color analyzer for estimating the needed amount of nitrogen fertilizer. *Comput. Electron. Agric.* **2015**, *116*, 228–233. [CrossRef]
25. Padilla, F.M.; de Souza, R.; Peña-Fleitas, M.T.; Gallardo, M.; Giménez, C.; Thompson, R.B. Different responses of various chlorophyll meters to increasing nitrogen supply in sweet pepper. *Front. Plant Sci.* **2018**, *9*, 1752. [CrossRef]
26. Daughtry, C.S.T.; Walthall, C.L.; Kim, M.S.; Brown De Colstoun, E.; McMurtrey, J.E., III. Estimating corn leaf chlorophyll concentration from leaf and canopy reflectance. *Remote Sens. Environ.* **2000**, *74*, 229–239. [CrossRef]
27. Navarro, G.; Caballero, I.; Silva, G.; Parra, P.C.; Vázquez, Á.; Caldeira, R. Evaluation of forest fire on Madeira Island using Sentinel-2A MSI imagery. *Int. J. Appl. Earth Obs. Geoinf.* **2017**, *58*, 97–106. [CrossRef]
28. Gong, P.; Pu, R.; Biging, G.S.; Larrieu, M.R. Estimation of forest leaf area index using vegetation indices derived from Hyperion hyperspectral data. *IEEE Trans. Geosci. Remote Sens.* **2003**, *41*, 1355–1362. [CrossRef]
29. Qi, J.; Chehbouni, A.; Huete, A.; Kerr, Y.; Sorooshian, S. A modified soil adjusted vegetation index. *Remote Sens. Environ.* **1994**, *48*, 119–126. [CrossRef]

30. Chen, J. Evaluation of vegetation indices and a modified simple ratio for boreal applications. *Can. J. Remote Sens.* **1996**, *22*, 229–242. [CrossRef]
31. Fieuzal, R.; Sicre, C.M.; Baup, F. Estimation of corn yield using multi-temporal optical and radar satellite data and artificial neural networks. *Int. J. Appl. Earth Obs. Geoinf.* **2017**, *57*, 14–23. [CrossRef]
32. Li, X.; Xu, X.; Bao, Y.; Huang, W.; Luo, J.; Dong, Y.; Song, X.; Wang, J. Retrieving LAI of winter wheat based on sensitive vegetation index by the segmentation method. *Sci. Agric. Sin.* **2012**, *45*, 3486–3496.
33. Rondeaux, G.; Steven, M.; Baret, F. Optimization of soil-adjusted vegetation indices. *Remote Sens. Environ.* **1996**, *55*, 95–107. [CrossRef]
34. Roujean, J.L.; Breon, F.M. Estimating PAR absorbed by vegetation from bidirectional reflectance measurements. *Remote Sens. Environ.* **1995**, *51*, 375–384. [CrossRef]
35. Jordan, C.F. Derivation of leaf-area index from quality of light on the forest floor. *Ecology* **1969**, *50*, 663–666. [CrossRef]
36. Huete, A.R. A soil-adjusted vegetation index (SAVI). *Remote Sens. Environ.* **1988**, *25*, 295–309. [CrossRef]
37. Baresel, J.P.; Rischbeck, P.; Hu, Y.; Kipp, S.; Barmeier, G.; Mistele, B.; Schmidhalter, U. Use of a digital camera as alternative method for non-destructive detection of the leaf chlorophyll content and the nitrogen nutrition status in wheat. *Comput. Electron. Agric.* **2017**, *140*, 25–33. [CrossRef]
38. Pagola, M.; Ortiz, R.; Irigoyen, I.; Bustince, H.; Barrenechea, E.; Aparicio-Tejo, P.; Lamsfus, C.; Lasa, B. New method to assess barley nitrogen nutrition status based on image colour analysis: Comparison with SPAD-502. *Comput. Electron. Agric.* **2009**, *65*, 213–218. [CrossRef]
39. Karcher, D.E.; Richardson, M.D. Quantifying turfgrass color using digital image analysis. *Crop Sci.* **2003**, *43*, 943–951. [CrossRef]
40. Guerrero, J.M.; Pajares, G.; Montalvo, M.; Romeo, J.; Guijarro, M. Support vector machines for crop/weeds identification in maize fields. *Expert Syst. Appl.* **2012**, *39*, 11149–11155. [CrossRef]
41. Tucker, C.J. Red and photographic infrared linear combinations for monitoring vegetation. *Remote Sens. Environ.* **1979**, *8*, 127–150. [CrossRef]
42. Wang, Y.; Wang, D.; Zhang, G.; Wang, C. Digital camera-based image segmentation of rice canopy and diagnosis of nitrogen nutrition. *Trans. Chin. Soc. Agric. Eng.* **2012**, *28*, 131–136.
43. Beniaich, A.; Naves Silva, M.L.; Avalos, F.A.P.; Menezes, M.D.; Candido, B.M. Determination of vegetation cover index under different soil management systems of cover plants by using an unmanned aerial vehicle with an onboard digital photographic camera. *Semin. Cienc. Agrar.* **2019**, *40*, 49–66. [CrossRef]
44. Wang, Y.; Wang, D.; Zhang, G.; Wang, J. Estimating nitrogen status of rice using the image segmentation of GR thresholding method. *Field Crops Res.* **2013**, *149*, 33–39. [CrossRef]
45. Li, Y.; Chen, D.; Walker, C.N.; Angus, J.F. Estimating the nitrogen status of crops using a digital camera. *Field Crops Res.* **2010**, *118*, 221–227. [CrossRef]
46. Gitelson, A.A.; Viña, A.; Arkebauer, T.J.; Rundquist, D.C.; Keydan, G.; Leavitt, B. Remote estimation of leaf area index and green leaf biomass in maize canopies. *Geophys. Res. Lett.* **2003**, *30*. [CrossRef]
47. Zhang, X. *Crop Roots and Utilization of Soil Water*; China Meteorological Press: Beijing, China, 1999.
48. Gao, L.; Yang, G.; Li, H.; Li, Z.; Feng, H.; Wang, L.; Dong, J.; He, P. Winter wheat LAI estimation using unmanned aerial vehicle RGB-imaging. *Chin. J. Eco Agric.* **2016**, *24*, 1254–1264.
49. Yu, C.; Qin, J.; Xu, J.; Nie, H.; Luo, Z.; Cen, K. Straw combustion in circulating fluidized bed at low-temperature: Transformation and distribution of potassium. *Can. J. Chem. Eng.* **2010**, *88*, 874–880. [CrossRef]
50. Gu, Y.; Wylie, B.K.; Howard, D.M.; Phuyal, K.P.; Ji, L. NDVI saturation adjustment: A new approach for improving cropland performance estimates in the Greater Platte River Basin, USA. *Ecol. Indic.* **2013**, *30*, 1–6. [CrossRef]
51. De Souza, R.; Peña-Fleitas, M.T.; Thompson, R.B.; Gallardo, M.; Grasso, R.; Padilla, F.M. The use of chlorophyll meters to assess crop N status and derivation of sufficiency values for sweet pepper. *Sensors* **2019**, *19*, 2949. [CrossRef]

remote sensing

MDPI

Article

Evaluation of RGB, Color-Infrared and Multispectral Images Acquired from Unmanned Aerial Systems for the Estimation of Nitrogen Accumulation in Rice

Hengbiao Zheng, Tao Cheng, Dong Li, Xiang Zhou, Xia Yao, Yongchao Tian, Weixing Cao and Yan Zhu *

National Engineering and Technology Center for Information Agriculture, Key Laboratory for Crop System Analysis and Decision Making, Ministry of Agriculture, Jiangsu Key Laboratory for Information Agriculture, Jiangsu Collaborative Innovation Center for Modern Crop Production, Nanjing Agricultural University, Nanjing 210095, Jiangsu, China; 2015201019@njau.edu.cn (H.Z.); tcheng@njau.edu.cn (T.C.); lidongmath@163.com (D.L.); x_zhou1989@163.com (X.Z.); yaoxia@njau.edu.cn (X.Y.); yctian@njau.edu.cn (Y.T.); caow@njau.edu.cn (W.C.)

* Correspondence: yanzhu@njau.edu.cn; Tel.: +86-8439-6598; Fax: +86-25-8439-6672

Received: 3 April 2018; Accepted: 23 May 2018; Published: 25 May 2018

Abstract: Unmanned aerial system (UAS)-based remote sensing is one promising technique for precision crop management, but few studies have reported the applications of such systems on nitrogen (N) estimation with multiple sensors in rice (*Oryza sativa* L.). This study aims to evaluate three sensors (RGB, color-infrared (CIR) and multispectral (MS) cameras) onboard UAS for the estimation of N status at individual stages and their combination with the field data collected from a two-year rice experiment. The experiments were conducted in 2015 and 2016, involving different N rates, planting densities and rice cultivars, with three replicates. An Oktokopter UAS was used to acquire aerial photography at early growth stages (from tillering to booting) and field samplings were taken at a near date. Two color indices (normalized excess green index (NExG), and normalized green red difference index (NGRDI)), two near infrared vegetation indices (green normalized difference vegetation index (GNDVI), and enhanced NDVI (ENDVI)) and two red edge vegetation indices (red edge chlorophyll index ($CI_{red\ edge}$), and DATT) were used to evaluate the capability of these three sensors in estimating leaf nitrogen accumulation (LNA) and plant nitrogen accumulation (PNA) in rice. The results demonstrated that the red edge vegetation indices derived from MS images produced the highest estimation accuracy for LNA (R^2: 0.79–0.81, root mean squared error (RMSE): 1.43–1.45 g m^{-2}) and PNA (R^2: 0.81–0.84, RMSE: 2.27–2.38 g m^{-2}). The GNDVI from CIR images yielded a moderate estimation accuracy with an all-stage model. Color indices from RGB images exhibited satisfactory performance for the pooled dataset of the tillering and jointing stages. Compared with the counterpart indices from the RGB and CIR images, the indices from the MS images performed better in most cases. These results may set strong foundations for the development of UAS-based rice growth monitoring systems, providing useful information for the real-time decision making on crop N management.

Keywords: UAS; multiple sensors; vegetation index; leaf nitrogen accumulation; plant nitrogen accumulation

1. Introduction

Nitrogen (N) plays a key role in crop growth and yield formation. Rice (*Oryza sativa* L.) is one of the largest consumers of N fertilizers [1]. Crop nitrogen accumulation (NA), as a product of nitrogen content (NC) and biomass includes not only information on the N status, but also canopy capacity during crop growth [2,3]. Therefore, the estimation of leaf nitrogen accumulation (LNA) and plant nitrogen accumulation (PNA) is not only useful for evaluating crop production capability and predicting grain quality, but also supporting diagnosis of N status in crop production.

Remote sensing (RS) techniques have been proved to be a promising approach for crop growth monitoring [4], nutrition diagnosis [5] and yield prediction [6] in different crops. RS data is widely used to monitor crop N status, including satellite imagery [7,8], aerial photographs [9] and canopy reflectance spectra [5,10]. For example, Yao et al. [5] used ground-based hyperspectral data to detect the leaf N concentration of various growth stages in winter wheat. Huang et al. [8] used FORMOSAT-2 satellite data to estimate aboveground biomass and plant N uptake at the panicle initiation stage of rice in northeast China. Boegh et al. [9] used airborne multispectral (MS) data to quantify canopy N concentration in agriculture. Although ground-based RS sensors are easy to operate and can be used to obtain high estimation accuracy for crop growth parameters, they are costly and labor intensive. Satellite imagery often provides insufficient spatial resolution to monitor crop growth status for smallholders and is easily affected by cloud conditions. Manned airborne platforms can be used to obtain imagery with high spatial and temporal resolutions, but they are limited by high operational complexity and costs.

Unmanned aerial systems (UAS) have become prevalent and offer numerous advantages in precision agriculture, such as the ultra-high spatial resolution (e.g., centimeters), relatively low operational costs and the near real-time image acquisition [11,12]. However, they also have a significantly low payload capacity, so light-weight compact sensors are required, and these sensors have been developed and mounted on various UAS for remote sensing applications [13–16]. The most affordable sensors are off-the-shelf digital cameras with red, green and blue bands. One of the successful applications of UAS in agricultural management was reported by Hunt et al. [11], who used a digital camera attached to a model aircraft to estimate the biomass of corn, alfalfa and soybean. They found a linear correlation between biomass and the normalized green–red difference index (NGRDI) derived from the RGB images. Zhu et al. [17] investigated the possibility of using digital imagery collected by a UAS to monitor N status in paddy fields. Córcoles et al. [18] measured canopy cover in an onion crop with RGB images from UAS and determined the relationship between canopy cover and leaf area index (LAI). However, reflectance in the near infrared (NIR) wavelengths varied most across the growing season due to increasing biomass at the canopy level [19], and numerous vegetation indices (VIs) proposed for biomass or LAI estimation included NIR bands [20–22]. Therefore, a few studies have attempted to modify RGB cameras by replacing the red or blue channel with a NIR filter to obtain color near-infrared (CIR) images. For instance, Hunt et al. [23] developed a digital CIR camera system and found a good correlation between LAI and green normalized difference vegetation index (GNDVI) from the imagery of winter wheat. This CIR camera system has also been used by Hunt et al. [24] to estimate the biomass of winter cover crops. Lebourgeois et al. [25] used a UAS equipped with two digital cameras (RGB and CIR cameras) to investigate sugarcane N status and determined the optimal VI to estimate leaf and canopy N content. Recently, mounting MS cameras has become a new alternative approach for precision agriculture. More complex analysis has been done using MS cameras with more spectral bands, allowing advanced vegetation studies. They were applied to water stress detection [26], disease detection [27] and vigor monitoring [28,29]. The UAS images in the aforementioned studies have been used to estimate the agronomic parameters LAI [30–32] and biomass [24,33]. N status, as one of the most important agronomic parameters in precision farming, needs to be addressed with UAS due to the low efficiency of other RS techniques.

Whether a UAS platform could be applied to monitor N status needs to be evaluated. Furthermore, the performance of multiple sensors onboard UAS for the same purpose has rarely been compared. Von Bueren et al. [34] evaluated the spectral quality of four different types of optical UAS-based sensors (RGB camera, CIR camera, six band MS camera and a high resolution spectrometer) and found that a MS camera can collect imagery with high radiometric quality in the red edge region. Although RGB and CIR cameras have been widely applied in precision agriculture, the stability of the models established with indices from RGB and CIR images needs to be validated with the test set. Most importantly, in order to better apply UAS to precision agriculture, we need to clarify how far these cameras can be used during the whole season or at a certain growth stage, and how to choose a proper camera for N status monitoring in terms of cost, data processing efficiency and study area size. Therefore, the objective of the present study was to evaluate the performance of different sensors onboard UAS on LNA and PNA estimation in rice. The anticipated results would provide guidance on how to choose a proper camera for N accumulation estimation, and lay foundations for developing non-destructive and rapid monitoring of N status with UAS in rice crops.

2. Materials and Methods

2.1. Experimental Design

Two field experiments were designed, involving different N rates, planting densities and rice cultivars. All the experiments were conducted in the experimental station of the National Engineering and Technology Center for Information Agriculture (NETCIA) located in Rugao city, Jiangsu province, China (120°45′E, 32°16′N) (Figure 1). In 2015, one japonica rice cultivar Wuyunjing24 (V1) and one indica rice cultivar Yliangyou1 (V2) were seeded at day of year (DOY) 136 and transplanted into the paddy field at DOY 166. Four N rates (0 (N0), 100 (N1), 200 (N2) and 300 (N3) kg N ha^{-1} as urea) were applied with one density (0.30 m × 0.15 m) for the minimum and maximum N rates and two densities (0.30 m × 0.15 m and 0.50 m × 0.15 m) for the intermediate N rates. The N fertilizers were applied in the form of urea: 40% as basal fertilizer before transplanting, 10% at the tillering stage, 30% at the jointing stage and 20% at the booting stage. The plot size was 30 m^2 with 6 m in length and 5 m in width. In 2016, the experiment was conducted with the same plot size and rice cultivars. The two rice cultivars were seeded at DOY 138 and transplanted with two planting densities (0.30 m × 0.15 m and 0.50 m × 0.15 m) into the paddy field at DOY 168. Three N rates (0 (N0), 150 (N1) and 300 (N2) kg N ha^{-1} as urea) were applied with 40% as basal fertilizer before transplanting, 20% at the tillering stage, 20% at the jointing stage and 20% at the booting stage.

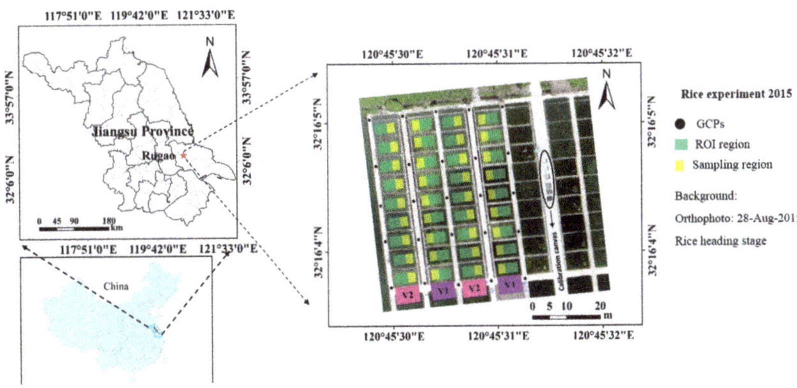

Figure 1. Study site: rice experiment at the NETCIA experimental station in 2015. GCPs, ground control points used for band to band registration.

2.2. Field Measurements

The UAS flight dates were adjusted to the field sampling dates as much as weather conditions allowed (Table 1). Due to the poor weather conditions, we could only take high quality UAS images on 5 August, 2015. Although this was three days after the fertilization application, we believe significant N uptake by the rice plants has not yet occurred [35]. For the field destructive sampling, 3 hills of rice plants were randomly selected from the sampling region of each plot and separated into leaves and stems. All the samples were oven-dried for 30 min at 105 °C and later at 80 °C to a constant weight, then weighted, ground and stored in plastic bags for chemical analysis. The total N content in the leaf and stem tissues was determined using the micro-Keldjahl method [36]. The N accumulation (g m^{-2}) of leaves (LNA) and plants (PNA) was calculated as the product of N content (%) and dry biomass (t ha^{-1}) of leaves and whole aboveground plants, respectively.

Table 1. Synthesis of field management and data acquisition calendar.

Year	Seedling	N Fertilizer	UAS Flight Date	Field Sampling Date	Growth Stage	Image Acquisition		
						MS	RGB	CIR
2015	16 May	23 June	21 July	21 July	Tillering	√	√	√
		2 August	5 August	31 July	Jointing	√	√	√
		14 August	14 August	15 August	Booting	√	√	√
2016	18 May	25 June	21 July	21 July	Tillering	-	√	√
		4 August	6 August	6 August	Jointing	√	√	√
		14 August	14 August	14 August	Booting	√	√	√

Note: MS, RGB and CIR represent multispectral, red–green–blue and color infrared.

2.3. *UAV Campaigns and Sensors*

The UAV used in this study was the Mikrokopter OktoXL [37], an eight-rotor aircraft with a maximum payload capacity of 2.5 kg. This UAV has a flight duration of 8–25 min, depending on the battery and actual payload. Three cameras were mounted onboard the UAV separately for image collection and their technical specifications were listed in Table 2. The Tetracam mini-MCA6 (Tetracam Inc., Chatsworth, CA, USA) MS camera has six channels and was evaluated in the literature for other purposes [32,38–40]. The camera has user configurable band pass filters (Andover Corporation, Salem, NH, USA) of 10-nm full-width at half-maximum and center wavelengths at blue (490 nm), green (550 nm), red (680 nm), red edge (720 nm), NIR1 (800 nm) and NIR2 (900 nm). The camera was run on a 3 s shutter release interval and collected images in a 10 bit RAW format. The other two cameras collected true color and CIR images in JPEG format, respectively. The true color camera was a Canon 5D Mark III (Canon Inc., Tokyo, Japan) with a 22.1 megapixel (MP) CMOS sensor. The CIR camera was a NIR–green–blue camera, which was a modified Canon PowerShot SX260 with a 12.1 MP CMOS sensor (www.maxmax.com). After filter modification, the red channel became near-infrared and the other two channels remained the same. These two cameras were set to continuous data capturing at 1 frame per second (fps), with an exposure time manually set for each flight according to light conditions.

Table 2. Technical specifications of the three cameras and image captures from the UAS.

Camera	Version	Field of View	Image Size	Altitude (m)	Coverage (Single Image) (ha)	Pixel Size (on the Ground) (mm)
RGB	Canon 5D Mark III	74° × 53°	3840 × 5760	50	0.75	13
Color infrared	Canon PowerShot SX260	72° × 52°	3000 × 4000	100	1.41	36
Multispectral	Mini-MCA	38° × 31°	1024 × 1280	100	0.38	54

The UAS was flown over the paddy field on three dates at the heights of 50 m and 100 m above ground level during the 2015 and 2016 growing seasons (Table 2). Flight speed and route planning were fixed during the whole season. Additionally, all flights were carried out in stable ambient light conditions between 11:00 a.m. and 1:30 p.m. when the sun elevation angle was in the range of 56–72°.

2.4. *Image Processing*

2.4.1. Image Preprocessing

After the flights, the MS images were downloaded from the memory cards and processed in the IDL/ENVI environment (Exelis Visual Information Solutions, Boulder, CO, USA). The image preprocessing workflows followed Kelcey and Lucieer [41]. In short, the following three corrections were applied: (i) noise reduction using dark current imagery; (ii) lens vignetting correction based on spatially dependent correction factors; and (iii) removal of lens distortion with a modified Brown–Conrady model. To reduce the band-to-band misalignment, the six bands were co-registered with a total of 25 ground control points marked on the concrete roads within the study area (Figure 1). Six spectrally flat calibration canvases (1.2 × 1.2 m) with reflectance intensities at 3%, 6%, 12%, 22%, 48% and 64% were placed within the UAS flight overpass. With these calibration canvases, the digital number (DN) values of the images were transformed into reflectance values by applying an empirical line correction method [37,42]. The empirical line correction coefficients established between the convoluted ASD-FieldSpec4 and the six acquired mini-MCA spectral bands were then applied per pixel to the MS images. The reflectance of each plot was represented by the average of the reflectance values over the non-sampling area of the plot.

RGB and CIR cameras are consumer-grade sensors and there is no rigorous pre-processing procedure for them. Dark offset imagery was generated for the two cameras in a totally dark environment and there is no need to do noise correction due to extremely low noise. The effect of lens vignetting and distortion could be ignored since the RGB and CIR images covered a large area and the photos with the sampling plots in the nadir area were selected for subsequent analysis. The DN values of the RGB and CIR images were not radiometrically corrected to surface reflectance using the calibration canvas due to the complexity of determining the spectral sensitivity for each channel. As the RGB and CIR images had higher spatial resolutions than the MS image, they were down-sampled to the same spatial resolution as that of the MS image (Figure 2). The DN value of each plot was represented by the average of DN values over the non-sampling area of the plot.

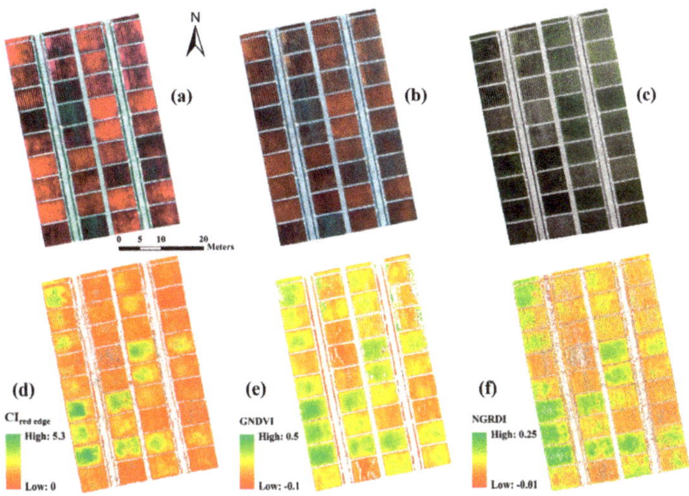

Figure 2. UAS images acquired from three cameras: MS image (**a**), CIR image (**b**) and RGB image (**c**) and their corresponding VI maps: red edge chlorophyll index ($CI_{red\ edge}$) map (**d**), GNDVI map (**e**) and NGRDI map (**f**). Study area (36 plots) was extracted from the three images. MS image is displayed with NIR, G and B bands.

2.4.2. Calculation of Vegetation Indices

Given the specific band configurations of the three cameras, six VIs were examined for the estimation of LNA and PNA (Table 3). From the RGB images, we calculated two color indices, the normalized excess green index (NExG) [43] and NGRDI [44], because NExG has been proved to possess the ability to quantify crop responses under different conditions [45], and NGRDI was originally proposed for vegetation fraction estimation and found to be related to crop biomass before canopy closure with RGB imagery from UAS [11]. From the CIR images, we chose two NIR VIs, GNDVI [46] and enhanced NDVI (ENDVI), because GNDVI was shown to be linked to chlorophyll concentration, and ENDVI was recommended by the company that converted the cameras (www.maxmax.com) and also shown to detect vegetation vigor well [45]. For the MS images, we had more options for VIs due to the multiple bands. Except for the color indices and NIR VIs, two red edge VIs ($CI_{red\ edge}$ [47] and DATT [48]) were employed for N accumulation estimation, because they both had a strong capability in chlorophyll contents estimation in higher plants. The VIs from the MS images were calculated with reflectance values and those from the RGB and CIR images were with DN values.

Table 3. Spectral vegetation indices evaluated in this study.

Index	Name	Formula	References	Camera
NExG	Normalized Excess green index	(2*G − R − B)/(G + R + B)	[43,45]	MS, RGB
NGRDI	Normalized green-red difference index	(G − R)/(G + R)	[21,44]	MS, RGB
GNDVI	Green normalized difference vegetation index	(NIR − G)/(NIR + G)	[46]	MS, CIR
ENDVI	Enhanced normalized difference vegetation index	(NIR + G − 2*B)/(NIR + G + 2*B)	(www.maxmax.com)	MS, CIR
$CI_{red\ edge}$	Red edge chlorophyll index	NIR/RE-1	[47]	MS
DATT	DATT	(NIR − RE)/(NIR − R)	[48]	MS

Note: For RGB and CIR images, NIR, R, G and B are the radiometric normalized pixel values of the NIR, red, green and blue band, respectively. For multispectral images, NIR, RE, R, G and B are the reflectance values of the near infrared, red edge, red, green and blue bands, respectively.

2.5. Data Analysis

The correlations between agronomic parameters and VIs were analyzed using MATLAB R2010b software (The MathWorks, Inc., Natick, MA, USA). The experimental data collected in 2015 were used to develop the regression models and data collected in 2016 were subsequently used to validate the regression models. Generally, the estimation performance of the model was evaluated by comparing the differences in coefficient of determination (R^2) and root mean square error (RMSE). The model performance was more accurate with an R^2 near to 1, and an RMSE near to 0.

3. Results

3.1. Variation of Agronomic Variables over the Three Growth Stages

Figure 3 shows the overall trend of leaf or plant N status and biomass from the 2015 field dataset. Leaf N concentration kept decreasing from the tillering stage (2.55%) to the booting stage (2.28%). Plant N concentration decreased from tillering (1.85%) to jointing (1.37%) and remained stable following the application of the second top-dressing fertilizer. Leaf biomass increased consistently as the rice plants developed (from 0.62 t ha^{-1} to 2.52 t ha^{-1}). Plant biomass increased from 1.27 t ha^{-1} at the tillering stage to 2.95 t ha^{-1} at the jointing stage and then doubled after the second top-dressing. LNA and PNA were close at the tillering and jointing stages, but became significantly different after the second fertilization.

3.2. Comparison between Counterpart Indices from MS and RGB Images

Two counterpart indices (NExG and NGRDI), computed from the MS and RGB images, were employed to evaluate the capability of these two cameras in LNA and PNA estimation. Figure 4 shows that NExG-RGB had comparable performance to NExG-MCA in LNA and PNA estimation at the tillering and jointing stages. With regard to the pooled datasets across all growth stages, NExG-RGB performed slightly better than NExG-MCA. NGRDI-RGB was superior to NGRDI-MCA at individual stages, explaining 19% and 15% more variability of LNA and PNA for all growth stages, respectively.

Obviously, the relationships between color indices and N accumulation exhibited a higher scattering of data points for the booting stage than for the tillering and jointing stages. Therefore, the correlation for the combination of the tillering and jointing stages (LNA: R^2 = 0.71; PNA: R^2 = 0.71) was significantly higher than that for all three stages (LNA: R^2 = 0.44; PNA: R^2 = 0.56). Additionally, color indices from both types of images exhibited the best performance in LNA and PNA estimation for the jointing stage (Figure 4).

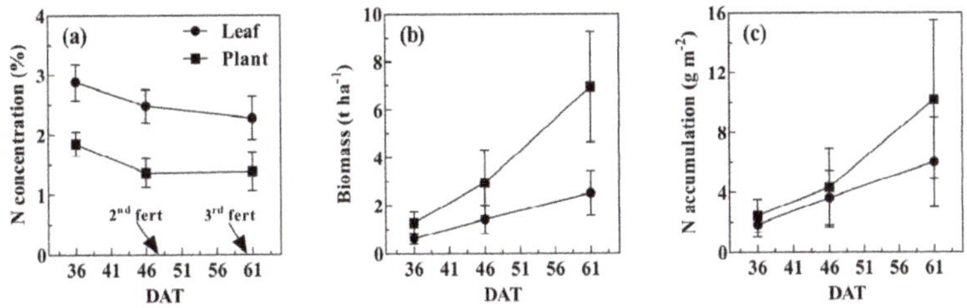

Figure 3. Average measurements of leaf or plant N concentration (**a**), biomass (**b**) and N accumulation (**c**) derived from the 2015 field dataset. Vertical bars denote standard deviation; arrows indicate dates of N top-dressing fertilization. DAT denotes days after transplantation.

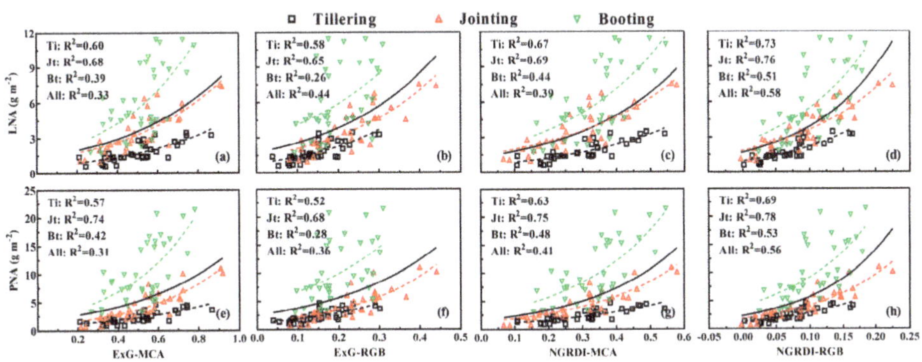

Figure 4. Leaf N accumulation (LNA, g m^{-2}) and plant N accumulation (PNA, g m^{-2}) plotted against vegetation indices from RGB and MS images: NExG-MCA (**a**,**e**), NExG-RGB (**b**,**f**), NGRDI-MCA (**c**,**g**) and NGRDI-RGB (**d**,**h**). The black solid line is for pooled data and the dashed lines are for individual stages. All regressions are statistically significant ($p < 0.01$).

3.3. Comparison between Counterpart Indices from MS and CIR Images

To compare the capability of the MS and CIR cameras in estimating N accumulation, two counterpart VIs (GNDVI and ENDVI) were computed from the MS and CIR images and their performance is shown in Figure 5. Compared with GNDVI-CIR, GNDVI-MCA performed slightly better in the LNA and PNA estimation for individual stages and exhibited more consistent relationships with LNA (R^2: 0.78–0.81) and PNA (R^2: 0.72–0.83) across the three stages (Figure 5a,e). In contrast, when the data from all stages were pooled, GNDVI-CIR explained 10% and 7% more variation in LNA and PNA, respectively (Figure 5b,f). ENDVI-MCA was superior to ENDVI-CIR in LNA and PNA estimation across all three growth stages. Among the three growth stages, the best performance of LNA and PNA estimation was observed for the jointing stage with all the evaluated VIs.

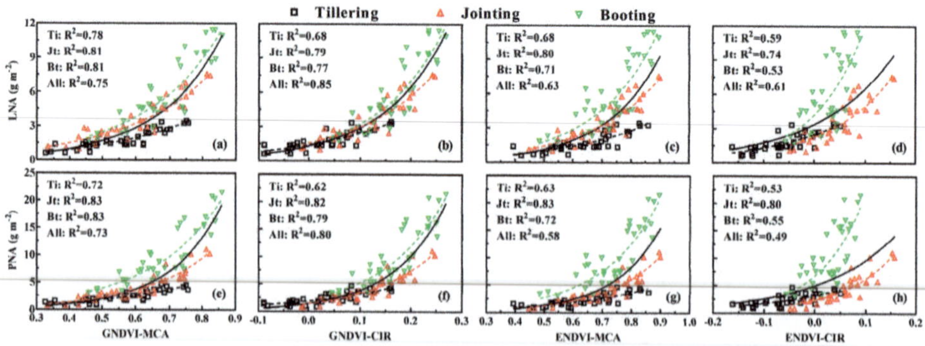

Figure 5. Leaf N accumulation (LNA, g m^{-2}) and plant N accumulation (PNA, g m^{-2}) plotted against vegetation indices from CIR and MS images: GNDVI-MCA (**a**,**e**), GNDVI-CIR (**b**,**f**), ENDVI-MCA (**c**,**g**) and ENDVI-CIR (**d**,**h**). The black solid line is for pooled data and the dashed lines are for individual stages. All regressions are statistically significant ($p < 0.01$).

3.4. Evaluation of Red Edge Indices from Multispectral Images

Figure 6 exhibits that $CI_{red\ edge}$ and DATT performed consistently well in LNA (R^2: 0.82–0.89) and PNA (R^2: 0.82–0.88) estimation across different growth stages. An all-stage model of DATT could explain 89% of LNA variability and 86% of PNA variability, which was slightly higher than that of $CI_{red\ edge}$. However, the relationship of LNA and PNA with $CI_{red\ edge}$ was linear, but that with DATT was nonlinear. Compared with the best performing index from the CIR images (GNDVI-CIR), DATT was slightly superior to GNDVI in LNA and PNA estimation. Specially, DATT could explain more than 30% of the variation in LNA and PNA compared to the best performing indices from RGB images (NGRDI-RGB).

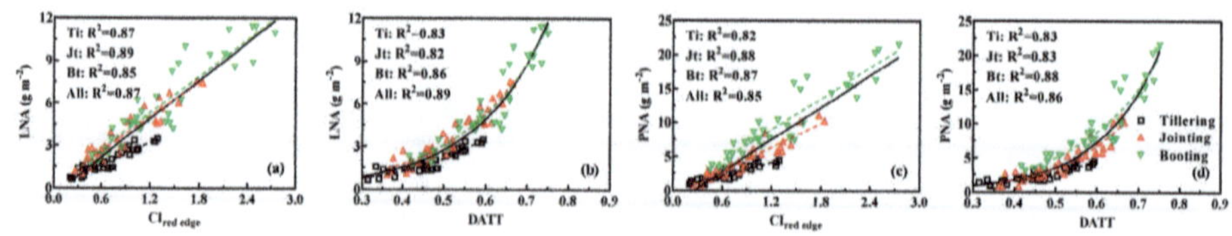

Figure 6. Leaf N accumulation (LNA, g m^{-2}) and plant N accumulation (PNA, g m^{-2}) plotted against red edge indices from MS images: $CI_{red\ edge}$ (**a**,**c**) and DATT (**b**,**d**). The black solid line is for pooled data and the dashed lines are for individual stages. All regressions are statistically significant ($p < 0.01$).

3.5. Model Validation

The derived regression models from the calibration datasets were validated with the test set for all models (Table 4). Color indices (NExG and NGRDI) from the MS and RGB images exhibited low estimation accuracies for LNA ($R^2 \leq 0.58$ and RMSE ≥ 0.92 g m^{-2}) and PNA ($R^2 \leq 0.66$ and RMSE $\geq$ 1.18 g m^{-2}) for both the individual stages and the all-stage group. The highest accuracy for LNA and PNA estimation was produced by NGRDI-RGB ($R^2 = 0.52$ and RMSE = 1.61 g m^{-2}) and NGRDI-MCA ($R^2 = 0.47$ and RMSE = 2.44 g m^{-2}) at the jointing stage, respectively. Compared with the counterpart indices from the RGB images, VI-MCA performed better in most cases (Figures 7 and 8). However, the estimation accuracy improved significantly when only the tillering and jointing stages were considered. The NGRDI and NExG yielded satisfactory performance in LNA ($R^2 = 0.64$ and RMSE = 1.16 g m^{-2}) and PNA ($R^2 = 0.69$ and RMSE = 1.95 g m^{-2}), respectively (Figure 9a,b).

Table 4. Accuracy assessment with the determination coefficient (R^2) and root mean squared error (RMSE) values for the estimation of rice leaf N accumulation (LNA, g m^{-2}) and plant N accumulation (PNA, g m^{-2}) with vegetation indices in 2016.

Indices	LNA (g m^{-2})										PNA (g m^{-2})									
	Tillering		Jointing		Booting		Jointing & Booting		All Three Stages		Tillering		Jointing		Booting		Jointing & Booting		All Three Stages	
	R^2	RMSE	R^2	RMSE	R^2	RMSE	R^2	RMSE	R^2	RMSE	R^2	RMSE	R^2	RMSE	R^2	RMSE	R^2	RMSE	R^2	RMSE
NExG-MCA	-	-	0.29	1.74	0.21	3.41	0.16	5.16	-	-	-	-	0.33	3.09	0.19	5.06	0.17	4.63	-	-
NExG-RGB	0.42	1.04	0.34	4.02	0.21	4.70	0.19	6.14	0.30	**2.04**	0.66	**1.18**	0.41	6.27	0.19	9.06	0.16	8.92	0.29	7.44
NGRDI-MCA	-	-	0.41	1.74	0.36	2.58	0.30	2.29	-	-	-	-	0.47	2.44	0.33	4.81	0.31	3.83	-	-
NGRDI-RGB	0.58	0.92	0.52	1.61	0.51	2.50	0.41	2.66	0.50	2.22	0.58	1.62	0.56	2.60	0.46	4.63	0.37	4.53	0.49	3.76
GNDVI-MCA	-	-	0.78	**1.25**	0.79	1.36	0.69	1.49	-	-	-	-	0.84	1.67	0.77	**2.56**	0.71	2.41	-	-
GNDVI-CIR	**0.71**	**0.75**	0.70	2.34	0.55	2.67	0.46	2.50	**0.65**	2.08	**0.74**	1.60	0.74	2.39	0.52	5.13	0.45	3.73	**0.65**	**3.12**
ENDVI-MCA	-	-	0.64	1.33	0.63	1.80	0.49	2.02	-	-	-	-	0.71	2.46	0.61	3.21	0.51	3.52	-	-
ENDVI-CIR	0.55	1.76	0.37	2.04	0.14	3.82	0.10	2.34	0.45	2.18	0.60	2.77	0.40	4.25	0.12	7.03	0.10	4.10	0.45	3.75
$CI_{red\ edge}$	-	-	0.85	1.53	0.82	**1.33**	0.79	**1.43**	-	-	-	-	0.88	**1.15**	**0.83**	2.72	0.81	2.38	-	-
DATT	-	-	**0.86**	1.90	**0.83**	1.38	**0.81**	1.45	-	-	-	-	**0.90**	1.69	0.82	2.58	**0.84**	**2.27**	-	-

Note: The number in bold denotes the maximal R^2 or minimal RMSE in each column.

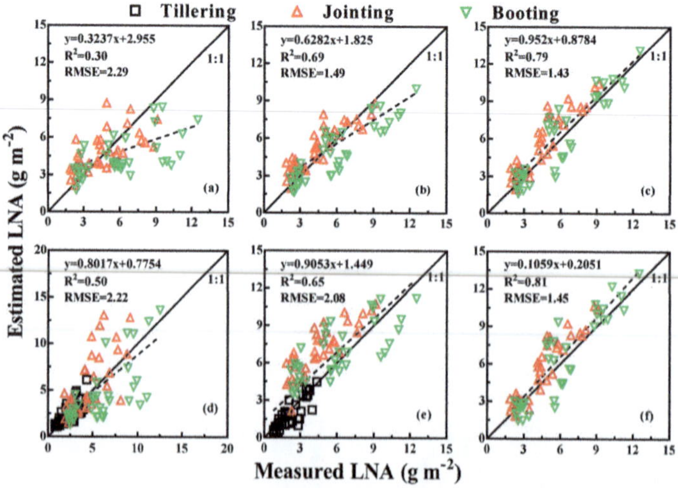

Figure 7. The 1:1 relationship between the estimated and measured leaf N accumulation (LNA, g m^{-2}) in rice based on NGRDI-MCA (**a**), GNDVI-MCA (**b**), $CI_{red\ edge}$ (**c**), NGRDI-RGB (**d**), GNDVI-CIR (**e**), and DATT (**f**). Only RGB and CIR images have tillering data. The solid line indicates the 1:1 line and the dashed line indicates the regression line.

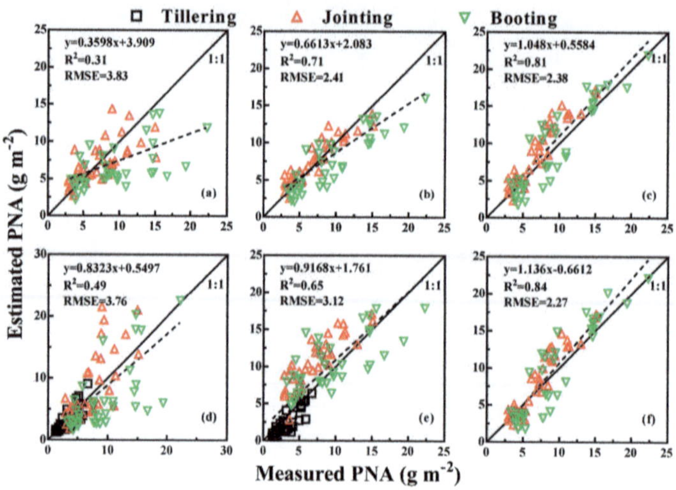

Figure 8. The 1:1 relationship between the estimated and measured plant N accumulation (PNA, g m^{-2}) in rice based on NGRDI-MCA (**a**), GNDVI-MCA (**b**), $CI_{red\ edge}$ (**c**), NGRDI-RGB (**d**), GNDVI-CIR (**e**), and DATT (**f**). Only RGB and CIR images have tillering data. The solid line indicates the 1:1 line and the dashed line indicates the regression line.

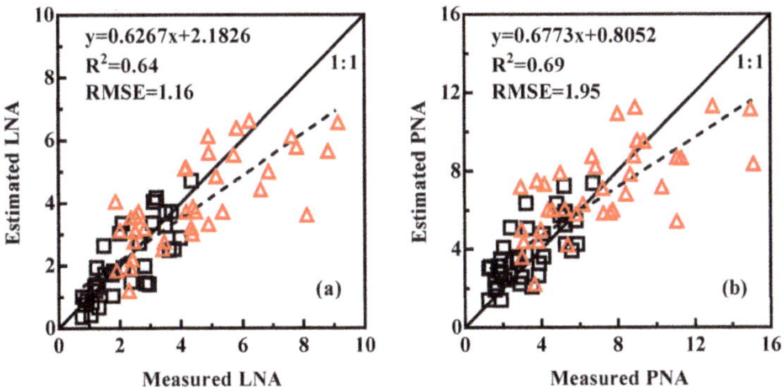

Figure 9. The 1:1 relationship between the estimated and measured leaf N accumulation (LNA, g m^{-2}) and plant N accumulation (PNA, g m^{-2}) in rice based on two-stage (tillering and jointing) model using color indices derived from the RGB images: NGRDI-LNA model (**a**) and NExG-PNA model (**b**). The solid line indicates the 1:1 line and the dashed line indicates the regression line.

The GNDVI-CIR and ENDVI-CIR models performed poorly for both individual stages and the all-stage group. Only GNDVI-CIR obtained moderate estimation accuracy for LNA (R^2 = 0.71 and RMSE = 0.75 g m^{-2}) and PNA (R^2 = 0.65 and RMSE = 3.12 g m^{-2}) at the jointing stage and in the all-stage group, respectively. However, compared with VI-CIR, the counterpart indices from the MS images performed significantly better. GNDVI-MCA produced the highest estimation accuracy of LNA (R^2 = 0.79 and RMSE = 1.36 g m^{-2}) and PNA (R^2 = 0.84 and RMSE = 1.67 g m^{-2}) for the booting and jointing stages, respectively. Moreover, the two-stage group (booting and jointing) model of GNDVI-MCA yielded satisfactory performance in LNA and PNA estimation (Figures 7b and 8b).

Obviously, the estimation accuracy of $CI_{red\ edge}$ and DATT was comparable and highest among all the evaluated indices across all growth stages except for the jointing stage in LNA estimation (Table 4). The two-stage model (jointing and booting) based on the $CI_{red\ edge}$ exhibited satisfactory performance in LNA (R^2 = 0.79 and RMSE = 1.43 g m^{-2}) and PNA (R^2 = 0.81 and RMSE = 2.38 g m^{-2}) estimation (Figures 7c and 8c). Figure 10 shows the spatial distribution of LNA and PNA estimated by $CI_{red\ edge}$ in the study area for the jointing and booting stages. These figures highlight an increase of N accumulation between the two dates, and the spatial variability in N accumulation across N treatments.

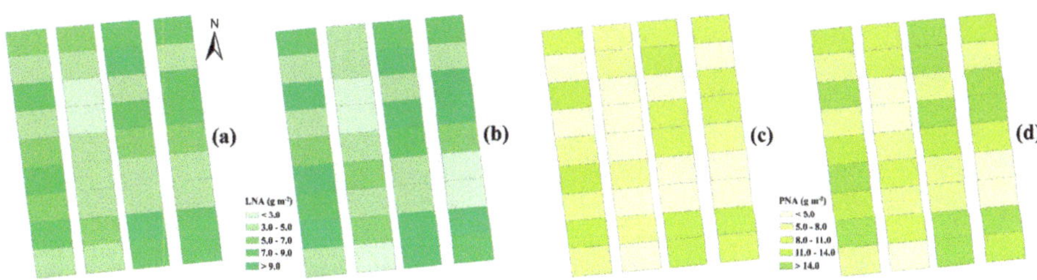

Figure 10. N accumulation maps using the derived generic relationship between $CI_{red\ edge}$ and N accumulation at different growth stages, LNA at jointing stage (**a**) and booting stage (**b**); PNA at jointing stage (**c**) and booting stage (**d**).

4. Discussion

4.1. Feasibility of Fitting a Single VI-LNA or VI-PNA Model for All Growth Stages

Agronomic parameters (e.g., biomass and plant N uptake) were often estimated at the individual growth stage or stage groups separated by pre-heading and post-heading, and the model for the early stages performed better than the latter or the whole season [13,49–51]. In this study, we established both stage groups and individual stage models for LNA and PNA estimation. Color indices (NExG and NGRDI) from RGB images performed poorly in N accumulation estimation with pooled data from different stages, but the performance improved significantly when excluding the booting stage (Figure 9). Similar results were found in wheat plant N uptake [52] and barley biomass [13] estimation, because the reflectance in the visible region was easily saturated when the chlorophyll content was relatively higher at the booting stage [19]. The NIR VIs from the CIR images yielded better performance with the all-stage model than the individual-stage model, especially for GNDVI-CIR (Figures 7c and 8c). Hunt et al. [24] also used GNDVI derived from CIR images to estimate winter cover crop biomass, but the correlations were not high due to much greater biomass variation within experimental strips than among strips.

For MS images, two red edge indices ($CI_{red\ edge}$ and DATT) produced quite satisfactory performance at the individual stage and stage groups. As shown in Figures 7 and 8, the estimates of the red edge indices with two-stage models were closest to the 1:1 line, because N accumulation was dominated by biomass and biomass could be more accurately estimated with red edge indices than the NIR or visible VIs at a high canopy density [49,53,54]. Moreover, this MiniMCA camera could deliver MS imagery of high radiometric quality in the red edge region [34]. Considering the linear relationship with LNA and PNA, $CI_{red\ edge}$ could be considered a suitable indicator for rice N accumulation estimation from the UAV platform.

4.2. Performance of Counterpart Indices from Different Sensors

In this study, the counterpart VIs from MS images performed significantly better than that from RGB or CIR images (Table 4). That is partly due to the lack of radiometric correction for RGB and CIR images. The established models from RGB and CIR images had poorer stability in DN values than reflectance between the two years, because the DN values were significantly affected by the exposure sets on the digital cameras based on overall light intensity [11].

Although the DN-based models were presented in this study, they were also compared with the reflectance-based models, and the latter did not show a consistent improvement with radiometric calibration (data not shown). This is probably due to the inaccuracy in the spectral response function and the quality of the RGB and CIR cameras, which were not as well calibrated in-house as the MS cameras. In addition, radiometric calibration was difficult for the RGB and CIR cameras in this study due to the complexity of obtaining their spectral response functions for each channel. The use of a calibration canvas may not be practical when multiple flights are needed to cover a large area and solar illumination varies between flights [24]. On the other hand, the RGB and CIR cameras possessed much broader bandwidth (over 50 nm) than that of the MS camera (10 nm). Previous studies have also shown that narrow-band vegetation indices were superior to broad bandwidth indices in LAI [55] and N [56] estimation. Future work could focus on improving the stability and transitivity of models established with RGB and CIR images through different fields and years.

These VIs are structural indices, which are more related to biomass than N concentration, thus we found the VIs were weakly correlated with N concentration (data not shown), which corresponds with the findings in other studies [2,52]. Because N concentration monitoring is affected by many factors, such as canopy structural parameters (e.g., LAI, biomass). Knyazikhin et al. [57] proposed directional area scattering factor (DASF) to eliminate the influence of canopy structural parameters with hyperspectral data. However, DASF could not be obtained due to the limited bands in these sensors. N accumulation composed of N concentration and biomass, contains information on canopy structure and N status, thus it can be taken as a good indicator for a N nutrition assessment [2,58,59]. However, the only stress for rice growth in this study was N deficiency, because other practical management was the same for all the plots. Treatments within N stress were labeled with low N concentration and biomass, thus the low N accumulation provides a real statement of the N deficit.

4.3. Choice of an Appropriate Camera for Precision Agriculture

UAS have proven to be an ideal technique in precision agriculture for crop growth mapping [60,61] and monitoring [13,14,33]. Simultaneously, many aspects of sensor selection should be taken into consideration, such as band configuration and price. Digital cameras (both RGB and CIR cameras) are easy to use and are affordable for most researchers. Additionally, RGB and CIR images could be used after being downloaded from a memory card without any preprocessing (e.g., band registration) and a number of sophisticated commercial softwares (e.g., Agisoft Photoscan and Pix4D) have been developed to process the RGB and CIR images. These advantages make it more efficient to process the image data and easier to cover a large experimental area [34]. Furthermore, digital cameras could collect images with ultra-high spatial resolution to construct a high density point cloud, which has been used to estimate plant height [13,62]. Therefore, digital cameras are widely used for precision agriculture, as stated in the review by Zhang and Kovacs [63].

Crop growth condition at the early stages is very important for the yield formation and N management is practiced at the tillering and jointing stages [64], thus it is essential to provide farmers with N status in order to supply appropriate rate of fertilizer [65]. Given the low cost of RGB and CIR cameras and satisfactory performance in N accumulation estimation, we believe RGB cameras could be used to monitor N status for the two early stages of rice growth. CIR cameras would be a good choice for N accumulation estimation for all three stages, if the demand on estimation accuracy is not so strict and financial budget is tight. Besides, the CIR camera was also suited for LAI estimation in winter wheat [23].

Compared with the digital cameras, the MS camera performed significantly better in LNA and PNA estimation. However, the miniMCA (700 g) is much heavier than the commercial digital camera (270 g), which is a great burden for UAS energy supply [66]. Though MS images could be processed with the Pix4D and Photoscan software packages, the spatial coverage is not as efficient as that of the RGB images due to the narrow field of view. In addition, the channels in new MS cameras such as RedEdge are still designed in arrays, and much effort is still needed to correct for the band-to-band misalignment. New products may have better performance in hardware design, but we still face similar challenges in image collection and preprocessing. In this study, only one image was used for the small study area and whether or not the result from the small plot experiments could be applied to a larger area needs to be investigated with more data in the future.

Although a limited number of VIs was selected to estimate rice N accumulation in this study, they were a proper representation of the function of these sensors in N status monitoring. We would like to extend our study to use more VIs with the three sensors on this topic. Moreover, the emerging compact hyperspectral sensors (e.g., UHD 185, Rikola) might provide a better choice for UAS to monitor crop growth status, because they have been successfully applied to estimate aboveground biomass [67] and leaf N content [68] in winter wheat. In addition, N accumulation is composed of N concentration and biomass, and biomass could be related to N stress but also to many other stresses. Our future work will focus on detecting the N concentration directly to find a real statement of the N deficit with VIs.

5. Conclusions

In this study, we evaluated three sensors (RGB, CIR and MS cameras) onboard UAS for the estimation of LNA and PNA in rice. UAS images were acquired in two continuous years at the early three growth stages (tillering, jointing and booting). Six VIs (NExG, NGRDI, GNDVI, ENDVI, $CI_{red\ edge}$ and DATT) computed from the corresponding images were employed to estimate LNA and PNA for individual stages and stage groups, and the established VI-LNA and VI-PNA models were evaluated with the test set. The results demonstrated that both NGRDI-RGB and GNDVI-CIR had satisfactory performance in model calibration for the early two (tillering and jointing) and all three growth stages, respectively. NGRDI-RGB and EXG-RGB exhibited moderate estimation accuracy in LNA and PNA model validation for the early two stages (tillering and jointing), respectively. GNDVI-CIR yielded moderate estimation accuracy for all three stages. Red edge indices ($CI_{red\ edge}$ and DATT) from the MS images performed consistently well in model calibration and validation. Therefore, RGB and CIR cameras are good avenues for the qualitative monitoring of N status, due to their satisfactory performance in model calibration. Additionally, in consideration of their low cost and highly efficient data processing they are also good alternatives for quantitative N estimation only for the early two stages (tillering and jointing), due to the moderate estimation accuracy. The MS camera could be taken as a proper sensor for quantitative analysis of N status monitoring with red edge indices. These results further prove that UAS is a reliable approach for monitoring N accumulation, which is a good indicator for N diagnosis and subsequent N management. Future work should be directed to examining these established models with multi-year datasets to improve robustness and applicability.

Author Contributions: T.C., X.Y., Y.T., W.C., and Y.Z. conceived and designed the experiments; D.L., X.Z., and H.Z. performed the experiments; H.Z., X.Z., and D.L. analyzed the data; H.Z., T.C. and Y.Z. wrote the paper.

Acknowledgments: This work was supported by grants from the National Key Research and Development Program of China (2016YFD0300601), the Special Fund for Agro-scientific Research in the Public Interest (201303109), the Award for Jiangsu Distinguished Professor and the Priority Academic Program Development of Jiangsu Higher Education Institutions (PAPD), China. The Natural Science Foundation of Jiangsu Province (BK20150663).

Conflicts of Interest: The authors declare no conflict of interest.

References

1. Janssen, B.H. Efficient use of nutrients: An art of balancing. *Field Crops Res.* **1998**, *56*, 197–201. [CrossRef]
2. Chu, X.; Guo, Y.; He, J.; Yao, X.; Zhu, Y.; Cao, W.; Cheng, T.; Tian, Y. Comparison of different hyperspectral vegetation indices for estimating canopy leaf nitrogen accumulation in rice. *Agron. J.* **2014**, *106*, 1911–1920. [CrossRef]
3. Wang, W.; Yao, X.; Tian, Y.C.; Liu, X.J.; Jun, N.I.; Cao, W.X.; Zhu, Y. Common Spectral Bands and Optimum Vegetation Indices for Monitoring Leaf Nitrogen Accumulation in Rice and Wheat. *J. Integr. Agric.* **2012**, *11*, 2001–2012. [CrossRef]
4. Zheng, H.; Cheng, T.; Yao, X.; Deng, X.; Tian, Y.; Cao, W.; Zhu, Y. Detection of rice phenology through time series analysis of ground-based spectral index data. *Field Crops Res.* **2016**, *198*, 131–139. [CrossRef]
5. Yao, X.; Ren, H.; Cao, Z.; Tian, Y.; Cao, W.; Zhu, Y.; Cheng, T. Detecting leaf nitrogen content in wheat with canopy hyperspectrum under different soil backgrounds. *Int. J. Appl. Earth Obs. Geoinf.* **2014**, *32*, 114–124. [CrossRef]
6. Serrano, L.; Filella, I.; Penuelas, J. Remote sensing of biomass and yield of winter wheat under different nitrogen supplies. *Crop Sci.* **2000**, *40*, 723–731. [CrossRef]
7. Blaes, X.; Chomé, G.; Lambert, M.-J.; Traoré, P.S.; Schut, A.G.; Defourny, P. Quantifying Fertilizer Application Response Variability with VHR Satellite NDVI Time Series in a Rainfed Smallholder Cropping System of Mali. *Remote Sens.* **2016**, *8*, 531. [CrossRef]
8. Huang, S.; Miao, Y.; Zhao, G.; Yuan, F.; Ma, X.; Tan, C.; Yu, W.; Gnyp, M.L.; Lenz-Wiedemann, V.I.; Rascher, U. Satellite remote sensing-based in-season diagnosis of rice nitrogen status in Northeast China. *Remote Sens.* **2015**, *7*, 10646–10667. [CrossRef]
9. Boegh, E.; Soegaard, H.; Broge, N.; Hasager, C.; Jensen, N.; Schelde, K.; Thomsen, A. Airborne multispectral data for quantifying leaf area index, nitrogen concentration, and photosynthetic efficiency in agriculture. *Remote Sens. Environ.* **2002**, *81*, 179–193. [CrossRef]
10. Inoue, Y.; Sakaiya, E.; Zhu, Y.; Takahashi, W. Diagnostic mapping of canopy nitrogen content in rice based on hyperspectral measurements. *Remote Sens. Environ.* **2012**, *126*, 210–221. [CrossRef]

11. Hunt, E.R.; Cavigelli, M.; Cst, D.; Mcmurtrey, J.I.; Walthall, C.L. Evaluation of digital photography from model aircraft for remote sensing of crop biomass and nitrogen status. *Precis. Agric.* **2005**, *6*, 359–378. [CrossRef]
12. Lelong, C.C.; Burger, P.; Jubelin, G.; Roux, B.; Labbé, S.; Baret, F. Assessment of unmanned aerial vehicles imagery for quantitative monitoring of wheat crop in small plots. *Sensors* **2008**, *8*, 3557–3585. [CrossRef] [PubMed]
13. Bendig, J.; Yu, K.; Aasen, H.; Bolten, A.; Bennertz, S.; Broscheit, J.; Gnyp, M.L.; Bareth, G. Combining UAV-based plant height from crop surface models, visible, and near infrared vegetation indices for biomass monitoring in barley. *Int. J. Appl. Earth Obs. Geoinf.* **2015**, *39*, 79–87. [CrossRef]
14. Hunt, E.R.; Daughtry, C.S.T.; Mirsky, S.B.; Hively, W.D. Remote sensing with unmanned aircraft systems for precision agriculture applications. In Proceedings of the Second International Conference on Agro-Geoinformatics, Fairfax, VA, USA, 12–16 August 2013.
15. Jin, X.; Liu, S.; Baret, F.; Hemerlé, M.; Comar, A. Estimates of plant density of wheat crops at emergence from very low altitude UAV imagery. *Remote Sens. Environ.* **2017**, *198*, 105–114. [CrossRef]
16. Lebourgeois, V.; Bégué, A.; Labbé, S.; Mallavan, B.; Prévot, L.; Roux, B. Can commercial digital cameras be used as multispectral sensors? A crop monitoring test. *Sensors* **2008**, *8*, 7300–7322. [CrossRef] [PubMed]
17. Zhu, J.; Wang, K.; Deng, J.; Harmon, T. Quantifying Nitrogen Status of Rice Using Low Altitude UAV-Mounted System and Object-Oriented Segmentation Methodology. In Proceedings of the ASME 2009 International Design Engineering Technical Conferences and Computers and Information in Engineering Conference, San Diego, CA, USA, 30 August–2 September 2009.
18. Córcoles, J.I.; Ortega, J.F.; Hernández, D.; Moreno, M.A. Estimation of leaf area index in onion (*Allium cepa* L.) using an unmanned aerial vehicle. *Biosyst. Eng.* **2013**, *115*, 31–42. [CrossRef]
19. Hatfield, J.L.; Gitelson, A.A.; Schepers, J.S.; Walthall, C.L. Application of Spectral Remote Sensing for Agronomic Decisions. *Agron. J.* **2008**, *100*, 117–131. [CrossRef]
20. Jordan, C.F. Derivation of leaf area index from quality of light on the forest floor. *Ecology* **1969**, *50*, 663–666. [CrossRef]
21. Tucker, C.J. Red and photographic infrared linear combinations for monitoring vegetation. *Remote Sens. Environ.* **1979**, *8*, 127–150. [CrossRef]
22. Huete, A.R. A soil-adjusted vegetation index (SAVI). *Remote Sens. Environ.* **1988**, *25*, 295–309. [CrossRef]
23. Hunt, E.R.; Hively, W.D.; Fujikawa, S.J.; Linden, D.S.; Daughtry, C.S.; McCarty, G.W. Acquisition of NIR-green-blue digital photographs from unmanned aircraft for crop monitoring. *Remote Sens.* **2010**, *2*, 290–305. [CrossRef]
24. Hunt, E.R.; Hively, W.D.; McCarty, G.W.; Daughtry, C.S.T.; Forrestal, P.J.; Kratochvil, R.J.; Carr, J.L.; Allen, N.F.; Fox-Rabinovitz, J.R.; Miller, C.D. NIR-green-blue high-resolution digital images for assessment of winter cover crop biomass. *GISci. Remote Sens.* **2011**, *48*, 86–98. [CrossRef]
25. Lebourgeois, V.; Bégué, A.; Labbé, S.; Houlès, M.; Martiné, J.F. A light-weight multi-spectral aerial imaging system for nitrogen crop monitoring. *Precis. Agric.* **2012**, *13*, 525–541. [CrossRef]
26. Suárez, L.; Zarco-Tejada, P.; González-Dugo, V.; Berni, J.; Fereres, E. Detecting water stress in orchard crops using PRI from airborne imagery. In Proceedings of the 6th EARSeL SIG IS Workshop Imaging Spectroscopy: Innovative Tool for Scientific and Commercial Environmental Applications, Tel-Aviv, Israel, 16–19 March 2009.
27. Calderón, R.; Navas-Cortés, J.A.; Lucena, C.; Zarco-Tejada, P.J. High-resolution airborne hyperspectral and thermal imagery for early detection of Verticillium wilt of olive using fluorescence, temperature and narrow-band spectral indices. *Remote Sens. Environ.* **2013**, *139*, 231–245. [CrossRef]
28. Rey, C.; Martín, M.P.; Lobo, A.; Luna, I.; Diago, M.P.; Millan, B.; Tardáguila, J. Multispectral imagery acquired from a UAV to assess the spatial variability of a Tempranillo vineyard. In *Precision Agriculture '13*; Wageningen Academic Publishers: Wageningen, The Netherlands, 2013.
29. Candiago, S.; Remondino, F.; de Giglio, M.; Dubbini, M.; Gattelli, M. Evaluating Multispectral Images and Vegetation Indices for Precision Farming Applications from UAV Images. *Remote Sens.* **2015**, *7*, 4026–4047. [CrossRef]
30. Hunt, E.; Hively, W.D.; Daughtry, C.S.; McCarty, G.W.; Fujikawa, S.J.; Ng, T.; Tranchitella, M.; Linden, D.S.; Yoel, D.W. Remote sensing of crop leaf area index using unmanned airborne vehicles. In Proceedings of the Pecora 17 Symposium, Denver, CO, USA, 18–20 November 2008.

31. Mathews, A.J.; Jensen, J.L.R. Visualizing and Quantifying Vineyard Canopy LAI Using an Unmanned Aerial Vehicle (UAV) Collected High Density Structure from Motion Point Cloud. *Remote Sens.* **2013**, *5*, 2164–2183. [CrossRef]
32. Yao, X.; Wang, N.; Liu, Y.; Cheng, T.; Tian, Y.; Chen, Q.; Zhu, Y. Estimation of Wheat LAI at Middle to High Levels Using Unmanned Aerial Vehicle Narrowband Multispectral Imagery. *Remote Sens.* **2017**, *9*, 1304. [CrossRef]
33. Bendig, J.; Bolten, A.; Bennertz, S.; Broscheit, J.; Eichfuss, S.; Bareth, G. Estimating Biomass of Barley Using Crop Surface Models (CSMs) Derived from UAV-Based RGB Imaging. *Remote Sens.* **2014**, *6*, 10395–10412. [CrossRef]
34. Von Bueren, S.; Burkart, A.; Hueni, A.; Rascher, U.; Tuohy, M.; Yule, I. Comparative validation of UAV based sensors for the use in vegetation monitoring. *Biogeosci. Discuss.* **2014**, *11*, 3837–3864. [CrossRef]
35. Gastal, F.; Lemaire, G.; Durand, J.-L.; Louarn, G. Quantifying crop responses to nitrogen and avenues to improve nitrogen-use efficiency. In *Crop Physiology (Second Edition)*; Academic Press: Cambridge, MA, USA, 2005.
36. Bremner, J.M.; Mulvancy, C.S. Nitrogen-total. In *Methods of Soil Analysis. Part 2 Chemical and Microbiological Properties*; American Society of Agronomy: Madison, WI, USA, 1982; pp. 595–624.
37. Zhou, X.; Zheng, H.B.; Xu, X.Q.; He, J.Y.; Ge, X.K.; Yao, X.; Cheng, T.; Zhu, Y.; Cao, W.X.; Tian, Y.C. Predicting grain yield in rice using multi-temporal vegetation indices from UAV-based multispectral and digital imagery. *ISPRS J. Photogram. Remote Sens.* **2017**, *130*, 246–255. [CrossRef]
38. Pérez-Ortiz, M.; Peña, J.; Gutiérrez, P.; Torres-Sánchez, J.; Hervás-Martínez, C.; López-Granados, F. A semi-supervised system for weed mapping in sunflower crops using unmanned aerial vehicles and a crop row detection method. *Appl. Soft Comput.* **2015**, *37*, 533–544. [CrossRef]
39. Severtson, D.; Callow, N.; Flower, K.; Neuhaus, A.; Olejnik, M.; Nansen, C. Unmanned aerial vehicle canopy reflectance data detects potassium deficiency and green peach aphid susceptibility in canola. *Precis. Agric.* **2016**, *17*, 659–677. [CrossRef]
40. Garcia-Ruiz, F.; Sankaran, S.; Maja, J.M.; Lee, W.S.; Rasmussen, J.; Ehsani, R. Comparison of two aerial imaging platforms for identification of Huanglongbing-infected citrus trees. *Comput. Electron. Agric.* **2013**, *91*, 106–115. [CrossRef]
41. Kelcey, J.; Lucieer, A. Sensor Correction of a 6-Band Multispectral Imaging Sensor for UAV Remote Sensing. *Remote Sens.* **2012**, *4*, 1462–1493. [CrossRef]
42. Smith, G.M.; Milton, E.J. The use of the empirical line method to calibrate remotely sensed data to reflectance. *Int. J. Remote Sens.* **1999**, *20*, 2653–2662. [CrossRef]
43. Woebbecke, D.; Meyer, G.; von Bargen, K.; Mortensen, D. Color indices for weed identification under various soil, residue, and lighting conditions. *Trans. ASAE* **1995**, *38*, 259–269. [CrossRef]
44. Gitelson, A.A.; Kaufman, Y.J.; Stark, R.; Rundquist, D. Novel algorithms for remote estimation of vegetation fraction. *Remote Sens. Environ.* **2002**, *80*, 76–87. [CrossRef]
45. Rasmussen, J.; Ntakos, G.; Nielsen, J.; Svensgaard, J.; Poulsen, R.N.; Christensen, S. Are vegetation indices derived from consumer-grade cameras mounted on UAVs sufficiently reliable for assessing experimental plots? *Eur. J. Agron.* **2016**, *74*, 75–92. [CrossRef]
46. Gitelson, A.A.; Kaufman, Y.J.; Merzlyak, M.N. Use of a green channel in remote sensing of global vegetation from EOS-MODIS. *Remote Sens. Environ.* **1996**, *58*, 289–298. [CrossRef]
47. Gitelson, A.A.; Gritz, Y.; Merzlyak, M.N. Relationships between leaf chlorophyll content and spectral reflectance and algorithms for non-destructive chlorophyll assessment in higher plant leaves. *J. Plant Physiol.* **2003**, *160*, 271–282. [CrossRef] [PubMed]
48. Datt, B. Visible/near infrared reflectance and chlorophyll content in Eucalyptus leaves. *Int. J. Remote Sens.* **1999**, *20*, 2741–2759. [CrossRef]
49. Cheng, T.; Song, R.; Li, D.; Zhou, K.; Zheng, H.; Yao, X.; Tian, Y.; Cao, W.; Zhu, Y. Spectroscopic Estimation of Biomass in Canopy Components of Paddy Rice Using Dry Matter and Chlorophyll Indices. *Remote Sens.* **2017**, *9*, 319. [CrossRef]
50. Li, F.; Gnyp, M.L.; Jia, L.; Miao, Y.; Yu, Z.; Koppe, W.; Bareth, G.; Chen, X.; Zhang, F. Estimating N status of winter wheat using a handheld spectrometer in the North China Plain. *Field Crops Res.* **2008**, *106*, 77–85. [CrossRef]

51. Tilly, N.; Aasen, H.; Bareth, G. Fusion of Plant Height and Vegetation Indices for the Estimation of Barley Biomass. *Remote Sens.* **2015**, *7*, 11449–11480. [CrossRef]
52. Cao, Q.; Miao, Y.; Feng, G.; Gao, X.; Li, F.; Liu, B.; Yue, S.; Cheng, S.; Ustin, S.L.; Khosla, R. Active canopy sensing of winter wheat nitrogen status: An evaluation of two sensor systems. *Comput. Electron. Agric.* **2015**, *112*, 54–67. [CrossRef]
53. Gitelson, A.A.; Andrés, V.; Arkebauer, T.J.; Rundquist, D.C.; Galina, K.; Bryan, L. Remote estimation of leaf area index and green leaf biomass in maize canopies. *Geophys. Res. Lett.* **2003**, *30*, 335–343. [CrossRef]
54. Mutanga, O.; Skidmore, A.K. Narrow band vegetation indices overcome the saturation problem in biomass estimation. *Int. J. Remote Sens.* **2004**, *25*, 3999–4014. [CrossRef]
55. Elvidge, C.D.; Chen, Z. Comparison of broad-band and narrow-band red and near-infrared vegetation indices. *Remote Sens. Environ.* **1995**, *54*, 38–48. [CrossRef]
56. Perry, E.M.; Roberts, D.A. Sensitivity of Narrow-Band and Broad-Band Indices for Assessing Nitrogen Availability and Water Stress in an Annual Crop. *Agron. J.* **2008**, *100*, 1211–1219. [CrossRef]
57. Knyazikhin, Y.; Schull, M.A.; Stenberg, P.; Mõttus, M.; Rautiainen, M.; Yang, Y.; Marshak, A.; Carmona, P.L.; Kaufmann, R.K.; Lewis, P.; et al. Hyperspectral remote sensing of foliar nitrogen content. *Proc. Natl. Acad. Sci. USA* **2013**, *110*, 185–192. [CrossRef] [PubMed]
58. Bajwa, S.; Mishra, A.; Norman, R. Canopy reflectance response to plant nitrogen accumulation in rice. *Precis. Agric.* **2010**, *11*, 488–506. [CrossRef]
59. Yu, K.; Li, F.; Gnyp, M.L.; Miao, Y.; Bareth, G.; Chen, X. Remotely detecting canopy nitrogen concentration and uptake of paddy rice in the Northeast China Plain. *ISPRS J. Photogramm. Remote Sens.* **2013**, *78*, 102–115. [CrossRef]
60. Luna, I.; Lobo, A. Mapping Crop Planting Quality in Sugarcane from UAV Imagery: A Pilot Study in Nicaragua. *Remote Sens.* **2016**, *8*, 500. [CrossRef]
61. Torres-Sánchez, J.; Peña, J.M.; Castro, A.I.D.; López-Granados, F. Multi-temporal mapping of the vegetation fraction in early-season wheat fields using images from UAV. *Comput. Electron. Agric.* **2014**, *103*, 104–113. [CrossRef]
62. Holman, F.H.; Riche, A.B.; Michalski, A.; Castle, M.; Wooster, M.J.; Hawkesford, M.J. High Throughput Field Phenotyping of Wheat Plant Height and Growth Rate in Field Plot Trials Using UAV Based Remote Sensing. *Remote Sens.* **2016**, *8*, 1031. [CrossRef]
63. Zhang, C.; Kovacs, J.M. The application of small unmanned aerial systems for precision agriculture: A review. *Precis. Agric.* **2012**, *13*, 693–712. [CrossRef]
64. Ding, Y.F.; Zhao, C.H.; Wang, Q.S. Effect of application stage of panicle fertilizer on rice grain yield and the utilization of nitrogen. *J. Nanjing Agric. Univ.* **2003**, *26*, 5–8.
65. Haboudane, D.; Miller, J.R.; Tremblay, N.; Zarco-Tejada, P.J.; Dextraze, L. Integrated narrow-band vegetation indices for prediction of crop chlorophyll content for application to precision agriculture. *Remote Sens. Environ.* **2002**, *81*, 416–426. [CrossRef]
66. Torres-Sánchez, J.; López-Granados, F.; de Castro, A.I.; Peña-Barragán, J.M. Configuration and specifications of an Unmanned Aerial Vehicle (UAV) for early site specific weed management. *PLoS ONE* **2013**, *8*, e58210. [CrossRef] [PubMed]
67. Yue, J.; Yang, G.; Li, C.; Li, Z.; Wang, Y.; Feng, H.; Xu, B. Estimation of Winter Wheat Above-Ground Biomass Using Unmanned Aerial Vehicle-Based Snapshot Hyperspectral Sensor and Crop Height Improved Models. *Remote Sens.* **2017**, *9*, 708. [CrossRef]
68. Liu, H.; Zhu, H.; Wang, P. Quantitative modelling for leaf nitrogen content of winter wheat using UAV-based hyperspectral data. *Int. J. Remote Sens.* **2017**, *38*, 2117–2134. [CrossRef]

remote sensing

MDPI

Article

Predicting Canopy Nitrogen Content in Citrus-Trees Using Random Forest Algorithm Associated to Spectral Vegetation Indices from UAV-Imagery

Lucas Prado Osco [1], Ana Paula Marques Ramos [2], Danilo Roberto Pereira [2], Érika Akemi Saito Moriya [3], Nilton Nobuhiro Imai [3], Edson Takashi Matsubara [4], Nayara Estrabis [1], Maurício de Souza [1], José Marcato Junior [1], Wesley Nunes Gonçalves [1,4], Jonathan Li [5], Veraldo Liesenberg [6,*] and José Eduardo Creste [7]

1 Faculty of Engineering, Architecture, and Urbanism and Geography, Federal University of Mato Grosso do Sul, Av. Costa e Silva, Campo Grande 79070-900, Brazil; pradoosco@gmail.com (L.P.O.); nayara.estrabis@ufms.br (N.E.); mauriciomdseng@gmail.com (M.d.S.); jose.marcato@ufms.br (J.M.J.); wesley.goncalves@ufms.br (W.N.G.)

2 Environmental and Regional Development, University of Western São Paulo, R. José Bongiovani, 700-Cidade Universitária, Presidente Prudente 19050-920, Brazil; anaramos@unoeste.br (A.P.M.R.); danilopereira@unoeste.br (D.R.P.)

3 Department of Cartographic Science, São Paulo State University, Presidente Prudente 19060-900, Brazil; eakemisaito@gmail.com (É.A.S.M.); nilton.imai@unesp.br (N.N.I.)

4 Faculty of Computer Science, Federal University of Mato Grosso do Sul, Av. Costa e Silva, Campo Grande 79070-900, Brazil; edsontm@facom.ufms.br

5 Department of Geography and Environmental Management and Department of Systems Design Engineering, University of Waterloo, Waterloo, ON N2L 3G1, Canada; junli@uwaterloo.ca

6 Forest Engineering Department, Santa Catarina State University (UDESC), Av. Luiz de Camões, 2090-Conta Dinheiro, Lages 88520-000, Brazil

7 Agronomy Development, University of Western São Paulo, R. José Bongiovani, 700-Cidade Universitária, Presidente Prudente 19050-920, Brazil; jcreste@unoeste.br

* Correspondence: veraldo.liesenberg@udesc.br; Tel.: +55-(49)-3289-9301

Received: 14 November 2019; Accepted: 5 December 2019; Published: 6 December 2019

Abstract: The traditional method of measuring nitrogen content in plants is a time-consuming and labor-intensive task. Spectral vegetation indices extracted from unmanned aerial vehicle (UAV) images and machine learning algorithms have been proved effective in assisting nutritional analysis in plants. Still, this analysis has not considered the combination of spectral indices and machine learning algorithms to predict nitrogen in tree-canopy structures. This paper proposes a new framework to infer the nitrogen content in citrus-tree at a canopy-level using spectral vegetation indices processed with the random forest algorithm. A total of 33 spectral indices were estimated from multispectral images acquired with a UAV-based sensor. Leaf samples were gathered from different planting-fields and the leaf nitrogen content (LNC) was measured in the laboratory, and later converted into the canopy nitrogen content (CNC). To evaluate the robustness of the proposed framework, we compared it with other machine learning algorithms. We used 33,600 citrus trees to evaluate the performance of the machine learning models. The random forest algorithm had higher performance in predicting CNC than all models tested, reaching an R^2 of 0.90, MAE of 0.341 g·kg^{-1} and MSE of 0.307 g·kg^{-1}. We demonstrated that our approach is able to reduce the need for chemical analysis of the leaf tissue and optimizes citrus orchard CNC monitoring.

Keywords: UAV multispectral imagery; spectral vegetation indices; machine learning; plant nutrition

1. Introduction

Remote sensing of agricultural fields is important to assist its management through a low-cost and non-destructive approach. The usage of remote sensing systems supports data acquisition in a more frequent and faster manner, being more valuable to evaluate plants than most traditional agronomic procedures [1,2]. In the nutritional analysis, different remote sensing techniques were evaluated recently [3–7]. Regardless of the conducted approach, the spectral analysis of the vegetation is viewed as a reasonable alternative to estimate plant health conditions.

One important issue to correctly manage agricultural fields is to know the nitrogen (N_2) content in plants. N_2 is one of the main nutrients required for foliar development and photosynthetic activity, influencing plant productivity [7]. However, applications of excessive amounts of fertilization in agricultural fields are still a common but erroneous practice [8]. This practice negatively impacts plants, provoking their intoxication, and the environment through the leaching and the volatilization processes of the non-absorbed part [9]. Consistent monitoring of the nutrient in leaf-tissue is essential to improve the management of crops and orchards.

The traditional agronomic methods to determine N_2 rely on the chemical analysis of the leaf-tissue. Those methods are normally labor-intensive, time-consuming, and highly costly and they produce environmentally dangerous residuals [10]. As a non-destructive, clean and fast approach, remote sensing data like multispectral imagery obtained from unmanned aerial vehicles (UAV)-based sensors are often being used to monitor the nitrogen content in plants [11,12]. The wide market availability of UAV, the high-spatial-resolution and the potential of multispectral imagery are some of the reasons behind it [13]. Yet, because of the amount of data produced, remote sensing techniques in combination with high-resolution images are requiring more robust techniques to be evaluated.

Recently, machine learning algorithms have been used for different remote sensing applications [14–18]. Algorithms like artificial neural networks (ANN), support vector machine (SVM), decision trees (DT), random forests (RF), and others are powerful tools in assisting in UAV-based image analysis [19]. These algorithms performed quite well in current approaches involving plant conditions such as nutritional status [20], water-quantity [21], biomass [19], and chlorophyll content [22]. These studies have also considered the contribution of individual bands and spectral vegetation indices in their evaluations.

To estimate N_2 in plants, many studies evaluated the potential of spectral vegetation indices in crops such as wheat, maize, rice, corn, and others [12,23–26]. They can be applied at different scales, such as leaf or canopy level [27] and mitigate anisotropy effects, background shadows, and soil brightness contributions [27–29]. Nevertheless, these advantages over individual spectral bands have yet to be further explored with machine learning algorithms.

Combining machine learning algorithms with spectral vegetation indices is a fairly new practice since these algorithms ensure good performance even with several variables as input features [20]. As the spectral indices are generally simple to be computed and may reduce interference from other non-plant surface targets, they can be considered a feasible practice into measuring N_2 in tree-canopy scales. Agricultural fields such as citrus orchards may benefit from this type of analysis, as spectral indices are known to mitigate anisotropic effects from the tree-canopies.

In citrus plants, few studies evaluated the canopy nitrogen content (CNC) with remotely sensed data, and no study was found, up to the moment, involving the use of spectral indices with machine learning models. Since machine learning models can use additional information obtained directly from spectral indices, our hypothesis is that this combined information may result in an interesting outcome for the N_2 prediction in tree-canopy structures. Although machine learning algorithms have been employed in the leaf nitrogen content (LNC) analysis, there are still few studies that have incorporated a large dataset of spectral indices into their data-set. To the best of our knowledge, these have not been evaluated at a canopy-structure level.

In this paper, we propose a new framework to infer nitrogen content in citrus-tree at a canopy level using spectral vegetation indices calculated from UAV-imagery and the RF algorithm. First, we investigate the individual spectral indices performances and their relation to the CNC. Second, we combined the spectral indices into an RF model and evaluated its performance. We compared the proposed framework with other machine learning methods to prove the robustness of our approach. This paper is organized as follows: Section 2 presents related works. Section 3 describes the method employed in the analysis. Sections 4 and 5 present and discuss the results, respectively. Finally, Section 6 concludes this research.

2. Related Work

With the high availability of UAVs, the evaluation of nutritional condition in plants with high-spatial-resolution images has turned into common practice. Often used statistical methods like the principal component analysis (PCA), partial least square regression (PLSE), stepwise multiple linear regression (SMLR), and others were already implemented in the nitrogen content analysis [30,31]. However, these methods presented different predictions accuracies in this task. This demands for more robust and intelligent algorithms, such as machine learning models. The use of machine learning into predicting nitrogen content is fairly new in remote sensing applications and already presented interesting findings. Though, none incorporated these models to evaluate nitrogen content at a canopy-level.

Regarding the LNC assessment, a study [32] calculated the nitrogen nutrition index (NNI) and evaluated it with machine learning models using RGB images. The authors used a potted pakchoi experiment in a greenhouse and compared the performance of different algorithms in two different stages. Random forest presented the best overall performance, reaching prediction accuracies of 0.82 and 0.94 in the seedling and harvest stages, respectively. Another study [33] evaluated LNC prediction in EO-1 Hyperion hyperspectral data and reached an R^2 of 0.67 for LNC in sugar-cane using the RF model.

The practice of combining several spectral vegetation indices with machine learning models is unusual. To estimate LNC, up until now, only two studies achieved this task, and both concluded that the RF model is a valid approach [20,34]. The first study [34] used 26 spectral indices of WorldView-2 images as input features into an RF model. In this study, red-edge based vegetation indices were the most significant variables for predicting LNC, and their combination with the algorithm returned an R^2 of 0.89 for grass LNC. The second study [20] evaluated LNC in wheat crops with 19 spectral vegetation indices derived from hyperspectral UAV-based images. First-derivative indices were better related to LNC and predicted it with an R^2 of 0.72 using the RF model.

In citrus plants, until this moment, no machine learning model was implemented to predict CNC. Still, one approximation was conducted with hyperspectral measurements in orange-leaves [31]. The authors applied a PLSR in a 350 to 2500 nm spectrum and discovered that the 448, 669, 719, 1377, 1773, and 2231 nm wavelengths were better correlated with LNC, and returned an R^2 of 0.83 and an RMSE of 0.122% for the validation dataset. In UAV-based images, a past study evaluated the performance of PLSR into predicting LNC and returned an R^2 of 0.647 [35]. The authors, however, indicated that new approaches should be conducted to improve the LNC prediction in citrus-trees trough UAV images.

In a previous study, we used spectral wavelengths recorded with a field spectroradiometer to classify a UAV-image with different spectral algorithms [36]. This study was conducted in the same experimental area as the study presented here and returned a classification accuracy of 85.7% and a kappa index of 0.75 for the spectral angle mapper (SAM) algorithm. Still, this method only returned a classified map with three N_2 classes (low, medium, and high), not being suitable to produce more detailed information. This approach also needed a field spectroradiometer to construct the spectral library used by the algorithm, which discourages its replication by low-budget models.

The use of spectral vegetation indices, derived from UAV-images, in conjunction with machine learning algorithms, has yet to be explored in the evaluation of tree-canopies such as citrus orchards. The random forest learner already demonstrates high potentials to predict LNC in other crops, and it indicated what spectral indices better contributed to its performance. Though, no machine learning method is universally appropriate, requiring comparison against others in order to test its robustness. For that, we present a methodological approach to consider the usage of the random forest algorithm into prediction CNC in citrus-trees at a canopy-level based on spectral vegetation indices data.

3. Materials and Method

The proposed method consisted of three phases (Figure 1). Phase 1 describes the survey method performed to collect field data from a Valencia-orange orchard. Here we acquired aerial images with a multispectral Parrot Sequoia camera embedded in a UAV platform and collected leaf samples in the experimental area. Phase 2 focuses on image processing procedures. They were conducted in commercial software named Pix4DMapper, and sampling points were created in a geographical information system (GIS) environment with the open-source software QGIS 3.4. Phase 3 was separated into two stages. Primarily, we selected the available spectral indices for the Parrot Sequoia bands and compared the CNC with them. After that, we used these spectral indices as input parameters for our RF approach and evaluated it with a cross-validation method.

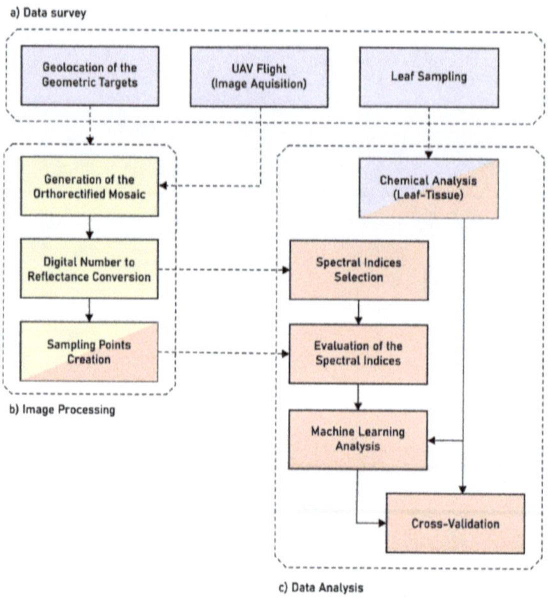

Figure 1. The workflow of the steps of the proposed method.

3.1. Data Survey

The study was conducted in a commercial orchard of Valencia orange (*Citrus sinensis* Valencia) trees planted on a Citrumelo swingle rootstock. The trees had reached their maturation stages at the period of the survey, with 5 years from their initial planting. The survey was carried out on 22 March 2018, when the plants were in their vegetative phase. The experiment was conducted in an area of 71.4 ha divided into 24 field-plots containing 752 plants per hectare, at 7 m × 1.9 m spacing (Figure 2). The area was previously fertilized with 250 kg.ha^{-1} of saturated nitrogen in ammonium nitrate.

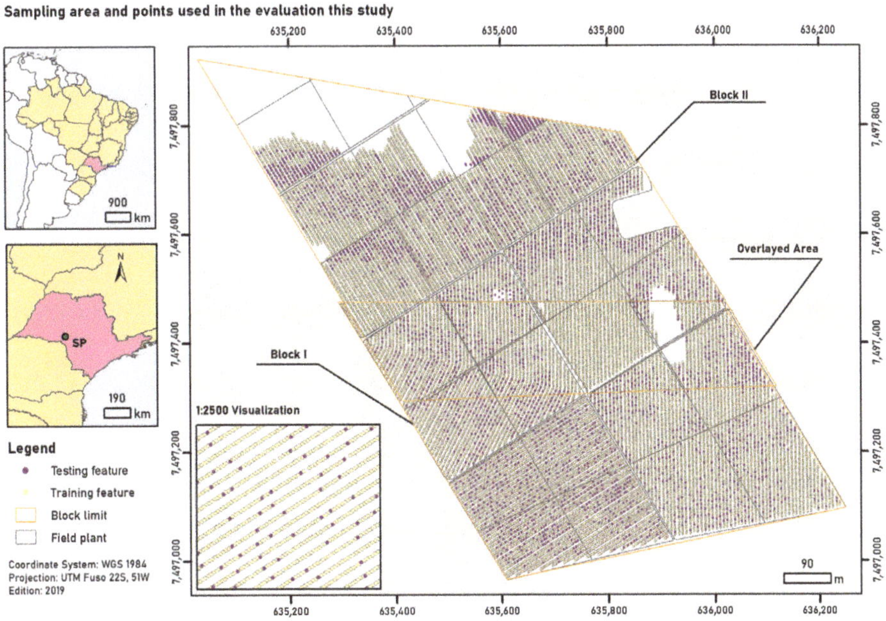

Figure 2. Study area and points used in the evaluation of the spectral indices.

The flight was conducted between 13:00 and 14:00. (local time), at 120 m high, resulting in a 12.9 cm ground sample distance (GSD) image. An eBee SenseFly UAV equipped with the Parrot Sequoia camera was used. This camera records images in the following spectral regions: green (530–570 nm), red (640–680 nm), red-edge (730–740 nm), and near-infrared (770–810 nm). Before the flight, a calibration plaque unique for the Sequoia equipment was recorded with the Parrot Sequoia camera to normalize the local illumination. The main characteristics of the UAV-mounted sensor used are summarized in Table 1.

Table 1. Parrot Sequoia camera details and flight conditions.

Band	Wavelength (nm)	Spectral Resolution	10 Bits	Flight High	120 m
Green	550 (± 40)	**Spatial Resolution**	12.9 cm	**Flight Time**	01:30 P.M.
Red	660 (± 40)	**HFOV**	70.6°	**Weather**	Partially cloudy
Red-edge	735 (± 10)	**VFOV**	52.6°	**Precipitation**	0 mm
Near-infrared	790 (± 40)	**DFOC**	89.6°	**Wind**	At 1–2 m/s

Horizontal field of view (HFOV); vertical field of view (VFOV); displayed field of view (DFOC).

The study area was separated into 27 field-plants (Figure 2). In each field, we collected an *n* amount of leaves from *n* number of trees. Both numbers variated according to plant-field size and the number of trees per field. In total, approximately 4000 leaf samplings were gathered in this area. The method applied here followed standard recommended procedures. For each citrus-tree sampled, we collected the 3rd or 4th leaf of a fruit-branch, at a medium canopy height. The leaves were all visually health with no signs of diseases or damage. They were separated and identified in plastic bags and submitted to the laboratory analysis.

In the laboratory, leaf area (cm^2) was measured for each sample using a digital analysis method [37]. Later, they were washed and dried in an oven at 60–65 °C for 48 h, and then crushed. The Kjeldahl titration method was applied to determine the LNC. This method is divided into three stages: (1) digestion, (2) distillation in a nitrogen distiller, and (3) titration with sulfuric acid (H_2SO_4) [38].

After this, the averaged LNC values were associated with their correspondent field-plant. In our studied area, we obtained LNC between 23.2 and 29.5 g·kg^{-1}, with a variance of 2.33 g·kg^{-1}.

3.2. *Image Pre-Processing and Sampling Points*

The image pre-processing was performed in the Pix4DMapper software, in which we divided two mosaic blocks to optimize it (Figure 2). We first optimized the interior and exterior parameters and generated the sparse dense cloud based on the structure-from-motion (SfM) method. Later, we generated the point clouds based on MVS (multi-view stereo) approach. For the SfM, we used a total of nine control points in cross-format with approximately 50 cm × 50 cm in size, distributed equally in the experimental area. We measured the coordinates of these targets with a Leica Plus GS15 Global Navigation Satellite System (GNSS), dual-frequency in real-time kinematic (RTK) mode, with a 3 mm precision.

The UAV flight was approved by the Department of Airspace Control (DECEA), responsible for the Brazilian airspace. The images were acquired with an 80% longitudinal and 60% lateral overlaps. The orthomosaic was composed of 2389 scenes altogether. We converted the digital number (DN) values to surface reflectance values using the calibration parameters described in the Parrot Sequoia manual [39]. Finally, an orthorectified surface reflectance image was generated for each band at each block (I and II). Both image blocks were used to create a unique mosaic.

In a GIS environment, we manually identified 33,600 citrus-trees in our experimental area (Figure 2). This was performed in the QGIS 3.4 open-source software and used photointerpretation techniques to delineate point-features to mark the location of each tree. We attributed a radius (1.1–1.4 m) for each tree-canopy in order to calculate their respective leaf area index (LAI) in relation to the ground area. The mentioned variation in radius was used because of the difference in canopy sizes.

Light interaction at the canopy level is dominantly affected by canopy structure, and the leaf properties may be canceled out. Since our image data is at a landscape level, spectral vegetation indices should only be linked to LNC after considering this detail. To correctly compare average LNC measured in the laboratory with data extracted from the UAV-based image, we scaled up the LNC from leaf to canopy level (CNC). To do this, we multiplied the averaged LNC of each plant-field with the calculated LAI. This process can be described by Equation (1) [40]. This procedure resulted in 33,600 trees with known scaled up nitrogen content.

$$CNC = \overline{LNC_l} \times LAI, \tag{1}$$

where LNC_l is the averaged value measured in the laboratory and LAI is given by the ratio between the leaf area and ground area. LNC was measured in g·kg^{-1}, while LAI is dimensionless. The resulting CNC was also in g·kg^{-1}. This equation is based on a canopy development model that considered the relationship between nitrogen content and the LAI at different growth stages [40,41]. In our study case, the citrus-trees were evaluated during their intermediate phase (i.e., with neither young or old leaves), so a linear relationship can be assumed [41].

We separated the trees into training and testing data-sets (Figure 2). The training points consisted of 90% (80% train and 10% validation) of the entire data-set, while the testing points were represented by the remaining 10%. We evaluated the distributions for the validation and testing data to determine if there was a significant difference between samples (Figure 3). This prior evaluation returned a normal distribution for both datasets (Shapiro–Wilk p-value equal to 0.77 and 0.79) and no statistical difference between both means (test t student p-value of 0.5144).

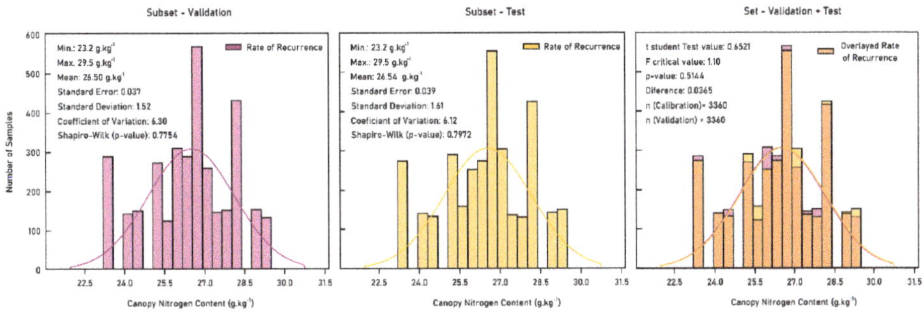

Figure 3. Distribution of spectral sampling points as a function of leaf nitrogen content (LNC).

3.3. Spectral Vegetation Indices

To gather the available spectral vegetation indices to this study, we based our selection on the Index Database [42]. We considered the spectral indices associates with the Parrot Sequoia band ranges (Table 2). We identified them according to their purpose (variable) and scale (canopy and/or leaf level). The values of parameters such as soil-lines (L), which are required by some spectral indices, were adopted based upon literature recommendation.

Table 2. The spectral vegetation indices associated with the Sequoia camera used in this study.

Index	Equation	Variable	Scale
ARVI2 (Atmospherically Resistant Vegetation Index 2)	$-0.18 + 1.17 * \left[\frac{(R\lambda_{nir} - R\lambda_{red})}{(R\lambda_{nir} + R\lambda_{red})}\right]$	Vitality	Canopy
CCCI (Canopy Chlorophyll Content Index)	$\frac{(R\lambda_{nir} - R\lambda_{rededge})/(R\lambda_{nir} + R\lambda_{rededge})}{(R\lambda_{nir} - R\lambda_{red})/(R\lambda_{nir} + R\lambda_{red})}$	Chlorophyll	Leaf/Canopy
CG (Chlorophyll Green)	$(R\lambda_{nir}/R\lambda_{green})^{-1}$	Chlorophyll	Leaf/Canopy
CIgreen (Chlorophyll Index Green)	$(R\lambda_{nir}/R\lambda_{green}) - 1$	Chlorophyll/LAI	Leaf/Canopy
CIrededge (Chlorophyll Index RedEdge)	$(R\lambda_{nir}/R\lambda_{rededge}) - 1$	Chlorophyll/LAI	Leaf/Canopy
Ctr2 (Simple Ratio 695/760 Carter2)	$R\lambda_{rededge}/R\lambda_{nir}$	Chlorophyll/Stress	Leaf
CTVI (Corrected Transformed Vegetation Index)	$\frac{NDVI+0.5}{\lvert NDVI+0.5\rvert} * \sqrt{NDVI + 0.5}$	Vegetation	Leaf/Canopy
CVI (Chlorophyll Vegetation Index)	$R\lambda_{nir} * (R\lambda_{red}/R\lambda_{green}{}^2)$	Chlorophyll	Canopy
GDVI (Difference NIR/Green Difference Vegetation Index)	$R\lambda_{nir} - R\lambda_{green}$	Vegetation	Leaf
GI (Simple Ratio 554/677 Greenness Index)	$R\lambda_{green}/R\lambda_{red}$	Chlorophyll	Leaf
GNDVI (Normalized Difference NIR/Green NDVI)	$(R\lambda_{nir} - R\lambda_{green})/(R\lambda_{nir} + R\lambda_{green})$	Chlorophyll	Leaf
GRNDVI (Green-Red NDVI)	$\frac{R\lambda_{nir}-(R\lambda_{green}+R\lambda_{red})}{R\lambda_{nir}+(R\lambda_{green}+R\lambda_{red})}$	Vegetation	Leaf/Canopy
GSAVI (Green Soil Adjusted Vegetation Index)	$\frac{(1+L)*(R\lambda_{nir}-R\lambda_{green})}{(R\lambda_{nir}+R\lambda_{green}+L)}$	Vegetation	Canopy
IPVI (Infrared Percentage Vegetation Index)	$\frac{R\lambda_{nir}}{\left(\frac{R\lambda_{nir}+R\lambda_{red}}{2}\right)*(NDVI+1)}$	Vegetation	Canopy
MCARI1 (Modified Chlorophyll Absorption in Reflectance Index 1)	$1.2 * [2.5 * (R\lambda_{nir} - R\lambda_{red}) - 1.3 * (R\lambda_{nir} - R\lambda_{green})]$	Chlorophyll	Leaf/Canopy
MSAVI (Modified Soil Adjusted Vegetation Index)	$\frac{2*R\lambda_{nir}+1-\sqrt{(2*R\lambda_{nir}+1)^2-8*(R\lambda_{nir}-R\lambda_{red})}}{2}$	Vegetation	Canopy
MSR (Modified Simple Ratio)	$(SR - 1)/\sqrt{(SR + 1)}$	Vegetation	Leaf

Table 2. *Cont.*

Index	Equation	Variable	Scale
MTVI (Modified Triangular Vegetation Index)	$1.2 * [1.2 * (R\lambda_{nir} - R\lambda_{green}) - 2.5 * (R\lambda_{red} - R\lambda_{green})]$	Vegetation	Leaf/Canopy
ND682/553 (Normalized Difference 682/553)	$(R\lambda_{red} - R\lambda_{green})/(R\lambda_{red} + R\lambda_{green})$	Vegetation	Leaf/Canopy
NDVI (Normalized Difference Vegetation Index)	$(R\lambda_{nir} - R\lambda_{red})/(R\lambda_{nir} + R\lambda_{red})$	Biomass/Others	Leaf/Canopy
Norm G (Normalized G)	$R\lambda_{green}/(R\lambda_{nir} + R\lambda_{red} + R\lambda_{green})$	Vegetation	Leaf/Canopy
Norm NIR (Normalized NIR)	$R\lambda_{nir}/(R\lambda_{nir} + R\lambda_{red} + R\lambda_{green})$	Vegetation	Leaf/Canopy
Norm R (Normalized R)	$R\lambda_{red}/(R\lambda_{nir} + R\lambda_{red} + R\lambda_{green})$	Vegetation	Leaf/Canopy
OSAVI (Optimized Soil Adjusted Vegetation Index)I	$\frac{(1+0.16)*(R\lambda_{nir} - R\lambda_{red})}{(R\lambda_{nir} + R\lambda_{red} + 0.16)}$	Vegetation	Canopy
RDVI (Renormalized Difference Vegetation Index)	$(R\lambda_{nir} - R\lambda_{red})/\sqrt{(R\lambda_{nir} + R\lambda_{red})}$	Chlorophyll	Leaf/Canopy
SAVI (Soil-Adjusted Vegetation Index)II	$\frac{(1+L)*(R\lambda_{nir} - R\lambda_{red})}{(R\lambda_{nir} + R\lambda_{red} + L)}$	Biomass	Canopy
SR672/550 (Simple Ratio 672/550 Datt5)	$R\lambda_{red}/R\lambda_{green}$	Chlorophyll	Leaf
SR750/550 (Simple Ratio 750/550 Gitelson and Merzlyak 1)	$R\lambda_{rededge}/R\lambda_{green}$	Chlorophyll	Leaf/Canopy
SR800/550 (Simple Ratio 800/550)	$R\lambda_{nir}/R\lambda_{green}$	Chlorophyll/Biomass	Leaf
TraVI (Transformed Vegetation Index)	$\sqrt{NDVI + 0.5}$	Vegetation	Leaf/Canopy
TriVI (Triangular Vegetation Index)	$0.5[120(R\lambda_{rededge} - R\lambda_{green}) - 200(R\lambda_{red} - R\lambda_{green})]$	Chlorophyll	Leaf/Canopy
SR (Simple Ratio)	$R\lambda_{nir}/R\lambda_{red}$	Vegetation	Leaf
WDRVI (Wide Dynamic Range Vegetation Index)	$(0.1*R\lambda_{nir} - R\lambda_{red})/(0.1*R\lambda_{nir} + R\lambda_{red})$	Biomass/LAI	Leaf/Canopy

L = 0.5; NDVI = normalized differential vegetation index; SR = simple ratio; LAI = leaf area index. The variable "Vegetation" indicates that the applicability of the index occurs in a general sense, and is not specific like the others; NIR = near-infrared. Central wavelengths are indicated in Table 1.

The relation between the spectral indices and the CNC was calculated using linear and exponential regressions. The metrics used to evaluate them were the regression coefficient (R^2), the root-mean-squared error (RMSE), and the correlation coefficient (r) values. This prediction returned a statistical comparison for the individual spectral indices and helped evaluate their performance. For both regression and correlation analysis, we adopted a coefficient interval of 95%.

3.4. Analysis

The RF algorithm is based on regression trees and relies on the hypothesis that the overall accuracy can be improved by implementing the prediction results of combined independent predictors [20]. As previously stated, we distributed the 33,600 trees into training (80% train and 10% validation) and testing (10%) datasets. In a computational environment, we needed to define the number of trees, the number of nodes and the stop criteria for the RF model. To avoid overfitting, we performed a hyperparametrization process.

To define the most appropriate hyperparameter, we used a cross-validation stochastic approach, where we separated our dataset into 10 folds. One-fold was used to validate the model performance, while the remaining nine folds were used to train the model. This test was repeated until all the 10 folds were evaluated individually. In our study, the number of nodes did not interfere with the prediction accuracy, so a fixed parameter was adopted after initial tests. We determined the number of trees to be equal to 200 since it did not result in any practical gains with higher quantities (Figure 4).

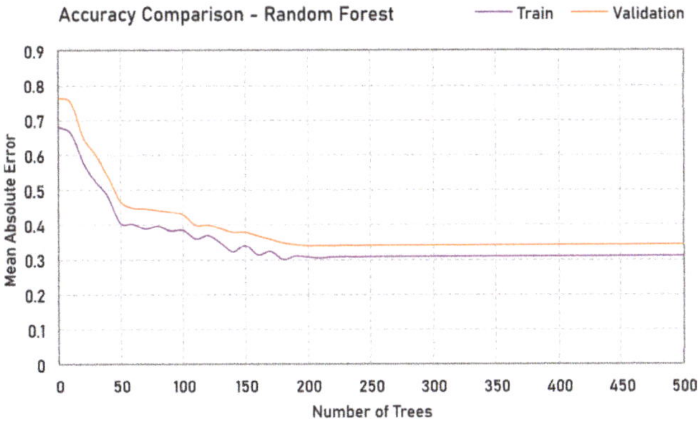

Figure 4. Training accuracy with the difference in accuracy for the random forest (RF) model.

We implemented an extreme gradient boosting (XGBoost) model to verify its impact on the RF performance. The XGBoost uses a forward-learning ensemble method to obtain predictive results in gradually improved estimations. This model computes second-order gradients of the loss function and an advanced regularization (L1 and L2) type [43]. We then performed an evaluation of the models' accuracy when implementing less spectral indices. For this, we selected the 5 and 10 spectral indices that presented a higher contribution to our model. Lastly, since our RF model consisted of 200 trees, we evaluated its trees by a Pythagorean plot. As shorter trees are better in contributing to prediction values, we plotted one example of its first five levels to ascertain the relationship between the spectral indices in the RF model.

The compiled dataset was also used in other machine learning algorithms, like ANN, SVM, DT and LR [44]. These algorithms are considered standard approaches in machine learning evaluation and therefore were compared against RF to test its robustness. We applied the same conjunction of training and test data and performed their hyperparameterization. For this, we used a cross-validation stochastic approach, and data folds implemented in the RF evaluation were also used for these algorithms. The RF and the remaining machine learning algorithms were processed with the open-source RapidMiner 9.5 software, which runs in its own Python library [45]. The parameters of all methods have been set to the library default values except those described in this section.

The hyperparameterization process considered the individual characteristics of the evaluated algorithms. The stop criteria were defined once it did not reduce the MAE since it only increased the processing time needed. For the ANN algorithm, we adopted one hidden layer with 500 neurons and applied a linear activation function to the output layer. We also adopted the Adam Optimizer with a regularization (α) equal to 0.0001. The SVM was applied with a radial basis function (RBF) kernel $\exp(-g|x - y|^2)$, where the gamma (g) value was set automatically, with a regression loss equal to 50.00, a tolerance of 0.001 and an interaction limit of 500. For the DT method, we determined the number of leaves to be equal or higher than 2 and adopted a maximum tree-depth of 100. Finally, we applied two regression models using the Ridge (L2) regularization and a Lasso (L1) regularization, with both strengths (α) equal to 0.015.

The proposed approach was evaluated with the following metrics: mean squared error (MSE); coefficient of variance of the root mean square error (CVRMSE); mean absolute error (MAE); and R^2. We then qualitatively evaluated the RF predictions by plotting it in a regression graphic and also calculating the individual contribution of each spectral vegetation index for the model. Finally, we loaded the prediction results in a map, indicating the nitrogen content at each tree-canopy. For that, we created a new column feature in our tree dataset. This map was used to evaluate qualitatively the CNC through the experimental area.

4. Results

Spectral indices were computed to evaluate the direct relation with CNC in citrus-tress (Table 3). At least 12 of the tested spectral indices (total of 33) presented a regression coefficient above 0.5 and only 05 had a correlation coefficient higher than 0.7 in linear and exponential regressions. In literature, lower correlation values (0.5) have been used to determine regions favorable to the estimation of nitrogen content in citrus trees [31]. Though, in our study, most of these spectral indices presented RMSE higher than 1 g·kg^{-1}, which is a considerable discrepancy for the CNC in our experimental area.

Table 3. Regression analysis between the spectral vegetation indices and the canopy nitrogen content (CNC).

Index	R^2	RMSE	Equation	r
ARVI2	0.12	2.014	y = 67.36x − 31.18	0.3504
CCCI	**0.57**	**1.145**	**y = 86.55x − 0.004121**	**0.6954**
CG	**0.57**	**1.123**	**y = 3.008x − 5.782**	**0.6796**
CI_{green}	0.26	1.853	y = 3.008x − 2.774	0.4796
$CI_{rededge}$	**0.57**	**1.223**	**y = 26.13x + 6.714**	**0.6072**
Ctr2	0.11	2.031	y = −125.5x + 34.09	−0.2282
CTVI	0.12	2.020	y = 178.9x − 184.1	0.2430
CVI	**0.51**	**1.359**	**y = 3.572x + 0.2191**	**0.6424**
GDVI	0.43	1.515	$y = -698.6x^2 + 607.1x - 104.9$	0.5996
GI	0.30	1.797	y = −23.09x + 62.69	−0.3493
GNDVI	0.42	1.431	y = 186x − 126.6	0.5853
GRNDVI	0.26	1.821	y = 82.78x − 33.62	0.3996
GSAVI	**0.52**	**1.279**	**$y = -1608x^2 + 1989x - 588.1$**	**0.6690**
IPVI	0.13	2.006	y = 87.83x − 51.58	0.2607
MCARI1	0.45	1.188	$y = -394.2x^2 + 523.7x - 46.9$	0.5731
MSAVI	**0.62**	**1.013**	**$y = -1748x^2 + 2431x - 817.1$**	**0.7626**
MSR	0.23	1.887	y = 6.52x + 2.101	0.3792
MTVI	0.45	1.288	$y = -394.2x^2 + 523.7x - 46.9$	0.5731
ND682/553	0.11	2.029	y = 37x + 33.79	0.2319
NDVI	0.12	2.014	y = 78.81x − 43.3	0.2504
Norm G	0.47	1.134	y = −438.4x + 63.11	−0.6188
Norm NIR	0.32	1.621	y = 165.6x − 116.4	0.4996
Norm R	0.11	2.030	y = −168.6x + 35.29	−0.2288
OSAVI	0.39	1.529	y = 68.43x − 25.89	0.5032
RDVI	**0.54**	**1.154**	**$y = -2168x^2 + 2671x - 795.3$**	**0.6028**
SAVI	**0.58**	**1.045**	**$y = -2123x^2 + 2747x - 861.5$**	**0.6813**
SR672/550	0.10	2.175	$y = -881.4x^2 + 1197x - 379.4$	0.0982
SR750/550	**0.61**	**1.022**	**y = 7.301x − 18.77**	**0.7991**
SR800/550	**0.57**	**1.083**	**y = 3.008x − 5.782**	**0.7296**
$T_{ra}VI$	0.12	2.020	y = 178.9x − 184.1	0.2430
$T_{ri}VI$	**0.63**	**1.001**	**$y = -0.782x^2 + 24.49x - 164.3$**	**0.8012**
VIN	0.27	1.832	y = 0.9866x + 9.754	0.3238
WDRVI	**0.58**	**1.076**	**$y = -466.2x^2 + 238.9x - 3.666$**	**0.7166**

Indices highlighted showed a regression coefficient (R^2) above 0.50. All spectral indices returned a *p*-value under 0.05.

In our machine learning analysis, the proposed framework considering the RF model performed better than most of the other machine learning algorithms (Table 4) and individual spectral indices (Table 3). The algorithm returned a prediction with an MSE of 0.307 g·kg^{-1}, and MAE of 0.341 g·kg^{-1} and an R^2 equal to 0.90. The XGBoost model implemented also returned similar metrics.

Table 4. Performance results of each selected algorithm prediction evaluated in this study.

Model	MSE	CVRMSE	MAE	R^2
Support Vector Machine	2.055	5.149	1.011	0.65
Decision Tree	0.347	2.225	0.462	0.85
Random Forest	0.307	2.098	0.341	0.90
Random Forest (XGBoost)	0.300	2.043	0.327	0.90
Artificial Neural Network	1.676	4.168	0.865	0.70
Linear Regression (Ridge)	2.041	5.895	0.984	0.63
Linear Regression (Lasso)	2.010	5.790	0.965	0.65

Other algorithms like DT presented interesting results, being closer to the predictions of the RF model. This is to be expected since the random forest is based on the idea of multiple decision trees. Algorithms such as SVM, ANN, and LR had lower performances. However, they were still better than the previous individual spectral indices analysis.

The CNC predicted by the proposed framework with the RF was evaluated in a plot considering the measured CNC in comparison with the returned CNC from the prediction (Figure 5). Although the graphic shows some predictions distancing from the 1:1 relationship (dashed line), the high amount of data used (n = 3360) helped to estimate the result in a regression coefficient of 0.90. A slight decline in the line suggests that our approach was better at predicting CNC between 26 and 28 g·kg^{-1}.

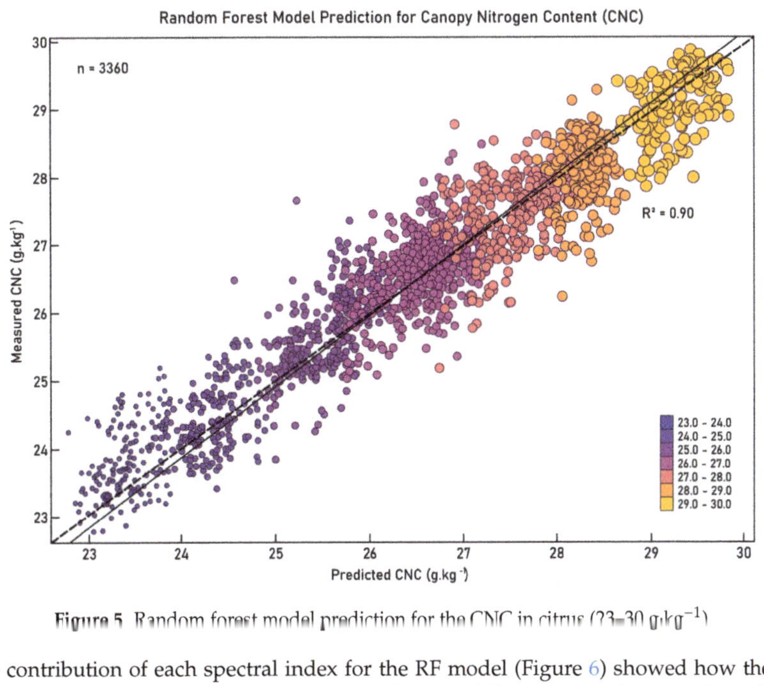

Figure 5. Random forest model prediction for the CNC in citrus (23–30 g·kg^{-1}).

The contribution of each spectral index for the RF model (Figure 6) showed how the highest related spectral indices with CNC (Table 2) assisted the model into the prediction. This information should be considered when analyzing the performance of the model, since it may help future research to reduce the amount of data incorporated into the algorithm. For the prediction of CNC in citrus-trees, the returned results indicate a higher contribution of spectral vegetation indices like SR750–550 (8.2%), TriVI (7.3%), and CI (Red-edge; 6.2%). The first 10 spectral indices composed more than 55% of the contribution value (Figure 6).

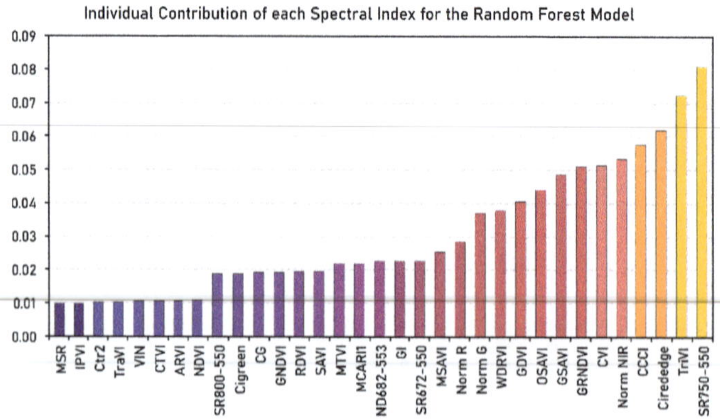

Figure 6. Individual contribution (in %) of each spectral vegetation index for the RF model.

By implementing the best five and 10 spectral indices into the RF model, we verified a slight decrease in its performance (Table 5). This is indicative of how even lower contributions may assist the model. Regardless, this decrease was relatively small, and a tradeoff between the number of spectral indices used and the obtained accuracy is something that should be considered. Another observation is that the XGBoost model presented better results, which may consist of an alternative to reduce the number of spectral indices while improving the performance of the algorithm.

Table 5. Performance results of the random forest model with less spectral indices as input.

Model	Indices (n)	MSE	CVRMSE	MAE	R^2
Random Forest	5	0.376	2.342	0.477	0.83
Random Forest (XGBoost)	5	0.350	2.253	0.412	0.85
Random Forest	10	0.345	2.215	0.401	0.85
Random Forest (XGBoost)	10	0.318	2.127	0.357	0.88

A qualitative evaluation of one of the most representative trees returned by the random forest model helped ascertained the relationship between the spectral indices (Figure 7). An evaluation of shorter trees demonstrated that the best individual spectral indices (Figure 6) returned the strongest contributions.

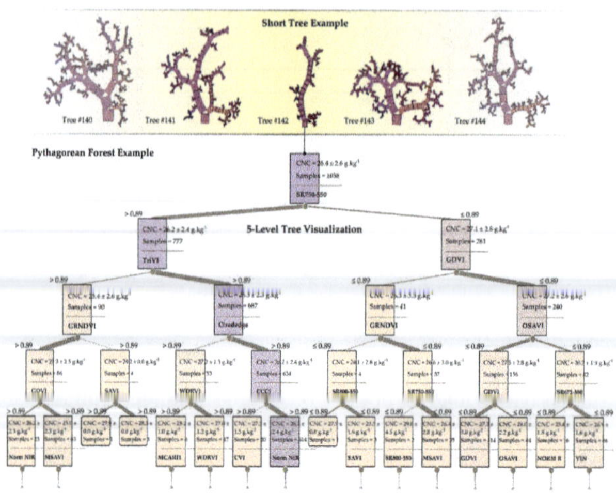

Figure 7. Example of one short tree (initial 5-levels) of the RF model applied.

The predictions returned by the proposed approach were incorporated into the map dataset. This procedure resulted in a qualitative map where a field technician can evaluate the CNC in each known citrus-tree with a prediction R^2 of 0.90 (Figure 8).

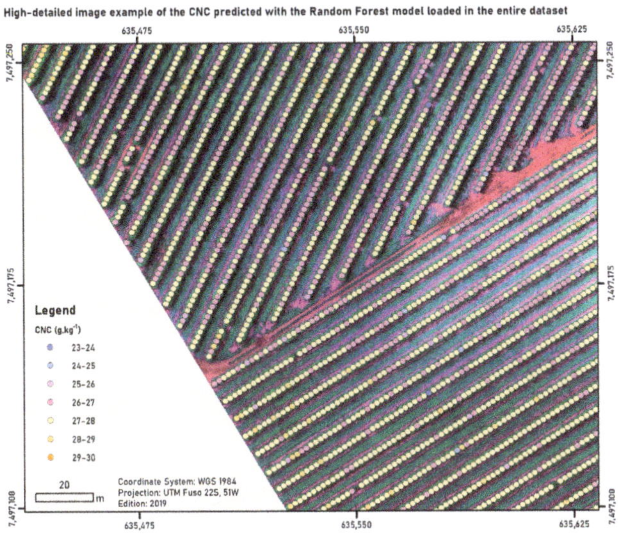

Figure 8. High-detailed image example of the CNC predicted with the RF model loaded in the entire dataset.

5. Discussion

The proposed framework of this study predicted, with high accuracy (R^2 of 0.90 and MSE of 0.307 $g \cdot kg^{-1}$), the amount of CNC present in citrus trees. In a previous study in the same area, we applied the SAM algorithm while submitting simulated spectral curves as input data, reaching an accuracy of 85.7% [36]. The method employed here allowed us to determine if machine learning models, specifically RF, were capable of performing this task. A contribution of this study is the use of spectral vegetation indices as input variables in a machine learning model to predict the amount of N_2 at a tree-canopy level. The method presented here can be replicated to other orchards and cultivars once specifics such as sensor type and available spectral indices are considered.

As related, spectral indices have proved to be an important mechanism in the evaluation of N_2 in other crops [22–25]. However, our results showed a low prediction accuracy (R^2 below 0.7) by relating them directly to the CNC in the citrus-trees. Regardless, in the RF model, spectral indices developed considering red-edge bands generally performed better than other spectral indices. This observation was also found in the evaluation of the LNC in another study [33]. The majority of the high-relation indices also observed here were developed at the canopy level (Table 2).

The combination of spectral vegetation indices with machine learning models proved to be a suitable approach to predict CNC. The RF algorithm performed better than others, with decision tree being closely related. Other algorithms such as SVM and LRs showed similar accuracy to the individual spectral indices' regression values. The ANN, although had a superior advantage over SVM and LRs, returned values far below the RF and DT algorithms. This comparison indicates how appropriate the regression analysis with RF is to predict CNC. This finding was also observed in other related studies [19,32–34].

The distribution of predicted values versus measured values (Figure 5) demonstrated how effective our approach was. Despite presenting errors in the extreme CNC ranges (23–24 and 29–30 $g \cdot kg^{-1}$), the model was able to reduce most errors in the intermediary values (26–28 $g \cdot kg^{-1}$). The high accuracy obtained can be explained because of the way the regression was calculated within the model since it considered the individual contribution of independent variables by combining its results. Still, it was recommended that future research always compare it against other machine learning algorithms.

We considered a total of 33 spectral indices, which was more than previous similar studies have used [19,33]. The individual contribution of each spectral index (Figure 6) demonstrated that the first 10 indices accounted for more than 55% of the total contribution for RF. Additionally, we performed evaluation tests considering the first five and 10 spectral indices (Table 5), in which also returned high accuracies. An evaluation of the shorter trees (Figure 7) corroborated the information shown by the individual spectral indices rank (Figure 6). Lastly, the XGBoost model helped to improve the accuracy of the algorithm when using a smaller number of spectral indices (Table 5). This indicates that it is possible to reduce the number of spectral indices used and still obtain highly accurate results. This information is important since it helps new researches to reduce the amount of processed data, which impacts training and testing time performances.

An adverse condition mentioned by other studies is the contribution of soil brightness to CNC evaluation. As spectral indices are directly obtained from reflectivity, it makes it difficult to avoid background effects [19]. However, in this study, since our data consisted only of delineated tree-canopies (Figures 2 and 8), the soil-brightness contribution was considered minimal. Another possible impact is the anisotropy effect of the trees. Although spectral indices are known to reduce these effects, the fact that we scaled the leaf-measured LNC in the laboratory to an N_2 in canopy-level should contribute to reducing this factor.

Finally, another contribution of our study was the construction of a map indicating the nitrogen content at a canopy level with an accurate prediction of 0.90 and an MSE of 0.307 g·kg^{-1}. This type of map is effective as it may offer farmers and agronomic technicians the opportunity to evaluate individual plants. The agronomical method of collecting leaf-tissue and performing its chemical analysis in the laboratory is spatially limited due to practical reasons. Since our approach returned a high prediction for each individual tree, it is safe to assume that remotely sensed data outperforms the traditional method in relation to trees covered per area. This can positively impact fertilization methods and promote better yield predictions.

6. Conclusions

This study proposed a new framework to infer the nitrogen content at canopy-level in citrus trees. We evaluated the performance of the RF algorithm associated with spectral indices and compared it with other machine learning algorithms. Our approach demonstrated that the combination of spectral vegetation indices and the random forest algorithm is a powerful tool for CNC estimation. While a regression between the spectral indices and the CNC returned low coefficients (R^2 at 0.10–0.63), the combined indices into the RF model resulted in an R^2 of 0.90 and an MAE of 0.341 g·kg^{-1}. This accuracy was higher than previous research, in which we evaluated spectral analysis algorithms for the same experimental field. In conclusion, we recommend the integration of spectral indices with machine learning algorithms like RF to assess CNC in citrus-trees.

Author Contributions: Conceptualization, L.P.O., A.P.M.R. and J.E.C.; methodology, L.P.O., É.A.S.M. and A.P.M.R.; formal analysis, L.P.O. and M.d.S.; resources, J.E.C., N.N.I., J.M.J. and E.T.M.; data curation, É.A.S.M. and L.P.O.; writing—original draft preparation, L.P.O. and A.P.M.R. writing—review and editing, J.M.J., D.R.P., J.L., V.L., W.N.G., N.E. and E.T.M.; supervision, A.P.M.R., J.M.J., N.N.I. and J.E.C.; project administration, L.P.O., A.P.M.R. and J.E.C.; funding acquisition, L.P.O. and J.E.C.

Funding: This research was partially funded by CAPES/Print (p: 88881.311850/2018-01). V. Liesenberg is supported by FAPESC (2017TR1762) and CNPq (313887/2018-7).

Conflicts of Interest: The authors declare no conflict of interest.

References

1. Huang, S.; Miao, Y.; Yuan, F.; Cao, Q.; Ye, H.; Lenz-Wiedemann, V.I.S.; Bareth, G. In-season diagnosis of rice nitrogen status using proximal fluorescence canopy sensor at different growth stages. *Remote Sens.* **2019**, *11*, 1847. [CrossRef]
2. Cui, B.; Zhao, Q.; Huang, W.; Song, X.; Ye, H.; Zhou, X. A new integrated vegetation index for the estimation of winter wheat leaf chlorophyll content. *Remote Sens.* **2019**, *11*, 974. [CrossRef]
3. Zhang, K.; Ge, X.; Shen, P.; Li, W.; Liu, X.; Cao, Q.; Tian, Y. Predicting rice grain yield based on dynamic changes in vegetation indexes during early to mid-growth stages. *Remote Sens.* **2019**, *11*, 387. [CrossRef]
4. Brinkhoff, J.; Dunn, B.W.; Robson, A.J.; Dunn, T.S.; Dehaan, R.L. Modeling mid-season rice nitrogen uptake using multispectral satellite data. *Remote Sens.* **2019**, *11*, 1837. [CrossRef]
5. Tilly, N.; Bareth, G. Estimating nitrogen from structural crop traits at field scale—A novel approach versus spectral vegetation indices. *Remote Sens.* **2019**, *11*, 2066. [CrossRef]
6. Li, Z.; Jin, X.; Yang, G.; Drummond, J.; Yang, H.; Clark, B.; Zhao, C. Remote sensing of leaf and canopy nitrogen status in winter wheat (Triticum aestivum L.) based on N-PROSAIL model. *Remote Sens.* **2018**, *10*, 1463. [CrossRef]
7. Song, Y.; Wang, J. Soybean canopy nitrogen monitoring and prediction using ground based multispectral remote sensors. In Proceedings of the IEEE International Geoscience and Remote Sensing Symposium (IGARSS), Beijing, China, 10–15 July 2016; pp. 6389–6392. [CrossRef]
8. Cilia, C.; Panigada, C.; Rossini, M.; Meroni, M.; Busetto, L.; Amaducci, S.; Boschetti, M.; Picchi, V.; Colombo, R. Nitrogen status assessment for variable rate fertilization in maize through hyperspectral imagery. *Remote Sens.* **2014**, *6*, 6549–6565. [CrossRef]
9. Ciampitti, I.A.; Salvagiotti, F. New insights into soybean biological nitrogen fixation. *Agron. J.* **2018**, *110*, 1185–1196. [CrossRef]
10. Chhabra, A.; Manjunath, K.R.; Panigraphy, S. Non-point source pollution in Indian agriculture: Estimation of nitrogen losses from rice crop using remote sensing and GIS. *Int. J. Appl. Earth Obs. Geoinf.* **2010**, *12*, 190–200. [CrossRef]
11. Zheng, H.; Cheng, T.; Li, D.; Zhou, X.; Yao, X.; Tian, Y.; Zhu, Y. Evaluation of RGB, color-infrared and multispectral images acquired from unmanned aerial systems for the estimation of nitrogen accumulation in rice. *Remote Sens.* **2018**, *10*, 824. [CrossRef]
12. Zheng, H.; Li, W.; Jiang, J.; Liu, Y.; Cheng, T.; Tian, Y.; Zhu, Y.; Cao, W.; Zhang, Y.; Yao, X. A Comparative assessment of different modeling algorithms for estimating LNC in winter wheat using multispectral images from an unmanned aerial vehicle. *Remote Sens.* **2018**, *10*, 2026. [CrossRef]
13. Al-Najjar, H.A.H.; Kalantar, B.; Pradhan, B.; Saeidi, V.; Halin, A.A.; Ueda, N.; Mansor, S. Land cover classification from fused DSM and UAV images using convolutional neural networks. *Remote Sens.* **2019**, *11*, 1461. [CrossRef]
14. Wu, L.; Zhu, X.; Lawes, R.; Dunkerley, D.; Zhang, H. Comparison of machine learning algorithms for classification of LiDAR points for characterization of canola canopy structure. *Int. J. Remote Sens.* **2019**, *40*, 5973–5991. [CrossRef]
15. Dyson, J.; Mancini, A.; Frontoni, E.; Zingaretti, P. Deep learning for soil and crop segmentation from remotely sensed data. *Remote Sens.* **2019**, *11*, 1859. [CrossRef]
16. Zhong, L.; Hu, L.; Zhou, H. Deep learning based multi-temporal crop classification. *Remote Sens. Environ.* **2019**, *221*, 430–443. [CrossRef]
17. Wolanin, A.; Camps-Valls, G.; Gómez-Chova, L.; Mateo-García, G.; van der Tol, C.; Zhang, Y.; Guanter, L. Estimating crop primary productivity with Sentinel-2 and Landsat 8 using machine learning methods trained with radiative transfer simulations. *Remote Sens. Environ.* **2019**, *225*, 441–457. [CrossRef]
18. Ashapure, A.; Oh, S.; Marconi, T.G.; Chang, A.; Jung, J.; Landivar, J.; Enciso, J. Unmanned aerial system based tomato yield estimation using machine learning. Proc. SPIE 11008, Autonomous Air and Ground Sensing Systems for Agricultural Optimization and Phenotyping IV. In Proceedings of the SPIE Defense + Commercial Sensing, Baltimore, MD, USA, 16–18 April 2019. [CrossRef]
19. Pham, T.D.; Yokoya, N.; Bui, D.T.; Yoshino, K.; Friess, D.A. Remote sensing approaches for monitoring mangrove species, structure, and biomass: Opportunities and challenges. *Remote Sens.* **2019**, *11*, 230. [CrossRef]

20. Liang, L.; Di, L.; Huang, T.; Wang, J.; Lin, L.; Wang, L.; Yang, M. Estimation of leaf nitrogen content in wheat using new hyperspectral indices and a random forest regression algorithm. *Remote Sens.* **2018**, *10*, 1940. [CrossRef]
21. Krishna, G.; Sahoo, R.N.; Singh, P.; Bajpai, V.; Patra, H.; Kumar, S.; Sahoo, P.M. Comparison of various modelling approaches for water deficit stress monitoring in rice crop through hyperspectral remote sensing. *Agric. Water Manag.* **2019**, *213*, 231–244. [CrossRef]
22. Shah, S.H.; Angel, Y.; Houborg, R.; Ali, S.; McCabe, M.F. A random forest machine learning approach for the retrieval of leaf chlorophyll content in wheat. *Remote Sens.* **2019**, *11*, 920. [CrossRef]
23. Schlemmer, M.; Gitelson, A.; Schepers, J.; Ferguson, R.; Peng, Y.; Shanahan, J.; Rundquist, D. Remote estimation of nitrogen and chlorophyll contents in maize at leaf and canopy levels. *Int. J. Appl. Earth Obs. Geoinf.* **2013**, *25*, 47–54. [CrossRef]
24. Huang, S.; Miao, Y.; Yuan, F.; Gnyp, M.L.; Yao, Y.; Cao, Q.; Wang, H.; Lenz-Wiedemann, V.I.S.; Bareth, G. Potential of RapidEye and WorldView-2 satellite data for improving rice nitrogen status monitoring at different growth stages. *Remote Sens.* **2017**, *9*, 227. [CrossRef]
25. Kalacska, M.; Lalonde, M.; Moore, T.R. Estimation of foliar chlorophyll and nitrogen content in an ombrotrophic bog from hyperspectral data: Scaling from leaf to image. *Remote Sens. Environ.* **2015**, *169*, 270–279. [CrossRef]
26. Chen, P.; Haboudane, D.; Tremblay, N.; Wang, J.; Vigneault, P.; Baoguo, L. New spectral indicator assessing the efficiency of crop nitrogen treatment in corn and wheat. *Remote Sens. Environ.* **2010**, *114*, 1987–1997. [CrossRef]
27. Cammarano, D.; Fitzgerald, G.J.; Casa, R.; Basso, B. Assessing the robustness of vegetation indices to estimate wheat N in Mediterranean environments. *Remote Sens.* **2014**, *6*, 2827–2844. [CrossRef]
28. Hunt, E.R.J.; Doraiswamy, P.C.; Mcmurtrey, J.E.; Daughtry, C.S.T.; Perry, E.M.; Akhmedov, B. A visible band index for remote sensing leaf chlorophyll content at the canopy scale. *Int. J. Appl. Earth Obs. Geoinf.* **2013**, *21*, 103–112. [CrossRef]
29. Kooistra, L.; Clevers, J.G.P.W. Estimating potato leaf chlorophyll content using ratio vegetation indices. *Remote Sens. Lett.* **2016**, *7*, 611–620. [CrossRef]
30. Zhai, Y.; Cui, L.; Zhou, X.; Gao, Y.; Fei, T.; Gao, W. Estimation of nitrogen, phosphorus, and potassium contents in the leaves of different plants using laboratory-based visible and near-infrared reflectance spectroscopy: Comparison of partial least-square regression and support vector machine regression methods. *Int. J. Remote Sens.* **2013**, *34*, 2502–2518. [CrossRef]
31. Min, M.; Lee, W. Determination of significant wavelengths and prediction of nitrogen content for citrus. *Am. Soc. Agric. Eng.* **2005**, *48*, 455–461. [CrossRef]
32. Xiong, X.; Zhang, J.; Guo, D.; Chang, L.; Huang, D. Non-Invasive Sensing of Nitrogen in Plant Using Digital Images and Machine Learning for *Brassica Campestris* ssp. *Chinensis* L. *Sensors* **2019**, *19*, 2448. [CrossRef]
33. Abdel-Rahman, E.M.; Ahmed, F.B.; Ismail, R. Random forest regression and spectral band selection for estimating sugarcane leaf nitrogen concentration using EO-1 Hyperion hyperspectral data. *Int. J. Remote Sens.* **2013**, *34*, 712–728. [CrossRef]
34. Ramoelo, A.; Cho, M.A.; Mathieu, R.; Madonsela, S.; van de Kerchove, R.; Kaszta, Z.; Wolff, E. Monitoring grass nutrients and biomass as indicators of rangeland quality and quantity using random forest modelling and WorldView-2 data. *Int. J. Appl. Earth Obs. Geoinf.* **2015**, *43*, 43–54. [CrossRef]
35. Liu, X.F.; Lyu, Q.; He, S.L.; Yi, S.L.; Hu, D.Y.; Wang, Z.T.; Deng, L. Estimation of carbon and nitrogen contents in citrus canopy by low-altitude remote sensing. *Int. J. Agric. Biol. Eng.* **2016**, *9*, 149–157. [CrossRef]
36. Osco, L.P.; Marques Ramos, A.P.; Saito Moriya, É.A.; de Souza, M.; Marcato Junior, J.; Matsubara, E.T.; Creste, J.E. Improvement of leaf nitrogen content inference in Valencia-orange trees applying spectral analysis algorithms in UAV mounted-sensor images. *Int. J. Appl. Earth Obs. Geoinf.* **2019**, *83*, 101907. [CrossRef]
37. Katabuchi, M. LeafArea: An R package for rapid digital image analysis of leaf area. *Ecol. Res.* **2015**, *30*, 1073–1077. [CrossRef]
38. Nitrogen Determination by Kjeldahl Method PanReac AppliChem ITW Reagents. Available online: https://www.itwreagents.com/uploads/20180114/A173_EN.pdf (accessed on 9 February 2019).
39. Sequoia, P. Parrot Sequoia Manual. © 2019 Parrot Drones SAS. Available online: https://parrotcontact.parrot.com/website/user-guides/sequoia/sequoia_user_guide.pdf (accessed on 19 November 2019).

40. Ling, B.; Goodin, D.G.; Mohler, R.L.; Laws, A.N.; Joern, A. Estimating canopy nitrogen content in a heterogeneous grassland with varying fire and grazing treatments: Konza Prairie, Kansas, USA. *Remote Sens.* **2014**, *6*, 4430–4453. [CrossRef]
41. Yin, X.; Lantinga, E.A.; Schapendonk, A.H.C.M.; Zhong, X. Some quantitative relationships between leaf area index and canopy nitrogen content and distribution. *Ann. Bot.* **2003**, *91*, 893–903. [CrossRef]
42. IDB. Index DataBase. A Database for Remote Sensing Indices. The IDB Project, 2011–2019. Available online: https://www.indexdatabase.de/ (accessed on 7 February 2019).
43. XGBoost. eXtreme Gradient Boosting. Available online: https://github.com/dmlc/xgboost (accessed on 23 November 2019).
44. Mitchell, T.M. *Machine Learning*, 1st ed.; McGraw-Hill, Inc.: New York, NY, USA, 1997.
45. RapidMiner. RapidMiner Python Package. Available online: https://github.com/rapidminer/python-rapidminer (accessed on 23 November 2019).

remote sensing

MDPI

Article

Improving Unmanned Aerial Vehicle Remote Sensing-Based Rice Nitrogen Nutrition Index Prediction with Machine Learning

Hainie Zha [1], Yuxin Miao [1,2,*], Tiantian Wang [1], Yue Li [1], Jing Zhang [1], Weichao Sun [1], Zhengqi Feng [1] and Krzysztof Kusnierek [3]

1 International Center for Agro-Informatics and Sustainable Development (ICASD), College of Resources and Environmental Sciences, China Agricultural University, Beijing 100193, China; zhahn@cau.edu.cn (H.Z.); tianw@cau.edu.cn (T.W.); yli@cau.edu.cn (Y.L.); jingz9468@gmail.com (J.Z.); sunweichao628@gmail.com (W.S.); fengzhengqi20@gmail.com (Z.F.)

2 Precision Agriculture Center, Department of Soil, Water and Climate, University of Minnesota, St. Paul, MN 55108, USA

3 Center for Precision Agriculture, Norwegian Institute of Bioeconomy Research (NIBIO), Nylinna 226, 2849 Kapp, Norway; krzysztof.kusnierek@nibio.no

* Correspondence: ymiao@umn.edu or ymiao@cau.edu.cn

Received: 13 December 2019; Accepted: 6 January 2020; Published: 8 January 2020

Abstract: Optimizing nitrogen (N) management in rice is crucial for China's food security and sustainable agricultural development. Nondestructive crop growth monitoring based on remote sensing technologies can accurately assess crop N status, which may be used to guide the in-season site-specific N recommendations. The fixed-wing unmanned aerial vehicle (UAV)-based remote sensing is a low-cost, easy-to-operate technology for collecting spectral reflectance imagery, an important data source for precision N management. The relationships between many vegetation indices (VIs) derived from spectral reflectance data and crop parameters are known to be nonlinear. As a result, nonlinear machine learning methods have the potential to improve the estimation accuracy. The objective of this study was to evaluate five different approaches for estimating rice (*Oryza sativa* L.) aboveground biomass (AGB), plant N uptake (PNU), and N nutrition index (NNI) at stem elongation (SE) and heading (HD) stages in Northeast China: (1) single VI (SVI); (2) stepwise multiple linear regression (SMLR); (3) random forest (RF); (4) support vector machine (SVM); and (5) artificial neural networks (ANN) regression. The results indicated that machine learning methods improved the NNI estimation compared to VI-SLR and SMLR methods. The RF algorithm performed the best for estimating NNI (R^2 = 0.94 (SE) and 0.96 (HD) for calibration and 0.61 (SE) and 0.79 (HD) for validation). The root mean square errors (RMSEs) were 0.09, and the relative errors were <10% in all the models. It is concluded that the RF machine learning regression can significantly improve the estimation of rice N status using UAV remote sensing. The application machine learning methods offers a new opportunity to better use remote sensing data for monitoring crop growth conditions and guiding precision crop management. More studies are needed to further improve these machine learning-based models by combining both remote sensing data and other related soil, weather, and management information for applications in precision N and crop management.

Keywords: fixed-wing UAV remote sensing; nitrogen status diagnosis; random forest; precision nitrogen management; machine learning

1. Introduction

Rice (*Oryza sativa* L.) is one of the most important crops in the world, consumed by more than 60% of China's population as a staple food. Rice production in China is a major consumer of nitrogen

(N) fertilizers, but the N use efficiency (NUE) is less than 30% [1]. Uniform fertilizer application across the fields according to experience or regional guidelines is the common practice and can lead to over-application of N at low yielding areas. The over-application of N fertilizers can result in enhanced reactive N losses to the environment, affecting human health, ecosystem services, biodiversity, climate change, and sustainability [1,2]. Precision N management (PNM) has the potential to effectively improve NUE, reduce soil and groundwater pollution, and increase farmers' income [2]. Efficient tools for rapid and in-season diagnosis of rice N status over large areas are essential for the practical implementation of the PNM strategies.

When N fertilizers are applied in the fields, they need to be converted to plant available forms (nitrate (NO_3^-) or ammonium (NH_4^+)) before they can be used by plants. The time they take for these conversions depends on the fertilizer type, soil temperature, soil moisture, soil pH, and soil aeration, etc. [3]. In rice production, N fertilizers are recommended to be applied in several splits to improve NUE, including application before planting or transplanting (called basal N fertilizer), at tillering stage (called tiller N fertilizer), at panicle initiation or stem elongation stage (called panicle N fertilizer), and at heading stage (called grain N fertilizer). It is important to diagnose rice N status during the growing season at different key N application stages, so topdressing N rates can be adjusted to better meet crop N needs. N nutrition index (NNI) is a reliable N status indicator and is defined as the ratio of plant N concentration (PNC) over critical N concentration (N_c), which is the minimum PNC that will achieve maximum aboveground biomass (AGB) production [4–6]. NNI > 1 indicates surplus N supply, while NNI < 1 indicates N deficiency, and NNI around 1 represents optimal N nutritional status [6]. However, NNI determination requires destructive sampling and chemical analysis, which limits its application in PNM. Therefore, the interests in technologies allowing nondestructive estimation of NNI over large areas are increasing.

Proximal and remote sensing technologies are commonly used for estimating crop N status nondestructively and at low cost [7–11]. A number of studies have used proximal canopy sensors to estimate NNI of various crops [12–17]. However, the usage of proximal sensors is not efficient for large production fields and mounting the sensors on the ground vehicles is not suitable for rice production. Satellite remote sensing has also been used for monitoring crop growth and N status in large areas. The FORMOSAT-2 satellite images were used to estimate rice NNI and diagnose N status and the results indicated that a practical approach was to use the satellite images to estimate rice AGB and plant N uptake (PNU), which were then used to calculate N_c and NNI (R^2 = 0.52) [2]. The potential of using FORMOSAT-2, RapidEye, and WorldView-2 satellite data to estimate rice NNI were also evaluated and the results indicated that WorldView-2 satellite data performed the best [18]. However, in rice production areas, complete overcast weather conditions are very common, and it is very challenging to obtain satellite image data at the growth stages needed for guiding in-season topdressing N recommendations.

In recent years, unmanned aerial vehicle (UAV)-based remote sensing has developed rapidly, due to its low cost, ease of operation, and wide field of view [19,20]. The advances in data processing software have followed, allowing for automated development of image products [21]. A number of studies have used UAV remote sensing for crop N status diagnosis in various crops [22–27]. Most of these studies focused on identifying the optimum vegetation index (VI) and used linear regression method to estimate NNI or other N status indicators. The research should advance towards including more significant VIs and using nonlinear methods to improve the N status diagnosis with UAV remote sensing.

Over the past decade, machine learning (ML) methods have been widely adopted in complex and data-intensive areas such as medicine, astronomy, biology, and precision agriculture, due to their capability to discover information hidden in the data [28]. One of the main advantages of ML is that they are capable of solving significant nonlinear problems using datasets from multiple sources [29]. Agricultural remote sensing inversion is a typical nonlinear problem, and ML has been applied to solve it with satisfactory results [30,31]. For example, Han et al. used UAV remote sensing data and ML to estimate maize (*Zea mays* L.) biomass (R^2 = 0.70) [32]. Ali et al. developed a model for the estimation of grassland biomass by using adaptive neuro-fuzzy inference system and multi-temporal remote sensing (R^2 = 0.85) [33]. Pantazi et al. developed an artificial neural network (ANN)-based wheat yield prediction model using normalized difference VI (NDVI) derived from satellite imagery and eight soil parameters [34]. Liu et al. estimated wheat leaf N content using a multilayer perceptron neural network model and hyperspectral image data [35]. Zheng et al. compared different ML methods for estimating winter wheat leaf N content using UAV multispectral images and found that the fast processing random forest (RF) algorithm performed the best among the tested methods (R^2 = 0.79, RMSE = 0.33) [36].

The literature of using ML on the UAV-borne reflectance data for rice crop N status is limited. Therefore, the objective of this study was to evaluate five different approaches for estimating rice aboveground biomass (AGB), plant N uptake (PNU), and N nutrition index (NNI) at stem elongation (SE) and heading (HD) stages in Northeast China: (1) single VI (SVI); (2) stepwise multiple linear regression (SMLR); (3) random forest (RF); (4) support vector machine (SVM); and (5) artificial neural networks (ANN) regression. Our hypothesis is that the machine learning methods can analyze both linear and nonlinear relationships between a dependent variable and multiple independent variables and can improve the prediction of rice N status indicators using multiple VIs than methods using single VI or multiple linear regression method using multiple VIs. This paper is organized in the following sections: Section 1 provides an introduction of the background and objective of this research. Section 2 describes the field experiments, data collection and analysis methods. Section 3 presents the results, and Section 4 discusses the results. Section 5 concludes this research.

2. Materials and Methods

2.1. Study Site

The study site is located at the Jiangsanjiang Experiment Station of the China Agricultural University (47.2°N, 132.6°E) in the Sanjiang Plain of Heilongjiang Province, Northeast China (Figure 1). Sanjiang Plain belongs to a typical cool-temperate sub-humid continental monsoon climate zone. Japonica rice is the main planting crop in this cold region. The average sunshine hours are about 2300–2600 per year, the frost-free period is only about 110–135 days per year. The mean annual temperature is about 2 °C, and the average daily temperature is 19.9 °C during the growing season. The average rainfall is 500–600 mm per year, about 72% of which occurs between June and September [4]. The primary soil type in the Sanjiang Plain is Albic soil, classified as Mollic Planosols in the FAO-UNESCO system and Typical Argialbolls in Soil Taxonomy [37].

2.2. Experimental Setup

Ten plot experiments were conducted in 2017 and 2018 involving two Japonica rice cultivars Longjing 31 (with 11 leaves) and Longjing 21 (with 12 leaves), five N rates (0, 40, 80, 120, and 160 kg N ha^{-1}), two different planting densities (27 and 33 hills m^{-2}). All of the experiments adopted the randomized complete block design with three replicates (Figure 1). The size of each plot was 7 × 9 m and did not change during the study period. The N fertilizer was applied with three splits in the N rate experiments: 40% as basal application before transplanting, 30% at the tillering stage, and 30% at the stem elongation (SE) stage. Phosphorus and potassium fertilizers were applied uniformly across the plot experiments at the rates of 50 kg P_2O_5 ha^{-1} and 105 kg K_2O ha^{-1}, respectively. Phosphorus was applied in a single rate before transplanting and potassium was applied in two equal splits before transplanting and at the SE growth stage.

In addition to the plot experiments, three on-farm experiments were conducted in cooperation with three selected farmers of Qixing Farm in 2017 and 2018 in order to compare different precision rice management systems (Figure 1). The soil organic matter (OM) content was 30.2, 37.5, and 43.2 g kg^{-1} for Fields 1, 2, and 3, respectively. Treatments in each experiment included (1) Farmer's Practice

(FP); (2) Regional Optimal Management (ROM); (3) Precision Rice Management 1 (PRM1) with remote sensing-based N recommendation at stem elongation stage; (4) PRM2; and (5) PRM3. PRM2 and 3 used two different rates of controlled-release fertilizer as basal fertilizer. The plot size for each treatment varied from 20 × 8 m to 30 × 10 m, depending on the farmer's field situation. The rice cultivar was Longjing 31 (an 11 leaf cultivar), and each treatment was performed in triplicate. The details of planting density and N application rates are given in Table 1.

These plot and on-farm experiments were conducted for other objectives, but this study took advantage of the variable N status in these experiments to evaluate different UAV remote sensing-based N status estimation methods.

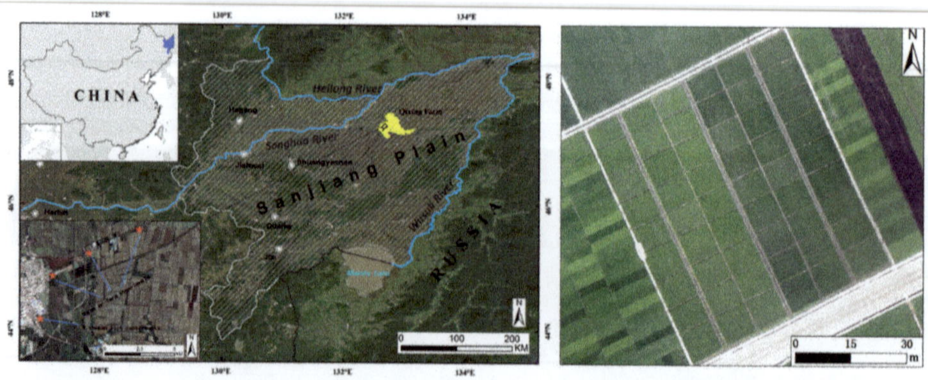

Figure 1. The location of the study sites (**left**) and a Red Green Blue (RGB) image of the N rate experimental plots (**right**).

Table 1. Fertilizer application rate and timing for different treatments in the on-farm experiments conducted in 2017 and 2018 at Qixing Farm.

Treatment *	Planting Density (plants m^{-2})	Total N Rate (kg ha^{-1})	Base N (kg ha^{-1})	Tiller N (kg ha^{-1})	Panicle N (kg ha^{-1})
FP	24	120	79	21	20
ROM	27	120	79	21	20
PRM1	27	?	71	21	?
PRM2	27	?	80	-	?
PRM3	27	?	80	-	?

Note: * FP: Farmer's Practice; ROM: Regional Optimum Management; PRM1–3: Precision Rice Management Strategy 1, 2, and 3. To be determined by sensor-based in-season N recommendation algorithm.

2.3. Field Data Collection and NNI Parametrization

After spectral data collection at the SE and heading (HD) growth stages, three hills of rice plants were randomly selected according to the average tillering numbers in each plot and removed with roots. They were washed with clear water, and the roots were removed with scissors. The cleaned samples were separated into leaves, stems, and panicles (at heading), put into the oven under 105 °C for 30 min to deactivate the enzymes, and then dried to a constant weight at about 80 °C to determine dry AGB. N concentrations for leaves, stems, and panicles were determined using the standard Kjeldahl method [38]. PNC was determined based on the weighted average of the N content of all rice organs. The PNU was determined by multiplying PNC with AGB.

The critical N dilution curve of rice in Northeast China developed by Huang et al. [8] shown in Equation (1) was used in this research for AGB larger than 1 t ha^{-1}:

$$Nc = 27.7W - 0.34, \quad (1)$$

where Nc is the critical N concentration (%) in the AGB, and W is the shoot dry weight expressed in t ha^{-1}. For AGB less than 1 t ha^{-1}, the Nc was set to a concentration of 2.77%.

The NNI was calculated using Equation (2)

$$\text{NNI} = \text{Na}/\text{Nc}, \tag{2}$$

where Na is the measured N concentration.

The NNI was also alternatively calculated using PNU, as given in Equation (3)

$$\text{NNI} = \text{PNU}/(\text{Nc} \times \text{AGB}), \tag{3}$$

where PNU is plant N uptake (kg ha^{-1}), and AGB is the aboveground biomass in t ha^{-1}.

2.4. UAV Image Acquisition and Preprocessing

This study utilized the eBee SQ fixed-wing UAV system (SenseFly, Cheseaux-sur-Lausanne, Switzerland) with Parrot Sequoia camera onboard. This camera includes a four-band multispectral camera (1.2 MP, 1280 × 960 pixels) with a green band (550 + 20 nm), red band (660 + 20 nm), Red edge band (735 + 5 nm), the near-infrared band (790 + 20 nm) and Red Green Blue (RGB) camera (16 MP, 4608 × 3456 pixels). The unit is equipped with the upwards-oriented irradiance sensor for automated control of the integration time on the detectors. The camera system was referenced for the current downwelling radiation before each flight mission using a white Spectralon® panel (Labsphere, Inc., North Sutton, NH, USA). The UAV missions were conducted between 10:00 and 14:00, under windless and clear-sky conditions.

The UAV mission control and image acquisition were performed by the flight control software eMotion Ag 3.5.0 (SenseFly, Cheseaux-sur-Lausanne, Switzerland). The flight altitude was 106 m, the ground sampling distance was about 0.1 m per pixel and the images were taken with the forward overlap and the side overlap of 85% and 75%, respectively [39]. After the data acquisition, the geotagged images were mosaicked using Pix4Dmapper Ag software (Pix4D SA, Prilly, Switzerland) to obtain the spectral reflectance image of the entire scene, covering the whole experimental area. The mosaic was later orthorectified in ENVI 5.1 software (ENVI, Harris Geospatial Solutions, Inc., Boulder, Colorado, USA), using the ground control points referenced by a survey-grade GNSS receiver (CHCNAV, LT500, Shanghai, China) [39]. A total of four UAV reflectance orthoimages were obtained at the SE and HD growth stages in 2017 and 2018. The plot boundaries were digitized and used as regions of interest to select and average image pixels at a given sampling point in order to relate them to the groundtruth data.

2.5. Data Analysis

In this study, the reflectance data from the four spectral bands were used to calculate 72 VIs (Table A1) and both raw reflectance data of the three wavebands and VIs were used in the analyses. The calculated VIs were ranked by R^2 for their relationships with AGB, PNU, and NNI and the top performing indices were further investigated.

The data collected in 2017 and 2018 were pulled together and then randomly divided into training dataset (70%) and test dataset (30%). A total of 381 observations were obtained in 2017 and 2018, 266 of which were used as training dataset and 115 as test dataset (Table 2). Among the analyzed crop properties, the AGB was the most variable parameter, with coefficient of variation (CV) being 37.54% for training and 42.37% for the test dataset, followed by PNU (CV% of 34.31% and 39.59%). PNC and NNI had similar variability, with CV of 6.03% and 16.42% in the training dataset and 15.47% and 17.86% in the test dataset, respectively. NNI ranged from 0.57 to 1.28 in the training dataset, and 0.58 to 1.21 in the test dataset. The data range of all training datasets encompassed the test dataset range, which ensured that the test data would not exceed the scope of the trained models. The training dataset was used to establish the simple regression models using linear, quadratic, power, exponential, and logarithmic functions or SMLR models between the VIs and AGB, PNU, and NNI. Established models were evaluated using the test dataset. The coefficient of determination (R^2), root mean square error (RMSE), and relative error (RE) were used to assess the models. The higher the R^2 and the lower the RMSE and RE, the higher was the precision and accuracy of the model for predicting the N status indicators. The scikit-learn [40,41], a Python machine learning library, was used in this study to establish models for estimation of AGB, PNU, and NNI using three conventional ML methods: RF, SVM, and ANN regressions. Tenfold cross-verification and grid search were used to find the optimal parameters during model development. The test dataset and R^2, RMSE, RE were used to evaluate the accuracy of the models.

Table 2. Descriptive statistics of rice aboveground biomass (AGB), plant nitrogen concentration (PNC), plant nitrogen uptake (PNU), and nitrogen nutrition index (NNI) across nitrogen treatments, varieties, and years.

	Minimum	Maximum	Mean	SD	CV (%)
		Training dataset (n = 266)			
AGB (t ha^{-1})	0.98	10.86	5.28	1.98	37.54
PNC (g kg^{-1})	8.75	20.99	15.65	2.51	16.03
PNU (kg ha^{-1})	15.73	154.10	80.60	27.74	34.41
NNI	0.57	1.28	0.97	0.16	16.42
		Test dataset (n = 115)			
AGB (t ha^{-1})	1.51	10.45	5.25	2.22	42.37
PNC (g kg^{-1})	9.36	20.04	15.53	2.40	15.47
PNU (kg ha^{-1})	23.83	154.09	79.62	31.53	39.59
NNI	0.58	1.21	0.95	0.17	17.86

Note: SD: standard deviation of the mean; CV: coefficient of variation (%).

3. Results

3.1. Single Spectral Band Analysis

The coefficient of determination for the relationships between the reflectance of each of the four wavebands and rice N status indicators at different growth stages are shown in Figure 2 for both training and testing datasets. The NIR band consistently had the highest R^2 for AGB and PNU, while for PNC and NNI, the sensitivity of different wavebands changed with growth stages. In general, the relationships between reflectance of different wavebands and PNC were weaker than other N status indicators.

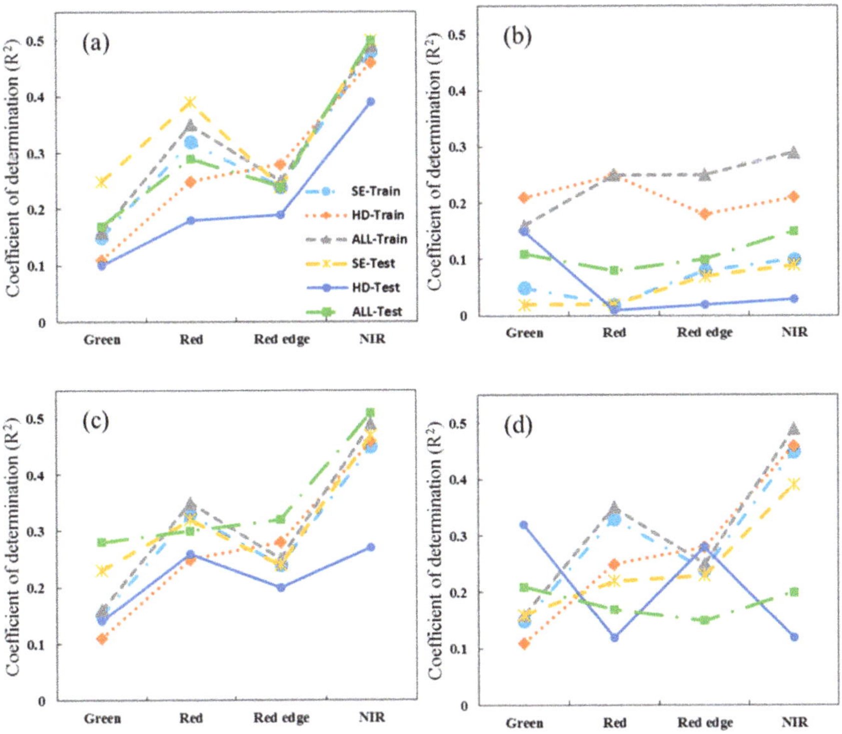

Figure 2. Coefficients of determination (R^2) for the relationships between reflectance of unmanned aerial vehicle (UAV) camera bands and aboveground biomass (**a**), plant N concentration (**b**), plant N uptake, (**c**) and nitrogen nutrition index (**d**) at different stages (SE, stem elongation stage; HD, heading stage; All, across stages) for both training and test datasets.

3.2. Vegetation Index Analysis

The three top performing VIs for estimating rice N status indicators based on the training dataset are given in Table 3. At best, a single VI could explain 65%, 65%, and 74% of AGB variation at the SE, HD, and across growth stages, respectively. The corresponding R^2 was 0.61, 0.69, and 0.73 for PNU at SE, HD, and across stages, respectively. For NNI, 43%, 63%, and 39% of the variabilities were explained by the best VI at SE, HD, and across growth stages, respectively. All these relationships were significant at $p < 0.01$.

The VIs with the highest R^2 were selected to establish the regression models for prediction of AGB, PNU, and NNI, which were validated using the test dataset and the results are shown in Figure 3. The models performed worse for AGB and PNU at SE and HD stages compared with calibration models, but slightly better across growth stages. For NNI, the models performed better with the test dataset. The indirect estimation of NNI performed slightly better than the direct approach at SE and HD stages, but somewhat worse across growth stages (Figure 4).

Table 3. The top three performing vegetation indices for estimating rice AGB, PNU, and NNI. All the relationships were significant at $p < 0.01$.

AGB (t ha^{-1})					PNU (kg ha^{-1})					NNI				
Index	Model	R^2	RMSE	RE (%)	Index	Model	R^2	RMSE	RE (%)	Index	Model	R^2	RMSE	RE (%)
						Stem elongation stage								
GOSAVI	E	0.65	0.58	16	GOSAVI	E	0.61	11.40	18	NNIR	Q	0.43	0.10	11
GRDVI	P	0.64	0.58	16	GRDVI	P	0.60	11.63	18	GOSAVI	E	0.42	0.10	11
GSAVI	E	0.63	0.59	16	NNIR	E	0.60	12.38	20	GRDVI	P	0.42	0.10	11
						Heading stage								
NLI	P	0.65	1.01	15	GOSAVI	P	0.69	15.84	16	GNDVI	P	0.63	0.11	12
WDRVI	P	0.61	1.08	16	NDVI	P	0.61	17.54	18	CIg	Q	0.63	0.11	11
GSAVI	P	0.59	1.03	15	WDRVI	P	0.60	17.57	18	GRVI	Q	0.63	0.11	11
						Across growth stages								
MGSAVI	E	0.74	1.10	21	GOSAVI	Q	0.73	14.95	19	CIg	Q	0.39	0.13	13
GRDVI	E	0.73	1.12	21	MGSAVI	E	0.69	16.30	20	GRVI	Q	0.38	0.13	13
GSAVI	E	0.73	1.13	21	GRDVI	E	0.69	16.48	21	GWDRVI	Q	0.38	0.13	13

Note: E, P, and Q: the exponential, power, and quadratic fit. The vegetation index abbreviations are explained in Table A1.

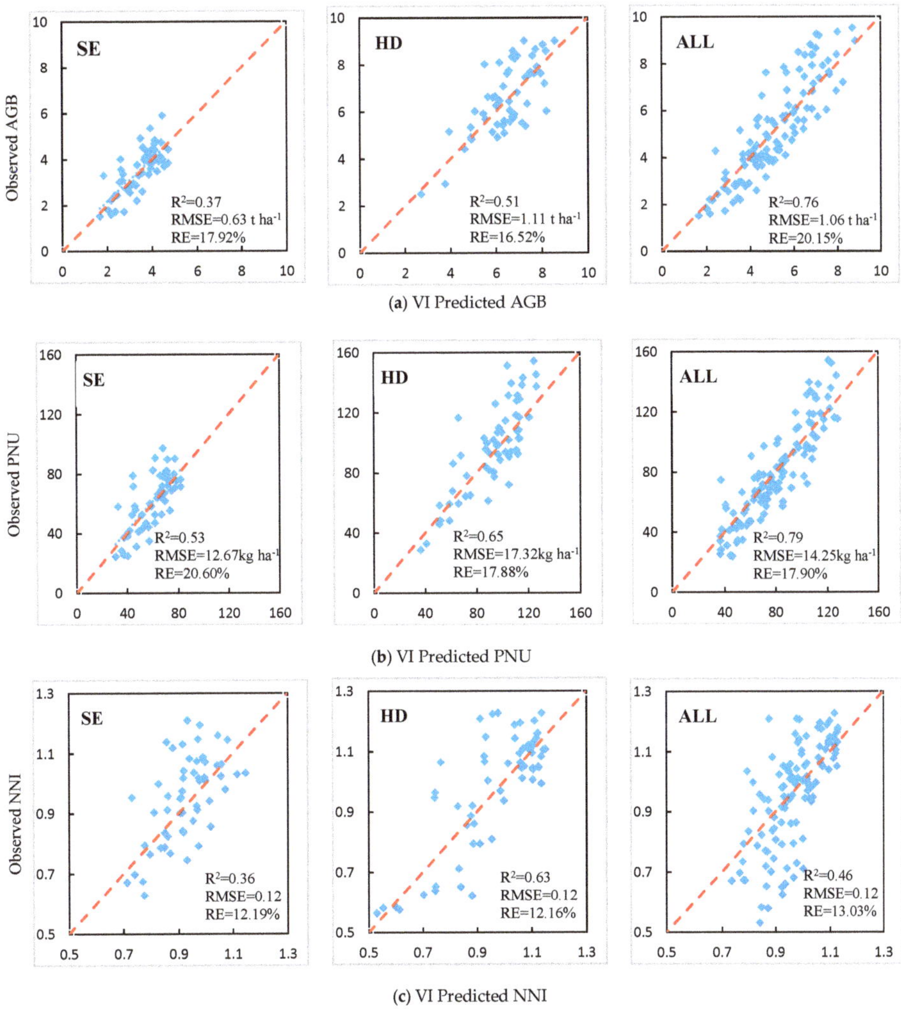

Figure 3. Relationships between predicted and observed AGB (**a**), PNU (**b**), and NNI (**c**) using the vegetation index approach within the test dataset at stem elongation (SE), heading (HD), and across growth stages (ALL). The red line is the 1:1 line.

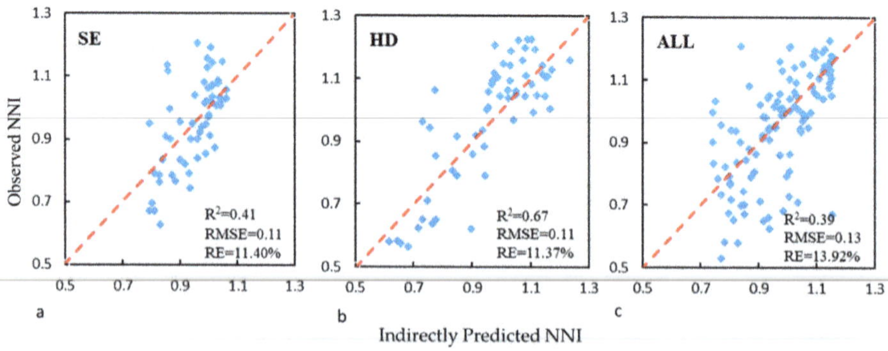

Figure 4. Relationships between observed and indirectly predicted NNI based on predicted AGB and PNU within the test dataset at SE (**a**), HD (**b**), and ALL (**c**). The red lines are 1:1 lines.

3.3. *Stepwise Multiple Linear Regression (SMLR) Analysis*

The SMLR analysis results indicated that the models could explain 69%, 62%, and 68% of AGB variation at SE, HD, and across stages using 2–4 VIs, respectively (Table 4). Similar results were obtained for PNU. These models explained 54%, 75%, and 40% of the NNI variability at the SE, HD, and across stages, respectively. These models performed better than models based on single VI in terms of R^2, RMSE, and RE.

The test results given in Table 5 indicate that the SMLR performed better at estimating AGB and NNI than models using single VI, while for PNU estimation, the two modeling methods performed similarly. Moreover, the results of indirect prediction of NNI were similar to the results of direct prediction.

Table 4. Stepwise multiple linear regression (SMLR) models based on unmanned aerial vehicle (UAV) data for estimation of rice AGB, PNU, and NNI at SE, HD, and ALL with data from training the dataset.

Stage	Regression Equation	R^2	RMSE	RE (%)
	AGB (kg ha^{-1})			
SE	−4.053 + 4.384*GNDVI + 0.211*RESR + 16.482*MTCAR/OSAVI	0.69	0.51	14
HD	−5.475 + 8.159*MCARI3 + 1.106*MSR	0.62	0.97	14
All	7.906 + 81.541*MGSAVI − 90.222*GSAVI − 3.516*MCARI2*OSAVI	0.68	1.11	21
	PNU (kg ha^{-1})			
SE	−198.601 + 353.387*GOSAVI + 132.397*MNDRE2 − 91.552*MCARI1	0.63	10.32	16
HD	−267.115 + 579.684*GOSAVI − 206.772*RE	0.69	15.18	16
All	6.614 + 613.62*MGSAVI − 1711.01*SAVI + 248.331*REDVI + 1237.866*RDVI	0.73	14.38	18
	NNI			
SE	−7.976 + 32.438*NNIR − 15.718*NNIRI + 16.493*RE − 7.852*MGSAVI + 0.038*SAVI*SR	0.54	0.09	9
HD	−36.417 + 39.501*GNDVI + 103.241*NGI − 2.601*MNDI	0.75	0.09	9
All	0.983 + 0.776*MNDRE2 − 7.632*NGI + 7.384*R	0.40	0.13	13

Note: R, RE, are the red and red edge bands, respectively. The vegetation index abbreviations are explained in Table A1.

Table 5. Validation results of the SMLR models for estimating rice AGB, PNU, and NNI at SE, HD, and ALL.

Parameter	SE			HD			ALL		
	R^2	RMSE	RE (%)	R^2	RMSE	RE (%)	R^2	RMSE	RE (%)
AGB (t ha^{-1})	0.61	0.51	14	0.52	1.09	16	0.77	1.05	20
PNU (kg ha^{-1})	0.60	11.60	19	0.65	16.96	18	0.80	13.76	17
NNI	0.52	0.10	10	0.74	0.09	10	0.53	0.11	12
NNI_Indirect	0.51	0.10	11	0.74	0.10	10	0.49	0.11	12

3.4. Performance of Machine Learning Models

For estimating AGB and PNU, the RF and ANN models consistently performed better than the SVM models, while for NNI, the RF model consistently performed the best at different growth stages, based on the calibration dataset (Table 6). The validation results indicated that the RF models performed consistently the best among the tested methods, including the indirect estimation of NNI (Table 7). Some models, especially those based on the ANN method, did not validate well with the test dataset, indicating the problem of overfitting.

Table 6. The calibration result of random forest (RF), support vector machine (SVM), and artificial neural networks (ANN) modeling algorithms at SE, HD, and ALL for rice AGB, PNU, and NNI.

Parameter		SE			HD Subset			ALL		
		R^2	RMSE	RE (%)	R^2	RMSE	RE (%)	R^2	RMSE	RE (%)
AGB (t ha^{-1})	RF	0.87	0.33	9	0.85	0.6	9	0.92	0.54	10
	SVM	0.74	0.47	13	0.62	0.79	11	0.88	0.69	17
	ANN	0.88	0.32	9	0.77	0.74	11	0.97	0.31	19
PNU (kg ha^{-1})	RF	0.93	4.59	7	0.93	7.05	7	0.90	8.59	16
	SVM	0.65	10.05	16	0.70	15.07	15	0.73	14.38	18
	ANN	0.71	9.1	14	0.73	13.53	14	0.95	6.47	8
NNI	RF	0.94	0.03	3	0.96	0.03	3	0.93	0.04	4
	SVM	0.65	0.08	8	0.79	0.08	8.52%	0.75	0.08	8.08%
	ANN	0.73	0.07	7	0.81	0.08	8.66%	0.55	0.11	10.61%

Table 7. The validation result of RF, SVM, and ANN model algorithms at SE, HD, and ALL for rice AGB, PNU, and NNI.

Parameter		SE			HD			ALL		
		R^2	RMSE	RE (%)	R^2	RMSE	RE [%]	R^2	RMSE	RE (%)
AGB (t ha^{-1})	RF	0.64	0.58	16	0.61	1.00	15	0.83	0.58	16
	SVM	0.38	0.76	22	0.59	1.01	15	0.81	0.95	18
	ANN	0.60	0.62	17	0.39	1.24	18	0.65	1.31	25
PNU (kg ha^{-1})	RF	0.62	11.52	19	0.69	16.45	17	0.83	12.81	16
	SVM	0.55	11.92	19	0.49	20.98	22	0.79	14.16	18
	ANN	0.57	12.13	9	0.63	17.89	18	0.74	15.88	20
NNI	RF	0.58	0.09	10	0.79	0.09	9	0.72	0.09	9.34
	SVM	0.46	0.10	11	0.70	0.11	11	0.62	0.11	11
	ANN	0.56	0.10	10	0.79	0.09	9	0.61	0.11	11
NNI_Indirect	RF	0.54	0.10	10	0.64	0.10	10	0.64	0.10	11
	SVM	0.37	0.11	12	0.49	0.15	16	0.50	0.12	13
	ANN	0.34	0.17	18	0.58	0.14	15	0.46	0.15	16

3.5. Random Forest Models Based on Selected Vegetation Indices

For practical applications, the RF models were optimized by removing VIs not important for the performance of the model. This resulted in simpler models, yet with comparable performance to the models based on all the tested VIs (Table 8). Models established at the SE stage, although outperforming the models based on single VI or SMLR models, performed worse in comparison with the models at the HD stage and across stages. The indirect NNI estimation approach gave worse results than the direct approach, which was similar to the results obtained with SMLR analysis.

Depending on the analyzed subset, from 17 to 23 VIs were selected by the RF models at different growth stages and the top five VIs are listed in Table 9. The relative importance of different VIs changed with growth stages or dependent variables. Green Optimized Soil Adjusted Vegetation Index (GOSAVI) was consistently selected among the top five indices at SE, HD, or across growth stages for both AGB and NNI prediction, and at SE and HD stages for PNU prediction. Normalized near-infrared (NNIR) and red edge difference vegetation index (REDVI) were among the top five indices for AGB and PNU at the SE stage, and for NNI at both SE and HD stages.

Table 8. The calibration and validation results of RF models based on selected vegetation indices at SE, HD, and ALL for rice AGB, PNU, and NNI.

Parameter		SE			HD			ALL		
		R^2	RMSE	RE	R^2	RMSE	RE	R^2	RMSE	RE
AGB	Calibration	0.91	0.28	7.66	0.95	0.35	5.06	0.97	0.36	6.90
(t ha^{-1})	Validation	0.66	0.58	16.45	0.69	0.88	13.15	0.83	0.92	17.54
PNU	Calibration	0.94	4.11	6.52	0.96	5.32	5.46	0.94	7.35	9.12
(kg ha^{-1})	Validation	0.66	11.13	18.10	0.69	16.39	16.94	0.85	12.37	15.55
NNI_direct	Calibration	0.94	0.03	3.33	0.96	0.04	3.65	0.93	0.04	4.45
	Validation	0.61	0.09	9.98	0.79	0.09	9.06	0.74	0.09	8.72
NNI_indirect	Validation	0.53	0.10	10.77	0.72	0.10	10.60	0.67	0.10	10.13

Table 9. The relative importance of top five vegetation indices selected by RF models at SE, HD, and ALL for rice AGB, PNU, and NNI.

AGB (t ha^{-1})		PNU (kg ha^{-1})		NNI	
SE	*N* = 21		*N* = 21		*N* = 22
NNIR	0.09	NNIR	0.22	REDVI	0.21
REDVI	0.09	REDVI	0.20	NNIR	0.13
MSR_G	0.0	GOSAVI	0.12	MERIS	0.06
GOSAVI	0.07	NLI	0.05	MTCARI/OSAVI	0.05
CIg	0.06	REOSAVI	0.04	GOSAVI	0.05
HD	*N* = 17		*N* = 20		*N* = 23
OSAVI	0.30	GOSAVI	0.49	GNDVI	0.53
MCARI3	0.23	GWDRVI	0.18	NNIR	0.09
VIopt	0.10	NRI2	0.04	GOSAVI	0.09
GOSAVI	0.07	NRI	0.04	NGI	0.04
MCARI1/MRETVI	0.05	Green	0.03	REDVI	0.02
ALL	*N* = 19		*N* = 23		*N* = 23
GRDVI	0.37	GRDVI	0.49	CIg	0.24
GOSAVI	0.30	GRVI	0.14	GOSAVI	0.10
NLI	0.06	NNIR	0.05	Red	0.06
MNDRE	0.04	SAVI*SR	0.05	RETVI	0.05
OSAVI	0.04	GSAVI	0.04	MDD	0.04

Note: N—number of variables selected by the respective models. The vegetation index abbreviations are explained in Table A1.

3.6. Nitrogen Status Diagnosis at the Farm Scale

The N status diagnosis maps for the study area were created based on the predicted NNI using the fixed wing UAV remote sensing images and the RF models at the SE (Figure 5) and HD (Figure 6) stages in 2017. At the SE stage, the majority fields had optimal or surplus N status, with less N deficient areas (Figure 3). At the HD stage, the majority fields had deficient or surplus N status, with less areas having optimal N status (Figure 4). For the N plot experiments, most of the plots receiving less than 120 kg N ha^{-1} were classified as N deficient, while most plots receiving 160 kg N ha^{-1} were classified as surplus N, whereas parts of these plots were also categorized as optimal N and parts of plots receiving 120 kg N ha^{-1} were also classified as N surplus.

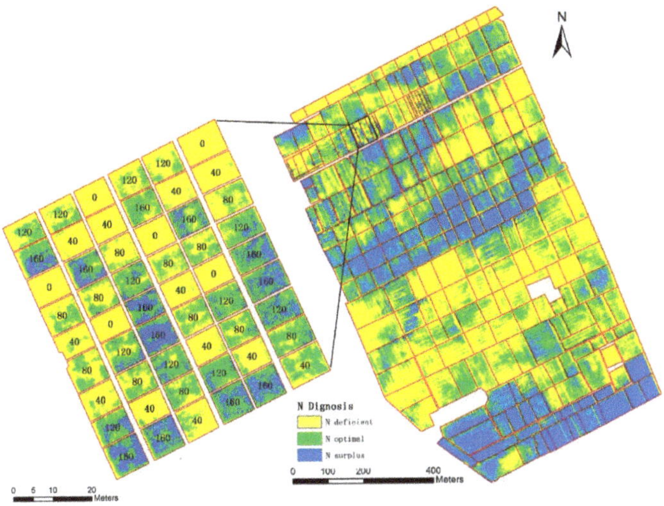

Figure 5. The N status diagnosis maps of the study area and N rate experimental plots based on NNI predicted by the RF model at SE stage of 2017.

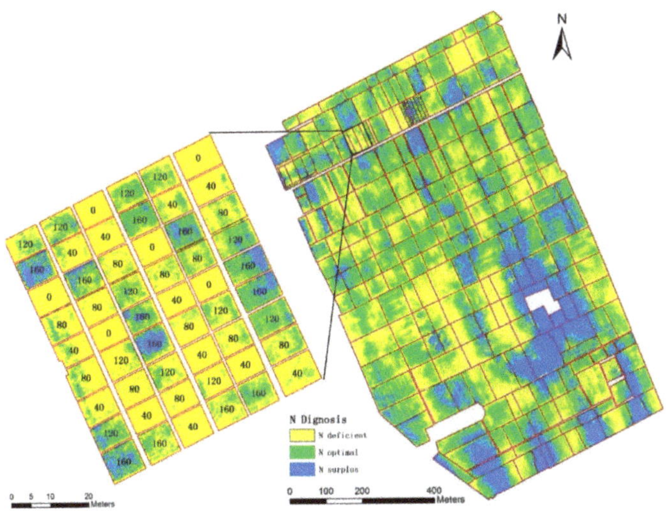

Figure 6. The N status diagnosis maps based on the predicted NNI using the RF model for the study area and N rate experimental plots at HD stage in 2017.

4. Discussion

4.1. Estimating Rice N Status Indicators Using Single Vegetation Index

Using UAV-based remote sensing for in-season crop N status diagnosis and guiding variable rate N application is very attractive. The reflectance of single spectral bands can be used to estimate crop N status, as indicated by the results of this study. However, this approach only uses the reflectance of only one spectral band. A common approach to use reflectance information from more than one spectral band is to develop VIs, which are mathematical combinations of reflectance from two or more spectral bands. VIs are expected to perform better than single spectral wavebands. Many different factors may influence the performance of VIs, including soil and water backgrounds, weeds, cover crops in the interrow, the types of plants, and the growth stages of crops, etc. [42]. Growth stage can have a strong influence on the sensitivity and performance of different wavelengths and VIs for estimating crop parameters [43,44]. For rice, soil and water background can have a strong influence on canopy reflectance at early growing season before rice canopy closure (e.g., tillering stage or SE stage). At later growth stages with canopy closure (e.g., HD stage), some VIs like normalized difference vegetation index (NDVI) can become saturated [44]. In addition, the emergence of panicles makes the canopy reflectance more complicated, increasing the reflectance in visible spectral region but decreasing reflectance in the NIR region [45]. As a result, many different VIs have been developed for different applications [42]. It is necessary to evaluate the published VIs and identify the best performing VIs for a particular application (e.g., estimation of rice N status indicators).

The results of this study indicated that GOSAVI, Nonlinear Index (NLI), and Modified Green Soil Adjusted Vegetation Index (MGSAVI) performed best, explaining 65%, 65%, and 74% of rice AGB variability at SE, HD, and across growth stages, respectively. The GOSAVI explained 61%, 69%, and 73% of PNU variability at the SE, HD, and across growth stages, respectively. However, at best 63% of the NNI variability could be explained by Green Normalized Difference Vegetation Index (GNDVI) at the HD stage, but only 43% and 39% of NNI variability could be explained at the SE and across stages. The results of Cao et al. using active canopy sensor Crop Circle ACS-470 indicated that 54%–79%, 59%–83%, and 59%–77% of rice AGB, PNU, and NNI variabilities could be explained by the best performing VIs, respectively [46]. This study gave similar results for AGB and PNU, however the NNI estimations at SE and across growth stages were worse than the results obtained by Cao et al. [46]. This may be due to the fact that the UAV image sampling included the entire areas of the plots. The soil and water background may have more influence on the reflectance when compared with handheld canopy sensor in the study of Cao et al. [46]. As a result, using UAV remote sensing and VI-based approach could not achieve acceptable NNI estimation at the SE stage before canopy closure. In a similar research with winter wheat across smallholder farmer fields, Chen et al. explained 72%, 64%, and 46% variation in winter wheat (*Triticum aestivum* L.) AGB, PNU, and NNI at the SE stage using single VI-based approach with eBee SQ UAV remote sensing [39]. Their results were comparable to our results, with AGB and PNU being better predicted than NNI.

4.2. The Performance of Different Machine Learning Modeling Methods

In addition to single VI, SMLR and three different ML algorithms were applied to predict rice N status indicators in this study. The SMLR model performed significantly better than models based on single VI. Our results are consistent with the results of previous studies with winter wheat [47]. SMLR models use more VIs with spectral information related to the variables of interest and are flexible and easy to perform [48–51].

The SMLR models can only model linear combination of predictors [52], while the ML models can also model nonlinear relationships. The RF regression algorithm is an ensemble-learning algorithm that combines a broad set of regression trees. A regression tree represents a set of conditions or restrictions that are hierarchically organized and successively applied from a root to a leaf of the tree [53–55]. The SVM algorithm is based on statistical learning theory and can be regarded as the same type of network, can also be used for both classification and regression problems [56]. ANN regression is based on the gradient learning method. It is a nonparametric nonlinear model that uses neural network spreading between layers and simulates human brain receivers and information processing [57,58]. All three ML models performed better than models based on single VI. The three ML models all achieved better results than SMLR models in calibration, but in the validation analysis, only the RF models performed consistently better than SMLR. The possible reason for such results is that ML modeling often results in an over-fitting phenomenon, and the robustness and generalization ability of RF are stronger than the other ML methods [31,36,58–60].

The results of NNI indirect estimation approach were generally worse than the direct estimation approach. This is possibly due to the estimation of AGB and PNU in the indirect estimation approach that led to the accumulation of errors.

In summary, the results of this study indicated that the RF algorithm could be used to predict NNI directly at different growth stages. It performed better than other evaluated approaches. The NNIR and REDVI indices were the most important predictors at the SE stage, GNDVI and NNIR were most important at the HD stage, while green chlorophyll index (CIg) and GOSAVI were the most important predictors across growth stages. The relative importance of different VIs varied with growth stages and N status indicators. NNIR, REDVI, and other VIs containing red edge and near-infrared bands were more important for NNI estimation. Some of the VIs were significantly correlated. When the model needs to select input parameters, if the correlation between two VIs is very high, the RF model tends to select only one VI and abandons the other. Many of the VIs with small weights were selected in the models, because these algorithms need to use more dimensions to explain the variation of the data.

4.3. Challenges and Future Research Needs

In this study, multispectral data and VIs were obtained using fixed-wing UAV remote sensing, and the rice NNI distribution maps at different growth stages were created based on RF model prediction. The NNI map at the SE stage can be used to guide farmers to apply N fertilizers at the variable rates. The use of fixed-wing UAV remote sensing can effectively overcome the limitations of satellite remote sensing and proximal crop canopy sensing, and provide a reliable data source for diagnosis of the rice N nutritional status and in-season variable rate recommendation.

At present, most UAVs for remote sensing are powered with batteries, and the operation time is still quite short. For example, the eBee SQ system can only fly about 40 min, which limits the data acquisition ability of a single UAV operator. If UAVs adopt larger battery capacities and more effective battery charging in the future, effectively solving the problem of insufficient power, it will greatly increase the operational efficiency and the area monitored by a single unit. In addition, the field preparation for setting up ground control points, reflectance panels, and flight design is also very time consuming. The advances in technology have made it possible to achieve similar precision without the use of ground control points, to get rid of ground reflectance panels by using incident light sensors and greatly simplify flight design [61]. UAV remote sensing is also significantly affected by weather conditions, like winds, rain, or clouds [61,62]. Mounting active canopy sensors on UAV may provide a practical solution to such weather limitations [63].

In this study, UAV image-based rice crop reflectance was the single data source used in the ML models. This alone showed that nonlinear ML models improved NNI estimation compared to the simple VI-based methods. In addition to the commonly used red, green, blue, red edge, and near-infrared bands, other spectral regions should be studied for N status diagnosis, like shortwave infrared (SWIR)-based indices [45,64] or using hyperspectral cameras [62]. Studies found that a combination of multispectral and thermal images using relevance vector machines improved the estimation of plant chlorophyll concentration [65] and has the potential to simultaneously identify N and water stress. In the future, meteorological data, soil data, terrain attributes, and the information about crop management can be used together with remote sensing data to improve the performance of the ML models and NNI estimation [66].

5. Conclusions

In this study, eBee SQ UAV images were used to evaluate the VI, SMLR, and three ML algorithms (RF, SVM, and ANN) to estimate rice AGB, PNU, and NNI at the SE, HD, and across stages, and the NNI maps were created to diagnose N nutritional status of rice fields at the study site in Northeast China. The results indicated that ML methods could significantly improve the estimation of rice NNI compared to single VI and SMLR models, especially using an optimized RF algorithm, with 94% and 96% of the NNI variability being explained for the calibration dataset at the SE and HD stages, respectively, and 61% and 79% of NNI variability being explained for the test dataset at the SE and HD stages, respectively. The RMSE was 0.09, and RE was less than 10%. It is concluded that the RF modeling method can significantly improve the prediction of rice NNI using UAV remote sensing. The application machine learning methods offers a new opportunity to better use remote sensing data for monitoring crop growth conditions and guiding precision crop management. More studies are needed to further improve these machine learning-based models by combining both remote sensing data and other related soil, weather, and management information for applications in precision N and crop management.

Author Contributions: Y.M. and H.Z. conceptualized and designed the research. H.Z., T.W., Y.L., J.Z., W.S., and Z.F. conducted the field experiments and performed data collection. H.Z. analyzed the data and wrote the original manuscript. Y.M. and K.K. reviewed and revised the paper. The research was supervised by Y.M. and K.K. secured the research funding and managed the projects. All authors have read and agreed to the published version of the manuscript.

Funding: This research was funded by the Norwegian Ministry of Foreign Affairs (SINOGRAIN II, CHN-17/0019), the National Key Research and Development Program of China (2016YFD0200600, 2016YFD0200602), National Basic Research Program of China (2015CB150405), and the Internationalization Training and Promotion Project of Graduate Students in China Agricultural University.

Acknowledgments: The authors thank Guojun Li, Yankai Niu, and Guangming Zhao at Jansanjiang Bureau of Agriculture in Heilongjiang Province, Beijing Aoteng Yanshi Technology Co., Ltd., and Anhui Yigang Information Technology Co., Ltd. for their assistance during this research. We also would like to thank all of the farmers for their cooperation in this research.

Conflicts of Interest: The authors declare no conflict of interest.

Appendix A

Table A1. The vegetation indices evaluated in this study. G, R, RE, and NIR indicate green red, red edge, and near infrared band reflectance.

Index	Formula	Reference
Green Ratio Vegetation Index (GRVI)	NIR/G	[67]
Green Difference Vegetation Index (GDVI)	NIR − G	[68]
Green Normalized Difference Vegetation Index (GNDVI)	(NIR − G)/(NIR + G)	[69]
Green Wide Dynamic Range Vegetation Index (GWDRVI)	(a*NIR − G)/(a*NIR + G) (a = 0.12)	[46]
Green Chlorophyll Index (CIg)	NIR/G − 1	[70]
Modified Green Simple Ratio (MSR_G)	(NIR/G − 1)/SQRT(NIR/G + 1)	[46]
Green Soil Adjusted Vegetation Index (GSAVI)	1.5*[(NIR − G)/(NIR + G + 0.5)]	[71]
Modified Soil Adjusted Vegetation Index (MSAVI)	0.5*[2*NIR + 1 − SQRT((2*NIR + 1)2 − 8*(NIR − G))]	[72]
Green Optimal Soil Adjusted Vegetation Index (GOSAVI)	(1 + 0.16)(NIR − G)/(NIR + G + 0.16)	[73]
Green Re-normalized Different Vegetation Index (GRDVI)	(NIR − G)/SQRT(NIR + G)	[46]
Normalized Green Index (NGI)	G/(NIR + RE + G)	[71]
Normalized Red Edge Index (NREI)	RE/(NIR + RE + G)	[46]
Normalized Red Index (NRI)	R/(NIR + RE + R)	[14]
Normalized NIR Index (NNIR)	NIR/(NIR + RE + G)	[71]
Modified Double Difference Index (MDD)	(NIR − RE) − (RE − G)	[14]
Modified Normalized Difference Index (MNDI)	(NIR − RE)/(NIR − G)	[46]
Modified Enhanced Vegetation Index (MEVI)	2.5*(NIR − RE)/(NIR + 6*RE − 7.5*G + 1)	[46]
Modified Normalized Difference Red Edge (MNDRE)	[NIR − (RE − 2*G)]/[NIR + (RE − 2*G)]	[46]
Modified Chlorophyll Absorption In Reflectance Index1 (MCARI1)	[(NIR − RE) − 0.2*(NIR − R)](NIR/RE)	[46]

Table A1. *Cont.*

Index	Formula	Reference
Modified Chlorophyll Absorption In Reflectance Index 2 (MCARI2)	$\frac{1.5[2.5(NIR-R)-1.3(NIR-RE)]}{\sqrt{(2NIR+1)^2-(6NIR-5\sqrt{R})-0.5}}$	[14]
Normalized Difference Vegetation Index (NDVI)	(NIR − R)/(NIR + R)	[74]
Ratio Vegetation Index(RVI)	NIR/R	[75]
Difference Vegetation Index (DVI)	NIR − R	[68]
Renormalized Difference Vegetation Index (RDVI)	(NIR − R)/SQRT(NIR + R)	[76]
Wide Dynamic Range Vegetation Index (WDRVI)	(a*NIR − R)/(a*NIR + R) (a = 0.12)	[77]
Soil-Adjusted Vegetation Index (SAVI)	1.5*(NIR − R)/(NIR + R + 0.5)	[78]
Optimized SAVI (OSAVI)	(1 + 0.16)*(NIR − R)/(NIR + R + 0.16)	[73]
Modified Soil-adjusted Vegetation Index (MSAVI)	0.5*[2*NIR + 1 − SQRT((2*NIR + 1)2 − 8*(NIR − R))]	[72]
Transformed Normalized Vegetation Index (TNDVI)	SQRT((NIR − R)/(NIR + R) + 0.5)	[79]
Modified Simple Ratio (MSR)	(NIR/R − 1)/SQRT(NIR/R + 1)	[80]
Optimal Vegetation Index (VIopt)	1.45*((NIR^2 + 1)/(R + 0.45))	[81]
MERIS Terrestrial Chlorophyll Index (MTCI)	(NIR − RE)/(RE − R)	[82]
Nonlinear Index (NLI)	(NIR^2 − R)/(NIR2 + R)	[83]
Modified Nonlinear Index (MNLI)	1.5*(NIR^2 − R)/(NIR^2 + R + 0.5)	[84]
NDVI*RVI	(NIR^2 − R)/(NIR + R^2)	[84]
SAVI*SR	(NIR^2 − R)/[(NIR + R + 0.5)*R]	[84]
Normalized Difference Red Edge (NDRE)	(NIR − RE)/(NIR + RE)	[85]
Red Edge Ratio Vegetation Index (RERVI)	NIR/RE	[86]
Red Edge Difference Vegetation Index (REDVI)	NIR − RE	[46]
Red Edge Re-normalized Different Vegetation Index (RERDVI)	(NIR − RE)/SQRT(NIR + RE)	[46]
Red Edge Wide Dynamic Range Vegetation Index (REWDRVI)	(a*NIR − RE)/(a*NIR + RE) (a = 0.12)	[46]
Red Edge Soil Adjusted Vegetation Index (RESAVI)	1.5*[(NIR − RE)/(NIR + RE + 0.5)]	[46]
Red Edge Optimal Soil Adjusted Vegetation Index (REOSAVI)	(1 + 0.16)(NIR − RE)/(NIR + RE + 0.16)	[46]
Modified Red Edge Soil Adjusted Vegetation Index (MRESAVI)	0.5*[2*NIR + 1 − SQRT((2*NIR + 1)2 − 8*(NIR − RE))]	[46]
Optimized Red Edge Vegetation Index (REVIopt)	100*(lnNIR − lnRE)	[87]
Red Edge Chlorophyll Index (CIre)	NIR/RE − 1	[88]
Modified Red Edge Simple Ratio (MSR_RE)	(NIR/RE − 1)/SQRT(NIR/RE + 1)	[14]
Red Edge Normalized Difference Vegetation Index (RENDVI)	(RE − R)/(RE + R)	[89]
Red Edge Simple Ratio (RESR)	RE/R	[90]
Modified Red Edge Difference Vegetation Index (MREDVI)	RE − R	[46]
MERIS Terrestrial Chlorophyll Index (MTCI)	(NIR − RE)/(RE − R)	[82]
DATT Index (DATT)	(NIR − RE)/(NIR − R)	[91]
Normalized Near Infrared Index (NNIRI)	NIR/(NIR + RE + R)	[14]
Normalized Red Edge Index (NREI)	RE/(NIR + RE + R)	[14]
Normalized Red Index (NRI)	R/(NIR + RE + R)	[14]
Modified Double Difference Index (MDD)	(NIR − RE) − (RE − R)	[14]
Modified Red Edge Simple Ratio (MRESR)	(NIR − R)/(RE − R)	[14]
Modified Normalized Difference Index (MNDI)	(NIR − RE)/(NIR + RE − 2R)	[14]
Modified Enhanced Vegetation Index (MEVI)	2.5*(NIR − R)/(NIR + 6*R − 7.5*RE + 1)	[14]
Modified Normalized Difference Red Edge (MNDRE2)	(NIR − RE + 2*R)/(NIR + RE − 2*R)	[14]
Red Edge Transformed Vegetation Index (RETVI)	0.5*[120*(NIR − R) − 200*(RE − R)]	[14]
Modified Chlorophyll Absorption In Reflectance Index 3 (MCARI3)	[(NIR − RE) − 0.2*(NIR − R)](NIR/RE)	[14]
Modified Chlorophyll Absorption In Reflectance Index 4 (MCARI4)	$\frac{1.5[2.5(NIR-G)-1.3(NIR-RE)]}{\sqrt{(2NIR+1)^2-(6NIR-5\sqrt{G})-0.5}}$	[14]
Modified Transformed Chlorophyll Absorption In Reflectance Index (MTCARI)	3*[(NIR − RE) − 0.2*(NIR − R)(NIR/RE)]	[14]
Modified Red Edge Transformed Vegetation Index (MRETVI)	1.2*[1.2*(NIR − R) − 2.5*(RE − R)]	[14]
Modified Canopy Chlorophyll Content Index (MCCCI)	NDRE/NDVI	[92]
MCARI1/OSAVI	MCARI1/OSAVI	[14]
MCARI2/OSAVI	MCARI2/OSAVI	[14]
MTCARI/OSAVI	MTCARI/OSAVI	[14]
MCARI1/MRETVI	MCARI1/MRETVI	[14]
MTCARI/MRETVI	MTCARI/MRETVI	[14]

References

1. Miao, Y.; Stewart, B.A.; Zhang, F. Long-term experiments for sustainable nutrient management in China. A review. *Agron. Sustain. Dev.* **2011**, *31*, 397–414. [CrossRef]
2. Cao, Q.; Miao, Y.; Feng, G.; Gao, X.; Liu, B.; Liu, Y.; Li, F.; Khosla, R.; Mulla, D.J.; Zhang, F. Improving nitrogen use efficiency with minimal environmental risks using an active canopy sensor in a wheat-maize cropping system. *Field Crop Res.* **2017**, *214*, 365–372. [CrossRef]

3. Havlin, J.L.; Tisdale, S.L.; Nelson, W.L.; Beaton, J.D. *Soil Fertility and Fertilizers: An Introduction to Nutrient Management*, 8th ed.; Pearson, Inc.: Upper Saddle River, NJ, USA, 2014; pp. 117–184.
4. Huang, S.; Miao, Y.; Cao, Q.; Yao, Y.; Zhao, G.; Yu, W.; Shen, J.; Yu, K.; Bareth, G. A new critical nitrogen dilution curve for rice nitrogen status diagnosis in Northeast China. *Pedosphere* **2018**, *28*, 814–822. [CrossRef]
5. Greenwood, D.; Gastal, F.; Lemaire, G.; Draycott, A.; Millard, P.; Neeteson, J. Growth rate and % N of field grown crops: Theory and experiments. *Ann. Bot.* **1991**, *67*, 181–190. [CrossRef]
6. Lemaire, G.; Jeuffroy, M.; Gastal, F. Diagnosis tool for plant and crop N status in vegetative stage. *Eur. J. Agron.* **2008**, *28*, 614–624. [CrossRef]
7. Inoue, Y.; Sakaiya, E.; Zhu, Y.; Takahashi, W. Diagnostic mapping of canopy nitrogen content in rice based on hyperspectral measurements. *Remote Sens. Environ.* **2012**, *126*, 210–221. [CrossRef]
8. Filella, I.; Serrano, L.; Serra, J.; Penuelas, J. Evaluating wheat nitrogen status with canopy reflectance indices and discriminant analysis. *Crop Sci.* **1995**, *35*, 1400–1405. [CrossRef]
9. Caturegli, L.; Corniglia, M.; Gaetani, M.; Grossi, N.; Magni, S.; Migliazzi, M.; Angelini, L.; Mazzoncini, M.; Silvestri, N.; Fontanelli, M.; et al. Unmanned aerial vehicle to estimate nitrogen status of turfgrasses. *PLoS ONE* **2016**, *11*, e158268. [CrossRef]
10. Jay, S.; Maupas, F.; Bendoula, R.; Gorretta, N. Retrieving LAI, chlorophyll and nitrogen contents in sugar beet crops from multi-angular optical remote sensing: Comparison of vegetation indices and PROSAIL inversion for field phenotyping. *Field Crop. Res.* **2017**, *210*, 33–46. [CrossRef]
11. Clevers, J.G.P.W.; Gitelson, A.A. Remote estimation of crop and grass chlorophyll and nitrogen content using red-edge bands on Sentinel-2 and -3. *Int. J. Appl. Earth Obs.* **2013**, *23*, 344–351. [CrossRef]
12. Cao, Q.; Miao, Y.; Feng, G.; Gao, X.; Li, F.; Liu, B.; Yue, S.; Cheng, S.; Ustin, S.L.; Khosla, R. Active canopy sensing of winter wheat nitrogen status: An evaluation of two sensor systems. *Comput. Electron. Agric.* **2015**, *112*, 54–67. [CrossRef]
13. Padilla, F.M.; Gallardo, M.; Pena-Fleitas, M.T.; Souza, R.D.; Thompson, R.B. Proximal optical sensors for nitrogen management of vegetable crops: A review. *Sensors* **2018**, *18*, 2083. [CrossRef]
14. Lu, J.; Miao, Y.; Shi, W.; Li, J.; Yuan, F. Evaluating different approaches to non-destructive nitrogen status diagnosis of rice using portable RapidSCAN active canopy sensor. *Sci. Rep.* **2017**, *7*, 14073. [CrossRef]
15. Morier, T.; Cambouris, A.N.; Chokmani, K. In-season nitrogen status assessment and yield estimation using hyperspectral vegetation indices in a potato crop. *Agron. J.* **2015**, *107*, 1295–1309. [CrossRef]
16. Xia, T.; Miao, Y.; Wu, D.; Shao, H.; Khosla, R.; Mi, G. Active optical sensing of spring maize for in-season diagnosis of nitrogen status based on nitrogen nutrition index. *Remote Sens.* **2016**, *8*, 605. [CrossRef]
17. Padilla, F.M.; Pena-Fleitas, M.T.; Gallardo, M.; Thompson, R.B. Evaluation of optical sensor measurements of canopy reflectance and of leaf flavonols and chlorophyll contents to assess crop nitrogen status of muskmelon. *Eur. J. Agron.* **2014**, *58*, 39–52. [CrossRef]
18. Huang, S.; Miao, Y.; Yuan, F.; Gnyp, M.; Yao, Y.; Cao, Q.; Wang, H.; Lenz-Wiedemann, V.; Bareth, G. Potential of RapidEye and WorldView-2 satellite data for improving rice nitrogen status monitoring at different growth stages. *Remote Sens.* **2017**, *9*, 227. [CrossRef]
19. Ham, Y.; Han, K.K.; Lin, J.J.; Golparvar-Fard, M. Visual monitoring of civil infrastructure systems via camera-equipped Unmanned Aerial Vehicles (UAVs): A review of related works. *Vis. Eng.* **2016**, *4*, 1. [CrossRef]
20. Yang, G.; Liu, J.; Zhao, C.; Li, Z.; Huang, Y.; Yu, H.; Xu, B.; Yang, X.; Zhu, D.; Zhang, X. Unmanned aerial vehicle remote sensing for field-based crop phenotyping: Current status and perspectives. *Front. Plant Sci.* **2017**, *8*, 1111. [CrossRef]
21. Yao, H.; Qin, R.; Chen, X. Unmanned aerial vehicle for remote sensing applications—A review. *Remote Sens.* **2019**, *11*, 1443. [CrossRef]
22. Barbedo, J.G.A. A review on the use of unmanned aerial vehicles and imaging sensors for monitoring and assessing plant stresses. *Drones* **2019**, *3*, 40. [CrossRef]
23. Deng, L.; Mao, Z.; Li, X.; Hu, Z.; Duan, F.; Yan, Y. UAV-based multispectral remote sensing for precision agriculture: A comparison between different cameras. *ISPRS J. Photogramm.* **2018**, *146*, 124–136. [CrossRef]
24. Pajares, G. Overview and current status of remote sensing applications based on Unmanned Aerial Vehicles (UAVs). *Photogramm. Eng. Remote Sens.* **2015**, *81*, 281–330. [CrossRef]

25. Severtson, D.; Callow, N.; Flower, K.; Neuhaus, A.; Olejnik, M.; Nansen, C. Unmanned aerial vehicle canopy reflectance data detects potassium deficiency and green peach aphid susceptibility in canola. *Precis. Agric.* **2016**, *17*, 659–677. [CrossRef]
26. Vega, F.A.; Ramírez, F.C.; Saiz, M.P.; Rosúa, F.O. Multi-temporal imaging using an unmanned aerial vehicle for monitoring a sunflower crop. *Biosyst. Eng.* **2015**, *132*, 19–27. [CrossRef]
27. Wang, H.; Mortensen, A.K.; Mao, P.; Boelt, B.; Gislum, R. Estimating the nitrogen nutrition index in grass seed crops using a UAV-mounted multispectral camera. *Int. J. Remote Sens.* **2019**, *40*, 2467–2482. [CrossRef]
28. Qiu, J.; Wu, Q.; Ding, G.; Xu, Y.; Feng, S. A survey of machine learning for big data processing. *EURASIP J. Adv. Signal Process.* **2016**, *2016*, 67. [CrossRef]
29. Chlingaryan, A.; Sukkarieh, S.; Whelan, B. Machine learning approaches for crop yield prediction and nitrogen status estimation in precision agriculture: A review. *Comput. Electron. Agric.* **2018**, *151*, 61–69. [CrossRef]
30. Liakos, K.; Busato, P.; Moshou, D.; Pearson, S.; Bochtis, D. Machine learning in agriculture: A review. *Sensors* **2018**, *18*, 2674. [CrossRef]
31. Ali, I.; Greifeneder, F.; Stamenkovic, J.; Neumann, M.; Notarnicola, C. Review of machine learning approaches for biomass and soil moisture retrievals from remote sensing data. *Remote Sens.* **2015**, *7*, 16398–16421. [CrossRef]
32. Han, L.; Yang, G.; Dai, H.; Xu, B.; Yang, H.; Feng, H.; Li, Z.; Yang, X. Modeling maize above-ground biomass based on machine learning approaches using UAV remote-sensing data. *Plant Methods* **2019**, *15*, 10. [CrossRef] [PubMed]
33. Ali, I.; Cawkwell, F.; Dwyer, E.; Green, S. Modeling managed grassland biomass estimation by using multitemporal remote sensing data—A machine learning approach. *IEEE J. Sel. Top. Earth Obs. Remote Sens.* **2017**, *10*, 3254–3264. [CrossRef]
34. Pantazi, X.E.; Moshou, D.; Alexandridis, T.; Whetton, R.L.; Mouazen, A.M. Wheat yield prediction using machine learning and advanced sensing techniques. *Comput. Electron. Agric.* **2016**, *121*, 57–65. [CrossRef]
35. Liu, H.; Zhu, H.; Wang, P. Quantitative modelling for leaf nitrogen content of winter wheat using UAV-based hyperspectral data. *Int. J. Remote Sens.* **2017**, *38*, 2117–2134. [CrossRef]
36. Zheng, H.; Li, W.; Jiang, J.; Liu, Y.; Cheng, T.; Tian, Y.; Zhu, Y.; Cao, W.; Zhang, Y.; Yao, X. A comparative assessment of different modeling algorithms for estimating leaf nitrogen content in winter wheat using multispectral images from an unmanned aerial vehicle. *Remote Sens.* **2018**, *10*, 2026. [CrossRef]
37. Xing, B.; Dudas, M.J.; Zhang, Z.; Xu, Q. Pedogenetic characteristics of albic soils in the three river plain, Heilongjiang Province. *Acta Pedol. Sin.* **1994**, *31*, 95–104.
38. Lv, W.; Ge, Y.; Wu, J.; Chang, J. Study on the method for the determination of nitric nitrogen, ammoniacal nitrogen and total nitrogen in plant. *Spectrosc. Spect. Anal.* **2004**, *24*, 204–206.
39. Chen, Z.; Miao, Y.; Lu, J.; Zhou, L.; Li, Y.; Zhang, H.; Lou, W.; Zhang, Z.; Kusnierek, K.; Liu, C. In-season diagnosis of winter wheat nitrogen status in smallholder farmer fields across a village using unmanned aerial vehicle-based remote sensing. *Agronomy* **2019**, *9*, 619. [CrossRef]
40. Pedregosa, F.; Varoquaux, G.; Gramfort, A.; Michel, V.; Thirion, B.; Grisel, O.; Blondel, M.; Prettenhofer, P.; Weiss, R.; Dubourg, V. Scikit-learn: Machine learning in Python. *J. Mach. Learn. Res.* **2011**, *12*, 2825–2830.
41. Abraham, A.; Pedregosa, F.; Eickenberg, M.; Gervais, P.; Mueller, A.; Kossaifi, J.; Gramfort, A.; Thirion, B.; Varoquaux, G. Machine learning for neuroimaging with scikit-learn. *Front. Neuroinform.* **2014**, *8*, 14. [CrossRef]
42. Xue, J.; Su, B. Significant remote sensing vegetation indices: A review of developments and applications. *J. Sens.* **2017**, *2017*, 1353691. [CrossRef]
43. Hatfield, J.L.; Prueger, J.H. Value of using different vegetative indices to quantify agricultural crop chateristics at different growth stages under varying management practices. *Remote Sens.* **2010**, *2*, 562–578. [CrossRef]
44. Gnyp, M.L.; Miao, Y.; Yuan, F.; Ustin, S.L.; Yu, K.; Yao, Y.; Huang, S.; Bareth, G. Hyperspectral canopy sensing of paddy rice aboveground biomass at different growth stages. *Field Crop Res.* **2014**, *155*, 42–55. [CrossRef]
45. Tang, Y.; Huang, J.; Wang, R. Change law of hyperspectral data in related with chlorophyll and carotenoid in rice at different developmental stages. *Rice Sci.* **2004**, *11*, 274–282.
46. Cao, Q.; Miao, Y.; Wang, H.; Huang, S.; Cheng, S.; Khosla, R.; Jiang, R. Non-destructive estimation of rice plant nitrogen status with Crop Circle multispectral active canopy sensor. *Field Crop. Res.* **2013**, *154*, 133–144. [CrossRef]

47. Jin, X.; Yang, G.; Xu, X.; Yang, H.; Feng, H.; Li, Z.; Shen, J.; Lan, Y.; Zhao, C. Combined multi-temporal optical and radar parameters for estimating LAI and biomass in winter wheat using HJ and RADARSAR-2 Data. *Remote Sens.* **2015**, *7*, 13251–13272. [CrossRef]
48. Miao, Y.; Mulla, D.J.; Randall, G.W.; Vetsch, J.A.; Vintila, R. Combining chlorophyll meter readings and high spatial resolution remote sensing images for in-season site-specific nitrogen management of corn. *Precis. Agric.* **2009**, *10*, 45–62. [CrossRef]
49. Faurtyot, T.; Baret, F. Vegetation water and dry matter contents estimated from top-of-the-atmosphere reflectance data: A simulation study. *Remote Sens. Environ.* **1997**, *61*, 34–45. [CrossRef]
50. Chen, W.; Zhao, J.; Cao, C.; Tian, H. Shrub biomass estimation in semi-arid sandland ecosystem based on remote sensing technology. *Glob. Ecol. Conserv.* **2018**, *16*, e479. [CrossRef]
51. Qin, H.; Wang, C.; Xi, X.; Tian, J.; Zhou, G. Estimation of coniferous forest aboveground biomass with aggregated airborne small-footprint LiDAR full-waveforms. *Opt. Express* **2017**, *25*, A851–A869. [CrossRef]
52. Forkuor, G.; Hounkpatin, O.K.L.; Welp, G.; Thiel, M. High Resolution Mapping of Soil Properties Using Remote Sensing Variables in South-Western Burkina Faso: A comparison of machine learning and multiple linear regression models. *PLoS ONE* **2017**, *12*, e170478. [CrossRef] [PubMed]
53. Breiman, L.; Friedman, J.; Olshen, R.; Stone, C. *Classification and Regression Trees (The Wadsworth Statistics/Probability Series)*; Chapman and Hall/CRC: New York, NY, USA, 1984; pp. 1–358.
54. Rodriguez-Galiano, V.; Mendes, M.P.; Garcia-Soldado, M.J.; Chica-Olmo, M.; Ribeiro, L. Predictive modeling of groundwater nitrate pollution using Random Forest and multisource variables related to intrinsic and specific vulnerability: A case study in an agricultural setting (Southern Spain). *Sci. Total Environ.* **2014**, *476*, 189–206. [CrossRef] [PubMed]
55. Wang, L.; Zhou, X.; Zhu, X.; Dong, Z.; Guo, W. Estimation of biomass in wheat using random forest regression algorithm and remote sensing data. *Crop J.* **2016**, *4*, 212–219. [CrossRef]
56. Durbha, S.S.; King, R.L.; Younan, N.H. Support vector machines regression for retrieval of leaf area index from multiangle imaging spectroradiometer. *Remote Sens. Environ.* **2007**, *107*, 348–361. [CrossRef]
57. He, Q. Neural Network and its Application in IR. In *Graduate School of Library and Information Science, University of Illinois at Urbana-Champaign*; Spring: Champaign, IL, USA, 1999.
58. Yuan, H.; Yang, G.; Li, C.; Wang, Y.; Liu, J.; Yu, H.; Feng, H.; Xu, B.; Zhao, X.; Yang, X. Retrieving soybean leaf area index from unmanned aerial vehicle hyperspectral remote sensing: Analysis of RF, ANN, and SVM Regression Models. *Remote Sens.* **2017**, *9*, 309. [CrossRef]
59. Yao, X.; Huang, Y.; Shang, G.; Zhou, C.; Cheng, T.; Tian, Y.; Cao, W.; Zhu, Y. Evaluation of six algorithms to monitor wheat leaf nitrogen concentration. *Remote Sens.* **2015**, *7*, 14939–14966. [CrossRef]
60. Yue, J.; Yang, G.; Li, C.; Li, Z.; Wang, Y.; Feng, H.; Xu, B. Estimation of winter wheat above-ground biomass using unmanned aerial vehicle-based snapshot hyperspectral sensor and crop height improved models. *Remote Sens.* **2017**, *9*, 708. [CrossRef]
61. Tsouros, D.C.; Bibi, S.; Sarigiannidis, P.G. A review on UAV-based applications for precision agriculture. *Information* **2019**, *10*, 349. [CrossRef]
62. Hunt, E.R., Jr.; Daughtry, C.S.T. What good are unmanned aircraft systems for agricultural remtoe sensing and precision agriculture? *Int. J. Remote Sens.* **2018**, *39*, 5345–5376. [CrossRef]
63. Li, S.; Ding, X.; Kuang, Q.; Ata-Ul-Karim, S.T.; Cheng, T.; Liu, X.; Tian, Y.; Zhu, Y.; Cao, W.; Cao, Q. Potential of UAV-based active sensing for monitoring rice leaf nitrogen status. *Front. Plant Sci.* **2018**, *9*, 1834. [CrossRef]
64. Herrmann, I.; Karnieli, A.; Bonfil, D.J.; Cohen, Y.; Alchanatis, V. SWIR-based spectral indices for assessing nitrogen content in potato fields. *Int. J. Remote Sens.* **2010**, *31*, 5127–5143. [CrossRef]
65. Elarab, M.; Ticlavilca, A.M.; Torres-Rua, A.F.; Maslova, I.; McKee, M. Estimating chlorophyll with thermal and broadband multispectral high resolution imagery from an unmanned aerial system using relevance vector machines for precision agriculture. *Int. J. Appl. Earth Observ. Geoinf.* **2015**, *43*, 32–42. [CrossRef]
66. Pullanagari, R.R.; Kereszturi, G.; Yule, I. Integrating airborne hyperspectral, topographic, and soil data for estimating pasture quality using recersive feature elimination with random forest regression. *Remote Sens.* **2018**, *10*, 1117. [CrossRef]
67. Buschmann, C.; Nagel, E. In vivo spectroscopy and internal optics of leaves as basis for remote sensing of vegetation. *Int. J. Remote Sens.* **1993**, *14*, 711–722. [CrossRef]

68. Tucker, C.J. Red and photographic infrared linear combinations for monitoring vegetation. *Remote Sens. Environ.* **1979**, *8*, 127–150. [CrossRef]
69. Gitelson, A.A.; Kaufman, Y.J.; Merzlyak, M.N. Use of a green channel in remote sensing of global vegetation from EOS-MODIS. *Remote Sens. Environ.* **1996**, *58*, 289–298. [CrossRef]
70. Gitelson, A.A. Remote estimation of canopy chlorophyll content in crops. *Geophys. Res. Lett.* **2005**, *32*, L08403. [CrossRef]
71. Sripada, R.P.; Heiniger, R.W.; White, J.G.; Meijer, A.D. Aerial color infrared photography for determining early in-season nitrogen requirements in corn. *Agron. J.* **2006**, *98*, 968. [CrossRef]
72. Qi, J.; Chehbouni, A.; Huete, A.R.; Kerr, Y.H.; Sorooshian, S. A modified soil adjusted vegetation index. *Remote Sens. Environ.* **1994**, *48*, 119–126. [CrossRef]
73. Rondeaux, G.; Steven, M.; Baret, F. Optimization of soil-adjusted vegetation indices. *Remote Sens. Environ.* **1996**, *55*, 95–107. [CrossRef]
74. Rouse, J.W., Jr.; Haas, R.H.; Schell, J.A.; Deering, D.W. Monitoring vegetation systems in the Great Plains with ERTS. In *NASA. Goddard Space Flight Center 3d ERTS-1 Symphony*; NASA: Washington DC, USA, 1974; pp. 309–317.
75. Jordan, C.F. Derivation of leaf-area index from quality of light on the forest floor. *Ecology* **1969**, *50*, 663–666. [CrossRef]
76. Roujean, J.; Breon, F. Estimating PAR absorbed by vegetation from bidirectional reflectance measurements. *Remote Sens. Environ.* **1995**, *51*, 375–384. [CrossRef]
77. Gitelson, A.A. Wide dynamic range vegetation index for remote quantification of biophysical characteristics of vegetation. *J. Plant Physiol.* **2004**, *161*, 165–173. [CrossRef] [PubMed]
78. Huete, A.R. A soil-adjusted vegetation index (SAVI). *Remote Sens. Environ.* **1988**, *25*, 295–309. [CrossRef]
79. Sandham, L. Surface temperature measurement from space: A case study in the south western cape of south Africa. *S. Afr. J. Enol. Vitic.* **1997**, *18*, 25–30. [CrossRef]
80. Chen, J.M. Evaluation of vegetation indices and a modified simple ratio for boreal applications. *Can. J. Remote Sens.* **1996**, *22*, 229–242. [CrossRef]
81. Reyniers, M.; Walvoort, D.J.; De Baardemaaker, J. A linear model to predict with a multi-spectral radiometer the amount of nitrogen in winter wheat. *Int. J. Remote Sens.* **2006**, *27*, 4159–4179. [CrossRef]
82. Dash, J.; Curran, P. The MERIS terrestrial chlorophyll index. *Int. J. Remote Sens.* **2004**, *25*, 5403–5413. [CrossRef]
83. Goel, N.S.; Qin, W. Influences of canopy architecture on relationships between various vegetation indices and LAI and FPAR: A computer simulation. *Remote Sens. Rev.* **1994**, *10*, 309–347. [CrossRef]
84. Gong, P.; Pu, R.; Biging, G.S.; Larrieu, M.R. Estimation of forest leaf area index using vegetation indices derived from Hyperion hyperspectral data. *IEEE Trans. Geosci. Remote Sens.* **2003**, *41*, 1355–1362. [CrossRef]
85. Barnes, E.; Clarke, T.; Richards, S.; Colaizzi, P.; Haberland, J.; Kostrzewski, M.; Waller, P.; Choi, C.; Riley, E.; Thompson, T.; et al. Coincident detection of crop water stress, nitrogen status and canopy density using ground based multispectral data. In Proceedings of the Fifth International Conference on Precision Agriculture, Bloomington, MN, USA, 16–19 July 2000.
86. Gitelson, A.A.; Merzlyak, M.N.; Lichtenthaler, H.K. Detection of red edge position and chlorophyll content by reflectance measurements near 700 nm. *J. Plant Physiol.* **1996**, *148*, 501–508. [CrossRef]
87. Jasper, J.; Reusch, S.; Link, A. Active sensing of the N status of wheat using optimized wavelength combination: Impact of seed rate, variety and growth stage. *Precis. Agric.* **2009**, *9*, 23–30.
88. Gitelson, A.A.; Gritz, Y.; Merzlyak, M.N. Relationships between leaf chlorophyll content and spectral reflectance and algorithms for non-destructive chlorophyll assessment in higher plant leaves. *J. Plant Physiol.* **2003**, *160*, 271–282. [CrossRef] [PubMed]
89. Elsayed, S.; Rischbeck, P.; Schmidhalter, U. Comparing the performance of active and passive reflectance sensors to assess the normalized relative canopy temperature and grain yield of drought-stressed barley cultivars. *Field Crop. Res.* **2015**, *177*, 148–160. [CrossRef]
90. Erdle, K.; Mistele, B.; Schmidhalter, U. Comparison of active and passive spectral sensors in discriminating biomass parameters and nitrogen status in wheat cultivars. *Field Crop. Res.* **2011**, *124*, 74–84. [CrossRef]

91. Datt, B. Visible/near infrared reflectance and chlorophyll content in Eucalyptus leaves. *Int. J. Remote Sens.* **1999**, *20*, 2741–2759. [CrossRef]
92. Long, D.S.; Eitel, J.U.; Huggins, D.R. Assessing nitrogen status of dryland wheat using the canopy chlorophyll content index. *Crop Manag.* **2009**, *8*. [CrossRef]

remote sensing

MDPI

Article

Leaf Nitrogen Concentration and Plant Height Prediction for Maize Using UAV-Based Multispectral Imagery and Machine Learning Techniques

Lucas Prado Osco [1,*], José Marcato Junior [2], Ana Paula Marques Ramos [3], Danielle Elis Garcia Furuya [3], Dthenifer Cordeiro Santana [4], Larissa Pereira Ribeiro Teodoro [5], Wesley Nunes Gonçalves [2,6], Fábio Henrique Rojo Baio [5], Hemerson Pistori [6,7], Carlos Antonio da Silva Junior [8] and Paulo Eduardo Teodoro [4,5]

1 Faculty of Engineering and Architecture and Urbanism, University of Western São Paulo (UNOESTE), Rodovia Raposo Tavares, km 572 - Limoeiro, Pres. Prudente, SP 19067-175, Brazil

2 Faculty of Engineering, Architecture and Urbanism and Geography, Federal University of Mato Grosso do Sul (UFMS), Cidade Universitária, Av. Costa e Silva, Pioneiros, MS 79070-900, Brazil; jose.marcato@ufms.br (J.M.J.); wesley.goncalves@ufms.br (W.N.G.)

3 Post-Graduate Program of Environment and Regional Development, University of Western São Paulo (UNOESTE), Rodovia Raposo Tavares, km 572 - Limoeiro, Pres. Prudente, SP 19067-175, Brazil; anaramos@unoeste.br (A.P.M.R.); daniellegarciafuruya@gmail.com (D.E.G.F.)

4 Programa de Pós-Graduação em Agronomia - Área de Concentração: Produção Vegetal, da UEMS, Unidade Universitária de Aquidauana, Aquidauana, MS 79200-000, Brazil; dthennyfer.santana@hotmail.com (D.C.S.); paulo.teodoro@ufms.br (P.E.T.)

5 Department of Agronomy, Federal University of Mato Grosso do Sul (UFMS), Rodovia MS 306, km. 305 Caixa Postal 112, Chapadão do Sul, MS 79560000, Brazil; larissa_ribeiro@ufms.br (L.P.R.T.); fabiobaio@ufms.br (F.H.R.B.)

6 Faculty of Computing, Federal University of Mato Grosso do Sul (UFMS), Cidade Universitária, Av. Costa e Silva, Pioneiros, MS 79070-900, Brazil; pistori@ucdb.br

7 Inovisão, Universidade Católica Dom Bosco, Av. Tamandaré, 6000, Campo Grande, MS 79117-900, Brazil

8 Department of Geography, State University of Mato Grosso (UNEMAT), Av. dos Ingas, 3001 - Jardim Imperial, Sinop, MT 78555-000, Brazil; carlosjr@unemat.br

* Correspondence: lucasosco@unoeste.br

Received: 14 September 2020; Accepted: 3 October 2020; Published: 5 October 2020

Abstract: Under ideal conditions of nitrogen (N), maize (*Zea mays* L.) can grow to its full potential, reaching maximum plant height (PH). As a rapid and nondestructive approach, the analysis of unmanned aerial vehicles (UAV)-based imagery may be of assistance to estimate N and height. The main objective of this study is to present an approach to predict leaf nitrogen concentration (LNC, g kg^{-1}) and PH (m) with machine learning techniques and UAV-based multispectral imagery in maize plants. An experiment with 11 maize cultivars under two rates of N fertilization was carried during the 2017/2018 and 2018/2019 crop seasons. The spectral vegetation indices (VI) normalized difference vegetation index (NDVI), normalized difference red-edge index (NDRE), green normalized difference vegetation (GNDVI), and the soil adjusted vegetation index (SAVI) were extracted from the images and, in a computational system, used alongside the spectral bands as input parameters for different machine learning models. A randomized 10-fold cross-validation strategy, with a total of 100 replicates, was used to evaluate the performance of 9 supervised machine learning (ML) models using the Pearson's correlation coefficient (r), mean absolute error (MAE), coefficient of regression (R^2), and root mean square error (RMSE) metrics. The results indicated that the random forest (RF) algorithm performed better, with r and RMSE, respectively, of 0.91 and 1.9 g.kg^{-1} for LNC, and 0.86 and 0.17 m for PH. It was also demonstrated that VIs contributed more to the algorithm's performances than individual spectral bands. This study concludes that the RF model is appropriate to predict both agronomic variables in maize and may help farmers to monitor their plants based upon their LNC and PH diagnosis and use this knowledge to improve their production rates in the subsequent seasons.

Keywords: UAV; random forest; nitrogen; maize

1. Introduction

Remote sensing techniques aligned with precision agriculture practices are being investigated in researches with different farmlands [1]. In recent years, the increase of market-availability of unmanned aerial vehicles (UAV) encouraged multiple applications in this field. Agriculture remote sensing is a promising field as it supports a multidisciplinary view of different problems related to crop mapping [2] and has been implemented in multiple subjects, such as environment control [3], temporal analysis [4], phenology [5], yield-prediction [6–9], and nutritional analysis [10–12]. These studies revealed the importance of evaluating techniques and sensing data to deal with such tasks.

A relevant topic for farmers and technicians is the correct monitoring of their farmlands, as nutrient absorption rates are connected with plant-growth and yield estimates. An important nutrient related to plant-growth is Nitrogen (N). N benefits leaf development and photosynthetic activity in plants, influencing their productivity [13]. Plants that have nutritional deficiencies related to N show visual symptoms in their leaves, known as chlorosis [14,15]. This nutrient is commonly applied in agricultural areas and it is one of the most contributive nutrients to global production. However, the incorrect diagnosis may be a problem from both economic and environmental point-of-views [16,17].

To circumvent the aforementioned problem, agronomic technicians rely on traditional methods of chemical leaf tissue analysis to determine the amount of N absorbed by the plant [18]. However, this practice is viewed as a destructive, time-consuming, and highly-priced approach. Thus, it is difficult to adopt the traditional analysis as a recurrent procedure to monitor multiple areas and stages [19]. As a rapid, nondestructive, and highly-replicable method, UAV-based image analysis may be of assistance to perform plant nutrient content and growth-status estimate [20–24].

As an alternative method, multispectral data analysis collected with sensor systems represents a promising approach to increase the precision in area monitoring [20]. Predicting nutrient content and plant height with remote systems and automated intelligent methods is gaining attention in agriculture practices. With multispectral sensors, at canopy or leaf levels, different studies predicted leaf nitrogen concentration (LNC) in maize (*Zea mays* L.) [16], winter-wheat (*Triticum aestivum*) [21], cotton (*Gossypium hirsutum*) [22], rice (*Oryza sativa*) [23], citrus (*Citrus sinensis*) [18,24], among others. Although hyperspectral sensors stand out in their ability to characterize the spectral response with high accuracies [25,26], multispectral sensors are used more frequently in agriculture remote sensing since they are economically viable and accessible to most of the front-end users.

Predicting agronomic variables with multispectral data is a common practice in remote sensing applications. However, performing this task with machine learning techniques is still a recent and relevant topic in agriculture remote sensing since it provides a robust and direct approach to evaluate different agronomic variables. Machine learning is considered a subgroup inside of the artificial intelligence area in which algorithms can learn from data and then discover patterns in the dataset, deciding on new and similar information. The algorithms have the potential to model several types of datasets using linear and parametric and nonlinear and nonparametric approaches [12,27,28], including multispectral images [29]. Different machine learning algorithms like random forests (RF), decision trees (DT), artificial neural network (ANN), support vector machines (SVM), among many others, have been adopted to attend various applications in agriculture remote sensing [5,30–32].

Machine learning has helped to increase not only the prediction's accuracy of some agronomic variables but also assisted in solving complex problems related to data heterogeneity. A revision study on yield and N content prediction [33] concluded that advances in remote sensing technologies and machine learning techniques will result in more cost-effective and comprehensive solutions for a better crop state assessment. The combination of machine learning techniques and vegetation indices

(VI) is also an important subject in agricultural applications and has been adopted in different studies, some of which are related to maize characteristics predictions [34,35].

Under ideal conditions of N, maize plants can grow to their full potential reaching maximum height [36,37]. Considering that, implementing different approaches to estimate height and N with UAV-based remote systems is essential to optimize the monitoring of areas with multiple varieties. Currently, one of the main objectives of maize breeding programs is to identify genotypes with high efficiency in N usage [38,39]. Obtaining rapid predictions with an alternative approach like machine learning and UAV-based image may enable programs, technicians, and farmers to evaluate multiple genotypes each year, allowing them to optimize the selection of the most promising plants concerning N use efficiency. In this matter, the main idea behind this proposal is to present a feasible alternative to monitor N and plant height (PH) with machine learning techniques in UAV based imagery.

By implementing the aforementioned approach, farmers can monitor their LNC in maize plants and select the areas or maize varieties (based upon their location or plots) that are most promising based upon their diagnosis and use this knowledge to improve their production rates in subsequent seasons. As machine learning has been proved [23–30] to be a robust approach to evaluate heterogeneous data, it could return important results when considering different genotypes of maize plants. In this paper the following questions are addressed: (1) which machine learning models are most suitable to predict LNC and PH in maize (*Zea mays* L.) plants with spectral data from UAV-based image? and (2) amongst all predictor variables (spectral indices, bands, and the combination of both), which one is the most useful for mapping LNC and PH based on the machine learning approach?

2. Materials and Methods

The proposed method was divided into 4 main phases: (1) the description of in-field experiments and how the experimental design was mounted, (2) the extraction of variables LNC and PH, (3) the image preprocessing and calculation of the VIs investigated, and (4) the experimental protocol implemented. Each main phase is described in detail in the following subsections.

2.1. Field Trials

The experiment was carried out in the municipality of Chapadão do Sul, State of Mato Grosso do Sul, Brazil (18°46′26′′ S, 52°37′28′′ W, and an average altitude of 810 m), during the 2017/2018 and 2018/2019 crop seasons. In this experiment, 11 maize cultivars cultivated under two rates of nitrogen fertilization in topdressing, 60 kg ha^{-1}—considered as low and 180 kg ha^{-1}—considered as high, were investigated, with four replicates of each plot. The cultivars used in the experiment were: Caimbé; CatiVerde; Gorotuba; AlAvaré; BRS106; BRS4103; BRS4104; Diratininga; SCS154; SCS155; and SCS156. The dimensions of each plot were five rows, spaced at 0.45 m each, with a 5 m length. Because it corresponds to a relatively small experimental area, the soil here presents similar conditions. This area is constantly monitored and soil corrections are conducted whenever necessary.

The corn cultivars and N rates were allocated in the same plots in both seasons. The use of several cultivars and two rates of N aimed to create different situations promoted by farmers in Brazil. Thus, the models tested can estimate the variables for these conditions in both seasons. The integration between multiple varieties also was important to provide enough samples for the machine learning models to learn the necessary features: LNC and PH. It was also necessary to build a dataset heterogeneous enough to demonstrate the feasibility of these techniques. The geographic location of the area, along with the experiment plots, is displayed in Figure 1.

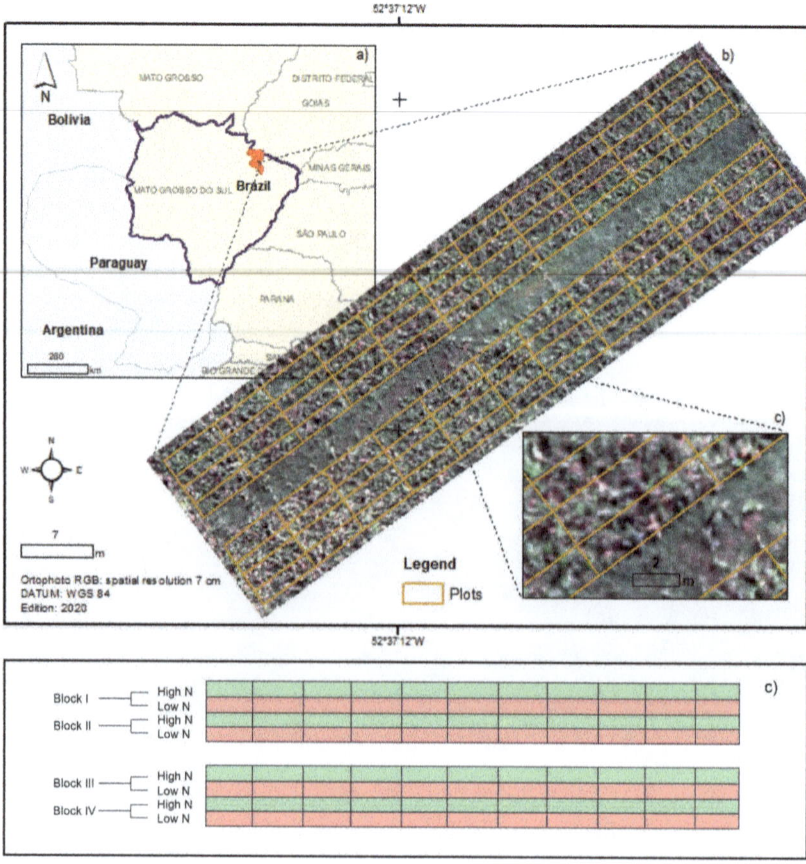

Figure 1. The geographic location of the study area. (**a**) Corresponds with the location of the municipality within the Mato Grosso do Sul State in Brazil; (**b**) represents the plots in the experimental site; and (**c**) is the scheme concerning the experimental fertilization rates used.

2.2. Evaluated Variables

Maize plants were evaluated at the V12 stage. The images were collected at this stage because plants have reached their full potential in terms of growth and nitrogen absorption in this phase. The average third of five leaves of maize plants were collected in each experimental unit. The LNC (g kg^{-1}) was obtained by the methodology described in [40]. In this regard, N was evaluated with in-field measurements and following agronomical standard procedures. For this, the Kjeldahl titration technique was applied, which is divided into (1) digestion, (2) distillation in an N distiller, and, (3) titration with sulfuric acid (H_2SO_4). On this same date, PH (m) was obtained with an average of five plants chosen at random in each plot. For this, a measuring tape was used, positioned from the base of the plant to its apex (i.e., the highest point of the plant; at the top of the canopy). A tracking GNSS with high precision accuracy was used to map the crop plots (yellow-grids in Figure 1), ensuring that the collected data was representative of each plot.

This provided a total of 176 in-field observations of LNC and PH. The measure mean-values of LNC and PH, for both seasons (2017/2018 and 2018/2019), did not result in statistical differences at a p-value under 0.05. For this, a Shapiro-Wilk test followed by a pairwise t student test was used. When calculating the variance for LNC, values of 21.33 g.kg^{-1} and 20.21 g.kg^{-1} in 2017/2018 and 2018/2019 crop seasons were obtained, respectively. As for PH, the variance obtained was 0.107 m and 0.101 m for 2017/2018 and 2018/2019 crop seasons, respectively. This information, alongside with the p-value under 0.05 for both LNC and PH in each year, indicated that the seasons returned similar conditions for both analyses.

2.3. Image Acquisition and Vegetation Indices

The following spectral regions were used for calculating the VIs: green (G), red (R), red-edge (RE), and near-infrared (NIR). The described wavelength (nm) is the bands' center on both sensors. The area was recorded during the first crop season (2017/2018) with a MicaSense Red-Edge multispectral sensor (G: 560 nm, R: 668 nm, RE: 717 nm, and NIR: 842 nm) embedded in a UAV-multirotor X800. For the second crop season (2018/2019), a Sensefly eBee RTK fixed-wing remotely piloted aircraft was used. The eBee was equipped with the Sensefly Parrot Sequoia multispectral sensor (G: 550 nm, R: 660 nm, RE: 735 nm, and NIR: 790 nm).

Both sensors acquired spectral data in the aforementioned wavelengths and used a luminosity sensor allowing the calibration of the acquired values. The two overflights were performed at 100 m altitude, returning a spatial image resolution (ground sample distance—GSD) of 0.10 m, and were conducted at 10:00 h (local time). Figure 2 summarizes the climatic and atmospheric conditions of each crop season (2017/2018 and 2018/2019). As described, both flights occurred at the V12 stage of each season.

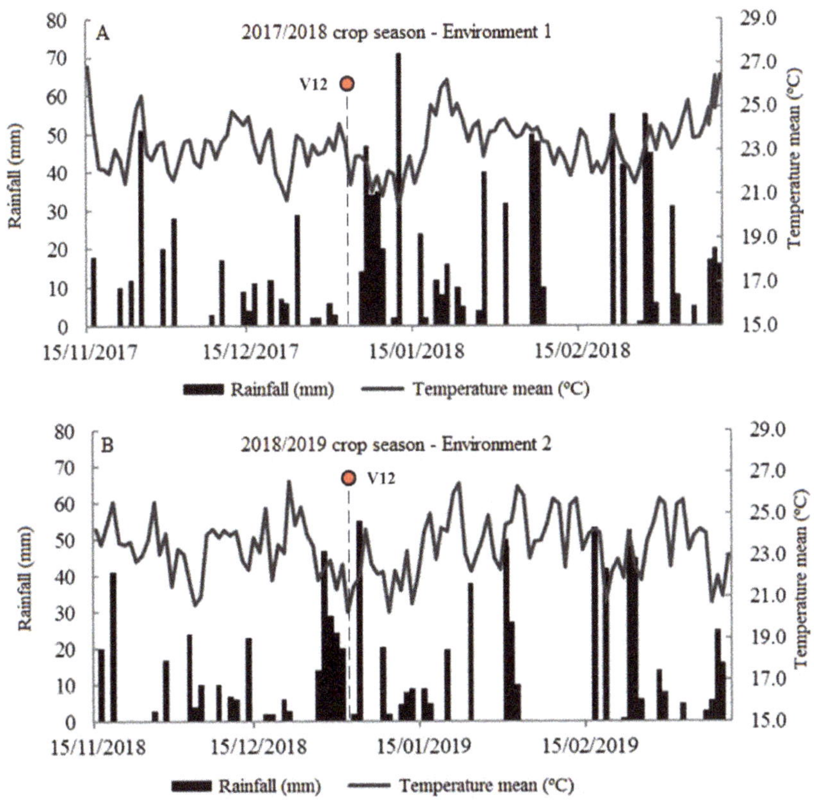

Figure 2. The climatic and atmospheric conditions registered during the crop seasons of 2017/2018 and 2018/2019.

For the image preprocessing, the Pix4DMapper was used, optimizing the interior and exterior parameters of the image. A sparse dense cloud based on the structure-from-motion (SfM) technique and point clouds based on the MVS (multi-view stereo) with multiple control points collected were used. These points were collected with a global navigation satellite system (GNSS), dual-frequency in real-time kinematic (RTK) mode. Images were acquired with 80% longitudinal and 60% lateral overlaps, and the digital number (DN) was converted to surface reflectance using the calibration parameters described in the manual of both sensors. The calibration and luminosity corrections were also necessary to minimize the influence of soil brightness. Because spectral indices were used in this study, this interference was also minimized. The plantation itself also was in a stage fully developed and covered most of the soil in the spatial resolution registered, making its contribution minimal to the spectral behavior of the plants.

During an experimental phase, multiple VIs were calculated with the aforementioned spectral bands. However, most of the indices did not return promising results and also presented redundancy over the tests. Because of that, only four main VIs were implemented in the machine learning models: normalized difference vegetation index (NDVI) [41], normalized difference red-edge index (NDRE) [42], green normalized difference vegetation (GNDVI) [43], and soil adjusted vegetation index (SAVI) [44]. These VIs are among the most commonly used indices to predict plant health and conditions. The equations arranged below demonstrate the spectral data used to obtain these VIs, respectively.

$$NDVI = (NIR - R)/(NIR + R) \tag{1}$$

$$NDRE = (NIR - RE)/(NIR + RE) \tag{2}$$

$$GNDVI = (NIR - G)/(NIR + G) \tag{3}$$

$$SAVI = (1 + 0.5)(NIR - R)/(NIR + R + 0.5) \tag{4}$$

2.4. Data Analysis

The pixel values for each plantation plot were extracted from the images. These values were used as input to estimate the measured in-field values of LNC and PH in their corresponding plot. A randomized 10-fold cross-validation sampling strategy, with a total of 10 repetitions, was used to evaluate the performance of 9 supervised machine learning models (Table 1). To evaluate the performance of each model as well as the relationship between the predicted and observed variables, the root mean squared error (RMSE) and mean absolute error (MAE) metrics were used.

The number of samples implemented was similar to others presented in previous research [12,28], which also discussed the required quantities of input data to train these types of algorithms. With the cross-validation approach, 90% of the 176 samples were used to train the models and 10% to test it. Because this process was repeated, 100 randomized test-sets were constructed. In summary, this type of validation is repeated sequentially, constantly changing the folder used for validating the algorithm [12,18,27,45]. In this manner, the algorithm is always validated with data not used at its training phase. In this experiment, the entire procedure was also repeated 100 times, which means that the models were built from scratch in every repetition.

Two decision tree-based machine learning algorithms were used here: the reduced error pruning tree with backfitting, and the random forest method with 100% of the training set as bagging size. A K-nearest neighbor was also used with three different K values: 1, 5, and 10. Support vector machines adopting sequential minimal optimization (SMO) have been tested under 2 different kernels: radial based functions and polynomial. Finally, a linear and a kernel-based regressor were also included for comparison. The linear regression uses a grid-search strategy for model selection based on the Akaike information criterion and the kernel-based regressor is a radial basis function network.

Table 1. Machine learning algorithms and input data variation used in this study.

Test Order	ML Model	Reference
#1	REPTree—REPT	Saha et al. [46]
#2	Random Forest—RF	Belgiu et al. [47]
#3	K-Nearest Neighbor (K=1)—1NN	Ali et al. [48]
#4	K-Nearest Neighbor (K=5)—5NN	Ali et al. [48]
#5	K-Nearest Neighbor (K=10)—10NN	Ali et al. [48]
#6	SVM-RBF—SVMR	Nalepa et al. [49]
#7	Support Vector Machine-Polynomial—SVMP	Nalepa et al. [49]
#8	Linear Regression—LR	Štepanovský et al. [50]
#9	RBF Regression—RBF	Cheshmberah et al. [51]

The library default values were adopted for the number and depth of trees, nodes, and leaves in the decision tree models, as well as a different number of neighbors (1, 5, and 10) for the KNN algorithm. As stated, two functions (RBF and polynomial) were considered for SVM, the exp(-gamma*|u-v|^2) and the (gamma*u′*v + coef0)^2, respectively. Each value regarding the described variables was set to be calculated automatically considering the overall best predictions with an epsilon loss curve equal to 0.1. Last, a grid search approach was used to fine-tune the linear regression model (RBF Regression), thus performing a hyperparametrization of this particular model.

All the models have been tested using three sets of variables: (#1) a set with spectral-bands only (SB), (#2) a set only with Vis, and (#3) a set including both SB and VIs together. During the experimental phase, different hybrid combinations of the adopted models were evaluated. However, the combinations are not discussed in this manuscript mainly because they did not result in interesting outcomes as well as the separation between SBs and VIs. After determining the best overall algorithm, an inference model was calculated to produce a prediction map over the UAV image. This map was used to ascertain the relationship between the predicted variables and help discuss the implications of the proposal of this study.

Additionally, based on the overall best algorithm, the most contributive input data used by the learner were also identified. For this, a classifier attribute evaluation that estimates the worth of an attribute by using this specified classifier (in our case, the overall best algorithm) was implemented with the rank search as a selection method. The evaluated rank metric was based on a merit score obtained with the ZeroR regressor. This merit corresponds with the relative increase in the performance of the model in relation to the ZeroR classifier. The ZeroR was used since it takes the average value of the target variable and uses this value as a prediction. In this regard, the rank method can return a merit number even greater than 1 (since relative increase may exceed 100%). This procedure was important to determine the significance of each SB or VI to infer LNC and PH in maize crops.

3. Results

3.1. Relationship among the Agronomic Variables

To present the relationship among the evaluated variables, Figure 3 was prepared to display the Pearson's correlations between LNC and PH with the SBs and VIs evaluated in this study. The magnitude of the correlations was different for each N fertilization rate (high and low).

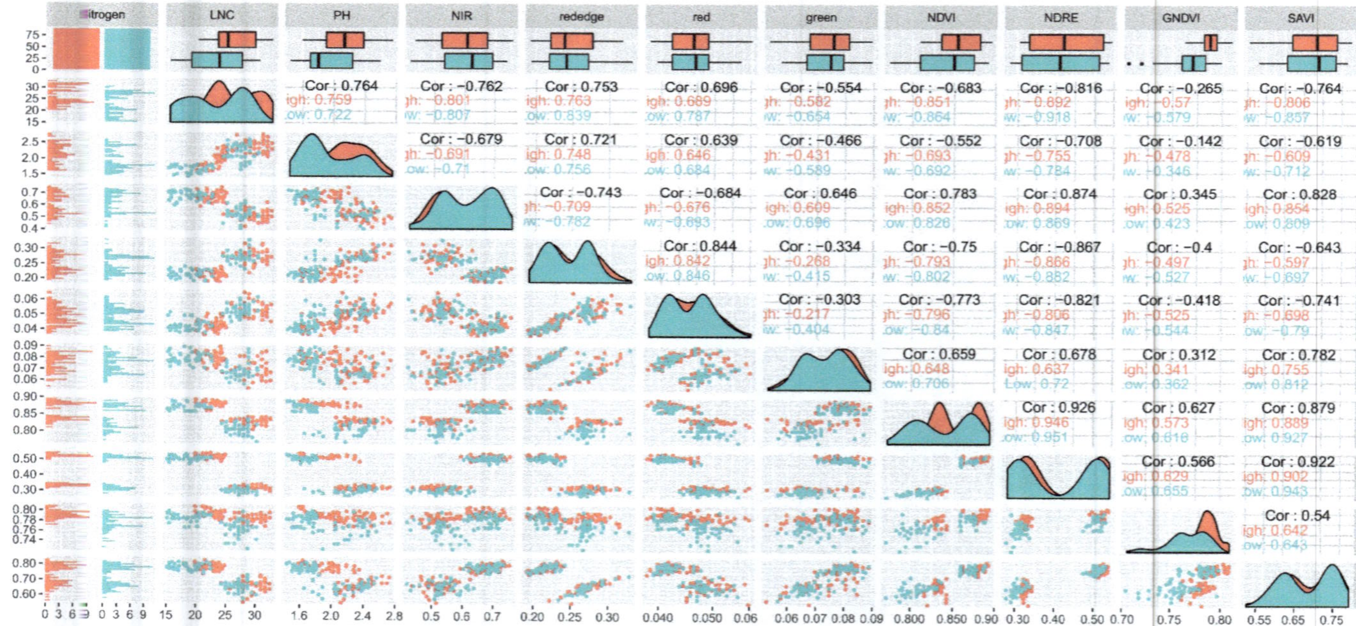

Figure 3. Pearson's correlation (r) between each variable implemented in this study. The lower triangle corresponds with the dispersion graphics of each correlation in a pairwise manner. The higher triangle displays the correlation value. The main diagonal illustrates a kernel density-based histogram between the two N rates and the distribution of it according to the respective variables. The top row displays box-plots of the values according to the associated N content, and the first column performs a comparison between the basic distributions of data across the two N conditions.

3.2. Models' Performances for LNC and PH Prediction

Figures 4 and 5 display the boxplots for the RMSE using 100 runs (10 repetitions of 10-fold cross-validation) of each machine learning algorithm under the 3 data input configurations: SB, VI, and SB+VI. Figure 4 displays the boxplot for LNC, whereas, in Figure 5, PH is displayed. Regarding the LNC estimate, the RMSE indicates a higher averaged performance of the RF model with a smaller interquartile range for the VI and SB+VI configuration. The performance of RF using only the SBs is lower than using the other configuration sets for both LNC and PH. The three KNN models also showed lower values of RMSE for VI alone than SB+VI, with a clear advantage for higher K sets (e.g., 5 and 10).

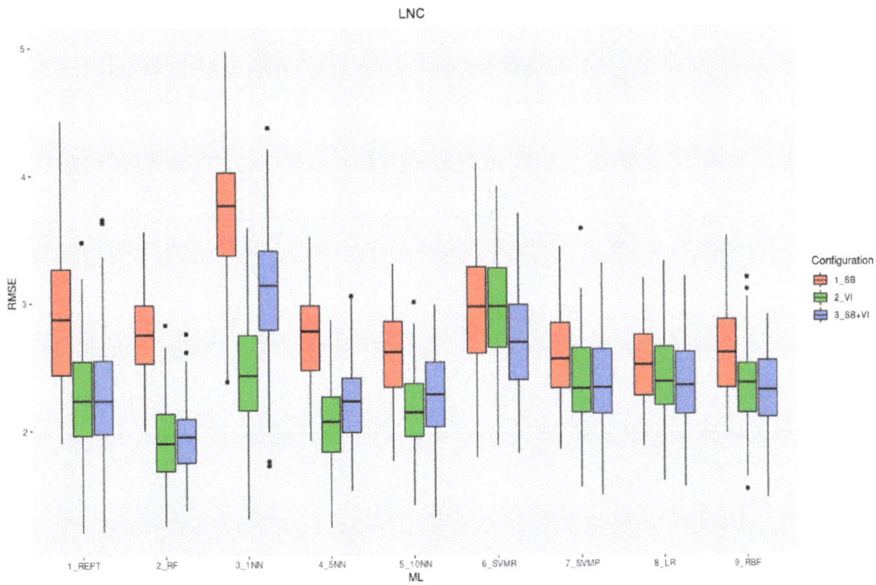

Figure 4. Boxplot for the root mean squared error (RMSE) for leaf nitrogen concentration (LNC) (g kg^{-1}) estimation using 9 machine learning algorithms, over 100 runs, and 3 input configurations: (1) spectral bands (SBs), (2) vegetation indices (VIs), and (3) SB+VI. The point plots indicate outliers encountered during the phase of the 100 different test repetitions.

In the boxplots for PH, there is a slightly lower averaged RMSE for RF when comparing it against the other models, but the combination of SBs + VIs seemed to lower the performance of the model. Beyond RF, the REPT and KNN models presented good results for the VIs dataset alone. Although some outliers were detected in the estimations, each box-plot was constructed over a 95% confidence interval. The overall performance of the best model (RF) presented an RMSE equal to 1.9 g.kg^{-1} and 0.17 m, for both LNC and PH, respectively.

To better ascertain the relationship between predictions and measured variables, the regression of the overall three best methods for each variable (PH and LNC) was plotted. It used the configuration set #2 containing only the VIs as input variables (Figure 6).

The PH scatterplot in Figure 6 demonstrates how consistent the RF model was when predicting this variable. As for the LNC prediction, it is possible to notice that the two topdressing conditions of N fertilization rates are separated by the model. This is an important observation since it demonstrates that the RF approach was able to separate distinctly the low and high rate levels.

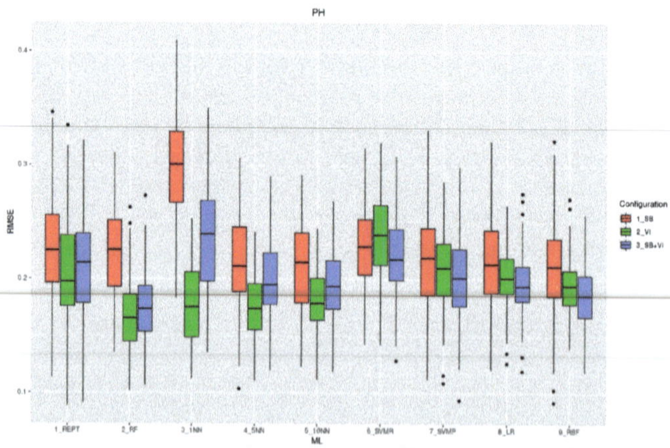

Figure 5. Boxplot for the root mean square error (RMSE) for plant height (PH) (m) estimation using 9 machine learning algorithms, over 100 runs, and 3 input configurations: 1) spectral bands (SBs), 2) vegetation indices (VIs), and 3) SB+VI. The point plots indicate outliers encountered during the phase of the 100 different test repetitions.

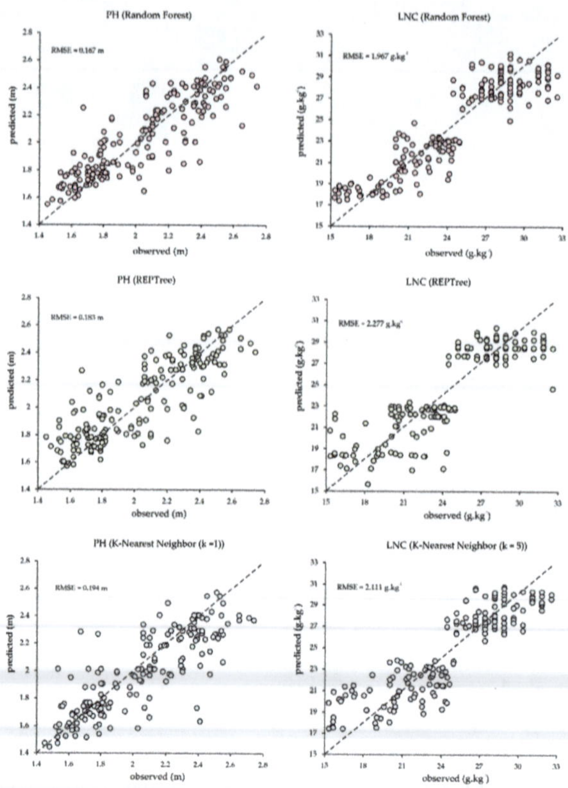

Figure 6. Relationship between the predicted and observed variables (LNC and PH) using the overall best models and configuration. The dashed-line corresponds with a 1:1 linear-fit.

To better demonstrate the feasibility of the proposal, a map of the predicting values for both LNC and PH was constructed with the RF model using the VIs as input parameters (Figure 7). This map can provide a qualitative approach for the result. Once trained, a machine learning model can calculate or perform inference over the image data, returning a visual representation of the area.

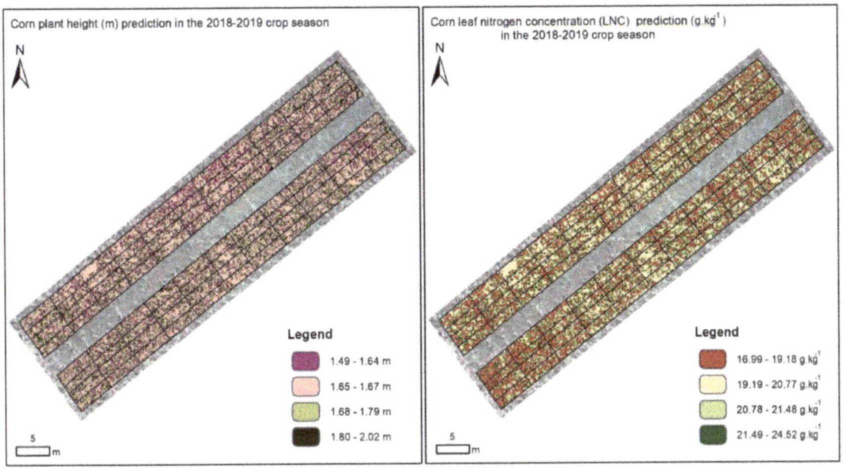

Figure 7. Prediction map over the 2018/2019 crop season indicating the LNC and PH related to every pixel in the unmanned aerial vehicle (UAV)-based image. The model used for inference was the random forest (RF) and only the VIs were used as input data to calculate it.

Table 2 ranks the contribution of individual attributes for RF for estimating LNC and PH. This metric was estimated with configuration #3 (VIs + SBs), considering all input data to confront both spectral bands and spectral indices' importance to the model. The merit for each attribute indicated that VIs like NDVI, NDRE, and SAVI were more contributive than SBs. This result supports the observation over previous analysis, that the VIs configuration returned better accuracy than the SBs configuration. The merit score was obtained from the ranking-based approach in the method section.

Table 2. Ranking order describing the merit of each machine learning model used in this study.

Order	Attribute	Merit (avg.) - LNC	Merit (avg.) - PH
1	NDVI	1.018 ± 0.043	0.939 ± 0.045
2	NDRE	1.004 ± 0.039	0.897 ± 0.048
3	SAVI	0.912 ± 0.047	0.862 ± 0.046
4	Red-Edge	0.88 ± 0.049	0.802 ± 0.046
5	Near-Infrared	0.842 ± 0.04	0.734 ± 0.046
6	Red	0.828 ± 0.047	0.719 ± 0.043
7	Green	0.714 ± 0.052	0.596 ± 0.046
8	GNDVI	0.454 ± 0.044	0.499 ± 0.045

4. Discussion

The evaluation of multiple cultivars and different quantities of N fertilizer was implemented to simulate the characteristics encountered in most maize-crops around Brazil. With this experimental design, using the spectral data in three distinct configurations, we investigated the performance of a set of machine learning algorithms, like the REPT, RF, KNN, SVM (with RBF and polynomial kernels), and LR. The leaners returned similar predictions according to the respective configuration set when predicting both LNC and PH. The adopted configurations were indicative of the importance of VIs in the prediction of these agronomic variables. In a direct analysis regarding the relationship between each variable (Figure 3), a negative correlation of LNC with NIR and VIs and a positive correlation with R was found. This observation differentiates from the literature [10–17] since the LNC is closely related to chlorophyll influence over these spectral variables. Still, this could configure a particular case related to field conditions raised in this study. The same observations were also noted in a previous study, conducted in a Citrus orchard, which also observed the same spectral bands from the Parrot Sequoia embedded sensor [11].

The RF algorithm may share a similar trend in the nutrient analysis since different and recent types of research concluded that this learner is obtaining optimal and balanced results with different spectral data. In similar research [52], an experiment conducted with maize and multispectral imagery from the orbital scale demonstrated that VIs showed strong performance. Another research [34], aiming to estimate maize, stated that the RF learner returned the highest accuracies among the evaluated algorithms. For N content, although not conducted in maize crops, multiple types of research [25,33,53–56] also concluded that the RF learner, as well as other types of regressors based on decision trees, were appropriate to model LNC. In the presented approach, the errors encountered with this model are relatively lower or similar when in comparison to the aforementioned studies.

RF is one of the most powerful methods in the current literature related to machine learning tasks [57–62]. The increase in data dimensionality is often seen as a problem for most traditional methods. In this study, the increase in dimensionality was also necessary to improve the overall accuracy of this model. As for limitations, the major difficulty associated with this method, as well as the other machine learning approaches, is the small amount of data [63]. However, as agronomic variables are onerous to obtain, tests with multiple repetitions and configurations sets were conducted to ensure the accuracy of this proposal. The strategy of adopting different configurations and repetitions should be further explored in future research, where the number of instances is relatively small.

The performance of each algorithm was, as discussed, evaluated with different configurations. This analysis returned interesting outcomes, as the accuracy of the learners were better with the VIs as attributes (Figures 4 and 5). The importance of VIs as estimators of N and PH was evaluated in previous papers [18,52,64–67]. This is mainly because the VIs enhance some characteristics related to biological variables, such as chlorophyll content and biomass, which are highly correlated with LNC and PH. Nonetheless, when implemented in machine learning methods, it is difficult to understand the exact function in the model's predictions. However, when considering different scenarios, as well as implementing the rank-based approach presented here, it is possible to shine some light onto this process. The rank demonstrated that most contributions are provided by the VIs, and, to a lesser extent, the SBs with their respective surface reflectance values. This type of evaluation is important since it provides a matter to indicate which input variables are more suitable to model the evaluated problem, which can reduce the amount of data input, resulting in an accurate and more rapid estimative.

As for the image itself, the major limitation of a UAV image data collection is the low capacity to compensate and analyze larger areas. However, this type of aerial remote sensing is important when considering the spatial resolution and highly detailed information obtained on the vegetation cover, permitting an analysis at a plant or crop-plot level [68–71]. Additionally, by evaluating crop at an aerial view, it is easier to ascertain the relationship between spectral data and biophysical variables, since the end-user can reduce the amount of noise introduced in the system by extracting only pixels corresponding with the canopy itself.

The approach presented here may also be implemented with different datasets over diverse areas, crops, and sensors. The approach of adopting multiple machine learning models and VIs could also be used to predict agronomic variables like other macronutrients and micronutrients. Previous experiments already suggest the possibility of inference other nutrients with spectral data from proximal sensors [12,62,72]. In this regard, additional experiments could consider multispectral data from sensors embedded in UAVs. Here, the particular objective was to investigate the contribution of multispectral data in machine learning methods to nutrient content (N) and height (PH). The advantage of LNC and PH prediction with UAV-based images is that it promotes a rapid and cost-efficient manner to the recurrent monitoring of the agricultural landscapes. However, the traditional agronomic method should not be substituted but assisted by remote sensing technologies and computational techniques such as the ones indicated here.

5. Conclusions

In this study, a machine learning approach was implemented to estimate LNC (g kg^{-1}) and PH (m) for maize plants. It was tested whether the models are impacted by data input regarding different combinations of SBs and VIs. It also demonstrated which one of the implemented learners is more suitable to predict both parameters (LNC and PH). The conducted experiment showed that the RF algorithm performed better, with RMSE equal to 1.9 g.kg^{-1} and 0.17 m, for LNC and PH, respectively. The VIs contributed more to the algorithm's performances than the SBs. This paper concludes that the proposed approach of machine learning models is appropriate to predict these agronomic variables. This method may be used in research that intends to evaluate different types of crops or applied in precision agriculture practices and assist in decision-making models. Regardless, future experiments should be conducted in more practical conditions.

Author Contributions: Conceptualization, H.P., J.M.J., and Y.Y.; methodology, P.E.T., F.H.R.B., C.A.d.S.J., H.P., J.M.J.; investigation, H.P.; writing—original draft preparation, H.P., L.P.O, J.M.J. A.P.M.R., D.E.G.F.; writing—review and editing, L.P.O., J.M.J., A.P.M.R., W.N.G.; project administration, P.E.T., F.H.R.B., L.P.R.T., D.C.S. All authors have read and agreed to the published version of the manuscript.

Funding: This research was funded by CNPq, grant number 303559/2019-5, 433783/2018-4, 314902/2018-0, and 304173/2016-9; CAPES, grant number 88881.311850/2018-01, and; Fundect, grand number 59/300.066/2015, and 59/300.095/2015.

Acknowledgments: The authors acknowledge the support of UFMS (Federal University of Mato Grosso do Sul), UCDB (Dom Bosco Catholic University), CNPq (National Council for Scientific and Technological), and CAPES (Coordination for the Improvement of Higher Education Personnel - Finance code 001).

Conflicts of Interest: The authors declare no conflict of interest. The funders had no role in the design of the study; in the collection, analyses, or interpretation of data; in the writing of the manuscript, or in the decision to publish the results.

References

1. Weiss, M.; Jacob, F.; Duveiller, G. Remote sensing for agricultural applications: A meta-review. *Remote Sens. Environ.* **2020**, *236*, 111402. [CrossRef]
2. Wang, S.; Azzari, G.; Lobell, D.B. Crop type mapping without field-level labels: Random forest transfer and unsupervised clustering techniques. *Remote Sens. Environ.* **2019**, *222*, 303–317. [CrossRef]
3. Hunt, E.R.; Daughtry, C.S.T. What good are unmanned aircraft systems for agricultural remote sensing and precision agriculture? *Int. J. Remote Sens.* **2018**, *39*, 5345–5376. [CrossRef]
4. Zhong, L.; Hu, L.; Zhou, H. Deep learning based multi-temporal crop classification. *Remote Sens. Environ.* **2019**, *221*, 430–443. [CrossRef]
5. Haghverdi, A.; Washington-Allen, R.A.; Leib, B.G. Prediction of cotton lint yield from phenology of crop indices using artificial neural networks. *Comput. Electron. Agric.* **2018**, *152*, 186–197. [CrossRef]
6. Chen, S.W.; Shivakumar, S.S.; Dcunha, S.; Das, J.; Okon, E.; Qu, C.; Taylor, C.J.; Kumar, V. Counting Apples and Oranges with Deep Learning: A Data-Driven Approach. *IEEE Robot. Autom. Lett.* **2017**, *2*, 781–788. [CrossRef]
7. Hunt, M.L.; Blackburn, G.A.; Carrasco, L.; Redhead, J.W.; Rowland, C.S. High resolution wheat yield mapping using Sentinel-2. *Remote Sens. Environ.* **2019**, *233*, 111410. [CrossRef]
8. Jin, Z.; Azzari, G.; You, C.; Di Tommaso, S.; Aston, S.; Burke, M.; Lobell, D.B. Smallholder maize area and yield mapping at national scales with Google Earth Engine. *Remote Sens. Environ.* **2019**, *228*, 115–128. [CrossRef]
9. Sun, J.; Di, L.; Sun, Z.; Shen, Y.; Lai, Z. County-level soybean yield prediction using deep CNN-LSTM model. *Sensors* **2019**, *19*, 4363. [CrossRef]

10. Delloye, C.; Weiss, M.; Defourny, P. Retrieval of the canopy chlorophyll content from Sentinel-2 spectral bands to estimate nitrogen uptake in intensive winter wheat cropping systems. *Remote Sens. Environ.* **2018**, *216*, 245–261. [CrossRef]
11. Osco, L.P.; Marques Ramos, A.P.; Saito Moriya, É.A.; de Souza, M.; Marcato Junior, J.; Matsubara, E.T.; Imai, N.N.; Creste, J.E. Improvement of leaf nitrogen content inference in Valencia-orange trees applying spectral analysis algorithms in UAV mounted-sensor images. *Int. J. Appl. Earth Obs. Geoinf.* **2019**, *83*, 101907. [CrossRef]
12. Osco, L.P.; Ramos, A.P.M.; Pinheiro, M.M.F.; Moriya, É.A.S.; Imai, N.N.; Estrabis, N.; Ianczyk, F.; de'Araújo, F.F.; Liesenberg, V.; de Castro Jorge, L.A.; et al. A machine learning approach to predict nutrient content in valencia-orange leaf hyperspectral measurements. *Remote Sens.* **2020**, *12*, 906. [CrossRef]
13. Song, Y.; Wang, J. Soybean canopy nitrogen monitoring and prediction using ground based multispectral remote sensors. In Proceedings of the 2016 IEEE International Geoscience and Remote Sensing Symposium (IGARSS), Beijing, China, 10–15 July 2016; pp. 6389–6392. [CrossRef]
14. Cammarano, D.; Fitzgerald, G.J.; Casa, R.; Basso, B. Assessing the robustness of vegetation indices to estimate wheat N in mediterranean environments. *Remote Sens.* **2014**, *6*, 2827–2844. [CrossRef]
15. Xue, J.; Gao, S.; Fan, Y.; Li, L.; Ming, B.; Wang, K.; Xie, R.; Hou, P.; Li, S. Traits of plant morphology, stalk mechanical strength, and biomass accumulation in the selection of lodging-resistant maize cultivars. *Eur. J. Agron.* **2020**, *117*, 126073. [CrossRef]
16. Cilia, C.; Panigada, C.; Rossini, M.; Meroni, M.; Busetto, L.; Amaducci, S.; Boschetti, M.; Picchi, V.; Colombo, R. Nitrogen status assessment for variable rate fertilization in maize through hyperspectral imagery. *Remote Sens.* **2014**, *6*, 6549–6565. [CrossRef]
17. Wang, J.; Shen, C.; Liu, N.; Jin, X.; Fan, X.; Dong, C.; Xu, Y. Non-Destructive evaluation of the leaf nitrogen concentration by In-Field visible/Near-Infrared spectroscopy in pear orchards. *Sensors* **2017**, *17*, 538. [CrossRef]
18. Prado Osco, L.; Marques Ramos, A.P.; Roberto Pereira, D.; Akemi Saito Moriya, É.; Nobuhiro Imai, N.; Takashi Matsubara, E.; Estrabis, N.; de Souza, M.; Marcato Junior, J.; Gonçalves, W.N.; et al. Predicting canopy nitrogen content in citrus-trees using random forest algorithm associated to spectral vegetation indices from UAV-imagery. *Remote Sens.* **2019**, *11*, 2925. [CrossRef]
19. Muñoz-Huerta, R.F.; Guevara-Gonzalez, R.G.; Contreras-Medina, L.M.; Torres-Pacheco, I.; Prado-Olivarez, J.; Ocampo-Velazquez, R.V. A review of methods for sensing the nitrogen status in plants: Advantages, disadvantages and recent advances. *Sensors* **2013**, *13*, 10823–10843. [CrossRef]
20. O'Connell, J.L.; Byrd, K.B.; Kelly, M. Remotely-sensed indicators of N-related biomass allocation in Schoenoplectus acutus. *PLoS ONE* **2014**, *9*. [CrossRef]
21. Zheng, H.; Li, W.; Jiang, J.; Liu, Y.; Cheng, T.; Tian, Y.; Zhu, Y.; Cao, W.; Zhang, Y.; Yao, X. A comparative assessment of different modeling algorithms for estimating leaf nitrogen content in winter wheat using multispectral images from an unmanned aerial vehicle. *Remote Sens.* **2018**, *10*, 2026. [CrossRef]
22. Raper, T.B.; Varco, J.J. Canopy-scale wavelength and vegetative index sensitivities to cotton growth parameters and nitrogen status. *Precis. Agric.* **2014**, *16*, 62–76. [CrossRef]
23. Brinkhoff, J.; Dunn, B.W.; Robson, A.J.; Dunn, T.S.; Dehaan, R.L. Modeling mid-season rice nitrogen uptake using multispectral satellite data. *Remote Sens.* **2019**, *11*, 1837. [CrossRef]
24. Liu, Y.L.; Lyu, Q.; He, S.L.; Yi, S.L.; Liu, X.F.; Xie, R.J.; Zheng, Y.; Deng, L. Prediction of nitrogen and phosphorus contents in citrus leaves based on hyperspectral imaging. *Int. J. Agric. Biol. Eng.* **2015**, *8*, 80–88. [CrossRef]
25. He, L.; Song, X.; Feng, W.; Guo, B.B.; Zhang, Y.S.; Wang, Y.H.; Wang, C.Y.; Guo, T.C. Improved remote sensing of leaf nitrogen concentration in winter wheat using multi-angular hyperspectral data. *Remote Sens. Environ.* **2016**, *174*, 122–133. [CrossRef]
26. Miyoshi, G.T.; Arruda, M.D.; Osco, L.P.; Marcato Junior, J.; Gonçalves, D.N.; Imai, N.N.; Tommaselli, A.M.; Honkavaara, E.; Gonçalves, W.N. A novel deep learning method to identify single tree species in UAV-based hyperspectral images. *Remote Sens.* **2020**, *12*, 1294. [CrossRef]
27. Maxwell, A.E.; Warner, T.A.; Fang, F. Implementation of machine learning classification in remote sensing: An applied review. *Int. J. Remote Sens.* **2018**, *39*, 2784–2817. [CrossRef]
28. Singh, A.; Ganapathysubramanian, B.; Singh, A.K.; Sarkar, S. Machine Learning for High-Throughput Stress Phenotyping in Plants. *Trends Plant Sci.* **2016**, *21*, 110–124. [CrossRef]

29. Ball, J.E.; Anderson, D.T.; Chan, C.S. Comprehensive survey of deep learning in remote sensing: Theories, tools, and challenges for the community. *J. Appl. Remote Sens.* **2017**, *11*, 1. [CrossRef]
30. Osco, L.P.; Ramos, A.P.M.; Moriya, É.A.S.; Bavaresco, L.G.; de Lima, B.C.; Estrabis, N.; Pereira, D.R.; Creste, J.E.; Júnior, J.M.; Gonçalves, W.N.; et al. Modeling hyperspectral response of water-stress induced lettuce plants using artificial neural networks. *Remote Sens.* **2019**, *11*, 2797. [CrossRef]
31. Berger, K.; Verrelst, J.; Féret, J.B.; Wang, Z.; Wocher, M.; Strathmann, M.; Danner, M.; Mauser, W.; Hank, T. Crop nitrogen monitoring: Recent progress and principal developments in the context of imaging spectroscopy missions. *Remote Sens. Environ.* **2020**, *242*, 111758. [CrossRef]
32. Zha, H.; Miao, Y.; Wang, T.; Li, Y.; Zhang, J.; Sun, W. Improving Unmanned Aerial Vehicle Remote Sensing-Based Rice Nitrogen Nutrition Index Prediction with Machine Learning. *Remote Sens.* **2020**, *12*, 215. [CrossRef]
33. Chlingaryan, A.; Sukkarieh, S.; Whelan, B. Machine learning approaches for crop yield prediction and nitrogen status estimation in precision agriculture: A review. *Comput. Electron. Agric.* **2018**, *151*, 61–69. [CrossRef]
34. Han, L.; Yang, G.; Dai, H.; Xu, B.; Yang, H.; Feng, H.; Li, Z.; Yang, X. Modeling maize above-ground biomass based on machine learning approaches using UAV remote-sensing data. *Plant Methods* **2019**, *15*, 1–19. [CrossRef] [PubMed]
35. Yang, N.; Liu, D.; Feng, Q.; Xiong, Q.; Zhang, L.; Ren, T.; Zhao, Y.; Zhu, D.; Huang, J. Large-scale crop mapping based on machine learning and parallel computation with grids. *Remote Sens.* **2019**, *11*, 1500. [CrossRef]
36. Sharma, L.K.; Bu, H.; Franzen, D.W.; Denton, A. Use of corn height measured with an acoustic sensor improves yield estimation with ground based active optical sensors. *Comput. Electron. Agric.* **2016**, *124*, 254–262. [CrossRef]
37. Jiang, T.; Liu, J.; Gao, Y.; Sun, Z.; Chen, S.; Yao, N.; Ma, H.; Feng, H.; Yu, Q.; He, J. Simulation of plant height of winter wheat under soil Water stress using modified growth functions. *Agric. Water Manag.* **2020**, *232*, 106066. [CrossRef]
38. Almeida, V.C.; Viana, J.M.S.; DeOliveira, H.M.; Risso, L.A.; Ribeiro, A.F.S.; DeLima, R.O. Genetic diversity and path analysis for nitrogen use efficiency of tropical popcorn (Zea mays ssp. everta) inbred lines in adult stage. *Plant Breed.* **2018**, *137*, 839–847. [CrossRef]
39. Torres, L.G.; Rodrigues, M.C.; Lima, N.L.; Horta Trindade, T.F.; Fonseca e Silva, F.; Azevedo, C.F.; DeLima, R.O. Multi-trait multi-environment Bayesian model reveals g x e interaction for nitrogen use efficiency components in tropical maize. *PLoS ONE* **2018**, *13*, e0199492. [CrossRef]
40. Silva, F.C. *Da Manual de Análises Químicas de Solos, Plantas e Fertilizantes; Fernando do Amaral Pereira*, 2nd ed.; Embrapa Informação Tecnológica: Brasilia, Brazil, 2009.
41. Rouse, J.W.; Hass, R.H.; Schell, J.A.; Deering, D.W. Monitoring vegetation systems in the great plains with ERTS. Third Earth Resour. *Technol. Satell. Symp.* **1973**, *1*, 309–317.
42. Gitelson, A.; Merzlyak, M.N. Quantitative estimation of chlorophyll-a using reflectance spectra: Experiments with autumn chestnut and maple leaves. *J. Photochem. Photobiol. B Biol.* **1994**, *22*, 247–252. [CrossRef]
43. Gitelson, A.A.; Kaufman, Y.J.; Merzlyak, M.N. Use of a green channel in remote sensing of global vegetation from EOS- MODIS. *Remote Sens. Environ.* **1996**, *58*, 289–298. [CrossRef]
44. Huete, A.R. A Soil-Adjusted Vegetation Index (SAVI). *Remote Sens. Environ.* **1988**, *25*, 295–309. [CrossRef]
45. Mitchell, T.M. *Machine Learning*, 1st ed.; McGraw-Hill, Inc.: New York, NY, USA, 1997.
46. Saha, S.; Saha, M.; Mukherjee, K.; Arabameri, A.; Ngo, P.T.T.; Paul, G.C. Predicting the deforestation probability using the binary logistic regression, random forest, ensemble rotational forest, REPTree: A case study at the Gumani River Basin, India. *Sci. Total Environ.* **2020**, *730*, 139197. [CrossRef] [PubMed]
47. Belgiu, M.; Drăgu, L. Random forest in remote sensing: A review of applications and future directions. ISPRS J. Photogramm. *Remote Sens.* **2016**, *114*, 24–31. [CrossRef]
48. Ali, N.; Neagu, D.; Trundle, P. Evaluation of k-nearest neighbour classifier performance for heterogeneous data sets. *SN Appl. Sci.* **2019**, *1*, 1–15. [CrossRef]
49. Nalepa, J.; Kawulok, M. Selecting training sets for support vector machines: A review. *Artif. Intell. Rev.* **2019**, *52*, 857–900. [CrossRef]
50. Štepanovský, M.; Ibrová, A.; Buk, Z.; Velemínská, J. Novel age estimation model based on development of permanent teeth compared with classical approach and other modern data mining methods. *Forensic Sci. Int.* **2017**, *279*, 72–82. [CrossRef]

51. Cheshmberah, F.; Fathizad, H.; Parad, G.A.; Shojaeifar, S. Comparison of RBF and MLP neural network performance and regression analysis to estimate carbon sequestration. *Int. J. Environ. Sci. Technol.* **2020**. [CrossRef]
52. Kross, A.; McNairn, H.; Lapen, D.; Sunohara, M.; Champagne, C. Assessment of RapidEye vegetation indices for estimation of leaf area index and biomass in corn and soybean crops. *Int. J. Appl. Earth Obs. Geoinf.* **2015**, *34*, 235–248. [CrossRef]
53. Varela, S.; Dhodda, P.R.; Hsu, W.H.; Prasad, P.V.V.; Assefa, Y.; Peralta, N.R.; Griffin, T.; Sharda, A.; Ferguson, A.; Ciampitti, I.A. Early-season stand count determination in Corn via integration of imagery from unmanned aerial systems (UAS) and supervised learning techniques. *Remote Sens.* **2018**, *10*, 343. [CrossRef]
54. Miphokasap, P.; Wannasiri, W. Estimations of Nitrogen Concentration in sugarcane using hyperspectral imagery. *Sustainability.* **2018**, *10*, 1266. [CrossRef]
55. Abdel-Rahman, E.M.; Ahmed, F.B.; Ismail, R. Random forest regression and spectral band selection for estimating sugarcane leaf nitrogen concentration using EO-1 Hyperion hyperspectral data. *Int. J. Remote Sens.* **2013**, *34*, 712–728. [CrossRef]
56. Liang, L.; Di, L.; Huang, T.; Wang, J.; Lin, L.; Wang, L.; Yang, M. Estimation of leaf nitrogen content in wheat using new hyperspectral indices and a random forest regression algorithm. *Remote Sens.* **2018**, *10*, 1940. [CrossRef]
57. Rocha, A.D.; Groen, T.A.; Skidmore, A.K. Spatially-explicit modelling with support of hyperspectral data can improve prediction of plant traits. *Remote Sens. Environ.* **2019**, *231*, 111200. [CrossRef]
58. Quanzhou, Y.; Shaoqiang, W.; Hao, S.; Kun, H.; Lei, Z. An Evaluation of Spaceborne Imaging Spectrometry for Estimation of Forest Canopy Nitrogen Concentration in a Subtropical Conifer Crop of Southern China. *J. Resour. Ecol.* **2014**, *5*, 1–10. [CrossRef]
59. Wang, T.; Thomasson, J.A.; Yang, C.; Isakeit, T.; Nichols, R.L. Automatic classification of cotton root rot disease based on UAV remote sensing. *Remote Sens.* **2020**, *12*, 1310. [CrossRef]
60. Bruning, B.; Liu, H.; Brien, C.; Berger, B.; Lewis, M.; Garnett, T. The Development of Hyperspectral Distribution Maps to Predict the Content and Distribution of Nitrogen and Water in Wheat (*Triticum aestivum*). *Front. Plant Sci.* **2019**, *10*, 1–16. [CrossRef]
61. Zhou, W.; Zhang, J.; Zou, M.; Liu, X.; Du, X.; Wang, Q.; Liu, Y.; Liu, Y.; Li, J. Prediction of cadmium concentration in brown rice before harvest by hyperspectral remote sensing. *Environ. Sci. Pollut. Res.* **2019**, *26*, 1848–1856. [CrossRef]
62. Gao, J.; Meng, B.; Liang, T.; Feng, Q.; Ge, J.; Yin, J.; Wu, C.; Cui, X.; Hou, M.; Liu, J.; et al. Modeling alpine grassland forage phosphorus based on hyperspectral remote sensing and a multi-factor machine learning algorithm in the east of Tibetan Plateau, China. ISPRS J. Photogramm. *Remote Sens.* **2019**, *147*, 104–117. [CrossRef]
63. Agarwal, S. Data Mining: Data Mining Concepts and Techniques. In Proceedings of the 2013 International Conference on Machine Intelligence and Research Advancement, Katra, JK, India, 21–23 December 2013; Available online: https://ieeexplore.ieee.org/abstract/document/6918822/ (accessed on 5 October 2020).
64. Knoblauch, C.; Watson, C.; Berendonk, C.; Becker, R.; Wrage-Mönnig, N.; Wichern, F. Relationship between remote sensing data, plant biomass and soil nitrogen dynamics in intensively managed grasslands under controlled conditions. *Sensors* **2017**, *17*, 1483. [CrossRef]
65. Ramoelo, A.; Cho, M.A.; Mathieu, R.; Madonsela, S.; van de Kerchove, R.; Kaszta, Z.; Wolff, E. Monitoring grass nutrients and biomass as indicators of rangeland quality and quantity using random forest modelling and WorldView-2 data. *Int. J. Appl. Earth Obs. Geoinf.* **2015**, *43*, 43–54. [CrossRef]
66. Chen, J.; Li, F.; Wang, R.; Fan, Y.; Raza, M.A.; Liu, Q.; Wang, Z.; Cheng, Y.; Wu, X.; Yang, F.; et al. Estimation of nitrogen and carbon content from soybean leaf reflectance spectra using wavelet analysis under shade stress. *Comput. Electron. Agric.* **2019**, *156*, 482–489. [CrossRef]
67. Wolanin, A.; Camps-Valls, G.; Gómez-Chova, L.; Mateo-García, G.; van der Tol, C.; Zhang, Y.; Guanter, L. Estimating crop primary productivity with Sentinel-2 and Landsat 8 using machine learning methods trained with radiative transfer simulations. *Remote Sens. Environ.* **2019**, *225*, 441–457. [CrossRef]
68. Gebremedhin, A.; Badenhorst, P.; Wang, J.; Giri, K.; Spangenberg, G.; Smith, K. Development and validation of a model to combine NDVI and plant height for high-throughput phenotyping of Herbage Yield in a perennial ryegrass breeding program. *Remote Sens.* **2019**, *11*, 2494. [CrossRef]

69. Zhang, Q.; Liu, Y.; Gong, C.; Chen, Y.; Yu, H. Applications of deep learning for dense scenes analysis in agriculture: A review. *Sensors.* **2020**, *20*, 1520. [CrossRef]
70. Li, Y.; Cao, Z.; Lu, H.; Xiao, Y.; Zhu, Y.; Cremers, A.B. In-field cotton detection via region-based semantic image segmentation. *Comput. Electron. Agric.* **2016**, *127*, 475–486. [CrossRef]
71. Xiong, X.; Zhang, J.; Guo, D.; Chang, L.; Huang, D. Non-invasive sensing of nitrogen in plant using digital images and machine learning for brassica campestris ssp. Chinensis, L. *Sensors* **2019**, *19*, 2448. [CrossRef]
72. Ling, B.; Goodin, D.G.; Raynor, E.J.; Joern, A. Hyperspectral analysis of leaf pigments and nutritional elements in tallgrass prairie vegetation. *Front. Plant Sci.* **2019**, *10*, 1–13. [CrossRef]

remote sensing

MDPI

Article

A Novel Machine Learning Approach to Estimate Grapevine Leaf Nitrogen Concentration Using Aerial Multispectral Imagery

Ali Moghimi [1], Alireza Pourreza [1,*], German Zuniga-Ramirez [1,2], Larry E. Williams [2,3] and Matthew W. Fidelibus [2,3]

1 Department of Biological and Agricultural Engineering, University of California, Davis, One Shields Ave, Davis, CA 95616, USA; amoghimi@ucdavis.edu (A.M.); gzunigaramirez@ucanr.edu (G.Z.-R.)
2 Kearney Agricultural Research and Extension Center, 9240 S. Riverbend Avenue, Parlier, CA 93648, USA; lewilliams@ucanr.edu (L.E.W.); mwfidelibus@ucanr.edu (M.W.F.)
3 Department of Viticulture and Enology, University of California, Davis, 595 Hilgard Ln, Davis, CA 95616, USA
* Correspondence: apourreza@ucdavis.edu

Received: 8 September 2020; Accepted: 21 October 2020; Published: 26 October 2020

Abstract: Assessment of the nitrogen status of grapevines with high spatial, temporal resolution offers benefits in fertilizer use efficiency, crop yield and quality, and vineyard uniformity. The primary objective of this study was to develop a robust predictive model for grapevine nitrogen estimation at bloom stage using high-resolution multispectral images captured by an unmanned aerial vehicle (UAV). Aerial imagery and leaf tissue sampling were conducted from 150 grapevines subjected to five rates of nitrogen applications. Subsequent to appropriate pre-processing steps, pixels representing the canopy were segmented from the background per each vine. First, we defined a binary classification problem using pixels of three vines with the minimum (low-N class) and two vines with the maximum (high-N class) nitrogen concentration. Following optimized hyperparameters configuration, we trained five machine learning classifiers, including support vector machine (SVM), random forest, XGBoost, quadratic discriminant analysis (QDA), and deep neural network (DNN) with fully-connected layers. Among the classifiers, SVM offered the highest F1-score (82.24%) on the test dataset at the cost of a very long training time compared to the other classifiers. Alternatively, QDA and XGBoost required the minimum training time with promising F1-score of 80.85% and 80.27%, respectively. Second, we transformed the classification into a regression problem by averaging the posterior probability of high-N class for all pixels within each of 150 vines. XGBoost exhibited a slightly larger coefficient of determination ($R^2 = 0.56$) and lower root mean square error (RMSE) (0.23%) compared to other learning methods in the prediction of nitrogen concentration of all vines. The proposed approach provides values in (i) leveraging high-resolution imagery, (ii) investigating spatial distribution of nitrogen across a vine's canopy, and (iii) defining spatial zones for nitrogen application and smart sampling.

Keywords: grapevine; hyperparameter optimization; machine learning; multispectral imaging; nitrogen; precision viticulture; UAV

1. Introduction

The United States of America is the seventh largest table grape (*Vitis vinifera* L.) producing country in the world with 100 million 9 kg-boxes produced annually since 2012. Approximately 99% of the USA's table grapes are produced in California, primarily in the southern San Joaquin Valley (SJV) [1]. One of the most common table grape cultivars in the SJV, and worldwide, is Flame Seedless [1]. Production of Flame Seedless, like other table grapes, is particularly capital-intensive, so it is imperative to maximize yield and quality to ensure profitability [2]. An important factor that contributes to grape yield and quality is vine mineral nutrition [3,4]. In SJV vineyards, nitrogen (N) is the mineral element that most often needs to be supplemented by fertilization [5]. Insufficient N limits vine vigor [6,7]. On the other hand, unnecessary application of N fertilizers needlessly increases the production cost, may cause unintended nutrient imbalances, contributes to environmental contamination [8]), stimulates excess vegetative growth, and reduces grape quality [9]. In fact, excessive application of nitrogen to crop plants has led to significant nitrate groundwater pollution in the SJV [10]. To help prevent further nitrate contamination to groundwater, growers in the SJV are required to implement nitrogen management plans.

Laboratory analyses of plant tissue samples can help determine if supplemental fertilizer treatments are necessary and are thus a fundamental component of N management plans [11]. Petioles are the most commonly sampled tissue of grapevines, but leaf blades may be used instead [11]. Tissue nutrient content changes markedly over the course of each season, hence samples are collected at particular growth stages to facilitate the comparison of data collected in other seasons or different locations. Bloom (anthesis) is the earliest standard sample collection time, and petioles or leaf blades collected at bloom are generally considered a reliable indicator of grapevine nutrient status and are thus widely used by commercial growers [12]. The collection of tissue samples is laborious, the analyses are relatively expensive, and care must be taken to ensure the samples collected are representative of the management area [4].

The increased availability and affordability of unmanned aerial and ground vehicles and advanced non-contact sensing technologies (e.g., hyper/multispectral sensors) has prompted research on remote sensing as a complimentary or alternative approach for estimating nitrogen status in specialty crops such as grape [4,13], apple [14], citrus [15], and almond [16]. An effective remote sensing protocol might offer some advantages over conventional testing, particularly with respect to spatial resolution and cost. For example, remote sensing could potentially reveal spatial variation of N status in grapevines, which can assist in identifying hot spots for smart sampling (i.e., directed sampling) and generating precise maps to develop a variable rate N fertilization program. However, the use of remote sensing for N estimation in crops has been constrained by issues associated with data analysis, such as overfitting, the curse of dimensionality, and developing robust, scalable, and generalizable predictive models. These large image-based datasets, often having high spatial, spectral, and temporal resolution, may require more sophisticated data analytics techniques to fully capitalize on the data, whereas conventional methods have previously relied on spectral indices [17].

Spectral indices have some significant limitations as analytical tools. For instance, normalized difference vegetation index (NDVI) saturates when vegetation coverage is dense [18], and atmospherically resistant vegetation index (ARVI) saturates when chlorophyll concentration reaches a certain level [19]. Previous studies have recommended a wide range of spectral indices and electromagnetic regions for predicting nitrogen content in crops. Pacheco-Labrador et al. [20] summarized more than 80 spectral indices from the literature for N content estimation. They achieved a strong coefficient of determination ($R^2 = 0.76$) between a spectral index composed of three spectral bands (1310, 1720, and 730 nm) and leaf nitrogen concentration in Holm oak. In addition, they identified red edge as a region with a high correlation to leaf N concentration. A number of studies reported electromagnetic regions suitable for N estimation including shortwave infrared (SWIR) [21], green, red edge, and near-infrared [20], visible region, mainly blue region [22], blue, green, and red edge [23], and red edge position [24]. While red edge has been frequently suggested as one of the most relevant spectral regions, Hansen and Schjoerring [22] questioned the importance of red edge as it was not identified as a significant band related to the N content of two winter wheat cultivars in their study. After calculating all possible normalized ratio indices for wavebands between 350 and 2200 nm aimed at predicting N content of five species, Ferwerda et al. [25] did not identify an optimal normalized ratio index that correlated well with the N content of five plant species. Considering the limitations of spectral indices, there is a need for advanced data analytics techniques to develop mathematical predictive models through harnessing the power of big data and computer processing.

The emergence of advanced machine learning techniques, along with high-performance computational power, has provided new opportunities to translate image-based datasets into novel insights. In agriculture, machine learning and deep learning have been recently implemented to analyze images captured for various applications, such as biotic stress detection [26,27], abiotic stress detection [28,29], nitrogen estimation [30,31], spectral features selection for high-throughput phenotyping [32], weed detection [33,34] and yield prediction [17,35].

Although the capability of machine learning has been verified in the agricultural domain, its full potential has been restricted by the limited number of ground truth data while developing a robust and generalizable predictive model demands a large and diverse dataset to capture the inherent large spatial and temporal variability in complex biological problems. Ground truth data collection is generally expensive, labor-intensive, and time-consuming. In addition, collecting ground truth data in agriculture often requires destructive experiments, like tissue sampling for nitrogen content measurements. A limited number of ground truth data confines the efficient use of advanced machine learning algorithms because it limits the tuning of the model's hyperparameter and learning of the model's parameters. Configuration of hyperparameters and parameters of a model is required for developing an underlying function that effectively maps the input variables to the desired output variables and maintains its performance on unseen datasets.

This study was motivated by the desire to develop a data-driven, decision-support tool to facilitate grapevine nitrogen management. The overarching goal of this study was to leverage high-resolution aerial multispectral imagery and advanced machine learning techniques to develop robust predictive models for grapevine nitrogen estimation at the bloom (anthesis) stage when tissue sampling is often performed. The specific objectives were to (i) develop predictive models through the optimized configuration of hyperparameters and parameters for accurate nitrogen concentration estimation at bloom, (ii) compare the performance of trained models in pixel-based binary classification and in vine-based prediction of nitrogen concentration, and (iii) investigate the spatial distribution of nitrogen across a canopy through pixel-based soft classification (i.e., calculating the posterior probabilities of classes).

2. Materials and Methods

2.1. Experimental Site

The research was conducted in a commercial table grape vineyard near Kingsburg, California. *Vitis vinifera* L. cv. Flame Seedless table grapes grafted onto Freedom rootstock were planted in 2012. Vine and row spacing was 1.83 and 3.66 m, respectively. The vines were trained to quadrilateral cordons and supported by a 3.05 m wide, Y-shaped, open gable trellis system.

All vines within the experimental portion of the vineyard were subjected to regular cultural practices performed by the grower with the exception of nitrogen fertilization. Vines in the study area were not fertilized or received different amounts of N as part of an N fertilization trial. Specifically, the effects of three levels of N (0, 19.2, or 48 g/vine) and two levels of split applications (2 versus 10 applications/season) were tested in a split-plot design where the level of N was the main plot factor (replicated five times), and the number of split applications was the subplot factor. This design required a total of 30 plots (3 levels of N × 2 split application treatments × 5 replications), each containing five individual vines, resulting in a study with 150 vines. Due to the combination of main plot factors (levels of N) and subplot factors (levels of split applications), at bloom, the time tissue samples were collected, the various plots had received a fraction of the total seasonal amount of N to be applied. Depending on the N and split application treatments to which the plots were assigned, the vines had received 0.0, 3.9, 9.6, 9.7, or 24.2 g N/vine (Figure 1A). The differential application of N affected leaf N concentration, which made the site an appropriate place to evaluate the potential of remote sensing to estimate leaf N concentration of table grapes.

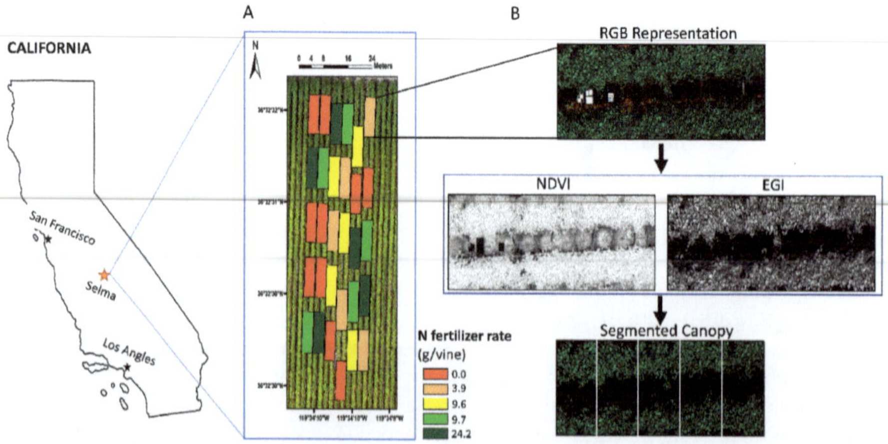

Figure 1. (**A**) Experimental site composed of 150 vines with five nitrogen application treatments at Selma, California. Each plot contained five consecutive vines in a row. Depending on the nitrogen and split application treatments assigned to the plots, the vines had received 0, 3.9, 9.6, 9.7, or 24.2 g N/vine at the time the data were collected (bloom). (**B**) Segmentation process of canopy pixels using excess green index (EGI) and normalized difference vegetation index (NDVI). The extent of each vine within a plot was determined based on the markers' location, spatial resolution (~1 cm/pixel), and vines spacing (1.83 m).

Leaf tissue samples were collected from each of the five vines within each experimental plot immediately after aerial imagery at bloom. We collected one shoot at random from each of the 150 vines. Entire shoots were thoroughly rinsed in deionized water, and all of the leaf blades on each shoot were clipped from their petioles with shears, dried in a forced-air oven, ground into a fine powder, and then submitted to a commercial laboratory (Dellavalle Laboratory, Inc., Fresno, CA, USA) for determination of total N using the automated combustion method [36]. Nitrogen concentration values are expressed as a percentage of leaf dry weight. We followed a similar protocol except that the vine that each shoot was collected from was noted and considered a separate sample.

2.2. *Airborne Multispectral Imaging System*

The images were collected with the RedEdge 3 camera system (MicaSense, Inc., Seattle, WA, USA), which simultaneously captures five discrete spectral bands, including blue, green, red, red edge, and near-infrared. The field of view of the lens for all bands was 47.2 degrees. The size of the images was 1280 × 960 pixels, and they were saved with 16-Bit TIFF. Table 1 summarizes the specification of the multispectral camera.

Table 1. Specification of MicaSense RedEdge 3 multispectral camera.

Bands	Center Wavelength (nm)	Bandwidth FWHM *
Blue	475	20
Green	560	20
Red	668	10
Red Edge	717	10
Near-infrared	840	40

* full width at half maximum.

2.3. Aerial Imagery Campaign

Multispectral images were collected from the experimental plots during the bloom stage on 8 May 2019. The multispectral camera system was mounted on the dual downward gimbal mount of DJI Matrice 210 (Shenzhen, China) UAV using a 3D-printed mounting bracket. The multispectral images were captured individually from each single plot in a hovering state at an altitude about 15 m above ground level using the camera's manual capture functionality through the WIFI interface to achieve sufficient spatial resolution (about 1 cm/pixel) for minimizing the number of mixed pixels. An automatic exposure setting was used during aerial multispectral imagery to leverage the sensor's full dynamic range while avoiding saturation. The images were captured around solar noon with clear sky conditions.

2.4. Pre-Processing of Multispectral Images

A python-based framework called Micasense_preprocessing (version 1.0.0) was developed to automate all pre-processing steps, including radiometric calibration, unwarping (removing lens distortion for better image alignment), and bands alignments [37].

At the first step of radiometric calibration, raw images were converted to radiance to account for sensor-dependent factors such as gain, exposure setting, and vignette effects [17]. The Micasense_preprocessing uses information embedded in the header file of images for radiance conversion. Afterwards, the radiance images were converted to reflectance to account for the time-dependent factor, which is mainly variation in the intensity of incident light. The Micasense_preprocessing converted the radiance images to reflectance using incoming irradiance measured by a five-band incident light sensor integrated with the camera. The incident light sensor was mounted on the top of the aircraft using a 3D-printed mounting bracket while pointing upward to measure downwelling irradiance at each band per each captured image. In addition, the Micasense_preprocessing unwarped the images, aligned, and cropped each image to the common frame among all five bands.

2.5. Grapevine's Canopy Segmentation

Two markers were placed on top of the first and last vine's trunk in each plot to facilitate the identification of each plot extent. Afterwards, the extent of each vine within a plot was determined based on the markers' location, spatial resolution (~1 cm/pixel), and vines spacing (1.83 m). Lastly, pixels representing the canopy were segmented from the background using a binary mask obtained from applying empirical thresholds on the excess green index (EGI) and NDVI (i.e., multiplying EGI and NDVI binary masks) [29] (Figure 1B). The canopy pixels located in the shaded area were removed during the segmentation process.

2.6. Analysis of Multispectral Dataset

Prediction of nitrogen at the vine-scale was essentially a regression problem in which we have a continuous output variable (i.e., nitrogen concentration). While we had a single value for nitrogen concentration per a given vine, there were thousands of pixels (i.e., samples) for each vine. One common approach is to average across all of the pixels for an individual vine to generate one feature vector (i.e., spectral response) per each output. Although this approach is widely used to address ill-matched problems, when there is only one output variable representing several samples, it can be a naïve solution because it ignores the variability of N concentrations within vines. This spatial variability, however, can be revealed by using an appropriate approach for analysis of high-resolution imagery.

In the first step of the analysis, a binary classification problem was defined at the pixel level to leverage the large number of pixels obtained through high-resolution imagery. In addition, this approach enabled us to investigate the distribution of nitrogen concentration across the top of the canopy. However, since the finest resolution at which nitrogen management decisions can be made is at the vine scale, we converted the pixel-based soft classification problem into a vine-based regression problem. The details of the proposed methodology are described hereinafter.

2.6.1. Training Machine Learning Classifiers

Based on the nitrogen concentration measured through tissue sampling for each vine, we defined two classes, low-N class and high-N class, such that a balanced dataset was generated with an approximately equal number of observations per each class. The low-N class entailed three vines (n = 42,604 pixels) with the minimum amount of nitrogen concentration, and the high-N class comprised two vines (n = 41,817 pixels) with the maximum amount of nitrogen concentration among all 150 vines (Figure 2A). Therefore, the matrix of features had 84,421 rows (i.e., pixels) and five columns (i.e., spectral bands), and the target array with the length of 84,421 contained the class labels (low-N or high-N) corresponding to each of the samples. The averaged spectral response of low-N and high-N classes are presented in Figure 2B.

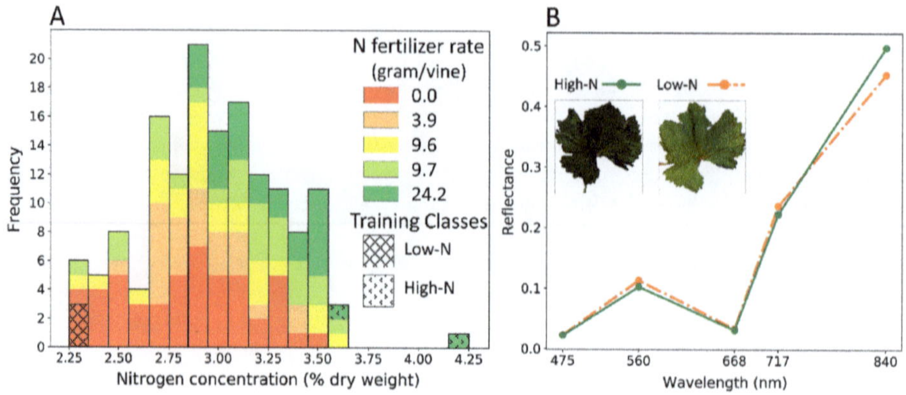

Figure 2. (**A**) Stacked histogram of measured leaf nitrogen concentration (% dry weight) showing the contribution of individual nitrogen fertilizer rate (the amount of nitrogen that vines received until bloom) in each bin. Three vines with minimum nitrogen concentration were used as the low-N class, and two vines with maximum nitrogen concentration were used as the high-N class. The mean and standard deviation of the measured nitrogen concentration for 150 vines were 2.98% and 0.34%, respectively. (**B**) Averaged spectral response of the low-N class (three vines with 42,604 pixels) and the high-N class (two vines with 41,817 pixels).

After shuffling the dataset, about 10% of the data from each class was held out as a test dataset (4190 pixels for high-N and 4253 pixels for low-N class) for an unbiased evaluation of the prediction models, and the rest of the dataset (90%) was used for hyperparameter tuning and learning the model's parameters.

The training dataset was standardized using the z-score technique such that each new feature had a zero-mean and unit-variance distribution. The mean and standard deviation of the training dataset was used for standardizing the test dataset [17].

In this study, we implemented five machine learning algorithms, including support vector machine (SVM), random forest, XGBoost, quadratic discriminant analysis (QDA), and deep neural network (DNN) with fully-connected layers. These classifiers were selected to examine various types of classifiers in terms of objective function, interpretability, scalability, sensitivity to outliers and noise, and ability to handle collinearity among the input features.

All models were developed in Python (version 3.7.3) using the scikit-learn library (version 0.22) [38], XGBoost library (version 0.90) [39], and TensorFlow library (version 2.1.0) [40].

2.6.2. Hyperparameter Optimization

The performance of machine learning models is largely influenced by the configuration of their internal parameters, so-called hyperparameters [41]. The aim of optimizing hyperparameters, which should be defined upfront by model developers, is to identify an optimal set of hyperparameters that offers the best performance on a validation dataset based on a desired validation metric such as F1-score or prediction error. Through hyperparameter optimization, as a key step in the development process of machine learning models, we attempt to develop a model that could generalize its performance on an unseen test dataset.

The most widely used approaches for optimizing hyperparameters have been grid and random search, among which random search tends to outperform [42]. However, grid and random search suffer from ignoring the past results observed during previous search iteration; hence they are inefficient in terms of required processing time to explore search space. Recently, the Bayesian optimization approach has been applied for hyperparameter tuning of machine learning models [43,44]. Using this approach, the history of previously evaluated trials is leveraged to strategically navigate the next hyperparameter search in search space in a more time-efficient manner. It has been shown that hyperparameter optimization techniques based on Bayesian optimization could significantly outstrip random search in terms of lower validation error and required computation time [43,45].

In this study, we utilized Tree Parzen Estimator [43], a similar method to Bayesian optimization, yet faster [46]. An optimal set of hyperparameters for each machine learning classifier was identified using the Optuna library (version 1.3.0) [47] in python. This library is a define-by-run application programming interface (API) that provides an automated framework to dynamically search hyperparameter space for the identification of an optimal set of hyperparameters through strategic searching and pruning method [47]. The list of hyperparameters optimized for each classifier is presented in Table A1.

2.7. *Comparing Classifiers on Test Dataset*

The performance of classifiers was evaluated on the test dataset using several metrics such as F1-score, precision, recall, and area under the curve (AUC) in the receiver operating characteristics curve [48]. In addition, the performance of the classifiers was also compared using a 10-fold cross-validation (CV) on the training dataset. The required time for training and testing the classifiers was also recorded to compare the scalability of the models.

2.8. *Transforming Classification into Regression*

The ultimate goal for developing machine learning predictive models was to predict the nitrogen concentration at the vine-scale at which fertilizer decision and application are performed. The classifiers were developed to return posterior class probability—a soft classification approach rather than a hard-binary classification. The posterior probability of high-N given spectral data $\left(P(high\_N|X)\right)$ for all pixels representing each of 150 vines was averaged to calculate the probability of high-N at the vine scale. Then, the coefficient of determination between $P(high\_N|X)$ and the measured nitrogen concentration was computed for each predictive model. Root mean square error (RMSE) was also calculated as another metric for prediction error. SVM was omitted for this step as the standard SVM does not offer per-class posterior probability [49]. In this approach, the performance of models trained on pixels of five vines was evaluated on a large dataset, including vines with N concentration distribution outside of the training dataset composed of pixels of five vines with extreme nitrogen concentration.

The entire process from data collection to data analysis and interpretation is summarized as a flowchart in Figure A1.

3. Results

3.1. Bands Pairwise Correlation

Patterns and relationships between the features (spectral bands) for the two classes (low-N and high-N) were investigated through exploratory data analysis (EDA) in which bands were mutually plotted against each other in the shape of a rectangular 5 × 5 matrix (Figure 3).

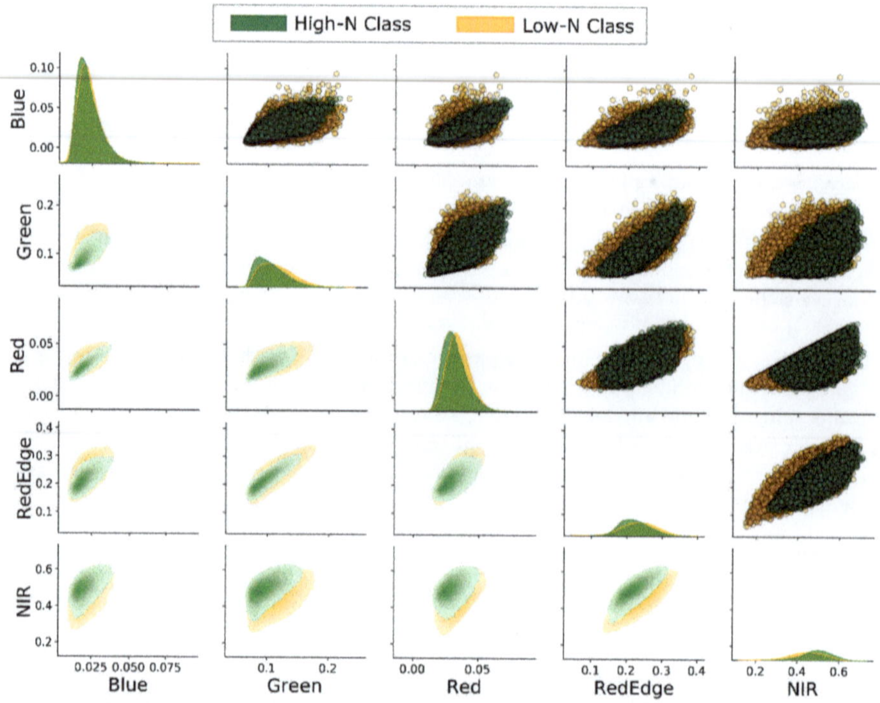

Figure 3. Multispectral bands pair plot in the shape of a rectangular 5 × 5 matrix. Upper diagonal plots show scatter plots of pixels in two-dimensional feature spaces; each spanned by a pair of bands. Diagonal plots illustrate the probability density function of two classes estimated with the univariate kernel density estimate (KDE) technique. Lower diagonal plots depict the probability density function of each class estimated by bivariate KDE across the 2-dimensional feature space spanned by a pair of bands. The regions shown with a darker color refer to a denser area.

The upper triangle of the grid in Figure 3 illustrates the scatter plots of pixels in two-dimensional feature spaces, each spanned by two bands. These scatter plots provide information about the correlation between bands, distribution of the pixels, and the extent of low-N class deviation from high-N class. In general, spectral bands exhibited positive correlation mutually. The scatter plots in Figure 3 indicate that red edge has a strong correlation with both near-infrared (NIR) and red bands, whereas green band exhibits relatively less correlation with other bands. In addition, scatter plots depict how different pair-band combinations differentiated low-N and high-N classes. For instance, NIR and green bands, which were less correlated, offered better discrimination of the two classes compared to the other pairs. It also appears that the NIR-RedEdge scatter plot is the other feature space in which the two classes are more separated while there is a strong correlation between the NIR and red edge bands.

Pixels representing low-N class tend to scatter more in the feature space, indicating more variation among pixels. Alternatively, pixels of the high-N class tend to cluster with less variation, generating denser area in the scatter plots. However, the regions with a higher density per each class are not noticeable in the scatter plots due to a large number of pixels and a high degree of overlapping. Therefore, the bivariate kernel density estimate (KDE) is shown in the lower triangle of the grid in Figure 3. The bivariate KDE plots illustrate the variability of the underlying probability density function of each class across the two-dimensional feature space spanned by two bands. While the scatter plots provide information about the distribution of the pixels from the two classes, the bivariate KDE plots demonstrate the probability density of the pixels per each class—regions with higher probability density are shown with a darker color in Figure 3. Similar to the scatter plots, bivariate KDE plots indicate the positive correlation between spectral bands as the major axis of the ellipsoid in pairwise plots has a positive slope.

The diagonal plots in Figure 3 show the underlying probability density function of two classes estimated with univariate KDE—a nonparametric technique with a Gaussian kernel function. The univariate KDE plots provide insight into the reflectance pattern of pixels representing the two classes. For instance, the low-N pixels tended to reflect more in green and red bands, which explains why the leaves with low N concentration appear to have a light green-yellow color.

3.2. Hyperparameter Tuning

The goal of hyperparameter tuning was to assure an optimal set of the model's hyperparameters that returns the best performance is used during the training step of the model. Figure 4 shows the variation of classification accuracy for SMV as a function of hyperparameter values for more than 100 trials, where the accuracy reached a plateau. For this dataset, the "radial basis function" tended to result in a higher accuracy; therefore, it was selected as the kernel for the SVM classifier. The optimal values for the regularization parameter (C = 52.07) and kernel coefficient (gamma = 0.21) were identified based on the SVM performance across the trials. The list of hyperparameters, search space domain per each hyperparameter, and the selected optimal hyperparameters are shown in Table A1 per each classifier.

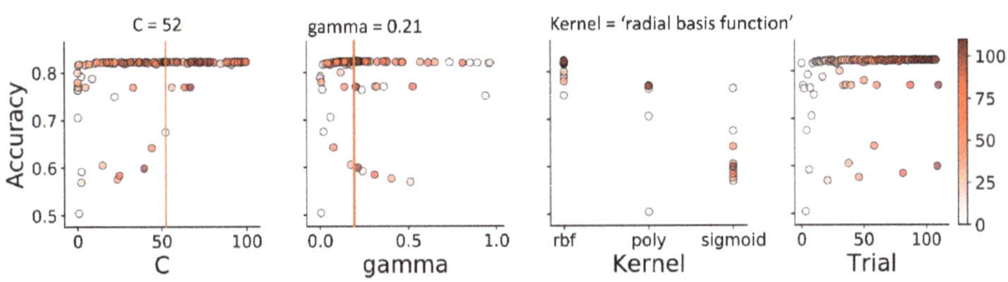

Figure 4. Hyperparameters optimization for support vector machine (SMV) classifier. The optimal hyperparameters set, which returns the maximum accuracy (0.82), includes: C = 52.07 (regularization parameter), gamma = 0.21 (kernel coefficient), Kernel = "radial basis function".

3.3. Performance of Classifiers

The performance of five classifiers was examined in this study. Once the classifiers were trained on the training dataset ($n = 75,978$) with the optimal hyperparameters, their performance was evaluated on the test dataset ($n = 8443$) using several metrics including F1-score, precision, recall, and area under the curve (AUC) in receiver operating characteristics (ROC) curve.

Table 2 presents the performance of classifiers on the test dataset as well as the training dataset using a 10-fold cross-validation. Among the classifiers, SVM offered the best performance at the cost of long training time compared to the other classifiers. Alternatively, QDA and XGBoost required the minimum training time. While F1-mean and AUC provide information about the ability of the classifiers in learning from the training dataset and generalization on an unseen test dataset, the required time for training and testing of classifiers indicates the scalability of the classifiers to larger datasets. QDA needed a small fraction of a second (0.02 s) to return a promising F1-mean (80.85%), which was about 1.4% lower than the highest F1-mean obtained by SVM (82.24%). It should be noted that AUC was not calculated for SVM because the SVM classifier does not return the probability of belonging to classes for a given sample.

Table 2. Comparing the performance of classifiers on test (n = 8443) and training datasets (n = 75,978). The result of the training dataset was obtained using 10-fold cross-validation.

Classifier	Test Dataset							Training Dataset	
	F1 [1] (%)	Precision (%)	Recall (%)	AUC [2] (%)	Threshold [3]	Training time (s)	Prediction time (s)	F1 -mean (%)	F1 StD [4]
SVM	82.24	82.35	82.25	–	–	128.59	3.45	82.25	0.32
QDA	80.85	80.96	80.86	89.01	0.54	0.02	0.01	80.40	0.30
Random Forest	81.81	81.81	81.81	90.25	0.51	91.39	0.62	81.55	0.23
XGBoost	80.27	80.29	80.27	88.29	0.51	7.29	0.02	80.94	0.31
DNN	81.68	81.79	81.69	89.90	0.52	316.09	0.57	81.34	0.52
Ensemble	82.24	82.31	82.25	90.31	–	543.37	4.66	–	–

[1] Average of F1-scores weighted by samples of each class; [2] Area under the curve in receiver operating characteristics curve; [3] Optimal decision threshold for classification; [4] Standard deviation of F1-score among 10 folds.

To aggregate the benefits of each classifier, an ensemble classifier was constructed by combining the prediction made by the classifiers. For a given test sample, all trained classifiers were used to predict its class label. Using the voting approach, a class label with the maximum of votes was assigned to the test sample. The ensemble classifier was not able to improve the performance of classification. This might suggest there are pixels in the test dataset that the majority of the classifiers were not able to classify them correctly; hence, the ensemble approach did not enhance the classification performance. For instance, some of the pixels in high-N class might present a very similar reflectivity to low-N class and vice versa.

The high F1-score averaged among the 10-folds of the training dataset and its low standard deviation across the folds for each classifier, indicated the performance of the classifiers was consistent regardless of which folds of samples were used as training and test datasets. In addition, it suggested a similar F1-score can be expected by deploying the trained classifiers using all training dataset to predict the class labels of the test dataset. This observation that the F1-scores obtained for 10-fold CV and test dataset were very close to one another might suggest the optimal hyperparameters were identified for each classifier.

3.4. Nitrogen Prediction of Vines

Once the classifiers had been trained on the training dataset using the optimal hyperparameters, they were utilized to predict the probability of high-N class for all pixels of other vines. The predicted probability values were then averaged across the pixels of a vine to compute one probability value ($P(high\_N|X)$) per each vine. Figure 5 shows the coefficient of determination (R^2) and root mean square error (RMSE) for nitrogen prediction of four classifiers capable of predicting the class label as a probability. XGBoost exhibited a slightly larger coefficient of determination and lower RMSE compared to other learning methods in the prediction of N concentration of vines.

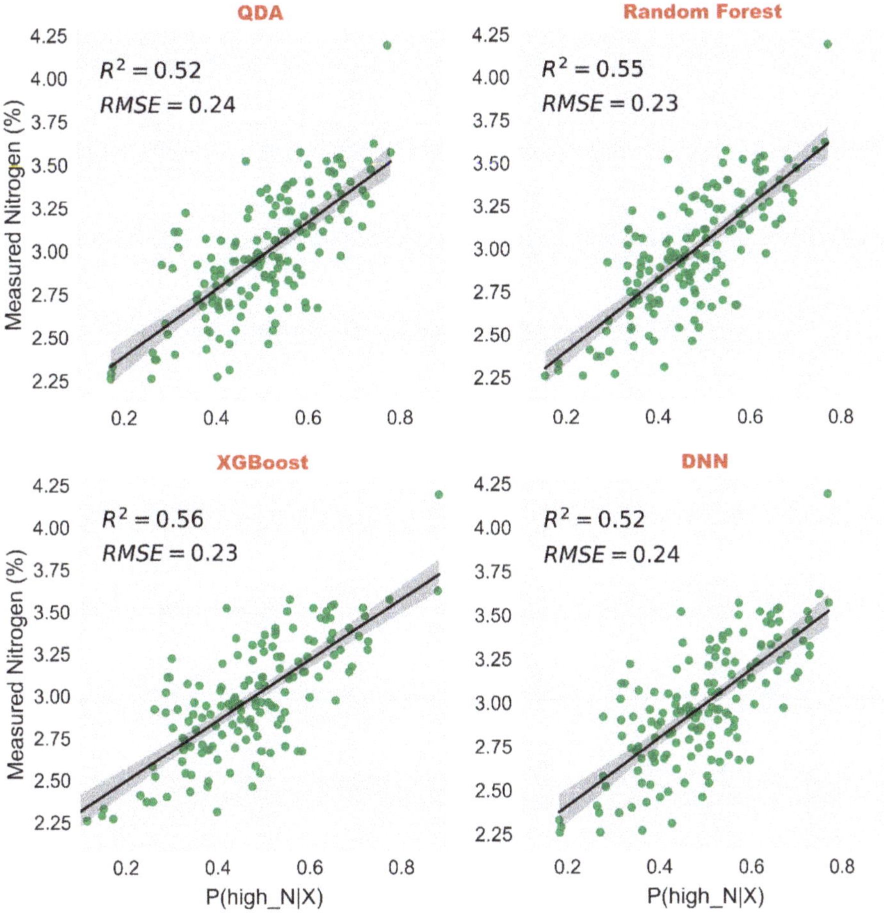

Figure 5. Performance of four learning models in prediction of nitrogen (N) concentration for all 150 vines. For a given vine, N concentration was measured through tissue sampling and predicted N was obtained by averaging the posterior probability of high_N class for all pixels of the vine. The coefficient of determination (R^2) and root mean square error (RMSE) are reported as two metrics representing the goodness of fit per each learning method.

4. Discussion

4.1. *Influence of Leaf Nitrogen Concentration on Spectral Characteristics*

A mechanistic understanding of how incident light interacts with leaves of different N concentrations is required for a meaningful interpretation of remote sensing data, such as aerial multispectral images. In this study, the observed differences between the spectral responses of low-N and high-N classes might be the result of physiological changes, such as leaf chlorophyll concentration, which is correlated with the N concentration of fully expanded leaf blades [13].

4.1.1. Visible Bands (Blue, Green, and Red)

A healthy (i.e., non-stressed) leaf has a tendency to absorb a larger extent of incident light in blue and red bands for photosynthesis activities. Therefore, the higher reflectivity of low-N pixels in blue and red bands can be an indication of a decline in chlorophyll content. Compared to red band, the reflectivity of low-N class at blue band was less affected by the N concentration—a larger alteration was observed in the red band (Figure 2B and diagonal plots in Figure 3). This observation may suggest that chlorophyll in grape leaves, with high absorption at blue and red bands, was more sensitive to N deficiency compared with carotenoids, which have high absorption in blue regions. The ratio between carotenoid and chlorophyll has been previously reported to increase in plants under stress or with senescing leaves [50]. Based on this alteration in the relative ratio between carotenoid and chlorophyll, spectral indices such as the normalized pigments chlorophyll ratio index (NPCI) [51] and normalized difference plant senescence index (NDPSI) [17] have been developed to assess N deficiency in leaves and to segment senescent leaves in aerial imagery, respectively. In the green band, pixels representing low-N class tended to show a greater reflectance (Figure 2B and diagonal plots in Figure 3) in response to a decrease in leaf chlorophyll concentration [52,53].

4.1.2. Red Edge Band

Red edge is historically known to be an indirect estimator of plant N status as red edge position was shown to shift to a shorter wavelength in response to a decline in chlorophyll content [54,55], and chlorophyll content itself was shown to be correlated with N content in grapes [13,50]. Although an accurate calculation of red edge position is not possible in multispectral imaging due to its coarse spectral resolution, the higher reflectivity of low-N class in red and red edge suggests the red edge position has shifted towards shorter wavelengths (Figure 2B).

4.1.3. Near-Infrared Band

Healthy leaves tend to have a high reflectivity in near-infrared due to their internal cellular structure [56]. Pixels of low-N class tended to have lower reflectance in near-infrared band compared to the pixels representing the high-N class (Figure 2B). The reduction in near-infrared reflectance of leaves with low nitrogen concentration may indicate that the internal cellular structure of the leaves sustained damages as previous studies demonstrated the reflectance of leaves in near-infrared is largely influenced by leaf structure [57,58].

In summary, the near-infrared band exhibited the largest absolute difference between the spectral reflectance of low-N and high-N class, followed by the red edge, green, red, and blue bands. This is in agreement with a previous study that showed that near-infrared, red edge, and green bands have the highest coefficient of determination between leaf reflectance from 400 to 2500 nm and leaf nitrogen concentration in Holm oak [20]. However, if we sort the multispectral bands based on the absolute values of percentage difference $\left(\left|\frac{\rho_{high_N}-\rho_{low_N}}{\rho_{high_N}}\right| \times 100\right)$, the order will become green (10.41%), near-infrared (9.02%), red (8.63%), red edge (6.09%), and blue (2.61%).

4.2. *Significance of the Proposed Data-Driven Method*

Prediction of leaf N concentration was inherently a regression problem, which requires the prediction of a continuous variable (i.e., N concentration) given multiple input variables (i.e., multispectral images). One of the novel contributions of this study was to deploy machine learning classifiers for this type of regression problem. The process started with training supervised binary classifiers on pixels of vines with extreme N concentration. Once a classifier was trained with its optimal set of hyperparameters, it was utilized to predict N concentration of a given vine in the form of probability of belonging to either of classes (i.e., low-N or high-N).

The alternative approach was to train a regression algorithm for the direct prediction of N concentration at the vine scale. In this approach, the spectral response of pixels representing vine canopy should be averaged per vine to obtain one input feature vector per ground truth data, which was the vine's N concentration measured through tissue sampling. Therefore, substantial spatial and spectral information is diminished through the averaging process across the thousands of pixels in a vine [17]. In addition, to train a robust regression algorithm for learning a complex problem such as nitrogen prediction, more substantial ground truth data are required in this approach since the total number of samples is limited to the number of vines, which was 150 in this study.

The proposed method in this study, on the other hand, offers several advantages which are discussed here.

4.2.1. Appropriate Use of High-Resolution Imagery

One of the benefits of the proposed framework is to leverage thousands of samples in the form of pixels attained through high-resolution aerial imagery. Although a dataset with a large number of samples, compared to the number of features, does not necessarily assure developing accurate learning models, it allows having (i) an adequate training dataset to develop a robust mapping function for mapping from an input space to an output space, (ii) a validation dataset with a sufficient number of samples to identify an optimal set of hyperparameters and better tuning the parameters of the learning model, and (iii) a test dataset with more diverse samples to reliably evaluate the generalization performance of the model.

The results demonstrated the efficiency in learning from large training data and the reliability of the trained models in predicting the class labels of unseen test samples. The high F1-mean and low F1-std achieved by the 10-fold CV on the training dataset indicated low bias and variance error, respectively (Table 2). Moreover, the F1-mean of CV can be used as an estimation for the expected model accuracy on unseen data. This was verified by the results obtained on the test dataset, indicating the generalization ability of the trained classifiers on new datasets. In addition, the promising regression results confirmed the validity of the proposed approach in predicting N concentration in grapevines.

4.2.2. Spatial Distribution of N Across the Vine's Canopy

Nitrogen is a significant determinant for the photosynthetic activity of plants, and its distribution within the canopy is a key element of photosynthesis and carbon gain at the canopy level [59]. The pixel-based probability, ($P(high\_N|X)$ or $P(low\_N|X)$), estimated by the trained predictive models can assist in the mechanistic understanding of the N distribution at the surface of the canopy in a non-invasive and quantitative manner.

Figure 6A shows the distribution of the posterior probability of high-N class ($P(high\_N|X)$) at the pixel level, obtained by a XGBoost classifier and two widely used spectral indices, normalized difference vegetation index (NDVI) and normalized difference red edge index (NDRE), for two vines one with the lowest N concentration (2.41%) and the other with the highest N concentration (4.19%) among all vines. While the distribution of $P(high\_N|X)$ for the two vines are well-separated, there is a high degree of overlap between the distribution of NDVI and NDRE for these two vines with low and high N concentrations. This highlights the ability of the predictive models in distinguishing vines with low N concentrations from vines with high N concentrations.

Figure 6B illustrates pixel-based probability ($P(high\_N|X)$) obtained by XGBoost classifier along with NDVI and NDRE, represented with a colormap, for two plots with zero and 24.2 g applied nitrogen per vine. Figure 6B demonstrates how the results achieved by XGBoost could capture the variability of N across the plots—a vine with a high N concentration exhibits more dark green pixels, whereas a vine with low N concentration displays more light yellow pixels. Similar to the histograms, there is a distinct difference between the vines with low and high N concentration for the $P(high\_N|X)$ compared to NDVI and NDRE. In essence, machine learning algorithms provide a wider dynamic range, which is visually more appealing for human sensory comparison. Furthermore, machine learning can be used as a practical tool to spot hot zones with low nitrogen concentration in a large commercial orchard. Among the two indices, NDRE performed better in discrimination of the vines with extreme N concentration.

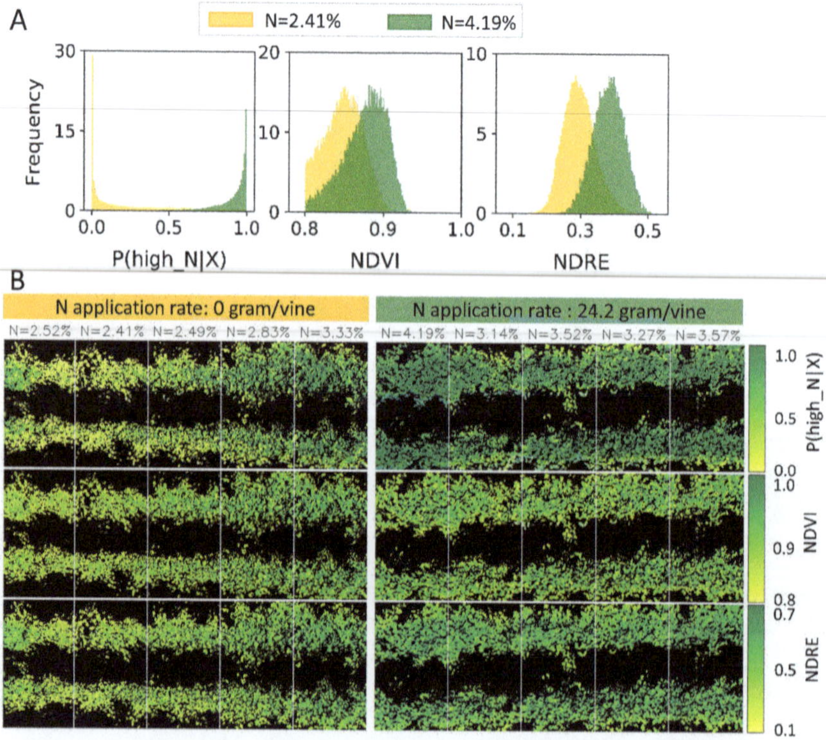

Figure 6. Comparing the results obtained by one of the machine learning techniques (posterior probability of the high-N class, $P(high\_N|X)$, obtained by XGBoost) with two widely used spectral indices, normalized difference vegetation index (NDVI) and normalized difference red edge index (NDRE). (**A**) Histogram of $P(high\_N|X)$ for two vines with minimum and maximum nitrogen concentration among all 150 vines compared with NDVI histogram and NDRE histogram. A distinct separation exists in $P(high\_N|X)$ histogram for the pixels of vines with the two extreme nitrogen concentrations compared with NDVI and NDRE. (**B**) Pixel-base $P(high\_N|X)$, NDVI, and NDRE represented with a colormap for two plots with zero, and 24.2 g of nitrogen applied per vine. Each plot contains five vines, and the nitrogen concentration measured through leaf sampling is shown at the top of each vine. $P(high\_N|X)$ obtained by XGBoost captures the spatial variability of N across the plots, offers a wider dynamic range, which is visually more appealing for human sensory comparison and is more useful in spotting hot zones suffering from low nitrogen in a vineyard.

According to $P(high\_N|X)$ calculated for vines in the plot with 24.2 g of N applied per vine, pixels representing the leaves at the apical portion of the shoots (edge of the canopy toward the middle of the row) were classified as low-N, most probably because these are young leaves with less chlorophyll content than more mature leaves. Grapevine leaves at the shoot apex have been shown to have significantly lower chlorophyll and N (when expressed per unit leaf area) than those located further down the shoot (mid and basal positions) at bloom, however, the N concentration (expressed per unit dry weight) of the apical leaves are significantly greater than those of the more mature leaves [60]. In this study, pixels representing apical leaves in the high N plot (i.e., 24.2 g/vine) exhibited similar spectral responses to leaves with decreased N in the zero applied N plots, even though their N concentration may be high. This agrees with the conclusions of Friedel et al. [13] who reported a disconnect between chlorophyll and N in young grape leaves. Consequently, the predictive models classified the pixels representing apical leaves as the low-N class (Figure 6B).

4.2.3. Adjustable Decision Threshold for Spatial Zoning of Nitrogen

In this study, the optimal decision threshold for classifying low-N and high-N classes was calculated based on the maximum weighted F1-score. However, the proposed framework in this study offers flexibility in the discretion of decision thresholds to define N management zones according to various vineyard-specific conditions and management strategies. For example, in a conservative N management approach, large threshold values for $P(high\_N|X)$ can be defined to determine N management zones, aimed to assure minimum N stress. Alternatively, in an environmentally friendly approach, smaller threshold values can be defined to reduce the risk of environmental contamination caused by excessive N application. Therefore, in the proposed method, agronomic expert knowledge can be integrated with machine learning to define optimal N management zones.

4.2.4. Directed Sampling from Hot Spot in Vineyard

The proposed data-driven method can help growers collect tissue samples in a more intelligent and efficient manner. The descriptive N status map generated by the proposed method (similar to Figure 6) can be used to efficiently identify vines with a various N status aimed at directed sampling. The conventional random or grid sampling techniques, or sampling based on vineyard history, can be replaced by the directed sampling technique, which can provide insights into N spatial variability.

4.3. *Limitation of Multispectral Imaging*

As discussed above, young leaves with less chlorophyll may exhibit a spectral response similar to leaves with low N concentration. In addition, many other stresses or diseases can lead to chlorophyll degradation in leaves, resulting in spectral patterns similar to leaves with low N concentration [61]. For instance, leaves with water stress in grapevine may exhibit similar spectral characteristics to leaves suffering from N deficiency as they tend to have a lower reflectivity in near-infrared and higher reflectivity in red edge and red bands compared to well-irrigated leaves [62]. Therefore, multispectral imaging with a limited number of typical bands (blue, green, red, red edge, and near-infrared) may not serve as the best tool for N assessment, in particular for a commercial vineyard where other stress/diseases inducing chlorophyll degradation might be prevalent. In such cases, hyperspectral imaging may be used to identify the most informative spectral bands aimed at developing a custom-designed multispectral sensor for particular stress/disease detection, such as N deficiency in grapevines [32,63].

5. Conclusions

This study proposed an innovative method for the analysis of high-resolution aerial multispectral images captured at the bloom stage to assess nitrogen concentration in vines. A supervised binary classification problem was defined to, (i) benefit from a training machine learning on a larger dataset obtained through high-resolution imagery, (ii) provide insight on spatial variability of nitrogen concentration within a single vine as well as across the grape vineyard, and (iii) accommodate diverse points of view, benefit-oriented or environmental-oriented perspectives, in defining an optimal threshold for fertilizer management decisions. For this purpose, five commonly used machine learning classifiers were trained with an optimal set of hyperparameters. The highest F1-score (82.24%) on test dataset was achieved by SVM with maximum training time, whereas QDA and XGBoost required the minimum training time with promising F1-scores of 80.85% and 80.27%, respectively. Afterwards, we transformed the classification problem into a regression problem to predict N concentration at vine scale. Through implementing a soft classification approach, the posterior probability of high-N class given spectral data ($P(high\_N|X)$) for all pixels of a vine was averaged to be used as an indication of nitrogen concentration at vine scale. Among the predictive models, XGBoost performed slightly better in terms of coefficient of determination and RMSE in the prediction of nitrogen concentration. The findings of this study can offer immediate practical applications for sustainable nitrogen management, such as (i) providing insights on nitrogen variability in vineyards, which could be useful for variable rate management, (ii) identifying hot zones with low nitrogen content for a more informed and efficient tissue sampling. In addition, we investigated the impact of low nitrogen concentration on the spectral characteristics of leaves in five bands. Based on the percentage difference between the averaged spectral response of low-N and high-N class, the largest difference was observed for green, near-infrared, red, red edge, and blue. To identify the most informative bands for nitrogen estimation, a sensor, like a hyperspectral camera, with a higher spectral resolution, is required along with advanced feature selection techniques.

Author Contributions: Conceptualization, A.M. and A.P.; methodology, A.M.; software, A.M.; formal analysis, A.M.; data curation, G.Z.-R., M.W.F., and L.E.W.; writing—original draft preparation, A.M.; writing—review and editing, A.P., G.Z.-R., L.E.W., and M.W.F.; visualization, A.M., A.P.; supervision, A.P.; funding acquisition, A.P., M.W.F., and L.E.W. All authors have read and agreed to the published version of the manuscript.

Funding: This research was funded by the California Table Grape Commission.

Acknowledgments: The authors would like to gratefully acknowledge the funding provided by the California Table Grape Commission. The authors also thank Mario Salinas and Jaclyn Stogbauer for their valuable assistance in data collection and tissue sampling.

Conflicts of Interest: The authors declare no conflict of interest. The funders had no role in the design of the study; in the collection, analyses, or interpretation of data; in the writing of the manuscript, or in the decision to publish the results.

Appendix A

Table A1. List of hyperparameters, search space, and optimal values per each classifier. Optimized configuration of hyperparameters was identified using five-fold cross validation in Optuna library (version 1.3.0) in python (version 3.7.3).

Classifier	Hyperparameters	Search Space Domain	Optimal Parameter
support vector machine (SMV)	"kernel"	("poly", "rbf", "sigmoid")	"rbf"
	"C"	(1×10^{-2}, 1×10^{2})	52.07
	"gamma"	(1×10^{-3}, 1×10^{0})	0.21
discriminant analysis	"classifier"	(QDA, LDA)	QDA
	"reg_param" (QDA)	(0, 1×10^{-3})	5.27×10^{-7}
random forest	"n_estimators"	(100, 1000)	496
	"criterion"	("gini", "entropy")	"entropy"
	"min_samples_split"	(2, 100)	38
	"min_samples_leaf"	(2, 50)	3
	"max_features"	(2, 5)	2
	"max_depth"	(10, 1000)	110
	"bootstrap"	(True, False)	TRUE
XGBoost	"num_boost_round"	(100, 1000)	292
	"learning_rate"	(0.01, 0.5)	0.45
	"feature_fraction"	(0.1, 1.0)	0.9
	"subsample"	(0.1, 1.0)	0.94
	"booster"	("gbtree", "gblinear", "dart")	"gbtree"
	"lambda"	(1×10^{-8}, 1.0)	1.90×10^{-4}
	"alpha"	(1×10^{-8}, 1.0)	3.60×10^{-7}
	"max_depth"	(1, 9)	9
	"eta"	(1×10^{-8}, 1.0)	0.02
	"gamma"	(1×10^{-8}, 1.0)	3.28×10^{-5}
	"grow_policy"	("depthwise", "lossguide")	"lossguide"
deep neural network (DNN)	"n_hidden_layers"	(1, 10)	6
	"weight_decay"	(1×10^{-10}, 1×10^{-3})	4.44×10^{-8}
	"activation"	("relu", "sigmoid", "tanh")	"tanh"
	"n_units_in_hidden_layers"	(4, 10)	(6, 6, 5, 6, 6, 5)
	"optimizer"	("RMSprop", "Adam", "SGD")	"Adam"
	"learning_rate"	(1×10^{-5}, 1×10^{-1})	4.38×10^{-3}
	"batch_size_power"	(5, 9)	2^{5}

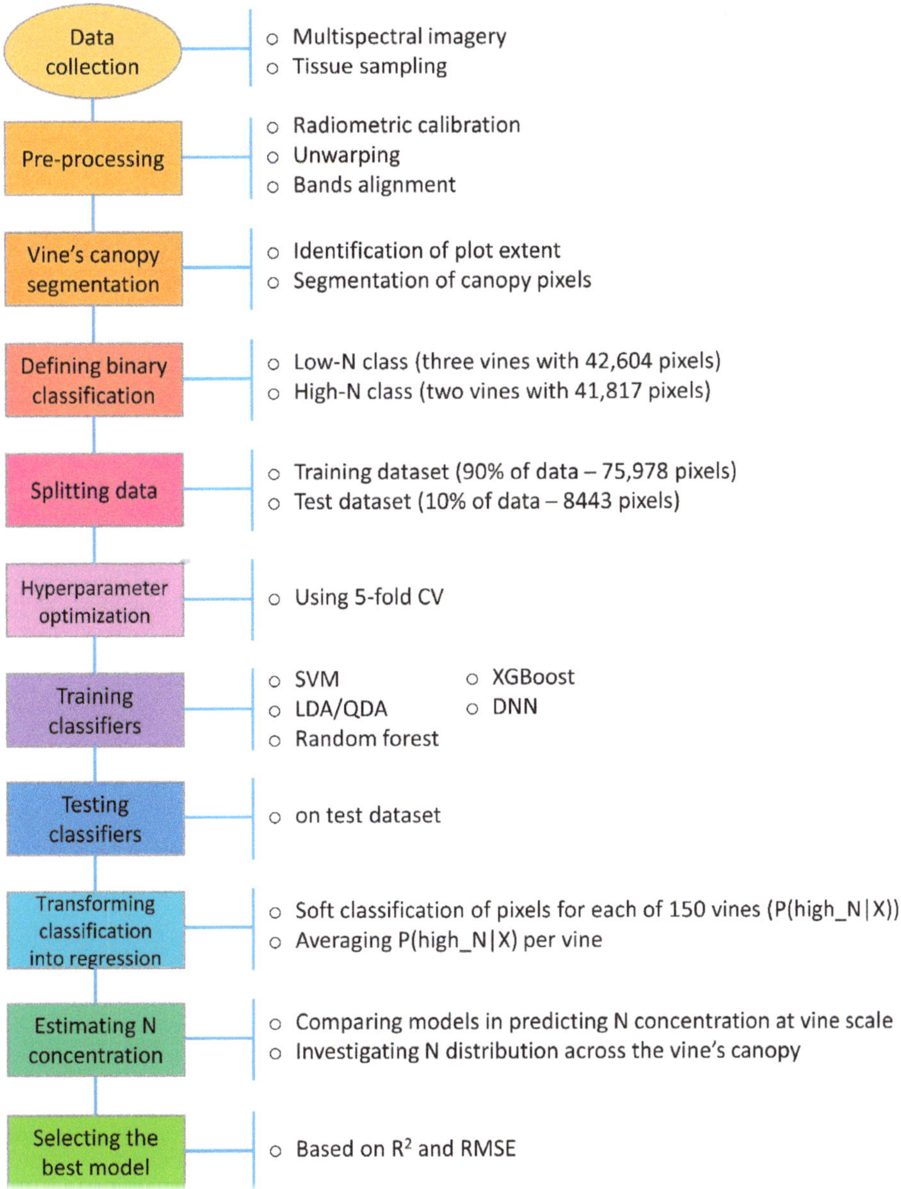

Figure A1. Workflow consisted of all steps from data collection to data analysis.

References

1. Fidelibus, M.W.; Hashim-Buckey, J.; Vasquez, S. Mondo et mercato: Stati Uniti. In *L'Uva da Tavola*; Angelini, R., Ed.; Bayer CorpScience S.r.l.: Milano, Italy, 2010; pp. 506–518.
2. Fidelibus, M.; El-kereamy, A.; Zhuang, G.; Haviland, D.; Hembree, K.; Stewart, D. *Sample Costs to Establish and Produce Table Grapes. San Joaquin Valley South. Flame Seedless, Early Maturing*; UC Agricultural Issues Center: Davis, CA, USA, 2018.

3. Christensen, L.P.; Peacock, W.L. Mineral nutrition and fertilization. In *Raisin Production Manual;* Christensen, L.P., Ed.; University of California, Agriculture Natural Resources, Communication Services: Oakland, CA, USA, 2000; pp. 102–114. ISBN 9781879906440.
4. Anderson, G.; Van Aardt, J.; Bajorski, P.; Heuvel, J.V. Detection of wine grape nutrient levels using visible and near infrared 1nm spectral resolution remote sensing. *Auton. Air Ground Sens. Syst. Agric. Optim. Phenotyping* **2016**, *9866*, 98660. [CrossRef]
5. Christensen, L.P.; Kasimatis, A.N.; Jensen, F.L. *Grapevine Nutrition and Fertilization in the San Joaquin Valley;* University of California: Berkeley, CA, USA, 1978; ISBN 9780931876257.
6. Conradie, W.J. Distribution and Translocation of Nitrogen Absorbed During Early Summer by Two-Year-Old Grapevines Grown in Sand Culture. *Am. J. Enol. Vitic.* **1991**, *42*, 180–190.
7. Grechi, I.; Vivin, P.; Hilbert, G.; Milin, S.; Robert, T.; Gaudillère, J.-P. Effect of light and nitrogen supply on internal C: N balance and control of root-to-shoot biomass allocation in grapevine. *Environ. Exp. Bot.* **2007**, *59*, 139–149. [CrossRef]
8. Keller, M.; Kummer, M.; Vasconcelos, C. Soil nitrogen utilisation for growth and gas exchange by grapevines in response to nitrogen supply and rootstock. *Aust. J. Grape Wine Res.* **2001**, *7*, 2–11. [CrossRef]
9. Ferrara, G.; Malerba, A.D.; Matarrese, A.M.S.; Mondelli, D.; Mazzeo, A. Nitrogen Distribution in Annual Growth of 'Italia' Table Grape Vines. *Front. Plant Sci.* **2018**, *9*, 1374. [CrossRef] [PubMed]
10. Harter, T.; Lund, J.R.; Darby, J.; Fogg, G.E.; Howitt, R.; Jessoe, K.; Pettygrove, S.G.; Quinn, J.F.; Viers, J.H.; Boyle, D.B.; et al. *Addressing Nitrate in California's Drinking Water with a Focus on Tulare Lake Basin and Salinas Valley Groundwater;* Report for the State Water Resources Control Board Report to the Legislature; UC Davis Center for Watershed Sciences: Davis, CA, USA, 2012.
11. Mills, H.A.; Jones, J.B. *Plant Analysis Handbook II;* MicroMacro: Athens, GA, USA, 1996; ISBN1 1878148052. ISBN2 9781878148056.
12. Iland, P.; Dry, P.; Proffitt, T.; Tyerman, S. *The Grapevine: From the Science to the Practice of Growing Vines for Wine;* Patrick Iland Wine Promotions Pty Ltd.: Adelaide, Australia, 2011; ISBN 9780958160551.
13. Friedel, M.; Hendgen, M.; Stoll, M.; Löhnertz, O. Performance of reflectance indices and of a handheld device for estimating in-field the nitrogen status of grapevine leaves. *Aust. J. Grape Wine Res.* **2020**, *26*, 110–120. [CrossRef]
14. Ye, X.; Abe, S.; Zhang, S. Estimation and mapping of nitrogen content in apple trees at leaf and canopy levels using hyperspectral imaging. *Precis. Agric.* **2019**, *21*, 198–225. [CrossRef]
15. Min, M.; Lee, W.S. Determination of Significant Wavelengths and Prediction of Nitrogen Content for Citrus. *Trans. ASAE* **2005**, *48*, 455–461. [CrossRef]
16. Zarate-Valdez, J.L.; Muhammad, S.; Saa, S.; Lampinen, B.D.; Brown, P.H. Light interception, leaf nitrogen and yield prediction in almonds: A case study. *Eur. J. Agron.* **2015**, *66*, 1–7. [CrossRef]
17. Moghimi, A.; Yang, C.; Anderson, J.A. Aerial hyperspectral imagery and deep neural networks for high-throughput yield phenotyping in wheat. *Comput. Electron. Agric.* **2020**, *172*, 105299. [CrossRef]
18. Gitelson, A.A. Wide Dynamic Range Vegetation Index for Remote Quantification of Biophysical Characteristics of Vegetation. *J. Plant Physiol.* **2004**, *161*, 165–173. [CrossRef]
19. Gitelson, A.A.; Kaufman, Y.J.; Merzlyak, M.N. Use of a green channel in remote sensing of global vegetation from EOS-MODIS. *Remote Sens. Environ.* **1996**, *58*, 289–298. [CrossRef]
20. Pacheco-Labrador, J.; Gonzalez-Cascon, R.; Martín, M.P.; Riaño, D. Understanding the optical responses of leaf nitrogen in Mediterranean Holm oak (*Quercus ilex*) using field spectroscopy. *Int. J. Appl. Earth Obs. Geoinf.* **2014**, *26*, 105–118. [CrossRef]
21. Kokaly, R. Spectroscopic Determination of Leaf Biochemistry Using Band-Depth Analysis of Absorption Features and Stepwise Multiple Linear Regression. *Remote Sens. Environ.* **1999**, *67*, 267–287. [CrossRef]
22. Hansen, P.M.; Schjoerring, J.K. Reflectance measurement of canopy biomass and nitrogen status in wheat crops using normalized difference vegetation indices and partial least squares regression. *Remote Sens. Environ.* **2003**, *86*, 542–553. [CrossRef]
23. Tian, Y.; Yao, X.; Yang, J.; Cao, W.; Hannaway, D.; Zhu, Y. Assessing newly developed and published vegetation indices for estimating rice leaf nitrogen concentration with ground- and space-based hyperspectral reflectance. *Field Crop. Res.* **2011**, *120*, 299–310. [CrossRef]
24. Cho, M.A.; Skidmore, A.K. A new technique for extracting the red edge position from hyperspectral data: The linear extrapolation method. *Remote Sens. Environ.* **2006**, *101*, 181–193. [CrossRef]

25. Ferwerda, J.G.; Skidmore, A.K.; Mutanga, O. Nitrogen detection with hyperspectral normalized ratio indices across multiple plant species. *Int. J. Remote Sens.* **2005**, *26*, 4083–4095. [CrossRef]
26. Esgario, J.G.; Krohling, R.A.; Ventura, J.A. Deep learning for classification and severity estimation of coffee leaf biotic stress. *Comput. Electron. Agric.* **2020**, *169*, 105162. [CrossRef]
27. Qiu, R.; Yang, C.; Moghimi, A.; Zhang, M.; Steffenson, B.J.; Hirsch, C.D. Detection of Fusarium Head Blight in Wheat Using a Deep Neural Network and Color Imaging. *Remote Sens.* **2019**, *11*, 2658. [CrossRef]
28. Tong, H.; Madison, I.; Long, T.; Williams, C.M. Computational solutions for modeling and controlling plant response to abiotic stresses: A review with focus on iron deficiency. *Curr. Opin. Plant Biol.* **2020**, *57*, 8–15. [CrossRef]
29. Moghimi, A.; Yang, C.; Miller, M.E.; Kianian, S.F.; Marchetto, P.M. A Novel Approach to Assess Salt Stress Tolerance in Wheat Using Hyperspectral Imaging. *Front. Plant Sci.* **2018**, *9*, 1182. [CrossRef]
30. Berger, K.; Verrelst, J.; Féret, J.-B.; Hank, T.; Wocher, M.; Mauser, W.; Camps-Valls, G. Retrieval of aboveground crop nitrogen content with a hybrid machine learning method. *Int. J. Appl. Earth Obs. Geoinf.* **2020**, *92*, 102174. [CrossRef]
31. Nigon, T.J.; Yang, C.; Paiao, G.D.; Mulla, D.J.; Knight, J.F.; Fernández, F.G. Prediction of Early Season Nitrogen Uptake in Maize Using High-Resolution Aerial Hyperspectral Imagery. *Remote Sens.* **2020**, *12*, 1234. [CrossRef]
32. Moghimi, A.; Yang, C.; Marchetto, P.M. Ensemble Feature Selection for Plant Phenotyping: A Journey from Hyperspectral to Multispectral Imaging. *IEEE Access* **2018**, *6*, 56870–56884. [CrossRef]
33. Espejo-Garcia, B.; Mylonas, N.; Athanasakos, L.; Fountas, S.; Vasilakoglou, I. Towards weeds identification assistance through transfer learning. *Comput. Electron. Agric.* **2020**, *171*, 105306. [CrossRef]
34. Pantazi, X.; Tamouridou, A.; Alexandridis, T.K.; Lagopodi, A.; Kashefi, J.; Moshou, D. Evaluation of hierarchical self-organising maps for weed mapping using UAS multispectral imagery. *Comput. Electron. Agric.* **2017**, *139*, 224–230. [CrossRef]
35. Feng, L.; Zhang, Z.; Ma, Y.; Du, Q.; Williams, P.; Drewry, J.; Luck, B. Alfalfa Yield Prediction Using UAV-Based Hyperspectral Imagery and Ensemble Learning. *Remote Sens.* **2020**, *12*, 2028. [CrossRef]
36. Gavlak, R.; Horneck, D.; Miller, R.O. *Soil, Plant and Water Reference Methods for the Western Region 1*, 3rd ed.; Western Rural Development Center (WREP-125): Logan, UT, USA, 2005.
37. Moghimi, A. Micasense_Preprocessing (Version 1.0.0). 2020. Available online: https://doi.org/10.5281/zenodo.3988680 (accessed on 20 June 2020).
38. Pedregosa, F.; Varoquaux, G.; Gramfort, A.; Michel, V.; Thirion, B.; Grisel, O.; Blondel, M.; Prettenhofer, P.; Weiss, R.; Dubourg, V.; et al. Scikit-learn: Machine Learning in Python. *J. Mach. Learn. Res.* **2011**, *12*, 2825–2830.
39. Chen, T.; Guestrin, C. XGBoost: A Scalable Tree Boosting System. In Proceedings of the 22nd ACM SIGKDD International Conference on Knowledge Discovery and Data Mining, San Francisco, CA, USA, 13–17 August 2016; ACM: New York, NY, USA, 2016; pp. 785–794.
40. Abadi, M.; Agarwal, A.; Paul Barham, E.B.; Chen, Z.; Citro, C.; Corrado, G.S.; Davis, A.; Dean, J.; Devin, M.; Ghemawat, S.; et al. TensorFlow: Large-scale machine learning on heterogeneous systems. *arXiv* **2015**, arXiv:1603.04467.
41. Li, L.; Jamieson, K.; Rostamizadeh, A.; Gonina, E.; Ben-tzur, J.; Hardt, M.; Recht, B.; Talwalkar, A. A System for Massively Parallel Hyperparameter Tuning. In Proceedings of the Machine Learning and Systems, Austin, TX, USA, 2–4 March 2020; pp. 230–246.
42. Bergstra, J.; Bengio, Y. Random Search for Hyper-Parameter Optimization. *J. Mach. Learn. Res.* **2012**, *13*, 281–305.
43. Bergstra, J.; Bardenet, R.; Bengio, Y.; Kégl, B. Algorithms for Hyper-Parameter Optimization. In *Advances in Neural Information Processing Systems 24*; Shawe-Taylor, J., Zemel, R.S., Bartlett, P.L., Pereira, F., Weinberger, K.Q., Eds.; Curran Associates, Inc.: New York, NY, USA, 2011; pp. 2546–2554.
44. Snoek, J.; Larochelle, H.; Adams, R.P. Practical Bayesian Optimization of Machine Learning Algorithms. In *Proceedings of the 25th International Conference on Neural Information Processing Systems—Volume 2*; Curran Associates Inc.: Red Hook, NY, USA, 2012; pp. 2951–2959.
45. Bergstra, J.; Yamins, D.; Cox, D.D. Making a Science of Model Search: Hyperparameter Optimization in Hundreds of Dimensions for Vision Architectures. In Proceedings of the 30th International Conference on Machine Learning, Atlanta, GA, USA, 16–21 June 2013; Volume 28.

46. Franceschi, L.; Donini, M.; Frasconi, P.; Pontil, M. Forward and Reverse Gradient-Based Hyperparameter Optimization. In Proceedings of the 34th International Conference on Machine Learning, Sydney, Australia, 6–11 August 2017.
47. Akiba, T.; Sano, S.; Yanase, T.; Ohta, T.; Koyama, M. Optuna: A Next-generation Hyperparameter Optimization Framework. In Proceedings of the 25rd ACM SIGKDD International Conference on Knowledge Discovery and Data Mining, Anchorage, AK, USA, 25 July 2019.
48. Tan, P.-N.; Steinbach, M.; Kumar, V. *Introduction to Data Mining*, 1st ed.; Addison-Wesley Longman Publishing Co., Inc.: Boston, MA, USA, 2005; ISBN 0321321367.
49. Platt, J.C.; Platt, J.C. Probabilistic Outputs for Support Vector Machines and Comparisons to Regularized Likelihood Methods. *Adv. Large Margin Classif.* **1999**, *10*, 61–74.
50. Peñuelas, J.; Filella, I. Visible and near-infrared reflectance techniques for diagnosing plant physiological status. *Trends Plant Sci.* **1998**, *3*, 151–156. [CrossRef]
51. Peñuelas, J.; Gamon, J.; Fredeen, A.; Merino, J.; Field, C. Reflectance indices associated with physiological changes in nitrogen- and water-limited sunflower leaves. *Remote Sens. Environ.* **1994**, *48*, 135–146. [CrossRef]
52. Daughtry, C. Estimating Corn Leaf Chlorophyll Concentration from Leaf and Canopy Reflectance. *Remote Sens. Environ.* **2000**, *74*, 229–239. [CrossRef]
53. Zhao, D.; Reddy, K.R.; Kakani, V.G.; Reddy, V. Nitrogen deficiency effects on plant growth, leaf photosynthesis, and hyperspectral reflectance properties of sorghum. *Eur. J. Agron.* **2005**, *22*, 391–403. [CrossRef]
54. Ayala-Silva, T.; Beyl, C.A. Changes in spectral reflectance of wheat leaves in response to specific macronutrient deficiency. *Adv. Space Res.* **2005**, *35*, 305–317. [CrossRef]
55. Filella, I.; Penuelas, J. The red edge position and shape as indicators of plant chlorophyll content, biomass and hydric status. *Int. J. Remote Sens.* **1994**, *15*, 1459–1470. [CrossRef]
56. Gates, D.M.; Keegan, H.J.; Schleter, J.C.; Weidner, V.R. Spectral Properties of Plants. *Appl. Opt.* **1965**, *4*, 11–20. [CrossRef]
57. Knipling, E.B. Physical and physiological basis for the reflectance of visible and near-infrared radiation from vegetation. *Remote Sens. Environ.* **1970**, *1*, 155–159. [CrossRef]
58. Slaton, M.R.; Hunt, E.R.; Smith, W.K. Estimating near-infrared leaf reflectance from leaf structural characteristics. *Am. J. Bot.* **2001**, *88*, 278–284. [CrossRef]
59. Hikosaka, K.; Anten, N.P.R.; Borjigidai, A.; Kamiyama, C.; Sakai, H.; Hasegawa, T.; Oikawa, S.; Iio, A.; Watanabe, M.; Koike, T.; et al. A meta-analysis of leaf nitrogen distribution within plant canopies. *Ann. Bot.* **2016**, *118*, 239–247. [CrossRef] [PubMed]
60. Poni, S.; Intrieri, C.; Silvestroni, O. Interactions of LeafAge, Fruiting, and Exogenous Cytokinins in Sangiovese Grapevines under Non-Irrigated Conditions. II. Chlorophyll and Nitrogen Content. *Am. J. Enol. Vitic.* **1994**, *45*, 278–284.
61. Huang, C.H.; Singh, G.P.; Park, S.H.; Chua, N.-H.; Ram, R.J.; Park, B.S. Early Diagnosis and Management of Nitrogen Deficiency in Plants Utilizing Raman Spectroscopy. *Front. Plant Sci.* **2020**, *11*, 663. [CrossRef]
62. Zarco-Tejada, P.J.; González-Dugo, V.; Williams, L.E.; Suárez, L.; Berni, J.A.J.; Goldhamer, D.A.; Fereres, E. A PRI-based water stress index combining structural and chlorophyll effects: Assessment using diurnal narrow-band airborne imagery and the CWSI thermal index. *Remote Sens. Environ.* **2013**, *138*, 38–50. [CrossRef]
63. Omidi, R.; Moghimi, A.; Pourreza, A.; Aly, M.E.-H.; Eddin, A.S. Ensemble Hyperspectral Band Selection for Detecting Nitrogen Status in Grape Leaves. *arXiv* **2020**, arXiv:2010.04225.

Publisher's Note: MDPI stays neutral with regard to jurisdictional claims in published maps and institutional affiliations.

remote sensing

MDPI

Article

Estimating Leaf Nitrogen Content in Corn Based on Information Fusion of Multiple-Sensor Imagery from UAV

Xingang Xu [1,2,*], Lingling Fan [1], Zhenhai Li [1], Yang Meng [1], Haikuan Feng [1], Hao Yang [1] and Bo Xu [1]

[1] Key Laboratory of Quantitative Remote Sensing in Agriculture of Ministry of Agriculture, Beijing Research Center for Information Technology in Agriculture, Beijing Academy of Agriculture and Forestry Sciences, Beijing 100097, China; p17301156@stu.ahu.edu.cn (L.F.); lizh@nercita.org.cn (Z.L.); mengy@nercita.org.cn (Y.M.); fenghk@nercita.org.cn (H.F.); yangh@nercita.org.cn (H.Y.); xub@nercita.org.cn (B.X.)

[2] School of Agricultural Engineering, Jiangsu University, Zhenjiang 212013, China

* Correspondence: xxg2007@aliyun.com; Tel.: +86-010-51503676

Abstract: With the rapid development of unmanned aerial vehicle (UAV) and sensor technology, UAVs that can simultaneously carry different sensors have been increasingly used to monitor nitrogen status in crops due to their flexibility and adaptability. This study aimed to explore how to use the image information combined from two different sensors mounted on an UAV to evaluate leaf nitrogen content (LNC) in corn. Field experiments with corn were conducted using different nitrogen rates and cultivars at the National Precision Agriculture Research and Demonstration Base in China in 2017. Digital RGB and multispectral images were obtained synchronously by UAV in the V12, R1, and R3 growth stages of corn, respectively. A novel family of modified vegetation indices, named coverage adjusted spectral indices (CASIs (CASI $= \mathrm{VI}/(1 + \mathrm{FV}_{cover})$, where VI denotes the reference vegetation index and FV_{cover} refers to the fraction of vegetation coverage), has been introduced to estimate LNC in corn. Thereby, typical VIs were extracted from multispectral images, which have the advantage of relatively higher spectral resolution, and FV_{cover} was calculated by RGB images that feature higher spatial resolution. Then, the PLS (partial least squares) method was employed to investigate the relationships between LNC and the optimal set of CASIs or VIs selected by the RFA (random frog algorithm) in different corn growth stages. The analysis results indicated that whether removing soil noise or not, CASIs guaranteed a better estimation of LNC than VIs for all of the three growth stages of corn, and the usage of CASIs in the R1 stage yielded the best R^2 value of 0.59, with a RMSE (root mean square error) of 22.02% and NRMSE (normalized root mean square error) of 8.37%. It was concluded that CASIs, based on the fusion of information acquired synchronously from both lower resolution multispectral and higher resolution RGB images, have a good potential for crop nitrogen monitoring by UAV. Furthermore, they could also serve as a useful way for assessing other physical and chemical parameters in further applications for crops.

Keywords: RGB; multispectral; coverage adjusted spectral index; vegetation index; vegetation coverage; random frog algorithm

Citation: Xu, X.; Fan, L.; Li, Z.; Meng, Y.; Feng, H.; Yang, H.; Xu, B. Estimating Leaf Nitrogen Content in Corn Based on Information Fusion of Multiple-Sensor Imagery from UAV. *Remote Sens.* **2021**, *13*, 340. https://doi.org/10.3390/rs13030340

Received: 16 December 2020
Accepted: 18 January 2021
Published: 20 January 2021

1. Introduction

Nitrogen is a critical nutrient element for crop growth. The proper application of nitrogen fertilizer has a significant influence on final crop yield and quality. Leaf nitrogen in crop canopies is an important indicator that characterizes the nitrogen nutrition status in crops. Accurate and dynamic estimation of crop leaf nitrogen is of significance for rationally managing nitrogen fertilization [1,2].

Conventional methods for detecting crop nitrogen usually involve outdoor sampling and indoor testing, which is not only time-consuming and laborious but also destructive and lagging. With the development of spectral detection technology, non-destructive remote sensing has become an attractive tool for crop nitrogen assessment. For regional-scale monitoring of crop nitrogen, satellite-based optical remote-sensing technology is widely used, but at a field scale it is often hindered by cloud cover and high cost [3]. In recent years, family farms have gradually become an important mode of operation in Chinese agriculture, and there is a great demand for new technologies. As a rapidly developing technology, with the advantages of high efficiency, low cost, and convenience of use [4–6], UAVs (unmanned aerial vehicle) have begun to play an active role in agricultural field management [7–9]. The spatial resolution of UAV images can reach centimeter or even millimeter scales, which has unparalleled advantages on the field scale for diagnosing crop nitrogen nutrition in family farms. Recently, there have been many reports on crop nitrogen assessment using UAVs. Some vegetation indices (VI) and spectral bands from UAV-based multispectral imagery have been used with machine learning methods to effectively predict leaf nitrogen content (LNC) in maize [10], and have also been employed to evaluate the nitrogen accumulation in corn canopies, and map the spatial nitrogen variation in corn fields [11]. UAV-based hyperspectral data with plenty of shallow bands had a good performance for evaluating LNC in the different growth stages of wheat [12]. The combined features of vegetation coverage and VIs extracted from UAV-based RGB images could show some potential for estimating LNC in maize [13]. In addition, UAV-based images from RGB, multispectral, and thermal sensors have been used to acquire vegetation and temperature parameters for assessing nitrogen use efficiency in crops [14].

It is noted that UAV images have a higher spatial resolution compared with conventional satellite images, and the problem of soil background affecting crop canopy spectra is more prominent. Knowing how to decrease the influence of soil noise as much as possible is vital to the accurate estimation of crop nitrogen by remote sensing [15]. A few vegetation indices (VIs) have been developed to eliminate or minimize the effects of soil background. Huete et al. [15] proposed the soil adjusted vegetation index (SAVI) to describe soil–vegetation systems, and lessen the sensitivity of vegetation indices (VIs) to soil background. Rondeaux et al. [16] recommended the optimized SAVI (OSAVI) to increase the response to vegetation, while decreasing the variability from the soil. Due to changes of soil humidity and solar incidence angle, Major et al. [17] suggested three modified SAVIs (SAVI2, SAVI3, and SAVI4) to alleviate the influence of soil background. In the above studies, these VIs were designed to generally decrease soil noise by adding an adjusting factor into the combination of both red and near-infrared reflectance bands, usually suitable for remote sensing images with relatively low spatial resolution or more mixed pixels. Liu et al. [18] used three threshold methods with RGB digital images from UAV to separate wheat from soil, and then evaluate nitrogen status in wheat with RGB-based VIs. Yao et al. [19] developed a coverage adjustment spectral index (CASI), into which fractional vegetation cover (FV_{cover}) was introduced to monitor LNC in winter wheat with ground hyperspectral measurements under different soil backgrounds, and obtained a better accuracy of LNC estimation. In the existing reports on crop nitrogen assessment, the selected VIs are usually calculated only using the UAV images from a single sensor, even if there are multi-sensor images acquired synchronously from an aerial platform, and it is seldom considered that, in terms of the complementary advantages of different sensors in spatial and spectral resolution, the multiple-sensor images could be simultaneously applied to extract some VIs for crop nitrogen evaluation. Possibly, VIs using a fusion of information will have a better performance in assessing nitrogen status in crops.

The above mentioned CASI ($CASI = VI/(1 + FV_{cover})$) could decrease the influence of soil noise, and has been proved to be effective in estimating LNC in wheat, but both the VI and FV_{cover} in CASI were calculated by ground canopy spectral measurements and digital pictures, respectively, not using UAV remote sensing images [19]. Hence, it is worth discussing whether CASI is also adaptable to estimate LNC in corn when using UAV-based multi-sensor images, and at present little research on this has been conducted. In this study, CASI was tested to evaluate LNC in corn using UAV multi-sensor images. Digital RGB and multispectral images were acquired synchronously from UAV sensors at three growth stages of corn. Considering that multispectral images have the advantage of relatively higher spectral resolution, and RGB images are characterized by higher spatial resolution, the two inputs, VI and FV_{cover}, into CASI were extracted from multispectral and RGB images, respectively. Thereby, CASI could make use of the complementary advantages of different sensors in spatial and spectral resolution, and integrated the useful information from two sensors. This might be helpful to effectively assess nitrogen status in crops.

This study focuses on the potential applications of both UAV multi-sensor images and the modified vegetation index, CASI, for LNC assessment. The objectives of this study were: (i) to evaluate the potential performance of CASI for LNC estimates using UAV-based data; (ii) to compare the ability of the selected existing VI and the corresponding CASI to reduce the influence of soil noise; (iii) to provide a new concept and method reference using UAV multi-sensor images to monitor nitrogen status in crops.

2. Materials and Methods

2.1. UAV System

In this present study, an eight-rotor electric UAV carrying digital RGB and multispectral cameras was used to collect images in the study area. A DJI-S1000+ (DJI, Guangdong, China) was chosen as the UAV flight platform, which weighs 4.4 kg, and has an automated flight control system allowing for user-defined waypoint operations and custom mission planning, a maximum flight time of 20 min, and a LiPo battery (6S, 10,000~20,000 mAh). The digital RGB imager carried by the UAV was equipped with a high-definition Cyber-shot DSC-QX100 camera (Sony, Minato, Tokyo, Japan) that weighs 179 g and uses an Exmor R CMOS sensor with an effective 20 megapixels. A Parrot Sequoia (Parrot, USA) multispectral camera which consists of four single-band photoreceptors with 1.2 megapixels was also mounted on the UAV; the Sequoia camera weighs 107 g and has built-in 64 GB storage and SD expansion slots. More details about the parameters of the sensors are in Table 1.

Table 1. Main parameters of the sensors mounted on the UAV in this study.

Sensor Name	Sensor Type	Spectral Band (nm)	Resolution (pixels)	FOV (H° × V°)	Shutter Type	Weight (g)
Parrot Sequoia	MS	550, 660, 735, 790 *	1280 × 960	62.2 × 48.7	Global	107
Cyber-shot DSC-QX100	RGB	N/A	5472×3648	82.0 × 64.5	Global	179

MS: multispectral, FOV: field of view, H: horizonal, V: vertical. * for convenience, the 550 nm band is denoted by GRE, 660 nm RED, 735 nm REG, and 790 nm NIR in the subsequent analyses.

2.2. Data Collection and Acquisition

2.2.1. Study Site and Experimental Design

The experiment was conducted at the National Precision Agriculture Research Demonstration Base in 2017. The base is located in the northeast of Xiaotangshan Town, Changping District, Beijing (40°00′–40°21′N, 116°34′–117°00′E, 36 m). There is a temperate continental monsoon climate in this area, and the soil is a moist and heavy loam, with a nitrate N (NO_3-N) content of 3.2–14.8 mg/kg, an ammonium nitrogen (NH_4-N) content of 10.2–12.3 mg·kg^{-1}, an available phosphorus content of 3.1–21.2 mg·kg^{-1}, an effective potassium content of 86.8–120.6 mg·kg^{-1}, and an organic matter content of 15.8–20.2 g·kg^{-1}.

The experiment involved 48 plots, among which 24 plots concerned different cultivars (P) in the north and the other 24 were linked with nitrogen (N) rates in the south, each plot was 6 × 10 m, with a 1 m isolation zone (Figure 1). Eight cultivars were selected for sowing, with three repetitions for each cultivar in the northern experiment field, and among the eight cultivars: four (JK9681, JK528, JK9689, and JK665) belong to the JingKe (JK) corn family in China; two (NK718 and JNK728) are widely cultivated in Northeast China and the northern regions of North China, and have some differences in disease resistance and maturity; and another two (JD58 and XY335) are high-yielding and early-maturing varieties in North China. For 24 plots in the north of the experiment field, N fertilizer treatment was the conventional N rate (N2): 193 kg·ha^{-1}. In the southern half of experiment field, the 24 plots were treated with the following N rates: no nitrogen application (N0), 97 kg·ha^{-1} (N1), 193 kg·ha^{-1} (N2), and 289 kg·ha^{-1} (N3), each N treatment was repeated six times, and the two cultivars (JK968 and ZD958) that are the most widely cultivated in North China were selected to sow in the 12 plots (Figure 1). Management procedures, such as pest management, weed control, and phosphate and potassium application in all plots were carried out according to local practical standards.

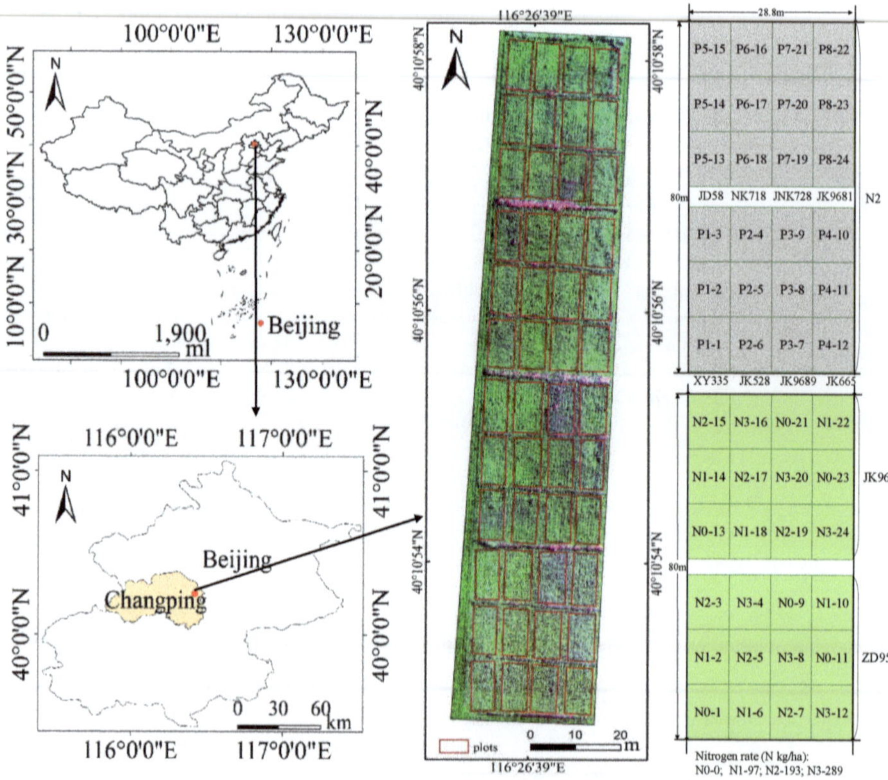

Figure 1. Location of the study area and experimental design.

2.2.2. Determination of Leaf Nitrogen Concentration

This experiment covered the three important growth stages, the V12, R1, and R3 stages of summer corn. V12 is a critical stage during corn vegetative growth, the twelfth leaf of the corn plant is fully unfolded, and the female ear begins floret differentiation in the V12 stage. R1 belongs to the transition period from vegetative growth to reproductive growth, and in this stage the filament of the female ear emerges from the bract. R3 is an important stage during corn reproductive growth, and the corn seeds turn yellow in the R3 stage [20]. In the three stages, the differences of corn canopy structure are relatively obvious and representative, which can result in different influences of the soil background on canopy spectra.

The collections of ground samples were conducted on 7 and 18 August, and 1 September in 2017. Three representative corn plants were selected in each of the forty-eight plots, and the aboveground corn plants were collected for subsequent analysis. In laboratory conditions, the green leaves of the three corn plants from each plot were separated from stems and de-enzymed at 105 °C for 30 min, and then dried at 80 °C to constant weight for chemical analysis. LNCs (g 100 g^{-1}, %) for the dried leaf samples were determined by a Kjeldahl meter (Buchi B–339, FOSS, Sweden). Table 2 shows the statistics for the measured LNC.

Table 2. Descriptive statistics for measured corn leaf nitrogen content (LNC) from three growth stages.

Growth Stage	Sample Number	Max	Min	Mean	SD	CV
V12	48	3.17	1.42	2.38	0.43	0.18
R1	48	3.49	1.17	2.63	0.35	0.13
R3	48	3.36	1.37	2.36	0.45	0.19

SD: standard deviation, CV: coefficient of variation.

2.3. Methods and Principles

2.3.1. Image Acquisition

UAV surveys with two camera sensors collecting digital RGB and multispectral imagery were conducted under clear and cloudless weather conditions from 10:00 a.m. to 14:00 p.m. on 7 and 18 August, and 1 September in 2017. The UAV missions investigating the 48 plots were performed at a flight height of 60 m and a flight speed of 6 m·s^{-1}, with a tilt angle of 0° for the two cameras, and a shortest interval of 2 s between imagery acquisitions. Both the front overlap and the side overlap of UAV flights were set to 80%.

2.3.2. Image Processing

It has been proven that high-definition digital images can be transformed into new color spaces by HSV (hue saturation value), and it is very easy to distinguish soil from vegetation in the HSV-based images [21,22]. In this study, the colors (red, green, blue) of digital RGB images from the UAV were converted into hue, saturation, and value by HSV, and then threshold segmentation was used to differentiate between vegetation and soil in the images. Figure 2 shows the spatial color transformation distribution of the study area. It can be seen that after the original RGB images have been transformed by HSV two times, the soil is able to be clearly distinguished from crop vegetation in the HSV2 images.

Some existing studies have demonstrated that random forest classification (RFC) can be well adopted for medical disease classification, habitat identification in ecological environments, and land cover classification based on multispectral images [23–25]. For the UAV multispectral images in this study, RFC was implemented to differentiate between vegetation and soil, and then created a vector including corn vegetation and soil.

2.3.3. Vegetation Coverage

Fractional vegetation coverage (FV_{cover}), which refers to the percentage of vertical projection area of vegetation per ground surface area, is a critical parameter for assessing crop density and soil factor. Ground-based RGB images have been used to effectively extract vegetation coverage by methods of color transformation and threshold segmentation [26,27]. By comparison, UAV-based RGB images have the distinct advantages of monitoring a large-scale area, as well as high spatial resolution, which facilitate the acquisition of vegetation coverage and the evaluation of crop growth in fields. In the present study, RGB images derived from UAV were converted into new image data via two HSV transformations, so as to identify background soil and corn canopy area more easily [21]. Identified corn pixels in each plot were then divided by the total number of pixels in the same plot to acquire FV_{cover}. Vegetation coverage is calculated as follows:

$$FV_{cover} = \frac{A_{veg}}{A_{total}} \quad (1)$$

where FV_{cover} is the coverage of vegetation, A_{veg} is the number of corn pixels in each plot, and A_{total} is the total number of pixels in the same plot.

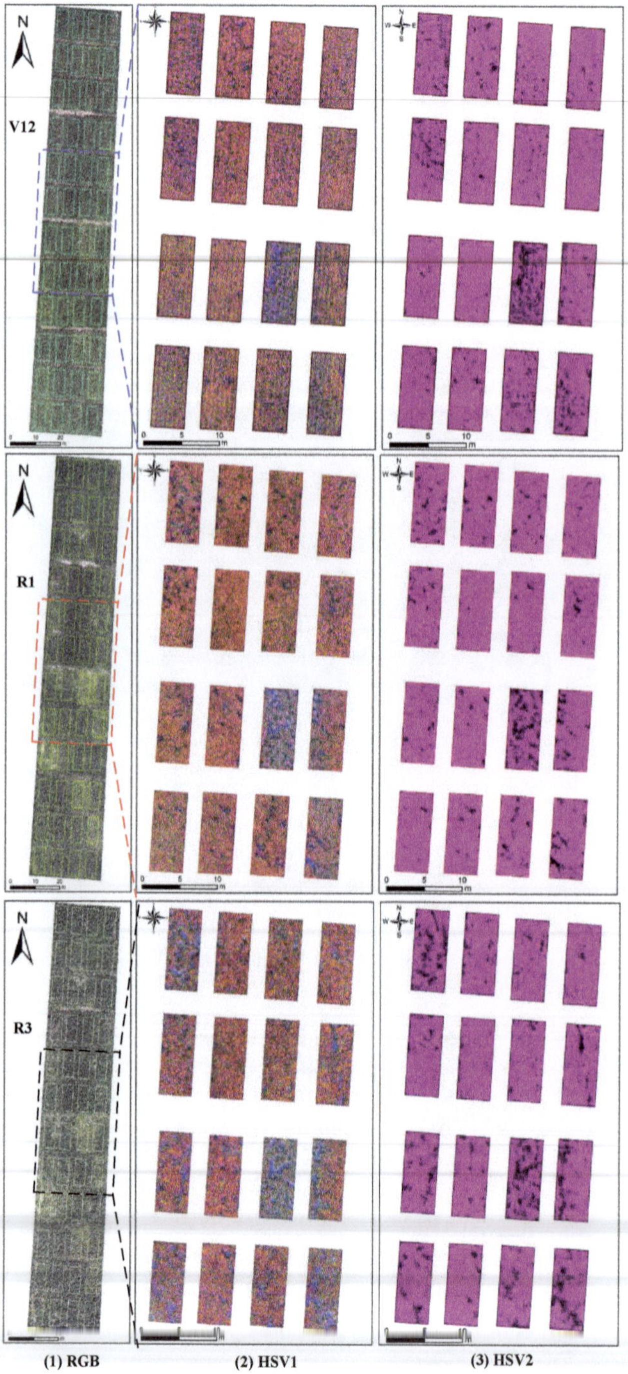

Figure 2. Spatial distribution maps of RGB images with color transformation in three different stages (V12, R1, and R3): (1) original RGB images in the study area, (2) hue saturation value (HSV) primary transformation, and (3) HSV secondary transformation, respectively.

Yao et al. [19] proposed the novel index, CASI, to quantitatively correct for the effect of changing vegetation density and soil cover on wheat LNC estimates, on the basis of understanding the L parameters of SAVI previous studied. Namely, FV_{cover} factor was added to VIs as follows: $CASI = VI/(1 + FV_{cover})$. However, the two inputs of CASI in the study of Yao et al., VI and FV_{cover}, were calculated by ground-based canopy spectra and digital RGB pictures in wheat fields, respectively. In the present study, the extraction of CASI was performed by using UAV-based images in the following framework (Figure 3).

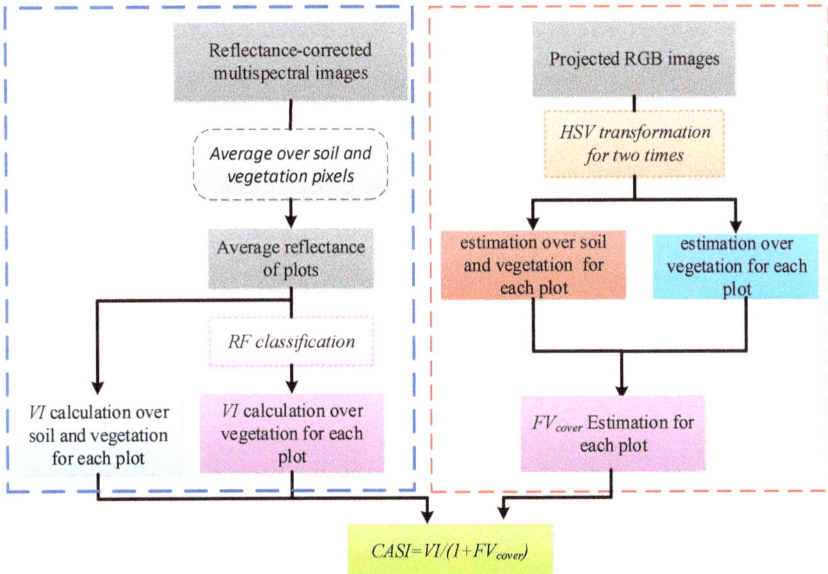

Figure 3. Flowchart illustrating the extraction of coverage adjusted spectral indices (CASI) with unmanned aerial vehicle (UAV) multispectral and RGB images.

2.3.4. Spectral Variables

Reflectance bands (green, red, red-edge, and near-infrared) from multispectral images derived from UAV were used as canopy spectral features, which were variably linked with plant activity, canopy structure, and nutrient situation [28–30]. Vegetation indices that can indicate crop nutrition and growth status have been widely used in agricultural fields. In this study, multispectral images were utilized to extract a set of VIs and the corresponding CASIs for LNC estimation. Table 3 lists the selected VIs that are related to the nitrogen status in crops.

2.3.5. Random Frog Algorithm

The random frog algorithm (RFA) is an effective tool for variable selection on the basis of the reversible jump Markov chain Monte Carlo method [44]. Generally, the RFA works in three steps: (1) randomly give an initial subset V_0 with Q variables; (2) propose a candidate subset V^* containing Q^* variables based on V_0, take V^* with a certain probability as V_1, and then replace V_0 with V_1. Loop this step until N iterations; (3) calculate a selection probability of each variable, which will be used as a measure of variable importance and a criterion of variable selection [44–46]. In the present study, RFA was used to select the better variables among both four bands and a set of VIs or CASIs from UAV-based images for LNC estimation. First, all spectral variables were ranked in descending order, according to the selection probabilities computed by RFA, and then the top five variables were selected as the optimal features for evaluating N status in corn.

Table 3. Vegetation indices from UAV-based multispectral images in the study.

VIs	Name	Formula	Reference
DVI	Difference Vegetation Index	$DVI = R_{NIR} - R_{RED}$	[31]
GNDVI	Green NDVI	$GNDVI = (R_{NIR} - R_{GRE})/(R_{NIR} + R_{GRE})$	[32]
MNLI	Modified Non-linear Vegetation Index	$MNLI = 1.5({R_{NIR}}^2 - R_{RED})/({R_{NIR}}^2 + R_{RED} + 0.5)$	[33]
MSAVI	Modified SAVI	$MSAVI = \left(2R_{NIR} + 1 - \sqrt{(2R_{NIR} + 1)^2 - 8(R_{NIR} - R_{RED})}\right)/2$	[34]
MSR	Modified Simple Ratio	$MSR = (R_{NIR}/R_{NIR} - 1)/\sqrt{R_{NIR}/R_{NIR} + 1}$	[35]
MTVI2	Modified Triangular Vegetation Index 2	$MTVI2 = \frac{1.5(1.2(R_{NIR} - R_{GRE}) - 2.5(R_{RED} - R_{GRE}))}{\sqrt{(2R_{NIR}+1)^2 - (6R_{NIR} - 5\sqrt{R_{RED}}) - 0.5}}$	[36]
NDVI	Normalized Difference Vegetation Index	$NDVI = (R_{NIR} - R_{RED})/(R_{NIR} + R_{RED})$	[37]
NLI	Non-linear Vegetation Index	$NLI = ({R_{NIR}}^2 - R_{RED})/({R_{NIR}}^2 + R_{RED})$	[38]
OSAVI	Optimization of Soil-Adjusted Vegetation Index	$OSAVI = 1.16(R_{NIR} - R_{RED})/(R_{NIR} + R_{RED} + 0.16)$	[16]
R_M	Red Model	$R\_M = R_{NIR}/R_{REG} - 1$	[39]
RDVI	Renormalized Difference Vegetation Index	$RDVI = (R_{NIR} - R_{RED})/\sqrt{R_{NIR} + R_{RED}}$	[40]
RVI	Ratio Vegetation Index	$RVI = R_{NIR}/R_{RED}$	[41]
RVIgreen	Green Ratio Vegetation Index	$RVI_{green} = R_{NIR}/R_{GRE}$	[42]
TVI	Triangular Vegetation Index	$TVI = 0.5(120(R_{NIR} - R_{GRE}) - 200(R_{RED} - R_{GRE}))$	[43]

Note: R_{GRE} denotes green band reflectance, R_{RED} red band reflectance, R_{REG} red-edge band reflectance, and R_{NIR} near-infrared reflectance.

2.3.6. Partial Least Squares Regression

Partial least squares regression (PLSR) is a widely used technique for studying the relationships between multiple dependent variables and multiple independent variables. PLSR integrates the merits of principal component analysis (PCA), canonical analysis (CCA), and linear regression analysis, and can efficiently obtain the dominant factors with the strongest explanatory power for dependent variables. Especially, it is very useful for solving problems such as with obvious multicollinearity between variables, or when the number of variables is greater than the number of samples [47].

2.3.7. Data Evaluation

In this study VIs were extracted from UAV-based multispectral images before and after removing soil noise, and their correlations with LNC were analyzed, as well as the corresponding CASIs. The coefficient of determination (R^2), root mean square error (RMSE), and normalized RMSE (NRMSE, %) were used to explain and quantify the relationship with LNC, and the three metrics can be expressed as follows:

$$R^2 = 1 - \Sigma_{i=1}^{n}(x_i - y_i)^2/\Sigma_{i=1}^{n}(x_i - \overline{x})^2 \tag{2}$$

$$RMSE = \sqrt{\Sigma_{i=1}^{n}(x_i - y_i)^2/n} \tag{3}$$

$$NRMSE = \frac{RMSE}{\overline{x}} * 100 \tag{4}$$

where x_i and y_i are the measured and predicted LNC in corn, $\overline{x}$ is the mean of measured LNC, and n is the total number of samples.

3. Results

3.1. Relationship between VIs, CASIs, and LNC

Table 4 shows the correlation coefficients between LNC in corn, VIs, and CASIs from the UAV images before and after removing background soil. Correspondingly, both larger fractions and darker colors of the pie charts within the red rectangle area indicate stronger correlations between LNC, Vis, and CASIs in Figure 4.

Table 4. Correlations between LNC, VIs, and CASIs in the three growth stages.

Indices	Growth Stages/VIs vs CASIs						Growth Stages (Removing Soil Noise)/VIs vs CASIs					
	V12		R1		R3		V12		R1		R3	
GRE	−0.377 **	−0.416 **	−0.690 **	−0.695 **	−0.454 **	−0.525 **	−0.262	−0.297	−0.036	−0.12	−0.607 **	−0.606 **
RED	−0.449 **	−0.510 **	−0.667 **	−0.662 **	−0.614 **	−0.628 **	−0.26	−0.295	0.158	0.072	−0.586 **	−0.590 **
REG	−0.494 **	−0.686 **	−0.672 **	−0.728 **	−0.436 **	−0.594 **	−0.304	−0.343	0.107	−0.017	−0.610 **	−0.610 **
NIR	0.152	−0.119	0.064	−0.394 **	0.450 **	−0.032	−0.211	−0.276	0.317	0.186	−0.580 **	−0.583 **
DVI	0.393 **	0.26	0.592 **	0.598 **	0.611 **	0.613 **	0.317	0.293	0.391 **	0.379 **	0.589 **	0.589 **
GNDVI	0.502 **	0.343	0.656 **	0.658 **	0.541 **	0.503 **	0.314	0.294	0.664 **	0.669 **	0.622 **	0.625 **
MNLI	0.540 **	0.577 **	0.677 **	0.676 **	0.527 **	0.550 **	0.462 **	0.460 **	0.677 **	0.705 **	0.175	0.237
MSAVI	0.466 **	0.33	0.592 **	0.676 **	0.616 **	0.619 **	0.319	0.296	0.406 **	0.395 **	0.592 **	0.594 **
MSR	0.462 **	0.387 **	0.569 **	0.571 **	0.589 **	0.588 **	0.34	0.321	0.424 **	0.417 **	0.578 **	0.579 **
MTVI2	0.351	0.242	0.22	0.131	0.473 **	0.450 **	0.309	0.284	−0.149	−0.233	0.488 **	0.473 **
NDVI	0.480 **	0.281	0.593 **	0.597 **	0.616 **	0.618 **	0.302	0.278	0.416 **	0.406 **	0.594 **	0.597 **
NLI	0.474 **	0.465 **	0.527 **	0.525 **	0.586 **	0.597 **	0.382 **	0.365	0.692 **	0.717 **	0.333	0.408 **
OSAVI	0.483 **	0.293	0.596 **	0.600 **	0.616 **	0.619 **	0.307	0.283	0.412 **	0.401 **	0.592 **	0.594 **
R_M	0.584 **	0.509 **	0.707 **	0.709 **	0.641 **	0.655 **	0.506 **	0.494 **	0.712 **	0.721 **	0.545 **	0.571 **
RDVI	0.477 **	0.307	0.595 **	0.601 **	0.614 **	0.616 **	0.311	0.287	0.406 **	0.395 **	0.592 **	0.593 **
RVI	0.445 **	0.375 **	0.548 **	0.540 **	0.566 **	0.546 **	0.366	0.335	0.643 **	0.613 **	0.605 **	0.589 **
RVIgreen	0.481 **	0.429 **	0.641 **	0.638 **	0.541 **	0.475 **	0.356	0.324	0.426 **	0.387 **	0.563 **	0.548 **
TVI	0.323	0.192	0.333	0.285	0.516 **	0.495 **	0.306	0.28	−0.066	−0.145	0.521 **	0.509 **

** Denotes significant at 0.01 level, for the two columns in each growth stage, the 1st column means the correlations between VIs and LNC, the 2nd column for those between CASIs and LNC.

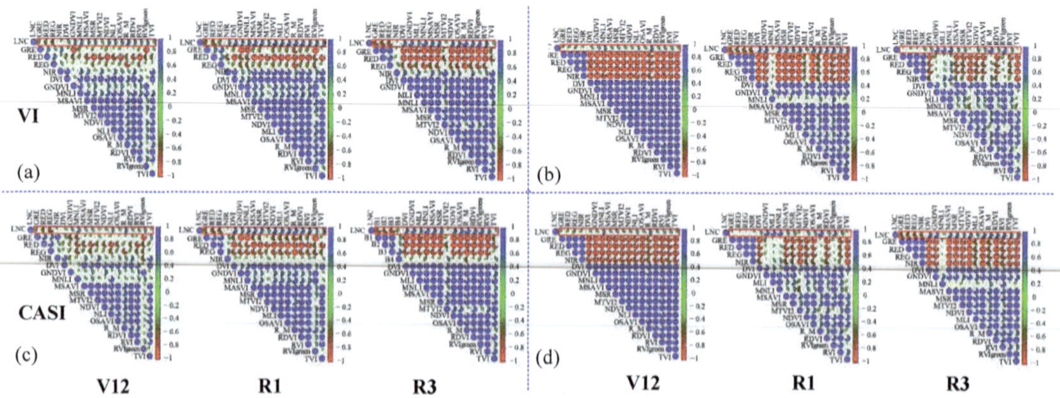

Figure 4. Relationships between VIs, CASIs, and LNC in the three stages of V12, R1, and R3. (**a**,**b**) denote the correlations between VIs and LNC before and after removing the soil background, respectively. (**c**,**d**) refer to the correlations between CASIs and LNC before and after soil removal, respectively.

It can be seen from Table 4 that whether eliminating soil noise or not, the four bands (GRE, RED, REG, and NIR) in the R3 stage are more stably and strongly correlated with LNC in comparison with the V12 and R1 stages. On the other hand, most of the VIs in the V12 stage show worse relationships with LNC, while some VIs in the R1 and R3 stages have significant correlations with LNC, especially GNDVI, R_M, and RVI. As a matter of fact, due to the relatively low vegetation cover in the corn fields in the V12 stage, the soil background had a great influence on the crop canopy reflectance, which directly impacts the sensitivity of spectral vegetation indices to nitrogen status to a greater or lesser extent. However, on account of increasing vegetation cover in the R1 and R3 stages, vegetation indices usually displayed a closer relationship with LNC.

From Figure 4, it can be noticed that CASIs have a similar behavior to VIs. Namely, CASIs in the V12 stage generally show a relatively weak correlation with LNC, and in the R1 and R3 stages the CASIs have closer relationships with LNC. Nevertheless, most of CASIs into which the factor of FV_{cover} was added have a slightly better correlation than the corresponding Vis, whether removing soil noise or not in the latter two stages. This indicates that CASI with the vegetation adjusting factor might have a better performance than VI under high cover conditions. From Table 4 and Figure 4, it can also be seen that in the V12 stage CASIs show no better correlations with LNC in comparison with VIs before removing the soil background, but the CASI's relationships with LNC are similar to VI's after eliminating soil noise.

3.2. Selection of Vegetation Features by RFA

To estimate LNC in corn, first the random frog algorithm (RFA) was used to search for optimal features among the selected VIs and CASIs, and then a PLSR analysis was carried out by using the five optimal features in each of the three growth stages. Table 5 lists the features selected by RFA under different soil removal conditions in the three growth stages of corn. It can be observed from Table 5 that the features selected by RFA from VIs or CASIs show stability, and have no changes after removing soil for the same growth stage, and that the selected features from CASIs also exhibit a greater consistency than those from VIs in the three stages of corn; for instance there are the same features in the V12 and R3 stages. In addition, Table 5 also shows that most of the optimal five features selected by RFA from VIs and CASIs are the same for the same growth stage, and only one or two features are different, which will make it comparable to estimate LNC with optimal VIs and CASIs in this way.

Table 5. Features selected by random frog algorithm (RFA) under different soil-removing conditions in the three stages.

Types	Conditions	Stages	Features
VIs	Not removing soil	V12	REG, NLI, DVI, MNLI, R_M
		R1	NLI, NIR, NDVI, RDVI, OSAVI
		R3	GNDVI, NIR, GRE, DVI, MNLI
	Soil removal	V12	REG, NLI, DVI, MNLI, R_M
		R1	NLI, NIR, NDVI, RDVI, OSAVI
		R3	GNDVI, NIR, GRE, DVI, MNLI
CASIs	Not removing soil	V12	REG, NIR, GRE, DVI, MNLI
		R1	REG, NIR, NDVI, RDVI, OSAVI
		R3	REG, NIR, GRE, DVI, MNLI
	Soil removal	V12	REG, NIR, GRE, DVI, MNLI
		R1	REG, NIR, NDVI, RDVI, OSAVI
		R3	REG, NIR, GRE, DVI, MNLI

It was also noticed that the red-edge (REG) and near infrared (NIR) band features are significantly selected for LNC estimates under different soil-removing conditions in the three stages. As a matter of fact, the famous red edge feature has been widely used to assess vegetational parameters, such as nitrogen and chlorophyll content [39,48–50], while the near infrared band is also the key component of most VIs [16,31–43].

3.3. Estimation of LNC in Corn Using CASIs and VIs

Based on the above optimal features selected by RFA in each growth stage, PLSR was used to estimate LNC in corn. Figure 5 shows the relationships between the predicted and measured LNC using VIs without and with removing soil noise in the different stages. It can be seen that the estimations of LNC with removal of soil noise are generally better than those without eliminating soil noise, which indicates it is necessary for assessing nitrogen status to take soil influence into account.

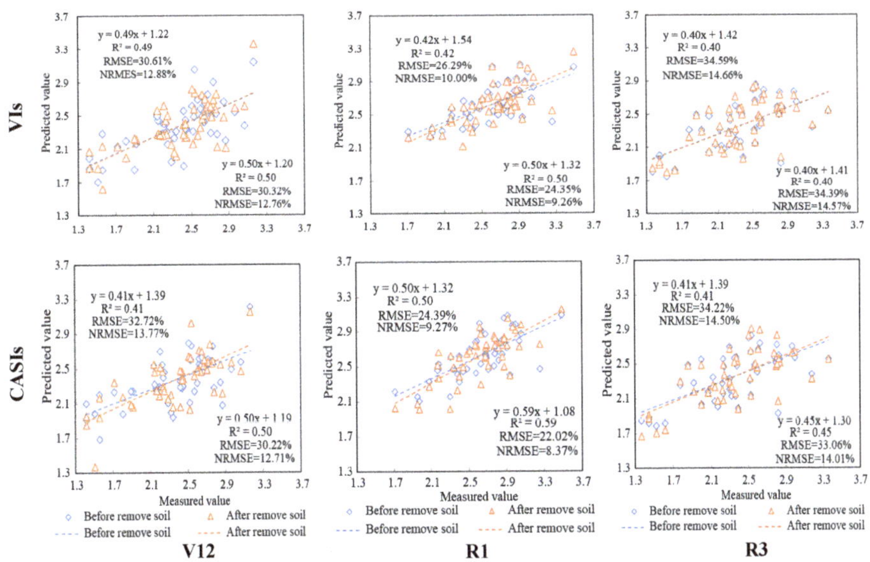

Figure 5. Relationships between the predicted and measured values of LNC using optimal VIs and CASIs selected by RFA before and after removing soil noise in the three stages of V12, R1, and R3.

Unlike VIs using only multispectral images from UAV, CASIs which were constructed by adding the factor, FV_{cover}, to VIs were extracted to estimate LNC in corn using UAV RGB and multispectral images in this study. Figure 5 shows that the CASIs had a better performance after soil elimination than before soil removal, with the best R^2 of 0.59, RMSE of 22.02%, and NRMSE of 8.37% in the R1 stage. What is more, it can be seen from Table 6 that the CASIs generally exhibited the better capability for estimating LNC than the Vis, whether removing background soil or not.

Table 6. Comparison using optimal CASIs and VIs to evaluate LNC in different growth stages, with different processing for the soil background.

Conditions	Stages	CASI			VI		
		R^2	RMSE	NRMSE	R^2	RMSE	NRMSE
No removal of soil	V12	0.41	32.72%	13.77%	0.49	30.61%	12.88%
	R1	0.50	24.39%	9.27%	0.42	26.29%	10.00%
	R3	0.41	34.22%	14.50%	0.40	34.59%	14.66%
Soil removal	V12	0.50	30.22%	12.71%	0.50	30.32%	12.76%
	R1	0.59	22.02%	8.37%	0.50	24.35%	9.26%
	R3	0.45	33.06%	14.01%	0.40	34.39%	14.57%

3.4. Mapping LNC Based on UAV Images at Plot Scale

VIs at different growth stages of corn were extracted from UAV multispectral images covering the 48 plots. The best five were selected by RFA from VIs to establish the PLSR model for LNC estimates, and then the LNC distributions at plot scale were mapped for each stage (Figure 6). The mapping analysis results show that although there was the same nitrogen treatment (N2) at the 24 plots in the northern study area, LNC in the three stages still ranges from 2.00% to 3.14% due to trait differences among cultivars, and this situation is similar whether soil noise is removed or not. In addition, it can also be noticed that the LNC at the same cultivar plots shows an obvious consistency in the R1 stage, which also happens in this stage after soil elimination. This may be because corn in the R1 stage is the most vigorous, vegetation coverage is also the largest, and soil has the least influence on the corn canopies.

At the 24 plots in the southern study area, the different nitrogen application rates resulted in distinct differences in corn growth. LNC ranged greatly from 1.71% to 2.75% in the V12 and R3 stage, and only from 2.00% to 2.75% in R1 stage. It can be seen from Figure 6 that after eliminating soil background, LNC changes at plot scale were more consistent with nitrogen treatments in comparison with no soil removal.

The five sensitive CASIs using UAV RGB and multispectral images were selected by RFA to construct the models for estimating LNC based on PLSR, and LNC distributions for each stage were mapped at plot scale (Figure 7). Like VI-based LNC distributions, LNC using CASI at the northern 24 plots in the three stages ranged greatly from 1.69% to 3.22% due to cultivar differences, even if there was the same nitrogen treatment; this was particularly evident in the V12 and R3 stages, and only in the R1 stage LNC shows some consistency at the same cultivar plots. Comparing with no soil removal, LNC with soil elimination at the northern plots slightly increased.

In the southern plots, corn growth in the three stages was closely related to nitrogen treatment. High nitrogen application rates (N2 and N3) indicated better growth with high LNC for corn, no nitrogen treatment (N0) showed the worst growth. From Figure 7, it can be seen that LNC changed at plot scale, consistent with nitrogen treatments generally, and it was more distinct in the R1 stage after soil removal. In addition, corn growth in the R1 stage was the most vigorous among the three stages, which was well reflected by the LNC distribution in the R1 stage.

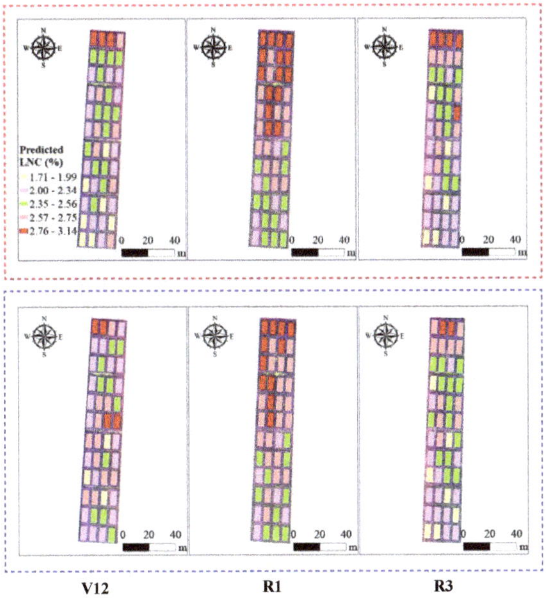

Figure 6. LNC distribution maps of corn using VIs from UAV multispectral images at plot scale in the three stages of V12, R1, and R3. The distribution maps of LNC without soil removal are shown in the red dotted box, those with eliminating soil are in the blue dotted box.

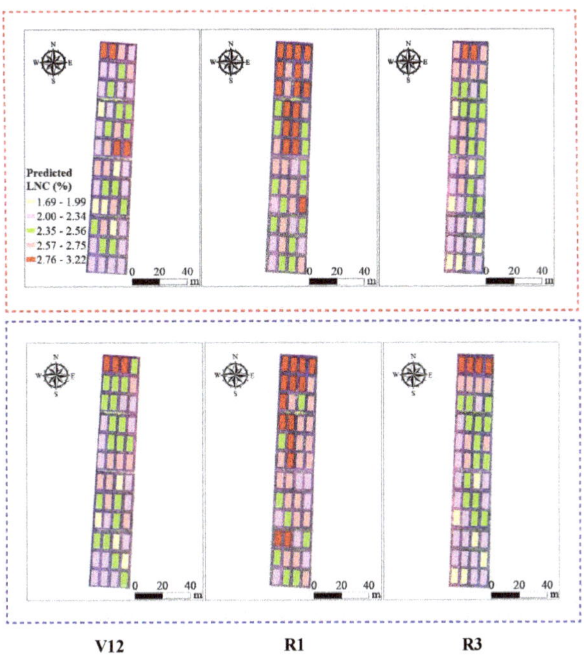

Figure 7. Distribution maps of LNC in corn based on CASIs using UAV RGB and multispectral images at plot scale in the three stages of V12, R1, and R3. The maps of LNC without soil removal are shown in the red dotted box, and those with soil removal are in the blue dotted box.

4. Discussion

4.1. Partial Soil or Shadow Removal on UAV Images

Compared with satellite remote-sensing images, the low-altitude UAV images have a higher spatial resolution, and the vegetation textures in images are also more prominent, however, there is also much more soil or shadow background in UAV-based images. Low-altitude UAV remote-sensing images are strongly affected by planting density, row distance, canopy structure, and other factors, but these factors are usually linked with the soil [16,51]. For instance, lower crop planting density will lead to lower vegetation coverage, and the soil background and shadow will certainly increase in images. Hence, the problem of how to alleviate the influence from soil or shadow noise must inevitably be considered in UAV applications. In previous studies, RGB images have been usually converted into HSV color space for soil and shadow detection [52–54]. In this study, HSV color transformation was applied for the elimination of soil and shadow background from UAV RGB images, and this indicates that it is very effective to remove soil or shadow by HSV. Particularly, HSV was consecutively used two times in this study, and the soil and shadow could be clearly identified in the HSV2 images (Figure 2), which may have a certain significance for the other related studies.

4.2. Vegetation Features for LNC Evaluates

Vegetation indices (VIs) derived from the combination of characteristic bands in multispectral data can highlight the properties of vegetation, and be widely used for estimating crop parameters, such as nitrogen and chlorophyll content [48,49,55]. However, VIs are usually affected by many factors, and exhibit different traits. Some have great potential for suppressing soil noise [16,34], and others are good at anti-saturation for vegetation spectra [36,41]. Therefore, effectively selecting sensitive features from VIs to assess nitrogen status should be considered [10,56,57]. In the present study, the random frog algorithm (RFA) was implemented to select a set of the sensitive features from UAV images to estimate LNC in the V12, R1, and R3 stages of corn. Among the sets of sensitive features for the three stages, the red-edge (REG) and near infrared (NIR) band features were significantly selected for LNC estimates; the two features had also been used to predict maize LNC in an existing report [10]. In practice, REG is one of the most widely used spectral features for evaluating crop parameters [39,55–59] and NIR is also the key component for most typical VIs [16,31–43].

CASI is actually a kind of VI, just adding the parameter of vegetation cover, FV_{cover}, to the conventional VI. Compared with VIs, CASIs had generally better correlations with LNC in this study, which is in line with a previous study [19]. The existing research has demonstrated that soil background can greatly influence the relationships between canopy spectra and N status [19,60–62]. The Addition of FV_{cover} to the basis of VI should be a good method for eliminating the influence of soil, which should make CASI more sensitive to nitrogen status in crops. In addition, it should be noted that in this study the elimination of soil noise was applied only for UAV multispectral images with relatively low spatial resolution. Despite soil removal from multispectral images by threshold segmentation, there were still mixed pixels with soil and vegetation cover derived from UAV RGB images with higher spatial resolution; CASIs integrating VI with FV_{cover} should be able to further eliminate some noise after removing soil noise.

4.3. Advantages and Disadvantages of UAV Data Applications

There have been some reports where UAV images from multiple sensors were simultaneously implemented to estimate crop parameters, and in these studies the images from different sensors were often used to individually extract a variety of spectral variables, then the variables were directly input into various models for evaluating crop growth status, while little attention has been paid to the application of a fusion of information from images of different sensors [10,14,63]. In this study, high-definition RGB images were used to efficiently extract vegetation coverage (FV_{cover}) in plots, and multispectral images for the calculation of typical vegetation indices (VIs), then CASIs that integrate VIs with FV_{cover} were acquired to estimate LNC in corn. CASIs can make good use of the complementary advantages from two types of UAV images, and are a simple and effective method of information fusion. In fact, it is a developing trend to use a fusion of information for crop growth assessment [64].

However, there are still some deficiencies that need to be improved further. Due to the influence of the shooting angles and solar-irradiance angles on low-altitude UAV, some soil and shadow could not be completely removed. The training models constructed in this study only used data in three stages of one year. Thus, further validation will be needed using data from many years or different areas in the future.

5. Conclusions

This study explored how to use digital RGB and multispectral images from two sensors mounted simultaneously on a low-altitude UAV to assess nitrogen status in three growth stages of corn. CASIs that connect VIs with vegetation cover (FV_{cover}) used the UAV images from two sensors to estimate LNC in corn based on the PLS with RFA methods. The results proved that CASIs exhibited the better estimation of LNC in comparison with VIs in all of the three growth stages of corn, regardless of soil removal. It is concluded that CASIs fusing useful information from both digital RGB and multispectral images acquired synchronously by UAV appear very promising, and have potential for remotely monitoring leaf nitrogen content in crops. The preliminary research of using CASIs can also provide new ideas and referential methods for UAV monitoring of other biochemical parameters.

Author Contributions: X.X. put forward the general idea and revised the manuscript. L.F. analyzed the data and wrote the draft of the manuscript. Z.L. and H.Y. gave comments and suggestions to the manuscript. Y.M., H.F. and B.X. provided data and data acquisition capacity. All authors have read and agreed to the published version of the manuscript.

Funding: This study was supported by National Key Research and Development Program (2019YFE0125300, 2017YFD0201501), National Natural Science Foundation of China (41571416).

Institutional Review Board Statement: Not applicable.

Informed Consent Statement: Not applicable.

Data Availability Statement: The data that support the findings of this study are available from the corresponding author, upon reasonable request.

Acknowledgments: The authors thank Weiguo Li and Hong Chang for their assistance in field data collection and farmland management.

Conflicts of Interest: The authors declare no conflict of interest.

References

1. Fitzgerald, G.; Rodriguez, D.; O'Leary, G. Measuring and predicting canopy nitrogen nutrition in wheat using a spectral index—The canopy chlorophyll content index (CCCI). *Field Crop. Res.* **2010**, *116*, 318–324. [CrossRef]
2. Xue, L.; Luo, W.; Cao, W.; Tian, Y. Research progress on the water and nitrogen detection using spectral reflectance. *Int. J. Remote Sens.* **2003**, *7*, 73–80.
3. Tremblay, N. Determining nitrogen requirements from crops characteristics. Benefits and challenges. In *Recent Research Development Agronomy and Horticultur*, 1st ed.; Pandalai, S.G., Ed.; Research Signpost: Kerala, India, 2004; pp. 157–182.
4. Mulla, D.J. Twenty five years of remote sensing in precision agriculture: Key advances and remaining knowledge gaps. *Biosyst. Eng.* **2013**, *114*, 358–371. [CrossRef]
5. Ballesteros, R.; Ortega, J.F.; Hernandez, D.; Moreno, M.A. Applications of georeferenced high-resolution images obtained with unmanned aerial vehicles. Part I: Description of image acquisition and processing. *Precis. Agric.* **2014**, *15*, 579–592. [CrossRef]
6. Peña, J.M.; Torres-Sánchez, J.; de Castro, A.I.; Kelly, M.; López-Granados, F. Weed mapping in early-season maize fields using object-based analysis of Unmanned Aerial Vehicle (UAV) images. *PLoS ONE* **2013**, *8*, e77151. [CrossRef]
7. Swain, K.C.; Thomson, S.J.; Jayasuriya, H.P.W. Adoption of an unmanned helicopter for low-altitude remote sensing to estimate yield and total biomass of a rice crop. *Trans. ASABE* **2010**, *53*, 21–27. [CrossRef]

8. Shamshiri, R.R.; Hameed, I.A.; Balasundram, S.K.; Ahmad, D.; Weltzien, C.; Yamin, M. Fundamental Research on Unmanned Aerial Vehicles to Support Precision Agriculture in Oil Palm Plantations. In *Agricultural Robots-Fundamentals and Applications*, 1st ed.; Zhou, J., Ed.; IntechOpen: Nanjing, China, 2019; pp. 91–116.
9. Swain, K.C.; Jayasuriya, H.P.W.; Salokhe, V.M. Suitability of low-altitude remote sensing images for estimating nitrogen treatment variations in rice cropping for precision agriculture adoption. *J. Appl. Remote Sens.* **2007**, *1*, 6656–6659. [CrossRef]
10. Osco, L.P.; Junior, J.M.; Ramos, A.P.M.; Furuya, D.E.G.; Santana, D.C.; Teodoro, L.P.R.; Gonçalves, W.N.; Baio, F.H.R.; Pistori, H.; Junior, C.A.D.S.; et al. Leaf nitrogen concentration and plant height prediction for maize using UAV-based multispectral imagery and machine learning techniques. *Remote Sens.* **2020**, *12*, 3237. [CrossRef]
11. Lee, H.; Wang, J.; Leblon, B. Using linear regression, Random Forests, and Support Vector Machine with unmanned aerial vehicle multispectral images to predict canopy nitrogen weight in corn. *Remote Sens.* **2020**, *12*, 2071. [CrossRef]
12. Liu, H.; Zhu, H.; Wang, P. Quantitative modelling for leaf nitrogen content of winter wheat using UAV-based hyperspectral data. *Int. J. Remote Sens.* **2017**, *38*, 2117–2134. [CrossRef]
13. Lu, J.; Cheng, D.; Geng, C.; Zhang, Z.; Xiang, Y.; Hu, T. Combining plant height, canopy coverage and vegetation index from UAV-based RGB images to estimate leaf nitrogen concentration of summer maize. *Biosyst. Eng.* **2021**, *202*, 42–54. [CrossRef]
14. Kefauver, S.C.; Vicente, R.; Vergara-Díaz, O.; Fernandez-Gallego, J.A.; Kerfal, S.; Lopez, A. Comparative UAV and field phenotyping to assess yield and nitrogen use efficiency in hybrid and conventional barley. *Front. Plant Sci.* **2017**, *8*, 1733. [CrossRef] [PubMed]
15. Huete, A.; Jackson, R.D.; Post, D.F. Spectral response of a plant canopy with different soil backgrounds. *Remote Sens. Environ.* **1985**, *17*, 37–53. [CrossRef]
16. Rondeaux, G.; Steven, M.D.; Baret, F. Optimization of soil adjusted vegetation indices. *Remote Sens. Environ.* **1996**, *55*, 95–107. [CrossRef]
17. Major, D.; Frederic, B.; Guyot, G. A ratio vegetation index adjusted for soil brightness. *Int. J. Remote Sens.* **1990**, *11*, 727–740. [CrossRef]
18. Liu, S.; Yang, G.; Jing, H.; Feng, H.; Li, H.; Chen, P.; Yang, W. Retrieval of winter wheat nitrogen content based on UAV digital image. *Trans. CSAE* **2019**, *35*, 75–85, (In Chinese with English Abstract).
19. Yao, X.; Ren, H.; Cao, Z.; Tian, Y.; Cao, W.; Zhu, Y.; Cheng, T. Detecting leaf nitrogen content in wheat with canopy hyperspectrum under different soil backgrounds. *Int. J. Appl. Earth Obs. Geoinf.* **2014**, *32*, 114–124. [CrossRef]
20. Ritchie, S.W.; Hanway, J.J. *How a Corn Plant Develops*; Special Report No. 48; Iowa State University of Science and Technology, Cooperative Extension Service: Ames, IA, USA, 1986; pp. 1–21.
21. Tsai, C.M.; Yeh, Z.M. Contrast enhancement by automatic and parameter-free piecewise linear transformation for color images. *IEEE Trans. Consum. Electron.* **2008**, *54*, 213–219. [CrossRef]
22. Guo, P.; Wu, F.; Dai, J.; Wang, H.; Xu, L.; Zhang, G. Comparison of farmland crop classification methods based on visible light images of unmanned aerial vehicles. *Trans. CSAE* **2017**, *33*, 112–119, (In Chinese with English Abstract).
23. Vimal, C.; Sathish, B. Random Forest Classifier Based ECG Arrhythmia Classification. *Int. J. Healthc. Inf. Syst. Inform.* **2010**, *5*, 1–10. [CrossRef]
24. Torres, M.; Qiu, G.P. Automatic habitat classification using image analysis and random forest. *Ecol. Inform.* **2014**, *23*, 126–136. [CrossRef]
25. Akar, Ö.; Güngör, O. Classication of multispectral images using Random Forest algorithm. *J. Geod. Geoinf.* **2012**, *1*, 105–112. [CrossRef]
26. Li, C.; Wang, J.; Liu, L.; Wang, R. Automated digital image analyses for estimating percent ground cover of winter wheat based on object features. *J. Zhejiang Univ. Agric. Life Sci.* **2004**, *30*, 650–656, (In Chinese with English Abstract).
27. Jia, J.; Hu, Y.; Liu, L. Useing Digital Photography to Measure Vegetation Coverage in Qinghai-Tibet Plateau. *J. Geoinf. Sci.* **2010**, *12*, 880–885, (In Chinese with English Abstract).
28. Zhang, J.; Pu, R.; Yuan, L. Integrating remotely sensed and meteorological observations to forecast wheat powdery mildew at a regional scale. *IEEE J. Sel. Top. Appl. Earth Obs. Remote Sens.* **2014**, *7*, 4328–4339. [CrossRef]
29. Fan, L.; Zhao, J.; Xu, X.; Liang, D.; Yang, G.; Feng, H.; Yang, H.; Yulong, W.; Chen, G.; Wei, P. Hyperspectral-based estimation of leaf nitrogen content in corn using optimal selection of multiple spectral variables. *Sensors* **2019**, *19*, 2898. [CrossRef]
30. Xu, X.; Zhao, C.; Wang, J.; Zhang, J.; Song, X. Using optimal combination method and in situ hyperspectral measurements to estimate leaf nitrogen concentration in barley. *Precis. Agric.* **2014**, *15*, 227–240. [CrossRef]
31. Richardson, A.J.; Wiegand, C.L. Distinguishing vegetation from soil background information. *Photogramm. Eng. Remote Sens.* **1977**, *43*, 1541–1552.
32. Gitelson, A.; Kaufman, Y.J.; Merzlyak, M.N. Use of a green channel in remote sensing of global vegetation from EOS-MODIS. *Remote Sens. Environ.* **1996**, *58*, 289–298. [CrossRef]
33. Gong, P.; Pu, R.; Biging, G.S.; Larrieu, M.R. Estimation of forest leaf area index using vegetation indices derived from Hyperion hyperspectral data. *IEEE Trans. Geosci. Remote Sens.* **2003**, *41*, 1355–1362. [CrossRef]
34. Qi, J.G.; Chehbouni, A.; Huete, A.; KERR, Y.; Sorooshian, S. A modified soil adjusted vegetation index. *Remote Sens. Environ.* **1994**, *48*, 119–126. [CrossRef]
35. Chen, J.M. Evaluation of vegetation indices and a modified simple ratio for boreal applications. *Can. J. Remote Sens.* **1996**, *22*, 229–242. [CrossRef]

36. Haboudane, D.; Miller, J.R.; Pattey, E.; Zarco-Tejada, P.J.; Strachan, I.B. Hyperspectral vegetation indices and novel algorithms for predicting green LAI of crop canopies: Modeling and validation in the context of precision agriculture. *Remote Sens. Environ.* **2004**, *90*, 337–352. [CrossRef]
37. Rouse, J.W.; Haas, R.H.; Schell, J.A.; Deering, D.W. Monitoring Vegetation Systems in the Great Plains with ERTS. In Proceedings of the Third ERTS Symosium, Washington, DC, USA, 10–14 December 1973; Volume 1, pp. 48–62.
38. Goel, N.S.; Qin, W. Influences of canopy architecture on relationships between various vegetation indices and LAI and Fpar: A computer simulation. *Remote Sens. Rev.* **1994**, *10*, 309–347. [CrossRef]
39. Gitelson, A.A.; Viña, A.; Ciganda, V.; Rundquist, D.C.; Arkebauer, T.J. Remote estimation of canopy chlorophyll content in crops. *Geophys. Res. Lett.* **2005**, *32*, L08403. [CrossRef]
40. Roujean, J.L.; Bŕeon, F.M. Estimating PAR absorbed by vegetation from bidirectional reflectance measurements. *Remote Sens. Environ.* **1995**, *51*, 375–384. [CrossRef]
41. Jordan, C.F. Derivation of Leaf Area Index from Quality of Light on the Forest Floor. *Ecology* **1969**, *50*, 663–666. [CrossRef]
42. Xue, L.H.; Cao, W.X.; Luo, W.H.; Dai, T.B.; Zhu, Y. Monitoring leaf nitrogen status in rice with canopy spectral reflectance. *Agron. J.* **2004**, *96*, 135–142. [CrossRef]
43. Broge, N.H.; Leblanc, E. Comparing prediction power and stability of broadband and hyperspectral vegetation indices for estimation of green leaf area index and canopy chlorophyll density. *Remote Sens. Environ.* **2001**, *76*, 156–172. [CrossRef]
44. Li, H.; Xu, Q.; Liang, Y. Random Frog: An efficient reversible jump Markov Chain Monte Carlo-like approach for variable selection with applications to gene selection and disease classification. *Anal Chim. Acta* **2012**, *740*, 20–26. [CrossRef]
45. Hu, M.H.; Zhai, G.T.; Zhao, Y.; Wang, Z.D. Uses of selection strategies in both spectral and sample spaces for classifying hard and soft blueberry using near infrared data. *Sci. Rep.* **2018**, *8*, 6671. [CrossRef] [PubMed]
46. Zhang, C.; Kong, W.; Liu, F.; He, Y. Measurement of aspartic acid in oilseed rape leaves under herbicide stress using near infrared spectroscopy and chemometrics. *Heliyon* **2016**, *2*, e00064. [CrossRef] [PubMed]
47. Wold, S.; Ruhe, A.; Wold, H.; Dunn, W.J., III. The collinearity problem in linear regression. The partial least squares (PLS) approach to generalized inverses. *SIAM J. Sci. Stat. Comput.* **1984**, *5*, 735–743. [CrossRef]
48. Li, F.; Mistele, B.; Hu, Y.; Chen, X.; Schmidhalter, U. Reflectance estimation of canopy nitrogen content in winter wheat using optimised hyperspectral spectral indices and partial least squares regression. *Eur. J. Agron.* **2014**, *52*, 198–209. [CrossRef]
49. Hunt, E.R., Jr.; Doraiswamy, P.C.; McMurtrey, J.E.; Daughtry, C.S.T.; Perry, E.M.; Akhmedov, B. A visible band index for remote sensing leaf chlorophyll content at the canopy scale. *Int. J. Appl. Earth Obs. Geoinf.* **2013**, *21*, 103–112. [CrossRef]
50. Cho, M.A.; Skidmore, A.K. A new technique for extracting the red edge position from hyperspectral data: The linear extrapolation method. *Remote Sens. Environ.* **2006**, *101*, 181–193. [CrossRef]
51. Rasmussen, J.; Ntakos, G.; Nielsen, J.; Svensgaard, J.; Poulsenb, R.N.; Christensena, S. Are vegetation indices derived from consumer-grade cameras mounted on UAVs sufficiently reliable for assessing experimental plots? *Eur. J. Agron.* **2016**, *74*, 75–92. [CrossRef]
52. Farou, B.; Rouabhia, H.E.; Seridi, H.; Akdag, H. Novel approach for detection and removal of moving cast shadows based on RGB, HSV and YUV color spaces. *Comput. Inform.* **2017**, *36*, 837–856. [CrossRef]
53. Surkutlawar, S.; Kulkarni, R.K. Shadow suppression using RGB and HSV color space in moving object detection. *Int. J. Adv. Comput. Sci. Appl.* **2013**, *4*, 164.
54. Zhang, Y.; Hartemink, A.E. A method for automated soil horizon delineation using digital images. *Geoderma* **2019**, *343*, 97–115. [CrossRef]
55. Li, F.; Miao, Y.; Feng, G.; Yuan, F.; Yue, S.; Gao, X.; Liu, Y.; Liu, B.; Ustin, S.L.; Chen, X. Improving estimation of summer maize nitrogen status with red edge-based spectral vegetation indices. *Field Crop. Res.* **2014**, *157*, 111–123. [CrossRef]
56. Osco, L.P.; Ramos, A.P.M.; Pinheiro, M.M.F.; Moriya, É.A.S.; Imai, N.N.; Estrabis, N.; Ianczyk, F.; de'Araújo, F.F.; Liesenberg, V.; de Castro Jorge, L.A.; et al. A machine learning approach to predict nutrient content in valencia-orange leaf hyperspectral measurements. *Remote Sens.* **2020**, *12*, 906. [CrossRef]
57. Zheng, H.; Li, W.; Jiang, J.; Liu, Y.; Cheng, T.; Tian, Y.; Zhu, Y.; Cao, W.; Zhang, Y.; Yao, X. A comparative assessment of different modeling algorithms for estimating leaf nitrogen content in winter wheat using multispectral images from an unmanned aerial vehicle. *Remote Sens.* **2018**, *10*, 2026. [CrossRef]
58. Dong, T.; Liu, J.; Shang, J.; Qian, B.; Ma, B.; Kovacs, J.M.; Walters, D.; Jiao, X.; Geng, X.; Shi, Y. Assessment of red-edge vegetation indices for crop leaf area index estimation. *Remote Sens. Environ.* **2019**, *222*, 133–143. [CrossRef]
59. Fu, Y.; Yang, G.; Li, Z.; Li, H.; Li, Z.; Xu, X.; Song, X.; Zhang, Y.; Duan, D.; Zhao, C.; et al. Progress of hyperspectral data processing and modelling for cereal crop nitrogen monitoring. *Comput. Electron. Agric.* **2020**, *172*, 105321. [CrossRef]
60. Sun, L.; Cheng, L. Anti-soil background capacity with vegetation biochemicalcomponent spectral model. *Acta Ecol. Sin.* **2011**, *31*, 1641–1652.
61. Gilabert, M.; González-Piqueras, J.; Garcııa-Haro, F.; Meliá, J. A generalized soil-adjusted vegetation index. *Remote Sens. Environ.* **2002**, *82*, 303–310. [CrossRef]
62. Chen, P.; Haboudane, D.; Tremblay, N.; Wang, J.; Vigneault, P.; Li, B. New spectral indicator assessing the efficiency of crop nitrogen treatment in corn and wheat. *Remote Sens. Environ.* **2010**, *114*, 1987–1997. [CrossRef]
63. Maimaitijiang, M.; Sagan, V.; Sidike, P.; Daloye, A.M.; Erkbol, H.; Fritschi, F.B. Crop Monitoring Using Satellite/UAV Data Fusion and Machine Learning. *Remote Sens.* **2020**, *12*, 1357. [CrossRef]
64. Maimaitijiang, M.; Sagan, V.; Sidike, P.; Hartling, S.; Esposito, F.; Fritschi, F.B. Soybean yield prediction from UAV using multimodal data fusion and deep learning. *Remote Sens. Environ.* **2020**, *237*, 111599. [CrossRef]

remote sensing

MDPI

Article

A Comparative Assessment of Different Modeling Algorithms for Estimating Leaf Nitrogen Content in Winter Wheat Using Multispectral Images from an Unmanned Aerial Vehicle

Hengbiao Zheng [1,2,3,4], Wei Li [1,2,3,4], Jiale Jiang [1,2,3,4], Yong Liu [1,2,3,4], Tao Cheng [1,2,3,4], Yongchao Tian [1,2,3,4], Yan Zhu [1,2,3,4], Weixing Cao [1,2,3,4], Yu Zhang [1,2,3,4] and Xia Yao [1,2,3,4,*]

1 National Engineering and Technology Center for Information Agriculture, Nanjing Agricultural University, Nanjing 210095, China; 2015201019@njau.edu.cn (H.Z.); 2017101043@njau.edu.cn (W.L.); jialejiang@njau.edu.cn (J.J.); 2013101027@njau.edu.cn (Y.L.); tcheng@njau.edu.cn (T.C.); yctian@njau.edu.cn (Y.T.); yanzhu@njau.edu.cn (Y.Z.); caow@njau.edu.cn (W.C.); zhangyu@njau.edu.cn (Y.Z.)

2 Key Laboratory for Crop System Analysis and Decision Making, Ministry of Agriculture, Nanjing Agricultural University, Nanjing 210095, China

3 Jiangsu Key Laboratory for Information Agriculture, Nanjing Agricultural University, Nanjing 210095, China

4 Jiangsu Collaborative Innovation Center for Modern Crop Production, Nanjing Agricultural University, Nanjing 210095, China

* Correspondence: yaoxia@njau.edu.cn; Tel.: +86-025-8439-6565

Received: 1 November 2018; Accepted: 10 December 2018; Published: 13 December 2018

Abstract: Unmanned aerial vehicle (UAV)-based remote sensing (RS) possesses the significant advantage of being able to efficiently collect images for precision agricultural applications. Although numerous methods have been proposed to monitor crop nitrogen (N) status in recent decades, just how to utilize an appropriate modeling algorithm to estimate crop leaf N content (LNC) remains poorly understood, especially based on UAV multispectral imagery. A comparative assessment of different modeling algorithms (i.e., simple and non-parametric modeling algorithms alongside the physical model retrieval method) for winter wheat LNC estimation is presented in this study. Experiments were conducted over two consecutive years and involved different winter wheat varieties, N rates, and planting densities. A five-band multispectral camera (i.e., 490 nm, 550 nm, 671 nm, 700 nm, and 800 nm) was mounted on a UAV to acquire canopy images across five critical growth stages. The results of this study showed that the best-performing vegetation index (VI) was the modified renormalized difference VI (RDVI), which had a determination coefficient (R^2) of 0.73 and a root mean square error (RMSE) of 0.38. This method was also characterized by a high processing speed (0.03 s) for model calibration and validation. Among the 13 non-parametric modeling algorithms evaluated here, the random forest (RF) approach performed best, characterized by R^2 and RMSE values of 0.79 and 0.33, respectively. This method also had the advantage of full optical spectrum utilization and enabled flexible, non-linear fitting with a fast processing speed (2.3 s). Compared to the other two methods assessed here, the use of a look up table (LUT)-based radiative transfer model (RTM) remained challenging with regard to LNC estimation because of low prediction accuracy (i.e., an R^2 value of 0.62 and an RMSE value of 0.46) and slow processing speed. The RF approach is a fast and accurate technique for N estimation based on UAV multispectral imagery.

Keywords: UAV; multispectral imagery; LNC; vegetation index; non-parametric regression; radiative transfer model

1. Introduction

Nitrogen (N) is one of the most important nutrients required for plant growth and is therefore critical for crop production. A deficiency in N significantly reduces crop photosynthetic yields while the excessive use of fertilizers for this element leads to both resource waste and environmental pollution [1,2]. Furthermore, leaf N content (LNC) at early growth stages (e.g., jointing and booting) is a good indicator for N fertilizer application [3], and LNC at late growth stages (e.g., after heading) is highly related to the final grain quality [4]. Quantification of LNC is therefore a prerequisite for the production of high-yield and good-quality crops while causing minimal environmental impact.

Remote sensing (RS) has become an attractive technique in precision agricultural assessment as it can be used to monitor crop growth status rapidly and nondestructively. The main RS platforms currently in use include satellite, manned airborne, and ground-based approaches, which can all be equipped with various kinds of sensors. Although satellite images can be used to monitor N status across large areas [5,6], they cannot provide sufficient accuracy because of their low spatio-temporal resolution. Even though manned airborne platforms are able to capture images at high spatio-temporal resolution, this approach is limited by both high operational complexity and cost [7].

In contrast, ground-based RS platforms are able to attain high N status estimation monitoring accuracy [8,9], but this approach remains inefficient when used over large areas, while unmanned aerial vehicle (UAV)-based RS platforms provide a low-cost alternative for collecting RS data at high spatio-temporal resolution [10,11]. This platform has been widely applied in precision agriculture and has been utilized for LAI [12] as well as biomass estimations [10,13], but few studies to date have discussed N status detection using this approach [14,15]. It therefore remains an open question whether, or not, UAV images can be used to monitor N status.

A range of methods have so far been proposed that use spectral data to model N content, including statistical and chemometric algorithms alongside physical models. The statistical method has been used most commonly to monitor N content based on optical measurements from different platforms [8,16]. Empirical relationships between LNC and canopy optical properties have also been calibrated using experimental datasets, an approach that has proven to be both efficient and accurate [8,9,17]. It is also the case, however, that retrieval algorithms based on vegetation indices (VIs) tend to exhibit poor model portability because they are easily influenced by band configuration, index formulation, and fitting function [18]. Besides, most VIs are easily saturated at high N content levels [8,19].

An additional set of techniques that have been commonly applied to identify variables for N modeling comprise non-parametric algorithms, including partial least square regression (PLSR), artificial neural networks (ANNs), random forest (RF), and support vector machines (SVMs) [3,20,21]. These approaches make full use of all spectral data and avoid multicollinearity that is inherent to multiple linear regressions [20]. As these methods have also been shown to be very efficient for processing nonlinear data, it is likely that they are also able to deal with high-dimensional data [21] although performance remains an issue [22,23]. In the earlier study, Verrelst et al. [23] investigated the efficiency of four machine learning regression algorithms at estimating leaf chlorophyll content (LCC), LAI, and fractional vegetation cover (FVC), specifically neural networks (NN), support vector regression (SVR), kernel ridge regression (KRR), and Gaussian processes regression (GPR). As the results of this study showed that the latter was more efficient compared to the other three [23], it will also be worthwhile to investigate the performance of different non-parametric algorithms for LNC estimation.

It remains challenging to quantify LNC differences based on small-plot experiments using several cultivars as well as N application levels and planting densities. As differences in N content under experimental conditions are generally limited, established models might be unstable in practical applications. It is also the case that a significant component of variations in canopy optical properties are also due to changes in sun zenith angles, canopy structures, and background. As these differences significantly affect the relationships between spectral parameters and N content, a model based on physical parameters should enable us to clearly explain these potentially confounding factors.

Although a PROSAIL radiative transfer model (RTM) [24] used in combination with hyperspectral reflectance has been shown to provide an effective method for estimating crop LAI [25,26] and LCC [27], it remains unclear how this approach can be utilized to offer enough LNC estimation accuracy with UAV multispectral imagery.

The different modeling algorithms discussed above were studied here using a range of species and sites. One key aim of this research was to comprehensively compare these approaches and determine the optimal retrieval method for a particular objective, especially when using UAV images. A range of questions remains to be addressed, including which VI is optimal for wheat LNC estimation? Which non-parametric algorithm provides the best estimates? How well do physical models perform for LNC retrieval when based on UAV multispectral images? Additionally, which is the best approach when all three modeling algorithms are compared in terms of processing efficiency, model simplification, and estimation accuracy? The objective of this study was therefore to evaluate the performance of these three different retrieval methods for winter wheat LNC estimation using UAV multispectral imagery.

2. Materials and Methods

2.1. Experimental Design

Three field experiments were conducted over two growing seasons (2013–2014 and 2014–2015) in Rugao City (120°45′E, 32°16′N) within Jiangsu Province in eastern China. The predominant soil type is loam and the soil organic matter was 18.9–24.6 g/kg, available N was 140.56–150.41 mg/kg, total nitrogen was 1.87–2.07 g/kg, available phosphorus was 50.12–57.84 mg/kg, and available potassium was 90.32–96.76 mg/kg. These experiments involved different N rates, planting densities, and wheat cultivars, and comprised a randomized complete block design with three replicates, thus there were 36, 30, and 36 treatments for Exp. 1, Exp. 2, and Exp. 3, respectively. A mixture of 120 kg/ha P_2O_5 and 120 kg/ha K_2O was applied to all treatments prior to seeding. Crop management followed local standard practices for wheat production; additional details regarding these three experiments are provided in Table 1.

Table 1. Details of the three field experiments.

Experiment	Year	Cultivar	N Rate (kg/ha)	Planting Density (plants/ha)	Sampling Date	Growth Stage	N
Exp. 1	2013–2014	Yangmai 18 Shengxuan 6	0, 100, 300	1.5×10^6 3.0×10^6	14 March 9/15/23 April 6 May	Jointing, Booting, Heading, Anthesis, Filling	159
Exp. 2	2013–2014	Xumai 30 Ningmai 13	0, 75, 150, 225, 300	2.4×10^6	14 March 9/15/23 April 6 May	Jointing, Booting, Heading, Anthesis, Filling	135
Exp. 3	2014–2015	Yangmai 18 Shengxuan 6	0, 100, 300	1.5×10^6 2.4×10^6	26 March 8/17/25 April 6 May	Jointing, Booting, Heading, Anthesis, Filling	164

2.2. Data Collection

2.2.1. UAV System and Image Acquisition

An eight-rotor MK-Oktokopter UAV (Mikrokopter Inc., Moormerland, Germany) was used to carry a six-channel multispectral Tetracam mini-MCA6 camera (Tetracam Inc., Chatsworth, CA, USA) to collect images in this study (Figure 1). The specific parameters of this UAV and camera are shown in Table 2. This multispectral camera was equipped with five spectral channels (i.e., 490 nm, 550 nm, 671 nm, 700 nm, and 800 nm) with a 10 nm bandwidth, and had an incident light sensor (ILS). All UAV campaigns were undertaken in stable ambient light conditions (between 11:00 and 13:30) at five critical growth stages (Table 1). The UAV was flown at a height of 150 m, and images were collected with spatial resolution of 8.125 cm. After each flight, only one image with high quality was selected for image analysis due to the small study area (50 m × 35 m).

Figure 1. The UAV equipped with multispectral camera used in this study.

Table 2. Specifications of UAV and Mini-MCA multispectral camera.

UAV		Camera	
Weight (g)	2050	Weight (g)	700
Battery weight (g)	520	Geometric resolution (pixel)	1280 × 1024
Maximum payload (g)	2500	Radiometric resolution (bit)	10
Flight duration (min)	8–41	Speed (frame/s)	1.3
Radius (m)	1000	Focal length (mm)	9.6

2.2.2. Ground Sampling

A total of 30 wheat plant samples were randomly collected from each plot subsequent to each UAV campaign in order to determine LNC values (%). All the green leaves from each sample were separated from stems, oven-dried at 80 °C to a constant weight, and then weighed. Dried leaf samples were ground to pass through a 1 mm screen and stored in plastic bags for subsequent chemical analysis. Total leaf N concentration was determined using the micro-Kjeldahl method. Leaf chlorophyll content (Cab) was measured using a soil and plant analyzer development (SPAD) 502 (Minolta Camera Co., Osaka, Japan) with sub-samples (five plants) randomly selected and the first, second, and third fully expanded leaves chosen from three layers encompassing the base, middle, and top parts of wheat leaves. Averaged SPAD readings were taken as sample values in each case. Absolute leaf chlorophyll content (LCC) was then obtained using an equation that expresses the relationship between SPAD readings and LCC values [28].

2.3. *Image Processing*

The pre-processing UAV image workflows used in this analysis followed those proposed by [12,29], and included noise reduction, veginetting, and lens distortion correction as well as band registration and radiometric calibration. Thirty ground control points (GCPs) were evenly distributed in the experimental area, and the geographic coordinates were determined by X900 GNSS (Huace Inc., Beijing, China). The GCPs were used for band registration and georeferencecing processed in the ENVI/IDL environment (Exelis Visual Information Solutions, Boulder, CO, USA). After that, radiometric calibration was conducted by the empirical line method [30] with four standard calibration canvas with different reflectance values (3%, 22%, 48%, and 82%). Reflectance was then extracted from each radiometrically corrected image using a region of interest (ROI) from each plot.

2.4. Retrieval Techniques

2.4.1. Parametric Modeling Algorithms

The parametric modeling algorithm used in this analysis was based on VI calculated with reflectance from UAV multispectral images. Thus, 19 kinds of VI formulations, including two-band, three-band, and four-band indices, encompassing all possible combinations were used to develop correlations versus wheat LNC (Table 3). Linear regression between LNC and all VIs was utilized to eliminate the impact of functions as opposed to band selection and index formulation.

Table 3. Commonly used vegetation indices.

Index	Formula	Reference
Two-band		
Ratio VI (RVI)	$R_{\lambda1}/R_{\lambda2}$	[31]
Difference VI (DVI)	$R_{\lambda1} - R_{\lambda2}$	[31]
NDVI	$(R_{\lambda1} - R_{\lambda2})/(R_{\lambda1} + R_{\lambda2})$	[32]
Renormalized difference VI (RDVI)	$(R_{\lambda1} - R_{\lambda2})/(R_{\lambda1} + R_{\lambda2})^{0.5}$	[33]
Soil adjusted VI (SAVI)	$1.5(R_{\lambda1} - R_{\lambda2})/(R_{\lambda1} + R_{\lambda2} + 0.5)$	[34]
Optimized soil adjusted VI (OSAVI)	$(1 + 0.16)(R_{\lambda1} - R_{\lambda2})/(R_{\lambda1} + R_{\lambda2} + 0.16)$	[35]
Optimized VI (VI_{opt})	$(1 + 0.45)(R_{\lambda1}{}^2 + 1)/(R_{\lambda2} + 0.45)$	[36]
Modified sample ratio (MSR)	$((R_{\lambda1}/R_{\lambda2}) - 1)/(SQRT((R_{\lambda1}/R_{\lambda2}) + 1))$	[37]
Three-band		
Enhanced VI (EVI)	$2.5(R_{\lambda1} - R_{\lambda2})/(R_{\lambda1} + 6R_{\lambda2} - 7.5R_{\lambda3} + 1)$	[38]
Modified normalized difference (mND)	$(R_{\lambda1} - R_{\lambda2})/(R_{\lambda1} + R_{\lambda2} - 2R_{\lambda3})$	[39]
Modified sample ratio (mSR)	$(R_{\lambda1} - R_{\lambda2})/(R_{\lambda3} - R_{\lambda2})$	[39]
Modified chlorophyll absorption in RI (MCARI)	$(R_{\lambda1} - R_{\lambda2} - 0.2(R_{\lambda1} - R_{\lambda3}))(R_{\lambda1}/R_{\lambda2})$	[40]
Transformed chlorophyll absorption in RI (TCARI)	$3((R_{\lambda1} - R_{\lambda2}) - 0.2(R_{\lambda1} - R_{\lambda3})(R_{\lambda1}/R_{\lambda2}))$	[41]
Three-band index 1 (TBI1)	$(R_{\lambda1} - R_{\lambda2} - R_{\lambda3})/(R_{\lambda1} + R_{\lambda2} + R_{\lambda3})$	[42]
Three-band index 2 (TBI2)	$(R_{\lambda1} - R_{\lambda2} + 2R_{\lambda3})/(R_{\lambda1} + R_{\lambda2} - 2R_{\lambda3})$	[17]
Four-band		
Vogelmann index (VOG)	$(R_{\lambda1} - R_{\lambda2})/(R_{\lambda3} + R_{\lambda4})$	[43]
MERIS terrestrial chlorophyll index (MTCI)	$(R_{\lambda1} - R_{\lambda2})/(R_{\lambda3} - R_{\lambda4})$	[44]
TCARI/OSAVI	TCARI/OSAVI	[41]
MCARI/OSAVI	MCARI/OSAVI	[40]

$R_{\lambda1}$, $R_{\lambda2}$, $R_{\lambda3}$, and $R_{\lambda4}$ denote the reflectance of spectral bands randomly selected from 490 nm, 550 nm, 671 nm, 700 nm, and 800 nm.

2.4.2. Non-Parametric Modeling Algorithms

The SimpleR toolbox [45] was used in this study to implement 13 non-parametric modeling algorithms and to develop models. A comprehensive description of these algorithms was presented in [46]. These non-parametric approaches can be further subdivided into linear and non-linear regressions; of these, three fall into the former category-least-squares linear regression (LSLR), principal component regression (PCR), and partial least-squares regression (PLSR)-while 10 fall into the latter-artificial neutral networks (ANN), decision trees (DT), regression trees (RT), bagging trees (BaT), and boosting trees (BoT) as well as random forest (RF), relevance vector machine (RVM), kernel ridge (KRR), and Gaussian processes regressions (GPR) alongside variational heteroscedastic GPR (VH-GPR) and extreme learning machines (ELM).

2.4.3. Physical Based Modeling

The widely used PROSAIL radiative transfer model comprises a combination of the SAIL canopy reflectance and PROSPECT leaf optical properties models. The combined approach was utilized here to retrieve canopy parameter data and was generated via both the latter two methods, PROSPECT-5 and 4SAIL. A look-up-table (LUT) was then applied; these efficient inversion algorithms are commonly used for agronomic parameter retrieval [46,47]. The imposed boundaries and distributions of PROSAIL input variables used in this study are summarized in Table 4; these values were obtained from field measurements and other studies that have utilized the same crops [47,48]. Thus, uniform distributions of LCC and normally distributed LAI were sampled 100 times, uniform carotenoid distributions were sampled 50 times, and all other variables were held constant. A resultant LUT dataset comprising 500,000 parameter combinations was chosen for this analysis; a total of 22 cost functions, including the insertion of up to 50% Gaussian noise into simulated data and multiple best solutions, were considered to optimize the LUT inversion strategy to address radiative transfer model issues [49]. After LCC was retrieved from the PROSAIL model, LNC was indirectly obtained on the empirically linear relationship between LCC and LNC.

Table 4. PROSAIL model input parameters.

Parameters	Units	Range	Distribution
Leaf: PROSPECT-5			
Leaf structure index (N)	Unitless	1.2–1.8	Gaussian
Leaf chlorophyll content (LCC)	[μg/cm^2]	25–75	Gaussian
Leaf dry matter content (C_m)	[g/cm^2]	0.013	
Leaf water content (C_w)	[cm]	0.018	
Canopy: 4SAIL			
Leaf area index (LAI)	[m^2/m^2]	0–7	Gaussian
Soil scaling factor (α_{soil})	Unitless	0.3	
Average leaf angle (ALA)	[°]	60	
Hotspot parameter (HotS)	[m/m]	0.2	
Diffuse incoming solar radiation (skyl)	[%]	10	
Sun zenith angle (θ_s)	[°]	25	
View zenith angle (θ_v)	[°]	0	
Sun-sensor azimuth angle (Φ)	[°]	0	

2.5. Model Calibration and Validation

Table 5 lists the calibration and validation of models on different methods. Data collected from all experiments were pooled to examine the relationship between VIs and LNC with linear regression, and then the optimal bands' configurations were determined. Both the LNC-VI model and non-parametric model were calibrated and validated with a k-fold (k = 10) cross-validation procedure. The whole dataset was randomly divided into 10 equal-sized sub-datasets. Nine sub-datasets were used as the calibration (training) dataset and the rest was used as the validation (test) dataset, then this procedure was repeated 10 times [48]. For the physical-based modeling method, predicted LNC values, after being retrieved from the empirical model, were compared with the field measured values. The predictive capability of those models with different methods was then assessed using the determination coefficient (R^2) and root mean square error (RMSE). All the above procedures were implemented using MATLAB 2014a (The MathWorks Inc., Natick, MA, USA).

Table 5. Calibration and validation of the models on different methods.

Method	Calibration	Validation
Parametric	10-fold cross validation, nine sub-datasets used for calibration (training), the rest for validation (test), repeated 10 times	
Non-parametric		
Physical-based model	LCC retrieved from PROSAIL, LNC obtained through the empirically linear model between LCC and LNC with measured data	All retrieved LNC values compared with measured LNC values

3. Results

3.1. Optimal VI Determination

Relationships between LNC and 19 different formulas with random bands were established, and the best-performing VIs in each case are listed in Table 6. Results show that both RDVI and SAVI performed equally well in the case of two-band indices (R^2 = 0.73 and RMSE = 0.38, respectively), outperforming other examples. In addition, optimal VI values for each formulation comprising two bands were constructed with a red edge (720 nm) and a near infrared band (800 nm); results show that EVI was superior to others in terms of LNC estimation in the case of three-band indices yielding an R^2 and RMSE of 0.73 and 0.38, respectively. Data show that all four-band indices exhibited similar LNC estimation efficiency even when encompassing different band combinations, but performed worse than optimal two-band and three-band VI variations. It is also clear that formulation type exerts a significant influence on VI performance even when the same bands are employed. In addition, this modeling method is characterized with an extremely fast speed (within 0.05 s) under the MATLAB. From the scatter plots shown in Figure 2, the saturation at high LNC values still exists despite a relatively high R^2, resulting in low estimation accuracy at high values.

Table 6. Cross-validation statistics and processing speed for the best-performing vegetation index (VI) under each formulation.

	VI	Optimal Bands	R^2	RMSE (%)	Processing Speed (s)
Two-band	RVI	λ1: 700; λ2: 800	0.49	0.52	0.029
	DVI	λ1: 800; λ2: 700	0.67	0.41	0.029
	NDVI	λ1: 800; λ2: 700	0.49	0.52	0.046
	RDVI	**λ1: 800; λ2: 700**	**0.73**	**0.38**	**0.029**
	SAVI	λ1: 800; λ2: 700	0.73	0.38	0.030
	OSAVI	λ1: 800; λ2: 671	0.70	0.40	0.029
	VI_{opt}	λ1: 800; λ2: 671	0.69	0.40	0.029
	MSR	λ1: 700; λ2: 800	0.48	0.52	0.028
Three-band	EVI	λ1: 800; λ2: 700; λ3: 490	0.73	0.38	0.031
	mND	λ1: 800; λ2: 700; λ3: 490	0.69	0.40	0.029
	mSR	λ1: 700; λ2: 490; λ3: 800	0.68	0.41	0.026
	MCARI	λ1: 550; λ2: 700; λ3: 800	0.69	0.41	0.029
	TCARI	λ1: 550; λ2: 700; λ3: 800	0.68	0.41	0.028
	TBI1	λ1: 671; λ2: 700; λ3: 550	0.56	0.48	0.028
	TBI2	λ1: 800; λ2: 490; λ3: 671	0.55	0.49	0.028
Four-band	VOG	λ1: 490; λ2: 700; λ3: 800; λ4: 671	0.70	0.40	0.027
	MTCI	λ1: 671; λ2: 800; λ3: 700; λ4: 490	0.69	0.40	0.027
	TCARI/OSAVI	λ1: 550; λ2: 700; λ3: 800; λ4: 490	0.66	0.42	0.028
	MCARI/OSAVI	λ1: 550; λ2: 700; λ3: 800; λ4: 490	0.66	0.42	0.028

The row in bold type denotes the best-performing VI.

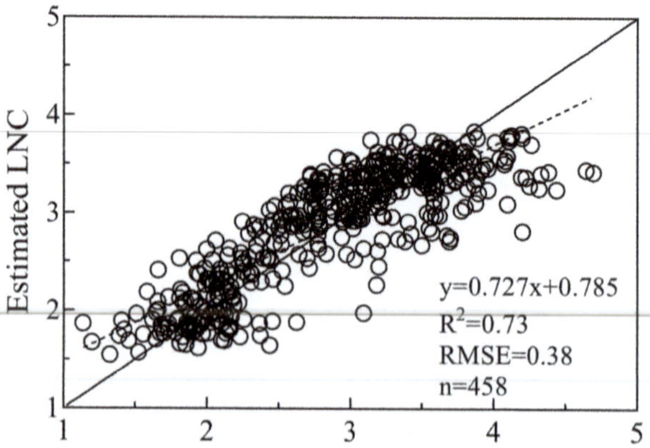

Figure 2. Comparison between measured and estimated LNC values with the best performing VI.

3.2. Optimal Non-Parametric Modeling Algorithm Determination

A total of 13 non-parametric modeling algorithms were utilized in this study to estimate wheat LNC (Table 7); data show that all outperformed optimal VI, with the exception of RT. Indeed, the best-performing regression method was RF, which yielded an R^2 of 0.79 and an RMSE of 0.33 and had a fast processing speed of 2.28 s. In addition, we found that the majority of nonlinear non-parametric modeling algorithms were superior to their linear counterparts. Albeit yielding accurate estimates, RVM, ELM, VH-GPR, and NN approaches all proceeded very slowly. In contrast, the linear non-parametric regression models of LSLR, PCR, and PLSR were all extremely fast, more rapid even than their parametric counterparts.

The data presented in Figure 3a comprise scatter plots of measured LNC values versus estimated ones derived from the optimal non-parametric RF algorithm. In this case, estimated values at the high level turned out to be closer to the 1:1 line than those generated from RDVI. Thus, after measuring the importance of predictor variables using the mean squared error (MSE) [50], it is clear that these values for NIR (800 nm) bands were the largest among the five, followed by the red (671 nm) band (Figure 3b). The red (671 nm) and NIR (800 nm) bands are therefore more important for LNC estimation than any of their counterparts.

Table 7. Performance of different non-parametric modeling algorithms in LNC estimation ranked according to RMSE values.

Non-Parametric Algorithm	R^2	RMSE (%)	Processing Speed (s)
Random Forest (RF)	0.79	0.33	2.284
Bagging Trees (BaT)	0.78	0.34	2.700
Kernel Ridge Regression (KRR)	0.78	0.35	1.934
Neural Network (NN)	0.77	0.35	10.406
VH Gaussian Process Regression (VH-GPR)	0.77	0.35	17.059
Gaussian Process Regression (GPR)	0.77	0.35	1.265
Extreme Learning Machine (ELM)	0.76	0.36	20.068
Least-Squares Linear Regression (LSLR)	0.75	0.36	0.007
Boosting Trees (BoT)	0.75	0.37	2.301
Relevance Vector Machine (RVM)	0.75	0.37	268.473
Partial Least-Squares Regression (PLSR)	0.74	0.37	0.016
Principal Component Regression (PCR)	0.73	0.38	0.009
Regression Trees (RT)	0.69	0.40	0.616

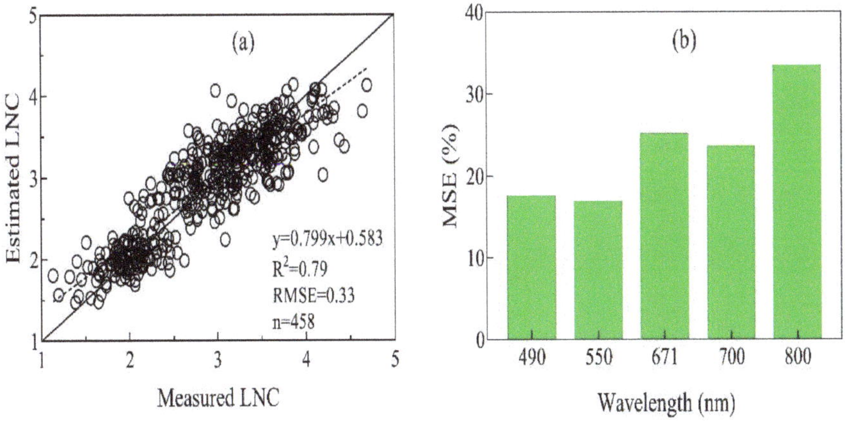

Figure 3. Comparison of measured and estimated LNC values derived using the RF modeling algorithm (**a**) and MSE values for this model at different spectral bands (**b**).

3.3. Performance of LUT-Based PROSAIL Inversion Performance

The data presented in Table 8 illustrate the performance of the LUT-based PROSAIL model with different cost functions, noise proportions, and multiple solutions. In this case, however, as LNC could not be retrieved directly from the PROSAIL model, LCC was initially estimated. The optimal inversion strategy for LCC retrieval used in this study was $K(x) = \log(x)^2$ with R^2 and RMSE values of 0.81 and 7.05, respectively. LNC was then indirectly estimated subsequent to LCC inversion via the relationship between LCC and LNC (Figure 4a). The PROSAIL model performance in LNC inversion was not particularly satisfactory with an R^2 value of 0.62 (Figure 4b); thus, compared to both VIs and non-parameter modeling methods, the LUT-based PROSAIL approach actually performed worse although the processing speed in this case was comparable with those of non-linear non-parametric regression algorithms.

Table 8. Performance of different regularization strategies used in the PROSAIL model ranked according to RMSE values.

Cost Function	Noise (%)	Multiple Solutions (%)	R^2	RMSE ($\mu g/cm^2$)	Processing Speed (s)
$K(x) = \log(x)^2$	29	9	0.81	7.05	2.04
$K(x) = x(\log(x)) - x$	41	41.5	0.75	8.24	1.85
Neyman chi-square	37	10.5	0.74	8.74	1.86
W Kagan	37	10.5	0.74	8.74	1.85
Kullback-Leibler	45	11.5	0.81	8.98	1.92
Jeffreys-Kullback-Leibler	45	19.5	0.80	9.17	1.76
Bhattacharyya divergence	45	19.5	0.81	9.26	2.03
Pearson chi-square	50	43	0.78	9.33	1.85
L-divergence Lin	47	20.5	0.81	9.35	2.16
Shannon (1948)	47	20.5	0.81	9.35	1.98
Shannon entropy	50	21.5	0.81	9.45	1.82
Harmonique toussaint	50	21	0.81	9.50	1.85
K-divergence Lin	50	30.5	0.80	9.54	1.96
Negative exponential disparity	48	20.5	0.79	9.65	1.92
Exponential	50	48	0.59	11.84	1.98
Normal distribution-LSE	50	50	0.47	13.10	1.74
Geman and McClure	50	50	0.46	13.16	1.79
$K(x) = -\log(x) + x$	39	50	0.79	13.19	1.98
Least absolute error	50	50	0.34	15.16	1.75
$K(x) = \log(x) + 1/x$	50	50	0.07	17.61	1.96

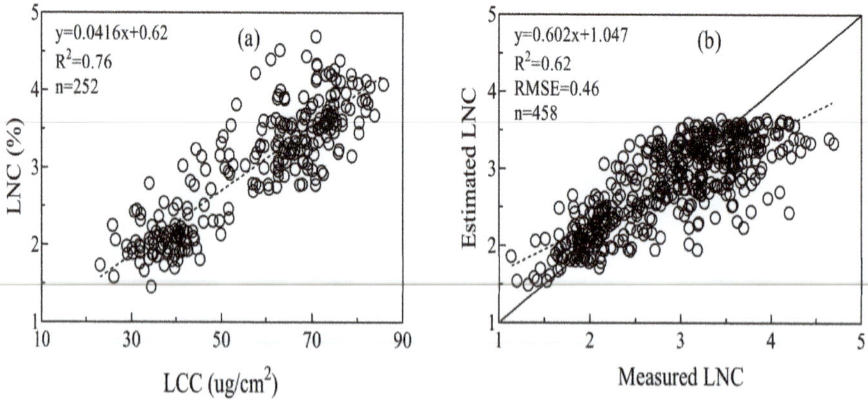

Figure 4. Empirical linear relationship between LCC and LNC values (**a**). Measured versus estimated LNC values derived from the most effective LUT-based inversion scheme (Table 8) (**b**).

3.4. Effects of Growth Stage, Cultivar, and Cultivation Factors on Estimation Accuracy

The data presented in Table 9 summarize the effects of growth stage, cultivar, planting density, and year on the estimation accuracy of different methods. These records show that for different growth stages, both RDVI and LUT-based methods performed better in the middle of the season (i.e., booting, heading, and anthesis) compared to either early (i.e., jointing) or late (filling) stages. An RF approach was able to obtain accurate estimates from jointing to anthesis stages alongside lower ones at the filling stage.

The results of this study reveal varied RDVI performance depending on the wheat cultivar; the most accurate estimates were recovered for Ningmai 13 (RRMSE = 10.4%) while the worst were seen for Shenxuan 6 (RRMSE = 14.0%). The RF approach also generated satisfactory and stable values for different cultivars with RRMSE ranging between 10.7% and 12.0%, while the LUT-based retrieval method also performed equally in all cases.

As planting density increased, LNC estimation accuracy gradually decreased based on RDVI and the best performance was obtained at the lowest density. At the same time, the LUT-based retrieval method yielded highest accuracies at the lowest density while the RF approach led to comparable performance at different planting densities. All three methods performed better for 2014 than for 2015.

Table 9. Relative RMSE (RRMSE, %) values for different wheat LNC estimation methods under different conditions.

Sub-Group	Treatment	Different Modeling Algorithms		
		RDVI	RF	LUT
Growth stage	Jointing	16.0	11.4	16.53
	Booting	8.8	8.8	12.60
	Heading	10.0	9.9	12.80
	Anthesis	11.7	11.7	14.03
	Filling	17.9	16.2	22.92
Variety	Yangmai 18	13.1	11.3	16.34
	Shengxuan 6	14.0	12.0	16.43
	Xumai 30	13.4	11.9	16.51
	Ningmai 13	10.4	10.7	15.41
Plant density	1.5×10^6 plants/ha	12.1	12.1	13.41
	2.4×10^6 plants/ha	12.4	11.7	16.30
	3×10^6 plants/ha	14.4	11.1	16.34
Year	2014	12.0	11.2	0.14
	2015	14.6	12.2	0.18

4. Discussion

Although ground-based spectral data and satellite images have been widely utilized to monitor the N status of crops [9,16,51], few studies to date have assessed the capabilities of UAV platforms. We evaluated the performance of UAV images using different modeling algorithms and demonstrate that this approach provides a reliable technique for winter wheat leaf N content estimation.

The results of this analysis show that in terms of parametric approaches, use of an RDVI modified with NIR and red edge bands provides optimal VI values for LNC estimation (i.e., R^2 = 0.73; RMSE = 0.38); this result is in close agreement with the previous findings of Inoue et al. [20] and Yao et al. [52], who noted that a combination of NIR and red edge bands provides an efficient approach for N status monitoring. The RDVI is also advantageous because it optimizes the vegetation signal and therefore has an improved degree of sensitivity in high-biomass regions; this approach is able to enhance vegetation monitoring via decoupling of the canopy background signal and reducing atmospheric influence [38].

However, even though results of sufficient accuracy were obtained in this analysis using a simple model, a number of drawbacks remain, including the fact that this approach becomes saturated at high N rates and canopy densities; it is easily affected by the growth stage, and information is lost at other spectral bands. Indeed, the RDVI performed poorly at both jointing and filling stages (Table 8), a result that might be explained by the fact that the canopy was mixed with soil background during the early stage and then panicles later in development. Furthermore, the accuracy of estimation decreased from the booting to filling stage, which might be due to the differences of the leaf biomass at varied stages [53]. The use of the VI incorporating more bands was also unable to generate higher accuracy than a two-band approach; furthermore, different formulas with the same bands performed significantly in LNC estimation, which indicated that both band configuration and VI formulation played an important role in LNC estimation. It is also crucial to consider the applicability of VI-LNC models as the performance of these approaches often depends on the ecological site, crop type, and growth stage [54]. The RDVI-LNC model should therefore be tested using additional datasets so as to extend its capability in the future.

It is well known that vegetation canopy spectral signatures are dominated by numerous biophysical and biochemical variables [55,56]. Thus, compared with parametric methods, most non-parametric algorithms tend to perform better because this regression family makes full use of all spectral information and so are able to better handle confounding factors when compared to VI values [20,22]. Although linear non-parametric algorithms performed lightly worse than their nonlinear counterparts in this analysis, these approaches possessed an extremely fast processing speed; this attribute indicates that these methods comprise a promising technique that can be integrated into crop monitoring systems.

Previous studies have also shown that linear non-parametric algorithms, such as PLSR, are able to generate satisfactory estimates for crop biomass [3] as well as N [20] and chlorophyll content [22]. The results of this study show that amongst non-parametric algorithms, the RF approach was both the most accurate and stable method under different conditions because RF provides a nonlinear regression with LNC and has the advantage of dealing with a large dataset with high speed and efficiency [50,57]. Furthermore, RF also has the ability to rank the importance of variables [50,57]. We therefore recommend that the RF approach would be a reliable technique for crop N estimation, even though many software packages do not yet include this algorithm.

The LUT-based retrieval method used in this study had the lowest LNC estimation accuracy of the three approaches tried, in contrast to previous research results [27,58]. Indeed, as some variables (e.g., LAI and chlorophyll content) could be retrieved directly from the PROSAIL model while LNC was generated indirectly from the empirical relationship between LNC and LCC [59,60], estimation accuracy was influenced by retrieval equation accuracy. We also note that the LUT-based retrieval method has a number of drawbacks, including the need for too many input parameters, large data size, and long processing times, and the fact that only parameters inherent to the model can be retrieved. However, a physical model has the advantage of offering uncertainty estimates, which provide information on model transplantation possibilities.

Although previous studies have attempted to employ UAV-based images to monitor crop N status [14,15], the datasets used was small and so estimation accuracy was unsatisfactory. In contrast, the results of this study show that the RDVI generated higher estimates while the non-parametric RF regression method led to a higher degree of accuracy under different conditions. These results suggest that UAV-based multispectral images provide a promising approach that can be applied to crop N status monitoring. However, even though a high predictive accuracy was obtained in this study, the established LNC model will still need to be tested with data from other ecological sites and crop types as the variables used here came from just one site. We also show that the PROSAIL model is not suitable for LNC retrieval because of its low predictive accuracy unless the relationship between this variable and LNC can be made more robust.

5. Conclusions

A range of modeling algorithms (i.e., parametric, non-parametric, and physical retrieval) were employed in this study to estimate winter wheat LNC using UAV-based multispectral images. Estimation models were then cross-validated with datasets from different growing seasons, including different stages, cultivars, N rates, and planting densities. In terms of parametric regressions, modified RDVI with a red edge and NIR bands turned out to comprise the best-performing index with the most accurate cross-validated result (i.e., R^2 = 0.73, RMSE = 0.38). This method was also characterized by an extremely high processing speed and a saturation effect at high LNC levels. In terms of non-parametric regression approaches, we showed that the RF method comprised the best-performing algorithm (i.e., R^2 = 0.79, RMSE = 0.33), also with a fast processing speed. The use of a physical retrieval method remains challenging for LNC estimations because of undeterminable input variables and low prediction accuracies.

Author Contributions: X.Y. and Y.Z. conceived and designed the experiments; W.L., J.J., Y.L. and Y.Z. performed the experiments; Y.L. analyzed the data; H.Z. and X.Y. wrote the paper. All authors contributed to the interpretation of results and editing of the manuscript.

Funding: This research was funded by the National Key Research and Development Program of China (2016YFD0300601), the National Natural Science Foundation of China (31671582), Jiangsu Qinglan Project, the 111 project (B16026), Jiangsu Collaborative Innovation Center for Modern Crop Production (JCICMCP), the Priority Academic Program Development of Jiangsu Higher Education Institutions (PAPD), Qinghai Project of Transformation of Scientific and Technological Achievements (2018-NK-126), and Jiangsu Province Key Technologies R&D Program (BE2016375).

Acknowledgments: The authors would like to thank all reviewers and editors for their comments on this paper.

Conflicts of Interest: The authors declare no conflict of interest.

Abbreviations

UAV	unmanned aerial vehicle
RS	remote sensing
LNC	leaf nitrogen content
LAI	leaf area index
LCC	leaf chlorophyll content
SPAD	soil and plant analyzer development
RVI	ratio vegetation index
DVI	difference vegetation index
NDVI	normalized difference vegetation index
RDVI	renormalized difference vegetation index
SAVI	soil adjusted vegetation index
OSAVI	optimized soil adjusted vegetation index
VI_{opt}	optimized vegetation index
MSR	modified sample ratio
EVI	enhanced vegetation index
MCARI	modified chlorophyll absorption in reflectance index
TCARI	transformed chlorophyll absorption in reflectance index
TBI	three-band index
VOG	Vogelmann index
MTCI	MERIS terrestrial chlorophyll index
LSLR	least-squares linear
PCR	principal component
PLSR	partial least-squares regression
ANN	artificial neutral networks
DT	decision trees
RT	regression trees
BaT	bagging trees
BoT	boosting trees
RF	random forest
RVM	relevance vector machine
KRR	kernel ridge
GPR	Gaussian processes regressions
VH-GPR	variational heteroscedastic GPR
ELM	extreme learning machines
RTM	radiative transfer model
LUT	look-up-table
R^2	determination coefficient
RMSE	root mean square error
RRMSE	relative root mean square error
ILS	incident light sensor
GCP	ground control point
ROI	region of interest

References

1. Hatfield, J.L.; Gitelson, A.A.; Schepers, J.S.; Walthall, C.L. Application of spectral remote sensing for agronomic decisions. *Agron. J.* **2008**, *100*, 117–131. [CrossRef]
2. Ju, X.T.; Xing, G.X.; Chen, X.P.; Zhang, S.L.; Zhang, L.J.; Liu, X.J.; Cui, Z.L.; Yin, B.; Christie, P.; Zhu, Z.L.; et al. Reducing environmental risk by improving N management in intensive Chinese agricultural systems. *Proc. Natl. Acad. Sci. USA* **2009**, *106*, 3041–3046. [CrossRef] [PubMed]
3. Hansen, P.M.; Schjoerring, J.K. Reflectance measurement of canopy biomass and nitrogen status in wheat crops using normalized difference vegetation indices and partial least squares regression. *Remote Sens. Environ.* **2003**, *86*, 542–553. [CrossRef]
4. Tan, C.; Guo, W.; Wang, J. Predicting grain protein content of winter wheat based on Landsat TM images and leaf nitrogen content. In Proceedings of the International Conference on Remote Sensing, Environment and Transportation Engineering, Nanjing, China, 24–26 June 2011.
5. Eitel, J.U.H.; Long, D.S.; Gessler, P.E.; Smith, A.M.S. Using in-situ measurements to evaluate the new RapidEye™ satellite series for prediction of wheat nitrogen status. *Int. J. Remote Sens.* **2007**, *28*, 4183–4190. [CrossRef]
6. Huang, S.; Miao, Y.; Yuan, F.; Gnyp, M.L.; Yao, Y.; Cao, Q.; Wang, H.; Lenz-Wiedemann, V.I.; Bareth, G. Potential of RapidEye and WorldView-2 satellite data for improving rice nitrogen status monitoring at different growth stages. *Remote Sens.* **2017**, *9*, 227. [CrossRef]
7. Boegh, E.; Soegaard, H.; Broge, N.; Hasager, C.B.; Jensen, N.O.; Schelde, K.; Thomsen, A. Airborne multispectral data for quantifying leaf area index, nitrogen concentration, and photosynthetic efficiency in agriculture. *Remote Sens. Environ.* **2002**, *81*, 179–193. [CrossRef]
8. Tian, Y.C.; Yao, X.; Yang, J.; Cao, W.X.; Hannaway, D.B.; Zhu, Y. Assessing newly developed and published vegetation indices for estimating rice leaf nitrogen concentration with ground-and space-based hyperspectral reflectance. *Field Crops Res.* **2011**, *120*, 299–310. [CrossRef]
9. Yao, X.; Ren, H.; Cao, Z.; Tian, Y.; Cao, W.; Zhu, Y.; Cheng, T. Detecting leaf nitrogen content in wheat with canopy hyperspectrum under different soil backgrounds. *Int. J. Appl. Earth Obs. Geoinf.* **2014**, *32*, 114–124. [CrossRef]
10. Bendig, J.; Yu, K.; Aasen, H.; Bolten, A.; Bennertz, S.; Broscheit, J.; Gnyp, M.L.; Bareth, G. Combining UAV-based plant height from crop surface models, visible, and near infrared vegetation indices for biomass monitoring in barley. *Int. J. Appl. Earth Obs. Geoinf.* **2015**, *39*, 79–87. [CrossRef]
11. Zheng, H.; Cheng, T.; Li, D.; Zhou, X.; Yao, X.; Tian, Y.; Cao, W.; Zhu, Y. Evaluation of RGB, color-infrared and multispectral images acquired from unmanned aerial systems for the estimation of nitrogen accumulation in rice. *Remote Sens.* **2018**, *10*, 824. [CrossRef]
12. Yao, X.; Wang, N.; Liu, Y.; Cheng, T.; Tian, Y.; Chen, Q.; Zhu, Y. Estimation of wheat LAI at middle to high levels using unmanned aerial vehicle narrowband multispectral imagery. *Remote Sens.* **2017**, *9*, 1304. [CrossRef]
13. Yue, J.; Yang, G.; Li, C.; Li, Z.; Wang, Y.; Feng, H.; Xu, B. Estimation of winter wheat above-ground biomass using unmanned aerial vehicle-based snapshot hyperspectral sensor and crop height improved models. *Remote Sens.* **2017**, *9*, 708. [CrossRef]
14. Gnyp, M.L.; Panitzki, M.; Reusch, S.; Jasper, J.; Bolten, A.; Bareth, G. Comparison between tractor -based and UAV-based spectrometer measurements in winter wheat. In Proceedings of the 13th International Conference on Precision Agriculture, Monticello, IL, USA, 31 July–3 August 2016.
15. Li, J.; Zhang, F.; Qian, X.; Zhu, Y.; Shen, G. Quantification of rice canopy nitrogen balance index with digital imagery from unmanned aerial vehicle. *Remote Sens. Lett.* **2015**, *6*, 183–189. [CrossRef]
16. Stroppiana, D.; Boschetti, M.; Brivio, P.A.; Bocchi, S. Plant nitrogen concentration in paddy rice from field canopy hyperspectral radiometry. *Field Crops Res.* **2009**, *111*, 119–129. [CrossRef]
17. Wang, W.; Yao, X.; Yao, X.; Tian, Y.; Liu, X.; Ni, J.; Cao, W.; Zhu, Y. Estimating leaf nitrogen concentration with three-band vegetation indices in rice and wheat. *Field Crops Res.* **2012**, *129*, 90–98. [CrossRef]
18. Rivera, J.P.; Verrelst, J.; Delegido, J.; Veroustraete, F.; Moreno, J. On the semi-automatic retrieval of biophysical parameters based on spectral index optimization. *Remote Sens.* **2014**, *6*, 4927–4951. [CrossRef]
19. Zhu, Y.; Tian, Y.; Yao, X.; Liu, X.; Cao, W. Analysis of common canopy reflectance spectra for indicating leaf nitrogen concentrations in wheat and rice. *Plant Prod. Sci.* **2007**, *10*, 400–411. [CrossRef]

20. Inoue, Y.; Sakaiya, E.; Zhu, Y.; Takahashi, W. Diagnostic mapping of canopy nitrogen content in rice based on hyperspectral measurements. *Remote Sens. Environ.* **2012**, *126*, 210–221. [CrossRef]
21. Yao, X.; Huang, Y.; Shang, G.; Zhou, C.; Cheng, T.; Tian, Y.; Cao, W.; Zhu, Y. Evaluation of six algorithms to monitor wheat leaf nitrogen concentration. *Remote Sens.* **2015**, *7*, 14939–14966. [CrossRef]
22. Atzberger, C.; Guérif, M.; Baret, F.; Werner, W. Comparative analysis of three chemometric techniques for the spectroradiometric assessment of canopy chlorophyll content in winter wheat. *Comput. Electron. Agric.* **2010**, *73*, 165–173. [CrossRef]
23. Verrelst, J.; Muñoz, J.; Alonso, L.; Delegido, J.; Rivera, J.P.; Camps-Valls, G.; Moreno, J. Machine learning regression algorithms for biophysical parameter retrieval: Opportunities for Sentinel-2 and-3. *Remote Sens. Environ.* **2012**, *118*, 127–139. [CrossRef]
24. Jacquemoud, S.; Baret, F.; Andrieu, B.; Danson, F.M.; Jaggard, K. Extraction of vegetation biophysical parameters by inversion of the PROSPECT + SAIL models on sugar beet canopy reflectance data. Application to TM and AVIRIS sensors. *Remote Sens. Environ.* **1995**, *52*, 163–172. [CrossRef]
25. Liang, L.; Di, L.; Zhang, L.; Deng, M.; Qin, Z.; Zhao, S.; Lin, H. Estimation of crop LAI using hyperspectral vegetation indices and a hybrid inversion method. *Remote Sens. Environ.* **2015**, *165*, 123–134. [CrossRef]
26. Yang, G.; Zhao, C.; Xing, Z.; Huang, W.; Wang, J. LAI Inversion of Spring Wheat Based on PROBA/CHRIS Hyperspectral Multi-Angular Data and PROSAIL Model. Available online: http://xueshu.baidu.com/usercenter/paper/show?paperid=7ff0cdf37366c37dd6406bf9aa80a99f&site=xueshu_se&hitarticle=1 (accessed on 11 December 2018).
27. Botha, E.J.; Leblon, B.; Zebarth, B.; Watmough, J. Non-destructive estimation of potato leaf chlorophyll from canopy hyperspectral reflectance using the inverted PROSAIL model. *Int. J. Appl. Earth Obs. Geoinf.* **2007**, *9*, 360–374. [CrossRef]
28. Uddling, J.; Gelang-Alfredsson, J.; Piikki, K.; Pleijel, H. Evaluating the relationship between leaf chlorophyll concentration and SPAD-502 chlorophyll meter readings. *Photosynth. Res.* **2007**, *91*, 37–46. [CrossRef]
29. Kelcey, J.; Lucieer, A. Sensor correction of a 6-band multispectral imaging sensor for UAV remote sensing. *Remote Sens.* **2012**, *4*, 1462–1493. [CrossRef]
30. Smith, G.M.; Milton, E.J. The use of the empirical line method to calibrate remotely sensed data to reflectance. *Int. J. Remote Sens.* **1999**, *20*, 2653–2662. [CrossRef]
31. Jordan, C.F. Derivation of leaf area index from quality of light on the forest floor. *Ecology* **1969**, *50*, 663–666. [CrossRef]
32. Tucker, C.J. Red and photographic infrared linear combinations for monitoring vegetation. *Remote Sens. Environ.* **1979**, *8*, 127–150. [CrossRef]
33. Roujean, J.L.; Breon, F.M. Estimating PAR absorbed by vegetation from bidirectional reflectance measurements. *Remote Sens. Environ.* **1995**, *51*, 375–384. [CrossRef]
34. Huete, A.R. A soil-adjusted vegetation index (SAVI). *Remote Sens. Environ.* **1988**, *25*, 295–309. [CrossRef]
35. Rondeaux, G.; Steven, M.; Baret, F. Optimization of soil-adjusted vegetation indices. *Remote Sens. Environ.* **1996**, *55*, 95–107. [CrossRef]
36. Reyniers, M.; Walvoort, D.J.; De Baardemaaker, J. A linear model to predict with a multi-spectral radiometer the amount of nitrogen in winter wheat. *Int. J. Remote Sens.* **2006**, *27*, 4159–4179. [CrossRef]
37. Chen, J.M. Evaluation of vegetation indices and a modified simple ratio for boreal applications. *Can. J. Remote Sens.* **1996**, *22*, 229–242. [CrossRef]
38. Huete, A.; Didan, K.; Miura, T.; Rodriguez, E.P.; Gao, X.; Ferreira, L.G. Overview of the radiometric and biophysical performance of the MODIS vegetation indices. *Remote Sens. Environ.* **2002**, *83*, 195–213. [CrossRef]
39. Sims, D.A.; Gamon, J.A. Relationships between leaf pigment content and spectral reflectance across a wide range of species, leaf structures and developmental stages. *Remote Sens. Environ.* **2002**, *81*, 337–354. [CrossRef]
40. Daughtry, C.S.T.; Walthall, C.L.; Kim, M.S.; De Colstoun, E.B.; McMurtrey Iii, J.E. Estimating corn leaf chlorophyll concentration from leaf and canopy reflectance. *Remote Sens. Environ.* **2000**, *74*, 229–239. [CrossRef]
41. Haboudane, D.; Miller, J.R.; Tremblay, N.; Zarco-Tejada, P.J.; Dextraze, L. Integrated narrow-band vegetation indices for prediction of crop chlorophyll content for application to precision agriculture. *Remote Sens. Environ.* **2002**, *81*, 416–426. [CrossRef]

42. Tian, Y.C.; Gu, K.J.; Chu, X.; Yao, X.; Cao, W.X.; Zhu, Y. Comparison of different hyperspectral vegetation indices for canopy leaf nitrogen concentration estimation in rice. *Plant Soil* **2014**, *376*, 193–209. [CrossRef]
43. Zarco-Tejada, P.J.; Miller, J.R.; Noland, T.L.; Mohammed, G.H.; Sampson, P.H. Scaling-up and model inversion methods with narrowband optical indices for chlorophyll content estimation in closed forest canopies with hyperspectral data. *IEEE Trans. Geosci. Remote Sens.* **2001**, *39*, 1491–1507. [CrossRef]
44. Dash, J.; Curran, P.J. The MERIS terrestrial chlorophyll index. *Int. J. Remote Sens.* **2004**, *25*, 5403–5413. [CrossRef]
45. Camps-Valls, G.; Gómez-Chova, L.; Muñoz-Marí, J.; Lázaro-Gredilla, M.; Verrelst, J. simpleR: A Simple Educational Matlab Toolbox for Statistical Regression. In: V2. Available online: https://www.uv.es/gcamps/software.html (accessed on 10 December 2018).
46. Verrelst, J.; Rivera, J.P.; Veroustraete, F.; Muñoz-Marí, J.; Clevers, J.G.; Camps-Valls, G.; Moreno, J. Experimental Sentinel-2 LAI estimation using parametric, non-parametric and physical retrieval methods-A comparison. *ISPRS J. Photogramm. Remote Sens.* **2015**, *108*, 260–272. [CrossRef]
47. Zhang, L.; Guo, C.L.; Zhao, L.Y.; Zhu, Y.; Cao, W.X.; Tian, Y.C.; Cheng, T.; Wang, X. Estimating wheat yield by integrating the WheatGrow and PROSAIL models. *Field Crops Res.* **2016**, *192*, 55–66. [CrossRef]
48. Li, H.; Liu, G.; Liu, Q.; Chen, Z.; Huang, C. Retrieval of winter wheat leaf area index from Chinese GF-1 satellite data using the PROSAIL model. *Sensors* **2018**, *18*, 1120. [CrossRef] [PubMed]
49. Rivera, J.P.; Verrelst, J.; Leonenko, G.; Moreno, J. Multiple cost functions and regularization options for improved retrieval of leaf chlorophyll content and LAI through inversion of the PROSAIL model. *Remote Sens.* **2013**, *5*, 3280–3304. [CrossRef]
50. Breiman, L. Random Forests. *Mach. Learn.* **2001**, *45*, 5–32. [CrossRef]
51. Huang, S.; Miao, Y.; Zhao, G.; Yuan, F.; Ma, X.; Tan, C.; Yu, W.; Gnyp, M.L.; Lenz-Wiedemann, V.I.; Rascher, U.; et al. Satellite remote sensing-based in-season diagnosis of rice nitrogen status in Northeast China. *Remote Sens.* **2015**, *7*, 10646–10667. [CrossRef]
52. Yao, X.; Zhu, Y.; Tian, Y.; Feng, W.; Cao, W. Exploring hyperspectral bands and estimation indices for leaf nitrogen accumulation in wheat. *Int. J. Appl. Earth Obs. Geoinf.* **2010**, *12*, 89–100. [CrossRef]
53. Li, F.; Gnyp, M.L.; Jia, L.; Miao, Y.; Yu, Z.; Koppe, W.; Zhang, F. Estimating N status of winter wheat using a handheld spectrometer in the North China Plain. *Field Crops Res.* **2008**, *106*, 77–85. [CrossRef]
54. Nigam, R.; Bhattacharya, B.K.; Vyas, S.; Oza, M.P. Retrieval of wheat leaf area index from AWiFS multispectral data using canopy radiative transfer simulation. *Int. J. Appl. Earth Obs. Geoinf.* **2014**, *32*, 173–185. [CrossRef]
55. Asner, G.P. Biophysical and biochemical sources of variability in canopy reflectance. *Remote Sens. Environ.* **1998**, *64*, 234–253. [CrossRef]
56. Ustin, S.L. Remote sensing of canopy chemistry. *Proc. Natl. Acad. Sci. USA* **2013**, *10*, 804–805. [CrossRef] [PubMed]
57. Belgiu, M.; Dragut, L. Random forest in remote sensing: A review of applications and future directions. *ISPRS J. Photogramm. Remote Sens.* **2016**, *114*, 24–31. [CrossRef]
58. Jiang, J.; Comar, A.; Burger, P.; Bancal, P.; Weiss, M.; Baret, F. Estimation of leaf traits from reflectance measurements: Comparison between methods based on vegetation indices and several versions of the PROSPECT model. *Plant Methods* **2018**, *14*, 23. [CrossRef] [PubMed]
59. Cammarano, D.; Fitzgerald, G.; Basso, B.; O'Leary, G.; Chen, D.; Grace, P.; Fiorentino, C. Use of the Canopy Chlorophyl Content Index (CCCI) for remote estimation of wheat nitrogen content in rainfed environments. *Agron. J.* **2011**, *103*, 1597–1603. [CrossRef]
60. Zhao, C.; Wang, Z.; Wang, J.; Huang, W.; Guo, T. Early detection of canopy nitrogen deficiency in winter wheat (Triticum aestivum L.) based on hyperspectral measurement of canopy chlorophyll status. *N. Z. J. Crop Hortic. Sci.* **2011**, *39*, 251–262. [CrossRef]

remote sensing

Article

Prediction of Early Season Nitrogen Uptake in Maize Using High-Resolution Aerial Hyperspectral Imagery

Tyler J. Nigon [1,2], Ce Yang [2,*], Gabriel Dias Paiao [1], David J. Mulla [1], Joseph F. Knight [3] and Fabián G. Fernández [1]

1 Department of Soil, Water and Climate, University of Minnesota, Saint Paul, MN 55108, USA; nigo0024@umn.edu (T.J.N.); gdiaspai@umn.edu (G.D.P.); mulla003@umn.edu (D.J.M.); fabiangf@umn.edu (F.G.F.)

2 Department of Bioproducts and Biosystems Engineering, University of Minnesota, Saint Paul, MN 55108, USA

3 Department of Forest Resources, University of Minnesota, Saint Paul, MN 55108, USA; jknight@umn.edu

* Correspondence: ceyang@umn.edu; Tel.: +1-612-626-6419

Received: 11 March 2020; Accepted: 9 April 2020; Published: 12 April 2020

Abstract: The ability to predict spatially explicit nitrogen uptake (NUP) in maize (*Zea mays* L.) during the early development stages provides clear value for making in-season nitrogen fertilizer applications that can improve NUP efficiency and reduce the risk of nitrogen loss to the environment. Aerial hyperspectral imaging is an attractive agronomic research tool for its ability to capture spectral data over relatively large areas, enabling its use for predicting NUP at the field scale. The overarching goal of this work was to use supervised learning regression algorithms—Lasso, support vector regression (SVR), random forest, and partial least squares regression (PLSR)—to predict early season (i.e., V6–V14) maize NUP at three experimental sites in Minnesota using high-resolution hyperspectral imagery. In addition to the spectral features offered by hyperspectral imaging, the 10th percentile Modified Chlorophyll Absorption Ratio Index Improved (MCARI2) was made available to the learning models as an auxiliary feature to assess its ability to improve NUP prediction accuracy. The trained models demonstrated robustness by maintaining satisfactory prediction accuracy across locations, pixel sizes, development stages, and a broad range of NUP values (4.8 to 182 kg ha^{-1}). Using the four most informative spectral features in addition to the auxiliary feature, the mean absolute error (MAE) of Lasso, SVR, and PLSR models (9.4, 9.7, and 9.5 kg ha^{-1}, respectively) was lower than that of random forest (11.2 kg ha^{-1}). The relative MAE for the Lasso, SVR, PLSR, and random forest models was 16.5%, 17.0%, 16.6%, and 19.6%, respectively. The inclusion of the auxiliary feature not only improved overall prediction accuracy by 1.6 kg ha^{-1} (14%) across all models, but it also reduced the number of input features required to reach optimal performance. The variance of predicted NUP increased as the measured NUP increased (MAE of the Lasso model increased from 4.0 to 12.1 kg ha^{-1} for measured NUP less than 25 kg ha^{-1} and greater than 100 kg ha^{-1}, respectively). The most influential spectral features were oftentimes adjacent to each other (i.e., within approximately 6 nm), indicating the importance of both spectral precision and derivative spectra around key wavelengths for explaining NUP. Finally, several challenges and opportunities are discussed regarding the use of these results in the context of improving nitrogen fertilizer management.

Keywords: cross-validation; feature selection; hyperparameter tuning; image processing; image segmentation; nitrogen fertilizer recommendation; supervised regression

1. Introduction

Nitrogen (N) fertilizer inputs are crucial for achieving high crop yields, but the loss of reactive N from agricultural systems leads to atmospheric, surface water, and groundwater pollution [1–3], ultimately diminishing environmental quality and human well-being [4,5]. Despite the potential environmental consequences to society, the pressure on producers to increase productivity oftentimes leads to N fertilizer applications in excess of crop requirement [6,7]. Without more efficient N fertilizer applications, the increasing global population and subsequent rising demand for food are expected to cause an increase in the loss of reactive N in the future [8].

A strategy to reduce the likelihood of reactive N loss is to apply part of the crop's N requirement after emergence, delaying the application until crop demand is near its maximum. Grain yields and crop N use are not uniform across seasons [9], so this delayed application also provides the opportunity to adapt N fertilizer rates in a dynamic manner based on the influence of weather and N cycle processes on early season crop growth or stress. Maize N requirement typically varies spatially [10], so this strategy can be more effective with variable rate/site-specific applications [11].

The fundamental barrier to our ability of making reliable site-specific N fertilizer recommendations is the lack of understanding and inability to accurately quantify soil N supply and how it is expected to change throughout the season [12]. This is particularly challenging because both natural N supply (i.e., non-fertilizer) and crop N requirement are dynamic in both space and time and are difficult to predict [13]. Determination of early season crop N uptake (NUP) can be helpful for making N fertilizer recommendations due to its connection with natural N supply (e.g., mineralization) and crop N requirement. Remote sensing offers the opportunity to capture near real-time information about crop N status [14], and it can be an efficient way to assess the spatial variability across an entire field or farm. With the availability of robust, reliable unmanned aircraft and low payload hyperspectral line-scanning imagers in recent years, there is now the opportunity to use high-resolution, aerial hyperspectral imagery to predict early season NUP in maize. As a research tool, aerial hyperspectral imaging is attractive for its ability to scale, especially compared to other methods that use ground-based hyperspectral point measurements for N fertilizer management [15].

There are tradeoffs in pixel size related to the approach for which remote sensing imagery is captured. It is typically preferred to capture images with finer spatial resolution, but this comes with a tradeoff of larger data storage and processing requirements, as well as longer acquisition time and battery/fuel requirement for a given spatial extent. Academic research is commonly conducted on small plot areas that do not cover a large extent, and a low flight altitude (e.g., approximately 20 m) may be practical for covering the extent of the experimental area. However, this is rarely suitable at the field scale (including on-farm research) because of the large areas that must be imaged. This leads to a likely dilemma where much of the data used for research and development is inherently different from the data used by practitioners (e.g., the inherent pixel size differs). Therefore, it is important to develop prediction models that are either scale-agnostic or that are specifically tailored to the conditions for which practitioners will be able to practically implement them.

The objectives of this work were to: (i) compare supervised learning techniques for their ability to predict maize NUP at the early- to mid-vegetative development stages (i.e., V6 to V14) using spectral features from high-resolution aerial hyperspectral imagery, (ii) quantify the potential model improvement by including an auxiliary feature derived before the segmentation process, and (iii) evaluate how the difference in pixel size at image capture affects prediction accuracy based on the inherent ability to segment pixels most influenced by soil and/or shadow.

2. Materials and Methods

2.1. Field Experiments

Data from three experiments in southern Minnesota were used to evaluate the objectives of this study. The Wells experiment (43.85437, −93.72077) was conducted near Wells, Minnesota in 2018, and the two Waseca experiments were conducted at the Agroecology Research Farm (44.063635, −93.540281) near Waseca, Minnesota in 2019 (hereby referred to as the Waseca "small plot" and "whole field" experiments). Weed, pest, and disease management were carried out by farm managers using approaches typical for the areas. The N fertilizer rate across all experiments ranged from 0 to 231 kg ha^{-1}. Experimental treatments included conventional and sidedress N applications, whereby the conventional treatments had all N applied around the time of planting, and the sidedress treatments had 45 kg ha^{-1} N applied at planting and the remainder applied as sidedress close to the V8 development stage.

2.1.1. Wells Experiment

The Wells, Minnesota dataset was acquired from a broader experiment that evaluated the effect of tillage (conventional tillage, strip-till, and no-till) and tile drainage (drained and undrained) on N response and contained 144 experimental plots across four replications. The previous crop (i.e., in 2017) was soybeans, and the Marna silty clay loam (fine, smectitic, mesic Vertic Endoaquolls) and Nicollet silty clay loam (fine-loamy, mixed, superactive, mesic Aquic Hapludolls) soils were the two predominant soil series at the site. The crop was planted (Pioneer hybrid P9929AMXT) on 17 May 2018 at a population of 86,400 plants ha^{-1}. Nitrogen was applied as urea + N-(n-butyl) thiophosphoric triamide (NBPT; urease inhibitor) on 21 May 2018. Phosphorus and potassium fertilizers were applied on 25 May 2018 at rates that ensure optimum soil fertility for maize production in Minnesota [16].

2.1.2. Waseca Experiments

The experimental boundaries for both the small plot and whole field experiments near Waseca, Minnesota are illustrated in Figure 1 [17]. The previous crop (i.e., in 2018) was soybeans, and the Nicollet clay loam (fine-loamy, mixed, superactive, mesic Aquic Hapludolls) and Webster clay loam (fine-loamy, mixed, superactive, mesic Typic Endoaquolls) soils were the two predominant soil series at the site. Both the small plot and whole field experiments were planted (Channel hybrid 199-1VT2P RIB) on 3 June 2019 at a population of 87,722 plants ha^{-1} immediately following a tillage pass. Preplant N was applied to the small-plot experiment as urea + NBPT on 17 May 2019. The preplant N treatment was applied to the whole field experiment in two phases: the first phase was applied as a urea + NBPT/ammonium sulfate blend (45 kg N ha^{-1}) on 16 May 2019, and the second phase was applied as urea + NBPT at a rate of 179 kg nitrogen ha^{-1} on 3 June 2019. Sidedress N was applied as urea + NBPT on 3 July 2019 (at the V7 development stage) with a 4.6 m applicator (Gandy Orbit-Air, Gandy Company; Owatonna, Minnesota; application width of eight crop rows) equipped with a variable rate controller (Viper version 3.2.0.0, Raven Industries, Inc.; Sioux Falls, South Dakota).

2.2. *Crop Nitrogen Uptake*

To convert biomass from a per-plant basis to a per-area basis, a plant emergence of 83,361 plants ha^{-1} was assumed for the Wells experiment (96.4% emergence), and plant emergence of 85,251 plants ha^{-1} was assumed for both Waseca experiments (97.0% emergence). These emergence assumptions were based on average stand count observations conducted shortly after emergence. Within two days of capturing hyperspectral imagery, tissue samples were collected to measure tissue N concentration in the above-ground biomass and above-ground total NUP. Plants were sampled by cutting the base of the stem just above the soil surface. Additional details regarding sample acquisition (i.e., date, development stage, number of samples, etc.) are listed in Table A1. Plant samples were oven-dried at 60 °C to constant weight, weighed, and ground to pass a 1 mm sieve. Total N concentration was determined for each sample using Kjeldahl digestion [18] (Wells experiment) or dry combustion [19] (Waseca experiments).

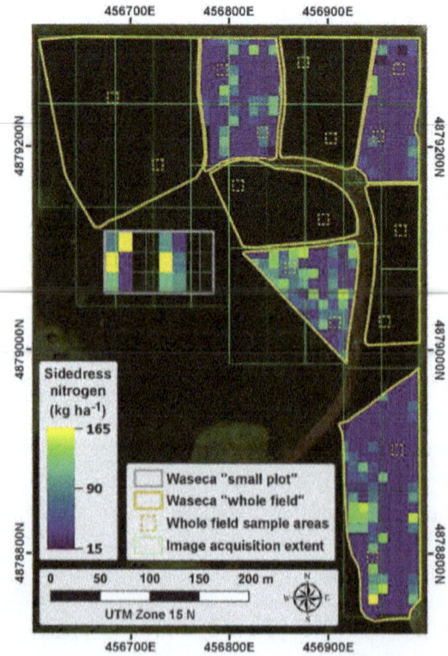

Figure 1. Waseca "small plot" and "whole field" experimental boundaries. The whole field sample areas, whole-field hyperspectral image extents, and sidedress N fertilizer rates are also illustrated.

2.3. *Airborne Spectral Imaging System*

Hyperspectral images were captured via an airborne spectral imaging system (Resonon, Inc., Bozeman, Montana). The airborne system included a flight computer (Resonon NUMI, flight code version 1.36) that integrated the following hardware for seamless data capture: (i) a pushbroom (i.e., line scanning) hyperspectral imaging spectrometer (Resonon Pika IIg VNIR), (ii) an inertial measurement unit (IMU; Ellipse2-N, SBG Systems, Carrières-sur-Seine, France), (iii) a single band GNSS (Global Navigation Satellite System) antenna (TW2410, Tallysman, Inc., Ottawa, Ontario) surface mounted and lifted 15 cm away from the unmanned aerial vehicle flight controller, and (iv) a 240 GB solid-state hard drive for data storage (Intel S3500 Series). The specifications of the spectral imager are presented in Table 1. The flight computer and spectral imager were powered by a designated lithium polymer battery (Ronin 4350 mAh, DJI, Inc., Shenzhen, China), and they were rigidly mounted to a three-axis gimbal (Ronin-MX, DJI, Inc., Shenzhen, China) on board an unmanned aerial vehicle (DJI Matrice 600 Pro, DJI, Inc., Shenzhen, China).

Table 1. Specifications of the hyperspectral imaging spectrometer.

Spectral Range (nm)	Spectral Resolution (nm)	Spectral Channels	Spatial Channels	Bit Depth	Field of View (Degrees)
400–900	2.1	240 [1]	640	12	33.0

[1] Although 240 spectral channels were available, many were clipped out because of high noise.

2.4. *Airborne Image Capture*

Resonon Ground Station software (version 3.123) was used to program the flight computer before each flight campaign. The "bright areas of interest" auto-exposure setting was used, which adjusted the gain and exposure time based on ambient lighting conditions and the general brightness of the target when each image began to be captured. A moderate framerate of 109 (frames per second) was used for each image campaign because it was suitable for the desired ground speeds and altitudes for achieving spatial integrity (i.e., pixels captured with a 1:1 aspect ratio in the cross-track and along-track directions).

A polygon that delineated the desired area for image capture, together with real-time location data from the GNSS receiver of the airborne system, dictated whether the spectral imager was to capture data at a particular location. Image capture commenced when the airborne system entered the polygon boundary and ceased when it exited the polygon boundary.

The gimbal was programmed to maintain the position of the spectral imager at nadir, which involved real-time gimbal adjustments (pitch and roll), due mostly to UAV movement and wind force. The gimbal maintained the heading of the spectral imager relative to the heading of the UAV so that the spectral image array (640 spatial channels) remained approximately perpendicular to the heading of the UAV flight direction. Autonomous flight missions were created and executed using DJI Ground Station Pro (iOS app version 1.8.2 + for 2018 flights and iOS app version 2.0.2 + for 2019 flights). The altitude, speed, and resulting image pixel size (i.e., ground-resolved distance) for each experimental site are presented in Table 2.

Table 2. Spatial description of aerial image campaigns at each experimental site. The cropped plot size refers to the spatial extent of cropped image at each sample location.

Site	Altitude	Ground Speed	Ground Swath	Pixel Size	Area Captured	Cropped Plot Size
	m	m s^{-1}	m	cm	ha	m
Wells	40	4.0	23.7	4.0	4.5	6.2 × 1.8
Waseca small-plot [1]	20–25	2.0–2.5	11.8–14.8	2.0–2.5	0.7	1.8 × 1.8
Waseca whole-field	80	8.0	47.4	8.0	11.2	10 × 10

[1] Altitude was 20 m on the V6 and V8 development stages, but it was 25 m at the V14 stage to increase the ground swath. The corresponding cropped plot size at V14 was 1.8 m × 2.3 m.

2.5. Reference Panels

Reference panels were constructed by applying a 50/50 mixture of gray paint and barium sulfate ($BaSO_4$) by weight to high-density fiberboard (60 cm × 60 cm × 3.2 mm). Painted $BaSO_4$ diffuses incoming solar irradiance in various directions to minimize specular reflection and has less than a 5% difference from a Lambertian surface for 0 to 55 degree zenith angles [20]. The spectral profile of the reference panels from approximately 400 to 900 nm was nearly flat, with an average reflectance of 42.1% (σ = 1.1%; measured with a spectroradiometer; ASD FieldSpec 4; Analytical Spectral Devices, Inc., Longmont, Colorado). Immediately before spectral image acquisition, the reference panels were strategically placed throughout the experimental area so that images containing reference panels were captured at least every 90 s to account for temporal variation in solar illumination.

2.6. Image Pre-Processing

All image pre-processing steps were carried out using Spectronon software (version 2.134; Resonon, Inc., Bozeman, Montana). Raw hyperspectral images were corrected to radiance ($\mu W\ sr^{-1}\ cm^{-2}\ \mu m^{-1}$) in a two-step process using the Radiance Conversion plugin (Equations (1) and (2)) and gain and offset files supplied by the manufacturer of the spectral imager. The *Radiance Conversion plugin* first scaled the gain file (denoted as a) to the integration time and gain of the data cube (denoted as x):

$$\bar{a} = a * \left(\frac{a_\tau * a_\rho}{x_\tau * x_\rho}\right) \tag{1}$$

where $\bar{a}$ is the scaled gain file, a_τ is the integration time of the *gain* file, a_ρ is the gain of the *gain* file, x_τ is the integration time of the data cube, and x_ρ is the gain of the data cube. The raw data cube was then corrected to radiance:

$$X_{rad} = \left(\frac{X_{dn}}{\bar{a}} - b\right) * R \tag{2}$$

where X_{rad} is the radiance corrected data cube in units of $\mu W\ sr^{-1}\ cm^{-2}\ \mu m^{-1}$, X_{dn} is the raw data cube, b is the offset file with the closet match to x_τ and x_ρ, and R is the calibrated radiance source that was used to illuminate the integrating sphere.

Following radiometric correction, images were georectified (using the *Georectification plugin*) based on time-synced data collected by the GNSS receiver and IMU of the airborne system (i.e., latitude, longitude, altitude, yaw, pitch, and roll). A 1.0 m digital elevation model image was used as the basis for projecting each image line to the appropriate elevation above mean sea level. The line-scanning imager sometimes fails to capture data when it passes over the target area too quickly, and it conversely captures duplicate data when it passes too slowly. The *Georectification plugin* used a linear interpolation method to adjust pixel values where there was missing or duplicate data.

Then, georectified radiance images were converted to reflectance by applying a single spectrum correction to each image [21] based on the relationship between the radiance of the reference panels from the imagery and the average measured reflectance. Radiance pixels were manually extracted, taking care to avoid pixels near the edges of panels. If there was no reference panel present in an image, radiance data extracted from the reference panel captured closest in time were used for reflectance conversion instead.

2.7. *Image Post-Processing*

All image post-processing was performed in Python (version 3.7.3) using the *hs_process* package (version 0.0.3) [22]; the *hs_process* package leverages *Spectral Python* (version 0.20) [23] and the *Geospatial Data Abstraction Library* (version 2.3.3) [24]. Post-processing steps included cropping, spectral clipping and smoothing, and segmentation.

2.7.1. Cropping

To allow for seamless batch processing on all subsequent processing and analysis steps, spectral images were spatially subset to exclude all areas outside the plot boundary by cropping the images to plot boundaries. The image size following the spatial cropping step for each experimental location is summarized in Table 2.

2.7.2. Spectral Clipping and Smoothing

Since there was low signal-to-noise, the spectral bands shorter than 430 nm (n = 13) and longer than 880 nm (n = 3) were clipped (i.e., removed) from every image cube. The spectral bands near the O_2 absorption region [25] (760–776 nm; n = 7) and H_2O absorption region [26] (813–827 nm; n = 7) were also clipped because of relatively high noise. In total, 210 spectral bands were kept for analysis. Following spectral clipping, the Savitzky–Golay smoothing procedure (11-band moving window, second-order polynomial fitting) was applied to the spectral domain of each image pixel [27]. Following these steps, at least one cropped spectral image existed for every plot used for analysis.

2.7.3. Choice of the Auxiliary Feature and Image Segmentation

The MCARI2 (Modified Chlorophyll Absorption Ratio Index Improved) spectral index (Equation (3)) was applied to each image to segment the vegetation pixels from those that represent soil and shadow. MCARI2 was chosen because it incorporates a soil adjustment factor (based on the concept developed by Huete [28]) that was optimized with the constraint of preserving the sensitivity of MCARI2 to leaf area index and insensitivity to chlorophyll influence [29].

$$MCARI2 = \frac{1.5[2.5(\lambda_{800} - \lambda_{670}) - 1.3(\lambda_{800} - \lambda_{550})]}{\sqrt{(2 * \lambda_{800} + 1)^2 - \left(6 * \lambda_{800} - 5 * \sqrt{\lambda_{670}}\right) - 0.5}} \quad (3)$$

Before segmentation was carried out, a preliminary analysis was conducted to identify an auxiliary feature that could be made available to the learning models to accomplish the objectives of this study. Specifically, the goal of the preliminary analysis was to find a metric sensitive to above-ground biomass and/or chlorophyll concentration across the range of development stages evaluated in this study. Many pre-segmentation MCARI2 descriptive statistics (e.g., mean, median, standard deviation, various percentiles, etc.) were calculated and evaluated for their relationship with above-ground biomass and chlorophyll concentration. Following the analysis, the 10th percentile MCARI2 was the auxiliary feature chosen because of its relatively strong linear relationship with above-ground biomass across development stages.

Since the leaf area index was expected to differ across N treatments at the development stages evaluated in this study, the most appropriate MCARI2 segmentation threshold was also expected to vary. To account for these differences, the upper MCARI2 threshold was dynamically calculated as the 90th percentile within each cropped image. All pixels in the data cube with MCARI2 values below the 90th percentile threshold were masked out and excluded from any subsequent analyses. The remaining unmasked pixels were assumed to be the pixels that were most likely to represent pure vegetation and were least influenced by soil and shadow (i.e., as mixed pixels).

2.8. Cross-Validation

There were 247 NUP observations across four development stages (V6, V8, V10, and V14), two growing seasons (2018 and 2019), and four pixel sizes available for testing the hypotheses of this experiment (Table 2; Table A1). This full dataset was used for model training and testing, and it represented a robust combination of sampling and acquisition conditions during the early- to mid-vegetative development stages.

A stratified sampling approach was used to assign observations to the training and test sets because of the varying number of observations for each experiment and the robust combination of sampling and acquisition conditions. Stratification ensured that the training and test sets followed the same approximate probability distribution, which reduced the likelihood of model overfitting. Stratification was performed within the experiment and development stage, whereby 60% of the observations were assigned to the training set (n = 148) and the remaining 40% were assigned to the test set (n = 99). The test set was held out during training and tuning and was only used to assess the final prediction performance of the regression models. A stratified random sample from the training set (75%; n = 111) was used for feature selection. For hyperparameter tuning, a repeated stratified k-fold cross-validation technique was applied to the full training set. The cross-validation used for hyperparameter tuning implemented three repetitions and four splits (a different randomization was applied in each repetition), and hyperparameter combinations were evaluated using the mean of the 12 validation scores.

A *Yeo-Johnson* power transformation was applied to the NUP response measurements to minimize the inherent heteroscedasticity of the dataset [30].

2.9. Feature Selection

When limited to a fixed, small number of observations to train a model, model accuracy tends to decrease as the dimensionality of the input dataset increases. For the small number of observations in this experiment relative to the 210 original input bands, it was wise to reduce the dimensionality on the hyperspectral dataset before training the model. The aim of dimensionality reduction was to minimize multicollinearity while preserving the critical information necessary to accurately predict the response variable.

Feature selection was used as a means of dimensionality reduction and was performed before and independent of training the prediction models using the Lasso (least shrinkage and selection operator) algorithm. Lasso is a supervised regression-based model that performs regularization and identifies the most informative, least redundant features for predicting the response variable [31]. Mathematically, Lasso solves the minimization of the least-squares penalty with an added ℓ_1 regularization term.

Feature selection is possible with Lasso because its ℓ_1 regularization term induces sparsity by forcing the sum of the absolute value of the feature coefficients to be less than some fixed constant (controlled by the α parameter), effectively forcing coefficients to be exactly zero for features that are collinear and otherwise uninformative. Upon model convergence, features with coefficients equal to zero are neglected, and features with non-zero coefficients are such because they contain the richest information and are therefore allowed to contribute to the prediction model.

Lasso was implemented in Python (version 3.7.3) using the *Lasso* function of the *sklearn.linear_model* module (*scikit-learn* version 0.22) [32]. A coordinate descent algorithm was implemented to fit the feature coefficients. Since the α parameter controls the degree of regularization on the cost function (i.e., the penalty term) and effectively controls the number of features whose coefficients are forced to zero, α was adjusted from $1.0e^{+00}$ to $1.0e^{-04}$ in a logarithmic manner to explore the effect of feature number on training and cross-validation scores. The root mean squared error (RMSE) of prediction (Equation (4)) was the objective function used for scoring:

$$RMSE = \sqrt{\frac{1}{n}\sum_{i=1}^{n}(y_i - \hat{y}_i)^2} \quad (4)$$

where n is the number of observations, y_i is measured NUP for the i-th observation, and $\hat{y}_i$ is the predicted NUP for the i-th observation. As α becomes smaller and approaches zero, the cost function becomes more similar to that of linear regression, resulting in more features having non-zero coefficients. It was common for multiple α values to yield a similar number of features. In such cases, the feature set with the lowest cross-validation error (i.e., the best score) was ultimately used for model testing.

2.10. Model Tuning and Prediction

Four supervised regression models were used to predict early season NUP: *Lasso*, *random forest*, *support vector*, and *partial least squares*. All models were fit and tuned using the *scikit-learn* Python package (version 0.22). Each of these models has hyperparameters that must be set before the learning process because they are not directly learned during model training. It was necessary to tune these hyperparameters to ensure that they were set to optimum values to avoid underfitting and to accurately test the final prediction performance of the trained model. Tuning was performed using a grid-search cross-validation technique that exhaustively searched over explicit hyperparameter ranges (implemented using the *GridSearchCV* function of the *sklearn.model_selection* module). The hyperparameter ranges evaluated for each regression model are described in the following sections. The hyperparameters that had the lowest cross-validation error (i.e., RMSE) were ultimately used for model testing.

Lasso regression is an attractive prediction model because it tends to perform well with a relatively small number of features, ultimately resulting in a simpler and more interpretable model. Lasso was tuned for its only hyperparameter, α (ranged from $1.0e^{+00}$ to $1.0e^{-04}$).

Support vector regression models are popular prediction models because they are typically effective in high-dimensional spaces (including those where the number of features is greater than the number of samples), they are memory efficient during training, and they are versatile in that different kernel functions can be specified as their decision function. SVR is unique in that it aims to find a function $f(x)$ that deviates from y_i by no more than ε, while simultaneously remaining as flat as possible [33,34]. The SVR model was tuned for both the *linear* and *radial basis* kernel functions. The *linear* kernel function was tuned on the ε (ranged from $1.0e^{+00}$ to $1.0e^{-03}$) and C (ranged from 200 to 800) hyperparameters, and the *radial basis* kernel function was tuned on the ε (ranged from $1.0e^{+00}$ to $1.0e^{-03}$), C (ranged from 10 to 70), and γ (ranged from 1 to 20) hyperparameters.

Random forests are attractive because they are invariant to scaling and transformations of feature values, they are robust to the inclusion of irrelevant features, and they produce models that can be inspected [35]. The random forest model uses an ensemble learning method in which many decision trees are learned in parallel (i.e., there is no interaction between the various trees during the learning process) [36]. The random forest model was tuned for the *maximum number of features to consider* (ranged from n_{features} to $n_{\text{features}}/20$) and the *minimum number of samples required to split an internal node* (ranged from 2 to 10).

Partial least squares regression (PLSR) transforms input features into a new feature space by forming a linear combination of features (i.e., components) that maximizes the covariance between its components and the response variable(s). As a result of this dimensionality reduction (i.e., feature extraction), PLSR can be used with any number of features [37]. It is especially useful when there are more features than observations, or when the features are highly collinear (e.g., in hyperspectral image analysis). The PLSR model was tuned for two hyperparameters: the *number of components* (ranged from 2 to 10) and whether *scaling* was implemented.

Model accuracy was evaluated using both the RMSE (Equation (4) and mean absolute error (MAE) of prediction (Equation (5))):

$$MAE = \frac{1}{n}\sum_{i=1}^{n}\left|y_i - \hat{y}_i\right|. \quad (5)$$

3. Results

3.1. Image Segmentation and MCARI2 Analysis

The MCARI2 histograms illustrate the approximate distribution of the vegetative fraction in each image before segmentation (Figure 2). The range of the MCARI2 distribution decreased as development stage progressed, and the 90th percentile MCARI2 generally increased. Images captured from the Waseca whole field experiment that had a coarser 8 cm pixel size (Figure 2d–f) showed a consistently increasing trend in the 90th percentile MCARI2 up to the V14 development stage, whereas images captured from the Waseca small-plot experiment that had a finer 2–2.5 cm pixel size (Figure 2a–c) peaked by the V8 development stage. The histograms generally shifted from a skewed right distribution at the V6 development stage to a skewed left distribution at the V14 development stage, indicating an increase in vegetative cover as the crop developed. This trend is especially apparent for the finer pixel size, providing clear evidence that the image pixel size affects the distribution of MCARI2 values.

In general, there is a higher likelihood that unmasked pixels represent pure vegetation as the 90th percentile MCARI2 increases. Imagery captured at a coarser pixel size showed a consistently lower 90th percentile MCARI2 across the mid- to late-vegetative development stages (Figure 3b). Alternatively, the larger pixel size showed a consistently higher 10th percentile MCARI2 (Figure 3a). Across all development stages, the mean difference between the smaller and larger pixel sizes was 0.12 and –0.09 for the MCARI2 10th and 90th percentiles, respectively.

The 90th percentile MCARI2 could adequately segment pure vegetation (Figure 2), but it had a weak relationship with above-ground biomass due to saturation after the V10 development stage (Figure 4b). The coefficient of determination (R^2) and root mean squared error (RMSE; smaller is better) improved moderately when considering only the V6, V8, and V10 development stages (the best fit from ordinary least squares regression is represented by the dashed line in Figure 4b), but it is not ideal to have a model that is constrained by development stage. However, the 10th percentile MCARI2 did not saturate out as extensively and had a strong relationship with above-ground biomass (Figure 4a; R^2 of 0.75 and RMSE of 907 kg ha^{-1}).

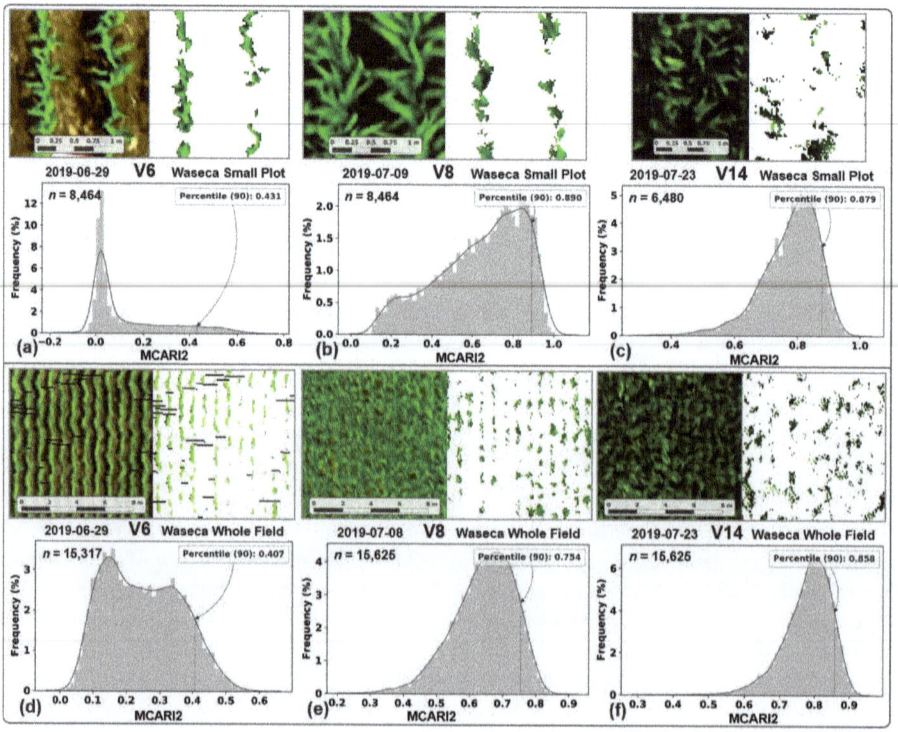

Figure 2. Aerial hyperspectral images (represented as a true color render) for the Waseca small plot (**a–c**) and Waseca whole field (**d–f**) experiments before and after segmentation at the V6 (a and d), V8 (b and e), and V14 (c and f) development stages. The histograms illustrate the 90th percentile MCARI2 values (Modified Chlorophyll Absorption Ratio Index Improved) for each image, which was used as the threshold for segmentation.

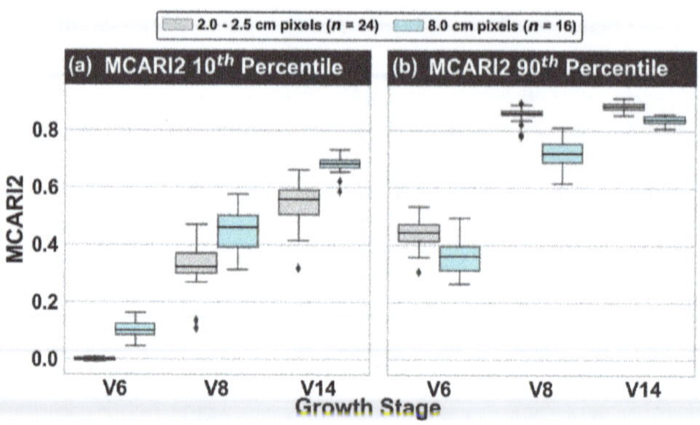

Figure 3. Effect of pixel size on (**a**) MCARI2 (Modified Chlorophyll Absorption Ratio Index Improved) 10th and (**b**) 90th percentile MCARI2 values at the V6, V8, and V14 development stages before image segmentation. The pixel size was 2.0 cm at the V6 and V8 development stage, but 2.5 cm at the V14 development stage.

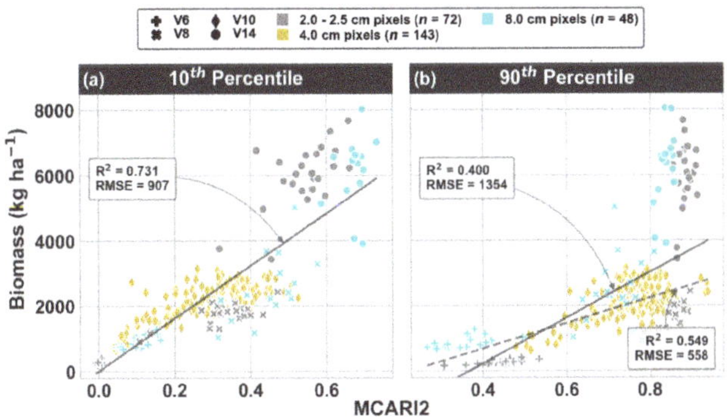

Figure 4. Relationship between above-ground biomass and MCARI2 (Modified Chlorophyll Absorption Ratio Index Improved) 10th (**a**) and 90th (**b**) percentile values. Best fit lines from ordinary least squares linear regression are illustrated, as well as their respective coefficient of determination (R^2) and root mean squared error (RMSE). The dashed line in (b) represents the best fit from ordinary least squares regression when considering only data from the V6, V8, and V10 development stages. The MCARI2 10th percentile was chosen as the auxiliary feature to complement the spectral features for model training because it did not saturate out as extensively as the MCARI2 90th percentile.

3.2. Feature Selection

Adjusting the α parameter from $1.0e^{+00}$ to $1.0e^{-04}$ in the Lasso algorithm governed the number of non-zero feature coefficients from 1 to 79, respectively (Figure 5). Few selected features were in the visible spectral region, and *all* features came from the red edge and near-infrared regions when less than 10 features were selected. The spectral areas of particular significance for NUP were from 735 to 744 nm and from 867 to 879 nm (as indicated by the features that were selected when the number of features was less than five). It is unclear why exactly adjacent features (i.e., within just a few spectral bands) were oftentimes selected by Lasso, but this may suggest that the change in spectral reflectance between adjacent bands (i.e., derivative spectra) is meaningful for explaining NUP.

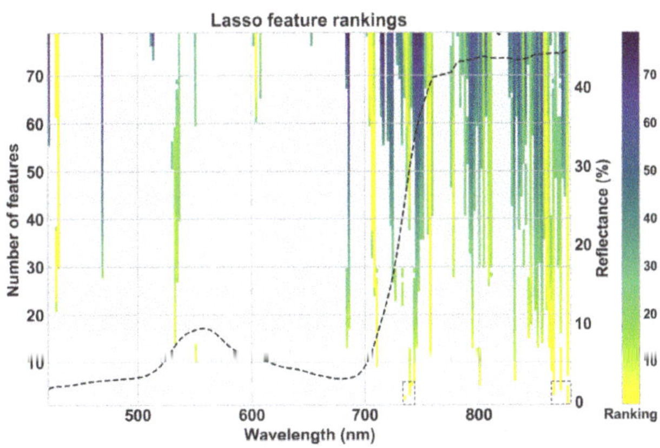

Figure 5. Hyperspectral features selected by the Lasso algorithm (the colored areas indicate selected features). The color indicates the ranking of the feature coefficients determined by the regularization term (yellow color corresponds to a ranking with higher feature coefficients). The top five features from 735 to 744 nm and from 867 to 879 nm are outlined to emphasize the most significant spectral regions for NUP. The mean hyperspectral reflectance across all segmented images used in this study (i.e., mean vegetation) is overlaid for reference.

3.3. Hyperparameter Tuning

The most common hyperparameter values across all training instances (i.e., for varying numbers of selected features) for each prediction model are listed in Table 3. The frequency of the mode indicates the proportion of training instances for which the listed hyperparameter value was optimal. Lower frequencies correspond to hyperparameters whose optimal value varied across the range of tuning values, whereas higher frequencies correspond to hyperparameters that were not particularly variable.

Table 3. The mode and frequency of the mode across all training instances for the tuned hyperparameters in each of the prediction models. Hyperparameter tuning results are listed for the analysis that used spectral features only, as well as the analysis that used spectral features in addition to the auxiliary feature. The hyperparameters and their values are listed according to the conventions of the *scikit-learn* function arguments [32].

Model Parameters	Spectral Features Only		With Auxiliary Feature	
	Modal Value	Frequency	Modal Value	Frequency
Lasso				
alpha	0.001	61%	0.0001	100%
Support vector regression				
kernel	"rbf"	82%	"linear"	92%
Gamma [1]	5	40%	-	-
C [1]	30	38%	200	84%
epsilon [1]	0.01	48%	0.01	86%
Random forest				
min_samples_split	2	82%	2	46%
max_features	0.3	70%	0.9	61%
Partial least squares				
n_components	7	39%	7	61%
scale	1	77%	1	79%

[1] The denoted support vector regression hyperparameters are summarized across only the training instances with the modal kernel function.

Lower Lasso α values (i.e., $1.0e^{-03}$ to $1.0e^{-04}$) were optimal, probably because lower α values offer less restriction on the number of selected features. In SVR, the *radial basis* kernel function was optimal when only spectral features were used, but the *linear* kernel function was optimal with the addition of the 10th percentile MCARI2 from the image segmentation step. In random forest, the *minimum number of samples required to split an internal node* was usually optimal at two, and the *maximum number of features to consider at each tree node* was typically either 30% or 90% of the total number of features. In PLSR, the optimal *number of components* generally increased as the number of features increased (*data not shown*). Although the optimal *number of components* was seven across all training instances, the number of components never exceeded five with less than 12 features (however, note that there cannot be more components than features in PLSR).

As the number selected features increased, the training and validation errors of the optimally tuned models generally decreased and the R^2 values increased (Figure 6). Most of the error reduction occurred in the addition of the first few features, and improvement in model fit usually plateaued by approximately 10 selected features (PLSR plateaued at approximately 30 features; data not shown). Of the four models evaluated, random forest showed more overfitting during hyperparameter tuning than Lasso, SVR, or PLSR as illustrated by the greater difference between the training and validation errors/R^2 values. This difference tended to increase as the number of selected features increased (although the magnitude and consistency of this observation varied among models), indicating overfitting with an increasing number of selected features. As a result of a more overfit model, the subsequent validation error in the random forest model was greater than that of Lasso, SVR, and PLSR, whose validation errors were all comparable. The standard deviation (σ) of error across the four folds and three repetitions of the repeated k-fold cross-validation (i.e., the shaded region of Figure 6) remained fairly constant as the number of selected features increased, and it was comparable across all four models (σ was approximately 2 kg ha^{-1}).

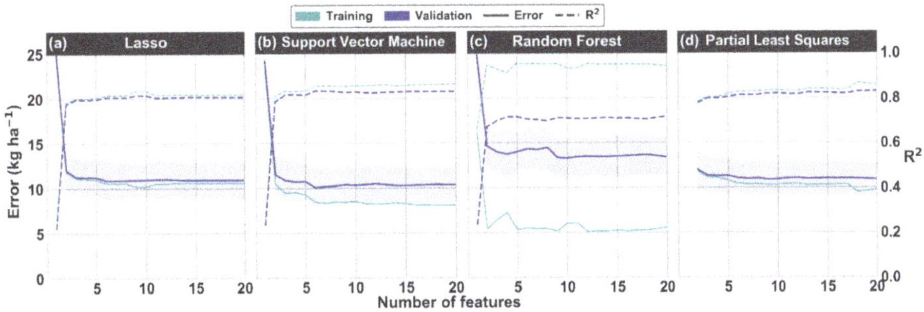

Figure 6. Influence of feature number on the mean absolute error (MAE) and coefficient of determination (R^2) during hyperparameter tuning. The MAE and R^2 are illustrated for both the training and validation datasets for nitrogen uptake predicted by the (**a**) Lasso, (**b**) support vector regression, (**c**) random forest, and (**d**) partial least squares models evaluated in this study. The shaded region surrounding the validation error represents the standard deviation (σ) of error across the four folds and three repetitions of the repeated k-fold cross-validation applied during hyperparameter tuning.

3.4. Nitrogen Uptake Predictions

Across the three experimental sites, NUP between the V6 and V14 development stages ranged from 4.8 to 181.5 kg ha^{-1} (mean measured NUP in the test set was 57.1 kg ha^{-1}; Figure 7). There was a clear improvement in NUP prediction performance when the 10th percentile MCARI2 was available as an auxiliary feature (Figure 7e–h). All predictions in Figure 7 used five input features (predictions from Figure 7a–d used only spectral features, whereas predictions from Figure 7e–h used the 10th percentile MCARI2 auxiliary feature in addition to four spectral features). By adding the auxiliary feature, the average MAE and RMSE across the four prediction models were reduced by 1.6 (14%) and 1.8 (11%) kg ha^{-1}, respectively. Overall, the performance among Lasso, SVR, and PLSR was comparable, while the performance of random forest was substantially inferior as indicated by the higher error values. Among the models that used only spectral features (Figure 7a–d), the MAE of SVR was marginally lower than Lasso and PLSR (however, the RMSE was comparable). Comparing among the models that used the 10th percentile MCARI2 (Figure 7e–h), the MAE of Lasso and PLSR were marginally lower than SVR. The best-fit lines of all models are remarkably similar to the 1:1 line, especially for the models that used only spectral features, indicating that there was no tendency to under- or overpredict across the observational range of NUP. Among the three image capture pixel sizes, there does not appear to be a trend in the contribution of error by pixel size (i.e., the larger pixel size does not appear to have higher error, per se). Across all models, the relative MAE improved from 20.3% to 17.5% and the relative RMSE improved from 28.0% to 24.8% by adding the auxiliary feature (Table A2).

The *Yeo-Johnson* power transform served its purpose, as the variance of predicted NUP increased proportionally to the increase in NUP itself. This is illustrated by the increasing deviation of the points from the 1:1 line as measured NUP increased, and it is apparent by observing the increase in stratified error values shown along the right axis of each subplot. The SVR model with spectral features only performed particularly well at predicting low and very low NUP values (i.e., NUP < 50 kg ha^{-1}) as indicated by the low stratified MAE (i.e., 3.5 kg ha^{-1}) compared to the other models (average MAE of 4.6 kg ha^{-1}). SVR was the only learning model among the four that did not show an improvement in prediction error at the very low NUP level (i.e., NUP < 25 kg ha^{-1}) by making use of the 10th percentile MCARI2 feature. The SVR stratified MAE actually increased from 3.5 to 3.9 kg ha^{-1} (all others decreased at the very low NUP stratification), but they were still the lowest among the other learning models. Although the random forest models had higher overall MAE, they performed satisfactory at the very low NUP level (i.e., fairly similar to the other models). Although there were small differences in MAE within the medium and high NUP levels (NUP ≥ 50 kg ha^{-1}), neither Lasso, SVR, or PLS clearly emerged as the best performer. The NUP stratifications reported in Figure 7 roughly coincide with the four observed development stages. In general, the MAE increased as the development stage progressed (data not shown), which was largely attributed to the increasing variance in measured NUP with the increase in NUP itself.

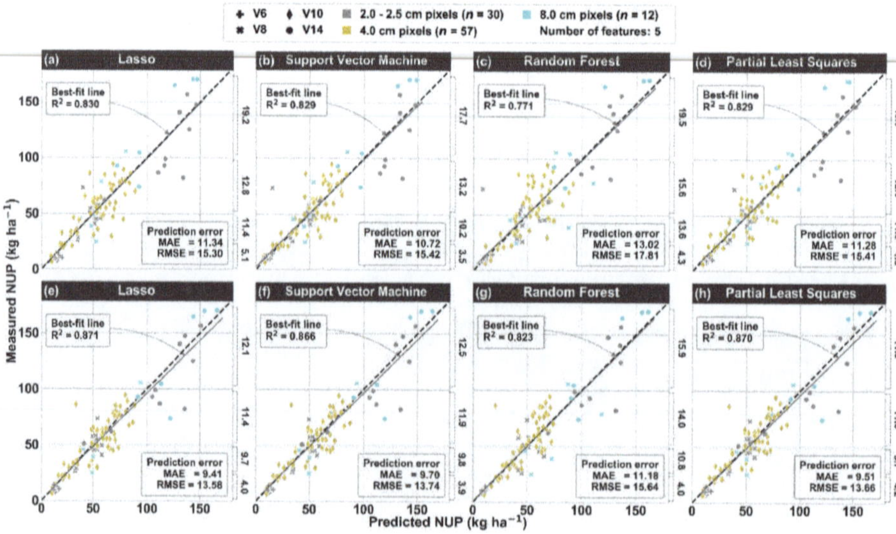

Figure 7. Measured and predicted nitrogen uptake (NUP) values from each of the four learning models using the test dataset (n = 99). Plots in the top row (**a–d**) used five spectral features (739, 741, 801, 867, and 873 nm), and plots in the bottom row (**e–h**) used four spectral features (737, 739, 867, and 871 nm) in addition to the 10th percentile MCARI2 (Modified Chlorophyll Absorption Ratio Index Improved) before image segmentation. The mean absolute error (MAE) and root mean squared error (RMSE) correspond to the prediction error (i.e., deviation from the 1:1 dashed line), whereas the coefficient of determination (R^2) corresponds to the best-fit line. MAE values stratified by measured NUP are shown along the right axis of each plot ("very low", "low", "medium", and "high" groups are stratified as: NUP < 25 kg ha^{-1}; 25 ≥ NUP < 50 kg ha^{-1}; 50 ≥ NUP < 100 kg ha^{-1}; and NUP ≥ 100 kg ha^{-1}, respectively).

For any number of input features, the NUP prediction performance improved when the 10th percentile MCARI2 was available as an auxiliary feature (Figure 8). Across all models, the MAE of prediction was reduced by 1.2 kg ha^{-1} when considering the 10th percentile MCARI2 (summarized data not shown). Average improvements within models ranged from 0.96 kg ha^{-1} (SVR model) to 1.4 kg ha^{-1} (Lasso model). The number of input features required to reach optimal performance was reduced with the 10th percentile MCARI2 feature available (optimal performance is indicated by the minimum MAE across all number of input features). The Lasso, SVR, and PLSR models reached optimal performance with less than 10 input features with the 10th percentile MCARI2 auxiliary feature available. If using spectral features only, Lasso, SVR, and PLSR did not usually reach optimal performance until at least 20 input features were used. This evidence indicates that the inclusion of the 10th percentile MCARI2 value before segmentation not only improved overall prediction accuracy, but it also reduced the number of input features required to reach optimal performance (optimal prediction errors were usually reached with only three features).

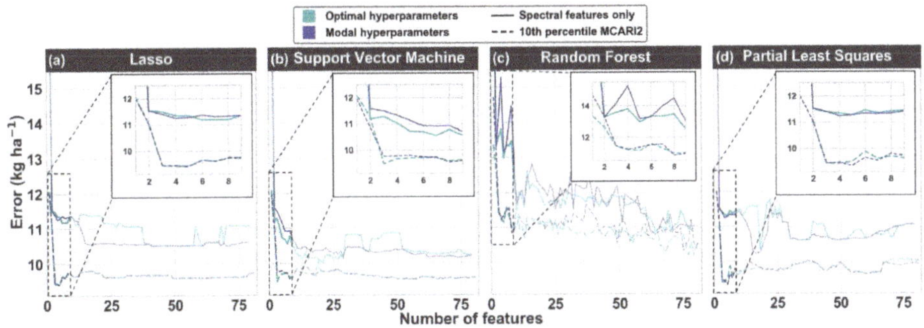

Figure 8. Influence of feature number on the mean absolute error (MAE) of the test dataset for Lasso (**a**), support vector regression (**b**), random forest (**c**), and partial least squares (**d**) models. The error designated by *optimal hyperparameters* indicates that the model hyperparameters were adjusted for each input feature number based on the results of the hyperparameter tuning, and the error designated by *modal hyperparameters* indicates that the hyperparameters were fixed for all number of features. Errors are presented for models that used spectral features only (solid lines), as well as well as for models that used the 10th percentile MCARI2 (Modified Chlorophyll Absorption Ratio Index Improved) auxiliary feature in addition to spectral features (dashed lines).

Unlike the other learning models, the random forest prediction error continued to decrease as the number of features increased, and it was not clear if the optimal performance was ever reached with the range of input features evaluated. Furthermore, random forest errors were noisy (i.e., they had a tendency to change abruptly as the number of features increased), and they were clearly higher than the other three models across all the features evaluated.

There was not a clear advantage in using the optimal hyperparameters compared to using the modal hyperparameters from Table 3 (note that the *optimal* hyperparameters were determined during hyperparameter tuning individually for each group of selected features). In general, as the number of features increased, the error trend was usually smoother when the modal hyperparameters were used.

4. Discussion

4.1. Model Comparison

Although all four supervised regression models (Lasso, SVR, random forest, and PLSR) had satisfactory prediction errors, errors from random forest were consistently higher (+10% average MAE across all input features). It was beyond the scope of this study to assess the computational efficiency of the models, but it may be worth noting that the SVR model took substantially more time to train than the other models. This was especially apparent during hyperparameter tuning, because the grid search method required that each model be trained for all possible hyperparameter combinations and all input features. If considering the computational requirement in addition to prediction accuracy, Lasso and PLSR generally emerged as the preferred models among those evaluated for predicting maize NUP using hyperspectral imagery.

There are many studies that correlate active or passive remote sensing with NUP in maize [14], but it seems rare that NUP predictions are evaluated with proper cross-validation techniques. The only report of cross-validated NUP prediction errors in maize was a study that used spectral indices from an active canopy sensor to predict NUP in the early vegetative development stages using simple linear regression [38]. Xia et al. [38] reported NUP RMSE as low as 16.6 kg ha^{-1} across development stages (V5 to V10). The current study showed NUP RMSE as low as 13.6 kg ha^{-1} across development stages (Figure 7), which was an 18% improvement compared to Xia et al. [38]. This improvement was observed despite the fact that predictions were made across a broader range of development stages

(V6 to V14) and NUP values (4.8 to 182 kg ha^{-1}) in the current study, for which larger errors can be expected due to increasing NUP variability for observations at later development stages.

Building cross-validated prediction models offers the advantage of interpreting remote sensing data in a way that is more extensible for practitioners. Rather than focusing on comparing interpretations of remote sensing information [14,39], remote sensing can instead be used to directly predict N status indicators familiar to agronomists (e.g., crop NUP, chlorophyll content, etc.). These N status indicators likely differ in their fundamental importance for making site-specific fertilizer recommendations, and thus the challenge of improving N recommendations can be broken down so research objectives focus instead on improving our fundamental understanding of N dynamics in maize (e.g., relating N status indicators to crop N demand). The value of comparing the accuracy and practicality of various remote sensing approaches [14] should not be diminished. Rather, we identify two independent requirements for making site-specific N recommendations using remote sensing: (i) the ability to reliably predict crop N status indicators familiar to agronomists, and (ii) having a clear understanding of how various N status indicators relate to soil N supply [12] and/or crop N requirement [13]. Thus, to improve the ability to provide reliable site-specific N fertilizer recommendations, we must not only be able to predict maize crop NUP, but we must understand how N supply and maize N requirement are expected to change throughout the season.

4.2. Segmentation

The goal for segmentation in this N study was to remove the pixels least likely to represent pure vegetation without having a bias against low chlorophyll plants. The MCARI2 spectral index was ideal for segmentation because it is sensitive to leaf area index and resistant to chlorophyll, which were important considering that plants with low N rates were expected to exhibit low chlorophyll. Shadowed pixels also tend to be problematic during segmentation, which MCARI2 handled especially well compared to other common spectral indices. Image pixel size slightly affected the range of MCARI2 histograms (Figure 2; a wider histogram corresponds to a higher coefficient of variation), even across the relatively small pixel sizes used in this study. The decreased variability observed for coarser pixel sizes tends to provide less specific biophysical crop information [40], but this issue was not severe for the small range in pixel sizes used in this study (i.e., 2.0–8.0 cm). Although early season weed control was not problematic at any of the sites used in this experiment, we do not expect basic segmentation methods (such as those used for this analysis) to discern differences between crop and weeds. Thus, from a practical perspective, the use of remote sensing for predicting biophysical parameters such as NUP requires suitable weed control, or perhaps a more sophisticated segmentation method that is capable of masking out weed pixels.

4.3. Inclusion of an Auxiliary Feature

A drawback of segmenting soil, shadow, and mixed pixels from the purest vegetation pixels is that information describing the canopy cover may be lost. It may be detrimental to segment generally unwanted pixels because their contribution may have a positive influence in explaining variability after averaging spectra from all pixels within a plot/region of interest and passing it to the regression model for training or validation. Therefore, it was desirable to identify a metric from the segmentation process that could be recorded and passed on to the subsequent steps for model evaluation so the benefits of segmentation could be exploited without sacrificing information about canopy cover. The 90th percentile MCARI2 was a suitable metric for segmenting pure vegetation, but it had a weak relationship with above-ground biomass due to saturation after the V10 development stage (Figure 4b). Since the 10th percentile MCARI2 did not saturate out as extensively (Figure 4a; i.e., there was a linear relationship across the range of biomass values), it was considered a more suitable metric for improving NUP prediction accuracy. This notion was supported by the improved overall prediction accuracy with the inclusion of the 10th percentile MCARI2 auxiliary feature. If segmentation of soil, shadow, or mixed pixels is possible (i.e., when using high-resolution aerial imagery), it is recommended to identify a metric before the segmentation process that can be used as an auxiliary input feature, because it may improve overall model accuracy.

At any particular development stage, it was not surprising that the 90th percentile MCARI2 was lower for the larger pixel size—this simply illustrates that the larger pixel sizes are more influenced by the nearby soil and have a greater degree of mixing. The more interesting question is perhaps whether this greater degree of mixing observed with the coarser pixels translated to a bias and/or increased NUP prediction error from the models. Unfortunately, there was an offset in timing between image acquisition and tissue sampling for the Waseca whole field experiment at the V6 development stage, so prediction data are limiting for the 8.0 cm pixel size. However, with the data available, there is no evidence of a bias or increased prediction error from plots that had larger inherent pixel sizes. As long as the models are trained using data that are representative and within the constraints of data that will be used for predictions, these results show that imagery with up to 8.0 cm pixel size can be used during the early development stages (i.e., up to V14) without sacrificing NUP prediction accuracy.

4.4. Spectral Feature Selection

Hyperspectral sensors are incredible research tools because they offer hundreds of spectral bands, but the results of this study show that optimal prediction accuracy was usually achieved with only a few features. Features in the red edge (near 740 nm) and near-infrared (near 870 nm) regions were particularly important spectral regions for NUP (Figure 5). Reflectance in the red edge region changes rapidly due to the transition from strong pigment absorption in the red region to light scattering in leaves in the near-infrared region [41]. The Lasso algorithm oftentimes selected spectral features that were adjacent to each other (i.e., within approximately 6 nm). This not only illustrates the apparent importance of spectral precision and bandwidth of the sensor, but it also suggests that the slope in reflectance (i.e., derivative spectra) between adjacent bands is especially meaningful for explaining NUP. The inflection point in the red edge region is known to shift to shorter wavelengths as a consequence of reduced chlorophyll content [42], and perhaps the selection of adjacent spectral features is capturing this phenomenon.

4.5. Ongoing Challenges

The ability to predict spatially explicit NUP during the early- to mid-vegetative development stages has a clear value to maize producers that wish to make in-season N applications. Hyperspectral imagery is undoubtedly a promising tool for such an endeavor, but several technical and logistical challenges remain as described more fully below.

4.5.1. Cost of Specialty Sensors

First, the cost of deploying an unmanned hyperspectral imaging unit is not likely to be practical for commodity crops such as maize. Until the cost of high-resolution, hyperspectral imaging substantially decreases, this technology is unlikely to be implemented. Only 2–5 features were typically required to achieve optimal or near-optimal NUP prediction performance regardless of the prediction model used in this study (Figure 8). Given the relatively few number of features needed for predicting NUP, it may be interesting for sensor engineers and manufacturers to explore the possibility of designing and building specialized narrowband sensors (i.e., up to five specific spectral bands with 2–3 nm bandwidth) that are useful for specific applications such as N management. Although hyperspectral imagery is a great research tool for identifying the most useful spectral bands, it certainly may not be required for satisfactory NUP prediction accuracy. Instead, more simple multispectral aerial sensors, satellite sensors, and active ground-based sensors may be sufficient for adequately predicting NUP. Considerable research has been conducted to predict early season maize NUP using these various sensors and/or platforms [14]. However, it is rare that prediction accuracy is reported using proper cross-validation techniques, making it challenging to compare among sensors, platforms, or prediction models to draw sound conclusions about the most satisfactory methods.

4.5.2. Timeliness

Second, it may be challenging sometimes to acquire, process, train, and analyze data in the time required to make a timely management recommendation. The burden of the acquisition process can be ameliorated with hardware that is fully integrated (e.g., spectral sensor that works well with an unmanned platform) and is efficient (e.g., battery capacity, flight time, read/write speeds, etc.). Pre-processing (e.g., radiometric calibration, reflectance conversion, georeferencing, etc.) and post-processing tasks can generally be improved with software, although customized solutions may have to be designed and built for specific applications and/or processing methodologies. Supervised models are only as good as their training and validation accuracies, and it would be wise to continually add new training data over time to incorporate data from different environments to improve the robustness of the models. While this study quantified the improvement in NUP prediction performance by including the 10th percentile MCARI2 as an input feature, it did not address the value of segmentation in the first place. This same notion could be said for many of the oftentimes subjective methods within the meticulous image processing workflow (e.g., spectral smoothing, spectral binning, segmentation, etc.). Furthermore, it is rare that a practitioner using the methods described in this study would be able to access the same resources used in this study (e.g., an imager with similar specifications and calibration protocol, processing software, etc.). For these reasons, more attention should be given toward quantifying the overall effect on prediction accuracy after mimicking the specifications of various popular image sensors or modifying thresholds used during processing (e.g., 90th percentile MCARI2 used as a segmentation threshold).

4.5.3. Making a Fertilizer Recommendation

Third, the scientific understanding of how early season crop NUP should be used to make a N fertilizer recommendation is poorly understood. Precision N management aims to match the N supply with crop demand in both space and time to ensure optimal grain yield while reducing the risk of environmental pollution from excess N. Thus, the prediction of early season crop NUP as demonstrated in this study is only useful for practical purposes when there exists a known relationship between observed NUP and future crop demand. Although there are cases when the relationship between plant-based spectral measurements and optimal N rate can be weak to non-existent [43], there is considerable evidence suggesting the contrary [44–48]. This indicates that remote sensing observations during the growing season can perhaps be used to determine crop N requirements, at least in part.

Future crop demand is itself subject to uncertainty because of the close relationship between the state of the soil environment (e.g., temperature, moisture, biological activity, etc.) and N additions (e.g., mineralization) or losses (e.g., leaching, denitrification, etc.). This is especially true in higher rainfall regions such as Minnesota and is largely driven by uncertainty in weather. Indeed, it is common that the economic optimum N rate (EONR) varies within a particular field and across years [10], but it is not always clear what drives these differences [39]. The lack of understanding of future crop demand together with uncertainty in the weather introduces a general distrust in any N recommendation that is derived from remote sensing [39]. Although there is always expected to be some degree of uncertainty in crop N demand, a planned in-season N fertilizer application can reduce the risk of N loss before the advent of increased demand and rapid crop uptake [39], ultimately improving NUP efficiency. However, the risk of an in-season application is that wet weather conditions can delay the planned application timing, which can subsequently lead to reduced grain yield and/or increased residual soil N in some environments [49].

Future research should investigate the relationships between early season NUP and the EONR. Specifically, we must critically assess how spatially explicit, early season NUP is helpful for understanding the contributions to various parts of the N mass balance. Our understanding of net mineralization in soils is especially limiting [39], perhaps most fundamentally because of challenges with our ability to consistently estimate mineralizable N that varies due to soil properties, weather, sample collection timing, etc. [50]. However, there is perhaps an opportunity to use soil temperature and moisture sensors [51], as well as other factors such as topography and soil physical and chemical properties, to estimate trends in mineralization somewhat reasonably throughout the season. The potential for incorporating process-based models to calculate EONR using readily available data also shows considerable promise [52] and should be explored for its ability to relate EONR with early season crop NUP.

5. Conclusions

As implemented in this study, supervised regression using hyperspectral remote sensing offers the opportunity to observe N supply spatially during the early- to mid-vegetative development stages, albeit in the form of crop N uptake rather than soil N supply. Introducing an auxiliary feature to the supervised regression models substantially improved N uptake prediction accuracy for all the models evaluated. Although hyperspectral imagery captured at a larger pixel size (i.e., up to 8.0 cm) demonstrated a greater degree of mixing (i.e., between soil, shadow, and vegetation), there did not appear to be evidence that this larger inherent pixel size translated to a bias or increased prediction error. The N uptake predictions from this study are superior to other reports in the literature that use various passive and active remote sensing techniques. As exciting and promising as this may be for making progress toward improved N management in agriculture, it has limited practical value for precision N management without additional efforts in three critical areas: (i) the cost of this specialty spectral data should be reduced, (ii) the timeliness from data acquisition to prediction should be improved, and (iii) the relationship between early season N uptake and upcoming crop N demand must be predicted with some degree of certainty.

To take the next step of making fertilizer recommendations, and to do so with greater confidence, it is imperative that remote sensing research in precision N management begins to encompass data-driven predictions of spatially explicit crop N demand and/or estimated optimum N rate. In continuing this journey, prediction accuracy must be reported habitually using proper cross-validation techniques, which is a practice that is rarely demonstrated in the relevant literature to date. Aside from preventing model overfitting, cross-validated results allow more authentic comparisons to be made among experiments that demonstrate the use of remote sensing for predicting N uptake and other biophysical crop parameters. Further research may conclude that narrowband or hyperspectral imagery is not a necessity for satisfactory prediction results (after all, optimal prediction accuracy was usually achieved with only a few spectral features). However, this must be properly quantified using various sensor configurations (e.g., spectral features, bandwidths, etc.), acquisition methods (e.g., altitude, platform, etc.), and/or processing methods.

Author Contributions: Conceptualization: T.J.N., C.Y., and D.J.M.; methodology: T.J.N. and G.D.P.; software: T.J.N.; formal analysis: T.J.N.; writing—original draft preparation: T.J.N.; writing—review and editing: C.Y., G.D.P., D.J.M., J.F.K., and F.G.F.; visualization: T.J.N.; project administration: D.J.M. and F.G.F.; funding acquisition: D.J.M., C.Y., T.J.N., and F.G.F. All authors have read and agreed to the published version of the manuscript.

Funding: This research was possible because of funding provided by the Minnesota Department of Agriculture through the Minnesota Clean Water Land and Legacy Act, grant number 153761 PO 3000031069 (Waseca experiments), and the Minnesota Soybean Research and Promotion Council, grant numbers 00079668 and 00071830 (Wells experiment). Minnesota's Discovery, Research, and InnoVation Economy (MnDRIVE) and the University of Minnesota also provided financial support in the form of student fellowships and/or research assistantships.

Acknowledgments: We thank Ali Moghimi for his contributions toward helping to integrate the hyperspectral imager with the UAV, as well as his valuable perspectives regarding image processing and model training. We thank the staff at the University of Minnesota Southern Research and Outreach Center, especially Jeff Vetsch, for help with field activities for the Waseca experiments. We also thank the Soil, Water, & Climate field crew, János Forgács, Xiaolei Deng, Abdullahi Abdullahi, and Leanna Leverich for assistance with sample collection, and Wenhao Su for assistance with image pre-processing.

Conflicts of Interest: The authors declare no conflict of interest. The funders had no role in the design of the study; in the collection, analyses, or interpretation of data; in the writing of the manuscript, or in the decision to publish the results.

Appendix A

Table A1. Number of experimental observations (n), tissue sampling date, image acquisition date and local time, sampling area, number of tissue subsamples, and nitrogen extraction method for each experimental dataset subset by vegetative development stage.

Year	Experiment	ID	Observation n	Stage	Sampling Date	Image Date	Image Time	Sample Area	Subsample n	Nitrogen Extraction
2018	Wells	1	142	V10	29 June 2018	28 June 2018	11:49–12:00	1.5 m × 5 m (2 rows)	6	Kjeldahl
2019	Waseca small-plot	2	24	V6	29 June 2019	29 June 2019	12:21–12:28	1.5 m × 2 m (2 rows)	10	Dry combustion
2019		3	24	V8	9/10 July 2019 [1]	09 July 2019	11:40–11:46	1.5 m × 2 m (2 rows)	10	Dry combustion
2019		4	24	V14	23 July 2019	23 July 2019	12:03–12:09	1.5 m × 2 m (2 rows)	6	Dry combustion
2019	Waseca whole-field	5	16	V8	10 July 2019	08 July 2019	13:06–13:17	5 m × 10 m (6 rows)	6	Dry combustion
2019		6	16	V14	23 July 2019	23 July 2019	12:32–12:42	5 m × 10 m (6 rows)	6	Dry combustion

[1] V8 tissue sampling for the Waseca small-plot experiment began on 9 July and finished on 10 July.

Table A2. Relative mean absolute error (MAE) and relative root mean squared error (RMSE) for each of the models evaluated in this study. The relative error values in the *Spectral Features Only* column used five spectral features (739, 741, 801, 867, and 873 nm), and the relative error values in the *With Auxiliary Feature* column used four spectral features (737, 739, 867, and 871 nm) in addition to the 10th percentile MCARI2 (Modified Chlorophyll Absorption Ratio Index Improved) before image segmentation.

Objective Function/Model	Spectral Features Only	With Auxiliary Feature
Relative [1] MAE		
Lasso	19.9%	16.5%
Support vector	18.8%	17.0%
Random forest	22.8%	19.6%
Partial least squares	19.8%	16.7%
Relative [1] RMSE		
Lasso	26.8%	23.8%
Support vector	27.0%	24.1%
Random forest	31.2%	27.4%
Partial least squares	27.0%	23.9%

[1] The relative MAE and RMSE values were calculated by dividing the error value by the mean measured nitrogen uptake across all observations in the test set (57.1 kg ha^{-1}; $n = 99$).

References

1. Keeler, B.L.; Polasky, S. Land-use change and costs to rural households: A case study in groundwater nitrate contamination. *Environ. Res. Lett.* **2014**, *9*, 074002. [CrossRef]
2. Shcherbak, I.; Millar, N.; Robertson, G.P. Global metaanalysis of the nonlinear response of soil nitrous oxide (N_2O) emissions to fertilizer nitrogen. *Proc. Natl. Acad. Sci. USA* **2014**, *111*, 9199–9204. [CrossRef] [PubMed]
3. Van Metre, P.C.; Frey, J.W.; Musgrove, M.; Nakagaki, N.; Qi, S.; Mahler, B.J.; Wieczorek, M.E.; Button, D.T. High Nitrate Concentrations in Some Midwest United States Streams in 2013 after the 2012 Drought. *J. Environ. Qual.* **2016**, *45*, 1696. [CrossRef]
4. Keeler, B.L.; Gourevitch, J.D.; Polasky, S.; Isbell, F.; Tessum, C.W.; Hill, J.D.; Marshall, J.D. The social costs of nitrogen. *Sci. Adv.* **2016**, *2*. [CrossRef] [PubMed]
5. Erisman, J.W.; Galloway, J.N.; Seitzinger, S.; Bleeker, A.; Dise, N.B.; Petrescu, A.M.R.; Leach, A.M.; de Vries, W. Consequences of human modification of the global nitrogen cycle. *Philos. Trans. R. Soc. B Biol. Sci.* **2013**, *368*, 20130116. [CrossRef] [PubMed]
6. Conant, R.T.; Berdanier, A.B.; Grace, P.R. Patterns and trends in nitrogen use and nitrogen recovery efficiency in world agriculture. *Global Biogeochem. Cycles* **2013**, *27*, 558–566. [CrossRef]
7. Pannell, D.J. Economic perspectives on nitrogen in farming systems: Managing trade-offs between production, risk and the environment. *Soil Res.* **2017**, *55*, 473. [CrossRef]
8. Tilman, D.; Balzer, C.; Hill, J.; Befort, B.L. Global food demand and the sustainable intensification of agriculture. *Proc. Natl. Acad. Sci. USA* **2011**, *108*, 20260–20264. [CrossRef]
9. Dhital, S.; Raun, W.R. Variability in optimum nitrogen rates for maize. *Agron. J.* **2016**, *108*, 2165–2173. [CrossRef]
10. Mamo, M.; Malzer, G.L.; Mulla, D.J.; Huggins, D.R.; Strock, J. Spatial and temporal variation in economically optimum nitrogen rate for corn. *Agron. J.* **2003**, *95*, 958–964. [CrossRef]
11. Basso, B.; Dumont, B.; Cammarano, D.; Pezzuolo, A.; Marinello, F.; Sartori, L. Environmental and economic benefits of variable rate nitrogen fertilization in a nitrate vulnerable zone. *Sci. Total Environ.* **2016**, *545–546*, 227–235. [CrossRef] [PubMed]
12. Robertson, G.P.; Vitousek, P.M. Nitrogen in Agriculture: Balancing the Cost of an Essential Resource. *Annu. Rev. Environ. Resour.* **2009**, *34*, 97–125. [CrossRef]
13. Cassman, K.G.; Dobermann, A.; Walters, D.T. Agroecosystems, Nitrogen-use Efficiency, and Nitrogen Management. *AMBIO A J. Hum. Environ.* **2002**, *31*, 132–140. [CrossRef]
14. Corti, M.; Cavalli, D.; Cabassi, G.; Marino Gallina, P.; Bechini, L. Does remote and proximal optical sensing successfully estimate maize variables? A review. *Eur. J. Agron.* **2018**, *99*, 37–50. [CrossRef]
15. Li, F.; Wang, L.; Liu, J.; Wang, Y.; Chang, Q. Evaluation of leaf N concentration in winter wheat based on discrete wavelet transform analysis. *Remote Sens.* **2019**, *11*, 1331. [CrossRef]
16. Kaiser, D.E.; Lamb, J.A.; Eliason, R. Fertilizer Guidelines for Agronomic Crops in Minnesota. Retrieved from the University of Minnesota Digital Conservancy. 2011. Available online: http://hdl.handle.net/11299/198924 (accessed on 10 April 2020).
17. Wilson, G.L.; Mulla, D.J.; Galzki, J.; Laacouri, A.; Vetsch, J.; Sands, G. Effects of fertilizer timing and variable rate N on nitrate–N losses from a tile drained corn-soybean rotation simulated using DRAINMOD-NII. *Precis. Agric.* **2020**, *21*, 311–323. [CrossRef]
18. Bradstreet, R.B. Kjeldahl Method for Organic Nitrogen. *Anal. Chem.* **1954**, *26*, 185–187. [CrossRef]
19. Matejovic, I. Total nitrogen in plant material determinated by means of dry combustion: A possible alternative to determination by kjeldahl digestion. *Commun. Soil Sci. Plant. Anal.* **1995**, *26*, 2217–2229. [CrossRef]
20. Robinson, B.F.; Biehl, L.L. Calibration procedures for measurement of reflectance factor in remote sensing field research. In Proceedings of the SPIE 0196, Measurements of Optical Radiations, San Diego, CA, USA, 27–30 August 1979; pp. 15–26.
21. Smith, G.M.; Milton, E.J. The use of the empirical line method to calibrate remotely sensed data to reflectance. *Int. J. Remote Sens.* **1999**, *20*, 2653–2662. [CrossRef]
22. Nigon, T.J. HS-Process. Anaconda Cloud 2020. Available online: https://hs-process.readthedocs.io/ (accessed on 10 April 2020).
23. Boggs, T. Spectral Python. Anaconda Cloud 2020. Available online: http://www.spectralpython.net/ (accessed on 10 April 2020).

24. GDAL/OGR contributors. GDAL/OGR Geospatial Data Abstraction Library. Open Source Geospatial Foundation 2020. Available online: https://gdal.org (accessed on 10 April 2020).
25. Greenblatt, G.D.; Orlando, J.J.; Burkholder, J.B.; Ravishankara, A.R. Absorption measurements of oxygen between 330 and 1140 nm. *J. Geophys. Res.* **1990**, *95*, 18577. [CrossRef]
26. Hill, C.; Jones, R.L. Absorption of solar radiation by water vapor in clear and cloudy skies: Implications for anomalous absorption. *J. Geophys. Res. Atmos.* **2000**, *105*, 9421–9428. [CrossRef]
27. Savitzky, A.; Golay, M.J.E. Smoothing and Differentiation of Data by Simplified Least Squares Procedures. *Anal. Chem.* **1964**, *36*, 1627–1639. [CrossRef]
28. Huete, A. A soil-adjusted vegetation index (SAVI). *Remote Sens. Environ.* **1988**, *25*, 295–309. [CrossRef]
29. Haboudane, D.; Miller, J.R.; Pattey, E.; ZarcoTejada, P.J.; Strachan, I.B. Hyperspectral vegetation indices and novel algorithms for predicting green LAI of crop canopies: Modeling and validation in the context of precision agriculture. *Remote Sens. Environ.* **2004**, *90*, 337–352. [CrossRef]
30. Yeo, I.-K.; Johnson, R.A. A new family of power transformations to improve normality or symmetry. *Biometrika* **2000**, *87*, 954–959. [CrossRef]
31. Tibshirani, R. Regression shrinkage and selection via the Lasso. *J. R. Stat. Soc. Ser. B* **1996**, *58*, 267–288. [CrossRef]
32. Pedregosa, F.; Varoquaux, G.; Gramfort, A.; Michel, V.; Thirion, B.; Grisel, O.; Blondel, M.; Müller, A.; Nothman, J.; Louppe, G.; et al. Scikit-learn: Machine Learning in Python. *J. Mach. Learn. Res.* **2012**, *12*, 2825–2830.
33. Vapnik, V. *The Nature of Statistical Learning Theory*; Springer: New York, NY, USA, 1995; ISBN 978-1-4757-2440-0.
34. Smola, A.J.; Schölkopf, B. A tutorial on support vector regression. *Stat. Comput.* **2004**, *14*, 199–222. [CrossRef]
35. Hastie, T.; Tibshirani, R.; Friedman, J. *The Elements of Statistical Learning: Data Mining, Inference, and Prediction*, 2nd ed.; Springer: New York, NY, USA, 2009; ISBN 0387848576.
36. Breiman, L. Random forests. *Random For.* **2001**, *45*, 5–32.
37. Helland, I.S. On the structure of partial least squares regression. *Commun. Stat. Simul. Comput.* **1988**, *17*, 581–607. [CrossRef]
38. Xia, T.; Miao, Y.; Wu, D.; Shao, H.; Khosla, R.; Mi, G. Active optical sensing of spring maize for in-season diagnosis of nitrogen status based on nitrogen nutrition index. *Remote Sens.* **2016**, *8*, 605. [CrossRef]
39. Morris, T.F.; Murrell, T.S.; Beegle, D.B.; Camberato, J.J.; Ferguson, R.B.; Grove, J.; Ketterings, Q.; Kyveryga, P.M.; Laboski, C.A.M.; Mcgrath, J.M.; et al. Strengths and limitations of nitrogen rate recommendations for corn and opportunities for improvement. *Publ. Agron. J.* **2018**, *110*, 1–37. [CrossRef]
40. Polinova, M.; Jarmer, T.; Brook, A. Spectral data source effect on crop state estimation by vegetation indices. *Environ. Earth Sci.* **2018**, *77*, 1–12. [CrossRef]
41. Scotford, I.M.; Miller, P.C.H. Applications of spectral reflectance techniques in Northern European cereal production: A review. *Biosyst. Eng.* **2005**, *90*, 235–250. [CrossRef]
42. Filella, I.; Penuelas, J. The red edge position and shape as indicators of plant chlorophyll content, biomass and hydric status. *Int. J. Remote Sens.* **1994**, *15*, 1459–1470. [CrossRef]
43. Kitchen, N.R.; Sudduth, K.A.; Drummond, S.T.; Scharf, P.C.; Palm, H.L.; Roberts, D.F.; Vories, E.D. Ground-based canopy reflectance sensing for variable-rate nitrogen corn fertilization. *Agron. J.* **2010**, *102*, 71–84. [CrossRef]
44. Ma, B.L.; Subedi, K.D.; Costa, C. Comparison of crop-based indicators with soil nitrate test for corn nitrogen requirement. *Agron. J.* **2005**, *97*, 462–471. [CrossRef]
45. Scharf, P.C. Soil and plant tests to predict optimum nitrogen rates for corn. *J. Plant. Nutr.* **2001**, *24*, 805–826. [CrossRef]
46. Scharf, P.C.; Brouder, S.M.; Hoeft, R.G. Chlorophyll meter readings can predict nitrogen need and yield response of corn in the north-central USA. *Agron. J.* **2006**, *98*, 655–665. [CrossRef]
47. Schmidt, J.; Beegle, D.; Zhu, Q.; Sripada, R. Improving in-season nitrogen recommendations for maize using an active sensor. *Field Crop. Res.* **2011**, *120*, 94–101. [CrossRef]
48. Schmidt, J.P.; Dellinger, A.E.; Beegle, D.B. Nitrogen recommendations for corn: An on-the-go sensor compared with current recommendation methods. *Agron. J.* **2009**, *101*, 916–924. [CrossRef]
49. Spackman, J.A.; Fernandez, F.G.; Coulter, J.A.; Kaiser, D.E.; Paiao, G. Soil texture and precipitation influence optimal time of nitrogen fertilization for corn. *Agron. J.* **2019**, *111*, 2018–2030. [CrossRef]
50. Clark, J.D.; Veum, K.S.; Fernández, F.G.; Kitchen, N.R.; Camberato, J.J.; Carter, P.R.; Ferguson, R.B.; Franzen, D.W.; Kaiser, D.E.; Laboski, C.A.M.; et al. Soil sample timing, nitrogen fertilization, and incubation length influence anaerobic potentially mineralizable nitrogen. *Soil Sci. Soc. Am. J.* **2020**. [CrossRef]

51. Leirós, M.C.; Trasar-Cepeda, C.; Seoane, S.; Gil-Sotres, F. Dependence of mineralization of soil organic matter on temperature and moisture. *Soil Biol. Biochem.* **1999**, *31*, 327–335. [CrossRef]
52. Puntel, L.A.; Pagani, A.; Archontoulis, S.V. Development of a nitrogen recommendation tool for corn considering static and dynamic variables. *Eur. J. Agron.* **2019**, *105*, 189–199. [CrossRef]

remote sensing

MDPI

Article

Integrating Airborne Hyperspectral, Topographic, and Soil Data for Estimating Pasture Quality Using Recursive Feature Elimination with Random Forest Regression

Rajasheker R. Pullanagari [1,*], Gabor Kereszturi [2] and Ian Yule [1]

[1] New Zealand Centre for Precision Agriculture (NZCPA), School of Agriculture and Environment, Massey University, Private Bag 11 222, Palmerston North 4442, New Zealand; i.j.yule@massey.ac.nz
[2] Geosciences, School of Agriculture and Environment, Massey University, Private Bag 11 222, Palmerston North 4442, New Zealand; G.Kereszturi@massey.ac.nz
* Correspondence: p.r.eddy@massey.ac.nz; Tel.: +64-6-356-9099

Received: 19 May 2018; Accepted: 11 July 2018; Published: 13 July 2018

Abstract: Accurate and efficient monitoring of pasture quality on hill country farm systems is crucial for pasture management and optimizing production. Hyperspectral imaging is a promising tool for mapping a wide range of biophysical and biochemical properties of vegetation from leaf to canopy scale. In this study, the potential of high spatial resolution and airborne hyperspectral imaging for predicting crude protein (CP) and metabolizable energy (ME) in heterogeneous hill country farm was investigated. Regression models were developed between measured pasture quality values and hyperspectral data using random forest regression (RF). The results proved that pasture quality could be predicted with hyperspectral data alone; however, accuracy was improved after combining the hyperspectral data with environmental data (elevation, slope angle, slope aspect, and soil type) where the prediction accuracy for CP was R^2_{CV} (cross-validated coefficient of determination) = 0.70, $RMSE_{CV}$ (cross-validated root mean square error) = 2.06%, RPD_{CV} (cross-validated ratio to prediction deviation) = 1.82 and ME: R^2_{CV} = 0.75, $RMSE_{CV}$ = 0.65 MJ/kg DM, RPD_{CV} = 2.11. Interestingly, the accuracy was further out-performed by considering important hyperspectral and environmental variables using RF combined with recursive feature elimination (RFE) (CP: R^2_{CV} = 0.80, $RMSE_{CV}$ = 1.68%, RPD_{CV} = 2.23; ME: R^2_{CV} = 0.78, $RMSE_{CV}$ = 0.61 MJ/kg DM, RPD_{CV} = 2.19). Similar performance trends were noticed with validation data. Utilizing the best model, spatial pasture quality maps were created across the farm. Overall, this study showed the potential of airborne hyperspectral data for producing accurate pasture quality maps, which will help farm managers to optimize decisions to improve environmental and economic benefits.

Keywords: pasture quality; airborne hyperspectral imaging; random forest regression

1. Introduction

Pasture quality is a growing concern because it is a critical constraint for achieving optimal growth and performance for animal production [1]. To meet the nutritional requirements of animals, high quality pasture needs to be maintained on farms. Therefore, being able to accurately assess pasture quality is essential to maintain high quality feed throughout the year. Typically, the assessment of pasture quality, crude protein (CP) and metabolisable energy (ME), is derived from laboratory analysis; however, this method takes significant time and expense, which means these variables are not often measured.

The wider application of remote sensing techniques for precision grassland management is restricted due to heterogeneous pasture [2]. However, substantial progress has been made with sensors and analytical techniques in recent years that can provide comprehensive, site-specific, quantitative information for grasslands. Full range hyperspectral sensors utilize contiguous narrow spectral measurements of reflected light, which has the potential to capture strong narrow absorptions features caused by chemical bonds present in biochemicals of interest [3,4]. Subsequently, it has proved a powerful tool to quantify a wide range of grassland biophysical; biomass [5], dead vegetation fraction [6], and biochemical attributes such as: biomass, leaf area index [7,8], nitrogen [9–11], phosphorus [12], fiber [13,14], polyphenols [15], and cellulose [16].

Although a great variety of vegetation indices (VI's) are widely used to estimate various vegetation properties from reflected light, their potential is limited in the quantification of biochemicals under heterogeneous grassland systems [17]. Compared to the full spectrum-based models, the performance of models based on VI's was inconsistent due to different canopy characteristics [18]. Subsequently, multivariate statistics have been proposed to extract comprehensive information. For instance, Biewer et al. [19] found that compared to VIs, a full spectrum approach had shown strong correlation with forage quality variables such as CP, ME, ash, and acid detergent fibre (ADF) in mixed swards. Utilizing full spectrum data yielded higher accuracy than optimal narrowband VIs for estimating N concentration and biomass yield in bioenergy cropping systems [18]. Partial least squares regression (PLSR) is a widely used chemometric approach for quantifying pasture macronutrients [20] and quality [14,19] from hyperspectral data because it effectively addresses the problems of overfitting and collinearity. However, canopy reflectance data may be confounded by factors such as soil background, canopy structure, illumination, and viewing geometry, which leads to non-linear and complex relationships [21]. Consequently, Ramoelo et al. [22] suggested the use of non-linear algorithms and showed that kernel-based PLS (KPLS) is more powerful than traditional/linear PLS for estimating N and P concentrations of grass in heterogeneous savannah ecosystems. Verrelst et al. [23] highlighted the importance of non-linear regression methods to retrieve vegetation properties from remote sensing data. In recent years, machine-learning approaches gained in popularity because of their flexibility in explaining non-linear complex relationships without considering any statistical assumptions. Among the machine learning algorithms, random forest (RF) gained more importance in hyperspectral remote sensing due to its capability to deal with complex relationships [24].

The relationships could be improved if the hyperspectral data was combined with the environmental variables, as pasture quality is influenced by environmental factors [22]. Soil fertility is a key driver for pasture growth and quality; therefore, adequate supplies of nutrients through regular and balanced fertilization is essential to maintain high quality pasture. In addition to soil fertility, pasture quality varies spatially on hill country farms due to the impact and interactions of multiple influencing conditions, including topography (elevation, slope angle, and slope aspect), environmental factors (temperature, solar irradiation, rainfall, soil type, and soil moisture), and botanical composition [25]. Pasture quality is also dependent on agronomic management practices, such as stocking rate, pasture cover at set stocking, the shearing policy, and weaning date [26].

Since hyperspectral data carries redundant information, selecting relevant spectral variables in the modelling process could improve prediction accuracy and model robustness [27,28]. Although several approaches have been proposed for selecting the best features, Grenitto et al. [29] highlighted that recursive feature elimination (RFE) combined with RF could provide unbiased and stable results with improved accuracy. However, to our knowledge, this method was not investigated for estimating pasture quality attributes, motivating the present study, which aims to test the potential of multiple source information combined with RF–RFE to describe pasture quality (CP and ME) information. Also, the important hyperspectral and environmental variables will be screened using RF–RFE.

2. Materials and Methods

The study area, Limestone Downs, was located at Port Waikato (37°28.665′S, 174°45.540′E) in the northwest of New Zealand where mixed pasture is grown throughout the year (Figure 1). The total study area comprises approximately 3148 ha which classified into 190 paddocks with different sizes ranging from

1.5 to 41 ha. Perennial ryegrass (*Lolium perenne* L.) and white clover (*Trifolium repens* L.) are the dominate species, and a small proportion of kikuyu grass (*Cenchrus clandestinus*), dandelion (*Taraxacum officinale*), and catsear (*Hypochaeris radicata*) are also present. The study area is conventionally used for sheep and beef production. The mean annual precipitation and temperature ranges between 1250–1500 mm and 14.1–16 °C for the period 1971–2000. This study was conducted during the spring season where optimal conditions (temperature, rainfall, and sunshine) prevail for pasture growth.

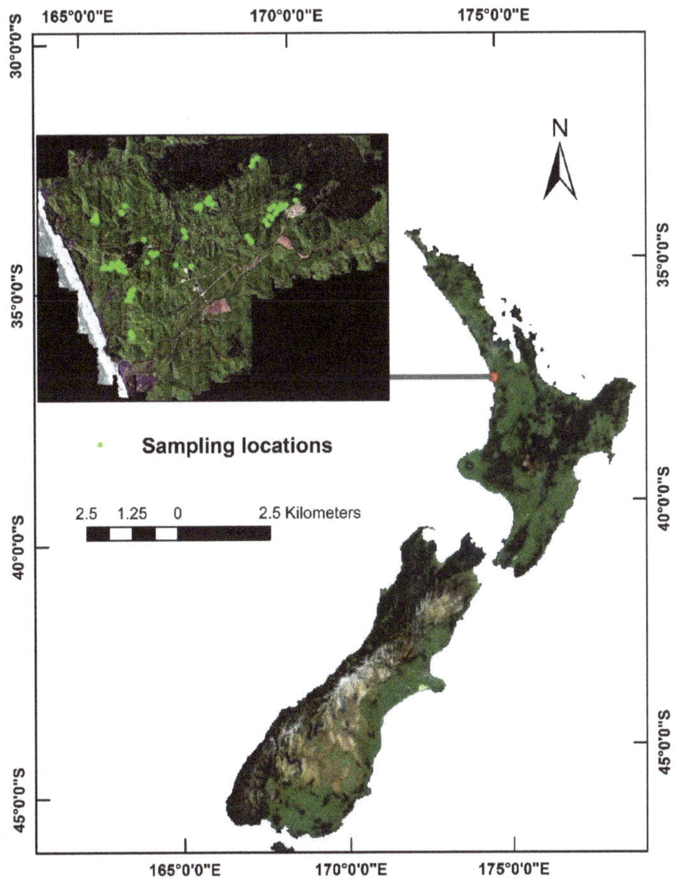

Figure 1. Location of the study area (Limestone Downs) in northwest New Zealand and the sampling plots. The inset map shows the RGB colour composite of the hyperspectral image over Limestone Downs.

A full-spectrum, pushbroom AisaFENIX (Specim Ltd., Oulu, Finland) hyperspectral imaging system was used in the study. The sensor measures upwelling radiance from 370 to 2500 nm as Digital Numbers (DN) with a spectral interval of 3.5–12.2 nm. The AisaFENIX sensor (Specim, Oulu, Finland) has a Field of View (FOV) of 32.2°, as well as an Instantaneous Field of View (IFOV) of 0.084°. The hyperspectral imaging system was mounted on a single-engine, fixed-wing aircraft which was flown at an elevation around 660 m to ensure ground sampling distance of approximately 1 m. To know the position of each pixel, the hyperspectral imaging system was coupled with an RT Oxford Survey+ Ltd., Global Navigation Satellite System (GNSS) and an Inertial Measurement Unit

(IMU). The image was collected between 10:30 and 12:00 New Zealand local time on 24th October 2014. The digital numbers (DN) were converted into radiance (W m^{-2} sr^{-1}) using factory provided radiometric calibration coefficients in CaliGeoPRO software (Specim, Finland). Surface reflectance values were obtained from radiance data using ATCOR4 (ReSe Ltd., Wil, Switzerland), which used geographic, temporal, and atmospheric parameters [30].

Within the study area, based on the access, 150 sites were selected for pasture using stratified random sampling. Elevation and slope angle variables were used as strata, and random sites were then selected from each strata. Since paddocks were not used as a basis for selecting the sites, the total sites came from 72 paddocks, where some of the sites fall under individual paddocks and the remainder from multiple locations of a paddock. Following the aerial campaign, at each of the 150 sites, a 0.5 × 0.5-m quadrat was placed on the ground and a pasture sample harvested to ground level using battery-powered hand shears. The cut samples were immediately placed in a labelled polythene bag, sealed, and stored in a chiller box. These boxes were then transported to an analytical laboratory (Analytical Research Laboratories Ltd., Napier, New Zealand) for immediate determination of CP and ME.

In addition to hyperspectral data, site elevation, slope angle, slope aspect, as well as soil type (Figure 2) were included in the analysis. Elevation, slope angle, and slope aspect maps were generated with a linear filter [31] on a low resolution (5 m) Light Detection And Ranging (LiDAR) Digital Terrain Model, captured in 2010. Soil type information was gathered from Massey University soil map archives. Based on the evolution of New Zealand soils, soil taxonomy, and local knowledge, a new classification was developed [32]. This study included 16 soil types: oxidic granular, allophanic brown, deep orthic, humic gley, humic organic, mottled orthic recent, orthic allophanic, orthic brown, orthic gley, orthic gley and sandy brown, rendzic melanic, sandy brown, sandy gley, sandy raw, sandy recent and typic oxidic granular (Figure 2). Allophanic soils are low density and low fertile which has ability to retain phosphorus in high. Brown soils dominated with clay minerals cover 43% of New Zealand. Gley soils are highly fertile and rich in organic matter. Similar to gley soils, organic soil is rich in organic matter and extremely acidic. Granular soils developed from andesitic to rhyolitic volcanic deposits with a moderate amount of weathering products, such as kaolinite. Melanic soils are highly fertile, with large populations of microorganisms. Oxidic soils are well-developed soils weathered from volcanic deposits, which are dominated with iron and aluminum oxides. Recent soils cover 6% of New Zealand, which is developed on volcanic tephra, and these soils are dominated by secondary illite minerals. The auxiliary data were resampled and co-registered with the hyperspectral image using a nearest neighbor interpolation method.

The spectral and environmental data in the corresponding sampling locations were extracted from a window size of 3 × 3 (9 m^2). The mean value from each window was considered as a response variable. The reflectance data was converted into first derivative reflectance (FDR) using Savitzky-Golay filter to highlight the subtle overlapping absorption peaks. Following the transformation, random forest regression (RFR) was applied to develop relationships between pasture quality and hyperspectral and environmental data. The soil data was converted into binary and stacked with the remaining variables. Since the full data contains different data types, scaling was performed, where each variable value was divided by its standard deviation. For the model development, 60% (n = 90) of total samples were selected, and the remaining 40% (n = 60) were used for validating the model performance. RF is an ensemble learning technique proposed by Brieman and Leo [32].

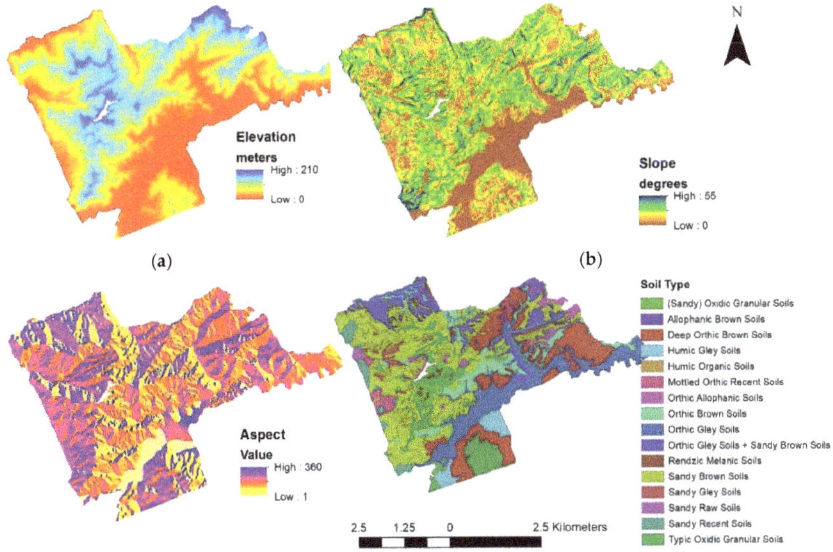

Figure 2. Topographic and soil variables used in this study: (**a**) elevation (**b**) slope angle (**c**) slope aspect (**d**) soil types of the Limestone Downs area.

RF is a collection of several decision trees where each tree is constructed independently with random samples (n) from the training data. Random samples were drawn with the replacement from the training data using a bootstrap aggregating (bagging) method, which was found to be a more robust method for obtaining a stable model and helped to avoid overfitting [33]. Usually, 64% of training data is selected as in-bag data, and the remaining 36% were referred to as out-of-bag (OOB) data. At each node, a random subset of variables were selected. RFR learns the behaviour of the selected m variables and finally selects best performing variables using least square error criteria. The final prediction results were obtained by averaging the predicted results from all trees.

For constructing the model, it is necessary to tune two important parameters: the number of variables at each split and the number of trees. Each split of the tree is determined using a randomized subset of the variables (the default is 1/3 of the total number of variables) at each node [34]. The number of trees was optimized using root mean square error (RMSE) and tested on different population of trees ranging from 50 to 448 using every 10th interval.

The RFE is a wrapper-based feature-ranking algorithm that searches within the space for optimal subset by performing optimization algorithms [35]. The construction of the model initiates with training data, and variables are then ranked according to their importance (Figure 3). While constructing the decision trees, each variable in the OOB data is randomly permuted. After this, RMSE values were calculated for OOB data, and the permuted variables were estimated (Figure 3). Based on the RMSE values, one variable was removed, and a new RF model was created using the remaining variables. This process was recursively applied until only one variable remained as input [36]. During the process of elimination, 10-fold cross validation was implemented to optimize the variable selection and to ascertain the standard deviation of error. In the recursion process, the model with minimum RMSE and with least standard deviation error was set as the optimum model; if it finds another model with a different subset of variables, it automatically updates and ranks. Finally, it selects the best variables yielding the smallest RMSE.

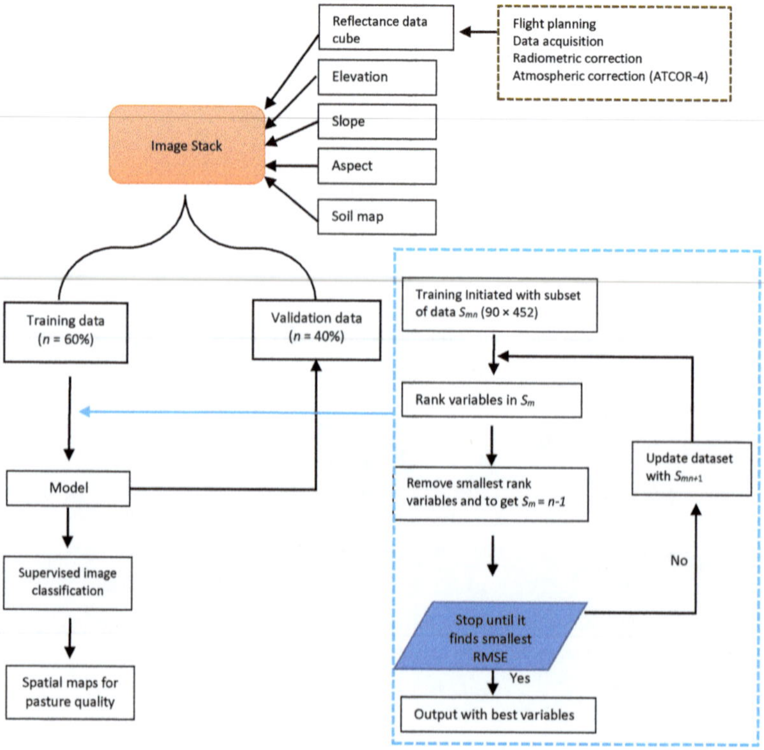

Figure 3. The workflow of the proposed procedure for creating pasture-quality maps.

The goodness of fit of the developed regression models were evaluated by calculating the cross-validated coefficient of determination (R^2_{CV}), cross-validated root mean square error ($RMSE_{CV}$), and cross-validated ratio to prediction deviation (RPD_{CV}).

$$RMSE_{CV} = \sqrt{\sum_{i=1}^{N} \frac{(\hat{y}_i - y)^2}{N}}$$

$$RPD_{CV} = \frac{SD(y)}{RMSE_{CV}}$$

where $\hat{y}$ is the predicted values, y is the observed value, and N is the number of samples. SD(y) refers to standard deviation of measured y. Models with RPD ≥ 2 predict well with reliable estimates [14]. Pasture quality maps were generated using the best model.

The final maps were created using masking. For this, the land surface cover maps were created by classifying the hyperspectral image. The land cover types included in this study are forest, bush, pasture, and non-vegetation areas. The classification was performed using supervised algorithm, support vector machine (SVM). SVM is a robust classification method widely used for hyperspectral imagery [37]. We have used radial basis function; their parameters (cost and gamma) were optimized using 5-fold cross-validation. High-resolution RGB images were used as a georeference for selecting the training pixels from the hyperspectral image. The training model was then used to extrapolate the hyperspectral image across the landscape. Model development, creating pasture-quality maps, and classification of hyperspectral image was performed in MATLAB® environment.

3. Results

The pasture quality attributes were estimated from the samples collected in the field. The descriptive statistics of pasture quality values for the calibration model are summarized in Table 1. The collected pasture samples had high variability for CP (CV = 23.05%) with a range from 6.06 to 25.64 and high standard deviation (std = 3.76). In contrast, ME exhibited low variability (CV = 12.55%) and a small range (6–12.50) of samples. As expected, a wide range of pasture variability, particularly with CP, exists on hill country farms due to diverse environmental conditions [38].

Table 1. Descriptive statistics: mean, standard deviation (std), minimum (min), maximum (max), and coefficient of variation (CV) pasture quality data was used to build a calibration model.

Pasture Quality	Mean	std	Min	Max	CV (%)
Crude protein (%)	16.31	3.76	6.06	25.64	23.05
Metabolisable energy (MJ/kg DM)	10.91	1.37	6.00	12.50	12.55

Note: std- standard deviation; CV- coefficient of variation.

FDR of corresponding pasture samples extracted from the AisaFENIX were presented as a function of pasture quality attributes (CP and ME) in Figure 4.

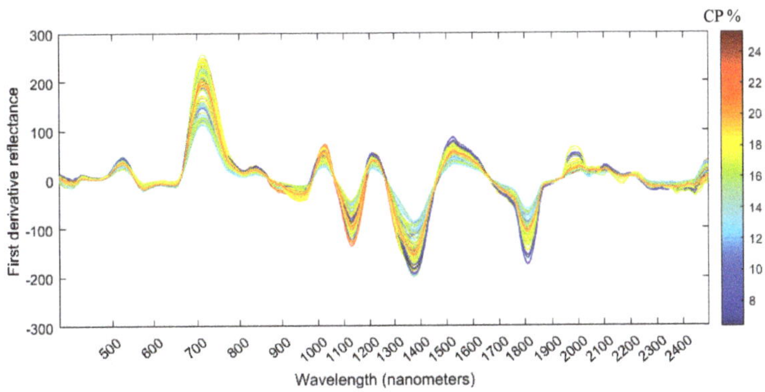

(a)

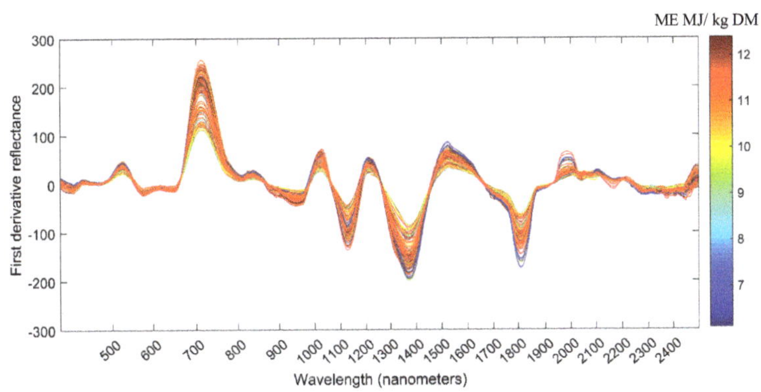

(b)

Figure 4. First derivative reflectance of pasture as a function of spectral wavelength. The color of the spectra indicates the content of CP (**a**) and ME (**b**).

The magnitude of absorption features are highly variable and complex with relevant pasture quality values. In Figure 4, there are a few distinctive spectral features of high and low pasture quality that can be seen around 1230 nm, 1340 nm, 1550 nm, and 1800 nm. The relationships between spectral data and measured values of CP and ME were shown in Table 2. The results from this study indicate that CP was predicted with high accuracy from hyperspectral data using RF technique (R^2_{CV} = 0.66, $RMSE_{CV}$ = 2.24, RPD_{CV} = 1.68). However, the accuracy was slightly improved (R^2_{CV} = 0.70, $RMSE_{CV}$ = 2.06, RPD_{CV} = 1.82) by adding environmental variables (elevation, slope angle, slope aspect, and soil type) (Table 2). The pasture ME was predicted with an R^2_{CV} of 0.61, $RMSE_{CV}$ = 0.85, and RPD_{CV} = 1.62 with hyperspectral data alone. The prediction accuracy of ME increased dramatically after including environmental variables (R^2_{CV} = 0.75, $RMSE_{CV}$ = 0.65, and RPD_{CV} = 2.11). Separate regression models were also created to assess the impact of only environmental variables on both CP and ME. This resulted in a relatively low accuracy models ($0.35 \geq R^2_{CV} \leq 0.31$).

When feature selection was performed using RF–RFE, the accuracy was further improved for both CP and ME. It is worth noting that the improvement was higher in the case of the CP model when compared to the ME model. The calibration model prediction results were consistent with validation results, though the validation results were slightly lower than the calibration results. Only 7–8% of hyperspectral variables were selected as important for describing CP and ME, which are present across the electromagnetic spectrum, though the majority of them are concentrated in the short wave infrared (SWIR) region. The selected important wavebands by RF–RFE for each pasture quality attribute are shown in Figure 5. For CP, the sensitive spectral bands are 505–554, 609, 612, 784, 787, 818, 822, 842, 932, 939, 946, 959, 1000, 1500, 1935, 2013, 2018, 2035, 2107, 2178, 2234, 2344, and 2420 nm. The sensitive spectral bands for ME are 517–520, 643, 653, 684, 691, 753, 849, 890, 939, 963, 1017, 1276, 1512–1520, 1618, 1785, 1796, 1802–1808, 1935, 1996, 2013, 2051, 2090, 2123, 2173, 2239, 2305, 2415, and 2420 nm. The pasture attributes CP and ME have moderate intercorrelation (R^2 = 0.38) and are found with few common bands (939, 1935, 2013, 2420 nm).

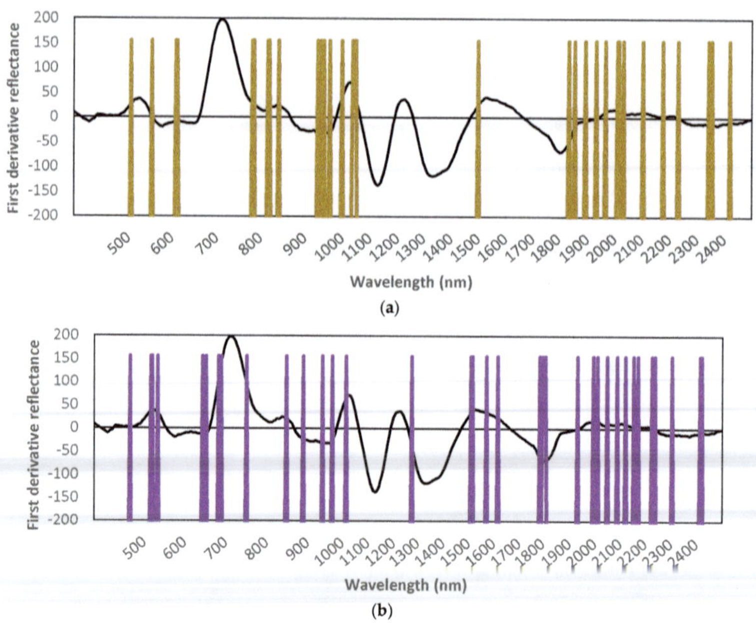

Figure 5. Important wavelengths selected using recursive feature selection-random forest regression for (a) crude protein (CP) (b) metabolizable energy (ME).

Table 2. Regression model results for crude protein (CP) and metabolizable energy (ME) using airborne hyperspectral imaging. Regression metrics are R^2_{CV}-cross-validated coefficient of determination, $RMSE_{CV}$-cross-validated root mean square error, and RPD_{CV}- cross-validated ratio to prediction deviation.

Pasture Quality	Variables	R^2_{CV}	$RMSE_{CV}$	RPD_{CV}	R^2_P	$RMSE_P$	RPD_P
CP	Full-spectrum	0.66	2.24	1.68	0.63	2.20	1.65
CP	Elevation, slope and aspect and soil type	0.31	3.13	1.20	0.27	3.81	0.95
CP	FS, elevation, slope, aspect and soil type	0.70	2.06	1.82	0.70	2.01	1.80
CP	Selected variables [a]	0.80	1.68	2.23	0.78	1.72	2.11
ME	FS	0.61	0.85	1.62	0.61	0.86	1.64
ME	Elevation, slope and aspect and soil type	0.35	1.10	1.24	0.32	1.22	1.15
ME	FS, elevation, slope, aspect and soil type	0.75	0.65	2.11	0.77	0.63	2.23
ME	Selected variables [b]	0.78	0.61	2.25	0.79	0.60	2.35

Note: FS—Full spectrum; [a] elevation, slope, soil type, and spectral features from Figure 5a. [b] slope, soil type and spectral features from Figure 5b.

In both CP and ME, the included environmental variables were found to have significant influence on model performance with improved accuracy.

The prediction models with the highest R^2_{CV} values were to create raster maps, depicting the spatial pattern of pasture quality (Figure 6). The spatial maps of CP and ME were masked using the land surface classification map from SVM, which is as accurate as 93.4% (overall accuracy). The pasture areas were highlighted with colored pixels and non-pasture areas left empty with a background of RGB image. The range of predicted CP is from 6.32 to 25.60% with high values in the east and west sides of the study area, while the south was dominated with low CP pasture (Figure 6). Compared with CP, ME was less variable across the area with the majority of the area dominated by moderate ME values (Figure 6).

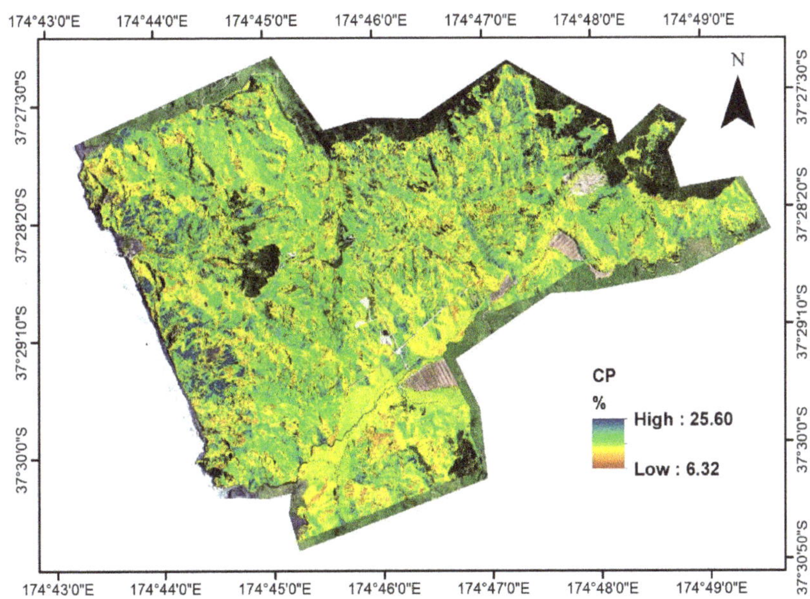

(a)

Figure 6. *Cont.*

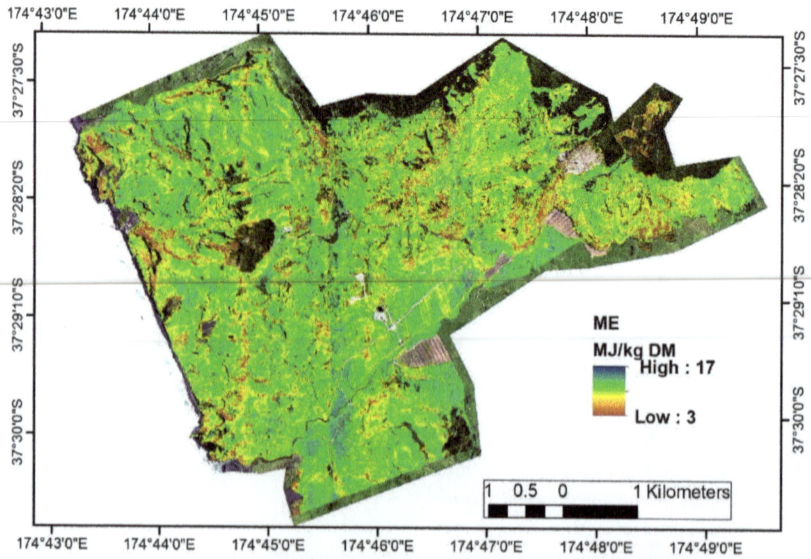

(b)

Figure 6. Pasture quality spatial maps produced from fusion of hyperspectral, topographic, and soil data using recursive feature selection–random forest regression (**a**) crude protein (CP) concentration map (**b**) metabolizable energy map. Predictions are only shown for pasture areas.

4. Discussion

Airborne hyperspectral imaging has potential for estimating CP and ME accurately and over large spatial extents which enables continuous spatial maps to be created. In this study, a hill country farm was imaged with an airborne hyperspectral system which produced accurate estimates for CP (RPD_{CV} = 2.23) and ME (RPD_{CV} = 2.25) of heterogeneous mixed pasture. The successful application of this technology in pasture quality is not surprising, as pointed out by previous studies [14,16,19,39]; however, the approach used in this study improved the prediction results by integrating the hyperspectral and environmental data-combined machine-learning algorithms. Such knowledge of the landscape could inform pasture and herd management decisions to improve animal production and assist in land stewardship efforts.

In Figure 4, the pasture quality relevant features were not very distinct. This is due to the fact that the variation in canopy reflectance was primarily influenced by direct and indirect confounding factors such as canopy structure, solar/viewing geometry, soil background, broad water absorptions, while the contribution from vegetation chemistry is very small (2–4%) [40]. These multiple factors also impede the attribute estimation to some extent, though the proposed approach produced accurate estimates. Further investigation is required to break down the individual influence from these factors on pasture quality.

Although the pasture was characterized by heterogeneity, RF has accounted for maximum explanation on pasture quality from the hyperspectral data. Many studies suggest that RF may be more powerful than the traditional multivariate regression methods as it extracts complex, non-linear information from the spectral data [24,27]. In this study, RF accounted for >70% of the variability in CP and ME from the hyperspectral data alone. When combined with the topographic and soil data, the RF–RFE approach showed an improvement in the prediction accuracy. The latter indicates the importance of considering environmental data for estimating pasture quality. Similarly, Ramoelo et al. [22] attempted to combine proximal hyperspectral data with environmental data to predict nitrogen and phosphorus concentration of grass in a savannah ecosystem using non-linear-PLSR, and the researchers found improved results over the hyperspectral data alone. From the results obtained in this study, we recommend the use of spectral and environmental variables together to provide improved prediction accuracy.

RF–RFE is capable of selecting important spectral and environmental features that are sensitive to pasture quality and improved the accuracy levels. Similarly, Granitto et al. [29] used RF–RFE for analysing high dimensional data and found it to be an efficient feature selection method, far better than traditional methods. Other researchers [21,41] found that considering noisy variables could interfere with model performance, which leads to over-fitting. Therefore, important relevant features need to be selected for robust estimates. As seen in Figure 5, the selected spectral bands are scattered over the whole spectrum indicating the importance of full-spectrum. The selected bands for each quality attribute are different because of the contrasting chemical composition of each attribute. However, bands at 939, 1935, 2013, and 2420 nm are mutually selected in both cases. This might be due to the presence of common functional bonds; hence, both quality attributes are noticed with correlation (Table 2). The band 1935 nm is influenced by broad water absorption centered at 1940 nm [40]. With both quality attributes, the spectral region from 500–770 nm related electronic transitions caused by pigment absorptions, such as chlorophyll, xanthophyll, and carotenoids. Clustering of sensitive bands are located in the SWIR region, which are mainly characterized by fundamental overtones and harmonics of O–H, C–C, C–H, and N–H [3,42]. In CP, the majority of the selected bands (932, 1000, 1500, 1935, 2035, 2178, 2234 and 2344 nm) are closely assigned with bond vibrations of protein and nitrogen molecules [3]. Fundamentally, ME is mainly composed of crude fibre [43]; therefore, the majority of the selected bands (1017, 1276, 1512–1520, 1618, 1785, 1796, 1802–1808, 1935, 2090, 2123, 2239, 2305, 2415 and 2420 nm) are associated with vibrations of lignin, cellulose, and hemicellulose [3].

Pasture quality appeared to be correlated with topographic variables and soil type, which indicates that these variables are also one of the key drivers to influence pasture quality (Table 2). This leads to improved accuracy by combining the hyperspectral and environmental data (Table 2). In this study, a wide range of soil types with different nutrient levels are present across the farm, which directly support changes of pasture quality (CP and ME). For example, in Figures 2 and 6, gley soils dominated paddocks distributed with high quality pasture (high ME and optimal CP values) because of high fertility of soils. In contrast, paddocks with allophanic soils show low quality pasture (low ME and CP). Slope variables also positively influenced the pasture quality. Generally, flat regions were associateded with high fertility soils, while hilly regions lost soil fertility due to surface run-off [22]. Although the current study proves the feasibility of mapping pasture quality at local scale, under large-scale environments, the influence of topography and soil type might be different due to the presence of different soil types and environments. Moreover, the relationships might change with seasons due to variable weather conditions and pasture response. Therefore, further investigation is required before utilizing this model for large-scale environments.

Understanding the spatial variability of pasture quality in hill country farms allows for more efficient use of natural resources and improving agronomic management. Both CP and ME exhibited different spatial patterns, reflecting the different factors that influence each attribute. Fertilizer is a key input in hill country farms, as it helps to maintain high quality pasture. Traditional blanket application of fertilizer ignores spatial variability and can result in the application of excessive fertilizer on high-fertility zones and vice-versa on low fertility which can result in the loss of fertilizer into the environment; with fertilizer being such a large investment for farmers, it is important to ensure that the value of that investment is realized as fully as possible. In 2006, Murray and Yule [44] conducted an experiment to test the performance variable rate fertilizer (VRF) over blanket application at Limestone Downs based on broad scale annual pasture production. The single super phosphate was applied through aerial top-dressing aircraft with a controlled ground resolution of 18 m^2. They reported that the annual pasture production could be increased between 6.5–24.4% by VRF. They also conducted an economic analysis on implementing VRF over the blanket application where they found that this technique could increase 26% annual cash returns per hectare [45]. These findings clearly indicate the potential to improve fertilizer use efficiency and the economic benefits, as well as to reduce the risk of fertilizer wastage contaminating the environment from VRF [38]. From the current research, we are able to quantify pasture quality attributes more accurately at fine scale, which are the better indicators of spatial variability within field over the annual pasture production. These pasture quality maps can provide the necessary inputs for VRF applications. In addition, these comprehensive spatial information quality maps enable the farm managers to ensure proper mineral nutrition of ruminant animals [46].

5. Conclusions

This study evaluated the feasibility of combing the hyperspectral, topography, and soil data to retrieve pasture quality attributes of heterogeneous pasture using RF–RFE. Since the pasture quality variation is driven by many environmental and management factors, our results showed that combining hyperspectral data with generally available environmental factors (elevation, slope angle, and soil type) improved the prediction accuracy in comparison with hyperspectral data alone. This result also confirmed that RF-RFE markedly improved the estimates of pasture quality (RPD = 2.11–2.35) by selecting the most sensitive variables across the spectrum and environmental data. Elevation, slope, and soil type were found as important variables to predict CP while for ME, the same variables with the exception of elevation were found to be significant. The applicability of the proposed approach needs to be investigated on a large-scale where highly variable soil types exist. Overall, the final accurate pasture quality spatial maps allow farmers to optimize their agronomic decisions.

Author Contributions: This experiment was conducted and designed by P.R.R., The hyperspectral image processing was performed by K.G. The data analysis performed by P.R.R. The manuscript written by P.R.R. and K.G. In addition, Y.I.J. has provided significant contributions for manuscript preparation.

Funding: The research was funded through Primary Growth Partnership (PGP) programme "Pioneering to Precision-Application of Fertiliser in Hill Country" jointly led by Ravensdown Limited and Ministry of Primary Growth Industries (MPI), New Zealand.

Acknowledgments: The authors gratefully acknowledge the funding bodies (MPI and Ravensdownn Limited) and the field team from Massey University and AgResearch.

Conflicts of Interest: The authors declare no conflict of interest.

References

1. Holmes, C.W.; Wilson, G.F.; Mackenzie, D.D.S.; Flux, D.S.; Brookes, I.M.; Davey, A.W.F. *Milk Production from Pasture*; Massey University: Palmerston North, New Zealand, 2007; p. 602.
2. Schellberg, J.; Hill, M.J.; Gerhards, R.; Rothmund, M.; Braun, M. Precision agriculture on grassland: Applications, perspectives and constraints. *Eur. J. Agron.* **2008**, *29*, 59–71. [CrossRef]
3. Curran, P.J. Remote sensing of foliar chemistry. *Remote Sens. Environ.* **1989**, *30*, 271–278. [CrossRef]
4. Kokaly, R.F.; Asner, G.P.; Ollinger, S.V.; Martin, M.E.; Wessman, C.A. Characterizing canopy biochemistry from imaging spectroscopy and its application to ecosystem studies. *Remote Sens. Environ.* **2009**, *113* (Suppl. 1), S78–S91. [CrossRef]
5. Mutanga, O.; Skidmore, A.K. Narrow band vegetation indices overcome the saturation problem in biomass estimation. *Int. J. Remote Sens.* **2004**, *25*, 3999–4014. [CrossRef]
6. Pullanagari, R.R.; Kereszturi, G.; Yule, I.J. Quantification of dead vegetation fraction in mixed pastures using aisafenix imaging spectroscopy data. *Int. J. Appl. Earth Obs. Geoinf.* **2017**, *58*, 26–35. [CrossRef]
7. Haboudane, D.; Miller, J.R.; Pattey, E.; Zarco-Tejada, P.J.; Strachan, I.B. Hyperspectral vegetation indices and novel algorithms for predicting green LAI of crop canopies: Modeling and validation in the context of precision agriculture. *Remote Sens. Environ.* **2004**, *90*, 337–352. [CrossRef]

8. Thenkabail, P.S.; Mariotto, I.; Gumma, M.K.; Middleton, E.M.; Landis, D.R.; Huemmrich, K.F. Selection of hyperspectral narrowbands (HNBs) and composition of hyperspectral twoband vegetation indices (HVIs) for biophysical characterization and discrimination of crop types using field reflectance and hyperion/eo-1 data. *IEEE J. Sel. Top. Appl. Earth Obs.* **2013**, *6*, 427–439. [CrossRef]
9. Li, F.; Miao, Y.; Hennig, S.; Gnyp, M.; Chen, X.; Jia, L.; Bareth, G. Evaluating hyperspectral vegetation indices for estimating nitrogen concentration of winter wheat at different growth stages. *Precis. Agric.* **2010**, *11*, 335–357. [CrossRef]
10. Mutanga, O.; Skidmore, A.K. Integrating imaging spectroscopy and neural networks to map grass quality in the kruger national park, South Africa. *Remote Sens. Environ.* **2004**, *90*, 104–115. [CrossRef]
11. Mutanga, O.; Skidmore, A.K. Red edge shift and biochemical content in grass canopies. *ISPRS J. Photogramm. Remote Sens.* **2007**, *62*, 34–42. [CrossRef]
12. Mutanga, O.; Kumar, L. Estimating and mapping grass phosphorus concentration in an African savanna using hyperspectral image data. *Int. J. Remote Sens.* **2007**, *28*, 4897–4911. [CrossRef]
13. Knox, N.M.; Skidmore, A.K.; Prins, H.H.T.; Asner, G.P.; van der Werff, H.M.A.; de Boer, W.F.; van der Waal, C.; de Knegt, H.J.; Kohi, E.M.; Slotow, R.; et al. Dry season mapping of savanna forage quality, using the hyperspectral carnegie airborne observatory sensor. *Remote Sens. Environ.* **2011**, *115*, 1478–1488. [CrossRef]
14. Pullanagari, R.; Yule, I.; Tuohy, M.; Hedley, M.; Dynes, R.; King, W. In-field hyperspectral proximal sensing for estimating quality parameters of mixed pasture. *Precis. Agric.* **2012**, *13*, 351–369. [CrossRef]
15. Skidmore, A.K.; Ferwerda, J.G.; Mutanga, O.; Van Wieren, S.E.; Peel, M.; Grant, R.C.; Prins, H.H.T.; Balcik, F.B.; Venus, V. Forage quality of savannas—Simultaneously mapping foliar protein and polyphenols for trees and grass using hyperspectral imagery. *Remote Sens. Environ.* **2010**, *114*, 64–72. [CrossRef]
16. Thulin, S.; Hill, M.J.; Held, A.; Jones, S.; Woodgate, P. Hyperspectral determination of feed quality constituents in temperate pastures: Effect of processing methods on predictive relationships from partial least squares regression. *Int. J. Appl. Earth Obs. Geoinf.* **2012**, *19*, 322–334. [CrossRef]
17. Pullanagari, R.; Yule, I.; Hedley, M.; Tuohy, M.; Dynes, R.; King, W. Multi-spectral radiometry to estimate pasture quality components. *Precis. Agric.* **2012**, *13*, 442–456. [CrossRef]
18. Foster, A.J.; Kakani, V.G.; Mosali, J. Estimation of bioenergy crop yield and n status by hyperspectral canopy reflectance and partial least square regression. *Precis. Agric.* **2017**, *18*, 192–209. [CrossRef]
19. Biewer, S.; Fricke, T.; Wachendorf, M. Development of canopy reflectance models to predict forage quality of legume-grass mixtures. *Crop Sci.* **2009**, *49*, 1917–1926. [CrossRef]
20. Sanches, I.D.; Tuohy, M.P.; Hedley, M.J.; Mackay, A.D. Seasonal prediction of in situ pasture macronutrients in new zealand pastoral systems using hyperspectral data. *Int. J. Remote Sens.* **2013**, *34*, 276–302. [CrossRef]
21. Pullanagari, R.R.; Kereszturi, G.; Yule, I.J. Mapping of macro and micro nutrients of mixed pastures using airborne aisafenix hyperspectral imagery. *ISPRS J. Photogramm. Remote Sens.* **2016**, *117*, 1–10. [CrossRef]
22. Ramoelo, A.; Skidmore, A.K.; Cho, M.A.; Mathieu, R.; Heitkönig, I.M.A.; Dudeni-Tlhone, N.; Schlerf, M.; Prins, H.H.T. Non-linear partial least square regression increases the estimation accuracy of grass nitrogen and phosphorus using in situ hyperspectral and environmental data. *ISPRS J. Photogramm. Remote Sens.* **2013**, *82*, 27–40. [CrossRef]
23. Verrelst, J.; Camps-Valls, G.; Muñoz-Marí, J.; Rivera, J.P.; Veroustraete, F.; Clevers, J.G.; Moreno, J. Optical remote sensing and the retrieval of terrestrial vegetation bio-geophysical properties—A review. *ISPRS J. Photogramm. Remote Sens.* **2015**, *108*, 273–290. [CrossRef]
24. Mutanga, O.; Adam, E.; Cho, M.A. High density biomass estimation for wetland vegetation using worldview-2 imagery and random forest regression algorithm. *Int. J. Appl. Earth Obs. Geoinf.* **2012**, *18*, 399–406. [CrossRef]
25. White, J.; Hodgson, J. *New Zealand Pasture and Crop Science*; Oxford University Press: Oxford, UK, 2000.
26. Gray, D.; Reid, J.; Kemp, P.; Brookes, I.; Horne, D.; Kenyon, P.; Matthew, C.; Morris, S.; Valentine, I. Controlling pasture quality on hill country-key decisions and techniques. *Agron. N. Z.* **2003**, *34*, 73–82.
27. Abdel-Rahman, E.M.; Ahmed, F.B.; Ismail, R. Random forest regression and spectral band selection for estimating sugarcane leaf nitrogen concentration using eo-1 Hyperion hyperspectral data. *Int. J. Remote Sens.* **2013**, *34*, 712–728. [CrossRef]
28. Andersen, C.M.; Bro, R. Variable selection in regression—A tutorial. *J. Chemom.* **2010**, *24*, 728–737. [CrossRef]
29. Granitto, P.M.; Furlanello, C.; Biasioli, F.; Gasperi, F. Recursive feature elimination with random forest for ptr-ms analysis of agroindustrial products. *Chemom. Intell. Lab. Syst.* **2006**, *83*, 83–90. [CrossRef]

30. Richter, R.; Schläpfer, D. Geo-atmospheric processing of airborne imaging spectrometry data. Part 2: Atmospheric/topographic correction. *Int. J. Remote Sens.* **2002**, *23*, 2631–2649. [CrossRef]
31. Burrough, P.A.; McDonnell, R.A.; Lloyd, C.D. *Principles of Geographical Information Systems*; Oxford University Press: Oxford, UK, 2015.
32. Breiman, L. Random forests. *Mach. Learn.* **2001**, *45*, 5–32. [CrossRef]
33. Fawagreh, K.; Gaber, M.M.; Elyan, E. Random forests: From early developments to recent advancements. *Syst. Sci. Control Eng.* **2014**, *2*, 602–609. [CrossRef]
34. Oliveira, S.; Oehler, F.; San-Miguel-Ayanz, J.; Camia, A.; Pereira, J.M.C. Modeling spatial patterns of fire occurrence in mediterranean europe using multiple regression and random forest. *For. Ecol. Manag.* **2012**, *275*, 117–129. [CrossRef]
35. Guyon, I.; Weston, J.; Barnhill, S.; Vapnik, V. Gene selection for cancer classification using support vector machines. *Mach. Learn.* **2002**, *46*, 389–422. [CrossRef]
36. Zhou, Q.; Zhou, H.; Zhou, Q.; Yang, F.; Luo, L. Structure damage detection based on random forest recursive feature elimination. *Mech. Syst. Signal Process.* **2014**, *46*, 82–90. [CrossRef]
37. Burai, P.; Deák, B.; Valkó, O.; Tomor, T. Classification of herbaceous vegetation using airborne hyperspectral imagery. *Remote Sens.* **2015**, *7*, 2046–2066. [CrossRef]
38. Scrimgeour, F.G. Pathways ahead for New Zealand hill country farming. *J. N. Z. Grassl.* **2016**, *78*, 73–82.
39. Kawamura, K.; Watanabe, N.; Sakanoue, S.; Inoue, Y. Estimating forage biomass and quality in a mixed sown pasture based on partial least squares regression with waveband selection. *Grassl. Sci.* **2008**, *54*, 131–145. [CrossRef]
40. Wang, Z.; Skidmore, A.K.; Wang, T.; Darvishzadeh, R.; Heiden, U.; Heurich, M.; Latifi, H.; Hearne, J. Canopy foliar nitrogen retrieved from airborne hyperspectral imagery by correcting for canopy structure effects. *Int. J. Appl. Earth Obs. Geoinf.* **2017**, *54*, 84–94. [CrossRef]
41. Zou, X.; Zhao, J.; Povey, M.J.W.; Holmes, M.; Mao, H. Variables selection methods in near-infrared spectroscopy. *Anal. Chim. Acta* **2010**, *667*, 14–32.
42. Kokaly, R.F. Investigating a physical basis for spectroscopic estimates of leaf nitrogen concentration. *Remote Sens. Environ.* **2001**, *75*, 153–161. [CrossRef]
43. Özelçam, H.; Kırkpınar, F.; Tan, K. Chemical composition, in vivo digestibility and metabolizable energy values of caramba (*Lolium multiflorum* cv. Caramba) fresh, silage and hay. *Asian Australas. J. Anim. Sci.* **2015**, *28*, 1427–1432. [CrossRef] [PubMed]
44. Murray, R.I.; Yule, I.J. Developing variable rate application technology: Scenario development and agronomic evaluation. *N. Z. J. Agric. Res.* **2007**, *50*, 53–63. [CrossRef]
45. Murray, R.I.; Yule, I.J. Developing variable rate application technology: Economic impact for farm owners and topdressing operators. *N. Z. J. Agric. Res.* **2007**, *50*, 65–72. [CrossRef]
46. Soder, K.J.; Stout, W.L. Effect of soil type and fertilization level on mineral concentration of pasture: Potential relationships to ruminant performance and health. *J. Anim. Sci.* **2003**, *81*, 1603–1610. [CrossRef] [PubMed]

remote sensing MDPI

Article

Modeling Mid-Season Rice Nitrogen Uptake Using Multispectral Satellite Data

James Brinkhoff [1,*], Brian W. Dunn [2], Andrew J. Robson [1], Tina S. Dunn [2] and Remy L. Dehaan [3]

1 Applied Agricultural Remote Sensing Centre, University of New England, Armidale, NSW 2351, Australia
2 NSW Department of Primary Industries, 2198 Irrigation Way, Yanco, NSW 2703, Australia
3 EH Graham Centre for Agricultural Innovation (NSW Department of Primary Industries and Charles Sturt University), Locked Bag 588, Wagga, NSW 2678, Australia
* Correspondence: james.brinkhoff@une.edu.au

Received: date; Accepted: date; Published: date

Abstract: Mid-season nitrogen (N) application in rice crops can maximize yield and profitability. This requires accurate and efficient methods of determining rice N uptake in order to prescribe optimal N amounts for topdressing. This study aims to determine the accuracy of using remotely sensed multispectral data from satellites to predict N uptake of rice at the panicle initiation (PI) growth stage, with a view to providing optimum variable-rate N topdressing prescriptions without needing physical sampling. Field experiments over 4 years, 4–6 N rates, 4 varieties and 2 sites were conducted, with at least 3 replicates of each plot. One WorldView satellite image for each year was acquired, close to the date of PI. Numerous single- and multi-variable models were investigated. Among single-variable models, the square of the NDRE vegetation index was shown to be a good predictor of N uptake (R^2 = 0.75, RMSE = 22.8 kg/ha for data pooled from all years and experiments). For multi-variable models, Lasso regularization was used to ensure an interpretable and compact model was chosen and to avoid over fitting. Combinations of remotely sensed reflectances and spectral indexes as well as variety, climate and management data as input variables for model training achieved $R^2 < 0.9$ and RMSE < 15 kg/ha for the pooled data set. The ability of remotely sensed data to predict N uptake in new seasons where no physical sample data has yet been obtained was tested. A methodology to extract models that generalize well to new seasons was developed, avoiding model overfitting. Lasso regularization selected four or less input variables, and yielded R^2 of better than 0.67 and RMSE better than 27.4 kg/ha over four test seasons that weren't used to train the models.

Keywords: rice; nitrogen management; remote sensing; multispectral imagery; reflectance index; multiple variable linear regression; Lasso model

---

1. Introduction

Nitrogen (N) is a key input for plant development due to its role in the production of chlorophyll, which is crucial for photosynthesis [1]. It has a significant role in attaining crop yield potential [2] and quality [3]. However, excessive N fertilization results in pollution, leading to negative environmental outcomes [4]. Further, N use efficiencies are often low, which leads to non-optimal production costs [5].

Rice N application can be optimized to meet desired targets [6], such as yield, financial return on N cost [7], quality and protein content [8]. If too little N is available to the plants, the crop will not reach its yield potential [9]. If too much N is applied, there is a risk that the crop will lodge, thus decreasing yield and increasing the time needed to harvest [10]. Excess N also increases the risk of low temperature induced floret sterility, thereby reducing yield [11] and also increases the risk of disease [10]. Rice protein is determined by N status, with high levels of N resulting in lower cooked rice quality [8].

A single pre-season N application is less effective than split application with mid-season topdressing [12]. The Oryza-0 crop model has been used in simulations to optimize N application splits to optimize yield [7]. The most effective time for a mid-season application is at the panicle initiation (PI) stage [13], which occurs at the start of the reproductive phase of rice development, when the panicle begins to form in the base of the stem [14]. Typically, a base N rate is applied at the start of the season, then tissue tests are done mid-season to determine additional N top-dressing rates [15].

Pre-season soil tests have not proven to be a reliable way to determine paddy soil N requirements [16,17]. Detailed recommendations of N top-dressing rates as a function of PI N uptake and rice variety have been developed, utilising physical samples of plants at PI [15,18]. Site-specific management of rice N application has been shown to both improve yield [19] and reduce total N applied [20]. This improves farming profitability (due to reduced input costs and increased revenue) and environmental outcomes (due to reduced N loss to the atmosphere and water systems) [21]. N requirements also vary spatially within paddies [22], which motivates the application of spatially varying N at PI to match requirements. Studies of the application of such precision agriculture variable-rate techniques in wheat [23] and sugarcane [24] have shown improvement in N use efficiency.

There are a number of methods to determine N uptake at PI, with varying degrees of cost-effectiveness, practicability, accuracy and spatial resolution. Plant tissue samples can be collected and analyzed, however growers have little time and motivation to perform this task [15]. When they do take samples, they are often only at a few points and so may not encompass the variability in their paddies. In addition, sample sizes are small in size (typically 0.2 m^2), so sampling location can have a large impact on the degree to which the samples accurately represent the variation of N uptake across a site. Point measurements of rice N status can also be made using in-field sensors, such as SPAD meters [1]. However, measuring individual leaves results in limited accuracy due to leaf-to-leaf variation, and many leaves must be sampled to represent plant N uptake adequately [12]. These point sampling methods are labour intensive and therefore have seen limited adoption. Therefore, developing an operational remote sensing based method to determine N uptake is of great interest, as it captures spatial variation and may eliminate the need for field sampling. Typically industry desires variable N prescription maps with rates separated into 30 kg/ha classes (as aerial spreaders are probably not much more accurate than this and there is minimal impact on yield at PI topdressing rates less than 30 kg/ha [15]). Therefore, N uptake prediction errors of less than 30 kg/ha are required [18].

Canopy chlorophyll content has strong relationships with remotely sensed reflectances, and canopy nitrogen (N) content is strongly related to chlorophyll content [25]. Thus, remotely sensed data can be used to predict canopy nitrogen content. Vegetation indexes using bands in the red-edge (740 nm), near infrared (790 nm) and green (550 nm) regions are particularly important to predict canopy N status [26]. Chlorophyll content is strongly related to the red-edge and green absorption of leaves, while near infrared characteristics are related to structure and leaf thickness [26].

Sensor types include optical cameras (3 bands—red, green blue) [27], multispectral sensors (often adding near-infrared and red-edge bands to optical bands) [21] and hyperspectral sensors [28], which capture reflectance at over many narrow bands. Hyperspectral data was used to investigate determining N content at the heading stage [29], and at the panicle formation stage [28,30]. Significantly different relationships between plant N status and hyperspectral data (measured in field with hyperspectral spectroradiometers) were found before and after heading [31], explained by rice plant morphological differences. They found linear combinations of reflectance bands resulted in better prediction than ratio indexes or normalized difference indexes. The derivatives of reflectance with respect to wavelength may give greater accuracy across varieties, regions and seasons as shown in [28]. Despite the accuracy and predictive power available from hyperspectral measurements, there are significant drawbacks including the cost of the sensors and data processing power required to derive useful models [32]. These limit the current applicability of hyperspectral sensors to monitor rice crops for industry-wide commercial applications. However, these studies using hyperspectral sensors have shown the most important wavelengths for sensing N status, and therefore what wavelengths would be desired in multispectral sensors for this application. Multispectral sensors are relatively cost-effective, and have been used to predict rice N uptake with reasonable accuracy [27], particularly when the sensor includes detection of the red-edge part of the spectrum.

Some studies have used hand-held or proximal sensors [31]. Mounting sensors on unmanned aerial vehicles (UAVs) allows collecting within-field variability data in much less time than using proximal sensors [21,33,34]. Some proximal sensing systems use active sensors, where the incident radiation is generated by the device. In contrast, most UAV systems utilise passive sensors, which rely on incident radiation from the sun and measure reflected radiation. Thus they require irradiance measurement or measurement of a known reflectance target in order to generate radiometrically corrected reflectance data [35]. Thus, to generate accurate data, it is crucial to ensure sensors are calibrated accurately for current radiance conditions [36], which may limit widespread grower adoption. The restrictive regulations on UAV operations, including line-of-sight and non-autonomous requirements, and the costs of acquiring images over larger areas currently render UAVs un-economical for large-area industry-wide application [37]. Satellite-based multispectral data for rice N sensing is an alternative [38,39].

To achieve the goal of an operational remote-sensing based model to predict N uptake, a crucial factor is the stability of models over seasons, locations, varieties and management practices such as planting dates. Significant seasonal variation in models has been observed, attributed to differences in environmental conditions [29]. These drivers, such as cumulative temperature and solar radiation, affect the dynamics of N uptake in rice [40]. Soil N is also dynamic, and models that seek to address this were discussed in [41]. Thus, it may be necessary to account for date differences between image captures and the dates that the actual N uptake is required.

Relatively few reports [29,30] train models on one or more seasons and test the model's prediction accuracy on a separate season. If a model is required to predict future seasons N uptake from past data, it is important to separate model training, validation and testing data based on season, rather than randomly selecting points from all available seasons data. Another important factor for models is that they be interpretable and easily integrated into geographic information system (GIS) platforms. These goals may not be achieved if black-box machine learning methods are used, such as support vector machine (SVM) models and neural networks. Similar to [21], we employed multi-variable linear regression using the Lasso regularization technique to select a subset of the most important predictors, while discarding less important predictors.

Our work is part of a project to provide an operational tool to Australian rice growers, to generate spatial maps of optimal N prescription rates to be applied at panicle initiation using remote sensing. It aims to provide a viable alternative to manual plant sampling, which has suffered limited industry adoption due to the time and effort required. Plant sampling also provides limited spatial resolution so is not useful for potential variable-rate applications. We assessed the stability of remotely sensed data over four seasons, two sites, four varieties and different sowing dates. Optimal combinations of data input variables to best predict N uptake were found, using data from WorldView satellite multispectral sensors, as well as climate, variety and management data. The WorldView sensors were chosen as they offer a combination of high spectral and spatial resolution. They are thus able to test our methodology of generating multi-season models of N uptake with fewer limitations from the sensors themselves. We draw conclusions on the best performance that can be expected in the study environment, and discuss potential transfer of the methodology to other, more cost effective, sensors.

2. Materials And Methods

2.1. Experiments

The study was based in Leeton, NSW, Australia. A list of all the experiments used in this work is given in Table 1. It includes data from four rice growing seasons (harvest in autumn each year from 2015–2018), four rice varieties (Reiziq™, Sherpa™, Langi and Topaz™), multiple sowing dates and two sites with differing soil types (Yanco Agriculture Institute (YAI) [146.425E, 34.618S] and Leeton Field Station (LFS) [146.361E, 34.606S]). The soil types of these sites according to the definitions in [42] are red-brown earth at YAI, and self-mulching clay at LFS. Further soil properties at these sites are described in [30]. Sowing date is usually a day before the date of first flush shown in Table 1. One image per season was acquired from the WorldView-2 (2015 and 2016) or WorldView-3 (2017 and 2018) satellites, with image dates close to PI dates. The study area is shown in Figure 1, together with the satellite image capture areas and field sites.

All experiments were drill sown at 0.19 m row spacing and flush irrigated until the 3 to 4 leaf growth stage (see growth stage definitions in [43]). Several nitrogen rates were applied at this stage with urea spread on the dry soil surface prior to the application of permanent water to the experiments. Table 1 shows the WorldView image dates, first flush (FF) date, permanent water (PW) date, and physical sampling dates. It also shows the number of plots, rice varieties (R = Reiziq™, S = Sherpa™, L = Langi, T = Topaz™) and N rates for each experiment. N rates were within the range 0 to 300 kg/ha, with rates per season and experiment indicated in Table 1. There are at least three replicates of each N × variety treatment in each experiment. Plot sizes ranged from 6.3 × 5.7 m^2 to 10.5 × 11.4 m^2. Each plot was sampled on a number of dates, only the sample date closest to the image date is shown in Table 1.

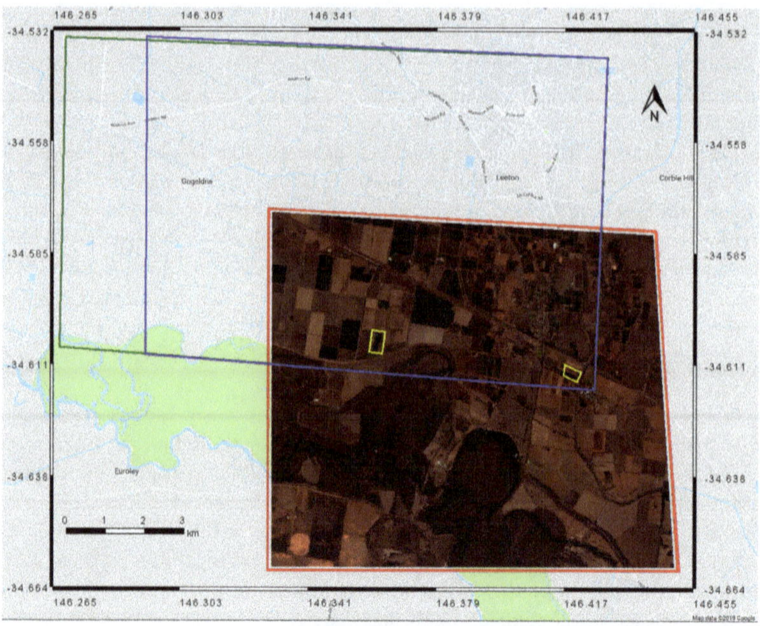

Figure 1. WorldView image areas. 21 December 2014 and 30 December 2015 (red), 7 January 2017 (green) and 9 January 2018 (blue). The red-green-blue image of the 2014 capture is also shown. The study sites are shown in yellow (LFS in the center and YAI on the right).

Table 1. Table of experiments, indicating management dates and N rates, rice varieties and number of plots. For experiments with multiple sowing dates, the early date is denoted SD1 and the later SD2.

Year	Site	Exp	N Rates (kg/ha)	Varieties	Plots	First Flush	Perm Water	Sample	Image
2015	LFS		0, 60, 120, 180, 240	R, S, L, T	80	16 Oct 2014	20 Nov 2014	29 Dec 2014	21 Dec 2014
2015	YAI	SD1	0, 60, 120, 180	R, S, L, T	48	10 Oct 2014	19 Nov 2014	30 Dec 2014	21 Dec 2014
2015	YAI	SD2	0, 60, 120, 180	R, S, L, T	48	27 Oct 2014	27 Nov 2014	5 Jan 2015	21 Dec 2014
2016	LFS	SD1	75, 150, 225, 300	R, S, L, T	48	21 Oct 2015	26 Nov 2015	30 Dec 2015	30 Dec 2015
2016	LFS	SD2	75, 150, 225, 300	R, S, L, T	48	5 Nov 2015	4 Dec 2015	31 Dec 2015	30 Dec 2015
2016	YAI		0, 60, 120, 180, 240	R, S, L, T	60	15 Oct 2015	20 Nov 2015	30 Dec 2015	30 Dec 2015
2017	LFS		0, 60, 120, 180, 240, 300	R, S, L, T	72	31 Oct 2016	2 Dec 2016	6 Jan 2017	7 Jan 2017
2017	YAI		0, 60, 120, 180, 240	R, S, L, T	60	20 Oct 2016	1 Dec 2016	9 Jan 2017	7 Jan 2017
2018	LFS		0, 60, 120, 180, 240, 300	R, L, T	54	30 Oct 2017	30 Nov 2017	6 Jan 2018	9 Jan 2018
2018	YAI		0, 60, 120, 180	R, L, T	36	27 Oct 2017	1 Dec 2017	6 Jan 2018	9 Jan 2018

2.2. PI N Sampling Methodology

Panicle initiation (PI) occurs at the start of the reproductive phase in rice growth. There are methods to predict the timing of PI using environmental variables [44]. Field sampling was used to determine the exact time of PI, defined as the time when three out of ten main stems have a panicle 1–3 mm long [14].

Above ground plant samples were collected from each plot at PI from an area of six plant rows (0.19 m row spacing) by one metre. Five whole plant grab samples including leaves and stems (approximately 20 g each) were collected from each sample to create a 100 g subsample which was dried in a microwave oven before being dried overnight in a conventional oven at 60 °C. This subsample was ground and analysed for N concentration by Dumas combustion [45] and multiplied by the measured dry matter weight to calculate PI nitrogen uptake.

2.3. Climate Data

In order to assess the possibility that climate variation from season-to-season could affect the relationship between rice N uptake at PI and remotely sensed data, weather data for each of the seasons was obtained and processed. Aggregated climate parameters between critical management and measurement days were calculated, these days being the first flush (FF), the date permanent water (PW) was applied, the date the N uptake was sampled and the date the satellite image was captured.

Growth degree days [46] was calculated. The base temperature was set to 10 °C, which is commonly used in the Australian rice growing system [44]:

$$\mathrm{GDD} = \sum_{t_0}^{t_N}(T_{\max}+T_{\min})/2 - 10, \tag{1}$$

where t_0 and t_N are the beginning and end days of the GDD summation, and $T_{\max}$ and $T_{\min}$ are the daily maximum and minimum temperatures, respectively.

The sum of solar radiation between dates was calculated by accumulating daily solar radiation. Reference evapotranspiration (ETo) in mm was calculated from weather observations using the standardized ASCE equation [47]. In this work, the sum of these parameters between two dates will be indicated using nomenclature such as GDD(image-FF) for the accumulated growth degree days between the image date and first flush date.

2.4. Worldview Satellite Data

The 21 December 2014 and 30 December 2015 images were from the WorldView-2 satellite, with 2 m resolution for the multispectral bands. The 7 Januar 2017 and 9 January 2018 images were from the WorldView-3 satellite, with 1.2 m multispectral resolution. All imagery was acquired under cloud-free conditions. The band definitions of the sensors on both platforms are shown in Table 2.

Table 2. WorldView-2 and WorldView-3 band definitions.

Abbreviation	Band	Band Edges (nm)
c	Coastal	400–450
b	Blue	450–510
g	Green	510–580
y	Yellow	585–625
r	Red	630–690
re	Red edge	705–745
nir	Near infrared	770–895
nir2	Near infrared 2	860–1040

The satellite images were imported into Google Earth Engine [48] where subsequent image analysis was performed. The at-sensor pixel data were converted to top-of-atmosphere reflectance using the metadata supplied with the images, and the procedure given by the image provider at https://www.digitalglobe.com/resources. Then, atmospheric haze correction was applied using the Dark Object Subtraction method [49].

Shapefiles of each of the trial plots for each season, site and experiment were created and overlaid on the satellite imagery. The averages of each of the eight band reflectances for each plot were calculated (with plot boundaries internally buffered by 1 m to avoid edge effects), and exported as a data table for further analysis.

2.5. Data Pre-Processing

Data analysis was performed using the Python programming language in Google's Colaboratory cloud-based environment. Extensive use was made of tools such as Pandas [50], and the scikit-learn [51] and StatsModels [52] packages for statistical modeling. Data from the various sources was merged:

- Satellite remote sensed data (average band reflectances per plot).
- Aggregated climate (GDD, Solar, ETo) data and number of days between dates of first flush, permanent water, sample collection, image capture.
- Field rice sample data (N uptake per plot). For most experiments in Table 1, the plots were sampled numerous times to correspond with dates of image capture and growth stage. So the dataset had a row for each of these samples (i.e., multiple rows per plot).

As well as the raw reflectances, ratio and normalized difference indexes were calculated. It is commonly understood that taking ratios of remotely-sensed reflectances provides some normalization against the effects of variable radiance and atmospheric absorption between image captures. Therefore, as well as investigating the relationship between the band reflectances and N uptake, 2-band ratio spectral indexes (RSIs) and normalized difference spectral indexes (NDSI) from the 8 reflectance bands were investigated, defined as [28]:

$$\mathrm{RSI}(b1, b2) = b1/b2 \tag{2}$$

$$\mathrm{NDSI}(b1, b2) = (b1 - b2)/(b1 + b2) \tag{3}$$

For example, the widely used normalized difference vegetation index (NDVI) is NDSI(nir,r), and the normalized difference red edge (NDRE) is NDSI(nir,re). All possible RSIs and NDSIs from all independent combinations of the eight reflectance bands in Table 2 were calculated and added to the dataset.

A number of additional vegetation indexes that aren't RSIs or NDSIs were also trialled. These were shown to be significant for predicting N uptake in previous works [25,28]:

$$\text{CIre} = nir/re - 1 \tag{4}$$

$$\text{CIg} = nir/g - 1 \tag{5}$$

$$\text{SAVI(nir,re)} = (1 - 0.08) \times (nir - re)/(nir + re - 0.08) \tag{6}$$

2.6. N Uptake Models

2.6.1. Single-Variable Models

Data from all the experiments was pooled. As the N uptake was sampled on multiple dates for each experiment, data from the sample dates closest to the image dates was chosen for preliminary analyses, which are the sample dates shown in Table 1.

Correlation analysis was performed of all RSI and NDSI bands against N uptake, to determine which are likely to be the dominant predictors. This indicates how well a linear model can describe the relationship between N uptake and spectral indexes:

$$y = a \times x + b \tag{7}$$

where a and b are the slope and intercept of the model, y is N uptake and x is the spectral index being assessed. This nomenclature is used throughout this paper. The coefficient of determination R^2 was used to quantify the power of each index to explain the variance in the N uptake:

$$R^2 = 1 - \frac{\sum_{i=0}^{N-1}(y_i - \hat{y}_i)^2}{\sum_{i=0}^{N-1}(y_i - \overline{y})^2} \tag{8}$$

where N is the number of observations, $\hat{y}_i$ is the i-th model prediction and y_i is the i-th observation, and $\overline{y}$ is the mean of the observations.

Models were further evaluated using the root mean squared error (RMSE):

$$\text{RMSE} = \sqrt{\sum_{i=0}^{N-1}(y_i - \hat{y}_i)^2/N} \tag{9}$$

RMSE is an easily interpretable measure of model performance as it retains the units of the quantity being observed and predicted. In our case, this is kg/ha of N uptake. Note, in other studies, N uptake or plant nitrogen accumulation (PNA) is reported in g/m^2, which can be converted to N uptake in kg/ha by multiplying by ten.

Model coefficients were tested for significance at the $p = 0.05$ level using the StatsModels Ordinary Least Squares methods (statsmodels.regression.linear_model.OLS).

Following this, nonlinear transformations were investigated to enable models to better fit N uptake. These included the exponential transformation:

$$y = b\exp^{a \times x} \tag{10}$$

$$\ln y = a \times x + \ln b \tag{11}$$

where a and b are the regression coefficients. N uptake is denoted by y in this and following equations, and x is the vegetation index being used (for example, NDRE). The second line indicates that by taking the logarithm of both sides of the equation, the model can be fit using linear regression. The disadvantage of this formulation is that the predicted variable, $\ln y$, no longer has units of kg/ha.

A second transformation involved squaring the unitless input vegetation index. It will be shown that this linearizes the N uptake vs. index relationship, when the selected index is NDRE, and also renders the y-intercept insignificant. This squared transformation is:

$$y = a \times x^2 + b \approx a \times x^2 \tag{12}$$

where a is the slope of the N uptake vs. x^2 line, and b is the intercept.

In order to assess the effect of secondary factors (such as season, variety, site, sowing date) on the models, Equation (12) was fit to subsets of the data. The regression lines for each subset were assessed for significant difference from the regression line of the whole dataset by first dividing the N uptake by $a \times \text{NDRE}^2$ (with a set to the extracted coefficient for the whole dataset), then applying the Tukey HSD test at the alpha = 0.05 level (using the statsmodels.stats.multicomp.pairwise_tukeyhsd component of the StatsModels package in Python [52]).

2.6.2. Multi-Variable Models

The analysis then proceeds to multi-variable linear regression, to investigate how much more accurate the N uptake model can be across seasons, varieties, sites, management with more complex models. The possible inputs to the model are: reflectances, RSIs, NDSIs, varieties, all combinations of dates between [first flush, permanent water, image, sample], as well as the growth degree days, solar radiation, rain and evapotranspiration over these days.

Relationships between N uptake and input variables often followed a squared rather than linear trend. In addition, interaction between many of the input variables was found to be significant, as evidenced by variable selection methods (described below) making use of these interaction terms. Therefore, the input variables were transformed with a second-order polynomial so that squared terms of all input variables, and cross terms between all combinations of input variables were generated. As a simple example, if the raw input variables were [NDRE, NDVI], the transformed variables would be [NDRE^2, NDRE $\times$ NDVI, NDVI^2]. In practice, we considered models with up to 33 variables (including remotely sensed bands, variety, climate parameters, important dates), giving 594 transformed variables. The input variables were then scaled using the standard scaler in scikit-learn (sklearn.preprocessing.StandardScaler). This centers and normalizes the variances of each of the input variable data, which is important for fitting models with regularization (described below).

In order to produce models that are interpretable and straightforward to implement in a variety of GIS platforms, black-box models such as SVM models and neural networks were avoided. Instead, we used linear regression with Lasso regularization from the scikit-learn package (sklearn.linear_model.Lasso). This has the advantage of selecting the most important input variables and setting all others to 0, because it uses L1 regularization [53]. The Lasso objective is to minimize the following, with w_j being the coefficient of each input variable x_j and $_i$ denoting observation number:

$$\sum_i (y_i - \sum_j x_{ij} w_j)^2 + \alpha \sum_j |w_j| \tag{13}$$

The Lasso α parameter multiplies the L1 term, so larger values of α result in more compact models (more coefficients w_j set to 0). The relative importance of the input variables can be assessed by examining the magnitude of the remaining coefficients.

Three-fold cross validation was used to select the best Lasso α using the GridSearchCV tool in scikit-learn (sklearn.model_selection.GridSearchCV [51]). This splits the training/validation data into three groups, training a model on two and validating (assessing the accuracy of model predictions using a score such as RMSE) on the third. The procedure is illustrated in Figure 2a. This is done three times, first with groups 1 & 2 as training data and group 3 as validation data, then groups 1 & 3 as training data and group 2 as validation data, then groups 2 & 3 as training data and group 1 as validation data. The average of the validation scores (RMSE of N uptake prediction) is taken. This process is repeated as the Lasso α was swept. The α that gives the best average RMSE score is selected. The model is then re-trained with this best α with all the training/validation data. Finally, this model was evaluated on it's ability to predict the held back test data.

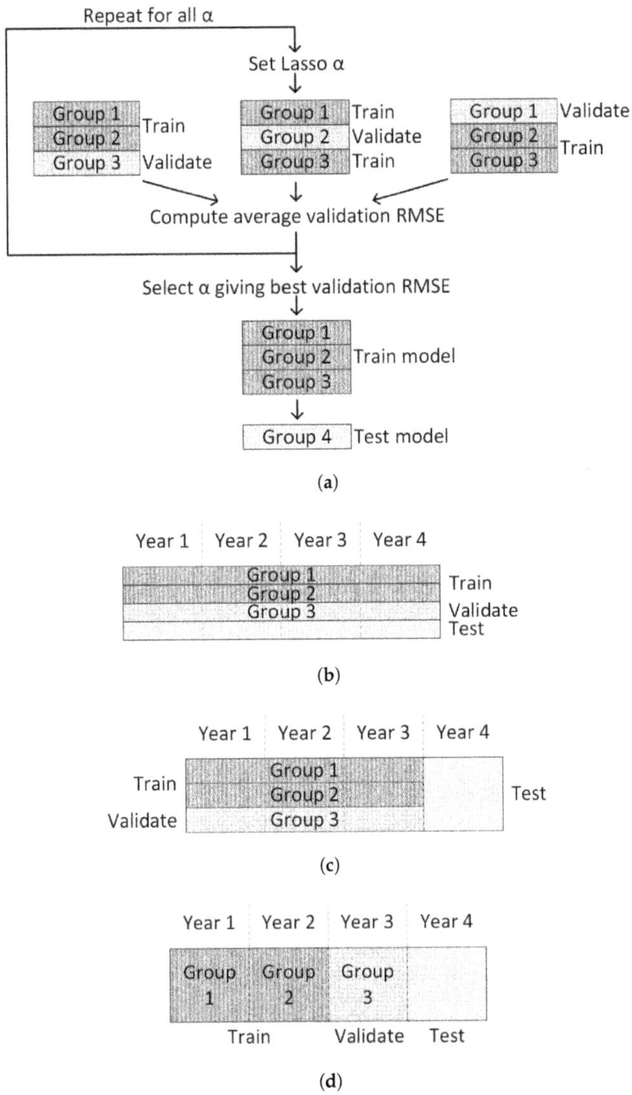

Figure 2. Training-validation-test model extraction methods with multi-season data. (**a**) The three-fold cross validation procedure used to select the best Lasso α and evaluate the model against held-back test data. (**b**) Randomly assigning training/validation and test data points from all seasons data. (**c**) Randomly assigning training/validation and test data points from three seasons data and testing the extracted model on the fourth season. (**d**) Training a model on two seasons data and validating on a third season (repeating this three times for all combinations of training/validation data), then testing the extracted model on the fourth season. Note, only the first of the three folds in the cross validation procedure of (**a**) is shown in (**b**,**d**).

2.6.3. Multi-Season Models

Three ways of splitting the multi-season data into training, validation and test sets were evaluated. The first is shown in Figure 2b. The entire data set was split randomly into training/validation and test data, with 30 % of the data held for test.

However, for operational use, models need to be trained on previous season data to predict N uptake at PI for a new season, so that topdressing can be applied soon after an image is captured. To assess the ability of the remotely sensed data to achieve this aim, a second method of splitting data took the training data from three seasons and the extracted model was tested on the fourth season data, as shown in Figure 2c. Three-fold cross validation was used with Lasso regression to select the best α, with randomly selected training and validation points from three season's data. Then this model's prediction was tested on the remaining season's data. The WorldView 8-band reflectances were used as the input variables, with the $NDRE^2$ band added, which was found to increase accuracy. It will be shown that this method overfits the model to the three training seasons, and so does not generalize well to the fourth test season.

To seek to solve this overfitting issue, a different cross-validation method was used to select the optimal Lasso α from the three training seasons, shown in Figure 2d. Rather than randomly assigning points from the three training seasons to model training and validation sets over three folds, the training and validation sets included two seasons and one season respectively. This was implemented using the Leave One Group Out method in scikit-learn (sklearn.model_selection.LeaveOneGroupOut) [51]. For example, for testing the model on the 2018 season, the cross validation for Lasso model training used [training, validation] folds of [2015 + 2016, 2017], [2015 + 2017, 2016], [2016 + 2017, 2015] to select α, before testing the final model on the 2018 data. Because the model training process is validating performance against a held out season data in each of the cross validation folds, the model can then generalize better to a totally new season, as overfitting to within-season data is avoided.

3. Results

3.1. Correlation between Remotely Sensed Data and N Uptake

The correlation between each of the RSIs and NDSIs and N uptake for the pooled data from all experiments was assessed using the coefficient of determination (R^2). The results are shown in Figure 3. The best ratio index is RSI(nir2,re) with $R^2 = 0.78$, closely followed by RSI(nir,re) with $R^2 = 0.74$. For the normalized difference indexes, the best index is NDRE = NDSI(nir,re), with $R^2 = 0.73$. These results indicate ratio indexes are at least as good as normalized difference indexes for predicting N uptake. The single-variable linear regression analysis that follows uses NDRE as the input variable. The performance using other combinations of bands will be described in following sections.

NDVI ($R^2 = 0.26$) does not perform as well as NDRE ($R^2 = 0.73$), as seen in Figure 3. The reason for this is indicated in Figure 4. NDVI saturates at large values of N uptake. In contrast, NDRE continues to increase at high values of N uptake, making it a better predictor.

There is a nonlinear relationship between NDRE and N uptake, evident in Figure 4b. Fitting a linear model leads to large residuals at high and low values of N uptake. Various equations were used to attempt to capture this relationship. Two are shown in Figure 5. An exponential relationship appears to fit the trend of the data closely, as seen in Figure 5b. This is described by Equation (10). The disadvantage of this formulation is that the predicted variable (logarithm of N uptake) no longer has units of kg/ha, making evaluation statistics such as RMSE difficult to interpret.

Another possibility to describe the nonlinear relationship between NDRE and N uptake is shown in Figure 5c, where instead of transforming the predicted variable (N uptake), we square the unitless input variable NDRE. This model is described by Equation (12). This transformation linearizes the N uptake vs. index relationship. It also gives a lower RMSE than the linear or ln(N) options and the predicted variable retains physical units of kg/ha.

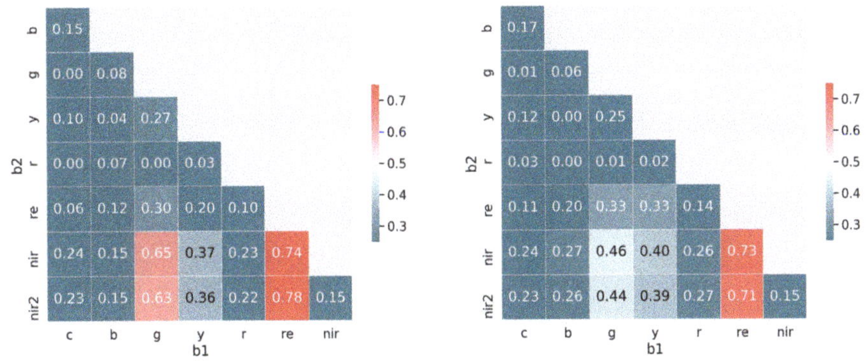

(a) (b)

Figure 3. Coefficient of determination R^2 between sampled nitrogen uptake and derived image indexes. (**a**) Ratios of image bands (b_1/b_2). (**b**) Normalized difference ratios of image bands ($(b_1 - b_2)/(b_1 + b_2)$.

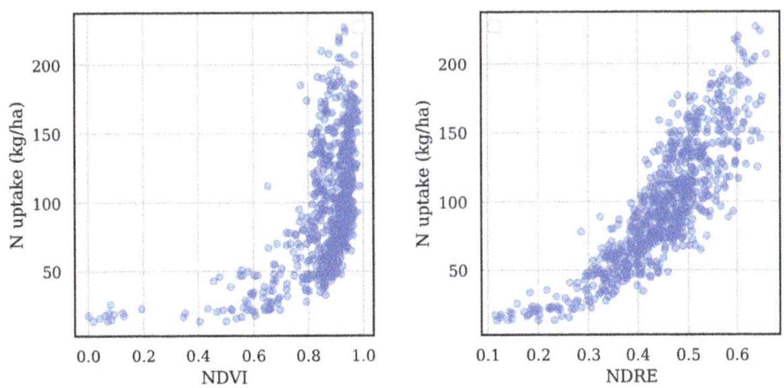

(a) (b)

Figure 4. Nitrogen uptake vs. NDVI (**a**) and NDRE (**b**).

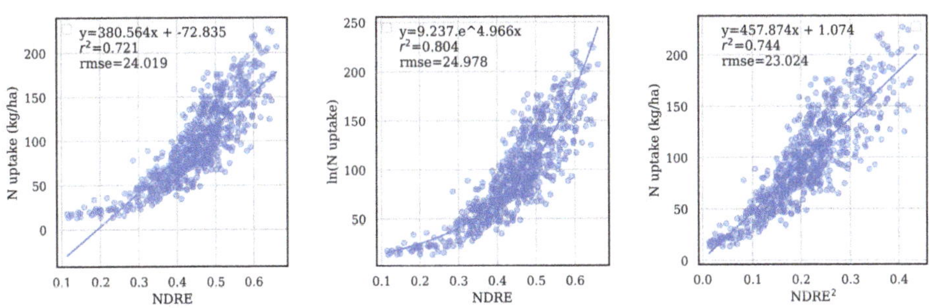

(a) (b) (c)

Figure 5. Comparison of fitting equations, (**a**) N vs. NDRE. (**b**) ln(N) vs. NDRE. (**c**) N vs. $NDRE^2$.

The y-intercept of the N uptake vs. $NDRE^2$ model was not significant at $p < 0.05$, so was omitted. The slope was significant at $p < 0.05$. The resulting model equation fit to the four-season dataset is $y = a \times x^2 = 457 \times NDRE^2$. The residuals were examined and followed a normal distribution. Maps of modeled N uptake were generated using this equation and NDRE computed from the satellite images. Figure 6 shows the measured and predicted N uptake map from the 2017 season at the LFS site. Good agreement is observed, with $R^2 = 0.88$ and RMSE = 21.3 kg/ha.

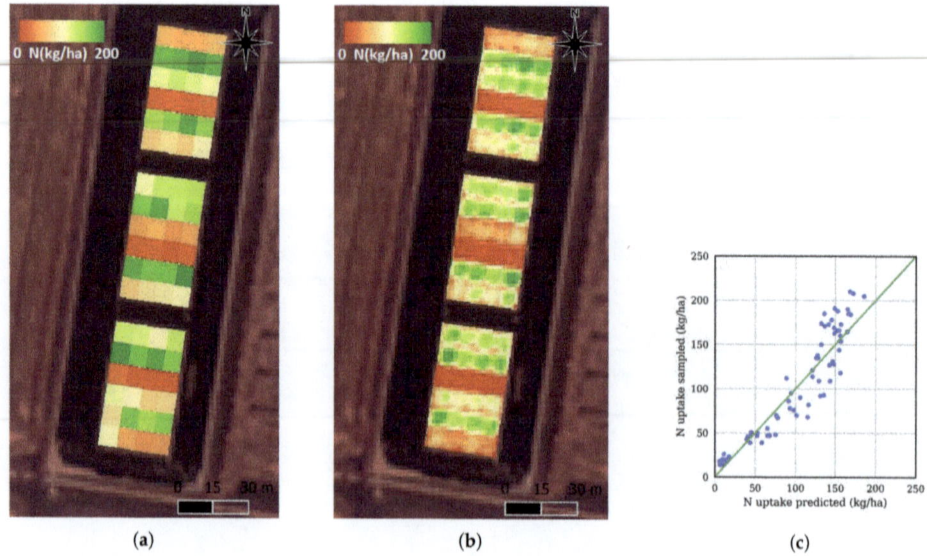

Figure 6. Comparison of sampled and predicted N uptake from the 2017 experiment at the LFS site. Red indicates 0 kg/ha, green indicates 200 kg/ha N uptake. (**a**) Sampled. (**b**) Predicted. (**c**) Graph showing sampled vs. predicted N uptake per plot.

3.2. Effect of Secondary Variables On Models

The dataset was split into batches, based on (i) year, (ii) variety, (iii) site and (iv) sowing date; in order to assess the effect of these factors on the relationship between remotely sensed data and N uptake. Equation (12) was fit to each of the subsets of data. The results are shown in Table 3. The statistically significant differences in the subsets are indicated in the Group column of the table. There are significant differences in the model slopes (a) between different years, from 387 (2018) to 528 (2016). The regressions per year are shown in Figure 7. There are significant differences between the Reiziq™model and the other variety models. There are no significant differences between sites or sowing time (before or after 25-October).

Table 3. Linear regression results for N uptake vs. $NDRE^2$, with separate regression coefficients for different groupings of experiments.

Variable	Value	Points	a	R^2	RMSE (kg/ha)	Group
ALL	ALL	554	457	0.75	22.8	
Year	2015	176	450	0.71	15.5	b
	2016	156	528	0.69	19.2	c
	2017	132	473	0.85	23.1	bc
	2018	90	387	0.80	22.7	a

Table 3. *Cont.*

Variable	Value	Points	*a*	R^2	RMSE (kg/ha)	Group
Variety	Sherpa™	116	491	0.81	19.5	b
	Reiziq™	146	409	0.76	22.9	a
	Langi	146	464	0.79	21.2	b
	Topaz™	146	499	0.79	20.5	b
Site	LFS	302	452	0.83	20.0	a
	YAI	252	463	0.61	25.6	a
Sow date	Early	296	485	0.77	21.0	a
	Late	258	431	0.77	22.9	a

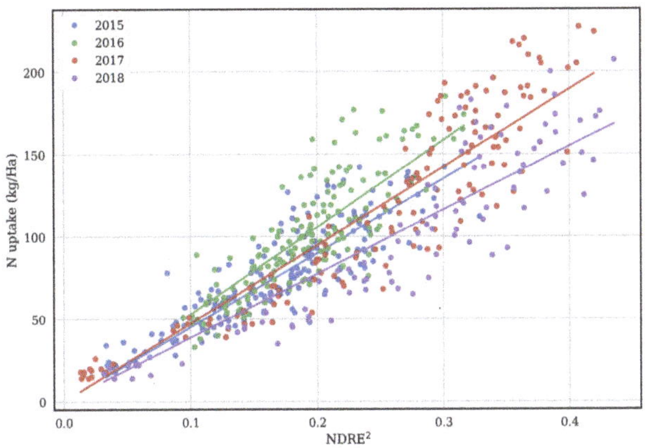

Figure 7. Regression of N uptake vs. $NDRE^2$ per year.

3.3. *Effect of Different Sampling and Image Dates*

There were multiple sample dates for each of the experiments (see Section 2.5), though the analysis up until now has only made use of the sample data closest to the date where the satellite image was captured. Nitrogen uptake is dynamic, so it is reasonable to expect the actual N uptake at the image date will be different to the sampled N uptake if the sample was taken on a different date. To test the effect of different sampling and image dates, sample data from all dates was used, and the data was separated into buckets of 10 days between image and sample date. The model of Equation (12) was fit to each bucket. The results are shown in Table 4. As expected, generally when the N uptake was sampled after the image capture (negative Days(image-sample)), the regression slope is steep, as the N uptake would have increased relative to $NDRE^2$. The best relationship is observed when the sampling date is close to the image date, as indicated by the highest $R^2 = 0.76$ with Days(image-sample) = 0.

Table 4. Linear regression results for N uptake vs. $NDRE^2$, with separate regression coefficients for different number of days between N uptake sampling and image capture, binned into 10 day buckets.

Days (Image-Sample)	Points	*a*	R^2
−30	176	586	0.59
−20	48	492	0.27
−10	284	534	0.64
0	600	451	0.76
10	36	336	0.63

3.4. Improving the Accuracy of The Model

Up until now, a single input variable model has been used (Equation (12) with x = NDRE). In this section, more complex multi-variable linear regression models are explored, using Lasso regression to avoid overfitting and to eliminate insignificant variables, as described in Section 2.6.2.

First, data for training/validating and testing the models are randomly chosen from all four seasons, as shown in Figure 2b. The results are shown in Table 5. Models with different numbers and combinations of input variables were tested. Many multispectral sensors have less bands than the 8 bands of the WorldView sensors (Table 2). Therefore, models were generated using only 4 out of the 8 possible bands to test degradation in accuracy expected with less capable sensors. These four bands are g, r, re, nir.

The first group of models at the top of Table 5 shows results from single input variable models (Lasso $\alpha = 0$ as no variable selection is needed). The second group of models in the middle of the table includes all RSIs and NDSIs formed by combinations of all reflectance bands. Results are shown for all combinations of 4 bands b, g, r and nir (RSI4 and NDSI4), as well as all combinations of WorldView sensor's 8 bands (RSI8 and NDSI8). The third group of models are for combinations of the raw reflectance bands without taking ratios or normalized differences (R4 and R8). The final models also add variety and climate data. The model in the last row adds in all sample data, with multiple sample dates per image capture (see Section 2.5).

For each combination of input variables, the table indicates the best Lasso α, which gave the lowest validation RMSE. It also shows the number of raw input variables (# X), the total number of input variables after processing with a second-order polynomial (# $(X)^2$), and the number of variables that were retained after Lasso regularization (# selected X). The last column in the table shows the 3 most significant input variables used in the model.

For the single-input variable models, $NDRE^2$ is the best predictor of N uptake, with test RMSE of 23.7 kg/ha.

Models using combinations of RSIs and NDSIs achieve similar performance, with 8 bands improving RMSE by 1–2 kg/ha relative to 4 bands. NDSIs using 8 bands yielded N uptake RMSE of 18.9 kg/ha.

Using the raw reflectances (R4 and R8 in the third section of Table 5) instead of RSIs or NDSIs yields lower errors by around 3 kg/ha. Adding variety data does not improve the model. Adding climate and management date variables (R8 $\times$ climate) does improve the model by around 1.5 kg/ha to 14.6 kg/ha. The most important bands in the models are re, nir and nir2.

The final row in Table 5 shows the results from the model extracted using N uptake data from all sample dates. The Lasso regularization selected 110 input variables and achieved an R^2 of 0.90 and RMSE of 14.2 kg/ha on the test data. An example of the training and test RMSE and the number of selected model features as a function of the Lasso alpha is shown in Figure 8, corresponding to this last row in Table 5. The figure demonstrates that reducing alpha increases the number of input variables, which gives lower RMSE on the training data, but results in overfitting so that RMSE on the test data is degraded. The optimum alpha is chosen as the one giving the lowest test RMSE, in this case, 0.004. This model is significantly better than the simple single input variable models. However, the model becomes difficult to interpret because of the large number of input variables. It is also questionable if the model will fit future season's data accurately, a consideration that will be explored in the following section.

Table 5. Results from multi-variable regression models with all season data pooled, and train/test points randomly selected (as shown in Figure 2b). #X is the number of raw input variables. $\#(X)^2$ is the number of input variables after the polynomial is applied. # selected X is the number of variables selected after Lasso regularization. Most important X shows the three most important variables.

Input Variables	Best α	# X	# $(X)^2$	# Selected X	Train R^2	Test R^2	Train RMSE	Test RMSE	Three Most Important X
NDVI	0	1	1	1	0.24	0.30	38.1	41.4	NDVI
NDRE	0	1	1	1	0.72	0.74	23.2	25.5	NDRE
$NDRE^2$	0	1	1	1	0.74	0.77	22.3	23.7	$NDRE^2$
CIre	0	1	1	1	0.73	0.77	22.8	23.8	CIre
CIg	0	1	1	1	0.63	0.67	26.7	28.5	CIg
SAVI	0	1	1	1	0.62	0.67	26.9	28.5	SAVI(nir,re)
RSI4	0.01	6	27	19	0.83	0.82	18.2	20.8	RSI(r,g) × RSI(re,g), RSI(r,g), RSI(re,g)
RSI8	0.1	28	434	24	0.86	0.85	16.2	19.5	RSI(nir2,re) × RSI(nir2,nir), RSI(b,c) × RSI(re,r), RSI(nir2,c) × RSI(nir2,re)
NDSI4	0.04	6	27	10	0.82	0.82	18.4	21.0	$NDSI(nir,re)^2$, NDSI(r,g) × NDSI(re,g), $NDSI(r,g)^2$
NDSI8	0.1	28	434	21	0.86	0.85	16.2	18.9	$NDSI(nir2,re)^2$, $NDSI(re,c)^2$, NDSI(y,c) × NDSI(r,b)
R4	0.004	4	14	13	0.87	0.87	15.7	17.7	nir^2, re × nir, r × re
R8	0.004	8	44	29	0.90	0.90	13.9	16.0	b × r, c × b, c^2
R8 × variety	0.02	12	90	36	0.91	0.89	13.3	16.2	$nir2^2$, g × re, g × nir
R8 × climate	0.02	29	464	39	0.92	0.91	12.7	14.6	$nir2^2$, re × nir, nir × Rain(image-PW)
R8 × variety × climate	0.020	33	594	61	0.93	0.91	11.7	14.9	re × r, nir × Sow_week, nir × Rain(image-PW)
As above, all sample dates	0.004	33	594	110	0.93	0.90	12.4	14.2	y × Sow_week, nir × Solar(image-PW), g × re

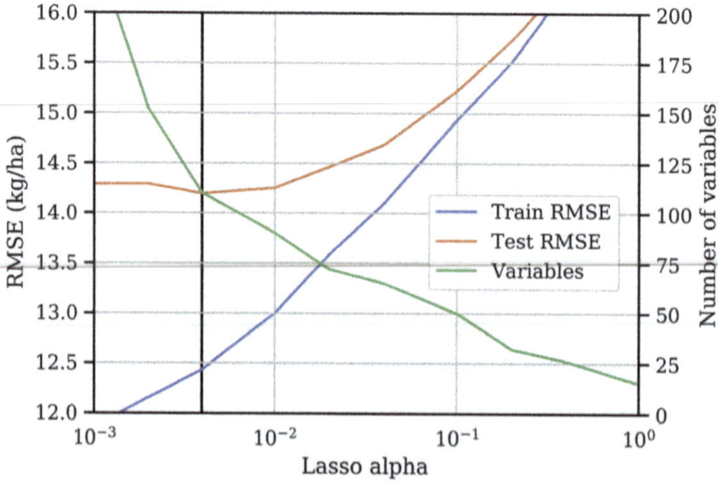

Figure 8. RMSE for the Lasso model as a function of α including all variables (last row of Table 5).

3.5. Model Consistency with Season

Previous sections have pooled data from all seasons, with models trained on randomly selected training and test data points from the pooled data. It was seen that using this method, where model training data includes points from seasons being predicted, quite good absolute accuracy can be achieved (RMSE < 20 kg/ha). Now we examine the ability of models trained on three known seasons to predict results in a fourth unknown season, using the methods described in Section 2.6.3.

The first method randomly selects training and validation data from three seasons, then tests on the fourth, as shown in Figure 2c. The results are shown in Table 6. The models use the eight raw bands and NDRE^2 as input variables. All combinations of training/validation and test years are evaluated. The resulting models are quite complex with small α and more than 29 input variables selected. When the model is trained on 2016–2018 data, and tested on the 2015 data, reasonable performance is achieved (RMSE = 17.2 kg/ha). However, for other test seasons, the predictions are poor. The models with 2017 and 2018 test seasons give a negative R^2. This is possible with R^2 defined as in Equation (8), and indicates the models are not generalizing to the test data. These results indicate that when models are tested on data from unseen seasons using the model training methodology in Figure 2c, generated models may overfit the training/validation data and so will be unlikely to predict future season N uptake accurately.

To seek to solve this overfitting issue, a different cross-validation method was used to select optimal Lasso α. This is shown in Figure 2d, and proceeds by holding out one of the training seasons for model validation, and repeating over three combinations of training/validation seasons in three-fold cross validation, before testing on the remaining season from which the model has not seen data. The results using this method are shown in Table 7. Now Lasso regularization has selected quite simple models, with relatively high α and only three or four input variables selected. The worst RMSE across the four test seasons is 27.4 kg/ha, and worst R^2 is 0.67. In all seasons the most important input variable is nir2 $\times$ NDRE^2 with strong similarity between models from all four groups of seasons.

In order to assess the simple NDRE^2 model of Equation (12) using this same multi-season model training strategy (Figure 2d), this model was fit to three years data, then tested against the remaining year. The worst case RMSE was 36.0 kg/ha for the 2018 test year. Thus, we confirm Lasso regression with multiple input variables produces better models (worst case RMSE 27.4 kg/ha for the 2018 test year).

Table 6. Lasso regression with one year held out as test data, and the other three years as training data with randomized cross validation (as shown in Figure 2c).

Test Year	Best α	# X	# $(X)^2$	# Selected X	Train R^2	Test R^2	Train RMSE	Test RMSE
2015	0.002	9	54	43	0.93	0.65	13.2	17.2
2016	0.001	9	54	53	0.92	0.38	13.5	27.2
2017	0.001	9	54	48	0.89	−27.04	12.6	311.1
2018	0.004	9	54	29	0.91	−6.47	13.5	140.4

Table 7. Lasso regression with one year held out as test data, and the other three years as training data with cross validation splits based on training data years (as shown in Figure 2d).

Test Year	Best α	# X	# $(X)^2$	# Selected X	Train R^2	Test R^2	Train RMSE	Test RMSE	Three Most Important X
2015	10.0	9	54	3	0.79	0.67	22.5	16.8	nir2 $\times$ NDRE^2, nir2^2, g
2016	4.0	9	54	4	0.85	0.67	19.0	19.9	nir2 $\times$ NDRE^2, r $\times$ NDRE^2, c $\times$ re
2017	4.0	9	54	4	0.80	0.78	17.2	27.4	nir2 $\times$ NDRE^2, nir2^2, g
2018	4.0	9	54	3	0.84	0.72	17.5	27.4	nir2 $\times$ NDRE^2, NDRE^2, g

4. Discussion

This study has demonstrated the accuracies that can be obtained using multi-spectral satellite data to predict the N uptake of rice at panicle initiation. This is an essential pre-requisite to developing an operational mid-season N topdressing recommendation tool for the rice industry, based on remote sensing.

Table 8 gives a summary of results from a selection of papers on predicting rice N uptake using remote sensing, for papers that report predicted N uptake root mean squared error (RMSE). The table assumes N uptake at jointing is similar to that at PI, and it shows the results for the best model formulation. The last column shows the worst case test season RMSE for the limited number of studies that test models on different seasons than those used to train the models.

Table 8. Published research investigating prediction of N uptake of rice. Root mean squared error (RMSE) of the N uptake is given for cases when training and test data come from the same season, and when the test data comes from a different season (worst case is shown). T = tillering, PI = panicle initiation, H = heading, F = filling, MS = multispectral, HS = hyperspectral, PLSR = partial least squares regression, IPLSR = interval PLSR, MLR = multiple linear regression, VI = vegetation index.

Ref	Stage	Platform	Sensor	Model	Seasons	RMSE (kg/ha) Same Seasons	RMSE (kg/ha) Test Season
[54]	T, PI, H, F	Handheld	MS	VI	2	7.07	-
[28]	PI	Handheld	HS	IPLSR	4	11.7	-
[30]	PI	Handheld	HS	PLSR	3	17.4	34.9
[27]	PI	UAV	MS	VI	2	11.5	-
[29]	H	Airborne	HS	PLSR, MLR	3	11.98	67.1
This	PI	Satellite	MS	MLR	4	14.6	27.4

Our work indicates that the model generation strategy critically depends on whether in-season sampling data is available. Complex multi-variable models predicting N uptake from remotely sensed data can be very accurate, providing in-season physical samples and remote sensing data is available for the season being predicted. In this case, previous studies have shown N uptake prediction errors as low as 7 kg/ha [54], and many others around 12 kg/ha [27–29]. However, these don't consider the practically important application of predicting N uptake in new seasons for which sampling data is not yet available. With this limitation, we found N uptake RMSEs of better than 19 kg/ha with remote sensing data from four seasons as shown in Table 5. Addition of management and climate data enabled prediction RMSE better than 15 kg/ha.

However, it is rarely the case that in-season data is available by the time predicted N uptake is needed to generate topdressing requirements, a factor which isn't considered in most previous studies.

The timing and logistics from image acquisition with simultaneous physical N uptake sampling, to model extraction from this data, to N uptake prediction, to N recommendation and finally to applying the top-dressing necessitates applying models extracted from previous data to new seasons. However, seasonal variability in models leading to inaccuracy in extrapolating models to new seasons was tested and noted by [29] and [30]. We likewise showed that simply extracting models using randomly selected training and validation data points from other seasons can result in very poor prediction of N uptake in a new season, with RMSEs greater than 100 kg/ha in some cases, as shown in Table 6.

To solve this issue, we proposed a cross-validation procedure that uses training and validation data sets from separate seasons, which results in better generalization of N uptake prediction to new seasons. Testing across the four seasons of this study, we obtained RMSEs between 16.8–27.4 kg/ha. Notably, this was using multispectral sensors and from satellite platforms, which makes the methodology suitable for industry adoption.

The results illustrated a number of additional challenges to producing accurate N uptake models. The days between image acquisition and N uptake sampling affects model parameters, shown in Table 4. Ideally, sampling should occur at the same time as image acquisition. This result shows the importance of having image capture close to the date at which N uptake is required if possible. It also motivates developing models for the dynamic change of N uptake with time and environmental conditions in case image acquisition must occur on a date significantly differently than the PI date. Ultimately, a model for the dynamics of N uptake with climate and time is desirable, so that the calculated N uptake from the image can be extrapolated to PI date. Models predicting soil N in flooded systems were reviewed in [41]. These could be coupled with rice crop models to determine N uptake. The time and climate-dependent behavior of remotely sensed vegetation indexes that predict N uptake were shown in [40]. Bringing remote sensing, environmental data and crop models together to improve accuracy of N uptake predictions may be an approach to improving N recommendations.

There are differences in the model of N uptake from remote sensed data from season to season that are difficult to model. More work is needed to determine the causes of these differences and how they can be incorporated into a model to improve the absolute accuracy to new seasons before sample data is available. Possible avenues for investigation include soil chemistry differences, water management differences, ponded water temperature differences, leaf area index differences at the image date (as ponded water may interact with canopy reflectance) and un-modelled climate effects [55].

We investigated different combinations of remote sensing bands and spectral indexes. An important band for predicting N uptake on the WorldView satellites is nir2, at 860–1040 nm. Most multispectral sensors do not have an nir band with wavelengths this large, and some accuracy is lost using the lower wavelength nir band (770–895 nm) which is more common in a variety of sensors. We found models generated using four multispectral bands commonly found in sensors (green, red, red edge, near infrared) produced prediction errors 1–2 kg/ha higher than if all eight multispectral bands of the WorldView sensors were used. We found, in agreement with other works, the NDVI is a poor predictor of high values of N uptake, and indexes including the red-edge band are able to predict a greater range of values [26]. We also showed that squaring the NDRE index follows the trend of N uptake closely.

We now consider the prospects of implementing a viable methodology of generating PI N uptake maps covering the rice industry in a given area. The cost of WorldView imagery is relatively high (though less in $/ha than airborne data), and the minimum capture area is large. It may be feasible for commercial use if a large number of paddies are grouped in a small area (so the minimum area capture covers many fields), but may not be feasible for more sparsely distributed fields. Therefore, extending our multi-season N uptake prediction methodology to more cost-effective platforms is desirable. An additional consideration is that image acquisition needs to be arranged well in advance, so models forecasting PI date need to be accurate [44]. There is also the risk of cloud cover on the selected acquisition date. Aerial platforms will remove limitations of image acquisitions on days with cloud cover and may offer more flexibility with late adjustments of image dates. UAVs offer an alternative for small areas, provided an experienced operator is available, who follows calibration protocols strictly. It would be useful to investigate the integration of models from various platforms and sensors in order to mitigate issues related to lack of availability of quality images from a single platform in a given season.

Following determination of N uptake at PI, recommendations of optimal N topdressing rates are needed. These recommendations may be generated using studies such as the one described in [15], where plots were sampled for PI N uptake, and various topdressing rates were applied across plots. Then final grain yield results are used to generate tables of optimal PI N application as a function of PI N uptake. Variable rate topdressing resolution using aerial application is around 30 kg/ha. The multi-season model RMSE is as high as 27.4 kg/ha when predicting N uptake in an unseen season. However, given the remote sensing approach removes the labour associated with sampling, and the fact that it is unfeasible to sample enough points to adequately represent spatial variability, the remote sensing approach remains an attractive solution.

5. Conclusions

This work has demonstrated the ability of multi-spectral satellite images to predict absolute N uptake in rice at the panicle initiation growth stage, using data from four seasons, two sites, four varieties, and multiple sowing dates. A simple single variable model using the $NDRE^2$ index was shown to give reasonable performance, with N uptake RMSE of 22.8 kg/ha and R^2 of 0.75 when all four season's data were pooled. More complex models incorporating combinations of input bands, climate and management date data achieved R^2 of 0.91 and RMSE of 14.6 kg/ha. However, this methodology requires physical samples to be available for model training for each season that the N uptake is to be predicted. We developed a model fitting methodology to predict future seasons, that avoids model overfitting providing multiple previous seasons data are available. Using this methodology, prediction performance ranged from R^2 of 0.67–0.78 and RMSE of 16.8–27.4 kg/ha over four test seasons for which the model had seen no data. Model regularization limited the complexity to four or less input variables. This accuracy is sufficient for operational use in prescribing variable-rate application of N at the panicle initiation growth stage.

Author Contributions: Conceptualization, B.W.D., R.L.D.; data curation, J.B.; methodology, B.W.D., R.L.D., A.J.R. and J.B.; software, J.B.; validation, B.W.D., A.J.R. and J.B.; formal analysis, J.B.; investigation, B.W.D.; writing—original draft preparation, J.B.; writing—review and editing, B.W.D., A.J.R., T.S.D., R.L.D.; visualization, J.B.; project administration, B.W.D.; funding acquisition, B.W.D.

Funding: This research was funded by AgriFutures Australia, project number PRJ-011058, 'Improving mid-season nitrogen management of rice', and project number PRJ-009772, 'Moving forward with NIR and remote sensing'.

Acknowledgments: The authors are grateful for the work of Chris Dawe and Craig Hodges in contributing to field work. AGScene Pty Ltd contributed to the previous acquisition and analysis of WV3 imagery that has served as a precursor to this paper. The map in Figure 1 was generated in Google Earth Engine, using some elements coded by Gennadii Donchyts.

Conflicts of Interest: The authors declare no conflict of interest.

References

1. Muñoz-Huerta, R.F.; Guevara-Gonzalez, R.G.; Contreras-Medina, L.M.; Torres-Pacheco, I.; Prado-Olivarez, J.; Ocampo-Velazquez, R.V. A Review of Methods for Sensing the Nitrogen Status in Plants: Advantages, Disadvantages and Recent Advances. *Sensors* **2013**, *13*, 10823–10843, doi:10.3390/s130810823. [CrossRef] [PubMed]
2. Tilman, D.; Cassman, K.G.; Matson, P.A.; Naylor, R.; Polasky, S. Agricultural sustainability and intensive production practices. *Nat. Lond.* **2002**, *418*, 671–677, doi:10.1038/nature01014. [CrossRef] [PubMed]

3. López-Bellido, L.; Fuentes, M.; Castillo, J.E.; López-Garrido, F.J. Effects of tillage, crop rotation and nitrogen fertilization on wheat-grain quality grown under rainfed Mediterranean conditions. *Field Crops Res.* **1998**, *57*, 265–276, doi:10.1016/S0378-4290(97)00137-8. [CrossRef]
4. Camargo, J.A.; Alonso, Á. Ecological and toxicological effects of inorganic nitrogen pollution in aquatic ecosystems: A global assessment. *Environ. Int.* **2006**, *32*, 831–849, doi:10.1016/j.envint.2006.05.002. [CrossRef] [PubMed]
5. Raun, W.R.; Johnson, G.V. Improving Nitrogen Use Efficiency for Cereal Production. *Agron. J.* **1999**, *91*, 357–363, doi:10.2134/agronj1999.00021962009100030001x. [CrossRef]
6. Ten Berge, H.F.M.; Thiyagarajan, T.M.; Shi, Q.; Wopereis, M.C.S.; Drenth, H.; Jansen, M.J.W. Numerical optimization of nitrogen application to rice. Part I. Description of MANAGE-N. *Field Crops Res.* **1997**, *51*, 29–42, doi:10.1016/S0378-4290(96)01042-8. [CrossRef]
7. Ten Berge, H.F.M.; Shi, Q.; Zheng, Z.; Rao, K.S.; Riethoven, J.J.M.; Zhong, X. Numerical optimization of nitrogen application to rice. Part II. Field evaluations. *Field Crops Res.* **1997**, *51*, 43–54, doi:10.1016/S0378-4290(96)01041-6. [CrossRef]
8. Lee, K.J.; Lee, B.W. Modeling for recommending panicle nitrogen topdressing rates for yield and milled-rice protein content. *J. Crop Sci. Biotechnol.* **2012**, *15*, 335–343, doi:10.1007/s12892-012-0117-8. [CrossRef]
9. Yao, Y.; Miao, Y.; Huang, S.; Gao, L.; Ma, X.; Zhao, G.; Jiang, R.; Chen, X.; Zhang, F.; Yu, K.; et al. Active canopy sensor-based precision N management strategy for rice. *Agron. Sustain. Dev.* **2012**, *32*, 925–933, doi:10.1007/s13593-012-0094-9. [CrossRef]
10. Samonte, S.O.P.; Wilson, L.T.; Medley, J.C.; Pinson, S.R.M.; McClung, A.M.; Lales, J.S. Nitrogen Utilization Efficiency. *Agron. J.* **2006**, *98*, 168–176, doi:10.2134/agronj2005.0180. [CrossRef]
11. Gunawardena, T.A.; Fukai, S.; Blamey, F.P.C. Low temperature induced spikelet sterility in rice. I. Nitrogen fertilisation and sensitive reproductive period. *Aust. J. Agric. Res.* **2003**, *54*, 937–946, doi:10.1071/ar03075. [CrossRef]
12. Xue, L.; Li, G.; Qin, X.; Yang, L.; Zhang, H. Topdressing nitrogen recommendation for early rice with an active sensor in south China. *Precis. Agric.* **2014**, *15*, 95–110, doi:10.1007/s11119-013-9326-5. [CrossRef]
13. Lee, Y.J.; Yang, C.M.; Chang, K.W.; Shen, Y. A Simple Spectral Index Using Reflectance of 735 nm to Assess Nitrogen Status of Rice Canopy. *Agron. J. Madison* **2008**, *100*, 205–212. [CrossRef]
14. Dunn, T.; Dunn, B. Identifying Panicle Initiation in Rice. Available online: https://www.dpi.nsw.gov.au/__data/assets/pdf_file/0003/449823/identifying-panicle-initiation-in-rice.pdf (accessed on 2 May 2019).
15. Dunn, B.W.; Dunn, T.S.; Orchard, B.A. Nitrogen rate and timing effects on growth and yield of drill-sown rice. *Crop Pasture Sci.* **2016**, *67*, 1149–1157, doi:10.1071/CP16331. [CrossRef]
16. Cassman, K.G.; Dobermann, A.; Cruz, P.C.S.; Gines, G.C.; Samson, M.I.; Descalsota, J.P.; Alcantara, J.M.; Dizon, M.A.; Olk, D.C. Soil organic matter and the indigenous nitrogen supply of intensive irrigated rice systems in the tropics. *Plant Soil* **1996**, *182*, 267–278, doi:10.1007/BF00029058. [CrossRef]
17. Russell, C.A.; Dunn, B.W.; Batten, G.D.; Williams, R.L.; Angus, J.F. Soil tests to predict optimum fertilizer nitrogen rate for rice. *Field Crops Res.* **2006**, *97*, 286–301, doi:10.1016/j.fcr.2005.10.007. [CrossRef]
18. Dunn, B. Improving Topdressing Recommendations for Rice. Available online: https://agrifuturesrice.squarespace.com/s/Improving-topdressing-recommendations-for-rice.pdf (accessed on 2 May 2019).
19. Haefele, S.M.; Wopereis, M.C.S. Spatial variability of indigenous supplies for N, P and K and its impact on fertilizer strategies for irrigated rice in West Africa. *Plant Soil* **2005**, *270*, 57–72, doi:10.1007/s11104-004-1131-5. [CrossRef]
20. Peng, S.; Buresh, R.J.; Huang, J.; Zhong, X.; Zou, Y.; Yang, J.; Wang, G.; Liu, Y.; Hu, R.; Tang, Q.; et al. Improving nitrogen fertilization in rice by sitespecific N management. A review. *Agron. Sustain. Dev.* **2010**, *30*, 649–656, doi:10.1051/agro/2010002. [CrossRef]
21. Stavrakoudis, D.; Katsantonis, D.; Kadoglidou, K.; Kalaitzidis, A.; Gitas, I.Z. Estimating Rice Agronomic Traits Using Drone-Collected Multispectral Imagery. *Remote Sens.* **2019**, *11*, 545, doi:10.3390/rs11050545. [CrossRef]
22. Moharana, S.; Dutta, S. Spatial variability of chlorophyll and nitrogen content of rice from hyperspectral imagery. *ISPRS J. Photogramm. Remote Sens.* **2016**, *122*, 17–29, doi:10.1016/j.isprsjprs.2016.09.002. [CrossRef]
23. Raun, W.R.; Solie, J.B.; Johnson, G.V.; Stone, M.L.; Mullen, R.W.; Freeman, K.W.; Thomason, W.E.; Lukin, E.V. Improving nitrogen use efficiency in cereal grain production with optical sensing and variable rate application. *Agron. J. Madison* **2002**, *94*, 815. [CrossRef]

24. Bramley, R.G.V.; Ouzman, J.; Gobbett, D.L. Regional scale application of the precision agriculture thought process to promote improved fertilizer management in the Australian sugar industry. *Precis. Agric.* **2019**, *20*, 362–378, doi:10.1007/s11119-018-9571-8. [CrossRef]
25. Schlemmer, M.; Gitelson, A.; Schepers, J.; Ferguson, R.; Peng, Y.; Shanahan, J.; Rundquist, D. Remote estimation of nitrogen and chlorophyll contents in maize at leaf and canopy levels. *Int. J. Appl. Earth Obs. Geoinf.* **2013**, *25*, 47–54, doi:10.1016/j.jag.2013.04.003. [CrossRef]
26. Hatfield, J.L.; Gitelson, A.A.; Schepers, J.S.; Walthall, C.L. Application of Spectral Remote Sensing for Agronomic Decisions. *Agron. J.* **2008**, *100*, S-117–S-131, doi:10.2134/agronj2006.0370c. [CrossRef]
27. Zheng, H.; Cheng, T.; Li, D.; Zhou, X.; Yao, X.; Tian, Y.; Cao, W.; Zhu, Y. Evaluation of RGB, Color-Infrared and Multispectral Images Acquired from Unmanned Aerial Systems for the Estimation of Nitrogen Accumulation in Rice. *Remote Sens.* **2018**, *10*, 824, doi:10.3390/rs10060824. [CrossRef]
28. Inoue, Y.; Sakaiya, E.; Zhu, Y.; Takahashi, W. Diagnostic mapping of canopy nitrogen content in rice based on hyperspectral measurements. *Remote Sens. Environ.* **2012**, *126*, 210–221, doi:10.1016/j.rse.2012.08.026. [CrossRef]
29. Ryu, C.; Suguri, M.; Umeda, M. Multivariate analysis of nitrogen content for rice at the heading stage using reflectance of airborne hyperspectral remote sensing. *Field Crops Res.* **2011**, *122*, 214–224, doi:10.1016/j.fcr.2011.03.013. [CrossRef]
30. Dunn, B.W.; Dehaan, R.; Schmidtke, L.M.; Dunn, T.S.; Meder, R. Using Field-Derived Hyperspectral Reflectance Measurement to Identify the Essential Wavelengths for Predicting Nitrogen Uptake of Rice at Panicle Initiation. *J. Near Infrared Spectrosc.* **2016**, *24*, 473–483. [CrossRef]
31. Yu, K.; Li, F.; Gnyp, M.L.; Miao, Y.; Bareth, G.; Chen, X. Remotely detecting canopy nitrogen concentration and uptake of paddy rice in the Northeast China Plain. *ISPRS J. Photogramm. Remote Sens.* **2013**, *78*, 102–115, doi:10.1016/j.isprsjprs.2013.01.008. [CrossRef]
32. Adão, T.; Hruška, J.; Pádua, L.; Bessa, J.; Peres, E.; Morais, R.; Sousa, J.J. Hyperspectral Imaging: A Review on UAV-Based Sensors, Data Processing and Applications for Agriculture and Forestry. *Remote Sens.* **2017**, *9*, 1110, doi:10.3390/rs9111110. [CrossRef]
33. Zheng, H.; Cheng, T.; Li, D.; Yao, X.; Tian, Y.; Cao, W.; Zhu, Y. Combining Unmanned Aerial Vehicle (UAV)-Based Multispectral Imagery and Ground-Based Hyperspectral Data for Plant Nitrogen Concentration Estimation in Rice. *Front. Plant Sci.* **2018**, *9*, 936, doi:10.3389/fpls.2018.00936. [CrossRef] [PubMed]
34. Guan, S.; Fukami, K.; Matsunaka, H.; Okami, M.; Tanaka, R.; Nakano, H.; Sakai, T.; Nakano, K.; Ohdan, H.; Takahashi, K. Assessing Correlation of High-Resolution NDVI with Fertilizer Application Level and Yield of Rice and Wheat Crops Using Small UAVs. *Remote Sens.* **2019**, *11*, 112, doi:10.3390/rs11020112. [CrossRef]
35. Tu, Y.H.; Phinn, S.; Johansen, K.; Robson, A. Assessing Radiometric Correction Approaches for Multi-Spectral UAS Imagery for Horticultural Applications. *Remote Sens.* **2018**, *10*, 1684, doi:10.3390/rs10111684. [CrossRef]
36. Von Bueren, S.K.; Burkart, A.; Hueni, A.; Rascher, U.; Tuohy, M.P.; Yule, I.J. Deploying four optical UAV-based sensors over grassland: Challenges and limitations. *Biogeosciences* **2015**, *12*, 163–175, doi:10.5194/bg-12-163-2015. [CrossRef]
37. Matese, A.; Toscano, P.; Gennaro, S.F.D.; Genesio, L.; Vaccari, F.P.; Primicerio, J.; Belli, C.; Zaldei, A.; Bianconi, R.; Gioli, B. Intercomparison of UAV, Aircraft and Satellite Remote Sensing Platforms for Precision Viticulture. *Remote Sens.* **2015**, *7*, 2971–2990, doi:10.3390/rs70302971. [CrossRef]
38. Huang, S.; Miao, Y.; Zhao, G.; Yuan, F.; Ma, X.; Tan, C.; Yu, W.; Gnyp, M.L.; Lenz-Wiedemann, V.I.S.; Rascher, U.; et al. Satellite Remote Sensing-Based In-Season Diagnosis of Rice Nitrogen Status in Northeast China. *Remote Sens.* **2015**, *7*, 10646–10667, doi:10.3390/rs70810646. [CrossRef]
39. Huang, S.; Miao, Y.; Yuan, F.; Gnyp, M.L.; Yao, Y.; Cao, Q.; Wang, H.; Lenz-Wiedemann, V.I.S.; Bareth, G. Potential of RapidEye and WorldView-2 Satellite Data for Improving Rice Nitrogen Status Monitoring at Different Growth Stages. *Remote Sens.* **2017**, *9*, 227, doi:10.3390/rs9030227. [CrossRef]
40. Zhang, K.; Ge, X.; Shen, P.; Li, W.; Liu, X.; Cao, Q.; Zhu, Y.; Cao, W.; Tian, Y. Predicting Rice Grain Yield Based on Dynamic Changes in Vegetation Indexes during Early to Mid-Growth Stages. *Remote Sens.* **2019**, *11*, 387, doi:10.3390/rs11040387. [CrossRef]
41. Nurulhuda, K.; Gaydon, D.S.; Jing, Q.; Zakaria, M.P.; Struik, P.C.; Keesman, K.J. Nitrogen dynamics in flooded soil systems: An overview on concepts and performance of models. *J. Sci. Food Agric.* **2018**, *98*, 865–871, doi:10.1002/jsfa.8683. [CrossRef]

42. Hornbuckle, J.W.; Christen, E.W. Physical Properties of Soils in the Murrumbidgee and Coleambally Irrigation Areas. Available online: https://publications.csiro.au/rpr/pub?list=BRO&pid=procite:76ba0c1d-8687-4114-847e-f2e6dac6c94b (accessed on 5 July 2019).
43. Troldahl, D.; Dunn, B.; Fowler, J.; Garnett, L.; Groat, M.; Mauger, T. Rice Growing Guide 2018. Available online: https://www.dpi.nsw.gov.au/__data/assets/pdf_file/0007/829330/RGG-accessible-22Aug2018.pdf (accessed on 7 May 2019).
44. Darbyshire, R.; Crean, E.; Dunn, T.; Dunn, B. Predicting panicle initiation timing in rice grown using water efficient systems. *Field Crops Res.* **2019**, *239*, 159–164, doi:10.1016/j.fcr.2019.05.018. [CrossRef]
45. Daun, J.K.; DeClercq, D.R. Comparison of combustion and Kjeldahl methods for determination of nitrogen in oilseeds. *J. Am. Oil Chem. Soc.* **1994**, *71*, 1069–1072, doi:10.1007/BF02675898. [CrossRef]
46. McMaster, G.S.; Wilhelm, W.W. Growing degree-days: One equation, two interpretations. *Agric. For. Meteorol.* **1997**, *87*, 291–300, doi:10.1016/S0168-1923(97)00027-0. [CrossRef]
47. Walter, I.A.; Allen, R.G.; Elliott, R.; Jensen, M.E.; Itenfisu, D.; Mecham, B.; Howell, T.A.; Snyder, R.; Brown, P.; Echings, S.; et al. ASCE's standardized reference evapotranspiration equation. In Proceedings of theWatershed Management 2000 and Operations Management 2000, Collins, CO, USA, 20–24 June 2000; doi:10.1061/40499(2000)126. [CrossRef]
48. Gorelick, N.; Hancher, M.; Dixon, M.; Ilyushchenko, S.; Thau, D.; Moore, R. Google Earth Engine: Planetary-scale geospatial analysis for everyone. *Remote Sens. Environ.* **2017**, *202*, 18–27, doi:10.1016/j.rse.2017.06.031. [CrossRef]
49. Chavez, P.S. An improved dark-object subtraction technique for atmospheric scattering correction of multispectral data. *Remote Sens. Environ.* **1988**, *24*, 459–479, doi:10.1016/0034-4257(88)90019-3. [CrossRef]
50. McKinney, W. Data Structures for Statistical Computing in Python. In Proceedings of the 9th Python in Science Conference, Austin, TX, USA, 28 June–3 July 2010; pp. 51–56.
51. Pedregosa, F.; Varoquaux, G.; Gramfort, A.; Michel, V.; Thirion, B.; Grisel, O.; Blondel, M.; Prettenhofer, P.; Weiss, R.; Dubourg, V.; et al. Scikit-learn: Machine Learning in Python. *J. Mach. Learn. Res.* **2011**, *12*, 2825–2830.
52. Seabold, S.; Perktold, J. Statsmodels: Econometric and Statistical Modeling with Python. In Proceedings of the 9th Python in Science Conference, Austin, TX, USA, 28 June–3 July 2010; p. 6.
53. Hastie, T.; Tibshirani, R.; Friedman, J. *The Elements of Statistical Learning: Data Mining, Inference, and Prediction*, 2nd ed.; Springer Series in Statistics; Springer: New York, NY, USA, 2009.
54. Xue, L.; Cao, W.; Luo, W.; Dai, T.; Zhu, Y. Monitoring Leaf Nitrogen Status in Rice with Canopy Spectral Reflectance. *Agron. J.* **2004**, *96*, 135–142, doi:10.2134/agronj2004.1350. [CrossRef]
55. Samborski, S.M.; Tremblay, N.; Fallon, E. Strategies to Make Use of Plant Sensors-Based Diagnostic Information for Nitrogen Recommendations. *Agron. J.* **2009**, *101*, 800–816, doi:10.2134/agronj2008.0162Rx. [CrossRef]

remote sensing

Article

Developing a Proximal Active Canopy Sensor-based Precision Nitrogen Management Strategy for High-Yielding Rice

Junjun Lu [1,2], Yuxin Miao [1,3,*], Wei Shi [1], Jingxin Li [1], Xiaoyi Hu [1,4], Zhichao Chen [2], Xinbing Wang [1] and Krzysztof Kusnierek [5]

1 International Center for Agro-Informatics and Sustainable Development (ICASD), College of Resources and Environmental Sciences, China Agricultural University, Beijing 100193, China; junjunlu@hpu.edu.cn (J.L.); shiwei@alu.cau.edu.cn (W.S.); lijingxin@alu.cau.edu.cn (J.L.); hxy1223@alu.cau.edu.cn (X.H.); xbwang2020@cau.edu (X.W.)

2 Agro-Geoinformatics Research Center (ARC), School of Surveying and Land Information Engineering, Henan Polytechnic University, Jiaozuo 454000, China; czc@hpu.edu.cn

3 Precision Agriculture Center, Department of Soil, Water and Climate, University of Minnesota, St. Paul, MN 55108, USA

4 Beijing Agriculture Technology Extension Station, Beijing 100029, China

5 Center for Precision Agriculture, Norwegian Institute of Bioeconomy Research (NIBIO), Nylinna 226, 2849 Kapp, Norway; Krzysztof.Kusnierek@nibio.no

* Correspondence: ymiao@umn.edu

Received: 16 February 2020; Accepted: 29 April 2020; Published: 2 May 2020

Abstract: RapidSCAN is a portable active canopy sensor with red, red-edge, and near infrared spectral bands. The objective of this study is to develop and evaluate a RapidSCAN sensor-based precision nitrogen (N) management (PNM) strategy for high-yielding rice in Northeast China. Six rice N rate experiments were conducted from 2014 to 2016 at Jiansanjiang Experiment Station of China Agricultural University in Northeast China. The results indicated that the sensor performed well for estimating rice yield potential (YP_0) and yield response to additional N application ($RI_{Harvest}$) at the stem elongation stage using normalized difference vegetation index (NDVI) (R^2 = 0.60–0.77 and relative error (REr) = 6.2–8.0%) and at the heading stage using normalized difference red edge (NDRE) (R^2 = 0.70–0.82 and REr = 7.3–8.7%). A new RapidSCAN sensor-based PNM strategy was developed that would make N recommendations at both stem elongation and heading growth stages, in contrast to previously developed strategy making N recommendation only at the stem elongation stage. This new PNM strategy could save 24% N fertilizers, and increase N use efficiencies by 29–35% as compared to Farmer N Management, without significantly affecting the rice grain yield and economic returns. Compared with regional optimum N management, the new PNM strategy increased 4% grain yield, 3–10% N use efficiencies and 148 \$ ha^{-1} economic returns across years and varieties. It is concluded that the new RapidSCAN sensor-based PNM strategy with two in-season N recommendations using NDVI and NDRE is suitable for guiding in-season N management in high-yield rice management systems. Future studies are needed to evaluate this RapidSCAN sensor-based PNM strategy under diverse on-farm conditions, as well as to integrate it into high-yield rice management systems for food security and sustainable development.

Keywords: RapidSCAN sensor; nitrogen recommendation algorithm; in-season nitrogen management; nitrogen use efficiency; yield potential; yield responsiveness

1. Introduction

As one of the major cereal crops in the world, more than half of the world's population takes rice (*Oryza sativa* L.) as the staple food [1]. The area under rice cultivation in Asia accounts for 90% of the world's total rice area [2]. At the same time, inappropriate nitrogen (N) fertilizer application rates and timing result in low N use efficiency (NUE) in this area [3]. Northeast China is a major rice production region in China and the abovementioned management problems are common [4–6]. Facing these challenges, Chinese agricultural scientists have developed regional optimum N management (RONM) systems, aiming to obtain higher yields with less resources and N losses suitable for different regions [4,6,7]. The RONM system using fixed N rates and timing optimum for a region may not be optimal for a specific site, year, and variety in that region [8,9]. Precision N management (PNM) strategies consider both spatial and temporal variability in soil N supply and crop N demand. They have the potential to further improve NUE over the RONM strategy [8].

Active crop canopy sensors have been increasingly used to develop in-season site-specific N management strategies, allowing non-destructive real-time diagnosis of crop N status and N recommendations. They have their own light sources and are not affected by environmental light conditions [8]. The GreenSeeker active canopy sensor (Trimble Navigation Limited, Sunnyvale, CA, USA) is a commonly used sensor for guiding in-season N management [8]. It has red (R) and near-infrared (NIR) spectral wavebands and two default vegetation indices (VI): normalized difference vegetation index (NDVI) and ratio vegetation index (RVI) [8]. A GreenSeeker sensor-based PNM strategy has been previously developed to improve NUE while maintaining rice yield in Northeast China [10]. In this strategy, basal and tillering N rates were the same as RONM, while panicle fertilizer rates at the stem elongation stage were adjusted based on N fertilizer optimization algorithm using the GreenSeeker sensor [10]. The key components of this strategy include in-season estimation of yield potential without additional topdressing N application (YP_0) and N response index ($RI_{Harvest}$) [10–12]. The potential yield with sufficient topdressing N application (YP_N) can be estimated by multiplying YP_0 and $RI_{Harvest}$ [11] and then the N topdressing requirement is estimated by dividing the yield increase (YP_N-YP_0) by the average NUE [10]. Because of the saturation problem of NDVI at moderate to high biomass conditions, the estimation of YP_0 and $RI_{Harvest}$ using GreenSeeker NDVI was not very satisfactory across all stages [10] or at later growth stages (e.g., the heading stage) [13]. A previous study using the Crop Circle ACS 470 sensor (Holland Scientific, Inc., Lincoln, NE, USA) indicated that red edge-based VIs had the potential to overcome the NDVI saturation problem and improve the estimation of YP_0 and $RI_{Harvest}$, especially at later rice growth stages [13].

The RapidSCAN CS-45 sensor (Holland Scientific Inc., Lincoln, Nebraska, USA) is a relatively new alternative active crop canopy sensor available in the market. It is a lightweight and convenient portable sensor with built-in global positioning system and red edge (RE) band in addition to red and near infrared bands. Another advantage of the RapidSCAN sensor is that the sensor data collection is not influenced by measurement height in the range of 0.3 to 3 m [14]. It provides NDVI and normalized difference red edge (NDRE) as two default VIs, in addition to the R, RE, and NIR waveband reflectance. Besides NDVI and NDRE, many different VIs can be calculated. This sensor was found to perform well for estimating rice N status indicators at different growth stages [14]. Zhang et al. [15] reported that NDRE had a better rice yield prediction accuracy than NDVI from stem elongation to booting stage using the RapidSCAN sensor. More studies are needed to develop RapidSCAN sensor-based PNM strategies for rice.

To ensure both food security and agricultural sustainable development, integrated precision rice management systems have been developed to increase rice yield and NUE simultaneously [16]. In such systems, in addition to N applications before transplanting, at the tillering and stem elongation stages, grain N fertilizer was also applied at the heading stage to better meet the N demand of high-yielding rice. The previously developed GreenSeeker-based PNM strategies did not perform well to guide grain N fertilizer application at the heading stage because of the NDVI saturation [13]. The RapidSCAN sensor has the potential to overcome the saturation problem of NDVI and research is needed to develop a RapidSCAN sensor-based PNM strategy for high-yielding rice management systems that will guide topdressing N applications at both stem elongation and heading stages.

Therefore, the objectives of this study are to (i) evaluate the potential of in-season estimation of the rice yield potential and the response to N application at different growth stages using the RapidSCAN CS-45 sensor, (ii) develop a RapidSCAN sensor-based PNM strategy for high-yielding rice, and (iii) evaluate the RapidSCAN sensor-based PNM strategy for different varieties, N status, and years in Northeast China.

2. Materials and Methods

2.1. Study Site

The study was conducted in Sanjiang Plain, Heilongjiang Province, Northeast China (47.2°N, 132.6°E). The main soil type in this area is Albic soil, classified as Mollic Planosols in the FAO-UNESCO system, and typical Argialbolls in the soil taxonomy [17]. The study site is located in a cool-temperate sub-humid continental monsoon climate zone. The temperature ranges from −41 °C in the winter to 38 °C in the summer, with a mean annual temperature of 1.9 °C. About 72% of its annual precipitation (500–600 mm) occurs from June to September. The annual frost-free period is about 120–140 days long [16].

2.2. Calibration and Validation Experiments

Six plot experiments (Exp.) were conducted from 2014 to 2016 at Jiansanjiang Experiment Station of China Agricultural University, involving two different varieties, N rates and sensor-based N management strategies (Table 1). Each experiment had the same five N rates (0, 40, 80, 120, and 160 kg N ha^{-1}). In addition, the experiments in 2015–2016 consisted of a sensor-based PNM treatment using the RapidSCAN sensor. The N fertilizer was applied in five N rate treatments (except the control treatment without N application) as three splits: 40% as basal N before transplanting, 30% at tillering stage, and the remaining 30% N at the stem elongation stage. According to previous studies [6,9], the N rate treatment for 120 kg N ha^{-1} was used as RONM system in this region. The sensor-based PNM treatments were also based on the RONM system, with the same basal (48 kg ha^{-1}) and tillering (36 kg ha^{-1}) N rates. The panicle and grain fertilizer rates were determined according to active canopy sensor-based N recommendation algorithm and applied at the stem elongation and heading stages, respectively.

Table 1. The details about the crop growth and crop sensing within the experiments performed in this study.

Experiment	Year	Variety	Transplanting Date	Harvest Date	Stem Elongation Stage		Heading Stage	
					Sensing Date	DAT	Sensing Date	DAT
Exp. 1	2014	Longjing 31	19 May	29 September	3 July	45	26 July	68
Exp. 2	2015		20 May	4 October	6 July	47	30 July	71
Exp. 3	2016		19 May	25 September	5 July	47	25 July	67
Exp. 4	2014	Longjing 21	19 May	29 September	7 July	49	26 July	68
Exp. 5	2015		20 May	4 October	7 July	48	2 August.	74
Exp. 6	2016		19 May	25 September	5 July	47	25 July	67

Note: DAT: the number of days from transplanting to sensing.

Exp. 1 to 3 used Longjing 31, which is an 11-leaf variety requiring about 130 days to reach maturity. Exp. 4 to 6 used Longjing 21, which is a 12-leaf variety that needs about 133 days to maturity. All plot experiments were replicated three times in a randomized complete block design. The N source was granular urea. To evaluate the potential of the crop canopy sensors to estimate rice YP_0 and $RI_{Harvest}$ at the stem elongation and heading stages, each plot of all experiments (except the 0 kg N ha^{-1} treatment) was divided into two parts: 4.5 × 9 m as the main plot and 2.5 × 9 m as the subplot without receiving the third N application. For all the treatments, 50 kg P_2O_5 ha^{-1} in the form of $Ca(H_2PO_4)_2$ was applied before transplanting and 105 kg K_2O ha^{-1} in the form of KCl was applied as two splits: 50% before transplanting and 50% at the stem elongation stage. Rice seedlings were prepared in a greenhouse and transplanted into the experimental fields in mid-May. The field and crop management in these experiments followed the regional recommendations.

2.3. Proximal Sensing Data Collection

The RapidSCAN CS-45 sensor was used to collect reflectance data in this study. The sensor with modulated light emitting diodes irradiates the crop canopy and determines a portion of the radiation reflected from the crop canopy, without being affected by ambient illumination. The internal polychromatic light source includes three spectral bands centered at R (670 nm), RE (730 nm), and NIR (780 nm) wavelengths. According to the manufacturer, the sensor has the unique feature of Pseudo Solar Reflectance measurements that are independent of height in the range of 0.3 m to 3 m. Considering the potential influence of viewing angle and measurement area on sensor readings, the sensor footprint was parallel to the plant rows with the beam of light being perpendicular to rice canopy about 0.7–0.9 m above the canopy. The sensor was carried at a consistent speed to collect sensor readings from four different rows (3 m per row) in the middle of each plot. The reflectance values were then averaged to represent the reflectance for each plot.

Following the methodology established in the previous study on the RapidSCAN sensor [14], fifty-one VIs were evaluated in this study for estimating YP_0 and $RI_{Harvest}$ and the best performing VIs for calibration and validation are listed in Table 2. NDVI and NDRE were provided as two default indices for this sensor (see Table 2). Reflectance data were collected at stem elongation and heading stages, which were the key stages for panicle and grain fertilizer applications.

Table 2. NDVI, NDRE, and the best performing vegetation indices for calibration and validation from RapidSCAN used in this study.

Index	Formula	Reference
Normalized difference vegetation index (NDVI)	$(NIR - R)/(NIR + R)$	Rouse et al. [18]
Normalized difference red edge (NDRE)	$(NIR - RE)/(NIR + RE)$	Barnes et al. [19]
Ratio vegetation index (RVI)	NIR/R	Jordan [20]
Modified soil-adjusted vegetation index (MSAVI)	$\left[(2 \times NIR + 1) - \sqrt{(2 \times NIR + 1)^2 - 8(NIR - R)}\right]/2$	Qi et al. [21]
Modified simple ratio (MSR)	$(NIR/R - 1)/\sqrt{NIR/R + 1}$	Chen [22]
Optimal vegetation index (VI_{opt})	$1.45 \times \left(NIR^2 + 1\right)/(R + 0.45)$	Reyniers et al. [23]
Nonlinear index (NLI)	$\left(NIR^2 - R\right)/\left(NIR^2 + R\right)$	Goel and Qin [24]
NDVI*RVI	$\left(NIR^2 - R\right)/\left(NIR + R^2\right)$	Gong et al. [25]
Red edge wide dynamic range vegetation index (REWDRVI)	$(0.12INIR - RE)/(0.12RNIR + RE)$	Cao et al. [26]
Red edge optimal soil adjusted vegetation index (REOSAVI)	$(1 + 0.16)(NIR - RE)/(NIR + RE + 0.16)$	Cao et al. [26]
Modified red edge soil adjusted vegetation index (MRESAVI)	$\left[(2 \times NIR + 1) - \sqrt{(2 \times NIR + 1)^2 - 8(NIR - RE)}\right]/2$	Cao et al. [26]
Optimized red edge vegetation index ($REVI_{opt}$)	$100 \times (\ln NIR - \ln RE)$	Jasper et al. [27]
Normalized near infrared index (NNIRI)	$NIR/(NIR + RE + R)$	Lu et al. [14]

2.4. Plant Sampling and Measurements

At the stem elongation, heading, and maturity stages, 3 hills with tillers representative of each plot were randomly selected for assessing the aboveground biomass. After cleaning with water, all roots were removed. The plant samples were then oven dried for 30 min at 105 °C and then at 70 °C until constant weight, and weighed to determine their biomass. They were later ground to pass a 0.5 mm sieve. Plant N concentration was determined using the Kjeldahl-N method.

Rice was harvested at the end of September or early October. Grain yield was determined by hand harvesting three 1 m^2 areas in each plot where spectral reflectance data were collected. Grains were separated from straw using a small grain thresher and then weighed. Grain moisture was determined immediately after weighing. The rice grain weight was adjusted to a moisture content of 140 g kg^{-1}.

Agronomic efficiency of N (AE_N) and partial factor productivity of N (PFP_N) were calculated using the following equations:

$$AE_N\ (kg\ kg^{-1}) = \frac{\text{Grain yield} - \text{Grain yield at control}}{\text{N rate}} \times 100 \quad (1)$$

$$PFP_N\left(kg\ kg^{-1}\right) = \frac{\text{Grain yield}}{\text{N rate}} \quad (2)$$

2.5. Development and Evaluation of RapidSCAN-Based Precision Nitrogen Management Strategies

Based on Yao et al. [10], the RapidSCAN-based PNM strategy in this study was developed by first establishing the models to estimate YP_0 and $RI_{Harvest}$ using in-season estimate of yield (INSEY) and in-season N response index based on VI (RI-VI), respectively. INSEY can be regarded as an estimate of average daily biomass production from the time of transplanting to the day of sensing [11]. It was calculated as NDVI divided by the number of growing degree days > 0 [10]. In this study, however, the number of days from transplanting to sensing was used instead of growing degree days to calculate INSEY, similarly to the method of Cao et al. [13]. With respect to their study, the selected RapidSCAN VIs were used here to replace the GreenSeeker NDVI or RVI. $RI_{Harvest}$ indicates the actual crop yield response to additional N within a given year [28,29] and was calculated as follows [13]:

$$RI_{Harvest} = \frac{Yield\_N_{rich}}{Yield\_CK} \quad (3)$$

where $Yield\_N_{rich}$ is the average yield of plots receiving sufficient N application (the 160 kg N ha^{-1} treatment in this study), and Yield_CK is the average yield of plot without receiving the third N application at the stem elongation stage or the fourth N application at the heading stage.

RI-VI was calculated in the same way as $RI_{Harvest}$, with the exception that VIs derived from RapidSCAN sensor were used instead of yield. YP_N was calculated by multiplying YP_0 and $RI_{Harvest}$. Finally, the N topdressing requirement is estimated by dividing the yield gap ($YP_N - YP_0$) by the average AE of topdressing N ($AE_{topdressing}$) [10]. The $AE_{topdressing}$ should be higher than the one for the whole season, and will be predicted during the growing season using the predicted $RI_{Harvest}$ [30].

To ensure sufficient N supply for grain filling and higher NUE in high-yield rice management systems, a strategy for in-season site-specific N management of rice using RapidSCAN at stem elongation and heading stages was developed in this study (Figure 1). First, the topdressing N application rate (N_{rate}) at stem elongation stage was determined as mentioned above, and then this rate was split in two doses, 2/3 as panicle fertilizer at stem elongation stage (SE_N_{rate}) and 1/3 for grain fertilizer at the heading stage. Second, the RapidSCAN sensor was used to estimate the potential yield with added N application at the heading stage (HD_YP_N). The difference between estimated YP_N at stem elongation and heading stages (HD_YP_N - SE_YP_N) was used to adjust the remaining 1/3 N_{rate} to match the crop N demand at the heading stage. Therefore, the recommended N topdressing application rate at the heading stage (HD_N_{rate}) can be determined as follows (Figure 1):

$$HD\_N_{rate} = \frac{HD\_YP_N - SE\_YP_N}{AE_{topdressing}} + \frac{1}{3}N_{rate} \quad (4)$$

where HD_N_{rate} is the topdressing N application rate at heading stage, HD_YP_N is the predicted yield potential with topdressing N application at heading stage, SE_YP_N is the predicted yield potential with topdressing N application at the stem elongation stage, and $AE_{topdressing}$ is the topdressing N agronomic efficiency.

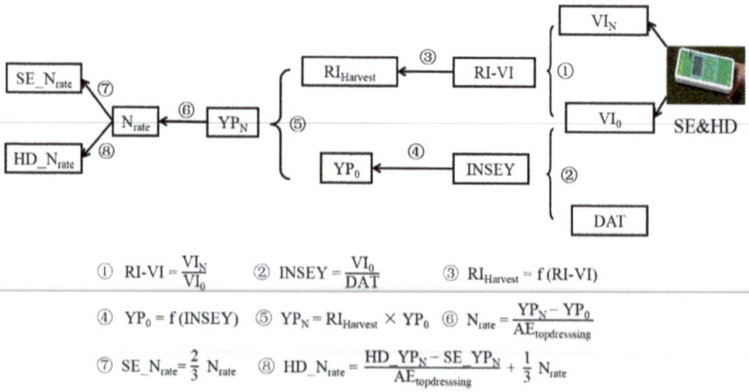

Figure 1. RapidSCAN sensor-based in-season N recommendation algorithm developed for determining topdressing N rates at stem elongation stage and heading stage of high-yielding rice in this study. VI_N: vegetation index at plots with sufficient N fertilization plots; VI_0: vegetation index at plots without additional N topdressing application; RI-VI: in-season N response index based on vegetation index; $RI_{Harvest}$: N response index based on yield; DAT: the number of days from planting to the date of sensing; INSEY: in-season estimate of yield; YP_0: the potential yield without additional topdressing N application; N_{rate}: recommended topdressing N application rate at stem elongation stage; $AE_{topdressing}$: agronomic efficiency of topdressing N; SE_YP_N or HD_YP_N: the potential yield with added topdressing N application at the stem elongation stage or heading stage, respectively; SE_N_{rate} or HD_N_{rate}: recommended topdressing N application rate at the stem elongation stage or heading stage, respectively.

In addition, two restrictions were applied after considering the rice production situation in Northeast China:

$$YP_N \le YP_{max}\left(12\ \text{t ha}^{-1}\right) \tag{5}$$

$$N_{min}\left(0\ \text{kg ha}^{-1}\right) \le N_{rate} \le N_{max}\left(48\ \text{kg ha}^{-1}\right) \tag{6}$$

where YP_{max} is the maximum obtainable yield, N_{min} and N_{max} are the minimum and maximum topdressing N rates.

According to the definition and the methods of Raun et al. [11] and [12], the YP_{max} in the study region was set to 12 t ha^{-1} based on previous studies and farmer survey data in this region [10,13,16,31]. The topdressing N application rates at the stem elongation stage was set to 0 to 48 kg ha^{-1} based on farmer surveys and previous studies in this region [10,16,31].

In order to determine whether the restrictions applied to the rice PNM strategies were suitable, three N rate treatments (80, 120, and 160 kg ha^{-1}) from Exp. 1–6 were chosen to represent three rice N status before topdressing (deficient, optimal, and surplus, respectively). They were used to evaluate the RONM and the developed PNM strategies by calculating the differences between economically optimum N rate (EONR) and N rates recommended by RONM or PNM strategies. In order to evaluate the potential of the developed PNM strategy, the RapidSCAN-based PNM treatment in Exp. 2–3 and 5–6 was compared for yield, N rate, and NUE with the control treatment (0 kg N ha^{-1}), the 160 kg N ha^{-1} treatment reflecting the farmer N management (FNM) and the 120 kg N ha^{-1} identical with the RONM. For the RapidSCAN sensor-based PNM treatment, the topdressing N rate was estimated based on the PNM strategy developed in this study using NDVI at the stem elongation stage and NDRE at the heading stage (Exp. 2 and 5 using data up to 2015; Exp. 3 and 6 using data up to 2016).

Economic return to N (E, \$ ha^{-1}) was used to evaluate the profitability of different N management systems, and was calculated as follows:

$$E = (Y_N - Y_0) \times P_Y - N_{total} \times P_N \tag{7}$$

where Y_N (kg ha^{-1}) is the rice grain yield with N application, Y_0 (kg ha^{-1}) is the rice grain yield of the check treatment without any N application, P_Y is rice grain price (0.44 \$ kg^{-1}). N_{total} is the total N fertilizer application rate (kg ha^{-1}). P_N is the N fertilizer price (0.54 \$ kg^{-1}).

2.6. Statistical Analysis

Data collected from the three-year experiments were pooled together and then randomly divided into calibration dataset (67% of the observations) and validation dataset (33% of the observations) for the estimation of YP_0 and $RI_{Harvest}$ using RapidSCAN sensor. The coefficients of determination (R^2) for the relationships between VIs and agronomic parameters were calculated using SPSS 18.0 (SPSS Inc., Chicago, Illinois, USA), and the models with the highest R^2 were selected. In addition to R^2, the performance of the models for predicting YP_0 and $RI_{Harvest}$ was also evaluated using the root mean square error (RMSE) and relative error (REr). Analysis of variance were conducted using the SAS software package Version 9.0 (SAS Institute Inc., Cary, NC, USA). The means for treatments were compared with least significant difference (LSD) test at the 0.05 probability level (at $p < 0.05$).

3. Results

3.1. Changes in NDVI vs. NDRE among Different N Rates, Varieties, Stages, and Years

Rice grain yield was significantly affected by the factors of N rates, varieties, and years, and RapidSCAN-based NDVI and NDRE also showed similar results (Table 3). Except for the variation of year, NDVI was significantly affected across different N rates, varieties, and growth stages. NDRE was significantly affected by these factors, except for variety. The changes in NDVI vs. NDRE were also shown in Figure 2. YP_0, NDVI, and NDRE all increased with N rates (Figure 3). The average NDVI showed significant difference between Longjing 31 (0.68) and Longjing 21 (0.72) across years, growth stages, and N levels.

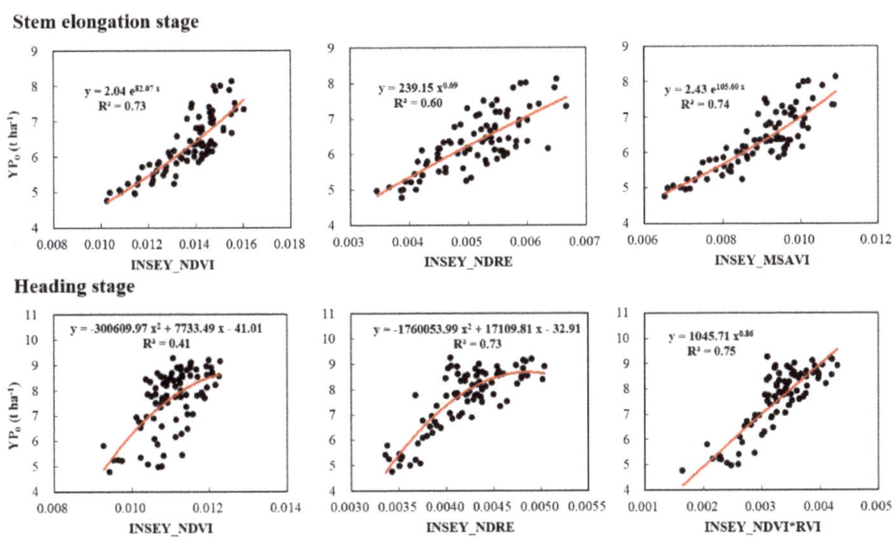

Figure 2. The relationships between yield without additional topdressing N application (YP_0) and in-season estimate of yield (INSEY) calculated with NDVI, NDRE, MSAVI, and NDVI*RVI across all varieties at stem elongation stage and heading stage.

Table 3. Significance of mean squares in the analysis of yield without additional topdressing N application (YP_0), RapidSCAN NDVI, and NDRE under five N rates (0, 40, 80, 120, 180 kg ha^{-1}) combined at two stages (stem elongation and heading stage), across three years (2014-2016) for two varieties (Longjing 31 and Longjing 21).

Source of Variation	Degree of Freedom	Significance of Mean Square		
		NDVI	NDRE	YP_0
N level	4	***	***	***
Variety	1	**	ns	***
Stage	1	***	***	***
Year	2	ns	***	***

Note: *, **, and *** indicate significance at 0.05, 0.01, and 0.001 probability levels, respectively. ns = non-significant.

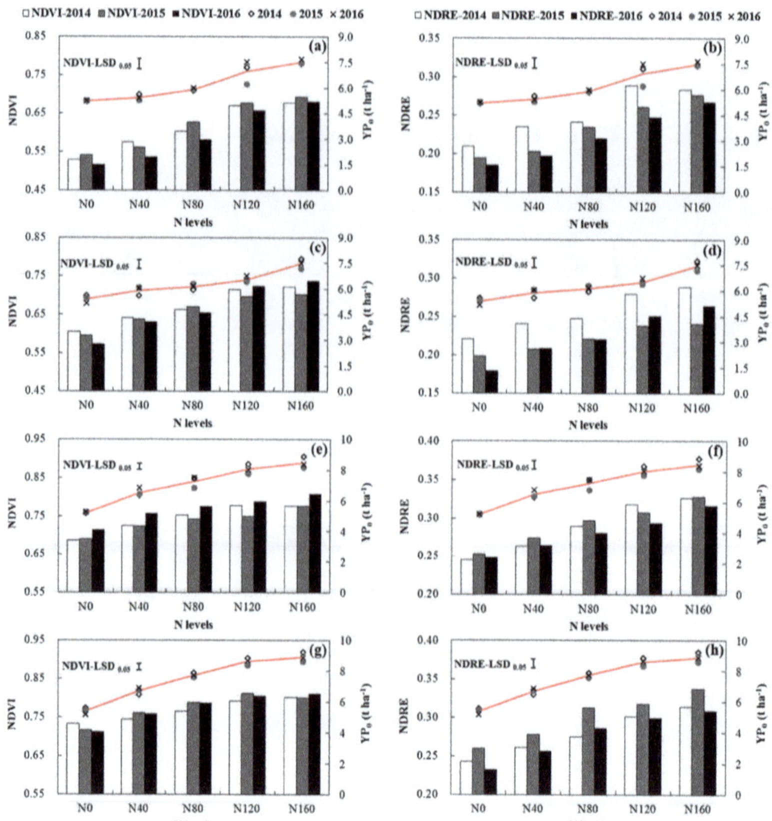

Figure 3. Rice yield without additional topdressing N application (YP_0) or NDVI (**a**; **c**; **e**; **g**) and YP_0 or NDRE (**b**; **d**; **f**; **h**) obtained by RapidSCAN at the stem elongation (**a–d**) and heading (**e–h**) stages as affected by different N rates for Longjing 31 (**a**; **b**; **e**; **f**) and Longjing 21 (**c**; **d**; **g**; **h**) in 2014–2016, respectively. Different color bars represent the value of NDVI or NDRE in different years. Different dots represent YP_0 in different years. The red curves were the curves of rice yield potential without additional topdressing N application (YP_0). Vertical bars represent the LSD value ($p = 0.05$) among different N levels.

3.2. Yield Without Additional Topdressing N Application

The performance of the INSEY calculated with NDVI, NDRE, and best performing VIs to estimate rice YP_0 varied with different growth stages across N rate treatments, sites, and years (Table 4 and Figure 2). At the stem elongation growth stage, two varieties performed similarly. INSEY-NDVI explained 72–76% of YP_0 variability, which was better than INSEY-NDRE (60–66%). The INSEY calculated with best performing VIs (nonlinear index (NLI) (INSEY_NLI), modified simple ratio (MSR) (INSEY_MSR), and modified soil-adjusted vegetation index (MSAVI) (INSEY_MSAVI)) performed similarly (R^2 = 0.74–0.78) to INSEY_NDVI.

Table 4. Calibration and validation results for predicting yield without additional topdressing N application (YP_0) using the in-season estimate of yield (INSEY) calculated with the RapidSCAN's default indices (NDVI and NDRE) and the best performing vegetation indices for two varieties at the stem elongation (SE) and heading (HD) stages in 2014–2016.

Variety	Stage	Calibration Results			Validation Results		
		Index	Model	R^2	R^2	RMSE	REr
Longjing 31	SE	INSEY_NDVI	E	0.76	0.70	0.49	7.7
		INSEY_NDRE	P	0.66	0.62	0.55	8.5
		INSEY_NLI	E	0.78	0.70	0.49	7.6
	HD	INSEY_NDVI	Q	0.59	0.54	0.74	9.9
		INSEY_NDRE	Q	0.89	0.76	0.54	7.3
		INSEY_REOSAVI	Q	0.89	0.76	0.55	7.3
Longjing 21	SE	INSEY_NDVI	E	0.72	0.66	0.39	6.3
		INSEY_NDRE	P	0.62	0.34	0.54	8.7
		INSEY_MSR	E	0.74	0.63	0.41	6.5
	HD	INSEY_NDVI	Q	0.28	0.64	0.93	12.1
		INSEY_NDRE	Q	0.77	0.75	0.66	8.5
		INSEY_MRESAVI	Q	0.78	0.72	0.69	9.0
Across varieties	SE	INSEY_NDVI	E	0.73	0.66	0.47	7.4
		INSEY_NDRE	P	0.60	0.49	0.56	8.8
		INSEY_MSAVI	E	0.74	0.65	0.47	7.4
	HD	INSEY_NDVI	Q	0.41	0.59	0.81	10.7
		INSEY_NDRE	Q	0.73	0.70	0.65	8.6
		INSEY_NDVI*RVI	P	0.75	0.70	0.66	8.7

Note: Q, E, and P: stand for quadratic, exponential, and power models. RMSE: root mean square error. REr: relative error (%).

At the heading stage, however, INSEY_NDVI did not perform very well, explaining only 28-59% of the YP_0 variability. Moreover, INSEY_NDRE performed consistently better than INSEY_NDVI, explaining 73–89% of the YP_0 variability (Table 4). The INSEY calculated with red edge optimal soil adjusted vegetation index (REOSAVI) (INSEY_REOSAVI), modified red edge soil adjusted vegetation index (MRESAVI) (INSEY_MRESAVI), and NDVI*RVI (INSEY_NDVI*RVI), did not perform significantly better than INSEY_NDRE (Table 4). Furthermore, the YP_0 was better estimated using INSEY for Longjing 31 than for Longjing 21.

The validation results were similar to the calibration results (Table 4). At the stem elongation stage, the INSEY_NDVI and INSEY calculated with best performing VI had similar performance for predicting YP_0, with R^2, RMSE, and REr of 0.65–0.66, 0.47, and 7.4% across both varieties. At the heading stage, INSEY_NDRE and INSEY calculated with best performing VI performed similarly for predicting YP_0, with R^2, RMSE, and REr of 0.70, 0.65–0.66, and 8.6–8.7%, respectively.

3.3. *The Responsiveness to Topdressing N Application*

The performance of the RI_VI calculated with NDVI (RI_NDVI), NDRE (RI_NDRE), and best performing VIs to estimate rice $RI_{Harvest}$ varied with growth stages across N rate treatments, sites, and years (Table 5 and Figure 4). At the stem elongation stage, RI_NDRE and RI calculated with best performing VIs (optimal vegetation index (VI_{opt}), red edge wide dynamic range vegetation index (REWDRVI) and RVI) did not perform significantly better than RI_NDVI (R^2 = 0.67–0.78). The $RI_{Harvest}$ was estimated better for Longjing 31 (R^2 = 0.71–0.79) than for Longjing 21 (R^2 = 0.68–0.71) or across varieties (R^2 = 0.64–0.68). The validation results showed similar pattern to the calibration results (Table 5). The prediction of $RI_{Harvest}$ for Longjing 31 (R^2 = 0.72–0.78, RMSE = 0.08–0.09 and REr = 6.1–6.8) was better than for Longjing 21 (R^2 = 0.60–0.67, RMSE = 0.10–0.11 and REr = 7.0–7.4) or across varieties (R^2 = 0.61–0.63, RMSE = 0.11 and REr = 7.8–8.0).

Table 5. Calibration and validation results for the response index calculated with yield ($RI_{Harvest}$) predicted by response index calculated with NDVI, NDRE, and the best performing vegetation indices for two varieties at stem elongation stage (SE) and heading stage (HD) in 2014–2016.

Variety	Stage	Calibration Results			Validation Results		
		Index	Model	R^2	R^2	RMSE	REr
Longjing 31	SE	RI_NDVI	Q	0.78	0.77	0.08	6.2
		RI_NDRE	Q	0.71	0.72	0.09	6.8
		RI_VI_{opt}	Q	0.79	0.78	0.08	6.1
	HD	RI_NDVI	Q	0.92	0.65	0.12	10.2
		RI_NDRE	Q	0.92	0.75	0.10	8.4
		RI_NNIRI	Q	0.95	0.73	0.10	8.7
Longjing 21	SE	RI_NDVI	Q	0.68	0.60	0.11	7.3
		RI_NDRE	Q	0.69	0.63	0.11	7.4
		RI_REWDRVI	Q	0.71	0.67	0.10	7.0
	HD	RI_NDVI	E	0.67	0.60	0.16	13.4
		RI_NDRE	Q	0.79	0.82	0.10	8.7
		RI_$REVI_{opt}$	Q	0.79	0.82	0.10	8.7
Across varieties	SE	RI_NDVI	Q	0.67	0.61	0.11	8.0
		RI_NDRE	Q	0.64	0.62	0.11	8.0
		RI_RVI	Q	0.68	0.63	0.11	7.8
	HD	RI_NDVI	Q	0.79	0.62	0.14	11.5
		RI_NDRE	Q	0.85	0.78	0.10	8.3
		RI_REOSAVI	Q	0.85	0.78	0.10	8.3

Note: Q, E, and P stand for the quadratic, exponential, and power models. RMSE: root mean square error. REr: relative error (%).

At the heading stage, RI_NDRE and RI calculated with best performing VIs (normalized NIR index (RI_NNIRI), red edge optimal VI (RI_$REVI_{opt}$), REOSAVI (RI_REOSAVI)) performed similarly for the prediction of $RI_{Harvest}$ (Table 5). They worked better than RI_NDVI (R^2 = 0.79–0.85 vs. R^2 = 0.67–0.79) for Longjing 21 or across varieties based on the calibration results. However, they performed similarly for Longjing 31 (R^2 = 0.92–0.95). With validation, RI_NDRE (R^2 = 0.75–0.82, RMSE = 0.10 and REr = 8.3–8.7) and RI calculated with best performing VIs (R^2 = 0.73–0.82, RMSE = 0.10 and REr = 8.3–8.7) were better than RI_NDVI (R^2 = 0.60–0.65, RMSE = 0.12–0.16 and REr = 10.2–13.4) for a specific variety or across varieties (Table 5). The RI calculated with best performing VIs performed better at the heading stage than the stem elongation stage for either calibration or validation.

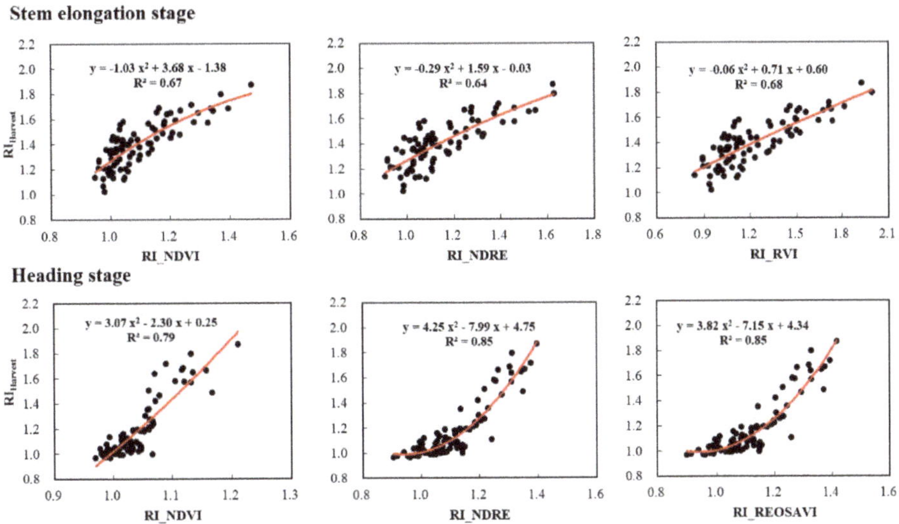

Figure 4. The relationships between response index calculated with yield ($RI_{Harvest}$) and response index calculated with NDVI, NDRE, RVI, and REOSAVI across all varieties at stem elongation stage and heading stage.

3.4. In-Season Prediction of Nitrogen Use Efficiency

$AE_{topdressing}$ varied significantly between Longjing 31 (39.5 kg kg^{-1}) and Longjing 21 (59.0 kg kg^{-1}) (Figure 5a). $RI_{Harvest}$ had a strong positive relationship with $AE_{topdressing}$, with the correlation being stronger for Longjing 21 (R^2 = 0.76) than for Longjing 31 (R^2 = 0.61) (Figure 5b). Therefore, in-season predicted $RI_{Harvest}$ was used to predict $AE_{topdressing}$ for side-dress N fertilizer recommendations in this study.

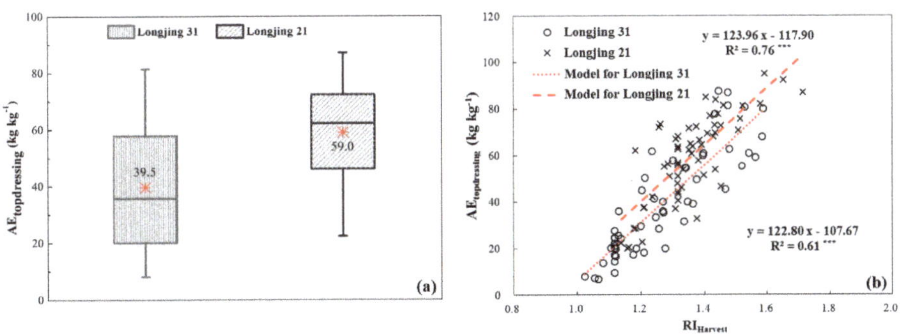

Figure 5. The difference in $AE_{topdressing}$ between Longjing 31 and Longjing 21 (**a**) and its relationship with $RI_{Harvest}$ (**b**). The red ✱ indicates the average of $AE_{topdressing}$ in Figure 5a. The red lines are the different regression models for Longjing 31 and Longjing 21 in Figure 5b; *** indicates significance at the level of $p < 0.001$.

3.5. Evaluating Different Precision Nitrogen Management Strategies Under Variable Nitrogen Status Using Scenario Analysis

Based on the abovementioned results (Tables 4 and 5), different PNM strategies were developed, as explained in Methods and Figure 3. To evaluate the performance of these strategies under different

N status, the N rate treatments (80, 120, and 160 kg ha^{-1}) in Exp.1–6 were selected for scenario analysis to determine the difference between recommended N rates (ΔN_{rate}) based on PNM strategies and EONR calculated using the N responses in each variety-year (Figures 6–8). The RONM strategy did not consider the variation of years, varieties, and rice N status, and used a fixed N topdressing rate (36 kg N ha^{-1}). Four RapidSCAN sensor-based PNM strategies were evaluated. The tested PNM strategies all had the same basal and tillering N application rates as RONM, but panicle fertilizer at the stem elongation stage was recommended using NDVI or the best performing VIs or panicle and grain fertilizer rates at the stem elongation and heading stages were recommended using NDVI and NDRE or the best performing VIs.

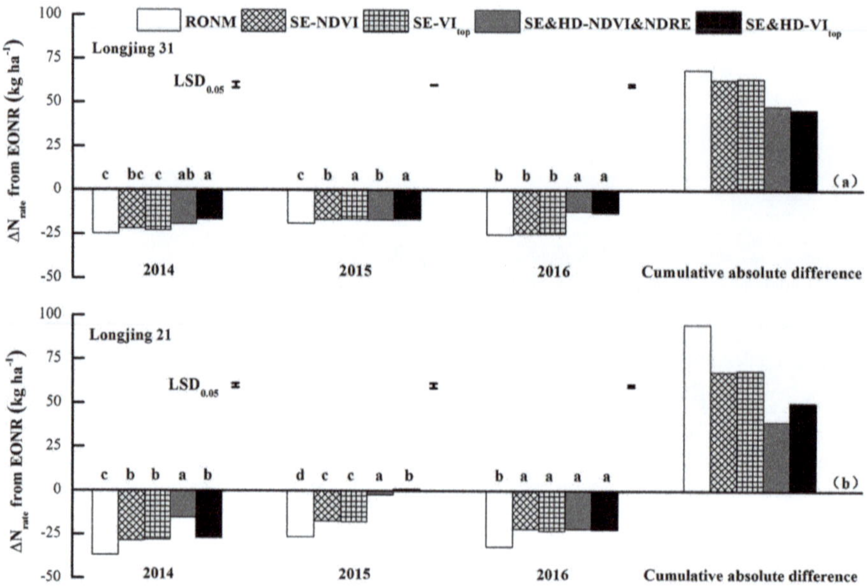

Figure 6. The difference of recommended topdressing N application rate (ΔN_{rate}) from economically optimal N rates (EONR) between RONM strategy and RapidSCAN sensor-based PNM strategies for the rice varieties of Longjing 31 (**a**) and Longjing 21 (**b**) under deficient N status before topdressing (N80) in 2014–2016. RONM: the regional optimum N management topdressing N rate (36 kg N ha^{-1} in this study). SE-NDVI and SE-VI_{top} indicated recommended topdressing N application rate calculated by the models based on NDVI and the top performing VIs (VI_{top}) for different varieties at stem elongation stage, respectively. SE&HD-NDVI&NDRE and SE&HD-VI_{top} indicated recommended topdressing N application rates calculated by the models with NDVI (at stem elongation stage) and NDRE (at heading stage), and VI_{top} at stem elongation and heading stage, respectively. Vertical bars represent the LSD value ($p = 0.05$). Different letters indicate significant difference at $p < 0.05$ level within the same year.

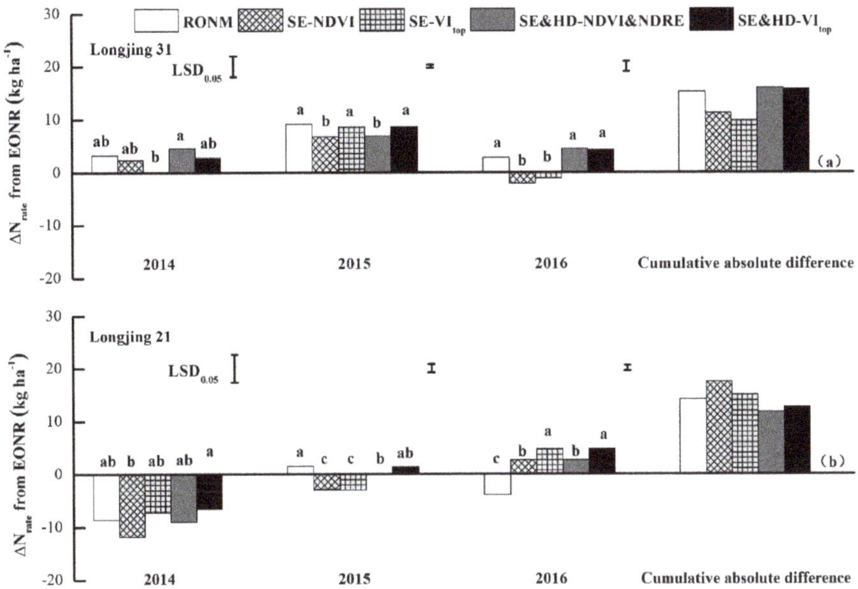

Figure 7. The difference of recommended topdressing N application rate (ΔN_{rate}) from economically optimal N rates (EONR) between RONM strategy and RapidSCAN sensor-based PNM strategies for the rice varieties of Longjing 31 (**a**) and Longjing 21 (**b**) under optimal N status before topdressing (N120) in 2014–2016. RONM: the regional optimum N management topdressing N rate (36 kg N ha^{-1} in this study). SE-NDVI and SE-VI_{top} indicated recommended topdressing N application rates calculated by the models using NDVI and the top performing VIs (VI_{top}) for different varieties at stem elongation stage, respectively. SE&HD-NDVI&NDRE and SE&HD-VI_{top} indicated recommended topdressing N application rates calculated by the models using NDVI (at stem elongation stage) and NDRE (at heading stage), and VI_{top} at stem elongation and heading stage, respectively. Vertical bars represent the LSD value (p = 0.05). Different letters indicate significant difference at $p < 0.05$ level within the same year.

There were significant differences between RONM and N rates recommended by different RapidSCAN sensor-based PNM strategies under different years, varieties, and rice N status. The N topdressing rate of RONM was consistently lower than EONR under deficient N status (Figure 6), but higher than EONR under surplus N conditions (Figure 8). Under relatively optimum N conditions, RONR was slightly higher than EONR for Longing 31, but lower for Longjing 21 (Figure 7).

All the four PNM strategies recommended higher N rates than RONM under deficient N conditions and lower N rates under surplus N conditions, while under relatively optimal N conditions, the PNM strategies would recommend slightly higher, lower or similar N rates depending on the year, variety, and PNM strategy. In a specific year across different N conditions, the ΔN_{rate} for the PNM strategies was in a range of 0–15 kg N ha^{-1}. Among all the four PNM strategies, the PNM strategies based on in-season N recommendations at both stem elongation and heading stages (with 3-year cumulative ΔN_{rate} of 40-50 kg N ha^{-1}) performed better than the PNM strategies making N recommendation only at the stem elongation stage for Longjing 21 under different N conditions (with 3-year cumulative ΔN_{rate} of 68-69 kg N ha^{-1}). For Longjing 31, the PNM strategies making N recommendations at both stem elongation and heading stages performed much better under deficient N conditions, with 3-year cumulative ΔN_{rate} (46-48 kg N ha^{-1}) being lower than the PNM strategies making N recommendations only at the stem elongation stage (63-64 kg N ha^{-1}). Irrespective of the varieties, using NDVI and

NDRE for panicle and grain N fertilizer recommendations had better or similar performance as the best performing VIs.

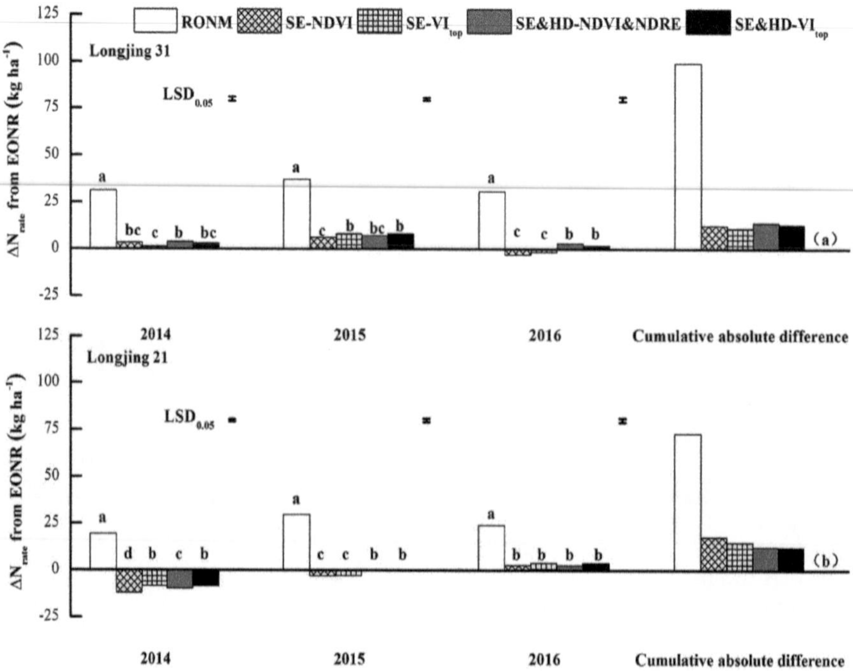

Figure 8. The difference of recommended N topdressing application rate (ΔN_{rate}) from economically optimal N rates (EONR) between RONM strategy and RapidSCAN sensor-based PNM strategies for the rice varieties of Longjing 31 (**a**) and Longjing 21 (**b**) under surplus N status before topdressing (N160) in 2014–2016. RONM: the regional optimum N management topdressing N rate (36 kg N ha^{-1} in this study). SE-NDVI and SE-VI$_{top}$ indicated recommended topdressing N application rates calculated by the models using NDVI and the top performing VIs (VI$_{top}$) for different varieties at stem elongation stage, respectively. SE&HD-NDVI&NDRE and SE&HD-VI$_{top}$ indicated recommended topdressing N application rates calculated by the models using NDVI (at stem elongation stage) and NDRE (at heading stage), and VI$_{top}$ at stem elongation and heading stage, respectively. Vertical bars represent the LSD value ($p = 0.05$). Different letters indicate significant difference at $p < 0.05$ level within the same year.

3.6. Evaluation Experiments

Based on the above results, we chose NDVI and NDRE for panicle and grain fertilizer recommendation to evaluate the potential of RapidSCAN sensor-based PNM strategy in the evaluation experiments. These experiments included the 0 kg N ha^{-1} treatment as control (CK), the 160 kg N ha^{-1} treatment as FNM treatment, the 120 kg N ha^{-1} treatment as RONM treatment, and RapidSCAN-based PNM treatment in Exp. 2–3 and 5–6 (Table 6). The RapidSCAN sensor-based PNM treatment in evaluation experiments had the same N rate as RONM (N120) before topdressing and made two in-season adjustments at the stem elongation and heading growth stages.

Table 6. N application rate (N_{rate}), grain yield, agronomy efficiency of N (AE_N), partial factor productivity of N (PFP_N), and economic returns of different rice N management strategies in the evaluation experiments conducted in 2015 and 2016.

	Treatment	2015		2016		Across Two Years		
		Longjing 31	Longjing 21	Longjing 31	Longjing 21	Longjing 31	Longjing 21	Average
N_{rate} (kg ha^{-1})	CK	0	0	0	0	0	0	0
	FNM	160	160	160	160	160	160	160
	RONM	120	120	120	120	120	120	120
	PNM	118	119	122	127	120	123	121
Yield (t ha^{-1})	CK	5.26 b	5.46 b	5.25 b	5.17 c	5.25 c	5.31 c	5.28 c
	FNM	8.19 a	8.60 a	8.40 a	8.90 ab	8.30 a	8.75 ab	8.52 a
	RONM	7.78 a	8.36 a	8.05 a	8.68 b	7.91 b	8.52 b	8.22 b
	PNM	8.14 a	8.59 a	8.43 a	9.06 a	8.29 a	8.83 a	8.56 a
AE_N (kg kg^{-1})	CK	-	-	-	-	-	-	-
	FNM	18.3 b	19.6 c	19.7 c	23.3 b	19.0 c	21.5 c	20.3 c
	RONM	21.0 ab	24.2 b	23.3 b	29.3 a	22.2 b	26.8 b	24.5 b
	PNM	24.4 a	26.4 a	26.1 a	30.8 a	25.3 a	28.6 a	26.9 a
PFP_N (kg kg^{-1})	CK	-	-	-	-	-	-	-
	FNM	51.2 c	53.7 c	52.5 c	55.6 b	51.9 c	54.7 b	53.3 c
	RONM	64.8 b	69.7 b	67.1 b	72.3 a	65.9 b	71.0 a	68.5 b
	PNM	68.9 a	72.4 a	69.1 a	71.6 a	69.0 a	72.0 a	70.5 a
Economic return ($ ha^{-1})	CK	-	-	-	-	-	-	-
	FNM	1205 a	1297 a	1300 a	1556 a	1253 a	1426 a	1339 a
	RONM	1045 a	1214 a	1166 a	1481 a	1106 b	1348 b	1227 b
	PNM	1205 a	1315 a	1334 a	1647 a	1269 a	1481 a	1375 a

Note: CK: the control treatment with no N application; RONM: regional optimum N management; FNM: farmer N management; PNM: RapidSCAN-based precision N management strategy. Within a column for parameter, values followed by different letters are significantly different ($p < 0.05$).

Making two in-season N recommendations to better meet rice N requirements, the RapidSCAN-based PNM strategy recommended different N rates for different years and varieties. For Longjing 31, the total recommended N rates by the PNM strategy were 118–122 kg N ha^{-1}, which were similar to the rates of the RONM strategy but 24–26% lower than the N rates of the FNM strategy. For Longjing 21, the total N rates recommended by the PNM strategy were similar to or higher than the rates given by the RONM strategy but 21–26% lower than the N rate of the FNM strategy (Table 6).

The RapidSCAN-based PNM strategy resulted in higher yield than RONM strategy across years and varieties. For Longjing 31, the yield of the PNM strategy was not significantly different from FNM and RONM strategies in a specific year, but it was significantly higher (5%) than the RONM strategy across two years. For Longjing 21, the PNM strategy significantly increased grain yield by 4% with respect to the RONM strategy in 2016 or across two years, but the increase was not significantly different from the FNM strategy.

The PNM strategy led to the highest AE_N and PFP_N across years and varieties. For Longjing 31, the PNM strategy increased AE_N and PFP_N by 14% and 5% over the RONM strategy and both by 33% over the FNM strategy across two years, respectively. For Longjing 21, the PNM strategy increased AE_N by 33% and PFP_N by 32% over the FNM strategy across two years. Compared with RONM, the PNM strategy increased AE_N by 7%, but did not improve PFP_N significantly. Across varieties and years, the PNM strategy increased AE_N and PFP_N by 32–33% over FNM and by 3–10% over RONM.

The PNM strategy consistently increased economic returns compared with RONM by 101–168 $ ha^{-1} for specific year-variety combination or across years and varieties. The PNM strategy increased economic returns by 0-91 $ ha^{-1} over the FNM strategy, but none of the increase was statistically significant.

4. Discussion

4.1. Selection of Vegetation Indices for Precision N Management Strategy Development

NDVI and NDRE are two widely used VIs in precision agriculture and have good relationships with aboveground biomass and plant N uptake [14,32]. They are the default VIs of the RapidSCAN sensor, and increased with N application rates in this study, which was in agreement with Aranguren et al. [33]. Besides NDVI and NDRE, many other VIs can be calculated from the three wavebands of RapidSCAN [13,34]. It is important to know if any other VI will perform consistently better than the default VIs and should be selected for PNM strategy development.

In this study, we found that the two default indices (NDVI and NDRE) were sufficient for estimating key parameters to develop the PNM strategy at different key growth stages and other VIs would not be needed.

The stem elongation stage is the key stage to apply panicle fertilizer for rice in Northeast China. The rice canopy was not closed at this stage, and the biomass was not very high yet. The NDVI values were not saturated and achieved similar performance for predicting YP_0 (R^2 = 0.66-0.70; REr = 6.3-7.7%) and $RI_{Harvest}$ (R^2 = 0.60–0.77; REr = 6.2–8.0%) as best performing VIs (R^2 = 0.63–0.70 and 0.63–0.78; REr = 6.5–7.6% and 6.1–7.8%, respectively). The recommended N rates using NDVI were also similar to the best performing VIs across both varieties based on the scenario analysis results. This agreed with many studies achieving good results with GreenSeeker NDVI-based PNM systems at early growth stages [10,12,34–38].

The heading stage is the key stage to apply grain fertilizer for rice in Northeast China in high-yield rice management systems [13,16]. GreenSeeker NDVI become saturated at this stage and cannot be used to make N recommendations [10,13,16,34,39]. As a result, the PNM strategy making two in-season N recommendations at stem elongation and heading stages only using NDVI had the worst performance among all the PNM strategies evaluated in the scenario analysis. One approach to overcome the NDVI saturation problem is to combine NDVI with relative plant height data [30]. Another approach is to replace NDVI with other VIs that can overcome the saturation effect [32]. The RapidSCAN sensor has a RE band in addition to the R and NIR bands, which can be used to calculate more RE-based VIs to overcome the saturation problems of NDVI [13,14,40]. In this study, good relationships between RE-based VIs and YP_0 or $RI_{Harvest}$ were found at the heading stage, with R^2 being 0.75–0.95 for calibration, and R^2 and REr being 0.70–0.82 and 7.3–9.0% for validation, respectively. Encouragingly, the NDRE index achieved similar performance for predicting YP_0 (R^2 = 0.70–0.76; REr = 7.3–8.6%) and $RI_{Harvest}$ (R^2 = 0.75–0.82; REr = 8.3–8.7%) as best performing VIs (R^2 = 0.70–0.76 and 0.73–0.82; REr = 7.3–9.0% and 8.3–8.7%, respectively). Furthermore, the N rates recommended by the PNM strategy using NDVI at the stem elongation stage and NDRE at heading stage were very similar to the N rates recommended by the PNM strategy using best performing VIs at the stem elongation and heading stages.

4.2. In-Season Prediction of Agronomic Efficiency of Topdressing Nitrogen

The NUE is an important parameter for developing crop sensing-based PNM strategies, and can directly influence the final recommended N rates. Previous crop sensor-based PNM strategies generally used a constant NUE value. Raun et al. [12] used the expected recovery efficiency to calculate the N topdressing rate for wheat with the range of 0.5 to 0.7 in America. Cao et al. [34] used 0.4 as the expected recovery efficiency in developing PNM strategy for winter wheat in North China Plain. According to the multi-site-year data in N plot experiments, Yao et al. [10] used fixed $AE_{topdressing}$ (26.79 kg kg^{-1}) for the development of rice PNM strategy.

Studies indicated that NUE is influenced by environmental factors (e.g., soil fertility, water supply, sunshine, accumulated temperature, etc.) and the plant genotypes (e.g., ability to absorb N, photosynthetic rate, stress resistance, root system structure, etc.) [41–44]. Wang et al. [30] considered the influence of early N application and soil types on AE_N and used in-season predicted $RI_{Harvest}$

with GreenSeeker sensor to predict AE_N for developing crop sensor-based PNM strategy for spring maize in Northeast China. Following their idea, this study found that early N application and rice variety were two key factors influencing the $AE_{topdressing}$ of rice in the study region. If early season N application rate is low, N deficiency will lead to a higher efficiency of N use, while a high N application rate during the early growth stage will result in N surplus conditions before topdressing, which will lead to a lower use efficiency of the topdressing N fertilizer. $RI_{Harvest}$ is an indicator of crop N response to additional N application. If early season N status is deficient, the crop will be more responsive to additional topdressing or side-dressing N application. On the other hand, if the early season N status is surplus, the crop will not be responsive to additional N application, and the NUE for the topdressing or side-dressing N application will be lower. Therefore, the $RI_{Harvest}$ and $AE_{topdressing}$ should be positively correlated, as found in this study (R^2 = 0.61 ~ 0.76). This finding was consistent with the results of Wang et al. [30] for spring maize (R^2 = 0.72 ~ 0.74). On the other hand, the big difference of $AE_{topdressing}$ between Longjing 31 (39.5 kg kg^{-1}) and Longjing 21 (59.0 kg kg^{-1}) is shown in Figure 5a. The variety with longer growing period will require more N supply and the photosynthetic utilization will be higher. Therefore, in-season prediction of $AE_{topdressing}$ can be based on crop N status using in-season active sensor predicted response index and variety- or variety group-specific models should be developed.

4.3. Evaluation of Different Precision N Management Strategies

In scenario analysis, the four PNM strategies for N rate recommendations performed much better than the RONM strategy, especially under deficient or surplus N status. The PNM strategies making two in-season N recommendations at both stem elongation and heading stages performed better than PNM strategies making only one in-season N recommendation at the stem elongation stage, and their recommended N rates were 15-28 kg N ha^{-1} closer to EONR under deficient N status across years and varieties. For the longer-growing variety (Longjing 21), the PNM strategy making two N recommendations performed consistently better under different N conditions, being 2-28 kg N ha^{-1} closer to EONR across years. However, no significant difference was found under optimal or surplus N status for Longjing 31, the shorter-growing variety. One of the reasons might be that Longjing 21 typically required a higher N rate and more time (3-7 days) to booting and grain filling than Longjing 31 [9]. The PNM strategy with two in-season N recommendations can give us two opportunities to adjust N application rates according to in-season crop growth and weather conditions, and can better meet rice N needs to increase grain yield. This is more important for the long-season variety Longjing 21 than the shorter-season variety of Longjing 31 [45].

In the evaluation experiments, the RapidSCAN-based PNM strategy could achieve similar rice yield as the FNM strategy, save 24% N fertilizers, and increase NUE by an average of 32–33% across years and varieties (Table 6). Because of the low N fertilizer prices under agricultural subsidies by the government in China, the difference of economic return between PNM and FNM was small (average 36 \$ ha^{-1} annually) across varieties. These results agreed with the results of Yao et al. [10], who reported that PNM strategies based on GreenSeeker or chlorophyll meter increased PFP_N by 48-65% without significantly increasing the yield in comparison to FNM.

Although the total N rates of PNM and RONM strategies in the evaluation experiments were not significantly different, the PNM strategy allowed us to better adjust N distribution for different varieties and plant N status, resulting in the increased grain yield, AE_N, PFP_N, and economic return by an average of 4%, 10%, 3%, and 148 \$ ha^{-1},respectively (Table 6). Zhao et al. [16] found that a chlorophyll meter-based PNM with two in-season N recommendations at the same growth stages as the PNM strategy developed in this study had higher N accumulation at later growth stages to achieve higher yield, and larger panicle size and grain fill percentage. Wang et al. [45] integrated a GreenSeeker sensor-based PNM strategy making one in-season N recommendation at stem elongation stage into a high-yield management system (with optimized transplanting density and water management) increased rice grain yield and NUE by 10% and 1–33% over regional optimum rice management, respectively. Future studies are needed to integrate the RapidSCAN sensor-based PNM strategy developed in this study into similar high-yield management systems to further evaluate its potential to simultaneously improve rice yield, NUE, and economic returns over the regional optimum management system under diverse on-farm conditions.

5. Conclusions

The RapidSCAN sensor has the potential to predict rice grain yield and response to additional N application in Northeast China at the stem elongation and heading stages during the growing season, with R^2 and REr being 0.63–0.82 and 6.1–9.0%, respectively. At the stem elongation stage, NDVI could be used to predict YP_0 and $RI_{Harvest}$ (R^2 = 0.66–0.70 and 0.60–0.77; REr = 6.3–7.7% and 6.2–8.0%) with similar performance as the best performing indices (R^2 = 0.63–0.70 and 0.63–0.78; REr = 6.5–7.6% and 6.1–7.8%, respectively). At the heading stage, NDRE could be used to predict YP_0 and $RI_{Harvest}$ (R^2 = 0.70–0.76 and 0.75–0.82; REr = 7.3–8.6% and 8.3–8.7%, respectively) as accurately as the best performing indices (R^2 = 0.70–0.76 and 0.73–0.82; REr = 7.3–9.0% and 8.3–8.7%, respectively). The combination of the two default indices (NDVI and NDRE) of the RapdiSCAN sensor was sufficient for rice PNM strategy development without the need of incorporating other indices. Across the two tested varieties under deficient N conditions or for Longjing 21 under different N conditions, the PNM strategy making two in-season N recommendations performed better than the PNM strategies making only one in-season N recommendation at the stem elongation stage. The RapidSCAN sensor-based PNM strategy with panicle and grain fertilizer recommendations could lead to similar yield as FNM, save 24% N application rate, and increase AE_N and PFP_N by an average of 33% and 32% across years and varieties, respectively. Compared with RONM, the PNM strategy increased grain yield, AE_N, PFP_N, and economic return by an average of 4%, 10%, 3%, and 148 \$ ha^{-1} across years and varieties, respectively. Future studies are needed to further evaluate the RapidSCAN sensor-based PNM strategy under diverse on-farm conditions, and integrate it into high-yield rice management systems.

Author Contributions: Conceptualization, J.L. (Junjun Lu) and Y.M.; methodology and validation, Y.M., J.L. (Junjun Lu), and X.W.; investigation, J.L. (Junjun Lu), W.S., J.L. (Jingxin Li), and X.H.; resources: Y.M., and J.L. (Junjun Lu); writing—original draft preparation, J.L. (Junjun Lu); writing—review and editing, Y.M., K.K. and Z.C.; supervision, Y.M.; project administration and funding acquisition, Y.M., J.L. (Junjun Lu), and K.K. All authors have read and agreed to the published version of the manuscript.

Funding: This research was funded by the Key National Research and Development Program (2016YFD0200600; 2016YFD0200602), Norwegian Ministry of Foreign Affairs (SINOGRAIN II, CHN-17/0019), Doctoral Foundation of Henan Polytechnic University (B2019-5), Key Scientific Research Projects of Universities Supported by Foundation of Henan Educational Committee (20A210013), and Key Scientific and Technological Project of Henan Province (192102310038).

Acknowledgments: The kind assistance and support provided for this research by Yuan Gao, Wen Yang, Yong Yang, Haisheng He, Honglin Li, Fengyan Liu, and Xiaolong Wang at Jiansanjiang Institute of Agricultural Research and Mr. Guojun Li from Jiansanjiang Department of Agriculture, Jiansanjiang Branch Bureau of Agricultural Reclamation are highly appreciated.

Conflicts of Interest: The authors declare no conflict of interest.

Abbreviations

Abbreviation	English Full Name	Unit
$AE_{topdressing}$	The topdressing nitrogen agronomic efficiency	%
AE_N	Agronomic efficiency of nitrogen	%
EONR	Economically optimum nitrogen rate	kg N ha^{-1}
FNM	Farmer nitrogen management	-
HD_N_{rate}	The recommended nitrogen topdressing application rate at the heading stage	kg N ha^{-1}
HD_YP_N	The potential yield with added nitrogen topdressing application at the heading stage	kg ha^{-1}
INSEY	In-season estimate of yield	-
N	Nitrogen	-
NDRE	Normalized difference red edge	-
NDVI	Normalized difference vegetation index	-
NIR	Near infrared	-
N_{rate}	The recommended N topdressing application rate	kg N ha^{-1}
NUE	Nitrogen use efficiency	-
PFP_N	Partial factor productivity of nitrogen	kg kg^{-1}
PNM	Precision nitrogen management	-
Q, E or P	The quadratic, exponential, or power fit	-
R	Red	-
R^2	The coefficients of determination	-
RE	Red edge	-
REr	Relative error	%
$RI_{Harvest}$	Yield responsiveness to additional nitrogen fertilizer applications	-
RI-VI	In-season nitrogen response index based on vegetation index	-
RMSE	The root mean square error	Same units as statistical data used
RONM	Regional optimum nitrogen management	-
RVI	Ratio vegetation index	-
SE_N_{rate}	The recommended nitrogen topdressing application rate at the stem elongation stage	kg N ha^{-1}
SE_YP_N	The potential yield with added nitrogen topdressing application at the stem elongation stage	kg ha^{-1}
SE-NDVI	The recommended nitrogen rate using normalized difference vegetation index at the stem elongation stage	kg N ha^{-1}
VI_0	Vegetation index at plots without additional nitrogen topdressing application	-
VI_N	Vegetation index at plots with sufficient nitrogen fertilization plots	-
VIs	Vegetation indices	-
YP_0	The yield potential without additional nitrogen application	kg ha^{-1}
YP_{max}	The maximum obtainable yield	kg ha^{-1}
YP_N	The potential yield with added nitrogen fertilization	kg ha^{-1}
ΔN_{rate}	The difference between recommended nitrogen rate and economically optimum nitrogen rate	kg N ha^{-1}

References

1. Cantrell, R.P.; Reeves, T.G. The cereal of the world's poor takes center stage. *Science* **2002**, *296*, 53. [CrossRef] [PubMed]
2. Normile, D. Reinventing rice to feed the world. *Science* **2008**, *321*, 330–333. [CrossRef] [PubMed]
3. Foley, J.A.; Ramankutty, N.; Brauman, K.A.; Cassidy, E.S.; Gerber, J.S.; Johnston, M.; Mueller, N.D.; O'Connell, C.; Ray, D.K.; West, P.C.; et al. Solutions for a cultivated planet. *Nature* **2011**, *478*, 337–342. [CrossRef] [PubMed]
4. Zhang, F.; Chen, X.; Vitousek, P. Chinese agriculture: An experiment for the world. *Nature* **2013**, *497*, 33–35. [CrossRef] [PubMed]
5. Peng, X.; Yang, Y.; Yu, C.; Chen, L.; Zhang, M.; Liu, Z.; Sun, Y.; Luo, S.; Liu, Y. Crop management for increasing rice yield and nitrogen use efficiency in Northeast China. *Agron. J.* **2015**, *107*, 1682–1690. [CrossRef]
6. Wu, L.; Chen, X.; Cui, Z.; Wang, G.; Zhang, W. Improving nitrogen management via a regional management plan for Chinese rice production. *Environ. Res. Lett.* **2015**, *10*, 095011. [CrossRef]
7. Chen, X.; Cui, Z.; Fan, M.; Vitousek, P.; Zhao, M.; Ma, W.; Wang, Z.; Zhang, W.; Yan, X.; Yang, J. Producing more grain with lower environmental costs. *Nature* **2014**, *514*, 486–489. [CrossRef]
8. Cao, Q.; Miao, Y.; Feng, G.; Gao, X.; Liu, B.; Liu, Y.; Li, F.; Khosla, R.; Mulla, D.J.; Zhang, F. Improving nitrogen use efficiency with minimal environmental risks using an active canopy sensor in a wheat-maize cropping system. *Field Crop. Res.* **2017**, *214*, 365–372. [CrossRef]
9. Zhang, J.; Miao, Y.; Batchelor, W.; Lu, J.; Wang, H.; Kang, S. Improving high-latitude rice nitrogen management with the CERES-rice crop model. *Agronomy* **2018**, *8*, 263. [CrossRef]

10. Yao, Y.; Miao, Y.; Huang, S.; Gao, L.; Ma, X.; Zhao, G.; Jiang, R.; Chen, X.; Zhang, F.; Yu, K.; et al. Active canopy sensor-based precision N management strategy for rice. *Agron. Sustain. Dev.* **2012**, *32*, 925–933. [CrossRef]
11. Raun, W.R.; Solie, J.B.; Johnson, G.V.; Stone, M.L.; Mullen, R.W.; Freeman, K.W.; Thomason, W.E.; Lukina, E.V. Improving nitrogen use efficiency in cereal grain production with optical sensing and variable rate application. *Agron. J.* **2002**, *94*, 815–820. [CrossRef]
12. Raun, W.R.; Solie, J.B.; Stone, M.L.; Martin, K.L.; Freeman, K.W.; Mullen, R.W.; Zhang, H.; Schepers, J.S.; Johnson, G.V. Optical sensor-based algorithm for crop nitrogen fertilization. *Commun. Soil Sci. Plan.* **2005**, *36*, 2759–2781. [CrossRef]
13. Cao, Q.; Miao, Y.; Shen, J.; Yu, W.; Yuan, F.; Cheng, S.; Huang, S.; Wang, H.; Yang, W.; Liu, F. Improving in-season estimation of rice yield potential and responsiveness to topdressing nitrogen application with Crop Circle active crop canopy sensor. *Precis. Agric.* **2016**, *17*, 136–154. [CrossRef]
14. Lu, J.; Miao, Y.; Shi, W.; Li, J.; Yuan, F. Evaluating different approaches to non-destructive nitrogen status diagnosis of rice using portable RapidSCAN active canopy sensor. *Sci. Rep.* **2017**, *7*, 14073. [CrossRef] [PubMed]
15. Zhang, K.; Ge, X.; Shen, P.; Li, W.; Liu, X.; Cao, Q.; Zhu, Y.; Cao, Q.; Tian, Y. Predicting rice grain yield based on dynamic changes in vegetation indexes during early to mid-growth stages. *Remote Sens.* **2019**, *11*, 387. [CrossRef]
16. Zhao, G.; Miao, Y.; Wang, H.; Su, M.; Fan, M.; Zhang, F.; Jiang, R.; Zhang, Z.; Liu, C.; Liu, P.; et al. A preliminary precision rice management system for increasing both grain yield and nitrogen use efficiency. *Field Crop. Res.* **2013**, *154*, 23–30. [CrossRef]
17. Xing, B.; Dudas, M.J.; Zhang, Z.; Xu, Q. Pedogenetic characteristics of albic soils in the Three River Plain, Heilongjiang province. *Acta. Pedolog. Sin.* **1994**, *31*, 95–104, (In Chinese with English Abstract).
18. Rouse, J.W.; Haas, J.R.H.; Schell, J.A.; Deering, D.W. Monitoring vegetation systems in the Great Plains with ERTS. In *Third Earth Resources Technology Satellite-1 Symposium*; NASA Special Publication: Washington, DC, USA, 1974; pp. 309–317.
19. Barnes, E.M.; Clarke, T.R.; Richards, S.E.; Colaizzi, P.D.; Haberland, J.; Kostrzewski, M.; Waller, P.; Choi, C.; Riley, E.; Thompson, T. Coincident detection of crop water stress, nitrogen status and canopy density using ground-based multispectral data. In Proceedings of the International Conference on Precision Agriculture and Other Resource Management, Bloomington, IN, USA, 16–19 July 2000; pp. 16–19.
20. Jordan, C.F. Derivation of leaf-area index from quality of light on the forest floor. *Ecology* **1969**, *50*, 663–666. [CrossRef]
21. Qi, J.; Chehbouni, A.; Huete, A.; Kerr, Y.; Sorooshian, S. A modified soil adjusted vegetation index. *Remote Sens. Environ.* **1994**, *48*, 119–126. [CrossRef]
22. Chen, J.M. Evaluation of vegetation indices and a modified simple ratio for boreal applications. *Can. J. Remote Sens.* **1996**, *22*, 229–242. [CrossRef]
23. Reyniers, M.; Walvoort, D.J.J.; Baardemaaker, J.D. A linear model to predict with a multi-spectral radiometer the amount of nitrogen in winter wheat. *Int. J. Remote Sens.* **2006**, *27*, 4159–4179. [CrossRef]
24. Goel, N.S.; Qin, W. Influences of canopy architecture on relationships between various vegetation indices and LAI and FPAR. *Remote Sens. Rev.* **1994**, *10*, 309–347. [CrossRef]
25. Gong, P.; Pu, R.; Biging, G.S.; Larrieu, M.R. Estimation of forest leaf area index using vegetation indices derived from Hyperion hyperspectral data. *IEEE Trans. Geosci. Remote* **2003**, *41*, 1355–1362. [CrossRef]
26. Cao, Q.; Miao, Y.; Wang, H.; Huang, S.; Cheng, S.; Khosla, R.; Jiang, R. Non-destructive estimation of rice plant nitrogen status with Crop Circle multispectral active canopy sensor. *Field Crop. Res.* **2013**, *154*, 133–144. [CrossRef]
27. Jasper, J.; Reusch, S.; Link, A. Active sensing of the N status of wheat using optimized wave-length combination: Impact of seed rate, variety and growth stage. In *Precision Agriculture'09*; Henten, E.J.V., Goense, D., Lokhorst, C., Eds.; Wageningen Academic Publishers: Wageningen, The Netherlands, 2009; pp. 23–30.
28. Johnson, G.V.; Raun, W.R. Nitrogen response index as a guide to fertilizer management. *J. Plant Nutr.* **2003**, *26*, 249–262. [CrossRef]
29. Mullen, R.W.; Freeman, K.W.; Raun, W.R.; Johnson, G.V.; Stone, M.L.; Solie, J.B. Identifying an in-season response index and then potential to increase wheat yield with nitrogen. *Agron. J.* **2003**, *95*, 347–351. [CrossRef]

30. Wang, X.; Miao, Y.; Dong, R.; Chen, Z.; Guan, Y.; Yue, X.; Fang, Z.; Mulla, D.J. Developing active canopy sensor-based precision nitrogen management strategies for maize in Northeast China. *Sustainability* **2019**, *11*, 706. [CrossRef]
31. Su, M.; Huang, S.; Zhao, G.; Zhao, Q.; Yao, Y.; Miao, Y. Current rice management practices of farmers in Heilongjiang land reclamation area and improvement strategies. *North. Rice* **2012**, *42*, 28–33, (In Chinese with English Abstract).
32. Li, S.; Ding, X.; Kuang, Q.; Ata-Ul-Karim, S.T.; Cheng, T.; Liu, X.; Tian, Y.; Zhu, Y.; Cao, W.; Cao, Q. Potential of UAV-based active sensing for monitoring rice leaf nitrogen status. *Front. Plant Sci.* **2018**, *9*, 1834. [CrossRef]
33. Aranguren, M.; Castellón, A.; Aizpurua, A. Crop sensor-based in-season nitrogen management of wheat with manure application. *Remote Sens.* **2019**, *11*, 1094. [CrossRef]
34. Cao, Q.; Miao, Y.; Li, F.; Gao, X.; Liu, B.; Lu, D.; Chen, X. Developing a new Crop Circle active canopy sensor-based precision nitrogen management strategy for winter wheat in North China Plain. *Precis. Agric.* **2017**, *18*, 2–18. [CrossRef]
35. Tubaña, B.S.; Arnall, D.B.; Walsh, O.; Chung, B.; Solie, J.B.; Girma, K.; Raun, W.R. Adjusting midseason nitrogen rate using a sensor-based optimization algorithm to increase use efficiency in corn. *J. Plant Nutr.* **2008**, *31*, 1393–1419. [CrossRef]
36. Macnack, N.; Khim, B.C.; Mullock, J.; Raun, W. In-season prediction of nitrogen use efficiency and grain protein in winter wheat (*Triticum aestivum* L.). Commun. *Soil Sci. Plan.* **2014**, *45*, 2480–2494. [CrossRef]
37. Purba, J.; Sharma, R.K.; Jat, M.L.; Thind, H.S.; Gupta, R.K.; Chaudhary, O.P.; Chandna, P.; Khurana, H.S.; Kumar, A.; Uppal, H.S.; et al. Site-specific fertilizer nitrogen management in irrigated transplanted rice (*Oryza sativa*) using an optical sensor. *Precis. Agric.* **2015**, *16*, 455–475.
38. Li, F.; Miao, Y.; Zhang, F.; Cui, Z.; Li, R.; Chen, X.; Zhang, H.; Schroder, J.; Raun, W.R.; Jia, L. In-season optical sensing improves nitrogen-use efficiency for winter wheat. *Soil Sci. Soc. Am. J.* **2009**, *73*, 1566–1574. [CrossRef]
39. Bonfil, D.J. Wheat phenomics in the field by RapidScan: NDVI vs. NDRE. *Isr. J. Plant Sci.* **2016**, 1–14. [CrossRef]
40. Lu, J.; Miao, Y.; Shen, J.; Cao, Q.; Huang, S.; Wang, H.; Wu, H.; Hu, S.; Hu, X. Improving estimation of rice yield potential using active canopy sensor Crop Circle ACS 430 in Northeast China. In *Precision Agriculture'15*; Stafford, J.V., Ed.; Wageningen Academic Publishers: Wageningen, The Netherlands, 2015; pp. 102–115. [CrossRef]
41. Fan, M.; Shen, J.; Yuan, L.; Jiang, R.; Chen, X.; Davies, W.J.; Zhang, F. Improving crop productivity and resource use efficiency to ensure food security and environmental quality in China. *J. Exp. Bot.* **2011**, *63*, 13–24. [CrossRef]
42. Ju, C.; Buresh, R.J.; Wang, Z.; Zhang, H.; Liu, L.; Yang, J.; Zhang, J. Root and shoot traits for rice varieties with higher grain yield and higher nitrogen use efficiency at lower nitrogen rates application. *Field Crop. Res.* **2015**, *175*, 47–55. [CrossRef]
43. Ladha, J.K.; Pathak, H.; Krupnik, T.J.; Six, J.; Kessel, C.V. Efficiency of fertilizer nitrogen in cereal production: Retrospects and prospects. *Adv. Agron.* **2005**, *87*, 85–156.
44. Li, S.; Wang, Z.; Hu, T.; Gao, Y.; Stewart, B.A. Nitrogen in dryland soils of China and its management. *Adv. Agron.* **2009**, *101*, 123–181.
45. Wang, H.; Miao, Y.; Zhao, G.; Yao, Y.; Khosla, R. Evaluating different integrated precision rice management strategies in Northeast China. In Proceedings of the Second International Conference on Agro-Geoinformatics (Agro-Geoinformatics 2013), Fairfax, VA, USA, 12–16 August 2013; pp. 419–422. [CrossRef]

remote sensing

MDPI

Article

Agronomic and Economic Potential of Vegetation Indices for Rice N Recommendations under Organic and Mineral Fertilization in Mediterranean Regions

Beatriz Moreno-García [1,2,*], Mª Auxiliadora Casterad [1], Mónica Guillén [1] and Dolores Quílez [1]

1 Unidad de Suelos y Riegos (asociada a EEAD-CSIC), Centro de Investigación y Tecnología Agroalimentaria de Aragón, Avda Montañana 930, 50059 Zaragoza, Spain; acasterad@aragon.es (M.A.C.); mguillenc@aragon.es (M.G.); dquilez@aragon.es (D.Q.)

2 231 ENGR Hall, Biological &Agricultural Engineering, Univ. of Arkansas, Fayetteville, AR 72701, USA

* Correspondence: bemorenog@gmail.com; Tel.: +1-479-575-2876

Received: 18 October 2018; Accepted: 27 November 2018; Published: 29 November 2018

Abstract: Rice (*Oryza sativa* L.) farmers in Mediterranean regions usually apply organic or mineral fertilizers before seeding that are supplemented with mineral nitrogen (N) later in the season. In general, the midseason N is applied without consideration of the actual crop N status, which may lead to over-fertilization and associated environmental problems. Thus, the purpose of this study was to design and evaluate a N recommendation approach using aerial images for Mediterranean paddy rice systems. A two-year rice field experiment was established in northeastern Spain, with different rates of pig slurry (PS) and mineral N fertilizer. Multispectral aerial images were taken at the rice booting stage, and several vegetation indices (VIs) were calculated. The VIs showed strong relationships with yield and the relations significantly differed between the PS and mineral fertilization treatments. The strongest relations with yield were obtained with $gMCARI_{NIR}$, proposed in this study, ($R^2 = 0.67$), GNDVI ($R^2 = 0.64$) and $MCARI_{NIR}$ ($R^2 = 0.64$), indicating the importance of including the green band information. The N recommendation approach generated using the VIs information showed a high success (87.5%) in the preliminary evaluation. The economic and environmental analysis showed that this approach provides a useful tool when compared to the usual farmer practices.

Keywords: flooded rice; pig slurry; aerial remote sensing; vegetation indices; N recommendation approach; Mediterranean conditions

1. Introduction

Rice (*Oryza sativa*, L.) is the staple food for nearly half of the world's seven billion people [1]. Nitrogen (N) is an essential element that is required to obtain high rice yields. Although N application increases rice productivity, poor N use efficiency is characteristic of irrigated rice systems, due to rapid losses of applied N; hence, adjusting N rates and the time of application to crop N requirements is crucial for optimizing N use efficiency and avoiding environmental problems, such as emission of greenhouse gases or surface and groundwater nitrate pollution [2].

In Europe, rice is mostly cultivated in Mediterranean countries, and 17% of the total area is cultivated in Spain [3]. The rice extension in northeastern Spain (Aragon and Catalonia regions) represents 24% [4] of the rice lands in the country. Besides, half of the national pig production is concentrated in this area [5]; thus, it is necessary to integrate pig slurry (PS) in the N fertilization schedule for rice fields. The most common practice of local farmers is to apply PS or chemical fertilizers before seeding (~70% of crop N requirements) and to complement it with mineral N-topdressing at the end of the tillering stage or during stem elongation. However, N is typically applied during mid-season without consideration of the crop N status at that time.

Although some studies have shown that a single optimum N application before flooding could allow maximum yield [6–8], estimation of the potential yield at the beginning of the season is difficult, because it is strongly influenced by yearly variation in weather conditions, mainly temperature and wind [9]; thus, this N management presents a high risk for over-fertilization. However, if N rates are reduced before seeding and later complemented with topdressing N rates that are adjusted to the crop N status at that time, over-fertilization and related environmental problems can be reduced or avoided and N use efficiency may be increased, as reported by Nishikawa et al. [10] and Xie et al. [11].

One of the most common tools for real-time N management in cereals is a chlorophyll meter (SPAD 502, Minolta Corp., Ramsey, NJ, USA; N-tester, Yara International ASA, Oslo, Norway). This tool has been used successfully to monitor leaf N status and guides N fertilization during the crop season in different cereals [12,13], including rice [14,15]. However, this method involves measurements at the leaf level; thus, it is inadequate for large-scale applications. Other approaches include the use of active crop canopy sensors, such as the GreenSeeker (Trimble Navigation Limited, Sunnyvale, CA, USA) or Crop Circle (Holland Scientific Inc., Lincoln, NE, USA), which measure the amount of light reflected by the canopy [16]. These sensors reduce the measurement time and are more suitable for field applications; nevertheless, it is still necessary to enter the field and perform the measurements in different areas to obtain a representative value. Aerial or satellite remote sensing images improve these measurements, because they can provide spectral information for whole fields and large areas [17], although they also have limitations [18]. Satellites have improved their spatial and temporal resolution over the last years, and cover larger areas at the same time, but are subject to fixed scheduling and strongly depend on cloud cover. Unmanned aerial vehicles (UAVs) offer very high spatial resolution, flexibility on scheduling, and its acquisition are independent on cloud cover conditions, however they are limited for the use in extensive areas and during wind conditions and the cost of a great effort for mosaicking and geocoding. The aircraft platform sits in between satellites and UAVs with more flexibility than satellite and better facility to cover the whole scene than UAVs [18].

Different authors have proposed different approaches for determining the rice N requirements in-season by using the relationships between canopy spectral information (vegetation indices; VIs) and different crop parameters [15,19–21]. The key to the success of these approaches is the prediction of yield responsiveness to additional N fertilization. Therefore, the establishment of strong relationships between spectral information and yield or other crop parameters is necessary for generating sound topdressing N recommendations.

The Normalized Difference Vegetation Index (NDVI), which integrates information in the red (R) and near-infrared (NIR) bands, is one of the most widely applied vegetation indices (VIs). The NDVI has been related to the leaf area index (LAI), biomass or yield in rice [22–24] and other crops [25–27], and it is the most widely used VI for N recommendation approaches [15,20,21,28]. However, for high chlorophyll (Chl) concentration or large vegetation coverage values, NDVI losses sensitivity and saturates [16,29]. Different authors have tried to handle this saturation phenomenon. Gitelson et al. [30] proposed the Green NDVI (GNDVI), that considers the green (G) instead the red (R) band in its formulation, and found that GNDVI was much more sensitive to Chl concentration, in a wide range of Chl variations, than the original NDVI. Lately, this index has been applied to rice [16]. Other authors have proposed three-band VIs to solve the saturation issue [16,31].

Other indices have been formulated to evaluate responses to variation in chlorophyll and N content. One example is the Modified Chlorophyll Absorption in Reflectance Index (MCARI) that was designed to be responsive to chlorophyll variation [32]; later this index was modified by different [illegible] changes [31,33] and was applied to rice [16,33].

Most studies focused on the use of canopy reflectance for estimating crop parameters or adjusting mid-season N in rice paddy fields have been conducted in Asia, and different agricultural practices are used in this region compared to the Mediterranean areas of Europe. Under Mediterranean conditions, Gilabert and Meliá [24] established yield prediction based on VIs obtained from satellite images in Valencia (Spain), and Casanova et al. [22] estimated LAI and biomass from VIs obtained from radiometer measurements in the Ebro Delta (Spain). Recently, efforts have been underway to integrate multispectral information into farm advisory systems. The Earth obseRvation Model based RicE information Service (ERMES) project is being developed in Spain, Italy, and Greece with the primary aim of rice yield prediction [34,35]. This project also includes the support of rice growers for fertilization or pest control and management, and two applications (PocketLAI and PocketN) have been developed for estimating in-season the crop status (LAI, leaf and plant N content) from digital photographs acquired with commercial smartphones [36,37]. However, although the perspectives based on spectral information to improve N management and increase N use efficiency in rice are promising, more studies focusing on the development of approaches to estimate the N topdressing needs in season are needed, with a special emphasis on the effects of organic fertilizers, such as PS on rice spectral information. Knowledge of the spectral response sensitivity to differences between mineral and organic fertilization is essential to develop sound approaches for N recommendations. Moreover, the use of aerial images, that cover large areas, should be studied, so that recommendations can be made to different farmers or for large farm areas.

Therefore, the main purpose of this study was to design and evaluate a N recommendation approach using multispectral images in a Mediterranean paddy rice system under organic and mineral fertilization. To achieve this goal, three sub-objectives were proposed:

1. To establish and compare the relationships between different VIs derived from aerial multispectral information and yield at the rice booting stage, and evaluate possible differences in these relationships due to organic and mineral fertilization.
2. To design and evaluate the agronomic performance of a N topdressing recommendation approach based on the information obtained in sub-objective 1.
3. To compare economically and environmentally different scenarios for N adjustment based on the recommendation approach defined in sub-objective 2.

The work has been developed as a pilot study on the japonica rice cultivar, Guadiamar, which accounts for 70% of the rice surface in the Aragon region in Spain.

2. Materials and Methods

2.1. Experimental Design and Agricultural Practices

The study was conducted in a flooded rice field (41°45′31.87″ N, 0°2′18.16″ W), with three fertilization strategies during two consecutive years (2012 and 2013) [7]. The climate of the region is semiarid continental Mediterranean with high temperatures during the summer and low precipitation (15.0 °C annual average temperature and 349 mm annual precipitation; average period 1980–2010).

The experimental design was a split plot with four repetitions (Figure 1). The main plots were assigned to three basal fertilization strategies consisting of two rates of pig slurry (PS) equivalent to 120 kg NH_4-N·ha^{-1} (PS120) and 170 kg NH_4-N·ha^{-1} (PS170) (Table 1, Pig slurry treatments, PS) and a mineral treatment (ammonium sulfate) at different rates (Table 1, Mineral treatments, M). Secondary plots included different topdressing N rates applied as ammonium sulfate. The experimental plots were 6 m wide by 12 m long for the PS treatments and 6 m wide by 5 m long for the M treatments (Figure 1). The total N rates ranged between 0 and 320 kg N·ha^{-1} and were applied as basal fertilization or as a combination of basal and topdressing fertilization for a total of 22 treatments (Table 1).

Pig slurry was band-spread on 15 May 2012 and 9 May 2013, and PS rates were established according to the ammonium N concentration of the PS, which was measured in situ by a Quantofix® N-volumeter [38] and by conductimetry [39]. On the same days, the basal mineral N fertilizer (ammonium sulfate) was applied to the plots of the M treatments at the corresponding rates together with P (100 kg P_2O_5·ha^{-1}) and K (100 kg K_2O·ha^{-1}) to avoid P or K deficiency.

The japonica rice cultivar, Guadiamar, was broadcast-seeded on 16 May 2012 and 15 May 2013 at a seed rate of 180 kg·ha^{-1}. In 2012, the field was immediately flooded after seeding; however, in 2013, flooding preceded rice seeding. Topdressing N was applied at the end of the tillering stage on 4 July 2012 and 29 July 2013 as ammonium sulfate. The field remained flooded until approximately one month before harvest, except for occasional drainages for the application of herbicides, pesticides and fungicides according to habitual practices in the area.

Rice was harvested from 15–17 October 2012 and 25 October 2013. Grain moisture (PM-600 grain moisture tester, Keller, Japan) was measured to adjust the yield to a moisture content of 140 g·kg^{-1}.

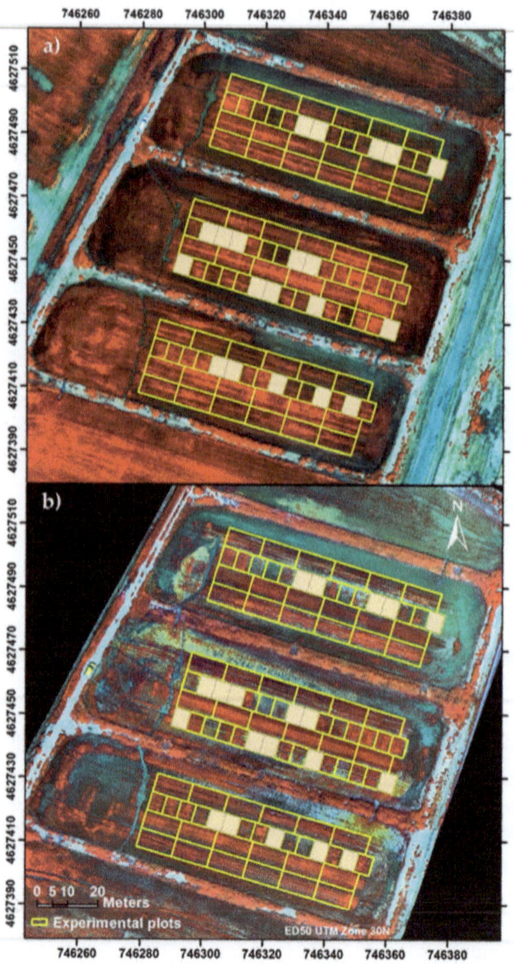

Figure 1. Experimental plots layout (yellow lines) superimposed on the images (false color RGB: Near-Infrared band/ Green band/ Red band) taken with a multispectral camera at the booting stage, 30 July 2012 (**a**) and 13 August 2013 (**b**). The shaded areas represent additional treatments excluded from the analysis in this paper.

Table 1. Amounts of N applied (kg N ha^{-1}) before seeding (BS) and topdressing (TS) in the different treatments. For the PS treatments, amount indicates target N rates.

Pig Slurry Treatments (PS)			Mineral Treatments (M)		
	BS kg NH_4-N·ha^{-1}	TP kg N·ha^{-1}		BS kg N·ha^{-1}	TP
PS120M0	120	-	M120M0	120	-
PS120M30	120	30	M120M30	120	30
PS120M60	120	60	M120M60	120	60
PS120M90	120	90	M120M90	120	90
PS120M120	120	120	M120M120	120	120
PS120M150	120	150			
PS170M0	170	-	Control (M0)	-	-
PS170M30	170	30	M30	30	-
PS170M60	170	60	M60	60	-
PS170M90	170	90	M90	90	-
PS170M120	170	120	M120 = M120M0	120	-
PS170M150	170	150	M150	150	-

2.2. Spectral Information

Spectral information was collected using a spectral camera from a manned plane (RS Servicios de Teledetección SL, Lleida, Spain). The spectral camera collected data at four wavelengths. The center wavelengths were as follows: Band 1 (blue-B): 450 nm; band 2 (green-G): 550 nm; band 3 (red-R): 675 nm; and band 4 (near infrared-NIR): 780 nm; the bandwidth was 20 nm (for the four bands). The images were collected at solar noon, to minimize shadows effects in the spectral response, on 30 July 2012 and 13 August 2013 at the rice booting stage, a few days before heading. The images were provided pre-processed by SpecTerra Services (WA, Australia) with the differential illumination effect corrected and were ortho-rectified and mosaicked. The spatial resolution was 0.1 m and the radiometric resolution was 16 bits. The spectral information was provided in digital values, thus adimensional VIs were chosen for the study.

Different VIs were evaluated for their relationship to rice yield (Table 2). The indices RVI, GRVI, NDVI and GNDVI were included because they were consistently shown to be related to agronomic parameters, such as yield, and have been used in the development of approaches to recommend N topdressing in different crops. Furthermore, we included three indices, MCARI1, $MCARI_{NIR}$ and $gMCARI_{NIR}$, which were derived from the Modified Chlorophyll Absorption in Reflectance Index, that was originally developed to be responsive to chlorophyll variation [32].

Haboudane et al. [31] modified the MCARI and proposed the MCARI1 with suppression of the R_{700}/R_{670} ratio to lower the sensitivity to chlorophyll effects and replacement of the red-edge wavelength (R_{700}) by the near-infrared wavelength (R_{800}) to increase the sensitivity to green LAI variation.

Cao et al. [33] modified the MCARI to work with the Crop Circle ACS-470 active sensor. The MCARI modified by Cao et al. [33], also denoted MCARI1 retained the formula of the MCARI, but the R_{700} and R_{670} were replaced by R_{NIR} (R_{760}) and $R_{red\ edge}$ (RE, R_{730}), respectively. This index showed consistent relations with rice aboveground biomass (R^2 = 0.79) and plant N uptake (R^2 = 0.83) across growth stages [33]. In our study, we have adapted this index to the available bands, NIR 780 nm and red 675 nm, and we have denoted the index as $MCARI_{NIR}$ (Table 2).

Table 2. Vegetation indices (VIs) evaluated in this study (Green-G: 550 nm, Red-R: 675 nm and Near Infrared-NIR: 780 nm).

Indices (VIs)	Formula	Reference
RVI Ratio Vegetation Index	NIR/R	[40]
GRVI Green Ratio Vegetation Index	NIR/G	[41]
NDVI Normalized Difference Vegetation Index	(NIR-R)/(NIR+R)	[42]
GNDVI Green Normalized Difference Vegetation Index	(NIR-G)/(NIR+G)	[30]
MCARI1 Modified Chlorophyll Absorption in Reflectance Index1	1.2[2.5(NIR-R)-1.3(NIR-G)]	[31]
$MCARI_{NIR}$ Modified Chlorophyll Absorption in Reflectance $Index_{NIR}$	[(NIR-R)-0.2(NIR-G)](NIR/R)	Adapted from Cao et al. [33]
$gMCARI_{NIR}$ Green peak Modified Chlorophyll Absorption in Reflectance $Index_{NIR}$	[(NIR-R)-(G-R)](NIR/R) = (NIR-G)(NIR/R)	Proposed in this study

The index $gMCARI_{NIR}$, proposed in this work, retains the MCARI structure, includes the near-infrared wavelength to increase the sensitivity to biomass changes as proposed by Haboudane et al. [31] and incorporates the green reflectance peak information (G-R) to account for the sensitivity to chlorophyll concentrations (Table 2). This index is the product of Green Difference Vegetation Index (NIR-G; [43]) and the Ratio Vegetation Index (NIR/R; [40]).

These vegetation indices have been successfully applied to monitor field crops at different spatial resolutions; spatial resolution of centimeters using field devices, as radiometers or Crop Circle sensors [16,21–23,33], spatial resolution of meters with aerial imagery [18,31], or spatial resolution of several meters with satellite images [17,18,24,25].

The VIs were calculated for each pixel and the average values for each plot were extracted. The extraction was performed using a mask for each plot that excluded a width of 1 m from the borders to avoid edge effects. In 2013, due to adverse meteorological conditions, rice seed germination was hindered and five plots were removed from the analyses because of bad emergence.

2.3. Relationship between Yield and Vegetation Indices

Many studies have reported improvements in yield estimation of irrigated rice after relativization of VIs with respect to overfertilized plots [21] to obtain sufficiency indices. Although some authors have obtained strong relationships between vegetation indices and yield in different years without converting the yield to relative value [20,21,24,44], in our study, it was not possible to obtain an equation to predict absolute yield values, due to high interannual yield variability. The variability in yield between years (7.8 $Mg{\cdot}ha^{-1}$ in 2012 and 5.9 $Mg{\cdot}ha^{-1}$ in 2013 in this study) in the study area is high for two reasons. Firstly, the area is at the low limit of temperature for adequate rice cultivation. Secondly, there is strong variability in meteorological conditions between years. Hence, relative yield (R_yield) and the relative VI (R_VI) were calculated for each year (Equations (1) and (2)).

$$R\_yield_{(plot)} = \frac{Yield_{(Plot)}}{Yield_{max}} \quad (1)$$

$$R\_VI_{(plot)} = \frac{VI\ value_{(Plot)}}{VI_{max}} \quad (2)$$

Linear-plateau equations ([7,45]; Equation (3)) were adjusted to model the response of yield to N rates in the PS and the M treatments for each of the two years in order to get $Yield_{max}$ and VI_{max}.

$$If\ N < C;\ Y = a + b{\cdot}N \quad (3)$$

[illegible]

where Y ($Mg{\cdot}ha^{-1}$) is the yield; N is the applied nitrogen rate ($kg\ N{\cdot}ha^{-1}$) (this rate is the sum of N applied before seeding and topdressing); a (intercept) is the yield at 0 $kg\ N{\cdot}ha^{-1}$; b is the increase in yield per unit increase in N; and C is the critical (or optimum) N rate, i.e., the minimum N rate above which the maximum yield is obtained. $Yield_{max}$ is the maximum yield or plateau of Equation (3), i.e., the yield for N $\geq$ than the critical N rate (C). The maximum VI (VI_{max}) for each year was calculated as the average VI value of treatments with N rates equal to, or higher than, the critical N rate (N $\geq$ C).

The relationships between relativized yield and relativized vegetation indices were established by regression analysis for the years 2012 and 2013 and the pooled data (using 100% of the data). The models tested were linear, multiplicative, logarithmic and exponential. The coefficient of determination and the root mean square error (RMSE) of the regressions were used to evaluate the performance of the seven VIs.

The differences between years and between PS and M treatments were evaluated by a test of equality for regression lines across groups [46].

2.4. N Topdressing Recommendation Approach

2.4.1. Design

The N topdressing recommendation approach was based on the calculation of a N sufficiency index (NSI) as indicator of the crop N status, i.e., as quantifiers of N nutrition index. This approach was defined by Denuit et al. [12] using the HydroAgri N-Tester sensor and corrects the readings of the sensor with the readings of the N over-fertilized plot to obtain sufficiency indices (in this case relativized VIs, i.e., NSI is R_VI, Equation (4)). Thus, a NSI value below 0 means the plot could present a nutritional deficit, and NSI values of 1 or above 1 means the plot presents 'a priori' a good nutritional status. This type of approach has been successfully tested in rice by Chen et al. [19] or Xue and Yang [20].

The approach is based on the relationship between the variables Delta NSI and Delta N (Equation (5)). Delta NSI (Equation (6)) is the difference between NSI and 1. Thus, conceptually if Delta NSI is 0 or positive, the plot should have a good nutritional status, and if it is negative, the plot might present a N deficit. Delta N (Equation (7)) is the difference between the N applied in the treatment (N_T) and the critical N rate (C) obtained in the yield response to N (Equation (3)).

This approach was selected because it does not require absolute yield predictions; instead, N deficit or additional N required for raising NSI from a current level to a target level is calculated based on the relationship of NSI with N rate [20].

$$NSI = R\_VI_{(Plot)} = \frac{VI\ value_{(Plot)}}{VI_{max}} \tag{4}$$

$$Delta\ N = f(Delta\ NSI) \tag{5}$$

$$Delta\ NSI = NSI - 1 \tag{6}$$

$$Delta\ N = N_T - C \tag{7}$$

If Delta N estimate from spectral information is positive, topdressing N fertilization is not necessary; and on the other hand, if Delta N estimate is negative, the plot must be fertilized. In this case, Delta N gives the N deficit, i.e., the amount of N that needs to be applied to obtain maximum yield.

Three of the four replicates (75% of the data) of the experiment were used for establishing the model (Equation (5)). The fourth replicate was used for performance evaluation. The replicate used in the evaluation process was not the same for all treatments and was determined by random draw in each treatment.

Among the seven VIs evaluated, the two indices that presented the strongest relation to yield were chosen for the establishment of the N recommendation model. These two indices were GNDVI and $gMCARI_{NIR}$, (see Section 3.1.2).

2.4.2. Validation Process

In the validation process, the values of R_GNDVI and R_gMCARI$_{NIR}$ for the validation plots (25% data) in each year (2012 and 2013) were obtained dividing the VIs values obtained in each plot by the average values of the over-fertilized plots of the corresponding year (2012 and 2013). Plots from PS120M150 and PS170M150 were considered overfertilized for PS treatments and plots from M120M90 and M120M120 for M treatments (Table 2). Then, the values of Delta N were obtained using Equation (5). The performance of the model was evaluated by the percent of success. The sign of Delta N and the actual R_yield were compared in the validation plots and the plots were assigned to one of the following three options.

- *Success:* Delta N was negative (i.e., the plot would have needed additional N fertilization) and the actual R_yield was below 1; or Delta N was positive (i.e., the plot would not have needed additional N fertilization) and the actual R_yield was equal or higher than 1.
- *Failure by excess*: Delta N was negative, but the actual R_yield was equal or above 1 (the approach would have recommended additional N fertilization, but the plot had reached the optimum yield).
- *Failure by defect*: Delta N was positive, but the actual R_yield was below 1, (the approach would not have recommended additional N fertilization, but the plot had not reached the optimum yield).

Success and error percentages were calculated considering four strategies:

- The use of GNDVI
- The use of gMCARI$_{NIR}$
- The combination GNDVI & gMCARI$_{NIR}$: The plot will only be fertilized if both VIs recommend additional N fertilization.
- The combination GNDVI or gMCARI$_{NIR}$: If one of the VIs recommends N fertilization, the plot will be fertilized even if the other index does not recommend N fertilization.

The fertilization treatments varied from 0 to 240 kg N·ha^{-1} and hence, some of the plots presented a high N deficiency. These plots with high N deficiency do not represent a real field situation and give an easy success in the validation; therefore, plots with R_yield below 0.7 were eliminated of the validation. Therefore, the set of 24 plots used for validation represents a real scenario.

2.4.3. Economic and Environmental Analysis

The net benefit due to N topdressing application was calculated according to Equation (8).

$$Net\ benefit = (Y_{Ntop} - Y_{real}) \cdot Price_{grain} - Price_{N} \cdot N_{rate} - Cost_{Napplicattion} \quad (8)$$

where Y_{Ntop} (t·ha^{-1}) is the yield hypothetically reached (according to response equation) if the plot had been fertilized; Y_{real} (t·ha^{-1}) is the actual yield in the plot, $Price_{grain}$ is the rice grain price ($·t^{-1}); $Price_N$ is the N (ammonium sulfate) price ($·kg^{-1}); N_{rate} (kg·ha^{-1}) is the N rate established for each scenario; and $Cost_{Napplication}$ ($·ha^{-1}) is the price for N application (fuel + man labor). Prices of the local market (year 2015) [47] were considered for the calculation. The net benefit did not include the costs of the image acquisition and further processing. The net benefit was calculated for each of the 24 plots used in the validation process.

Different scenarios were economically compared to a reference scenario.

- *Reference*: All plots are fertilized with a fixed predefined N rate (N_{ref}) without using any recommendation approach. This practice is currently used by farmers in the study area and will be considered as the reference scenario.

- *Scenario 1*: Plots are fertilized according to Delta N estimates. Topdressing N is applied to the plots when Delta N is negative. The N rate is given by Delta N, (i.e., if Delta N = −30, N rate will be 30 kg N·ha^{-1}) (Equation (9)).

$$\begin{array}{ll} If\ Delta\ N \geq 0; & N\ rate = 0 \\ If\ Delta\ N < 0; & N\ rate = |Delta\ N| \end{array} \quad (9)$$

- *Scenario 2*: Is a variation of scenario 1, but a minimum topdressing N rate (N_m) is established (Equation (10)). This option was considered since machinery is not prepared to apply fertilizers at low rates and farmers do not usually go inside the field to apply small N rates.

$$\begin{array}{ll} If\ Delta\ N \geq 0; & N\ rate = 0 \\ If\ Delta\ N < 0 \text{ and } |Delta\ N| \geq N_m; & N\ rate = |Delta\ N| \\ If\ Delta\ N < 0 \text{ and } |Delta\ N| < N_m; & N\ rate = N_m \end{array} \quad (10)$$

- *Scenario 3*: Is a variation of scenario 2. Plots are fertilized according to Delta N approach establishing a fixed predefined N rate (N_{fix}), i.e., if Delta N estimate is positive, the plots will not be fertilized, if the Delta N estimate is negative, the plots will be fertilized with N_{fix} (Equation (11)).

$$\begin{array}{ll} If\ Delta\ N \geq 0; & N\ rate = 0 \\ If\ Delta\ N < 0; & N\ rate = N_{fix} \end{array} \quad (11)$$

To consider the environmental impact of overfertilization, N excess was also evaluated. For the plots that would be fertilized according to the different scenarios, N excess was calculated as the difference between the N rate applied and the N rate that would be necessary according to yield response to N equations (Equation (3)).

For all scenarios, different predefined N rates (N_{fix}) and minimum N rates (N_m) were evaluated in the range 0–200 kg N·ha^{-1} and the net benefit and the N excess were represented graphically.

3. Results

3.1. Relationships between Yield and Vegetation Indices

3.1.1. Influence of the Year and TYPE of fertilizer

Significant relationships were observed between R_yield and the seven R_VIs and those relationships did not significantly differ between the two years (Table 3).

Table 3. Best models for the relationships between R_yield (y) (relativized to maximum value of each year) and R_VIs (x) (relativized to maximum value of each year) for years 2012 and 2013, for the PS and M treatments. The last two rows give the significance of the comparison between years and the type of fertilizer.

	R_RVI	R_GRVI	R_NDVI	R_GNDVI	R_MCARI1	R_MCARI$_{NIR}$	R_gMCARI$_{NIR}$
2012	y = 0.22 + 0.74x	y = 0.04 + 0.93x	y = −0.18 + 1.13x	y = 0.29 + 0.68x	y = −0.22 + 1.18x	$y = 0.98x^{0.43}$	$y = 1.00x^{0.33}$
2013	y = 0.22 + 0.75x	y = 0.14 + 0.84x	y = −0.39 + 1.36x	y = 0.22 + 0.76x	y = −0.17 + 1.12x	$y = 0.99x^{0.48}$	$y = 1.01x^{0.39}$
PS	y = 0.25 + 0.75x	y = 0.20 + 0.80x	y = −0.26 + 1.26x	y = 0.31 + 0.70x	y = −0.12 + 1.12x	$y = 1.03x^{0.44}$	$y = 1.04x^{0.35}$
M	y = 0.19 + 0.72x	y = −0.01 + 0.94x	y = −0.26 + 1.17x	y = 0.22 + 0.71x	y = −0.33 + 1.21x	$y = 0.93x^{0.46}$	$y = 0.95x^{0.37}$
Year	n.s.	n.s.	n.s.	n.s.	n.s.	n.s.	n.s
Fertilizer	***	***	***	**	***	***	***

n.s = not significant; ** $p < 0.01$; *** $p < 0.001$.

Thus, information from years with different crop development and yield potential could be joined if both yield and the indices are relativized to the maximum values of each year. However, the relationship between R_yield and R_VIs significantly differed between the PS and M treatments (Table 3; Figure 2a for NDVI). For example, for the NDVI, for a fixed value of the R_NDVI, the PS treatments had a higher expected yield than the M treatments (Figure 2a), i.e., the PS treatments reached that maximum yield with a lower value of the R_NDVI than in the M treatments.

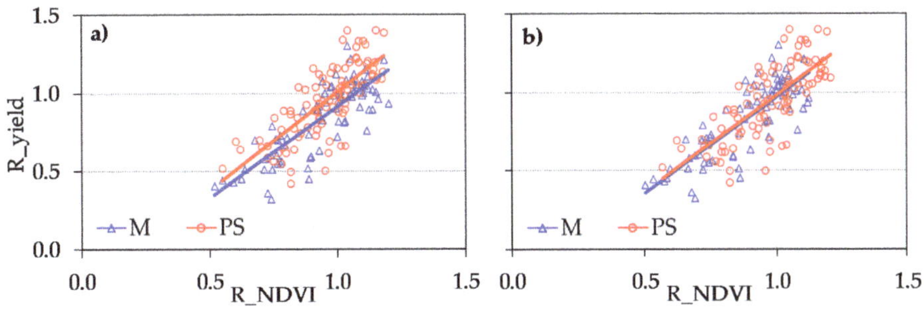

Figure 2. Relationship between R_yield and R_NDVI (relativized to maximum values of each year) in the PS and M treatments (**a**) and relationship between R_yield (relativized to maximum value of each year) and R_NDVI (relativized to maximum value of each year and within each year individually for the PS and M treatments) for PS and M treatments (**b**); 2012 and 2013 pooled data.

These results suggest that the VIs need to be relativized considering the PS and M treatments separately. Therefore, the R_VIs were recalculated using maximum VI values separately for the PS and M treatments within each year. After this process, the relationship between R_yield and R_VIs did not significantly differ between the PS and M treatments (Figure 2b).

3.1.2. Performance and Comparison of the Indices

The R_yield and the seven R_VIs for each year individually and for the pooled data over the two years (2012&2013) were significantly related (Table 4). Linear models fitted well the relationships between R_yield and the R_RVI, R_GRVI, R_NDVI, R_GNDVI and R_MCARI1 (Figure 3a–d) (Table 4). For the R_$MCARI_{NIR}$ and R_$gMCARI_{NIR}$ (Figure 3e,f), the increase in R_yield per unit increased in the indices decreased as the value of the indices increased, indicating that the relationship was not linear. For these two indices, the best fit was obtained with the multiplicative model (R_yield = a·R_VI^b). The residuals from all regressions were independent and normally distributed.

Table 4. Coefficients of determination (R^2) of the relationships between R_yield and R_VIs for years 2012 and 2013 and the pooled data, and RMSE for the pooled data.

	Model Type †	2012 n = 88	2013 n = 83	Pooled 2012+2013 n = 171	
		R^2	R^2	R^2	RMSE
R_RVI	L	0.70 ***	0.56 ***	0.62 ***	0.149
R_GRVI	L	0.74 ***	0.53 ***	0.61 ***	0.151
R_NDVI	L	0.74 ***	0.56 ***	0.63 ***	0.148
R_GNDVI	L	0.77 ***	0.56 ***	0.64 ***	0.144
R_MCARI1	L	0.69 ***	0.40 ***	0.52 ***	0.168
R_$MCARI_{NIR}$	M	0.74 ***	0.58 ***	0.64 ***	0.145
R_$gMCARI_{NIR}$	M	0.77 ***	0.61 ***	0.67 ***	0.139

*** $p < 0.001$. † L = linear, M = multiplicative.

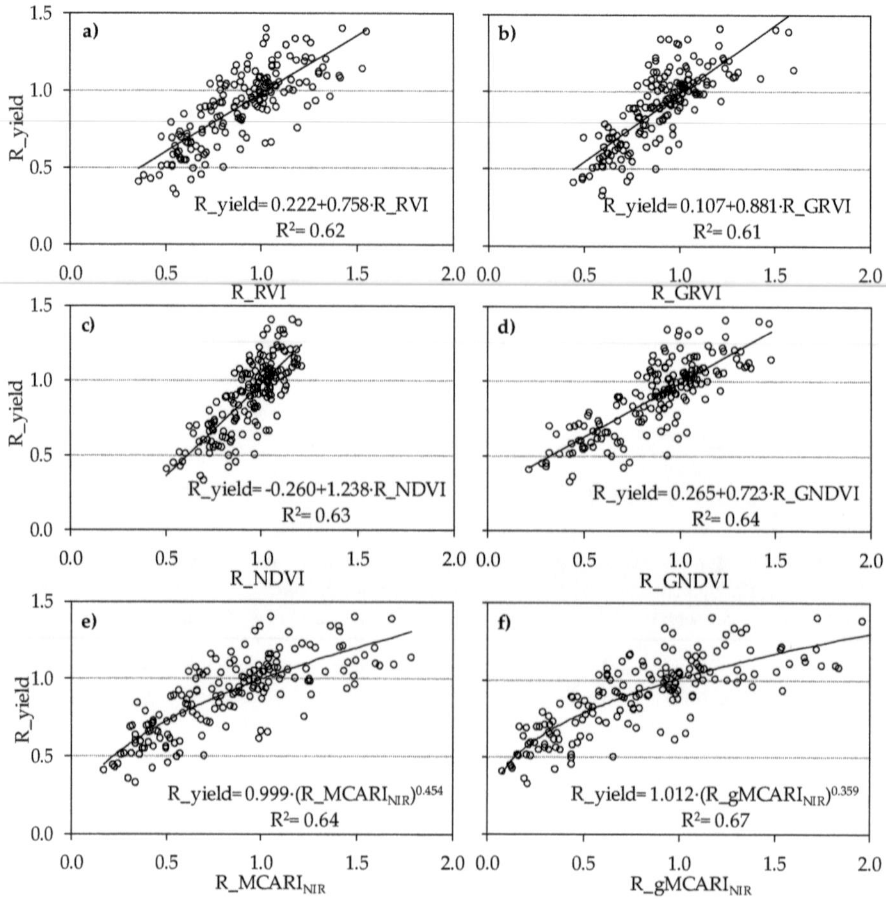

Figure 3. Relationship between R yield and R_VIs (R_RVI (a), R_GRVI (b), R_NDVI (c), R_GNDVI (d), R_MCARI$_{NIR}$ (e), and R_gMCARI$_{NIR}$ (f)) for the pooled data over the two years.

The coefficients of determination ranged between 0.40 and 0.77 (Table 4); in general, the R_NDVI and R_GNDVI improved the relationships in comparison to the R_RVI and R_GRVI (all of them including two bands in their definition). R_gMCARI$_{NIR}$, which includes information on three bands, slightly improved the relationship in comparison to the two-band VIs.

In 2012, R_GNDVI and R_gMCARI$_{NIR}$ explained 77% of the R_yield variability and performed better than the other indices. For 2013, R_gMCARI$_{NIR}$ was the index that explained the highest percentage (61%) of the R_yield variability.

When data from both years were pooled, R_gMCARI$_{NIR}$, R_MCARI$_{NIR}$ and R_GNDVI showed the highest coefficients of determination (R^2 = 0.67, 0.64 and 0.64 respectively) and the lowest RMSE (0.139, 0.145, 0.144) and all of them included the green band information in their definition.

3.2. N Topdressing Recommendation Approach

3.2.1. Design of the N Topdressing Recommendation Approach

The indices that presented the strongest relation to yield were GNDVI, MCARI$_{NIR}$ and gMCARI$_{NIR}$. The index gMCARI$_{NIR}$ was selected for the design of the model because it showed the best relation to yield. GNDVI was selected versus $MCARI_{NIR}$ (both with the same strength in their relation to yield) with the aim to test two different functions (linear and multiplicative) and two different indices (two and three bands VIs).

The equations adjusted (Figure 4) to estimate Delta N from Delta NSI (GNDVI) and Delta NSI ($gMCARI_{NIR}$) were Equations (12) and (13):

$$Delta\ N = 2.843 + 211.491 \cdot Delta\ NNI(GNDVI) \quad (12)$$

$$Delta\ N + 200 = 207.291 \cdot (Delta\ NNI(gMCARI_{NIR}) + 1)^{0.517} \quad (13)$$

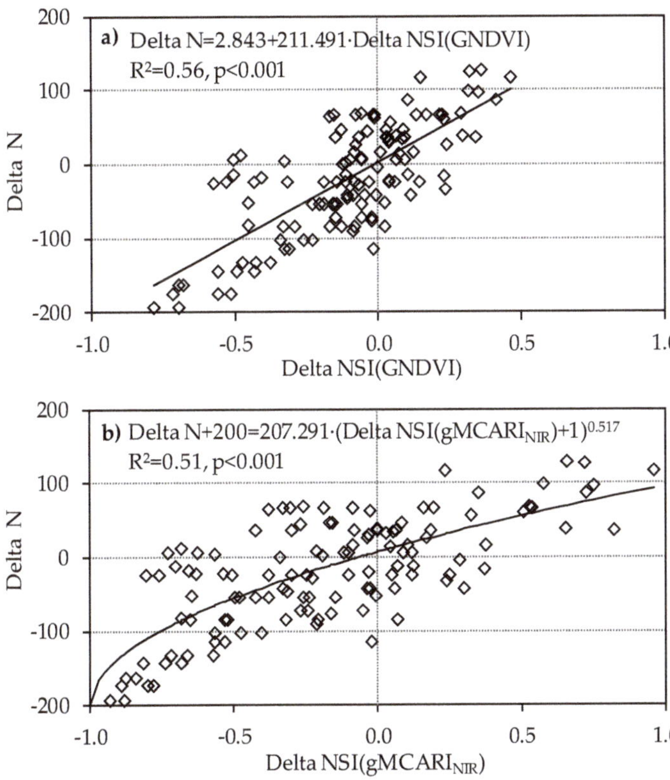

Figure 4. Relationship between Delta N (N increased or decreased compared with the optimum treatment) and Delta NSI(GNDVI) (**a**) or Delta NSI($gMCARI_{NIR}$) (**b**) (R_GNDVI-1 or R_$gMCARI_{NIR}$-1) (pooled data of the two years, 75% data).

3.2.2. Assessment of the N Topdressing Recommendation Approach

The N recommendation approach designed had a high success rate (higher than 83%) and performed better with $gMCARI_{NIR}$, with an 87.5% of success, than with the GNDVI with an 83.3% success (Table 5). The percentage of failure by excess was lower for $gMCARI_{NIR}$ than for GNDVI; however, the percentage of failure by defect was higher for $gMCARI_{NIR}$ than for GNDVI. The combination of both VIs did not improve the ability for N recommendation prediction (Table 5). The GNDVI & $gMCARI_{NIR}$ strategy showed the same percentages of success and failure than $gMCARI_{NIR}$, and the GNDVI or $gMCARI_{NIR}$ possibility showed the same results than GNDVI.

Since the combination of VIs did not improve the ability of N recommendation in comparison to using only one VI, $gMCARI_{NIR}$ (the VI with the highest percentage of success) was considered the best option and it was evaluated in the economic and environmental analysis.

Table 5. Percentage of success and failure by excess and defect using the indices GNDVI, $gMCARI_{NIR}$ and the combinations GNDVI & $gMCARI_{NIR}$ and GNDVI or $gMCARI_{NIR}$ (25% data, excluded plots with R_yield below 0.7, total number of plots used for validation: 24).

Strategy	SUCCESS	EXCESS	DEFECT
R_GNDVI	83.3	12.5	4.2
R_$gMCARI_{NIR}$	87.5	4.2	8.3
R_GNDVI&R_$gMCARI_{NIR}$	87.5	4.2	8.3
R_GNDVI or R_$gMCARI_{NIR}$	83.3	12.5	4.2

In the framework of the four scenarios analyzed in this work, the net benefit due to topdressing N fertilization ranged between −50 $\$\cdot ha^{-1}$ and 132 $\$\cdot ha^{-1}$ (the additional benefit if these plots had been fertilized with additional N) (Figure 5).

Figure 5 shows the net benefit (a) and N excess (b) according to the fixed predefined N rate (N_{fix}) for the reference scenario and scenario 3 or to the minimum N rate (N_m) in the case of scenario 2.

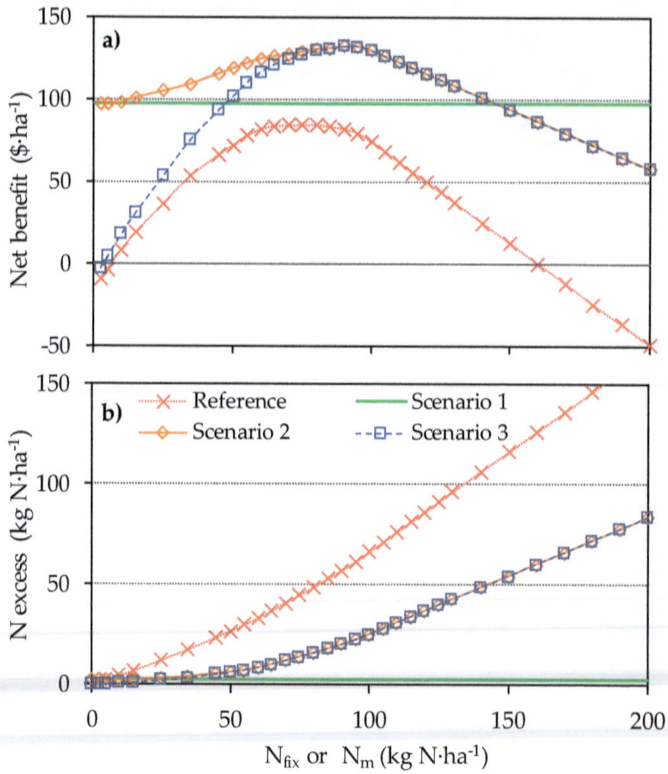

Figure 5. Net benefit ($\$\cdot ha^{-1}$) (a) and N excess (kg $N\cdot ha^{-1}$) (b) according to the fixed predefined N rate (N_{fix}, kg $N\cdot ha^{-1}$) or minimum N rate (N_m, kg $N\cdot ha^{-1}$). Each point is the average of the 24 plots used in the economic analysis.

The *reference scenario* (current farmers' practice), where all plots are fertilized, obtains the lowest net benefit, with a maximum of 84 $\$\cdot ha^{-1}$ for a rate of 75 kg $N\cdot ha^{-1}$ (Figure 5a), and the highest N excess of the four scenarios reaching 45 kg N excess$\cdot ha^{-1}$ for the N rate of maximum net benefit (Figure 5b).

The use of the N recommendation approach (*Scenario 1*) shows a net benefit of 97 $\$\cdot ha^{-1}$ and a N excess of 2.3 kg $N\cdot ha^{-1}$. The net benefit and N excess associated with this strategy have been represented as a horizontal green line in Figure 5 for comparative purposes with the rest of scenarios.

The establishment of a threshold minimum N rate (N_m) in *Scenario 2*, shows a higher net benefit than *Scenario 1*, reaching 132 $\$\cdot ha^{-1}$ for a threshold N rate of 90 kg $N\cdot ha^{-1}$ (Figure 5a). However, the increase of the net benefit (35 $\$\cdot ha^{-1}$) is counterbalanced with an increase in N excess (20 $kg\cdot N\cdot ha^{-1}$ for the highest benefit versus 2.3 kg $N\cdot ha^{-1}$ in *Scenario 1*) (Figure 5b).

In the case of fixing a predefined N rate (*Scenario 3*), the maximum net benefit (132 $\$\cdot ha^{-1}$) is equal to that in Scenario *2* for the rate of 90 $kg\cdot ha^{-1}$ (Figure 5a). In this scenario, the net benefit dramatically decreases as the N_{fix} decreases; therefore, the election of the optimum N_{fix} rate is a key factor, i.e., if N_{fix} is too low, the net benefit could be dramatically reduced. For the optimum N_{fix} rate of this study (90 $kg\cdot ha^{-1}$), the N excess was 20 $kg\cdot N\cdot ha^{-1}$, the same as in *Scenario 2* (Figure 5b).

4. Discussion

4.1. Relationships between Yield and Vegetation Indices

4.1.1. Influence of the Type of Fertilizer

The relationship between R_yield and R_VIs showed differential effects for the PS and M fertilization treatments for the seven VIs analyzed (Figure 2a). Although studies focusig on the differences in spectral information between different types of fertilization in rice fields are lacking, there are some studies in other crops that support these results. Zhao et al. [48] measured the canopy apparent photosynthesis (CAP), the photosynthetic rate of flag leaves, LAI and yield in a winter wheat experiment in plots fertilized with cow manure, urea and a mixed application. The results showed that during the early growth period, a single application of urea promoted better crop development and resulted in higher values of CAP or LAI; however, during the late growth stage, a single application of cow manure and the mixed application delayed the leaf senescence process compared to a single application of urea. The results suggested that mixed application of organic and inorganic fertilizers delayed leaf senescence and maintained better canopy structure and higher photosynthetic capability at the late grain filling stage, which resulted in a higher grain yield. In our study, we found the same type of behavior as Zhao et al. [48], i.e., for the same value of the R_VIs at the booting stage, the PS treatments reached a higher yield compared to the M plots. This better development during the late stage is usually related to higher availability of micronutrients provided by the organic fertilizers [49–51].

The observed differences between the type of fertilization is an important result for the development and use of N recommendation approaches. This result implies that it is necessary to establish over-fertilized plots for each type of fertilization.

4.1.2. Performance and Comparison of the Indices

The R_VIs showed good relationships with R_yield (Table 4). Other studies have also found good relationships between these indices and rice yield [15,16,24,29,52]. The coefficients of determination are in the range reported by the abovementioned studies at the same growth stage (booting) [15,16].

One common problem with the two-band VIs is that they become saturated under high values of biomass [16,29] and different authors have proposed three-band VIs to handle this saturation phenomenon. In our study, the R_MCARI1 (three-band VI) did not improve the R_yield prediction in comparison with the traditional two-band indices and the relationship even worsened; this result contrasts the study reported by Haboudane et al. [31] in which the MCARI1 was less sensitive to the saturation phenomenon, although these findings were for different crops than rice (i.e., corn, wheat and soybean). However, the R_MCARI$_{NIR}$ and R_gMCARI$_{NIR}$ had a strong relationship with R_yield and performed equal to or better than the traditional R_NDVI and R_GNDVI (Table 4). The index proposed in this study (R_gMCARI$_{NIR}$) was the best for R_yield prediction and improved the relationship obtained with the R_MCARI$_{NIR}$.

Although the relationships between R_yield and the R_MCARI$_{NIR}$ or R_gMCARI$_{NIR}$ (three-band VIs) were stronger than for the two-band VIs, there was only a 1–7% increase in the variability explained by these indices in comparison to two-band VIs, such as the R_NDVI and R_GNDVI (Table 4). These results contrast with those from the study reported by Cao et al. [16] in which the yield potential variability explained by the three-band indices was between 21 and 26% higher than the variability explained by the NDVI or RVI at the rice booting stage.

These differences in yield prediction improvement with three-band indices may be related to potential yield. The highest yield in this study was approximately 8000 kg·ha^{-1}; however, yields reached 10,000 kg·ha^{-1} in the study conducted by Cao et al. [16] and 14,000 kg·ha^{-1} for the study conducted by Harrell et al. [29] in which the saturation phenomenon was also observed. Therefore, the saturation phenomenon can be attributed to high biomass conditions in high-yield systems. In these systems, two-band VIs do not perform well, especially at later stages with high biomass conditions; however, three-band VIs can overcome this saturation problem.

Thus, in low-yielding rice systems similar to our system, two-band indices, such as the NDVI and GNDVI, are not subject to the saturation phenomenon, due to low crop biomass. This hypothesis is in agreement with the results of Xue et al. [21], who found strong relationships between the NDVI and rice yield potential across the growing season from tillering to grain filling, with yield levels lower than 8000 kg·ha^{-1} (similar to this study); thus, the saturation phenomenon, due to high biomass conditions was not observed.

Thus, in our study, the best indices to predict R_yield were the R_GNDVI, R_MCARI$_{NIR}$ and R_gMCARI$_{NIR}$, both incorporate the green band. This result supports the findings of Cao et al. [16], which demonstrated that the best VIs for estimation of the response index at harvest (R_yield) were green band-based VIs.

4.2. Assessment of the N Topdressing Recommendation Approach

The approach evaluated in this study showed success percentages higher than 80% and reaching 87.5% for the best options (Table 5). Thus, these options confer an advantage over the application of N topdressing without any advice. Nevertheless, it is important to point out that this validation process means only a preliminary evaluation of the approach, since it is based on the comparison of the approach's recommendation and the actual yield harvested in the same experiment where the models (equations) were obtained. The approach should be tested further in real field situations, fertilizing the fields following the recommendations of the models and evaluating the yield obtained [15,20,21] compared to other fields with a standard fixed N rate.

In contrast to other studies where only one index was used in the development of these approaches, in our study, the combination of two indices at the same time was tested (the plot is fertilized when both VIs recommend N topdressing or when one of the VIs recommends N fertilization even if the other index does not recommend N fertilization). Although these options did not increase the percentage of success under the studied conditions, it should be considered in further studies, because in other situations it could increase the percentage of success.

The results show that the Reference Scenario, in which all plots are fertilized without spectral information contribution, is the least recommended scenario of the four evaluated scenarios (Figure 5). This scenario shows the lowest net benefit and the highest N excess, suggesting that this option is neither economically nor environmentally viable when compared to the other scenarios. Therefore, the use of a decision tool for in season topdressing N recommendation is clearly advantageous.

The N recommendation approach evaluated in this study seems to be a good strategy. Three possibilities were evaluated: Fertilizing with the N rate according to the approach, Delta N, (Scenario 1), establishing a minimum N rate, N_m, (Scenario 2) or establishing a fixed predefined N rate, N_{fix}, (Scenario 3). When a minimum N rate is established (Scenario 2), the maximum net benefit increases in comparison to Scenario 1, the maximum difference is 35 $\$\cdot ha^{-1}$, for a minimum N rate of 90 kg $N\cdot ha^{-1}$, but the N excess increases in 18 kg $N\cdot ha^{-1}$ (Figure 5). Combining net benefit and environmental impact, we would recommend a N rate of 60 kg $N\cdot ha^{-1}$ with an increase in net benefit of 27 $\$\cdot ha^{-1}$ and an increase in N excess of 6.1 kg $N\cdot ha^{-1}$ in comparison to Scenario 1. On the other hand, when a fixed predefined N rate is established (Scenario 3), the maximum net benefit obtained is the same as that obtained in Scenario 2. However, in Scenario 3, a careful selection of the N_{fix} rate is crucial, since the net benefit can dramatically lower with small modifications of the N rate (Figure 5); this N rate will depend on the year and it is difficult to estimate. Hence, Scenario 1 and Scenario 2 seem to be better than Scenario 3. Scenario 2 was considered because farmers do not usually enter the field to apply small N rates or even machinery is not able to apply tiny amounts of fertilizer, thus a minimum N rate should be considered. This option allows increasing the net benefit under a large interval of N_m rates (from 0 to 150 kg $N\cdot ha^{-1}$) (Figure 5), in which the net benefit is the same or higher than in Scenario 1. Therefore, in Scenario 2 the net benefit is not at risk; however, the N excess increases as N_m increases. The cost of the risk for contamination, the N excess, has not been considered in the economic analysis and should be incorporated in future works. As a first approach in Scenario 2, N_m rates between 50–60 kg $N\cdot ha^{-1}$ are believed to be optimal considering both economic return and N excess.

Thus, the economic analysis suggests that the best option is the use of the N recommendation approach fertilizing with the recommended N rate (Scenario 1) or establishing a minimum N rate (Scenario 2) in order to increase the net benefit, although the increase of N excess should be economically evaluated.

These findings should be validated in a different set of field experiments, where some of the plots were fertilized according to the approach's N recommendation and other plots according to the local practices in order to evaluate the real benefits.

The inclusion of the cost of the images' capture and further processing is crucial to evaluate whether these tools are economically feasible, although multispectral information arises as a useful tool for increasing net benefit and decreasing N excess in the agro-systems of Northeast Spain.

5. Conclusions

The seven studied VIs showed good relationships with yield. However, differences between the PS and M treatments were detected, i.e., fields fertilized with pig slurry had a different spectral response than those fertilized with synthetic fertilizers. This is an important result for the development and use of N recommendation approaches since it is necessary to establish over-fertilized plots for each type of fertilization.

The best relationships between yield and the VIs were obtained with indices including the green band. Additionally, three-band VIs performed better than the traditional two-band VIs, but the improvement was minor compared to other studies with higher yields when the saturation of two-band indices was observed. The best relationships were obtained with GNDVI, $MCARI_{NIR}$ and $gMCARI_{NIR}$; the latter was proposed in this study, all three included the green band, and the $MCARI_{NIR}$ and $gMCARI_{NIR}$ included 3 bands.

The N recommendation approach evaluated was useful for N recommendation. The best option was the use of the index $gMCARI_{NIR}$ achieving an 87.5% of success and the combination of VIs did not improve the ability for N recommendation prediction. The economic analysis showed that the use of a N recommendation approach clearly increases the net benefit and lowers the N excess in comparison to fertilization without any recommendation. The best option in this study was the use of the recommended N rate (Delta N) or establishing a minimum N rate (optimum minimum N rate between 50 and 60 kg N·ha^{-1}).

Therefore, the use of aerial remote sensing is a promising tool for developing strategies for advising rice farmers. However, more research is needed, including a validation of this approach to field level, the inclusion of the cost of the recommendation system in the evaluation of the net benefit and the response of yield to spectral information in earlier crop development stages to adjust the N topdressing as much as possible to the usual practices in the area.

Author Contributions: B.M-G. was responsible for field data acquisition, data analysis, and design and evaluation of the N recommendation approach and wrote the manuscript draft. M.A.C. was responsible for remote sensing activities, analysis and interpretation of images information. M.G. collected and processed field and aerial images data. D.Q. was responsible for funding acquisition and directed the study. All the authors contributed to the discussion and revisions of the manuscript.

Funding: This study was funded by the National Institute for Agricultural and Food Scientific Research and Technology of Spain (INIA) and FEDER funds (RTA2010-0126-C02-01 and RTA2013-0057-C05-04). B. Moreno-García was granted with an FPI-INIA fellowship.

Acknowledgments: We thank Rosa Gómez for her assistance in image processing analysis, Ramón Isla for the critical review of the manuscript and the field and laboratory personnel of the Soils and Irrigation Department of CITA.

Conflicts of Interest: The authors declare no conflict of interest.

References

1. Mohanty, S. Trends in global rice consumption. *Rice Today* **2013**, *12*, 44–45.
2. Cassman, K.G.; Peng, S.; Olk, D.C.; Ladha, J.K.; Reichardt, W.; Dobermann, A.; Singh, U. Opportunities for increased nitrogen-use efficiency from improved resource management in irrigated rice systems. *Field Crop. Res.* **1998**, *56*, 7–39. [CrossRef]
3. FAOSTAT. Crops Production. Available online: http://faostat3.fao.org/home/E (accessed on 26 March 2015).
4. MAPAMA. Superficies y Producciones de Cereales. Año 2016. Ministerio de Agricultura y Pesca, Alimentación y Medio Ambiente. Available online: http://www.magrama.gob.es/es/estadistica/temas/estadisticas-agrarias/agricultura/superficies-producciones-anuales-cultivos/ (accessed on 27 January 2018).
5. MAPAMA. Encuestas ganaderas 2016. Ganado Porcino. Ministerio de Agricultura y Pesca, Alimentación y Medio Ambiente. Available online: http://www.magrama.gob.es/es/estadistica/temas/estadisticas-agrarias/ganaderia/encuestas-ganaderas/#para4 (accessed on 27 January 2018).
6. Bond, J.A.; Bollich, P.K. Yield and quality response to rice cultivars to preflood and late-season nitrogen. *Crop Manag.* **2007**. Available online: www.plantmanagementnetwork.org/cm/ (accessed on 5 July 2016). [CrossRef]
7. Moreno-García, B.; Guillén, M.; Quílez, D. Response of paddy rice to fertilisation with pig slurry in northeast Spain: Strategies to optimise nitrogen use efficiency. *Field Crop. Res.* **2017**, *208*, 44–54. [CrossRef]
8. Wilson, C.; Slaton, N.; Norman, R.; Miller, D. Efficient use of fertilizer. In *Rice Production Handbook*; University of Arkansas Division of Agriculture Cooperative Extension Service: Little Rock, AR, USA, 2001; pp. 51–75.
9. Schlegel, A.J.; Grant, C.A.; Havlin, J.L. Challenging approaches to nitrogen fertilizer recommendations in continuous cropping systems in the Great Plains. *Agron. J.* **2005**, *97*, 391–398. [CrossRef]
10. Nishikawa, T.; Li, K.Z.; Inoue, H.; Umeda, M.; Hirooka, H.; Inamura, T. Effects of the Long-Term Application of Anaerobically-Digested Cattle Manure on Growth, Yield and Nitrogen Uptake of Paddy Rice (*Oryza sativa* L.), and Soil Fertility in Warmer Region of Japan. *Plant. Prod. Sci.* **2012**, *15*, 284–292. [CrossRef]
11. Xie, W.X.; Wang, G.H.; Zhang, Q.C.; Guo, H.C. Effects of nitrogen fertilization strategies on nitrogen use efficiency in physiology, recovery, and agronomy and redistribution of dry matter accumulation and nitrogen accumulation in two typical rice cultivars in Zhejiang, China. *J. Zhejiang Univ. Sci. B* **2007**, *8*, 208–216. [CrossRef] [PubMed]
12. Denuit, J.P.; Olivier, M.; Goffaux, M.J.; Herman, J.L.; Goffart, J.P.; Destain, J.P.; Frankinet, M. Management of nitrogen fertilization of winter wheat and potato crops using the chlorophyll meter for crop nitrogen status assessment. *Agronomie* **2002**, *22*, 847–853. [CrossRef]

13. Isla, R.; Valentín, F.; Quílez, D.; Guillén, M.; Aibar, J.; Maturano, M. Comparison of decision tools to improve the nitrogen management in irrigated maize under mediterranean conditions in Spain. In Proceedings of the 16th ASA Conference, Armidale, Australia, 14–18 October 2012.
14. Peng, S.; Garcia, F.V.; Laza, R.C.; Sanico, A.L.; Visperas, R.M.; Cassman, K.G. Increased N-use efficiency using a chlorophyll meter on high-yielding irrigated rice. *Field Crop. Res.* **1996**, *47*, 243–252. [CrossRef]
15. Yao, Y.; Miao, Y.; Huang, S.; Gao, L.; Ma, X.; Zhao, G.; Jiang, R.; Chen, X.; Zhang, F.; Yu, K.; Gnyp, M.L.; Bareth, G.; Liu, C.; Zhao, L.; Yang, W.; Zhu, H. Active canopy sensor-based precision N management strategy for rice. *Agron. Sustain. Dev.* **2012**, *32*, 925–933. [CrossRef]
16. Cao, Q.; Miao, Y.; Shen, J.; Yu, W.; Yuan, F.; Cheng, S.; Huang, S.; Wang, H.; Yang, W.; Liu, F. Improving in-season estimation of rice yield potential and responsiveness to topdressing nitrogen application with Crop Circle active crop canopy sensor. *Precis. Agric.* **2015**. [CrossRef]
17. Huang, S.; Miao, Y.; Zhao, G.; Yuan, F.; Ma, X.; Tan, C.; Yu, W.; Gnyp, M.L.; Lenz-Wiedemann, V.I.S.; Rascher, U.; Bareth, G. Satellite remote sensing-based in-season diagnosis of rice nitrogen status in Northeast China. *Remote Sens.* **2015**, *7*, 10646–10667. [CrossRef]
18. Matese, A.; Toscano, P.; Di Gennaro, S.F.; Genesio, L.; Vaccari, F.P.; Primicerio, J.; Belli, C.; Zaldei, A.; Bianconi, R.; Gioli, B. Intercomparison of UAV, aircraft and satellite remote sensing platforms for precision viticulture. *Remote Sens.* **2015**, *7*, 2971–2990. [CrossRef]
19. Chen, Q.; Tian, Y.; Yao, X.; Cao, W.; Zhu, Y. Comparison of five nitrogen dressing methods to optimize rice growth. *Plant. Prod. Sci.* **2014**, *17*, 66–80. [CrossRef]
20. Xue, L.; Yang, L. Recommendations for nitrogen fertiliser topdressing rates in rice using canopy reflectance spectra. *Biosyst. Eng.* **2008**, *100*, 524–534. [CrossRef]
21. Xue, L.; Li, G.; Qin, X.; Yang, L.; Zhang, H. Topdressing nitrogen recommendation for early rice with an active sensor in south China. *Precis. Agric.* **2014**, *15*, 95–110. [CrossRef]
22. Casanova, D.; Epema, G.F.; Goudriaan, J. Monitoring rice reflectance at field level for estimating biomass and LAI. *Field Crop. Res.* **1998**, *55*, 83–92. [CrossRef]
23. Gnyp, M.L.; Miao, Y.; Yuan, F.; Ustin, S.L.; Yu, K.; Yao, Y.; Huang, S.; Bareth, G. Hyperspectral canopy sensing of paddy rice aboveground biomass at different growth stages. *Field Crop. Res.* **2014**, *155*, 42–55. [CrossRef]
24. Gilabert, M.A.; Meliá, J. Usefulness of the temporal analysis and the normalized difference in the study of rice by means of landsat-5 TM images: Establishment of Relationships for Yield Prediction Purpose. *Geocarto Int.* **1990**, *5*, 27–32. [CrossRef]
25. Li, W.; Niu, Z.; Wang, C.; Huang, W.; Chen, H.; Gao, S.; Li, D.; Muhammad, S. Combined Use of Airborne LiDAR and Satellite GF-1 Data to Estimate Leaf Area Index, Height, and Aboveground Biomass of Maize during Peak Growing Season. *IEEE J. Sel. Top. Appl. Earth Obs. Remote Sens.* **2015**, *8*, 4489–4501. [CrossRef]
26. Sultana, S.R.; Ali, A.; Ahmad, A.; Mubeen, M.; Zia-Ul-Haq, M.; Ahmad, S.; Ercisli, S.; Jaafar, H.Z.E. Normalized difference vegetation index as a tool for wheat yield estimation: A case study from Faisalabad, Pakistan. *Sci. World J.* **2014**, *2014*. [CrossRef] [PubMed]
27. Quarmby, N.A.; Milnes, M.; Hindle, T.L.; Silleos, N. The use of multi-temporal NDVI measurements from AVHRR data for crop yield estimation and prediction. *Int. J. Remote Sens.* **1993**, *14*, 199–210. [CrossRef]
28. Raun, W.R.; Solie, J.B.; Johnson, G.V.; Stone, M.L.; Mutten, R.W.; Freeman, K.W.; Thomason, W.E.; Lukina, E.V. Improving nitrogen use efficiency in cereal grain production with optical sensing and variable rate application. *Agron. J.* **2002**, *94*, 815–820. [CrossRef]
29. Harrell, D.L.; Tubana, B.S.; Walker, T.W.; Phillips, S.B. Estimating Rice Grain Yield Potential Using Normalized Difference Vegetation Index. *Agron. J.* **2011**, *103*, 1717–1723. [CrossRef]
30. Gitelson, A.A.; Kaufman, Y.J.; Merzlyak, M.N. Use of a green channel in remote sensing of global vegetation from EOS-MODIS. *Remote Sens. Environ.* **1996**, *58*, 289–298. [CrossRef]
31. Haboudane, D.; Miller, J.R.; Pattey, E.; Zarco-Tejada, P.J.; Strachan, I.B. Hyperspectral vegetation indices and novel algorithms for predicting green LAI of crop canopies: Modeling and validation in the context of precision agriculture. *Remote Sens. Environ.* **2004**, *90*, 337–352. [CrossRef]
32. Daughtry, C.S.T.; Walthall, C.L.; Kim, M.S.; De Colstoun, E.B.; McMurtrey, J.E., III. Estimating corn leaf chlorophyll concentration from leaf and canopy reflectance. *Remote Sens. Environ.* **2000**, *74*, 229–239. [CrossRef]

33. Cao, Q.; Miao, Y.; Wang, H.; Huang, S.; Cheng, S.; Khosla, R.; Jiang, R. Non-destructive estimation of rice plant nitrogen status with Crop Circle multispectral active canopy sensor. *Field Crop. Res.* **2013**, *154*, 133–144. [CrossRef]
34. Busetto, L.; Casteleyn, S.; Granell, C.; Pepe, M.; Barbieri, M.; Campos-Taberner, M.; Casa, R.; Collivignarelli, F.; Confalonieri, R.; Crema, A. Downstream services for rice crop monitoring in Europe: From regional to local scale. *IEEE J. Sel. Top. Appl. Earth Obs. Remote Sens.* **2017**, *10*, 5423–5441. [CrossRef]
35. Nutini, F.; Confalonieri, R.; Crema, A.; Movedi, E.; Paleari, L.; Stavrakoudis, D.; Boschetti, M. An operational workflow to assess rice nutritional status based on satellite imagery and smartphone apps. *Comput. Electron. Agric.* **2018**, *154*, 80–92. [CrossRef]
36. Confalonieri, R.; Foi, M.; Casa, R.; Aquaro, S.; Tona, E.; Peterle, M.; Boldini, A.; De Carli, G.; Ferrari, A.; Finotto, G.; Guarneri, T.; Manzoni, V.; Movedi, E.; Nisoli, A.; Paleari, L.; Radici, I.; Suardi, M.; Veronesi, D.; Bregaglio, S.; Cappelli, G.; Chiodini, M.E.; Dominoni, P.; Francone, C.; Frasso, N.; Stella, T.; Acutis, M. Development of an app for estimating leaf area index using a smartphone. Trueness and precision determination and comparison with other indirect methods. *Comput. Electron. Agric.* **2013**, *96*, 67–74. [CrossRef]
37. Confalonieri, R.; Paleari, L.; Movedi, E.; Pagani, V.; Orlando, F.; Foi, M.; Barbieri, M.; Pesenti, M.; Cairati, O.; La Sala, M.S.; Besana, R.; Minoli, S.; Bellocchio, E.; Croci, S.; Mocchi, S.; Lampugnani, F.; Lubatti, A.; Quarteroni, A.; De Min, D.; Signorelli, A.; Ferri, A.; Ruggeri, G.; Locatelli, S.; Bertoglio, M.; Dominoni, P.; Bocchi, S.; Sacchi, G.A.; Acutis, M. Improving invivo plant nitrogen content estimates from digital images: Trueness and precision of a new approach as compared to other methods and commercial devices. *Biosyst. Eng.* **2015**, *135*, 21–30. [CrossRef]
38. Piccinini, S.; Bortone, G. The fertilizer value of agricultural manure: Simple rapid methods of assessment. *J. Agric. Eng. Res.* **1991**, *49*, 197–208. [CrossRef]
39. Yagüe, M.R.; Quílez, D. On-farm Measurement of Electrical Conductivity for the Estimation of Ammonium Nitrogen Concentration in Pig Slurry. *J. Environ. Qual.* **2012**, *41*, 893–900. [CrossRef] [PubMed]
40. Jordan, C.F. Derivation of Leaf-Area Index from Quality of Light on the Forest Floor. *Ecology* **1969**, *50*, 663–666. [CrossRef]
41. Inada, K. Spectral ratio of reflectance for estimating chlorophyll content of leaf. *Jpn. J. Crop Sci.* **1985**, *54*, 261–265. [CrossRef]
42. Rouse, J.W.J.; Hass, R.H.; Schell, J.A.; Deering, D.W. Monitoring vegetation systems in the great plains with ERTS. In Proceedings of the Third Earth Resources Technology Satellite-1 Symposium, Washington, DC, USA, 10–14 December 1973; pp. 309–317.
43. Tucker, C.J.; Elgin, J.H.; McMurtrey, J.E.; Fan, C.J. Monitoring corn and soybean crop development with hand-held radiometer spectral data. *Remote Sens. Environ.* **1979**, *8*, 237–248. [CrossRef]
44. Wang, Y.P.; Chang, K.W.; Chen, R.K.; Lo, J.C.; Shen, Y. Large-area rice yield forecasting using satellite imageries. *Int. J. Appl. Earth Obs. Geoinf.* **2010**, *12*, 27–35. [CrossRef]
45. Cerrato, M.E.; Blackmer, A.M. Comparison of models for describing corn yield response to nitrogen-fertilizer. *Agron. J.* **1990**, *82*, 138–143. [CrossRef]
46. Dixon, W.J. *BMDP Statistical Software Manual*; University of California Press: Berkeley, CA, USA, 1985.
47. IAEST. Precios Pagados por los Agricultores. Coyuntura Agraria de Aragón. Departamento de Desarrollo Rural y Sostenibilidad. Available online: http://www.aragon.es/DepartamentosOrganismosPublicos/Institutos/InstitutoAragonesEstadistica/AreasTematicas/10_Precios/ci.08_Precios_agrarios.detalleDepartamento (accessed on 8 March 2016).
48. Zhao, J.; Dong, S.T.; Liu, P.; Zhang, J.W.; Zhao, B. Effects of long-term mixed application of organic and inorganic fertilizers on canopy apparent photosynthesis and yield of winter wheat. *Chin. J. Appl. Ecol.* **2015**, *26*, 2362–2370.
49. Shahid, M.; Shukla, A.K.; Bhattacharyya, P.; Tripathi, R.; Mohanty, S.; Kumar, A.; Lal, B.; Gautam, P.; Raja, R.; Panda, B.B.; Das, B.; Nayak, A.K. Micronutrients (Fe, Mn, Zn and Cu) balance under long-term application of fertilizer and manure in a tropical rice-rice system. *J. Soils Sed.* **2016**, *16*, 737–747. [CrossRef]
50. Dong, W.; Zhang, X.; Wang, H.; Dai, X.; Sun, X.; Qiu, W.; Yang, F. Effect of Different Fertilizer Application on the Soil Fertility of Paddy Soils in Red Soil Region of Southern China. *PLoS ONE* **2012**, *7*. [CrossRef] [PubMed]

51. Huang, Q.R.; Hu, F.; Huang, S.; Li, H.X.; Yuan, Y.H.; Pan, G.X.; Zhang, W.J. Effect of Long-Term Fertilization on Organic Carbon and Nitrogen in a Subtropical Paddy Soil. *Pedosphere* **2009**, *19*, 727–734. [CrossRef]
52. Chang, K.W.; Shen, Y.; Lo, J.C. Predicting rice yield using canopy reflectance measured at booting stage. *Agron. J.* **2005**, *97*, 872–878. [CrossRef]

remote sensing

MDPI

Article

Crop Sensor-Based In-Season Nitrogen Management of Wheat with Manure Application

Marta Aranguren *, Ander Castellón and Ana Aizpurua

NEIKER-Basque Institute for Agricultural Research and Development, Berreaga 1, 48160 Derio, Biscay, Spain; acastellon@neiker.eus (A.C.); aaizpurua@neiker.eus (A.A.)

* Correspondence: maranguren@neiker.eus

Received: 4 April 2019; Accepted: 6 May 2019; Published: 8 May 2019

Abstract: It is difficult to predict the crop-available nitrogen (N) from farmyard manures applied to soil. The aim of this study was to assess the usefulness of the proximal sensors, Yara N-Tester^{TM} and RapidScan CS-45, for diagnosing the N nutritional status of wheat after the application of manures at sowing. Three annual field trials were established (2014–2015, 2015–2016 and 2016–2017) with three types of fertilizer treatments: dairy slurry (40 t ha^{-1} before sowing), sheep manure (40 t ha^{-1} before sowing) and conventional treatment (40 kg N ha^{-1} at tillering). For each treatment, five different mineral N fertilization doses were applied at stem elongation: 0, 40, 80, 120, and 160 kg N ha^{-1}. The proximal sensing tools were used at stem elongation before the application of mineral N. Normalized values of the proximal sensing look promising for adjusting mineral N application rates at stem elongation. For dairy slurry, when either proximal sensor readings were 60–65% of the reference plants with non-limiting N, the optimum N rate for maximizing yield was 118–128 kg N ha^{-1}. When the readings were 85–90%, the optimum N rate dropped to 100–110 kg N ha^{-1} for both dairy slurry and conventional treatments. It was difficult to find a clear relationship between sensor readings and yield for sheep manure treatments. Measurements taken with RapidScan C-45 were less time consuming and better represent the spatial variation, as they are taken on the plant canopy. Routine measurements throughout the growing season are particularly needed in climates with variable rainfall. The application of 40 kg N ha^{-1} at the end of winter is necessary to ensure an optimal N status from the beginning of wheat crop development. These research findings could be used in applicator-mounted sensors to make variable-rate N applications.

Keywords: precision N fertilization; chlorophyll meter; NDVI; NDRE; NNI; canopy reflectance sensing; N mineralization; farmyard manures; *Triticum aestivum*

1. Introduction

Few cereal agroecosystems supply enough nitrogen (N) to sustain satisfactory crop production without fertilizers. To ensure that the potential yield is reached each year, fertilizers are often applied in excessive quantities, causing N loss to the atmosphere and water. In agricultural systems, mineral N is mainly lost through ammonia volatilization, denitrification and leaching [1]. Correct dose and application timing of N fertilizer is important so that crops make best use of the N applied with minimum risk of losses and adverse environmental impacts.

Agronomic decisions in cereals are implemented by using a growth-stage key which provides a common reference for describing crop development. Management by growth stage is critical to optimize N fertilization strategies. The Zadoks Cereal Growth-Stage Key [2] is the most commonly used growth-stage key for cereals, in which the development of the cereal plant is divided into 10 distinct development phases covering 100 individual growth stages. Individual growth stages are denoted by the prefix GS (growth stage). The principle Zadoks growth stages used in relation to N management are the beginning of tillering (GS21) and beginning of stem elongation (GS30). GS21 is the stage when tillers (lateral shoots emerging at the base of the main stem of the plant) start emerging. Each tiller has the potential to produce a spike. GS30 is the stage in which the final spikelet can be observed within the stem of the main tiller.

The application of fertilizers consists of two methods, basal application and topdressing application. In the former, fertilizers are distributed over the field and mixed with soil before sowing. In the latter, fertilizers are applied in the soil surface. In cereal crops, local farmers commonly apply organic manures as basal fertilizers rotationally every two or three years. Organic manures are typically applied as a basal dressing before sowing due to the humid climate in winter and spring that usually hinders the entrance of machinery to fields. The application of organic fertilizers is generally combined with the application of mineral fertilizers. Conversely, when organics are not applied as initial fertilizers, mineral N is applied in two topdressing applications: 40 kg N ha^{-1} at GS21 to encourage tillering and a second and greater application at GS30.

Organic manure is a heterogeneous material collected from livestock raising facilities. The manure's physical status depends on farm practice and storage conditions. Farmyard manures are solids that are commonly mixed with lignocellulosic bedding materials. Slurries are liquids generated by mixing the solid manures with wastewater, washing water or urine. In recent decades, with the rise in livestock industry, the disposal of these farmyard manures and slurries has increased [3]. There is an excess of manure in many regions around the world, and the amount of manure available for land application is increasing [4]. The application of these manures into agricultural soil allows the recycling of their nutrient value for fertilizing crops and increasing soil fertility, making better use of resources and economic sense [5]. However, organic manures can pose a considerable environmental risk if they are not carefully treated and applied. It is important to point out that the N applied with the organic manures is generally less available to crops than N in mineral fertilizers [5]. In addition, it is difficult to predict the amount of available N from organic manures for crops since nutrient mineralization from manures is influenced by manure characteristics, soil and environmental temperature, soil moisture, soil properties and microbial activity [1]. The capacity of the soil itself must be taken into account in order to provide N to crops and to improve cereal N fertilizer recommendations. In this sense, soil organic matter (SOM) is one of the most relevant soil compounds, as it has been reported that SOM accounted for 78% of the variation in mineralizable N, whereas other soil properties only accounted for 8% [6]. N mineralization in SOM is enhanced by increasing temperature [7] and when soil moisture is near field capacity [8]. It must not be forgotten that the weather varies considerably from year to year, causing large differences in yield potential in the same site [9,10]; consequently, crop N fertilizer demand widely varies.

In Western Europe, a soil test for estimating available mineral N (ammonium plus nitrate) in the soil profile has been widely used to measure the quantity of N fertilizer which needs to be applied. The technique is called N_{min}-method. Generally, for winter wheat (as for many other crops), crop N need is calculated at the end of winter or early spring based on a target yield. The recommended rate of N fertilizer is calculated by the predicted N demand for the target yield minus the measured soil N_{min} value at the end of winter, where the rapid period of crop growth starts. Even if soil analysis should give reliable information, it is often perceived as imprecise for several reasons [11]. In fact, translating a few values of the soil samplings to a heterogeneous field makes the method imprecise. Apart from that, both the sampling and the determination of N_{min} require time-consuming procedures.

The plant itself is considered a relevant indicator of N availability from any origin (organic manure, soil N supply or mineral fertilizer) within the growing season. Strategies based on plant indicators have been developed for the adjustment of N fertilizer application during the growing season. Ravier et al. [11] showed some decision rules for determining N fertilizer application on the basis of the crop Nitrogen Nutrition Index (NNI). The NNI is calculated relative to the critical N concentration of the aerial parts of the crop defined as the minimal concentration required for the maximum production of aerial dry matter [12]. However, NNI determination requires destructive and time-consuming procedures of the plant N content and crop biomass which make it impractical for farmers.

Various types of optical sensing tools have been developed for assessing the N status of plants within the growing season [13,14]. However, these tools are unable to measure N content of the crop directly [15], therefore they are based on the measurement of compounds, such as chlorophyll [12]. Optical crop sensing is relatively easy to perform, and a range of sensing techniques and sensors are commercially available [13]. Regarding transmission or absorption sensors, hand-held chlorophyll meters such as Yara N-Tester^{TM} can provide rapid results for diagnostic purposes [14,16]. Chlorophyll meter readings have been widely proven to be well correlated with leaf chlorophyll and N concentrations in wheat; therefore, chlorophyll content can be used to diagnose the N status of plants [17–19], making them interesting tools for modulating the N rate. Furthermore, they can be used to decide whether a supplementary dose is to be applied in order to increase grain N content [8,20]. The use of chlorophyll meter measurements as an alternative to the NNI has been tested [21–23].

When considering reflectance sensors, ground-based active crop canopy reflectance sensors have been identified as potentially valuable tools for site-specific N management in cereals [24–26], because these sensors are not affected by clouds, unlike aerial or satellite sensing. Ground-based sensors have been developed to assess the plant nutritional status and guide variable-rate N application for different grain crops [15,23,27]. Spectral data collected by RapidScan CS-45 is converted into canopy green area measurements by calculating vegetation indices such as the Normalized Difference Vegetation Index (NDVI) or Normalized Difference Red Edge (NDRE). Marti et al. [28] found significant correlations between the NDVI, yield and biomass in wheat. Lu et al. [29] determined that RapidScan CS-45 was useful for non-destructively estimating the NNI of rice and Li et al., 2018 [30] showed a great potential for monitoring rice leaf N status. Similarly, Bonfil et al., 2017 [31] showed that the use of RapidScan CS-45 allows rapid and accurate crop monitoring and yield estimation. Zhang et al., 2019 [32] predicted grain yield based on RapidScan CS-45 measurements. In a previous study, we demonstrated that the normalized readings of the proximal sensing tools, Yara N-Tester^{TM} (chlorophyll meter) and RapidScan CS-45 (ground-based active-light proximal sensor), were good indicators of the N nutritional status of the plant, as the NNI or N_{min}. The mineral fertilizer rate applied at stem elongation (GS30, following Zadoks et al., 1994 [2]) could be modulated with the use of those tools [33].

In this study, we annex a third year when the weather conditions in spring after the second and greater N application at GS30 were very dry (it did not rain in the following 20 days after application) in comparison with the first two years. The climate of the area is Temperate–Mediterranean according to the temperature regime of the Papadakis classification and humid Mediterranean according to the water regime [34], with an average rainfall of 779 mm year^{-1} and an average annual temperature of 11.5 °C. This climate covers a great portion of the territory of Araba (The Basque Country, norther Spain) and, although it is classified as Mediterranean, maritime and temperate characteristics are very evident [35]. In this area, three of the ten years have dry periods around the second and greater application at GS30 (media 1978–2017; period 15 March–1 May), as in the third year of this study. Weather conditions are variable from year to year [33] and the mineralization of nutrients from SOM and farmyard manures is influenced by many factors such as manure characteristics, soil and air temperature, soil moisture, soil properties and microbial activity [8]. Therefore, making the prediction of the N needed by the crop in different situations is necessary but complicated. Yara N-TesterTM and RapidScan CS-45 have shown promise for making N recommendations [17,19,20,29,30,32] and provide a practical and affordable option for on-farm implementation. The aim of this study was to assess the usefulness of the proximal sensing tools, Yara N-Tester^{TM} and RapidScan CS-45, for adjusting the optimum N rate at GS30 when farmyard manures are applied before sowing in the variable humid Mediterranean climate conditions. We also aimed to decide whether the 40 kg N ha^{-1} rate at GS21 is necessary when organic manures are applied as basal fertilizers. In addition, the utility of a reference area on the field with non-limited N supply is questioned [14], making the diagnosis more complicated for farmers.

2. Materials and Methods

Three field trials were established in Arkaute (Araba, Basque Country, Spain) at NEIKER-Tecnalia facilities (42 °85′N, 2 °62′W; elevation 515 m above sea level) in three consecutive wheat-growing seasons: 2014–2015, 2015–2016 and 2016–2017 (defined as 2015, 2016 and 2017) in different fields under rainfed conditions. There was a 130-meter (m) distance between the three field trials. In the field trial carried out in 2015, a soil pit was made and after describing and analyzing its horizons, the soil was classified as Hypercalcic Kastanozem [36]. The mineralogical properties of the soil were analyzed by X-ray diffraction. The soil contained 40% clay (clay < 2 μm: Illite (%) = 30; Kaolinite (%) = 12; Smectite (%) = 58). In the field trials carried out in 2016 and 2017, several prospective holes were observed, verifying that the three soils had similar characteristics. The three fields were flat.

Representative soil samples were taken from each field trial to analyze the physical and chemical properties before wheat sowing from depths of 0–30 and 30–60 centimeter (cm). Soil texture was analyzed by the Pippete method [37] and classified (0–30 cm, sandy clay loam and 30–60cm, clay loam) [38]. Soils that had high pH values (1:2.5 soil:water using a pH-Meter CG840; 8.0–8.5), were calcareous (3.6–58% according to soil depth) [39] and had moderate organic matter content [40] in the upper layer (2–2.5%).

2.1. Experimental Setup and Treatments

In the area (Figure 1) where this study was carried out (Araba, Basque Country, norther Spain), the application of organic fertilizers is generally combined with the application of mineral fertilizers. Farmyard manures are typically applied as a basal dressing before sowing. For mineral fertilizers, an application is made at GS21 [2] and the last and greater mineral N dressing application is made at GS30 [2].

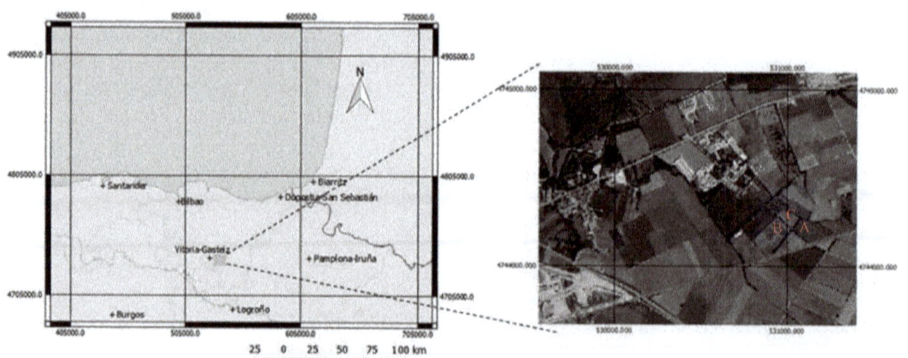

Figure 1. Location of the study (Arkaute, Araba) in the Basque Country, northern Spain. Letters "A", "B" and "C" represent the field trials in the 2015, 2016 and 2017 wheat-growing seasons, respectively.

Three kinds of initial fertilization were applied: dairy slurry (40 tons per hectare t ha^{-1}), sheep farmyard manure (40 t ha^{-1}) and conventional treatment (no organic fertilizer basal dressing and 40 kg N ha^{-1} at tillering). These three types of fertilization were combined with five N rates (calcium ammonium nitrate, NAC 27%) in the topdressings applied at GS30 (0, 40, 80, 120 and 160 kg N ha^{-1}). Apart from the treatments, two controls were established: a control without N fertilization (0N) and an overfertilized control plot (80 kg N ha^{-1}; 80 kg N ha^{-1} applied at tillering and 200 kg N ha^{-1} applied at GS30; Table 1). The experiment was a factorial randomized complete block design with three factors (year, initial fertilization and N rate at GS30) and four replicates. The area of each plot was 4 m in width and 8 m in length.

Table 1. N application rates and timing for three initial fertilization treatments for field trial (2015, 2016 and 2017) and control (0N) and overfertilized (280N) plots. GS21 is the beginning of tillering (end of winter [2]); GS30 is the beginning of stem elongation [2].

Initial Fertilization	2015		2016		2017		2015–2016–2017		Treatment Identification
	Total N[a] (kg ha^{-1})	N–NH_4^{+b} (kg ha^{-1})	Total N[a] (kg ha^{-1})	N–NH_4^{+b} (kg ha^{-1})	Total N[a] (kg ha^{-1})	N–NH_4^{+b} (kg ha^{-1})	Topdressing at GS21 (kg N ha^{-1})	Topdressing at GS30 (kg N ha^{-1})	
Conventional [–]	–	–	–	–	–	–	40	0	40N + 0N
								40	40N + 40N
								80	40N + 80N
								120	40N + 120N
								160	40N + 160N
Dairy Slurry (DS) [40 t ha^{-1}]	192	104	144	80	120	68	–	0	DS + 0N
								40	DS + 4N
								80	DS + 80 N
								120	DS + 120N
								160	DS + 160N
Sheep manure (SM) [40 t ha^{-1}]	336	0	592	200	448	–	–	0	SM + 0N
								40	SM + 40N
								80	SM + 80N
								120	SM + 120N
								160	SM + 120N
Control	–	–	–	–			–	–	0N
Overfertilized	–	–	–	–			80	200	280N

Organic amendments were applied on 13 November 2014, 4 November 2015 and 17 November 2016. Slurry and manure were sampled and analyzed for total N and $NH_4{}^+$–N (Table 1). We decided to apply manure at 40 t ha^{-1} and slurry at 40 t ha^{-1} because this is the usual rate at which organic amendments are applied as initial fertilizers in Araba, Basque Country, Spain.

Wheat (*Triticum aestivum* var. Cezanne) was sown on 24 November 2014, 06 November 2015 and 18 November 2016, and was harvested on 21 July 2015, 2 August 2016 and 2 August 2017. The sowing rate was 220 kg seed ha^{-1}. The preceding crops were flax (*Linum usitatissimum*), rapeseed (*Brassica napus*) and wheat (*Triticum aestivum*), respectively.

2.2. Mineral N Samples (N_{min}) and Biomass Samples for the Nitrogen Nutrition Index (NNI)

Three soil samples from replicates I, II and III (there were four replicates per treatment) and three rows of aboveground biomass of wheat of one meter in length (four replications) were taken for each kind of initial fertilization treatment (dairy slurry, sheep farmyard manure and conventional treatment), in 40N + 0N, DS + 0N and SM + 0N treatments (Table 1) at 0–30 at two times: (1) At GS21 (09 March 2015, 19 January 2016 and 02 March 2017) just before mineral N fertilization in the conventional treatment; (2) at GS30 (04 April 2015, 17 March 2016 and 06 April 2017) just before mineral N fertilization. The soil samples were analyzed for soil mineral nitrogen ($NH_4{}^+$–N plus $NO_3{}^-$–N) by spectrophotometry [41,42].

Fresh biomass samples were weighed and oven dried, and the dried biomass samples were again weighted for dry matter content determination. Biomass was estimated and N concentration was determined by Kjeldahl's method [43] to calculate the NNI (1) following Lemaire et al. [17]:

$$NNI = Na/Nc \tag{1}$$

The NNI was calculated as a ratio of plant N concentration (Na) and the critical N concentration (Nc) in aerial biomass (shoots). Nc (2) is defined as the minimum concentration of N necessary to achieve maximum aerial biomass at any stage of vegetative growth [44]. For the wheat critical N concentration, dilution curves were developed for winter wheat in France [43] as a universal relationship. In 2012, another was built in China [45] with the aim of adjusting the original to the conditions in China. Since we did not have sufficient data to develop a specific function for our region, we used the original dilution curve developed by Justes et al. [46] due to the proximity to France.

$$Nc = 5.35 * DM^{-0.442} \tag{2}$$

where DM is the amount of dry matter accumulated in the aerial biomass expressed in t ha^{-1}. Nc is expressed in % DM.

2.3. Proximal Sensing Tools for Adjusting the Optimum N Rate at GS30

The proximal sensing tools for the diagnosis of the N nutritional status tested were Yara N-Tester^{TM} (Yara International ASA, Oslo, Norway) and RapidScan CS-45 (Holland Scientific, Lincoln, NE, USA). Yara N-Tester^{TM} is a clip-on hand-held chlorophyll meter which measures light transmitted by the plant leaf at two different wavelengths, 650 (red light) and 940 nm (near infrared light, NIR). The ratio of the light transmitted at these wavelengths, in addition to the ratio determined with no sample, is processed by the instrument to produce a digital reading. The measurement point should be in the middle of the blade of the youngest, fully developed leaf. The values obtained are unitless and they express relative chlorophyll content. Thirty random measurements are recorded to get the representative value in each sampling point. RapidScan CS-45 is a portable entirely self-contained ground-based active crop canopy sensor that integrates a data logger, graphical display, GPS, active crop sensor and power source into a small and compact instrument. It measures crop reflectance at 670, 730 and 780 nm and provides the NDVI and NDRE. The measurements with RapidScan CS-45 were taken as the sensor was passed over the crop surface at approximately 1 m at constant walking speed. The sensor´s unit was handheld and two rows per elemental plot were scanned. NDVI and NDRE values were averaged to generate a value for that plot.

Measurements with the proximal sensing tools were taken in four replications in each kind of initial fertilization treatment (dairy slurry, sheep farmyard manure and conventional treatment) just before applying the topdressing at GS30. Measurements were taken in 40N + 0N, DS + 0N and SM + 0N treatments (Table 1). In addition, samples were taken in the control (0N) and overfertilized treatments (280N). We will refer to the Yara N-Tester^{TM} absolute values as abs_NTester and the RapidScan CS-45 absolute values as abs_NDVI and abs_NDRE. The measurements with Yara N-Tester^{TM} and RapidScan CS-45 were taken as described in Aranguren et al. [33].

Normalized values for the Yara N-Tester^{TM} and RapidScan CS-45 measures were calculated to avoid the noise encountered by variables other than N fertilizer. These values were calculated as a percentage by assigning the 100% value to the overfertilized treatment (280N) described previously, similar to the technique suggested by Follett and Follett [47]. We will refer to the Yara N-Tester^{TM} normalized values as nor_NTester and the RapidScan CS-45 normalized values as nor_NDVI and nor_NDRE.

2.4. Grain Yield

Yields were recorded at crop maturity using a plot harvester (1.5 × 8 m; Wintersteiger AG, Ried, Austria). For comparisons between fields, yields were converted to a 12% dry matter basis.

To show grain yield response to N fertilization at GS30, the grain yield was determined by fitting a Quadratic Plateau Function. It has been shown to best describe the yield response to N fertilization in humid Mediterranean climate conditions of Araba (Basque Country, northern Spain) after comparison with other models (quadratic and square root) as was shown by Ortuzar-Iragorri et al. [48], who selected those models because other authors reported that they were frequently used for studying the relationship between yield and N fertilization. Ortuzar-Iragorri et al. [48] used the NLIN procedure (SAS Institute 1998 [49]) to adjust the data to the proposed models to obtain statistical parameters. The most important attribute of the function is where yield becomes relatively insensitive to increases in N fertilizer addition at GS30. The optimum N rate was determined following Ortuzar-Iragorri et al. [48]. In the neighborhood region of Navarra (northern Spain), with similar climatic conditions, Arregui et al. [9] also used a Quadratic Plateau Function for wheat yield response to N fertilization. In the case of corn, a Quadratic Plateau Function was also shown to be the best model for yield response to N fertilization [49–52]. A Quadratic Plateau Function (3) was used to indicate the optimum N rate at GS30 to achieve the maximum yield (yield vs. N fertilizer rate at GS30):

$$y = a + bN + cN^2 \tag{3}$$

where y is the dependent variable (yield, kg ha^{-1}), N is the N rate applied at GS30 (kg N ha^{-1}) and a, b and c are coefficients.

The equations obtained with the Quadratic Plateau Function were used to determine the economically optimal dose of N (4), following the technique suggested by Aizpurua et al. [53]. According to their findings, the revenues obtained can be calculated as:

$$Revenue\ yield = w\,Y - fN \tag{4}$$

where w is the wheat price (€ kg^{-1}), f is the fertilizer price (€ kg^{-1}), N is the nitrogen rate (kg ha^{-1}) and Y is the quadratic plateau function.

When the revenue yield is derived with respect to the N rate and equals to zero, maximum revenues would be obtained, and that would be the economical optimum rate based on yield. The wheat grain price used was 0.18 € kg^{-1}, and the fertilizer price was 1.19 € kg N [54].

2.5. Statistical Analysis

The Quadratic Plateau Function was used for the wheat yield response. A nonlinear regression procedure was carried out using *R 3.2.5'* software [55] to plot curves that best described the yield response to N fertilizer application.

Both soil mineral nitrogen (N_{min}) and NNI measurements were conducted before applying the N fertilization rates at GS21 and at GS30. The proximal sensing tool (Yara N-Tester^{TM} and RapidScan CS-45) measurements (abs_NTester, abs_NDVI and abs_NDRE) were conducted before applying the N fertilization rate at GS30. Overfertilized plots were used to normalize absolute values (nor_NTester, nor_NDVI and nor_NDRE). The factors analyzed for statistical analysis in all cases were growing season and initial fertilization by analyses of variance (ANOVA) using '*R 3.2.5'* software [55]. There was a significant interaction between growing season and initial fertilization for all these measurements. Therefore, an ANOVA was performed to analyze differences among initial fertilization treatments in each growing season. Another ANOVA was performed to analyze the differences among the wheat-growing seasons in each initial treatment. To separate the means, the Tukey test was used ($p \leq 0.05$), utilizing the *R package agricolae* [56].

3. Results

3.1. Grain Yield

Optimum N Rate at GS30

The optimum N rate at GS30 was different for each kind of initial fertilizer (conventional treatment, slurry or sheep manure) in each wheat-growing season. In 2015 (Figure 2a), the optimum N rate at GS30 was 98 kg N ha^{-1} in the conventional treatment (plus 40 kg N ha^{-1} at tillering). In the organic treatments, the optimum N rate at GS30 was approximately 118 kg N ha^{-1}. Maximum wheat grain yields were 8456, 8240 and 8356 kg ha^{-1} for conventional treatments, slurry and manure, respectively.

In 2016 (Figure 2b), the optimum N rate at GS30 in the conventional treatment was 109 kg N ha^{-1} (plus 40 kg N ha^{-1} at tillering). In the organic treatments, the optimum N rate in the slurry treatment was 98 kg N ha^{-1} and in the manure treatment was 147 kg N ha^{-1}. Maximum wheat grain yields were 10,227, 10,271 and 10,723 kg ha^{-1} for the conventional, slurry and manure treatments, respectively.

In 2017 (Figure 2c), the optimum N rate at GS30 in the slurry treatment was 128 kg N ha^{-1} and in the manure treatment, it was 156 kg N ha^{-1}. In the conventional treatment, the maximum yield was not achieved. Therefore, the maximum rate applied at GS30 (160 kg N ha^{-1}) was taken as the optimum N rate at GS30. Maximum wheat grain yields were 5841 and 6205 kg ha^{-1} for slurry and manure treatments, respectively.

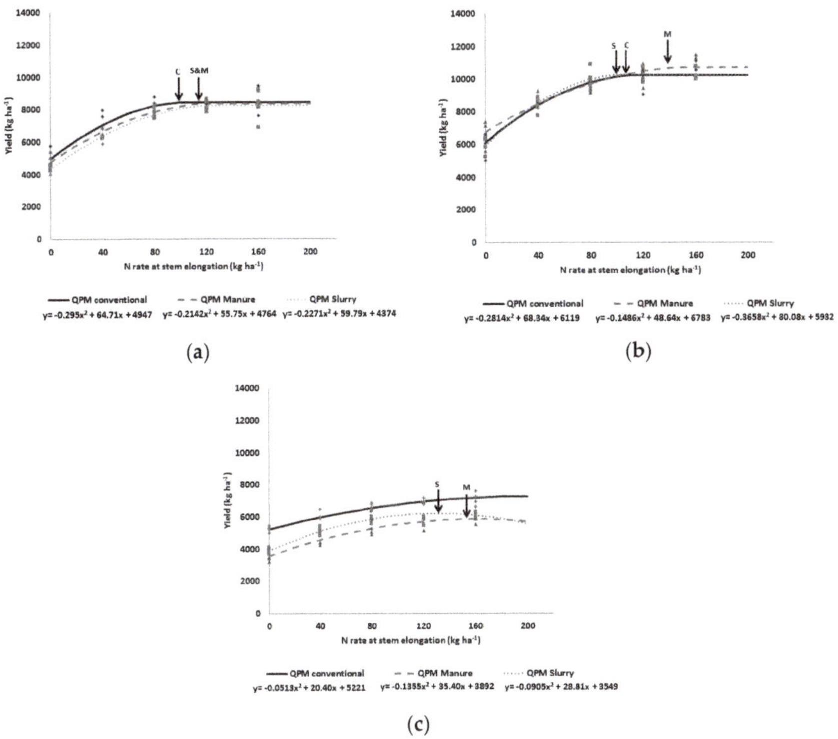

Figure 2. Effect of the N fertilization rate at stem elongation (GS30) on yield (kg ha^{-1}) in 2015 (**a**), 2016 (**b**) and 2017 (**c**); wheat-growing seasons with respect to initial fertilization: conventional treatment, dairy slurry, sheep manure. A quadratic plateau model was used to study the yield response. The arrow marked with C represents the N fertilization rate when the maximum yield was achieved by conventional treatment. S represents the N fertilization rate when the maximum yield was achieved by dairy slurry treatment, and M represents the N fertilization rate when the maximum yield was achieved by sheep manure treatment.

3.2. Economically Optimal Dose

In 2015 and 2016, the economically optimal dose at GS30 was almost the same as the optimum N rate at GS30. In 2015, the economically optimal dose was 98 kg N ha^{-1} for conventional treatment and 117 kg N ha^{-1} for organic amendments. In 2016, the economically optimal dose was 110, 100 and 141 kg N ha^{-1} for conventional, slurry and manure initial treatments, respectively. However, in 2017 the economically optimal dose at GS30 was lower than the optimum N rate at GS30 (131, 111 and 130 kg N ha^{-1} for conventional, slurry and manure initial treatments, respectively).

3.3. Soil Mineral Nitrogen (N_{min}) and Total Rainfall

Before sowing, soil N_{min} in 2015 was 50 kg N ha^{-1}; in 2016, it was 42 kg N ha^{-1} and in 2017, it was 34 kg N ha^{-1} (Table 2). At GS21, N_{min} values (Table 2) were lower in 2015 than in 2016 and 2017 in organic fertilization treatments. During the 2016 and 2017 wheat-growing seasons, the lower rainfall (Figure 3) allowed higher N_{min} values (around 30 kg N ha^{-1}) at GS21 (Table 2). In 2015 at GS21, conventional treatment had significantly higher N_{min} values than dairy slurry and sheep manure (Table 2). At GS30, 0N and conventional treatment presented higher values in 2017 than in 2015 and in 2016. N_{min} values in 2016 were extremely low in all cases (Table 2). No differences among treatments were detected at GS30 in any growing cycle.

Table 2. Soil N_{min} content (kg N ha^{-1}; 0–30 cm) at the beginning of the three wheat-growing seasons (2015, 2016 and 2017), beginning of tillering (GS21) and beginning of stem elongation (GS30) in Arkaute.

Growing Season	Treatments	Nmin (0–30cm; kg N ha^{-1})				
		Initial	GS21		GS30	
		Mean Values	Mean Values	sd	Mean Values	sd
2015	0N	50	22 A	5	12 b	5
	40 + 0N		22 A	5	13 b	4
	DS + 0N		4 B b	1	13	9
	SM + 0N		9 B b	3	12	5
2016	0N	42	30	9	1 c	1
	40 + 0N		30	9	3 b	1
	DS + 0N		32 a	4	4	0
	SM + 0N		30 a	2	1	4
2017	0N	34	33	12	26 a	3
	40 + 0N		33	12	32 a	13
	DS + 0N		36 a	11	16	3
	SM + 0N		16 ab	12	14	11

Means followed by a different capital letter indicate significant differences among initial treatments for each year (Tukey, $p \leq 0.05$). Means followed by a different lower-case letter indicate differences among each initial treatment for different years (Tukey, $p \leq 0.05$). Both 0N and 40 + 0N have the same value at GS21 because the measurement was taken before topdressing application in 40 + 0N at GS21.

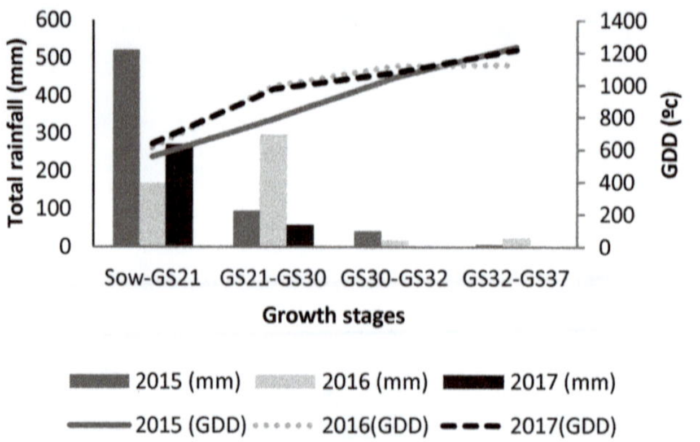

Figure 3. Total rainfall (mm) and growing degree days (GDD, °C) between wheat growth stages [2] in three wheat-growing seasons (2015, 2016 and 2017) in Arkaute. Sow, sowing.

3.4. NNI

NNI values at GS21 (Table 3) in 2016 were significantly higher than in 2015 and 2017, and dairy slurry and sheep manure treatments had significantly higher values than the treatments without organics as initial fertilizer. However, there were no differences in 2015 and 2017 among initial fertilization treatments.

At GS30 (Table 3), in treatments with organics as initial fertilizer and in 0N, NNI values were significantly higher in 2016 than in 2015 and 2017. However, conventional treatment presented higher values in 2016 and 2017 than in 2015. There were significant differences in 2016 and 2017 among initial fertilization treatments. In 2017, the overfertilized plot presented the highest NNI followed by the conventional treatment and 0N and organic treatments. There were no differences between 0N and organic treatments. In 2016, the overfertilized plot presented the highest NNI and there were no differences among the rest of the treatments. In 2015, there were no differences among initial fertilization treatments.

Table 3. Nitrogen Nutrition Index (NNI) at the beginning of tillering (GS21) and at stem elongation (GS30) for three initial fertilization treatments, as well as control (0N) and overfertilized (280), in three field trials (2015, 2016 and 2017) in Arkaute.

Growing Season	Treatments	NNI			
		GS21		GS30	
		Mean Values	sd	Mean Values	sd
2015	OverFert	0.37 b	0.02	ND	ND
	0N	0.37 b	0.02	0.26 b	0.03
	Conventional	0.37 b	0.02	0.35 b	0.09
	Dairy Slurry	0.34 b	0.05	0.37	0.08
	Sheep manure	0.38 b	0.06	0.29 b	0.02
2016	OverFert	0.60 B a	0.08	0.77 A a	0.10
	0N	0.60 B a	0.08	0.42 B a	0.04
	Conventional	0.60 B a	0.08	0.55 B a	0.01
	Dairy Slurry	0.67 A a	0.03	0.51 B	0.03
	Sheep manure	0.68 A a	0.05	0.51 B a	0.07
2017	OverFert	0.23 c	0.02	0.80 A	0.05
	0N	0.23 c	0.02	0.33 C b	0.03
	Conventional	0.23 c	0.02	0.53 B a	0.05
	Dairy Slurry	0.26 c	0.02	0.38 C	0.04
	Sheep manure	0.23 c	0.04	0.37 C b	0.05

Means followed by a different capital letter indicate significant differences among initial treatments for each year (Tukey, $p < 0.05$). Means followed by a different lower-case letter indicate differences among each initial treatment for different years (Tukey, $p < 0.05$). ND means no data. sd means standard deviation. 0N, 40 + 0N and OverFert treatments have the same value at GS21 because the measurement was taken before topdressing application in 40 + 0N and OverFert at GS21.

3.5. Proximal Sensing Tools

3.5.1. Absolute Values

The absolute values of RapidScan CS-45 (abs_NDVI and abs_NDRE) at GS30 were significantly higher in 2016 than in 2015 and 2017 in all treatments (Table 4). The absolute values of Yara N-Tester^{TM} (abs_NTester) were significantly higher in 2016 than in 2015 (Table 4) in dairy slurry and sheep manure treatments. In the conventional treatment, there were no significant differences among years (Table 4). Finding values to adjust the optimum N rate with absolute values was complicated because variability among years was high.

Means with different lower-case letters (a, b) represent significant differences among initial treatments in Yara N-TesterTM (abs_NTester) measurements for each year (Tukey, $p \leq 0.05$); means with different capital letters in italics (*A*, *B*) represent significant differences among initial treatments in NDVI RapidScan CS-45 measurements (abs_NDVI) for each year (Tukey, $p \leq 0.05$); means with different underlined capital letters (A, B) represent significant differences among initial treatments in NDRE RapidScan CS-45 measurements (abs_NDRE) for each year (Tukey, $p \leq 0.05$); means with different lower-case letters (y, z) represent significant differences among the three years in Yara N-TesterTM measurements (abs_NTester) for each treatment (Tukey, $p \leq 0.05$); means with different capital letters in italics (*Y*, *Z*) represent significant differences among the three years for NDVI RapidScan CS-45 measurements (abs_NDVI) in each treatment (Tukey, $p \leq 0.05$) and means with different underlined capital letters (Y, Z) represent significant differences among the three years for NDRE RapidScan CS-45 measurements (abs_NDRE) in each treatment (Tukey, $p \leq 0.05$).

Table 4. Absolute values obtained with tools for the diagnosis of the N nutritional status (Yara N-Tester^{TM} and RapidScan CS-45) at stem elongation (GS30) for three initial fertilization treatments and control (0 N) in three wheat-growing seasons (2015, 2016 and 2017) in Arkaute.

Growing Season	Treatments	abs_Ntester		RapidScan CS-45			
				abs_NDVI		abs_NDRE	
		Mean Values	sd	Mean Values	sd	Mean Values	sd
2015	0N	304b z	34	0.35 B Z	0.04	0.12 B Z	0.01
	40N	460 a	22	0.54 A Z	0.04	0.20 A Z	0.02
	DS	334 b z	16	0.39 B Z	0.04	0.14 B Z	0.02
	SM	332 b z	10	0.39 B Z	0.05	0.13 B Z	0.02
2016	0N	403 c y	8	0.55 B Y	0.05	0.18 B Y	0.02
	40N	477 a	14	0.67 AY	0.02	0.24 A Y	0.15
	DS	438 b y	15	0.65 AY	0.04	0.23 A Y	0.02
	SM	442 b y	28	0.66 AY	0.03	0.23 A Y	0.02
2017	0N	377 b z	13	0.39 C Y	0.06	0.12 B	0.01
	40N	507 a	54	0.59 A Z	0.03	0.19 A	0.01
	DS	382 b z	53	0.43 B Z	0.04	0.13 B	0.01
	SM	389 b z	12	0.41 C Z	0.04	0.13 B	0.01

3.5.2. Normalized Values

The normalized values of RapidScan CS-45 and Yara N-Tester^{TM} (as a percentage compared to the overfertilized (280 N) treatment) were significantly higher in 2016 (Figure 4b) than in 2015 and 2017 (Figure 4a,c) in all treatments with the exception of nor_NTester values in the conventional treatment. In 2015 (Figure 4a), the normalized measurements of both tools agreed with each other for all treatments, showing differences between the conventional treatment (88%), initial organic fertilization treatments (65%) and 0N (57%). In 2016 (Figure 4b), Yara N-Tester^{TM} showed differences between the conventional treatment (92%), control (78%) and both organic treatments (85%). RapidScan CS-45 showed differences between the conventional treatment (88%) and 0N (72%), but the tool did not detect differences between the conventional treatment and organic treatment (84–87%). In 2017 (Figure 4c), the measurements of both proximal tools detected differences between the conventional treatment (83–87%) and the remaining treatments (control and organics). In 2017, only nor_NDRE detected differences between control and organic treatments.

It was possible to find a relationship between normalized values of RapidScan CS-45 and Yara N-Tester^{TM} and optimum N rate at GS30 to achieve the maximum yield in the conventional treatment and in the slurry treatment (Table 5). When both proximal sensing tool values were 60–65%, the optimum N rate at GS30 to achieve the maximum yield was 118–128 kg N ha^{-1} in dairy slurry treatment (Table 5). When the readings were 85–90%, the optimum N rate was 100–110 kg N ha^{-1} in dairy slurry and conventional treatment (Table 5). In the sheep manure treatment, there was no clear relationship between sensor values and the optimum N rate. When values were 60–65%, the N recommendation was 117 or 155 kg N ha^{-1} and 147 kg N ha^{-1} for readings around 89%.

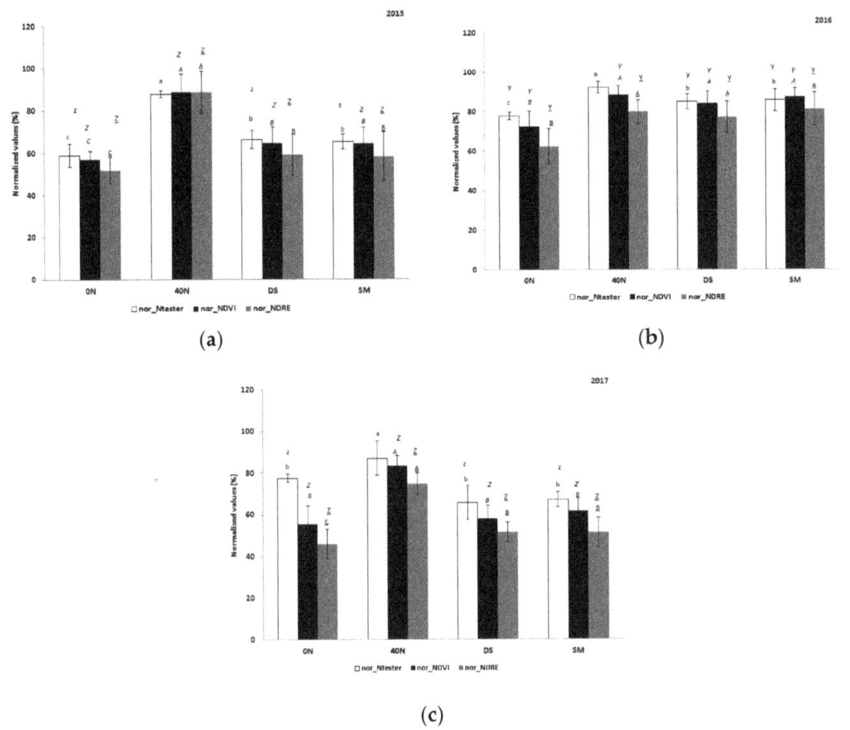

Figure 4. Relation between initial fertilization (conventional treatment (40N), dairy slurry (DS) and sheep manure (SM)) and control (0N) and values obtained with tools for the diagnosis of the N nutritional status (RapidScan CS-45, Yara N-Tester^{TM}) at stem elongation (GS30) in the 2015 (**a**), 2016 (**b**) and 2017 (**c**) wheat-growing seasons. Values were normalized assigning the 100% value to the overfertilized (280N) plot. Means with different lower-case letters (a, b) represent significant differences among initial treatments in Yara N-Tester^{TM} (nor_NTester) measurements for each year (Tukey, $p \leq 0.05$); means with different capital letters in italics (*A*, *B*) represent significant differences among initial treatments in NDVI RapidScan CS-45 measurements (nor_NDVI) for each year (Tukey, $p \leq 0.05$); means with different underlined capital letters (A, B) represent significant differences among initial treatments in NDRE RapidScan CS-45 measurements (nor_NDRE) for each year (Tukey, $p \leq 0.05$); means with different lower-case letters (y, z) represent significant differences among the three years in Yara N-Tester^{TM} measurements (nor_NTester) for each treatment (Tukey, $p \leq 0.05$); means with different capital letters in italics (*Y*, *Z*) represent significant differences among the three years for NDVI RapidScan CS-45 measurements (nor_NDVI) in each treatment (Tukey, $p \leq 0.05$) and means with different underlined capital letters (Y, Z) represent significant differences among the three years for NDRE RapidScan CS-45 measurements (nor_NDRE) in each treatment (Tukey, $p \leq 0.05$).

Table 5. Normalized values (nor_Ntester, nor_NDVI, nor_NDRE) obtained with tools for the diagnosis of the N nutritional status (Yara N-Tester and RapidScan CS-45) at stem elongation (GS30) and their corresponding optimal N application at GS30 for dairy slurry and conventional treatments.

Initial Fertilization	Proximal Tool Readings at GS30 (%)	Optimal N Application at GS30 (kg N ha^{-1})
Dairy Slurry	60–65	118–128
Dairy Slurry/Conventional	85–90	100–110

4. Discussion

Studies focused on the use of proximal sensing tools for adjusting topdressing N in wheat have been conducted in other climates. In this region with variable rainfall, it was necessary to study the usefulness of proximal sensing tools for an optimum N mineral rate application when organic manures had been applied as basal fertilizers.

4.1. Mineral N Fertilizer Reduction When Organic Fertilizer Was Applied before Sowing

In this study, maximum wheat grain yields were different in each growing season with 2016 being the year with the highest yields. Within each year for 2015 and 2016, maximum yields between the conventional treatment and organics as initial fertilizers treatment were comparable (Figure 2a,b). However, in 2017, there were differences among initial fertilization treatments, with the conventional treatment being 1300–1700 kg ha^{-1} more productive than the organic treatments depending on each individual N rate at GS30 (Figure 2c). In this regard, the optimum N rate at GS30 was different for each growing season and for each initial fertilization treatment. In fact, the application of organic manures as initial fertilizers reduced the mineral N rate used at GS30 in 2015 and 2016. In 2015, when organic manures were used as initial fertilizers, approximately 20 kg N mineral ha^{-1} less than in the conventional treatment was necessary to achieve the maximum wheat grain yield. In 2016, when applying slurry as an initial fertilizer, 51 kg N mineral ha^{-1} less than in the conventional treatment was necessary to achieve the maximum wheat grain yield. In 2016, when using manure as initial fertilizer, the same mineral N dose as in the conventional treatment was necessary to achieve the maximum wheat grain yield. However, in the manure treatment, the maximum yield was 500 kg ha^{-1} higher than in the conventional treatment. According to results shown in 2015 and 2016, the 40 kg N ha^{-1} application at the GS21 could be avoided when organic manures were applied even when N_{min} at GS21 was very low (4–9 kg N ha^{-1}) as in 2015 (Table 2). In 2017, following mineral N application at GS30, it did not rain until leaf flag emergence (GS37 [2]; Figure 3). In 2017, yields were higher in the conventional treatment where 40 kg N ha^{-1} had been applied at GS21 even if N_{min} values at GS21 were high (Table 2).

Both economic and environmental costs are key considerations when N fertilization management strategies are being developed [53]. The economically optimum N rate at GS30 was similar to the optimum N rate in 2015 and 2016. However, in 2017, when mineral fertilizer price and wheat prices are considered, the N rate is lower than the optimum N rate. The low precipitation after the mineral N application at GS30 did not allow N absorption by the wheat crop until one month later, negatively affecting N use efficiency, hence reducing the economic N rate and the economic benefit of the fertilizer.

4.2. Soil N Availability

As the N_{min} values showed, the soil N supply in soils amended with animal manure may be very variable (Table 2). The recommended rate of N fertilizer is calculated by the predicted N demand for the target yield minus the measured N_{min} at the end of winter. However, some factors make this technique imprecise. On the one hand, using a target yield is not feasible because, each growing season, yields vary depending on the weather (Figures 2 and 3). On the other hand, soil N_{min} at GS21 depends not only on the mineralization (SOM, applied organic compounds and weather conditions) but also on crop uptake. Although N_{min} values at GS21 in 2016 and 2017 were similar, yields achieved in 2016 (10,200–10,700) were higher than yields in 2017 (5800–6200; Figure 2). As the results showed (Tables 2 and 3), it is possible to have high N availability in soil and plants with low N nutritional status (as in 2017), as well as low N availability in soil and acceptable N nutritional statuses in plants (as in 2016).

The weather has a significant effect on yield. Since the weather can be difficult to predict in the long term, predicting the mineralization of organic forms of N and other nutrients into plant available mineral forms and the wheat demand remains a challenge [57]. Many factors control the decomposition of SOM and the mineralization of nutrients, rendering the prediction difficult [58].

Arregui and Quemada [59], in similar climate conditions (humid Mediterranean), showed that due to the high rainfall, low evapotranspiration and low crop demand, from sowing to mid-tillering (GS25), most of the mineral N present in the soil before sowing was lost by nitrate leaching, as could have happened in 2015. The lower N_{min} values in the treatments where organics were applied in 2015 could be explained by immobilization [57]. Basso et al. [10] showed that wheat yield production in the Mediterranean environment is highly affected by spring rainfall and the amount of soil water stored in soil before and during the growing season. In the area where the study was carried out (humid Mediterranean) [34], three of the ten years have dry periods (media 1978–2017; period 15 March–1 May) when N fertilizer topdressing is applied at GS30 at the stage of highest N uptake by the wheat crop. It should be mentioned that applying organics as initial fertilizers will probably not synchronize with the highest N demand by the crop because a long period of time passed between application and plant N uptake. However, the wet conditions in winter in Araba hinder the entrance of machinery to fields in spring. The unique alternative being their application in autumn, likely causing N loss by leaching.

Soil characteristics also have an effect on N mineralization. In our study, the soil presented 40% clay (58% Smectite). Chantigny et al. [60] suggested that clay fixation may have a negative effect on N availability during the crop growing season. Other studies also demonstrated that there is a significant interaction between manure and soil regarding the net mineralization and that the net N mineralization of cattle manures, cattle slurry and plant recovery is lower in clay soils is associated with the clay fixation of NH_4^+–N [61].

It is also remarkable that, in 2016, the preceding crop was rapeseed (*Brassica napus*). It has been shown that in the humid Mediterranean region of Spain, including rapeseed in the crop rotation increases wheat yields by about 10% [62]. Gallejones et al. [63] concluded that rapeseed as the preceding crop for wheat probably caused higher N mineralization.

Many factors affect soil N mineralization and soil N availability, making it difficult to estimate the amount of N mineralized from SOM to adjust the rate of N fertilizer required to optimize crop yield and also quality, even if soil N_{min} is known at the end of winter. Otherwise, fields are heterogeneous, and it implies laborious and expensive sampling and analysis [64]. Ravier et al. [11] highlighted many sources of uncertainty in soil N analysis (lack of standardized procedure for soil sampling or the choice of the sampling zone and extrapolation) that led to the exclusion of decision rules based on the monitoring of soil mineral content. Recommendations for the start of the crop cycle are usually based on soil parameters, and the recommendations for later stages in the crop cycle are based on plant indicators. Plants integrate soil variability, climate, crop management and other environmental influences, which are good indicators of the nutrient needs.

4.3. Nitrogen Nutrition Index (NNI)

The topdressing N fertilizer recommendation rate can be adjusted according to the NNI, which can indicate the magnitude of the N surplus or deficiency [17]. In our experiment, all treatments were under N deficiency (NNI < 0.9), even the overfertilized ones. However, the NNI was able to detect differences among treatments in the same growing season and differences among the wheat-growing seasons for the same treatment. Different thresholds of the NNI have been proposed in on-farm applications to diagnose the N status of plants [12,65–67]. Ravier et al. [23] determined the NNI threshold that the wheat crop can tolerate in different growing stages without reducing the maximum yields. Thus, they consider NNI values <0.4 as situations that should be avoided, and when the NNI is between 0.4 and 0.7 before ear has reached 1 cm in length (before GS30, following Zadoks et al., 1994 [2]), they recommend the application of 40 kg N ha^{-1} at that stage. In our case, in 2016 at GS21, NNI values were 0.60–0.68. In 2016, yields were comparable between the conventional treatment (where 40 kg N ha^{-1} was applied at GS21) and treatments where organics were applied as initial fertilizers without N application at GS21. This fact suggests that increasing the N rate at GS30 can reverse the N deficiency at GS21 and organics could match the conventional treatment (where 40 kg N ha^{-1} was applied at GS21). In 2015 at GS21, NNI values were lower than 0.4 (0.34–0.38), which means that the crop had a severe deficiency and that the yields will be reduced. In 2015, yields were lower than in 2016 but as in 2016, yields between the conventional treatment (where 40 kg N ha^{-1} was applied at GS21) and organics treatment were comparable. In 2017, NNI values were 0.23–0.26 at GS21 and it did not rain after N application at GS30 until GS37, causing low yields. In 2017, the application of 40 kg N ha^{-1} at GS21 in the conventional treatment had a significant effect on the yield, allowing higher yields in the conventional treatment than in the organic treatment. Those events in 2015 and 2016 suggested that the tolerable N deficiency in GS21 could be lower (0.3) than the one identified by Ravier et al. [23] and that crop yield can be restored with the topdressing N application at GS30 when the N deficiency is not very severe, as in 2017. However, they also concluded that the extent of tolerable N deficiencies in the early stages is less clear than in more advanced stages. In fact, a low NNI occurring during tillering can lead to a low number of stems, but a higher NNI during GS30 can lead to an increase in the number of grains per ear [17]. It is important to highlight that tolerating N deficiency gives the crop the opportunity to absorb the available soil mineral N before applying fertilizer [11]. The determination of the NNI requires representative samples to be taken, the determination of dry matter content and the determination of N concentration by laboratory analytics, making the determination of the NNI difficult and time-consuming. Moreover, as with soil sampling, farmers cannot instantly know the N status of plants.

4.4. Proximal Sensing Tools and Vegetation Indices

Abs_NTester readings increased as NNI values increased. For NNI = 1 values, Ziadi et al. [68] determined an absolute chlorophyll meter reading with Minolta SPAD-502 of 42.5 at GS30 in wheat. Similar absolute chlorophyll meter readings (ranging from 39 to 45, depending on the site-year) for NNI = 1 in wheat were also reported by Peltonen et al. [69] and Bundy and Andraski [70]. The crop sensor performance for N status diagnosis is influenced by many other variables apart from N: seasonal variation, plant water status, diseases and pests, plant growth stage, genotype, etc. [31]. Normalized readings have been recommended to reduce the year effect [21,23] and the noise promoted by other variables other than N fertilization [18].

In normalized values, chlorophyll meter readings have been correlated with the NNI [21,22]. Ziadi et al. [68] established the critical relativized chlorophyll meter readings ranging from 0.89 to 0.95 for NNI = 1. However, the normalization of the chlorophyll meter readings does not entirely remove the year effect [71], but these readings better predict the NNI [68]. Prost and Jeuffroy [21] suggested that the overfertilized plot should have an NNI higher than 1, indicating that N is not limiting. Ravier et al. [72] concluded that the use of normalized values is problematic because it is essential to ensure that a well-fertilized strip is not N deficient. They determined that using relativized values from an N-deficient overfertilized plot (NNI < 0.9) may lead to an error of 0.34%. The overfertilized treatments of our dataset did not obtain NNI ≥ 0.9 (Table 3) even after the application of large amounts of N. Yao et al. [73] showed that normalized values were more correlated with the NNI at later stages of development than at early stages. In our results, both absolute values and normalized values were able to similarly detect differences among treatments. However, finding values to adjust the optimum N rate was complicated with absolute values because variability among years was higher than with relativized values. Regarding normalized values, in dairy slurry treatment, when both proximal sensing tool values were 60–65%, the optimum N rate at GS30 to achieve the maximum yield was 118–128 kg N ha^{-1} (Table 5). In dairy slurry and conventional treatments, when the readings were 85–90%, the optimum N rate was 100–110 kg N ha^{-1} (Table 5). In the sheep manure treatment, there was no clear relationship between sensor values and the optimum N rate. When values were 60–65%, the N recommendation was 117 or 155 kg N ha^{-1} and 147 kg N ha^{-1} for readings around 90%. That may be due to the heterogeneous nature of manures that makes it difficult to predict how quickly and how much N will be transformed in plant-available N during the wheat-growing season. The composition of manures depends on animal diet, the amount of bedding, water and nutrient loss during storage and land application [4]. However, in this study, the main factor that brings variability is the maturation (curing) phase. Thus, sheep manure characteristics were different depending on the year: in 2016, the sheep manure was "fresh", containing NH_4^+ (0.33% of total N; Table 1), whereas in 2015 and 2017, it was "old" and the manures did not present any NH_4^+. In dairy slurry treatments, the NH_4^+ content in the three wheat-growing seasons was around 0.55% of total N, having lower variability compared to sheep manure. In 2016, the N availability in sheep manure treatment was high when the crop N demand was high, allowing greater tool values than in 2015 and 2017 (Table 4 and Figure 4). However, the N recommendation at GS30 was much higher in sheep manure treatments than in treatments with dairy slurry as initial fertilizer, even if tools' values were similar at GS30 (Table 4 and Figure 4). Manures also contain organic N that becomes available to crops after enzymatic hydrolysis. The amount of N mineralized from organic N depends on its physical–chemical characteristics, soil properties and climatic conditions [74]. Typically, slurries are more susceptible to enzymatic hydrolysis than solid manures that have bedding materials made of lignified compounds [4]. This can also justify the higher and variable N required by the crop when sheep manure was applied than when slurry was applied, especially when crop N requirements were high, as in 2016. Moreno-García et al. [75] and Zhao et al. [76] found that the higher availability of micronutrients provided by the organic fertilizers allowed higher yield than in conventional treatment and suggested that it is necessary to establish overfertilized plots for each type of fertilization.

These results look promising in order to adjust the N application rate at GS30 with the Yara N-Tester^{TM} and RapidScan CS-45 normalized readings when applying dairy slurry before sowing. However, in the sheep manure treatment, it is more difficult to find a relationship between sensor readings and yield. Both handheld N diagnostic tools were able to detect differences among N nutritional status in plants. Siband et al. [77] suggested that the NNI responded faster to N applications than chlorophyll meters, which is probably related to the time between N uptake and chlorophyll synthesis [22]. Ziadi et al. [68] saw that the whole plant can be less N deficient than the uppermost collared leaf. In this sense, evaluating the whole plant would be more precise than just a leaf. At early growth stages, evaluating the NDVI would be more accurate, although some authors such as Broge and Mortensen [78] highlighted that, when the wheat canopy is not closed, soil background exposure reduces the reliability of using reflectance for the estimation of crop N status. However, at the end of the growing cycle, the use of a chlorophyll meter is suitable [20], as 35% of N in the grain comes from the last developed leaf [79]. The utility of proximal tools to detect the N nutritional status of plants was supported by Arregui et al. [9] and Ortuzar-Iragorri et al. [18] who observed that chlorophyll meters enabled the prediction of the N status of plants. Furthermore, Mullen et al. [80] showed that in-season N demand for added N fertilizer in winter wheat could be detected using NDVI readings collected at GS30. Moreover, algorithms using crop canopy reflectance sensing to make N recommendation for wheat have been identified, and it has been shown that active canopy sensors could be used in determining variable N rate applications in wheat from the mid-growing season [81,82]. Marti et al. [28] found significant correlations between the NDVI, yield and biomass in wheat. On the other hand, Sylvester-Bradley et al. [64] developed an alternative strategy for signaling soil N status and observed that the NDVI of young canopies can signal soil N status where N_{min} is lower than 120–140 kg N ha^{-1}. In our conditions, generally N_{min} values were lower than 120–140 kg N ha^{-1}. In our case, both proximal tools showed a greater sensitivity than N_{min} when differentiating the initial fertilization treatments applied in the field trial. All of these findings were very promising because the type of fertilization could be changed to achieve a more precise N rate adjusted to the wheat crop demand. It should be noted that measurements taken with RapidScan CS-45 were less time consuming than with Yara N-Tester^{TM} and were taken on the plant canopy and not just in the uppermost fully expanded leaf, as with chlorophyll meters. Thus, samples taken with RapidScan CS-45 can better represent the spatial variation of the crop N status.

These hand-held tools are used during the wheat growth period (GS30), but many environmental variables (rainfall, temperature or relative humidity) affect crop growth and development after this stage until harvest [83]. However, any method of diagnosing N nutritional status at a particular stage has the same limitation. Furthermore, it has been shown that it is always possible to correct N deficiency until the end of the cereal growth season (GS65) [23] if soil is wet [84]. Since remote sensing measurements are not invasive and can be repeated several times during the growth period, information obtained on N status dynamics of plants can be used for decision making in N fertilizer management. What is more, ground-based remote sensing tools can be applied to satellite or airborne remote sensing.

5. Conclusions

Experimental findings to date have shown that Yara N-Tester^{TM} or Rapidscan CS-45 normalized readings look promising in order to adjust the N application rate at GS30 under rainfed conditions in humid Mediterranean climate conditions. For dairy slurry, when either proximal sensor readings were 60–65% of the reference plants with non-limiting N, the optimum N rate for maximizing yield was 118–128 kg N ha^{-1}. When the readings were 85–90%, the optimum N rate dropped to 100–110 kg N ha^{-1} for both dairy slurry and conventional treatments. However, in the sheep manure treatment, it is more difficult to find a relationship between sensor readings and yield due to the variable composition of the manure and subsequently, the available N.

When rainfall conditions between mid-March and April are as usual (seven out of ten years), a low N status (60–65%) can be recovered with no adverse effect on yield when mineral N is applied at GS30. However, during those years in which dry periods occur (three of ten), when mineral N is applied at GS30, the N deficiency is not recovered because the lack of soil moisture prevents proper N uptake by the crop. In these cases, the application of 40 kg N ha^{-1} at GS21 leads to higher yields. As the N rate at GS30 can be modulated with the proximal tools, the application of 40 kg N ha^{-1} at GS21 is necessary to ensure an optimal N status from the beginning of wheat crop development.

These hand-held tools are used during the wheat-growing period (GS30), but many environmental variables may affect crop development until harvest. In order to address this, given that remote sensing measurements are not invasive, these measurements should be taken periodically to monitor crop N status in an effective way. Routine measurements throughout the growing season are particularly needed in climates with variable rainfall. Measurements taken with RapidScan C-45 were less time consuming and better represent the spatial variation as they are taken on the plant canopy.

While Yara N-Tester^{TM} or Rapidscan CS-45 look promising for adjusting N application rates, further research is needed to improve the use of these sensors. These research findings could be used in applicator-mounted sensors to make variable-rate N applications.

Author Contributions: M.A., A.C. and A.A. contributed to the conception and design of the study. M.A and A.C. carried out fieldwork for acquisition of the data analysed. M.A., A.C. and A.A. contributed to different aspects of analysis and interpretation of data. M.A. and A.A. wrote the paper, and M.A., A.C. and A.A. authors revised it critically or approved of the final version.

Funding: This study was funded by the National Institute of Agricultural and Food Research and Technology (RTA2013-00057-01 and RTA2017-00088-C03-00) and by the Department for Economic Development and Infrastructures of the Basque Government. M. Aranguren is the recipient of a predoctoral fellowship from the Department for Economic Development and Infrastructures of the Basque Government.

Conflicts of Interest: The author declares no conflict of interest.

References

1. Cameron, K.C.; Di, H.J.; Moir, J.L. Nitrogen losses from the soil/plant system: a review. *Ann. Appl. Biol.* **2013**, *162*, 145–173. [CrossRef]
2. Zadoks, J.C.; Chang, T.T.; Konzak, C.F. A decimal code for the growth stages of cereals. *Weed Res.* **1974**, *14*, 415–421. [CrossRef]
3. FAO. *Pollution from Industrialized Livestock Production*; Agriculture and Consumer Protection Department: Rome, Italy, 2005.

4. Whalen, J.K.; Thomas, B.W.; Sharifi, M. Novel Practices and Smart Technologies to Maximize the Nitrogen Fertilizer Value of Manure for Crop Production in Cold Humid Temperate Regions. *Adv. Agron.* **2019**, *153*, 1–85.
5. Defra. Nutrient Management Guide (RB209). Available online: https://ahdb.org.uk/documents/RB209/RB209_Section2_WEB_2001.pdf (accessed on 11 January 2017).
6. Ros, G.H.; Hanegraaf, M.C.; Hoffland, E.; Van Riemsdijk, W.H. Predicting soil N mineralization: Relevance of organic matter fractions and soil properties. *Soil Boil. Biochem.* **2011**, *43*, 1714–1722. [CrossRef]
7. Eghball, B. Nitrogen Mineralization from Field-Applied Beef Cattle Feedlot Manure or Compost. *Soil Sci. Soc. J.* **2000**, *64*, 2024–2030. [CrossRef]
8. Eghball, B.; Wienhold, B.J.; Gilley, J.E.; Eigenberg, R.A. Mineralization of Manure Nutrients. *Biol. Syst. Eng. Pap. Publ.* **2002**, *139*.
9. Arregui, L.; Lasa, B.; Lafarga, A.; Irañeta, I.; Baroja, E.; Quemada, M. Evaluation of chlorophyll meters as tools for N fertilization in winter wheat under humid Mediterranean conditions. *Eur. J. Agron.* **2006**, *24*, 140–148. [CrossRef]
10. Basso, B.; Fiorentino, C.; Cammarano, D.; Cafiero, G.; Dardanelli, J. Analysis of rainfall distribution on spatial and temporal patters of wheat yield in Mediterranean environment. *Eur. J. Agron.* **2012**, *41*, 52–65. [CrossRef]
11. Ravier, C.; Jeuffroy, M.-H.; Meynard, J.-M. Mismatch between a science-based decision tool and its use: The case of the balance-sheet method for nitrogen fertilization in France. *NJAS-Wagening. J. Life Sci.* **2016**, *79*, 31–40. [CrossRef]
12. Ravier, C. Conception innovante d'une méthode de fertilization azotée: Ariculation entre diagnostic des usages, ateliers participatifs et modélisation. Ph.D. Thesis, University of Paris-Saclay, Saint-Aubin, France, 2017.
13. Antille, D.L.; Lobsey, C.R.; McCarthy, C.L.; Thomasson, J.A.; Baillie, C.P. A review of the state of the art in agricultural automation. Part IV: Sensor-based nitrogen management technologies. *ASABE Annu. Int. Meet.* **2018**. [CrossRef]
14. Diacono, M.; Rubino, P.; Montemurro, F. Precision nitrogen management of wheat. *Agron. Sustain. Dev.* **2013**, *33*, 219–241. [CrossRef]
15. Raun, W.R.; Solie, J.B.; Johnson, G.V.; Stone, M.L.; Mullen, R.W.; Freeman, K.W.; Thomason, W.E.; Lukina, E.V. Improving Nitrogen Use Efficiency in Cereal Grain Production with Optical Sensing and Variable Rate Application. *Agron. J.* **2002**, *94*, 815. [CrossRef]
16. Samborski, S.M.; Tremblay, N.; Fallon, E. Strategies to Make Use of Plant Sensors-Based Diagnostic Information for Nitrogen Recommendations. *Agron. J.* **2009**, *101*, 800. [CrossRef]
17. Lemaire, G.; Jeufroy, M.H.; Gastal, F. Diagnosis tool for plant and crop N status in vegetative stage: Teory and practices for crop N management. *Eur. J. Agron.* **2008**, *28*, 614–624. [CrossRef]
18. Ortuzar-Iragorri, M.A.; Alonso, A.; Castellón, A.; Besga, G.; Estavillo, J.M.; Aizpurua, A. N-Tester use in soft winter wheat: evaluation of nitrogen status and grain yield prediction. *Agron. J.* **2005**, *97*, 1380–1389. [CrossRef]
19. Piekelek, W.P.; Fox, R.H. Use of a chlorophyll meter to predict sidedress nitrogen requirements for maize. *Agron. J.* **1992**, *84*, 59–65. [CrossRef]
20. Ortuzar-Iragorri, M.A.; Aizpurua, A.; Castellón, A.; Alonso, A.; José, M.; Estavillo, J.M.; Besga, G. Use of an N-Tester chlorophyll meter to tune a late third nitrogen application to wheat under humid Mediterranean conditions. *J. Plant Nutr.* **2017**, *41*, 6635. [CrossRef]
21. Prost, L.; Jeuffroy, M.-H. Replacing the nitrogen nutrition index by the chlorophyll meter to assess wheat N status. *Agron. Sustain. Dev.* **2007**, *27*, 321–330. [CrossRef]
22. Debaeke, P.; Rouet, P.; Justes, E. Relationship Between the Normalized SPAD Index and the Nitrogen Nutrition Index: Application to Durum Wheat. *J. Plant Nutr.* **2006**, *29*, 75–92. [CrossRef]
23. Ravier, C.; Meynard, J.-M.; Cohan, J.-P.; Gate, P.; Jeuffroy, M.-H. Early nitrogen deficiencies favor high yield, grain protein content and N use efficiency in wheat. *Eur. J. Agron.* **2017**, *89*, 16–24. [CrossRef]
24. Shanahan, J.; Kitchen, N.; Raun, W.; Schepers, J.; Raun, W. Responsive in-season nitrogen management for cereals. *Comput. Electron. Agric.* **2008**, *61*, 51–62. [CrossRef]
25. Ali, A.M.; Thind, H.S.; Varinderpal-Singh, B.-S. A framework for refining nitrogen management in dry direct-seeded rice using GreenSeeker™ optical sensor. *Comput. Electron. Agric.* **2015**, *110*, 114–120. [CrossRef]

26. Mulla, D.J. Twenty five years of remote sensing in precision agriculture: Key advances and remaining knowledge gaps. *Biosyst. Eng.* **2013**, *114*, 358–371. [CrossRef]
27. Cao, Q.; Miao, Y.; Feng, G.; Gao, X.; Li, F.; Liu, B.; Yue, S.; Cheng, S.; Ustin, S.L.; Khosla, R. Active canopy sensing of winter wheat nitrogen status: An evaluation of two sensor systems. *Comput. Electron. Agric.* **2015**, *112*, 54–67. [CrossRef]
28. Marti, J.; Bort, J.; Slafer, G.A.; Araus, J.L. Can wheat yield be assessed by early measurements of Normalized Difference Vegetation Index? *Ann. Appl. Biol.* **2007**, *150*, 225. [CrossRef]
29. Lu, J.; Miao, Y.; Shi, W.; Li, J.; Yuan, F. Evaluating different approaches to non-destructive nitrogen status diagnosis of rice using portable RapidSCAN active canopy sensor. *Sci. Rep.* **2017**, *7*, 14073. [CrossRef] [PubMed]
30. Li, S.; Ding, X.; Kuang, Q.; Ata-Ui-Karim, S.T.; Cheng, T.; Liu, X.; Tian, Y.; Zhu, Y.; Cao, W.; Cao, Q. Potential of UAV-Based Active Sensing for Monitoring Rice Leaf Nitrogen Status. *Front. Plant Sci.* **2018**, *9*, 1834. [CrossRef] [PubMed]
31. Bonfil, D.J. Monitoring wheat fields by RapidScan: Accuracy and limitations. In Proceedings of the Conference on Precision Agriculture (ECPA 2017), Edinburgh, UK, 16–20 July 2017.
32. Zhang, K.; Ge, X.; Shen, P.; Li, W.; Liu, X.; Cao, Q.; Zhu, Y.; Cao, W.; Tian, Y. Predicting Rice Grain Yield Based on Dynamic Changes in Vegetation Indexes during Early to Mid-Growth Stages. *Remote Sens.* **2019**, *11*, 387. [CrossRef]
33. Aranguren, M.; Castellon, A.; Aizpurua, A. Topdressing nitrogen recommendation in wheat after applying organic manures: the use of field diagnostic tools. *Nutr. Cycl. Agroecosyst.* **2018**, *110*, 89–103. [CrossRef]
34. Papadakis, J. *Climates of the World and Their Agricultural Potentialities*; Libro de Edicion Argentina: Buenos Aires, Argentina, 1966.
35. Euskalmet. Euskal Meteorologia Agentzia. 2018. Available online: http://www.euskalmet.euskadi.eus/ (accessed on 21 November 2018).
36. IUSS Working Group, WRB. Base referencial mundial del recurso suelo 2014, Actualización 2015. Sistema internacional de clasificación de suelos para la nomenclatura de suelos y la creación de leyendas de mapas de suelos. In *Informes Sobre Recursos Mundiales de Suelos 106*; FAO: Roma, Italy, 2014.
37. Gee, G.W.; Bauder, J.W. Particle-size analysis. In *Methods of Soil Analysis: Part 1. Physical and Mineralogical Methods*; Klute, A., Ed.; SSSA: Madison, WI, USA, 1986; pp. 383–411.
38. Soil Survey Staff. *Soil Taxonomy: A Basic System of Soil Classification for Making and Interpreting Soil Surveys*, 2nd ed.; Natural Resources Conservation Service. U.S. Department of Agriculture: Washington, DC, USA, 1999.
39. MAPA. Métodos oficiales de análisis. In *Tomo III*; Ministerio de Agricultura, Pesca y Alimentación: Madrid, Spain, 1994.
40. Walkey, A.; Black, I.A. An examination of Degtjareff method for determining soil organic matter and a proposed modification of the chromic and titration method. *Soil Sci.* **1934**, *37*, 29–37. [CrossRef]
41. Cawse, P.A. The determination of nitrate in soil solutions by ultraviolet spectrophotometry. *Analyst* **1967**, *92*, 311–315. [CrossRef]
42. Nelson, D.W. Determination of ammonium in KCl extracts of soils by the salicylate method. *Commun. Soil Sci. Plant Anal.* **1983**, *14*, 1051–1062. [CrossRef]
43. AOAC, Association of Official Analytical Chemists International. Plants, 24, 127. In *Official Methods of AOAC International*, 16th ed.; Patricia, C., Ed.; AOAC International: Gaithersburg, MD, USA, 1999.
44. Lemaire, G.; Salette, J.; Sigogne, M.; Terrasson, J.P. Relation entre dynamique de croissance et dynamique de prélèvement d'azote pour un peuplement de graminées fourragéres. I: Etude de l'effet du milieu. *Agronomie* **1984**, *4*, 423–430. [CrossRef]
45. Yue, S.; Zhao, R.; Zhang, F.; Cui, Z.; Meng, Q.; Li, F.; Chen, X. Critical Nitrogen Dilution Curve for Optimizing Nitrogen Management of Winter Wheat Production in the North China Plain. *Agron. J.* **2012**, *104*, 523–529. [CrossRef]
46. Justes, E.; Mary, B.; Meynard, J.-M.; Machet, J.-M.; Thelier-Huche, L. Determination of a Critical Nitrogen Dilution Curve for Winter Wheat Crops. *Ann. Bot.* **1994**, *74*, 397–407. [CrossRef]
47. Follett, R.; Follett, R.; Halvorson, A. Use of a chlorophyll meter to evaluate the nitrogen status of dryland winter wheat. *Commun. Soil Sci. Plant Anal.* **1992**, *23*, 687–697. [CrossRef]

48. Ortuzar-Iragorri, M.A.; Castellón, A.; Alonso, A.; Besga, G.; Estavillo, J.M.; Aizpurua, A. Estimation of optimum nitrogen fertilizer rates in winter wheat in humid Mediterranean conditions, I: Selection of yield and protein response models. *Commun. Soil Sci. Plant Anal.* **2010**, *41*, 2293–2300. [CrossRef]
49. SAS Institute. *SAS version 8*; SAS Institute: Cary, NC, USA, 1998.
50. Cerrato, M.E.; Blackmer, A.M. Comparison of Models for Describing; Corn Yield Response to Nitrogen Fertilizer. *Agron. J.* **1990**, *82*, 138. [CrossRef]
51. Roberts, D.F.; Ferguson, R.B.; Kitchen, N.R.; Adamchuk, V.I.; Shanahan, J.F. Relationships between Soil-Based Management Zones and Canopy Sensing for Corn Nitrogen Management. *Agron. J.* **2012**, *104*, 119. [CrossRef]
52. Scharf, P.C.; Kitchen, N.R.; Sudduth, K.A.; Davis, J.G. Spatially Variable Corn Yield is a Weak Predictor of Optimal Nitrogen Rate Peter C. *Soil Sci. Soc. Am. J.* **2006**, *70*, 2154–2160. [CrossRef]
53. Aizpurua, A.; Estavillo, J.M.; Castellón, A.; Alonso, A.; Besga, G.; Ortuzar-Iragorri, M.A. Estimation of optimum nitrogen fertilizer rates in winter wheat in humid mediterranean conditions. II: Economically optimal dose nitrogen. *Commun. Soil Sci. Plant Anal.* **2010**, *41*, 301–307. [CrossRef]
54. MAPAMA. *Publicación de indicadores de precios y salarios agrarios*; Ministerio de Agricultura y Pesca, Alimentación y Medio Ambiente: Madrid, Spain, 2017.
55. R Core Team. *R: A Language and Environment for Statistical Computing*; R Foundation for Statistical Computing: Vienna, Austria, 2013.
56. De Mendiburu, F. Una herramienta de análisis estadístico para la investigación agrícola. Ph.D. Thesis, Universidad Nacional de Ingeniería, Lima, Peru, 2009.
57. Mohanty, M.; Sinha, N.K.; Reddy, K.S.; Chaudhary, R.S.; Rao, A.S.; Dalal, R.C.; Menzies, N.W. How important is the quality of organic amendments in relation to mineral N availability in soils? *Agric. Res.* **2013**, *2*, 99–110. [CrossRef]
58. Walley, F.; Yates, T.; Van Groenigen, J.-W.; Van Kessel, C. Relationships between Soil Nitrogen Availability Indices, Yield, and Nitrogen Accumulation of Wheat. *Soil Sci. Soc. J.* **2002**, *66*, 1549. [CrossRef]
59. Arregui, L.M.; Quemada, M. Drainage and nitrate leaching in a crop rotation under different N-fertilizer strategies: application of capacitance probes. *Plant Soil* **2006**, *288*, 57–69. [CrossRef]
60. Chantigny, M.H.; Angers, D.A.; Morvan, T.; Pomar, C. Dynamics of Pig Slurry Nitrogen in Soil and Plant as Determined with N. *Soil Sci. Soc. J.* **2004**, *68*, 637. [CrossRef]
61. Shah, G.M.; Rashid, M.I.; Shah, G.A.; Groot, J.C.J.; Lantinga, E.A. Mineralization and herbage recovery of animal manure nitrogen after application to various soil types. *Plant Soil* **2013**, *365*, 69–79. [CrossRef]
62. INTIA. Manual del cultivo de colza de otoño. Instituto Navarro de Tecnologías e Infraestructuras agroalimentarias. Available online: https://intiasa.es/repositorio/images/docs/ManualCOLZA2012.pdf (accessed on 11 January 2017).
63. Gallejones, P.; Castellón, A.; del Prado, A.; Unamunzaga, O.; Aizpurua, A. Nitrogen and sulphur fertilization effect on leaching losses, nutrient balance and plant quality in a wheat–rapeseed rotation under a humid Mediterranean climate. *Nutr. Cycl. Agroecosyst.* **2012**, *93*, 3355. [CrossRef]
64. Sylvester-Bradley, R.; Wiltshire, J.J.J.; Kindred, D.R.; Hatley, D.L.J.; Clarke, S. *Detecting Soil Nitrogen Supplies by Canopy Sensing—Project Report 460*; HGCA: London, UK, 2009.
65. Cilia, C.; Panigada, C.; Rossini, M.; Meroni, M.; Busetto, L.; Amaducci, S.; Boschetti, M.; Picchi, V.; Colombo, R. Nitrogen Status Assessment for Variable Rate Fertilization in Maize through Hyperspectral Imagery. *Remote Sens.* **2014**, *6*, 6549–6565. [CrossRef]
66. Huang, S.; Miao, Y.; Zhao, G.; Yuan, F.; Ma, X.; Tan, C.; Yu, W.; Gnyp, M.L.; Lenz-Wiedemann, V.I.; Rascher, U.; et al. Satellite Remote Sensing-Based In-Season Diagnosis of Rice Nitrogen Status in Northeast China. *Remote Sens.* **2015**, *7*, 10646–10667. [CrossRef]
67. Xia, T.; Miao, Y.; Wu, D.; Shao, H.; Khosla, R.; Mi, G. Active Optical Sensing of Spring Maize for In-Season Diagnosis of Nitrogen Status Based on Nitrogen Nutrition Index. *Remote Sens.* **2016**, *8*, 605.
68. Ziadi, N.; Brassard, M.; Bélanger, G.; Claessens, A.; Tremblay, N.; Cambouris, A.N.; Nolin, M.C.; Parent, L.-É. Chlorophyll Measurements and Nitrogen Nutrition Index for the Evaluation of Corn Nitrogen Status. *Agron. J.* **2008**, *100*, 1264. [CrossRef]
69. Peltonen, J.; Virtanen, A.; Haggrèn, E. Using a Chlorophyll Meter to Optimize Nitrogen Fertilizer Application for Intensively-Managed Small-Grain Cereals. *J. Agron. Sci.* **1995**, *174*, 309–318. [CrossRef]
70. Bundy, L.G.; Andraski, T.W. Diagnostic Tests for Site-Specific Nitrogen Recommendations for Winter Wheat. *Agron. J.* **2004**, *96*, 608–614. [CrossRef]

71. Ravier, C.; Quemada, M.; Jeuffroy, M.-H. Use of a chlorophyll meter to assess nitrogen nutrition index during the growth cycle in winter wheat. *Field Crop. Res.* **2017**, *214*, 73–82. [CrossRef]
72. Ravier, C.; Jeuffroy, M.H.; Gate, P.; Cohan, J.P.; Meynard, J.M. Combining user involvement with innovative design to develop a radical new method for managing N fertilization. *Nutr. Cycl. Agroecosyst.* **2018**, *110*, 117. [CrossRef]
73. Yao, Y.; Miao, Y.; Cao, Q.; Wang, H.; Gnyp, M.L.; Bareth, G.; Khosla, R.; Yang, W.; Liu, F.; Liu, C. In-Season Estimation of Rice Nitrogen Status with an Active Crop Canopy Sensor. *IEEE J. Sel. Top. Appl. Earth Obs. Sens.* **2014**, *7*, 4403–4413. [CrossRef]
74. Sharifi, M.; Zebarth, B.J.; Burton, D.L.; Rodd, V.; Grant, C.A. Long-Term Effects of Semisolid Beef Manure Application to Forage Grass on Soil Mineralizable Nitrogen. *Soil Sci. Soc. J.* **2011**, *75*, 649. [CrossRef]
75. Moreno-García, B.; Casterad, M.A.; Guillén, M.; Quílez, D. Agronomic and Economic Potential of Vegetation Indices for Rice N Recommendations under Organic and Mineral Fertilization in Mediterranean Regions. *Remote Sens.* **2018**, *10*, 1908. [CrossRef]
76. Zhao, J.; Dong, S.T.; Liu, P.; Zhang, J.W.; Zhao, B. Effects of long-term mixed application of organic and inorganic fertilizers on canopy apparent photosynthesis and yield of winter wheat. *Chin. J. Appl. Ecol.* **2015**, *26*, 2362–2370.
77. Siband, P.; Loyce, C.; Witt, C.; Dingkuhn, M. Evaluer le statut azote du riz irrigué. In *Modélisation des Agroécosystémes et Aide á la Decision*; Malézieux, E., Trébuil, G., Jaeger, M., Eds.; CIRAD: Montpellier, France, 2001; pp. 95–106.
78. Broge, N.; Mortensen, J. Deriving green crop area index and canopy chlorophyll density of winter wheat from spectral reflectance data. *Remote Sens. Environ.* **2002**, *81*, 45–57. [CrossRef]
79. Fuertes-Mendizábal, T.; Gonzalez-Murua, C.; González-Moro, M.B.; Estavillo, J.M. Late nitrogen fertilization affects nitrogen remobilization in wheat. *J. Plant Nutr. Soil Sci.* **2012**, *175*, 115–124. [CrossRef]
80. Mullen, R.W.; Freeman, K.W.; Raun, W.R.; Johnson, G.V.; Stone, M.L.; Solie, J.B. Identifying an In-Season Response Index and the Potential to Increase Wheat Yield with Nitrogen. *Agron. J.* **2003**, *95*, 347. [CrossRef]
81. Raun, W.R.; Solie, J.B.; Stone, M.L.; Martin, K.L.; Freeman, K.W.; Mullen, R.W.; Zhang, H.; Schepers, J.S.; Johnson, G.V.; Raun, W.; et al. Optical Sensor-Based Algorithm for Crop Nitrogen Fertilization. *Commun. Soil Sci. Plant Anal.* **2005**, *36*, 2759–2781. [CrossRef]
82. Calvo, N.I.R.; Rozas, H.S.; Echeverría, H.; Diovisalvi, N. Using Canopy Indices to Quantify the Economic Optimum Nitrogen Rate in Spring Wheat. *Agron. J.* **2015**, *107*, 459–465. [CrossRef]
83. Aranguren, M.; Castellón, A.; Aizpurua, A. Use of field diagnostic tools for top dressing nitrogen recommendation when organic manures are applied in humid Mediterranean conditions. In Proceedings of the 14th International Conference on Precision Agriculture, Montreal, QC, Canada, 24–27 June 2018.
84. Soenen, B.; Cohan, J.P.; Jeuffroy, M.H.; Meynard, J.M.; Ravier, C. Fertilisation azotée du blé: raisonner sans objectif de rendement? *Perspect. Agric.* **2017**, *445*, 40–42.

MDPI
St. Alban-Anlage 66
4052 Basel
Switzerland
Tel. +41 61 683 77 34
Fax +41 61 302 89 18
www.mdpi.com

Remote Sensing Editorial Office
E-mail: remotesensing@mdpi.com
www.mdpi.com/journal/remotesensing

www.ingramcontent.com/pod-product-compliance
Lightning Source LLC
Chambersburg PA
CBHW061452220926
51850CB00004B/74